U.S. MARINES IN THE KOREAN WAR

Edited by
Charles R. Smith

History Division
United States Marine Corps
Washington, D.C.
2007

Second Printing (2010)

PCN 106 0000 0100

Foreword

The anthology of articles that follows was compiled by the History and Museums Division during the 50th Anniversary commemoration of the Korean Conflict, 1950–1953. The focus of the various authors who wrote these historically related works on Korea did so to remember those Marines who fought and died in what some historians sometimes characterized as the "forgotten war." Forgotten or not, the Korean conflict was without parallel in Marine Corps history and no one who experienced it or lived through this era could ever forget the difficulties that they would encounter there.

The Korean War also represented a milestone in the developmental history of the Marine Corps. For perhaps what could very well be the last time, the Marine Corps made an opposed World War II style amphibious landing against a dedicated enemy. Korea was also the opening salvo in what became known as the Cold War. In reality, Korea represented the beginning of a series of "limited wars" that would be fought by the United States with the express political purpose of keeping such conflicts from developing into full blown world wars. Frustratingly for the men and women in uniform during the Cold War, political considerations frequently overrode military exigencies and logic. Having just successfully concluded a total war against an enemy whose objectives were clearly identifiable, the Korean conflict proved fraught with political twists and turns that made the military's job immensely more difficult. This was especially evident during the "stalemate" phase of the war, 1952–1953. No less bloody or violent, this period of the conflict saw the Marine Corps incur a significant number of casualties.

The Korean conflict was also important for operational reasons. It was clear that from 1950 on, limited wars fought by U.S. forces would be largely "come as you are affairs." During the summer and early fall of 1950, the Marine Corps learned a valuable lesson when it had to scramble to assemble its landing force for the Inchon operation, getting the 1st Marine Division into theater in the nick of time. No longer would the United States have the luxury of time in getting forces ready for limited wars. Next, for the first time, the advent of the helicopter would play a significant role in the combat plans of Marine units in the field. Experimentation with the concept of vertical assault, using this new technology took place during the conflict. Korea would also be the first time Marines would be given personal body armor or "flak jackets" to wear in combat. Such body armor would come in handy as the war settled into a stalemate along the 38th Parallel. While Marine elements had deployed to extremely cold locations in the past such as the occupation of Iceland by the 1st Marine Brigade (Provisional) in 1941, Korea would be the first time in the modern era where the Marine Corps would have to fight in extremely cold conditions. During Korea, the Corps came away with a new appreciation for the necessity of having the proper environmental gear tested and available for use by its combat and combat support troops. In sum, Korea set the operational tone that the Marine Corps would follow for the rest of the Cold War.

My special thanks is extended to Charles R. Smith, senior historian and editor of the Korean conflict commemorative history series and to Charles D. Melson, Chief Historian. These gentlemen, under the direction of then History and Museums Division Director Colonel John W. Ripley oversaw the painstaking process of editing the eleven separate historical pamphlets produced by the division for the 50th Anniversary of the Korean Conflict, 1950–1953.

Finally, let us pledge to never forget those Marines who served and sacrificed in this supposedly "forgotten war." It was entirely due to their military achievements at Inchon, at Chosin Reservoir, or in foxholes and bunkers along the 38th Parallel that we have a free and thriving South Korea today.

Dr. C. P. Neimeyer
Director of Marine Corps History

Preface

The Korean War was the first major armed clash between Free World and Communist forces, in what was to be called the Cold War. It was waged on land, on sea, and in the air over, and near the Korean peninsula, for more than three years. Among the U.S. forces committed on this far-off battlefront, it was once again the Marine Corps component that stood out in its sacrifice, military skills, and devotion to duty. When rushed into the battle during the first desperate weeks and months of the war, the quickly-organized and rapidly deployed 1st Provisional Marine Brigade and Marine Aircraft Group 33 helped to restore stability to the shattered U.S. Eighth Army front line around Pusan. It would be the first time that Marine air and ground elements, task organized under a single commander, had engaged in combat.

During the daringly conceived and executed United Nations counterstroke at Inchon, Marines accomplished this incredibly complex amphibious operation and the subsequent recapture of the South Korean capital, Seoul, with their customary spirit and precision, delivering a tactical blow that broke the backbone of the North Korean People's Army 1950 offensive. Never was Marine heroism and perseverance more conspicuous than during the bitter days of the Chosin Reservoir campaign, following the intervention of large-scale Chinese Communist Forces. Integrated ground and air action enabled more than 14,000 Marine, Army, and Royal Marine troops to break out of the entrapment and move south. The 1st Marine Division, considered by many to have been lost, properly evacuated its dead and wounded, brought out all operable equipment, and completed the withdrawal with tactical integrity, all the while dealing a savage blow to the enemy.

As the war of fire and movement turned into one of positional warfare that marked the final operations in Korea, the 1st Marine Division and 1st Marine Aircraft Wing again executed their respective missions with professional skill and dispatch, regardless of tactical problems and the dreary monotony that characterized the fighting around the Inje River and Hwachon Reservoir in the Punchbowl area, and the critical 35-mile front in West Korea near Panmunjom.

The more than half-century that now separates us from the Korean conflict has dimmed our collective memory. Many Korean War veterans considered themselves forgotten, their place in history sandwiched between World War II and the Vietnam War. This compilation grew out a joint endeavor by the Marine Corps History and Museums Division and Marine Corps Heritage Foundation to remedy that perceived oversight by highlighting the contributions and honoring the service of those Marines for today's Marines and the American people. The well-researched and highly-illustrated monographs, and now chapters, were written by Colonel Joseph H. Alexander, Captain John C. Chapin, Brigadier General Edwin H. Simmons, Colonel Allan R. Millet, Bernard C. Nalty, Major General John P. Condon, Commander Peter B. Mersky, USNR (Ret), and Lieutenant Colonel Ronald J. Brown, all distinguished historians, experts, or respected participants.

The authors and editor gratefully recognize and thank all who assisted in the preparation of the original monographs and of this volume by aiding in research and supplying photographs. Among them are the professional staffs of the former Marine Corps History and Museums Division, now the History Division, Quantico, Virginia; the Marine Corps Heritage Foundation, Quantico, Virginia; the Naval Historical Center, Washington, D.C.; the Archives and Special Collections Branch, Library of the Marine Corps, Quantico, Virginia; and the Modern Military Records and Still Pictures Branches of the National Archives and Records Administration, College Park, Maryland. And to the more than 150,000 Marines who served in Korea, and the 4,267 killed and 23,748 wounded in action, we say thank you.

Charles R. Smith
Editor

Contents

FIRE BRIGADE
U.S. Marines in the Pusan Perimeter

by Captain John C. Chapin, USMCR (Ret)

"The Marines have landed." How familiar the phrase, how extraordinary the circumstances on 2 August 1950. Instead of a beach saturated with enemy fire, the scene was a dock in the port of Pusan in the far southeast corner of Korea. The landing force was the 1st Provisional Marine Brigade; the situation it would soon face was one of desperate crisis.

The men arriving on board the transport ships that day knew they were going into battle, and their brigade commander, Brigadier General Edward A. Craig, had made his combat standards very clear in a meeting with his officers before the ships had sailed from San Diego: "It has been necessary for troops now fighting in Korea to pull back at times, but I am stating now that no unit of this brigade will retreat except on orders from an authority higher than the 1st Marine Brigade. You will never receive an order to retreat from me. All I ask is that you fight as Marines have always fought."

At sea, no one knew where the brigade would be committed to action, and the men knew nothing about the forthcoming enemy except it was called the North Korean People's Army (NKPA). On board their ships they had seen the situation maps which daily showed the steadily retreating line of defense, as the enemy drove irresistably farther and farther into South Korea. The regular physical fitness drills and weapons target practice took on an urgent new sense of purpose for the Marines.

AT LEFT: *Moving to attack: A corporal leads his automatic-rifle man through a rice paddy, oblivious to the dead enemy soldier underfoot.* Photo by David Douglas Duncan

Captain Francis I. "Ike" Fenton, Jr., then executive officer of Company B, later recalled:

> While on board ship our training area was limited. It was an impossibility to get the whole company together at one location. Consequently, we used passageways, boat decks, holds—any space we could find to lecture to the men and give them the little information that we had as to what was happening in Korea.
>
> We lectured on the characteristics of the T-34 tank and told the men about the kind of land mines we might expect. A lot of time was spent on blackboard tactics for the fire team, platoon, and company. We had the 3.5 rocket launcher, but no one present had ever fired one.

A variety of old World War II ships had brought the brigade. Task Force 53.7 had 10 ships. Two transports and a light carrier, the *Badoeng Strait* (CVE-116), transported the air arm, Marine Aircraft Group 33 (MAG-33). Two LSDs (landing ships, dock), two AKAs (cargo ships, attack), and three APAs (transports, attack) provided for the ground units. Pulling up alongside the dock at Pusan, the men of the brigade were split into three main units: the 2d Battalion, 5th Marines, on the *George Clymer* (APA-27), known to its passengers as the "Greasy George"; the 3d Battalion on the *Pickaway* (APA-222), with the regimental commander of the 5th Marines, Lieutenant Colonel Raymond L. Murray, on board; and the 1st Battalion on the *Henrico* (APA-45), which came limping into port last after a series of mechanical problems (even though it was known as the "Happy Hank").

Standing on the pier to meet the men was a disparate group of people: General Craig; Marines who had guarded the U.S. Embassy staff in its perilous journey all the way from the South Korean capital of Seoul to refuge in Pusan; some U.S. Army soldiers; a local band giving an earnest but painfully amateurish rendition of *The Marine Corps Hymn*; crowds of curious South Korean on-lookers; and undoubtedly some North Korean spies.

Craig was shocked to see the Marines watching the docking, as they casually leaned over the rails of their ships. He had previously sent an order through Army channels for the brigade to be prepared to march off the ships, combat ready, with weapons loaded. His

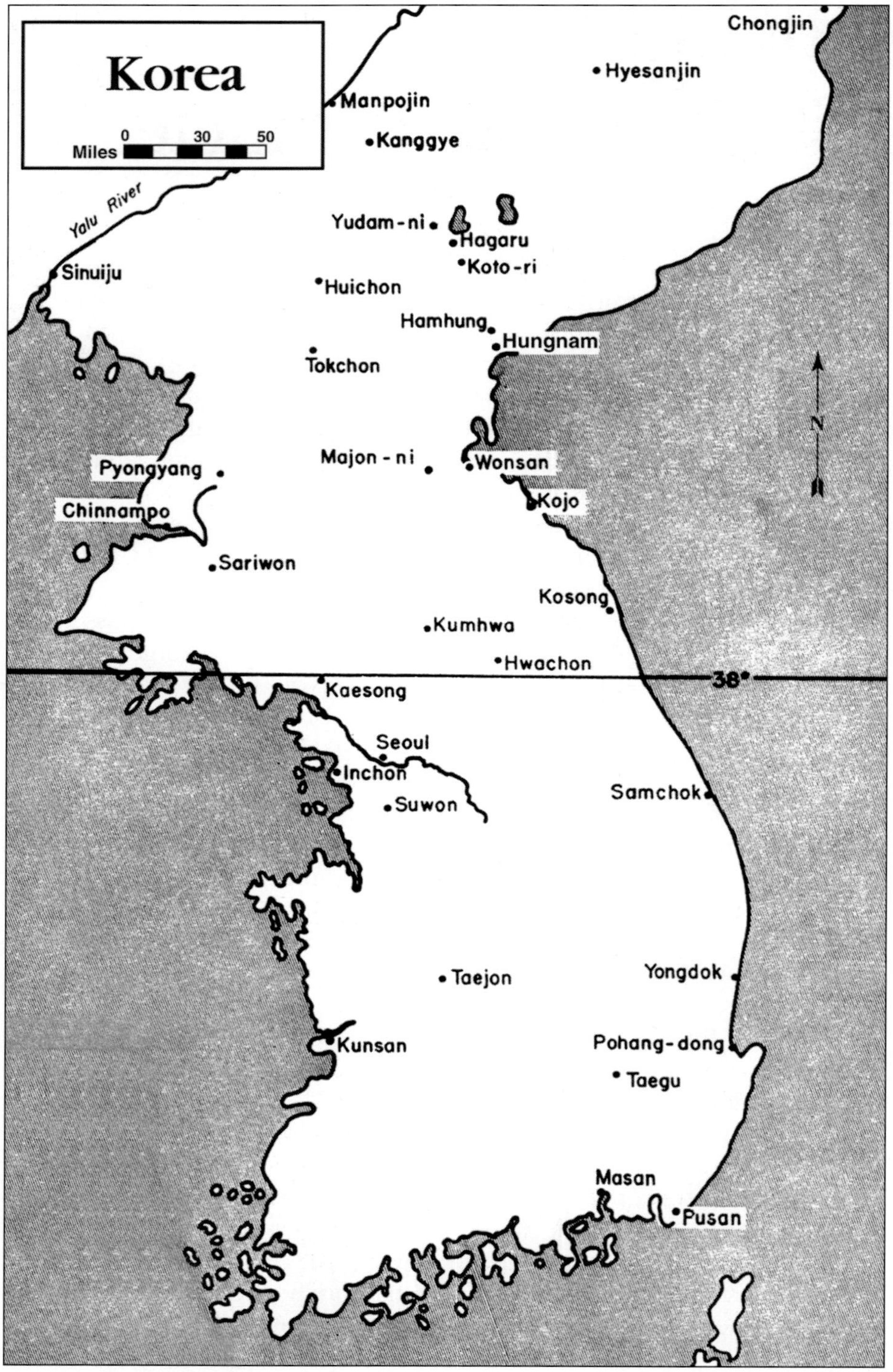

immediate, sharp inquiry to an officer on board revealed that his orders had never been received at sea. Accordingly, Craig immediately convened an officers' conference on the *Clymer*. His G-3, Lieutenant Colonel Joseph L. Stewart, announced that the brigade would move out at 0600 the following morning. This meant the men would spend the whole night unloading the ships and issuing full supplies of ammunition and rations, so that the brigade could move out on time. After making clear that he did not yet know where the brigade would be sent by Lieutenant General Walton H. Walker, the commanding officer of the U.S. Eighth Army in Korea, Craig concluded: "The Pusan perimeter is like a weakened dike and we will be used to plug holes in it as they open. We're a brigade, a fire brigade. It will be costly fighting against a numerically superior enemy. Marines have never yet lost a battle; this brigade will not be the first to establish such a precedent."

After a night of bedlam on the waterfront, 9,400 tons of supplies had been unloaded, but the brigade was to travel light, so most of these supplies and all personal baggage had to be left behind. Thus it was that the brigade was ready to move out on the morning of 3 August.

There was still uncertainty as to exactly where the men would enter combat. Walker's headquarters had telephoned Craig at midnight and told him to move the brigade to a town called Changwon, where Walker would temporarily hold the Marines in Eighth Army reserve. This would position the brigade strategically if Walker decided that his most pressing danger was an enemy breakthrough threat by the NKPA *6th Infantry Division* and the *83d Motorcycle Regiment*. The division was a highly professional, well-trained unit of Chinese Civil War veterans, and it had won a series of smashing victories since the invasion of South Korea a month earlier. Now these units had seized the town of Chinju and were poised to strike at the far southwestern corner of Walker's defense lines. Masan was their next probable target, and that was only 35 miles from Pusan.

The scene on the waterfront that morning was a study in contrasts. On one hand was the panicky atmosphere of the city of Pusan. A Marine officer felt it immediately: "A tension and excitement that was palpable . . . you could sense—

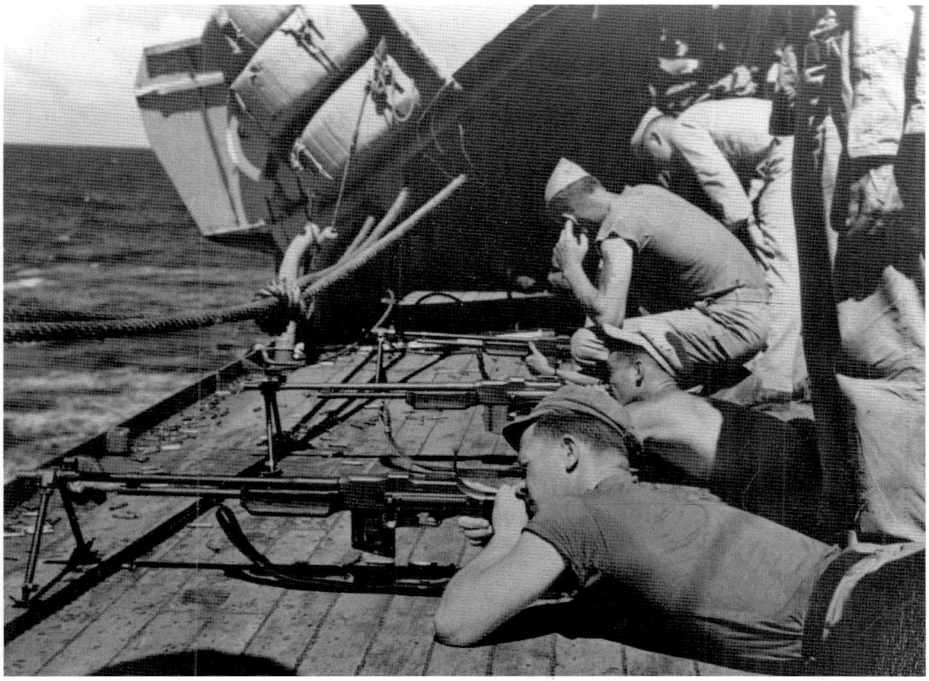

Department of Defense Photo (USMC) A1229

Life on board ship was busy as the Marines prepared for battle. Here three of them are test-firing their Browning Automatic Rifles.

expecting no nonsense, allowing none. And Marine leaders had never lost sight of their primary—their only—mission, which was to fight. The Marine Corps was not made pleasant for men who served in it. It remained the same hard, brutal way of life it had always been.

In 1950 . . . these men walked with a certain confidence and swagger. They were only young men like those about them in Korea, but they were conscious of a standard to live up to, because they had had good training, and it had been impressed upon them that they were United States Marines.

almost feel—fear. The people were scared to death. The North Koreans were very close."

On the other hand, there stood the solid, poised brigade which, with its aviation components, totaled 6,534 men. The three rifle battalions each had only two rifle companies, but, taken from the skeleton 1st Marine Division at Camp Pendleton, was a wide range of auxiliary units: a company each from the division's Signal, Motor Transport, Medical, Shore Party, Engineer, Ordnance, and Tank Battalions; detachments from the Service Battalion, Combat Service Group, Reconnaissance, and Military Police Companies; the 1st Amphibian Tractor Company; and Amphibian Truck Platoon. The 1st Battalion, 11th Marines, with three firing batteries, was also attached to provide the vital artillery support.

These units were permeated with an *esprit de corps* that was unique to the Marines. Author T. R. Fehrenbach had this analysis in his book, *This Kind of War*:

> In 1950 a Marine Corps officer was still an officer, and a sergeant behaved the way good sergeants had behaved since the time of Caesar,

Those young men of 1950 undoubtedly did not know that their predecessors had been to Korea before—four times, in fact. There had been a brief skirmish in 1871 (where the Marines were fired upon by a cannon dated 1313!). Subsequent landings took

Other Marines, in this case members of Company E, 2d Battalion, 5th Marines, huddled intently over instructions in the use of their light machine guns.

National Archives Photo (USMC) 127-N-A1291

National Archives Photo (USMC) 127-N-A1257

A color guard from the South Korean Army, carrying the colors of the United States, the United Nations, and South Korea, joins with a Korean band to greet 1st Provisional Brigade Marines on the dock at Pusan.

place in 1888 and 1894, and in 1905 Marines served as the Legation Guard in Seoul—little dreaming of the ordeal their successors there would undergo 45 years later.

Two things that were prominently visible on the pier were the 3.5-inch rocket launchers ("bazookas") and the M-26 Pershing tanks which equipped the Marines—new weapons that the battered Army divisions lacked.

Invisible, but fundamental to the action that lay ahead, were the qualities that had been ingrained into the Marines themselves. Joseph C. Goulden in *Korea: The Untold Story of the War* described the men this way: "They had been in combat training in the United States; they arrived in cohesive units in which officers and men had served together for months . . . . They insisted on controlling their own air support in coordinated actions based upon years of experience." Another writer, Clay Blair, in *The Forgotten War*, pointed out that "the ranks were filled with physically tough young men who had joined the corps to fight, not to sightsee. The Marines had superior firepower in squads, platoons, and companies."

However, amongst all the units in the Pusan Perimeter there was one point of similarity. Except for senior generals, no one—soldier or Marine—had more than a vague idea of how or why they came to be there in a life-or-death situation in a country of which they had never heard five weeks before.

High-Level Decisions

The actual events that had led up to the brigade being poised on that dock were a tangled skein of high-level meetings, flurries of orders, and long-distance airplane trips that spanned half the globe from New York to Washington, D.C., to California, to Honolulu, and to Tokyo.

It all began when alarm bells went off in the pre-dawn of 25 June 1950 at the United Nations in New York and the U.S. State Department in Washington. There had been a violent, surprise attack across the 38th Parallel, an invasion of South Korea by some 90,000 well-trained, heavily armed soldiers of the North Korean Peoples Army (NKPA). As the star-

Brigadier General Edward A. Craig

Edward A Craig was born on 22 November 1896 in Danbury, Connecticut, and attended St. Johns Military Academy in Wisconsin. After being commissioned in the Marine Corps in August 1917 he served in a wide range of posts: in Washington as aide to Major General Commandant John A. Lejeune in 1926, and in Haiti, the Dominican Republic, China, Nicaragua, and the Philippines, combined with tours on board the aircraft carriers USS *Yorktown* and USS *Enterprise*.

By May of 1942 he had been promoted to colonel, and this brought him command of the 9th Marines. He led his regiment in combat on Guadalcanal in July 1943, then that November on Bougainville where he was awarded the Bronze Star. In December 1943, he was given a temporary promotion to brigadier general. In July-August 1944 his regiment led the attack on Guam. Craig's valiant conduct there brought him a Navy Cross.

Moving to a staff assignment, he served as operations officer, V Amphibious Corps, in the assault on Iwo Jima in February 1945. A Legion of Merit was presented to him for that service.

Duty as assistant division commander, 1st Marine Division, in China in 1947, came with his promotion to permanent brigadier general. Craig then assumed command of the 1st Provisional Marine Brigade in June of that year. This brought him back to Guam, almost three years after he had participated in its recapture.

In 1949, he was transferred to Camp Pendleton as assistant division commander, 1st Marine Division. Very soon thereafter came the attack on South Korea, which led to his designation, for a second time, as Commanding General, 1st Provisional Marine Brigade. This time, however, it moved quickly to combat. When the brigade, after its victories in the Pusan Perimeter, was deactivated in September 1950, its troops were merged into a reformed 1st Marine Division. Craig reverted to his former billet as assistant division commander. For his noteworthy performance of duty during 1950 operations in Korea, he received an Air Medal with gold star, a Silver Star, and a Distinguished Service Medal.

January 1951 brought his promotion to lieutenant general, and a few months later, in June, he retired with 33 years of distinguished service. He died in December 1994.

tling news continued to pour in, it quickly became apparent that the half-trained, lightly armed Republic of Korea (ROK) troops defending South Korea were being smashed and overrun.

At an emergency meeting of the U.N. Security Council that same afternoon, followed by another two days later (27 June), there was a decision to call for armed force to repel the invaders. That was all that was needed for General Clifton B. Cates, Commandant of the Marine Corps, to seize the initiative. The next day, 28 June, he arranged a meeting with Admiral Forrest P. Sherman, Chief of Naval Operations, and Francis P. Matthews, Secretary of the Navy, and he recommended that the Fleet Marine Force (FMF) "be employed." Their reaction was noncommittal, since U.S. ground forces were not yet involved in South Korea. Cates, however, had learned over a long career a few things about combat—whether it was called by a euphemism such as "police action," as in this case, or was realistically a "war," in several of which he had distinguished himself. Acting upon instinct, when he returned from the indecisive meeting, he sent a warning order to the 1st Marine Division at Camp Pendleton, California, to get ready to go to war. (Due to peacetime cuts, the strength of the entire Marine Corps at that time was only 74,279, with 11,853 in the division and its accompanying aircraft wing. Thus, in reality, the "division" was little more than a reinforced regimental combat team in fighting strength, and the "wing" little more than a Marine air group.

Cates' gut reaction was confirmed the following day, on 29 June, when President Harry S. Truman authorized General of the Army Douglas MacArthur, Commander in Chief, Far East, in

Tokyo, to use the U.S. naval, ground, and air units he had available to support the desperate Republic of Korea forces.

Now there ensued examples of the arcane complexities of high-level decision-making at a time of great stress. At Cates' urging, Admiral Sherman asked Admiral Arthur W. Radford, Commander in Chief, Pacific Fleet, how long it would take to ship out a Marine regimental combat team (RCT). Radford replied on 2 July: "load in six days, sail in ten." Then, in a time-honored communications procedure for top-ranking officers, Sherman sent a private message for the eyes of MacArthur via his naval commander in the Far East, Vice Admiral C. Turner Joy, asking if the general would like a Marine RCT. Swamped with bad news from South Korea, MacArthur accepted immediately with "unusual enthusiasm."

Accordingly, he fired off to Sherman in Washington, D.C., that same day (2 July) an urgent radio request for a Marine RCT and a supporting Marine aircraft group (MAG).

A Brigade is Born

Sherman took the request to a meeting of the Joint Chiefs of Staff (JCS) for their decision. Although the Commandant of the Marine Corps was not, at that time, a member of the JCS, Cates felt that, since the decision directly affected "his Marines," he should be involved in it. Showing up uninvited at the meeting, he was allowed to join it in view of the disastrous news from the Korean front.

The JCS voted to commit the Marine RCT and MAG, and with Truman's concurrence, gave MacArthur the good news on 3 July. (Cates later asked Sherman how it had all come about, and the

M-26 Pershing Medium Tank and Its North Korean Counterpart

The M-26 Pershing, shown above, was the backbone of Marine armor during the first half of the Korean War. The 1st Tank Battalion, Fleet Marine Force, at Camp Pendleton, replaced its M-4A3 Sherman tanks with Pershings during the summer of 1950, shortly after the invasion of South Korea.

Company A, 1st Tank Battalion, sailed for Korea with the 1st Provisional Marine Brigade after having been able to test drive and fire only two of its new tanks. While enroute, 14 tanks were damaged when the cargo hold of a ship flooded. Landing at Pusan in August 1950, the tank crews had a brief familiarization period before going into action. In concert with the close-air support of Corsairs, 75mm recoilless rifles, and 3.5-inch rocket launchers, the tanks gave the brigade a level of firepower that proved very effective against the North Korean enemy.

Technical Data

Engine:	Ford V-8 gasoline, liquid-cooled, 500HP
Dimensions:	Length: 20 feet 8 inches Width: 11 feet 4 inches Height: 9 feet 1 inches
Weight:	46 tons
Maximum Speed:	On roads: 30mph Cross country: 18mph
Radius of Action:	On roads: 92 miles Cross country: 62 miles
Crew:	5
Armament:	One 90mm M3 gun Two .30-caliber machine guns M1919A4 (bow & coaxial) One .50-caliber M2 machine gun on turret

T-34 North Korean Medium Tank

After great success early in the war and acquiring a fearsome reputation, the T-34, not shown, met its nemesis in the Marines' anti-tank weapons. Supplied from Russian stocks, it weighed 32,000 kilograms and carried a crew of five men. A V-12 diesel engine gave it an off-road speed of 30 kilometers per hour. Armament was an 85mm gun, supplemented by two 7.62mm machine guns.

admiral replied in a baseball metaphor: "From Cates to Sherman, to Joy, to MacArthur, to JCS!")

Now Cates (and the Marine Corps) had to deliver. On 7 July he had the 1st Provisional Marine Brigade activated, but then a monumental effort, carried out at a frantic pace, was needed to assemble the essential manpower, equipment, and weapons—and to do all that in one week flat!

The initial building blocks were there, the 5th Marines at Camp Pendleton and MAG-33 at the nearby El Toro Marine Air Station. The critical manpower problem was to flesh out these units from their peacetime reductions so that they could fight with maximum effectiveness. By telegraph and telephone orders went out to regular Marines all over the country: "get to Pendleton NOW!" And so they came pouring in day and night by bus and plane and train, a flood of men from 105 posts and stations.

Marine Corps Historical Center Photo

General Clifton B. Cates

Captain "Ike" Fenton long remembered the ensuing problems:

> These men were shipped from the posts and stations by air, most of them arriving with just a handbag. Their seabags were to be forwarded at a later date. They didn't have dog tags and had no health records to tell us how many shots they needed. Their clothing generally consisted of khaki only, although a few had greens.
>
> They had no weapons and their 782 equipment was incomplete. We had a problem of trying to organize these men into a platoon and getting them all squared away before our departure date.

Other officers recalled odd aspects of those hectic days: no one got any sleep; some men were detailed to help in the filming of

The 3.5-Inch Rocket Launcher

The 3.5-inch rocket launcher (also known as the "Super Bazooka") offered the infantryman a portable rocket weapon, designed to be used as an anti-tank defense. Introduced in early 1950, the 3.5-inch launcher gave the Marine Corps the means to pierce any armored vehicle from a greater distance than previous launchers, and with improved accuracy.

The M20 3.5-inch launcher appeared after its predecessor, the 2.36-inch launcher (developed in World War II), proved ineffective against the Russian tanks in Korea. The M20 was a two-piece, smooth-bore weapon formed by connecting the front and rear barrels together. Weighing only 12 pounds with an assembled length of 60 inches, the 3.5-inch launcher was easily transportable and could be fired from a standing, sitting, or prone position. The M20 rocket had a "shaped charge" that concentrated the force of the explosion on a very small area, thus allowing the projectile to penetrate armor plate as thick as 11 inches. In addition to the weapon's deadly force, a unique gunsight that allowed for various ranges and speeds provided the 3.5 an accuracy up to 900 yards.

Hollywood's "Halls of Montezuma"; other men were detailed to fight fires in the Santa Margarita Mountains. The supply crisis was overcome by a precipitous change from an attitude of "counting shoelaces" to "take whatever you need." Acquisition of the new 3.5-inch rocket launcher was made possible by shipments from all over the country. There was an influx of senior staff noncommissioned officers, some of whom were physically unfit, and this led to some sergeants major being assigned to ride shotgun on ambulance jeeps. A number of World War II officers with no infantry experience also arrived.

The 5th Marines was beefed up by an emergency authorization to add a third platoon to each of the two rifle companies, but it proved impossible to get a third rifle company for each of the three battalions. This was a serious shortage. It meant that the battalions would have to go into battle without a company they could use for maneuver or have in reserve. And that would cause extra casualties in the weeks ahead. There were, however, two compensating factors for the shortages. First, the regiment was a well-trained, cohesive unit. Murray put it this way: "We had been extremely lucky, in the previous year we had virtually no turnover . . . so that we had a regiment which, for all intents and purposes, had been together for a full year, training."

Second, 90 percent of the brigade's officers had seen combat before on the bloody beaches and in the jungles of the Pacific. This was also the case for two-thirds of the staff noncommissioned officers. Here was a group of leaders well prepared for the rigors of combat.

With the addition of supporting units hastily assembled, a rein-

Lieutenant Colonel Raymond L. Murray

Born 30 January 1913 in Los Angeles, California, Murray grew up to attend Texas A&M College. While there he was enrolled in the Reserve Officer Training Course. Graduating in 1935 with a bachelor of arts degree, he did a short stint in the Texas National Guard and then was commissioned in the Marine Corps on 1 July. After Basic School, he was ordered to duty in China, 1937-1940. A radical change of scenery led to an assignment as a captain with the 1st Provisional Marine Brigade in Iceland, 1941-1942.

Moving overseas in November 1942 with the 2d Division, he was awarded the Silver Star in January 1943 for his service as Commanding Officer, 2d Battalion, 6th Marines, on Guadalcanal. Now a lieutenant colonel, he took his battalion on to Tarawa in November 1943, where he received a second Silver Star. This was followed by exploits on Saipan that brought him a Navy Cross and a Purple Heart in June 1944.

The years after World War II saw Murray in a variety of peacetime Marine Corps duties, leading to his taking over in July 1950 as Commanding Officer, 5th Marines (a billet normally reserved for a full colonel). When his regiment became the core of the 1st Provisional Marine Brigade in Korea, and then was a key unit in the Inchon-Seoul battles, he again distinguished himself in combat and was awarded his third Silver Star, a fourth one from the Army, and a Legion of Merit with Combat "V" in August and September 1950.

Further combat at the Naktong River, Inchon, Seoul, and the Chosin Reservoir brought a second Navy Cross and an Army Distinguished Service Cross.

In January 1951, after nearly eight years as a lieutenant colonel, he was promoted to full colonel, and then, after a sequence of duties in Washington, Quantico, and Camp Pendleton, to the rank of brigadier general in June 1959. This led to his assignment as Assistant Division Commander, 3d Marine Division, on Okinawa. Promoted to major general in February 1963, he saw duty as Deputy Commander, III Marine Amphibious Force, in Vietnam in October 1967.

After 33 years of highly decorated active duty, Murray retired in August 1968 as a major general.

forced RCT came into being. Such a unit normally would be commanded by a full colonel, but this case was different. As then-Lieutenant Colonel Murray later recalled:

> I was sure that a colonel would be brought in. It wasn't until sometime later when I was talking to [Major] General [Graves B.] Erskine [commanding the 1st Marine Division] . . . and he told me that when this broke, General Cates told him, "I'll get you a colonel as soon as I can to get out and take the regiment," and General Erskine said he told General Cates, "Don't need one. I've got somebody who can take the regiment."

Along with the manpower problem came materiel problems. The peacetime economies forced on all the military Services by political decisions in Washington had hit hard the resources of equipment, supplies, and weapons. The Marine Corps, however, had an ace up its sleeve for just such a high pressure, short-deadline situation as this.

Tucked away in the California desert was the huge Marine Supply Depot at Barstow. It had been filled five years earlier by following a prudent, far-seeing policy that countless past emergencies had taught the Marine Corps: "When you get a chance to stock up, do it, because you'll never know when you'll really need it!"

Thus, at the end of World War II, Marine salvage teams had looked around the Pacific islands for abandoned equipment. Then they brought it back to Barstow, repainted it "Marine green," stenciled "USMC" on it, and "mothballed" it for future use. From this treasure trove came the old jeeps, the old trucks, and the old amphibian tractors that would be so vital to the brigade's operations. Brand new, however, were the M-26 tanks with their 90mm guns. The Marines in the 1st Tank Battalion had trained in a different, older tank with different armament, and their race to switch over, train, and prepare for embarkation was typical of the pressure to which all hands were subjected. (Each tanker got to fire exactly two rounds before departure.)

When a brigade goes to war, it needs a lot of supplies and equipment. Here Marines labor on the dock at San Diego to load up the ships that will take them to war.

National Archives Photo (USMC) 127-N-A1063

It was the same frantic scene at El Toro as MAG-33 struggled to get its aviators and planes up to combat readiness. As with the ground troops, the organizational units were mere peacetime skeletons. Thus the 1st Marine Aircraft Wing (1st MAW) was a wing in name only, and had to be stripped bare just to give MAG-33 what it needed.

Adding a wholly new resource were "the first helicopter pilots of the United States Armed Forces to be formed into a unit for overseas combat service." They came from Quantico, Virginia, where, since 1947, the Marine Corps had pioneered helicopter combat techniques. On their arrival, there were just 48 hours to join up the four HO3S-1 helicopters with the four usable OY-2 observation planes, and have Marine Observation Squadron 6 (VMO-6) ready to ship out.

Somehow, it was done under the unbelievable pressure of time, and the brigade air-ground team was ready to sail on schedule. There was one final vignette that exemplified the morale of the men. A reporter-photographer, David Douglas Duncan, in his book, *This is War!*, described a scene where General Craig had spoken to a mass meeting of his men just before they went on board ship. When they heard they were headed for Korea and Craig referred to the traditional Marine role, "the men were dead-panned . . . expressionless." But Duncan continued:

> Then Craig, with his Brigade Surgeon standing at his side, told his men that as long as there were any Marines alive in Korea who could still fire a rifle, or toss a grenade, no other Marines would be left behind upon the battlefield, either wounded or dead. Over four thousand men shouted in unison as his Leathernecks gleefully slugged each other in the ribs, grinned happily and wanted to know when the hell they were going aboard ship.

On 14 July the ships left San Diego, taking Marines to combat once more.

Preparing the Way

With the troops enroute at sea, General Craig and Brigadier General Thomas J. Cushman had boarded an airplane and flown to Hawaii. There they met with Lieutenant General Lemuel C. Shepherd, Jr., Commanding General, Fleet Marine Force, Pacific (FMFPac). Craig underscored the painful shortage of rifle companies; the missing 105mm howitzers in his artillery, the 1st Battalion, 11th Marines; and his lack of motor transport.

Flying on to Tokyo on 19 July, the two Marine generals went quickly to meet General MacArthur. Craig made his feelings very clear:

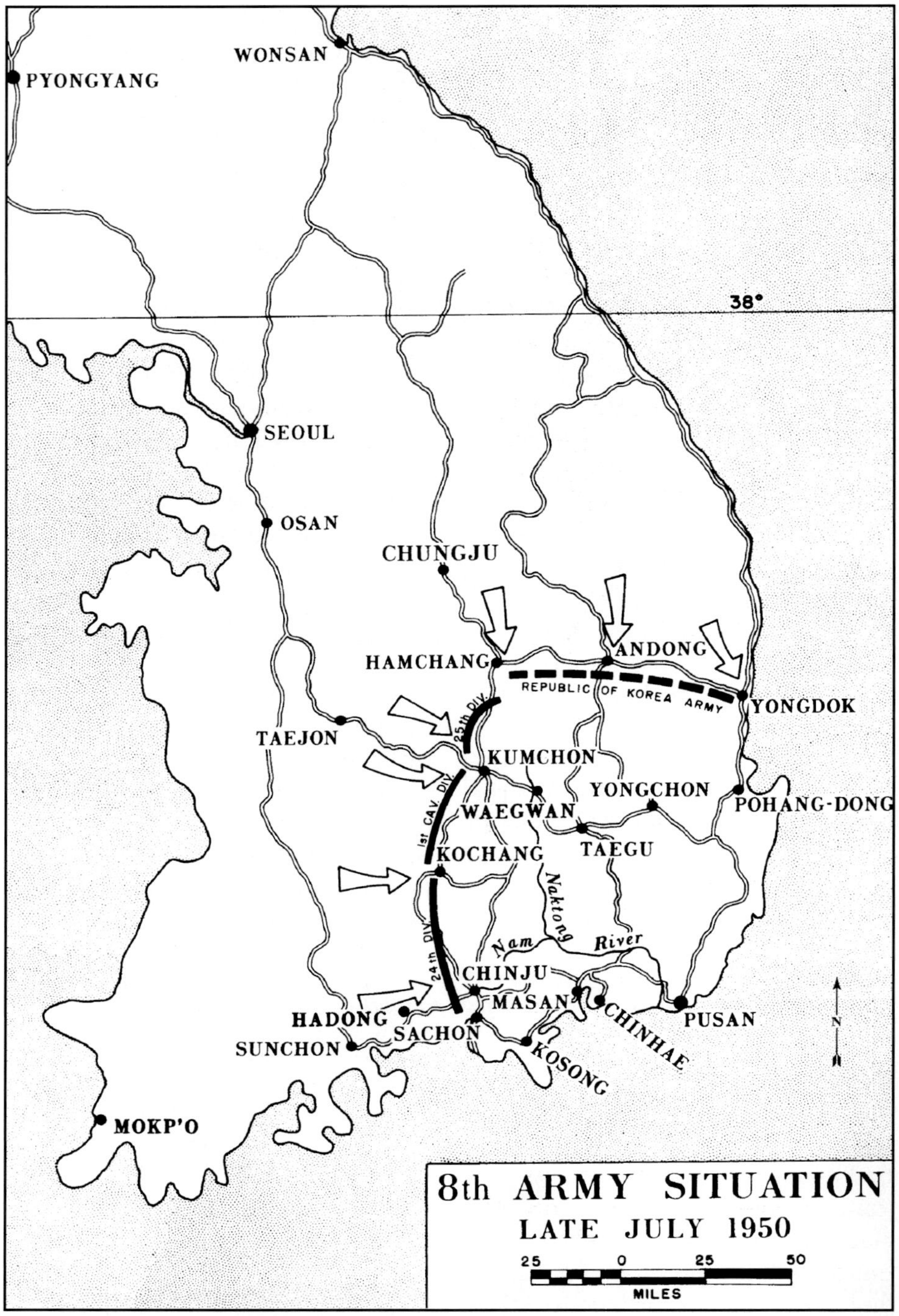

8th ARMY SITUATION
LATE JULY 1950

> While talking to General MacArthur, I informed him that we were on a peace-strength basis, that we were an air-ground team and had trained as such at Pendleton and would be very effective if left intact. However, I told him that if they took our air force away from us, our fighting potential would be cut about 99 percent as far as I was concerned.

MacArthur went on to assure Craig that the Marines could retain their planes, and he so informed Lieutenant General George E. Stratemeyer, commander of the U.S. Far East Air Forces. This was a great relief to Craig, who later stated that "Stratemeyer was very anxious to get Marine air under his command as soon as they arrived in that area."

The discussion continued on a harmonious note, with MacArthur saying, "I'm very glad to have you here with the 1st Brigade." When he learned of Craig's manpower shortages, he directed that a dispatch go to the JCS requesting full war strength for the brigade. (During this time, messages continued to fly back and forth regarding

National Archives Photo (USMC) 127-N-A1185

The men of the 1st Provisional Marine Brigade have landed in Pusan, Korea, and are marched off to combat.

the mobilization of the full 1st Marine Division for a future campaign that MacArthur was already planning. This led to the call-up of Marine Reserves on 19 July.)

The meeting ended with a directive from MacArthur to set up billets for the brigade in Japan. It was not to be. The situation in Korea had degenerated to a near-collapse. U.S. Army troops had been rushed from comfortable occupation duty in Japan to bolster the reeling ROK divisions. Things had gone badly—very badly. The official Army history recounts a continuous series of problems: tanks ambushed, sentries asleep, soldiers killed while riding in trucks instead of marching, repeated retreats, communication breakdowns, etc. The history characterizes the situation at the time the Marine brigade arrived by stating: "Walker was concerned about the failure of his troops to carry out orders to maintain contact with the enemy." Overall, it summarized the crisis in stark language: "Never afterwards were conditions as critical . . . . Never again did the North Koreans come as close to victory."

Faced with this situation, Walker, as ground commander, had withdrawn all the troops into a last-stand enclave called the Pusan Perimeter.

This was a 60-by-90-mile rectangle with the Sea of Japan on the east, the Korean Strait on the south, the Naktong River on the west, and a line of mountains on the north. It did have one advantage crucial to Walker. This was his ability, in this constricted area, to use his interior lines of movement to set up a final defensive perimeter with the capacity to rush emergency reinforcements to quell any serious enemy threat where a breakthrough seemed imminent.

With the whole beachhead on the Korean Peninsula now in such peril, Craig received new, urgent orders on 25 July: the brigade would go straight to Korea to serve as Walker's "fire brigade" where most needed. The next day Craig was in Taegu, Walker's headquarters in South Korea. He used his stay there to absorb all possible information on the fluid situation on the front lines—including a careful aerial survey he made of sites where his brigade might be thrown into action.

On 30 July, Craig headed for Pusan, set up a temporary command post, and wrote out a preliminary operations order for the brigade as the NKPA tide rolled over Chinju and headed for nearby Masan. Arrangements were made with MAG-33 in Japan to be ready for action the moment the brigade arrived on board its transports.

The next day, still without a decision by Walker on the deployment of the brigade, Craig sensed the threat to Masan, looming such a short distance from Pusan, as a probable priority. Accordingly, he decided to supplement his previous aerial view with a ground reconnaissance by jeep. Then he waited tensely for his brigade to arrive.

It came 2 August; it moved out 3 August. One historian, Donald Knox, crystallized that moment in

National Archives Photo (USN) 80-G-416920

The F4U Corsairs of VFM-214, VMF-323, and VMF(N)-513, with a helicopter of VMO-6, jam-pack the deck of the USS Badoeng Strait *enroute to Pusan.*

his book, *The Korean War*:

> The fluid situation the brigade would encounter in the Pusan Perimeter would demand the very elements the Marines had in abundance—courage, initiative, élan . . . . morale in the rifle companies was extremely high. In spite of what they'd heard, the Marines knew the North Koreans could be beaten. The Marine Corps was sending to Korea the best it had.

The Fire Brigade Goes to War: Crisis Number One

It was an early start; at 0600 on 3 August the "fire brigade" moved out to meet head-on the most urgent enemy threat. It went with a ringing message from Cates: "The proud battle streamers of our Corps go with you in combat. The pride and honor of many generations of Marines is entrusted to you today. You are the old breed. With you moves the heart and the soul and the spirit of all whoever bore the title United States Marine. Good luck and Godspeed."

Part of the men (1st Battalion) went by truck to the staging area of Changwon. Since the Marines had been forced by a shortage of shipping to leave their heavy equipment back in the United States, the transportation was made possible by borrowing two Army truck companies, with an additional bonus in the form of a loan of communication jeeps and reconnaissance company jeeps with .50-caliber machine guns. Going by train were the precious tanks and some of the men. Duncan, the reporter, described what those kind of trips were like:

> The first stage of moving up to the front was no prob-

Loading up the old Korean railroad cars, the men have their packs full, and carry entrenching tools, as they head for their Changwon staging area.

National Archives Photo (USMC) 127-N-A1181

lem, but it was slow. The troop trains were sturdy, wooden-bodied old coaches, leftovers from the days when the Japanese had run the country. . . . The Marines inside showed almost no interest in the slowly passing scenery. They ate their rations, oiled their weapons, slept in the vestibules between the cars with their rifles held close. They were professional men riding to work.

The planes of MAG-33 had a busy time that same day of 3 August. Under the command of General Cushman were the fighter squadrons VMF-323 ("Death Rattlers") under Major Arnold A. Lund and VMF-214 ("Black Sheep") commanded by Lieutenant Colonel Walter E. Lischeid. They were equipped with 60 of the gull-winged Corsair F4Us. One of their partners was Marine Night Fighter Squadron 513 (VMF[N]-513) ("Flying Nightmares") under Major Joseph H. Reinburg. This was a squadron specially trained for night fighting with its F4U-5N Corsairs and new twin-engine F7F Tigercats. The other partner was VMO-6, commanded by Major Vincent J. Gottschalk, with its four usable OY-2 light observation planes and, for the first time in real combat for any U.S. Service, four Sikorsky HO3S-1 helicopters.

When the ground elements of the brigade were unloading in Pusan, MAG-33 had been in Kobe, Japan. From there, VMF-323 had gone on board the *Badoeng Strait*, while VMF-214 was based on the *Sicily* (CVE-118). VMF(N)-513 was based at Itazuke Airfield on Kyushu Island. Marine Tactical Air Control Squadron 2 (MTACS-2) traveled by ship to Pusan. VMO-6 amazed the Japanese citizens

Brigadier General Thomas J. Cushman

Born on 27 June 1895 in Saint Louis, Cushman graduated from the University of Washington in Seattle and subsequently enlisted in the Marine Corps in July 1917. Commissioned in October 1918, he received his naval aviator wings the following year. Duty in Guam, Nicaragua, and Haiti followed the diverse Marine aviation pattern of the 1920s.

Next, in June 1933, came a tour in the Bureau of Aeronautics of the Navy Department, and then, broadening his interservice experience, he attended the Army Air Corps Tactical School in 1935. With the commitment of Marine aviation in World War II, Cushman was appointed Chief of Staff, Marine Aircraft Wings, Pacific. With a temporary rank of brigadier general in January 1944, he was next assigned as Air Defense Commander, Marianas Islands. For these services he was awarded a Bronze Star and the Legion of Merit with Combat "V."

When his rank was made permanent in 1947, he became Commanding General, Aircraft, Fleet Marine Force, Pacific, the following year. With the outbreak of the Korean War, he was assigned as Assistant Wing Commander, 1st Marine Aircraft Wing, in June 1950. With the forward echelon of the wing, he provided the air support for the Marine Brigade when it went to Korea. In 1951 he took command of the wing. His leadership there brought him his second Legion of Merit and a Distinguished Service Medal.

His final billet was Deputy Commander, Fleet Marine Force, Pacific, as a major general in 1953, and, after promotion to lieutenant general in 1954, he retired. He died in July 1972.

when it simply took off in its light observation planes and helicopters from the streets of Kobe. Four of its helicopters and four of its OY planes made the short hop to Pusan on 2 August, so they were there, ready to go with the brigade, even though they had not been visible in that memorable scene on the waterfront.

VMF-214 launched an eight-plane flight from the *Sicily* on 3 August and pummeled Chinju with incendiary bombs, rockets, and strafing, a small preview of what the Marines had in store for the NKPA *6th Division*. This attack took place less than a month after the receipt of official orders sending the planes to the Far East. (An even earlier mission—the first for Marine planes—had been on 4 July when two F4U Corsair photographic planes from MAG-12 on the carrier *Valley Forge* (CV-45) had joined in a Navy air strike against the North Korean capital of Pyongyang.)

On a succession of those early August days, all three of the Marine fighter squadrons kept up a steady pattern of bombing, strafing, and rocketing attacks on NKPA targets. On 5 August, for instance, Major Kenneth L. Reusser led a four-plane division of Corsairs to Inchon, the port of the South Korean capital of Seoul. There he was responsible for the discovery and the destruction of an enemy tank assembly plant, an oil refinery, and an oil tanker ship. The two Corsairs which Reusser flew on two successive strikes during his attacks of that day were severely damaged by enemy fire. He was awarded a gold star in lieu of a second Navy Cross for his heroism on this mission.

VMO-6 was also busy. The squadron had moved west from Pusan to Chinhae, a base close to the threatened city of Masan and

Vought F4U-4B Corsair

The familiar Vought F4U Corsair emerged out of World War II synonymous with American victory in the Pacific and became the aircraft most closely associated with Marine Corps aviation. The Corsair was a versatile, tough, and heavy fighter-bomber and night-fighter, and was easily recognized by its distinctive inverted gull wings. At the conclusion of the war, Vought's concentration was in the limited production of the F4U-4 models, producing 2,356 up to 1947.

The 4B model was equipped with a 2,100 horsepower engine of the Pratt and Whitney R-2800-18W type. The aircraft had a top speed of almost 450 mph, a climb rate of 3,870 feet per minute, and a range of more than 1,000 miles. Operational altitudes could be reached as high as 41,500 feet. Standard armament for the 4B were the awesome six .50-caliber machine guns, and a payload capability of eight 5-inch rockets and up to 4,000 pounds of ordnance.

When the 1st Provisional Marine Brigade made its entry into the Korean War, supporting the Marines on the ground were both Navy squadrons and, in particular, the Marine units from the carriers *Badoeng Strait* and *Sicily*,VMF-214, better known as the Black Sheep Squadron, and VMF-323, the Death Rattlers Squadron.

Starting on 7 August 1950, VMF-214 and -323, both of which had effectively absorbed the lessons of close air support during WWII, provided the brigade support by having four to 10 Corsairs continuously overhead. Flying a total of 6,575 combat support missions, the favorite ordnance carried for close air support missions was napalm, deadly jellied gasoline that was most effective against NKPA armor. The Corsairs usually carried two 150-gallon napalm bombs that weighed approximately 1,400 pounds apiece.

During the month of August, the close air support missions from *Badoeng Strait* and *Sicily* gave everyone a lasting impression. Observing from the ground, said an Army soldier of the Marine aviators: "The effectiveness of Marine close air support astonished Army troops fighting alongside the Leathernecks." On 18 August, several hundred NKPA fell under the Death Rattler's Corsairs' merciless air assault that pounded their retreat across the Naktong River.

National Archives Photo (USMC) 127-N-A130913

The eyes of the brigade: the OY light observation plane was invaluable in the rugged terrain and endless hills.

the brigade's forthcoming zone of action. This location had been a South Korean naval base and ammunition depot, but it had a 2,600-foot airstrip with two completed hangars and quonset huts for housing. So VMO-6 set up quickly for business.

Craig took off early on 3 August in one of its helicopters and put in a remarkable day that demonstrated the amazing versatility and usefulness of the new aircraft. He stopped to give instructions to the lead battalion on the march; he then selected a site for his forward command post (CP); and he then flew to Masan to confer with Walker and Major General William B. Kean, USA, commander of the 25th Infantry Division, to which the brigade would be attached. Finally, on his return trip, Craig landed three more times to meet with his unit commanders.

Craig's own later evaluation of this mobility was very specific. After noting that fast travel by jeep was often impossible due to traffic-clogged roads, considerable distances to his objectives, and frequent tactical moves, he contrasted these impediments with his obligations. These included conferences with Army generals, the need to return to his CP to issue orders, then to observe his Marines in the field, as well as the requirement that he reconnoiter the terrain before operations began. He then commented: "My staff faced the same problems. Time was always pressing. Fortunately, Marine helicopters attached to VMO-6 were always available for observation, communications, and control. These aircraft made my day! Without them I do not believe we would have had the success we did."

The squadron's OY-2 light planes were equally useful on that day as they flew convoy for the brigade and made reconnaissance flights over the staging area, looking for any signs of enemy infiltration. This proved so successful that VMO-6 set up a regular procedure to have an OY over the brigade area at all times during daylight hours. To provide this non-stop support, there were shifts with a new plane, new pilot, and new observer coming in relays every two hours. Similarly, two helicopters went every morning to the brigade CP, to be relieved at noon by two others.

This new element of air mobility proved to be a vital asset to the ground troops. Craig pointed out that "maps were poor, and no one in the brigade had personal knowledge of the terrain over which we were to fight. Helicopters were a life saver in this connection, as they provided the means for even commanders of small units to get into the air quickly from almost any point and identify roads, villages and key points prior to moving their troops." The helicopters soon were employed for a wide variety of additional missions: evacuating the wounded; transporting supplies to inaccessible hill peaks; scouting enemy locations; and rescuing downed fighter pilots.

Of course, the NKPA was quick to open fire whenever it spotted one of the helicopters on the ground. Duncan, the reporter, was again on the spot for one typical episode. He was cutting across one of the rice paddies to where an aircraft sat with rotor blades kept spinning for a fast take-off. General Craig emerged from that helicopter, checking the disposition of his troops. As the reporter looked closely at him, a conviction grew: "I knew that [he] could take anything that Korea could hand out."

Duncan's account continues: "Suddenly that old familiar bucket-swinging swoosh cut out all other sounds and two mortar bombs dropped into the riverbed. Great geysers of mud and gravel mixed with red-hot fragments shot into the sky. So did the helicopter. Before another bracket of bombs could fall the aircraft was halfway down the valley, General Craig was in his jeep headed for his CP on the mountainside."

With the full brigade concentrated at Changwon by the late afternoon of 3 August, Craig faced a very uncertain situation. Although he had been ordered into a "bivouac" status as Eighth Army reserve, he was wary, for his Changwon location was very close to a vital road junction at Chindong-ni where heavy fighting was taking place. With the perimeter shrinking at an alarming rate and an NKPA envelopment from the west headed straight for Pusan, Craig decided:

> We felt that going into bivouac would leave us wide open for surprise. To ensure our security and be prepared for any eventuality, I deployed the brigade tactically.
>
> Although a little trigger-happy, we were ready for combat, even though situated behind the so-called front-lines. During the few days we were at Changwon, we knew we were observed by enemy observation posts and patrols off on the flank. They did not bother us. A major penetration of the U.S. Army lines at Chindong-ni could have been fatal to us if we had been caught in bivouac.

The general's reference to "a little trigger-happy" was an understatement made some time later, for the first night they were anything but professional. In pitch darkness, with thoughts of enemy infiltration making some of the men tense, nervous firing broke out among the Marines.

Although there were varying opinions of how widespread the firing was, one private first class named Fred F. Davidson, later recalled:

> I raised my carbine and squeezed the trigger. The muzzle flash blinded me. For the next few seconds I saw

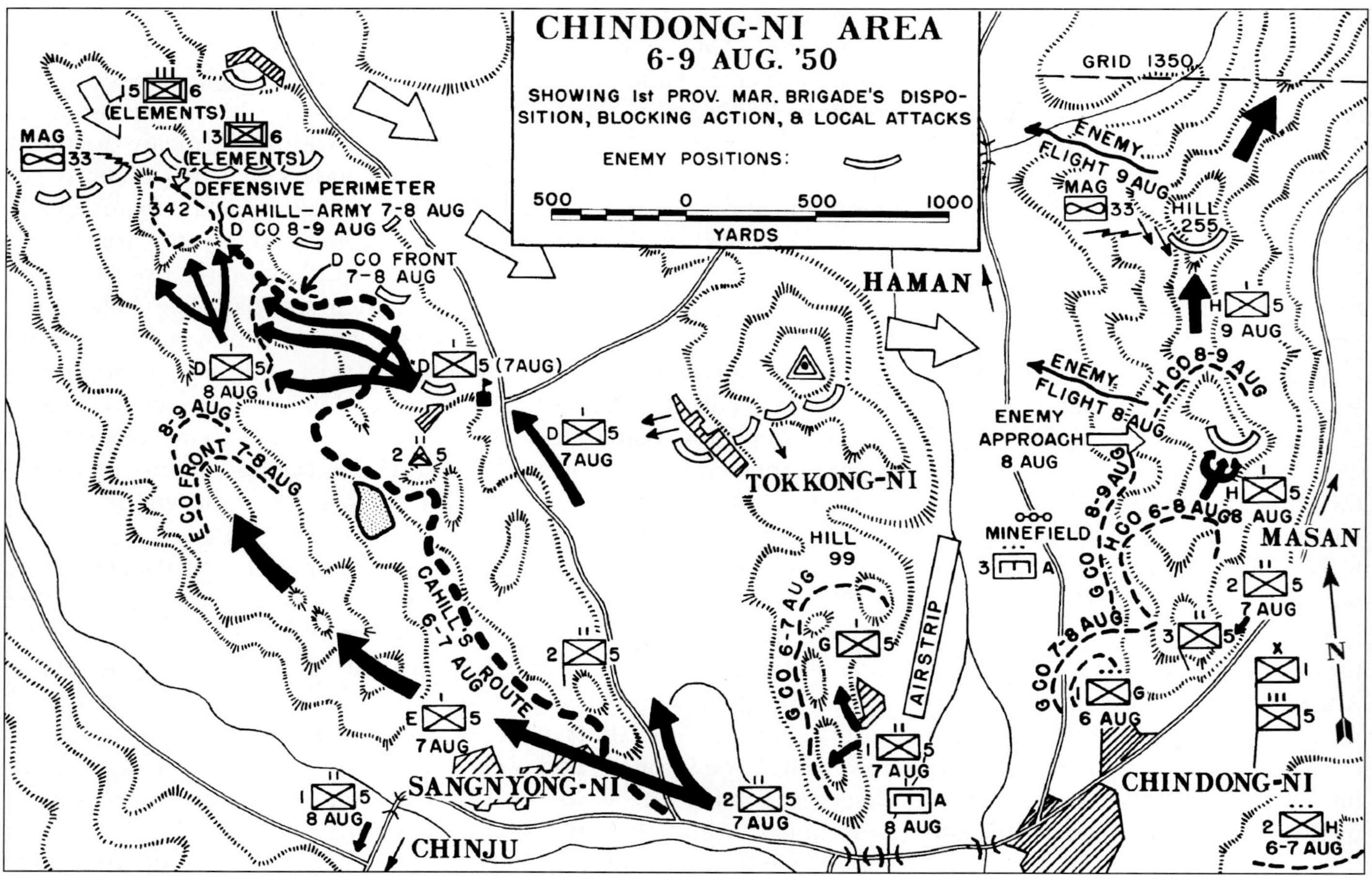

Weapons

A Marine in a rifle company had a wide variety of weapons that he could use himself, or that were available in other units to support him. As always, his basic weapon was his rifle.

U.S. Rifle, .30-Caliber, M1

The .30-caliber M1 rifle was a gas-operated, clip-fed, air-cooled, semi-automatic weapon. It weighed 9.5 pounds, had an average rate of aimed fire of 30 rounds per minute, a muzzle velocity of 2,600-2,800 feet per second, and a bullet clip capacity of eight rounds. Inherited from World War II, the M1 provided strong and accurate firepower for the rifleman.

U.S. Carbine, .30-Caliber, M1

The .30-caliber M1 carbine was a gas-operated, magazine-fed, air-cooled, semi-automatic shoulder weapon. The weight was only 5.75 pounds. Eight inches shorter than the M1 rifle, it had a muzzle velocity of just 2,000 feet per second, and a magazine capacity of 15 rounds. This size and weight led to its issuance to officers, although it lacked the hitting power of the M1.

Automatic Pistol, .45-Caliber, M1911A1

Regardless of what was officially prescribed, a number of Marines carried a .45-caliber automatic pistol in Korea. This was a time-honored weapon featured in the lore of the Corps. Described as a recoil-operated, magazine-fed, self-loading hand weapon, the .45-caliber weighed 2.76 pounds when fully loaded, was 8.59 inches in length, and had a capacity of seven rounds. The muzzle velocity was 802 feet per second, while the maximum effective range for the troops using it was only 25 yards. In close combat, it often proved invaluable.

To furnish a high volume of direct fire in support of the rifle platoons, there were three types of automatic weapons.

Browning, .30-Caliber, M1919A4

The .30-caliber "Browning" light machine gun was a recoil-operated, belt-fed, air-cooled weapon. It weighed 31 pounds, but with its tripod mount that rose to 49.75 pounds. While the "cyclical" rate of fire was 400-550 rounds per minute, the "usable" rate was really 150 rounds per minute. Muzzle velocity varied between 2,600 to 2,800 feet per second, depending on the cartridge used.

Browning, .30-Caliber, M1917A1

The "Browning" .30-caliber water-cooled "heavy" machine gun was extensively used in the battle for Seoul and in the trenches at the end of the war. Its effective rate of fire was 350-450 rounds per minute. With a muzzle velocity of 2,800 feet per minute, in direct fire its maximum effective range was 3,000 yards. This dropped to 300 yards for indirect fire. Its length was 38.5 inches. "Heavy" was an accurate term, since the gun alone weighed 41 pounds, and its tripod added another 53 pounds. Each ammunition belt contained 250 rounds.

Browning Automatic-Rifle, .30-Caliber, M1918A2

As a mainstay of the rifle squad, the "B-A-R" (as it was always called) was an air-cooled, gas-operated, magazine-fed, shoulder weapon. Weighing 20 pounds, it had a magazine capacity of 20 rounds. The man using it carried still more weight in the magazine pouches on his web belt. Although maximum range could be 5,500 yards, its effective range was 500 yards. There were two cyclical rates of fire for the BAR-man to choose: slow, 350, and normal, 550.

Two other specialized weapons were invaluable for the Marines during Pusan and the subsequent street fighting in Seoul. Against North Korean tanks, strong points, and snipers in buildings, they were deadly.

3.5-Inch Rocket Launcher

Familiarly call the "bazooka," the rocket launcher fired an 8.5-pound rocket with a hollow-shaped charge in its head. It weighed 15 pounds and was usually handled by a two-man team.

75mm Recoilless Rifle

The 75mm recoilless rifle fired conventional shells in a flat trajectory, weighed 105 pounds, and required a tripod in use. Its effective range was 1,000 to 2,000 yards.

Mortars

The 60mm mortar was a smooth-bore, muzzle-loading, high-angle-fire weapon used by a rifle company. With its base plate and bipod support, it weighed 42 pounds. Normal rate of fire was 18 rounds per minute, using either high explosive, white phosphorous, or illuminating shells. These had ranges varying from 1,075 to 1,985 yards.

The 81mm mortar could fire a 6.8 pound high explosive shell up to 3,290 yards. Its weight, combining barrel, base plate, and tripod totaled 136 pounds. Elevation could be varied from 40 to 85 degrees.

The 4.2-inch mortar, affectionately referred to as the "four-deuce," fired a round with more explosive power than a 105mm howitzer. Equivalent to 107mm in caliber, it could fire a 25.5-pound shell up to 4,400 yards. Total weight was 330 pounds.

105mm Howitzer M101A1

As a light, towed field artillery weapon, this was used in direct support of infantry units. A battalion had three batteries with six howitzers each. Weighing 4,950 pounds, the cannon fired a 33-pound shell to a maximum range of 11,000 meters. While usually moved by a truck, a heavy helicopter could also carry it. The 105 needed only three minutes to emplace and could sustain a rate of fire of three rounds per minute.

lights and stars. Andy shouted, "Hey, you almost hit me!" Oh, God, I didn't know I was aiming in that direction. It was so dark I couldn't see my front sight. I said to myself, "You better take it easy, ol' buddy, before you kill some Marine." Over to my rear someone else pulled off a round. Next it was someone to my front.

Then the firing pinballed from place to place all over the hill and back down toward the railroad track . . . . Finally . . . all firing ceased . . . . The rest of the night I lay awake, scared, my finger on the trigger.

The brigade's stay at Changwon was brief but useful. The rifle units got a pithy lecture about fire discipline and conducted patrols to the high ground beside them—a foretaste of the endless hill climbs ahead. The tank and artillery units had a opportunity at last to do some training in firing their weapons, and the Reconnaissance Company started its probing operations. Firm communications were set up with the fighter squadrons afloat.

Craig made two trips to Masan for planning meetings with Walker and Kean, and late on 5 August the brigade got the word to be prepared to move out by truck the next day to Chindong-ni with action to come immediately thereafter. The town was eight miles southwest of Masan on the road to Chinju. It was the point now subject to imminent NKPA attack.

Walker had assigned three units to this first offensive: the Marine brigade, two regiments of the 25th Infantry Division, and the Army's 5th RCT. They would be called Task Force Kean.

For the brigade, the 3d Battalion, commanded by Lieutenant Colonel Robert D. Taplett, was designated to move first on 6 August. Arriving at Chindong-ni, Taplett had to scout out the situation, since his battalion was due to be temporarily under the operational control of an Army colonel commanding the Army's 27th RCT there. When he got to the Army regimental command post (CP), the colonel was not there, and his operations officer did not know where he could be found, and neither could Taplett contact the commanding officer of the battalion in Chindong-ni. Its CP was there, right in the middle of the road, so Taplett quickly chose a very different location for his CP—on the reverse slope of a ridgeline.

As 7 August began, Task Force Kean was ready to jump off on the first real American offensive of the Korean War. Looking back on this day, Craig later felt that the fundamental requirement was for combat readiness. He had seen this in a brigade which was activated at Camp Pendleton on 7 July and was in combat by 7 August—only one month later.

It was in truth a memorable date for the brigade: exactly eight years earlier, to the day, Marines had opened the first American ground offensive of World War II at Guadalcanal. The plan now called for a three-pronged attack, with the brigade on the left following the south (roundabout) fork of the main road, the 5th RCT moving straight ahead on the road in the center (the direct line west to Chinju), and a regiment of the 25th Infantry Division swinging around in an arc on the right to join up with the 5th RCT halfway to Chinju.

It looked good on paper, but the NKPA refused to cooperate. The *6th Division* fully expected to continue its unbroken string of victo-

A Marine guides a work detail of South Korean carriers, bringing ammunition and water to the front lines.

National Archives Photo (USMC) 127-N-A1303

ries. Its commander, under orders to roll into Pusan forthwith, had issued this stirring proclamation to his men:

> Comrades, the enemy is demoralized. The task given to us is the liberation of Masan and Chinju and the annihilation of the remnants of the enemy. We have . . . accelerated the liberation of all Korea. However, the liberation of Chinju and Masan means the final battle to cut off the windpipe of the enemy. Comrades, this glorious task has fallen to our division! Men of the 6th Division, let us annihilate the enemy and distinguish ourselves!

Thus, just as Task Force Kean launched its attack, so did the *6th Division*. The Army's 5th RCT led off on the 7th with its 1st Battalion. When it got to the road junction west of Chindong-ni, for some unknown reason it took the left (south) fork that was assigned to the Marines instead of going straight ahead (west) on the road that led to Chinju. Advancing three miles on the wrong road, it left open to enemy control Hill 342 which overlooked and commanded the main supply route that the task force would need. Kean had ordered that this was to be held "at all costs."

A company of the 2d Battalion, 5th RCT, had earlier been on the hill, but it was now quickly surrounded and cut off. To help break the siege, a midnight order came from the 25th Division, via the commanding officer of the Army's 27th Infantry Regiment, to send a Marine platoon to help the beleaguered Army company on Hill 342. It would be the first infantry action for the brigade.

Department of Defense Photo (USMC) A413601

The pin-point attacks of their Corsairs gave the Marines invaluable close air support. Here the F4Us of VMF-323 are being loaded with rockets on board the USS Badoeng Strait *before a mission.*

Second Lieutenant John J. H. "Blackie" Cahill from Company G got the job that night of 6-7 August. Reinforced with a machine gun squad and a radio operator, he set out for the CP of the Army's 27th Infantry and then the CP of the 2d Battalion, 5th RCT. There he received the astonishing order that his one platoon was to relieve the Army's besieged company and hold the hill by itself. Moving out through the night of 6 August, the Marines suffered two wounded from fire that proved to be from the 2d Battalion, 5th RCT. There followed the next morning (7 August), the beginning of a hot day, an agonizing series of hill climbs in untempered sun which led to heat prostration and empty canteens, and then enemy fire on the platoon as it staggered upwards to the hilltop, urged on by Cahill and his noncommissioned officers. Only 37 of the original 52 men reached the top. Once there, Cahill used his radio to call his own 3d Battalion for badly needed supporting artillery fire and air drops of water and ammunition.

When the severity of the problems on Hill 342 became clearer, Company D from Lieutenant Colonel Harold S. Roise's 2d Battalion was sent into action on 7 August. As the NKPA continued to reinforce its troops, the rest of the 2d Battalion became heavily engaged nearby. In air temperature of 112 degrees men continually collapsed from nausea and heat exhaustion. Water was scarce and the slopes of the hill seemed to go on straight up forever. Finally, at the end of the day (7 August), Company D had nearly reached the crest, but, exhausted, dug in where it was for the night.

Meanwhile, the Army company and Cahill's platoon on the crest had had a brutal day. Parched for water and completely surrounded by enemy fire, they managed to hang on with reinforcements now near at hand. And so the day for the 2d Battalion ended in a stalemate with the enemy on and around Hill 342.

There were problems everywhere else. The 1st Battalion, 5th Marines, under Lieutenant Colonel George R. Newton, was backed up in Chindong-ni because the Army battalion had taken the wrong road. Taplett's 3d Battalion had relieved a battalion of the Army's

27th RCT the day before, but now the latter found itself attacked as it tried to move into reserve in the rear. The 5th RCT was stalled.

The official Army history describes this day of 7 August perfectly when it refers to "a general melee" amid "confusion." The problems were compounded when the NKPA slipped around Chindong-ni and occupied a commanding height, Hill 255, that dominated the task force's supply road to Masan in the rear.

Hearing of the stalled attack of his 5th RCT, Kean was exasperated and took prompt action. He contacted Craig, who never forgot the day. His men were relieving the Army's 27th RCT, with Chindong-ni to be the jump-off point for the Marines' attack once the Army's 5th RCT had cleared the road intersection just ahead. Craig remembered: "At Chindong-ni I found the most confused situation that I've encountered in the Marine Corps . . . . Finally, due to the inability of the Army to clear the road junction and the hold-up of our offensive, General Kean put all troops in that area under the Marine brigade commander, and I was given the brigade plus the [Army's] 24th Regiment and the 5th RCT."

This took place on 7 August, and now Craig would have to sort things out and get the task force moving forward. To do this, he acted in a typical way: he went straight to the front lines to observe the situation first-hand. This kind of on-the-spot leadership immediately struck Second Lieutenant Patrick G. Sivert, an observer overhead in an OY. He was "amazed" on that very first day at how close the brigade CP was to the front lines. In contrast, he noted that "with the other outfits in the surrounding area, it was just the converse. Consequently, our communications, for the most part, with the Marine units on the ground were almost always very good, and with the other units almost always very bad."

When Craig went forward, he found that the 5th RCT, under Colonel Godwin L. Ordway, USA, was still held up, even though "enemy resistance was light." It was clear to Craig that, to break the deadlock, he would need to launch a series of aggressive attacks by all his ground units, with heavy artillery and air support.

Thus, early the next morning, 8 August, Company D pushed to the crest of Hill 342. Cahill and the battered survivors greeted them with enormous relief. It remained, however, a touch-and-go situation. Enemy fire was sweeping the encircled position, Marine officers were going down, and NKPA riflemen

Exhausted due to the strenuous climb and scorching heat, Marines establish a hasty perimeter on a hillcrest west of Chindong-ni. Chingdong-ni would be where they got their "first taste of the enemy," whom they found to be "spirited, tenacious, and well trained."

Department of Defense Photo (USMC) A12036

were slowly and steadily worming their way up the approaches. A private in Company D, Douglas Koch, felt the pressure: "I felt pretty bad. This was a very hectic time. There'd been a lot of climbing, we were under fire . . . . Someone hollered that the lieutenant was dead . . . . Firing was hot and heavy. Guys fell around me." It grew worse. NKPA soldiers came right up to the Marine lines. The firefight continued to grow in intensity. When word was shouted that there was a new commanding officer, First Lieutenant Robert T. Hanifin, Jr., it was soon followed by the depressing news that he had collapsed in the heat. This passed command of the company to a veteran gunnery sergeant. Koch knew that there was only one thing for him and the surrounded men to do: hang on.

One of the reasons that they could "hang on" was that the Marines called on a weapon that the enemy had not previously experienced: air strikes that were not only immediate but also gave truly close air support. Panels were laid out to mark the ground positions, a radio call went to the forward air controller at battalion headquarters (who personally knew the pilots) and then to the control plane in the Corsairs already orbiting overhead. Down they screeched, strafing and rocketing. They came close in—very close in—to the defender's lines. Empty shell casings from their machine guns fell into the laps of the men below. This was more than the previously all-victorious NKPA troops had bargained for. Their firing slacked off, and the crucial hilltop held. Some 600 enemy attackers had failed in their attempt to cut the task force's main supply route.

These strikes were part of Craig's plan to push his men ahead with continuous close air support. In the first three days of combat, the two Marine fighter squadrons flew well over 100 sorties. The squadrons had tailored their flight schedules so that one or the other was always overhead, ready instantly to respond to calls for strikes during the daylight hours.

The other planes of MAG-33 were also daily demonstrating their worth. The OYs had bomb racks attached to their wing struts, thus enabling them to carry rations or cans of water to the ground troops panting in the heat and struggling up the ever-present hill slopes. This was supplemented by "daisy chains" of South Korean laborers who would pass up five-gallon cans of water, along with ammunition, to the men on the hilltops. The observation planes also became expert at spotting artillery fire for the 11th Marines. The OYs slow speed proved to be a big advantage. Sivert explained:

> In this type of terrain the enemy was so adept at camouflage that most of the time high-performance aircraft were just too fast to get down and search out a target. We in the slower moving aircraft were able to get down much lower, take our time in spotting a target, and then to stand off to one side or the other of the [bombing] runs, and make sure the aircraft were hitting the correct targets.
>
> Too, we were using the same maps that the ground commanders were using. They were able to give us targets and pinpoint the targets with exact coordinates.

Another advantage of the OYs was the ability to look down on hills (particularly reverse slopes) where the forward air controller (FAC) with the infantry on the ground was blocked from seeing the enemy target. Sivert found that a pattern of effective teamwork developed: the FAC would call on an OY to spot a target and give him the direction in which the bombing runs should be made. Sometimes the OY would even give the type of ammunition to be used on the target. Then, when the bombing runs had been completed, the OY would furnish damage estimates to the FAC. Teamwork was essential, since the OY could only communicate with the aircraft by relaying all directions through the FAC.

Helicopters also carried precious supply cargoes to isolated areas. In addition, they became invaluable in evacuating wounded riflemen. The fighter pilots developed an enthusiastic appreciation of these new "birds" when they similarly proved adept at rescuing pilots who had been shot down.

The full 2d Battalion was consolidating its control of Hill 342 on 8 August, much to the relief of Cahill (who received a Silver Star for his leadership). Meanwhile, the other rifle units of Murray's 5th Marines were also very busy. Taplett's 3d Battalion drew the assignment on 7 August of driving the enemy off the strategic Hill 255, which overlooked and blocked the main supply route (MSR) to the rear. The first small-scale attack on 8 August was directed at a lower hill that would give access to 255. It was repulsed. The commander of Company H, Captain Joseph C. Fegan, Jr., was later awarded a gold star in lieu of a second Silver Star for his bold actions when he personally led the next assault, after a platoon leader refused to move (Fegan relieved him for that). It came down to the messy business of cleaning out

Department of Defense Photo (USMC) A2262

Supporting fire from the howitzers of the 11th Marines was a crucial prelude to every attack of the riflemen.

each enemy foxhole, one at a time, for the NKPA troops fought to the death. Fegan was ably assisted by the heroics of such men as Corporal Melvin James (Army Distinguished Service Cross and Silver Star), and Technical Sergeant Ray Morgan and Private First Class Donald Terrio (Silver Stars).

When Company H hastily dug in for the night, Staff Sergeant James C. Davis had his platoon in a forward position only 75 yards from the enemy. While repairing a defective hand grenade, it slipped out of his grasp and dropped in the midst of his men. A posthumous award of a Navy Cross described his immediate reaction: "Without a moment's hesitation, he chose to sacrifice himself, rather than endanger his companions, and threw himself upon the live grenade."

In parallel action by Company G that day, Sergeant Jack E. Macy would later be awarded a Distinguished Service Cross for his perilous rescue trips to bring wounded men into safety. By the end of the day, the Marines were securely in possession of the first hill, with 255 looming ahead. The company had advanced more than 1,400 yards in the teeth of a fiercely resisting enemy. It had taken nine gruelling hours with great suffering from lack of water, heat exhaustion, and overexertion in the stifling weather. One man in the battalion later admitted: "Guys almost went mad for water. I never felt the kind of heat I felt in Korea. I just burned up. My hands went numb. I couldn't help myself; I began crying like a baby. I was ashamed. I felt I could crawl into a mouse hole and die, but I couldn't help what was happening to me."

This kind of water-deprivation and dehydration in the midst of blinding heat seriously affected the combat strength of all of the battalions. Murray, the regimental commander, admitted: "One time I figured I had about at least a third of my regiment lying at the side of the road with heat prostration."

In spite of the gruelling physical problems—and the fanatical resistance by the enemy—the battalion had now successfully positioned itself for the final lunge at Hill 255.

As Craig jockeyed his forces to meet the NKPA thrusts and launch his own attacks, Newton's 1st Battalion was finally able to move out of Chindong-ni early on 8 August. Its orders were to proceed to the now-famous road fork and take the left (south) route, while the Army's 5th RCT was to take the straight-ahead (west) route. Trying to approach the junction, Newton found that the 5th RCT was still stalled there. The road to the fork was jammed with soldiers and Army vehicles; it was a scene of "congestion and confusion." With the advance of the Marine battalion thus blocked, the solution for progress came in an order from Kean to Murray: send your 1st Battalion on a night march to Hill 308 to relieve the Army battalion that took your south road in error. It was expected to be a dangerous maneuver. The commander of the Army battalion felt that his companies were "cut off" by the NKPA; the Marines were to veer off the main road short of the clogged junction and file in column along narrow dikes in a wide rice paddy, totally exposed if fired upon; two South Korean civilians of unknown trustworthiness were to guide them through the pitch black night (since the assigned Army guide never appeared). Newton was deeply upset when the Army battalion prematurely withdrew from its position without waiting for the Marine relief force. As Andrew Geer described this unfortunate development in *The New Breed*, "there was a display of temper" between the two battalion commanders.

By midnight the Army troops had cleared the rice paddy paths, and the Marines quickly moved out. To the gratified surprise of the men, they encountered no enemy, and by dawn on 9 August they were safely assembled at the base of Hill 308. The battalion had been on the move, afoot, for 22 consecutive hours; the men were thirsty and dog-tired, but they had carried out the relief as ordered.

Kean, meanwhile, had not limited himself to his orders to Murray. He had come up to the deadlock at the junction, and his next orders were short and to the point. Indicating the hill that controlled the junction to one of his battalion commanders (who had earlier failed to capture the hill), Kean barked, "I want that hill tonight!" It was finally done.

The events of 8 August were not decisive in themselves, and did not appear to represent any real progress for the task force. Nevertheless, the groundwork had been laid, and Craig now had his troops where they were in position not only to crush the enemy's offensive, but also finally to make real progress of their own toward the ultimate objective of Chinju.

Two of the opposing forces, NKPA and Marine, had learned something about each other in these first clashes. Colonel Robert D. Heinl, Jr., in *Soldiers of the Sea* summed it up:

> The Marines got their first taste of the enemy. They found him spirited, tenacious, well trained, and generously equipped with Russian gear. Used to having the campaign their own way, the North Koreans fought confidently, but reacted with considerable surprise when they found themselves facing troops who gave no ground, hung on to their weapons, and brought in their wounded and dead.

A subsequent article in the *Marine Corps Gazette* by historian Lynn Montross analyzed the battle skills of the NKPA this way:

> The Marines learned to respect a hardy enemy for his skill at camouflage, ambush, infiltration, and use of cover. They learned that supporting air and artillery fires often had limited effect on a foe making clever use of reverse slope defenses to offset Marine concentrations. Thus a ridge might protect and conceal an enemy strong point until attackers were too close for supporting fires.

When this situation developed, with the heavy firepower of the Marines neutralized, their attack was reduced to the familiar basic essential of small arms fire fights. In these circumstances, the NKPA was able to meet them on even terms, man-to-man.

Just as the Marines had sized up the enemy, so, too, they had formed their own opinion of the

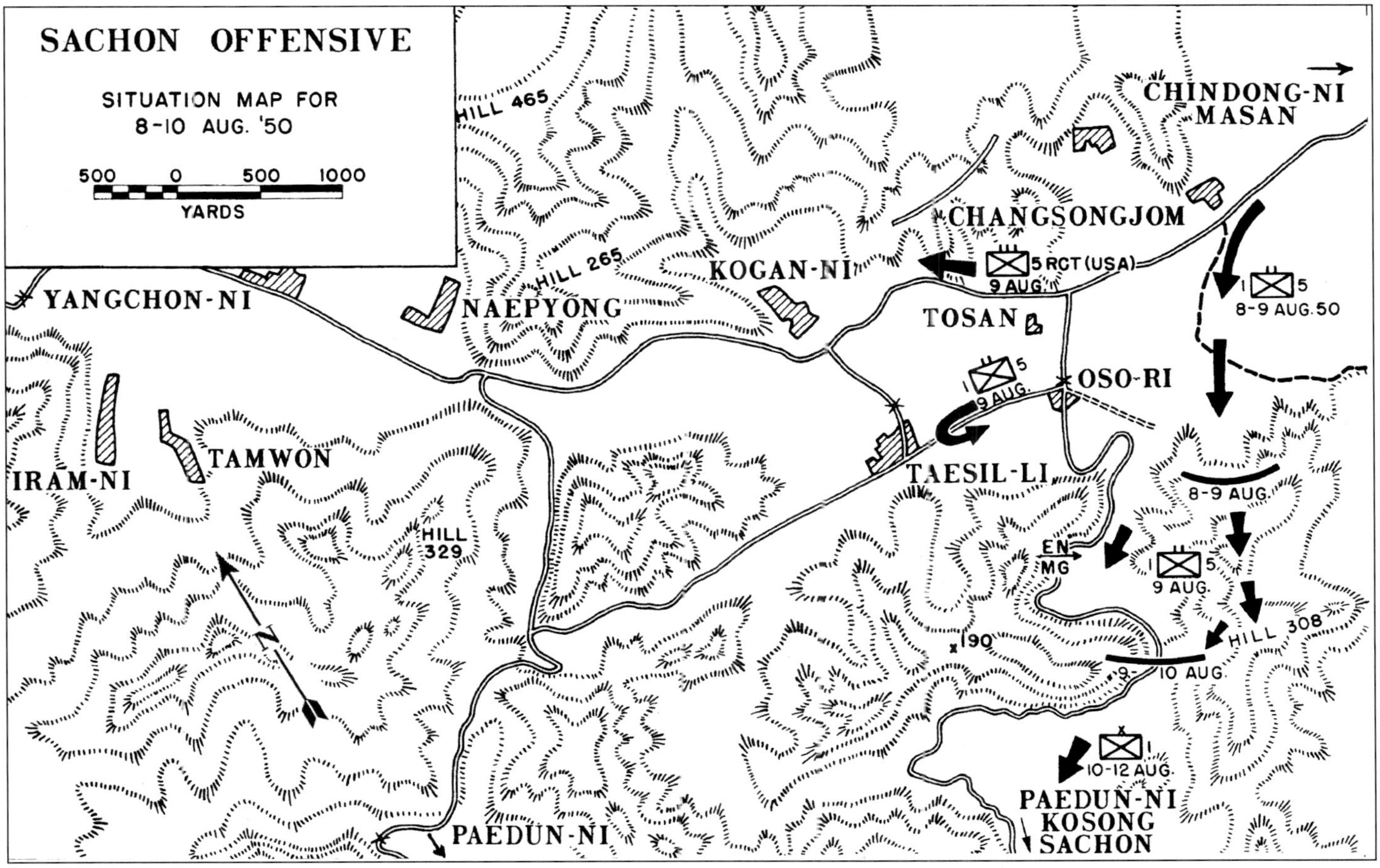

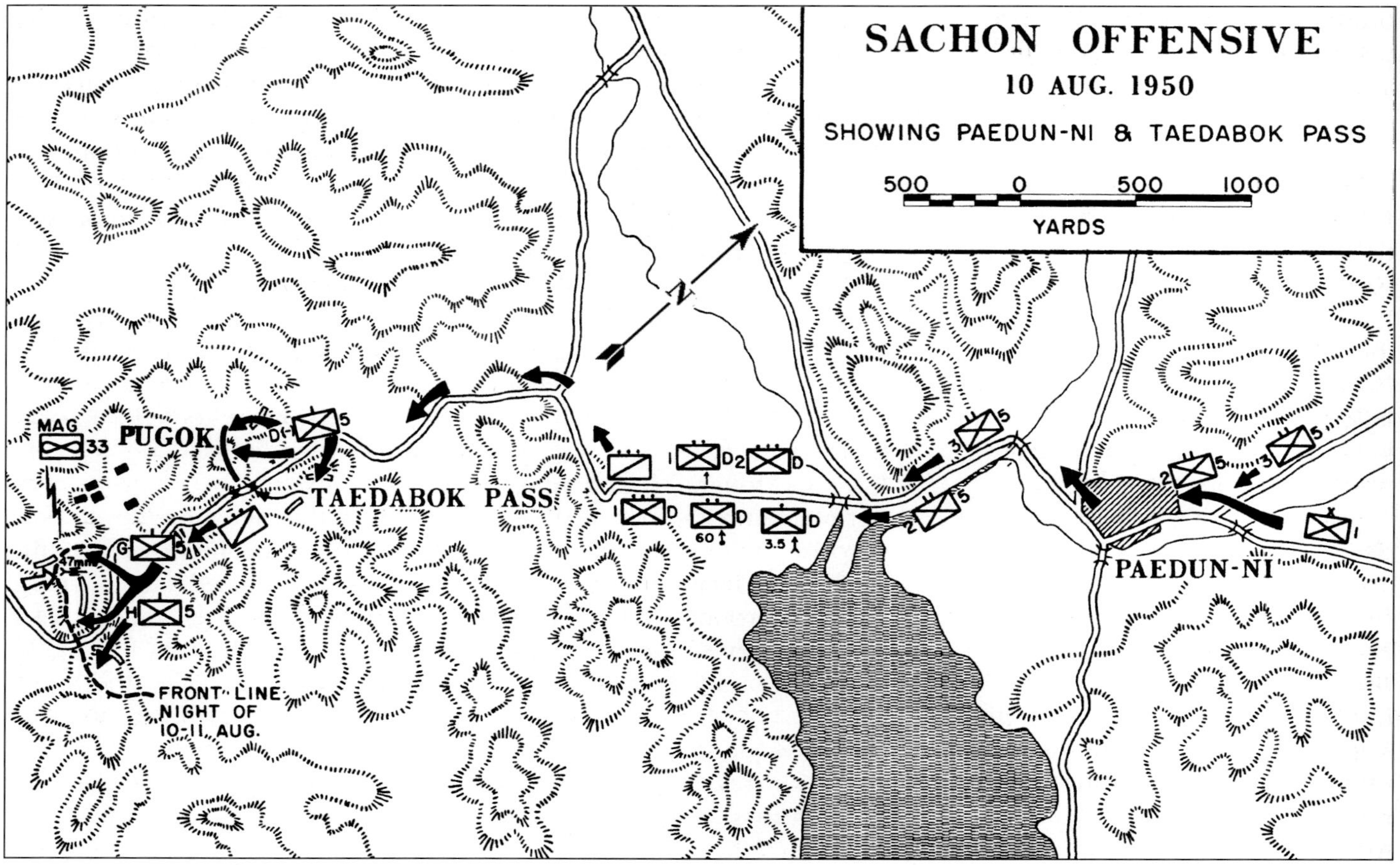

Army units with whom they were in contact. Other judgements were also being made at this time. An Army colonel had been sent by General Mark Clark's Army Field Forces Headquarters to evaluate the units of the Eighth Army in late July and early August. On 9 August he made his report to Lieutenant General Matthew B. Ridgway, whose aide prepared a detailed 12-point memorandum on the findings.

The report was very harsh. It is quoted at length in a recent book by Brigadier General Uzal W. Ent, USA (Ret), entitled *Fighting on the Brink: Defense of the Pusan Perimeter*. The book has Ent's summary, saying that the report "verbally ripped the officers and enlisted men of Eighth Army apart." It underscored three "principal deficiencies": lack of knowledge of infantry fundamentals; lack of leadership in combat echelons; and the absence of an aggressive fighting spirit.

Regardless of Army problems and wary of a tough enemy, but confident it could smash ahead, the 5th Marines made real progress on 9 August. Murray was a driver who knew that aggressive attacks would, in the end, reduce his casualties. Even though his 1st Battalion had barely arrived at the base of Hill 308, Murray radioed an order to attack immediately. Once again it was the familiar story of over-tired, thirsty men staggering up one more hill—this time after 27 hours of continuous, tense exertion. Fortunately, there was only sniper fire and the crest was secured, as the men collapsed on the broiling ground.

There was to be no let-up, however, for the beat-up 1st Battalion. Murray kept pushing. He ordered Newton to take his men back down from the hill they had just climbed so laboriously and to move along the south road towards the next objective, a village called Paedun-ni. It was a pathetic remnant that was able to come down that hill. There were only 30 men and two officers out of the whole company who were able to make it down without collapsing. Captain John L. Tobin, in bad shape himself, stayed with the rest of the men on the hilltop. Fenton painfully recalled the scene:

> The troops that had passed out had to be left where they had fallen, since no one had the strength to move them. The men who had heat prostration, but weren't out, tried to place themselves along the ridge where they could cover their fallen buddies in case of an enemy attack. The heat reached 114 degrees, and I

personally don't believe that our men on the hill could have repulsed 10 enemy troops.

Once Newton finally was able to get his survivors down to the Paedun-ni road, they were joined by his Headquarters Company, his Weapons Company, and a platoon of tanks. But Newton's troubles continued. He was stuck with obsolete Japanese maps which frequently used different names for towns, had no contour lines for the hills, and were undependable as to roads. This resulted in his taking the wrong fork in the road shortly after starting. Not one to be out of touch with his troops, Murray appeared shortly to correct the problem. It developed that the maps Newton and Murray had were each different. The upshot was that Murray decided that the whole column had to turn around on the primitive narrow road, retrace its steps, and take the other fork. Amidst the milling in this reversal, Newton was probably dismayed to see Craig appear on the confused scene. The general was not pleased, and without knowing the background, he expressed his thoughts in vivid language. When the battalion finally got restarted on the proper fork, Craig—another officer who kept in close touch with his troops—went with them to supervise the further attack he was planning. As evening fell, the 1st Battalion had come two miles from its jump-off and was ordered to dig in for the night.

Back in the zone of the 3d Battalion, the payoff came on 9 August for the hard fight the day before. The day began with a thorough saturation of Hill 255 by the artillery of the 1st Battalion, 11th Marines, under Lieutenant Colonel Ranson M. Wood.

The artillery batteries had to improvise their tactics during these early days in Korea. Ironically, they had suffered more casualties than the riflemen when the task force had begun its attack. Then, to counter the skillful infiltration of the NKPA, the three batteries would try to set up with one aiming to the north, one to the east, and one to the west, with protective foxholes around them. (Because the brigade was moving so fast, and with the penchant of the enemy for lightning hit-and-run tactics, the 11th Marines often would be able to set up only one battery for action.)

After the artillery had plastered the enemy positions on Hill 255, the battalion's forward air controller, First Lieutenant Daniel Greene, got on his radio, and the Corsairs then came wheeling in, this time with napalm's first scourge of the NKPA. It was a near-classic demonstration of the Marine concept of an air-ground team. When the riflemen scaled the final crest of the hill, there was little opposition. Nevertheless, the battles that led to the conquest of Hill 255 had cost Company H the loss of 25 percent of its men. When the 3d Battalion then joined up with part of the Army's 24th Infantry, the threat to the rear supply route (Masan to Pusan) had been eliminated.

With these hill captures by the three Marine battalions, the errant Army battalion of the 5th RCT, which had earlier taken the wrong fork at the junction, could now retrace its steps and rejoin its regiment. At last the 5th RCT moved out west towards a new objective on the road to Chinju.

This breakup of the log jam enabled Kean to relieve Craig of overall command of the task force and allowed the general to return to his own men on the afternoon of 9 August.

With his brigade now moving along its designated south road, Craig planned to exert maximum pressure on the NKPA by having the Marine battalions leap-frog

Marines carefully check individual huts to successfully drive North Korean defenders out of this village.

Department of Defense Photo (USMC) A15986

each other, pushing forward hard. The same procedures would be used by the companies and platoons. Whether it was advance guard, flankers out on the sides, or in the main column of the brigade, all the units would rotate. This enabled Craig to keep driving.

He had Murray pull Roise's 2d Battalion off Hill 342, and put it on trucks which brought it to an assembly point near Hill 308, a spot familiar to the 1st Battalion. Arriving there nearly at midnight on 9 August, Roise contemplated his situation. He had had 9 killed, 44 wounded, and a shocking 94 cases of heat prostration, the loss of key officers, and now his tired men were due to lead the attack in two hours—after the past 69 hours of climbing, fighting, and marching. Despite all this, he was relieved to see that the morale of his men appeared high. Furthermore, his riflemen had been reinforced by the attachment of a battery of artillery, a platoon of the powerful Pershing tanks, and a 75mm recoilless rifle platoon.

The attack on Paedun-ni was only the first objective enroute to the towns of Kosong and Sachon, the keys to the final goal of Chinju. Craig later described his reasoning:

> This night attack was in addition to an attack during the day, and, although the men were very tired and I hesitated to carry out the night movement, I considered that, if we could surprise the North Koreans and keep moving when the other American troops had already stopped for the night, that we might gain some added advantage—and this proved to be the case. We marched throughout the night and gained quite a bit of distance with only occasional shots being fired.

Moving through the 1st Battalion, the 2d Battalion had pressed forward through the night of 9-10 August, grateful that there was no opposition. There was an episode with a couple of tanks that got stuck, bringing both Craig and Murray to the spot with some strong words to move the rest of the column forward. By 0800 on 10 August, Roise and his men were in Paedun-ni.

And so 9 August ended with the Marine brigade finally all together as a unit and really starting to roll in high gear down the south road. The next day (10 August) brought some brisk action when the retreating enemy forces picked strategic places to delay the rapid advance of the Marine column. As usual, Craig had arrived at Paedun-ni by helicopter, and his refrain to the troops was to move ahead with "all speed." Accordingly, the 2d Battalion, even though it had just arrived, got ready to move out quickly for Kosong. The 3d Battalion followed.

With only a few trucks available, part of Company D was put on board, with the rest of the troops marching behind. As the trucks rolled down the road, they were preceded by a four-jeep reconnaissance team. Some 2 1/2 miles from Paedun-ni there was a section of the road where it made a sharp turn and narrowed along a defile 1,000 yards long underneath a large hill. It was called the Taedabok Pass, and 300 of the NKPA were dug in and carefully camouflaged waiting there in ambush. Their mortars, antitank guns, and artillery were ready to inflict heavy casualties on any troops who moved blindly into the pass.

However, the advance guard of the Marines was not moving blindly. Craig was well aware of the skill of the NKPA in ambushes and envelopments. He therefore had a policy of using his helicopters and OY planes to the maximum for reconnaissance of his front and flanks. In addition, he deployed a reconnaissance platoon in jeeps to scout ahead of the lead battalion. These men, Craig commented, "on two occasions uncovered very strong ambushes and suffered some casualties in getting out, but they did protect the main column."

One of those riding in a reconnaissance jeep was a young private first class. They were rolling happily down the road, thinking how quiet it was, when suddenly:

> The North Koreans opened up. [They] cut up the first couple of jeeps pretty bad. My group tumbled and ran for the ditch. I landed calf-deep in warm water. I heard machine guns chattering around me. Dirt kicked up along the road that was now lined with abandoned jeeps.
>
> Sergeant Dickerson shouted over the noise, "Those hills, the little low ones, over to the right, we gotta get over there. Gotta return fire from there." I picked up my BAR [Browning Automatic Rifle], and, crouched over, ran down the ditch.

At the same time an OY observation plane, flying less than 50 feet off the ground, spotted the ambush. With all hope of trapping the main column of Marines now gone, the NKPA poured on the fire. An antitank gun smashed a jeep. Now, coming up fast and deploying in counterattack on both sides of the road, the men of Company D went after the high ground at

1500 that afternoon. Their 60mm mortar fire silenced the antitank gun, and, when two Marine Pershing tanks arrived at 1630, their 90mm guns, combined with Corsair attacks, beat down the enemy fire.

The fact that there had been any surprise was on Murrays mind. He said later: "We moved pretty well along this road for a day, I guess, when we ran into an ambush. Shouldn't have been ambushed, we should have discovered it, but didn't. The advance guard failed to spot these people and got hit. Fortunately, though, the bulk of the regiment didn't get involved initially."

The ambush had delayed the brigade, but not for long and at a cost to the NKPA of hundreds of dead, wrecked vehicles, and large losses in weapons. Now the Marines were poised to sweep into Kosong.

Reinforcements arrived: the rest of the 2d Battalion on foot and the 3d Battalion by truck. Murray, of course, was there waiting for them. He took Taplett up to the top of one of the hills and they could see Kosong five miles away. The regimental commander, in his usual style, told Taplett to move his 3d Battalion through Roise's men at 1715 and attack immediately to clean out the pass and clear the way to Kosong. It was an unusual "pass through," since neither Murray or Taplett could locate Roise or his command post.

This order came as music to the ears of 2d Battalion Marines. Roise had had them moving and fighting for 88 hours over a distance of almost 50 miles. In spite of the never-ending hills and oppressive heat, the battalion had won each of its battles and inflicted more than 600 casualties on the enemy. Now it could actually relax for the moment. For the first time since going into action, there was enough water to drink and the men could eat their field rations in peace. Perspiration-soaked socks had brought on ulcer sores on their feet and ankles, so it was a blessed relief to be issued clean, dry socks.

As the 3d Battalion moved into position for its attack, the men were naturally concerned about enemy fire, but the first thing to hit them was friendly fire. One enlisted man later recounted his reaction:

> We passed through one of the other battalions. About 5:00 in the afternoon two American fighters [U.S. Air Force F-51s] zoomed down the road around 150 feet above our heads . . . . No matter where I ran, I couldn't seem to find an escape. Their .50-caliber bullets hit that hard, dry road and it sounded as if each was exploding. There was just nowhere to go to get out of the line of fire. Someone screamed, "Break out the air panels! Get the air panels!" The fighters left as suddenly as they had arrived.

By 1830 on 10 August, the lead platoons had jumped off in the attack, but they soon received heavy fire from two NKPA machine guns hidden at the far end of the pass. During this encounter, some Marines at the point were wounded, and platoon leader First Lieutenant Jack "Big Jack" Westerman made a daring rescue for which he was later presented a Navy Cross. Neutralizing those guns took the last of daylight, and so Murray had the battalion dig in for the night, sending men up the dominating hills for security. First Lieutenant Robert D. Bohn, the commander of Company G, was not very happy about that order: "It was just contrary to everything you're taught, to go up into enemy-held territory at night, no reconnaissance, nothing like that, and hold it."

Things got worse at dawn. The NKPA hit Bohn's company. Because he had had to feel his way up there in darkness, he really did not know exactly where he and his men were, but the enemy attack revealed:

> I was on the front line. I was on the forward slope of this hill, and my command group got hit. I got wounded, my mortar section chief got killed, and I had a couple of other casualties. But we were a well-trained outfit, so we immediately returned fire—I think there were maybe eight or ten of them, probably a delaying party—and we killed them all.
>
> It was very close. It was hand-grenade range and hand-to-hand in a couple of instances. I took hand-grenade fragments in the neck and shoulder, but they weren't too serious. It was the same hand grenade that killed a Marine right next to me. I killed the guy that threw it.

By the time the attack was finally beaten off, Bohn's cool and decisive handling of his men would result in the award of a Silver Star. However, Company G, which was due to lead the brigade's advance the morning of 11 August, was a half hour late getting to the appointed line of departure. John Toland's history, *In Mortal Combat*, records a remark to Bohn: "Murray was furious, 'When I say 0800, I don't mean 0801!' "

Company G then moved out at a

Sikorsky HO3S-1 Helicopter

Rotary-wing aircraft had come too late to have any effect on the tactics of World War II, although a few Sikorsky aircraft were used experimentally in the European and Pacific theaters near the end of the conflict. Following the war, it was the Marine Corps that took the lead in developing techniques and procedures for this new combat aircraft.

In February 1948, the first Sikorsky HO3S-1 helicopter was delivered to the first Marine helicopter squadron, experimental Marine Helicopter Squadron 1 (HMX-1), at Quantico, Virginia. Three months later, the squadron made the first helicopter troop lift in history.

Shortly after the Korean War broke out on 25 June 1950, 7 pilots, 30 enlisted men, and 4 HO3S-1 helicopters were detached from HMX-1 for service with the 1st Provisional Marine Brigade. Upon arrival at Marine Corps Air Station, El Toro, California, these elements were combined with 8 fixed-wing pilots, 33 enlisted men, and 4 "usable" OY-2 light observation planes to form the brigade's air observation squadron, Marine Observation Squadron 6 (VMO-6). The squadron's commanding officer, Major Vincent J. Gottschalk, was given just 48 hours to weld these two elements together before being shipped overseas.

Upon arrival in the Pusan Perimeter, VMO-6 set up its base at Chinhae on 2 August, ready for business. There was not long to wait. The next day, the brigade commander, Brigadier General Edward A. Craig, took off in one of the helicopters and gave a vivid demonstration of its versatility. In one day, he stopped to instruct a battalion, picked out the location for his forward command post, held a conference with U.S. Army commanders, and held three more meetings with his ground commanders.

Besides this role in command, the squadron's helicopters were "always available for observation, communications, and control." In addition, there were a wide variety of other missions: evacuating the wounded, rescuing downed fixed-wing pilots, transporting supplies, artillery spotting, and, scouting enemy dispositions. During the month of August 1950, VMO-6 helicopters amassed a total of 580 flights and the HO3S-1s chalked up the first successful combat missions. These missions were a harbinger of the large-scale deployments that would come.

Aircraft Data

Manufacturer: Sikorsky Aircraft Division of United Aircraft Corporation
Power Plant: Pratt and Whitney R 985 AN-7 Wasp Jr. Engine 9 Cylinder; Radial; Fan-Cooled; 450 Horsepower
Rotor Diameter: 48': 3 Blade Composite Construction
Tail Rotor Diameter: 8'5"; All Wood; 3 Blades
Length: 41' 13/4" Without Rotor Blades
Overall Length: 57' 1/2"
Height Overall: 12' 11"
Weight Empty: 3,795 Pounds
Maximum Gross Weight: 5,300 Pounds
Cruising Speed: 85 Miles Per Hour
Maximum Speed: 103 Miles Per Hour at Sea Level
Range: 260 Miles
Service Ceiling: 13,000'
Fuel Capacity: 108 U.S. Gallons
Seating: Four including Pilot

fast clip. It would be the pace of the point platoon which would govern the speed of the entire brigade. Accordingly, the advance flankers moved at a run to keep up with their platoon leader on the road. He, in turn, relieved them with fresh men as often as possible. The fast pace they set proved invaluable when they came upon any of the enemy. The Marines came to the first machine gun emplacement lurking on the route, and they hit it so hard and so unexpectedly that the five NKPA gunners were killed before they could fire a shot. Three more enemy positions fell to the same aggressive tactics of the point platoon.

With this kind of speed and skill up front, and with two Corsairs and an OY cruising overhead looking for any trouble, the brigade came wheeling down the road to reach the outskirts of Kosong by 1000. Softening up any potential defenders, the 105mm howitzers of the 11th Marines began raining high explosives on Kosong. This barrage and the onrushing brigade forced the opposing *83d Motorcycle Regiment* to pack up and seek safety in a hasty departure.

With the flight of the main body of the enemy, only a few snipers remained in Kosong. Company H passed through G and pushed rapidly into the town. On its heels came Taplett and Craig, with their hands on the helm, always close to the action. Meanwhile, Company G raced to seize control of Hill 88 southwest of the town and dominating the road to Sachon. The enemy was waiting there, but not for long. The Corsairs swooped in low with napalm, tank fire poured in, the howitzers of the 11th Marines blanketed the position, and the crest was quickly taken.

It was at Kosong that there was a clear example of the payoff from the long years of Navy-Marine cooperation: support of the brigade by Landing Ship Tanks. Craig fully realized their great value, for they proved a ready solution to the problem of getting supplies by truck on primitive, congested roads. Accordingly, he had had his helicopters make a reconnaissance of usable harbors on the nearby coast. Then the LSTs would move into a harbor that matched the brigade's advance. Craig described the pay-off:

> When we reached Kosong, we had an LST within six miles of that place on a covered road where we could unload and push forward supplies and build up a brigade dump at Kosong. Wounded could be evacuated immediately to the LST . . . . We always felt that we had a mobile base of supplies which we could bring in as necessary and that, even though we were separated by long distance or cut off from our rear base, we could always depend on these LSTs for supplies.

With Hill 88 secured, Craig had Taplett pull the men of Company G back, disregard other hills, and concentrate for an immediate drive by the brigade to Sachon. A pair of NKPA antitank guns were waiting on the route, but were discovered when an ambulance jeep was hit (killing a Navy corpsman). With its location disclosed the pair was quickly knocked out and the column surged forward, led by Company H with the forward air controller right up with the point men.

A few hours later the the marching men came upon an astonishing sight. When the *83d Motorcycle Regiment* hurriedly decamped from Kosong, its timing proved disastrous, for, just at that juncture, a flight of Corsairs from VMF-323 appeared on the scene. The pilots could hardly believe the tempting targets arrayed before their eyes, and the slaughter began; it came to be known as the "Kosong Turkey Shoot." The Corsairs swung low up and down the frantic NKPA column, raining death and destruction in a hail of fire from rockets and 20mm cannon. With the vehicles at the front and rear ends of its column destroyed, the enemy regiment was trapped. It was a scene of wild chaos: vehicles crashing into each other, overturned in ditches, afire, and exploding; troops fleeing for safety in every direction. Another flight from VMF-323 arrived, and, joined by U.S. Air Force F-51s, finished off the destruction of the trucks, jeeps, and motorcycles. Accounts of this NKPA debacle vary widely in their tallies of the number of vehicles destroyed: 100-200.

One thing was certain: when the ground troops reached the scene, the usable vehicles were quickly appropriated for the transportation-starved brigade. There was, in fact, a momentary slowdown in the fast advance of the Marines to stare. Joseph C. Goulden's *Korea: The Untold Story of the War* pictures the scene: "Black Soviet Army jeeps and motorcycles with sidecars, most of which had gone into battle in mint condition. Looking under the hoods, the Marines found the jeeps powered by familiar Ford Motor Company engines—apparent relics of American lend-lease aid to the Soviet Union during the Second World War."

The Marines found other things, too. Included in the wreckage were American jeeps the NKPA had captured earlier from U.S.

Photo by David Douglas Duncan

Across the rice paddies and up the endless hills, it was always on foot

Army troops, and Toland asserts that there were duffel bags containing Russian officers' uniforms.

There was another colorful episode which happened on the road that led back to Sachon in the rear of the motorcycle regiment. Andrew Geer's *The New Breed: The Story of the U.S. Marines in Korea* describes how Master Sergeant Herbert Valentine and Second Lieutenant Patrick G. Sivert were in an OY skimming the route when they observed a jeep making a high-speed getaway from the battle site. Sitting rigidly erect, arms folded, eyes never wavering from straight ahead, a high-ranking NKPA officer sat unmoving in the rear seat. The Marines in the OY came down close to the jeep and began firing their revolvers (the plane's only armament) at the fleeing target. Rifle fire came back from the jeep's front seat, but the officer remained rigid. This continued for a 20-mile stretch with no results. Finally, the terrified driver took one too many looks at the plane so close overhead, and the jeep hurtled over a cliff. The officer never budged from his fixed position as he plunged to his death.

Cruising the rest of the day in advance of the brigade, Marine air found other targets of opportunity. Geer totalled up the results:

> Score for the day to Marine Air: vehicles (all types) destroyed, 118; supply dumps destroyed, 2; ammunition dumps left burning, 2; buildings housing troops destroyed, 8; southeast section of Sachon set on fire; concentrations of troops south of Sachon, north of Kogan-ni and along route of withdrawal neutralized and dispersed with heavy casualties; one jeep presumed to be carrying a Very Important Person, destroyed.

There was, as always, a price the Marine aviators had to pay for these dramatic achievements. One pilot, Captain Vivian M. Moses, had his Corsair shot down by ground fire. When a helicopter from VMO-6 arrived to rescue him behind enemy lines, he was dead, the first death for MAG-33.

Another pilot, Lieutenant Doyle H. Cole, was luckier. Hit, his plane made a forced landing in the nearby ocean. He climbed out onto his emergency raft, and almost immediately a rescue helicopter appeared overhead and dropped to a position close above him. A rope was lowered and he was pulled up to safety. Glancing at the white hair of his rescuer, Cole slapped the old timer on the shoulder and said, "Thanks for the lift, buddy!" A second glance gave Cole a start. He saw the star on the dungarees and realized that it was Craig. An embarrassed, "Thank you, sir," blurted out, followed by a relaxed reply, "Glad to be of service, Lieutenant."

Down on the road, the brigade sped forward. Taplett and his air controller were up front with the lead platoon, and any time enemy resistance developed, in came the Corsairs. This immediacy of support was due to three factors. First, the Marines had been able to keep control of their own aviation, as MacArthur had promised Craig. Secondly, there were no upper echelons of command to delay strike requests. Each battalion and the regiment had its own tactical air control party. These control parties each consisted of an officer and six enlisted men; they each used a radio jeep and portable radios for direct orders to the planes. They worked with pilots who had had infantry training and had been carefully briefed on the ground situation. In addition, the brigade staff had an air section using four different radio networks for overall coordination, plus an observation section which used the OYs and helicopters of VMO-6 to pinpoint enemy targets for the Corsairs and control parties. Thirdly, the Marine fighter squadrons were very close by, based on the jeep carriers just offshore. Thus they could be overhead in minutes, rather than finally arriving from bases in Japan with only enough fuel for 15 minutes' support, which was the predicament of the U.S. Air Force.

As 11 August drew to a close, Taplett, after nearly being shot by an enemy soldier "playing possum," deployed his 3d Battalion on two hills by the road and had them dig in for the night. Sachon lay ahead, only a day's march away.

The men felt good. They were making rapid progress. As the official Marine history noted: "the enemy seemed to be disorganized if not actually demoralized. For the first time since the invasion began, a sustained Eighth Army counterattack had not only stopped the Red Korean steamroller but sent it into reverse."

In this happy frame of mind, the brigade got moving again early on the morning of 12 August. Enemy opposition was light, and the 1st Battalion in the lead quickly leaped forward 11 miles. Fenton noted that "the boys took quite a bit of pride in the fact that we had done all this moving on foot, while Army units moved mostly by motor. Morale was very high . . . There was evidence of considerable enemy disorganization . . . . We had them on the run and wanted to finish them off." By noon the brigade was only four miles from Sachon, and Chinju lay just eight miles beyond that. According to Geer, when a NKPA major was captured, he confessed, "Panic sweeps my men when they see the Marines with the yellow leggings coming at them."

Things looked good—too good. The old hands knew that something unpleasant always followed the good times. And so it did. With men from the Reconnaissance Company on the alert out front, Company B of the 1st Battalion poked its nose into a valley with a small village called Changchon. The Marines took a few shots at a pair of disappearing enemy soldiers, the first they had seen all day. The reply was thunderous. From the hills ahead and on either side of the road all hell broke loose, as 500 of the NKPA poured in fire from carefully camouflaged positions above the Marines. The enemy had brought up reinforcements from Sachon during the preceding night and set up an ambush here with the surviving members of the *83d Motorcycle Regiment* and part of the *2d Battalion, 15th Regiment*. The reconnaissance men had caused the trap to be sprung prematurely, before the whole Marine column could be caught in the heavy crossfire. Company B immediately rushed to help its reconnaissance men, but it was quickly pinned down by the avalanche of fire. An article by Fenton in the November 1951 *Marine Corps Gazette* told how its commander, Captain John L. Tobin, took his runners and headed forward, but halfway there:

> An enemy machine gun took them under fire, pinning them down in the rice paddy. Things were pretty hot, and Tobin noticed one of the runners shaking like an old Model-T Ford. He asked the Marine what was wrong and the boy replied that he was scared. Tobin put a big scowl on his face and replied, "Lad, Marines are never scared." Just then the enemy machine gunner got the range and was really kicking up the water and mud around them. Tobin turned to the runner and quickly added, "I see your point now. Let's get the hell out of here!"

The Corsairs and their napalm were called in, and, with their support, then fire from the tanks' 90mm guns, 4.2-inch mortars, and battalion artillery, the rest of the battalion cleaned the enemy off one hill after another in a hard four-hour battle. There was aggressive action by the rest of the

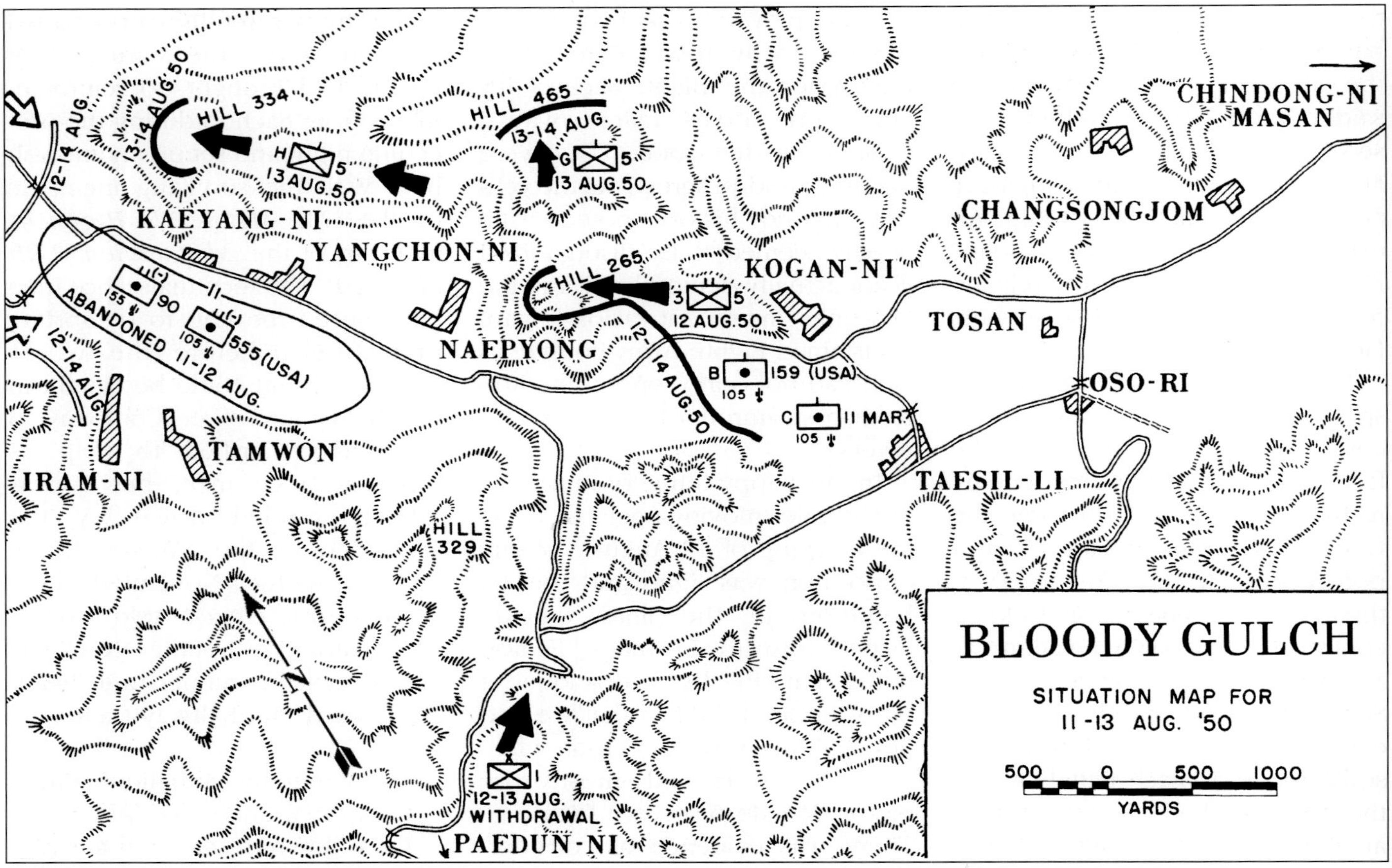

Marine column, and a squad leader in the 3d Battalion, Corporal Donald D. Sowl, was later awarded the Army's Distinguished Service Cross by order of General MacArthur.

There was a final flourish at the end of the day. A number of the enemy was spotted sneaking up the reverse slope of one of the hills. A veteran noncommissioned officer took a squad, deployed them along the ridgeline, and told them to wait silently. When the NKPA soldiers got within 75 feet, the sergeant gave his men the signal, and they poured out a sheet of fire. All 39 of the attackers were killed instantly, except for the officer leading them who was wounded and captured. Turned over to South Korean police to take back to the battalion CP for interrogation, the enemy officer did not survive the trip. As Geer wryly observed: "In the future they [the Marines] would conduct their own prisoners to the rear."

With all units dug in for the night, a rice paddy area of 1,000 yards between the two companies of the 1st Battalion was covered by the preregistered fire of mortars and artillery in case the enemy had any thoughts of a night attack. The brigade had now covered 29 miles of road (and much more counting the interminable distances up and down hills) in four short days. It had defeated the NKPA in every encounter, and here it was poised for the short step into Sachon. Next stop after that was the final objective, Chinju, now within easy reach of the hard-hitting brigade. Again, things looked good—too good.

This time the surprise came not from the NKPA in front but from the U.S. Army in the rear. Craig had received orders from Kean late in the morning of the day just ended, 12 August, to send without delay a reinforced battalion all the way back to the original starting point of the task force's drive, Chindong-ni. The Army's 5th RCT was in trouble again; its "push" towards Chinju had totally bogged down in what one account called "an epic disaster." With only two battalions left, Craig noted in his understated way that "the consequence was that our right flank . . . was exposed. There were many North Korean troops in that area, and we were, more or less, out on a limb at Sachon." Now the NKPA was cutting the main supply route behind the 5th RCT, and three batteries of the 555th and 90th Field Artillery Battalions had been completely overrun by the enemy. The Marine battalion was urgently needed to rescue the survivors from the shambles and restore the tactical situation.

The call from Kean began a hectic afternoon for Craig. Lynn

Montross in his book, *Cavalry of the Sky*, stressed the crucial mobility Craig enjoyed by repeated use of the helicopter. In a single afternoon, he took off from his CP at Kosong, then made two landings to give orders to his regimental commander, Murray, and to Taplett for the roadlift of the 3d Battalion to the crisis spot. Montross continued the story:

> Next, he spotted two columns of Marine trucks from the air and landed twice more to direct them to dump their loads and provide transportation for the troops. His G-3 [operations officer] and the battalion commander had meanwhile been sent ahead by helicopter to reconnoiter the objective area and plan for the Marines to deploy and attack upon arrival. Owing to these preparations, the assault troops seized part of the enemy position before darkness.

This fluid movement of Craig's enabled him, as a finale, to observe the start of the sunset attack enroute to a conference with Kean at Masan. While there he got the disheartening news that Walker wanted him to withdraw the brigade at daybreak. It was a gloomy ride for Craig back to his CP where he landed in early darkness.

The meeting with Kean not only confirmed the overwhelming problems of the 5th RCT, but also brought still more ominous news. The operations of Task Force Kean had been in the far southwestern sector of the Pusan Perimeter. Now the NKPA had crossed the Naktong River in the west center of the perimeter, broken the Army's lines, and were threatening to unhinge the entire defense of the peninsula. It was a time of real crisis, and Walker was calling on his battle-proven "fire brigade" to save the situation. This presented Craig with an even bleaker picture: he had to pull the rest of his brigade out of its successful drive toward Sachon and rush it north to stem the enemy breakthrough.

Withdrawal in the face of an aggressive enemy is one of the more difficult military operations. Newton, commander of the 1st Battalion, had gotten the word from Murray at midnight on 12 August to withdraw his men from their hilltop positions and form up on the road below at 0630 the following morning. There trucks would move them to their next combat assignment—unknown, as usual, to the men who would do the fighting.

Before it could get to the road, as the 1st Battalion was preparing to evacuate its positions on Hill 202, it was hit by a heavy assault. The veteran soldiers of the *6th Infantry Division* were experts at night attacks, and at 0450 they struck. It was close-in work. For a while, the outcome was in doubt. Separated from Company A, Company B was on its own. Its entire left flank was overrun, the communications wire was cut, and two Marine machine guns were captured and turned on the company. Fighting back face-to-face, the Marines called in fire from their 81mm and 4.2-inch mortars, together with artillery and 3.5-inch rocket rounds that pinpointed the enemy with fire barely in front of the defenders. Finally, at dawn, the situation was stabilized.

There now occurred "one of the most demoralizing incidents in Company B's experience for the entire campaign," as Fenton later commented. Tobin was ready at first light to move back and recover the wounded and missing men, just as Marine tradition (and Craig) had promised. It was not to be. Iron-clad orders from Walker to Craig to Murray to Newton forced an immediate withdrawal, in spite of Tobin's pleadings.

Fenton summarized the unanimous feeling:

> Twenty-nine bloody, sweating miles down the drain . . . . The men couldn't believe it. I couldn't believe it. It didn't seem possible, with all the lives we'd lost taking this ground, that we'd now just walk off and leave it. Baker Company's casualties for the morning's counterattack alone were 12 dead, 16 wounded, and 9 missing in action. And I'm certain those last nine were dead, too.
>
> I found it difficult to see men, veterans of the last war, older guys, sitting by the side of the road crying. They just didn't give a hoot. They were tired, disgusted. People just couldn't understand this part of the war.

A Relief Force

Leaving the 1st and 2d Battalions temporarily in the positions they had won in the Changchon area, Craig moved quickly on 12 August to organize the deployment of his 3d Battalion as a relief force for the overrun Army field artillery battalions. The orders from Kean had come at 1130 and by 1300 the riflemen and an artillery battery were in the trucks, on their way. A half hour later Taplett and the brigade operations officer, Lieutenant Colonel Stewart, were airborne to scout the disaster area by helicopter. They saw plenty of trouble: artillery pieces in disarray; jeeps on fire; American bodies lying in a stream bed; and, incongruously, one white table set in the midst of it all. The Army had "estimated" that 2,000 to 2,500 NKPA troops had infiltrated the area, smashed the Army artillery units, and were threatening the main supply route, so Taplett had originally presumed that there would be heavy combat for his battalion when it arrived. At the scene he saw no evidence of any such quantity of NKPA, and he strongly doubted the estimate.

The chaotic situation the Marines now saw had its roots in the events of the preceding day, 11 August. Without opposition, the 5th RCT had advanced just five miles from where it had started at the infamous road junction to a small village called Pongam-ni. The 555th "Triple Nickel" and 90th Field Artillery Battalions were in support, but were not protected or prepared for an enemy attack.

Marine procedures were much different. Craig later commented on this:

> The artillery had been trained in Pendleton in the methods of security. They were armed with bazookas, .50 calibers, and everything that the infantrymen would need to defend a position, and they were well trained in defense of their artillery positions. And they from that [first] day on took up defensive positions wherever they moved.
>
> As a result we never had a gun overrun. There were attempts at sniping and so forth, but we never had a gun taken or overrun; whereas I notice that the Army on a number of occasions in the perimeter lost whole batteries. It was simply, I think, because the artillerymen were not trained along the same lines as the Marines.

At this time, Kean was under heavy pressure from Walker to get the 5th RCT to leap ahead. So the division commander ordered his regimental commander (Colonel Godwin L. Ordway) to move part of his units quickly forward through the pass near Pongam-ni. Then there was indecision, delay, conflicting orders, and repeated failures in radio communications. As a result, part of the regiment went through the pass that night, and part stopped at Pongam-ni. With his command thus split up, and with enemy fire falling on the supply route to his rear, Ordway was in a difficult situation. It got worse after midnight on 11 August when telephone and radio communications with the artillery battalions was lost and the sounds of battle came from their direction. With the NKPA now on the high ground above him, Ordway decided at 0400 on 12 August to try to move the rest of his troops through the pass. A massive traffic jam ensued. As the official Army history noted: "During the hour or more before daylight, no vehicle in Ordway's range of vision moved more than 10 or 20 feet at a time."

As the infantry slowly moved out, the enemy quickly moved into the valley. Now the Army artillery, stalled behind the traffic jam, was a sitting duck. NKPA tanks and self-propelled guns were able to "approach undetected and unopposed, almost to point-blank range, and with completely disastrous effects." Enemy infantry from the *13th Regiment* of the *6th Division* closed in and added its firepower. It was a slaughter, and the artillery was completely overrun. A traumatic phone call from Brigadier General George B. Barth, USA, the 25th Division artillery commander, to Kean revealed the scope of the disaster and led Kean to order the rescue mission by the 3d Battalion, 5th Marines.

Kean also ordered a battalion of the 24th Infantry to bring relief by an attack towards Pongam-ni. This effort went nowhere on 12 August, and by the next day it was still two and a half miles from the artillery positions. The 555th had lost six of its 105mm howitzers, and the 90th had lost all six of the 155mm howitzers in one of its batteries. Along with some 300 men, probably 100 vehicles had been captured or destroyed (although the NKPA claimed an inflated 157 vehicles and 13 tanks). The Army had given the site the name "Bloody Gulch."

This was the grim situation that Taplett faced when his helicopter arrived on 12 August. He immediately had the aircraft land and he looked for the liaison officer who was supposed to meet him, now that he was coming under the operational control of the Army's 25th Infantry Division. No sign of any such person.

To try to get some information, Taplett was finally able to tap into a telephone line to the division headquarters in the rear and ask for orders. The reply was to "do what he thought was proper." That vague verbal order was all the leeway Taplett needed for immediate action. A helicopter reconnaissance was followed by a juncture with his troops. Then he led them by air to the valley from where he planned to attack the commanding ridges.

Less than three hours after boarding their trucks, the men of the 3d Battalion were at their assembly area, ready to jump off in an attack on a cold, rainy, miserable day. Taplett aggressively delayed only 15 minutes for an artillery preparation and some napalm runs by Marine Corsairs, and then moved out the riflemen. Without a single casualty, they soon reached the top of the first ridge. There they found signs that a substantial body of enemy troops had made a hasty departure, but this was a far cry from the resistance they had expected from the "2,000" or so enemy troops that Ordway had estimated had wreaked such havoc.

At 1900 Barth arrived to take command. Not knowing Taplett's style, he asked when the Marines would be ready to attack. Taplett presumably enjoyed a response one can easily imagine, "Sir, we've already done that, and my men are now digging in on top of the ridge." Barth graciously congratulated him.

The next morning, 13 August, the 3d Battalion attacked to secure the final ridges overlooking the pitiful remnants of the lost artillery. Again, there was no opposition, and by 1000 they were on top of their objectives. Craig later commented: "We found quite a number of Army artillerymen scattered through the area, hiding in various places." Besides those rescued by the Marines, some had fled and struggled back to safety with the 25th Division.

Taplett's men were now ready to go down, clean out any enemy, and retrieve the artillery pieces in the valley, but the Marines once more got orders that they could not take the objective they were poised to seize, but must, instead, move to the rear to meet the new enemy threat along the Naktong.

That marked the final episode in the Marine mission to aid the Army's 5th RCT. With all troops, Marine and Army, now pulled back to their starting point at Chindong-ni, it was the end of the offensive to occupy Chinju and, on 16 August, Task Force Kean was dissolved.

First Week's Results

Things had gone badly for the 5th RCT and its artillery, and the commanding officers of the regiment and the "Triple Nickel" battalion were both relieved of duty. Higher Army echelons were not pleased with their leadership or the morale and combat effectiveness of their men.

Craig, on the other hand was pleased. He had seen his brigade drive forward with vigor and professional skill. His officers were constantly aggressive, and the riflemen had done very well under fire. He noted that his men were "well trained and well led" by outstanding noncommissioned officers and "professional" officers who "knew their stuff." The reason for the brigade's achievements were clear to Craig:

> We were a generation of officers who grew up with the Marine Corps' standing operating procedures (SOPs) for amphibious operations. These were my "Bible" when I organized and trained an earlier Marine brigade on Guam during the period 1947-1949. During World War II we had repeatedly tested and refined our organization and techniques in landings all over the Pacific. These same SOPs enabled us to deploy to Korea quickly and fight effectively when we got there.

Equally important, the supporting arms had coordinated well with the infantrymen, with the close air support of MAG-33 demonstrating a wholly new element in the Korean War, flying more than 400 sorties in support of the brigade and other units of the Eighth Army. The Marines had twice been on the verge of seizing their objectives—first at Sachon-Chinju and then the recovery of the Army artillery—only to be pulled back by the strategic needs of the Eighth Army. Geer in his account concluded:

> The brigade came out of Changallon [Changchon] physically tough and psychologically hard . . . . They knew the enemy to be a vicious, skillfully led and well-equipped foe that could inflict heavy casualties in any action. They were prepared to meet with heavy losses and to carry on the attack, and were openly scornful of units unable to face these hard facts of war.

There had been a price, however. The brigade had had a total of 315 casualties, with 66 killed or died of wounds, 240 wounded, and 9 missing in action (when the 1st Battalion had not been allowed to recover them).

The action of that week had brought results on a wider, strategic scale. While there had been a failure to occupy Chinju, Task Force Kean had nevertheless been the first real American offensive of the Korean War. In a report to the United Nations, General MacArthur stated that "this attack not only secured the southern approaches to the beachhead, but also showed that the North Korean forces will not hold under attack."

The official Army history acknowledged in summary that "the task force had not accomplished what Eighth Army had believed to be easily possible—the winning and holding of the Chinju pass line," and, omitting any reference to the dramatic advance of the Marine brigade, admitted that the rest of the task force, "after a week of fighting, . . . was back

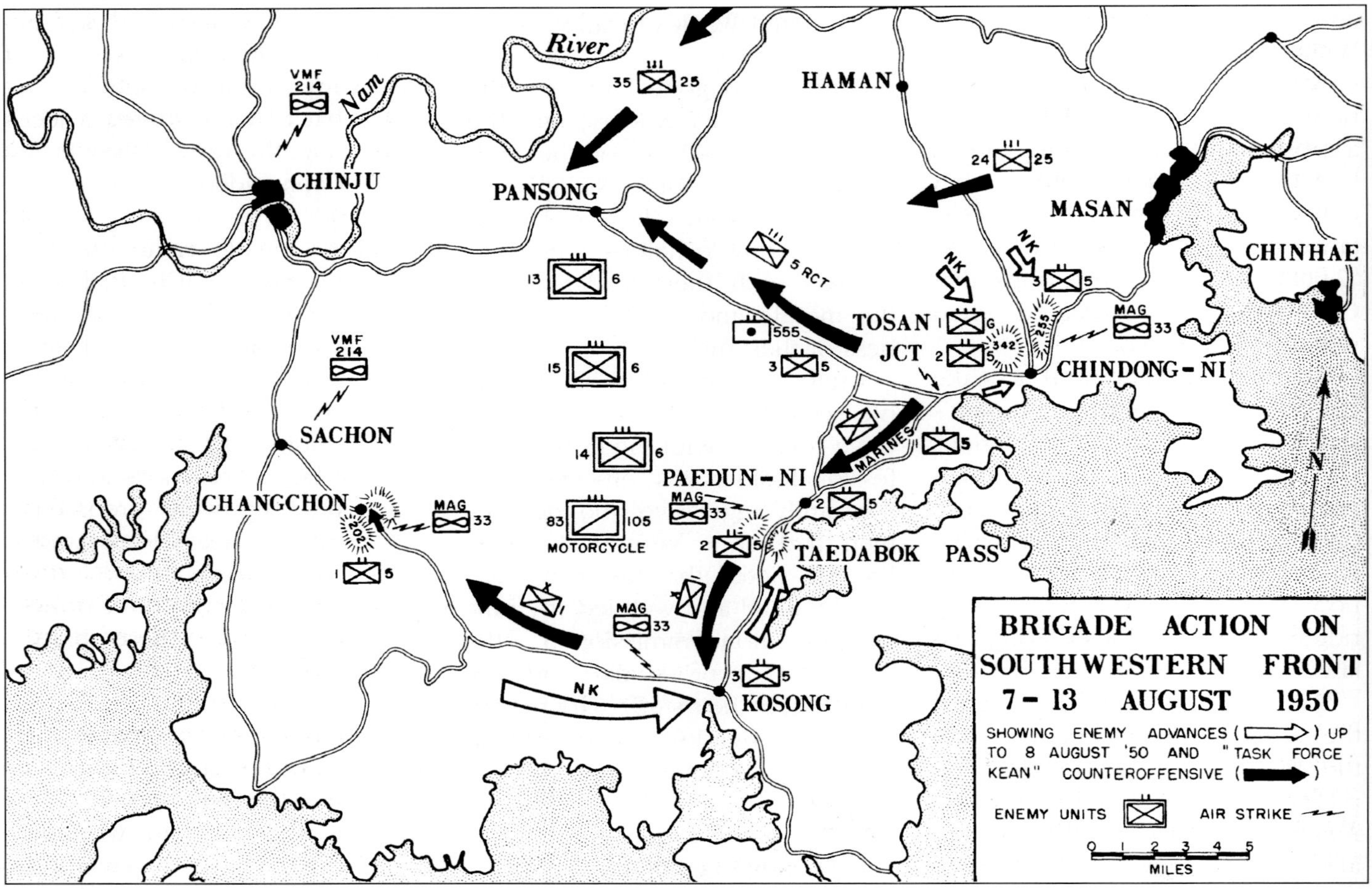

approximately in the positions from which it had started its attack." That history, however, went on to note "certain beneficial results . . . . It chanced to meet head-on the North Korean *6th Division* attack against the Masan position, and first stopped it and then hurled it back . . . . Task Force Kean also gained the time needed to organize and wire in the defenses that were to hold the enemy out of Masan during the critical period ahead."

The official Marine history could afford to be positive about the brigade's achievements:

> The Communist drive in this sensitive area came closest of all NKPA thrusts to the vital UN supply port of Pusan. Up to that time the NKPA units spearheading the advance—the *6th Infantry Division* and the *83d Motorcycle Regiment*—had never suffered a reverse worth mentioning since the outset of the invasion. Then the counterattack by the 1st Provisional Marine Brigade hurled the enemy back 26 miles in 4 days from the Chindong-ni area to Sachon.

It was estimated that the Marine air-ground team killed and wounded 1,900 of the enemy while destroying nearly all the vehicles of an NKPA motorized battalion in addition to infantry armament and equipment. The enemy threat in this critical area was nullified for the time being, and

A Marine skirmish line attacking over exposed ground to a nearby treeless hillcrest.

Photo by David Douglas Duncan

never again became so serious. Marine efforts assisted Army units of Task Force Kean in taking new defensive positions and defending them with fewer troops, thus freeing some elements for employment on other fronts. Finally, the Marines earned more time and space for the building up of Eighth Army forces in preparation for a decisive UN counteroffensive.

Interlude

With the conclusion of the drive towards Sachon, the Marines hoped for a respite before the next call to combat, which they knew was sure to come. Craig, however, had received orders at 0130, 14 August, to move his brigade as soon as possible to a place called Miryang. Using rail, trucks, and even an LST, his battalions made the trip of 75 miles in 26 hours. When the "Fire Brigade" arrived there, it was desperately needed in a new crisis.

Before the men moved out for combat, there was one blessed—though brief—interlude of relaxation: Marines from the rear, from staff positions, even tankers and artillerymen, were fed into the depleted rifle companies. (Another of the many times when there was a vital payoff for the Marine maxim, roughly: "No matter what your ultimate assignment may be, you will be trained first as a rifleman!") There was a pleasant grove of trees at Miryang, and the men could rest in the shade, get their first-ever bath in the river there, eat their first hot food, and exchange filthy, rotted uniforms for a fresh issue. Fegan commented: "Not only did I smell to high heaven, I also had dried blood all over my jacket."

That rest period was soon over. Upon arrival at Miryang, the brigade was placed under the operational control of the Army's 24th Infantry Division to meet a new threat. The situation was indeed critical. Ten days before, author Russell Spurr asserts, General Kim Chaek, front commander of the NKPA, had addressed his staff. Moving from the past (Sachon) battle to the forthcoming (Naktong) attack, he reputedly acknowledged that losses had been heavy, with the *6th Division* "reduced by half in the past week." He then went on to issue a clarion call for victory:

> The situation is not irretrievable. We have committed only a portion of our strength. I am therefore ordering the *4th Guards Division* to cross the Naktong River north of the present battlefield, capture Yongsan, and drive on to Miryang. This as you can see from the map, will sever the main supply route between Pusan and U.S. headquarters in Taegu; if we succeed, and I trust we shall, the northern part of the perimeter will collapse. It is defended largely by puppet troops and we know how they react when outflanked.

Enemy Breakthrough

The commander of the *4th Division* was Major General Lee Kwon Mu, a hardened professional who had fought with the Communists in China and served as a lieutenant in the Russian Army. Awarded North Korea's highest military decorations, the Order of Hero of the Korean Democratic People's Republic, and the National Flag First Class, for his earlier triumphs in South Korea, Mu had moved his 7,000 men into position on 4 August for a crucial attack across the Naktong River. The *4th Division* was a crack unit, given the honorary title of the *"Seoul" Division* for its triumphant earlier capture of the capital of South Korea. Leading the way were the *4th, 16th, and 18th Infantry Regiments*. They had moved stealthily into action the night of 5 August, wading across the Naktong under cover of darkness, while machine guns were pulled along on crude rafts. By the morning of 6 August, 1,000 of them had established a position on the east side, soon beefed up by artillery brought across the Naktong on a hidden, underwater bridge the NKPA had secretly constructed. This assault had meant the breaching of the last natural barrier which was counted on to protect the vital lifeline from Taegu to Pusan. It was at Taegu that General Walker had his headquarters for direction of the defense of the Pusan Perimeter.

This attack had come as a surprise to Brigadier General John H. Church, commander of the 24th Infantry Division. The subsequent threat was obvious. From the hills the NKPA had seized it dominated the road to Yongsan, five miles away. Twenty-five miles beyond that lay Miryang, and then the vital Pusan-Taegu main supply route (MSR). As Toland recorded: "Panic reached the government offices in Taegu." Walker, however, had remained cool, and the Army had entered a period of continuous battle. Some units were overrun and some soldiers had fled as NKPA soldiers appeared on flanks and rear. In a confusing period of separate confrontations, Army troops had been unable to push the NKPA here, and at another point in the north, back across the Naktong.

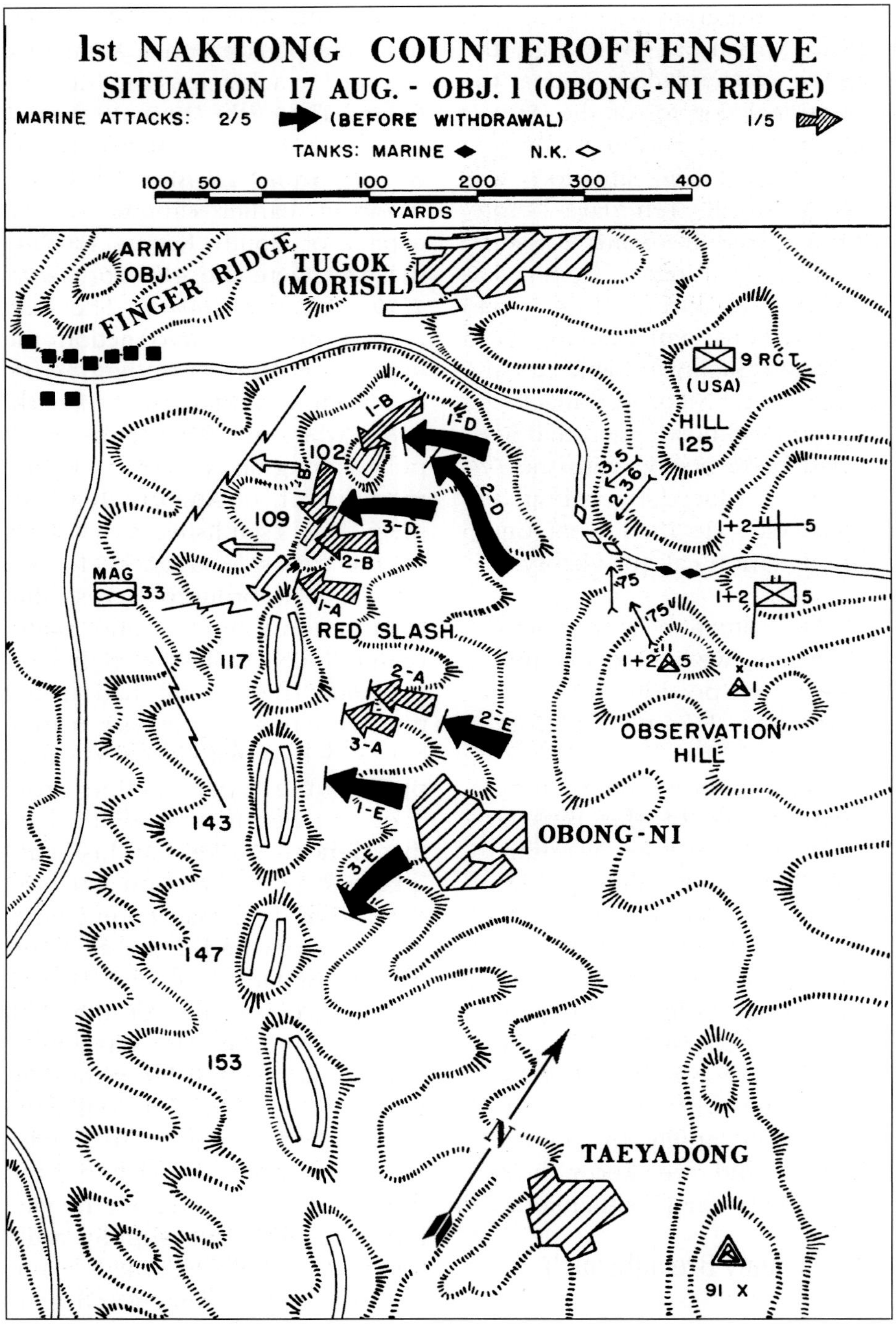

'Fire Brigade': Crisis Number Two

That was when Walker called in the Marines. Thus, on 15 August, Craig met with Church. Walker had earlier told Church, "I am going to give you the Marine Brigade. I want this situation cleaned up—and quick!" Craig made his plans following his meeting with Church. The brigade would move out of Miryang on 16 August to go on the attack. Geer records a British military observer who saw them getting started and sent a dispatch to Tokyo. He emphasized a "critical" situation in which Miryang could well be lost, then Taegu would become untenable, and "we will be faced with a withdrawal from Korea." In spite of these grim prospects, he got a premonition about the brigade. In spite of the "impossible odds" that he felt it faced, he described his gut feeling that it would check the *NKPA* advance:

> I realize my expression of hope is unsound, but these Marines have the swagger, confidence and hardness that must have been in Stonewall Jackson's Army of the Shenandoah. They remind me of the Coldstreams at Dunkerque. Upon this thin line of reasoning, I cling to the hope of victory.

That night, the tone of the attack was set when Murray told Newton: "You must take that ground tomorrow! You have to get on that ridge and take it! Understood?" Newton replied: "Understood! Understood! This battalion goes only one way—straight ahead!"

The brigade was to jump off at 0800 on 17 August as part of a planned full-scale effort by the Army's 24th Division, reinforced by the 9th Infantry Regiment. There was a happy history of linkage between the Marines and the 9th Infantry. They had served together in the battle for Tientsin during the Boxer Rebellion in China at the turn of the century, and again in the 2d Infantry Division in France during World War I. Now the 9th would operate on the brigade's right, with the Marines as the left wing of the attack. Three objective lines were assigned to the brigade, with the first being Obong-ni Ridge. Craig and Murray made an on-the-spot reconnaissance of the terrain which was a jumbled mass of hills and gullies. Because of the type of terrain to the left and the presence of the Army's 9th Regiment to the right, the only, reluctant choice was a frontal attack.

Department of Defense Photo (USMC) A2246

Marches had the men well spaced out to avoid unnecessary casualties. As this column heads toward its objective, the man second from the rear carries a flame thrower, while the man in front of him has a 3.5-inch rocket launcher. This picture was taken during a Naktong battle, showing a burning enemy tank, with the Marines carefully circling around it to avoid any explosion of its ammunition.

The shift to the new crisis area was a pressure-laden one for the Marines. Stewart, Craig's operations officer (G-3), remembered in later years that he was advised that the Naktong River line had been broken through, threatening the Pusan-Taegu MSR, and the brigade had to move there immediately to restore the front. He recalled:

> Things were so hectic that Roise, who was commanding the 2d Battalion, which was going on the line below Masan in a defensive position, received minimum orders to move. In fact, our radio contact was out and I wrote on a little piece of brown paper, "These are your trucks, move to Naktong at once."

Those were the only orders Roise ever got to move to the Naktong front. But they were all he needed in the hectic situation in which the Marines found themselves, for, when only a portion of the promised trucks showed up, many men in the battalion had to march until 0130 the next morning to reach the jump-off point for their attack a few hours later.

Waiting for the Marines, well dug-in and confident of victory, were the *18th Regiment* and a battalion of the *16th Regiment* of the NKPA *4th Division*. Geer quotes a speech by Colonel Chang Ky Dok, the regiment's veteran commanding officer:

> Intelligence says we are to expect an attack by American Marines. To us comes the honor of being the first to defeat these Marines soldiers. We will win where others have failed. I consider our positions impregnable. We occupy the high ground and they must attack up a steep slope. Go to your men and tell them there will be no retreat. I will take instant action against anyone who shows weakness.

Preparation by supporting units for the Marine riflemen's attack was inadequate. Artillery fire was ineffective. When the enemy positions were later examined, the foxholes were found to be very deep, sited along the length of the ridge slightly on the reverse slope. Thus shellfire on the forward slopes caused few casualties, nor could artillery get a trajectory to reach the enemy on the reverse slopes. Adding to the problem, there was only one air strike. Moreover, there would be little or no natural cover for the men who had to climb toward the six hills of Obong-ni, called by the news correspondents "No Name Ridge."

The 2d Battalion Attacks

Murray had an agreement with the Army's 9th Infantry on the right flank that the Marines would attack first, supported by fire from the 9th. He picked Roise's 2d Battalion to lead off. It was a very thin front line for such a crucial moment: four understrength platoons totaling only 130 men from Companies D and E to lead the assault (with two platoons as reserves). "Red Slash Hill" was to be their dividing line.

One platoon of Company E, led by Second Lieutenant Nickolas D. Arkadis, hit the village of Obong-ni at the foot of two of the company objectives: Hills 143 and 147. Driving ahead through heavy fire, the platoon fought its way to the slopes beyond. Arkadis' leadership was later recognized by the award of a Silver Star.

Now both companies were out in the open, sometimes forced to crawl upwards, met with a continuous hail of enemy machine gun and mortar fire with barrages of grenades. Casualties mounted rapidly. Joseph C. Goulden tells of a correspondent who was watching and described the bloody scene: "Hell burst around the Leathernecks as they moved up the barren face of the ridge. Everywhere along the assault line, men dropped. To continue looked impossible. But, all glory forever to the bravest men I ever saw, the line did not break. The casualties were unthinkable, but the assault force never turned back. It moved, fell down, got up and moved again."

One platoon of Company D, with only 15 men remaining, did claw its way to the top of Hill 109 on Obong-ni Ridge, but it was too weak and too isolated when reinforcements simply could not reach it, so it had to pull back off the crest. Second Lieutenant Michael J. Shinka, the platoon leader, later gave the details of that perilous struggle:

> Running short of ammo and taking casualties, with the shallow enemy slit trenches for cover, I decided to fall back until some of the fire on my left flank could be silenced. I gave the word to withdraw and take all wounded and weapons. About three-quarters of the way down, I had the men set up where cover was available. I had six men who were able to fight.
>
> I decided to go forward to find out if we had left any of our wounded. As I crawled along our former position (on the crest of Hill 109), I came across a wounded Marine between two dead. As I grabbed him under the arms and pulled him from the foxhole, a bullet shattered my chin. Blood ran into my throat and I couldn't breathe.

Shinka, after being hit again, did manage to survive, and was later awarded a Bronze Star Medal. Another Company D Marine, Staff Sergeant T. Albert Crowson, single-handedly silenced two deadly machine gun emplacements and was awarded the Army's Distinguished Service Cross by order of General MacArthur.

By now, it had become clear that many of the casualties were caused by heavy enemy fire coming from the zone in front of the Army's 9th Regiment to hit the flank and rear of the Marines, and there had been no supporting fire from the 9th. Other problems arose when some men of Company E were nearing the crest which was their objective and they were hit by white phosphorus shells from "friendly" artillery fire. Then, later, some Marines were hit in a strafing attack by their own Corsairs.

By mid-day the men of the 2d Battalion, halfway up the hills, could do no more, having suffered 142 casualties, 60 percent of their original 240 riflemen. Murray ordered it to pull back, undoubtedly lamenting the fact that he did not have a third rifle company in the battalion, for it might well have seized the top of the ridge and held it. Craig stressed this point in a later interview, noting that "without a third company, or maneuver element, the battalion commanders were at a tactical disadvantage in every engagement. They lacked flexibility in the attack. On defense they had to scrape up whatever they could in order to have a reserve."

Pinpointing an example, Craig recalled:

> This condition became critical in the First Battle of the Naktong. 2d Battalion, 5th Marines' assault companies took heavy losses in the initial attack against Obong-ni Ridge, the strongpoint of the enemy's bridgehead over the Naktong. Since Roise had nothing left to use, the attack stalled.
>
> Murray then had to commit 1/5 [the 1st Battalion] prematurely to continue the attack. This took time, giving the enemy a breather right at the height of the battle. That night, when the enemy hit our positions on the ridge with a heavy counterattack,

A Marine tank bulldozer clears a destroyed NKPA T-34 tank from the road.

National Archives Photo (USMC) 127-N-A1338

Newton certainly could have used another company on line or in reserve. We were spread pretty thin, and it was nip and tuck on that ridge for several more hours.

The original battle plan had called for an attack in a column of battalions, with each battalion taking successively one of the series of three ridge lines (objectives 1, 2, and 3) that shielded the NKPA river crossing It was now painfully obvious that a sharp change must be made. Accordingly, Newton's 1st Battalion relieved the battered 2d on the hillsides at 1600 (17 August), with Company A replacing E, and B replacing D.

While the *18th Regiment* had hit the 2d Battalion hard, the bravery, skill, and determination of those Marines had caused serious losses in the enemy's ranks: 600 casualties and severe reductions in serviceable weapons. With his ammunition running low and no medical supplies so that most of his wounded men were dying, the NKPA commander's situation was critical, as described by Fehrenbach:

> He knew he could not withstand another day of American air and artillery pounding and a fresh Marine assault up the ridge. Because he had a captured American SCR-300 radio, tuned in on Marine frequencies, he knew that the 1st Battalion had relieved 2/5 along the front of Obong-ni, and he knew approximately where the companies of 1/5 were located, for the Marines talked a great deal over the air.

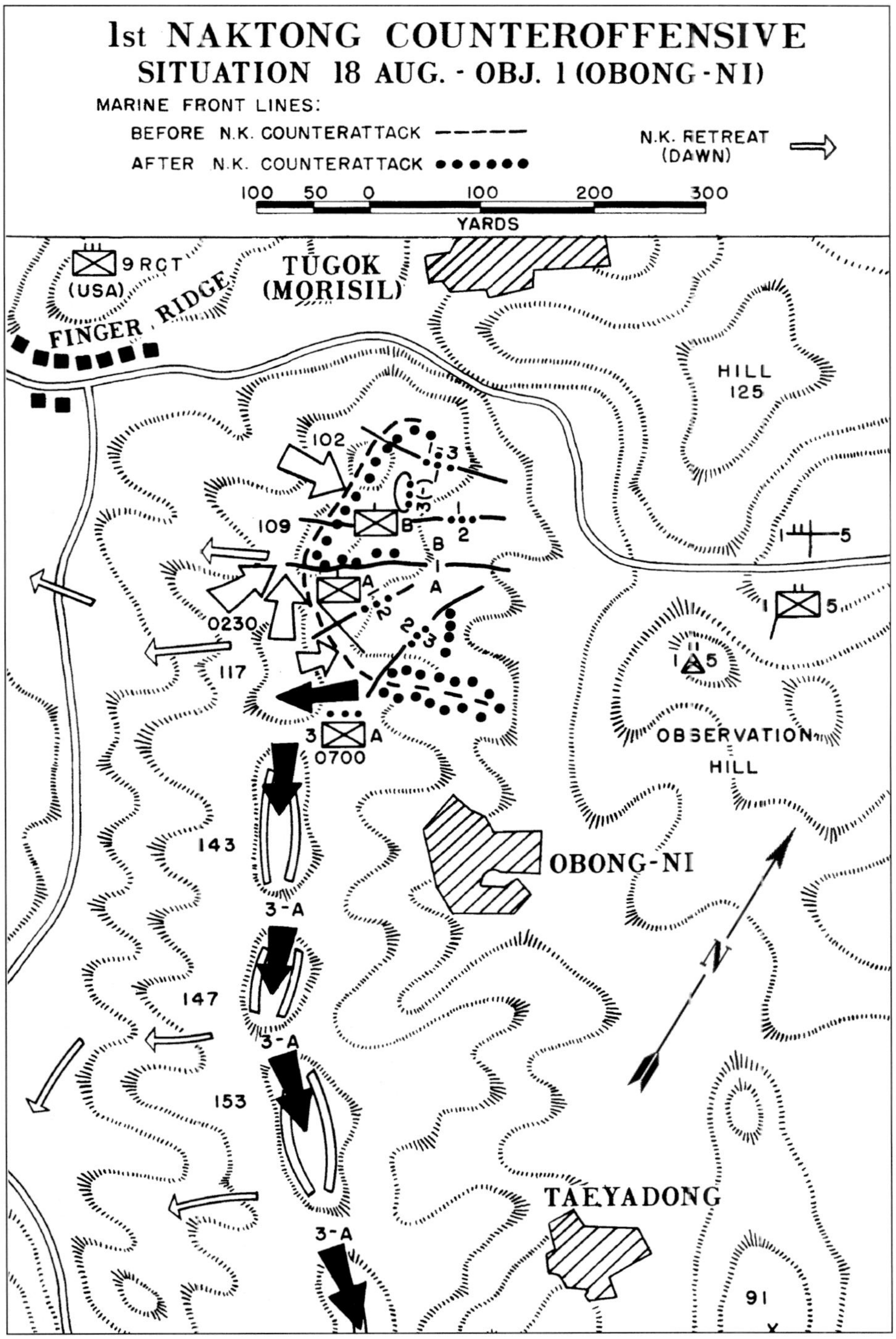

The 1st Battalion Attacks

The relief movement of the two battalions was covered by what the official Marine history described as "devastating fires" from the planes of MAG-33, the artillery of the 1st Battalion, 11th Marines, and the brigade's tank battalion. Then Companies A and B attacked up the daunting slopes. Simultaneously, after Murray had gone to see Church to request a change in the previously agreed-upon plan, the 9th Infantry jumped off in an attack. This eliminated the previous flanking fire on the Marines.

Helped by the advance bombardment, the two Marine companies were able to make slow (and costly) progress towards the crests. Company A attacked repeatedly, trying to reach the battalion's objective on the left: the tops of Hills 117 and 143. It proved impossible, in spite of very aggressive leadership by the officers (and gunnery sergeants who replaced them as they fell). The company could get only part way up the slopes when it was "pinned down by a solid sheet of Communist fire . . . casualties bled [the] skirmish line white and finally brought it to a stop."

Herbert R. Luster, a private first class in Company A, remembered his own searing experience in this brutal battle:

> It was evident no one saw the enemy but me . . . . I pulled back the bolt to cock the action of the BAR, pushed off the safety, settled back on my right foot, and opened fire. The flying dirt and tracers told me where my rounds were going. I emptied the rifle . . . . So I pushed the release with my right thumb and pulled the empty magazine out, stuck it in my jacket pocket, loaded and raised my BAR to my shoulder. Before I got it all the way up, red dirt kicked up in my face. A big jerk at my right arm told me I was hit. I looked down and saw blood squirting onto my broken BAR stock.

As always, there were gory episodes. Second Lieutenant Francis W. Muetzel in Company A was in an abandoned machine gun emplacement with his company executive officer and a rifleman from the 3d Platoon. He later recalled:

> The use of the abandoned machine gun emplacement proved to be a mistake. Enemy mortars and artillery had already registered on it . . . . Without registration of any kind, four rounds of enemy 82mm mortar fire landed around it. The blast lifted me off the ground, my helmet flew off. A human body to my left disintegrated. Being rather shook up and unable to hear, I crawled back to the CP . . . . About the time my hearing and stability returned . . . I thought of the 3d Platoon rifleman . . . . I returned to look for him. One of the mortar rounds must have landed in the small of his back. Only a pelvis and legs were left. The stretcher-bearers gathered up the remains with a shovel.

On the other side of "Red Slash Hill" that was the dividing line, Company B made some progress until it was pinned down by heavy fire from a nearby village on its flank. Captain John L. Tobin was wounded, so Fenton took over as company commander. Calling in an

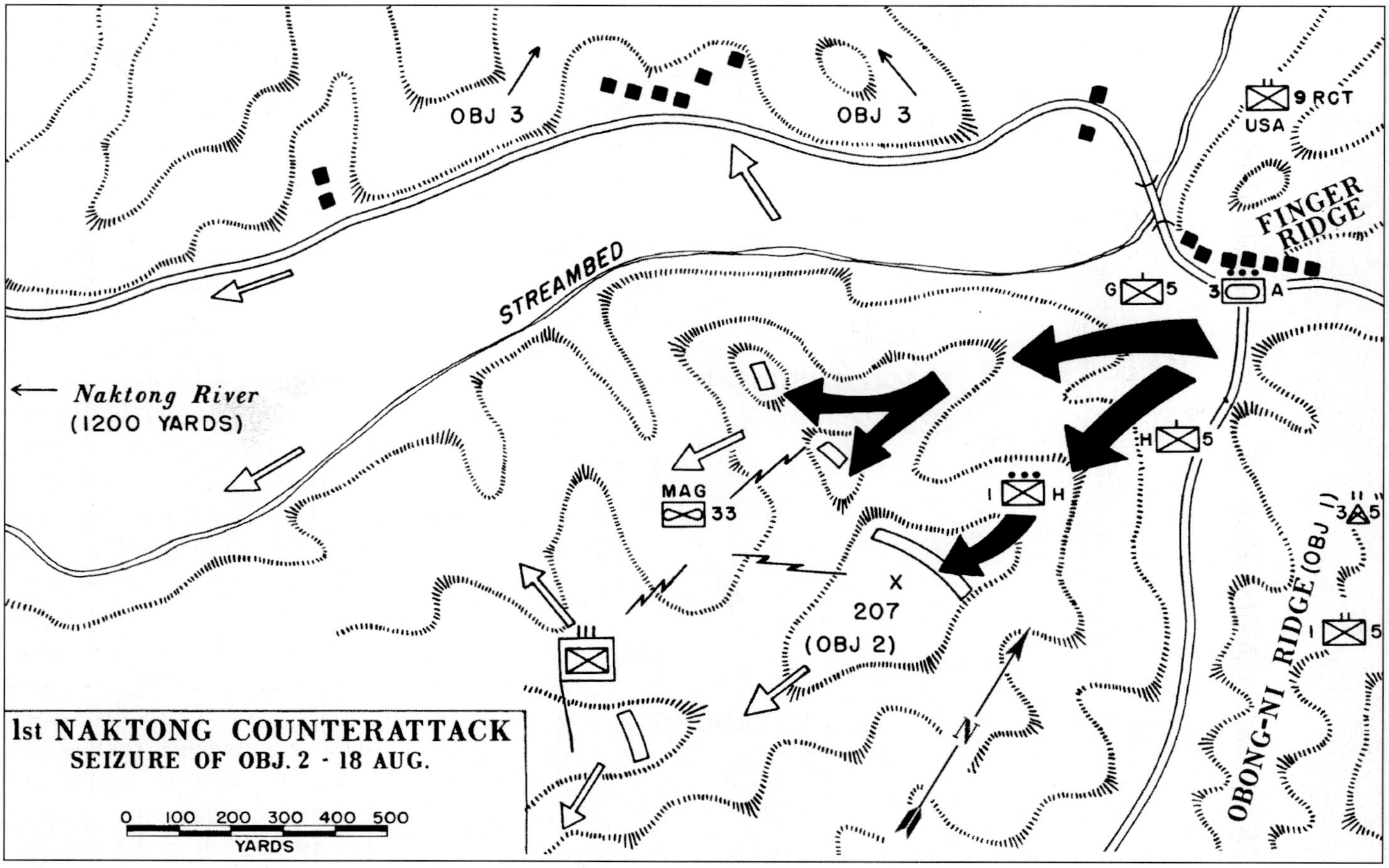

81mm mortar barrage from the battalion's weapons company, the riflemen were then able to lunge forward and seize the crests of Hills 102 and 109 by late afternoon (17 August).

The two battered companies settled down where they were and tied into each other to dig in night defensive positions. With the flood of casualties, the resulting manpower shortage caused the far left flank to dangle dangerously in the air. Newton threw together an improvised unit of men from his headquarters and service company personnel to cover that flank. The mortars and artillery were registered on probable enemy approach routes, including the crossing point on the Naktong River. Then their harassing fire missions went on all night to try to disrupt the enemy.

National Archives Photo (USMC) 127-N-A1160

Fire from supporting weapons was a crucial element as the rifle companies attacked. Above: 81mm mortars lay down a barrage. Below: a 75mm recoilless rifle blasts a specific target.

Smashing Enemy Tanks

At 2000 that night (17 August) the Marines had their first confrontation with the T-34 tanks of the NKPA. These were the tanks that had had such a fearsome reputation earlier in the war. The men of Company B from their hilltop perch saw four of them coming with a column of infantry, aimed to bypass the Marine riflemen, and, in a typical enemy tactic, probe to sow confusion amongst rear elements.

The Corsairs of MAG-33 were called in. They came roaring down, knocked out one tank, and scattered the accompanying enemy infantry. With a determination typ-

National Archives Photo (USMC) 127-N-A1461

ical of the hardy NKPA, the other three tanks came on.

When the news was flashed to Newton in his CP, he told Fenton to "let them go and they'd be dealt with in the rear." Back at Craig's brigade CP, there were two opposite reactions when the news arrived. A correspondent witnessed the scene: "Naval Captain Eugene Hering, brigade surgeon, jumped to his feet. "God Almighty!" he said. "The aid station's just a quarter of a mile from there! [Lieutenant (junior grade) Bentley] Nelson [one of the battalion's medical officers] won't leave his wounded! If those tanks break through . . . ." "They won't," the general said. "Newton will know what to do."

And he did. Summoning the Marine M-26 tanks and antitank weapons, Newton left the NKPA armor up to them. Fenton and the men of Company B had a ringside seat for the clash that followed. He later wrote:

> As the first tank rounded the corner down toward the 1st Battalion CP, it was met by 3.5" rocket fire from the antitank assault section, and fire from our 75mm recoilless weapons in position on the high ground on either side of the road. The tank was knocked out, and the second tank immediately came up and tried to go around it. The second, too, was hit in the track and skidded off the road. Our M-26 tanks finished him off [after a 2.36" white phosphorus rocket had ricocheted inside it, creating a fiery cauldron]. The third tank made the same mistake that the second tank made. He, too, tried to go around the other two tanks. One of our M-26 tanks hit this third tank with a direct hit. All three of these tanks were finished off by our M-26 tank platoon.

Back on the hills, the men of the 1st Battalion spent the midnight hours on the alert. The attacks that day had cost the brigade 205 casualties, and, to avoid the punishing Marine air strikes in daylight, the enemy was sure to counterattack during the darkness.

The Enemy Reaction

And it did. At 0230 a green signal flare soared into the sky, and the enemy hit—and hit hard. With their captured U.S. Army radio tuned to the Marines' frequency, the attackers knew the exact place where the two Marine companies were tenuously tied together, and they sought to drive a wedge in there and then envelop each company separately. With Company A only part way up Hill 117, machine gun fire from the crest and grenades rolling downhill covered the assault troops of the NKPA, as they ran down throwing more grenades and spraying submachine gun fire. A rifle platoon was in deep trouble, the mortar platoon was decimated, the Marine defense line was penetrated, the company was split in half, the battalion was assaulted, and the enemy forced Company A to make a partial withdrawal back to a spot near Hill 109.

Things were not much better in the Company B zone. With the two Marine companies split by the NKPA, the enemy assault smashed hard into Fenton's men. A platoon was overrun under the eerie light of mortar illuminating shells. The attackers charged into the CP, where hastily assembled stray Marines met them in bitter hand-to-hand combat. Possession of the two hard-won hills and, in fact, the outcome of the whole brigade attack hung in a delicate, trembling balance.

Just at that precarious moment, the phone rang in the CP of

On top of Objective Number 3, Marines look down on the Naktong River.
National Archives Photo (USMC) 127-N-A1401

Company B. It was Newton, calling to say that the position must be held "at all costs," and that he was pouring in all the supporting mortar and artillery fire he could muster. (This apparently prevented the NKPA from feeding in reinforcements to exploit the breakthroughs.) Newton's main message was a brutally frank reminder that, if the Marines retreated, they would simply have to grind their way back to the lost positions in the forthcoming days. Then Newton asked if they could hold on until daylight could bring relief. Fenton's reply has been variously reported: "We have gooks all around us"; "They've turned my left flank"; "Don't worry, Colonel. The only Marines that will be leaving this ridge tonight will be dead ones."

The supporting fire from the 4.2-inch mortar company proved to be an invaluable asset. With its high angle of fire, it was able to search out and wreak havoc on NKPA units shielded in gullies which Marine artillery fire could not reach. The company's commanding officer, First Lieutenant Robert M. Lucy, later recalled:

> The 1st Battalion was receiving a terrifically heavy counterattack. Our company was zeroed in on the hill and the valley in front of the battalion. When notified of this attack, we began firing our prearranged barrages. Later, where only one of these barrages had fallen, they counted 120 dead North Koreans with 12 cart-mounted machine guns, who had been massed in this little gulley behind the hill, a ridge in front of the battalion that would have caused them considerable trouble.

With many officers down and aided by the supporting fire, the noncommissioned officers took the lead in regrouping their units, and so the men of the depleted Companies A and B stood, and fought, and died, and finally held their ground. Typical of the unyielding defense were the examples of two platoon leaders, Second Lieutenant Hugh C. Schryver, Jr., in Company B, and Second Lieutenant Francis W. Muetzel in Company A. Both officers, although severely wounded, continued to lead their men with the "fierce determination" described in their citations for awards of the Silver Star. Slowly, toward dawn on 18 August, the enemy attacks weakened. But the Marines had paid a fearful price. Company B had begun the night with 190 enlisted men and five officers; the next morning there were only 110 left, with one officer

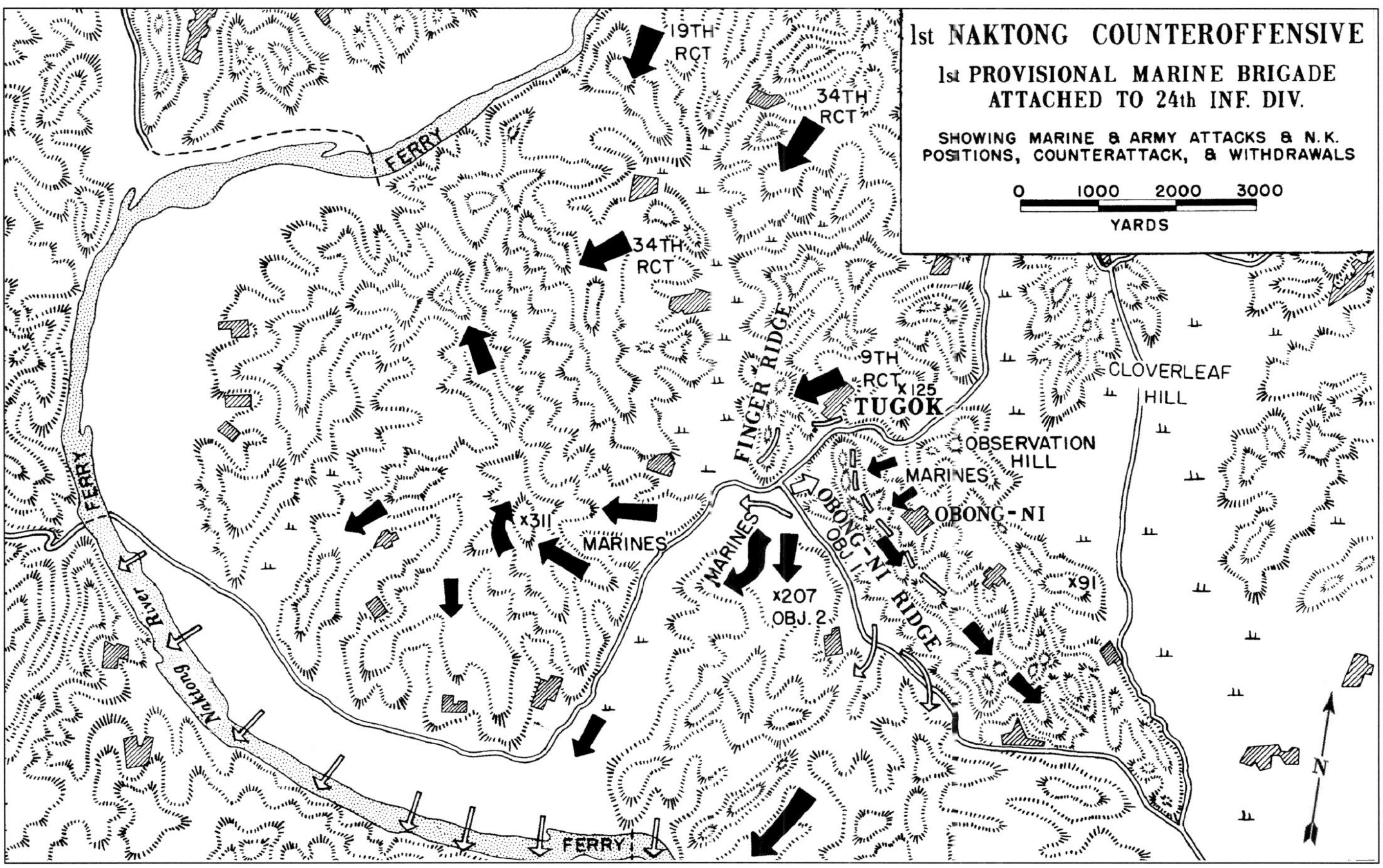

Department of Defense Photo (USMC) A-157140

With an enemy who was adept at infiltration and night attacks, Marines strung barbed wire whenever they had an opportunity.

still standing. Company A was in worse shape with just 90 men remaining from the 185 at the start of the night.

But the enemy had also paid a heavy price. The sequential attacks of the 2d and then the 1st Battalions and the dogged nighttime defense had caused hundreds of NKPA casualties so that, in Fehrenbach's words, "the *18th Regiment* was shattered beyond repair."

Craig ordered a resumption of the attack at 0700 the next morning, 18 August. None of the men on Obong-ni had had any sleep during the night past, but the Corsairs were back on station overhead, the enemy was weakening, and both Companies A and B moved once more into the assault. Company B worked men to its left to coordinate with Company A's effort to seize Hill 117. Four determined NKPA machine gunners there held up the advance, so the company commander, Captain John R. Stevens, got in touch with Newton to call in an air strike. There was legitimate concern about the fact that his Marines were too close, only 100 yards from the target, but a smoke rocket was fired into the emplacements from the control Corsair, and the next Corsair put a 500-pound bomb right onto the center of the target. The Marines lost one man killed, but the enemy was totally wiped out, and Company A's follow-up rush quickly took control of the crest. Time: 0734, request air strike; 0743, bomb delivered; 0748, on the crest.

There was a brief pause—well remembered by Muetzel:

> In an effort to calm the men after all they'd been through, I told them to break out rations and eat while they had a chance. I sat on the side of a hole and dangled my feet. On the other side of the hole lay a dead North Korean. He had caught one through the top of the head and looked pretty ugly. I was 23 years old and to reassure the men I tried to pull off a John Wayne stunt. When I was halfway through my can of meat and beans, decomposing gases caused the cadaver to belch. Black blood foamed out of its mouth and nose. I promptly lost my entire lunch. By the time the platoon got through laughing, the tension was broken and they were ready to go back to work.

And back to work the company went, moving aggressively to take the remaining hilltops. Resistance was minimal now, and soon all the heights of bloody Obong-ni Ridge were in Marine hands. As the men looked down the reverse slope of one of the hills, an unusual sight greeted their eyes. A clump of scrub pines lay below them, and, as they watched, astonished, the "clump" turned out to be a group of camouflaged enemy soldiers who arose and rushed downward in headlong flight.

The 1st Battalion now counted up the enemy weapons destroyed or abandoned: 18 heavy machine guns, 25 light machine guns, 63 submachine guns, 8 antitank rifles, 1 rocket launcher, and large stocks of ammunition and grenades.

The seizure of Obong-ni Ridge was crucial to the elimination of the threatening salient which had been driven into the Army's lines. As Geer summed it up, "it was evident the enemy had staked the defense of the Naktong Bulge on their ability to hold that key ridge."

Next: Objective 2

With Objective 1 now secured and the enemy in bad shape, Murray kept the pressure on. Taplett's 3d Battalion moved out that same morning of 18 August, bound for Objective 2, Hill 207 (the next rise west of Obong-ni). It was preceded by an intensive barrage from air, artillery, tanks, and mortars, including now supporting

National Archives Photo (USMC) 127-N-A2189

A "turkey shoot" ended the First Battle of the Naktong. Here a BAR man draws a bead on the fleeing enemy.

fire from the 9th Infantry on the right flank.

A correspondent in the rear was awed:

> The 155s began to roar and the snub-nosed 105s, and to one side the mortars were barking, and in front the squat tanks were slamming away with the 90mm guns whose muzzle blast can knock a man down at thirty feet, and above the hill, swooping low, the planes were diving in.
>
> You would see the smoke and fire flash of the rockets leaving the wings, and then would come the great tearing sound the rocket made in flight, and then the roar of its bursting against the hill. And after the rockets had gone, you would see the little round dots of smoke in the sky as the wing guns fired, and all the crest of the hill in front of How Company was a roaring, jumping hell of smoke and flame and dust and noise.

With this kind of preparation, "Objective 2 was not much of a fight," as an officer in Company G said. There was a grenade flurry near the crest of Hill 207, but a platoon of Company H was then able to rush the enemy positions, and it was all over by 1237.

There had been a tide of NKPA troops running for safety. Now it became a flood, increased by men driven from Hill 207. Everywhere the soldiers of the NKPA's "crack" *4th Division* on Obong-ni had themselves cracked and were fleeing westward in a disorganized, panic-stricken rout. It became a field day for the Marine artillery and planes—a thunderous hammering that caused massive waves of enemy deaths. There were "all kinds of bodies floating in the Naktong."

Final Victory: Objective 3

Taplett kept driving. Next target: "Objective 3," Hill 311, the last barrier before the Naktong River. There was another round of preparatory fire, this one featuring a dose of napalm, and one more time the riflemen moved out.

Things went fairly smoothly for Company G which, "brushing aside light resistance," was on the crest by 1730. Not so for Company H. It was badly hindered by difficult terrain and an obdurate enemy, and by 1825 was pinned down and unable to advance. Supporting fire from Company G and an attempted flanking maneuver by its Cahill platoon (which, 10 days earlier, had had that relief mission on Hill 342) were not

Department of Defense Photo (USMC) A2238

BGen Edward A. Craig receives the thanks of South Korean President Syngman Rhee at the Purple Heart ceremony.

enough help for H to be able to advance.

It was the last gasp of the NKPA, however. A heavy round of battalion mortar fire early the next morning, 19 August, was followed by a triumphant sweep of the hilltop by Company H, and the Marine battle to seize the three key ridges in the Naktong Bulge was over. One battalion stood on each of the three objectives, and the men of the brigade met the Army troops at the river's edge. Marine aviators reported, "the enemy was killed in such numbers that the river was definitely discolored with blood."

During the attack on Objective 3, the 3d Battalion surgeon came across a horrendous sight, demonstrating the savage brutality of the NKPA. A U.S. Army aid station had been overrun a week earlier, the wounded and bed-ridden men shot and bayoneted, their bodies then mutilated. Medics who bravely had stayed there to tend their men had had the hands wired behind their backs and then were murdered.

An incident occurred on one of these final nights that is very revealing of how personnel problems could be expeditiously dealt with in the "Old Corps"—particularly in a combat environment. Bohn had told his machine gun platoon lieutenant to check carefully on the positioning and coordination of the weapons' sites. When the company commander decided to inspect personally, he found wholly unsatisfactory results and crews who had not even seen their lieutenant. Steaming, he returned to his CP and hauled the lieutenant in for a very brief conversation:

> I said, "Have you put in the machine gun sections? Did you get around to check each section?" He said, "Yes, sir." So I relieved him. I called Taplett and said, "I don't even want this guy here tonight." I made him go back on his own, back to the battalion, and wrote an unsat report, un-officer-like conduct.
>
> It went up to Craig, and the guy was out of country in two or three days. It was so easy to do things like that then. And he was out of the Marine Corps. You can't do that today. You have to have a General Court-martial and everything else. There wasn't even a Court of Inquiry. Everyone agreed that he was a coward, and he was gone.

The brigade was now relieved by Army units—not always smoothly, but at least the tired Marines would get some rest.

The victory price for the Naktong Marines was clear: 66 dead, 278 wounded, but only 1 man missing in action. That last figure was the clearest indication of the value of Marine training and morale; there had been other units with distressingly high percentages of missing-in-action, but, as Edwin P. Hoyt summarizes in his history, *The Pusan Perimeter—Korea 1950*: "The Marines stood and fought, and they took care of their own."

The final results for the NKPA *4th Division* were shattering. Fewer than 3,000 men were able to get back across the Naktong, leaving more than 1,200 dead behind. The Marine brigade recovered a large amount of enemy equipment, including 34 artillery pieces (with five of them being captured Army 105mm howitzers), hundreds of automatic weapons, and thousands of rifles. The Army's official history sums it up: "The destruction . . . of the NKPA *4th Division* . . . was the greatest setback suffered thus far by the North Korean Army. The

4th Division never recovered from this battle."

After the brigade was pulled back off the hills it had won, Fenton described what he felt was the key reason for the Marine victory: "the finest batch of noncommissioned officers ever assembled in any Marine regiment. Not only were 75 percent of them combat veterans, he believed, but they had often stepped in as platoon leaders and were "outstanding." Fenton expanded on that:

> Squad leaders knew their job to the last detail. Many times I ended up with sergeants as platoon leaders after a big fire fight, and they did an excellent job. I just can't be too high in my praise.
>
> In some cases, it wasn't just noncoms. It was the PFCs and privates holding the job of a fire team leader or squad leader. It was their fine leadership, outstanding initiative, and control of the men that turned a possible defeat into a sweet victory.

On 20 August Craig learned from Church that the brigade had been detached from the 24th Division and was now part of Walker's Eighth Army reserve. There were letters of praise from both Walker and Church. The former wrote that the brigade's "excellence in leadership and grit and determination . . . upheld the fine tradition of the Marines in a glorious manner." Church graciously commented to the Marines that their "decisive and valiant offensive actions . . . predominantly contributed to the total destruction of the Naktong pocket." Perhaps the recognition the men of the brigade appreciated most came from their own Commandant.

National Archives Photo (USMC) 127-N-A1469

A high point of the brief interlude in the rest area: mail call!

General Cates' message said: "I am very proud of the performance of your air-ground team. Keep hitting them front, flanks, rear, and topside. Well done."

Another Brief Interlude

Thus the men of the brigade moved back into bivouac in an area near Masan known forever after as "The Bean Patch." Craig set up his CP there on 21 August and reported back again to Kean of the Army's 25th Division. The news was discouraging: all the land won in the brigade's drive to Sachon was now lost or under heavy enemy pressure, and the 11th Marines was needed to go back immediately to the original starting point two weeks earlier, Chindong-ni, to fire missions in support of the 25th Division.

But for the other Marines it was a wonderful, restorative change. Some 800 replacements arrived to fill in the painful gaps in the ranks; VMO-6 helicopters flew in hot food; letters from home and beer miraculously appeared; and new equipment was issued. But not enough of it. Fenton frankly noted that the equipment they had arrived with in the Bean Patch was in "terrible condition." It had deteriorated badly from exposure to heat, rain, and frequent immersion in rice paddies.

In addition, he commented:

> We were having a hard time getting Browning automatic rifles. Many of our BAR men had been casualties, and we were down to about three or four per platoon. You just couldn't get a BAR belt in Korea.
>
> Shoes were another big problem . . . . We reached the point where we had men running around in tennis shoes. Dungarees were in bad shape . . . . Our packs, which had been dropped at Pusan and were supposed to have been brought to us by the rear echelon, never arrived. The only way we could get a clean suit

Department of Defense Photo (USMC) A1441

Commanders and staff of the 5th Marines assembled for a photograph during a lull in the battle. Pictured in the front row, from left, are: LtCol George R. Newton, LtCol Harold S. Roise, and LtCol Robert D. Taplett. Second row, LtCol Raymond L. Murray and LtCol Lawrence C. Hays, Jr. Third row, Maj R. M. Colland, LtCol George F. Walters, Jr., Capt John V. Huff, Maj Kenneth B. Boyd, Maj Harold Wallace, Maj William C. Esterline. Fourth row, Capt Ralph M. Sudnick, Lt Robert M. Lucy, Lt Almarion S. Bailey, Lt Leo R. Jillisky, Lt Alton C. Weed, Capt Gearl M. English, WO Harold J. Michael, and CWO Bill E. Parrish

of dungarees was to wash them or survey the supply at the laundry unit when we took a shower.

There were no shelter halves either, so the men slept out in the open. A memorable event was a ceremony for the award of 87 Purple Heart medals, with South Korean President Syngman Rhee in attendance. The attrition rate among the officers had been fearful: five of the six company commanders were wounded and nine of the 18 rifle platoon leaders were wounded and four of the killed in action.

One platoon leader, Second Lieutenant Muetzel, received two Purple Hearts (with a Silver Star Medal to come later for his heroic actions on Obong-ni), while the gunnery sergeant of the Reconnaissance Company, a veteran of World War II wounds, received his fifth Purple Heart. It was a strain to try to look presentable for the ceremony, as Muetzel later remarked:

My leggings had been thrown away, my trousers were out at both knees, my right boot had two bullet holes in it, and my dungaree jacket had corporal's stripes stenciled on the sleeves. I grabbed a fast shave with cold water, hard soap, and a dull blade. Gene Davis loaned me a clean set of dungarees, Tom Gibson loaned me his second lieutenant's bars, and off I went with my troops.

Department of Defense Photo (USMC) A1507

After the First Battle of the Naktong, there was a welcome moment to clean up—with a South Korean boy holding the mirror.

Future Plans

While the troops were enjoying this temporary lull, some of Craig's senior staff officers were sent to Tokyo to confer on plans for the future use of the Marines. MacArthur had made bold—very bold—plans for a daring end-run around the NKPA besieging the Pusan Perimeter by making a surprise amphibious landing far to the rear, at Inchon. For this purpose he had urgently requested the full 1st Marine Division. Elements of it began arriving in Japan on 28 August, but there were massive problems to be overcome. The 1st Marines was on hand, but the 7th Marines would not arrive at Inchon until a week after D-day, with one battalion coming halfway around the world from the Mediterranean. The crucial unit for the forthcoming assault was supposed to be the battled-tested 5th Marines. It had already begun shipping its heavy equipment back down to Pusan, as plans were drawn to have it join the 1st Marine Division, even though it was now fully committed in combat. Morale soared in the brigade as the men looked forward to fighting side-by-side with fellow Marines.

Meanwhile, in Tokyo there were very tense moments. Time was critically short to mount an operation as complex as an amphibious assault. There were vigorous differences of opinion in Army-Navy-Marine meetings as to when or even whether the brigade should join the 1st Marine Division. On one hand, the Eighth Army staff felt, as the official Marine history bluntly put it, "Army morale would be hurt by taking the brigade away at a critical moment." And Walker placed an "extremely excited" telephone call to MacArthur's headquarters in Tokyo, saying in effect, "If I lose the 5th Marine Regiment, I will not be responsible for the safety of the front!" Thus there was strong Army pressure to substitute an Army regiment for the 5th Marines at the Inchon landing.

On the other hand, Major General Oliver P. Smith, Commanding General of the 1st Marine Division, supported by the three Navy admirals most closely involved, was equally adamant that, for a tricky amphibious landing, he had to have the 5th Marines which was trained for just such an operation. There was a deadlock.

The NKPA Attacks Again

Then, amidst these planning meetings, harsh reality came crashing down to complicate further decisions on the use of the brigade. The NKPA, realizing that time was running out for it, launched a final, convulsive attack to eradicate the Pusan Perimeter. Some 98,000 men in 13 divisions hit five separate points on the perimeter. Walker faced a brutal series of simultaneous problems. Where should he commit his limited reserves—in particular his proven Marine brigade? The two thrusts closest to Pusan were one against the Army's 25th Division in the same area of the far southwest, and another against the Army's 2d Division in the west central

Photo by David Douglas Duncan

The price of victory: a jeep takes two wounded men back to an aid station.

(Naktong) area. A breakthrough to capture Pusan would mean total disaster. (Military analysts in later years would speculate that that might well have happened if General Kim Chaek had ordered only diversionary attacks at four of the points, massed overwhelming strength at one point, and crashed through there.)

The NKPA assigned the *2d, 4th, 9th*, and *10th Divisions* to destroy the U.S. 2d Infantry Division before Miryang and drive through to the vital Pusan-Taegu MSR by way of Yongsan. Smashing into that division on 1 September, the North Korean assault quickly made a 4,000-yard penetration. The commanding general of the 2d Division, Major General Lawrence B. Keiser, USA, saw his division sliced in half, with his companies cut off or totally overrun, his defensive lines hustled back almost to Yongsan, and enemy infiltration in his rear. Neither Keiser nor his three regimental commanders had ever led troops in battle, and now the NKPA had punched a hole six miles wide and eight miles deep into their division. Obong-ni Ridge, so dearly bought, was back in enemy hands.

Now Walker made up his mind: the new Naktong Bulge had returned as his priority threat. Blair's book pointed out the logical, but painful, next step: "Walker came to a difficult and drastic decision: Once again he would have to call on Eddie Craig's Marines for help. The decision was drastic both because of the humiliation it would again cause the Army, and because Craig's Marines were a vital element in the Inchon invasion plan."

'Fire Brigade': Crisis Number Three

That was it. In the morning of 1 September, the orders came for the brigade, including the 11th Marines, to move by train and truck back once more to the Miryang assembly area. The reaction of the men was predictable: going back to regain what they had already won once.

When Craig had set up his CP in Miryang, his brigade came under the operational control of the Army's 2d Division. To old timers in the Marine Corps it surely brought back vivid and ironic memories of another time and place, when a Marine brigade had been teamed once before with the 9th and 23d Infantry as a proud part of the Army's superb 2d Division, 32 years earlier in France.

On 2 September, Craig had a conference with Keiser and the Eighth Army's Chief of Staff. General Shepherd later made a comment on this meeting which revealed the inherently gracious nature of Craig: "The Army division commander . . . went to Eddie, who was a brigadier, and said, 'General Craig, I'm horribly embarrassed that you have to do this. My men lost the ground that you took in a severe fight.' And Eddie, in his very gallant manner, said, 'General, it might have happened to me.' "

The Army officers at the meeting felt the situation was so desperate that the brigade should immediately be dribbled piecemeal into action, even though one of its battalions and its air control section had not yet arrived. Craig, who also knew when to make a stand, later remembered, "This was the only heated discussion I had in Korea with the Army." His stubborn view that the whole brigade should go into action as a unified air-ground team was finally accepted. Its attack would be down the Yongsan-Naktong road toward an all-too-familiar objective, Obong-ni Ridge. The 9th Infantry Regiment would be on its right, and other Army units on its left. Now the brigade, for the first time, appeared to have flanks that were secure enough to allow it to attack with two battalions abreast, Roise's 2d on the right and Newton's 1st

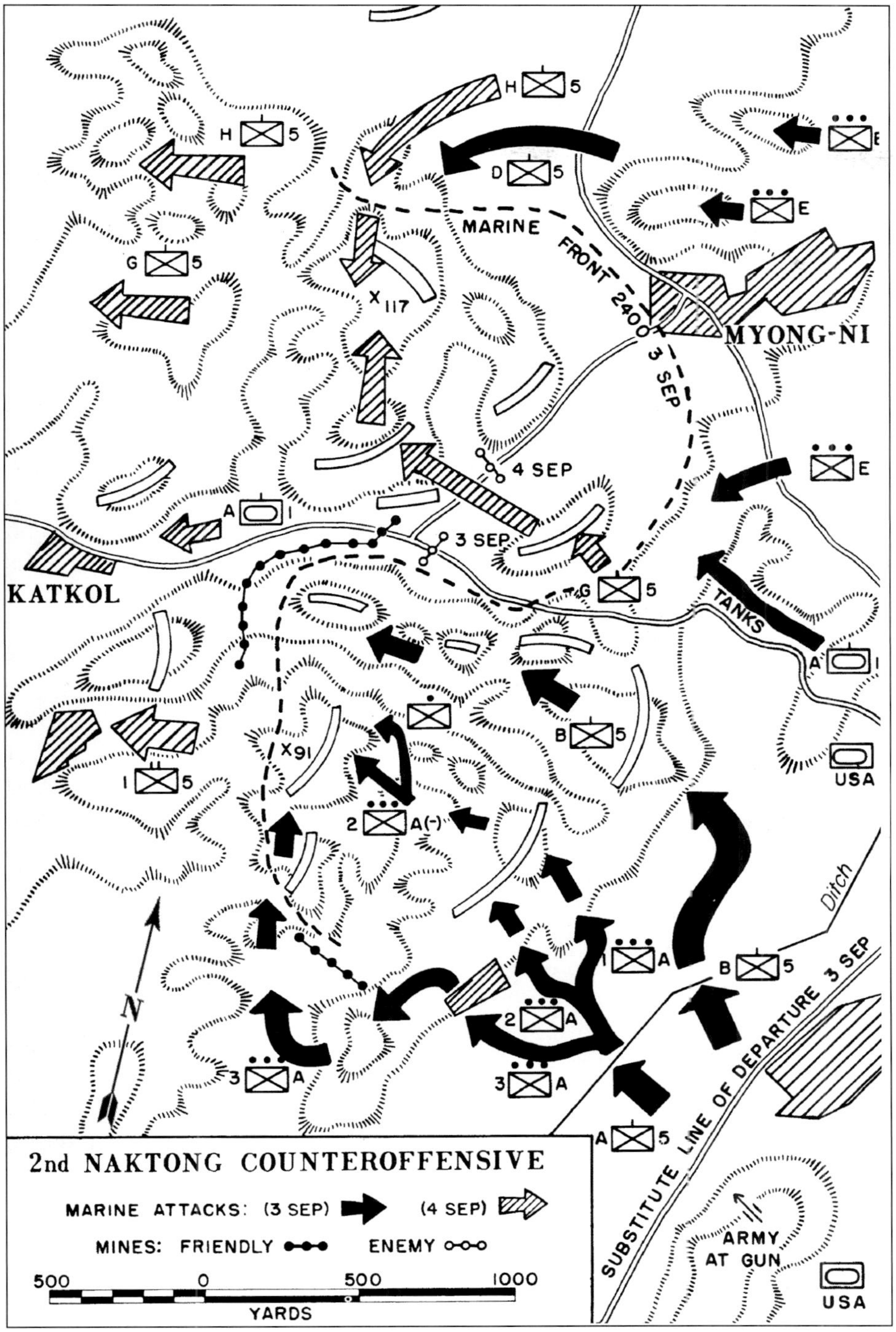

on the left. Taplett's 3d Battalion would block any enemy push along the southwest approaches to Yongsan.

Meanwhile, between 0300 and 0430, 3 September, the 2d Battalion moved into its forward assembly area north of Yongsan, with the 1st Battalion south of the town.

Opposite them, driving hard for Yongsan, were the NKPA divisions which had successfully advanced this far in the new Naktong Bulge. Immediately in front of the brigade was the *9th Division*. This was not a seasoned, professional outfit, such as the one the brigade had previously broken; rather, it had up to now been doing guard duty at Seoul. Behind it, in reserve, came a reconstituted *4th Division*, filled with new recruits after the massive casualties the brigade had inflicted on it in the first battle of the Naktong.

The Marines Attack

The Marine attack was to be launched early on the morning of 3 September. There were problems getting things started. Moving through Yongsan, the Marines were hit by small arms fire from snipers, but by 0630 they had worked their way to the western end of the town, and thought they were then headed forward to the agreed-upon line of departure for their main attack. Not so! During the night the Army troops on the ridgeline had "collapsed" and had been pushed back 1,000 yards. At 0645 Roise called Second Lieutenant Robert M. Winter to bring his tanks forward and lay down fire to cover the withdrawal of the Army troops. The original planned line of departure thus became the first objective when the Marines attacked.

The 2d Battalion jumped off at 0715, securing the right flank of the brigade's attack. To soften up his main objective, Roise called down a massive sheet of fire from tanks, air, mortars, artillery, and machine guns. The Marines pushed doggedly toward it wading through a rice paddy. Now the enemy's 9th Division quickly found its previous pattern of steady advances had ground to a screeching halt.

Craig, as was his wont, came up to check on the action. His observation post (OP) was between the tanks and Roise's OP. Enemy fire pounded the area, and Winter was wounded and had to be evacuated—but not before he offered Craig a bottle of whiskey from his tank. Winter was later awarded a Silver Star Medal for his leadership of his tank platoon that day.

Meanwhile, the 1st Battalion also moved out. Its attack route forced the men knee-deep into their own huge rice paddy. There they came under fire, but their supporting arms searched out the enemy positions. In particular, the Corsairs were able to engulf the NKPA with balls of napalm fire. A typical time of response was seven minutes from a strike request to execution.

This kind of seamless coordination in the Marine air-ground team was a source of great envy by the Army commanders who saw its decisive results. As Colonel Paul L. Freeman, USA, commander of the 23d Infantry (well off to the right of the brigade), wrote to General Matthew Ridgway in Washington:

> The Marines on our left were a sight to behold. Not only was their equipment superior or equal to ours, but they had squadrons of air in direct support. They used it like artillery. It was, "Hey, Joe, this is Smitty, knock off the left of that ridge in front of Item Company." They had it day and night . . . General, we just have to have air support like that, or we might as well disband the Infantry and join the Marines.

Photo by David Douglas Duncan

A distraught Cpl Leonard Hayworth pleads for more grenades, finds none, and must return empty-handed to his hard-pressed men.

By 1100 the 1st Battalion was at the base of its ridgeline objective. Working its way upwards under the protection of supporting 81mm mortar fire, Company A poised for a final charge. As soon as the fire lifted, the men sprang forward, screaming, shouting, firing every available weapon. To their amazement, a whole company of NKPA soldiers in front of them, shaken by the noise and the sight of charging Marines, leaped in a panic out of their concealed foxholes on the forward slope and fled back towards the crest of the hill. Then the long hours of practice on the rifle range really paid off: Marine marksmen coolly picked off most of the enemy as they ran. Company A immediately rushed to the crest. It was noon. In Company B, Fenton later observed:

> The 1st Battalion was able to move and seize the ridge line without encountering heavy opposition. I don't believe the enemy realized that we had a battalion to the left of the road, because he was prepared to take that high ground himself. We beat him there by a good 10 or 20 minutes and caught him coming across another rice paddy field. We really had a "turkey shoot."

Firing now from the heights, the Marine riflemen put on another display of precise marksmanship that must have stunned the simple peasant soldiers of the NKPA: the "yellow leggings" could kill with aimed fire at 400-500 yards. (Just as the Marines in that earlier brigade in France had stunned the Germans at Belleau Wood with trained rifle fire that killed at long range.)

What the 1st Battalion did not finish off, the 105s of the 11th Marines did. Those of the enemy who were left withdrew to Hill 117 in front of the 2d Battalion, but an artillery barrage was called down on them in transit, and wreaked more havoc.

In the 2d Battalion zone of attack there were some hard moments. When Company D was getting started in its assault, a tragic episode occurred. (Today, it is called "friendly fire" and results in great publicity. Fifty years ago, in the early days of the Korean War, it was regarded as just one of those unfortunate things that happened because close combat is always unpredictable.) The official Marine history did not even mention it, but it was seared into the memory of Private First Class Douglas Koch in Company D:

> Down the road from the north rolled four or five American tanks . . . . All of a sudden a machine gun stitched a stream of fire across the company's rear. I rolled over on one elbow and looked behind me. Someone yelled, "God, they're shooting at us." Instead of firing on the top of the hill, the tanks chose to fire at the bottom of the hill. I saw a puff of smoke. Just that quick a shell landed near me. It rolled me over into a little gully. I lay dazed. God, I thought, we're gonna get done in by our own goddamn outfit. While I lay with my head down, three or four more shells hit nearby . . . . A lot of men had been hit.

Naturally, this kind of ghastly mistake was temporarily shattering to the company, until the officers finally got their men moving again. But Koch and the others went back to their attack "still in shock."

This occurrence was, fortunately, a rarity. Elsewhere that morning of 3 September, Marine tanks were

Capt "Ike" Fenton, caught by surprise, described the grim moment: "We had been in one hell of a big battle. It was raining. The radio had gone out and we were low on ammunition."

Photo by David Douglas Duncan

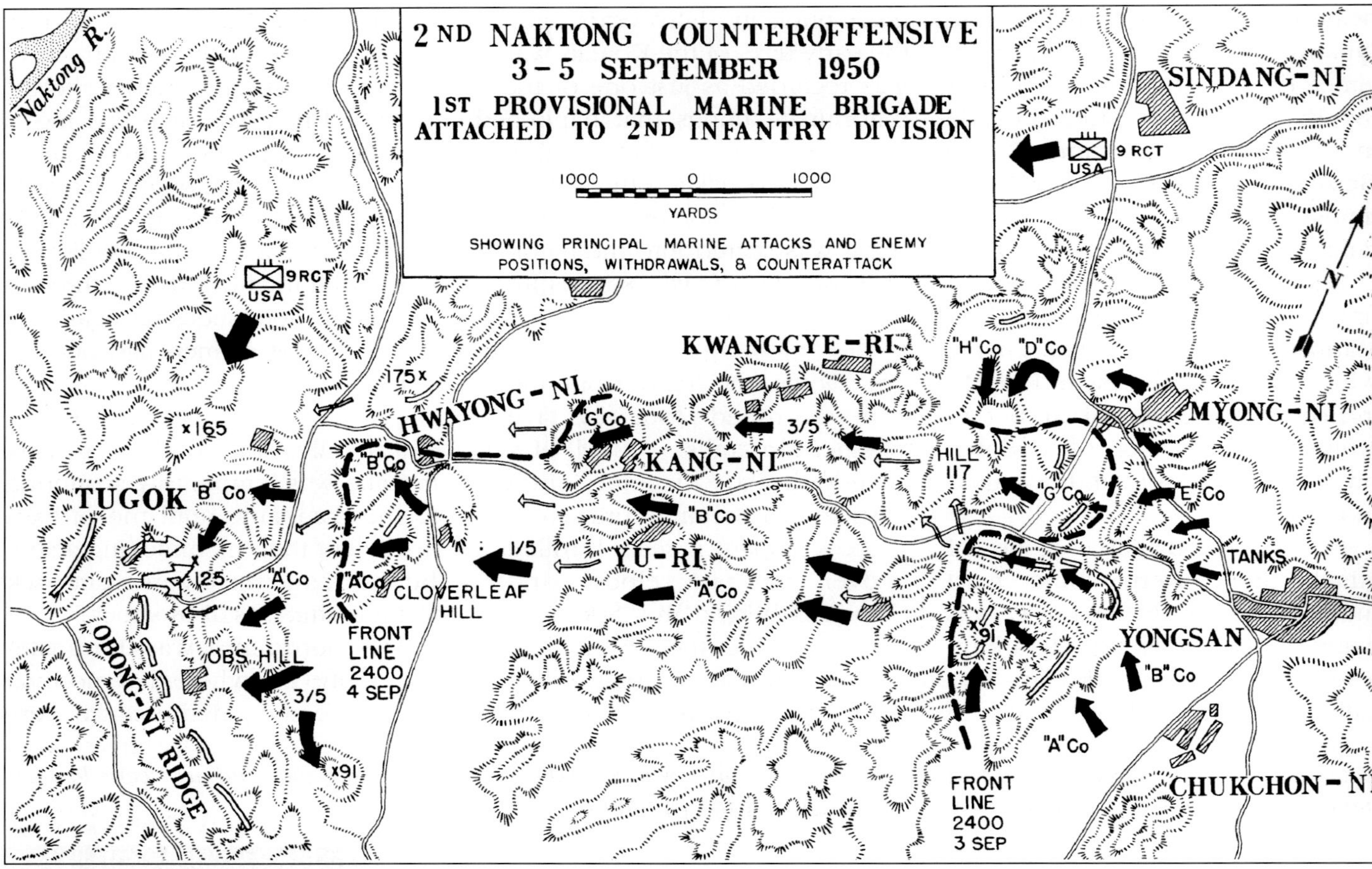

doing yeoman's work. They took on NKPA antitank weapons, surprised three T-34 tanks and wiped them out, then eliminated two more in the afternoon. This cleanup enabled the M-26s to concentrate their fire to good effect on enemy weapons and troop positions confronting the riflemen.

Marine air was also very active. With the squadrons shuttling so that one was always on hand to help, seven close air support missions were flown for the two assault battalions. Other Marine planes, guided by OYs, strafed and bombed, knocking out, among other things, 16 enemy gun and mortar positions.

Back on the ground, Company D's first objective was Hill 116, to try to cut off the enemy reinforcements coming over from the 1st Battalion's zone. Facing two NKPA battalions, the company found itself in a bloody battle. It was finally able to gain the crest of the northern spur of Hill 117, and there it dug in, isolated, some 500 yards from the rest of the 2d Battalion.

As the enemy troops filtered into the zone of the 2d Battalion, the men of the 1st Battalion were able to make good progress in the afternoon, with Company B reaching its part of Objective 2, a peak across the MSR from Hill 117. Company A, using a fancy triple envelopment seized its part, Hill 91, by 1630, and so all hands prepared for the usual night counterattack. Well they might. The 1st Battalion's right flank was dangling in air; it was trying to cover a front of nearly a mile; and its two rifle companies were 200 yards apart. The 2d Battalion was in an equally dangerous position, stretched over a 2,000-yard front, bent in a right angle, with Company D completely isolated.

Three things saved the Marines' precarious position. First, a bevy of their engineers moved in to sow a belt of antipersonnel mines, wired hand grenades, and blocks of TNT along the flanks. Secondly, VMF(N)-513 came on station with its F4U-5N Corsair and F7F Tigercat night fighters. Equipped with sophisticated radar, it was the only squadron to fly single-engine planes over Korea at night. Flying more than 2,000 hours of night missions in one month, it delivered this particular night six close air support strikes controlled from the two infantry battalions. Thirdly, a deluge of rain, accompanied by icy winds, further hindered any plans the battered NKPA troops might have contemplated for a counterattack.

As the Marines waited though the miserable, rainy night, even though they had driven two victorious miles west of Yongsan, their thoughts must have turned to the casualties of the past day: 34 killed and 157 wounded. Muetzel in the 1st Battalion later voiced what must have been a common sentiment after almost a month of grinding combat:

> [Men] came, were killed, and were carried away . . . . I knew this couldn't keep up . . . . We, me, all of us were eventually going to get it; it was just a matter of when and how bad . . . . It was just a god-awful mess—inadequate replacements, insufficient ammo, worn-out clothes and boots. No one much gave a rap about anything. Outside discipline was no longer a threat. What could the brass do to us that was worse than what we were doing? Each of us withdrew into our family—the squad, the platoon, the company, the regiment, the brigade, the Corps. Everyone else, bug off!

This same day, 3 September, witnessed a final showdown in the Tokyo planning meetings. A compromise solution to the deadlock emerged. Walker would get Army reinforcements and could temporarily use the Marine Brigade to meet his Naktong crisis. But it would have to be withdrawn by midnight 5 September to join the 1st Marine Division for the Inchon landing.

Continuing the Assault

Back with troops, in order to keep the pressure on the next morning (4 September), Murray had ordered Taplett's 3d Battalion to pass through the depleted 2d Battalion and resume the attack at 0800 with the 1st Battalion on its left. In 20 minutes, Taplett's men reached their first objective, then quickly took Hill 116 with almost no enemy resistance. Next, the battalion's main objective, Hill 117, was overrun by a pincer movement of Companies G and H. Incredibly, it was all over by 0840. No real enemy resistance had turned into a withdrawal, and now there were signs that was turning into a disorderly rout—a weird contrast to the bruising encounters the Marines had had the day before.

The 1st Battalion was simultaneously moving with equal rapidity. Shortly after starting, it occupied what appeared to have been a CP of the NKPA *9th Division*. Tents were still up, equipment was strewn around, and two abandoned T-34 tanks in perfect oper-

Marines meet almost no opposition as they top this hill in the Naktong River area on 4 September 1950

Department of Defense Photo (USMC) A8175

ating condition were captured (the first such to be taken and turned over to U.S. Army Ordnance for examination). The men in the battalion's steady advance saw the bodies of many dead NKPA soldiers and piles of abandoned or destroyed equipment, souvenirs of low-flying Corsair strikes and accurate fire from the 11th Marines poured on the retreating enemy. Among the litter were captured American guns, tanks, mortars, and vehicles which were returned to the 2d Division. The official Marine history described "a picture of devastation unequaled even by the earlier defeat of the NKPA *4th Division*." This time it was the *9th Division*'s turn to be hammered by the brigade.

By 1515 Newton's companies stood atop their first objectives, now less than 2,000 yards from the old killing ground on Obong-ni Ridge. Moving in coordinated tandem with them were Taplett's companies, which had pivoted to the west after seizing Hill 117.

Learning of the Marines' progress, Keiser gave Craig the go-ahead to have his brigade push on further toward Objective Two. Moving aggressively, using air strikes when held up, the 1st Battalion worked its way to the designated area (between Hill 125 and Observation Hill), securing Cloverleaf Hill by 1800.

Thus the brigade had advanced 3,000 yards and gained its objectives. Hoyt summarized the strategic importance of this: "The Marines had stopped the enemy's advance, saved Yongsan and the [MSR] road beyond [it], and put the North Korean *9th Division* into retreat."

As the Marine battalions dug in for the night they were in exposed positions similar to the preceding evening. Newton's men were 1,000 yards in front on the left, stretched paper-thin along a line almost a mile long. Taplett's men were no better off. Out of contact with the 1st Battalion on their left and the Army's 9th Infantry on their right, they curled up in a perimeter defense.

Expecting the usual NKPA night counterattack, the Marines again had their engineers put out a protective shell of mines, booby traps, and trip flares. There was heavy incoming shelling during the night, but that slacked off after a visit from the night fighter planes of VMF(N)-513. The rain poured down, but the enemy infantry apparently had been hit too hard during the day, and there was no assault.

When men are under heavy pressure in close combat little things can loom large in their minds. Fenton gave an example: "It had been raining all night, and the battalion had managed to get some hot coffee up to us, but just when the coffee arrived, we got the word to move out. We weren't able to distribute any of the coffee. This turn of events didn't do the morale any good. The men were soaking wet."

A more fundamental event took place that same night. Reluctantly following instructions from MacArthur, Walker issued an order that the Eighth Army would have to release all of its Marines at the end of the following day.

The Final Day

To finish off what the brigade had so successfully begun, Craig ordered both battalions to move out in a final attack the morning of 5 September. Before the 1st Battalion could get started, there was an unpleasant moment. Two U.S. Air Force F-51 fighters came screaming in over the Marines, strafing them. Miraculously, only one man was wounded.

The 3d Battalion started the day by showering a rain of fire from its high ground down on an NKPA attack on the 9th Infantry off to its right flank and rear. The 105s from the 11th Marines joined in, and the attack was shattered.

Now both battalions were ready to charge. And they did. The 1st Battalion jumped off at 0820 with the objective of capturing Hill 125 and Observation Hill, the brigade's segment of Phase Line Two. Obong-ni Ridge was then to be a special objective. Moving fast against light resistance, Newton had his men on his two target hills by 1100, and there Murray halted them until the 9th Infantry could come up to tie in on their right.

Meanwhile, the 3d Battalion was also moving ahead. Bohn had suggested that Company H, now commanded by Captain Patrick E. Wildman, serve as a base of fire to pin down the enemy, while he took Company G around the extreme left flank in an enveloping maneuver. "It worked beautifully," as he later reported, but then:

> As we were coming up, getting assembled . . . the North Koreans picked up on what we were doing. They had one of those old Russian [Maxim] wheeled machine guns, and I could see their officer. He was wheeling it up with his people. Jones saw him at the same time and he blew it up with the first round of 75 recoilless . . . . It was sheer luck.
>
> As soon as that happened, of course, we went smoking up, got over the top, and once we got to the top . . . we just rolled them up. It was outstanding.

So it was, that Company G was

in good shape on Hill 91, expecting to race ahead. Not so. Orders from Taplett at 1230 directed it to withdraw to Observation Hill and hold up there. The convergence of the 1st Battalion and the 9th Infantry had pinched out the 3d Battalion's area, so Company H joined in a sideslip behind the 1st Battalion to put the 3d Battalion on the left flank of the 1st, preparatory to a combined attack on Obong-ni Ridge. It, too, was told to stay in place; there would be a delay before any assault on Obong-ni.

With the heavy rain and ensuing fog Marine close air support was grounded, and this gave the NKPA an opportunity to launch a vicious daylight counterattack on the 1st Battalion. Company B, after an advance of 3,000 yards, was now located on a ridge line of Hill 125, parallel to and only 400 yards from Obong-ni. At 1420 an avalanche of enemy fire hit it. It was enfilade fire, mortars and machine guns, smothering both the reverse slope and the forward slope of the company's position. Fenton's comment was curt: "We were pinned down, and we couldn't move."

At 1430 the enemy infantry came on, some 300 strong. Fenton needed help, supporting fire and lots of it, but at this precise moment of peril all five of his radios, as well as the battalion's tactical radio, went dead in the downpour of rain. An enlisted runner, 22-year-old Private First Class William A. Wilson, was rushed off to the 9th Infantry, which had now come abreast on Company B's right flank. His message was urgent: "We need maximum supporting fire from your artillery, and we need it right now!" Meeting up with the adjacent Army company commander, Wilson was pointing out the target areas when the Army officer was struck down by machine gun fire and had to be evacuated. So the Marine coolly picked up his radio and directed the Army artillery fire to plaster Obong-ni and the adjoining enemy targets.

A runner had also been sent down to the MSR to warn the Marine tanks there that three NKPA T-34 tanks supporting the attack were coming towards them around

Marines assist wounded North Korean prisoners into jeep which took them to medical aid on 4 September 1950

Department of Defense Photo (USMC) A2130

the same bend that had been the scene of the previous tank battle two weeks earlier. The message was not in time. The lead enemy tank surprised the first Marine tank with its gun aimed left at Obong-ni. Several 85mm rounds knocked out the Marine tank. Its mate, trying to edge around the first tank, was also knocked out.

Then, out of the blue, a 3.5-inch rocket team, dispatched by Fenton, arrived at the carnage, soon joined by the battalion rocket team. In short order, they destroyed the first two enemy tanks, and then the third attacker, which turned out to be an armored personnel carrier. This made a total of eight steel hulks littering "The Bend."

While this dramatic tank confrontation was taking place, Fenton's infantry confrontation was also reaching a climax. He later described the tense situation:

> I found it necessary to place every man I had in the company on line. Rocket men, corpsmen, mortarmen, every available man went on line to stop this counterattack. To make matters worse, I began running low on ammunition. I was practically out of hand grenades, and things didn't look too rosy for us.
>
> Just at this time LtCol George Newton, my battalion commander, who had probably guessed my situation, sent a much-welcome platoon from A Company with five boxes of hand grenades. The enemy had closed so rapidly that we just took the hand grenades out of the case and tossed them to the men on the line. They would pull the pins and throw them. The enemy closed to less than 100 yards.

Adding to the intense pressure, the radios had not been functioning. Finally, at this crucial juncture, one radio was coaxed into service.

Marines examine two Soviet-made T-34 tanks destroyed by 3.5-inch rocket teams of the 1st Battalion 5th Marines, at "The Bend." The enemy tanks had supported the vicious North Korean daylight counterattack to blunt the battalion's assault on Obong-ni Ridge.

Department of Defense Photo (USMC)

Fenton quickly gave it to his forward observer for the 81mm mortars who called for immediate "fire for effect." When the mortars had finished deluging the NKPA attackers, there were only 18 rounds of ammunition left.

Duncan was with Company B during its wild battle and saw Master Sergeant Leonard R. Young positioning the men along the crest. (The later citation for a Silver Star described Young as "exposed to withering fire, [he] walked upright back and forth . . . placing men.") Then, Duncan wrote:

> He was shot. A machine gun bullet went right through his chest, knocking him into the mud. But not before he had given Ike Fenton the best that an old sergeant could give his company commander. He was still alive when they dragged him in across the slope.
>
> When they placed him upon a rough poncho-litter he looked up at Fenton, who stood with his hand touching the dripping canvas, and whispered, "God, I'm sorry Captain! I'm really sorry! But don't let them fall back! Please don't let them fall back." Fenton still had not said a word when the litter-bearers disappeared into the rain, and out of sight down the hill.

A crucial factor in the final, successful outcome of this struggle were reinforcements which came over from Company A: two platoons of riflemen, plus machine gunners, and mortarmen. Together with the combination of Army artillery fire and Marine 81mm mortar fire (which finally came within 50 yards of Company B), this broke the back of the NKPA attack, and secured the Marine positions.

Now, from their vantage point, the Marines could see the NKPA withdrawing from Obong-ni. It was an obvious signal that the enemy was thoroughly defeated, and the door was open for a quick and easy push all the way through to the Naktong River.

But the withdrawal deadline dictated by MacArthur had nearly arrived. All units were held up in position. The brigade counted up its casualties for that final day of battle, 5 September: 35 killed, 91 wounded, and, proudly, none missing in action.

At 1600 the battalion commanders all met with Murray to get the official word. Craig's directive was concise: "Commencing at 2400 5 September Brig moves by rail and motor to staging area Pusan for further operations against the enemy."

Relief and withdrawal at night from enemy contact is not as easy in practice as it is on paper. Hours after they were due, two Army lieutenants finally showed up to relieve the two companies of the 1st Battalion. Each had only a handful of men and very few weapons. As Muetzel recalled:

> An Army first lieutenant appeared with about 30 men who'd been scraped together from a headquarters unit . . . . I took the lieutenant to the very crest of the hill and had him dig in in a circle. He asked me to leave him our ammo for a 57mm recoilless rifle he had. Marines didn't have 57s, so he had a weapon and no ammo. He asked his sergeant to bring up their one machine gun. The sergeant told him it had been left back at the CP. I left behind about four cases of hand grenades.

So the battle-worn Marines slogged wearily through the mud and driving rain for three and a half miles to the rear. West of Yongsan, they finally boarded trucks, and by dawn 6 September they were on their way to Pusan, bone-tired but glad finally to leave those cruel hills of the Perimeter behind them.

Operational Results

As the truck riders' thoughts turned to their fellow Marines, they mourned the loss of good men and close friends. Those hills had cost the brigade 148 killed in action, 15 died of wounds, 9 missing in action (7 of these were later found to have been killed in action), and 730 wounded in action, for a casualty total of 902. Included in this total was a special category of men who had moved side by side with the Marines in combat, earning their undying admiration: the Navy corpsmen who had 22 casualties.

Looking back at what they had achieved in one short month, however, the men of the brigade could legitimately feel a sense of pride. They had traveled some 380 miles and mounted three difficult operations, each time facing and overwhelming heretofore successful enemy forces who had numerical superiority.

The initial brigade drive to Sachon had represented the first crisis in which a unit of the Eighth Army had been able to stop cold and then push back an enemy offensive: 26 miles in four days. Enemy casualties: 1,900.

The second crisis was a call for the "Fire Brigade" to stem the NKPA's dangerous breakthrough in the Naktong Bulge. There it literally destroyed the enemy's *4th Division*, with the Marine air and artillery arms contributing greatly to the slaughter. In addition, large quantities of captured U.S. Army weapons were seized and returned. MacArthur spoke of the enemy division as "decisively defeated . . . suffering very heavy losses in both personnel and equipment."

In the third crisis, the Second Battle of the Naktong, the brigade had again been rushed in to meet the swift advance of the NKPA *9th* and (a reconstituted) *4th Divisions*. When its counterattack smashed the enemy units in a mere three days, in conjunction with important U.S. Army attacks, the official Army history quoted prisoners as saying that this was "one of the bloodiest and most terrifying debacles of the war for a North Korean division." As a result, "the 9th and 4th enemy divisions were not able to resume the offensive."

Over the period of that single month, the enemy had paid a devastating price, an estimated 9,900 total casualties, and massive losses of equipment at the hands of the Marines.

The achievements of the brigade went far beyond dramatic tactical victories in the Pusan Perimeter. It had demonstrated in its mobilization a remarkable ability to pull together and ship out a large Marine combat unit in a pressure-laden, short time frame (six days).

It had also demonstrated a variety of other lessons in Korea: the crucial efforts of previous combat training on noncommissioned officers and officers; the value of the intangible, psychological factor of Marine *esprit de corps*; and the dazzling effectiveness of a tightly integrated aviation component. Called "the best close air support in the history of the Marine Corps," the operational statistics of MAG-33 showed a total of 1,511 sorties flown by the three squadrons, with 995 missions being close air support not only for the brigade, but also for U.S. Army and South Korean units. In addition, the OY light planes and the Sikorsky HO3S helicopters of VMO-6 had tallied 318 and 580 flights respectively in just the month of August. Moreover, the helicopters' successful first combat role had proven the certainty of their large scale use in the years to come.

An evaluation of all these factors led the official Marine history to summarize the overall, operational results of the brigade: "A careful examination of any of these operations in which Marines engaged discloses that a single failure would have a profound effect upon the entire UN effort."

The individual unit commanders who had led the brigade in its battles had a more forceful conviction. They felt that they had "saved the beachhead."

From Pusan to Inchon

The final chapter in the story of the 1st Provisional Marine Brigade is one that is less dramatic than its battles, but one which illustrates its organizational flexibility and skill. Again, as at Camp Pendleton previously, it had too much to do in too little time. Arriving in Pusan on 7 September with over-tired men, worn-out equipment, and understrength from casualties, the brigade had to cope with a thousand details to get ready to move out in a very few days for its next demanding combat assignment.

Sleeping in the open on the docks, the men ate on board the transports upon which they soon would sail. (Although Craig and his officers later recalled the troops sleeping in the adjoining warehouses.) Bohn remembered the human side of this return to "civilization." The ship that had brought him and his men to Pusan was once again there at dockside. The Navy officers came ashore and invited Bohn and all his officers and men to come on board, and then welcomed them with "steaks, hot food" and "all the PX stuff" the Marines had not seen in a long time and badly needed now.

Bohn went on to describe another way that their deficiencies were remedied:

> I'm probably not being sufficiently critical of the Marine Corps supply system because, if it hadn't been for the Army, we'd have been in trouble. We stole everything, including jeeps . . . . We saw some rail cars on the siding. My Marines just went in there and looked. Whatever the hell they wanted, they took. The Army didn't seem to mind that. Stole beer, too. And it worked.

It worked to such a degree that all the jeep trailers in another battalion were emptied, then stacked full with beer on ice—perfect for the hot, humid, summer weather. First, a big party for its own men, then for the sailors on the ship upon which they would embark, the *Henrico*. Then, however, things sort of got out of hand. Muetzel saw a jeep driven by two Marines race by, closely pursued by two MPs. The jeep went off the end of the dock into the water.

Then two other Marines, who had climbed over the fence around the dock area, returned in impressive style. They were driving a huge Brockway bridge transporter which they had "acquired." They quickly abandoned it at the MP checkpoint—leaving it nicely plugging the entrance to the dock until a qualified driver was later found. Muetzel went on to say:

> While we were waiting to

National Archives Photo (USMC) 127-GR-A2193

After an all-night ride from the front lines, over-tired Marine mortar crews assemble at the Pusan docks and prepare to board a U.S. Navy transport. In a matter of days they would linkup with the 1st Marine Division for the upcoming assault on Inchon

board the *Henrico*, we were required to turn in all the captured vehicles we were driving . . . . This left us unacceptably short of motor transportation. Consequently, vehicles were purloined from the Army. The worst offense I saw was the theft of the MP company commander's jeep. After a fast coat of green paint and phony numbers were slapped on, it was presented to Lieutenant Colonel George Newton, our battalion CO.

These shenanigans were, of course, only a counterpoint to the serious business at hand. To fill the gaps in the rifle unit, a large batch of replacements was on hand. These 1,135 officers and men would provide the manpower to give each battalion the third company which had been so sorely missed in the past battles. Now, for the first time in Korea, the 5th Marines reached full strength: 3,611 men. Although the fresh replacements' shiny new utility uniforms contrasted sharply with the bedraggled veterans, they soon fit in. Craig later commented that the new men "were integrated into the battalions without difficulty." Some of them were regular Marines and some were trained reservists, and Craig went on to say:

> Their [future] performance of duty was comparable in many ways, outside of, per-

haps, their weapons training and their tactical training in the field . . . . It speaks very well for the type of training and the adaptability of the Marines, both as individuals and as units, that such companies could be formed in the United States, join an active battalion just before landing, take part in that landing, and operate efficiently throughout the following campaigns.

In addition, a complete fourth regiment was attached to the brigade at this time. This was the 1st Korean Marine Regiment, 3,000 strong. The manpower was welcome, but there was just one problem. Craig explained: "These Korean Marines had never been issued arms, although they had been trained in their nomenclature and upkeep. They were, however, well drilled and had good discipline and spirit . . . arms were immediately issued."

For the brigade's well-used supporting arms, there was an intensive drive to clean up and service all the heavy equipment—tanks, trucks, and artillery pieces. For the infantry battalions, one critical need was new weapons. Many rifles, BARs, and particularly machine guns had been fired so much that the barrels were burned out, so replacements had to be issued.

Clothing was a disaster. Dungarees were rotted all the way through from rain and sweat, with the camouflage design faded out. Boots were "falling apart."

This kind of urgent need led Muetzel to strong measures. He badly wanted a new pair of boots, for the ones he wore had two bullet holes in the uppers and soles completely worn through. With none available from Marine supplies, he headed for the Eighth Army quartermaster. There he found a group of "scruffy" Marines being sharply told off for begging by an immaculate (rear echelon) Army major. The Marine group gave up and left, but Muetzel, looking like a "refugee" he admitted, persisted.

When the neatly-dressed major turned to go back to his office, Muetzel pushed into the building wearing his steel helmet, a dungaree jacket and pants with gaping holes, and tattered boots, and carrying a submachine gun and a .45 pistol on his hip. Now standing face to face, the major saw the lieutenant's bars on Muetzel's collar, glanced at his disreputable uniform, and started to say that he could not issue any boots. That did it! Muetzel burst out:

> I told him, simply, that I was just off the line, I was going right back onto the line, I was an infantry platoon leader, I didn't have a hell of a lot to lose, and I wanted a pair of boots right then and there! When he looked at my boots and noticed the bullet holes, he went right back into his stock and brought out a new pair of Army parachute jump boots . . . . I was ready to fight for those boots and that major knew it.

All during this time, the senior officers were involved in a different type of activity. They were closeted, preparing the after action reports, organizing the issue of supplies for re-equipment, thrashing out an embarkation plan, and familiarizing themselves with every planned detail that pertained to their unit's role in the forthcoming landing. Craig pushed them hard and soon—all too soon—the few days allotted had rushed by, and it was time to ship out. Starting the afternoon of 11 September, the troops began filing on board ship. The next day, the convoy sailed. Then, at 0001, 13 September, the brigade was deactivated and became part of the 1st Marine Division, bound for the historic amphibious assault at Inchon.

The brigade was now gone, but not forgotten. There was formal recognition of its achievements by two governments. The first was a Korean Presidential Unit Citation which recorded "outstanding and heroic performance of duty on the field of battle." Referring to the Naktong victories, the citation said: "The brigade attacked with such determination and skill as to earn the admiration of all . . . . The gallant Marine forces were instrumental in preventing the enemy from capturing their objective and cutting the north-south lines of communication. . . ."

The second award was a U.S. Presidential Unit Citation. This was a lengthy paean of praise for both the ground forces and the aviation units. It commended "extraordinary heroism in action . . . relentless determination . . . sheer resolution and esprit de corps . . . the brilliant record achieved. . . ." The award covered not only the brigade's ground units, but also MAG-33 and its squadrons.

They were fitting tributes to a special group of men who had truly earned a remarkable series of triumphs.

It would be a long war for the Marines in Korea, and there would be other much more famous battles to come, but the die was cast in those crucial first weeks of combat in August and September 1950. The Marine Corps had again decisively demonstrated that it was truly a "force in readiness," and that its rugged training and traditional *esprit de corps* could lead it to victory in "every clime and place."

About the Author

Captain John C. Chapin earned a bachelor of arts degree with honors in history from Yale University in 1942 and was commissioned later that year. He served as a rifle-platoon leader in the 24th Marines, 4th Marine Division, and was wounded in action in World War II during assault landings on Roi-Namur and Saipan.

Transferred to duty at the Historical Division, Headquarters Marine Corps, he wrote the first official histories of the 4th and 5th Marine Divisions. Moving to Reserve status at the end of the war, he earned a master's degree in history at George Washington University with a thesis on "The Marine Occupation of Haiti, 1915-1922."

Now a captain in retired status, he served for many years, starting in 1983, as a volunteer at the Marine Corps Historical Center. During that time he wrote the history of Marine Fighter-Attack (VMFA) Squadron 115. With support from the Historical Center and the Marine Corps Historical Foundation, he then spent some years researching and interviewing for the writing of a new book, *Uncommon Men—The Sergeants Major of the Marine Corps.* This was published in 1992 by the White Mane Publishing Company.

As part of the Historical Center's series of pamphlets commemorating the 50th anniversary of World War II, Captain Chapin wrote accounts of Marine operations in the Marshall Islands, on Saipan and Bougainville, and Marine aviation in the Philippines.

Sources

There are two basic official sources for the story of these early days in the Pusan Perimeter. For the Marine Corps, Lynn Montross and Capt Nicholas A. Canzona, USMC, *The Pusan Perimeter*, vol. 1 of *U.S. Marine Operations in Korea* (Washington: Historical Branch, G-3 Division, HQMC, 1954). For the Army, Roy E. Appleman, *South to the Naktong, North to the Yalu (June-November 1950), United States Army in The Korean War* (Washington: Office of the Chief of Military History, Department of the Army, 1961).

Also published officially is: LtCol Gary W. Parker, USMC, and Maj Frank M. Batha, Jr., USMC, *A History of Marine Observation Squadron Six* (Washington: History and Museums Division, HQMC, 1982).

Three articles that are very helpful and share a common publisher are: Ernest H. Giusti, "Marine Air Over the Pusan Perimeter," *Marine Corps Gazette*, May 1952; Lynn Montross, "The Pusan Perimeter: Fight for a Foothold," *Marine Corps Gazette*, June 1951; Col Nicholas A. Canzona, USMCR, "Marines Land at Pusan, Korea-August 1950," *Marine Corps Gazette*, August 1985; and Maj Francis I. Fenton, Jr., USMC, "Changallon Valley," *Marine Corps Gazette*, November 1951.

There is a wide range of commercially published books. Among those that contained useful material are: John Toland, *In Mortal Combat-Korea, 1950-1953* (New York: William Morrow and Company, 1991); Lynn Montross, *Cavalry of the Sky-The Story of U.S. Marine Helicopters* (New York: Harper and Brothers, 1954); Edwin P. Hoyt, *The Pusan Perimeter-Korea 1950* (New York: Stein and Day, 1984); David Douglas Duncan, *This Is War! A Photo-Narrative in Three Parts* (New York: Harper and Brothers, 1950-51); Clay Blair, *The Forgotten War* (New York: *Times* Books, 1987); Max Hastings, *The Korean War* (New York: Simon and Schuster, 1987) T. R. Fehrenbach, *This Kind of War* (New York: Macmillan, 1963); Joseph C. Goulden, *Korea: The Untold Story of the War* (New York: Times Books, 1982); Robert D. Heinl, Jr., *Soldiers of the Sea: The United States Marine Corps*, 1775-1962 (Annapolis: *United States Naval Institute*, 1962); and Robert Leckie, *Conflict* (New York: Putman, 1962).

A recent publication is by BGen Uzal W. Ent, USA (Ret), *Fighting On The Brink: Defense of the Pusan Perimeter* (Paducah, Ky: Turner Publishing Company, 1996).

Particular acknowledgment is made of the valuable quotations in Donald Knox, *The Korean War-Pusan to Chosin-An Oral History* (New York: Harcourt Brace Jovanovich, 1985) and Andrew Geer, *The New Breed-The Story of the U.S. Marines in Korea* (Nashville: The Battery Press, 1989).

In a different category is a book which the author affirms is based on actual Chinese sources: Russell Spurr, *Enter the Dragon-China's Undeclared War Against the U.S. in Korea, 1950-51* (New York: Henry Holt, 1988).

Personal interviews were helpful in meetings with: MajGen Robert D. Bohn, USMC (Ret); MajGen Charles D. Mize, USMC (Ret); and Col Robert D. Taplett, USMC (Ret).

Information also was contributed by MajGen Raymond L. Murray, USMC (Ret), and Col Francis I. Fenton, Jr., USMC (Ret), who kindly furnished postwar copies of *The Guidon*, a newsletter of B Company, with personal memoirs of combat.

The oral history transcripts at the Marine Corps Historical Center focus mainly on later events in Korea, but do have some observations by Craig, Murray, Bohn, Stewart, Sivert, Lucy, and LtCol Charles H. Brush, Jr., USMC, relating to the early days. In the Personal Papers Collection [now located at the Marine Corps University] there is a long memoir by PFC Herbert R. Luster who was a BAR-man in Company A (#1918-1A44). The files of the Reference Section contain much information pertinent to individual biographies and unit histories. Acknowledgement also is made to Col David Douglas Duncan, USMCR (Ret), for the use of his dramatic images.

OVER THE SEAWALL
U.S. Marines at Inchon

by Brigadier General Edwin H. Simmons, USMC (Ret)

Just three weeks away and there was still no approval from Washington for the Marines to land at Inchon on 15 September 1950. General of the Army Douglas MacArthur, determined to beat down the opposition to the landing, called a conference for late in the day, 23 August, at his headquarters in the Dai Ichi building in Tokyo.

Planning

As Commander in Chief, Far East (CinCFE), MacArthur considered himself empowered to conduct military operations more-or-less as he saw fit. But for an operation of the magnitude of Inchon and the resources it would require he needed approval from the highest level.

The Joint Chiefs of Staff (JCS), doubtful of the landing's chances of success, had sent out the Army Chief of Staff, General J. Lawton Collins, and the Chief of Naval Operations, Admiral Forrest P. Sherman, to review the situation directly with MacArthur. Now he would have to overcome their skeptical resistance. Collins was the JCS executive agent for the Far East Command and nominally higher in the chain-of-command than MacArthur—but only nominally. In World War I MacArthur was already a brigadier general when Collins was barely a captain. Now MacArthur had five stars and Collins four.

On this afternoon, First Lieutenant Alexander M. Haig's task was to lay out the pads of paper, pencils, and water glasses on the table of the sixth floor conference room. This done, he took his post seated in a straight-backed chair just outside the door. Haig, then the junior aide-de-camp to MacArthur's chief of staff, was destined to become, many years later, the Secretary of State.

The Marine Corps would have no voice at the meeting. The Corps had neither membership nor representation on the JCS. Admiral Sherman, not a strong champion of Marine Corps interests, was the service chief most directly concerned with the amphibious phase of the still tentative operation.

Opening Moves

Only two months before the meeting of MacArthur with Collins and Sherman, in the pre-dawn hours of 25 June, 25-year-old Lieutenant Haig, as duty officer at MacArthur's headquarters in Tokyo, received a phone call from the American ambassador in Seoul, John J. Muccio, that large formations of North Korean infantry had crossed the 38th Parallel. Haig informed his boss, Major General Edward M. "Ned" Almond, chief of staff of the Far East Command, who awakened MacArthur with the news. The United States was going to war.

Four days later, and a day after the fall of Seoul, MacArthur flew to Korea in the *Bataan*, to make a personal reconnaissance, taking with him Major General Almond. Korea stretched beneath them like a giant relief map. To the east of the Korean peninsula lay the Sea of Japan; to the west the Yellow Sea. The vulnerability of these two watery sides of the peninsula to a dominant naval power was not lost on a master strategist such as MacArthur. The *Bataan* landed at Suwon, 20 miles south of Seoul. MacArthur commandeered a jeep and headed north through, in his words, "the dreadful backwash of a defeated and dispersed army."

"Seoul was already in enemy hands," he wrote in his *Reminiscences* some years later. "The scene along the Han was enough to convince me that the defensive potential of South Korea had already been exhausted. The answer I had come to seek was there. I would throw occupation troops into this breach. I would rely upon strategic maneuver to overcome the great odds against me."

MacArthur returned to what he liked to call his "GHQ" in Tokyo, convinced that to regain the initiative the United States must use its amphibious capability and land behind the advancing North

AT LEFT: *The mop-up at Inchon turned up a group of young North Koreans, left as harassing forces.* Photo courtesy of *Leatherneck* Magazine

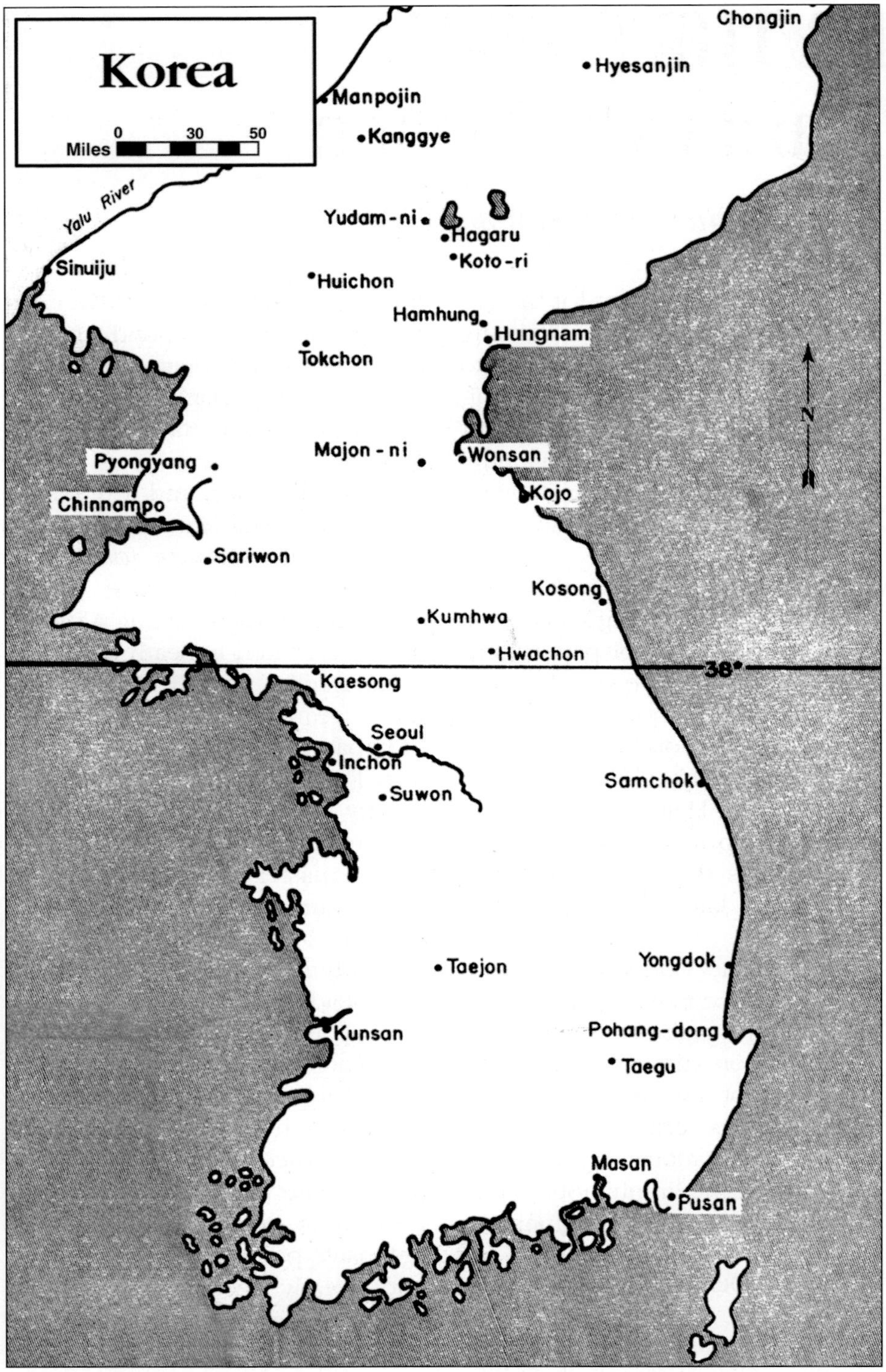

Koreans. He put his staff to work on a broad operational plan: two U.S. divisions would be thrown into the battle to slow the onrush of the North Korean People's Army (NKPA). A third division would land behind the NKPA and in a flanking attack liberate Seoul, the lost capital.

Unready Eighth Army

MacArthur had at his disposal in Japan the Eighth Army consisting of four divisions—the 7th, 24th, 25th, and the 1st Cavalry—all four at half-strength and under-trained. He began to move pieces of the 24th Division, rated at 65 percent combat-ready, to South Korea. His aim, he later said, was to trade space for time until a base could be developed at Pusan at the southern tip of the peninsula as a springboard for future operations.

Approval came from President Harry S. Truman for the imposition of a naval blockade and limited air operations. "The Air Force was under Lieutenant General George E. Stratemeyer, and the Navy under Vice Admiral C. Turner Joy, both able and efficient veterans of the war," wrote MacArthur.

But Vice Admiral Joy, as Commander Naval Forces, Far East, commanded virtually nothing. Vice Admiral Arthur D. "Rip" Struble, commander of the Seventh Fleet, a naval officer of considerable amphibious experience, reported not to Joy but to Admiral Arthur W. Radford who was both Commander in Chief, Pacific, and Commander in Chief, Pacific Fleet.

Lieutenant General George E. Stratemeyer commanded "FEAF" or Far East Air Forces. Subordinate to him were the Fifth Air Force in Japan, the Twentieth Air Force on Okinawa, and the Thirteenth Air Force in the Philippines.

Cates Offers Marines

Back in Washington, D.C., during the first hectic days after the North Korean invasion, the Commandant of the Marine Corps, General Clifton B. Cates, was not invited to attend the high-level meetings being held in the Pentagon. After four days of waiting, Cates drove to the Pentagon and, in his words, "kind of forced my way in."

"We were fighting for our existence," said General Lemuel C. Shepherd, Jr., who followed Cates as Commandant. "Sherman and the rest of these fellows wanted to keep us seagoing Marines, with a battalion landing team being the biggest unit we were supposed to have . . . . Everybody was against the Marine Corps at that time. Secretary of Defense Louis A. Johnson, always nagging, Truman hostile, and Cates carried that load all by himself and did it well."

Cates saw Admiral Sherman and told him the Marines could immediately deploy to Korea a brigade consisting of a regimental combat team and an aircraft group.

"How soon can you have them ready?" Sherman asked dubiously.

"As quickly as the Navy gets the ships," shot back Cates.

Sherman, overwhelmed perhaps by higher priorities, dallied two days before sending a back-channel message to Admiral Joy, asking him to suggest to MacArthur that he request a Marine air-ground brigade. MacArthur promptly made the request and on 3 July the JCS approved the deployment.

Cates did not wait for JCS approval. Formation of the 1st Provisonal Marine Brigade had already begun with troops stripped out of the half-strength 1st Marine Division. In four days' time—on 6 July—the brigade began to load out at San Diego for the Far East.

Several months before the breakout of war, MacArthur had requested amphibious training for his occupation troops. Troop Training Unit, Amphibious Training Command, Pacific Fleet, had been formed in 1943 for just such a purpose. Colonel Edward H. Forney, with Mobile Training Team Able and accompanied by an Air and Naval Gunfire Liaison Company (ANGLICO) training team, arrived in April 1950. A regiment in each of MacArthur's four divisions was to be amphibiously trained. Navy partner in the training would be Amphibious Group One (PhibGruOne) under Rear Admiral James H. Doyle.

A few days before the outbreak of the war Brigadier General William S. Fellers, commanding general of the Troop Training Unit, came out to Japan to inspect the progress being made by Forney and his team. Fellers and Forney were at a Fourth of July party being given by the American colony in Tokyo when an urgent message required their immediate presence at "GHQ." They arrived at the Dai Ichi—a tall building that had escaped the World War II bombing because the Imperial Palace was immediately across the way—to find a planning conference in progress with Almond at the helm. They learned that MacArthur had advanced the concept of a landing at Inchon, to be called Operation Bluehearts and to be executed on 22 July by the 1st Cavalry Division—and the 1st Provisional Marine Brigade, if the latter could be gotten there in time. Next day Colonel Forney became the G-5 (Plans) of the 1st Cavalry, one of MacArthur's favorite divisions.

Shepherd Meets with MacArthur

Three days after the interrupted Fourth of July party, Lemuel Shepherd, just promoted to lieutenant general and installed as Commanding General, Fleet Marine Force, Pacific, left Hawaii for Tokyo, accompanied by his operations officer, Colonel Victor H. Krulak. Shepherd had been urged to go to Tokyo by Admiral Radford, a good friend of the Marines, "to see MacArthur and find out what all this thing is about."

Shepherd saw his mission as being first to ensure that the 1st Provisional Marine Brigade was used as an integrated air-ground team and, second, to explore prospects for the use of additional Marine Corps forces.

"Having been with the 4th Brigade in France, I had learned that a Marine unit in an Army division is not good for the Corps," said Shepherd years later. Enroute to Tokyo he made up his mind that he was going to push for a Marine division to be sent to Korea.

General Shepherd met with Admiral Joy and General Almond on 9 July, and next day, accompanied by Colonel Krulak, saw MacArthur himself. He told them that the only hope for an early reversal of the disastrous situation was an amphibious assault against the enemy's rear.

"Here I was," said Shepherd later, "recommending that a Marine division be sent to Korea, and the Commandant didn't know anything about what I was doing."

MacArthur recalled to Shepherd the competence of the 1st Marine Division when it had been under his command during the Cape Gloucester operation at Christmas time in 1943. Shepherd had then been the assistant division commander. MacArthur went to his wall map, stabbed at the port of Inchon with the stem of his corncob pipe, and said: "If I only had the 1st Marine Division under my command again, I would land them here and cut off the North Korean armies from their logistic support and cause their withdrawal and annihilation."

Shepherd answered that if MacArthur could get JCS approval for the assignment of the 1st Marine Division, he could have it ready by mid-September. MacArthur told Shepherd to draft for his release a message to the JCS asking for the division.

Bluehearts, which would have used the 1st Cavalry Division, was abruptly cancelled. Planning in Tokyo, under Brigadier General Edwin K. "Pinky" Wright, USA, and his Joint Strategic Plans and Operations Group (JSPOG), shifted to an amphibious operation in September.

Under the U.N. Flag

On that same busy 10 July, MacArthur's mantle of authority was embroidered with a new title—Commander in Chief, United Nations Command or "CinCUNC." From then on operations in Korea and surrounding waters would be fought under the light-blue-and-white flag of the United Nations.

The sailing of the 1st Provisional Marine Brigade from San Diego began on 12 July. Core of the ground element was the 5th Marines; the air element was Marine Aircraft Group 33. Filling the brigade had gutted both the 1st Marine Division and the 1st Marine Aircraft Wing.

General Cates was in San Diego to see the Marines off. His long cigarette holder was famous; not many Marines knew that he used it because gas in World War I had weakened his lungs. General Shepherd was also on the dock and it gave him the opportunity to discuss with Cates his promise to MacArthur of a full division. Could the 1st Marine Division be assembled and made ready in such a short time?

"I don't know," said Cates dubiously; it would drain the Marine Corps completely.

"Clifton," said Shepherd simply, "you can't let me down."

Visitors from Washington

In Tokyo, where it was already 13 July, MacArthur was meeting with visitors from Washington—Army General Collins and General Hoyt S. Vandenberg, chief of staff of the newly independent Air Force. Also present were Admiral Radford, General Almond, and Lieutenant General Walton H. Walker. It had just been announced that Walker was shifting his flag from Japan to Korea, and the Eighth Army would become the Eighth U.S. Army in Korea, which yielded the acronym "EUSAK." MacArthur explained his reasons for cancelling Bluehearts and said that he had not yet chosen a new target date or location for an amphibious strike, but favored Inchon.

As soon as the meeting was over, Collins and Walker flew to Korea, where Walker opened a field headquarters at Taegu for his Eighth Army. Collins spent only an hour on the ground and did not leave the airport before returning to Tokyo.

Next day, the 14th, he was briefed by General Almond and Admiral Doyle, who had commanded Amphibious Group One since January. Before that for two years Doyle had headed the Amphibious Training Command, Pacific Fleet. During World War II he served on the staff of Amphibious Force, South Pacific.

Collins questioned the feasibility of landing at Inchon. Doyle said that it would be difficult but could be done. Before leaving Tokyo, Collins assured MacArthur that he would endorse the sending of a full-strength Marine division.

Earlier, during the planning for Operation Bluehearts, Doyle had expressed reservations over the use of the 1st Cavalry Division because it was not amphibiously trained. His relations with Almond were strained. He thought Almond arrogant and dictatorial and a person who "often confused himself with his boss."

Lieutenant Haig, Almond's aide and the keeper of his war diary, found his chief "volcanic" in personality, "brilliant" but "irascible," and, with all that, a "phenomenally gifted soldier." Almond, like his idol, General George S. Patton, Jr., designed his own uniforms and wore a pistol on a leather belt adorned with a huge crested buckle. He did this, he said, so as to be easily recognized by his troops.

General Walker, a tenacious man who deserved his nickname "Bulldog" (although he was "Johnnie" Walker to his friends), continued the piecemeal buildup of the Eighth Army. All of the 24th Division was committed by 7 July. The 25th Division completed its move from Japan on 14 July.

Tactical Air Control Problems

The 1st Cavalry Division was in process of loading out from Japan in Doyle's PhibGruOne when Bluehearts was cancelled in favor of an unopposed landing on 18 July at Pohang-dong, a port some 60 air miles northeast of Pusan. Plans developed for Bluehearts by both PhibGruOne and 1st Cavalry Division were used for the operation. For this non-hostile landing the Navy insisted on control of an air space 100 miles in diameter circling the landing site. This Navy requirement for control of air traffic over the objective area conflicted with Air Force doctrine which called for Air Force control of all tactical aircraft in the theater of operations.

Lieutenant General Earle E. "Pat" Partridge, whose Fifth Air Force Joint Operations Center was in Taegu side-by-side with Walker's Eighth Army headquarters, protested the Navy requirement that would have caused him to vacate the control of air over virtually all of the Pusan Perimeter. This began a doctrinal dispute involving the tactical control of air that would continue for the rest of the war.

Major General Oliver P. Smith

Oliver Prince Smith did not fill the Marine Corps "warrior" image. He was deeply religious, did not drink, seldom raised his voice in anger, and almost never swore. Tall, slender, and white-haired, he looked like a college professor is supposed to look and seldom does. Some of his contemporaries thought him pedantic and a bit slow. He smoked a pipe in a meditative way, but when his mind was made up he could be as resolute as a rock. He always commanded respect and, with the passage of years, that respect became love and devotion on the part of those Marines who served under him in Korea. They came to know that he would never waste their lives needlessly.

As commanding general of the 1st Marine Division, Smith's feud with the mercurial commander of X Corps, Major General Edward M. Almond, USA, would become the stuff of legends.

No one is ever known to have called him "Ollie." To his family he was "Oliver." To his contemporaries and eventually to the press, which at first tended to confuse him with the controversial Holland M. "Howlin' Mad" Smith of World War II, he was always "O. P." Smith. Some called him "the Professor" because of his studious ways and deep reading in military history.

Born in Menard, Texas, in 1893, he had by the time of America's entry into the First World War worked his way through the University of California at Berkeley, Class of 1916. While a student at Berkeley he qualified for a commission in the Army Reserve which he exchanged, a week after America's entry into the war on 6 April 1917, for the gold bars of a Marine Corps second lieutenant.

Department of Defense Photo (USMC) A8377

The war in Europe, where the Marines gained international fame, passed him by; he spent the war years in lonely exile with the garrison on Guam. Afterward, in the 1920s, he followed an unremarkable sequence of duty, much like that of most lieutenants and captains of the time: barracks duty at Mare Island, sea duty in the *Texas*, staff duty at Headquarters Marine Corps, and a tour with the Gendarmerie d'Haiti.

From June 1931 to June 1932, he attended the Field Officer's Course at Fort Benning. Next came a year at Quantico, most of it spent as an instructor at the Company Officer's Course. He was assigned in 1934 to a two-year course at the Ecole Superieur la de Guerre in Paris, then considered the world's premier school for rising young officers. Afterwards he returned to Quantico for more duty as an instructor.

The outbreak of World War II in 1939 found him at San Diego. As commanding officer of the 1st Battalion, 6th Marines, he went to Iceland in the summer of 1941. He left the regiment after its return to the States, for duty once again at Headquarters in Washington. He went to the Pacific in January 1944 in time to command the 5th Marines during the Talasea phase of the Cape Gloucester operation. He was the assistant commander of the 1st Marine Division during Peleliu and for Okinawa was the Marine Deputy Chief of Staff of the Tenth Army.

After the war he was the commandant of Marine Corps Schools and base commander at Quantico until the spring of 1948 when he became the assistant commandant and chief of staff at Headquarters. In late July 1950, he received command of the 1st Marine Division, destined for Korea, and held that command until May 1951.

After Inchon and Seoul, a larger, more desperate fight at Chosin Reservoir was ahead of him. In early 1951, the 1st Marine Division was switched from Almond's X Corps to Major General Bryant E. Moore's IX Corps. Moore died of a heart attack on 24 February 1951 and, by seniority, O. P. Smith became the corps commander. Despite his experience and qualifications, he held that command only so long as it took the Army to rush a more senior general to Korea.

O. P. Smith's myriad of medals included the Army Distinguished Service Cross and both the Army and the Navy Distinguished Service Cross for his Korean War Service.

On his return to the United States, he became the commanding general of the base at Camp Pendleton. Then in July 1953, with a promotion to lieutenant general, moved to the East Coast to the command of Fleet Marine Force, Atlantic, with headquarters at Norfolk, Virginia. He retired on 1 September 1955 and for his many combat awards was promoted to four-star general. He died on Christmas Day 1977 at his home in Los Altos Hills, California, at age 81.

Joint Chiefs Reluctant

Returning to Washington, Collins briefed his fellow chiefs on 15 July. He gave them the broad outlines of MacArthur's planned amphibious assault, but expressed his own doubts based on his experience in the South Pacific and at Normandy.

The JCS chairman, General of the Army Omar N. Bradley, thought it "the riskiest military proposal I ever heard of." In his opinion, MacArthur should be concentrating on the dismal immediate situation in South Korea rather than dreaming up "a blue sky scheme like Inchon." Bradley wrote later: "because Truman was relying on us to an extraordinary degree for military counsel, we determined to keep a close eye on the Inchon plan and, if we felt so compelled, finally cancel it."

The JCS agreed that the 1st Marine Division should be brought up to strength, but stopped short of committing it to the Far East. On 20 July, the Joint Chiefs informed MacArthur that the 1st Marine Division could not be combat ready until December. MacArthur erupted: the 1st Marine Division was "absolutely vital" to the plan being developed, under the codename Chromite, by General Wright's group. A draft, circulated at CinCFE headquarters on 23 July, offered three alternatives:

> Plan 100-B: A landing at Inchon on the west coast.
> Plan 100-C: A landing at Kunsan on the west coast.
> Plan 100-D: A landing at Chunmunjin-up on the east coast.

MacArthur's mind was now fully set on Inchon. He informed Collins, in his capacity as executive agent for the JCS, that lacking the Marine division, he had scheduled an amphibious assault at Inchon in mid-September to be executed by the 5th Marines and the 2d Infantry Division in conjunction with an attack northward by the Eighth Army. His message caused the chiefs to initiate a hurried teletype conference with MacArthur on 24 July. MacArthur prevailed and on the following day, 25 July, the chiefs finally approved MacArthur's repeated requests for the 1st Marine Division.

Gen Oliver P. Smith Collection, Marine Corps Research Center

MajGen Oliver P. Smith, left, assumed command of the 1st Marine Division at Camp Pendleton on 26 July 1950. Col Homer L. Litzenberg, Jr., right, arrived on 16 August with orders to reactivate the 7th Marines and have it ready for sailing by 3 September.

A New CG

Late in the afternoon of 25 July, Major General Oliver P. Smith arrived from Washington and checked in at the Carlsbad Hotel in Carlsbad, California. He was to take command of the 1st Marine Division at nearby Camp Pendleton on the following day. He phoned Brigadier General Harry B. Liversedge, the base commander and acting division commander, to let him know that he had arrived. Liversedge said that

Colonel Lewis B. "Chesty" Puller

Gen Oliver P. Smith Collection

The younger Marines in the 1st Marines were ecstatic when they learned their regiment was going to be commanded by the legendary "Chesty" Puller. Older officers and non-commissioned officers in the regiment were less enthusiastic. They remembered the long casualty list the 1st Marines had suffered at Peleliu while under Colonel Puller's command. His style was to lead from the front, and, when he went into Korea, he already had an unprecedented four Navy Crosses.

Born in 1898, Puller had grown up in Tidewater Virginia where the scars of the Civil War were still unhealed and where many Confederate veterans were still alive to tell a young boy how it was to go to war. Lewis (which is what his family always called him) went briefly to Virginia Military Institute but dropped out in August 1918 to enlist in the Marines. To his disappointment, the war ended before he could get to France. In June 1919, he was promoted to second lieutenant and then, 10 days later, with demobilization was placed on inactive duty. Before the month was out he had reenlisted in the Marines specifically to serve as a second lieutenant in the Gendarmerie d'Haiti. Most of the officers in the Gendarmerie were white Marines; the rank and file were black Haitians. Puller spent five years in Haiti fighting "Caco" rebels and making a reputation as a bush fighter.

He returned to the States in March 1924 and received his regular commission in the Marine Corps. During the next two years he did barracks duty in Norfolk, attended Basic School in Philadelphia, served in the 10th Marines at Quantico, and had an unsuccessful try at aviation at Pensacola. Barracks duty for two years at Pearl Harbor followed Pensacola. Then in 1928 he was assigned to the Guardia Nacional of Nicaragua. Here in 1930 he won his first Navy Cross. First Lieutenant Puller, his citation reads, "led his forces into five successive engagements against superior numbers of armed bandit forces."

He came home in July 1931 to the year-long Company Officers Course at Fort Benning. That taken, he returned to Nicaragua for more bandit fighting and a second Navy Cross, this time for taking his patrol of 40 Nicaraguans through a series of ambushes, in partnership with the almost equally legendary Gunnery Sergeant William A. "Iron Man" Lee.

Now a captain, Puller came back to the West Coast in January 1933, stayed a month, and then left to join the Legation Guard at Peiping. This included command of the fabled "Horse Marines." In September 1934, he left Peiping to become the commanding officer of the Marine detachment on board the *Augusta*, flagship of the Asiatic Fleet.

In June 1936, he came to Philadelphia to instruct at the Basic School. His performance as a tactics instructor and on the parade ground left its mark on the lieutenants who would be the captains, majors, and lieutenant colonels in the world war that was coming.

In June 1939, he went back to China, returning to the *Augusta* to command its Marines once again. A year later he left the ship to join the 4th Marines in Shanghai. He returned to the United States in August 1941, four months before the war began, and was given command of the 1st Battalion, 7th Marines, at Camp Lejeune. He commanded (he would say "led") this battalion at Guadalcanal and won his third Navy Cross for his successful defense of a mile-long line on the night of 24 October 1942. The fourth Navy Cross came for overall performance, from 26 December 1943 to 19 January 1944, at Cape Gloucester as executive officer of the 7th Marines. In February 1944, he took command of the 1st Marines and led it in the terrible fight at Peleliu in September and October.

Afterwards, he came back to command the Infantry Training Regiment at Camp Lejeune. Next he was Director of the 8th Marine Corps Reserve District with headquarters in New Orleans, and then took command of the Marine Barracks at Pearl Harbor. From here he hammered Headquarters to be given command, once again, of his old regiment, the 1st Marines.

After Inchon, there was to be a fifth Navy Cross, earned at the Chosin Reservoir. In January 1951, he received a brigadier general's stars and assignment as the assistant division commander. In May, he came back to Camp Pendleton to command the newly activated 3d Marine Brigade which became the 3d Marine Division. He moved to the Troop Training Unit, Pacific, on Coronado in June 1952 and from there moved east, now with the two stars of a major general, to Camp Lejeune to take command of the 2d Marine Division in July 1954. His health began to fail and he was retired for disability on 1 November 1955. From then until his death on 11 October 1971 at age 73 he lived in the little town of Saluda in Tidewater Virginia.

he had just received a tip from Washington that the division was to be brought to war strength and sail to the Far East by mid-August. Both Liversedge and Smith knew that what was left of the division was nothing more than a shell.

Smith took command the next day, 26 July. He had served in the division during World War II, commanding the 5th Marines in its Talasea landing at New Britain and was the assistant division commander at Peleliu. Only 3,459 Marines remained in the division at Camp Pendleton, fewer men than in a single full-strength regiment.

When the Joint Chiefs asked General Cates how he planned to bring the 1st Marine Division up to war strength, he had ready a two-pronged plan. Plan A would provide three rifle companies and replacements to the brigade already deployed. Plan B would use Reserves to fill up the division. Essential to the filling out of the 1st Marine Division—and the 1st Marine Aircraft Wing as well—was the mobilization of the Marine Corps Reserve. "Behind every Marine regular, figuratively speaking," wrote official historians Lynn Montross and Captain Nicholas A. Canzona, "stood two reservists who were ready to step forward and fill the gaps in the ranks."

The 33,527 Marines in the Organized Reserve in 1950 were scattered across the country in units that included 21 infantry battalions and 30 fighter squadrons. Virtually all the officers and noncommissioned officers had World War II experience, but the ranks had been filled out with youngsters, many of whom did not get to boot camp. Subsequent reserve training had included both weekly armory "drills" and summer active duty. Someone wryly decided they could be classified as "almost combat ready."

Behind the Organized Reserve was the Volunteer Marine Corps Reserve—90,044 men and women, most of them veterans, but with no further training after their return to civilian life. President Truman, with the sanction of Congress, authorized the call-up of the Marine Corps Reserve on 19 July. An inspired public information officer coined the phrase, "Minute Men of 1950."

On 26 July, the day following JCS approval of the 1st Marine Division's deployment, a courier arrived at Camp Pendleton from Washington with instructions for Smith in his fleshing-out of the 1st Marine Division: ground elements of the 1st Provisional Marine Brigade would re-combine with the division upon its arrival in the Far East; units of the half-strength 2d Marine Division at Camp Lejeune, North Carolina, would be ordered to Camp Pendleton and re-designated as 1st Marine Division units; all possible regulars would be stripped out of posts and stations and ordered to the division; and gaps in the ranks would be filled with individual Reserves considered to be at least minimally combat-ready.

Eighth Army Withdraws to Pusan

In Korea, at the end of July, Walker ordered the Eighth Army to fall back behind the Naktong River, the new defensive line forming the so-called "Pusan Perimeter." Both flanks of the Eighth Army were threatened. In light of this deteriorating situation, the Joint Chiefs asked MacArthur if he still planned an amphibious operation in September. An unperturbed MacArthur replied that "if the full Marine Division is provided, the chances to launch the movement in September would be excellent."

Reinforcements for Walker's Eighth Army began arriving directly from the United States, including the 1st Provisional Marine Brigade which debarked at Pusan on 3 August.

In Tokyo, General Stratemeyer became agitated when he learned that the 1st Provisional Marine Brigade, as an integrated air-ground team, intended to retain mission control of its aircraft. An uneasy compromise was reached by which the Marines were to operate their two squadrons of carrier-based Vought F4U Corsairs with their own controllers under the general coordination of Partridge's Fifth Air Force.

Reserve Comes to Active Duty

The first reservists to reach Pendleton—the 13th Infantry Company from Los Angeles, the 12th Amphibian Tractor Company of San Francisco, and the 3d Engineer Company from Phoenix—arrived on 31 July. Elements of the 2d Marine Division from Camp Lejeune began their train journey the same day. In that first week, 13,703 Marines joined the division.

On 4 August, the Commandant ordered the reactivation of the 1st Marines and 7th Marines. Both regiments had been part of the 1st Marine Division in all its World War II campaigns. The 1st Marines was activated that same day under command of the redoubtable Colonel Lewis B. "Chesty" Puller, who, stationed at Pearl Harbor as commanding officer of the Marine Barracks, had pestered Headquarters Marine Corps and General Smith with demands that he be returned to the command of the regiment he had led at Peleliu. By 7 August, the strength of the 1st Marine Division stood at 17,162.

The experiences of Lieutenant Colonel Thomas L. Ridge's 1st

Battalion, 6th Marines, were typical of the buildup being done at a dead run. Ridge had just taken command of the battalion. A crack rifle and pistol shot, he had spent most of World War II in intelligence assignments in Latin America, but in late 1944 was transferred to Fleet Marine Force, Pacific, in time for staff duty for Iwo Jima and Okinawa. As an observer at Okinawa he was twice wounded.

Ridge's battalion, barely returned to Camp Lejeune from six months deployment to the Mediterranean, traveled by ancient troop train to Camp Pendleton where it became the 3d Battalion of the reactivated 1st Marines. In about 10 days, the two-element, half-strength battalion expanded into a three-element, full-strength battalion. The two rifle companies

Major General Field Harris

During the course of the Korean War, Major General Field Harris would suffer a grievous personal loss. While he served as Commanding General, 1st Marine Aircraft Wing, his son, Lieutenant Colonel William F. Harris, was with the 1st Marine Division, as commanding officer of 3d Battalion, 7th Marines, at the Chosin Reservoir. The younger Harris' battalion was the rear guard for the breakout from Yudam-ni. Later, between Hagaru-ri and Koto-ri, Harris disappeared and was posted as missing in action. Later it was determined that he had been killed.

Field Harris—and he was almost always called that, "Field-Harris," as though it were one word—belonged to the open cockpit and silk scarf era of Marine Corps aviation. Born in 1895 in Versailles, Kentucky, he received his wings at Pensacola in 1929. But before that he had 12 years seasoning in the Marine Corps.

He graduated from the Naval Academy in March 1917 just before America's entry into World War I. He spent that war at sea in the *Nevada* and ashore with the 3d Provisional Brigade at Guantanamo, Cuba.

In 1919 he went to Cavite in the Philippines. After three years there, he returned for three years in the office of the Judge Advocate General in Washington. While so assigned he graduated from the George Washington University School of Law. Then came another tour of sea duty, this time in the *Wyoming*, then a year as a student at Quantico, and flight training at Pensacola. His new gold wings took him to San Diego where he served in a squadron of the West Coast Expeditionary Force.

He attended the Air Corps Tactical School at Langley Field, Virginia, after which came shore duty in Haiti and sea duty in the carrier *Lexington*. In 1935, he joined the Aviation Section at Headquarters, followed by a year in the Senior Course at the Naval War College in Newport, Rhode Island. In August 1941, he was sent to Egypt from where, as assistant naval attache, he could study the Royal Air Force's support of Britain's Eighth Army in its desert operations.

After Egypt and United States entry into the war, he was sent to the South Pacific. In the Solomons, he served successively as Chief of Staff, Aircraft, Guadalanal; Commander, Aircraft, Northern Solomons; and commander of air for the Green Island operation. Each of these

Department of Defense Photo (USMC) A310952

three steps up the chain of islands earned him a Legion of Merit. After World War II, he became Director of Marine Aviation in the Office of the Chief of Naval Operations (and received a fourth Legion of Merit). In 1948 he was given command of Aircraft, Fleet Marine Force, Atlantic. A year later he moved to El Toro, California, for command of Aircraft, Fleet Marine, Pacific, with concomitant command of the 1st Marine Aircraft Wing.

His Korean War service was rewarded with both the Army's and the Navy's Distinguished Service Medal. On his return to the United States in the summer of 1951, he again became the commanding general of Air, Fleet Marine Force, Atlantic. He retired in July 1953 with an advancement to lieutenant general because of his combat decorations, a practice which is no longer followed. He died in 1967 at age 72.

in the battalion each numbering about 100 men were doubled in size with a third rifle platoon added. A third rifle company was activated. The weapons company had no heavy machine gun platoon and only two sections in its antitank assault and 81mm mortar platoons. A heavy machine gun platoon was created and third sections were added to the antitank assault and 81mm mortar platoons. World War II vintage supplies and equipment came in from the mobilization stocks stashed away at the supply depot at Barstow, California—sufficient in quantity, poor in quality. The pressure of the unknown D-Day gave almost no time for unit shake-down and training.

Simultaneously with the ground unit buildup, Reserve fighter and ground control squadrons were arriving at El Toro, California, to fill out the skeleton 1st Marine Aircraft Wing. The wing commander, Major General Field Harris, Naval Academy 1917, and a naval aviator since 1929, had served in the South Pacific in World War II. More recently he had been Director of Aviation at Marine Corps Headquarters. He was one of those prescient senior Marines who foresaw a future for helicopters in amphibious operations.

Department of Defense Photo (USMC) A20115

MajGen Field Harris and a portion of the 1st Marine Aircraft Wing staff arrive at Barber's Point in Hawaii in early September enroute to the Far East. Leaving the Marine transport are, from left, Col Edward C. Dyer, Col Boeker C. Batterton, Col William G. Manley, and Gen Harris.

7th Infantry Division and KATUSA

In parallel actions, MacArthur on 4 August ordered Walker to rebuild the Army's 7th Infantry Division—the last division remaining in Japan—to full strength by 15 September. The division had been reduced to less than half-strength by being repeatedly culled for fillers for the three divisions already deployed to Korea. Until MacArthur's directive, the division was not scheduled to be up to strength until 1 October and not ready for amphibious operations until 1951. Now, the division was to get 30 percent of all replacements arriving from the United States. Moreover, a week later, on 11 August, MacArthur directed Walker to send 8,000 South Korean recruits to fill out the division.

The first of 8,600 Korean replacements, straight out of the rice paddies of South Korea and off the streets of Pusan, began arriving by ship at Yokohama a few days later. This infusion of raw untrained manpower, called "KATUSA"—Korean Augmentation of the U.S. Army—arrived for the most part in baggy white pants, white jackets, and rubber shoes. In three weeks they had to be clothed, equipped, and made into soldiers, including the learning of rudimentary field sanitation as well as rifle practice. The "buddy system" was employed—each Korean recruit was paired off with an American counterpart.

Major General David G. Barr, the 7th Infantry Division's commander, had been chief of staff of several commands in Europe during World War II. After the war he had headed the Army Advisory Mission in Nanking, China. He now seemed a bit old and slow, but he knew Chinese and the Chinese army.

1st Marine Division Loads Out

Loading out of the 1st Marine Division from San Diego began on 8 August. That same day, General Fellers, back from Japan, told Smith that the division would be employed in Korea between 15

Colonel Homer L. Litzenberg, Jr.

His troops called him "Litz the Blitz" for no particular reason except the alliteration of sound. He had come up from the ranks and was extraordinarily proud of it. Immediately before the Korean War began he was in command of the 6th Marines at Camp Lejeune, very much interested in his regimental baseball teams, and about to turn over the command to another colonel. When war came he was restored to command of the regiment and sadly watched his skeleton battalions depart for Camp Pendleton to form the cadre for the re-activated 1st Marines. This was scarcely done when he received orders to re-activate the 7th Marines on the West Coast.

Litzenberg was a "Pennsylvania Dutchman," born in Steelton, Pennsylvania, in 1903. His family moved to Philadelphia and, after graduating from high school and two years in the National Guard, he enlisted in the Marine Corps in 1922. Subsequent to recruit training at Parris Island, he was sent to Haiti. In 1925 he became a second lieutenant. East Coast duty was followed by expeditionary service in Nicaragua in 1928 and 1929, and then by sea service in a string of battleships—*Idaho*, *Arkansas*, *Arizona*, *New Mexico*—and the cruiser *Augusta*. After graduating from the Infantry School at Fort Benning in 1933, he had two years with a Marine Reserve battalion in Philadelphia. Next came two years on Guam as aide to the governor and inspector-instructor of the local militia. He came home in 1938 to serve at several levels as a war planner.

When World War II came, he was sent, as a major, to England to serve with a combined planning staff. This took him to North Africa for the amphibious assault of Casablanca in November 1942. He came home to form and command the 3d Battalion, 24th Marines, in the new 4th Marine Division, moving up to regimental executive officer for the assault of Roi-Namur in the Marshalls. He then went to the planning staff of the V Amphibious Corps for Saipan and Tinian.

After the war he went to China for duty with the Seventh Fleet and stayed on with Naval Forces Western Pacific. He came home in 1948 and was given command of the 6th Marines in 1949.

Department of Defense Photo (USMC) A4718

After Inchon, he continued in command of the 7th Marines through the battles of Seoul, Chosin Reservoir, and the Spring Offensive, coming home in April 1951. Soon promoted to brigadier general and subsequently to major general, he had many responsible assignments including assistant command of the 3d Marine Division in Japan, Inspector General of the Marine Corps, command of Camp Pendleton, and command of Parris Island. He returned to Korea in 1957 to serve as senior member of the United Nations component negotiating at Panmunjom. At the end of the year he came back for what would be his last assignment, another tour of duty as Inspector General.

He retired in 1959, with an elevation to lieutenant general because of his combat awards that included a Navy Cross, a Distinguished Service Cross, and three Silver Stars. He died in the Bethesda Naval Hospital on 27 June 1963 at age 68 and was buried in Arlington National Cemetery with full military honors.

and 25 September.

Much of the heavy equipment to be loaded arrived at dockside from the Barstow supply depot with no time for inspection. General Shepherd arrived on 13 August to observe and encourage, joined next day by General Cates. Puller's 1st Marines sailed from San Diego on 14 August, 10 days after activation. The Navy had very little amphibious shipping on the West Coast, and much of the division and its gear had to be lifted by commercial shipping.

Among the pressing matters discussed by Smith with his superiors Cates and Shepherd was the reactivation of the 7th Marines. Nucleus of the 7th Marines would be the skeleton 6th Marines, which had already lost two battalions to the 1st Marines. The 3d Battalion, 6th Marines, a half-strength peacetime battalion with pieces scattered around the Mediterranean, became the 3d Battalion, 7th Marines, with orders to proceed to Japan by way of the Suez Canal. Fillers for the battalion and a completely new third rifle company would have to come from Camp Pendleton.

What was left of the 6th Marines arrived at Pendleton on 16 August. The 7th Marines activated the next day. Colonel Homer L. "Litz the Blitz" Litzenberg, Jr., a mercurial man who had commanded the 6th Marines at Camp Lejeune, continued as commanding officer of the 7th Marines with orders to embark his regiment not later than 3 September.

Joint Chiefs Have a Problem

Although the National Defense Act of 1947 was in effect, the relationship of the Joint Chiefs to the theater commanders was not too clear. As a theater commander MacArthur had broad leeway in his actions. The JCS faced the Hobson's choice of asking MacArthur no questions and making no challenges, or exerting their capacity as the principal advisors to President Truman in his role.

The Joint Chiefs held an intensive series of briefings in the White House on 10 August, culminating in an afternoon meeting with the National Security Council. President Truman was told that a war-strength Marine division was being assembled for service in Korea. Admiral Sherman assured the President, however, that the JCS would have to pass on MacArthur's plans for an amphibious operation.

On 12 August, MacArthur issued CinCFE Operations Plan 100-B, specifically naming Inchon-Seoul as the objective area. No copy of this plan was sent to the JCS.

O. P. Smith Departs Pendleton

General Smith sent off the first echelon of his division headquarters by air on 16 August. Two days later he closed his command post at Camp Pendleton and left by air for Japan. Delayed by shipping shortages, outloading of a third of Smith's division—essentially the reinforced 1st Marines—was completed on 22 August. In all, 19 ships were employed.

Following close behind, Litzenberg beat by two days the embarkation date given him by Smith. The 7th Marines, filled up with regulars pulled away from posts and stations and reservists, sailed from San Diego on 1 September.

Marine Versus Air Force Close Support

General Stratemeyer, MacArthur's Air Force component commander, apparently first heard of the possibility of an Inchon landing on 20 July. His first action was to instruct his staff to prepare a small command group with which he could accompany MacArthur on the operation. Almost a month later, on 14 August, MacArthur discussed the proposed landing with Stratemeyer, pointing out that Kimpo Airfield, just west of the Han River from Seoul, was the best in Korea. MacArthur emphasized that the airfield must be quickly rehabilitated from any battle damage and put to use.

By then news stories were appearing that compared Fifth Air Force support of the Eighth Army unfavorably with the close air support being provided the Marine

USS Mount McKinley *(AGC7) was the command center afloat for the Inchon landing. It also served as a floating hotel for the large number of VIPs who were in Gen Douglas MacArthur's official party or were simply passing through.*

National Archives Photo (USN) 80-G-424523

brigade by its organic squadrons. On 23 August, Stratemeyer sent a memorandum to MacArthur stating that the news stories were another step "in a planned program to discredit the Air Force and the Army and at the same time to unwarrantedly enhance the prestige of the Marines." He pointed out that the Marine squadrons, operating from two aircraft carriers, were supporting a brigade of about 3,000 Marines on a front that could be measured in yards as compared to the Fifth Air Force which had to supply close air support for a front of 160 miles.

General Walker, collocated at Taegu with General Partridge, pulled the rug out from under General Stratemeyer's doctrinal concerns and contentions of unfairness, by commenting officially: "Without the slightest intent of disparaging the support of the Air Force, I must say that I, in common with the vast majority of officers of the Army, feel strongly that the Marine system of close air support has much to commend it . . . . I feel strongly that the Army would be well advised to emulate the Marine Corps and have its own tactical aviation."

Top Brass Gathers in Tokyo

General Collins and Admiral Sherman—the latter had not been to Korea before—made a quick visit on 22 August to Walker's Eighth Army headquarters at Taegu. Collins found Walker "too involved in plugging holes in his leaky front to give much thought to a later breakout." On the morning of 23 August, Collins accompanied Walker on a visit to all U. S. division commanders and the Marine brigade commander, Brigadier General Edward A. Craig. Collins found these field commanders confident but weary. Collins

National Archives Photo (USN) 80-G-422492

Gen Douglas MacArthur, center, greets Gen J. Lawton Collins, Chief of Staff, U.S. Army, and Adm Forrest P. Sherman, Chief of Naval Operations, upon their arrival in Tokyo on 21 August 1950. A critical conference would be held two days later at which MacArthur would have to convince these two members of the Joint Chiefs of Staff that a landing at Inchon was feasible.

then returned to Tokyo for the crucial conference at which MacArthur must overcome JCS reservations concerning the Inchon landing.

Major General Smith arrived at Haneda airport in Japan on 22 August and was met by his old friend, Admiral Doyle, the prospective Attack Force Commander. Smith later remembered that Doyle "was not very happy about the whole affair." They proceeded to Doyle's command ship, USS *Mount McKinley* (AGC 7). Smith's orders were to report his division directly to Commander in Chief, Far East, for operational control. His appointment with General MacArthur was set for 1730 that evening at the Dai Ichi building. Colonel Alpha L. Bowser, Jr., the division G-3, who had come out with the first echelon of Smith's staff, gave him a hurried briefing on the tentative plans for the division. "For the first time I learned that the division was to land at Inchon on 15 September," Smith wrote later.

On arriving at GHQ comfortably before the appointed time of 1730, Smith found that he was to meet first with Almond, who kept him waiting until 1900. Almond called most soldiers and officers "son," but when 58-year-old Almond addressed 57-year-old Smith as "son," it infuriated Smith. Almond further aggravated Smith by dismissing the difficulties of an amphibious operation as being "purely mechanical."

Having had his say, Almond ushered Smith into MacArthur's office. MacArthur, in a cordial and expansive mood, confidently told Smith that the 1st Marine Division would win the war by its landing at Inchon. The North Koreans had committed all their troops against the Pusan Perimeter, and he did not expect heavy opposition at Inchon. The operation would be somewhat "helter-skelter," but it would be successful. It was MacArthur's feeling that all hands would be home for Christmas, if not to the United States, at least to Japan.

Smith reported to Doyle his conviction that MacArthur was firm in his decision to land at Inchon on 15 September. Doyle replied that he thought there was still a chance to substitute Posung-Myun, a few miles to the south of Inchon, as a more likely landing site. Doyle was having his underwater demolition teams reconnoiter those beaches.

Next day, 23 August, Smith met again with Almond, this time accompanied by General Barr, commander of the 7th Infantry Division. When Smith raised the possibility of Posung-Myun as a landing site, Almond brushed him off, saying that any landing at Posung-Myun would be no more than a subsidiary landing.

Critical 23 August Conference Convenes

Smith was not invited to the 23 August conference. Nor was Shepherd. The all-important summit conference began with brief opening remarks by MacArthur. General Wright then outlined the basic plan which called for an assault landing by the 1st Marine Division directly into the port of Inchon. After the capture of Inchon, the division was to advance and seize, as rapidly as possible, Kimpo Airfield, the town of Yongdung-po, and the south bank of the Han River. The division was then to cross the river, capture Seoul, and seize the dominant ground to the north. Meanwhile, the 7th Infantry Division was to land behind the Marines, advance on the right flank, secure the south bank of the Han southeast of Seoul and the high ground north of Suwon. Thereafter, X Corps—1st Marine and 7th Infantry Divisions— would form the anvil against which the Eighth Army, breaking out of the Pusan Perimeter, would deliver the hammer blows that would destroy the North Korean Army.

After Wright's briefing, Doyle, as the prospective Attack Force commander, gave a thorough analysis of the naval aspects of the landing. Of greatest concern to Doyle were the tides. A point of contention was the length of the naval gunfire preparation. Doyle argued for three to four days of pre-landing bombardment by air and naval gunfire, particularly to take out the shore batteries. MacArthur's staff disputed this on the basis of the loss of tactical surprise. Admiral Sherman was asked his opinion and replied, "I wouldn't hesitate to take a ship up there."

"Spoken like a Farragut," said MacArthur.

With his concerns brushed aside, Doyle concluded his briefing with "the best that I can say is that Inchon is not impossible."

Collins questioned the ability of the Eighth Army to link up quickly with X Corps. He suggested Kunsan, to the south, as an alternate landing site. Sherman, in general terms, supported Collins' reservations. General MacArthur sat silently, puffing his pipe, for several moments. He then spoke and all agree that his exposition was brilliant. He dazzled and possibly confused his audience with an analogy from the French and Indian War, Wolfe's victory at Quebec: "Like Montcalm, the North Koreans will regard the Inchon landing as impossible. Like Wolfe I [can] take them by surprise."

As he himself remembered his summation years later in his memoirs:

> The Navy's objections as to tides, hydrography, terrain, and physical handicaps are indeed substantial and pertinent. But they are not insuperable. My confidence in the

MajGen David G. Barr, left, Commanding General of the U. S. Army's 7th Infantry Division meets with MajGen Edward M. Almond, Commanding General, X Corps, to discuss the Inchon landing. The 7th Division would land behind the Marines, advance on their right flank, and seize the commanding ground south of Seoul. National Archives Photo (USA) 111-SC349013

> Navy is complete, and in fact I seem to have more confidence in the Navy than the Navy has in itself . . . . As to the proposal for a landing at Kunsan, it would indeed eliminate many of the hazards of Inchon, but it would be largely ineffective and indecisive. It would be an attempted envelopment which would not envelop. It would not sever or destroy the enemy's supply lines or distribution center, and would therefore serve little purpose. It would be a "short envelopment," and nothing in war is more futile. But seizure of Inchon and Seoul will cut the enemy's supply line and seal off the entire southern peninsula . . . . This in turn will paralyze the fighting power of the troops that now face Walker . . . . If my estimate is inaccurate and should I run into a defense with which I cannot cope, I will be there personally and will immediately withdraw our forces before they are committed to a bloody setback. The only loss then will be my professional reputation. But Inchon will not fail. Inchon will succeed. And it will save 100,000 lives.

Others at the conference recalled MacArthur's closing words at the conference as being: "*We* shall land at Inchon, and *I* shall crush them." This said, MacArthur knocked the ashes of his pipe out into a glass ashtray, making it ring, and stalked majestically out of the room.

General Collins still harbored reservations. He thought a main point had been missed: what was the strength of the enemy at Inchon and what was his capability to concentrate there?

Admiral Sherman was momentarily carried away by MacArthur's oratory, but once removed from MacArthur's personal magnetism he too had second thoughts. Next morning, 24 August, he gathered together in Admiral Joy's office the principal Navy and Marine Corps commanders. Present, in addition to Sherman and Joy, were Admirals Radford and Doyle and Generals Shepherd and Smith. Despite general indignation over MacArthur's failure to give due weight to naval considerations, it was now abundantly clear that the landing would have to made at or near Inchon. But perhaps there was still room for argument for another landing site with fewer hydrographic problems. Shepherd announced that he was going to see MacArthur once again before returning to Pearl Harbor and that he would make a

Command Structure for Inchon

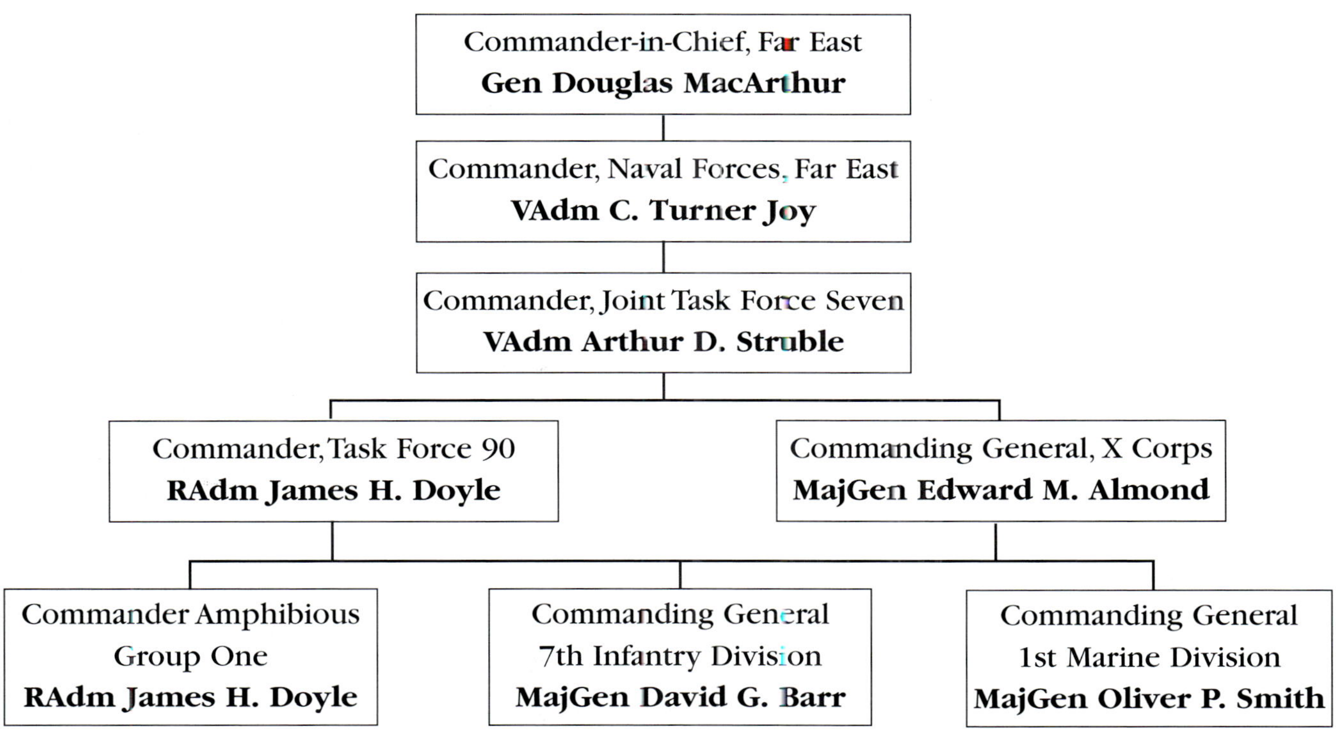

final plea for a landing south of Inchon in the vicinity of Posung-Myun.

Disappointment for General Shepherd

Shepherd, accompanied by Krulak, arrived at GHQ for his scheduled visit with MacArthur but was short-stopped by Almond who dismissed the Posung-Myun site, saying that Inchon had been decided upon and that was where the landing would be. The discussion became heated. Fortunately, MacArthur entered the room and waved Shepherd and Krulak into his office.

Shepherd had some expectation of being named the landing force commander. Admiral Sherman had recommended, without any great amount of enthusiasm, that Shepherd command X Corps for the operation because of his great amphibious experience and the expertise of his Fleet Marine Force, Pacific staff. General Wright on MacArthur's staff also recommended it, but a rumor was prevalent that Almond would get X Corps. MacArthur confirmed this intention, saying he would liked to have had Shepherd as commander, but that he had promised it to Almond. He asked if Shepherd would go along as his amphibious advisor. Shepherd hedged slightly. He said he would gladly go along as an observer.

Shepherd showed no rancor, then or later, at not getting command. He and Almond were both Virginians and both had gone to Virginia Military Institute—Almond, class of 1915 and Shepherd, class of 1917. Their personal relations were good but not close. Shepherd later characterized Almond as "an excellent corps commander. He was energetic, forceful, brave, and in many ways did a good job under difficult conditions." O. P. Smith would not come to share Shepherd's good opinion of Almond.

Plans Progress

The day following the 23 August conference, General Stratemeyer directed his staff to develop a FEAF plan to support the landing. The plan was to be separate from the CinCFE plan and was to provide mission direction for all combat aircraft not essential to the close support of the Eighth Army.

MacArthur, on 26 August, formally announced Almond's assignment as commanding general of X Corps. MacArthur had told him that he would continue, at the same time, to be the chief of staff of Far East Command. MacArthur's prediction was that Almond would soon be able to return to Tokyo. The landing at Inchon and subsequent capture of Seoul would end the war.

General Bradley's assessment of Almond was less than enthusiastic:

> Ned Almond had never commanded a corps—or troops in an amphibious assault. However, he and his staff, mostly recruited from MacArthur's headquarters, were ably backstopped by the expertise of the Navy and Marines, notably that of Oliver P. Smith, who commanded the 1st Marine Division, which would spearhead the assault.

MacArthur had not asked Collins and Sherman to approve his plan nor would they have had the authority to do so. The best they had to take back with them to Washington was a fairly clear concept of MacArthur's intended operations.

Collins and Sherman reported to Bradley and the other chiefs what they had learned about the Inchon plan, repeating their own misgivings. On 26 August, Bradley briefed President Truman and Secretary Johnson. The President was more optimistic than the chiefs.

'Conditional' Approval

On 28 August, the Joint Chiefs sent MacArthur a "conditional" approval, concurring in an amphibious turning movement, either at Inchon or across a favorable beach to the south. Chief "conditions" were that MacArthur was to provide amplifying details and keep them abreast of any modification of his plans. The Joint Chiefs specifically suggested preparation of an alternate plan for a landing at Kunsan.

X Corps dated its Operation Order No. 1, written largely by the facile pen of Colonel Forney, as 28 August; distribution was a day or so later. The 1st Marine Division "was charged with the responsibility as the Landing Force to assault INCHON, conduct beachhead operations, seize and protect KIMPO airfield, then advance to the HAN River line west of SEOUL. This achieved, the Division was further directed to seize SEOUL, and the commanding ground north of SEOUL, on order."

O. P. Smith's division staff, then on the *Mount McKinley*, was at half strength. Part of the remainder was enroute from the United States; part was with Craig's 1st Marine Brigade in the south of Korea. The brigade, although an organic part of the division, was still under the operational control of General Walker. Smith's staff, directed by Colonel Gregon A. Williams as chief of staff, worked well with Doyle's PhibGruOne staff. Above this harmonious relationship, the exact status of the more senior commands was indistinct and vaguely defined. From amidst a welter of paper, misunderstanding, ragged tempers, and sleep deprivation, Division Order 2-50, expanding on the corps order, emerged on 4 September.

Smith wrote later in the *Marine Corps Gazette*:

> By dedicated work on the part of the Division staff, with the wholehearted support of Adm Doyle's PhibGruOne staff, within three days a detailed plan for the Inchon Landing was drawn up, and two days later an advance planning draft of 1stMarDiv OpO 2-50 (Inchon Landing) was issued.

Time available for planning was so short that the assault regiments, contrary to amphibious doctrine, would get rigid landing plans drawn up completely by division.

The always dapper General Stratemeyer, seeking to solidify his contention that he was General MacArthur's tactical air commander, conferred with Joy, Struble, and Almond at CinCFE headquarters on 30 August. All that he could get was a general agreement on the adequacy of a CinCFE 8 July directive, "Coordination of Air Effort of Far East Air Forces and United States Naval Forces, Far East." Building on that, Stratemeyer sent a message to MacArthur, the gist of it being: "It is recognized that ComNavFE must have control of air operations within the objective area during the amphibious phase. Air operations outside of the objective area are part of the

Junior officers and enlisted Marines did not get a briefing on their unit's role in the landing until embarked in amphibious shipping enroute to the objective area. However, by then, because of leakage to the press, it was an open secret that the Marines were going to land at Inchon.

Department of Defense Photo (USMC) A2681

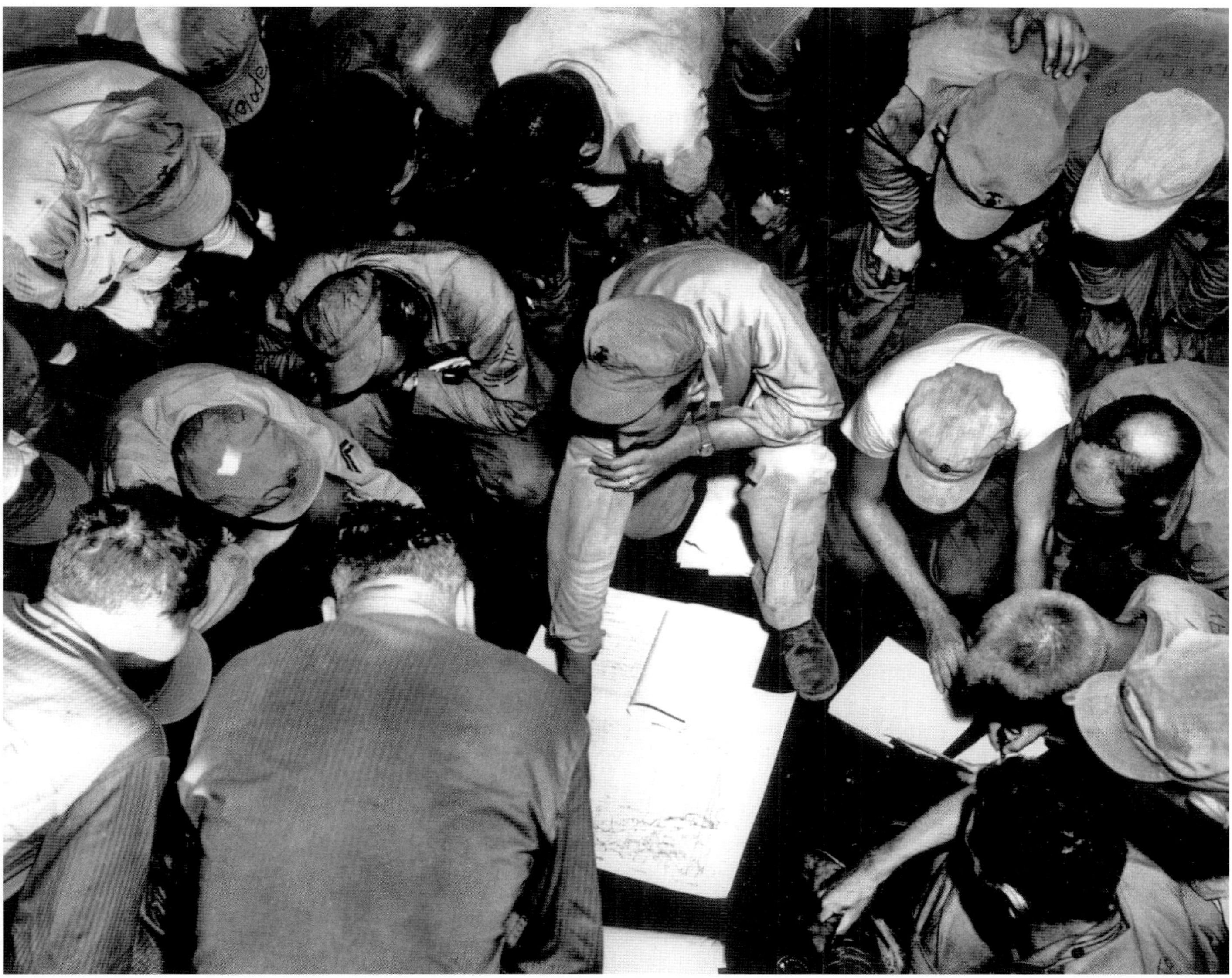

overall air campaign, and during the amphibious phase contribute to the success of the amphibious operation."

MacArthur's headquarters issued Operation Order No. 1 on 30 August, but neither a copy of this order nor any other amplifying detail had reached Washington by 5 September. On that date the chiefs sent a further request for details to MacArthur. Choosing to consider the 28 August JCS message to be sufficient approval, MacArthur dismissed the request with a brief message, stating "the general outline of the plan remains as described to you."

Later he would write that his plan "was opposed by powerful military in Washington." He knew that Omar Bradley, the JCS chairman, had recently testified to Congress that large-scale amphibious operations were obsolete. He disliked Bradley personally and derisively referred to him as a "farmer."

Both Bradley and Truman came from Missouri working-class families and were proud of it. A routine had been established under which the Joint Chiefs kept Truman informed, usually by a personal briefing by Bradley, of the current situation in Korea.

On 7 September, MacArthur received a JCS message which he said chilled him to the marrow of his bones. The message asked for an "estimate as to the feasibility and chance of success of projected operation if initiated on planned schedule."

The offending message reminded MacArthur that all reserves in the Far East had been committed to the Eighth Army and all available general reserves in the United States—except for the 82d Airborne Division—had been committed to the Far East Command. No further reinforcement was in prospect for at least four months. In light of this situation, a fresh evaluation of Inchon was requested.

MacArthur Protests

An indignant MacArthur fired back an answer, the concluding paragraph of which said: "The embarkation of the troops and the preliminary air and naval preparations are proceeding according to schedule. I repeat that I and all my commanders and staff officers are enthusiastic for and confident of the success of the enveloping movement."

The last sentence was manifestly not true. Lack of enthusiasm was readily apparent at all levels of command.

Next day, 8 September, the JCS sent MacArthur a short, contrite message: "We approve your plan and the President has been informed." The phrase "the President has been informed" annoyed MacArthur. To him it implied something less than presidential approval and he interpreted it as a threat on President Truman's part to overrule the Joint Chiefs. General Collins, for one, had no recollection of Truman ever expressing any doubt about the success of the Inchon landing or any inclination to override the actions of the JCS with respect to the operation.

Beach Reconnaissance

According to the intelligence available to General Smith, the enemy had about 2,500 troops in the Inchon-Kimpo region, including at least two battalions of the *226th Independent Marine Regiment* and two companies of the *918th Artillery Regiment*. The North Koreans had apparently prepared strong defensive positions. Reconnaissance reports indicated 106 hard targets, such as gun emplacements, along the Inchon beaches.

Some of the best beach intelligence was obtained by Navy offshore reconnaissance. Best known are the exploits of Lieutenant Eugene F. Clark, ex-enlisted man and an experienced amphibious sailor. He and two South Koreans left Sasebo on 31 August on board the British destroyer HMS *Charity*, transferred the next morning to a South Korean frigate, and landed that evening on Yong-hong-do, 14 miles off Inchon and one of the hundreds of islands that dotted Korea's west coast. The islanders were friendly. Clark organized the island's teenagers into coastwatching parties and commandeered the island's only motorized sampan. For two weeks he fought a nocturnal war, capturing more sampans, sending agents into Inchon, and testing the mud flats for himself. His greatest accomplishment was discovering that one of the main navigation lights for Flying Fish Channel was still operable. GHQ at Tokyo instructed him to turn on the light at midnight on 14 September. This he would do.

Anticipated hydrographic conditions were much more frightening than the quality of expected enemy resistance. Doyle's Attack Force would have to thread its way from the Yellow Sea through the tortuous Flying Fish Channel. As had already been determined, the 15th of September was the best day of the month because of the height and spacing of the tides. The morning high tide—an incredible 31.5 feet—would be at 0659 and the evening high tide at 1919. In between these times, as the tide fell, the currents would rip out of the channel at seven or eight knots, exposing mud flats across which even amphibian tractors

Terrain Handbook No. 65: Seoul and Vicinity (GHQ, Far East Command, 16 August 1950)

This pre-landing aerial photograph shows clearly the convoluted nature of the Inchon "beachhead." MajGen Oliver P. Smith, commanding the landing force, considered Wolmi-do, the island at the lower left of the photo, the key to the whole situation. Seizure of Wolmi-do would precede the main landings on Inchon itself.

could not be expected to crawl.

Wolmi-do: Key to Operation

Wolmi-do ("Moon Tip Island"), the long narrow island that formed the northern arm of Inchon's inner harbor, was thought to have about 500 defenders. Wolmi-do harbor was connected to the Inchon dock area by a 600-yard-long causeway. "Wolmi-do," wrote Smith, was "the key to the whole operation."

Brigade staff officers, headed by their chief of staff, Colonel Edward W. Snedeker, were called to Japan from Pusan. They recommended that the 3d Battalion, 5th Marines, be used for the assault of Wolmi-do.

Smith's plan, as it emerged, was to take Wolmi-do on the morning tide by landing the 3d Battalion, 5th Marines, across Green Beach. Then would come a long wait of 12 hours until the evening tide came in and the remainder of the division could continue the landing. The rest of the 5th Marines would cross Red Beach to the north of Wolmi-do, while Puller's 1st Marines landed over Blue Beach in the inner harbor to the south. Designation of the landing sites as "beaches" was misleading; the harbor was edged with cut-granite sea walls that would have to be scaled or penetrated.

Colonel Snedeker recommended that the new 1st Korean Marine Corps Regiment be added to the troop list. The assignment of the Republic of Korea (ROK) Marines to the division was approved by GHQ on 3 September. The Eighth Army was instructed to provide them weapons.

Almond asked Smith to take part in a war-gaming of the operation. Smith saw it as nothing more than a "CPX" or command post exercise and a waste of precious time. He sent a major in his place.

Almond inspected units of Barr's 7th Division at their camps—Fuji, McNair, McGill, Drake, and Whittington—between 31 August through 3 September. His aide,

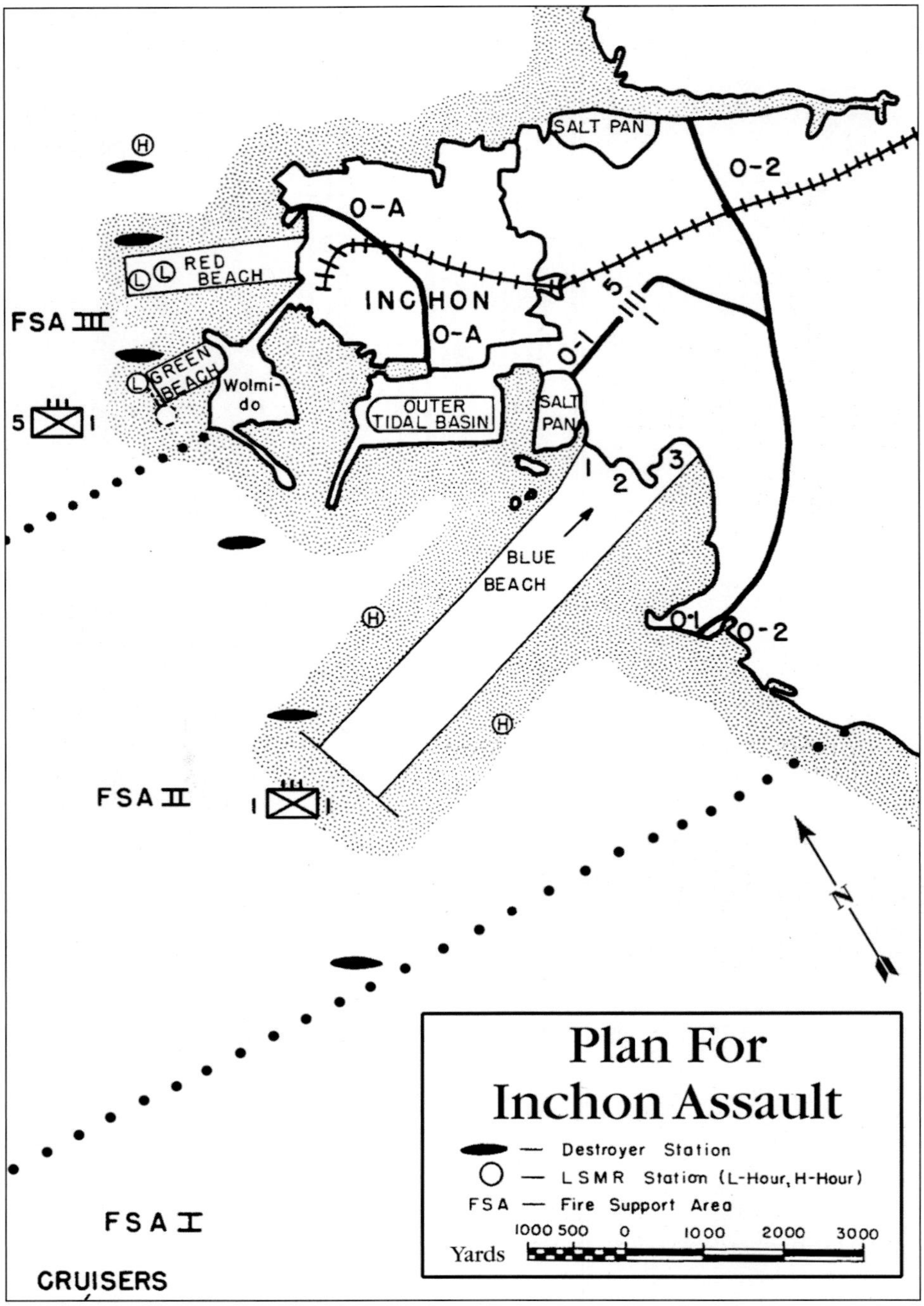

First Lieutenant Haig, accompanied him and took extensive notes. With few exceptions, Almond gained a "good" to "excellent" impression of the units he visited.

On the morning of 2 September Almond met with the officers of his Corps staff who were involved in his war game. He pointed out the necessity for frequent visits to subordinate units by commanding officers and the need for strong, well-organized, defenses for Corps headquarters. "The front line is the perimeter of the place where you happen to be," said Almond.

Meanwhile, the main body of the 1st Marine Division arrived at Kobe, Japan—except for the 5th Marines, which was still at Pusan, and the 7th Marines, which was still at sea.

Typhoon Jane Disrupts Embarkation

Typhoon Jane, with winds up to 74 miles an hour, struck Kobe on 3 September. Two feet of water covered the docks. One ship, with all the division's signal gear, settled to the bottom at her pier. All unloading and loading stopped for 24 hours. Property sergeants, called in from the outlying battalions, worked frantically to sort out their units' gear.

Adding to General Smith's worries, the availability of the 5th Marines was now challenged. General Walker, deeply involved in the bitter defense of the Naktong Bulge, strongly opposed the release of this now-seasoned regiment from his Eighth Army. To meet Walker's objections, and influenced by his own favorable impression of the 7th Division, Almond sent Colonel Forney, now the Marine Deputy Chief of Staff, X Corps, to ask O. P. Smith whether the 7th Marines would arrive in time to be substituted for the 5th Marines, or alternatively, if not, would the 32d Infantry be acceptable?

A conference on the proposed substitution was held on the evening of 3 September. Present, among others, were Generals Almond and Smith and Admirals Joy, Struble, and Doyle. Strangely, General Barr, the 7th Division's commander, was not there. The discussion became heated. Smith argued that the proposal went beyond a considered risk. If the substitution were made, he declared, he would change his scheme of maneuver. He would call off the landing of the 1st Marines over Blue Beach and give them the 5th Marines' mission of landing on Red Beach with the 32d Infantry following behind.

Admiral Struble (Shepherd thought him "slippery") resolved the contretemps by suggesting that a regiment of Barr's 7th Division be immediately embarked to stand off Pusan as a floating reserve, allowing the release of the 1st Provisional Marine Brigade. In General Smith's mind, Almond's proposal exemplified the wide gulf separating Army and Marine Corps thinking. As Colonel Bowser, General Smith's operations officer, remembered it, Doyle and Smith "came back about 11 o'clock having won their point, that the [Marine] brigade must come out of the Pusan perimeter and be part of our landing force."

The *Mount McKinley*, flagship of the Attack Force—with Smith on board so as to be in a better position to supervise the out-loading—set sail from Tokyo for Kobe on 4 September, arriving there early the next afternoon. That evening Smith called a conference of all available Marine Corps commanders to stress the urgency of the operation.

Almond Inspects Marines

A day later 6 September, General Almond came to Kobe to inspect 1st Marine Division units. He lunched with the staff noncommissioned officers at Camp Otsu accompanied by General Smith and Lieutenant Colonel Allan Sutter, then visited the 2d and 3d Battalions of the 1st Marines. Afterwards he went to Camp Sakai near Osaka to see the 11th Marines, the division's artillery regiment commanded by Colonel James H. Brower, and was favorably impressed." He commented in his diary: "A large percentage of the troops were drawn from active Marine reserve units . . . . The Army should have done likewise but did not."

In the evening Smith and his staff briefed him on the division's operation plan. Again Almond was favorably impressed, but he thought Smith's planned subsequent moves ashore too slow and deliberate. He stressed to Smith the need for speed in capturing Kimpo Airfield and Seoul itself. Smith was less impressed with Almond, saying: "The inspection consisted [of Almond] primarily questioning men, I suppose for the purpose of finding out what made Marines tick."

In the 1st Marine Division, operational planning trickled down to the battalion level. The 3d Battalion, 1st Marines, under Lieutenant Colonel Thomas L. Ridge, had steamed comfortably to Japan in the *General Simon B. Buckner* (AP 123) and was ensconced in what had been the barracks for a battalion of the 24th Infantry Division at Otsu on the south shore of Lake Biwa. There was no room for field training and the best the battalion could do was road-bound conditioning marches. The commanding officer and the three majors in the battalion were summoned to a meeting on board the regimental command ship berthed in Kobe. There had been a plethora of rumors, but now for the first time they learned officially that they were to land at Inchon. The regimental S-2, Captain Stone W. Quillian, went over the beach defenses, tapping a large map

President Harry S. Truman and Marine Commandant Gen Clifton B. Cates exchange warm greetings at a Marine Corps field demonstration at Quantico in June 1950, 10 days before the outbreak of the Korean War. This friendly relationship dissolved when Truman, in an ill-advised note, called the Marine Corps "the Navy's police force."

Department of Defense Photo (USMC) A407260

studded with suspected weapons emplacements. The S-3, Major Robert E. Lorigan, then briefed the scheme of maneuver. The 3d Battalion would be the right flank unit of the main landing. These were the D-Day objectives. *Tap, tap.* This piece of high ground was the battalion's objective. *Tap, tap.* This hook of land on the extreme right flank had to be taken. *Tap, tap.* The landing would be at 1730; it would be dark at 1900. There were no enthusiastic cheers from the listeners.

Then the regimental commander, Chesty Puller, got to his feet. "You people are lucky," he growled. "We used to have to wait every 10 or 15 years for a war. You get one every five years. You people have been living by the sword. By God, you better be prepared to die by the sword."

The troop list for the landing force totalled 29,731 persons, to be loaded out in six embarkation groups. Four groups would load out of Kobe, one group out of Pusan, and one group—made up of the Army's 2d Engineer Special Brigade—out of Yokohama. Not all units could be combat loaded; some compromises had to be accepted.

One Marine Corps unit that was not ready to go was the 1st Armored Amphibian Tractor Battalion, activated but not yet combat ready. The Army's Company A, 56th Amphibian Tractor Battalion, was substituted.

President Writes Letter

As the Marines combat loaded their amphibious ships at Kobe, the Pacific edition of *Stars and Stripes* reached them with a story that President Truman had called them "the Navy's police force." This compounded a previously perceived insult when the President labeled the United Nations intervention in Korea a "police action." The enraged Marines chalked on the tarpaulins covering their trucks and tanks, "Horrible Harry's Police Force" and similar epithets.

What had happened was that on 21 August, Congressman Gordon L. McDonough of California had written President Truman a well-intentioned letter urging that the Marines be given a voice on the Joint Chiefs of Staff. The President fired back a feisty note: "For your information the Marine Corps is the Navy's police force and as long as I am President that is what it will remain. They have a propaganda machine that is almost equal to Stalin's . . . . The Chief of Naval Operations is the Chief of Staff of the Navy of which the Marines are a part."

He had dictated the letter to his secretary, Rose Conway, and sent it without any member of his staff seeing it.

McDonough inserted the letter into the Congressional Record where it appeared on 1 September. The story reached the newspapers four days later and a great public outcry went up. By five o'clock the next afternoon Truman's advisors had prevailed upon him to send an apology to General Cates: "I sincerely regret the unfortunate choice of language which I used." Truman, in further fence-mending, in company with Cates, made a surprise visit two mornings later at a Marine Corps League convention coincidentally being held in Washington's Statler Hotel and charmed his audience.

Pulling Together the Landing Force

General Craig's 1st Provisional Marine Brigade was relieved of its combat commitment at midnight, 5 September. The brigade had done most of its fighting with a peacetime structure, that is, at about two-thirds its authorized wartime strength: two rifle companies to a battalion instead of three, four guns to an artillery battery instead of six. The 5th Marines did not get a third company for its three infantry battalions until just before mounting out for Inchon.

The Korean 1st Marine Regiment, some 3,000 men, commanded by Lieutenant Colonel Kim Sung Eun, arrived in Pusan on 5 September to join the 1st Marine Division. They were in khaki uniforms including cloth caps, and equipped with Japanese rifles and machine guns. The South Korean Marines were issued American uniforms—including helmets—and each was given one day on the rifle range to fire his new American weapons.

Built around a cadre drawn from the ROK Navy, the Korean Marine Corps ("KMCs" to the U.S. Marines) had been activated 15 April 1949. Company-size units had first deployed to southern Korea, and then to Cheju Island, to rout out Communist-bent guerrillas. After the North Korean invasion, the KMCs, growing to regimental size, had made small-scale hit-and-run raids along the west coast against the flank of the invaders.

Craig assigned Lieutenant Colonel Charles W. Harrison, until recently the executive officer of the 6th Marines at Camp Lejeune, as liaison officer to the KMCs. His party, given a radio jeep, was made up of three corporal radiomen, and a corporal driver. Harrison was well-chosen. His parents had been missionaries in Korea. He himself had graduated from the foreign high school in Pyongyang in 1928 and he had a working knowledge of Korean.

While the 5th Marines were

loading out, a paper, marked "Confidential" and giving specifics on a landing beach at Kaesong, was widely distributed and one or more copies were purposely "lost." Perhaps the word got back to the North Koreans.

The amphibious assault transport *Henrico* (APA 45) known to the fleet as "Happy Hank," had brought the 1st Battalion, 5th Marines, to Pusan. Now the ship received the same battalion, its numbers, thinned by the fighting in the Pusan Perimeter, now brought up to war strength. The Navy crew did their best to provide a little extra for their Marine passengers. The wardroom was made available to the officers 24 hours a day.

Marguerite "Maggie" Higgins, a movie-star-pretty blonde reporting on the war for the New York *Herald-Tribune* occupied one of the few staterooms. She had been a war correspondent in Europe during the last years of World War II and had been in Korea since the beginning of the new war. Ribald rumors as to her imagined nocturnal associations inevitably circulated throughout the ship.

Major General Field Harris, Commanding General, 1st Marine Aircraft Wing—O. P. Smith's aviator counterpart—arrived in Tokyo on 3 September. His forward echelon of the 1st Marine Aircraft Wing, was informed of the Inchon-Seoul operation three days later. Planning for the employment of Marine air was completed on 9 September. Marine Aircraft Group 33, relieved of its close support role in the Pusan Perimeter, would be the operating element. Harris and his forward echelon embarked at Kobe on 10 September as Tactical Air Command, X Corps.

Meanwhile, Almond continued his restless visits and inspections. On 9 September, General Barr briefed him on the 7th Infantry Division's plan of operations. Almond thought the plan adequate, but was concerned over possible problems of liaison and coordination with the 1st Marine Division. Events would prove him right

Almond's Good Ideas

A restive General Almond formed, for commando work, a Special Operations Company, X Corps, sometimes called a "Raider Group," under command of Colonel Louis B. Ely, Jr., USA. With Almond's encouragement, Ely proposed a raid to seize Kimpo Airfield. Almond asked Smith for 100 Marine volunteers to join the Special Operations Company; Smith, skeptical of the mission and unimpressed by Ely, stalled in providing Marines and the request was cancelled. As it turned out, Ely and his company would make an approach to the beach, but the distance from ship to shore proved too great for rubber boats.

Brigadier General Henry I. Hodes, USA, the assistant division commander of the 7th Infantry Division, visited Smith on the *Mount McKinley* on 9 September. Almond, still concerned by Smith's deliberate manner, had come up with yet another idea for the swift seizure of Kimpo. Almond's new plan called for landing a battalion of the 32d Infantry on Wolmi-do the evening of D-Day. It would "barrel" down the road to Seoul in trucks and tanks provided by the Marines. Smith, horrified by a plan he considered tactically impossible, told Hodes that he had no tanks to lend him.

The Secretary of the Navy, alerted by parents' complaints that underage sons were being sent to Korea, on 8 September sent a last-minute order to remove Marines under 18 before sailing, reducing the landing force by about 500 men. Those who were close to being 18 were held in Japan on other duties and eventually found their way to the division as replacements.

Second Typhoon

Weathermen said that a second typhoon, "Kezia," was following close behind "Jane." Rear Admiral Arleigh A. Burke, USN, had arrived in Tokyo from Washington to be Admiral Joy's deputy chief of staff. Burke attempted to make an office call on MacArthur to express his concerns regarding the coming typhoon and was blocked by Almond. Burke refused to discuss the matter with Almond and went back to his office. By the time he got there, a message was waiting that MacArthur would see him. Burke hurried back to GHQ and explained to MacArthur that if the typhoon came up and blew west there could be no landing on the 15th or 16th.

"What do we do, Admiral?" asked MacArthur.

"We sail early," said Burke. MacArthur agreed.

Navy meteorologists had first picked up signs of Kezia off the Mariana Islands on 6 September. Whipping up winds of 100 miles per hour, the typhoon moved steadily toward Japan and the East China Sea. Most endangered were the amphibious ships of Admiral Doyle's Attack Force. The route for all six transport groups to Inchon placed them squarely in the path of the on-coming oriental hurricane.

Both Doyle and O.P. Smith, the two who would bear the burden of directing the actual landing, were painfully aware that all the normal steps of preparing for an amphibious operation were either

Photo courtesy of LtCol Leo J. Ihli, USMC (Ret)

Marines prime an F4U-4B of VMF-323 for take-off from the deck of the light aircraft carrier Badoeng Strait *(CVE 116) standing off Korea. VMF-214, embarked in sister carrier* Sicily *(CVE 118), played a companion role in close support of the assault. The bent-wing Corsairs would prove once again to be ideal close support aircraft.*

being compressed or ignored completely in order to squeeze the operation into an impossibly short time frame. During World War II, at least three months would have been spent in planning and training for an operation of this magnitude. Beginning with Guadalcanal, a rehearsal—or rehearsals—was considered essential. For Inchon there would be no rehearsal. Doyle wryly concluded that a good deal would depend upon how skillfully the individual coxswains could perform in finding their way to the beaches.

Captain Martin J. "Stormy" Sexton, a World War II Raider and now aide-de-camp to General Smith, said later: "There was not even time for landing exercises by the LVTs. Some of the LVT crews had not even had the opportunity to try their engines out in the water and paddle around."

Execution

Marine aircraft squadrons VMF-214 and VMF-323 began the softening-up of Wolmi-do on 10 September with the delivery of napalm. Operating from the decks of the light carriers *Sicily* (CVE 118) and *Badoeng Strait* (CVE 116) ("Bing-Ding" to the Marines and sailors), the Marine fliers burned out most of the buildings on the island. Strikes by Navy aircraft from the big carriers *Valley Forge* (CV 45), *Philippine Sea* (CV 47), and *Boxer* (CV 21) continued for the next two days.

Joint Task Force 7 (JTF 7) was officially activated under Admiral Struble the following day, 11 September. Almond and X Corps would be subordinate to Struble and JTF 7 until Almond assumed command ashore and JTF 7 was dissolved.

Preliminary and diversionary air and naval gunfire strikes were roughly divided into 30 percent delivered north of Inchon, 30 percent south, and 40 percent against Inchon itself. Except for a few gunnery ships held back to protect the flanks of the Pusan Perimeter, JTF 7—in its other guise, the Seventh

Joint Task Force Seven

VAdm Arthur D. Struble

Task Force 90	Attack Force	RAdm James H. Doyle
Task Force 91	Blockade and Covering Force	RAdm Sir William G. Andrewes
Task Force 92	X Corps	MajGen Edward M. Almond
Task Force 99	Control and Reconnaissance Force	RAdm George R. Henderson
Task Force 77	Fast Carrier Group	RAdm Edward C. Ewen
Task Force 79	Service Squadron	Capt Bernard L. Austin
ROK Naval Forces		Cdr Michael L. Luosey[1]

[1] Liaison and Advisor

Fleet—included all the combatant ships in the Far East. Among them were three fast carriers, two escort carriers, and a British light carrier. In the final count, the force numbered some 230 ships, including 34 Japanese vessels, mostly ex-U.S. Navy LSTs (landing ships, tank) with Japanese crews. The French contributed one tropical frigate, *La Grandiere*, which arrived at Sasebo with a five-month supply of wine and a pin-up picture of Esther Williams, but no coding machine.

Mount McKinley, with Doyle, Smith, and their staffs on board, got underway from Kobe the morning of 11 September—a day ahead of schedule because of the approach of Typhoon Kezia—and steamed for Sasebo. Winds of the typhoon whipped up to 125 miles per hour. Doyle was gambling that Kezia would veer off to the north.

Almond held a last meeting at GHQ on 12 September to deal with the urgency for an early sailing because of the threat of Kezia. General Shepherd, General Wright, and Admiral Burke attended. That afternoon General MacArthur and his party left Haneda airport to fly to Itazuke air base. From there they would go by automobile to Sasebo.

MacArthur Goes to Sea

Because of the storm the *Mount McKinley* was late in reaching port. MacArthur's party waited in the Bachelor Officers Quarters, passing the time having sandwiches. It was close to midnight before the *Mount McKinley* rounded the southern tip of Kyushu and docked at Sasebo. MacArthur and his party boarded the ship and she was underway again within an hour. With General Shepherd came his G-3, Colonel Victor H. Krulak, and his aide and future son-in-law, Major James B. Ord, Jr.

MacArthur had five generals in his party—Shepherd, Almond, and Wright, and two others: Major General Courtney Whitney—his deputy chief of staff for civil affairs, but more importantly his press officer—and Major General Alonzo P. Fox. Fox was chief of staff to MacArthur in his capacity as "SCAP" (Supreme Commander Allied Powers) and Lieutenant Haig's father-in law. Absent from the group was Lieutenant General George Stratemeyer, USAF, who had had some expectation of accompanying MacArthur as his air boss. In assignment of spaces, MacArthur grandly ignored traditional ship protocol and took over Doyle's cabin. Doyle moved to his sea cabin off the flag bridge. Almond appropriated the ship's captain's cabin. O. P. Smith managed to keep his stateroom.

After breakfast on the morning of the 13th, Admiral Doyle led the embarked flag officers in a tour of the *Mount McKinley*, hoping to impress the Army generals that amphibious operations required specialization. MacArthur did not go along.

The absence of General Stratemeyer from MacArthur's party was a clear signal that the Navy had been successful in keeping the Air Force from operating within the amphibious objective area—a circle with a 100-mile radius drawn around Inchon. There would be no FEAF operations within this radius unless specifically requested by Struble. MacArthur remained above these doctrinal squabbles.

Operation 'Common Knowledge'

Neither General MacArthur nor Admiral Struble favored extensive air and naval gunfire preparation of the objective area, primarily because it would cause a loss of tactical surprise. Their concern was largely academic. All sorts of leakage circulated in Japan—and even reached the media in the United States—that an amphibious operation was being mounted out with a probable target of Inchon. At the Tokyo Press Club the impending landing was derisively called "Operation Common Knowledge." The North Korean command almost certainly heard these rumors and almost equally certain had tide tables for Inchon. Mao Tse Tung is supposed to have pointed at Inchon on a map of Korea and have said, "The Americans will land here."

American intelligence knew that the Russians had supplied mines, but how many had been sown in Flying Fish Channel? The lack of time and sufficient minesweepers made orderly mine-sweeping operations impossible.

'Sitting Ducks'

The pre-landing naval gunfire bombardment began at 0700 on 13 September with a column of cruisers and destroyers coming up the channel. The weather was good, the sea calm. Four cruisers—*Toledo* (CA 133), *Rochester* (CA 124), HMS *Kenya*, and HMS *Jamaica*—found their bombardment stations several miles south of Inchon and dropped anchor. Six destroyers—*Mansfield* (DD 728), *DeHaven* (DD 727), *Lyman K. Swenson* (DD 729), *Collett* (DD 730), *Gurke* (DD 783), and *Henderson* (DD 785)—continued on past the cruisers and were about to earn for themselves the rueful title of "Sitting Ducks."

What appeared to be a string of mines was sighted in the vicinity of Palmi-do. The destroyers opened fire with their 40mm guns and the mines began to explode. Leaving the *Henderson* behind to continue shooting at the mines, the five other destroyers steamed closer to their objectives. *Gurke* anchored 800 yards off Wolmi-do, which was being pounded by carrier air.

The remaining four destroyers took station behind *Gurke*. Just before 1300 they opened fire. Within minutes return fire came blazing back from hidden shore batteries. *Collett* took five hits, knocking out her fire direction system; her guns switched to individual control. *Gurke* took two light hits. *DeHaven* was slightly damaged. *Lyman K. Swenson* felt a near miss that caused two casualties. After an hour's bombardment the destroyers withdrew. One man had been killed—ironically Lieutenant (Junior Grade) David Swenson, nephew of the admiral for whom the destroyer was named—and eight were wounded.

From their more distant anchorage, the cruisers picked up the bombardment with 6-inch and 8-inch salvos. After that the carrier aircraft resumed their attack.

Next day, 14 September, five of the destroyers came back (the damaged *Collett* was left behind) and banged away again. At first the destroyers drew feeble return fire. By the time they withdrew 75 minutes later, having delivered 1,700 5-inch shells, there was no return fire at all. The Navy, with considerable satisfaction, reported Wolmi-do now ready for capture.

Attack Force Gathers

Admiral Doyle had won his gamble against the typhoon. The Yellow Sea was quiet and all elements of the Attack Force were in place off Inchon. General Craig's embarked 1st Provisional Marine Brigade, having arrived from Pusan, was formally dissolved on 13 September and its parts returned to the control of the parent division. Craig became the assistant division commander.

The Attack Force eased its way up Flying Fish Channel so as to be in the transport area before daylight on 15 September. General MacArthur spent a restless night. Standing at the rail of the *Mount McKinley* in the darkness, he entertained certain morbid thoughts, at least as he remembered them later in his *Reminiscences*: "Within five hours 40,000 men would act boldly, in the hope that 100,000 others manning the thin defense lines in South Korea would not die. I alone was responsible for tomorrow, and if I failed, the dreadful results would rest on judgment day against my soul."

George Gilman, an ensign in the *Mount McKinley*, had less lofty thoughts: "None of us boat group officers had ever had any experience operating under such tidal conditions before, let alone ever having been involved in an amphibious landing . . . . As the morning of September 15 approached, we realized we had all the ingredients for a disaster on our hands."

Destination Wolmi-do

L-hour was to be 0630. At 0545, the pre-landing shore bombardment began. Lieutenant Colonel Robert D. "Tap" Taplett's 3d Battalion, 5th Marines, was boated by 0600. The carrier-based Marine Corsairs completed their last sweep of the beach 15 minutes later.

"G Company was to land to the right of Green Beach in the assault, wheel right, and seize the domi-

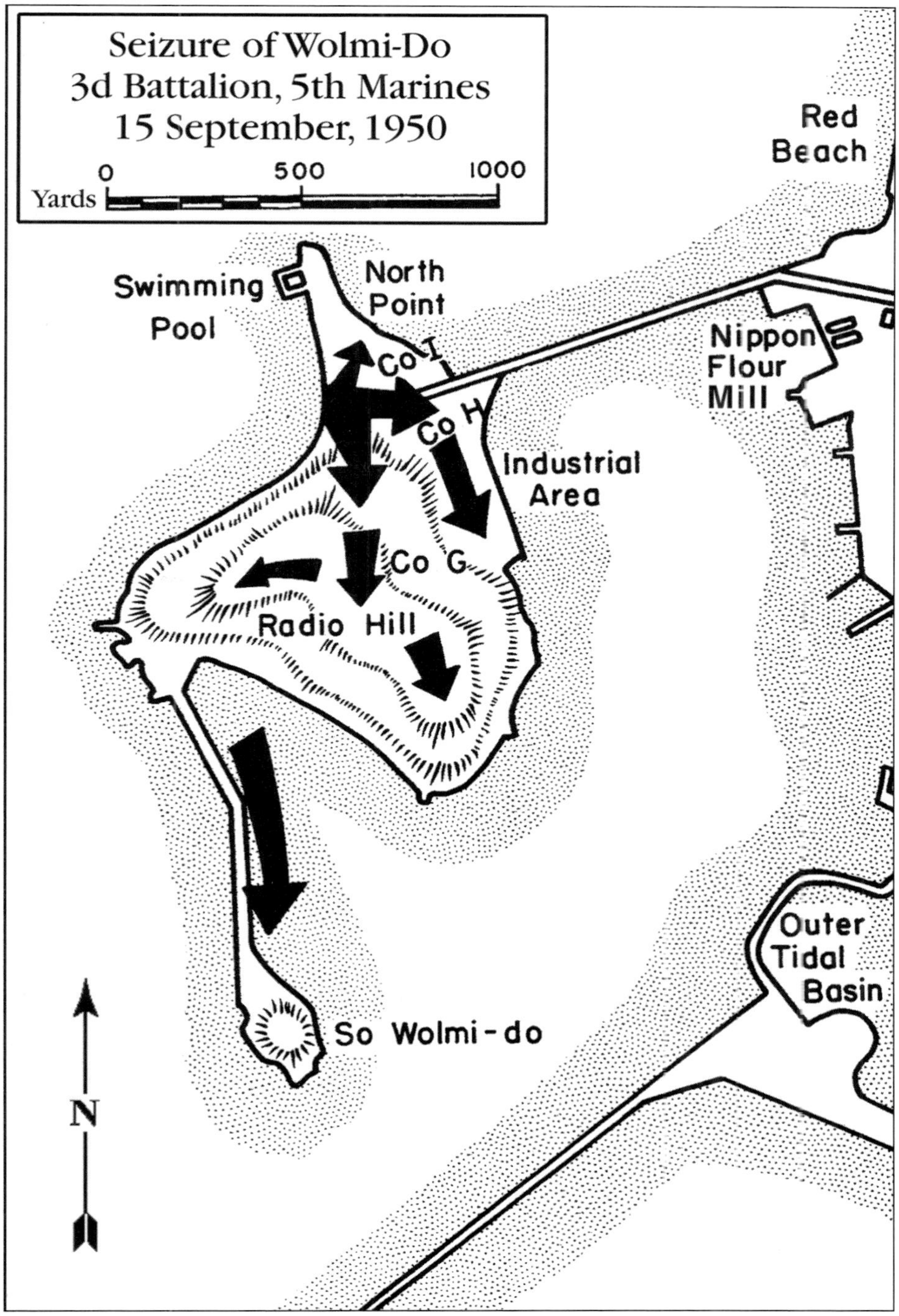

nant hill mass on the island, Radio Hill," remembered Robert D. "Dewey" Bohn (then a first lieutenant; he would retire a major general). His company was embarked in the fast destroyer transport *Diachenko* (APD 123). She stopped her engines at about 0300, the troop compartment lights came on, and reveille sounded over the public address system.

Most of the Marines were already awake. They hoped for the traditional "steak and eggs" pre-landing breakfast of World War II; instead they got scrambled powdered eggs, dry toast, and canned apricots. At about first light, Company G went over the side and down the cargo nets into the bobbing LCVPs, which then cleared the ship and began to circle.

Three LSMRs—medium landing ships converted to rocket ships—sent their loads of thousands of 5-inch rockets screeching shoreward toward Wolmi-do. The island seemed to explode under the impact. Then the landing craft began the run to Green Beach. MacArthur, Shepherd, Almond, Smith, Whitney, and Doyle all watched from the flag bridge of the *Mount McKinley*.

Seven LCVPs brought in the first wave, one platoon of Company G on the right and three platoons of Company H on the left. The landing craft converged on the narrow beach—scarcely 50 yards wide—and grounded at 0633, three minutes behind schedule. The remainder of the two assault companies came in as the second wave two minutes later. Resistance was limited to a few scattered shots.

Captain Patrick E. Wildman, commanding Company H, left a small detachment to clear North Point and then plunged across the island toward his objectives—the northern nose of Radio Hill and the shoreline of the burning industrial area facing Inchon. After a short pause to reorganize, Bohn took Company G towards the southern half of Radio Hill, 105 meters high. Resistance was half-hearted. At 0655, Sergeant Alvin E. Smith, guide of the 3d Platoon, secured an American flag to the trunk of a shattered tree. MacArthur, watching the action ashore from his swivel chair on the bridge of the *Mount McKinley*, saw the flag go up and said, "That's it. Let's get a cup of coffee."

Ten tanks—six M-26 Pershings and four modified M-4A3 Shermans, all under Second Lieutenant Granville G. Sweet—landed in the third wave at 0646 from three utility landing ships (LSUs). They crunched their way inland, poised to help the infantry.

Lieutenant Colonel Taplett land-

Department of Defense Photo (USMC) A2686

Reveille in the amphibious ships went at 0300 on the morning of 15 September. Marines hoped for the traditional "steak-and-eggs" D-day breakfast of World War II, but most transports fed simpler fare, such as powdered eggs and canned apricots. Breakfast on board the landing ships was even more spartan.

ed from his free boat a few minutes later. At almost the same time, Captain Robert A. McMullen brought in the fourth wave bearing Company I, the battalion reserve. His company, following behind Company H, encountered an angry nest of about a platoon of bypassed North Koreans. A flurry of hand grenades was exchanged. McMullen signaled Sweet's tanks to come forward. A Sherman with a dozer blade sealed the die-hard North Koreans in their holes.

Moving on to the near end of the causeway that stretched to Inchon itself, McMullen found more North Korean defenders hiding in a cave. One of Sweet's tanks fired a 90mm round into the mouth of the cave. There was a muffled explosion and 30 dazed and deafened North Koreans came staggering out with their hands above their heads. "Captured forty-five prisoners . . . meeting light resistance," radioed Taplett at 0745 to the *Mount McKinley*.

Wildman's Marines were finding it slow going in the ruins of the industrial area. Taplett ordered Bohn to take the rest of Radio Hill and by 0800 the high ground was Marine Corps property.

'Wolmi-do Secured'

Once again Taplett radioed the *Mount McKinley*, this time: "Wolmi-do secured."

With the success of the Marine landing blaring over the loudspeakers, MacArthur left the bridge

By 0655, the 3d Battalion, 5th Marines, had landed on Wolmi-do and had an American flag flying at the top of a shell-blasted tree. An hour later the battalion commander reported resistance as light and 45 dazed prisoners taken.

National Archives Photo (USMC) 127-GK-234I-A2694

Department of Defense Photo (USMC) A2798

Some North Korean defenders of Wolmi-do stubbornly remained in their cave-like positions and had to be burned out by flamethrowers. U.S. Marines were readily distinguishable at this stage of the war by their wear of camouflage helmet covers and leggings.

to pen a message to Admiral Struble in his flagship *Rochester*: "The Navy and Marines have never shone more brightly than this morning."

Ashore, Taplett consolidated his gains. His three rifle companies, by prearranged plan, took up defensive positions facing Inchon. The empty swimming pool at the tip of North Point became a stockade for prisoners.

At about 10 o'clock Taplett ordered Bohn to take Sowolmi-do, an islet dangling to the south of Wolmi-do with a lighthouse at the end of the causeway. Bohn sent Second Lieutenant John D. Counselman, leader of his 3d Platoon, with a rifle squad and a section of tanks. As a prelude to the assault, a flight of Corsairs drenched Sowolmi-do with napalm. Covered by the two tanks and a curtain of 81mm mortar fire, Counselman's riflemen crossed the narrow causeway, taking fire from a hill honey-combed with emplacements. Flamethrowers and 3.5-inch rocket launchers burned and blasted the dug-in enemy. Seventeen were killed, 19 surrendered, and eight or more managed to hide out. The lighthouse was taken and the job completed in less than two hours. Three Marines were wounded, bringing Taplett's casualties for the day to none killed, 17 wounded.

Word was passed that some of the North Koreans who had escaped were trying to swim for Inchon. A number of Bohn's Marines lined up rifle-range fashion and shot at what they saw as heads bobbing in the water. Others dismissed the targets as imaginary. Mopping up of the island was completed by noon.

Taplett, growing restless and seeing no sign of enemy activity, proposed to division that he make an assault on the city from his present position or at least a reconnaissance in force. Smith responded to his proposal with a firm negative.

Waiting for Evening Tide

The remainder of the division was steaming toward the inner transport area. There would now be a long wait until the evening tide swept in and the assault regi-

M-26 Pershing tanks, new to the Marines, began to land in the third wave at Wolmi-do and were soon put to use against North Korean fortified positions. A tank-infantry patrol assaulted and took Sowolmi-do, an islet dangling at the end of a causeway from the main island.

Department of Defense Photo (USMC)

be found in the way of targets within a 25-mile radius of Inchon. (The D-Day action for the aircraft on board the carrier *Boxer* was labeled "Event 15" and consisted of a strike with 12 F4U Corsairs and five AD Skyraiders.) The smoke of the bombardment and from burning buildings mixed with the rain so that a gray-green pall hung over the city.

H-Hour for the main landing was 1730. Lieutenant Colonel Raymond L. Murray's 5th Marines, minus the 3d Battalion already ashore on Wolmi-do, was to land over Red Beach, to the left and north of Wolmi-do. Murray's regiment was to seize the O-A line, a blue arc on the overlay to the division's attack order. On the ground O-A line swung 3,000 yards from Cemetery Hill on the north or left flank, through Observatory Hill in the center, and then through a maze of buildings, including the British Consulate,

Department of Defense Photo (USMC) A2723

Marines from the 3d Battalion, 5th Marines, escorted a steady stream of prisoners back to Green Beach on the seaward side of Wolmi-do. Landing ships and craft could beach as long as the tide was high, but once the tide receded they would be left high and dry on the mud flats.

ments could be landed. Marines, standing at the rail of their transports, strained their eyes looking for their intended beaches but could see nothing but smoke. The bombardment, alternating between naval gunfire and air strikes, continued.

During the course of the afternoon, Admiral Struble had his admiral's barge lowered into the water from the *Rochester* ("Roach-Catcher"). He swung by the *Mount McKinley* to pick up General MacArthur for a personal reconnaissance from close offshore of Wolmi-do and the harbor. Almond and Shepherd went with them.

They swung close to the seawall fronting the harbor. "General," said Shepherd, "You're getting in mighty close to the beach. They're shooting at us." MacArthur ignored the caution.

Naval gunfire and carrier air sought to hit everything that could

A corpsman bandages the forearm of a wounded North Korean prisoner on Wolmi-do. He and other prisoners were moved to one of the several prison stockades that were set up on the landing beaches.

Department of Defense Photo (USMC) A2802

Department of Defense Photo (USA) SC348839

Gen MacArthur indulged his passion for visiting the "front." During the interval between the morning and evening landings he personally "reconnoitered" the Inchon beaches in VAdm Arthur D. Struble's barge. Struble sits to MacArthur's right. On his left is Army MajGen Courtney Whitney, often-called MacArthur's "press secretary."

until it reached the inner tidal basin.

The 1st and 2d Battalions, 5th Marines, under Lieutenant Colonels George R. Newton and Harold S. Roise respectively, would land abreast across Red Beach. The new 1st ROK Marine Regiment would follow them ashore.

Newton and Roise had the Pusan Perimeter behind them, but not much other infantry experience. Newton, commissioned in 1938 from the Naval Academy, was with the Embassy Guard at Peking when World War II came on 7 December 1941 and spent the war as a prisoner of the Japanese. Roise, commissioned from the University of Idaho in 1939, had served at sea during the war.

In the assault, Newton's 1st Battalion and Roise's 2d Battalion would come away from the attack transports *Henrico* and *Cavalier* (APA 37) in landing craft. Both battalions would land in column of companies across the seawall onto narrow Red Beach. Newton, on the left, was to take Cemetery Hill and the northern half of Observatory Hill. Roise, on the right, was to take his half of Observatory Hill, the British Consulate, and the inner tidal basin.

"Two things scared me to death," said Roise of the landing plan. "One, we were not landing on a beach; we were landing against a seawall. Each LCVP had two ladders, which would be used to climb up and over the wall. This was risky . . . . Two, the landing was scheduled for 5:30 p.m. This would give us only about two hours of daylight to clear the city and set up for the night."

Captain Francis I. "Ike" Fenton, Jr., commander of Company B in Newton's battalion, sharply

Lieutenant Colonel Raymond L. Murray

Seldom does a Marine Corps regiment go into combat with a lesser grade than full colonel in command. But when Brigadier General Edward Craig arrived at Camp Pendleton in July 1950 to form the 1st Provisional Marine Brigade for service in Korea he found no reason to supplant the commanding officer of the 5th Marines, Lieutenant Colonel Raymond Murray. The tall, rangy Texan was an exception to the general rule. He had already made his reputation as a fighter and of being a step ahead of his grade in his assignments. As a major at Guadalcanal he had commanded the 2d Battalion, 6th Marines, and for his conspicuous gallantry had earned his first Silver Star.

After Guadalcanal, came Tarawa for the battalion and a second Silver Star for Murray, now a lieutenant colonel. Finally, at Saipan, although he was painfully wounded, Murray's control of his battalion was such that it brought him a Navy Cross.

Novelist Leon Uris served in Murray's battalion. Later, when he wrote his book *Battle Cry*, he used Murray as his model for "High Pockets" Huxley, his hard-charging fictional battalion commander.

Born in Alhambra, California, in 1913, Murray grew up in Harlingen, Texas. When he accepted his commission in July 1935, after graduating from Texas A&M College, then the incubator of many Army and Marine officers, he had behind him four years of the Army's Reserve Officers Training Corps and two years of the Texas National Guard. He had also starred at football and basketball. After attending Basic School, then in the Philadelphia Navy Yard, he was detailed to the 2d Marine Brigade in San Diego. The brigade went to troubled China a year later. Murray served for a short time in Shanghai, then moved to a prized slot in the Embassy Guard in Peking. He came back to San Diego in 1940 and returned to the 2d Marine Brigade which within months expanded into the 2d Marine Division. A 1st Provisional Marine Brigade was pulled out of the 2d Division in the summer of 1941 for service in Iceland. Murray, now a captain and soon to be a major, went with it. He was back in San Diego in April 1942 and in October sailed with the 6th Marines for the war in the Pacific.

He came home in August 1944 and served at Quantico, Camp Lejeune, Hawaii, and Camp Pendleton. Promotions were slow after 1945 and Murray was still a lieutenant colonel when the Korean War began in 1950. As commander of the infantry element of the later-day 1st Provisional Marine Brigade in the "fire brigade" defense of the Pusan Perimeter, he received his third and fourth Silver Stars for his staunch leadership.

At Inchon, Major General O. P. Smith gave Murray and his now-seasoned regiment the more complicated northern half of the landing. After Inchon and Seoul, Murray would continue in command through the Chosin Reservoir campaign. That battle in subzero weather brought him the Army's Distinguished Service Cross as well as his second Navy Cross. Finally, in January 1951 he was promoted to colonel.

Coming home from Korea in April 1951, he attended the National War College and then was handpicked to command The Basic School, since World War II at Quantico. Next he served at Camp Pendleton and Camp Lejeune. A promotion to brigadier general came in June 1959. Assignments in Okinawa, then Pendleton again, and Parris Island followed. Serving at Headquarters Marine Corps in 1967 as a major general, he was ordered to Vietnam as Deputy Commander, III Marine Amphibious Force. His strong physique finally failed him. He was invalided home in February 1968 to Bethesda Naval Hospital where he remained until his retirement on 1 August 1968. He now lives in Oceanside, California, close to Camp Pendleton.

Department of Defense Photo (USMC) A42922

Department of Defense Photo (USMC) A2865

Navy transports stand off Inchon and Wolmi-do before the landing. Amphibious lift for Inchon, some of it literally borrowed back from the Japanese, was a rusty travesty of the great amphibious armadas of World War II.

remembered the characteristics of Red Beach:

> Once on the beach there was an open area of about 200 yards. The left flank was marked by Cemetery Hill. From the sea it looked like a sheer cliff. To the right of Cemetery Hill was a brewery, some work shops, and a cotton mill. Further to the right and about 600 yards in from the beach was Observatory Hill, overlooking the entire landing area and considered critical; it was the regimental objective. Further to the right was a five-story office building built of concrete and reinforced steel.

Captain John R. Stevens' Company A was to land on the right flank. In the assault would be the 2d Platoon under Second Lieutenant Francis W. Muetzel and the 1st Platoon under Gunnery Sergeant Orval F. McMullen. In reserve was the 3d Platoon under First Lieutenant Baldomero Lopez, who had joined the company as it loaded out from Pusan.

Three miles to the south of the 5th Marines, Chesty Puller's 1st Marines was to land across Blue Beach. Puller's mission was to secure the O-1 line, a 4,000-yard arc that went inland as deep as 3,000 yards, and then hooked around to the left to cut off Inchon from Seoul.

Blue Beach One, 500 yards wide, had its left flank marked by a salt evaporator. What looked to be a road formed the boundary to the south with Blue Beach Two.

The 2d Battalion, 1st Marines, was under affable, white-haired Lieutenant Colonel Allan Sutter. After landing over Blue Beach One, he was to take a critical road junction about 1,000 yards northeast of the beach, and Hill 117, nearly two miles inland, which commanded Inchon's "back door" and the highway to Seoul, 22 miles away.

Sutter, a graduate of Valley Forge Military Academy and Dartmouth College, had gained his Marine Corps commission in 1937 through the Platoon Leaders Course, a program under which college students spent two summers at Quantico to qualify as second lieutenants. He then spent a year at the Basic School in Philadelphia before being assigned troop duties. During World War II, Sutter was a signal officer at Guadalcanal, Guam, and Okinawa.

Blue Beach Two, also 500 yards wide, had its left flank marked by the supposed road and its right flank by a narrow ramp jutting seaward. A cove, further to the right, named at the last minute "Blue Beach Three," offered an alternate or supplementary landing site. Ridge, with the 3d Battalion, was to cross the seawall girdling Blue Beach Two and take Hill 233, a mile southeast of the beach, and, on the extreme right, a small cape, flanking Blue Beach and topped by Hill 94.

At best, the four assault battalions coming across Red and Blue Beaches would have but two hours of high tide and daylight to turn the plan into reality. Smith, after fully committing his two regiments, would have nothing left as a division reserve except two half-trained Korean Marine battalions.

Assaulting Red Beach

It would be a long ride to Red Beach for the 1st and 2d Battalions of the 5th Marines. Troops began debarking from the transports at about 1530. "As you climb down that net into the LCVP you're scared," remembered Private First Class Doug Koch of Company D, 5th Marines. "What keeps you

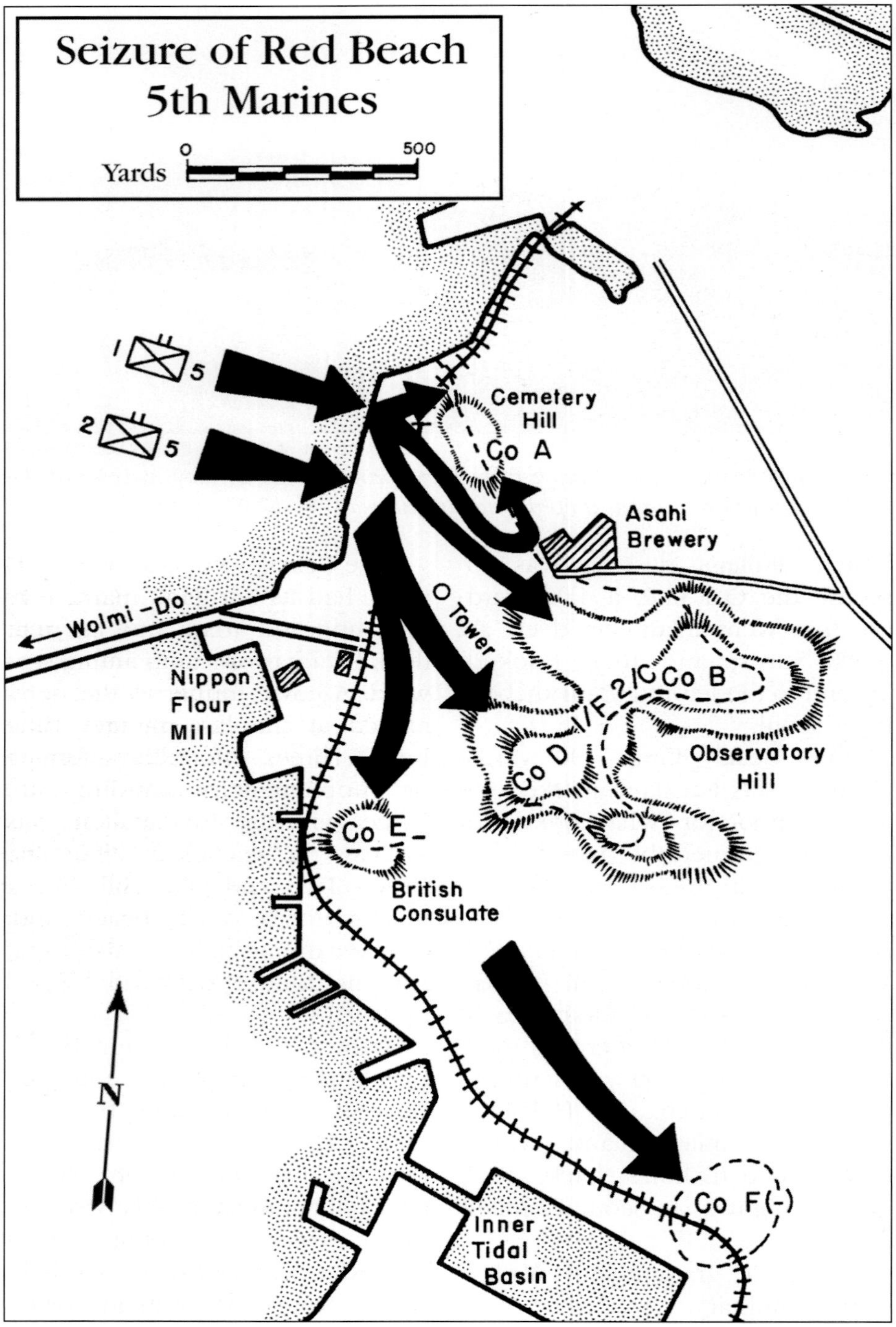

going is knowing this is what you have to do."

The *Horace A. Bass* (APD 124), the Red Beach control vessel, slowly steamed ahead with a long file of landing craft "trailing behind like a brood of ducklings."

The supporting rocket ships let go with a final fusillade of some 6,500 5-inch rockets. The resulting cloud of dust and smoke completely masked the beach area. The *Horace A. Bass*, an escort destroyer converted into a high-speed transport and anxious to get into the fight, banged away with her 5-inch guns. She then dipped her signal flag and the first wave headed for Red Beach.

The eight LCVPs in the first wave crossed the line of departure at H-8 with 2,200 yards to go. The four boats on the left carried the two assault platoons of Company A. Captain Steven's mission was to take Cemetery Hill and to secure the left flank of the beachhead. The four boats on the right carried the assault elements of Captain Samuel Jaskilka's Company E, which was to clear the right flank of the beach and then capture the hill that held the British Consulate.

As the first wave passed the mid-way point, two squadrons of Marine Corps Corsairs—VMF-214 under Lieutenant Colonel Walter E. Lischied and VMF-323 under Major Arnold A. Lund—came in to strafe both Red and Blue Beaches. They exhausted their loads and flew away. Not satisfied, Captain Stevens called for further air strikes against Red Beach. Four Navy A-4D Skyraiders made strafing passes until the wave had only 30 yards to go.

On the right, First Lieutenant Edwin A. Deptula's 1st Platoon, Company E, hit the seawall at 1731, one minute behind schedule. Designated Marines threw grenades up over the seawall, and after they exploded, Deptula took his platoon up the scaling ladders. A few stray rounds whined overhead.

Deptula pushed inland about 100 yards to the railroad tracks against no resistance. The rest of Company E landed about 10 minutes later. Captain Jaskilka (who would retire as a four-star general) quickly re-organized his company near the Nippon Flour Company building just south of the beachhead. Deptula's platoon continued down the railroad tracks to the British Consulate. Jaskilka sent another platoon to cross the railroad tracks and then move up the slope of 200-foot-high Observatory Hill.

On the left flank it was not quite that easy. One of the four landing

Gen Oliver P. Smith Collection, Marine Corps Research Center

Aerial photo of Red Beach shows the pounding it took in the pre-landing naval gunfire and air attacks. The 1st and 2d Battalions, 5th Marines, landed across this beach immediately north of the causeway leading to Wolmi-do.

craft, with half the 1st Platoon, Company A, on board, lagged behind with engine trouble. The remaining three boats reached the seawall at 1733. Sergeant Charles D. Allen took his half of the 1st Platoon over the wall and received fire from his north flank and from a bunker directly to his front. Several Marines went down.

To Allen's right, Second Lieutenant Frank Muetzel found a breach in the seawall and brought his 2d Platoon ashore. Facing them was a pillbox. Two Marines threw grenades and six bloody North Korean soldiers came out. Cemetery Hill loomed ahead, but Muetzel's immediate objective was Asahi Brewery. He slipped south of Cemetery Hill and marched unopposed down a street to the brewery. There was a brief indulgence in green beer.

Sergeant Allen, with his half-platoon, was making no progress against the bunker to his front. The second wave landed, bringing in the 3d Platoon under Baldomero Lopez and the missing half of the 1st Platoon. Too many Marines were now crowded into too small a space.

Lopez charged forward alone. He took out the bunker with a grenade and moved forward against a second bunker, pulling the pin from another grenade. Before he could throw it, he was hit. The grenade dropped by his side. He smothered the explosion with his body. This gained him a posthumous Medal of Honor. Two Marines went against the bunker with flamethrowers. They were shot down but the bunker was taken.

Captain Stevens's boat landed him in Company E's zone of action. Unable to get to his own company, he radioed his executive officer, First Lieutenant Fred F. Eubanks, Jr., to take charge. Stevens then radioed Muetzel to leave the brewery and get back to the beach where he could help out.

On the way back, Muetzel found a route up the southern slope of Cemetery Hill and launched an assault. The summit was alive with North Koreans, but there was no fight left in them. Dazed and spiritless from the pounding they had taken from the air and sea, they threw up their hands and surrendered. Muetzel sent them down to the base of the hill under guard.

Eubanks' Company E Marines meanwhile had bested the obstructing bunker with grenades and a flamethrower. His 1st and 2d Platoons pushed through and joined Muetzel's 2d Platoon. At 1755, 25 minutes after H-Hour, Captain Stevens fired an amber flare, signaling that Cemetery Hill was secure. It had cost his company eight Marines killed and 28 wounded.

Coming in on the third and fourth waves, Company C, 1st Battalion, was to take the northern half of Observatory Hill, and Company D, 2d Battalion, was to take the southern half. It did not work out quite that way. Parts of Companies C and D were landed on the wrong beaches. Company C, once ashore, had to wait 12 minutes for its commander, Captain Poul F. Pedersen. In Pedersen's boat was the fifth wave commander who had decided to tow a stalled LCVP. Once ashore, Pedersen had trouble sorting out his company from amongst the jumble of Marines that had gathered in the center of the beach.

Maggie Higgins, the *Herald-Tribune* correspondent, came off the *Henrico* in Wave 5 along with John Davies of the Newark *Daily News*, Lionel Crane of the London *Daily Press*, and a photographer. As their landing craft hit the seawall, the wave commander, First Lieutenant Richard J. "Spike" Schening, urged on his Marines with, "Come on you big, brave Marines. Let's get the hell out of here."

The photographer decided he had had enough and that he would go back to the *Henrico*. Maggie considered doing the same, but then, juggling her typewriter, she, along with Davies and Crane, followed Schening over the seawall.

Eight LSTs crossed the line of departure, as scheduled, at 1830 and were headed for the seawall. Seeing the congestion on Red Beach, the skippers of the LSTs concluded that the Marines were held up and could not advance.

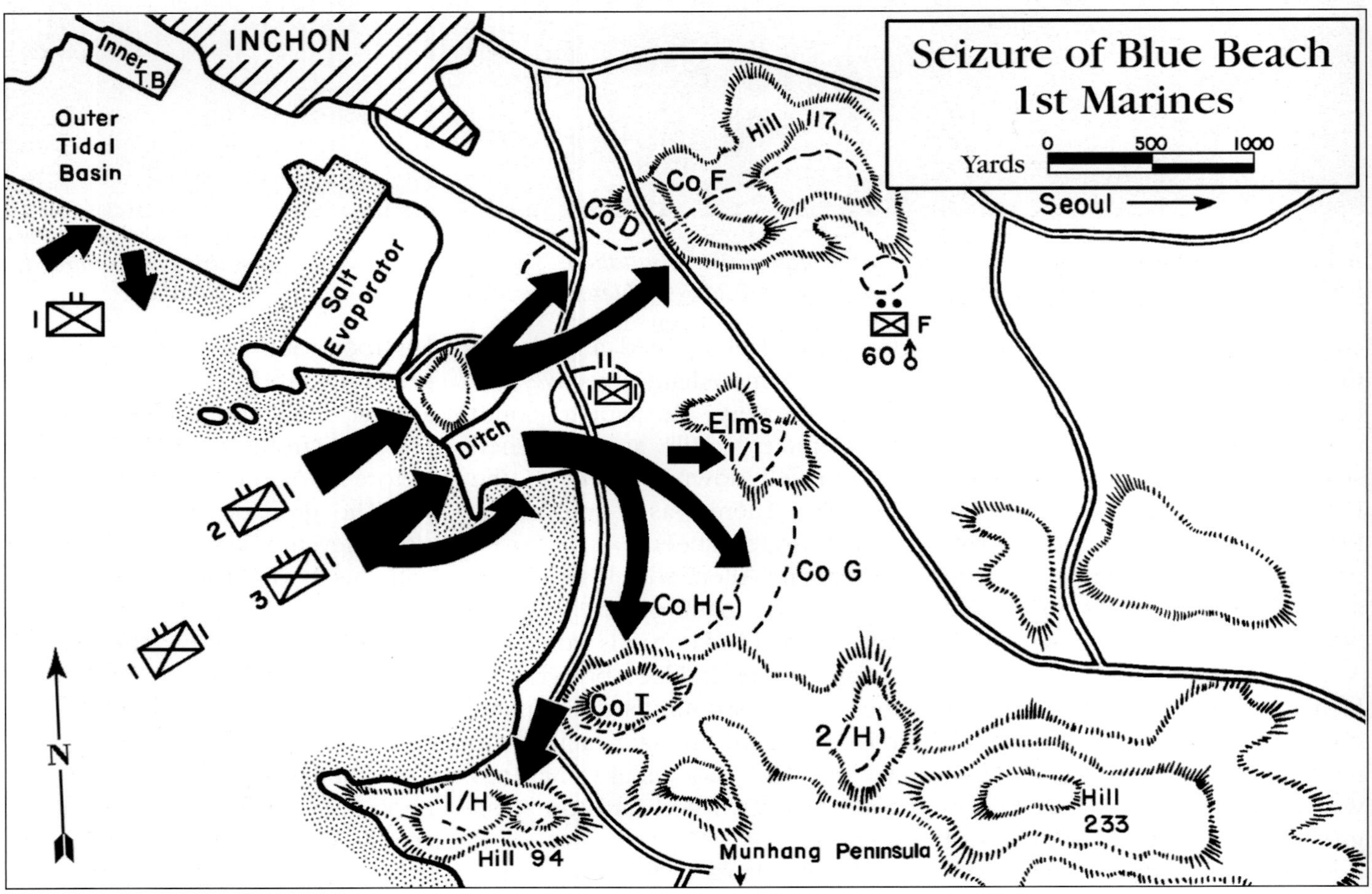

Department of Defense Photo (USMC) A3701

Marines enroute to Red Beach go over the side of their assault transport, down the cargo net hand-over-hand, and into the waiting LCVP, a version of the famous "Higgins boat" of World War II.

The lead LST received some mortar and machine-gun fire and fired back with its own 20mm and 40mm guns. Two other LSTs joined in. Unfortunately, they were spraying ground already occupied by the Marines.

The LST fire showered Muetzel's platoon, holding the crest of Cemetery Hill. Muetzel pulled back his platoon. As his Marines slid down the hill, they came under fire from a North Korean machine gun in a building on Observatory Hill. A chance 40mm shell from one of the LSTs knocked out the gun. Weapons Company and Headquarters and Service Company of Roise's 2d Battalion landed about 1830 and came under LST fire that killed one Marine and wounded 23 others.

By 1900, all eight LSTs had stopped firing and were nestled against the seawall. By then Second Lieutenant Byron L. Magness had taken his 2d Platoon, Company C, reinforced by Second Lieutenant Max A. Merritt's 60mm mortar section, up to the saddle that divided the crest of Observatory Hill. Their radios were not working and they had no flares. They had to inform the beach of their success by sending back a runner.

Meanwhile, the 1st Battalion's reserve company—Company B under Captain "Ike" Fenton—had landed in the 2d Battalion's zone. Lieutenant Colonel Newton ordered Fenton to assume Company C's mission and take the northern half of Observatory Hill. Six Marines were wounded along the way, but by about 2000 Fenton was at the top and tied in with the Magness-Merritt platoon.

In the right half of the regimental zone of action, Roise was getting the congestion on the beach straightened out. Company D, commanded by First Lieutenant H. J. Smith, had followed Company E ashore, but had landed to the left in the 1st Battalion zone. Smith (called "Hog Jaw" to make up for his non-existent first and second names) understood that Jaskilka's Company E was already on the crest of Observatory Hill. Under that assumption he started his company in route column up the street leading to the top of the hill. An enemy machine gun interrupted his march. After a brisk firefight that caused several Marine casualties, the enemy was driven off and Company D began to dig in for the night. A platoon from Company F, the battalion reserve, filled in the gap between Company D and the Magness-Merritt positions. The only part of the O-A line that was not now under control was the extreme right flank where the line ended at the inner tidal basin.

Maggie Higgins, after seeing the war, such as it was, found a boat

Department of Defense Photo (USMC) A3190

Marines go over the seawall forming the sharp edge of Red Beach. The Marine on the ladder has been identified as 1stLt Baldomero Lopez. Moments later he would give his life and earn a posthumous Medal of Honor.

on Red Beach that was returning to the *Mount McKinley*, where, after the personal intercession of Admiral Doyle, she was allowed to stay for the night. She slept on a stretcher in the sick bay. Next day, Admiral Doyle specified that in the future women would be allowed on board only between the hours of nine in the morning and nine at night. (About a month later, Maggie's transportation orders were modified. She would still be allowed on board any Navy ship but would have to be chaperoned by a female nurse.)

Murray, the regimental commander, came ashore at about 1830 and set up his command post at the end of the causeway that led from the mainland to Wolmi-do. Roise wished to stay where he was for the night, but Murray ordered him to reach the tidal basin. Company F, under Captain Uel D. Peters, faced around in the dark and plunged forward. Shortly after midnight, Roise reported that his half of the O-A line was complete.

Assaulting Blue Beach

The confusion was greater on Blue Beach than on Red Beach. Amphibian tractors, rather than landing craft, were used for the assault. The seawall was in disrepair with numerous breaks up which it was presumed the amphibian tractors could crawl. The 18 Army armored amphibians (LVT[A]s) forming the first wave crossed the line of departure at 1645 and headed toward Inchon. At four knots they needed three-quarters of an hour to hit the beach at H-Hour.

The soldiers had the compasses and seamanship to pierce the smoke and reached the beach on time. The second and following

First Lieutenant Baldomero Lopez

Baldomero Lopez was always eager. During World War II, he was 17 when he enlisted in the Navy in July 1943. Most thought him a Mexican American, but his father, also named Baldomero, as a young man had come to Tampa from the Asturias region of Spain. Los Asturianos, the men of Asturias, are known for their valor and honor.

He was appointed from the fleet to the Naval Academy in July 1944. His class, 1948A, was hurried through in three years. Lucky Bag, his class book, called him "one of the biggest hearted, best natured fellows in the brigade." Otherwise he does not seem to have been exceptional. His nickname at the Academy was "Lobo." This changed to "Punchy" after he came into the Marine Corps in June 1947, because it was generally believed that he had boxed while at Annapolis. After Basic School he stayed on at Quantico as a platoon commander in the Platoon Leaders Class. In 1948, he went to North China as part of a Marine presence that was in its last days. He served first as a mortar section leader and then as a rifle platoon commander at Tsingtao and Shanghai.

When the Marines closed out in China, he came back to Camp Pendleton. In the early summer of 1950, when the formation of the 1st Provisional Marine Brigade stripped the 1st Marine Division dry, he asked to be included but was left behind. He went out, however, to Korea in the draft that was sent to Pusan to fill the 5th Marines to war strength before embarking for Inchon. He was given the 3d Platoon, Company A, 1st Battalion.

Secretary of the Navy Dan Kimball presented the posthumous Medal of Honor to his father and mother at ceremonies in Washington on 30 August 1951.

Citation:

> For conspicuous gallantry and intrepidity at the risk of his life above and beyond the call of duty as a Rifle Platoon Commander of Company A, First Battalion, Fifth Marines, First Marine Division (Reinforced), in action against enemy aggressor forces during the Inchon invasion in Korea on 15 September 1950. With his platoon, First Lieutenant Lopez was engaged in the reduction of immediate enemy beach defenses after landing with the assault waves. Exposing himself to hostile fire, he moved forward alongside a bunker and prepared to throw a hand grenade into the next pillbox whose fire was pinning down that sector of the beach. Taken under fire by an enemy automatic weapon and hit in the right shoulder and chest as he lifted his arm to throw, he fell backward and dropped the deadly missile. After a moment, he turned and dragged his body forward in an effort to retrieve the grenade and throw it. In critical condition from pain and loss of blood, and unable to grasp the hand grenade firmly enough to hurl it, he chose to sacrifice himself rather than endanger the lives of his men and, with a sweeping motion of his wounded right arm, cradled the grenade under him and absorbed the full impact of the explosion. His exceptional courage, fortitude and devotion to duty reflect the highest credit upon First Lieutenant Lopez and the United States Naval Service. He gallantly gave his life for his country.

Department of Defense Photo (USMC) A43985

waves did not do so well. Rain and smoke had completely blotted out any view of the beach. From the bridge of his ship, the Blue Beach control officer watched the first two or three waves disappear into the smoke. He requested permission to stop sending any further waves ashore until he could see what was happening to them. Permission was denied.

As Major Edwin H. Simmons, the commander of Weapons Company, 3d Battalion, 1st Marines, remembered it:

> We had been told that a wave guide would pick us up and lead us to the line of departure . . . . Two LCVPs did come alongside our wave.

Department of Defense Photo (USMC) A2816

A key objective for the 5th Marines was the 200-foot-high Observatory Hill. Both the 1st and 2d Battalions converged on the hill with Marines from Company B taking the weather station on its top.

The first was filled with photographers. The second was loaded with Korean interpreters. Two of these were hastily dumped into my LVT, apparently under the mistaken notion that I was a battalion commander. Both interpreters spoke Korean and Japanese; neither spoke English. Time was passing, and we were feeling faintly desperate when we came alongside the central control vessel. I asked the bridge for instructions. A naval officer with a bullhorn pointed out the direction of Blue Two, but nothing could be seen in that direction except mustard-colored haze and black smoke. We were on our way when our path crossed that of

Marines setup a temporary barricade on the causeway to Inchon, after mopping up and consolidating their positions on Wolmi-do. Although not expecting a counterattack, they position a 3.5-inch rocket launcher and a machine gun just in case. The 3.5-inch rocket launcher proved itself adequate against the vaunted T-34 tank.

National Archives Photo (USMC) 127-N-A2747

Department of Defense Photo (USMC)

Amphibian tractors (LVTs) churn away from the landing ships (LSTs) that brought them to Inchon. "Amtracks" were used chiefly for the assault of Blue Beach within the inner harbor.

another wave. I asked if they were headed for Blue Two. Their wave commander answered, "Hell, no. We're the 2d Battalion headed for Blue One." We then veered off to the right. I broke out my map and asked my LVT driver if he had a compass. He looked at his instrument panel and said, "Search me. Six weeks ago I was driving a truck in San Francisco."

The nine Army LVT(A)s making up the first wave for Blue Beach One got ashore on schedule, but found themselves boxed in by an earth slide that blocked the exit road. The remaining nine Army armored amphibian tractors, forming Wave 1 for Blue Beach Two, made it to the seawall shortly after H-Hour but were less successful in getting ashore. The "road" separating Blue One and Two turned out to be a muck-filled drainage ditch. After exchanging fire with scattered defenders in factory buildings behind the seawall, the Army vehicles backed off and milled around, getting intermixed with the incoming troop-carrying Waves 2 and 3.

From his seat on the bridge of the *Mount McKinley*, MacArthur, surrounded by his gaggle of generals and admirals, peered through the gathering gloom of smoke, rain, and darkness and listened to the reports crackling over the loudspeaker. From his perspective, all seemed to be going well.

Lieutenant Colonel Sutter's second wave landed elements of both his two assault companies, Company D, under Captain Welby W. Cronk, and Company F, under Captain Goodwin C. Groff, across Blue Beach One shortly after H-Hour. Some of his amphibian tractors hung up on a mud bank about 300 yards offshore and their occupants had to wade the rest of the way. Most of Sutter's last three waves, bringing in his reserve, Company E, drifted to the right. As Sutter reported it: "For some unknown reason the third, fourth, and fifth waves were diverted from landing either on Beach BLUE-1 or along the rock causeway by a control boat. Instead they were directed to the right of the two beaches prescribed for the regiment and landed at Beach BLUE-3."

Wave 2 for Blue Beach Two, with Ridge's assault companies, passed through the Army tractors, Company G under Captain George C. Westover on the left, Company I under First Lieutenant Joseph R. "Bull" Fisher on the right. They reached the seawall about 10 minutes after H-Hour. The tractors bearing Company G formed up in column and muddled their way up the drainage ditch. Company I went over the seawall using aluminum ladders, some of which buckled. Assault engineers from Captain Lester G. Harmon's Company C, 1st Engineer Battalion, reached the wall and rigged cargo nets to help the later waves climb ashore.

Ridge, the 3d Battalion commander, accompanied by his executive office, Major Reginald R. Myers, seeing the congestion on Blue

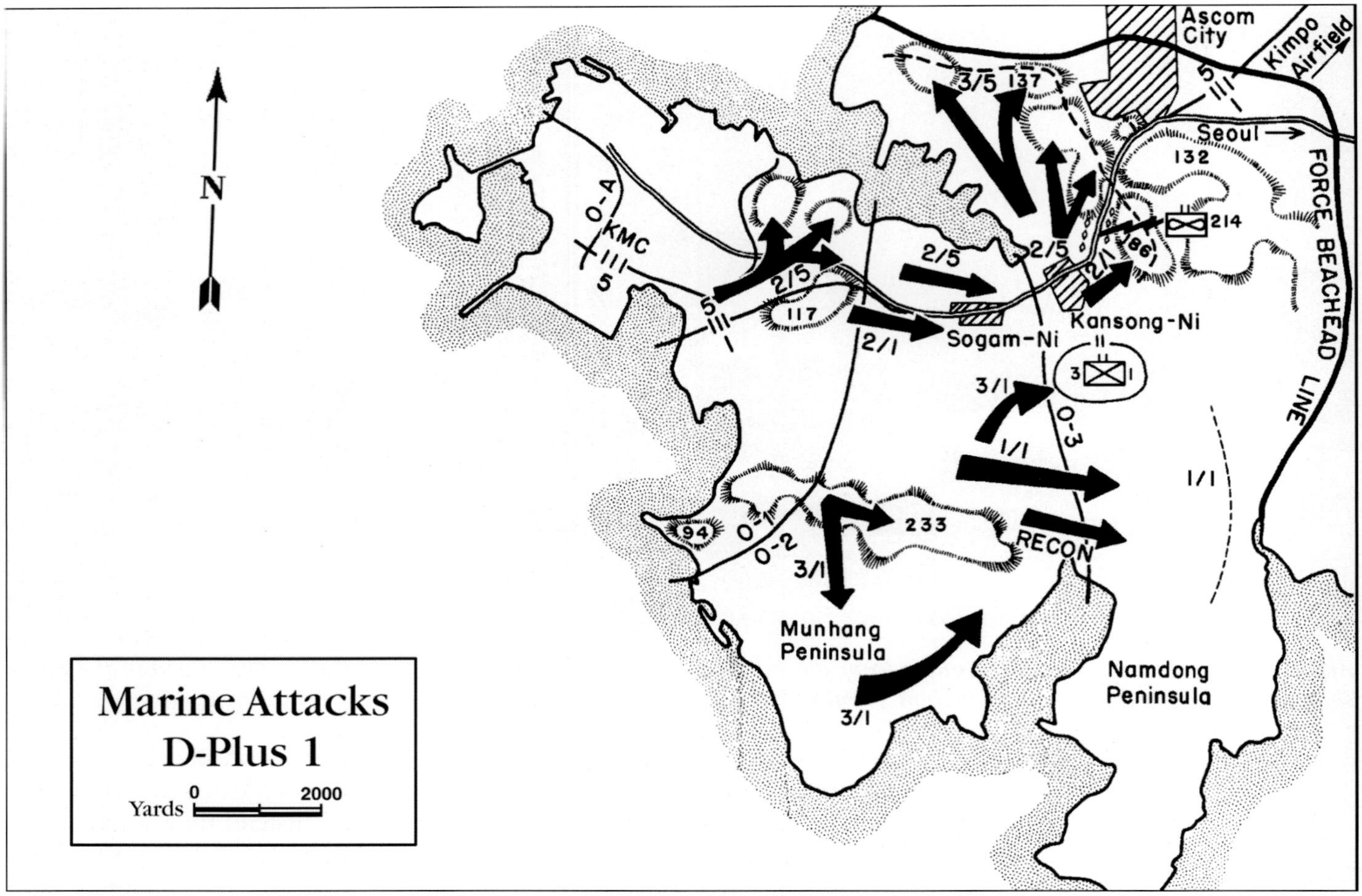

Marine Attacks D-Plus 1

Beach Two, moved in his free boat to explore the possibilities of Blue Beach Three. He found a mud ramp broken through the seawall and some of his battalion was diverted to this landing point. An enemy machine gun in a tower about 500 yards inland caused a few casualties before it was knocked out by fire from the Army's armored tractors.

More serious problems confronted Lieutenant Colonel Jack Hawkins' 1st Battalion, which was in regimental reserve. Boated in LCVP landing craft, he was ordered by Puller, who was already ashore, to land his battalion. If things had gone well Hawkins should have beached at about H+45 minutes or 1815. Veering off far to the left in the gloom, his leading waves mistook the wall of the tidal basin for the seawall of Blue Beach Two. Most of Company B and some of Company A had landed before Hawkins could correct the error. Most of those who landed were reboated and sent on to Blue Beach Two. Because of a shortage of boats, however, one platoon was left behind. Marching overland to Blue Beach Two this orphan platoon gilded the lily by picking up a bag of prisoners enroute.

The 3d Battalion, 1st Marines' reserve—Company H under Captain Clarence E. Corley, Jr.—landed across Blue Beaches Two and Three. The 1st Platoon, led by First Lieutenant William Swanson, had the mission of securing the right flank of the bridgehead. Swanson slid his platoon behind Company I and moved against a platoon-sized enemy dug in on Hill 94, which topped the fishhook cape bounding the beachhead on the south. The North Koreans were driven out, but at a cost. Swanson himself was severely wounded in the thigh and evacuated. (Swanson returned to the 3d Battalion in late winter 1951, was wounded in the hand at the end of March, and killed by one of our own mines on 15 May 1951.)

Corley's Company H, less its 1st Platoon, moved into the gap between Companies G and I. The 2d Platoon, Company H, was sent forward at midnight to outpost Hill 233, a mile to the front, got halfway there, to Hill 180, and received permission to stay put for the night.

Generals Almond and Shepherd came in with the ninth wave, along with Admiral Struble, for a looksee at how events were progressing on Blue Beach. Almond's aide, Lieutenant Haig, had come in to

National Archives Photo (USMC) 127-GK-234I-A409339

On the morning following the landing, the Marines marched through Inchon itself against no resistance. Initially, the Marines enjoyed a 10 to one numerical advantage over the mediocre defense force. North Korean resistance stiffened in both numbers and quality as the attack moved inland toward Seoul.

Red Beach on board one of the LSTs. He had with him Almond's personal baggage and the wherewithal to establish a mobile command post including a van fitted out as sleeping quarters and an office. In transit Haig had lost two of the general's five jeeps, swept over the side of the LST in the typhoon. When Haig met up with his boss, Almond's first question was whether Haig had gotten his baggage ashore without getting it wet.

While the 5th Marines were assaulting Red Beach, Brigadier General Craig—with his brigade dissolved and now the assistant division commander—came ashore at Wolmi-do and, joining Taplett's 3d Battalion, established an advance division command post. Craig had brought his brigade staff ashore intact to function as an interim division staff. Since his arrival in the objective area, Craig had had no opportunity to meet with O. P. Smith face-to-face.

During the night, Taplett's battalion crossed the causeway from Wolmi-do and rejoined the main body of the 5th Marines on Red Beach. Before morning the 1st Marine Division had all its first day's objectives. Resistance had been scattered—of the sort that goes down in the situation report as "light to moderate." Total Marine casualties for the first day's fighting were 20 killed, 1 died of wounds, 1 missing in action, and 174 wounded.

Assault Continues

At about midnight Puller and Murray received the division's attack order for the next day. Murray was to bring the 5th Marines up on line abreast of Puller's 1st Marines. The axis for the advance on Seoul would be the intertwined highway and railroad. The Korean Marine regiment

Department of Defense Photo (USA) SC348506

Once ashore at Inchon, the Marines see for themselves that naval gunfire had destroyed much of the city. Once ashore, the rule-of-thumb was that each assault battalion would have a cruiser or destroyer available for on-call missions.

was initially left behind in Inchon to mop up.

The day, 16 September, was clear and pleasant. The climate was about the same as our northeastern states at this time of year, warm during the day, a bit cool at night.

Murray elected to advance in column of battalions, leading off with Roise's 2d Battalion, followed by the 1st and 3d Battalions in that order. The 2d Battalion's advance through Inchon was strangely quiet. The enemy had vanished during the night.

Corsairs Against T-34s

Five miles to Murray's front, six of the vaunted Soviet-built T-34 tanks, without infantry escort, were rumbling down the Seoul highway toward him. Near the village of Kansong-ni, eight Corsairs from VMF-214 swept down on the advancing tanks with rockets and napalm. One Corsair, flown by Captain William F. Simpson, Jr., failed to come out of its dive, killing Simpson, but the tank attack was halted. One T-34 was engulfed in flames, a second had its tracks knocked off, and a third stood motionless on the road. A second flight of Corsairs came over to finish off the disabled T-34s. The pilots pulled away, thinking incorrectly that all six tanks were dead.

On the ground, Roise's 2d Battalion, 5th Marines, made solid contact with Sutter's 2d Battalion, 1st Marines, on Hill 117. The two battalions continued the advance against nothing heavier than sniper fire. By 1100 elements of both battalions were just short of Kansong-ni where they could see the smoke still rising from the fires set by the battle of T-34s and Corsairs.

Meanwhile, General Craig had moved his command group into Inchon itself. On the outskirts of the city, he found what he thought would be a good location for the division command post including a site close by where a landing strip could be bulldozed. He ordered his temporary command post moved forward.

Thirty "SCAJAP" LSTs, manned for the most part with Japanese crews, had been collected for the Inchon landing. Those that were carrying troops did not beach, but sent their passengers off in amphibian tractors. After the assault waves had swept ahead they did beach, when the tides permitted, for general unloading. Beach conditions and the mixed

The rubber-tired amphibious DUKW pulls a trailer about a mile outside of Inchon on the first morning after the landing. These "ducks" were used primarily to move guns, ammunition, and supplies for the artillery.

Department of Defense Photo (USA) SC348502

Department of Defense Photo (USA) SC348504

A curious Marine passes three knocked-out T-34 tanks. The vaunted Soviet-built tank proved no match for the array of weapons that the Marines could bring to bear, ranging from a Corsair fighter-bomber's rockets to the 3.5-inch rocket launchers in the rifle and weapons companies.

quality of the Japanese crews threw the planned schedule for unloading completely out of balance.

The landing and employment of tanks presented problems. The Marines had just received M-26 Pershings as replacements for their M-4A3 Shermans. Few of the members of Lieutenant Colonel Harry T. Milne's 1st Tank Battalion—except for Company A, which had been with the 5th Marines and had the M-26 at Pusan—were familiar with the Pershing. The tankers received their instruction on the new tanks on board ship—not the best place for tank training.

Major Vincent J. Gottschalk's VMO-6, the division's observation squadron, began flying reconnaissance missions at first light on D+1, 16 September. VMO-6 possessed eight Sikorsky HO3S-1 helicopters

M-26 Pershing tanks emerge from the maws of beached LSTs ("landing ships, tank") at Inchon. Marine tankers, previously equipped with the obsolescent M-4 Sherman tank, were re-equipped with the Pershings literally while on their way to the objective area.

Department of Defense Photo (USMC)

Department of Defense Photo (USMC) A130235

Maj Vincent J. Gottschalk, commanding officer of VMO-6, prepares to take off in an OY light observation aircraft. Among the varied missions of the squadron was spotting and adjusting artillery fire on the retreating North Koreans for the ground troops.

and eight OY airplanes and had been with the 1st Brigade at Pusan where for the first time Marines used helicopters in combat. That day, First Lieutenant Max Nebergall pulled a ditched Navy pilot out of Inchon harbor in the first of many rescue operations.

In the 1st Marines' zone of action Puller sent Ridge's 3d Battalion to make a sweep of Munhang Peninsula. Ridge used amphibian tractors as personnel carriers—a bold but dangerous practice—and advanced on a broad front, Companies G and I abreast with Company H following in reserve. Prisoners and materiel were taken, but there was almost no fighting. By noon the division held the 0-3 line, a front three miles long, secured on both flanks by water. Smith ordered Murray and Puller to move on forward and seize the Force Beachhead Line (FBHL) which would conclude the assault phase of the amphibious operation.

Murray chose to advance in two prongs. Roise with the 2d Battalion would continue to advance with his right flank tied to the Seoul highway. Taplett, coming up from behind with the 3d Battalion was to swing wide to the left. Newton, with the 1st Battalion, would follow in reserve.

Roise's battalion, escorted by Lieutenant Sweet's five M-26 Pershing tanks, moved up the road and at about 1330 rounded the bend into Kansong-ni. Two of Sweet's tanks crawled up a knoll from which they could cover the advancing riflemen. From this vantage point the Marine tankers saw three T-34 tanks, not dead as supposed, but ready for battle with hatches buttoned up and 85mm guns leveled on the bend in the road. Sweet's tanks smacked the T-34s with 20 rounds of armor-piercing shells. The T-34s went up in flames. Company D led the advance past the three burning hulks. Nearby the Marines found the two tanks knocked out earlier by the Corsairs. The sixth tank had vanished.

Company D continued for another thousand yards and then climbed a high hill on the west side of the road. Company F joined Company D on their left. They were still two miles from the Force Beachhead Line, but it looked like a good time and place to dig in for the night.

On Roise's left, Taplett's 3d

Department of Defense Photo (USA) SC349015

For much of the advance up the axis of the Inchon-Seoul highway, and even sometimes traveling cross-country, Marines used amphibian tractors as personnel carriers protected along the way by M-26 tanks. The North Koreans, in turn, tried to choke off these advances with ambushes and antitank mines.

Battalion advanced uneventfully and now held high ground overlooking the FBHL. His patrols reached the edge of Ascom City—once the village of Taejong-ni and now the remnants of a huge service command that had been used by the U.S. Army during the occupation—to his front. The sea was to his left.

A Korean civilian eager to assist the advancing forces, shows one of the division's reconnaissance Marines a large cache of dynamite and ammunition hidden in a storage cave. It was one of several caches uncovered by Capt Kenneth Houghton's Marines on the division's right flank.

Department of Defense Photo (USMC) A2821

South of the 5th Marines, Puller's 1st Marines, having spent most of the day pulling together its scattered parts, did not jump off in the new attack until about 1600. Sutter's 2d Battalion went forward on the right of the road past Kansong-ni for a thousand yards and then tied in with Roise's battalion for the night. Hawkins' 1st Battalion filled in between Sutter and Ridge. Ridge's 3d Battalion had done more hiking than fighting and at the end of the day was relieved by the Division Reconnaissance Company, under pugnacious Captain Kenneth J. Houghton, attached to the 1st Marines as the division's right flank element. Ridge's Marines went into regimental reserve. Houghton's reconnaissance Marines engaged no enemy but found huge caches of arms and ammunition.

O. P. Smith Opens His Command Post

General Craig had just gotten back from his search for a site for the division command post, when he learned that O. P. Smith, accompanied by Admiral Struble and General Shepherd, had landed. Smith was satisfied with Craig's recommended site. Craig then took him for a quick tour of the troop dispositions and at 1800 Smith officially assumed command ashore. During the day, General Almond visited Red Beach and the 5th Marines.

Smith was joined later that evening by Major General Frank E. Lowe, an Army Reserve officer and President Truman's personal observer, who had arrived unannounced. Lowe moved into the division command post. He and Smith got along famously. "His frank and disarming manner made him welcome throughout the division," remembered Smith.

Photo by Frank Noel, Associated Press

Front-line Marines found that stripping a prisoner bare took all the fight out of him and also eliminated the possibility of hidden weapons. Rear-echelon authorities found the practice distasteful and ordered the Marines to desist.

More Enemy T-34 Tanks

The night of 16-17 September was quiet, so quiet, the official history remarks, that a truck coming down the highway from Seoul drove unimpeded through the Marine front lines, until finally stopped by a line of M-26 tanks several hundred yards to the rear. The tankers, the 1st Platoon, Company A, under First Lieutenant William D. Pomeroy, took a surprised NKPA officer and four enlisted men prisoner.

Lieutenant "Hog Jaw" Smith, commander of Company D, 5th Marines, from his observation post overlooking the highway was sufficiently apprehensive, however, about a sharp bend in the road to the left front of his position to outpost it. He dispatched his 2d Platoon with machine guns and rocket launchers attached, all under Second Lieutenant Lee R. Howard, for that purpose.

During the night the North Koreans formed up a tank-infantry column—six T-34s from the *42d NKPA Mechanized Regiment* and about 200 infantry from the *18th NKPA Division* in Seoul—some miles east of Ascom City. Howard saw the lead tank at about dawn, reported its approach to "Hog Jaw" Smith, who reported it to Roise, who could not quite believe it. Obviously the North Koreans did not know the Marines were waiting for them. Howard let the column come abreast of his knoll-top position and then opened up. Official historians Montross and Canzona say: "The Red infantry went down under the hail of lead like wheat under the sickle." Corporal Oley J. Douglas, still armed with the 2.36-inch rocket launcher and not the new 3.5-inch, slid down the hill to get a better shot at the tanks. At a range of 75 yards he killed the first T-34 and damaged the second. The remaining four tanks continued to plow forward to be met by a cacophony of 90mm fire from Pomeroy's M-26 tanks at 600 yards range, 75mm recoilless rifle fire at 500 yards, and more rockets, some coming from Sutter's battalion on the other side of the road. Private First Class Walter C. Monegan, Jr., from Company F, 1st Marines, fired his 3.5-inch rocket launcher at point-blank range. Just which weapons killed which tanks would be argued, but the essentials were that all six T-34s were knocked out and their crews killed.

MacArthur Comes Ashore

MacArthur, instantly recognizable in his braided cap, sunglasses, well-worn khakis, and leather flight jacket, came grandly ashore that same morning, 17 September. His large accompanying party included Struble, Almond,

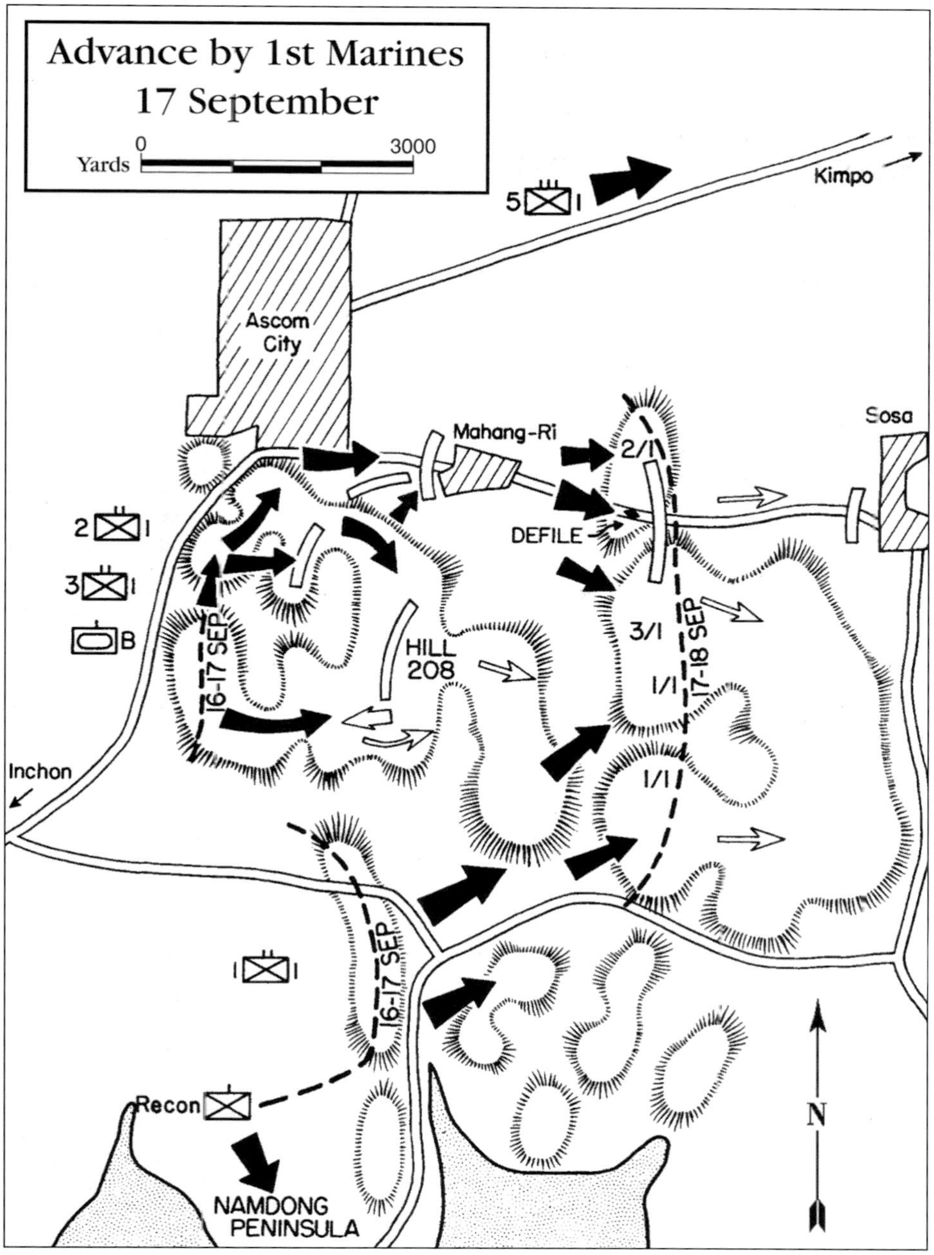

Shepherd, Whitney, Wright, and Fox; a bodyguard bristling with weapons; and a large number of the press corps. A train of jeeps was hastily assembled and the party proceeded to the 1st Marine Division headquarters in a dirt-floored Quonset hut where Smith joined the party. MacArthur presented him a Silver Star medal.

MacArthur and his entourage then visited Puller at the 1st Marines' observation post. MacArthur climbed the hill. Puller put down his binoculars and the two great actors shook hands. MacArthur gave Puller a Silver Star.

MacArthur's cavalcade next drove to the site of the still-smoking hulls of the dreaded North Korean T-34 tanks that had counterattacked at dawn. Shepherd, looking at the still-burning T-34s, commented to Almond that they proved that "bazookas" could destroy tanks.

"You damned Marines!" snorted Almond. "You always seem to be in the right spot at the right time . . . . MacArthur would arrive just as the Marines knocked out five tanks." Shepherd replied, "Well, Ned, we're just doing our job, that's all."

MacArthur climbed back into his jeep and the star-studded party drove on. Seven dazed North Korean soldiers crawled out from the culvert over which MacArthur's jeep had parked and meekly surrendered.

Next stop for MacArthur was the 5th Marines command post. MacArthur went to award Silver Stars to General Craig and Colonel Murray only to learn that his supply of medals was exhausted. "Make a note," he told his aide. The medals were delivered later.

MacArthur finished his tour with a visit to Green Beach at Wolmi-do, where unloading from the LSTs was progressing, and to see the occupants of the prisoner of war stockade—671 of them under guard of the 1st Marine Division's military police.

Ashore at Wolmi-do, MacArthur found evidence, to his great satis-

At a temporary aid station at Pier No. 2, designated Yellow Beach, a wounded Marine is given whole blood by a Navy corpsman. From this station, the wounded were evacuated to hospital ships off shore.

National Archives Photo (USA) 111-SC349024

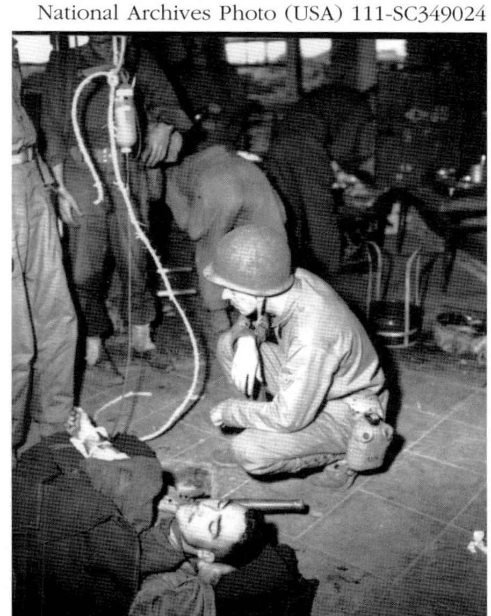

Department of Defense Photo (USA) SC348526

On the morning of 17 September, Gen MacArthur, surrounded by subordinates, bodyguards, and photographers, made a grand and much publicized tour of the Inchon beachhead. MacArthur is unmistakable in his crushed cap, sunglasses, and leather jacket. LtGen Lemuel C. Shepherd, Jr., on the left, is in his usual khakis and carrying his trademark cocomacaque, or Haitian walking stick. MajGen O. P. Smith, in khaki fore-n-aft cap and canvas leggings, trudges along behind Shepherd.

Almost obscured by the jeep's windshield, a photographer peers through his lens at the command echelons of the Inchon landing during the 17 September visit. Gen MacArthur in hawk-like profile stares straight ahead. MajGen O. P. Smith sits smiling in the middle of the rear seat, flanked on his right by MajGen Edward M. Almond and on his left by VAdm Arthur D. Struble. The unidentified Marine driver awaits instructions.

Department of Defense Photo (USA) SC348522

faction, that the enemy had begun an intensive fortification of the island. Later he pontificated: "Had I listened to those who wanted to delay the landing until the next high tides, nearly a month later, Wolmi-do would have been an impregnable fortress."

Almond, just before leaving with his boss to return to the *Mount McKinley*, informed Smith that Barr's 7th Infantry Division would begin landing the next day, coming in on the 1st Marine Division's right flank. Smith, returning to his command post, learned that Major General James M. Gavin, USA, of World War II airborne fame, had arrived to study the Marine Corps' use of close air support.

An airstrip was set up next to the division command post that same day, 17 September. After that, Gottschalk's VMO-6 flew a full schedule of observation, evacuation, liaison, and reconnaissance flights.

Department of Defense Photo (USA) SC348516

FMFPac commander LtGen Lemuel C. Shepherd, VMI 1917, on the right, points out something significant to the X Corps commander, MajGen Edward M. Almond, VMI 1915, as they move by motor launch from the Mount McKinley *to the beach. Shepherd, although relegated to the position of observer instead of corps commander, held no grudge against Almond.*

Marine helicopters, fragile and few in number, were found useful in evacuating severely wounded Marines to hospital facilities to the rear or at sea. As the war progressed, more suitable helicopters arrived and the practice became standard.

Photo by Frank Noel, Associated Press

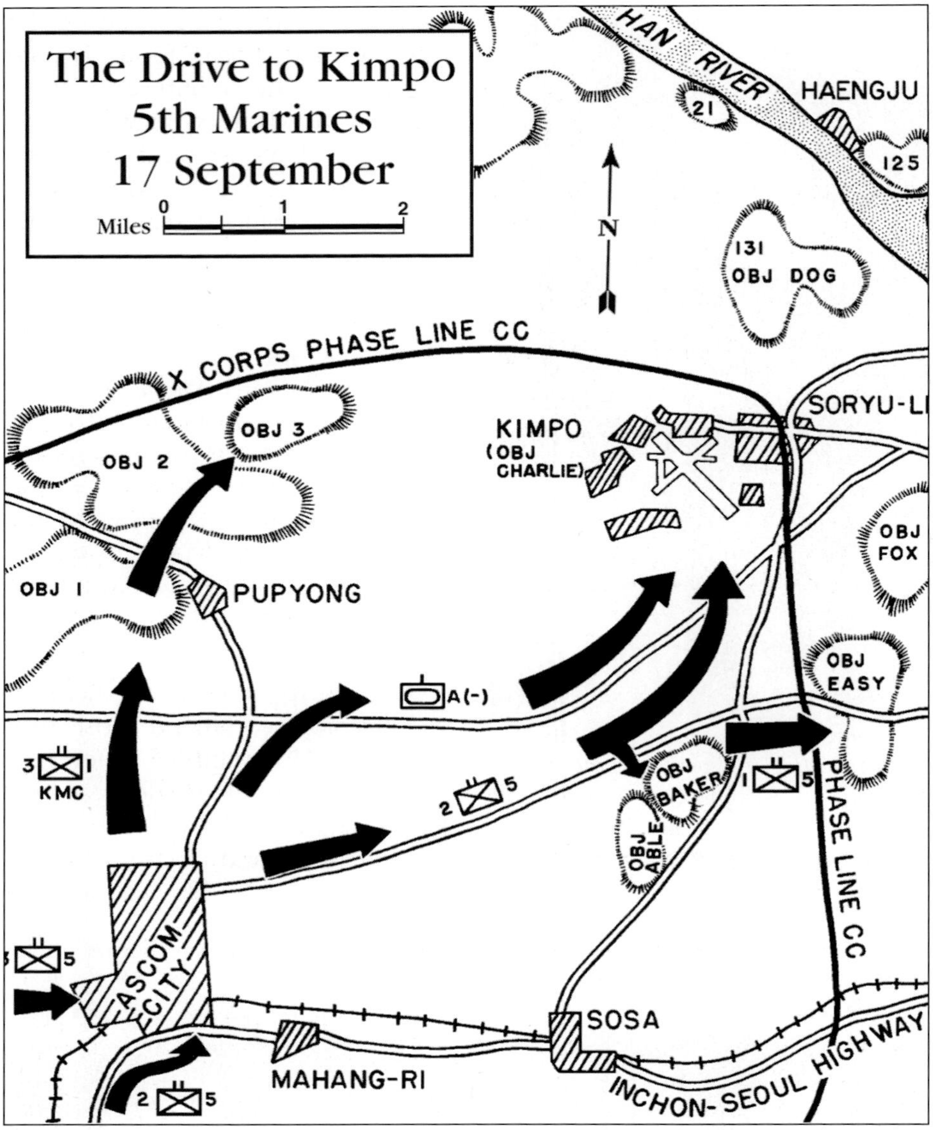

Infantry Advances

The battle with the T-34s delayed for an hour the jump-off for the day's attacks. The next phase line was 19 miles long and Murray's 5th Marines had two-thirds of it. At 0700, the Korean Marines' 3d Battalion had passed through Roise's 2d Battalion to clean up the outskirts of Ascom City. Roise himself jumped off two hours later, Captain Jaskilka's Company E in the lead. The advance was to be in column and then a left turn into Ascom City.

Company E, joined by 2d Platoon, Company F, spent the morning in a methodical clearing of the densely built-up area of little pockets of resistance. Roise found that the road on the map that was supposed to lead to his next objective, four miles distant, was nonexistent on the ground. The renewed advance did not get off until mid-afternoon.

The inexperienced 3d Battalion of the Korean Marines ran into trouble on the other side of Ascom City. Taplett's 3d Battalion, 5th Marines, in regimental reserve, moved in to help and efficiently knocked out the moderate resistance. Pomeroy came up with his platoon of M-26 tanks. Looking for Roise's 2d Battalion and, not finding the mythical road, he instead found Taplett's 3d Battalion. Eventually, Pomeroy reached the 2d Battalion and a road that would lead to Kimpo Airfield now about five miles away. He was joined by his company commander, Captain Gearl M. English, and another platoon of tanks.

Meanwhile, Roise advanced to two high hills some 4,000 yards south of Kimpo. He launched his attack against the airfield with Companies D and E in the assault. They moved rapidly against nothing but light small arms fire. Captain English brought up his tanks to help, assigning a tank platoon to support each of the assault companies. By 1800, Roise's Marines were at the southern end of the main runway. Each of his three rifle companies curled into separate perimeters for the night. Lieutenant Deptula's 1st Platoon, Company E, was positioned well out to the front in the hamlet of Soryu-li as an outpost.

During the afternoon, Newton and the 1st Battalion, 5th Marines, had moved up on Roise's right against no resistance. Taplett's 3d Battalion, having eased the situation for the Korean Marines, was two miles to the rear, again in regimental reserve.

With 1st Marines

Throughout the day, 17 September, Puller's 1st Marines had continued its advance. On the left flank Sutter, with the 2d Battalion, straddled the highway and moved forward behind an intermittent curtain of howitzer fire delivered by the 11th Marines. Essentially, Sutter was attacking due east from Mahang-ri to Sosa, two fair-sized villages. He deployed Company E on the left of the road, Company F on the right, and kept Company D in reserve. As the 5th Marines moved to the northeast toward

Department of Defense Photo (USMC)

When not moving from hill to hill, the Marines frequently found themselves attacking across flat rice paddies. Ironically, Kimpo, in addition to having the best airfield in Korea, was also known for growing the best rice.

Kimpo, a considerable gap widened between the two regiments.

North Korean resistance thickened as Sutter neared Sosa. Puller ordered Ridge to move the 3d Battalion up on Sutter's right flank. Ridge decided again to use amphibian tractors as personnel carriers. Westover's Company G clanked up the road behind the 2d Platoon, Company B tanks, under Second Lieutenant Brian J. Cummings. In a defile, some brave North Koreans tried to stop Cummings' M-26s with grenades. The advance on the road stalled. Company G got up on the high side of the defile to the right of the road. With Sutter's battalion on the left, the Marines had a converging "turkey shoot" and broke up the North Korean attack. Sutter and Ridge dug in for the night, each battalion on its own side of the defile. To their south, Hawkins' 1st Battalion and Houghton's Reconnaissance Company had cleared up Namdong Peninsula. The night would pass quietly for the 1st Marines.

To the rear, Inchon was in a shambles. Most of the city officials had fled before the North Korean capture of the city. Fortunately, Admiral Sohn Won Yil, the chief of naval operations of the ROK Navy, had come ashore to observe the operations of the Korean Marine Regiment. (He also received a MacArthur Silver Star.) Sohn picked a temporary mayor who was installed on the morning of 18 September by authority of a 1st Marine Division proclamation.

5th Marines Takes Kimpo

The night of 17-18 September was tense for the 5th Marines. Murray was certain that the North Koreans would not give up Kimpo, the best airfield in Korea, without a fight, and he was right. The airfield was under the apparent command of a Chinese-trained brigadier general, Wan Yong. The garrison, nominally the NKPA *1st Air Force Division*, was in truth a patchwork of bits and pieces of several regiments, with not more than a few hundred effectives.

The North Koreans went against Roise's well dug-in battalion in

Shore party operations followed close behind the assault waves and within a few days, stocks of ammunition, rations, and other supplies had reached the level needed for the drive to Seoul and its capture.

Photo by Frank Noel, Associated Press

Department of Defense Photo (USN) 420271

Fumigation and bath platoons would arrive later, but during the assault phase of the Inchon operation Marines seized the opportunity to clean up when and where they could. Helmets made convenient washbasins.

three badly coordinated attacks. The first hit Deptula's outpost at about 0300 in the morning, the Communists using rifles and machine pistols, backed by a T-34 tank. Deptula skillfully fought off four half-hearted assaults and by 0500 had withdrawn successfully to Company E's main line of resistance.

The second attack came from both the west and east against Jaskilka's Company E. The third attack hit Harrell's Company F further to the south. Both attacks were easily contained. The routed enemy fled toward the Han River.

At daylight Roise jumped off in pursuit. His Marines swept across the airfield, securing it and its surrounding villages by 1000. Companies E and F mopped up and Company D went on to take Hill 131 overlooking the Han. In 24 hours of fighting, Roise had lost four Marines killed and 19 wounded. His Marines had taken 10 prisoners and had counted about 100 enemy dead.

Correspondents and photographers examine a Russian-built Yak fighter in a destroyed hanger at Kimpo Airfield. Captured by the 2d Battalion, 5th Marines, Marine engineers quickly made the airfield operative and ready to receive elements of MAG-33.

National Archives Photo (USA) 111-SC349036

1st Marines Advances

In the 1st Marines' zone of action, Ridge, with the 3d Battalion outside of Sosa, decided that the center of North Korean resistance must be on Hill 123. During the night he called for naval gunfire. HMS *Kenya*, Captain P. W. Brock commanding, delivered some 300 rounds of 6-inch shells somewhere between Sosa and Hill 123. Ridge's naval gunfire spotter was not sure where they impacted, but Ridge, in the interest of inter-allied cordiality, sent Captain Brock a "well done."

At dawn Sutter charged ahead astride the Seoul highway, Company E on the left of the road and Company D on the right. Premature airbursts on the part of his artillery preparatory fires cost him

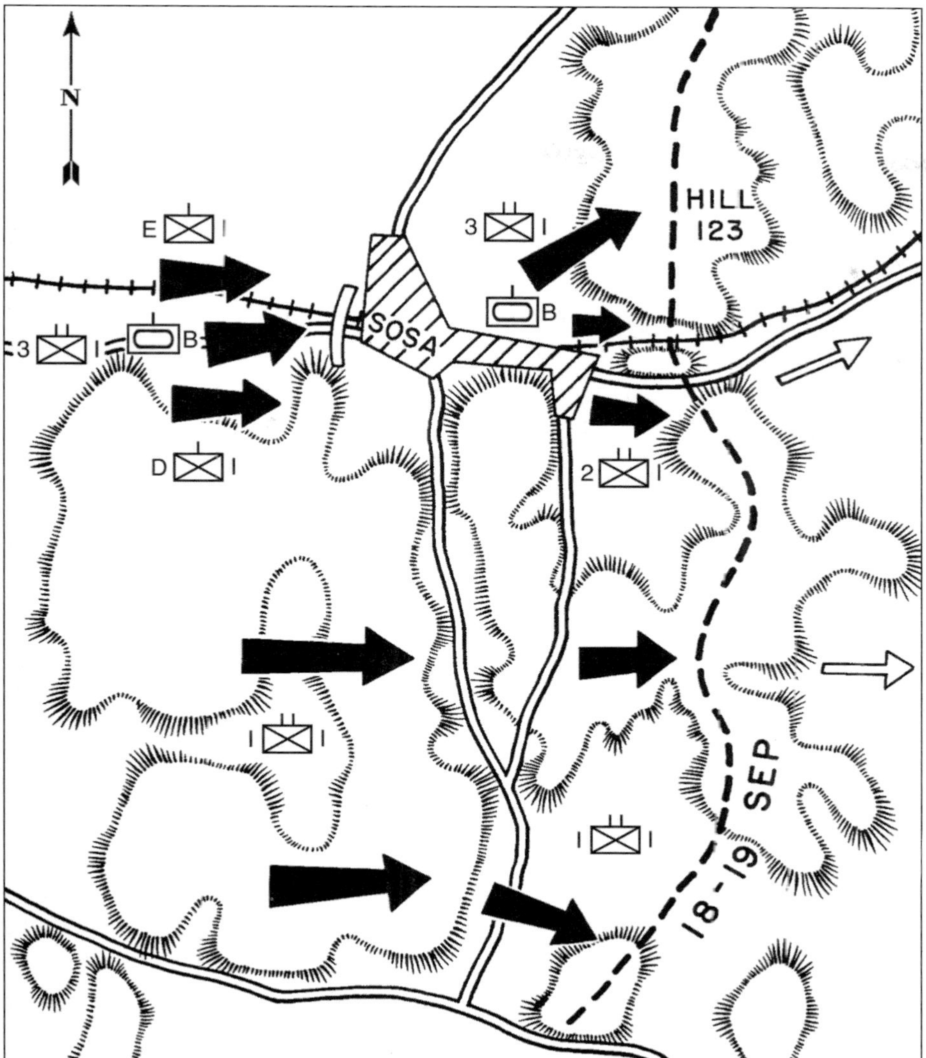

Capture of Sosa
1st Marines
18 September

0 3000
Yards

two killed and three wounded.

Behind the 2d Battalion, Ridge mounted up the 3d Battalion in a motorized column made up of a mixture of jeeps, amphibian tractors (LVTs), and amphibious trucks (DUKWs). Corsairs from VMF-214 worked over Sosa, sighted six T-34s beyond the town, and knocked out two of them. Ridge thundered ahead in a cloud of dust behind the tanks of Company B, 1st Tank Battalion. Together they brushed aside some light resistance, including an antitank roadblock. By noon Ridge had cleared the town. His battalion then swung to the left off the road and moved up Hill 123 while his naval gunfire spotter continued to look for some evidence as to where the *Kenya's* shells might have hit. The 3d Battalion was barely on the hill and not yet dug in when a barrage of North Korean 120mm mortar shells drenched their position causing 30 casualties. The romp over the green hills, marred as they were with the red-orange scars of shell holes and trench lines, was over. The war was getting serious.

Sutter's 2d Battalion, meanwhile, went straight ahead, left flank on the railroad tracks, into a defensive position about a mile beyond Sosa. A barrage of mortar shells cost him 14 casualties. Hawkins' 1st Battalion continued advancing on the right and for the third straight day encountered nothing but a few rifle shots.

Kimpo Airfield Becomes Operational

Murray displaced his command post forward from Ascom City to Kimpo. His regiment spent a quiet day sending patrols around the airfield. The field was in relatively good shape. A North Korean Soviet-built Yakovlev Yak-3 fighter and two Ilyushin "Shturmovik" attack aircraft were found in nearflyable condition.

The first aircraft to land at Kimpo was a Marine H03S-1 helicopter. It arrived at 1000 that morning, 18 September, piloted by Captain Victor A. Armstrong of VMO-6 and with General Shepherd and Colonel Krulak as passengers. General Craig who had just arrived by jeep met them.

Captain George W. King's Company A, 1st Engineer Battalion, made the field operational with temporary repairs. Generals Harris and Cushman came in by helicopter that afternoon. On their advice, General Almond authorized the establishment of Marine Aircraft Group 33 (MAG-33) on the field.

Corsairs began to arrive the next day. Harris set up the headquarters of his Tactical Air Command. Two Corsair squadrons, VMF-312 and VMF-212, came in. Night fighter squadron VMF(N)-542, under Lieutenant Colonel Max J.

National Archives Photo (USMC) 127-N-A3727

1stLt John V. Hanes flew in first Marine Corsair to land at Kimpo Airfield. Having taken hits while on a bombing mission, Hanes was grateful that there was a friendly airfield on which to land. BGen Thomas J. Cushman, Assistant Wing Commander of the 1st Marine Aircraft Wing, greets him.

Volcansek, Jr., arrived from Japan. There was a paper shuffle of squadrons between Marine Aircraft Groups 12 and 33. Marine Aircraft Group 33 under Brigadier General Thomas J. Cushman was now in business ashore. MAG-12 picked up the squadrons afloat. VMFs 214 and 323 continued to operate from the *Sicily* and *Badoeng Strait*, and the night-fighters of VMF(N)-513 from their base at Itazuke in Japan.

Reinforcements Arrive

On Murray's left the 2d KMC Battalion joined the 1st KMC Battalion. The ROK Army's 17th Regiment landed at Inchon and, temporarily under 1st Marine Division control was given an initial mission of completing the clean-up of the unswept area between Ascom City and the sea.

Almond, pressing forward, conferred with Smith on the morning of 18 September concerning the readiness of the 1st Marine Division to cross the Han. Smith pointed out that the 7th Division must take over its zone of action and free his right flank so he could concentrate his forces to cross the river. Smith already had it in his mind that the 5th Marines would go over first to be followed by the 1st Marines. His 7th Marines was still at sea. He went forward to Kimpo to discuss the matter with Murray.

The first unit of the 7th Division, the 32d Infantry, landed, as promised on the 18th, was attached temporarily to the 1st Marine Division. Smith relayed Almond's orders to the 32d to relieve the 1st Marines on the right flank and then to operate in the zone of action assigned to the 7th Division.

7th Division Becomes Operational

On the morning of 19 September, General Barr established his 7th Division's command post ashore. Almond called Barr and Smith together at the 1st Marine Division command post to discuss the 7th Division's immediate assumption of what had been the 1st Marines' zone of action south of the Inchon-Seoul highway.

The 31st Infantry had begun landing. The 32d Infantry would be detached from the 1st Marine Division at 1800. With these two regiments Barr was to begin operations. Smith would then be able to side-slip Puller's regiment fully to the left of the Seoul highway.

Almond's aide, Lieutenant Haig, who was a fly on the tent wall at these meetings, observed that "the Marines' respect for the 7th Division at this stage of the war was ostentatiously low."

Advancing to the Han

After that meeting, the peripatetic Almond went on to visit the command posts of both the 32d Infantry and the 1st Marines. He then proceeded to the 5th Marines command post on Kimpo Airfield to discuss with Murray the crossing of the Han that was scheduled for the following day. Murray told him that he planned to cross in column of battalions using amphibian tractors, amphibious trucks, and pontoon floats at a ferry crossing site northeast of Kimpo.

A significant range of hills separated the 5th Marines on Kimpo from Yongdung-po and the Han. During the night of 18-19 September, Murray had ordered Newton forward with the 1st Battalion to seize Hill 118 and then Hills 80 and 85, overlooking the Kalchon River near where it joined the Han.

At dawn, before Newton could move out, a company-sized North Korean force attacked Company C behind a shower of mortar shells. While Company C slaughtered the North Koreans, "Ike" Fenton's Company B moved against Hill 118. There was the usual air and

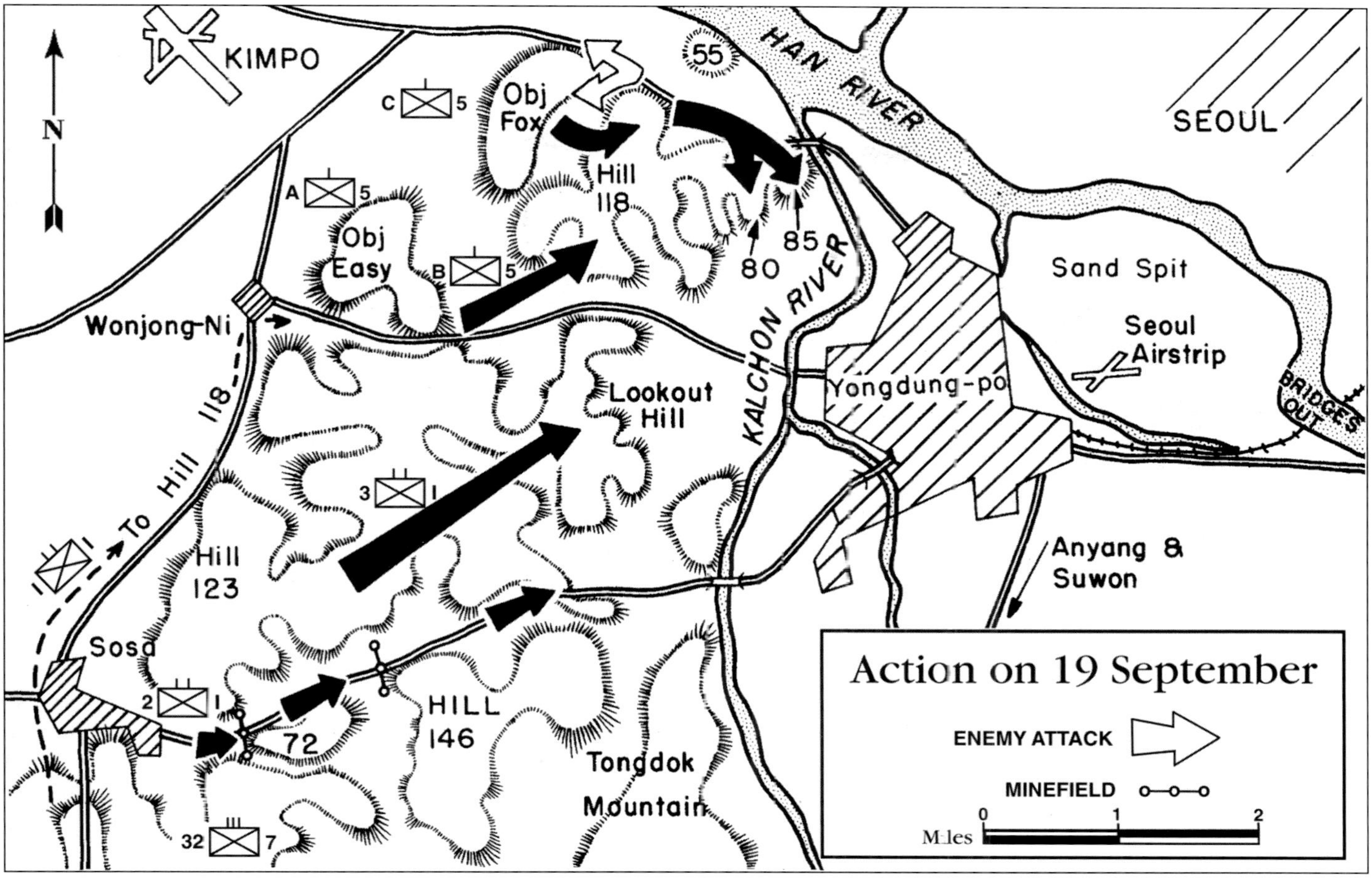

Action on 19 September

artillery preparation before the jump-off, and Company B took the peak of Hill 118 without suffering a single casualty. The trapped attacking North Koreans lost perhaps 300 dead (there is always optimism in the count of enemy dead) and 100 prisoners. Company C lost two killed and six wounded.

To the 5th Marines' right, Ridge's 3d Battalion, 1st Marines, with Companies H and I in the assault, moved off Hill 123 toward Lookout Hill, so-called because it gave a good view of the Kalchon and the town of Yongdung-po beyond. Official historians Montross and Canzona called the attack, which cost two killed and 15 wounded, "too successful," because it put Ridge's battalion well out in front of the 5th Marines on his left and Sutter's 2d Battalion on his right.

Sutter's battalion was advancing along the Seoul highway behind Captain Richard M. Taylor's Company C tanks and had gone little more than a quarter-mile when the lead M-26 hit a box mine that blew off a track and two road wheels. The antitank barrier of mines was formidable. The whole column came to a stop. Small-arms fire smashed in from neighboring Hill 72. The 11th Marines, the division's artillery regiment, took Hill 72 under howitzer fire. Corsairs from ever-ready VMF-214 came to help. A platoon of engineers under First Lieutenant George A. Babe blew up the box mines with "snowball" charges of C-3 plastic explosive. Sutter used all three of his rifle companies to uncover the minefield and force his way through. His infantry went forward a mile into heavy fighting around Hill 146 while the tanks waited on the side of the road. A second minefield was encountered, and more work by the engineers was needed. At 1900, Sutter ordered his battalion to dig in. His Marines had advanced nearly three miles at a cost of four killed and 18 wounded. Yongdung-po was still more than two miles in front of him.

Smith moved his command post forward the afternoon of 19 September to a site Craig selected about a mile and a half southeast of Kimpo; it had been used for U.S. dependents housing during the occupation. From here Smith was within easy jeep or helicopter distance of his front-line units. The abandoned Quonset huts were near ideal except for occasional harassment apparently by a single NKPA gun. The backbone for the perimeter defense around the command post was provided by a section of the Division Band trained as a machine gun platoon.

The 32d Infantry, now detached

from the division, was somewhere to Sutter's right rear. The Army battalion that relieved Hawkins' battalion had spent the day mopping up rather than continuing the attack.

Hawkins' 1st Battalion, 1st Marines, was on its way to relieve Newton's 1st Battalion, 5th Marines, an 11-mile motor march from the division's right flank. Captain Robert H. Barrow's Company A, 1st Marines, was the first to reach Hill 118 and relieve Fenton's Company B, 5th Marines.

Company C, 1st Marines, was to replace Company C, 5th Marines, on Hills 80 and 85. Newton was anxious to pull back his 1st Battalion, 5th Marines, to Kimpo to get ready for the river crossing the next day, and it was almost dark when Hawkins reached him. Company C, 1st Marines, under Captain Robert P. Wray, had not yet arrived.

Barrow, a tall Louisianian and a future Marine Commandant, realized the tactical importance of Hills 80 and 85 and radioed for permission to move Company A forward to the two hills. Permission was denied. Newton made it known that he would pull Company C off the hills no later than 2100. Wray's Company C did not reach Hill 118 until 2200; Hills 80 and 85 were left empty.

Confused Day

Before dawn the next day, 20 September, Hawkins' Marines on Hill 118 heard the North Koreans assault the empty hills. Then they came on in company-sized strength in a futile attack against the entrenched Marines on Hill 118.

Meanwhile, shortly before dawn a battalion-sized North Korean force, led by five T-34 tanks followed by an ammunition truck, came down the Seoul highway against Sutter's 2d Battalion, 1st Marines. Companies D and E held positions on each side of the road. The column roared through the gap between them and hit head-on against Company F's support position. The North Koreans were caught in a sleeve. Companies D and E poured fire into their flanks. Howitzer fire by the 2d and 4th Battalions, 11th Marines, sealed in the entrapped North Korean column. "A fortunate grenade was dropped in the enemy ammunition truck and offered some illumination," noted the 2d Battalion's Special Action Report, "enabling two tanks to be destroyed by 3.5" rocket fire."

The rocket gunner was Private First Class Monegan, the tank-killer

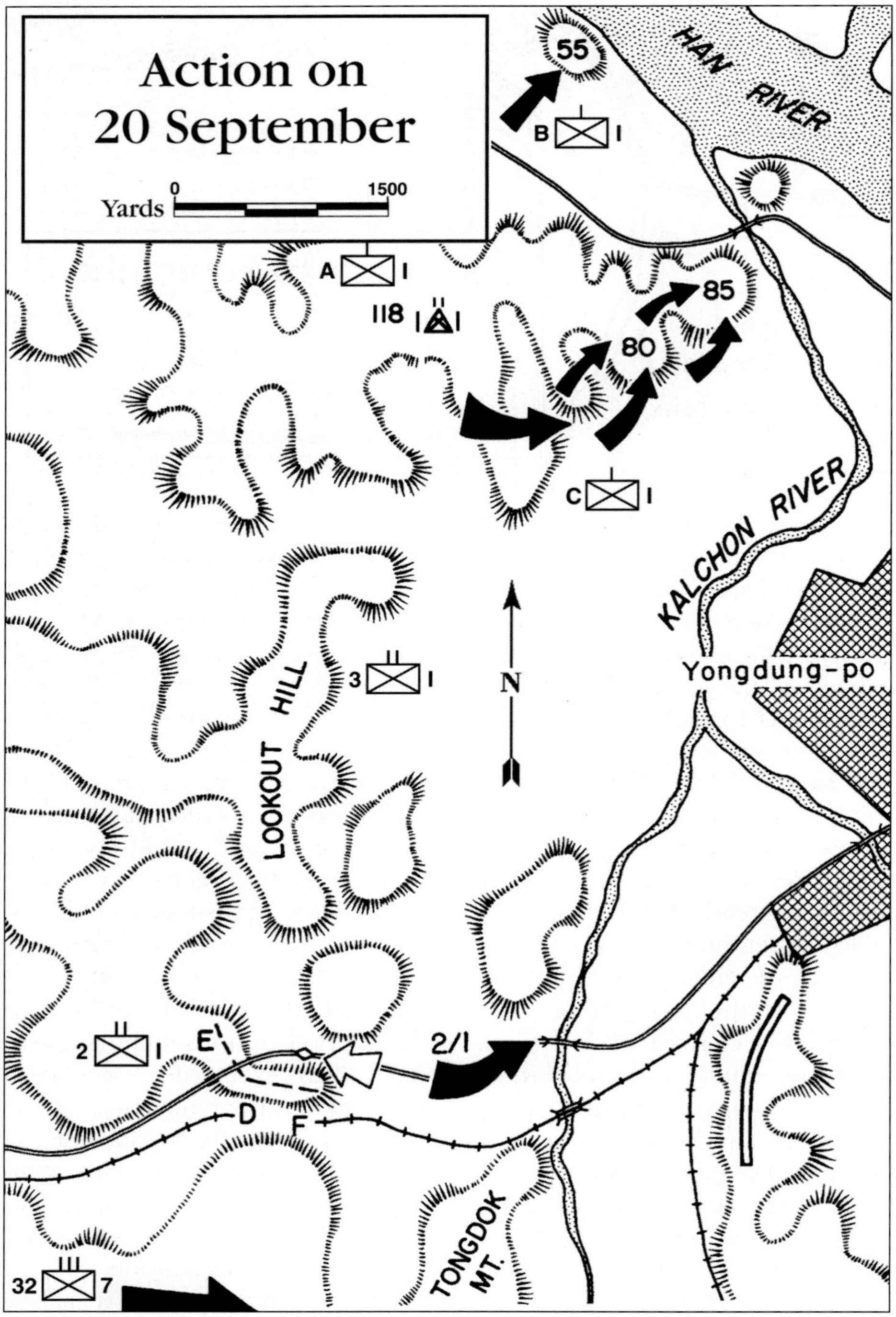

Private First Class Walter C. Monegan, Jr.

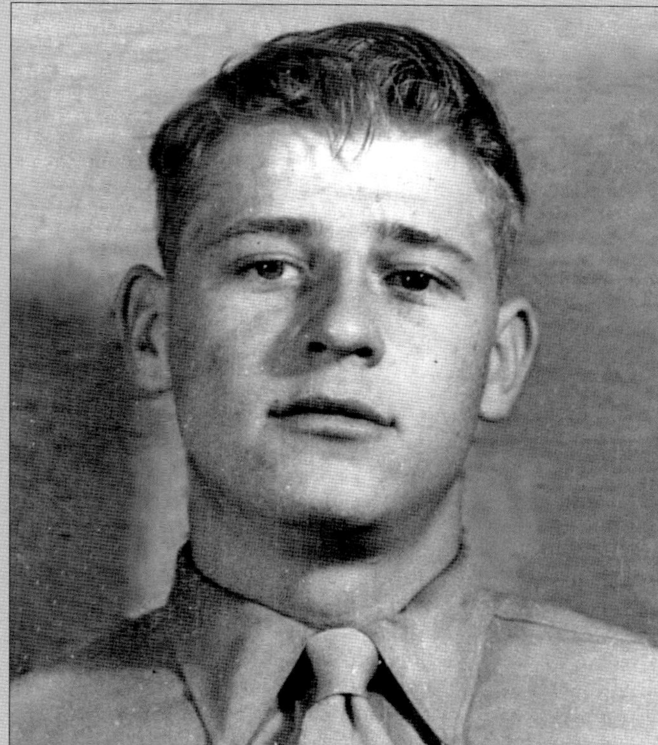

Department of Defense Photo (USMC) A45432

Nineteen-year-old Walter Monegan in five days of action fought two battles against North Korean T-34 tanks, won them both, and lost his own life.

Born on Christmas Day 1930, he could not wait until his 17th birthday, enlisting in the Army in November 1947. The Army discovered he was underage and promptly sent him home. He tried again on 22 March 1948, enlisting in the Marine Corps. After recruit training at Parris Island in June he was sent to China to join the 3d Marines at Tsingtao. After a year in China he came home, was stationed at Camp Pendleton for a year, and then was sent to Marine Barracks, Naval Air Station, Seattle. He had barely re-enlisted in July 1950 when he was ordered to return to Camp Pendleton to join the 2d Battalion, 1st Marines, then being formed.

His remains, buried temporarily at Inchon, were returned home and re-interred in Arlington National Cemetery on 19 July 1951. His wife, Elizabeth C. Monegan, holding their infant child, Walter III, received his posthumous Medal of Honor from Secretary of the Navy Dan Kimball, on 8 February 1952.

Citation:

> For conspicuous gallantry and intrepidity at the risk of his life above and beyond the call of duty while serving as a Rocket Gunner attached to Company F, Second Battalion, First Marines, First Marine Division (Reinforced), in action against enemy aggressor forces near Sosa-ri, Korea, on 17 and 20 September 1950. Dug in a hill overlooking the main Seoul highway when six enemy tanks threatened to break through the Battalion position during a pre- dawn attack on 17 September, Private First Class Monegan promptly moved forward with his bazooka under heavy hostile automatic weapons fire and engaged the lead tank at a range of less than 50 yards. After scoring a direct hit and killing the sole surviving tankman with his carbine as he came through the escape hatch, he boldly fired two more rounds of ammunition at the oncoming tanks, disorganizing the attack and enabling our tank crews to continue blasting with their 90-mm guns. With his own and an adjacent company's position threatened by annihilation when an overwhelming enemy tank-infantry force by-passed the area and proceeded toward the battalion Command Post during the early morning of September 20, he seized his rocket launcher and, in total darkness, charged down the slope of the hill where the tanks had broken through. Quick to act when an illuminating shell hit the area, he scored a direct hit on one of the tanks as hostile rifle and automatic weapons fire raked the area at close range. Again exposing himself he fired another round to destroy a second tank and, as the rear tank turned to retreat, stood upright to fire and was fatally struck down by hostile machine-gun fire when another illuminating shell silhouetted him against the sky. Private First Class Monegan's daring initiative, gallant fighting spirit and courageous devotion to duty were contributing factors in the success of his company in repelling the enemy and his self-sacrificing efforts throughout sustain and enhance the highest traditions of the United States Naval Service. He gallantly gave his life for his country.

of Soryu-li. He slid down the slope from Company F with his 3.5-inch rocket launcher and knocked out the first and second tanks. Machine gun fire killed him as he took aim on the third T-34. His family would receive a posthumous Medal of Honor. A third T-34 was captured intact. Sutter's battalion claimed 300 enemy dead. Half an hour after breaking up the North Korean attack, the 2d Battalion moved forward in its own attack.

Yongdung-po was drenched that day with shell-fire. Puller moved to align his regiment for the assault of the town. Hawkins was to take Hills 80 and 85. Sutter was to advance to the first of two highway bridges crossing the Kalchon. Ridge was to stay in reserve on Lookout Hill.

Hawkins sent out Captain Wray with Company C to capture Hills 80 and 85, that had been free for the taking, the day before. Wray, covered by the 81mm mortars and Browning water-cooled machine guns of Major William L. Bates, Jr.'s Weapons Company, made a text-

Second Lieutenant Henry A. Commiskey

Lieutenant Commiskey was no stranger to war. As an enlisted Marine he had been wounded at Iwo Jima and received a letter of commendation for "exhibiting high qualities of leadership and courage in the face of a stubborn and fanatical enemy."

Born in Hattiesburg, Mississippi, in 1927, he had joined the Marine Corps two days after his 17th birthday. He served more than five years as an enlisted man and was a staff sergeant drill instructor at Parris Island when he was selected for officer training in 1949. He completed this training in June 1950. Two months later he was with the 1st Marines and on his way to Korea.

He came from a family of fighters. His father had been a machine gun instructor in World War I. One brother had been with the Marine Raiders in World War II. Another brother was badly wounded while with the 187th Airborne Infantry in Korea.

In the action on 20 September, that gained Henry Commiskey the nation's highest award for valor, he escaped unscathed, but a week later he was slightly wounded in the fight for Seoul and on 8 December seriously wounded in the knee at the Chosin Reservoir. Sent home for hospitalization, he recovered and went to Pensacola in September 1951 for flight training, receiving his wings in June 1953 and then qualifying as a jet pilot.

He returned to Korea in April 1954 as a pilot with VMA-212. Coming home in September, he returned to line duty at his own request and was assigned once more to the 1st Marine Division. Next assignment was in 1956 to Jackson, Mississippi, close to his birthplace, for three years duty as a recruiter. In 1959, now a major, he went to the Amphibious Warfare School, Junior Course, at Quantico, and stayed on as an instructor at the Basic School. He retired from active duty in 1966 to Meridian, Mississippi, and died of a self-inflicted gunshot wound on 15 August 1971.

Department of Defense Photo (USMC) A43766

Citation:

> For conspicuous gallantry and intrepidity at the risk of his life above and beyond the call of duty while serving as a Platoon Leader in Company C, First Battalion, First Marines, First Marine Division (Reinforced), in action against enemy aggressor forces near Yongdungp'o, Korea, on 20 September 1950. Directed to attack hostile forces well dug in on Hill 85, First Lieutenant Commiskey, then Second Lieutenant, spearheaded the assault, charging up the steep slopes on the run. Coolly disregarding the heavy enemy machine-gun and small-arms fire, he plunged on well forward of the rest of his platoon and was the first man to reach the crest of the objective. Armed only with a pistol, he jumped into a hostile machine-gun emplacement occupied by five enemy troops and quickly disposed of four of the soldiers with his automatic pistol. Grappling with the fifth, First Lieutenant Commiskey knocked him to the ground and held him until he could obtain a weapon from another member of his platoon and kill the last of the enemy gun crew. Countinuing his bold assault, he moved to the next emplacement, killed two or more of the enemy and then led his platoon toward the rear nose of the hill to rout the remainder of the hostile troops and destroy them as they fled from their positions. His valiant leadership and courageous fighting spirit served to inspire the men of his company to heroic endeavor in seizing the objective and reflect the highest credit upon First Lieutenant Commiskey and the United States Naval Service.

Department of Defense Photo (USN) 80-G-426159

On 20 September, as the loading continues, an LST, beached until the next high tide comes in, has discharged its cargo. The small landing craft to the right are a 36-foot LCVP and two 50-foot LCMs.

book double envelopment of Hill 80 against stubborn resistance. The 1st Platoon, under Second Lieutenant William A. Craven, came in on the left. Second Lieutenant Henry A. Commiskey came in on the right with the 3d Platoon. Together they took Hill 80. The day was almost done but Wray went on against Hill 85, repeating his double envelopment. Craven set up a base of fire with his platoon on the northern slope of Hill 80. Second Lieutenant John N. Guild went forward on the left

Amphibious trucks, "ducks" to Marines, are readied at Inchon to be moved up for use in crossing the Han River. The division was well supported by the versatile trucks of the 1st Amphibian Truck Company, an element of the 1st Motor Transport Battalion.

National Archives Photo (USA) 111-SC348700

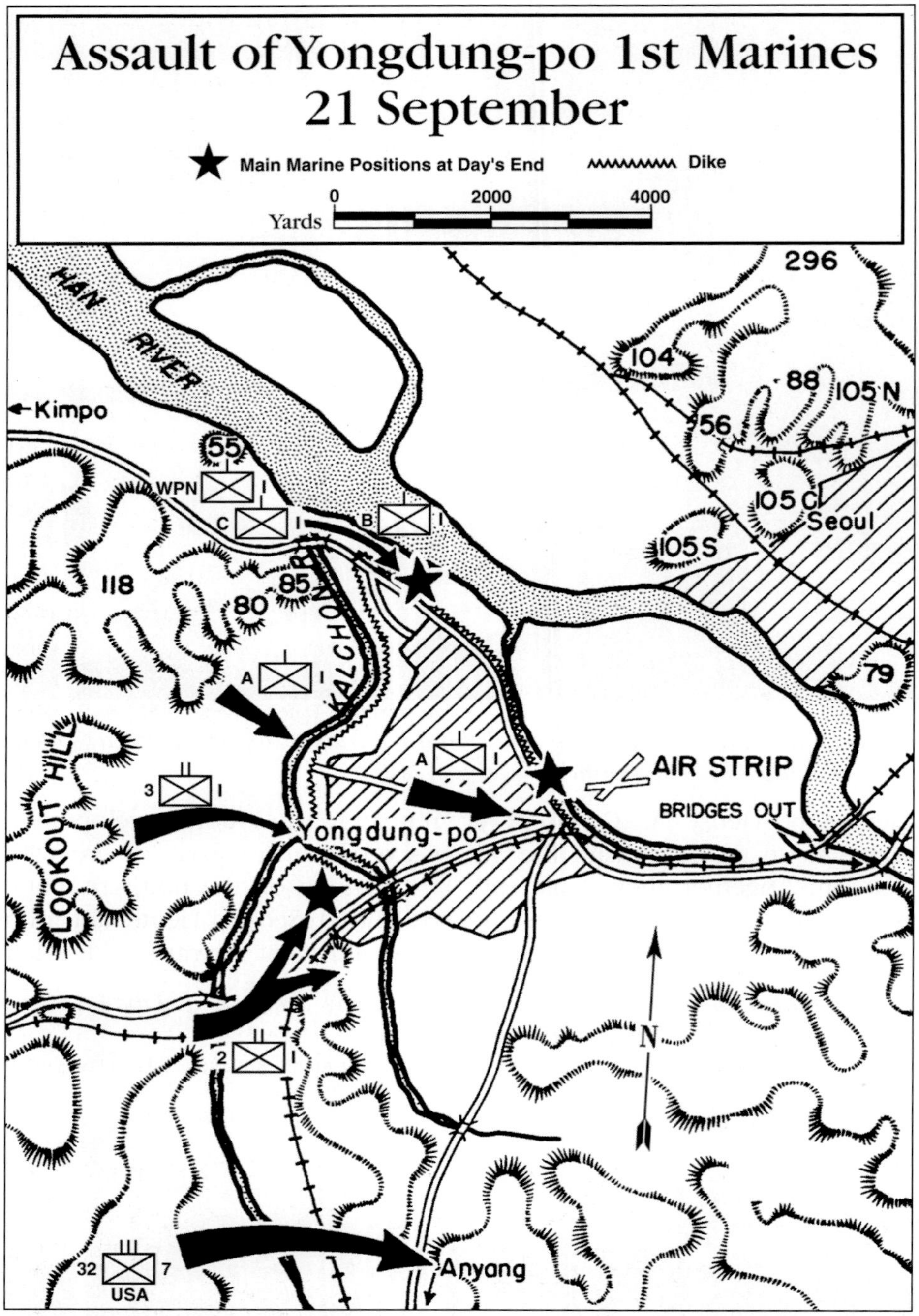

with his 2d Platoon and got almost to the top of the hill before being mortally wounded. Commiskey went out in front of his 3d Platoon in a one-man assault that earned him a Medal of Honor.

While Wray worked at capturing Hills 80 and 85, Hawkins' command group and Barrow's Marines watched as spectators from Hill 118. They saw to their left front, to their horror, a tracked "Weasel" with a wire party from the 1st Signal Battalion hit a mine on the approach to a bridge across the Kalchon near where it joined the Han. In full sight of Hill 118, two Marine wiremen were taken prisoner. A truck from Company A, 1st Engineers, with a driver and three passengers, unaware of the fate of the communicators, now came along the road. Barrow tried to catch their attention with rifle fire over their heads, but the truck continued into the ambush. Three engineers got away; one, Private First Class Clayton O. Edwards, was captured. (He would later escape from a train taking prisoners into North Korea.)

Meanwhile, Sutter's 2d Battalion, having begun the day by breaking up the T-34 tank-led North Korean attack, had moved forward uneventfully, except for harassing fire from their open right flank. They reached their day's objective, the highway bridge over the Kalchon, shortly after noon. The bridge was a long concrete span. The engineers inspected it and certified it strong enough to bear M-26 Pershing tanks for next day's attack into Yongdung-po itself. The second bridge, crossing a tributary of the Kalchon, lay 2,000 yards ahead. A high ridge, seemingly teeming with North Korean defenders, to the right of the road dominated the bridge. Sutter's neighbor on his right was Lieutenant Colonel Charles M. Mount, USA, with the 2d Battalion, 32d Infantry. The ridge commanding the second bridge was technically in Mount's zone of action. At 1300, Sutter asked Mount for permission to fire against the ridge. Mount readily agreed, but it took seven hours to get the fire mission cleared through the layers of regimental and division staffs and approved by X Corps. It was dark before Colonel Brower's 11th Marines was allowed to fire.

During the day, General Almond visited Colonel Puller at the 1st Marines' command post. Almond admired Puller's aggressive tactics and there was also a Virginia Military Institute connection. Puller, saying he could not reach Smith either by wire or radio, asked permission to burn Yongdung-po before committing his troops to its capture. Almond authorized its burning.

Almond's habit of visiting the

Marine regiments and issuing orders directly to subordinate commanders had become a serious aggravation to Smith. A division order went out that any direct order received from Almond would be immediately relayed to division headquarters for ratification.

Ready to Cross the Han

The shelling of Yongdung-po, now blazing with fires, continued throughout the night. Puller's plan of attack for the 1st Marines on 21 September was to have the 2d Battalion continue its advance astride the Inchon-Seoul highway. The 1st Battalion on the left would attack across country and the 3d Battalion, occupying Lookout Hill, would initially stay in reserve.

During the previous day, Captain Richard L. Bland had occupied Hill 55 overlooking the Han with Company B, 1st Marines. Now, shortly after dawn, he took his company across the bridge that had been the site of the ambush of the Marine communicators and engineers. In late afternoon, Hawkins sent Company C and Weapons Company across the bridge to join with Company B to form a perimeter for the night.

During the day, Ridge's 3d Battalion, in reserve on Lookout Hill, had grown impatient and had come forward prematurely, getting out in front of both the 1st and 2d Battalions. Its prospective assault companies, Companies G and I, reached and huddled behind the dike on the western bank of the Kalchon close to a water gate where a tributary entered into the main stream. This put them in good position to watch the approach march of Barrow's Company A to the Kalchon.

With Bland's Company B stalled on the opposite bank of the Kalchon, Hawkins had committed Company A to an attack from its positions on Hill 80 across a mile of rice paddies to the river. Barrow

Following a burst of sniper fire, Marines quickly take cover along a dike near the Han River. So far, the Marines had suffered only light casualties, while the North Koreans had lost heavily. National Archives Photo (USA) 111-SC349026

National Archives Photo (USMC) 127-N-A3610

Marines of Capt Robert H. Barrow's Company A, 1st Battalion, 1st Marines, move into Yongdung-po. Although the town seemed empty and dead, they carefully searched each building and side street but failed to uncover a flicker of enemy resistance.

deployed his platoons in a classic two-up one-back formation. As they came forward through the waist-high rice straw, a 3d Battalion officer, watching from his position behind the dike, was reminded of the stories he had been told of the Marines advancing through the wheat into Belleau Wood. Without a shot being fired, Company A waded the stream and marched into Yongdung-po. Barrow radioed Hawkins for instructions. Hawkins told him to keep on going.

The crossing of the Kalchon by Ridge's 3d Battalion was less easy. Going over the dike was eerily like going "over the top" of the trenches in the First World War. Second Lieutenant Spencer H. Jarnagin of Company G formed his platoon in line on the near side of the dike close to the water gate. At his whistle signal they started across. As they came out of the defilade provided by the dike, Maxim heavy machine guns on the opposite dike, perhaps 50 yards distant, opened up. Jarnagin fell back dead. His platoon recoiled, some of them wounded. Denied artillery support and with his 81mm mortars lacking ammunition, the battalion's Weapons Company commander called up his platoon of six water-cooled Browning machine guns.

During the rapid cross-country movement toward Seoul the heavy machine guns were initially attached by section to the rifle companies. They could not keep up with the light machine guns nor did the rifle company commanders fully understand their capabilities. Consequently they were pulled back to company control and employed in battery for overhead fire in the attack. Now, in this situation so much like the Western Front, they would come into their own.

With their barrels just clearing the top of the dike, the Brownings engaged the Maxims, just as they had done in 1918, and it was the Brownings that won. The 3d Battalion then crossed the Kalchon at the water gate, Westover's Company G to the left of the tributary, First Lieutenant Joseph Fisher's Company I to the right.

Early that morning Sutter's battalion crossed the second bridge without incident except for fire that continued to come in from across the boundary separating the 1st Marine Division from the 7th Infantry Division. Frustrated by the lack of artillery support, Sutter seized the bit in his teeth and shelled the offending ridge with his attached 4.2-inch mortars before sending up Companies E and F to take the high ground. While they were so engaged, Captain Welby W. Cronk took Company D along the highway and ran into another section of heavily fortified dike. Heavy fighting, supported by the ever-willing Corsairs of VMF-214, continued in Sutter's zone until late in the evening, when Sutter recalled Companies E and F to tuck them into a battalion perimeter for the night.

In Yongdung-po, Barrow could hear the furious firefight being waged by Sutter's battalion somewhere to his right. Crossing the town against scattered opposition Barrow reached yet another dike. Beyond it was a sandy flat reaching about a mile to the Han. To his left rear was Bland's Company B. Barrow dug in on the dike in a sausage-shaped perimeter. At nightfall, the Marines of Company A heard the characteristic chugging clatter of advancing tanks. Five T-34s, without infantry escort, came up the Inchon-Seoul highway and pumped steel into the western face of Company A's position. Barrow's 3.5-inch rocket gunners knocked out one and damaged two others.

Almond had been returning each evening to the *Mount McKinley*, but on the morning of 21 September he moved the headquarters of X Corps ashore and opened his command post in Inchon.

MacArthur came ashore again that afternoon enroute to Japan. A pride of generals—Almond, Shepherd, Smith, Barr, Harris, and Lowe—had gathered at Kimpo Airfield to see him off. Mutual congratulations were exchanged, and MacArthur flew to Tokyo. "He was, in my opinion, the greatest military leader of our century," mused General Shepherd, the Virginia gentleman, in 1967.

Later that day, in a ceremony at X Corps headquarters in Inchon and in accordance with established amphibious doctrine, overall command of the operation passed from Admiral Struble to General Almond.

By midnight, five infantry assaults against Barrow's position had followed the attack by the T-34s. All were beaten back, the heaviest fighting being in front of Second Lieutenant John J. Swords' 3d Platoon.

Pause in the Fighting

When the morning of 22 September came, Barrow's Marines

Gen Douglas MacArthur and MajGen Edward M. Almond examine a map at Kimpo Airfield shortly before the general's departure for Japan. MacArthur would tell the Joint Chiefs of Staff that "his forces were pounding at the gates of Seoul."

National Archives Photo (USA) 111-SC349084

National Archives Photo (USA) 111-SC349054

Navy Hospital Corpsmen Richard E. Rosegoom and Frank J. Yasso, assigned to the 1st Marine Division, give first aid to a wounded North Korean while another prisoner is marched to the rear. While always there for Marines, corpsmen also were available to treat prisoners of war and Korean civilians; the latter were second in number only to the Marines themselves.

were able to count 275 enemy dead. The four remaining T-34s, two damaged, two intact, were found abandoned nearby. The 1st and 3d Battalions renewed their attack and converged on Barrow's position against negligible resistance.

Sutter was not the only commander to complain about the fire control problems along the boundary between the two divisions. The 7th Division reported Marine Corps fire falling in its zone. Almond met with Barr and Smith and then told his aide, Lieutenant Haig, to telephone Corps headquarters and straighten out the situation.

Almond, continuing his critique of the Marines' performance, expressed his concern over Smith's "open" left flank. Smith explained to Almond his use of the 1st Korean Marine Corps Regiment, also that he had formed a Kimpo Airfield defense force, using combat support and service units. Almond appeared somewhat mollified.

The Korean Marines, leaving one battalion behind in Inchon, had followed the 5th Marines to Kimpo Airfield, and made its first attack northwest of the airfield on 19 September against light resistance. That same day the battalion from Inchon rejoined its parent regiment. Now, with one battalion to be left behind to cover the northwest flank, the KMC regiment prepared to follow the 5th Marines across the Han.

Smith's third organic infantry regiment, the 7th Marines, including the battalion that had come from the Mediterranean by way of the Suez Canal, had arrived in the harbor. Colonel Homer Litzenberg asked General Smith what element he wanted landed first. "An infantry battalion," said Smith. "And what next?" "Another infantry battalion."

Litzenberg opened his command

post two miles south of Kimpo. His 3d Battalion, under Major Maurice E. Roach, moved into an assembly area nearby. The 1st Battalion, under Lieutenant Colonel Thornton M. Hinkle, reached Hill 131 a mile north of the airfield sometime after midnight. The 1st Battalion, under Lieutenant Colonel Raymond G. Davis, stayed in the harbor to unload the ships that had brought in the regiment.

Smith made a note in his journal that Almond's concerns over open flanks had increased now that X Corps' command post was ashore. With the arrival of the 7th Marines, Smith himself could rest more easily concerning the security of his northwest flank.

Coordination between the 1st Marine Division and the 7th Division continued to be poor An extensive minefield delayed the 32d Infantry as it attacked along the Seoul-Suwon highway on 20 September, but on that same day the 32d did take T'ongdok mountain and a part of Copper Mine Hill. The rest of Copper Mine Hill was taken the next day and, as night fell, the Army regiment held a line two miles south of Anyang-ni. The big event of 22 September for the 32d Infantry was the capture of Suwon Airfield and opening it to friendly traffic.

Sutter's 2d Battalion reverted to regimental reserve the afternoon of 22 September after seven days in the assault. His grimy Marines gathered together in a bivouac area where they could wash and rest. The 22 September entry in Almond's war diary, dutifully kept by Haig, noted that Sutter's battalion had taken 116 casualties as "the result of aggressive forward movement without the required artillery preparation." That evening, Almond, after a busy day, entertained Admiral Doyle and selected staff officers at dinner at his newly established mess in Inchon.

Almond and Smith Disagree

By 23 September, the 32d Infantry had secured its objectives

Loaded in amphibious tractors and trucks, Korean Marines prepare to follow the 5th Marines across the Han River. A major portion of the 1st Korean Marine Corps Regiment had followed the 5th Marines to Kimpo Airfield, made its first attack northwest of the field, and were now poised to liberate the Korean capital.

National Archives Photo (USA) 111-SC348702

1st Marine Division Casualties

15-23 September 1950

Date	KIA[1]	DOW[2]	MIA[3]	WIA[4]	Total
15 Sept	20	1	1	174	196
16 Sept	2	1	1	22	26
17 Sept	6	0	0	70	76
18 Sept	7	3	0	92	102
19 Sept	10	1	0	61	72
20 Sept	24	1	3	119	147
21 Sept	30	3	0	198	231
22 Sept	27	3	0	135	165
23 Sept	19	7	0	117	143
Total	145	20	5	988	1,158

[1] KIA Killed in Action
[2] DOW Died in Action
[3] MIA Missing in Action
[4] WIA Wounded in Action

overlooking the Han, south and southeast of Yongdung-po. The 3d Battalion of the Army's highly regarded 187th Airborne Regiment, with Almond's "GHQ Raider Group" attached, arrived at Kimpo and temporarily came under 1st Marine Division control. Smith gave it the mission of covering his northwest flank, freeing the 7th Marines for a crossing of the Han.

Almond ordered his command post displaced forward from Inchon to Ascom City. During the day he visited Barr's command post and passed out a liberal number of Silver Stars, Bronze Stars, and Purple Hearts. Smith found Almond's practice of presenting on-the-spot awards disruptive and a cause for hurt feelings and misunderstandings. He thought Almond was inspired by Napoleon, but MacArthur was a more immediate practitioner. Smith had, it will be remembered, himself received a Silver Star from MacArthur as had Barr and Admiral Doyle. MacArthur was even more generous to Admiral Struble, giving him the Army's Distinguished Service Cross.

The 5th Marines was now firmly across the Han but was having difficulty in expanding its bridgehead. Mid-morning on the 23d, Almond met with Smith and urged him to put the 1st Marines across the river. He again complained that the Marines were not pressing the attack vigorously enough. Almond suggested that Smith cross the Han southeast of Seoul with the 1st Marines and then attack frontally into the city. Smith countered with a less-rash plan to have the 1st Marines cross at the 5th Marines' bridgehead. Almond reluctantly concurred.

From 15 through 23 September, the 1st Marine Division had suffered 165 men killed in action or died of wounds, 5 Marines still missing in action, and 988 men wounded. In turn the division had taken, by fairly accurate count, 1,873 prisoners, and claimed 6,500 enemy casualties.

During the day, 23 September, Smith visited the observation post of the 3d Battalion, 1st Marines, which had just taken Hill 108 overlooking the rail and highway bridges, their spans broken, into Seoul. A Marine major, who knew of O. P. Smith's study of the Civil War, presumed to remark that the position was similar to that of Burnside at Falmouth on the north bank of the Rappahannock across from Fredericksburg in December 1862. General Smith looked with amusement at the major and patiently explained that he would not make the same mistake as Burnside. There would be no frontal assault across the river into Seoul.

About the Author

Edwin Howard Simmons, a retired Marine brigadier general, was, as a major, the commanding officer of Weapons Company, 3d Battalion, 1st Marines, in the landing across Blue Beach Two at Inchon. His active service spanned 30 years—1942 to 1972—and included combat in World War II and Vietnam as well as Korea. A writer and historian all his adult life, he was the Director of Marine Corps History and Museums from 1972 until 1996 and is now the Director Emeritus.

He was born in Billingsport, New Jersey, the site of a battle along the Delaware River in the American Revolution, and received his commission in the Marine Corps through the Army ROTC at Lehigh University. He also has a master's degree from Ohio State University and is a graduate of the National War College. A onetime managing editor of the *Marine Corps Gazette*, he has been published widely, including more than 300 articles and essays. His most recent books are *The United States Marines: A History* (1998), *The Marines* (1998), and *Dog Company Six* (2000).

He is married, has four grown children, and lives with his wife, Frances, at their residence, "Dunmarchin," two miles up the Potomac from Mount Vernon.

Sources

The official history, *The Inchon-Seoul Operation* by Lynn Montross and Capt Nicholas A. Canzona, Volume II in the series, *U.S. Marine Operations in Korea, 1950-1953* (Washington, D.C.: Historical Branch, G-3 Division, HQMC, 1955), provided a centerline for this account.

Other official histories of great use were Roy E. Appleman, *South to the Naktong, North to the Yalu* (Washington, D.C.: Office of the Chief of Military History, Depart-ment of the Army, 1961); James A. Field, Jr., *History of United States Naval Operations: Korea* (Washington, D.C.: Government Printing Office, 1962); and James F. Schnabel, *Policy and Direction: The First Year* (Washington: Office of the Chief of Military History, Department of the Army, 1972).

Victory at High Tide (Philadelphia: Lippincott, 1968) by Col Robert D. Heinl, Jr., remains the best single-volume account of Inchon.

Among the other useful secondary sources were Alexander Haig, Jr., *Inner Circles* (New York: Warner Books, 1992); Clay Blair, *The Forgotten War: America in Korea, 1950-1953* (New York: Times Books, 1987); Gen Douglas MacArthur, *Reminiscences* (New York: McGraw Hill, 1964); Gen Omar N. Bradley and Clay Blair, *A General's Life: An Autobiography* (New York: Simon and Schuster, 1983); Donald Knox, *The Korean War: Pusan to Chosin* (San Diego: Harcourt Brace Jovanovich, 1985); Marguerite Higgins, *War in Korea: The Report of a Women Combat Correspondent* (Garden City: Doubleday & Company, 1951); Gen J. Lawton Collins, *Lightning Joe: An Autobiography* (Baton Rouge: Louisiana State University Press, 1979); and Cdr Malcolm W. Cagle and Cdr Frank A. Manson, *The Sea War in Korea* (Annapolis: U.S. Naval Institute, 1957).

Valuable insights were provided by an Inchon war game developed at the Marine Corps Historical Center (MCHC) in 1987, which examined the operation from the viewpoint of its principal commanders, using their reports, writings, and memoirs. Among the primary sources used, the most important were the unit files and records held by MCHC of the 1st Marine Division and its subordinate regiments and battalions. Also important were the biographic files held by Reference Section.

Other primary sources of great use were the oral histories, diaries, and memoirs of many of the participants. The most important of these were those of Generals Stratemeyer, Almond, Cates, Shepherd, O. P. Smith, Craig, V. H. Krulak, and Bowser, and Admirals Burke and Doyle. A fully annotated draft of the text is on file at the Marine Corps Historical Center. As is their tradition, the members of the staff at the Center were fully supportive in the production of this anniversary pamphlet. Photographs by Frank Noel are used with the permission of Associated Press/World Wide Photos.

BATTLE OF THE BARRICADES
U.S. Marines in the Recapture of Seoul

by Colonel Joseph H. Alexander, USMC (Ret)

Late on the afternoon of 24 September 1950, Captain Robert H. Barrow's Company A, 1st Battalion, 1st Marines, secured the military crest of Hill 79 in the southwest corner of Seoul, the enemy-occupied capital of the Republic of South Korea.

This momentous day for Barrow and his men began with a nerve-wracking crossing of the Han River in open-hatched DUKWs, the ubiquitous amphibious trucks of World War II. Debarkation on the north shore had been followed by an unorthodox passage of lines "on the fly" of the regiment's lead battalion and the subsequent high-tempo attack on Hill 79. Now the rifle company assumed defensive positions on the objective, the men gazing in awe at the capital city arrayed to their north and east, sprawling virtually to the horizon. Thousands of North Korean Peoples' Army (NKPA) troops lay waiting for them behind barricades or among countless courtyards and rooftops. Tens of thousands of civilians still clung to life in the battered city. The Marines were a very long way from the barren beaches of Tarawa or Peleliu. Even smoking Inchon, their amphibious objective 10 days earlier seemed far distant. Seoul would represent the largest objective the Marines ever assailed.

Earlier that day Colonel Lewis B. "Chesty" Puller, commanding the 1st Marines, issued a folded American flag to be raised on the regiment's first objective within the city limits. Barrow's battalion commander gave him the honor as the point company in the assault. The time was right. Barrow's men attached the national colors to a pole and raised them proudly on a rooftop on Hill 79. *Life* magazine photographer David Douglas Duncan, himself a Marine combat veteran, captured the moment on film. The photograph proved unremarkable—Hill 79 was no Mount Suribachi—but it reflected an indelible moment in Marine Corps history. Seven weeks earlier the 1st Marine Division was a division in name only. This afternoon a rifle company from that hastily reconstituted division had seized the first hill within occupied Seoul while all three regiments converged inexorably on the capital's rambling perimeter.

Barrow's flag-raising initiative enraged the neighboring 5th Marines, still slugging its way through the last of the bitterly defended ridges protecting the city's northwest approaches. Chang Dok Palace, the Republic of Korea's government center, lay within the 5th Marines' assigned zone. There, the 5th Marines insisted, should be the rightful place for the triumphant flag-raising. Barrow brushed aside the complaints. "Putting the flag on a bamboo pole over a peasant's house on the edge of Seoul does not constitute retaking the city," he said. Whether premature or appropriate, the flag raising on Hill 79 was an exuberant boost to morale at a good time. Chang Dok Palace lay just two miles north of Barrow's current position, but getting there in force would take the Marines three more days of extremely hard fighting.

By the night of 19 September Major General Oliver P. Smith, commanding the 1st Marine Division, had grounds for caution.

At left: *Lead elements of a Marine rifle squad pause by a captured North Korean barricade in Seoul to assign the next objective.* Photo by David Douglas Duncan

Capt Robert H. Barrow, commanding Company A, 1st Battalion, 1st Marines, pauses to raise the first American flag within the city limits of Seoul on Hill 79.

Photo by David Douglas Duncan

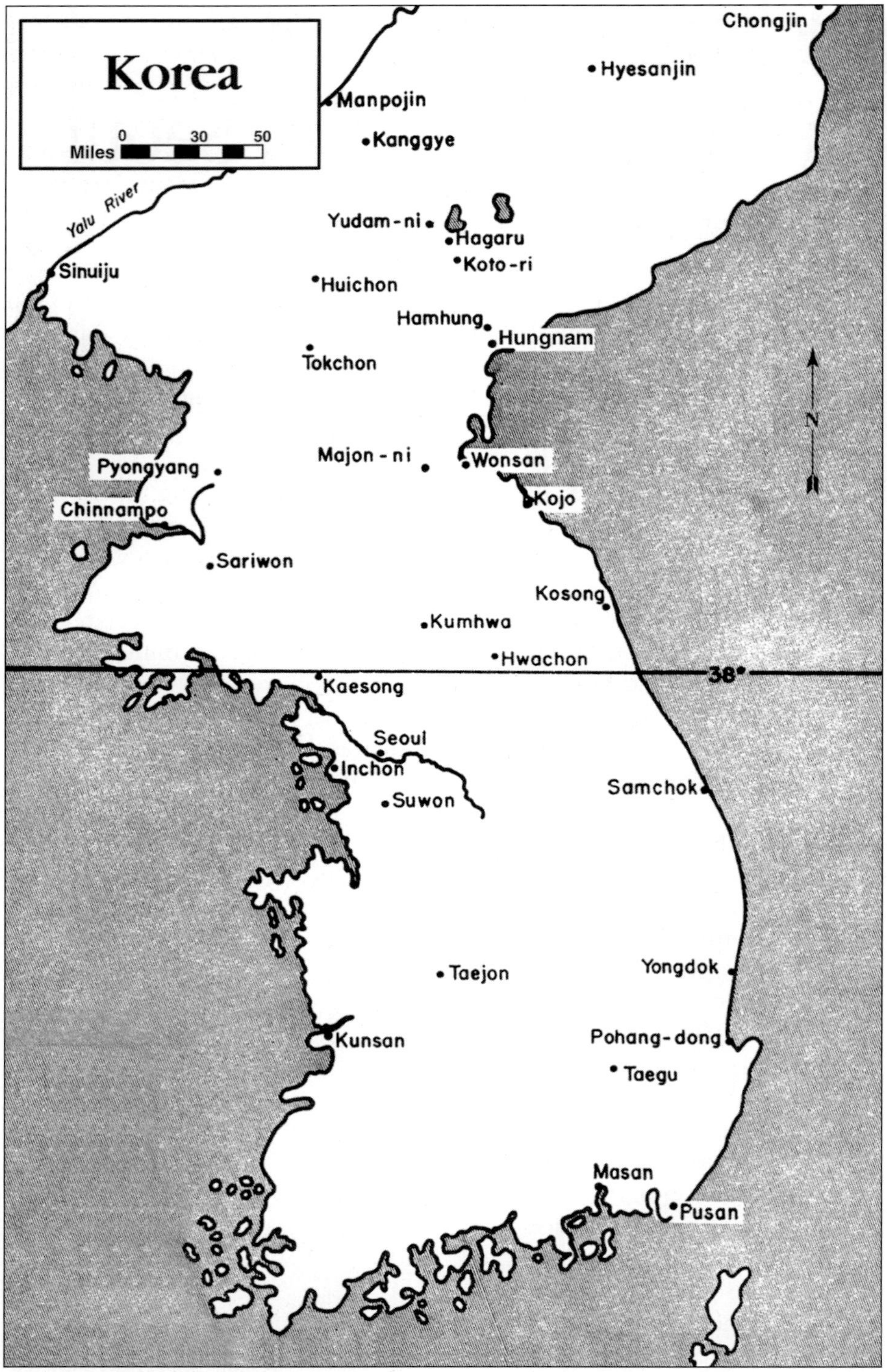

National Archives Photo (USA) 111-SC348519

MajGen Oliver P. Smith, a veteran of the Cape Gloucester, Peleliu, and Okinawa campaigns in the Pacific during World War II, commanded the 1st Marine Division throughout the Inchon-Seoul-Chosin campaigns.

Despite the impatient insistence on speed of advance by the X Corps commander, Major General Edward S. "Ned" Almond, USA, Smith knew he led a two-regiment division against an unknown enemy defending an enormous urban center.

On one hand, the pace of the allied build-up encouraged Smith. Two new Marine fighter squadrons had commenced flying into Kimpo Airfield since the 5th Marines captured it intact on the 18th, and they would launch their first Vought F4U Corsair strikes in support of the X Corps advance the morning of the 20th. The 32d Infantry Regiment of Major General David G. Barr's 7th Infantry Division had landed at Inchon and moved rapidly to cover the exposed right flank of Smith's approach to Seoul, south of Chesty Puller's 1st Marines. The 7th Marines' long, global journey to Inchon was about to end. Meanwhile, General Almond had strengthened Smith's light division by attaching two battalions of the 1st Republic of Korea (ROK) Marine Regiment, green but spirited sea soldiers.

Against these positive developments, O. P. Smith worried about his lack of a significant reserve, the absence of bridging material throughout X Corps, the morning's requirement to split his division on both sides of a tidal river, and the realization that the landing force would henceforth pass beyond the effective range of the guns of the fleet. He could also sense that North Korean resistance was stiffening and the quality of the opposition was improving. All signs pointed to a major clash in the week ahead.

Intelligence analysts on both division and corps staffs had diffi-

Principal Commanders, 1st Marine Division, Seoul

1st Marine Division

Commanding General: Major General Oliver P. Smith
Assistant Division Commander: Brigadier General Edward A. Craig
G-3: Colonel Alpha L. Bowser, Jr.

1st Marines

Commanding Officer: Colonel Lewis B. Puller
1st Battalion: Lieutenant Colonel Jack Hawkins
2d Battalion: Lieutenant Colonel Alan Sutter
3d Battalion: Lieutenant Colonel Thomas L. Ridge

5th Marines

Commanding Officer: Lieutenant Colonel Raymond L. Murray
1st Battalion: Lieutenant Colonel George R. Newton
2d Battalion: Lieutenant Colonel Harold S. Roise
3d Battalion: Lieutenant Colonel Robert D. Taplett

7th Marines

Commanding Officer: Colonel Homer L. Litzenberg, Jr.
1st Battalion: Lieutenant Colonel Raymond G. Davis
2d Battalion: Lieutenant Colonel Thornton M. Hinkle (Wounded in Action-Evacuated, September 28)
Major Webb D. Sawyer (from September 28)
3d Battalion: Major Maurice E. Roach

11th Marines

Commanding Officer: Colonel James H. Brower
1st Battalion: Lieutenant Colonel Ransom M. Wood
2d Battalion: Lieutenant Colonel Merritt Adelman
3d Battalion: Major Francis F. Parry
4th Battalion: Major William McReynolds

Other Division Units

Commanding Officer, 1st Shore Party Battalion: Lieutenant Colonel Henry P. Crowe
Commanding Officer, 1st Engineer Battalion: Lieutenant Colonel John H. Partridge
Commanding Officer, 1st Tank Battalion: Lieutenant Colonel Harry T. Milne
Commanding Officer, 1st Amphibian Tractor Battalion: Lieutenant Colonel Erwin F. Wann, Jr.
Commanding Officer, VMO-6: Major Vincent J. Gottschalk
Commanding Officer, 1st Service Battalion: Lieutenant Colonel Charles L. Banks
Commanding Officer, 1st Ordnance Battalion: Major Lloyd O. Williams
Commanding Officer, 1st Motor Transport Battalion: Lieutenant Colonel Olin L. Beall
Commanding Officer, 1st Medical Battalion: Commander H. B. Johnson, Jr., USN
Commanding Officer, 1st Signal Battalion: Major Robert L. Schreier
Commanding Officer, Reconnaissance Company: Captain Kenneth J. Houghton

culty defining an enemy order of battle after the Inchon landing because of the chaos the landing created in the headquarters of the NKPA in Pyongyang, the North Korean capital. Ignoring dozens of telltale indicators, the NKPA seemed astonished that the Commander in Chief, Far East, General of the Army Douglas MacArthur, could have landed such a large force amid Inchon's narrow channels and formidable mudflats. The Marines' quick seizure of the port, Ascom City, and Kimpo Airfield further disoriented the North Koreans.

By the night of the 19th-20th, however, the North Korean high command finally had major troop units on the move to defend the South Korean capital. They turned around the untested *18th NKPA Division*, bound from Seoul to the Pusan Perimeter, and recalled a veteran regiment of the *9th NKPA Division* from the southwest corner of the Naktong River. Most of these troops would defend the industrial suburb of Yongdungpo, directly south of the Han from central Seoul, against the 1st Marines.

On 20 September, while Lieutenant Colonel Raymond L. Murray led his 5th Marines across the Han River, two significant enemy units reached Seoul from assembly areas in North Korea to man the northwest defenses against this new American threat above the Han. From Sariwon came Colonel Pak Han Lin at the head of his *78th Independent Infantry Regiment*, some 1,500-2,000 untested troops in three infantry battalions. From nearby Chorwon came Colonel Wol Ki Chan's *25th NKPA Brigade*, more than 4,000 strong. Colonel Wol had received "postgraduate" tactical training in the Soviet Union and had trained his green troops well. His newly formed brigade contained an unusual concentration of crew-served weapons, including four heavy weapons battalions providing a proliferation of antitank and antiaircraft guns, plus heavy machine guns. Wol led the two units west of town to prepare last-ditch defenses along the same jumbled ridges where the Japanese had formerly conducted infantry-training exercises. General Smith's intuition had been correct. His North Korean enemy would shortly change from delaying tactics to hard-nosed, stand-and-deliver defense to the death.

North Korean Order Of Battle: Seoul/Wonsan Campaign

Defending the Northwest Approaches (Hill 296 Complex and beyond):

25th Brigade: Colonel Wol Ki Chan
78th Independent Infantry Regiment: Colonel Pak Han Lin
Seoul City Regiment

Defending Yongdungpo:

Elements of *3d Regiment, 9th Division*
Elements of *18th and 87th Divisions*

Defending Seoul:

Surviving components of the above forces
17th Rifle Division
43d Tank Regiment
19th Antiaircraft Regiment
513th Artillery Regiment
10th Railroad Regiment

Defending Uijongbu:

31st Regiment, 31st Division
75th Independent Regiment

Opposing 1st Battalion, 1st Marines, at Kojo:

10th Regiment, 5th Division: Colonel Cho Il Kwon

Opposing 3d Battalion, 1st Marines, at Majon-ni:

Elements of *15th Division:* Major General Pak Sun Chol

Two Rough Roads To Seoul

Few things could faze Lieutenant Colonel Murray, the 5th Marines' commander, after his month-long experience as the Eighth Army's "Fire Brigade" in the Pusan Perimeter, but preparing his veteran regiment for an opposed crossing of the Han River on 20 September proved a daunting task. To begin with, Murray found his command post crowded with high-ranking observers and correspondents. Each wondered how Murray would execute a crossing of such a broad river without heavy bridging material; all offered free advice. Murray abided these kibitzers for awhile, then cast them out.

LtCol Raymond L. Murray, a tall Texan who had earned a Silver Star on Guadalcanal, a second Silver Star on Tawara, and a Navy Cross on Saipan, commanded the 5th Marines.
Department of Defense Photo (USMC) A5850

A second situation proved more troublesome. While Murray felt confident the 1st Amphibian Tractor Battalion could shuttle his riflemen across in their tracked landing vehicles (LVTs then, AAVs now), and while he was reasonably sure Lieutenant Colonel John H. Partridge, the division engineer, could ferry his attached tanks across by using 50-foot pontoon sections, he still knew nothing of the river—its current, shoreline gradients, exit points. Nor did Murray know anything of the enemy's strength and capabilities in the vicinity of the abandoned ferry site at Haengju. Mile-long Hill 125 on the north bank dominated the crossing. Six years earlier Murray had led his 2d Battalion, 6th Marines, ashore at Saipan under direct fire from Japanese guns occupying the coastal hills, and he had no intention of repeating that experience here.

Murray asked General Smith to assign Captain Kenneth R. Houghton's division Reconnaissance Company to the crossing operation. Murray wanted an

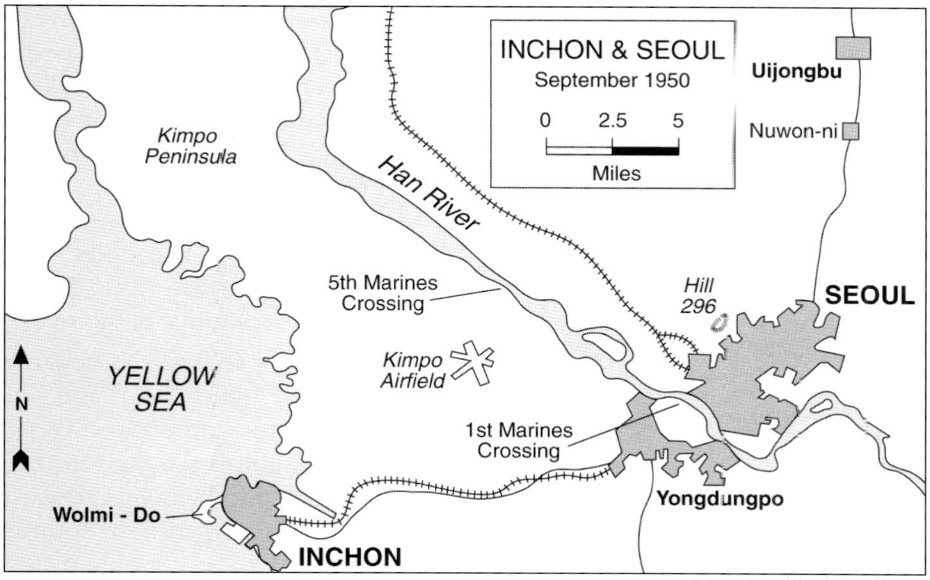

Mary Craddock Hoffman

advance party of reconnaissance Marines to swim the Han after dark on 19 September, stealthily determine any enemy presence, and then signal the remainder of the company to cross in LVTs. Murray then expected the company to man a defensive perimeter to cover the predawn crossing of Lieutenant Colonel Robert D. Taplett's 3d Battalion, 5th Marines.

Taplett considered the plan too ambitious. The Reconnaissance Company had the heart, he believed, but not the numbers (127 strong) to cover the sprawling high ground along the river. No one knew anything in advance about the possibility of enemy presence in strength along the far bank. Taplett quietly ordered his staff to draw up contingency plans for the crossing.

The North Koreans had not ignored the former ferry site. Aware that the Marines would likely cross the Han soon, the NKPA deployed an infantry battalion in the underbrush along Hill 125. Their camouflage discipline proved excellent. The Marines did not detect their presence throughout the afternoon and evening of the 19th.

After dark, Captain Houghton led 14 swimmers across the 400-yard-wide river. An ill-timed artillery mission set fire to a house in Haengju village, exposing the

Marine Corps amphibian tractors and DUKWs ferry troops across the Han River after the assault waves.
Photo by Frank Noel, Associated Press

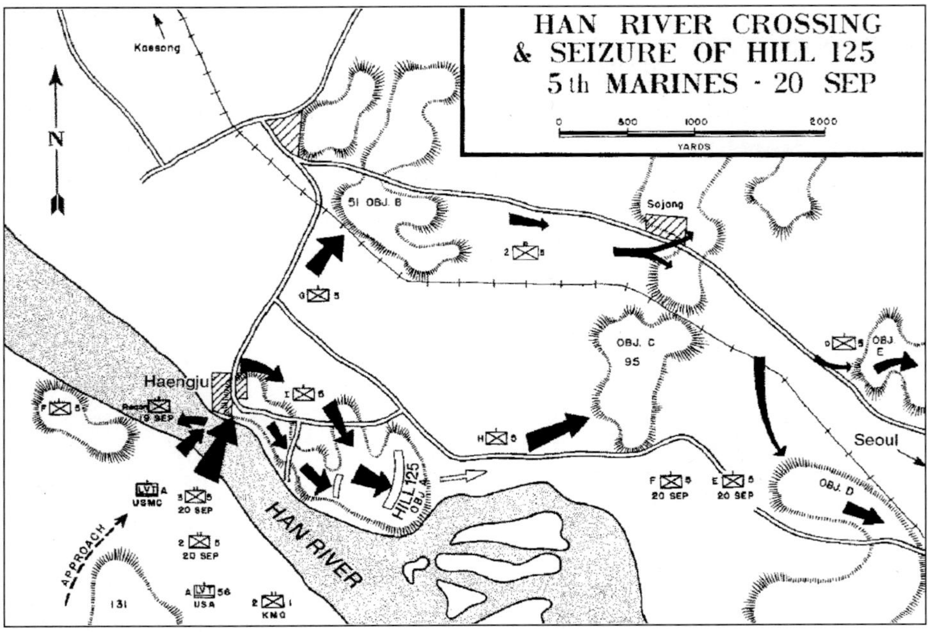

men in their final approach to the north bank. Technical Sergeant Ernest L. Defazio complained the blaze "lit up the place like a Christmas tree," but nothing stirred. Houghton dispatched four men to check for signs of the enemy on Hill 125, then sent an exultant but premature message to Murray: "The Marines have landed and the situation is well in hand." Houghton also radioed his executive officer to launch the balance of the company in its nine LVTs.

So far, so good. But few sounds attract more attention on a quiet night than the sudden revving up of nine pairs of Cadillac V-8 Amtrac engines. The noise seemed enough to wake the dead, and abruptly the NKPA battalion on Hill 125 opened a vicious fire against the approaching LVTs and Houghton's small group, now dangerously backlit by the burning building.

Second Lieutenant Philip D. Shutler commanded the second platoon of the Reconnaissance Company, his men divided between two LVTs that nosed into the river in column. Young as he was, Shutler had already been in tight spots. He had spent the month of August making night raids from USS *Horace A. Bass* (APD 124) in the Sea of Japan against the North Korean coastline, his Marines teamed with Underwater Demolition Team 1. Crossing the Han was a dissimilar experience, he later recounted. "Amphibian tractors were hardly stealthy vehicles," Shutler recalled. "We received enemy fire as soon as the vehicles entered the water. You could hear machine gun rounds plinking against the armored cab. Mortar rounds, possibly from our own 'four-deuce' tubes, were exploding in the river."

In the chaos some LVTs became stuck in the mud near the far shore, others veered away. Captain Houghton sprang into the river to rally the vehicles toward the landing site. Mortar rounds landed in the water near him; the concussion from one near miss knocked him out.

Lieutenant Shutler could see none of this from the crowded troop compartment of his lurching LVT. He scrambled topside, discovered to his horror that the vehicle had turned upstream, broadside to the NKPA gunners on Hill 125. He whacked the driver,

An LVT-3C of the 1st Amphibian Tractor Battalion takes off from the south bank of the Han with a load of American and Korean Marines, while Marine engineers prepare a pontoon bridge to carry equipment.

Photo by Frank Noel, Associated Press

jumped into the waist-deep water, and attempted to guide the vehicle directly ashore. He saw no sign of the advance swimmers.

At this point someone passed the word to abort the mission and return to the south bank. Five LVTs returned, leaving four stuck in the mud along the far shore. One of these contained Captain Houghton's unconscious body. Other Marines were missing. Shutler found one of his troops had died of wounds in the confused melee. The crossing had failed.

When Technical Sergeant Ernie DeFazio discovered his captain missing he promptly led a swimmer team back across the river. They rescued Houghton and his radio operator, retrieved two of the stuck vehicles and restored more than a bit of the company's honor.

But the night was nearly spent, the enemy occupied the crossing site in considerable strength, and every VIP in the theater—including General Douglas MacArthur—had announced their intentions of observing the morning crossing. As assistant division commander, Brigadier General Edward A. Craig frankly observed: "The eyes of the world were upon us. It would have looked bad for the Marines, of all people, to reach a river and not be able to cross."

The 5th Marines calmly decided to approach the crossing as an amphibious assault mission—tightly coordinated preliminary fires on the objective, an intermediate and final objective assigned, and troops organized into boat teams configured to each LVT. Taplett's 3d Battalion, 5th Marines, would lead the landing in assault waves, followed by Lieutenant Colonel Harold S. Roise's 2d Battalion, 5th Marines, to expand the beachhead; the entire regiment with its attached tank company to cross before dark. Marine Corsairs would arrive soon after sunrise to pound Hill 125 and scorch the Seoul-Kaesong highway to discourage any NKPA reinforcements.

Only a veteran force like the 5th Marines could have made such last-minute adaptations and passed the word to all hands in the remaining minutes before dawn. Taplett's original skepticism about the Reconnaissance Company's ability to hold an opposed bridgehead had served 3d Battalion, 5th Marines well; the battalion had already prepared worst-case alternative plans. By the time General Almond, Vice Admiral Arthur D. Struble, USN (Commander, Seventh Fleet), and Lieutenant General Lemuel C. Shepherd, Jr., USMC (Commanding General, Fleet Marine Force, Pacific) arrived they found Lieutenant Colonel Murray as unflappable as ever and the crossing well underway. Lieutenant Colonel Ransom M. Wood's 1st Battalion, 11th Marines, pounded the far bank with 105mm howitzers; Murray's own 81mm and 4.2-inch mortars joined the chorus. Taplett's first wave of six LVTs chugged resolutely on line towards the far bank.

At this point the NKPA battalion on Hill 125 opened a disciplined fire on the LVTs, scoring more than 200 hits on the vehicles as they trundled ashore. Fortunately their one antitank gun proved less accurate than their small arms fire. Taplett pressed on. His LVTs discharged Captain Robert A. McMullen's Company I, then pulled away for the return transit. McMullen quickly deployed his platoons up the open slopes of Hill 125 in a double envelopment. The fighting became point-blank and deadly.

With most NKPA gunners now taking aim at McMullen's Marines, the remaining companies of 3d Battalion, 5th Marines, crossed the river with relative ease. Corporal Larry V. Brom, a Company H squad leader, worried more about the claustrophobia his men experienced in their LVT's cramped troop compartment than "the occasional splat of bullets against the armor plate." Company H's LVTs lurched out of the river and continued rolling north, crossing the railroad and highway to secure distant Hill 51. Corporal Brom led his men in a mad dash up the rise as soon as the rear ramp dropped, vastly relieved to discover the crest undefended.

By contrast, Company I had its hands full taking Hill 125. The lower approaches contained scant cover. Well-sited NKPA gunners scythed down Captain McMullen's exposed 60mm mortar section and two sections of light machine guns.

The situation improved dramatically with the appearance overhead of four Corsairs from Lieutenant Colonel Walter E. Lischeid's Marine Fighter Squadron 214 (VMF-214). The Black Sheep pilots launched at 0551 from the escort carrier USS *Sicily* (CVE 118) in the Yellow Sea, southwest of Inchon, arriving over the river just in time to even the odds against Company I's arduous assault with a series of ear-splitting rocket and napalm attacks against the North Koreans defending the high ground. McMullen spurred his men forward, upward amid the bedlam. Their difficult double envelopment converged on the crest, culminating in a vicious flurry of hand-to-hand combat. An abrupt silence followed, broken only by the Marines gasping for breath.

Taking Hill 125 cost Company I 43 casualties; it inflicted at least 200 upon the enemy. It had been a beautifully executed tactical assault, highlighted by the high-

Department of Defense Photo (USMC) A409336

Advancing Marines examine the smoking ruin of a North Korean T-34 tank recently destroyed in an ambush.

speed, low-level strikes of the Corsairs. General Almond, observing this conflict from barely 500 yards away, admitted it was "one of the finest small-unit actions I've ever witnessed."

The forcible taking of Hill 125 meant the remainder of the 5th Marines could cross the river unimpeded. By the time General MacArthur arrived the crossing seemed routine. "You've done a perfect job," he told Lieutenant Colonel Murray, unaware of the all-night flail that preceded the perfection. Murray by then had his eye on the main objective, and he pointed upstream to the convoluted ridges that protected the approaches to Seoul from the northwest, the regimental route of advance. "They'll all evaporate very shortly," MacArthur assured Murray.

At a glance from long distance it seemed that the Supreme Allied Commander might have been right. Only eight miles separated Hill 125 at the Haengju crossing site from downtown Seoul. Murray's advance elements covered half that distance on the afternoon of the 20th, raising false hopes. Then NKPA resistance stiffened abruptly. It would take the 5th Marines a full week of desperate fighting to advance the final four miles into Seoul.

The 20th of September also began very early for Chesty Puller's 1st Marines on their final approach to Yongdungpo. The *87th NKPA Regiment* launched two predawn spoiling attacks against both flanks. The southern attack, led by five T-34 tanks, posed the greatest threat. The veteran NKPA troops endeavored to repeat their high-speed, straight-down-the-highway armored tactics that had proven wildly successful in the initial invasion, but their tanks had now lost their invulnerability. The armored column barreled blindly into a lethal L-shaped ambush set by Lieutenant Colonel Alan Sutter's 2d Battalion, 1st Marines. Short-range fire from Marine 3.5-inch bazookas knocked out the first two enemy tanks; a storm of direct and indirect fire cut down the supporting infantry, killing 300 men. The surviving North Koreans withdrew to their prepared defenses within Yongdungpo.

Puller pressed the advance, his 2d Battalion still astride the Inchon-Seoul highway, the 1st Battalion attacking through the hilly countryside below the Han. Sutter's lop-sided success in thwarting the NKPA tank attack pleased Puller, but the initial view of sprawling Yongdungpo from his observation post brought forth Puller's trademark scowl. The prospect of forcing a crossing of the high-banked Kalchon Canal, then fighting door-to-door through this large industrial suburb did not appeal to the veteran jungle fighter. When General Almond appeared from observing Murray's river crossing, Puller asked him for authorization to employ unrestricted firepower in taking the city. The corps commander agreed. Puller unleashed two battalions of supporting artillery (Lieutenant Colonel Merritt Adelman's 2d Battalion, 11th Marines, in direct support, and Major William McReynolds' 4th Battalion, 11th Marines, in general support) plus air strikes by Marine Corsairs. The *Sicily*-based Black Sheep followed their early-morning assistance to the 5th Marines with two dozen sorties against Yongdungpo, dropping 500-pound bombs and strafing with 20mm cannon and rockets. The city began to burn.

The 1st Marines commenced its main assault on Yongdungpo at 0630 the next morning. Neither Sutter's 2d Battalion or Lieutenant Colonel Jack Hawkins' 1st Battalion could sustain much headway. Crossing the Kalchon was like crossing a medieval castle moat; clambering over the dikes was akin to "going over the top" in the trenches of World War I. Sutter's outfit in particular took heavy casualties. The division's Special Action Report recorded the loss of 17 officers and 200 men by the 2d Battalion along the canal-like river by 21 September.

Puller committed elements of Lieutenant Colonel Thomas L. Ridge's 3d Battalion in the center, but a half dozen NKPA Maxim heavy machine guns took a grim toll of every attempt to cross the water gate sector of the Kalchon.

Ridge ordered Major Edwin H. Simmons, his Weapons Company commander, to suppress the fire. With his 81mm mortars temporarily out of ammunition and no artillery support immediately available, Simmons chose his Browning M1917A1 watercooled .30-caliber heavy machine guns for the mission. Proven veterans of the World War, the heavy Brownings were unsurpassed in providing rock-steady, sustained fire at a rate of 450-600 rounds per minute. Simmons massed these weapons with their barrels "just clearing the top of the dike." A fierce duel ensued—"heavies against heavies"—at an interval no greater than half a football field. The exchange was deafening, but Simmons' sturdy Brownings prevailed, allowing 3d Battalion, 1st Marines, to cross the Kalchon intact.

The Kalchon proved a barrier to the entire regiment on 21 September—with one memorable exception. While the battle raged on both sides—and shortly before Major Simmons' machine gun duel—Captain Robert H. Barrow, the future 27th Commandant, led his Company A, 1st Marines, through a rice field towards an uncommonly quiet sector of the Yongdungpo defenses. The North Koreans may have vacated this sector in order to more effectively contest the adjacent water gate fronting the 3d Battalion, an obvious crossing site. Barrow, however, expected to be hit at any moment. Simmons watched approvingly as Company A, 1st Marines, advanced past his immediate left flank, each platoon on line. "They were beautifully deployed," said Simmons. "As they came through the dry rice paddy I thought of the Marines coming through the wheat fields at Belleau Wood in 1918."

Private First Class Morgan Brainard of Barrow's company, though apprehensive about the spooky quiet, experienced similar thoughts as he crossed through the waist-high rice stalks. As he later described the advance:

> Somewhere off to our left, beyond the road and out of sight, beyond a line of trees we could hear the rattle of rifle and machine gun fire where Baker Company was going in . . . . To our immediate front, however, there was nothing but silence, as we continued to move forward

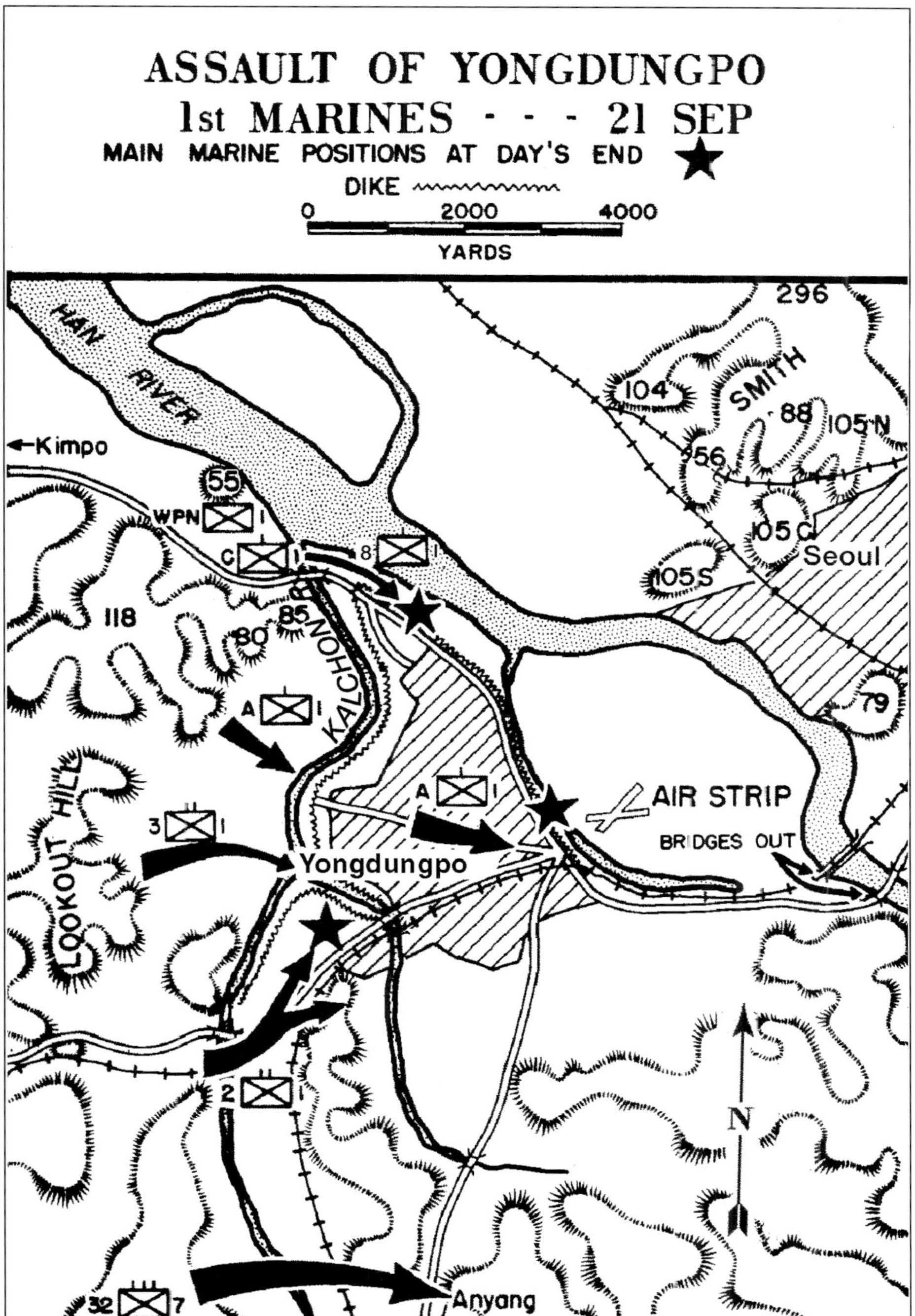

Department of Defense Photo (USMC) A3200

A column of M-26 Pershings and a bulldozer-configured M-4 Sherman advance towards Yongdungpo. The threat comes from the right flank, and firing has already been vigorous, judging from the spent 90mm shell casings alongside the road.

> through the field in perfect order. It was a classic-type infantry advance . . . but my mind kept racing back toward the stories I had read as a boy of the Marines attacking through the wheat fields of Belleau Wood . . . and I expected our peaceful scene would be shattered in a similar manner at any moment.

Captain Barrow acknowledged his serendipity. "We just happened to experience one of those rare fortunes of war . . . a momentary opportunity."

"We passed over the top of the dike quickly, slithered down the other side," recalled Brainard, "then inexplicably and stupidly stopped facing a stream [the Kalchon]. I mean the whole line stopped." The company gunnery sergeant quickly ended their hesitation: "Get in that goddamned water!"

Company A found itself entering the main street of Yongdungpo totally unopposed. "It was eerie," said Barrow. "We simply slithered into town undetected."

The 87th NKPA Regiment, desperately attempting to patch together a defense in depth, had accidentally left this critical approach unguarded, and Barrow took full advantage of the opening. His 200-man company flowed rapidly into the heart of the city, sweeping up surprised bands of

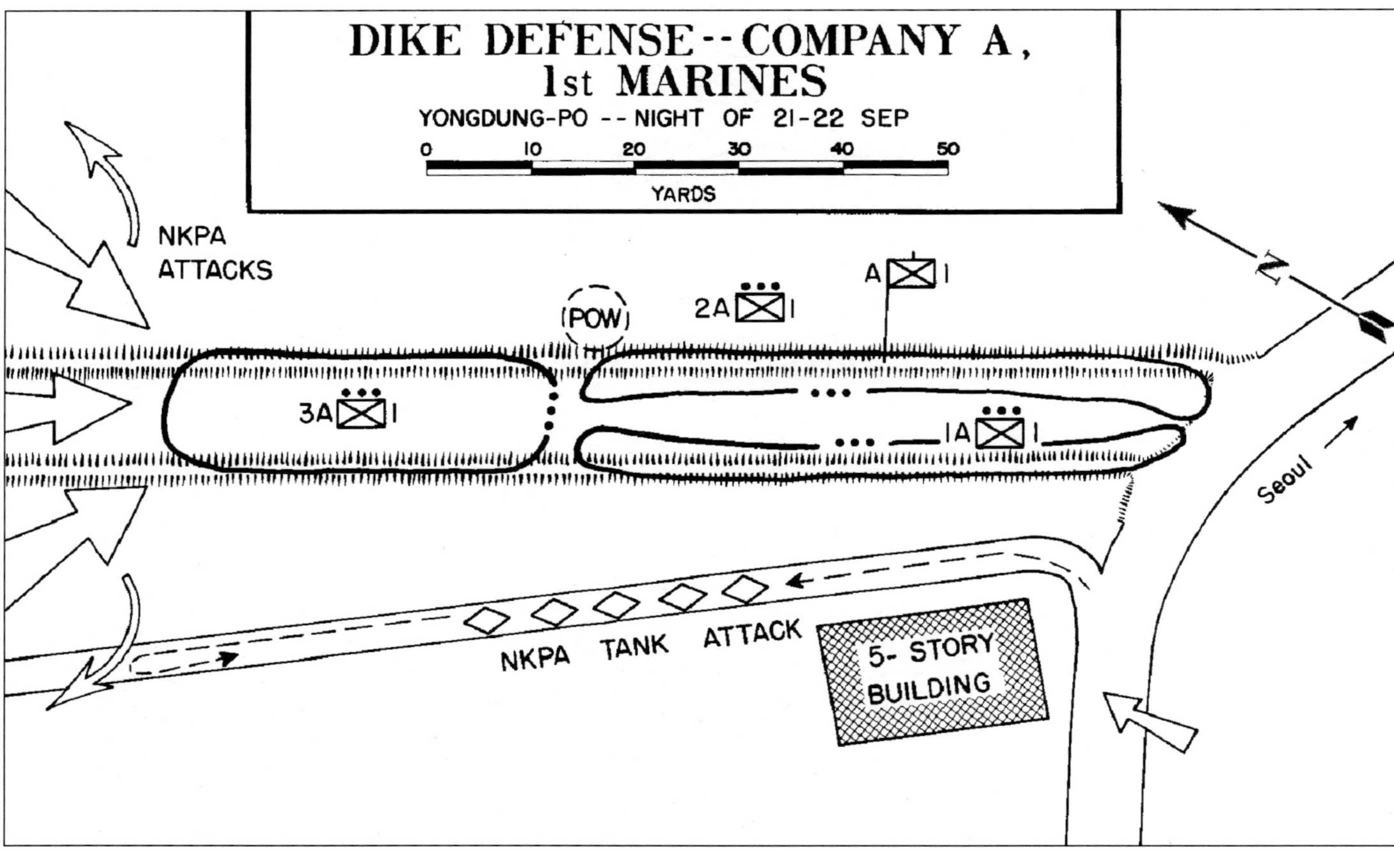

the enemy in the process.

Before dark they had cut the city in two. Barrow selected a sausage-shaped dike, 30 feet high and 150 yards long, as the place to make a stand for the night. "We immediately recognized that we had a valuable piece of real estate," he said. From the dike his Marines could interdict the intersection of the highways from Inchon and Kimpo.

Through this intersection at one point marched a large formation of unsuspecting NKPA infantry, singing political songs as they hurried to reinforce Yongdungpo's northwestern defenses. Barrow's interlocking machine guns and 60mm mortars cut down many and scattered the rest.

As darkness fell, Lieutenant Colonel Hawkins knew Barrow had executed a major penetration, but he could not reinforce this unexpected success. Barrow and Company A would be on their own—which was fine with Barrow. "We felt strong," he said. "We were not 'The Lost Company.'" "What followed," observed historian Jon T. Hoffman, "would become one of the great small-unit epics in the history of the Corps, to rank with Hunt's Point and Pope's Ridge [at Peleliu]."

Navy surgeon and corpsmen attached to the 1st Marine Division treat a badly injured two-year-old boy on the outskirts of Seoul.

Photo by Frank Noel, Associated Press

The NKPA attacked Company A shortly after dark with five Soviet-built T-34 tanks. The rattle and roar of their tracks as they approached almost unnerved Private First Class Morgan Brainard. "The squeaking and engine humming was drawing much closer, and as I crouched in my hole, I felt the ice-like shiver of pure fear." The tanks reached the intersection, then proceeded in column along a road parallel and extremely close to the Marines' positions dug into the side of the dike. The lead vehicle appeared enormous to Brainard: "In the moonlight I could see its turret with the long gun on it slowly circling back and forth, like some prehistoric, steel-backed monster sniffing for prey. I pressed tightly against the side of my hole, and waited for the flash and fire of its gun."

The tanks made five deliberate passes along that parallel track, firing their 85mm guns directly into the crowded dike from an ungodly short range of 25 yards. This was a terrifying experience for the Marines on the receiving end, but the dike's soft sand absorbed the base-detonated, armor-piercing shells, and there were few casualties. Meanwhile, Barrow's 3.5-inch rocket launcher teams stung the tanks repeatedly. "One of the most courageous acts that I ever witnessed was those brave young Marines with the 3.5s," he said. The first bazooka round Corporal Francis Devine ever fired in anger blasted a T-34 turret off its ring. Other gunners knocked out a second tank and damaged two more. The attached heavy machine gun section kept the vehicles buttoned up and peppered their vision blocks and periscopes. The surviving vehicles withdrew in disarray.

Department of Defense Photo (USA) SC348715

Elements of the 5th Marines advance through a burning village after crossing the Han River. The days of high mobility ended as the Marines reached the enemy main line of resistance in the high ridges on the outskirts of Seoul.

The enemy tanks may have been more successful had infantry accompanied them, but the NKPA riflemen did not appear until 0100. Four separate ground assaults followed, each beaten back by disciplined fire. "I expected to have a lot of promiscuous firing," said Barrow, but "my people didn't lose their fire discipline and go bananas and shoot randomly."

The enemy assembly area was so close to the Marines' defensive position that they could hear the voice of the local commander, unmistakably haranguing his troops into launching another attack. Corporal Billy D. Webb, an Oklahoma reservist "with fire in his eye," decided to even the odds. Slipping out of his foxhole—"for God's sake don't shoot me when I come back!"—Webb dashed through the adjoining maze of buildings, spotted an extremely animated officer trying to rally his troops for yet another attack, took careful aim, and shot him dead. Webb escaped in the resultant confusion, and the night assaults ceased before the Marines ran out of ammunition.

At dawn, Barrow counted 210 dead North Koreans around his beleaguered dike. "Yongdungpo did for A Company," said Barrow, "what no other thing could have done in terms of unifying it and giving it its own spirit, a spirit that said 'We can do anything.'"

If Barrow's company had "slithered" into Yongdungpo on the 21st, it was now the turn of the *87th NKPA Regiment*, having failed to oust the Marines throughout the night, to slither out of town the next morning. Barrow had skinned the cat, helping Puller capture a very difficult intermediate objective in two days of fighting. The road to Seoul for the 1st Marines now lay open, once the 5th Marines could advance eastward enough to cover their tactical crossing of the Han.

Back at Inchon, now well to the west of Puller's regiment at Yongdungpo, the offloading of fresh troops and combat cargo continued around the clock. By D+6, 21 September, 50,000 troops had landed, including Colonel Homer L. Litzenberg, Jr.'s 7th Marines, supported by Lieutenant Colonel Francis F. Parry's 3d Battalion, 11th Marines, a 105mm howitzer outfit.

The 7th Marines initially assumed security duties in the Inchon vicinity. General O. P. Smith critically needed them for the recapture of Seoul, but the newly formed outfit first required a day or two to shake itself down from the long deployment by sea. This did not take long. Lieutenant Colonel Raymond G. Davis' 1st Battalion, 7th Marines, for example, had conducted field firing from the fantail of their attack transport each day enroute. "We fired machine guns, rifles, mortars, and bullets, rocket launchers, and threw hand grenades at every piece of trash, orange crates, or whatever the ship's crew would toss overboard for us," said Davis. Within 48 hours the regiment moved out tactically, crossed the Han River, and began its own path towards Seoul's northern suburbs, somewhat northwest of the route of the 5th Marines. On the third day Parry's gunners fired their first rounds down range.

By the fortunes of war, the 5th Marines would pay the stiffest price of admission to enter Seoul. General MacArthur's beguiling assurance to Lieutenant Colonel Murray that the hills guarding the northwestern approaches to the capital "would all evaporate" proved famously false. The regiment would suffer a casualty rate more reflective of its recent history at Peleliu and Okinawa than the Korean peninsula.

Part of the difficulty came from the convoluted terrain, a sprawling series of hill masses, ridges, and draws extending from the Kaesong-Seoul highway in the north to the Han River in the south. "As an exercise in map reading," observed Marine historian Colonel Robert D. Heinl, Jr., "this ground is confusing and deceptive; for the tactician, it is a nightmare." Massive Hill 296 dominated the landscape; indeed, many of the other numbered peaks and knobs were in reality only protuberances of the hill's bony fingers extending to the Han and eastward into downtown Seoul itself. Confusingly, there were three Hill 105s in this complex (just as there had been three Hill 362s at Iwo Jima). Regimental planners nicknamed them for their linear sequence—Hills 105 North, Center, and South. All three would prove prickly objectives to seize and hold.

The North Koreans found the jumbled terrain around the Hill 296 complex to be ideal defensive ground. The fact that the Japanese had long used the same ridges for tactical training meant the preexisting availability of firing positions, command posts, and observation sites. Colonel Wol Ki Chan reached this preferred ground with his *25th NKPA Brigade* and Colonel Pak Han Lin's *78th Independent Infantry Regiment* just in time. Had the North Koreans been held up one more day passing through Seoul, the Marines might have seized Hill 296 and all of its deadly fingers with hardly a fight.

Colonels Wol and Pak deployed at least 6,000 troops into the hill complex. While yet to be tested in battle, the combined force was both well-led and well-trained. Wol's brigade also contained an abundance of heavy weapons units. Their crews spent the 20th and 21st digging in their weapons and registering their fire along the Marines' likely avenues of approach. Additional troops in odd-lot specialty organizations reinforced Wol during the battle for the hills, increasing his total force to nearly 10,000 men. The 5th Marines, even reinforced by their attachments and the ROK Marine battalion, could not match those numbers.

The 5th Marines had fought against highly experienced NKPA regiments in the Pusan Perimeter, units whose officers and non-commissioned officers had years of combat experience in China. The North Koreans they now faced lacked that background but made up for it with tenacity and firepower, including well-served high-velocity 76mm guns and 120mm heavy mortars. "Their mortar fire was very accurate," said veteran company commander Captain

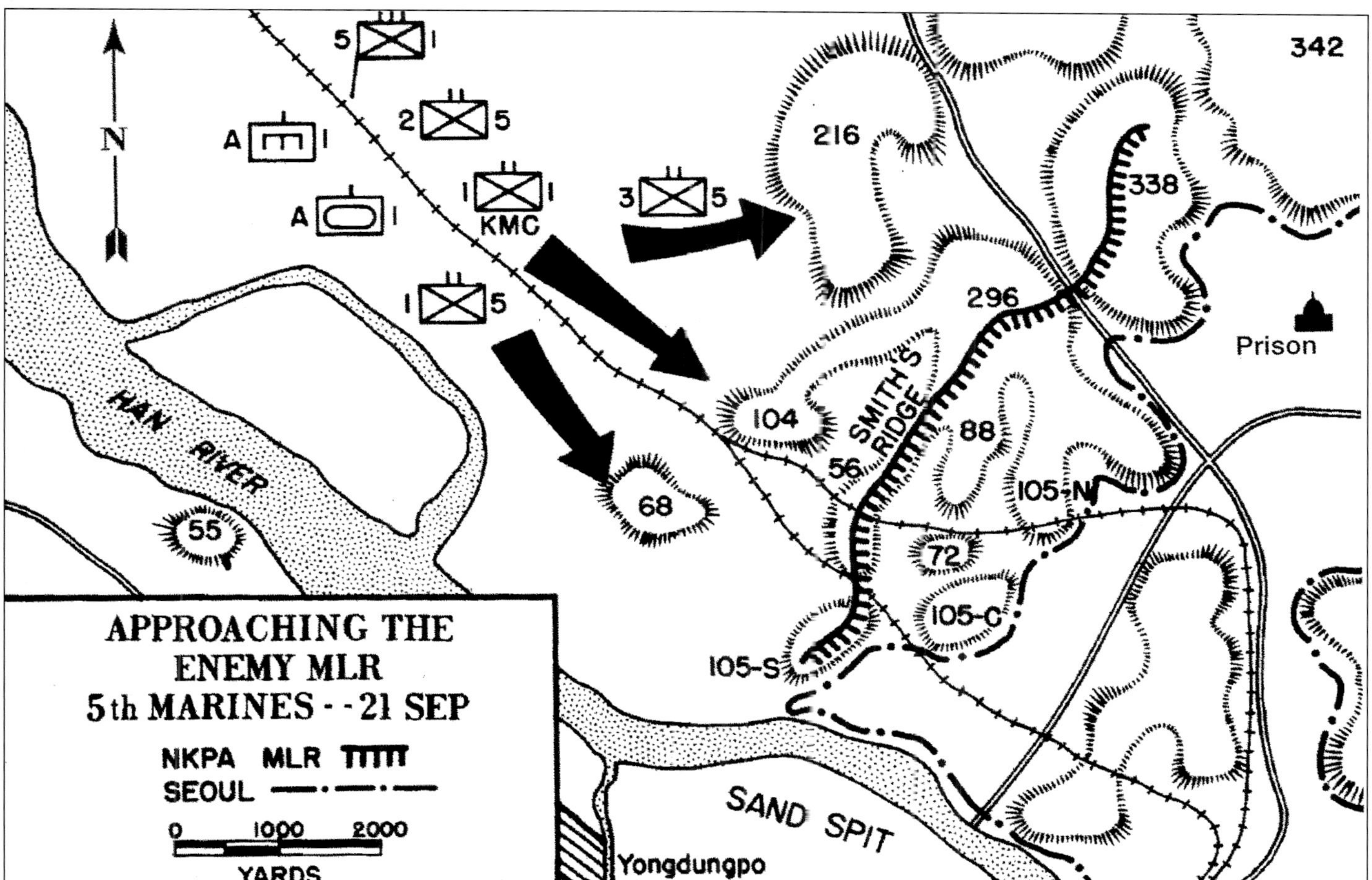

APPROACHING THE ENEMY MLR
5th MARINES - - 21 SEP

National Archives Photo (USMC) 127-N-A3860

Capt Francis I. "Ike" Fenton, Jr., center, enjoys a lighter moment with the Commandant, Gen Clifton B. Cates. Fenton's company experienced unremitting fire taking and holding Hill 105-South just outside the city limits.

Francis I. "Ike" Fenton, Jr. "They could really drop it in your lap."

Lieutenant Colonel Raymond Murray began the 22d of September with three of his four battalions on line: Taplett's 3d Battalion on the left, facing the main crest of Hill 296; Major Ko's 1st ROK Marine Battalion in the center, facing an exposed slope towards its objective, Hill 56; and Lieutenant Colonel George R. Newton's 1st Battalion on the right, aimed towards Hill 105-South. Lieutenant Colonel Harold S. Roise's 2d Battalion remained in reserve.

The battle for the hills got off to a bad start for Murray. During the night a North Korean shell exploded in his command post, causing many casualties. Murray survived with a small cut, but Lieutenant Colonel Lawrence C. Hays, his executive officer and fellow Tarawa veteran (1st Battalion, 8th Marines commanding officer at Red Beach Two), was badly hit and required emergency evacuation.

Murray nevertheless kicked off his regimental attack at 0700 on the 22d as planned. Taplett's 3d Battalion, 5th Marines, clawed its way steadily towards the steep crest of Hill 296, shaking off plunging fire from Communist positions north of the Kaesong Highway (the 7th Marines would not draw abreast to clear these positions along the left flank for another three days). Taplett's Marines maintained a steady rate of advance, the most promising of the week, halting only to resist company-sized counterattacks that boiled out of the draws and defiles along the shoulders of the hill mass.

Company H, 5th Marines, reached the hill's geographic crest by the end of the day. Corporal Larry Brom's platoon commander directed him to deploy his squad in a defensive sector along a grove of pine trees, and Brom supervised his men as they dug night positions and selected interlocking fields of fire. Satisfied with their preparations, he took off his pack and unfolded his e-tool (entrenching tool) to dig his own hole for the night. The squad had been uncommonly fortunate, Brom reflected, having lost only one man to enemy fire throughout the fighting along the Naktong, at Wolmi-do Island, and the advance east of Inchon. Here on Hill 296 their luck abruptly soured. A North Korean sniper shot Brom through the foot just after he knelt to unsling his pack. More fire sprayed the ridge crest. A gray-headed Korean "papa-san" scurried to Brom's side, scooped him up, and carried him piggyback down the reverse slope under intermittent fire to the battalion aid station. Brom gave him a fresh pack of cigarettes, all he possessed at the time. The old man bowed in gratitude, then returned back up the hill. For Corporal Brom, a two-year veteran of the 5th Marines, the war was over.

The incident of a Marine squad leader being picked off from long range at dusk by a North Korean sniper signified two developments. The NKPA had deployed front-line troops west of Seoul. Secondly, although the Marines had seized the crest of Hill 296, the North Koreans occupied defenses in depth throughout its massive fingers descending to the east and south.

The situation south of 3d Battalion, 5th Marines' advance validated these serious developments. On the 22d, the Korean Marine battalion encountered a furious fire from masked guns in every adjoining declivity each time it mounted an attack. Its objective was deceptive. Captain Fenton, operating on the Koreans' right flank, described Hill 56 as "a very insignificant looking low ridge that extended from 296 to 105-South." But the Koreans were advancing from low ground, through rice fields, exposed every step of the way to unrelenting artillery and mortar fire.

Murray directed Lieutenant Colonel Ransom M. Wood's supporting 1st Battalion, 11th Marines, to give the Koreans priority of fires. He also asked General Smith for more air support. This was forthcoming—the 1st Marines were mopping up Yongdungpo and the 7th Marines were not yet engaged. Major Arnold A. Lund led his Death Rattlers of VMF-323 off the escort carrier *Badoeng Strait* (CVE 116), which the aviators lovingly nicknamed "The Bing-Ding," in 42 sorties in support of the 5th Marines, the heaviest operational rate since D-Day at Inchon. Lieutenant Colonel Norman J. Anderson, the airborne tactical air controller for Marine Aircraft Group 33 (MAG-33), directed the strikes, then led one himself, a spectacular direct hit on Hill 72 (by now "Nellie's Tit" to the 5th Marines) that knocked out one of Colonel Wol's few tanks. Additional air strikes came from the newly arrived, Kimpo-based Lancers of VMF-212, commanded by Lieutenant Colonel Richard W. Wyczawski and Lieutenant Colonel Max J. Volcansek Jr.'s night-fighting Tigers of VMF(N)-542.

This was spectacular close air support—unerringly directed and delivered—and many North Koreans met their deaths from the skies, but their withering crossfire never ceased. The Korean Marines were literally stopped in their tracks. The advance of Newton's 1st Battalion, 5th Marines, on the right flank fared better, but only in relative terms. Attacking across 2,000 yards of open terrain cost Companies A and C dearly. The Marines found that one particularly deadly NKPA outpost contained a U.S. Browning .50-caliber heavy machine gun, captured during the first week of the war. Company A lost its last two officer platoon commanders in the assault. The cost was endemic with the 5th Marines. Seventeen of the regiment's original 18 platoon commanders had been killed or wounded in the first 50 days of combat in Korea, along with five of the six company commanders. Experienced non-commissioned officers took command of the platoons in Company A and continued the attack on Hill 105-South.

Captain "Ike" Fenton led Company B through Company A late in the day, then, leaning into a furious barrage from 1st Battalion, 11th Marines, joined Company C's dash for the crest of 105-South. It was a hollow victory. The battalion had suffered more than 40 casualties, and the enemy had mysteriously disappeared—"there were no bodies, not even any cartridge cases lying around," reported Fenton. Only later would the Marines discover the existence of a large cave on the hill's reverse slope, now a sanctuary for the former defenders, living and dead. In the meantime, punishing fire from the hills to the northeast began to rake the Marines exposed on the crest. As Heinl described Hill 105-South:

> [The hill] was no vacation spot. Before the sun set, enemy heavy machine guns began to scythe back and forth over the hilltop, while antitank guns, accurate as a sniper's rifle and a lot deadlier, flash-banged in with high-velocity rounds that left no time for a man to duck.

This was an unwelcome development to Fenton, who had lost only one killed and six wounded in his assault on the hill. Now, despite digging new foxholes along the military crest, his men would suffer stiff casualties from their hostile neighbors. "We were pinned down by day and counter-attacked by night," he said. To make matters worse, the Korean Marines' lack of progress left 1st

Marine Close Air Support in the Recapture of Seoul

"I believe the modern 'Marine Air-Ground Team' truly takes its departure from the crucible of the Korean War," reflected retired Lieutenant General Robert P. Keller, USMC, in a recent interview. Keller took command of the VMF-214 Black Sheep after North Korean antiaircraft gunners shot down Lieutenant Colonel Walter E. Lischeid over Seoul on 25 September 1950. Comparing this experience with his World War II service as a fighter pilot and squadron commander in the northern Solomons, Keller pointed to the emergence of close air support in the Korean War—"by Marines, for Marines"—as the principal difference. While ground Marines had enjoyed Marine air support at Peleliu, Iwo Jima, and Okinawa, it was never delivered more closely, nor more responsively than that provided by the F4-U Corsairs of the 1st Marine Aircraft Wing throughout the final four months of 1950, from the Pusan Perimeter through Inchon-Seoul to the Chosin Reservoir.

Major General Norman J. Anderson credited the success of this air support coordination to the hard work performed by Marine air and ground officers in the short interwar period. "The Marine Corps, having learned valuable lessons late in World War II, went to extremes in the late '40s to school its air and ground officers together and to structure its deployments as air-ground teams under a single command," he said. "This new structure served us well, then and ever since, beginning with the air-ground composition of the 1st Provisional Marine Brigade."

Of the four Marine fighter squadrons and two night fighter squadrons supporting the 1st Marine Division during the 33-day period from 7 September to 9 October, the Death Rattlers of VMF-323, commanded by Major Arnold A. Lund, saw more days in action and flew the most combat sorties (784, according to the official Marine Corps history of the Seoul campaign). The record comes with a bittersweet irony. The squadron had been in the process of a mandated deactivation when the war erupted, its pilots reassigned, its planes transferred for preservation. Saved at the last moment from the draconian cutbacks of the Truman Administration, the Death Rattlers reassembled in record time. During the Seoul campaign they launched from the escort carrier *Badoeng Strait* (CVE 116) in the Sea of Japan on missions ranging from reconnaissance to propaganda leaflet drops, but their most frequent mission by an order of magnitude was close air support.

The Black Sheep pilots of VMF-214 flew off the escort carrier *Sicily* (CVE 118), commanded by the legendary naval aviator Captain John S. Thach, USN, a World War II ace who in 1941 invented the "Thach Weave" to counter the Japanese Zero's technical superiority over

Photo Courtesy of LtCol Leo J. Ihli, USMC (Ret)

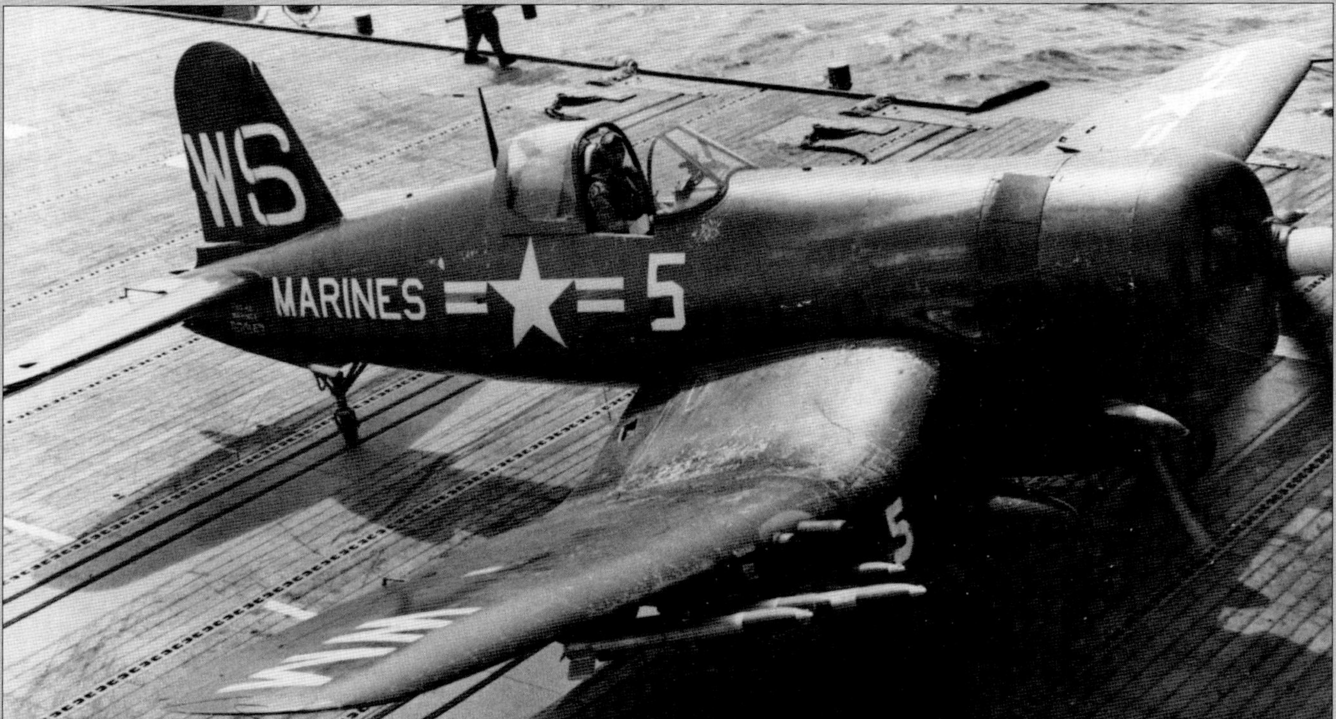

the F4F Wildcats. Thach became an enthusiastic advocate of Marine close air support. "It's like having artillery right over your shoulder!" he said. During the Seoul Campaign, Thach would often leave the bridge to attend the Black Sheep post-mission debriefings. "They took their work seriously. They really were the top pros in the business, I think, in the whole world. I had tremendous admiration for them."

So did the commanding general of the 1st Marine Division. "The effectiveness of the Marine air-ground team and close air support doctrine were reaffirmed with outstanding success," wrote Major General Oliver P. Smith after the liberation of Seoul.

For the troops on the ground, struggling to prevail against a well-armed enemy they could rarely see in the open, the firepower delivered by their fellow Marines overhead seemed awesome. Lieutenant Joseph R. Owen, the mortar platoon commander in Company B, 1st Battalion, 7th Marines, described his first experience with a close air strike during the battalion's battle for a ridge south of Uijongbu:

> The first of the gull-winged, dark blue Corsairs peeled from the circle and dove at the white smoke. Red tracers from its guns poured from the forward edges of the wings. The plane leveled off only yards above the ridgeline. We could see the pilot in the cockpit and the big, white Marine Corps emblem on the fuselage. . . . Then the [next] plane came in, this one dropping a pod of napalm. The black, coffin-shaped canister hit the ground, skipped a few feet above the surface, and exploded into a wall of flame that extended the length of the North Koreans' position. Two hundred yards below, we felt the shock of its explosion and a wave of searing heat.

While equally appreciative of the aviators' precision and valor, veteran infantry officer Captain Francis I. "Ike" Fenton, Jr., commanding Company B, 1st Battalion, 5th Marines, suggested even deadlier aerial firepower that could uproot North Koreans who took shelter in caves or railroad tunnels, as the 5th Marines experienced in the extended battle for Hill 105-South. "The close air support in Korea by the Marine Corps was outstanding," Fenton said. "However, I would like to see Marine aviation come up with a rocket with a napalm head. This rocket would be great for getting into tunnels, or into caves....The Koreans showed great fear for fire bombs....I believe a big rocket, maybe a Tiny Tim, that could carry a fairly good quantity of napalm, would be an excellent weapon."

Major General Field Harris' 1st Marine Aircraft Wing also provided close air support to the 7th Infantry Division, the other major component in X Corps during the Seoul campaign. Superbly assisted by Marine Captain Charles E. Crew's Far East Detachment, Air and Naval Gunfire Liaison Company, Fleet Marine Force, Pacific, the 1st Marine Aircraft Wing flew 1,024 sorties in support of the Army division in 57 days without a single casualty to front-line friendly troops, despite bombing and strafing runs as close as 200 yards. Brigadier General Homer W. Kiefer, USA, commanding the 7th Division's artillery, wrote an appreciative letter to the Commandant, stating: "The Marine system of control, in my estimation, approaches the ideal, and I firmly believe that a similar system should be adopted as standard for Army Divisions."

The Korean War as a whole would advance military aviation fully into the Jet Age, and soon the U.S. Air Force would wage epic air-to-air battles between its F-86 Sabers and the Soviet-built (and often Soviet-flown) MIG-15 fighters. Eventually the Marines would introduce in the skies over Korea their own jet fighter, the Grumman F9F-2 Panther, well armed for both air-to-air and air-to-ground missions. It was also the dawn of the Helicopter Age, and VMO-6 made military aviation history when it deployed to Pusan with the 1st Marine Brigade in August 1950 with four Sikorsky HO3S-1 helicopters.

By contrast the propeller-driven Corsair was now considered old and slow, hampered by a light payload capacity and too small a fuel tank. Landing the high-rise "U-birds" on the pitching deck of an escort carrier remained "adventurous," especially with the ship steaming westerly into a setting sun. "That bright red ball seemed to be sitting right on the fan-tail," General Keller recalled, "and it was difficult to make out the Landing Signal Officer, his signals, or even the deck." General Anderson cited another common hazard when trying to land an F4U into a setting sun: "The Corsair frequently managed to splatter the windshield with oil!"

Yet the Corsair in good hands proved highly reliable and durable for its age and the operating conditions. The hard-working maintenance crews of VMF-214 somehow averaged 95 percent availability of the Black Sheep Corsairs throughout the Pusan-Inchon-Seoul campaigns. And in the absence of a jet-propelled enemy air threat during those two months, the Corsair proved an invaluable contributor to the allied victories.

Certainly the ground Marines fighting towards Seoul or Uijongbu in the autumn of 1950 were very comfortable with the presence overhead of their protective Corsair, their familiar old "bent-wing widow-maker," the attack aircraft the Japanese in the previous war allegedly nicknamed "The Whistling Death." There is no record of what nickname the North Koreans may have used, but judging from the ever-increasing intensity of their ground fire the moment the F4Us swept into view, it was probable the Corsairs held their highest respect, as well.

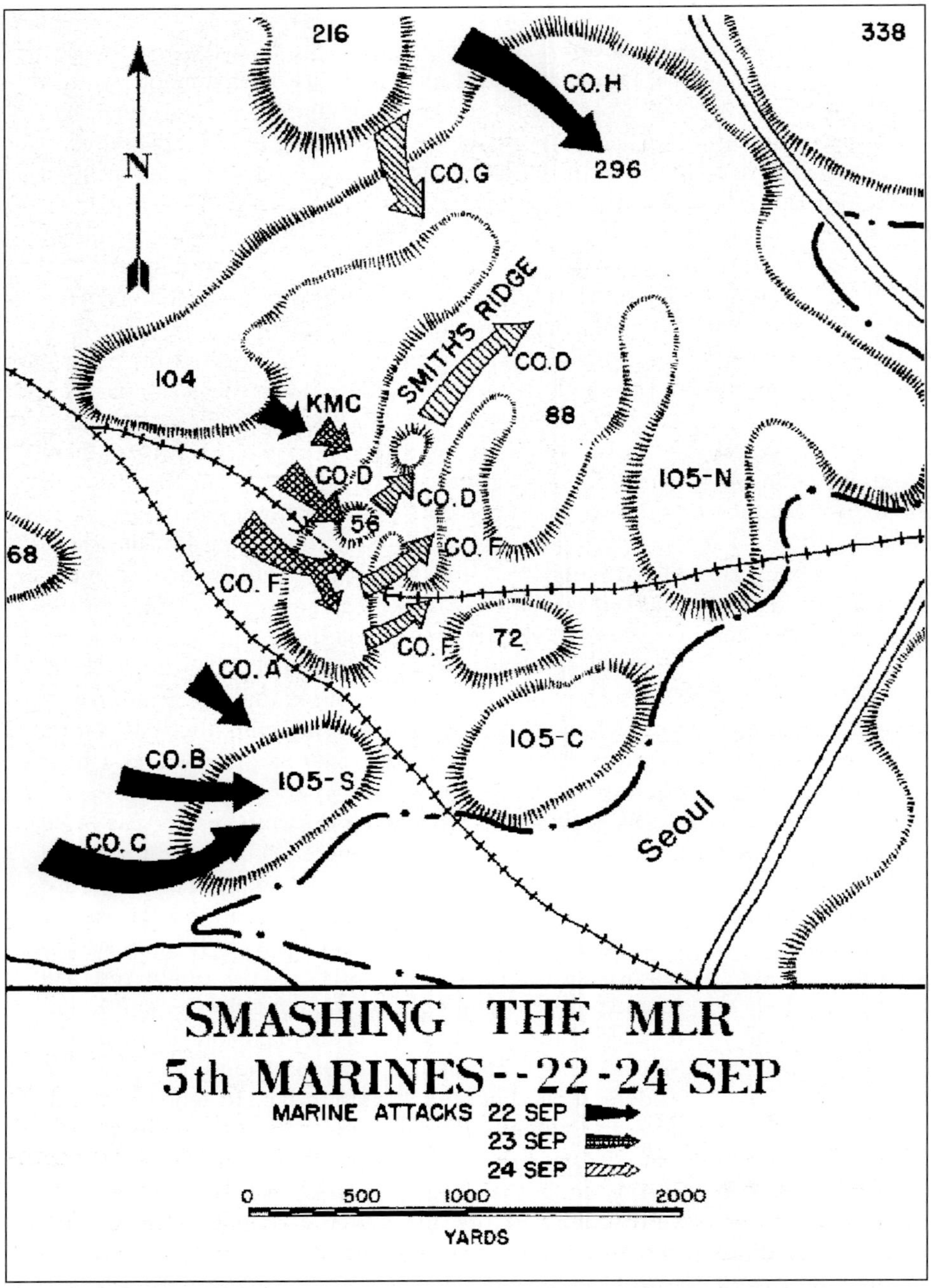

SMASHING THE MLR
5th MARINES--22-24 SEP

Battalion, 5th Marines' left flank fully exposed. Newton had to peel a company back to the starting position, and the day ended on that sour note.

Lieutenant Colonel Murray ordered the Korean Marines to resume their assault on Hill 56 the morning of 23 September, but try as they might the ROK troops were stopped cold by heavy fire. No one then realized that Colonel Wol had established his main line of resistance along the low ridge that passed through Hill 56. The insignificant-looking rise would become known as Smith's Ridge the following day.

Murray committed his reserve, ordering Lieutenant Colonel Roise to pass through the Koreans with 2d Battalion, 5th Marines, and continue the attack. Roise deployed Captain Uel D. Peters' Company F on the right and First Lieutenant H. J. "Hog Jaw" Smith's Company D on the left. Hugging the terrain and advancing by squad rushes, both companies were able in time to approach the higher ground with acceptable casualties, yet both suffered heavily in the close-in fighting that followed. This took the balance of the afternoon.

George Newton's 1st Battalion, 5th Marines, had all it could handle that day and night just maintaining its exposed forward position on Hill 105-South. In two days spent clinging to the hill's fire-swept crest, Companies B and C suffered 24 casualties. "All these men were hit in their foxholes," said Captain Fenton. "There was no way to keep the enemy from delivering plunging fire right in on top of us."

Robert Taplett's 3d Battalion, 5th Marines, also had its hands full throughout the 23d in repelling NKPA counterattacks against the crest of Hill 296 and trying to establish fire superiority against the enemy on a half-dozen circling hills. Clearly visible at one of these Communist strongpoints was a tall, fair-skinned officer with a charmed life, "Fireproof Phil." He may have been a Soviet military advisor, but whoever he was, Fireproof Phil exhibited unflagging disdain for Marine marksmanship. When riflemen, mortarmen, and artillerymen failed to knock him down, Taplett ordered up an M-26 Pershing tank. Sniping at Phil with a 90mm gun proved equally futile. The man dodged every round and kept exhorting his gunners to return fire until darkness shrouded the scene. The Marines never saw him again.

The 2d Battalion held Hill 56 throughout the night, but only by its collective fingernails. The assault companies were scattered and vulnerable. Lieutenant Colonel Max Volcansek's faithful night fighters circling overhead helped even the odds, but Marine artillery provided the greatest assistance. Wood's 1st Battalion, 11th Marines, fired all night long, illuminating the scorched battlefield and interdicting potential NKPA assembly areas. "I can't say enough about the artillery support we received that night," said Second Lieutenant Tilton A. Anderson, whose platoon had been reduced to seven men in the afternoon's fighting. "It was magnificent."

Major General Almond, the X Corps commander, grew impatient with the 1st Marine Division's slow progress north of the Han. Pressured by MacArthur to recapture Seoul by the third-month anniversary of the invasion, and mindful that the North Koreans would be fortifying the capital to a greater extent each day, Almond urged O. P. Smith to deploy the 1st Marines well beyond Yongdungpo to attack Seoul from the southeast. Almond's operations officer reflected his commander's impatience, saying: "The Marines were exasperatingly deliberate at a time when rapid maneuver was imperative."

Smith disagreed. Seizing Inchon against rear-echelon troops had been a relative cakewalk. Things had changed. The tenacity and firepower of the North Koreans battling the 5th Marines reminded Smith more of the Japanese at Peleliu or Okinawa. Seizing Seoul would therefore not be quick and easy, Smith argued, and the last thing he wanted was to wage that battle with his major components divided by the Han and attacking towards each other. Almond acquiesced to this logic, but he also decided to bring in Colonel Charles E. Beauchamp's 32d Infantry of the Army's 7th Infantry Division to attack the city from the southeast. Seoul would no longer be the sole province of the 1st Marine Division. Smith agreed to move Puller's 1st Marines across the Han the next morning, then loan the 1st Amphibian Tractor Battalion to X Corps to transport the 32d Infantry and the 17th ROK Regiment across the tidal river the following day.

Sketch by John DeGrasse

The 5th Marines learned tank-infantry coordination under intense pressure in the Pusan Perimeter. Here, attacking North Korean positions along the ridges outside Seoul, a fire team keeps up with its assigned Pershing tank.

Smith knew that Almond on his daily visits to the front-line regiments had taken to giving operational orders directly to Murray and Chesty Puller. In a heated private session, Smith asked Almond to knock it off. "If you'll give your orders to me," Smith said icily, "I'll see that they are carried out." Neither of Almond's division commanders, however, would successfully cure the commanding general of his impetuosity.

General Smith directed Puller to make his crossing slightly west of Yongdungpo, turn right, enter the city along the north bank, then execute a difficult pivot movement, wheeling the regiment north. Smith planned for Murray's 5th Marines to fight their way into the northwest sector of the city while Litzenberg's 7th Marines sealed off the NKPA access routes along the entire northern boundary. It was an ambitious and complicated plan. But the first order of business remained the destruction of the *25th NKPA Brigade* in the fortified barrier ridges to the northwest.

The battle for these ridges reached its climax on 24 September. The day broke with a low-lying mist, as Companies D and F arrayed themselves for the assault. Artillery preparations began at 0610. Company F jumped off 20 minutes later, seized the eastern end of the troublesome railroad tunnel, paused to allow a Corsair strike by the Lancers of VMF-212 (who would establish a 1st Marine Aircraft Wing record of 46 sorties this date), then dashed across the low ground to capture the heavily fortified eastern finger. This represented an encouraging start, but Company F was spent, having suffered more than a hundred casualties around the south edge of Hill 56 in the past 24 hours. Among the dead was Corporal Welden D. Harris, who had killed three North Korean soldiers in hand-to-hand fighting and been twice wounded the day before. Company F had given its all. Now it was all up to "Hog Jaw" Smith and Company D.

The recapture of Seoul would obviously require a team effort—Marines and Army, ground forces and air squadrons. But the keys to Seoul's access really came from two Marine rifle companies, Captain Robert Barrow's Company A, 1st Marines, at Yongdungpo during 21-22 September, and Captain H. J. Smith's Company D, 2d Battalion, 5th Marines, during the 23d-24th.

Company D faced the greater challenge. Captain Smith had to attack about 750 yards to the northeast across an open saddle, seize an extremely well-defended knoll, and continue beyond along an increasingly wooded ridge. This contested real estate became Smith's Ridge. Easily a thousand NKPA troops defended this terrain, well covered by the same sharpshooting gunners who had been making life so miserable for the 1st Battalion, 5th Marines, on Hill 105-South.

Smith began the day with a good-sized rifle company, but the mission required a battalion. Lieutenant Colonel Roise—who would join the ranks of the wounded this day but refuse evacuation—withheld Captain Samuel Jaskilka's Company E to exploit Smith's expected breakthrough and roll up the last hills to the east.

Captain Smith sensed what he faced and relied heavily on supporting arms, adding to the artillery fire missions and air strikes his own machine guns, mortars, and rocket launchers. Twice he punched ahead; twice he had to withdraw with heavy casualties. Nor did a flank attack succeed. An 11-man squad worked east then attacked north. The North Koreans shot them down to a man. Abruptly Smith's company was down to 44 Marines, including the 60mm mortar section, now out of ammunition and doubling as riflemen.

All it took was one North Korean prisoner of war to whip a pistol or grenade from under his loose clothing and attack his captor. Thereafter the Marines took no chances. Naked prisoners proceed under armed guard past a destroyed T-34 tank to a prison camp.

National Archives Photo (USA) 111-SC349027

By this time, the 11th Marines had been bombarding the ridgelines and reverse slopes of the objective for more than 24 hours. Ten Marine Corsairs from the Death Rattlers had rotated on station since sunrise, bombing, strafing, and dropping napalm canisters along the objective. Yet Colonel Wol's antiaircraft gunners had taken a toll: five of the Corsairs received extensive damage. Smith knew he was down to his final opportunity.

Smith called for a four-plane firing run, asking that the fourth Corsair execute a low but dummy pass to keep the enemy in their holes until the last possible moment. Major Lund's Corsair pilots flew this mission beautifully. As the third plane roared overhead Smith leapt to his feet screaming "follow me!" His Marines swept forward just beneath the last Corsair's low-level, ear-splitting run.

"Over they went," described Captain Nicholas A. Canzona of the engineer battalion in a 1956 *Marine Corps Gazette* account, "yelling wildly and firing their rifles, carbines, and BARs [Browning automatic rifles]. They entered upon a scene of carnage stretching out in every direction. Driving forward through the human wreckage, they shot and bayoneted anything that moved."

"Hog Jaw" Smith died at the bitter end, becoming Company D's

36th fatality of the assault. Seizing Smith's Ridge in fact cost the company 178 casualties of the 206 men who had advanced across the valley the previous day. But the reverse slopes of the complex looked like a charnel house. The surviving Marines began to count the windrows of NKPA bodies, most blasted hideously by Marine 105mm howitzers, Corsairs, and mortars. They reached 1,500 and had to stop counting; the task was too gruesome.

Company D had knocked down the center door to the *25th NKPA Brigade's* defenses, but more savage fighting remained to clear the final path to Seoul. Captain Jaskilka's fresh Company E moved through the gap between the remnants of Companies D and F, but encountered an extensive minefield and stubborn resistance on Nelly's Tit and Hill 105-Central beyond. The division engineers cleared the mines, but ridding the last hills of their die-hard defenders took Jaskilka another 24 hours. Lieutenant Colonel Taplett's 3d Battalion, 5th Marines, had a correspondingly difficult time snuffing out Hill 105-North. In close combat reminiscent of the Central Pacific in World War II, most of the enemy chose to die in place. Colonel Wol's fate remained unknown, presumed dead.

The 1st Battalion, 5th Marines, managed to maintain its precarious hold on the crest of 105-South while at the same time dispatching a large combat patrol down to the river to cover the crossing of Puller's 1st Marines. The nefarious hill would still represent a hornet's nest to all would-be occupants. It would take a combined assault by the 1st and 5th Marines and an armored column to close the cave and cut down the final defenders later that day.

National Archives Photo (USA) 111-SC349090

A wounded Marine is carried down from the front lines on the ridges northwest of Seoul. The 5th Marines' three-day battle for the ridges made the uneventful crossing of the Han by the 1st Marines possible.

The 5th Marines' three-day battle for the northwestern ridges made possible a surprisingly uneventful tactical crossing of the Han by the 1st Marines. The 2d and 3d Battalions crossed by LVTs; the 1st Battalion and Puller's command group made the crossing in DUKW amphibious trucks. NKPA opposition proved negligible. Lieutenant Colonel Henry P. "Jim" Crowe, who had created order out of chaos seven years earlier on Tarawa's Red Beach Three, swiftly deployed his 1st Shore Party Battalion along the landing site to keep troops and cargo moving inland, avoiding a dangerous bottleneck. Puller hustled his battalions eastward into the city, growling at the long time it would take his Pershing tanks to cross at the Haengju ferry further downstream and work their way back along the north bank.

General Smith finally had all three of his infantry regiments north of the Han and roughly in line. This same day, Lieutenant Colonel Litzenberg's 7th Marines experienced its first significant combat against an NKPA outpost to the northwest of Seoul. For Second Lieutenant Joseph R. Owen, commanding the 60mm mortar section in Company B, 1st Battalion, 7th Marines, the moment

Marine Combat Vehicles in the Seoul Campaign

The Marines mostly fought the first months of the Korean War with hand-me-down weapons and equipment from World War II stockpiles. In the case of combat vehicles, however, the Corps invested in two critical upgrades that provided a tactical edge in the recapture of Seoul: the M-26 Pershing medium tank and the LVT-3C amphibian tractor.

The sturdy M-4 Sherman tank had served the Marines well in the Pacific War from Tarawa through Okinawa, and by 1950 the tank battalions in the Fleet Marine Force were still equipped with the M-4A3-E8 "Easy Eight" version, featuring a 105mm gun. Yet the Sherman's success in the Pacific War was deceptive. Japanese tanks had provided no particular threat, the vehicle's narrow track width and high ground pressure had posed mobility problems in marginal terrain, and the Sherman's notoriously thin side and rear armor protection had proven inadequate against the enemy's 47mm antitank guns. The Sherman's prospects did not look favorable against the battle-proven T-34 medium tanks that the Soviet Union exported to client states like North Korea at the onset of the Cold War.

The Marines had foresightedly invested in the Army's acquisition of the M-26 Pershing 90mm-gun tank late in World War II. Their vehicles did not arrive in time for combat validation in Okinawa; nor could the postwar Corps afford to place them into operation, so the Pershings sat for several years in contingency reserve at the Marine supply base in Barstow, California.

When the Korean War erupted, the Commandant ordered the 1st Tank Battalion to deploy with the new Pershings in lieu of its Sherman "Easy Eights." The hasty transition was not pretty, especially in the case of the reinforced company assigned to the 1st Brigade for its early-July deployment. Few tankers had the opportunity for hands-on operation and maintenance training. The gunners were lucky to be able to fire two rounds each—and they had to use the more abundantly available 90mm antiaircraft rounds instead of the new but scarce high-velocity armor-piercing munitions. And since none of the new Marine Pershings were configured as flamethrowers or dozer-blade variants, the battalion sailed with an awkward mixture of old Shermans along with the M-26s, the making of a logistical nightmare.

The ragged transition made for an inauspicious combat debut for the Marine M-26s in Korea. Operating in the Pusan Perimeter southwest corner, one Pershing broke through the planking of a critical bridge, heightening fears that its 46-ton weight would prove too heavy for Korea's road network. A second vehicle threw a track while fording a stream, blocking the crossing. Things improved. The Marine Pershings established their dominance in a head-to-head engagement against T-34s in the first battle of the Naktong Bulge, then continued to sweep the field as the 1st Marine Division advanced on Seoul. The Sherman blade and flame variants also contributed materially, especially in the close engagement waged by Baker Company's tanks against cave-infested Hill 105-South on 25 September.

A Marine LVT-3C Bushmaster from the 1st Amphibian Tractor Battalion transfers troops to an LCVP.

Department of Defense Photo (USMC) A162956

National Archives Photo (USA) 111-SC348713

In the battle of downtown Seoul, the Pershings of Lieutenant Colonel Harry T. Milne's 1st Tank Battalion provided the crucial edge, time and again crashing through the North Korean barricades despite intense fire from the enemy's ubiquitous 45mm antitank guns. The battalion's War Diary for September reported the destruction of 13 NKPA tanks (which may have included several 76mm self-propelled guns) and 56 antitank guns or antiaircraft guns being fired horizontally at the approaching Pershings. The battalion lost five Pershings and one each of the flame and dozer Shermans in the recapture of Seoul.

The LVT-3C Bushmaster proved to be another smart investment for the Marines. Borg Warner's original LVT-3 had developed slowly during World War II, reaching the Fleet Marine Force out of numerical sequence and more than a year behind rival Food Machinery Corporation's LVT-4. Borg Warner built nearly 3,000 Bushmasters for the Marine Corps. The first vehicles arrived in time for the Okinawa invasion in the spring of 1945.

The Bushmaster was a welcome addition to the Marines' ship-to-shore team. Like its FMC predecessor, the Bushmaster came with a hinged rear ramp and sufficient cargo space to accommodate either a jeep or a 105mm howitzer. By mounting its twin Cadillac V-8 engines along the sides of both hulls, the Borg Warner engineers provided the Bushmaster with a cargo capacity that exceeded the LVT-4's by 3,000 pounds.

Faced with the need to upgrade their amphibian tractor fleet during the austere late 1940s, the Marines opted to modernize 1,200 low-mileage LVT-3s by raising the sides, installing aluminum covers over the troop/cargo compartment, and installing a small machine gun turret atop the cab. The Marines designated their newly modified vehicle the LVT-3C, and it proved remarkably well suited for both salt-water and fresh-water operations throughout the Korean peninsula. (The Republic of China Marine Corps employed American-built LVT-3Cs on Taiwan for a quarter of a century after the Korean War).

The Bushmasters of Lieutenant Colonel Erwin F. Wann, Jr.'s 1st Amphibian Tractor Battalion delivered Marines ashore at Inchon, transported each regiment plus the Army and ROK regiments across the Han under fire, and served as armored personnel carriers and cargo vehicles overland.

The 1st Marine Division was similarly well-supported by the versatile DUKWs of the 1st Amphibian Truck Company, an element of Lieutenant Colonel Olin L. Beall's 1st Motor Transport Battalion. (DUKW is not an acronym but an arcane industrial code used in World War II meaning an all-wheel-drive utility vehicle with twin rear wheel axles manufactured in 1942—"DUCKS" to Marines!)

Unfortunately the Marines fought the Inchon-Seoul campaign without the 1st Armored Amphibian Battalion. General Smith left the battalion with the division's rear echelon in Kobe as a temporary repository for the 500-plus, 17-year-olds ruled ineligible for combat by the Secretary of the Navy on the eve of the Inchon landing. The X Corps commander partially offset this lost capability by attaching the Army's Company A, 56th Amphibian Tractor Battalion, to the Marines. The Army company's 18 LVTA-5s equipped with snub-nosed 75mm howitzers spearheaded each river crossing, thereby proving themselves worthy recipients of the Presidential Unit Citation subsequently awarded the 1st Marine Division.

Painting by Col Charles H. Waterhouse, USMCR (Ret)

"First Firefight Above Seoul, B/1/7" portrays the intensity of night action that greeted the 7th Marines as they advanced to cut the roads leading north from Seoul.

was unforgettable:

> The North Korean mortars came. Spouts of earth and black smoke leaped about us, laced with flame and screaming shrapnel. The leaves from the bean plants spun in flurries, and the ground shook. I was suddenly in the midst of a frenzied storm of noise.

By the nature of their northern mission the 7th Marines would have scant contact with the other elements of the 1st Marine Division in the fight for Seoul. The other two regiments, however, would experience a dangerous interface, the 1st Marines attacking north through the heart of the city, the 5th Marines coming in from the northwest.

Concerned with the inherent risks facing these converging forces, Lieutenant Colonel Raymond Murray boarded a helicopter late in the afternoon of 24 September and flew to Chesty Puller's command post to coordinate the final assault. It was the first time the two commanders had ever met. Characteristically, Puller inquired of Murray the extent of the casualties he had sustained fighting for the northwest ridges. "He determined how good a fighter you were by how many casualties you had," Murray recalled. Murray's grim accounting of the 5th Marines' losses during the preceding three days made even Chesty Puller blink. The men then got down to work.

This was the time and setting when Captain Robert Barrow's Company A, 1st Battalion, 1st Marines, seized Hill 79 and raised the first flag in Seoul proper. The 1st Marine Division had entered the capital.

The Fight for Seoul

Seoul in 1950 was home to more than a million people, the fifth largest city in the Orient. While several hundred thousand civilian residents had fled the capital at the outbreak of the North Korean invasion, tens of thousands remained. Chesty Puller had ruefully predicted to a news correspondent that the North Koreans would defend the city in such a manner as to force the attacking Marines to destroy it. The ensuing three days would validate Puller's prediction. British correspondent Reginald Thompson would write despairingly: "Few people can have suffered so terrible a liberation."

X Corps launched its assault on Seoul proper the morning of 25 September. Lieutenant Colonel Erwin F. Wann, Jr.'s 1st Amphibian Tractor Battalion displaced during the night to Sansa-ri, a former ferry crossing 5,000 yards east of Yongdungpo. There, reinforced by Army LVTs of Company A, 56th Amphibian Tractor Battalion, the Marines embarked the 2d Battalion, 32d Infantry. Following a brief artillery and mortar barrage, the Amtracs plunged into the Han, shook off a few 76mm rounds, and at 0630 disembarked the soldiers on the far bank. Four Corsairs from Lieutenant Colonel Lischeid's VMF-214 Black Sheep squadron off the *Sicily* worked just ahead of the beachhead, coordinated by Marine tactical air control parties provided the 7th Division for the occasion.

The Army regiment completed the crossing by mid-afternoon and seized South Mountain, the 900-foot eminence (the Koreans call Nam-san) dominating southeastern Seoul. Late in the day, the 1st Amphibian Tractor Battalion delivered the 17th ROK Infantry across in trace, an exposed crossing that attracted considerably more NKPA long-range fires. Yet by nightfall all of General Almond's maneuver elements were in place north of the river.

General O. P. Smith worried that

the presence of the two additional regiments on his right flank would create dangerous crossfires and accidental meeting engagements, but the Army units maintained their positions on and around Nam-san, defending against major counterattacks, and later assaulted towards the east, well clear of the Marines' zone of action. No significant control problems developed.

At 0700 on the 25th, the 1st Marine Division kicked off its assault on Seoul. The plan of attack developed by Smith and his operations officer, Colonel Alpha L. Bowser, Jr., placed the biggest burden on the 1st Marines. Puller's regiment would attack to the north through the heart of the city on a mile-and-a-half front, bordered by Nam-san on the right and the Duk Soo Palace of the ancient rulers of Korea, on the left. Smith assigned the 1st Marines Objective Able, the high ground just beyond the city's northeastern limit, about six miles from Captain Barrow's forward position on Hill 79. Murray's 5th Marines would attack the northwest section of the capital, likewise on a mile-and-a-half front, seize Government House and Objective Baker, the high ground overlooking the Seoul-Uijongbu road from their dearly won positions along the Hill 296 complex. Litzenberg's 7th Marines would seize Objective Charlie, the high ground along the Seoul-Kaesong road six miles outside the city center. Smith continued his reinforcement of the 1st and 5th Marines with one battalion each of Korean Marines and assigned the balance of the Korean regiment as division reserve. Smith also attached the Division Reconnaissance Company to the 5th Marines to screen the high ground along its left flank. The 3d Battalion, 187th Airborne, under the operational control of the 1st Marine Division, would protect the Marines' western flank below the Han.

Colonel James H. Brower concentrated most of the howitzers of his 11th Marines in firing positions on the south bank of the Han near Yongdungpo. The big 155mm howitzers of the Army's 96th Field Artillery deployed nearby, ready to support either the Marines or the Army, as needed.

The action for the 5th Marines on 25 September was largely *deja*

The Marines fought two enemies in downtown Seoul—those who defended behind the barricades and the snipers seemingly hidden in every other window.

Photo by Frank Noel, Associated Press

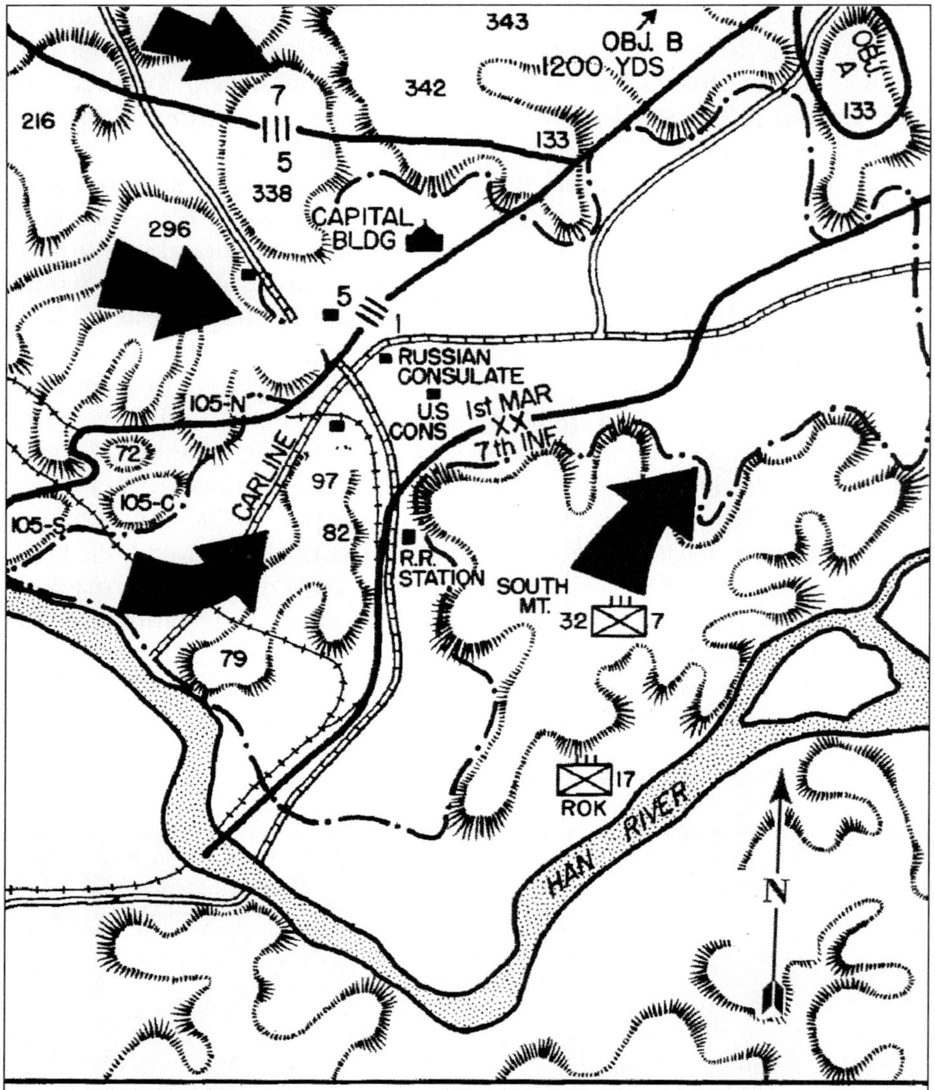

SEOUL ASSAULT PLAN

26 SEP 50

SHOWING DIVISION & REGTL BOUNDARIES

0 4000

YARDS

vu, the unfinished and still costly business of eliminating the residual positions of the *25th NKPA Brigade* along the eastern fingers of Hill 296 as described earlier. Here on two adjoining knobs, Company E, 2d Battalion, 5th Marines, and Companies H and I of 3d Battalion, 5th Marines, engaged the North Koreans in bloody close combat, again most ably supported by Marine Corsairs.

By now the *19th NKPA Antiaircraft Artillery Regiment* had learned how to deal with the terrifying strafing runs by Marine Corsairs. Increasingly, those antiaircraft gunners who survived the northwest ridge battles would turn Seoul into a "flak trap." September 25th reflected this new lethality, a particularly costly day for Marine Corsair squadron commanders. With the escort carrier *Sicily* and its embarked VMF-214 Black Sheep scheduled to rotate back to Inchon for repairs and resupply that afternoon, Lieutenant Colonel Walter Lischeid led the final sorties in support of the Army's river crossings. A North Korean gunner hit his Corsair over Seoul. Lischeid tried to nurse his crippled plane to Kimpo field but crashed in flames two miles shy of the airstrip.

In other aerial action on the 25th, Lieutenant Colonel Richard Wyczawski, commanding the Lancers of VMF-212, was wounded and shot down by hostile fire. So was Lieutenant Colonel Max Volcansek, commanding the night-fighting Tigers of VMF(N)-542, who barely bailed out before his plane crashed near Kimpo. Marines flying Sikorsky HO3S-1 helicopters from Marine Observation Squadron 6 (VMO-6) rescued both officers—Volcansek's rescue helicopter pulled him out of a rice paddy in a record six minutes elapsed time following notification—but all hands regretted the death of Lieutenant Colonel Lischeid.

Major Robert P. Keller, who had commanded three squadrons in the Pacific War, took over the Black Sheep. When a fellow aviator remarked, "Now you are the acting commanding officer," Keller retorted, "Acting, hell—I'm serious." Keller maintained the VMF-214 commitment to launching five-plane strikes every two hours. The Black Sheep pilots first plastered the ridge from which the antiaircraft battery had fired on Lischeid, then spent the remainder of the day delivering ordnance against targets ranging from railroad yards in the North Korean capital of Pyongyang to enemy troop concentrations in downtown Seoul, the other capital.

The nature of Marine close air support changed as the campaign entered the streets of Seoul. As Lieutenant Colonel Norman Anderson subsequently noted: "Bombing by its very nature gave way to the more easily accurate techniques of rocketing and strafing. . . . I feel we became increasingly aware of the need to avoid what we now call collateral damage." The Corsair's 20mm cannon could deliver a hellacious strafing run, but the "bent-wing U-Birds" could only carry 800 rounds, limiting the extent of this application. Anderson wistfully recalled his days of flying Marine Corps B-25s in the Philippines late in World War II, "a memorable strafer with 14 forward-firing, .50-caliber machine guns. Many's the time we might have put them to good use supporting Marines in the streets of Yongdungpo and Seoul. Alas, they were not carrier suitable."

On the ground in Seoul on 25 September progress came grudgingly to the 1st Marines despite its early start. Puller passed Ridge's 3d Battalion, 1st Marines, through Sutter's 2d Battalion, while, to Ridge's right, Hawkins adjusted the 1st Battalion's positions along Hill 79 to accommodate the 90-degree pivot to the northeast. This done, the regiment advanced methodically, Ridge and Hawkins abreast, Sutter in close reserve. The North Koreans resisted savagely, and Puller looked often for his missing tanks, still completing their long run east from the Haengju ferry crossing the previous afternoon.

Fresh minefields and sudden ambushes slowed Captain Bruce F. Williams' tank company, reinforced by a platoon each of infantry and combat engineers, once they crossed the river. As the armored column approached Seoul they drew fire from the southeast corner of Hill 105-South, still unconquered despite Captain Fenton's seizure of the crest three days earlier. This time, finally, the Marines had a force on the ground with the firepower, mobility, and shock action to finish the job. The tankers and engineers blew away a line of shacks blocking the base of the hill, thereby discovering the hidden cave mouth, and moved a flame tank up to the opening. Sensibly, the North Koreans began to surrender, one or two at first, then more than 100, outnumbering their captors.

The Marines to this point routinely made each prisoner of war strip buck naked, but they were shocked to find two women among this crew. Someone helpfully provided two pairs of long johns for the occasion, but the American press had a field day with the matter later, once the women got to the rear and complained. But it was a no-win situation for the Marines. The NKPA occupants of that cave had killed Marines from five different battalions; they were quite fortunate to escape the flame tank's horrors. As it was, other NKPA troops nearby had no intention of surrendering to the Marines. As Staff Sergeant Arthur Farrington reported:

> The enemy wounded were hoisted on board the tanks, 129 bare asses were lined up three abreast [between the vehicles] . . . when about 40-50 [North] Koreans jumped up to the left of the railroad tracks. They had been lying their doggo behind us all the time. We killed them with rifle, machine gun, and 90mm fire as they went across the paddies.

Captain Williams was understandably exultant as he led his column with its rich prizes into Seoul, but when he tried to recount the unit's success at 105-South to Chesty Puller, the colonel cut him short, saying, "I'm not

A Corsair flight on a close air support mission against targets in North Korea and around the South Korean capital.

Photo courtesy of LtCol Leo J. Ihli, USMC (Ret)

Department of Defense Photo (USMC) A3380

Marine riflemen and tanks advance north under fire along Seoul's principal boulevard.

interested in your sea stories young man. You're late. We've got fish to fry."

Puller sorely needed the tanks. The North Koreans defending Seoul lacked the numbers to occupy every building or side street, so they concentrated instead on the major avenues and thoroughfares. By now each significant intersection in the city featured an improvised barricade, typically protected by rice bags filled with sand or rubble, piled eight feet high by five feet wide, and defended by antitank guns, heavy machine guns, and mines. Marine historian Colonel Robert D. Heinl, Jr., likened the scene to 19th century France: "Every intersection was barricaded after the fashion of the Paris Commune: carts, earth-filled rice bags . . . furniture, and rubble." The Soviet Union's official newspaper *Pravda* compared the situation in Seoul to the Russian defense of Stalingrad in World War II: "There is firing behind every stone."

The axis of advance of Lieutenant Colonel Ridge's 3d Battalion, 1st Marines, was directly up Ma Po Boulevard towards the embassies and principal government buildings. Major Edwin Simmons later compared his company's advance to "attacking up Pennsylvania Avenue towards the Capitol in Washington, D.C." The boulevard was straight and wide—"once a busy, pleasant avenue lined with sycamores, groceries, wine and tea shops," according to Heinl. Trolley car tracks ran down the middle. Now NKPA barricades mushroomed at each intersection. Enemy snipers fired from blown out windows. Other NKPA troops lobbed Molotov cocktails from the rooftops onto the Marine tanks in the street below. And throughout all this mayhem fled thousands and thousands of terrified Korean refugees. Mines accounted for appalling casualties among them.

At one point Captain Robert Barrow halted his company along a particularly advantageous rise of ground overlooking the railroad yards and passenger station. For once he could clearly see the enemy troops moving into new positions, building fresh barricades, and preparing future ambushes. He called in artillery and mortar fire, employed his machine guns and rocket launchers, enjoying his dominant position. Strangely, he said, Lieutenant Colonel Hawkins kept urging him to advance. "We thought we were having a turkey shoot," Barrow recalled. "Nobody getting hurt and [us] knocking the hell out of them," but Hawkins said, "What's holding you up—move out!" When Barrow tried to explain his favorable position, Hawkins replied bluntly: "Unless you want a new

Street Fighting, 1950

Marines battling their way through the contested boulevards and back alleys of Seoul in September 1950 did so without benefit of the modern-day doctrine and training for "military operations in urban terrain." Street fighting at that time was an uncommon Marine experience. There had been a bloody two-day fight in downtown Vera Cruz, Mexico, in 1914, where Major Smedley D. Butler and Lieutenant Colonel Wendell C. "Whispering Buck" Neville led their men with axes and bayonets in attacking through the walls of the row-houses. Thirty years later, a different generation of Marines fought the Japanese through the burning streets of Garapan, Saipan, and again on a larger scale in the spring of 1945 amid the rubble of Naha, Okinawa.

But Seoul dwarfed Vera Cruz, Garapan, and Naha combined. An enormous, sprawling city dominated by steep hills, awash with terrified refugees, and stoutly defended by more than 20,000 North Koreans, Seoul constituted the largest, single objective ever assigned the Marines. Hue City in 1968 would take the Marines longer to recapture, but the casualties incurred at Hue, bad as they were, would not total half those sustained by the 1st Marine Division at Seoul.

National Archives Photo (USA) 111-SC-351392

Street fighting in Seoul involved forcibly uprooting the NKPA troops from either their roadblock barricades or their isolated strongpoints within or atop the buildings. Both required teamwork: engineers, tanks, and infantry for the barricades (often supported by artillery or Corsair strikes), and rifle squads supported by rocket launchers and scout-sniper teams against the strongpoints.

Door-to-door fighting proved to be as tense and exhausting in 1950 as it had been in Vera Cruz in 1914. As Private First Class Morgan Brainard of the 1st Marines recalled the action: "The tension from these little forays whittled us pretty keen . . . . I think if one's own mother had suddenly leapt out in front of us she would have been cut down immediately, and we all would probably have cheered with the break in tension." Brainard's company commander, Captain Robert H. Barrow, told a Headquarters Marine Corps tactics review board in 1951 that he quickly came to value the 3.5-inch rocket launcher in applications other than antitank defense. "We employed it in a very effective manner in Yongdungpo and in Seoul in the destruction of houses that had enemy in them. In many instances [our] 3.5 [gunners] simply shot at some of these fragile houses killing all the occupants."

The presence of so many civilian refugees in the streets and rubble vastly complicated the battle and necessitated extraordinary measures to ensure target identification and limit indiscriminate firing. Whenever troops stopped to reorganize "children appeared among them," observed the *Life* magazine photographer David Douglas Duncan. "Children gentle and tiny and wide-eyed as they fastened themselves to the men who first ignored them . . . then dug them their own little foxholes and expertly adapted helmets to fit their baby heads." Enemy snipers, mines, and long-range, heavy caliber antitank rifles took a toll among Marines and civilians alike. The ancient city became a ghastly killing ground.

battalion commander, you will attack at once." Barrow managed to convince Hawkins to come and see the situation for himself. Hawkins marveled at the abundance of targets under direct observation: "Get more mortars in there—get more artillery."

Yet Hawkins remained agitated, and Barrow soon saddled up his gunners and forward observers and plunged forward downhill into the maze of streets and railroad tracks (3d Battalion, 1st Marines, had Ma Po Boulevard; 1st Battalion, 1st Marines' axis of advance was less straightforward). Barrow and other junior officers in the 1st Marines later concluded that the pressure to advance had come down several echelons, possibly from the Tokyo headquarters of General MacArthur in his desire to recapture the capital by the symbolic third-month anniversary of its loss. "Who knows?" Barrow asked rhetorically. "Puller was

Photo by Frank Noel, Associated Press

The North Koreans built their barricades with burlap bags filled with dirt, rubble, or rice. Each position took the Marines an average of 45 to 60 minutes to overcome. Here a Marine rifleman scampers through a recently abandoned barricade during heavy fighting in Seoul's downtown business district.

being pushed by somebody in division. The division was being pushed by someone in Tenth Corps, and the corps was being pushed by the man himself, or someone speaking for him, back in Tokyo."

Top-level pressure notwithstanding, the two lead battalions of the 1st Marines could advance only 2,000 yards on the 25th. "Our advance this day was a foot-by-foot basis," said Lieutenant Colonel Ridge. North Korean mines knocked out two of Captain Williams' Pershing tanks; other vehicles sustained multiple hits from direct fire weapons. Ridge hunkered in for the night along Hill 97; Hawkins occupied Hill 82 to Ridge's immediate right rear. Company G and Weapons Company of 3d Battalion, 1st Marines, occupied the forward position, a roadblock protecting a key bridge on Ma Po Boulevard. The front lines were jagged; the North Koreans occupied several worrisome salients in close proximity.

Ridge directed Major Edwin H. Simmons, commanding Weapons Company, to coordinate the battalion's forward defenses. Simmons fortified the roadblock with two rifle squads, a section of his Browning heavy machine guns, a rocket squad, and a 75mm recoilless rifle section borrowed from the regimental antitank company. After supervising his attached engineers as they laid a series of antitank mines on the bridge, Simmons established his observation post (OP) in the cellar of an abandoned house on a rise to the left rear of the roadblock, protected by four additional heavy machine guns. His 81mm mortar platoon occupied uncommonly close firing positions 150 yards rearward, connected by phone

Under the watchful gaze of Joseph Stalin and Kim Il Sung, Marines crouch behind a barricade as enemy snipers resist their advance.

Photo courtesy of *Leatherneck* Magazine

wire to the OP. These were reasonable precautions given the volatile nature of the street fighting during the day and the nearby reentrants occupied by the North Koreans. Parts of the city still burned from the day's fighting, but the streets seemed quiet.

Then, shortly after 2000, a flash message from X Corps arrived in the division command post. Aerial observers had just reported "enemy fleeing city of Seoul on road north of Uijongbu." General Almond, sensing a great opportunity to crush the North Koreans, ordered an immediate advance by the 1st Marine Division, stating: "You will push attack now to the limit of your objectives in order to insure maximum destruction of enemy forces. Signed Almond."

The flash message stunned Colonel Bowser. The order was rife with unanswered questions—did Almond envision a five-mile night attack through the heart of the city by converging regiments out of direct contact with each other? And, by the way, how could an aerial observer distinguish at night between a column of retreating troops and a column of fleeing refugees? Bowser called his counterpart at X Corps with these questions but got nowhere. Neither did General Smith a moment later in a call to Almond's chief of staff. Smith shook his head and ordered his regimental commanders to comply—carefully. Throughout their smoking third of the city, the 1st Marine Division stirred and bitched. As one company commander queried: "A night attack without a reconnaissance or rehearsal? What are our objectives?" Private First Class Morgan Brainard recalled the grousing in the ranks that night: "We were all rousted out and mustered down on the darkened street by platoons. Scuttlebutt said we were going into the heart of Seoul in a surprise night attack."

After allowing his regimental commanders plenty of time to coordinate their plans, General Smith ordered the advance to kick off at 0145 following a 15-minute artillery preparation. The enemy moved first. Before midnight a sizable NKPA force hit Lieutenant Colonel Taplett's 3d Battalion, 5th Marines, on Hill 105-North. Lieutenant Colonel Murray and his executive officer attempted to make sense of the situation: "I'm afraid we'll have to delay pursuit of the 'fleeing enemy' until we see if Tap can beat off the counterattack."

As Major Simmons listened uneasily to the sounds of Taplett's firefight, less than 1,000 yards west, he received a call from Lieutenant Colonel Ridge ordering

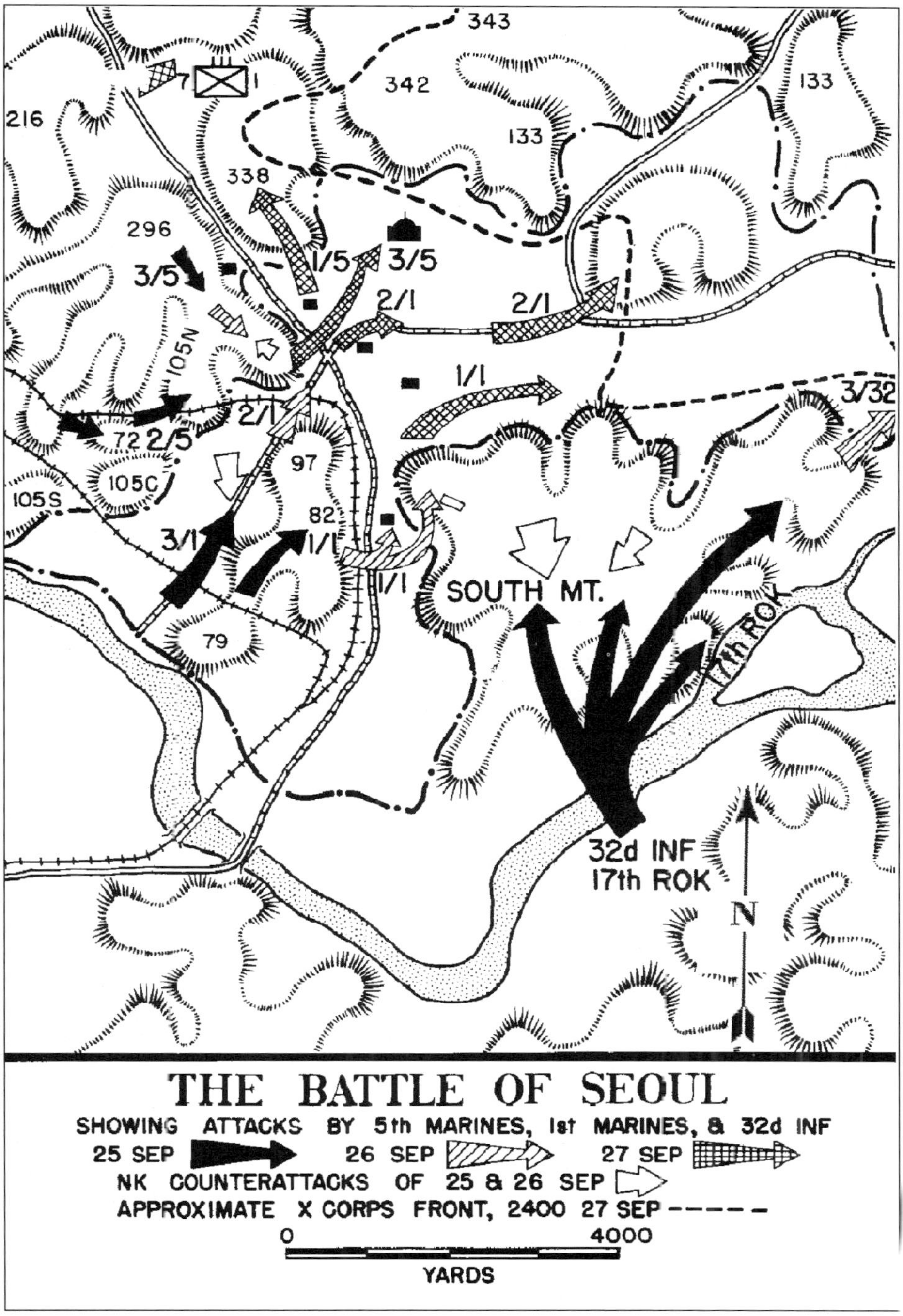

THE BATTLE OF SEOUL
SHOWING ATTACKS BY 5th MARINES, 1st MARINES, & 32d INF
25 SEP 26 SEP 27 SEP
NK COUNTERATTACKS OF 25 & 26 SEP
APPROXIMATE X CORPS FRONT, 2400 27 SEP
0 4000
YARDS

Sketch by Col Charles H. Waterhouse, USMCR (Ret)

A Marine artillery forward observer team adjusts supporting fires on enemy barricades in downtown Seoul. The firing batteries were south of the Han River. Forward observer teams had to adjust their fire with utmost precision in the crowded city.

him to dispatch a patrol to link with a similar patrol from the 5th Marines to facilitate the forthcoming night attack. Simmons protested the order. From the volume of fire to the west, a considerable NKPA force had moved between the two regiments. "I doubted a patrol could get through," said Simmons. Ridge repeated the order. Simmons assembled a patrol of Company G riflemen, led by Corporal Charles E. Collins. They departed about 1245. "I felt like I was kissing them goodbye," Simmons admitted.

The onset of the artillery preparatory fires heightened Simmons' concern for his patrol. Colonel Puller worried that the fire was inadequate for a general assault. At 0138, he asked Smith for a second fire mission, delaying the jump-off time to 0200. Fifteen minutes later the whole issue became moot.

Major Simmons first heard sounds of a nearby firefight and realized Collins' patrol had been intercepted. A moment later, at 0153, he heard the unmistakable sounds of tracked vehicles approaching the roadblock from the north, along with an almost instantaneous *crack!* of a Soviet T-34 85mm tank gun. The shell missed Simmons by inches and killed his radio operator at his side. Shaken, Simmons sounded the alarm. Far from fleeing the city, the enemy—at least this particular battalion of the *25th NKPA Brigade*—was charging due south down Ma Po Boulevard with six to 12 tanks and self-propelled guns, accompanied by infantry. As his roadblock defenders cut loose on the enemy tanks, Simmons called for artillery and 81mm mortar concentrations along the bridge, and the battle raged. General Smith, sobered by the ferocity of the NKPA assaults, postponed the division's night attack indefinitely.

The Marines would soon call the northwestern nose of Hill 97 "Slaughterhouse Hill," and from its slopes this night they inflicted a killing zone of epic proportions against the attacking armored column. Three battalions of the 11th Marines fired incessantly the next 90 minutes. At that point the tubes became so hot the howitzers had to ceasefire until they could cool down. In the lull, the NKPA tanks surged forward again. Simmons unleashed his beloved heavy Browning machine guns. "In the light of the burning buildings," he said, "I could see three [tanks} clearly, rolling forward on [the] boulevard about 500 yards to my front." Simmons saw the tracers from the Brownings whanging off the faceplates of the tanks. He asked for 155mm howitzer fire from the Army. The 31st Field Artillery Battalion responded with awesome firepower—360 rounds along 3d Battalion, 1st Marines' direct front.

Chesty Puller did not recognize

the radio call sign of the Army artillery liaison officer coordinating the 155mm howitzer missions that night, but he knew first-class fire support when he saw it. "This is Blade," he growled into his handset, "I don't know who in the hell you are, but thank God! Out."

The Army fire mission destroyed or disabled the last of the NKPA tanks threatening the 3d Battalion's roadblock, but several immobilized vehicles maintained a stubborn fire. One self-propelled gun continued to fire at Simmons' observation post, each shell screeching overhead barely a degree in elevation too high. Simmons feared the coming dawn would make his position terribly exposed, so he moved one of the 75mm recoilless rifles from the roadblock to the rubble-strewn front yard of the abandoned house. The crew stared anxiously into the darkness just north of the bridge, hoping to get off the first shot at dawn. Finally, in the gray half-light, the gunner spotted the enemy vehicle and squeezed his trigger. The round was a pinwheel hit—the self-propelled gun burst into flames. But the Marines had forgotten to consider the back blast of the recoilless rifle. "It bounced off the mud-and-wattle side of the house behind us and knocked us head-over-heels," Simmons said, adding "we thought it very funny at the time."

Sunrise brought Simmons more welcome news. Corporal Collins, having ordered the rest of his patrol back to the roadblock at their first encounter with the approaching NKPA armored column, covered its retreat with rifle fire, and then took refuge for the night in a cellar. Somehow he found a set of white robes commonly worn by the Korean civilians. Thus attired, he made his way through the still-dangerous streets to the 3d Battalion, 1st Marines' lines and safety.

The North Koreans executed a third major spoiling attack at 0500, launching a reinforced battalion against the 32d Infantry's positions on Nam-san. The Army regiment stood its ground and did not get rattled when one company was overrun. Making good use of his supporting arms, Colonel Charles Beauchamp organized a counterattack that drove the enemy out of the position and inflicted several hundred casualties.

Marine riflemen evacuate their wounded buddy under heavy enemy fire.
Department of Defense Photo (USA) SC351385

At daybreak, Colonel Puller arrived at Lieutenant Colonel Ridge's position. "You had better show me some results of this alleged battle you had last night," he warned. Ridge was unperturbed. He showed Puller the wreckage of the NKPA vehicles north of the bridge, the ruins of seven tanks, two self-propelled guns, and eight 45mm antitank guns. At least 250 dead North Koreans lay in clots along the boulevard (the official figure of 475 may have included those slain by Lieutenant Colonel Taplett's 3d Battalion, 5th Marines, that same night), and there were more than 80 prisoners in hand. The Marines' side of the battlefield seemed covered with a river of spent brass shell casings. Major Simmons' 10 Browning heavy machine guns had fired a phenomenal 120 boxes of ammunition during the night—30,000 rounds, a feat that even surpassed the volume fired by the legendary Sergeant "Manila John" Basilone at Guadalcanal in 1942 in Puller's old battalion. Colonel Puller flashed a rare grin.

Time magazine's combat correspondent Dwight Martin described the battlefield the morning of the 26th, as Sutter's 2d Battalion, 1st Marines, passed through Ridge's 1st Battalion:

> This morning Ma-Po wore a different look. The burned and blackened remains of the boulevard's shops and homes sent clouds of acrid smoke billowing over the city. Buildings still ablaze showered sparks and ashes high into the air to cascade down

This page and the next, street fighting in Seoul as captured by Life *magazine photographer David Douglas Duncan.*

프린트
34
34

on red-eyed, soot-faced Marines.

Given these circumstances, it is not surprising that the Marines greeted with hoots of derision the communique by General Almond that Seoul had been liberated at 1400 the previous afternoon, the 25th of September. "Three months to the day after the North Koreans launched their surprise attack south of the 38th Parallel," the message proclaimed, "the combat troops of X Corps recaptured the capital city of Seoul." To their astonishment, the Marines learned that their corps commander considered the military defenses of Seoul to be broken. "The enemy is fleeing the city to the northeast," the communique concluded. An Associated Press correspondent reflected the infantry's skepticism: "If the city had been liberated, the remaining North Koreans did not know it."

In truth the Marines and soldiers would still be fighting for full possession of the capital 48 hours past General Almond's announced liberation date, but the issue was insignificant. The troops viewed the battle from purely a tactical perspective; their corps commander sensed the political ramifications. Of far greater significance at this point was the fact that five infantry regiments with a total lack of experience waging coalition warfare with combined arms in an enormous urban center were nevertheless prevailing against a well-armed, disciplined enemy. General MacArthur's visionary stroke at Inchon had succeeded in investing the city of Seoul in just 11 days. In view of the allies' disheartening performance in the Korean War to date, MacArthur, and Almond, had earned the right to boast.

Further, although the Marines might not like to admit it, General Almond was essentially correct in his flash message the night of the 25th—the main body of the North Korean defenders, the remnants of a division, was indeed retreating north. What surprised all components of X Corps was the NKPA decision to expend the equivalent of an armored brigade in suicidal night attacks and die-hard defense of the main barricades to keep the Americans ensnared in the city.

Analyzing the NKPA decision to evacuate the main body of their defenders from Seoul is always risky, but there is evidence that the pullback resulted as much from their surprise at the unexpected crossing of the 32d Infantry and the 17th ROK Infantry from the southeast on the 25th—paired with the rapid advance of the 7th Marines, threatening the northern escape routes—as from the steady but predictable advance of the 5th and 1st Marines. Regardless, it was obvious to Almond and O. P. Smith that seizing such a mammoth objective as Seoul would require uncommon teamwork among Services, nations, and combat arms. Allied teamwork throughout the night attacks of 25-26 September had proven exemplary.

The Marines employed Corsairs and artillery to soften the barricades, then switched to 4.2-inch and 81mm mortars. The assault companies delivered machine gun and rocket fire on the fortifications to cover the deliberate minesweeping operations by combat engineers. Then came the M-26 Pershing tanks, often with other tanks modified as flamethrowers or bulldozers. On the heels of the tanks came the infantry with fixed bayonets. The process was unavoidably time-consuming—each barricade required 45-60 minutes to overrun—and each of these intermediate objectives took its toll in Marine and civilian casualties. The city smoked and burned.

As Lieutenant Colonel Jack Hawkins' 1st Battalion, 1st Marines, fought its way clear of the railroad yards and entered a parallel thoroughfare his riflemen stared in horror at the rampant destruction. As it appeared to Private First Class Morgan Brainard, the scene was one of "great gaping skeletons of blackened buildings with their windows blown out...telephone wires hanging down loosely from their drunken, leaning poles; glass and bricks everywhere; literally a town shot to hell."

Not all the fighting took place around the barricaded intersections. There were plenty of other NKPA soldiers holed up in the buildings and rooftops. Many of these soldiers became the prey of Marine scout-sniper teams, some armed with old Springfield '03 bolt action rifles fitted with scopes, others favoring the much newer M1-C semi-automatic rifles, match-conditioned weapons graced with cheek pads, flash suppressors, leather slings, and 2.2x telescopic sights. The snipers often worked in teams of two. One man used binoculars or a spotting scope to find targets for the shooter.

Many of the buildings in the city center were multi-story, and, according to Private First Class Brainard, "it meant going up the stairs and kicking open the doors of each room, and searching the balconies and backyard gardens as well." Often the Marines had to fight their way through the buildings, smashing their way through the walls like Smedley Butler's Marines in Vera Cruz in 1914.

Colonel Puller led his regiment from very near the forward elements. On this day he dismounted from his jeep and stalked up Ma Po Boulevard shortly behind Lieutenant Colonel Sutter's 2d

Combat Engineers in the Seoul Campaign

Among the many unsung heroes who provided ongoing combat support to the infantry regiments of the 1st Marine Division in the recapture of Seoul were the dauntless practitioners of Lieutenant Colonel John H. Partridge's 1st Engineer Battalion. As did the division as a whole, the engineers represented an amalgam of World War II veterans, new recruits, and a spirited group of reservists, including members of the 3d Engineer Company, United States Marine Corps Reserve, from Phoenix, Arizona.

Fortunately, the Inchon landing caught the NKPA forces in the region off guard, and the battalion had time to shake itself down in non-urgent missions before breaking into small units to tackle enemy minefields. The engineers at first cleared beach exits and assembly areas in the Inchon area, then moved out to help reconnoiter the roads leading east to Seoul. Of immediate concern to Major General Oliver P. Smith and his operations officer, Colonel Alpha L. Bowser, Jr., was whether the numerous bridges along the highways and secondary roads were sturdy enough to support the Marines' new M-26 Pershing tank with its 46 tons of combat-loaded weight.

The Marines encountered the first serious NKPA minefields (both antitank and anti-personnel) in the vicinity of Kimpo Airfield. The subsequent arrival of highly-trained, first-line North Korean reinforcements in defensive positions guarding the approaches to Seoul led to minefields of increasing size and sophistication. Soviet Red Army advisors had trained the NKPA in mine warfare, and many of the mines encountered by the Marines were made in Russia. These mines slowed the advance of the 1st Marine Division as it reached the outer defenses of Seoul along the Kalchon west of Yongdungpo or the avenues of approach to Hill 296 and its many subordinate peaks and ridgelines. Partridge's engineer teams performed their high-stress mine-clearing missions with progressive efficiency. This helped sustain the division's momentum and limited the time available to the enemy to more fully develop defensive positions within the city.

In Seoul, the Marines encountered barricaded roadblocks every 200 to 300 yards along the main boulevards. The North Koreans seeded most approaches with mines. The Marines formulated the necessary teamwork on the spot. The rifle company commander would shower the obstacle with fire, including smoke or white phosphorus mortar shells. Under this cover the engineer squad would hustle forward to clear the mines. Behind them would come the tanks, followed by the infantry. It was dangerous, often costly work. Sometimes a mine would detonate among the engineers. Sometimes they would miss a string and a tank would be lost. Most often, however, this painstaking process worked. Each barricade took an average of 45 minutes to clear. Utilizing this well-coordinated and increasingly proficient approach, the infantry battalions of the 1st Marines advanced an average of 1,200 yards each day—a small gain on a map, but an inexorable advance to the North Koreans.

The 1st Engineers provided another exceptional service to the division in the Seoul campaign—river cross-

Gen Oliver P. Smith Collection, Marine Corps Research Center

ing. The Han was a broad tidal river, and its few bridges had been blown up during the first week of the war. The Marines had amphibian tractors and DUKWs on hand to transport riflemen and small vehicles across against the current, but the campaign would never succeed without the means of ferrying, first, tanks and artillery, then heavy trucks and trailers. Major General Edward A. Almond, USA, commanding X Corps, had proven disingenuous in his repeated assurances that heavy bridging material would arrive in plenty of time to support the Marines' crossing in force. "General Almond promises bridge material," General O. P. Smith recorded in his journal. "This is an empty promise."

Thinking ahead, and acting on his own initiative, Lieutenant Colonel Partridge had obtained a pair of 50-ton pontoons in San Diego and had zealously protected them from the enraged embarkation officer who had to somehow load these unwelcome monsters on ships already stuffed with "essential" combat cargo. The pontoons proved priceless. Partridge had at least one section on hand to support the crossing of the 5th Marines at Haengju on 20 September. Lieutenant Colonel Henry P. Crowe's 1st Shore Party Battalion quickly established a smooth functioning ferry service, doubling their productivity with the arrival of Partridge's second pontoon. Here the 7th Marines crossed, as well as the company of M-26 tanks needed so direly by Colonel Puller in his first full day of street fighting in Seoul. General Almond's bridging material arrived in time to support the crossing of General MacArthur's official party as they arrived in Seoul on 29 September.

Greater glories awaited the 1st Engineer Battalion in the forthcoming Chosin Reservoir campaign, where they cleared an expeditionary airfield at Hagaru-ri and assembled the air-dropped Treadway Bridge in Funchilin Pass below Koto-ri, but their yeoman performance in close support of the 1st Marine Division's assault on Seoul set a standard of combined arms operations and greatly facilitated the timely recapture of the capital.

Battalion, 1st Marines, as it clawed its way along each city block. Sergeant Orville Jones, Puller's hand-picked driver throughout 1950-55, followed his colonel in the jeep, a short distance to the rear. Sometimes the Marines fighting door-to-door along the street would be appalled to see Chesty Puller walking fully exposed and abreast of the action, Jones recalled. "'Holy Jeez,' they would yell to each other—'Don't let Chesty get ahead of us—move it!'"

Yet even the famously aggressive Chesty Puller could not expedite the methodical reduction of the barricades. Puller admitted, "progress was agonizingly slow." Said the engineer Captain Nicholas A. Canzona: "It was a dirty, frustrating fight every yard of the way."

Army Lieutenant Robert L. Strickland, a World War II veteran now assigned as a cameraman for X Corps, got caught up in the street fighting. He sought shelter in an open courtyard behind a burning building, but the enemy fire came from all directions. "We got so much fire of all kinds that I lost count," he said. "I have seen a lot of men get hit both in this war and in World War II, but I think I have never seen so many get hit so fast in such a small area."

David Douglas Duncan, veteran Marine and extraordinary combat photographer for *Life* magazine, accompanied the 1st Battalion, 1st Marines, during their advance through the rail yards towards the station. Describing the action in his subsequent photo book *This is War*, Duncan highlighted the timely arrival of Marine Pershing tanks that "growled up across the railroad tracks, into the plaza—and met the enemy fire head on." Then, Duncan continued: "The tanks traded round for round with the heavily-armed, barricaded enemy—and chunks of armor and bits of barricade were blown high into the air. They were killing themselves at point-blank range."

Private First Class Brainard of Company A, 1st Marines, described a barricade that had just been demolished by a pair of M-26 Pershing tanks:

> We pass by the barricade which had been constructed with large-sized rice bags . . . and also with odd bits of furniture, such as tables, chairs, and wooden doors, all piled up together. There were about ten dead gooks sprawled in and around the obstruction, and the blackened antitank gun was tipped over on its side with lots of unused shells scattered around it.

The 1st and 5th Marines were now converging close enough that Colonel Puller's men could clearly see Lieutenant Colonel Murray's troops still fighting to clear the final eastern finger of Hill 296, the ridge that extended into the heart of Seoul. Certain riflemen in 1st Battalion, 1st Marines, spoke admiringly of the 5th Marines being "once more on top of the highest hill in the local vicinity—born billy goats."

The 5th Marines may have appreciated the compliment, but by 26 September they were sick and tired of the steep northwest approaches and the stubbornly defending remnants of the *25th*

Photo by Frank Noel, Associated Press

While the Pershing tank "Dead Eye Dick" advances beyond a captured North Korean barricade, a Marine sniper team waits for the 45mm antitank rounds to abate before moving into new firing positions.

NKPA Brigade. Captain Robert A. McMullen's Company I, the men who had spearheaded Lieutenant Colonel Taplett's crossing of the Han back at Haengju and earned the praise of General Almond by their double envelopment of Hill 125, would again be in the spotlight on the 26th. Taplett assigned McMullen the mission of sweeping the eastern terminus of the huge lower spur of Hill 296 that extended very near the major intersection of the Kaesong-Seoul highway and Ma Po Boulevard. Ahead, less than a mile to the northeast lay Government House. And not far beyond the palace was the boundary between the 1st and 7th Marines. By design, Murray's regiment, which had sustained the highest casualties the preceding week, was close to being pinched out and assigned a reserve role.

A brief helping hand from a Marine amid a day of great terror for the civilians—high explosives, burning buildings, downed power lines, and scattered families.

Photo by David Douglas Duncan

Private First Class Eugene A. Obregon

Born in November 1930, Private First Class Obregon enlisted in the Marine Corps in June 1948. Assigned to the 5th Marines, he was part of the 1st Provisional Marine Brigade, which was rushed to Korea in August 1950. He participated in the bloody battles at the Naktong River—crucial victories, which helped save the Pusan Perimeter from collapse.

When the 5th Marines re-embarked to join the 1st Marine Division for the assault landing at Inchon on 15 September, Obregon again took part. On 26 September, during the battle to recapture the South Korean capital, his heroic actions were recognized by a posthumous award of the Medal of Honor. The official citation reads, in part:

> For conspicuous gallantry and intrepidity at the risk of his life above and beyond the call of duty while serving with Company G, Third Battalion, Fifth Marines, First Marine Division (Reinforced), in action against enemy aggressor forces at Seoul, Korea, on 26 September 1950. While serving as an ammunition carrier of a machine-gun squad in a Marine rifle company which was temporarily pinned down by hostile fire, Private First Class Obregon observed a fellow Marine fall wounded in the line of fire. Armed only with a pistol, he unhesitatingly dashed from his covered position to the side of the casualty. Firing his pistol with one hand as he ran, he grasped his comrade by the arm with his other hand and, despite the great peril to himself, dragged him to the side of the road.
>
> Still under enemy fire, he was bandaging the man's wounds when hostile troops of approximately platoon strength began advancing toward his position. Quickly seizing the wounded Marine's carbine, he placed his own body as a shield in front of him and lay there firing accurately and effectively into the hostile group until he himself was fatally wounded by enemy machine-gun-fire. Private First Class Obregon enabled his fellow Marines to rescue the wounded man and aided essentially in repelling the attack.

Department of Defense Photo (USMC) A43987-B

The fellow Marine, whose life Obregon had saved, was Private First Class Bert M. Johnson. He recovered from his wounds and was returned to active duty. Obregon's sacrifice was memorialized when a building at Camp Pendleton, a ship, and a high school in the Los Angeles area were named after him.

—Captain John C. Chapin, USMCR (Ret)

But Hill 296 and Colonel Wol's hard-core survivors were not through with the 5th Marines. Company I's stouthearted advance encountered fierce opposition from the start. At one point McMullen led his troops into a maze of trenches manned by 200 North Koreans and forced them out by the sheer velocity of the assault—only to lose the position to a vicious counterattack. The two forces struggled across this contested ground the balance of the afternoon. At day's end the Marines held the field but were too depleted to exploit their advantage. Captain McMullen fell wounded and was evacuated. He had qualified for his seventh Purple Heart in two wars. Two Marines in Company G, 3d Battalion, 5th Marines, fighting in somewhat lower ground adjacent to Company I's battlefield, each received their fifth wound since the regiment's arrival in Pusan 53 days earlier.

Elsewhere during Company G's day-long fight, Corporal Bert Johnson, a machine-gunner, and Private First Class Eugene A. Obregon, his ammunition humper, tried to set up their weapon in an advanced position. The North Koreans charged, wounding Johnson with submachine gun fire. Obregon emptied his pistol at the shadows closing in, then dragged

Photo by David Douglas Duncan

Frightened civilians scurry to shelter while three Marine Corps Pershing tanks duel with North Korean antitank guns at a barricaded intersection along Ma Po Boulevard.

Johnson to a defilade position to dress his wounds. When the enemy swarmed too close, Obregon picked up a carbine and emptied the clip, always shielding Johnson with his body. There were too many of them, and in the end the North Koreans shot him to pieces. But Obregon had delayed their attack long enough for other Marines to hustle down the slope and rescue Johnson. Private First Class Obregon's family would receive his posthumous Medal of Honor.

The two rifle companies had fought their damnedest, but the 5th Marines still could not fight their way clear of the highlands. Stymied, Lieutenant Colonel Raymond Murray marshaled his forces for the final breakthrough on the morrow.

Nor was Murray in a position to maximize his supporting arms, as he had been able to do in the earlier assault on Smith's Ridge. He noted with some envy the volume of heavy-caliber indirect fire and the frequent Corsair missions being delivered in support of Puller's advance to his right front. "Chesty used a lot of artillery," Murray said later. "And you could almost see a boundary line between the two of [us], the smoke coming up from his sector and very little smoke coming up from mine." Lieutenant Colonel Jon Hoffman, author of Puller's definitive biography, noted the irony that Puller had been criticized six years earlier at Peleliu for abjuring supporting arms while his infantry elements shattered themselves in direct assaults against Bloody Nose Ridge. By comparison, Colonel Harold D. "Bucky" Harris, commander of the 5th Marines at Peleliu, had received praise for his policy of being "lavish with ordnance and stingy with the lives of my men." Now, in the streets of Seoul, it was Puller's turn to be "lavish with ordnance."

Another bitter lesson learned by the 1st Marine Division at Peleliu was how to protect its tanks from suicide sapper attacks. The "Old Breed" was the only division in the subsequent battle of Okinawa not to lose a tank to Japanese sappers. In downtown Seoul on 26 September, however, this distinctive streak ended. A nimble-footed North Korean darted out from the rubble, caught 2d Battalion, 1st Marines' riflemen by surprise, and flung a satchel charge atop a passing flame tank, then vanished in the blast and smoke. The crew escaped unscathed, but the tank was destroyed. Angered and embarrassed by this bad luck, 2d Battalion's NCOs forcibly reminded their men to watch the adjacent alleys and rubble piles, not the tanks. This paid off. The NKPA launched a dozen more sapper attacks against Marine tanks operating in the center of the boulevard; Lieutenant Colonel Sutter's troops cut each one of them down.

There was no real "school solution" that applied to the kaleidoscopic action taking place on the streets of Seoul on the 26th of September. Captain Norman R. Stanford, commanding Company E, 2d Battalion, 1st Marines, had as much tactical experience as anyone on the scene, having served as a company commander in the 1st Marines throughout Peleliu and Okinawa. Sutter ordered Stanford to follow Company F up the boulevard in trace, then take the right fork while Company F took the left at a designated intersection ahead. Sutter's closing guidance was succinct: "Move out fast and keep going." But Company F encountered a particularly nasty barricade just past the intersection and could not advance up the left fork.

Captain Stanford went forward to assess the delay. From 200 yards away the NKPA barricade looked unassailable:

> I took one look at the AT [antitank] muzzle blasts kicking aside the pall of smoke over the roadblock, and I

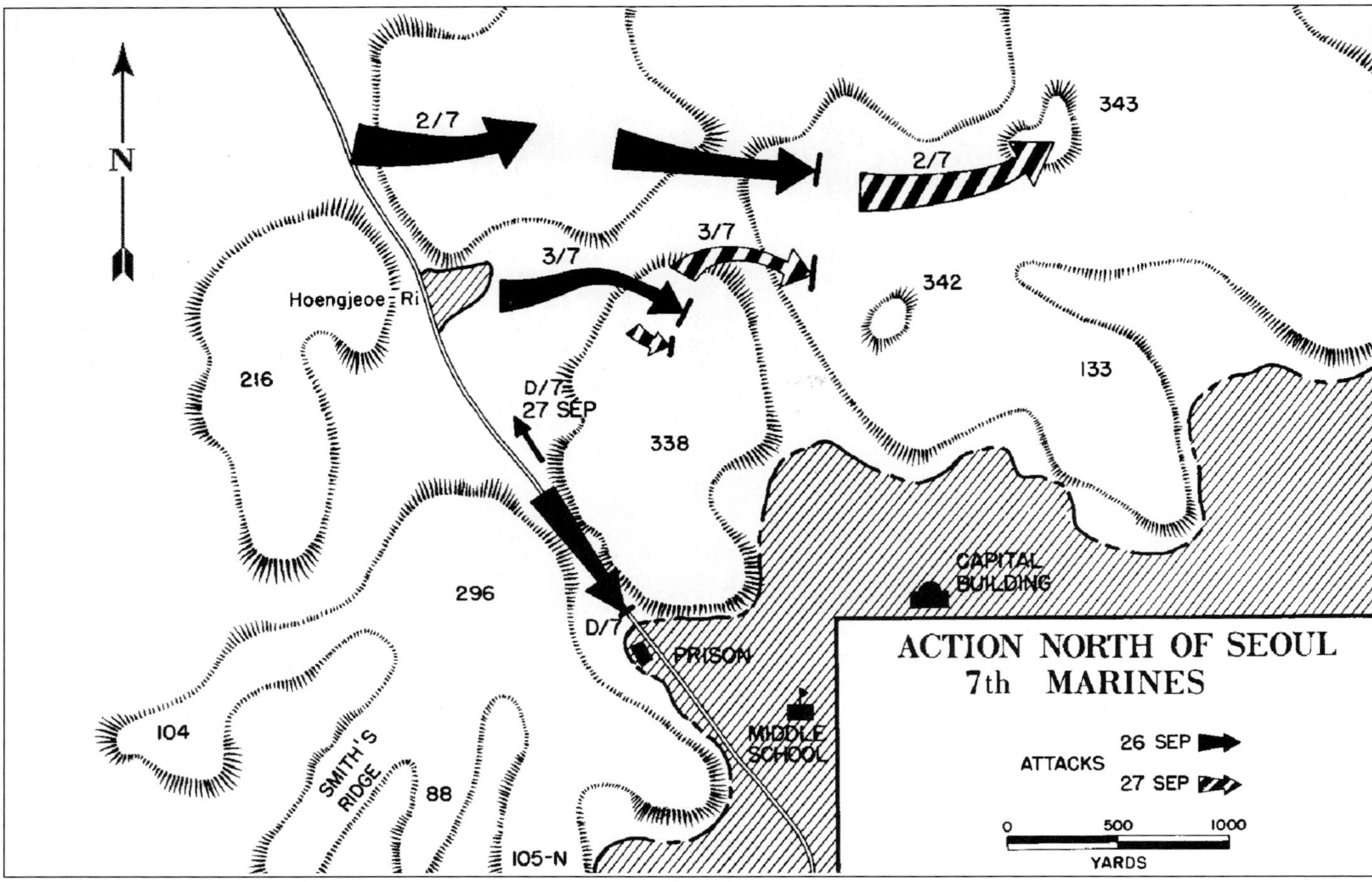

glanced at the thin flicker of automatic fire running across the barricade like a single line of flame and dived off the sidewalk into an alley.

Stanford's radio failed at this critical juncture. He had the option of bypassing Company F and the barricade and carrying out his assigned mission along the right fork, notwithstanding his naked left flank, or bowling straight ahead through Company F, smashing the barricade, and attacking with Company E up the left fork. He had the firepower—four tanks, an engineer platoon, rocket squads, and a 75mm recoilless rifle section attached. "I knew that we could go through anything for 250 yards," he said, risking the second option. He hurled his forces forward, towards the barricade. "We had it hot and heavy among the burning buildings and the crumbled sandbags of the barricade, and then they broke and ran . . . and we butchered them among the Russian AT [antitank] guns and the Japanese Nambu machine guns." Company E lost two officers and 18 men in their headlong assault. Captain Stanford was one of the wounded.

Sutter's battalion, like Taplett's along the ridge just to the west, had fought their best, but "the fleeing enemy" had limited his advance to 1,200 yards. Seoul would not fall this day.

Further to the northwest, and now not very far away, the 7th Marines veered towards the capital in keeping with O. P. Smith's orders to pinch out the 5th Marines. Company D, 2d Battalion, 7th Marines, led the advance along the Kaesong Highway as it threaded through two towering hills, the now-infamous Hill 296 on the right and Hill 338 on the left. First Lieutenant William F. Goggin, the machine gun officer, led the advance party.

Compared to all the grief being experienced by the other two regiments on the 26th, the 7th Marines enjoyed what at first appeared to be a cakewalk. Thousands of grateful civilians thronged the right-of-ways and hillsides, cheering the approaching Marines. It was an uncommon experience for Marines of any war to date, a welcome grace note to serve as a partial offset for the horrors to come. The North Koreans, of course, took prompt advantage of this opportunity.

The dense crowds prevented Company D from maintaining its own outriding flank protection along the ridges on both sides of the road and caused the van to overshoot the intended linkup point with the 5th Marines. The company unwittingly entered the city and the final defenses of one of the sacrificial battalions left behind by the departed *25th NKPA Brigade*.

Sudden machine gun fire from the front felled First Lieutenant William F. Goggin, halted the column, and created panic among the well wishers. Then other machine guns opened up at close range along the high ground on both sides. Another enemy force scrambled downhill to establish a blocking position in the rear. Company D was abruptly encircled and cut off.

Captain Richard R. Breen, though wounded early in the fighting, maintained his presence of mind. He still commanded a large, fresh, well-armed company. Once the civilians vanished and his Marines went to ground in good firing positions, he figured his men could hold their own, despite the danger. When Colonel Litzenberg called to see what kind of help he needed, Breen answered calmly, "We're okay, Colonel."

Had Company D's entrapment occurred two days earlier the ensuing darkness might have proven catastrophic, but by now the NKPA forces lacked the punch to finish the job. Additionally, Captain Breen received some spectacular help. Two U.S. Air Force C-47s dropped ammunition, rations, and medicine to the surrounded Marines just before dusk (one plane, badly shot up by North Korean antiaircraft gunners, had to crash-land at Kimpo). During the night Lieutenant (junior grade) Edward Burns, USN, the regimental surgeon, led a high-balling convoy of jeep ambulances through the enemy perimeter to retrieve 40 of Company D's most seriously wounded men.

Lieutenant Colonel Parry's 3d Battalion, 11th Marines, still in direct support of the 7th Marines, was the first artillery unit to cross the Han. At that point the infantry regiment extended from the north bank of the Haengju ferry crossing

By the third full day of the battle for Seoul, the city lay in ruins. Shell holes buckled the streets, rubble lay strewn everywhere, and a thousand fires blazed furiously. Exhausted Marines regroup for the final barricade assault.

Photo by Frank Noel, Associated Press

to the edge of Seoul, "a sector of 18 miles," said Parry, which required him to deploy "three batteries on three separate azimuths." Company D's encirclement on the edge of Seoul on the 26th caused a predicament. The company had crossed into the 5th Marines zone, and "it was several hours before we were able to obtain clearance to shoot." But Parry's gunners made up for the delays with pinpoint defensive fires around the Company D perimeter throughout the night. "We were credited by the company commander with saving their bacon," Parry said. The anticipated NKPA night attack did not materialize.

By now all of Colonel Homer L. Litzenberg's 7th Marines had received their separate baptisms of fire. One member of Company B, 1st Battalion, 7th Marines, recalled his own first combat encounter:

> The company was above Seoul when we ran into a firefight. We were moving at night. [There were] green tracers coming in, red tracers going out. It was confusing . . . I was scared [and] pretty much hugged the ground. I didn't even know how to dig a foxhole, but the Gunnery Sergeant told me how: "Make it like a grave."

The 26th of September, though devoid of major tactical gains in the fight for Seoul, ended with a significant operational breakthrough. At Suwon, 27 miles south of Seoul, three U.S. Army tanks of the 7th Cavalry raced into the perimeter of the 7th Division shortly before midnight. The Eighth Army had fought its way clear of the Pusan Perimeter, and its leading elements had linked up with X Corps.

For the 1st Marine Division, the climax of the Inchon-Seoul campaign came on 27 September, and most of O. P. Smith's disheveled troops seemed to sense the opportunity as soon as the new day dawned. Sunrise brought a special relief to Company D, 7th Marines, after its all-night vigil in the steep pass at the city limits. Litzenberg's relief column of tanks, infantry, and engineers fought their way into the position against negligible opposition. Captain Breen received his second wound during the extraction of his company, but the volume of enemy fire had diminished sharply from the previous day. While no one enjoyed being cut off, surrounded, and pinned down for 18 hours, Company D had acquitted itself well and learned lessons that would prove valuable in the hill fights ahead.

On this day, the 5th Marines finally fought their way clear of Hill 296 and into the city streets. By 0930, Taplett's 3d Battalion had linked up with Sutter's 2d Battalion, 1st Marines. Taplett wheeled northeast, grimly aiming for the huge red banners still flying over Government House and Chang Dok Palace.

As the lead battalions of both regiments lengthened their strides, a sense of friendly rivalry spurred them into a race to raise the national colors over key landmarks. The 1st Marines fought their way into several embassies, led by Company E, pausing to raise the flag over first the French, then the Soviet (with great irony), and finally the United States residences. Growled one gunnery sergeant: "It looks like the 4th of July around here."

Lieutenant Colonel Taplett's 3d Battalion, 5th Marines, had a brief but fierce fight on its final approach to the palace. Die-hard North Koreans, bolstered by a pair of self-propelled guns, fought to

Marines of Company G, 5th Marines, jubilantly yank down the Communist flag at Government House and run up the American and United Nations flags.

Photo courtesy of *Leatherneck* Magazine

Private First Class Stanley R. Christianson

Private First Class Christianson was born in January 1925 in Mindore, Wisconsin. After he enlisted in the Marine Corps in October 1942, he served with the 2d Division in three World War II campaigns. For his services, he was awarded a Letter of Commendation. Following duty during the occupation of Japan, he had a variety of assignments, including drill instructor at Parris Island.

When the Korean War broke out, he was a member of Company E, 2d Battalion, lst Marines, and took part in the Inchon assault. For his actions at Inchon, he received a Bronze Star Medal. During the subsequent battle for Seoul, he gave his life on 29 September, at the age of 25, on Hill 132. Private First Class Chistianson's citation for the Medal of Honor awarded him reads, in part:

> Manning one of the several listening posts covering approaches to the platoon area when the enemy commenced the attack, Private First Class Christianson quickly sent another Marine to alert the rest of the platoon. Without orders, he remained in his position and, with full knowledge that he would have slight chance of escape, fired relentlessly at oncoming hostile troops attacking furiously with rifles, automatic weapons and incendiary grenades. Accounting for seven enemy dead in the immediate vicinity before his position was overrun and he himself fatally struck down, Private First Class Christianson was responsible for allowing the rest of the platoon time to man positions, build up a stronger defense on that flank and repel the attack with 41 of the enemy destroyed, and many more wounded and three taken prisoner.

Department of Defense Photo (USMC) A-43986

After the war, his sacrifice was recognized by the dedication of a statue in his honor at Camp Lejeune, North Carolina. —Captain John C. Chapin, USMCR (Ret)

the end. Taplett's tank-infantry teams carried the day. Colonel Robert D. Heinl preserved the dramatic climax: "Moving at the high port up Kwangwhamun Boulevard, Company G, 5th Marines burst into the Court of the Lions at Government House, ripped down the red flag, and Gunnery Sergeant Harold Beaver ran up those same colors his forebears had hoisted 103 years earlier atop the Palace of the Montezumas." Two Korean Marines raised their national colors at the National Palace.

The fight for Seoul continued, especially along the towering ridgelines to the north, but by dusk in the city the NKPA had ceased to offer organized resistance. Twelve days after the surprise landing at Inchon (and two days after General Almond's victory communique), X Corps had seized sole possession of the capital city of the Republic of South Korea.

The 7th Marines continued to advance through the high ground north of the city, cutting the highway from Seoul to Uijongbu on the 28th. In this fighting Lieutenant Colonel Thornton M. Hinkle, commanding 2d Battalion, 7th Marines, was wounded and evacuated. Major Webb D. Sawyer took command. Meanwhile, the 31st Infantry and 17th ROK Infantry attacked to the east, successfully sealing off the last NKPA escape routes. There were still small bands of North Korean troops loose within the city—two of these struck the 2d Battalion, 1st Marines, in predawn counterattacks as late as the 29th. The first occurred at 0445, when an observation post on Hill 132 was infiltrated by an estimated 70 to 100 North Korean troops. A second attack hit the left flank of the battalion a short time later. Both attacks were repulsed with a loss of 28 Marines wounded and four killed, among them Private First Class Stanley R. Christianson, who subsequently received the Medal of Honor for his actions. Despite these counterattacks, the war was moving north, well above Seoul. Indeed, South Korean troops were about to cross the 38th Parallel.

On 29 September, General MacArthur and South Korean President Syngman Rhee and their wives returned to Seoul for a triumphant ceremony, accompanied by a large official retinue. The concentration of so many VIPs within the smoldering city so soon after the heavy fighting made General O. P. Smith nervous. Isolated NKPA antiaircraft gunners still exacted a price against allied planes flying over the city's northern suburbs, especially the slow flying observation aircraft of VMO-6, which lost a single-engine OY and an HO3S-1 helicopter on the day of the ceremony. Smith positioned Lieutenant Colonel Taplett's 3d Battalion, 5th Marines, on the hill overlooking the palace and Lieutenant Colonel Ridge's 3d Battalion, 1st Marines, along the route to be taken by the dignitaries—out of sight, but loaded for bear.

Despite the cost of more than 700 Marine casualties in seizing most of Seoul during the climactic three days of 25-28 September, only a handful of Marines attended the commemorative ceremony. Generals Smith and Craig, Colonel Puller, and Lieutenant Colonel Murray were there (Puller barely so; when a Military Police officer barred his jeep from the sedan entrance he ordered Sergeant Jones to drive over the officious major), but Colonels Litzenberg and Brower were still fighting the war north of the city and the 1st Marine Aircraft Wing senior officers were gainfully employed elsewhere. In retrospect it is unfortunate that more of those who had battled so hard for the victory—Marines, Navy corpsmen, soldiers, ROK troops, men of all ranks and specialties, grunts and aviators alike—could not have shared this special occasion. For a moment on the afternoon of the 27th, Seoul had seemed their dearly-won city. Two days later they were being told to remain out of sight of the official celebrants.

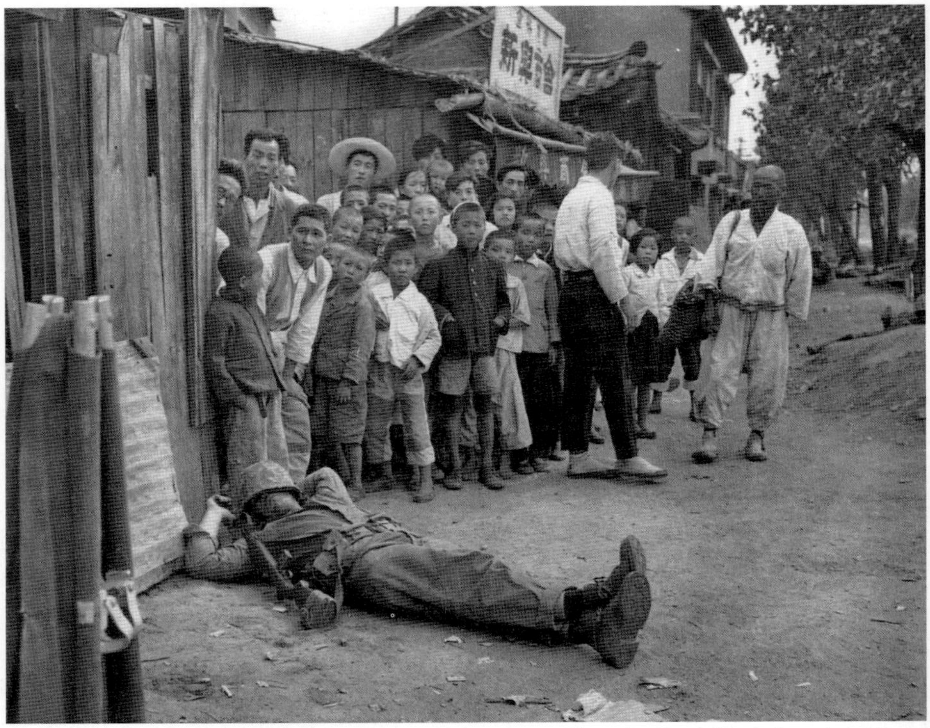

Photo by Frank Noel, Associated Press

South Korean civilians curiously observe one of their exhausted Marine liberators.

MacArthur conducted the special ceremony at high noon in the National Palace, ignoring the tinkle of broken glass that fell from the ceiling dome windows with every concussive rumble of distant artillery. "Mr. President," he intoned in his marvelous baritone voice, "By the grace of a merciful Providence our forces fighting under the standard of that greatest hope and inspiration of mankind, the United Nations, have liberated this ancient capital city of Korea . . . . I am happy to restore to you, Mr. President, the seat of your government that from it you may better fulfill your constitutional responsibilities." With tears running down his cheeks, MacArthur led the dignitaries in the Lord's Prayer. President Rhee was nearly overcome with emotion. To MacArthur he said: "We love you as the savior of our race."

The ceremony at the national capital represented Douglas MacArthur at his legendary finest. In the best of all worlds the Korean War would have ended on this felicitous note. In reality, however, the blazing speed with which MacArthur had reversed the seeming disaster in South Korea contained the seeds of a greater disaster to come in the north. The United States and the United Nations, flush with September's great victories, were fatally modifying their war aims to include the complete subjugation of North Korea and the forcible reunification of the entire peninsula. Already there were plans afoot to deploy the Marines north of the 38th Parallel. General Almond took O. P. Smith aside as they were leaving the ceremony and issued a warning order. The 1st Marine Division would soon be making another "end-run" amphibious landing on the northeast coast.

Other threats materialized. On the day following the Seoul ceremony, Chinese Premier Chou En-Lai warned the world that his nation "will not supinely tolerate" the invasion of North Korea. Few people in the West took him seriously.

In the meantime, Almond ordered Smith to seize and defend a series of blocking positions north of Seoul. The 5th Marines attacked northwest. The 3d Battalion, 5th Marines, executed an aggressive reconnaissance in force as far as the town of Suyuhyon against what the division special action report described as "moderate enemy resistance."

The 7th Marines drew the shortest straw, the division objective of Uijongbu, a vital road junction in the mountains 16 miles due north of Seoul. Here the highway and railroad tracks veer northeast towards the port of Wonsan and beyond, an important escape route for NKPA forces fleeing the "hammer and anvil" of the now converging Eighth Army with X Corps.

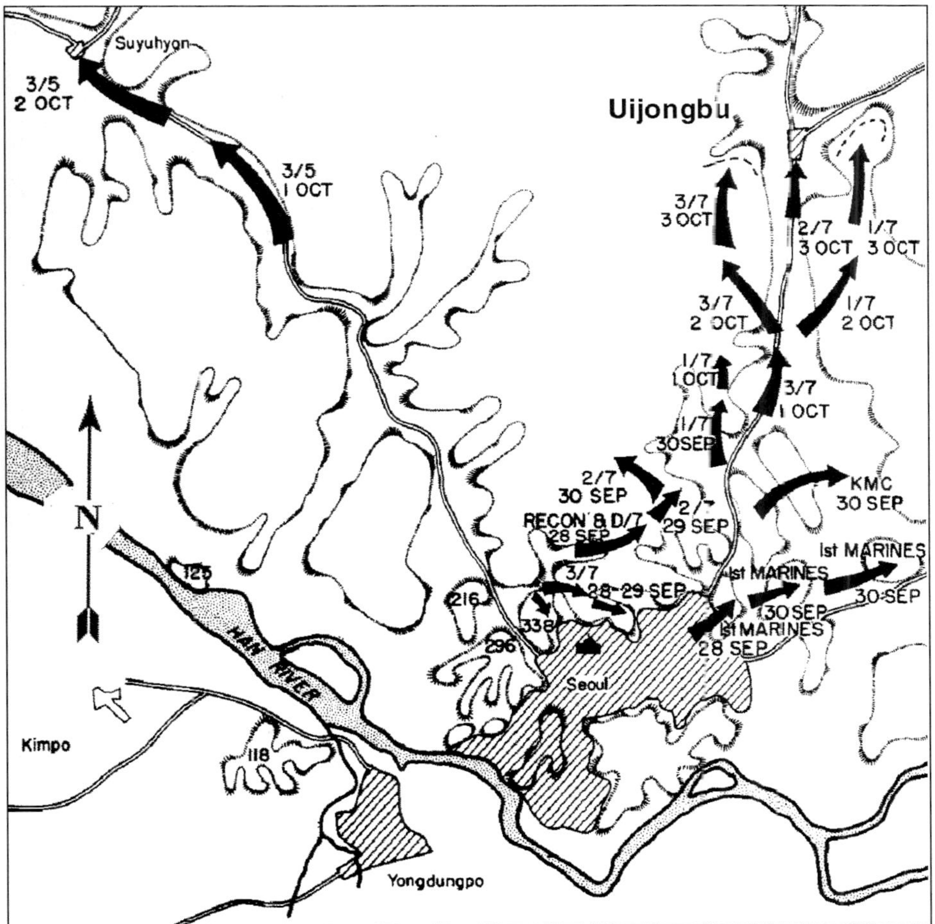

PURSUIT OF THE NKPA

&

CAPTURE OF UIJONGBU

28 SEP -- 3 OCT 1950

0 5 10

MILES

Smith reinforced the 7th Marines by attaching Major Parry's 3d Battalion, 11th Marines (reinforced with a battery of 155mm howitzers from 4th Battalion, 11th Marines), plus one company each of Pershing tanks, combat engineers, and Korean Marines, and an Army antiaircraft battery. This constituted a sizable force, virtually a small brigade, but Colonel Litzenberg would need every man in his three-and-a-half day battle for the road junction. Intelligence reports available to Litzenberg indicated he would be opposed by an amalgamation of NKPA units, including the remnants of the *Seoul City Regiment*; the *2d Regiment, 17th Rifle Division*, withdrawn from the Pusan Perimeter after the Inchon landing; and the fresh *75th Independent Regiment*, which reached Uijongbu from Hamhung the day before the 7th Marines attacked.

Principal air support for Litzenberg's advance would come from the Corsair pilots of Lieutenant Colonel Frank J. Cole's VMF-312, the Checkerboard squadron, newly arrived at Kimpo from Itami, Japan. Cole had commanded the same squadron as a major at the end of World War II and had trained his new aviators exceptionally well.

On 1 October, Colonel Litzenberg led his well-armed force northward. Advance aerial and map reconnaissance led him to conclude that the NKPA would most likely make a stand at

Photo by Frank Noel, Associated Press

A Marine mortar section moves north of Seoul past a grateful band of South Koreans.

Nuwon-ni where the highway passed through a narrow defile—a veritable "Apache Pass." Litzenberg planned for Lieutenant Colonel Raymond G. Davis' 1st Battalion, 7th Marines, to execute a tactical feint along the high ground on both sides, while Major Maurice E. Roach's tank-heavy 3d Battalion, 7th Marines, barreled straight through the pass during the distraction. The plan ran awry when Roach encountered a thick minefield in the pass. Litzenberg shifted both battalions to the high ground, and the Checkerboards of VMF-312 appeared at dawn on the 2d with a vengeance, bombing, strafing, and dropping napalm canisters. Davis and Roach scratched forward slowly along both ridges; the engineers labored in the minefields. But the North Koreans contested every yard, shooting down three Corsairs, disrupting the engineers, and limiting the Marines to less than a quarter-mile gain that day. During this fighting, Second Lieutenant Joseph R. Owen, the mortar officer in Company B, 1st Battalion, 7th Marines, learned bitter lessons in tactical communications. "The North Koreans," he said, "used whistles and bugles for battlefield command, more effective by far than our walkie-talkies." In addition, Lieutenant Lloyd J. Englehardt of VMO-6 flew his glassy-nosed HO3S-1 helicopter through heavy fire to rescue downed Checkerboard pilot Captain Wilbur D. Wilcox near the village of Chun-chon.

On 3 October, the regiment unveiled a good-luck piece, General Clifton B. Cates, Commandant of the Marine Corps, nicknamed "Lucky" Cates for his survival amid the First World War's bloodiest battlefields. Cates had flown to Korea to observe his Marines in action. Litzenberg's force put on a stellar show. The engineers having at considerable cost cleared the minefield in the defile, Major Webb D. Sawyer's 2d Battalion, 7th Marines, pounded straight up the middle. Soon they began overrunning enemy field pieces and had the enemy on the run. The NKPA had staked everything on holding the pass at Nuwon-ni and had little left to defend Uijongbu. Litzenberg unleashed all his forces. Sawyer's men stormed through the ruined town by late afternoon, the major pausing to telephone Litzenberg—widely known by his nickname "Litz the Blitz"—saying, "This is the Mayor of Blitz!"

The Uijongbu drive cost the 7th Marines 13 killed and 111 wounded, but the combat experience was worth the price to the newly formed regiment. Observed Lieutenant Joseph Owen: "For Baker-One-Seven it was combat training under fire; in those five days we became a good Marine rifle company."

The battle for the Nuwon-ni Pass marked the end of significant fighting in the Inchon-Seoul campaign. Almost immediately the 1st Marine Division turned over its assigned sector to the 1st Cavalry Division of the Eighth Army and began returning by regiments to the vicinity of Inchon for re-embarkation.

The leading elements of the division and other X Corps components assembled at a United Nations cemetery near Inchon on 6 October to honor their dead. Division Chaplain Robert M. Schwyhart led the spiritual salute.

From left, BGen Edward A. Craig; the Commandant of the Marine Corps, Gen Clifton B. Cates; and MajGen Oliver P. Smith, inspect the North Korean flag that recently was hauled down by Marines at Government House in Seoul.

Photo courtesy of *Leatherneck* Magazine

Major General Oliver P. Smith laid a wreath on the grave of Corporal Richard C. Matheny, a stalwart squadleader of the 5th Marines who before his death qualified in swift succession for the Bronze Star, Silver Star, and Navy Cross.

The combined Inchon-Seoul campaign cost the 1st Marine Division 2,450 casualties, according to the official history (415 killed or died of wounds; 2,029 wounded in action; 6 missing in action). North Korean gunners shot down 11 fighters of the 1st Marine Aircraft Wing. For their part, the Marines destroyed or captured 47 Russian-built tanks and sufficient heavy mortars, field guns, antitank guns, machine guns, and rifles to equip a good-sized brigade. A preponderance of the 14,000 NKPA fatalities claimed by X Corps in the campaign resulted from the combined air-ground fire of the Marines.

Such statistics had more relevance in World War II than in the murky political and psychological nature of limited warfare in the Atomic Age. The Cold War between the United States and the Soviet Union and their respective allies and surrogates was fully underway by 1950. In Seoul in September of that year, the United Nations for the first time restored the freedom of a democratic capital captured by Communist force of arms. The fact that all of X Corps' hard-fought gains would be swept away by the Chinese Communist counter-offensive three months later added to the bittersweet irony of this protracted war. In the final accounting, the 1953 ceasefire left Seoul firmly established as the capital of the Republic. Seoul's flourishing growth and development over the ensuing half century remain a tribute to the sacrifices of all those who fought and died to recapture

1st Marine Division Daily Casualties, Seoul Campaign

Date	Killed in Action	Died of Wounds	Missing in Action	Wounded in Action	Total Battle Casualties
20 Sep	24	1	3	119	147
21 Sep	30	3	0	198	231
22 Sep	27	3	0	135	165
23 Sep	19	7	0	117	143
24 Sep	68	4	0	217	289
25 Sep	33	4	1	238	276
26 Sep	29	7	0	167	203
27 Sep	33	3	0	153	189
28 Sep	8	4	0	31	43
29 Sep	19	1	0	49	69
30 Sep	11	2	0	48	61
1 Oct	2	1	0	16	19
2 Oct	15	1	0	81	97
3 Oct	2	1	0	35	38
4 Oct	0	0	0	3	3
5 Oct	1	1	0	3	5
Total	321	43	4	1,510	1,878

Appendix J. Montross & Canzana. *The Inchon-Seoul Operation*, 1955

and protect the ancient city.

Operation Yo-Yo The Wonsan Landing

General MacArthur ordered General Almond to re-embark X Corps and execute a series of amphibious landings along the east coast of North Korea. The 1st Marine Division would board designated shipping at Inchon and land tactically at Wonsan, the main event. The 7th Division would proceed south to Pusan to board its ships for a subsequent landing north of Wonsan. The original D-Day for the Marines at Wonsan was 15 October. The actual landing date was not even close.

Operation Chromite was the codename for the Inchon landing. The troops would nickname the Wonsan campaign "Operation Yo-Yo."

Inchon and Wonsan serve as book-end examples of amphibious warfare's risks and rewards. By all rights it should have been Inchon, with its legion of tactical and hydrographic dangers, that sputtered in execution. Wonsan, scheduled for attack by a larger and, by now, more experienced landing force against a sharply diminished enemy threat, should have been a snap. But in the irony of war, Inchon stands as a masterpiece, Wonsan as a laughingstock, as ill-conceived a landing as the United States ever conducted.

In late September 1950, there was nothing particularly wrong with the concept of a long-distance "Right Hook" amphibious landing from the Sea of Japan to seize Wonsan and other smaller ports along the North Korean coast. Wonsan at the time represented a reasonable objective, and the 1st Marine Division had proven its amphibious prowess in the difficult landing at Inchon and was expected to be available for the new mission in early October.

Wonsan had the best natural harbor in the Korean peninsula. Located 80 miles north of the 38th Parallel, the port's bulwark-like Kalma Peninsula provided an enormous sheltered harbor, a seven-inch tidal range, weak currents, rare fog, and a moderate beach gradient—all incomparably more favorable than Inchon. Wonsan's near-shore topography also offered a decent lodgement area, suitable for a division beachhead, before the Taebaek

Mountains—eastern North Korea's long, towering spine—reared upwards from the coastal plain. The port's strategic appeal centered on the combination of its accessible harbor with a high-capacity airfield, petroleum refining facilities, and its location astride major railroads and highways leading west to the North Korean capital of Pyongyang, north to Hungnam, and southwest to Seoul.

On 4 October, General Almond formally assigned the 1st Marine Division the mission of seizing and securing the X Corps base of operations at Wonsan, protecting the airfield, and continuing inland operations as assigned. Three unforeseen developments almost immediately knocked the Wonsan plans into a cocked hat: massive port congestion; a drastically accelerated invasion of North Korea; and the successful mining of their coastal ports by the North Koreans. As a consequence, MacArthur's celebrated "Right Hook" became suspended in mid-air, leaving the Marines (and all of X Corps) hanging in limbo—out of action—throughout a critical three-week period. The Wonsan landing, when it finally occurred, has been aptly described by military historians as "the most anticlimactic a landing as Marines have ever made."

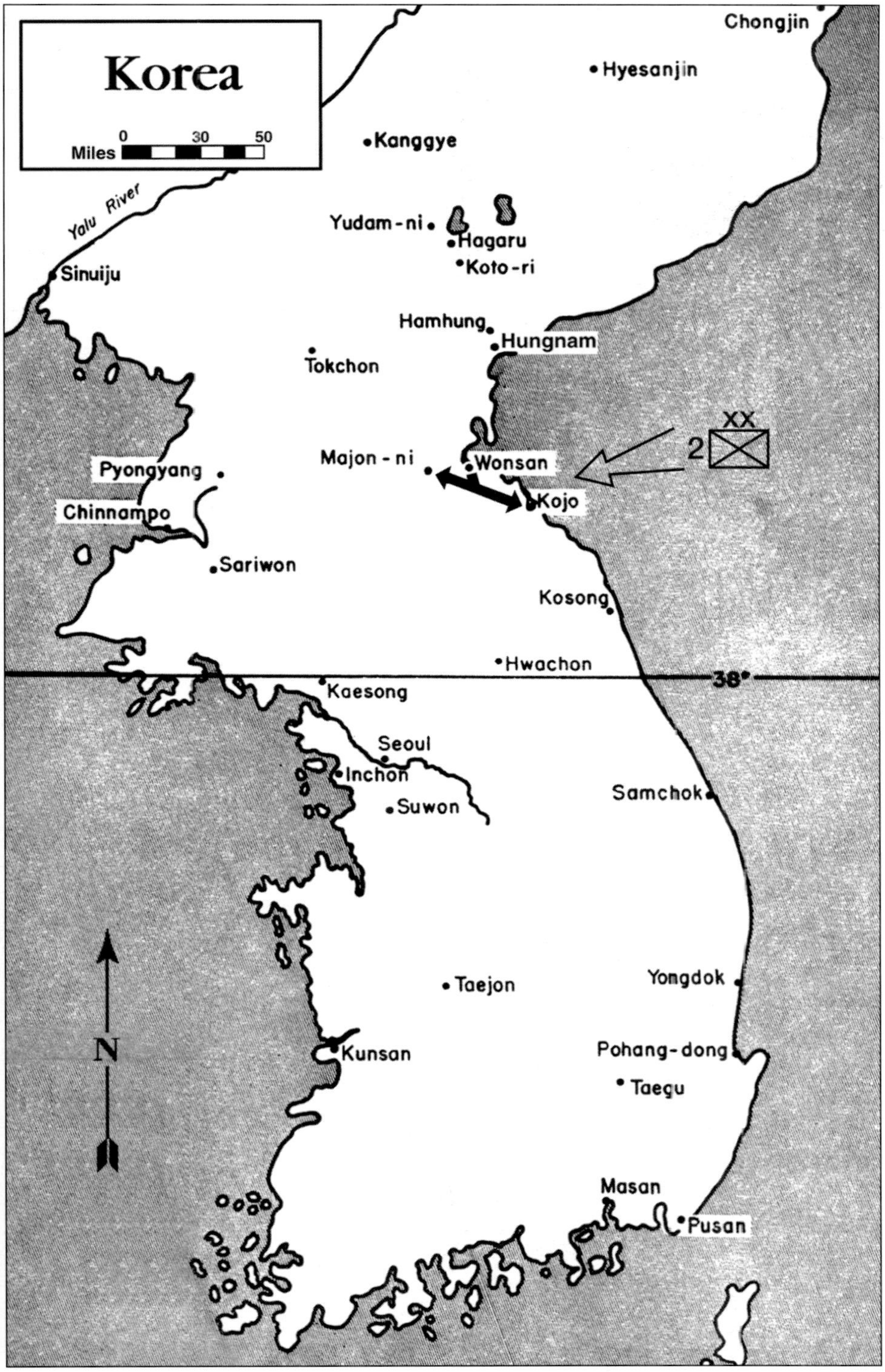

The 1st Marine Division operations order directed a simultaneous landing of the 1st and 7th Marines abreast on the eastern shore of the Kalma Peninsula, each supported by an artillery battalion and a battalion of Korean Marines. Wonsan airfield lay directly inland, as close to the landing beaches as Kadena and Yontan had been to the Hagushi beaches at Okinawa.

On 7 October, the day following the cemetery ceremony in Inchon, Major General O. P. Smith reported as landing force commander for the Wonsan expedition to Rear Admiral James H. Doyle, USN, commanding the Attack Force, U.S. Seventh Fleet. The division began embarking at Inchon the next day. It would take a week.

Here MacArthur's plans began to unwind. No one, it seems, had foreseen the tremendous strain about to be placed on the only two medium-capacity ports available, Inchon and Pusan, during a time of conflicting requirements to offload the mammoth supplies needed for the Eighth Army's invasion of western North Korea while simultaneously backloading two large divi-

Gen Oliver P. Smith Collection, Marine Corps Research Center

From left, MajGen Oliver P. Smith, MajGen Edward M. Almond, and RAdm James H. Doyle discuss plans for the Wonsan landing on board the Mount McKinley.

sions for the X Corps' "Right Hook." The piers, staging areas, and access roads in both ports became impossibly congested. Chaos reigned.

Combat loading for an opposed amphibious assault is an exact and time-consuming science. The 1st Marine Division, now fully fleshed out with the 7th Marines and other missing components, had 25,840 officers and men on the rolls for the Wonsan expedition, easily the largest division of any nation fighting in the Korean War. Admiral Doyle's Attack Force contained 66 amphibious ships plus six commercial cargo ships, but many of the vessels arrived late in the crowded port, few contained the preloaded 10-day levels of Class I, II, and IV supplies as promised, and the Attack Force still provided insufficient total lift capacity for all the division's rolling stock. The precise art of combat loading became the improvised "art of the possible," but each compromise cost time. As the division's special action report dryly noted, General Almond's prescribed D-day of 15 October "was moved progressively

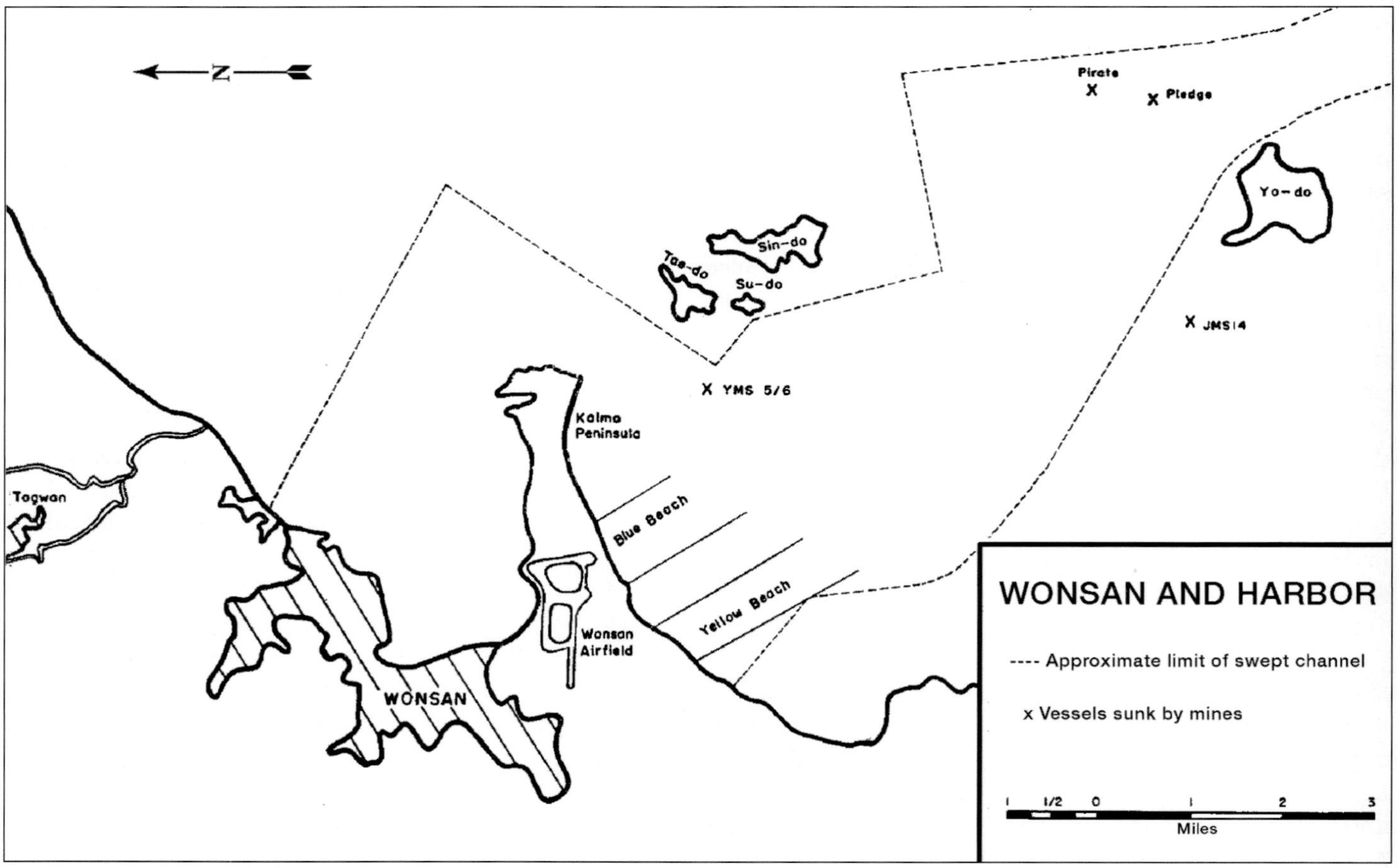

back to a tentative date of 20 October."

As junior officer in Company B, 1st Battalion, 7th Marines, Second Lieutenant Joseph R. Owen assumed the demanding duties of company embarkation officer. "We were assigned an old LST that our Navy had used in World War II," he said, "but which was now leased to Japan for use as a cargo ship." The Japanese captain spoke no English but conveyed to Owen by angry gestures his displeasure at what seemed to him to be gross overloading of his ship's safe lift capacity. When Owen's runner charged the bolt on his carbine the skipper abruptly acquiesced. "There was a shortage of shipping," Owen rationalized, "and, we were informed, we would be afloat for only a few days."

Private First Class Morgan Brainard of Company A, 1st Battalion, 1st Marines, boarded his assigned LST without bitching: "All we knew at that moment was that we were steaming south; that we were in dry clothes with a roof over our heads, and assured of two hot meals a day and the chance to take salt water showers . . . . Our slice of life seemed to be improving."

Most of Admiral Doyle's Wonsan Attack Force completed loading the 1st Marine Division and sortied from Inchon on 15 October, the original D-Day. By that time the other two factors that would render the planned assault meaningless had materialized. Five days earlier, Republic of Korea's I Corps had seized Wonsan by overland advance from the south. On 13 October, Major General Field Harris, commander of the 1st Marine Aircraft Wing, flew into Wonsan airfield, followed the next day by the Checkerboards of Lieutenant Colonel Cole's VMF-312 and other elements of Marine Aircraft Group 12.

In the meantime, Admiral Doyle's advance force commander discovered that the North Koreans had sewn the approaches to Wonsan with more than 2,000 anti-ship mines, both contact and magnetic. The U.S. Navy had only 12 minesweepers available in theater—as compared to the 100 employed in support of the Okinawa landing five years earlier—and even when reinforced by Korean and Japanese craft, the mission proved overwhelming. Two U.S. minesweepers hit mines and sank on 12 October. Heavy fire from North Korean coast defense guns hampered rescue operations. A Japanese sweeper sank on the 18th; the next day a huge mine practically vaporized a South Korean craft. Doyle's experiments in dropping 1,000-pound bombs and anti-submarine depth charges to create enough overpressure to detonate nearby mines failed. Even the fact that a linear, tactical landing had been replaced by a simpler administrative offload from amphibian tractors and landing craft in column did not help expedite the problem. Rear Admiral Allan E. Smith, commanding Task Force 95, the advance force, voiced the frustration of all hands when he reported: "we have lost control of the seas to a nation without a navy, using pre-World War I weapons, laid by vessels that were utilized at the time of the birth of Christ."

General Almond's frustration knew no bounds. On 20 October, with the war fast shifting away from his active influence (the Eighth Army entered the North Korean capital Pyongyang the previous day), and with no end in sight to the tedious minesweeping, Almond departed the flagship by helicopter and established his command post ashore at Wonsan.

As for the embarked Marines, rampant rumors swept the transports, especially beginning the afternoon of 19 October when the task force suddenly got underway, heading south. "War's over!" exclaimed many a Marine, "We're going back to Pusan and then heading home!" But the ships were only taking new precautions to protect themselves in hostile waters. For the next week—and a week is a very long time at sea on board transports as claustrophobically crowded as these—the ships reversed course every 12 hours, first heading south, then north, then starting over. Here emerged the sarcastic nickname "Operation Yo-Yo." As voiced by Marine Corps historians Lynn Montross and Nicholas A. Canzona in 1957: "Never did time die a harder death, and never did the grumblers have so much to grouse about."

The Japanese-crewed LST transporting Lieutenant Owen's company soon ran low on provisions and fresh water. As Owen recalled: "a three-week ordeal of misery and sickness . . . . The stench belowdecks made the air unbreathable."

Before long sickness swept the embarked landing force. Long lines of Marines suffering from dysentery and gastroenteritis overwhelmed poorly-equipped sick bays. The "Binnacle List" on board the converted civilian transport *Marine Phoenix* ran to 750 Marines at the height of the epidemic. The attack transport *Bayfield* (APA 33) reported a confirmed case of smallpox. As a final insult to the division's pride, a traveling USO show featuring Bob Hope and Marilyn Maxwell beat them to Wonsan, performing for an appreciative audience of Marine aviators and ROK soldiers while the fierce "spearhead" assault troops rocked in misery among the offshore swells.

At long last, on 26 October, the 1st Marine Division landed on the

Department of Defense Photo (USMC) A4313

Finally off the ships, the 1st Marine Division, which ended its interminable "Operation Yo-Yo" on 26 October, chugs ashore by Navy LCVP towards Wonsan, North Korea.

In the anticlimactic landing of the 1st Marine Division at Wonsan, troops dismount from a column of LVT-3Cs and their escorting LVTA-5 armored amphibians along the Wonsan airfield. A chill wind blows in from the looming Taebaek Mountains.

Department of Defense Photo (USN) 421366

Department of Defense Photo (USMC) A130364

The 1st Marine Division suffered the ignominy of landing at Wonsan weeks after South Korean forces had seized the port and the 1st Marine Aircraft Wing had arrived. Here a column of soaked infantrymen straggles ashore among good-hearted catcalls by VMF-312, the "Checkerboard" squadron.

Kalma Peninsula below Wonsan. "The day was bright and cold," recalled Private First Class Brainard of Company A, 1st Marines, "and the sea had a real chop to it as our [LVT] slid down the ramp and nosed forward into the water." The captain of Brainard's ship wished the departing Marines luck over the public address system, adding that MacArthur's headquarters had just announced that the troops should be "home by Christmas."

The airmen of the Checkerboard squadron hooted in derision as the infantry streamed ashore, puffing with exertion after three weeks of enforced inactivity. Lieutenant Owen encountered more sarcastic insults from the ROK troops who had captured the town 16 days before: "They had learned the middle-finger salute, which they rendered to us with great enthusiasm."

Colonel Puller bristled at the ignominy of his regiment being categorized as rear echelon troops due to no fault of their own. Then Brigadier General Edward A. Craig met Puller on the beach with the welcome news that he had just been selected for promotion to brigadier general. Puller's trademark scowl vanished momentarily. Then he turned to the job at hand. His regiment was about to be dispersed over a huge area of enemy territory, beginning with the deployment of Lieutenant Colonel Jack Hawkins' 1st Battalion, 1st Marines, on a special mission to relieve a ROK force guarding a supply depot at the coastal town of Kojo, 40 miles south.

Puller's dilemma reflected the drastic changes in United Nations' strategic objectives being formulated it seemed, day-by-day. The war in North Korea had evolved from the establishment of coastal operating bases and the methodical elimination of residual NKPA forces to a wide-open race to the Yalu River, the Chinese border. General O. P. Smith, accustomed to a relatively narrow zone of action in the Inchon-Seoul campaign, suddenly found himself responsible for a zone measuring 300 miles long by 50 miles deep. With General Almond already calling for two infantry regiments to advance as far north as Hamhung, Smith knew Puller's 1st Marines would be hard-pressed to cover an uncommonly large piece of real estate around the port of Wonsan. The Kojo assignment was but the first of several far-flung missions Puller would have to handle.

Half of Hawkins' battalion departed within hours after their landing on the 26th. The troops were still shaking out their sea legs when they clambered into a long line of empty gondola cars of a coal train bound for Kojo. It was an uncomfortable and singularly dirty ride. Captain Barrow noted that the residual coal "left a mark on all of us." The train had to make two trips to deliver the entire battalion, and those units traveling by road, like the attached artillery battery, did not arrive until the second night.

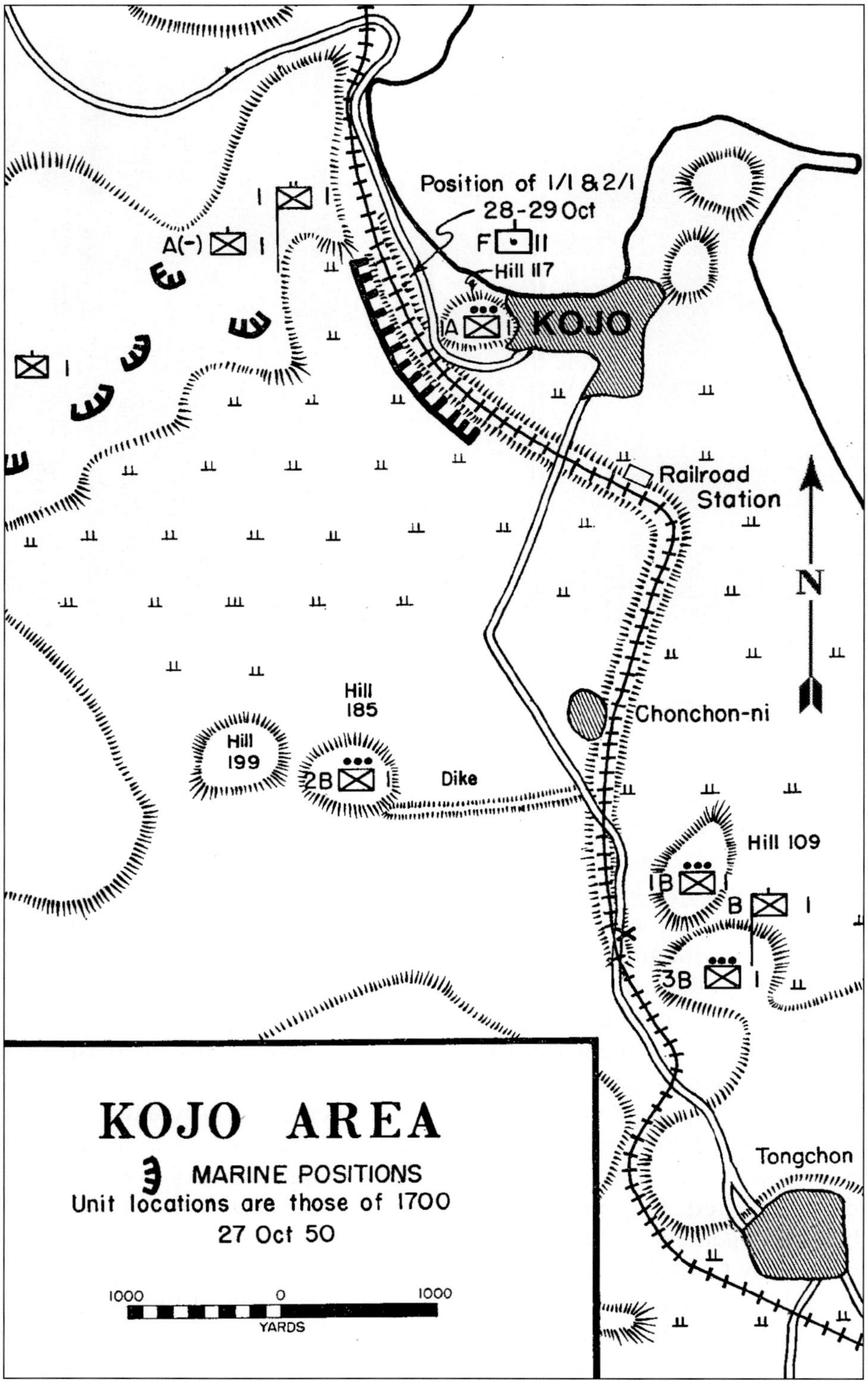

The troops dismounted from the train at Kojo stiff and disoriented. The town itself proved picturesque, but the supply dump had been largely emptied by the departing ROK garrison, too many hills dominated the town, and there was a critical lack of intelligence about a North Korean "guerrilla force" reportedly lurking in the area. "Quite candidly," admitted Barrow, "I never understood our mission."

The situation bothered Hawkins acutely. The late-afternoon approach of 3,000 refugees towards Kojo made him more uneasy. These Hawkins diverted into an assembly area outside the seaport, but their unimpeded approach reflected the vulnerability of his position. The largely depleted supply dump lay in low ground, difficult to defend. A well-defined avenue of approach into the seaport lay open from the south and southwest. "Therefore," Hawkins wrote shortly after the Kojo action, "I decided to place Company B in outpost positions to cover these approaches . . . . The remainder of the battalion would be deployed on the hill massif west of Kojo."

Accordingly, Captain Wesley C. Noren deployed Company B on outpost duty along three scattered hills two miles south of town. As night fell, Noren placed his men

Gen Oliver F. Smith Collection, Marine Corps Research Center

Men of the 1st Marines sweep through the village of Kojo following the sudden, violent, and well-coordinated North Korean night attacks of 27 October on the 1st Battalion's positions.

on 50 percent alert: each foxhole to contain one man awake, the other halfway zipped-up in his sleeping bag. The night was chilly; that morning the Marines had discovered the first ice of the season in the rice paddies. Their last fire-fight in burning Seoul a month ago had left them gasping in the heat. Now they began to shiver.

The security measures prescribed by Hawkins and Noren were normal under the assumed threat—light probing attacks by small bands of guerrillas. No one then knew that Noren's dispersed platoons had taken their night positions within direct observation of a significant organized force, the *10th Regiment, 5th NKPA Division*. Colonel Cho Il Kwon commanded this regiment, one of the highly disciplined forces led by veterans of the fighting in China that had spearheaded the invasion of South Korea four months earlier. Cho and his men had successfully evaded the "hammer and anvil" trap set by the United Nations forces after Inchon and returned essentially intact across the border. The regiment had left its tanks and artillery along the Naktong River, but still possessed plenty of mortars and machine guns. With more than a thousand assault troops at hand, Cho had the numbers and leadership to overwhelm Noren's outposts and simultaneously attack the flank of Hawkins' main positions west of Kojo.

Rarely in their long history had the 1st Marines been in such mortal danger. Cho's veterans moved out of their staging areas at nightfall and approached each outpost with disciplined stealth. These men were superb night-fighters. Some infiltrated undetected to within 10 feet of the nearest Marine foxholes. At precisely 2200, they attacked with submachine guns, grenades, and shrill screams.

Noren's rifle squads never had a chance. Seven Marines died in one platoon before they could even scramble out of their sleeping bags. Vicious hand-to-hand fighting swept the hilltops. Some units were cut off and scattered. Well-drilled junior non-commissioned officers grabbed disorganized indi-

Sketch by Col Charles H. Waterhouse, USMCR (Ret)

A Navy corpsman drags a wounded Marine out of the line of fire. The artist added the Red Cross band on the corpsman's arm to provide color to the composition, admitting in his original caption that most corpsmen discarded the brassard as too tempting a target to enemy snipers.

viduals and formed small counterattacks. When the 1st Platoon, abruptly missing 30 men, had to abandon Hill 109 just north of Noren's command post, Sergeant Clayton Roberts singlehandedly covered their withdrawal with his light machine gun until the North Koreans slipped in close and killed him.

Captain Noren kept his head, swiftly calling in mortar fire and trying to make sense of the pandemonium. One thing was sure—this was no guerrilla band. Judged by their night-fighting skills alone, Noren knew he was under attack by one of the original *Inmin Gun* outfits, supposedly destroyed by the United Nations' breakout.

Captain Noren held the scattered pieces of his beleaguered company together until 2350, no small achievement in the confusion, then radioed Hawkins for permission to withdraw. Hawkins assented. The battalion commander was much more in the dark than Noren, but he knew that Captain Robert P. Wray's Company C on the right rear flank had been hit hard by a violent surprise attack. Noren then executed a masterful point-by-point withdrawal under extreme enemy pressure. By 0200, he organized the surviving members of Baker Company into a 360-degree defensive circle along the railroad tracks about 2,500 yards below Kojo. At 0300, Noren established radio contact with the 4.2-inch mortar platoon whose steady fire then helped disrupt the NKPA forces converging on the small band.

Noren held his new position throughout the rest of the night. At the first grayness of dawn he began evacuating his wounded, dragging them north in ponchos through the thin ice and thick mud of the rice paddies. Suddenly a force of 200 of Colonel Cho's night fighters appeared out of the gloom, heading south, like vampires trying to outrun the sun. There was just enough light for the entire battalion to enjoy a "Turkey Shoot," including the newly arrived Battery F, 2d Battalion, 11th Marines. Seventy-five of this group of North Koreans never made it back to their sanctuary. Perhaps twice again as many NKPA bodies lay within the original Marine positions.

Fragmented reports of a major attack against 1st Battalion, 1st Marines, at Kojo began to arrive at the division command post around 0700 the next morning. Coincidentally, the first three helicopters of VMO-6 were just being ferried from Kimpo to Wonsan airfield. Captain Gene W. Morrison recalled landing at Wonsan during the emergency and not even shutting down his helicopter. He received an urgent cockpit briefing, then lifted off immediately for Kojo on a medical evacuation mission.

The sudden violence of the well-coordinated NKPA night attacks had shocked Lieutenant Colonel Hawkins deeply. His reports to division throughout the 28th reflected the concerns of an isolated commander under protracted stress: "Received determined attack . . . from sunset to sunrise by large enemy force," he reported in one message that reached General Smith about 1230. "One company still heavily engaged . . . . Have suffered 9 KIA, 39 WIA, 34 MIA probably dead . . . . If this position is to be held a regiment is required . . . . Shall we hold here or withdraw to North? Send all available helicopters for wounded."

Smith directed Colonel Bowser to send Puller and an additional battalion of the 1st Marines by immediate train to reinforce Hawkins. Smith also arranged for air strikes, destroyer bombard-

Aerial Medical Evacuation

A dramatic improvement in medical care for combat casualties became evident by the end of the Seoul campaign. According to historian J. Robert Moskin, Navy surgeons operated on 2,484 patients during the fighting for Inchon and Seoul. Only nine of these men died, a remarkable advance in the survival rate of those casualties who made it back to an aid station. Several factors contributed to this breakthrough, but one notable newcomer was the increased use of organic observation aircraft—principally the helicopter—for medical evacuation of severely wounded men.

The use of Marine Corps aircraft to evacuate casualties under fire began as early as 1928 in Nicaragua when First Lieutenant Christian F. Schilt landed his O2U Corsair biplane in the dusty streets of Quilali, a bravura performance, repeated 10 times, that resulted in the rescue of 20 men and a Medal of Honor for the intrepid pilot. Later, during the 1945 battle for Okinawa's Kunishi Ridge, the Marines evacuated hundreds of their casualties to a rear hospital by experimental use of their OY-1 single-engine observation aircraft, which trundled aloft from a dirt road just behind the front lines.

Marine Observation Squadron 6 (VMO-6) had been one of the observation squadrons that evacuated wounded men from the Kunishi Ridge battlefield. In Korea five years later the squadron again supported the 1st Marine Division. While the OY-1s occasionally transported wounded men in 1950, the mission increasingly became the province of the squadron's helicopters, the nation's first wartime use of the new technology.

The VMO-6 pilots flew Sikorsky HO3S-1 observation helicopters during the Seoul campaign. A bench seat behind the pilot could accommodate three passengers, but there was insufficient room in the cabin for a stretcher. To evacuate a non-ambulatory patient, according to historians (and helicopter pilots) Lieutenant Colonel Gary W. Parker and Major Frank M. Batha, Jr., the crew had to remove the right rear window and load the stretcher headfirst through the gap. The casualty's feet jutted out the open window.

Primitive as this arrangement may have been, the pilots of VMO-6 safely evacuated 139 critically wounded Marines during the Seoul campaign. Most of these men owed their lives to this timely evacuation. An unspoken but significant side benefit to these missions of mercy was their impact on the morale of the Marines still engaged in combat. Simply knowing that this marvelous new flying machine was on call to evacuate their buddies or themselves should the need arise was greatly reassuring.

The proficiency of the VMO-6 helicopter pilots proved reassuring to the fixed-wing pilots as well. A Marine helicopter had rescued Marine Aircraft Group 33's first downed aviator as early as the second day of the 1st Brigade's commitment to the Pusan Peninsula. Included in Major Robert P. Keller's post-Seoul campaign evaluation of his Black Sheep squadron's role in close air support operations were these comments: "The helicopters have done a wonderful service in rescuing downed pilots under the very guns of the enemy. The pilot should not start out cross country unless he is sure that helicopters are available."

Major Vincent J. Gottschalk's VMO-6 lost two OY-1s and two helicopters to enemy fire during the Seoul campaign. Fortunately, at least, none of the aircraft were transporting casualties at the time. Three months later, the squadron would transition to the Bell HTL-4 helicopter which could carry a stretcher mounted on each skid, in effect doubling the medical evacuation payload. In August 1951, in one of the high points of Marine Corps aviation history, Major General Christian F. Schilt, the hero of long-ago Quilali, took command of the 1st Marine Aircraft Wing in Korea.

National Archives Photo (USN) 80-A20546

ment, and a hospital-configured LST to be dispatched to Hawkins' aid.

Hawkins, convinced that the NKPA would return that night in great force, continued to send alarming messages to General Smith, but things had calmed down when Puller arrived with Sutter's 2d Battalion, 1st Marines, at 2230. There were no further attacks by the *10th NKPA Regiment*. Puller was decidedly unsympathetic to Hawkins' concerns (and in fact would replace him in command of 1st Battalion, 1st Marines, with Lieutenant Colonel Donald M. Schmuck a week later).

The next day, Captain Noren led a patrol south to recover a number of his missing troops who had gone to ground after being cut off during the night. Similarly, Captain George B. Farish, a VMO-6 helicopter pilot on a reconnaissance mission below Kojo, spotted the word "HELP" spelled out in rice straw in an open field, landed warily, and promptly retrieved smart-thinking Private First Class William H. Meister, one of Noren's lost sheep, from his nearby hiding place. The battalion's final count for that bloody night came to 23 killed, 47 wounded, and 4 missing.

On the same day, Captain Barrow led Company A south on a reconnaissance in force, accompanied by a destroyer offshore and a section of Corsairs overhead. Just as he reached his assigned turn-around point, a Corsair pilot advised him that a large number of enemy troops were digging in several miles farther south. Barrow directed the pilot to expend his ordnance on the target. He did so. Barrow then asked him if he could adjust naval gunfire. "Yeah, I can do that," came the reply. For the next half-hour the destroyer delivered a brisk fire, expertly adjusted by the pilot, who at the end reported many casualties and fleeing remnants. Barrow returned to Kojo without firing a shot, but fully convinced he had avenged Baker Company and taught his unknown opponent a lesson in

With the Kojo area secured, the 1st Marines, in coordination with Marine Corsairs, move out in reconnaissance in force.
Gen Oliver P. Smith Collection, Marine Corps Research Center

National Archives Photo (USMC) 127-N-A4553

The 5th Marines load on board a patched-up train for the move north to Hamhung in pursuit of the retreating North Koreans

combined arms firepower.

O. P. Smith could illafford to keep Puller and two battalions so far below Wonsan. With the 7th and 5th Marines already on their way far northward towards Hamhung, and an urgent requirement at hand for 3d Battalion, 1st Marines to proceed to Majon-ni, Smith had no Marine infantry units left to cover Wonsan's port, airfield, and road junctions. Members of the 1st Amphibian Tractor Battalion and other combat support Marines doubled as riflemen to fill the gaps until Puller's force returned in early November. The 1st Marine Division, however, would remain widely dispersed, a continuing concern to Smith. Indeed, his three infantry regiments would not be linked up for six weeks, from the administrative landing at Wonsan until the withdrawing columns from the Chosin Reservoir fought their way down to Chinhung-ni during the second week in December.

Other changes were in the wind. On 29 October, General Barr's 7th Division commenced its administrative landing at the small port of Iwon, another 60 miles above Hungnam. In the dramatic but strategically unsound "Race to the Yalu," two of Barr's units would become the only U.S. forces to actually reach the river. Meanwhile General Almond continued to look for opportunities to exploit the 1st Marine Division's amphibious capabilities. When Puller returned from Kojo, Almond warned him to be ready for an amphibious landing 220 miles northeast of Wonsan. The target was Chongjin, a seaport dangerously near the North Korean border with the Soviet Union.

Colonel Puller wasted little time worrying about another "End Run." Of more immediate concern to him was the commitment of Lieutenant Colonel Ridge's 3d Battalion, 1st Marines, to the defense of the key road junction at Majon-ni, a deployment that would last 17 days and provoke a dozen sharp firefights.

The mountain town of Majon-ni occupied a bowl-like plateau, encircled by higher ground, about 26 miles west-southwest of Wonsan. The roads from Wonsan, Pyongyang, and Seoul intersect here, and the highlands contain the headwaters of the Imjin River. Ridge's battalion arrived on 28 October to provide a screening and blocking force.

The terrain around Majon-ni lent itself more readily to the defense than Kojo, but Kojo had been much more accessible for the Marines. There were no rail lines, and the "highway" from Wonsan was a single-lane road that twisted through mountain passes, switchbacks, hairpin turns, and precipitous dropoffs. The troops called it "Ambush Alley."

In 1978, two Marine generals recalled their experiences as young officers involved in the defense and resupply of Majon-ni. To

Brigadier General Edwin H. Simmons, the Majon-ni mission was a difficult defensive line that had to be covered with a thin perimeter, thus providing 3d Battalion, 1st Marines, with good practice for their similar challenge ahead at Hagaru-ri, the crossroads mountain village situated at the southeastern edge of the Chosin Reservoir. "Majon-ni was a dress rehearsal for what was going to come up for us at Hagaru-ri," he said. For his part, General Robert H. Barrow considered Majon-ni more a precursor for Khe Sanh in 1968, a remote plateau in the mountains "at the end of a long, tenuous supply route in no-man's land."

Elements of the *15th NKPA Division* opposed Ridge's battalion in the mountains around Majon-ni. While more disorganized and much less proficient than the *10th Regiment* that had stung Hawkins' 1st Battalion, 1st Marines, so painfully at Kojo, these North Koreans were sufficiently trained fighters to threaten Ridge's perimeter each night and readily interdict the Marine convoys trundling carefully through Ambush Alley.

When Ridge went a week without resupply convoys being able to get through to Majon-ni, he requested an air drop of ammunition, gasoline, and rations. The 1st Air Delivery Platoon packaged 21 tons of these critical supplies into 152 parachutes. These were dropped over the Marine perimeter with uncommon accuracy by Air Force C-47s.

With ambushes occurring more frequently, Puller assigned Captain Barrow's Company A, 1st Battalion, 1st Marines, to escort a 34-vehicle convoy from Wonsan to Majon-ni at mid-afternoon on 4 November. Barrow was uncomfortable with both the late start and the slow progress. The North Koreans struck the column with heavy fire late in the afternoon. "They picked a good spot," said Barrow. He called for air strikes through a patchwork network and tried to work his infantry up the steep slopes towards the ambushers. "It soon became apparent that we were not going to be successful . . . and the bad thing was nightfall was approaching." Stymied, and embarrassed by the failure, Barrow ordered the huge 6x6s and Jeeps with trailers to turn around, a harrowing experience under automatic weapons fire. One vehicle went over the side. The Marines formed "a bucket-brigade" to retrieve the injured men.

Back at Wonsan, Captain Barrow dreaded having to report

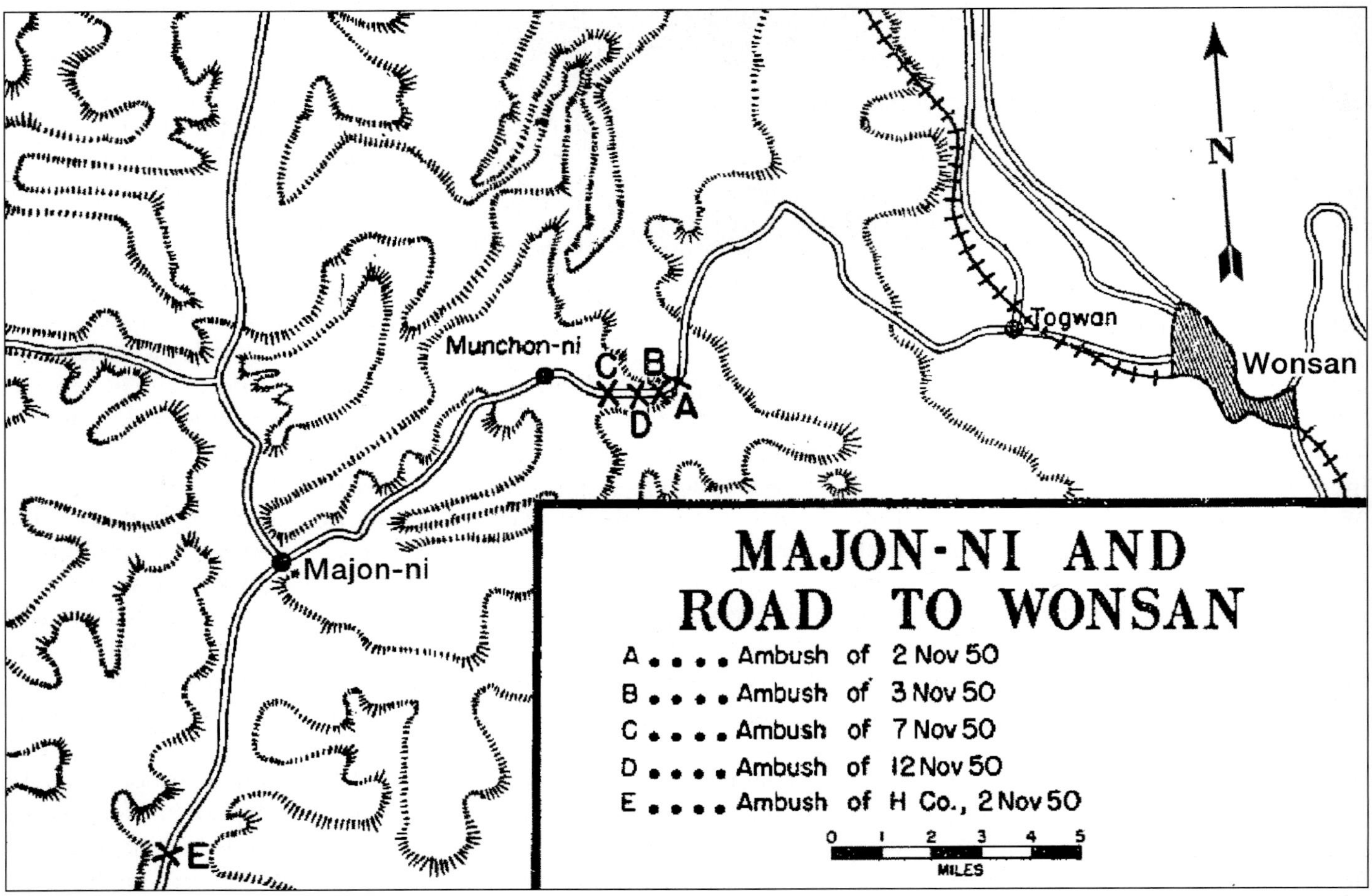

National Archives Photo (USMC) 127-N-A4492

Battery H, 11th Marines, runs its guns forward and prepares to go to work in support of the 7th Marines as the regimental combat team moves north from Hambung into the mountains of North Korea.

his lack of success to Chesty Puller. He found Puller in a school classroom, appropriately seated at the teacher's desk. "Colonel, I have failed you," he said. "No you didn't, old man," Puller growled, not unkindly. "Have a seat." Puller offered Barrow a drink of bourbon, then asked him what he needed to get the convoy through the next day. "More daylight and a forward air controller," Barrow replied.

Barrow departed Wonsan early on the 5th, inspired by an innovative tactic he had devised during the night. The North Koreans, he realized, could hear the trucks laboring up the pass long before they hove into view. He would therefore detach a reinforced platoon to precede the convoy on foot by several thousand yards—comforting for the convoy, although spooky for Second Lieutenant Donald R Jones' point platoon.

Private First Class Morgan Brainard was a member of the second fire team in Jones' dismounted advance patrol. After four bends in the road he looked back and saw the far-distant trucks begin to move. "We were then so far ahead, that I couldn't hear their engines, only our labored breathing," he said. "It was a lonely, eerie feeling, forty-two of us plodding up a bleak mountain road by ourselves."

"It worked!" said Barrow. Jones' point team caught the North Koreans cooking rice along the road, totally unaware and non-tactical. "We literally shot our way forward," said Brainard. More than 50 of the off-duty ambushers died in the surprise attack. "We just laid them out," Barrow recalled with obvious pride, adding, "sometimes the simplest solutions are the most successful."

Barrow delivered his convoy to Majon-ni, stayed to help 3d Battalion, 1st Marines, defend the perimeter against a large-scale night attack, then returned to Wonsan the next day, the emptied trucks now laden with more than 600 NKPA prisoners captured by Ridge's battalion. Yet Barrow's success did not end North Korean interdiction of the Marine convoys. They had learned their own lessons from their surprise defeat on 5 November and would fight smarter in two additional ambushes the following week.

On 10 November, the 3d Korean Marine Corps Battalion reinforced 3d Battalion, 1st Marines, at Majon-ni. The Korean Marines joined their American counterparts in a brief but heartfelt celebration of the 175th Birthday of the U.S. Marine Corps. Ridge's bakers outdid themselves with the resources at hand—an uneven yet ambitiously large cake, smeared with C-Ration jam—but what the hell!

The North Koreans struck the Marine perimeter once more in force the night of 11-12 November, then faded back into the mountains. On the 14th, Lieutenant Colonel Ridge turned over defense of the village to an Army battalion from the newly arrived 3d Infantry Division and led his men back to Wonsan, pleased that 3d Battalion, 1st Marines, had acquitted itself well on such an isolated mission. The battalion sustained 65 casualties defending Majon-ni; another 90 Marines became casualties in the series of convoy fights along Ambush Alley.

The Majon-ni mission ended three straight months of significant fighting between the Marines and main line elements of the North Korean Peoples' Army. Admirably supported by the 1st Marine Aircraft Wing, the Marines had fought well from the start, expanded effortlessly from a small brigade to a full-strength division, executed one of the most difficult amphibious landings in history, and —with the help of their allies and Army elements of X Corps—recaptured an enormous capital city. The resurgence of the Marines' standing within the national security community in Washington was downright dramatic.

But that phase of the Korean War had ended. A new, starkly different, and more troublesome phase had begun. The deceptive promise of "Home by Christmas" seemed abruptly swept away by a bone-chilling wind out of the Taebaek Mountains, out of Manchuria, a harbinger of an early winter—and perhaps something more ominous. By the time General Smith moved the division command post from Wonsan to Hungnam on 4 November, he had been receiving reports of Red Chinese troops south of the Yalu for 10 days. A patrol from 1st Battalion, 7th Marines, visited the headquarters of the 17th ROK Infantry near Sudong-ni on 31 October and confirmed the presence of prisoners of war from the *124th Division*, Chinese Communist Forces.

Colonel Homer Litzenberg, whose 7th Marines would lead the way into the Taebaek Mountains, warned his troops about the likelihood of a Third World War. "We can expect to meet Chinese Communist troops," he told them, "and it is important that we win the first battle. The results of that action will reverberate around the world, and we want to make sure that the outcome has an adverse effect in Moscow as well as Peiping."

The 7th Marines, wearing and carrying cold weather equipment, press north into the Taebaek Mountains in pursuit of North Korean forces. A burden now, they would come to depend on this gear in the coming month.

National Archives Photo (USMC) 127-N-A4524

About the Author

Colonel Joseph H. Alexander, USMC (Ret), served 29 years on active duty as an assault amphibian officer, including two tours in Vietnam and service as Chief of Staff, 3d Marine Division. He is a distinguished graduate of the Naval War College and holds degrees in history and national security from North Carolina, Jacksonville, and Georgetown Universities.

Colonel Alexander wrote the History and Museum Division's World War II 50th anniversary commemorative pamphlets on Tarawa, Iwo Jima, and Okinawa. His books include *A Fellowship of Valor: The Battle History of the U.S. Marines*; *Storm Landings: Epic Amphibious Battles of the Central Pacific*; *Utmost Savagery: The Three Days of Tarawa*; *Edson's Raiders: The 1st Marine Raider Battalion in WW II*; and (with Lieutenant Colonel Merrill L. Bartlett) *Sea Soldiers in the Cold War.* As chief military historian for Lou Reda Productions he has appeared in 15 documentaries for The History Channel and the Arts & Entertainment Network, including a four-part mini-series on the Korean War, "Fire and Ice."

Sources

Primary sources included the 1st Marine Division Special Action Reports for 29 August-7 October 1950, the war diaries of several ground and aviation units, and Gen Oliver P. Smith's official letters and memoir concerning the Seoul/Wonsan campaigns. Of the official history series, *U.S. Marine Operations in Korea,* the volumes by Lynn Montross and Nicholas A. Canzona (II: *The Inchon-Seoul Operation* [Washington, D.C., Historical Branch, G-3 Division, HQMC, 1955] and III: *The Chosin Reservoir Campaign* [Washing-ton, D.C., Historical Branch, G-3 Division, HQMC, 1957]), provide well-researched coverage of the recapture of Seoul and the Wonsan, Kojo, and Majon-ni operations. Among the Marine Corps Oral History Collection, I found most useful the interviews with Gen Robert H. Barrow, Col Francis I. Fenton, Jr., Maj Gen Raymond L. Murray, and LtCol Francis F. Parry. The interview with Adm John S. Thach, USN (Ret), in the U.S. Naval Institute's Oral History Collection, was consulted. I also benefited from direct interviews with MajGen Norman J. Anderson, Gen Robert H. Barrow, former SSgt Larry V. Brom, MGySgt Orville Jones, LtGen Robert P. Keller, LtGen Philip D. Shutler, and BGen Edwin H. Simmons. Contemporary quotations by PFC Morgan Brainard and Lt Joseph R. Owen are from their autobiographic books, Brainard's *Then They Called for the Marines* (formerly *Men in Low Cut Shoes* [Todd & Honeywell, 1986]) and Owen's *Colder Than Hell: A Marine Rifle Company at Chosin Reservoir* (Annapolis: Naval Institute Press, 1996).

Two official monographs proved helpful: LtCol Gary W. Parker and Maj Frank M. Batha, *A History of Marine Observation Squadron Six* (Washington, D.C.: History and Museums Division, HQMC, 1982), and Curtis A. Utz, *Assault from the Sea: The Amphibious Landing at Inchon* (Washington, D.C.: Naval Historical Center, 1994), which also includes the Seoul campaign.

Robert D. Heinl's stirring *Victory at High Tide: The Inchon-Seoul Campaign* (Philadelphia: Lippincott, 1968) leads the list of recommended books. I also suggest Bevin Alexander, *Korea: The First War We Lost* (New York: Hippocrene Books, 1986); Roy E. Appleman, *South to the Naktong, North to the Yalu* (Washington, D.C.: Office of the Chief of Military History, Department of the Army, 1961); Clay Blair, *The Forgotten War: America in Korea* (New York: *Times* Books, 1987) David Douglas Duncan's superb photo essay, *This is War! A Photo-Narrative in Three Parts* (New York: Harper & Brothers, 1951); George F. Hofmann and Donn A. Starry, *Camp Colt to Desert Storm. The History of U.S. Armored Forces* (Lexington: University of Kentucky Press, 1999); J. Robert Moskin, *The U.S. Marine Corps Story*, 3d ed. (Boston, Little Brown & Co., 1992); and Rod Paschall, *Witness to War: Korea* (New York: Perigree Books, 1995). Special thanks to LtCol Jon T. Hoffman, USMCR, for sharing advance copies of the Seoul/Wonsan chapters of his forthcoming biography of LtGen Lewis B. "Chesty" Puller.

I recommend these four vintage magazine essays: Nicholas A. Canzona, "Dog Company's Charge," *U.S. Naval Institute Proceedings* (Nov56); Ernest H. Giusti and Kenneth W. Condit, "Marine Air Over Inchon-Seoul," *Marine Corps Gazette*, June 1952; Lynn Montross, "The Capture of Seoul: Battle of the Barricades," *Marine Corps Gazette*, August 1951; and Norman R. Stanford, "Road Junction," *Marine Corps Gazette*, September 1951. For a more recent account, see Al Hemingway, "Marines' Battle for Seoul," *Military History*, August 1996.

The author acknowledges Mary Craddock Hoffman who designed the map of the overall Inchon-Seoul area, and Col David Douglas Duncan, USMCR (Ret), for allowing the use of his historical photographs of Seoul. Photographs by Frank Noel are used with permission of the Associated Press/Wide World Photos.

FROZEN CHOSIN
U.S. Marines at the Changjin Reservoir

by Brigadier General Edwin H. Simmons, USMC (Ret)

The race to the Yalu was on. General of the Army Douglas MacArthur's strategic triumph at Inchon and the subsequent breakout of the U.S. Eighth Army from the Pusan Perimeter and the recapture of Seoul had changed the direction of the war. Only the finishing touches needed to be done to complete the destruction of the North Korean People's Army. Moving up the east coast was the independent X Corps, commanded by Major General Edward M. Almond, USA. The 1st Marine Division, under Major General Oliver P. Smith, was part of X Corps and had been so since the 15 September 1950 landing at Inchon.

After Seoul the 1st Marine Division had reloaded into its amphibious ships and had swung around the Korean peninsula to land at Wonsan on the east coast. The landing on 26 October 1950 met no opposition; the port had been taken from the land side by the resurgent South Korean army. The date was General Smith's 57th birthday, but he let it pass unnoticed. Two days later he ordered Colonel Homer L. Litzenberg, Jr., 47, to move his 7th Marine Regimental Combat Team north from Wonsan to Hamhung. Smith was then to prepare for an advance to the Manchurian border, 135 miles distant. And so began one of the Marine Corps' greatest battles—or, as the Corps would call it, the "Chosin Reservoir Campaign." The Marines called it the "Chosin" Reservoir because that is what their Japanese-based maps called it. The South Koreans, nationalistic sensibilities disturbed, preferred—and, indeed, would come to insist—that it be called the "Changjin" Reservoir.

AT LEFT: *In this poignant photograph by the peerless Marine and* Life *photographer, David Douglas Duncan, the dead ride in trucks, legs bound together with pack straps.* Photo by David Douglas Duncan

General Smith, commander of the Marines—a quiet man and inveterate pipe-smoker (his favorite brand of tobacco was Sir Walter Raleigh)—was not the sort of personality to attract a nickname. His contemporaries sometimes referred to him as "the Professor" but, for the most part, to distinguish him from two more senior and better known General Smiths in the World War II Marine Corps—Holland M. "Howlin' Mad" Smith of famous temper and mild-mannered Julian C. Smith of Tarawa—he was known by his initials "O. P."

Across the Taebaek (Nangnim) Mountains, the Eighth Army, under Lieutenant General Walton H. Walker, was advancing up the west coast of the Korean peninsula. Walker, a short, stubby man, was "Johnnie" to his friends, "Bulldog" to the press. In World War II he had commanded XX Corps in General George S. Patton's Third Army and had been a Patton favorite. But these credentials held little weight with General Douglas MacArthur. He had come close to relieving Walker in August during the worst of the situation in the Pusan Perimeter. Relations between Almond and Walker were cool at best.

Department of Defense Photo (USMC) A88898

MajGen Oliver P. "O.P." Smith commanded the 1st Marine Division throughout the Chosin Reservoir campaign. A studious man, his quiet demeanor belied his extensive combat experience. His seemingly cautious style of leadership brought him into frequent conflict with MajGen Edward M. Almond, USA, the impetuous commanding general of X Corps.

MacArthur had given Almond command of X Corps for the Inchon landing while he continued, at least in name, as MacArthur's chief of staff at Far East Command. Almond, an ener-

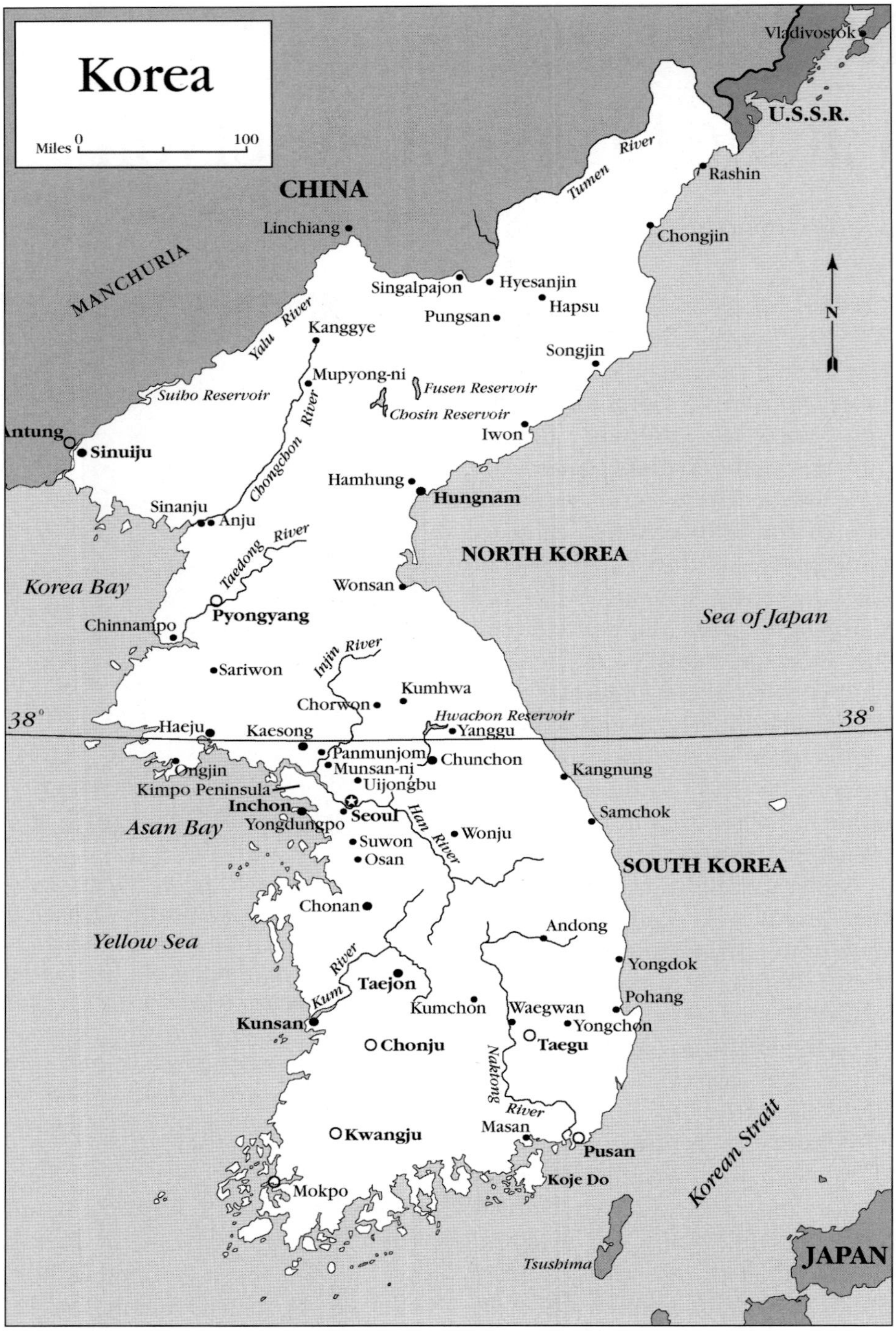

getic, ambitious, and abrasive man, still nominally wore both hats although his X Corps command post in Korea was a long distance from MacArthur's headquarters in Tokyo.

"General Almond in 1950 and 1951 in Korea had several nicknames," wrote Roy E. Appleman in his *Escaping the Trap: The U.S. Army X Corps in Northeast Korea.* "Generally, he was known to his friends and close associates as Ned. Other names were 'Ned, the Anointed,' which meant he was a favorite of General MacArthur's, and 'Ned, the Dread,' which referred to his power, his brusque manner, and sometimes arbitrary actions."

Many persons, both then and later, thought that X Corps should have now been subordinated to the Eighth Army. But on the 28th of October, "O. P." Smith was less concerned with these higher-level command considerations than he was with events closer to his headquarters at Wonsan. The 1st Battalion, 1st Marines, sent south of Wonsan to the coastal town of Kojo-ri, at the direction of X Corps, to protect a Republic of Korea supply dump had been roughly handled by a surprisingly strong North Korean attack. Smith thought that the battalion commander "was in a funk and it would be wise for Puller to go down and take charge." Colonel Lewis B. "Chesty" Puller, the regimental commander of the 1st Marines, left late that afternoon by rail with the 2d Battalion, 1st Marines, "to clear up the situation."

On the following day, Smith was annoyed by an order from Almond that removed the 1st Korean Marine Corps (KMC) Regiment from his operational control. The two commanders conferred on Monday, 30 October. Almond agreed to the return of one KMC battalion in order to expedite the move of the 5th and 7th Marines to Hamhung. After meeting with Almond, Smith flew by helicopter down to Kojo-ri and found that Puller indeed had the situation well in hand.

Tensions and differences between Almond and O. P. Smith were no secret. Almond had first met the Marine commander on Smith's arrival in Japan on 22 August 1950. As Almond still asserted a quarter-century later: "I got the impression initially (and it was fortified constantly later) that General Smith always had excuses for not performing at the required time the tasks he was requested to do."

With the 1st Marine Division

Department of Defense Photo (USN) 421392

Six weeks after the successful assault of Inchon, the 1st Marine Division made a delayed but unopposed landing at Wonsan on 26 October 1950. Heavy mining of the sea approaches with Soviet-made mines caused the delay.

After landing at Wonsan, the 1st Battalion, 1st Marines, was sent south to the coastal town of Kojo-ri. Here it was savaged by an unexpectedly strong North Korean attack. MajGen Smith sent Col Lewis B. "Chesty" Puller south with the 2d Battalion, 1st Marines, to "take charge."

Photo by Cpl W T. Wolfe, Department of Defense Photo (USMC) A4323

assigned as part of X Corps, Almond was Smith's operational commander. Smith's administrative commander continued to be Lieutenant General Lemuel C. Shepherd, Jr., commanding general of Fleet Marine Force, Pacific, with headquarters in Hawaii. While not in the operational chain-of-command, Shepherd was responsible for the personnel and logistical support of the 1st Marine Division. This gave Shepherd, an old war-horse, an excuse for frequent visits to the battlefield. Fifty-four-year-old Shepherd and 57-year-old Almond got along well, perhaps because they were both Virginians and both graduates of close-knit Virginia Military Institute—Almond, Class of 1915 and Shepherd, Class of 1917. "I liked him," said Shepherd of

LtGen Lemuel C. Shepherd, Jr., left, arrives at Wonsan Airfield on 31 October for one of his frequent visits and is greeted by MajGen Edward M. Almond. The two generals got along very well, perhaps because they were both Virginia Military Institute graduates. Both are wearing cuffed Army combat boots, a footgear favored by all those who could obtain them.

Photo by Cpl Jack Nash, National Archives Photo (USA) 111-SC351739

Photo by Cpl Jack Nash, National Archives Photo (USA) 111-SC351740

Using a map spread out on the hood of a jeep, MajGen Edward M. Almond briefs LtGen Lemuel C. Shepherd on his arrival at Wonsan on 31 October. At the extreme left is MajGen William J. Wallace, Director of Aviation at Headquarters U. S. Marine Corps, who accompanied Shepherd. MajGen Oliver P. Smith stands behind Shepherd and Almond. MajGen Field Harris is on the extreme right.

Almond at VMI, "but I really can't say he was one of my closer friends." Shepherd had had reason to expect command of X Corps for the Inchon landing, but MacArthur had given command to his chief of staff, Almond. Shepherd exhibited no visible grudge and Almond, in turn, always made Shepherd welcome on his visits and he often stayed with Almond in his mess. Years later, Shepherd, who considered Almond an excellent corps commander, said of him: "He was energetic, forceful, brave, and in many ways did a good job under most difficult conditions."

Concerning Almond's relations with Smith, Shepherd said: "He and O. P. just didn't get along, from the very first. They're two entirely different personalities. . . . O. P. [was] a cautious individual, a fine staff officer who considered every contingency before taking action. On the other hand Almond was aggressive and anxious for the X Corps to push ahead faster than Smith thought his division should. Smith wisely took every precaution to protect his flanks during his division's advance into North Korea, which slowed him down considerably. I'm sure Almond got into Smith's hair—just like I'm sure that I did too."

As a glance at a map will confirm, North Korea is shaped like a funnel, with a narrow neck—roughly a line from Wonsan west to Pyongyang—and a very wide mouth, the boundary with Red China and a bit with the Soviet Union on the north, formed by the Yalu and Tumen Rivers. Because of this geographic conformation, any force moving from the north to south had the advantage of a converging action. Conversely, forces moving from south to north must diverge. As Walker and Almond advanced to the north, the gap between Eighth Army and X Corps would grow wider and wider. This may have concerned Walker, but it does not seem to have bothered Almond—nor their common commander, General Douglas MacArthur, many miles away in Tokyo in what he liked to call his "GHQ."

General Shepherd arrived at Wonsan for one of his periodic visits on Tuesday, 31 October. Next morning Shepherd flew down to Kojo-ri to visit Puller and on his return to Wonsan he and Smith flew to Hamhung to see Litzenberg. That night Smith entered in his log:

> Litzenberg is concerned over the situation. He has moved up behind the 26th

Col Homer L. Litzenberg, Jr., commanding officer of the 7th Marines, was known to his troops as "Litz the Blitz," more for the alliteration than his command style. At the outbreak of the war, Litzenberg was in command of the 6th Marines at Camp Lejeune. In August 1950, the 7th Marines was hurriedly re-activated at Camp Pendleton using cadres drawn from the 6th Marines.

National Archives Photo (USMC) 127-N-A4719

ROK Regiment and will relieve them tomorrow. Two Chinese regiments have been identified to the front. The ROK regiment is very glad to be relieved by the Marines. The ROKs apparently have no stomach for fighting Chinese.

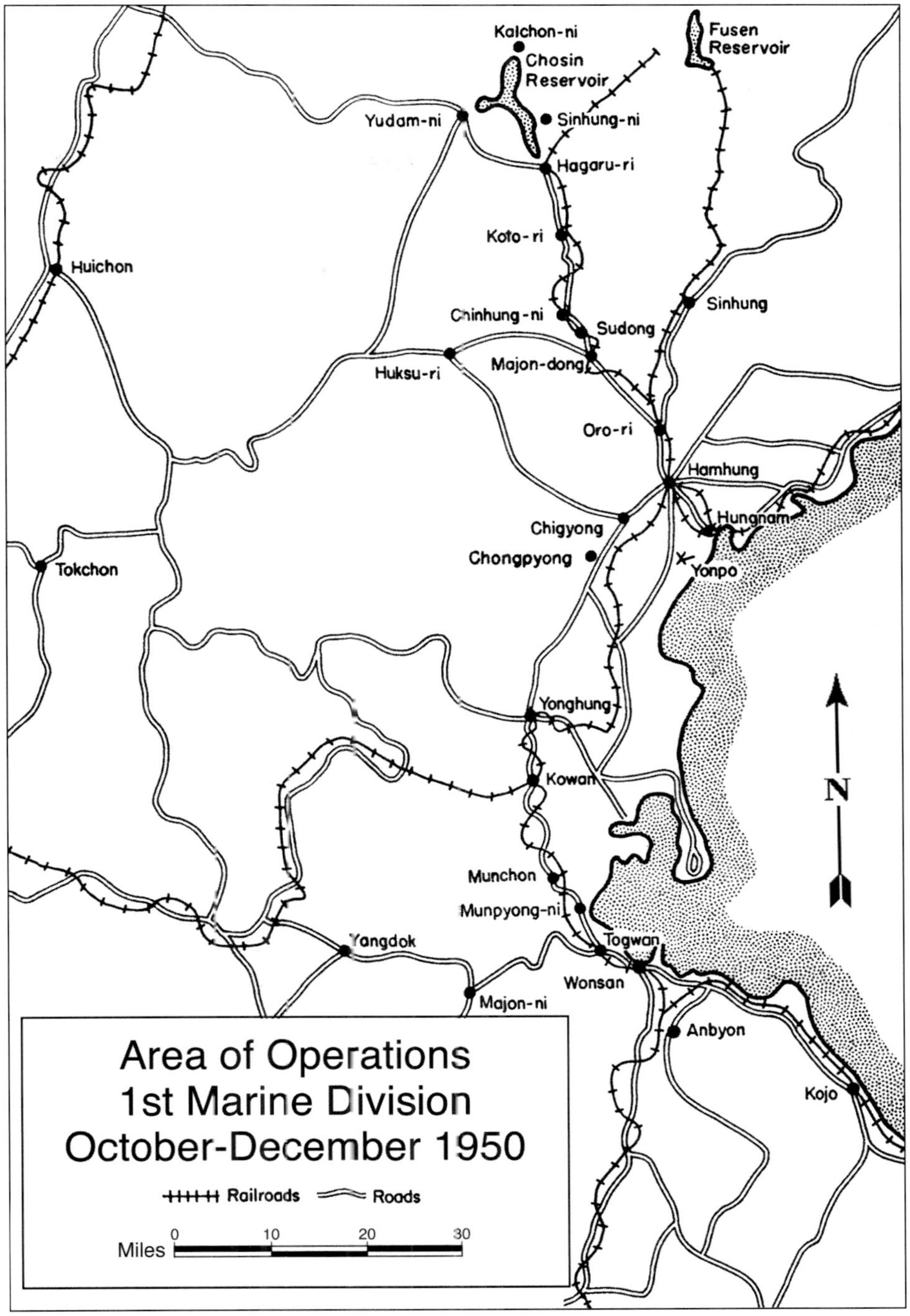

Area of Operations
1st Marine Division
October-December 1950

China Enters the War

Most Chinese historians now assert, and most Western historians are now ready to believe, that China entered the Korean War reluctantly.

In 1948, Kim Il Sung, then 37, had emerged under the patronage of the Soviet occupation as the leader of the so-called Democratic People's Republic of Korea with its capital in Pyongyang. In the West he remained a shadowy figure. Reputedly he had been a successful guerrilla fighter against the Japanese. He had returned to North Korea at World War II's end as a hero.

With the civil war against the Chinese Nationalists at a successful close, Mao in late 1949 and early 1950 released four divisions made up of soldiers of Korean origin to return to Korea. These Kim, under Soviet tutelage, reorganized into mirror images of Soviet rifle divisions in equipment and training. Early on Kim Il Sung learned how to play Mao Tse-tung against Stalin. For more than a year, Kim Il Sung zigzagged back and forth between Moscow and Peiping (not yet known in the West as "Beijing") seeking Stalin's and then Mao's support for an overt invasion of the South.

Both Stalin and Mao were at first skeptical of Kim's ambitions. Stalin cautioned Kim that he should cross the 38th Parallel only in a counteroffensive to a South Korea invasion of the north. Mao advised Kim to be prepared for protracted guerrilla warfare and not to attempt to reunify Korea by force.

For Mao, Kim's ambitious plans were a distraction. He was much more interested in completing his victory against the Chinese Nationalists by "liberating" Xingjiang, Tibet, and, most importantly, Taiwan. But in the spring of 1950 Stalin, playing his own game, gave Kim a qualified promise of Soviet support with the proviso that the North Korean leader consult with Mao. Accordingly, Kim went again to Peiping in mid-May 1950, put on a bold front and told Mao that Stalin had agreed with his plan to invade South Korea. A cautious Mao asked the Soviet ambassador to confirm Kim's assertion. A sly Stalin replied that while he approved Kim's plans, the deci-

Marine Corps Historical Center Photo Collection

Peng Dehuai, left, commander of the Chinese Communist Forces meets with Kim Il Sung, premier of the Democratic People's Republic of Korea. Peng later was China's Minister of National Defense but because of his criticism of the Great Leap Forward he was dismissed. During the Cultural Revolution he was arrested, imprisoned, tortured, and died in 1974 from a lack of medical care. He was posthumously rehabilitated as a "great revolutionary fighter" and loyal party member.

sion to invade South Korea was a decision to be made by China and North Korea. Until then Mao's position had been that Taiwan's liberation must have priority over Korea's unification. Reluctantly, Mao reversed these priorities but cautioned Kim on the strong possibility of intervention by the United States on the side of South Korea. Mao had almost no regular troops in northeast China. He told Kim that he would move troops into Manchuria, poised to cross into Korea, but that they would not enter the war unless American troops crossed the 38th Parallel.

The concentration of Chinese forces along the Yalu did not really begin until mid-July 1950 with the formation of the *Northeast China Border Defense Force*, about 260,000 troops. By mid-August Mao was certain that the United Nations forces would land at Inchon. On 23 August, the same day that MacArthur was wresting final approval for Inchon at a conference in Tokyo, Mao was meeting with his political and military leaders in Peiping. They were ordered to complete all preparations by the *Northeast China Border Defense Force* for war. The *Ninth Army Group*, which had been poised near Shanghai for the invasion of Taiwan (still known to the Western world as "Formosa"), was one of the major units ordered to move north.

The Inchon landing gave urgency to Chinese preparations to enter the war. Two days after the landing on 15 September 1950, a liaison party was sent to Pyongyang. Meanwhile, Kim Il Sung had asked Stalin for help, including putting pressure on China to send troops. Stalin considered the most acceptable assistance by Chinese armed forces would be in the form of "people's volunteers." In a telegram to Mao on 1 October, Stalin advised: "The Chinese soldiers may be considered as volunteers and of course will be commanded by the Chinese." Mao responded the next day that this was his intention.

For many years Western historians supposed that Lin Piao, a legendary Chinese Communist leader, commanded Chinese forces in Korea. They were wrong. At a 4 October conference in Peiping, Lin Piao argued strongly against sending troops into Korea to fight the Americans and refused to lead the intervention, using the subterfuge of poor health. Lin went off to Moscow for medical treatment and Mao named Peng Dehuai, a tough old revolutionary, to take his place. Peng, born in Hunan province of peasant stock, had emerged as a senior commander in Mao Tse-tung's famed Long March in 1934-1935. Peng arrived in Peiping too late for the 4 October meeting but met the next day with Mao who directed him to be ready to enter Korea by 15 October. On 8 October, Mao officially ordered the creation of the *Chinese People's Volunteers*, which would be the expeditionary element of the *Northeast China Border Defense Force*, with Peng as both military commander and political commissar.

That same day, 8 October, Mao sent his adroit vice-chairman and foreign minister, Zhou Enlai, to the Soviet Union to discuss with Stalin the provision of air assistance and military equipment. Zhou met with Stalin at his Black Sea resort. Stalin was noncommittal and said that he was not yet ready to provide air support. In Manchuria, Peng Dehuai was furious when he learned this. He stormed back to Peiping to meet again with Mao.

Meanwhile Kim Il Sung was pressing for immediate Chinese help.

In early October, Zhou Enlai informed the Indian ambassador in Peiping, Kavalam M. Panikkar, that if the United Nations forces crossed the 38th Parallel, China would send troops to defend North Korea. This warning reached Washington through diplomatic channels in New Delhi and Lon-don. Substantiating reports came through Moscow and Stockholm. The warnings were forwarded to MacArthur's GHQ in Tokyo.

On 15 October, the famous Wake Island meeting of President Harry S. Truman with General MacArthur took place. Truman's later blunt, but inadequate, explanation for the conference was "I wanted to have a personal talk with the General." A wary MacArthur perceived the meeting as a presidential ambush primarily designed to reinforce the Democratic Party's chances of success in the upcoming congressional elections. According to MacArthur, the possibility of Chinese intervention came up almost casually. He stated in his *Reminiscences* that the general consensus was that China had no intention of intervening. Truman would later say in his *Memoirs* that the threatened intervention in Korea was a prime reason for the meeting. He wanted MacArthur's "firsthand information and judgment."

What Truman took away from Wake Island was that the war in Korea was won and that the Chinese Communists would not attack. Asked about the chances of Chinese intervention, MacArthur, according to Truman, replied that there was very little chance that the Chinese would come in. At the most they might be able to get fifty or sixty thousand men into Korea, but since they had no air force, "if the Chinese tried to get down to Pyongyang, there would be the greatest slaughter."

MacArthur remembered the conversation quite differently. He would later say that it was a "prevarication" that he had predicted, "that under no circumstances would Chinese Communists enter the war." He characterized his Wake Island view on the possibility of Chinese intervention as "speculative." His own local intelligence, filtered through to him by his longtime G-2, Major General Charles A. Willoughby, USA, told him that large numbers of Chinese troops were massed across the Yalu, but his estimate was that America's virtually unopposed air power would make large-scale intervention impossible.

Four nights after the Wake Island meeting, on 19 October, the Chinese in massive numbers began crossing the Yalu.

General Almond's Ambitions

The entry of Chinese troops in force into North Korea was not picked up by United Nations intelligence, neither visually by aerial reconnaissance nor audibly by intercepts of radio signals. In northeast Korea, Almond continued his advance with great confidence. Almond's over-riding ambition was to beat his rival, General Walker, to the Yalu. His X Corps included two strong U.S. divisions—the 1st Marine and the 7th Infantry—and two Republic of Korea or "ROK" divisions—the Capital and 3d—and there were more troops on the way. With the expected arrival of the U.S. 3d Infantry Division the total would come up to 102,000, about two-thirds as many troops as Walker had in his Eighth Army.

The two ROK divisions, organized into the ROK I Corps, could

From left to right: MajGen David G. Barr, Commanding General, 7th Infantry Division; MajGen Edward M. Almond, Commanding General, X Corps; and Col Herbert B. Powell, Commanding Officer, 17th Infantry. After Barr's division made an unopposed landing at Iwon on 29 October, Almond pushed Barr to get to the Yalu. Powell's regiment was the spearhead for the advance.

Photo by Cpl Alex Klein, National Archives Photo (USA) 111-SC351957

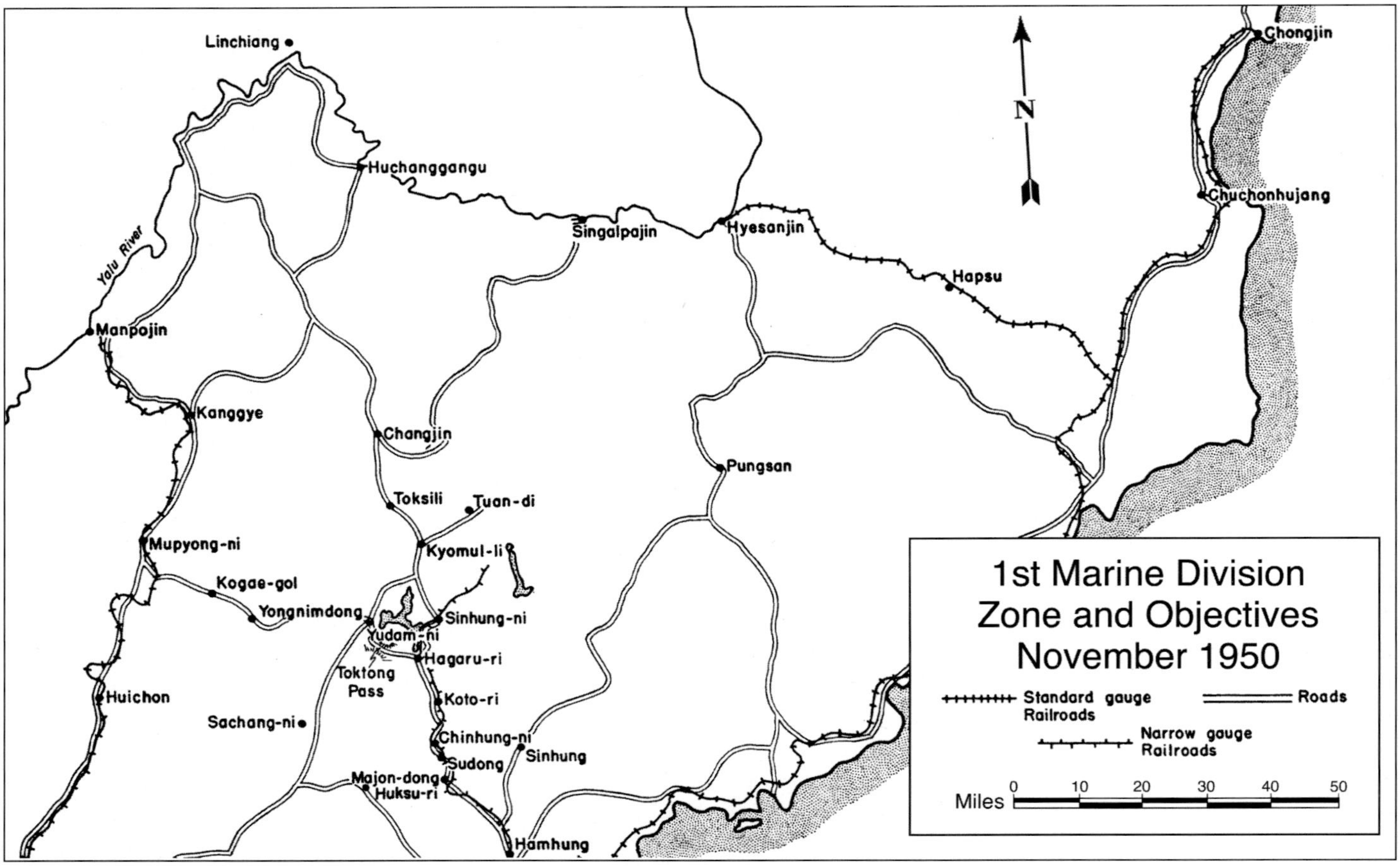

best be described as light infantry. They had no tanks and their only artillery were obsolescent 75mm howitzers. Almond's optimistic assessment of the ROK corps' fighting capabilities was "that they were a good deal better than the people they were chasing, the disorganized, disabled North Korean force."

"I realized," said Almond years later, "that we were scattered all over the landscape, but the general deployment was controlled by the terrain of the area in which the [X] corps was to operate." Almond should also have realized that there were strings tied to his employment of the 1st Marine Division and its companion 1st Marine Aircraft Wing. Shepherd, in Hawaii, was watching the use of the Marines very closely and so was the Commandant of the Marine Corps, General Clifton B. Cates, in Washington.

Army Major General David G. "Dave" Barr's 7th Division had loaded out from Pusan on 19 October as a follow-on to the 1st Marine Division at Wonsan. Barr had been the chief of staff of several commands in Europe during World War II. After the war he had headed the Army Advisory Mission in Nanking. At the war's beginning his 7th Division had been stripped to provide fillers for the 24th and 25th Divisions, the first divisions to be deployed to Korea. The 7th Division, before following the Marines ashore at Inchon, had been hurriedly brought to war strength with untrained South Korean recruits—the so-called KATUSA or "Korean Augmentees to the United States Army."

Now, when the Marine landing at Wonsan was delayed, Barr's destination was changed to Iwon, 75 miles northeast of Hungnam. Iwon was still theoretically in "enemy" territory, but it had good beaches and was known to be free of mines. Barr reloaded the 17th Regimental Combat Team under Colonel Herbert B. Powell into seven LSTs (tank landing ships) to be used in an amphibious assault in the event that the beaches were defended. They were not. As at Wonsan, the South Koreans had already taken the port from the land side. Powell's RCT-17 landed unopposed on 29 October and plunged ahead in a dash for Hyesanjin on the Yalu River. By the end of the month Powell's lead battalion was in a bitter four-day fight with the beaten, but still stubborn, North Koreans at Pungsan. In the days that followed, the remainder of the 7th Division came ashore. The 31st Infantry began landing on 3 November with the mission of moving in on the left flank of the 17th Infantry. The 32d Infantry followed on 4 November and went into bivouac northeast of Hungnam. On 8 November, the 31st Infantry ran into Chinese troops on the slopes of Paek-san, a 7,700-foot peak. In what was the 7th Division's first contact with the Chinese, the regiment reported at least 50 enemy killed.

Almond considered his control over the ROK I Corps to be no different than that he had over the 1st Marine Division and the 7th Infantry Division. Not all would agree, then or now, that his command of the ROK units was that complete. Their more binding orders came from President Syngman Rhee. While the landings at Wonsan and Iwon were still in prospect, the ROK Capital Division had marched steadily up the coast road to Iwon. The ROK 3d Division, meanwhile, had moved northwest from Hamhung toward the Chosin Reservoir.

On the Eighth Army front, the Chinese, whose presence in Korea had been doggedly denied at GHQ Tokyo, had by late October suddenly surfaced in formidable numbers. By the end of the month, the Chinese had defeated the ROK II Corps on the right flank of the Eighth Army to the point of disintegration, exposing the next unit to the left, the U.S. I Corps. General Walker ordered a general withdrawal to the Chungchon River. The Chinese did not pursue but broke off their offensive as suddenly as it began.

Separating the right flank of Eighth Army from X Corps was the Taebaek mountain range, the spine of the Korean peninsula and supposedly impassable to any significant number of troops. East of the Taebaek Mountains things seemed to continue to go well for General Almond and his X Corps. Almond did not appear to be perturbed by General Walker's problems.

Smith's Commanders and Staff

Because of the widely dispersed missions assigned his 1st Marine Division, General Smith had divided his command into regimental combat teams built around his three infantry regiments. RCT-5, under Lieutenant Colonel Raymond L. Murray, was assigned a zone behind Litzenberg's RCT-7. RCT-1, commanded by the legendary "Chesty" Puller, already the holder of four Navy Crosses, would remain for the time being in the vicinity of Wonsan fighting the remnants of one or more broken North Korean divisions struggling to get north.

All three regimental commanders had been successful battalion commanders in World War II. Moreover, Puller had commanded the 1st Marines at Peleliu. Now in this new war, Murray, 37, had brought the 5th Marines to Korea in a pell-mell rush to play a fire-brigade role in the defense of the Pusan Perimeter. Puller, 52, had arrived from Camp Pendleton with the 1st Marines in time for the Inchon landing. Litzenberg had formed the 7th Marines at Camp Pendleton, California, in a matter of days and had gotten to Korea in time to join in the battle for Seoul. Litzenberg was called "Litz the Blitz" by some, but this was more an alliteration—and maybe a little derisive at that—rather than a description of his command style, which tended to be cautious and buttoned-up. Because of his closely cropped prematurely white hair, some of his irreverent young lieutenants, and perhaps a few of his captains and majors often referred him to, as the "Great White Father."

Murray, the junior regimental commander, was simply known as "Ray." Among Marines, who like to argue over such things, Murray's 5th Marines, with the highest percentage of regulars and the longest time in the fight, would probably have rated highest in combat effectiveness. The 1st Marines, with Chesty Puller as its commander, most likely would have rated second. The 7th Marines, last to arrive and with the highest percentage of reserves, still had to prove itself and would have come in third.

Smith had a strong division staff: some members were already serving with the division when he took command and some that he had subsequently asked for. At the outbreak of the war Colonel Alpha L. Bowser, Jr., 40, had just been assigned as Force Inspector, Fleet Marine Force, Pacific. General Smith asked General Shepherd for his services as G-3 of the division. Shepherd assented and Bowser had become Smith's operational right-hand man. Bowser, Naval Academy 1932, had an enviable reputation in the Corps as a Quantico instructor, staff officer, expert in naval gunfire, and artillery battalion commander at Iwo Jima. He said later: "One of the major problems of the entire operation to the north of Wonsan was our open flank to the west. We never established contact with the right flank of the Eighth Army, and we had a great void out there which I at one time estimated to be somewhere in the neighborhood of 85 to 100 miles."

In a 1971 interview by D. Clayton James, noted historian and biographer of MacArthur, Bowser gave his considered opinion of Almond's leadership:

> General Almond was probably one of the most aggressive corps commanders I have ever seen in action. He was aggressive almost to a fault in my estimation. From

National Archives Photo (USMC) 127-N-A5895

For his brilliant staff work as division G-3, Col Alpha L. Bowser would later receive a Legion of Merit from MajGen Smith. Standing to Bowser's right is Col Bankson T. Holcomb, Jr., the division G-2, who received a similar decoration. These awards were made in January 1951.

> the standpoint of his own personal comfort and safety, he never gave it a thought. He was up at the front a great deal. He was what we referred to as a "hard charger."

However, Bowser went on to say:

> I questioned his judgment on many occasions. . . . I think that General Almond pictured this [campaign] in his mind's eye as a sweeping victory that was in his grasp. But he gambled and he lost. . . . A rather vain man in many ways. Ambitious. Could be a very warm personality as a personal friend. If he had one glaring fault, I would say it was inconsistency.

Bowser was not the only star player on Smith's team. Virtually all of the senior members of Smith's general and special staff were combat-tested veterans of considerable reputation.

Brigadier General Edward A. Craig, Smith's assistant division commander, now 54, had been commissioned in 1917, the same year as Shepherd and Smith. During the World War I years, while Smith was in garrison on Guam and Shepherd was winning laurels in France, Craig was fighting a kind of "cowboys and Indians" bush war against bandits in Haiti and Santo Domingo. In World War II he commanded the 9th Marines, first in training on Guadalcanal and then in combat on Bougain-ville and Guam—for the last he had a Navy Cross. He left the division in the summer of 1950 to command the 1st Provisional Marine Brigade in its adventures in the Pusan Perimeter, but rejoined the parent division in time for the Inchon landing. Like Smith, Craig was tall, slim, and prematurely white-haired. Smith, perhaps re-calling his own troubled period as the division's assistant division commander at Peleliu and being virtually ignored by the division commander, Major General William H. Rupertus, used Craig's services wisely and well, particularly as a roaming extension of his own eyes and ears. But Craig would be at home on emergency leave at a critical time in the campaign.

Smith's chief of staff, Colonel Gregon A. Williams, 54, had joined the division at Camp Pendleton in July. Before that he had been chief of staff of Fleet Marine Force, Pacific. A short, erect man, Williams had the reputation of being a "mean SOB." He had had a remarkable—but not unique for an officer of his vintage—career. He had enlisted in San Diego at the outbreak of World War I, but did not get to France, serving, in due time, in

Col Gregon A. Williams, shown here as a brigadier general, was Smith's chief of staff. Seldom seen in the field by the troops, Williams ran the division staff and headquarters with an iron hand. Like many senior Marine officers, he had had considerable service in China, both before and during World War II.

Department of Defense Photo (USMC) A42833

Nicaragua. As a sergeant he had been in the Dominican *Guardia Nacional* as a local lieutenant. A young Dominican lieutenant, Rafael Leonidas Trujillo, who would be the country's long-time dictator, became Williams' life-long friend. (The writer recalls that at the 1959 New Years reception at the presidential palace, in what was then Ciudad Trujillo, the single ornament on the grand piano in the ballroom was a silver-framed photograph of Colonel Williams.) Service in Nicaragua brought him a Navy Cross. At the beginning of the Pacific War, as an assistant naval attaché in Shang-hai, he was taken into custody by the Japanese and held as a prisoner until August 1942 when he was repatriated because of his diplomatic status. Undaunted, he returned to China to the new Nationalist capital in Chungking where he became involved in the support of guerrilla operations. In the summer of 1944 he was sent to the Pacific to take command of the 6th Marines, then involved in mopping-up on Saipan. He was a consummate chief of staff al-though not much loved by those who had to work for him. He got along famously with contemporaries such as Craig, but he terrified junior officers. Bowser, after some rough spells, said that, "He and I came to a perfect relationship." According to Bowser, Williams "took no guff" from Almond or Almond's chief of staff. Williams stayed close to the command post. Murray recalled seeing him only five or six times during their respective tours in Korea and then only at the division headquarters.

Much more visible to the command than Williams—and much better liked—was the deputy chief of staff, Colonel Edward W. Snedeker, 47. By training a communications officer, Snedeker had been Craig's chief of staff during the fighting by the 1st Provisional Marine Brigade in the Pusan Perimeter. Snedeker, Naval Academy 1926, had distinguished himself in World War II with a Silver Star from Guadalcanal and a Navy Cross from Okinawa where he commanded the 7th Marines. Smith's personal relationship with Snedeker was much closer than it was with Williams.

The G-1 (Personnel), Lieutenant Colonel Harvey S. Walseth, 39, Naval Academy 1935, had served in China before World War II and was a tank officer at Guadalcanal and Iwo Jima.

The G-2 (Intelligence), Colonel Bankson T. Holcomb, Jr., 42, movie-star handsome with his thin, clipped mustache and something of a bon vivant, was an old China hand. The cousin of Thomas Holcomb, who was the Commandant of the Marine Corps during World War II and the first Marine to reach the grade of four-star general, Bankson T. Holcomb, Jr., had graduated from high school in Peiping in 1925 and served two years as an enlisted man before going to the Naval Academy, Class of 1931. He returned to China in 1934 as an assistant to the naval attaché and Chinese language student, followed by two years as a Japanese language student in Tokyo. His speaking and reading ability in both Chinese and Japanese was rated as "excellent." He was at Pearl Harbor as an intelligence officer in December 1941 when the Japanese struck. In 1943 he returned to China once again, this time to operate out of the Nationalist capital of Chungking with Chinese guerrillas.

The G-4 (Logistics) Colonel

In January 1951, Col Edward W. Snedeker would receive a second Legion of Merit from MajGen Smith for his outstanding performance of duty as the division's deputy chief of staff. Earlier, in the Pusan Perimeter, he had been chief of staff of the 1st Provisional Marine Brigade.

National Archives Photo (USMC) 127-N-A5897

Francis M. McAlister, 45 and Naval Academy 1927, had fought as an engineer at Bougainville, Guam, and Okinawa. Like most Marine officers of his generation, he had served in Nicaragua and China.

Walseth, Holcomb, Bowser, and McAlister were the four pillars of the division's general staff. Bowser, the operations officer, and McAlister, the logistics chief, had a particularly close working partnership.

The much larger special staff ranged in grade from second lieutenant to colonel, from Second Lieutenant John M. Patrick, the historical officer, to Colonel James H. Brower, the division artillery officer. Most of the special staff were double-hatted; they also commanded the unit composed of their specialty. Brower commanded the division's artillery regiment, the 11th Marines, with three organic battalions of 105mm howitzers, a battalion of 155mm howitzers, and a battery of 4.5-inch multiple rockets. Brower, 42, was a Virginia Military Institute graduate, Class of 1931, who trained as an artilleryman at Fort Sill. During World War II he served as a staff officer with amphibious forces in the invasions of Sicily and Italy, but arrived in the Pacific in time for Okinawa.

Forty-four-year old Lieutenant Colonel John H. Partridge, as the commander of the 1st Engineer Battalion, was the division engineer. His engineers would work prodigies during the campaign. Partridge, Naval Academy 1936, had been with the 4th Marine Division at Roi-Namur, Saipan, Tinian, and Iwo Jima.

Smith and his staff, and his subordinate commanders and their respective staffs, worked their way through new sets of maps, analyzing the terrain and divining lines

Col James H. Brower was the division artillery officer and commanding officer of the 11th Marines. He fell ill at Hagaru-ri and was replaced in command by his executive officer, LtCol Carl A. Youngdale.

National Archives Photo (USMC) 127-N-A4696

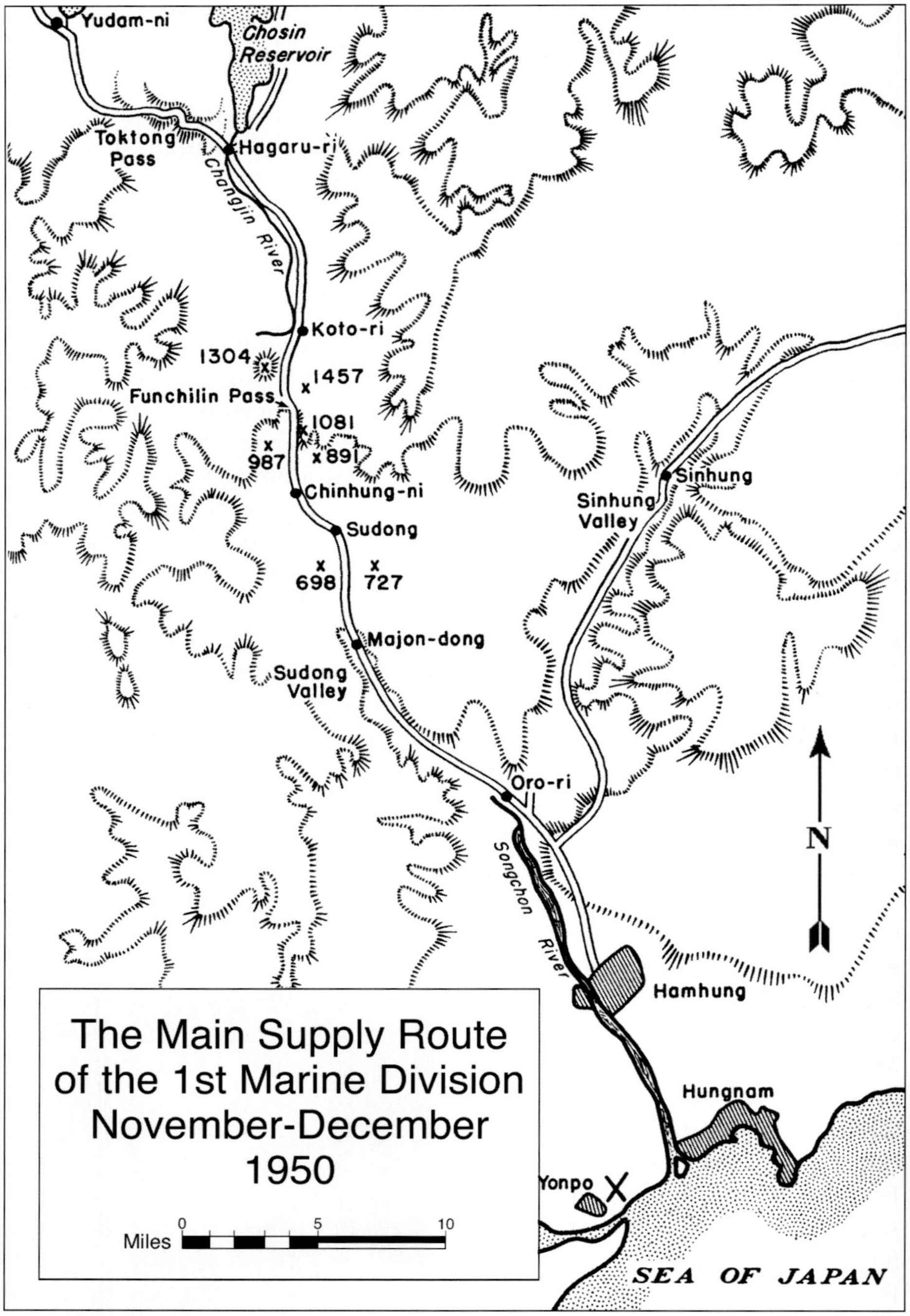

Distances in Road Miles	
Hungnam to Hamhung	8
Hamhung to Oro-ri	8
Oro-ri to Majon-dong	14
Majon-dong to Sudong	7
Sudong to Chinhung-ni	6
Chinhung-ni to Koto-ri	10
Koto-ri to Hagaru-ri	11
Hagaru-ri to Yudam-ni	14
TOTAL	78

of communication. Hamhung and Hungnam, then and now, are often confused. Hungnam is the port; it lies on the north side of the Songchon River where it empties into the Japanese Sea. Yonpo, with its airfield, which would prove critical, is on the south side of the estuary. Hamhung is the inland rail and highway nexus, straddling the main rail line from Wonsan north. A narrow-gauge (2' 6") line also started at Hamhung to Chinhung-ni. From there it climbed by cable car, now inoperative, to a plateau in the shadow of the Taebek Mountains. A road paralleled the narrow-gauge line. This would be the main supply route or "MSR" for the Marines' advance. The dirt-and-gravel road stretched 78 miles from Hamhung to Yudam-ni, which, as yet, was just a name on a map. "In only a few weeks," says the Corps' official history, "it would be known to thousands of Marines as the MSR, as if there never had been another."

RCT-7—with the 1st Motor Transport Battalion and Division Reconnaissance Company attached—received a partial issue of cold weather clothing before making the move north by truck and rail during the last three days of October.

Major Henry J. Woessner, 30, Naval Academy 1941, and operations officer of the 7th Marines, was at X Corps command post in Wonsan on 30 October when General Almond, standing before the Corps situation map, briefed General Barr on the upcoming operation. Barr's division was to push north to Hyesanjin on the Yalu. The 1st Marine Division was to reach the border by way of Chinhung-ni, Koto-ri, and Hagaru-ri. After describing this twin-pronged thrust to the Yalu, Almond again turned to the situation map. "When we have cleared all this out," he said with a broad sweep of his hand, "the ROKs will take over, and we will pull our divisions out of Korea."

Before leaving the command post, Woessner talked to an Army liaison officer who had just returned from the ROK 26th Regiment up near Sudong. The Army officer told him that the ROKs had collided with a Chinese force and had been driven back. Colonel Edward H. Forney, Almond's Marine Corps deputy chief of staff, arranged for Woessner to fly over the objective area in an Air Force North American T-6 Texan. Woessner saw no enemy on the flight to and over Hagaru-ri, but the rugged nature of the terrain impressed him.

Woessner, on his return that afternoon to the 7th Marines command post, made his report to Colonel Litzenberg who in turn called in his officers and noncommissioned officers and told them that they might soon be fighting the first battle of World War III. "We can expect to meet Chinese Communist troops," he said, "and it is important that we win the first battle."

Tuesday, 31 October

RCT-7 was scheduled to relieve the ROK 26th Regiment, 3d Division, in the vicinity of Sudong on 2 November. Litzenberg on 31 October cautiously sent out reconnaissance patrols from Hamhung to explore the route northward. One of the patrols from the 1st Battalion, 7th Marines—Captain Myron E. Wilcox, Jr., two lieutenants, three jeeps, and a fire team—reached the command post of the 26th Regiment near Sudong. Wilcox reported to Litzenberg that while there they had seen a Chinese prisoner. The South Koreans told Wilcox and his patrol that they had taken 16 Chinese prisoners and had identified them as belonging to the *124th Chinese Communist Force (CCF) Division*. The prisoners said they had crossed the Yalu in mid-October.

Further interrogation of the 16 prisoners had yielded that they were members of the *370th Regiment* of the *124th CCF Division,* which along with the *125th* and *126th Divisions*, made up the *42d CCF Army*. Roughly speaking, a Chinese army was the equivalent of a U.S. corps. On arriving from the Yalu, the *124th* had deployed in the center to defend the Chosin Reservoir, the *126th* had moved east to the vicinity of the Fusen Reservoir, and the *125th* to the western flank on the

Photo by Cpl Alex Klein, National Archives Photo (USA) 111-SC351718

Marines on 31 October saw for themselves the first Chinese prisoners taken by ROK I Corps. These prisoners, identified as belonging to the 124th CCF Division, *are wearing padded winter uniforms that offered little or no protection for the feet and hands, and the weather was about to turn cold.*

right of the *124th Division*.

Continuing the northward movement of his division from Wonsan, Smith ordered Murray to advance a battalion of the 5th Marines to Chigyong, eight miles southwest of Hamhung. Murray sent his 1st Battalion under Lieutenant Colonel George R. Newton. Newton, 35, Naval Academy 1938, had been a company commander in the Embassy Guard at Peiping in 1941 and spent World War II as a prisoner of war. Now, as a battalion commander, he had done well at Pusan and Inchon. One of Newton's companies was detached to relieve a company of the 7th Marines that was guarding Yonpo Airfield, five miles southwest of Hungnam.

Wednesday, 1 November

As yet the Marines had encountered no enemy anywhere along the MSR from Wonsan north to Hamhung; but Litzenberg was certain that he soon would be facing Chinese adversaries. On the following day, 1 November, he sent a stronger patrol from the attached division Reconnaissance Company to reconnoiter the Huksu-ri area about 45 miles northwest of Hungnam. This patrol, mounted in 21 jeeps and under First Lieutenant Ralph B. Crossman, after running into a small North Korean guerrilla force about three miles short of its objective, dug in for the night.

Meanwhile, the Marines began hearing rumors that the Eighth Army's 1st Cavalry Division—which they had last seen when the 1st Cavalry passed through the Marine lines north of Seoul headed for the successful capture of Pyongyang, the North Korean capital—was in serious trouble. If division headquarters had more

A 3.5-inch rocket section with Company C, 1st Battalion, 7th Marines, holds a position outside Hamhung on 31 October. The next day the 7th Marines would begin its march northward to relieve the 26th ROK Regiment near Sudong-ni.

Photo by Cpl Peter W. McDonald, National Archives Photo (USMC) 127-N-A5129

Photo by Cpl Peter W. McDonald, Department of Defense Photo (USMC) A4525

The 7th Marines began its motor march north from Hungnam on 1 November. A cautious Col Homer Litzenberg ordered LtCol Raymond G. Davis, commanding officer of the 1st Battalion, to reconnoiter to his front. Davis' Marines moved forward on foot. A crackle of small arms fire caused the column to halt momentarily. A full issue of cold-weather clothing has not yet been received. Only one Marine is wearing a parka.

definitive information, it did not filter down to the troops. Some 60 miles to the Marines' west in the Eighth Army zone of action, the Chinese had roughly handled the 8th U.S. Cavalry Regiment and the ROK 6th Division during the last days of October, but this had little or no effect on Almond's plans. Litzenberg's orders to advance remained unchanged. His first objective was to be Koto-ri.

During the day on 1 November, the 7th Marines made a motor march from Hungnam to an assembly area behind the ROK 26th Regiment, midway between Oro-ri and Majon-dong, without incident. Nevertheless, a cautious Litzenberg ordered Lieutenant Colonel Raymond G. Davis, 35, Georgia Tech 1938, to make a reconnaissance-in-force to South Korean positions north of Majon-dong with his 1st Battalion, 7th Marines. In World War II, Davis had commanded a heavy weapons company at Guadalcanal and an infantry battalion at Peleliu. The latter battle brought him a Navy Cross.

Late in the afternoon the regimental combat team curled up into a tight perimeter for the night. As part of RCT-7, Litzenberg had Major Francis F. "Fox" Parry's 3d Battalion, 11th Marines; the division Reconnaissance Company under Lieutenant Crossman; Company D, 1st Engineer Battalion, Captain Byron C, Turner; Company E, 1st Medical Battalion, under Lieutenant Commander Charles K. Holloway; detachments from the division's Signal Battalion, Service Battalion, and Military Police Company; and most of the 1st Motor Transport Battalion, Lieutenant Colonel Olin L. Beall.

Thursday, 2 November

The ROK 26th Regiment, awaiting relief, had withdrawn to a position about four miles south of Sudong. Early on the morning of 2 November, the South Koreans were probed by a CCF combat patrol estimated to be about two platoons in strength. Later that morning, Davis' 1st Battalion led the way out of the 7th Marines' perimeter toward the ROK lines at Majon-dong. Major Webb D. "Buzz" Sawyer, 32, followed with the 2d Battalion. A graduate of the University of Toledo, Sawyer had been commissioned in 1941. During World War II, as a captain and major he served with the 4th Marine Division at Roi-Namur, Saipan, Tinian, and Iwo Jima. Afterward, as an instructor at Quantico, he was known as an expert in the reduction of fortified positions.

Corsairs from Marine fighter squadron VMF-312 flew cover for what was essentially a parade northward. The passage of lines with the ROKs was over by 1030. The point, Company A, under Captain David W. Banks, took some scattered long-range fire and suffered a few casualties. Resistance thickened.

Major "Fox" Parry, 32, commanding the 3d Battalion, 11th Marines, was a Naval Academy graduate, Class of 1941. He had been the executive officer of an artillery battalion during Okinawa and immediately before Korea he had taken the yearlong Advanced Artillery Course at Fort Sill. At noon Battery I of Parry's artillery battalion fired the first of 26 fire missions covering the advance that would be shot during the day.

VMF-312, commanded by Lieutenant Colonel J. Frank Cole, flew 12 close air support missions and, as the light failed, night-fighter squadron VMF(N)-513, under Major J. Hunter Reinburg, delivered a few more.

Both Cole and Reinburg were experienced squadron commanders. Cole, 35, had entered the

Photo by Cpl Peter W. McDonald, Department of Defense Photo (USMC) A4549

Guns of Maj Francis F. "Fox" Parry's 3d Battalion, 11th Marines, covered the advance of the 7th Marines to Sudong. Here a 105mm howitzer from Capt Samuel A. Hannah's Battery G, laid for high-angle fire, awaits a fire mission. First artillery missions, shot on 2 November, alternated with close air support strikes by Marine Corsairs.

Marine Corps from the University of Nebraska in 1939 and had commanded fighter squadron VMF-111 in the Central Pacific and VMF-312 at Okinawa. Reinburg, 32, was the stepson of Marine Corps aviation great Lieutenant General Clayton C. Jerome. Reinburg had enlisted in the Naval Reserve in 1936 and transferred to the Marine Corps as an aviation cadet in 1940. Flying as a captain in the Solomons he became an ace, shooting down seven Japanese planes and destroying seven more on the ground. Before taking command of VMF(N)-513 he had spent a year as an exchange pilot flying night fighters with the British Royal Air Force. Technically, Marine fighter-bomber squadrons designated as "VMF(N)" were all-weather squadrons, but the "N" universally caused them to be called "night fighters." Reinburg's squadron flew twin-engine Grumman F7F-3Ns Tigercats

The main body of RCT-7, moving along the road in what Litzenberg called a "walking perimeter," had advanced just short of a mile by nightfall. Davis' 1st Battalion's nighttime positions were less than a mile south of Sudong, stretching across the valley from high ground to high ground. Behind him was Sawyer 's battalion similarly disposed. Sawyer was responsible for the high ground on both sides of the line of march. Captain Milton A. Hull, 30, commanding Company D, had some problems going up Hill 698 on the left hand side of the road. (Hills and mountains—both always called "hills"—were designated by their height in meters above sea level. Thus Hill 698 would be 698 meters or 2,290 feet in height.) A ROK company had precipitously given up its hillside position. The South Koreans, as they passed hurriedly southward, pointed back over their shoulders exclaiming "Chinese!"

Hull, a University of Florida graduate, had been commissioned in 1942 and had spent a good part of the war in China with the guerrillas, possibly with some of the same Chinese soldiers he was now fighting. Easy Company, under Captain Walter D. Phillips, Jr., passed through Hull's Dog Company to complete the fight, getting almost to the crest just

Col Homer Litzenberg called his road march a "walking perimeter." Here a part of the column pauses off the road, while Marine artillery and air pound the hills ahead. At nightfall on 2 November, LtCol Ramond Davis' 1st Battalion, lead element of the main body, halted one mile short of Sudong.

Department of Defense Photo (USMC) A4498

Photo by Sgt Frank C. Kerr, Department of Defense Photo (USMC) A4438

These Marines are members of Company D, 2d Battalion, 5th Marines. By 2 November, LtCol Raymond L. Murray's regiment had moved up by train to Hamhung with orders to patrol between Hamhung and Chigyong. Two days later the 3d Battalion was positioned near Oro-ri and the 2d Battalion was sent into Sinhung Valley. The 1st Battalion remained at Chigyong.

before midnight. (Rifle companies were almost invariably called by their name in the phonetic alphabet of the time: "Dog," "Easy," "Fox," and so on.) Farther to the rear was Major Maurice E. Roach's 3d Battalion, in a perimeter of its own, protecting the regimental train.

That morning, 2 November, Smith had met again with Almond. The 2d and 3d Battalions of the 5th Marines were moving by train to Hamhung. The 1st Battalion had already gone northward. Smith pointed out that the main supply route from Wonsan would be left exposed to guerrilla attack. Almond was not disturbed. He said that patrols could handle the guerrilla situation. Puller's 1st Marines, supported by elements of the 1st Tank Battalion, was given the responsibility from Wonsan northward to as far as Munchon. Murray's 5th Marines would patrol south from Hungnam to Chigyong. This left 54 miles from Chigyong south to Munchon uncovered except for light patrolling by Almond's Special Operations Company and a handful of South Korean counterintelligence agents.

Puller returned to Wonsan at midday on 2 November with his 1st Battalion. The 2d Battalion came back from Kojo-ri the following day. The 3d Battalion, 1st Marines, was still heavily engaged at Majon-ni, 26 miles west of Wonsan.

Friday, 3 November

Litzenberg did not know it, but he was two-thirds surrounded by the *124th CCF Division*. The *371st Regiment* was in the hills to his north and west. The *370th Regiment* was to his east. Somewhere behind these assault regiments, the *372d Regiment* stood ready in reserve.

By midnight on 2 November the 1st and 2d Battalions of the 7th Marines were being probed. An hour later both battalions were bending back from the weight of assaults on their flanks and Marines became acquainted with the Chinese habit of using flares

By 3 November, the 7th Marines was surrounded on three sides by the Chinese 124th Division. *Fighting grew fierce and casualties mounted. A sturdy masonry building in the shadow of high-tension lines coming down from the hydroelectric plant on the Changjin plateau became a battalion aid station.*

Photo by Cpl Peter W. McDonald, National Archives Photo (USMC) 127-N-A4541

National Archives Photo (USMC) 127-N-A4489

MajGen Smith used his assistant division commander, BGen Edward A. Craig, as his roving eyes and ears. Here, on 3 November, Craig, left, checks with Col Litzenberg whose 7th Marines had just made its first solid contact with the Chinese. Litzenberg is digging away at a half-size can of C-ration, probably canned fruit, a favorite.

Department of Defense Photo (USMC) A4501

Chinese prisoners, taken by Col Litzenberg's 7th Marines in the 3-4 November fighting, were identified as being members of the 370th *and* 371st Regiments *of the* 124th Division. *Chinese Communist "volunteers" fought well, but were surprisingly docile and uncomplaining once captured.*

and bugle calls to signal their attacks.

On the MSR, the roadblock in front of Able Company let a T-34 tank go by, thinking it was a friendly bulldozer. The single tank pushed through the company headquarters area and on through the battalion's 81mm mortar position, reaching Davis' command post. The startled Marines engaged the tank with rocket launchers and recoilless rifles; the tank took one or two hits and then turned around and headed north.

All three of Davis' rifle companies suffered heavy casualties as the night went on. The Chinese attackers got down to the road and wedged their way between the 2d and 3d Battalions. The regiment's 4.2-inch Mortar Company was overrun and lost one of its tubes. When morning came a confused situation faced the Marines. The Chinese were still in the valley. Getting rid of them would be an all-day effort. At first light Cole's VMF-312 came overhead with its Corsairs and was joined in mid-morning by Reinburg's Tigercats, pounding away with rockets, fragmentation bombs, and cannon fire. Parry's howitzers rendered yeoman service; before the end of the day his 18 guns had fired 49 missions delivering 1,431 105mm rounds. At closer range, Marine riflemen flushed out the Chinese enemy, fragmented now into individuals and small groups.

This would be Reinburg's last show. On 4 November he relinquished command of VMF(N)-513 to Ohio-born Lieutenant Colonel David C. Wolfe, 33, Naval Academy, Class of 1940. A big, athletic man, Wolfe had taken flight training as a captain and had commanded scout-bomber squadron VMB-433 in the Southwest Pacific during World War II.

The 1st Battalion counted 662 enemy dead in its zone of action. The 2d Battalion did not make a precise count but could not have been far behind. When Marine trucks came up with resupply, they carried back to Hungnam about 100 wounded Marines. Total Marine casualties for the two days—2 and 3 November—were 44 killed, 5 died of wounds, 1 missing, and 162 wounded, most of them in the 7th Marines.

As recorded in the official history by Lynn Montross and Nicholas A. Canzona, a tactical principle was emerging: "To nullify Chinese night attacks, regardless of large-scale penetrations and infiltration, defending units had only to maintain position until daybreak. With observation restored, Marine firepower invariably would melt down the Chinese mass to impotency." It was a principle that would serve the Marines well, time after time, in the coming several weeks.

Photo by Cpl L. B. Snyder, National Archives Photo (USMC) 127-N-A4463

BGen Edward Craig, on the left and wearing a Marine "fore-and-aft" cap, and MajGen Edward Almond, on the right with map case and wearing an Army cap, went forward to the 7th Marines command post to see for themselves the Chinese prisoners that had been taken. Col Homer Litzenberg, the regimental commander, is talking with BGen Craig.

Saturday, 4 November

The 7th Marines' positions remained essentially the same during the night of 3-4 November. The perimeters were peppered lightly, but there were no further Chinese assaults. Later it was learned that the *370th* and *371st CCF Regiments* were withdrawing to a defensive line, established by the *372d Regiment* about two miles north of Chinhung-ni, stretching from Hill 987 to Hill 891.

Litzenberg ordered increased patrolling to the north to begin at dawn on 4 November. Marines from Davis' 1st Battalion patrolled to the edge of Sudong, met no resistance, and returned to their perimeter. Crossman's Reconnaissance Company moved out in its jeeps at 0800. First Lieutenant Ernest C. Hargett took the point into Sudong and met a party of Chinese in the middle of the town. Hargett's men killed three and took 20 more as willing prisoners. Crossman now put Second Lieutenant Donald W. Sharon's 2d platoon into the point with the 1st Battalion coming behind them into Sudong.

The North Korean People's Army (NKPA) skeleton *344th Tank Regiment*, down to five Soviet-built T-34 tanks and apparently unable to negotiate Funchilin Pass, had been left on the low ground to fend for itself. One T-34 was abandoned after being damaged in its wild one-tank attack against the 7th Marines command post. The remaining four tanks took covered positions off the road. Sharon passed by the first hidden T-34 but bumped into the second. He and two of his Marines damaged the tank with hand grenades. Charlie Company, 7th Marines, with its own 3.5-inch rockets and reinforced with a section of 75mm recoilless rifles, came on the scene and finished off the second tank. A third tank emerged from a thatched hut. Engaged by both rocket launchers and recoilless rifles, the tank continued to move until stopped by the 5-inch rockets of a flight of Corsairs. The Marines now found the bypassed first tank.

Wounded Marines arrive at the battalion aid station of Maj Webb D. "Buzz" Sawyer's 2d Battalion, 7th Marines. Helicopter evacuation was still the exception rather than the rule. Most wounded Marines were hand-carried down to the closest road and then moved by jeep to the nearest aid station. Here they would be sorted out ("triage") and sent to the rear for more definitive treatment.

Photo by Cpl Peter W. McDonald, National Archives Photo (USMC) 127-N-A4546

Photo by Cpl Peter W. McDonald National Archives Photo (USMC) 127-N-A4496

Bodies of dead Marines, covered with ponchos and shelter halves, await further disposition. The 7th Marines fight at Sudong lasted three days, 2-4 November. Division casualties for the period, nearly all of them in the 7th Marines, totaled 61 killed in action, 9 died of wounds, 162 wounded, and 1 missing in action.

After receiving fire, the crew surrendered their tank and themselves. The fourth tank, now alone, surrendered without a fight. The *344th NKPA Tank Regiment* was no more. Litzenberg, having advanced almost four miles by mid-afternoon, ordered his regiment to halt for the night in a tight perimeter at Chinhung-ni.

For the first 43 miles north from Hungnam the 1st Marine Division's MSR was a two-lane highway passing through relatively flat terrain. At Chinhung-ni the road narrowed to one lane as it went up Funchilin Pass, climbing 2,500 feet in eight miles of zigzagging single-lane road clinging to the sides of the mountains; "a cliff on one side and a chasm on the other" as the official history described it. The narrow gauge railroad was operable as far as Chinhung-ni and it was decided to establish a railhead there.

The division Reconnaissance Company was ordered to move forward another mile, on up into Funchilin Pass, and outpost the southern tip of Hill 891. With Second Lieutenant Charles R. Puckett's 3d Platoon out in front, the reconnaissance Marines moved almost into the saddle separating Hills 987 and 891, already inconveniently occupied by the Chinese. A firefight developed. The company held its ground but lost two Marines killed, five wounded, and two jeeps destroyed.

On 4 November, Smith shifted his command post from Wonsan to Hungnam, occupying an abandoned engineering college on the outskirts of the city. In reconnoitering for the site, Smith's assistant division commander, Brigadier General Craig had been treated to the sight of 200 dead Koreans laid out in a row, executed by the Communists for no apparent reason. Smith flew to Hungnam by helicopter and occupied the new command post at about 1100. Most of his headquarters arrived by rail that evening, an uneventful trip except for a few scattered rifle

The 7th Marines entered Sudong on 4 November. Beyond Sudong the main supply route began its climb into Funchilin Pass. Here, a Marine patrol, troubled by a sniper, searches out a hamlet of thatched-roofed, mud-wattle huts.

Department of Defense Photo (USMC) A4537

Sketch by Sgt Ralph Schofield, USMCR

A Marine sniper draws a bead on a distant Chinese enemy. This sketch, and numerous others that follow, are by Cpl (later Sgt) Ralph H. Schofield, a talented Marine Corps reservist from Salt Lake City, who served as a Leatherneck magazine combat artist. A seasoned veteran of World War II, Schofield had fought as an infantryman in the South Pacific.

shots. The larger part of his headquarters would remain in place in Hungnam for the duration of the operation.

To the south of the 7th Marines, the battalions of Murray's RCT-5 were having their own adventures. "Our first assignment was to go to the east side of the reservoir," remembered Murray. "I wondered, why are they splitting us up like this?" By 4 November, the 1st Battalion (Lieutenant Colonel George R. Newton) had been left behind at Chigyong and detached to division control. The 3d Battalion (Lieutenant Colonel Robert D. "Tap" Taplett) was positioned near Oro-ri. The 2d Battalion (Lieutenant Colonel Harold S. Roise) had been sent into Sinhung Valley, five miles north and 15 miles east of the 7th Marines, to relieve the ROK 18th Regiment. The relief was accomplished without incident. Roise, 34, from Idaho, had spent World War II in the battleships *Maryland* (BB 46) and *Alabama* (BB 60). The 2d Battalion, 5th Marines, was his first infantry command and he had done well with it at Pusan, Inchon, and Seoul. His mission now was to block the Sinhung corridor and to find a northerly route to either the Chosin Reservoir or to the reservoir known to the Marines by its Japanese name "Fusen." The Korean name was "Pujon." Roise's mission carried him away from 1st Marine Division's axis of advance and into the zone of the 7th Infantry Division.

SSgt Meyer Rossum triumphantly displays a poster of Soviet dictator Joseph Stalin found in a by-passed Chinese bunker in the vicinity of Funchilin Pass. Marines would learn that the Chinese, despite problems of weather and terrain, were avid diggers and experts at field fortification.

National Archives Photo (USMC) 127-N-A4807

With 7th Marines, 5-6 November

Early on Sunday morning, 5 November, Major Roach's 3d Battalion, 7th Marines, passed through the 1st Battalion to continue the advance up Funchilin Pass. Hargett's reconnaissance platoon led the way. Rounding a hairpin turn, Hargett ran into Chinese fire and had four more Marines wounded. The 3d Battalion moved into the attack. Item Company was given Hill 987 and George Company Hill 891 as their objectives. Both were stopped by mid-morning by heavy small arms and machine gun fire. For the rest of the day the battle continued as a duel between Parry's 105mm howitzers and Chinese 120mm heavy mortars. From overhead, the Corsairs of VMF-312 delivered 37 close air support sorties.

At the top of the pass the road flattened onto a plateau and ran for two miles until it reached the village of Koto-ri where it rejoined the now-abandoned narrow gauge railroad. During the day General Smith gave Litzenberg the objective of reaching Koto-ri.

Roach's 3d Battalion continued the attack the next morning. How Company, under First Lieutenant Howard H. Harris, was to pass through George Company and move up the southern tip of Hill 891. Item Company, under First Lieutenant William E. Johnson, was to continue its attack against Hill 987. Both attacks went slowly, with the assaults not getting underway until mid-afternoon.

Second Lieutenant Robert D. Reem, leading one of How Company's platoons in the final assault, threw himself on a Chinese grenade and was killed. Harris radioed Roach that his company was exhausted. Roach relayed the report to Litzenberg who ordered the company to disengage and withdraw. Next morning, 7 November, Roach's battalion again moved up the slopes of both Hills 891 and 987, and this time found them empty of enemy. The Chinese had disappeared during the night. For most of the next three weeks traffic northward on the MSR would be unimpeded.

Operations North of Wonsan, 4-9 November

Meanwhile, in accordance with Almond's decision on 3 November, X Corps troops and the 1st Marine Division continued to share the responsibility for the Wonsan-Hungnam MSR. Operation of the Wonsan-Hamhung rail line came under X Corps Railway Transportation Section. The division began sending supply trains north daily from Wonsan. For two days they got through unmolested, but on the third day, 6 November, the train was halted at Kowan by torn-up rails. North Korean guerrillas then attacked the train, which was guarded by 39 Marines from Charlie Company of the 1st Amphibian Tractor Battalion. Taken by surprise, eight Marines were killed, two wounded, at the outset. Six more Marines were wounded in the ensuing firefight. The guard then broke off action and found protection within the perimeter of an Army artillery battalion.

Smith was promised the use of the Army's newly arrived 65th Regimental Combat Team to guard bridges and other key points along the route. Rail service from Wonsan to Hamhung was resumed on 9 November with the caution that passengers were to ride only in open gondola cars. Their steel sides promised some order of protection from small arms fire and mortar fragments. While Marines rattled northward in gondola cars, MacArthur on 9 November informed the Joint Chiefs of Staff that complete victory was still possible and reiterated his belief that U.S. air power would prevent the Chinese from crossing the Yalu in decisive numbers.

Arrival of 3d Infantry Division

The Army's 3d Infantry Division, its ranks hastily filled out with South Koreans, began arriving at Wonsan in early November. Major General Robert H. "Shorty" Soule, the division commander, was a paratrooper who had fought with the 11th Airborne Division under MacArthur in the Southwest Pacific. The first regiment of Soule's division to land was the 65th Infantry, made up largely of Puerto Ricans, on 5 November. Almond came, looked, and said he "didn't have much confidence in these colored troops."

During World War II, Almond had commanded the U.S. 92d Infantry Division which had almost all white officers but black rank-and-file. The division had turned in a mixed performance in Italy. Almond's prejudices were typical of his generation and Southern background. The regimental commander, Colonel Wil-liam W. Harris, West Point 1930, protested that most of his men were not "colored," but "white." Almond, unconvinced of the 65th RCT's reliability, told Harris that he was going to send the regiment north to Yonghung and then west across the mountains to make contact with the Eighth Army's right flank. Harris was appalled by these orders.

The 1st Shore Party Battalion, under command of legendary Lieutenant Colonel Henry P. "Jim" Crowe, 51, stayed behind at Wonsan to help the 3d Infantry Division land and unload. Crowe had enlisted during World War I and had a fabled career as football player, team shot, and bandit fighter, reaching the highly prized warrant grade of Marine gunner in

1934. He had a Silver Star from Guadalcanal and a Navy Cross from Tarawa where he commanded a battalion as a major. He thought Soule "one of the finest men" he ever met, but he found Almond "haughty."

5th Marines Operations, 5-8 November

There was now clear evidence that the Chinese, and some North Koreans, were out in front of Roise's 2d Battalion, 5th Marines, in Sinhung Valley but keeping their distance. Northwest of Sinhung itself and about 10 miles due east of Koto-ri, Dog Company captured a stray Chinese soldier found sleeping in a house. He proved to be a wealth of information. He said that he belonged to the *126th CCF Division*. He asserted that six CCF armies had arrived in North Korea and that a total of 24 divisions had been committed to the intervention. He had learned this in a series of lectures given by political officers to his regiment after it had crossed the border.

Smith conferred with Almond on the afternoon of 7 November. "He apparently has been somewhat sobered by the situation on the 8th Army front, which is not very good," Smith entered into his log. Almond promised Smith that he would let him concentrate the 1st Marine Division.

The 1st Battalion, 5th Marines, remained at Chigyong. On 7 November, Major Merlin R. Olson, 32, the battalion's executive officer, with Companies A and B reconnoitered in force west of Oro-ri to Huksu-ri. On the 8th, still short of his objective, Olson ran into a North Korean force, estimated at 2,000, and was recalled.

Meanwhile, Roise's patrols had found no useable road to either Chosin or Fusen Reservoirs but had learned that the road leading northeast to the Manchurian border, into the zone assigned to the U.S. 7th Infantry Division, could bear military traffic. One of his patrols touched a patrol from the 31st Infantry on 8 November. Smith had an understanding with Almond that if the 5th Marines could not get to the Fusen Reservoir by road, Barr's 7th Infantry Division would attempt to reach it from the east.

Photo by Cpl L. B. Snyder, National Archives Photo (USMC) 127-N-A4620

LtCol Raymond Davis' 1st Battalion, 7th Marines, occupied Koto-ri on the Marine Corps Birthday, 10 November, against no resistance. A day later needed replacements arrived. One or two battalion-sized replacement drafts arrived in Korea each month to keep the division's ranks—particularly the infantry units—at fighting strength.

7th Marines Operations, 8-11 November

On 8 November General Almond visited the 7th Marines. On learning that Captain Thomas E. Cooney, commander of George Company, had been twice slightly wounded on Hill 891, he awarded Cooney an on-the-spot Silver Star. His aide was caught without a supply of medals. Almond scribbled a note on a piece of paper—"Silver Star for Gallantry in Action"—and pinned it to Cooney's jacket.

A patrol of 15 Marines under First Lieutenant William F. Goggin of the 2d Battalion left Chinhung-ni at noon on 8 November, reached Koto-ri, and next evening returned unscathed to the lines of the 3d Battalion. Next day, 10 November and the Marine Corps Birthday, the 1st Battalion passed through the 3d Battalion and an hour-and-a-half later entered Koto-ri.

X Corps issued an order attaching the 65th Infantry and the ROK 26th Regiment to the 1st Marine Division. Two battalions of South Korean Marines were also to be attached. On receiving the order Smith learned that he was responsible for making contact with the Eighth Army. He gave orders to that effect to the 65th Infantry and was annoyed to find that Almond had already given the regiment's commander, Colonel Harris, detailed instructions down to the company level as to what to do. Something of the same happened with regards to the mission of the ROK 26th Regiment. "Such a procedure, of course, only creates confusion," Smith fussed in his log. "It was this type of procedure

Photo by Sgt John Babyak, National Archives Photo (USMC) 127-N-A4562

At division headquarters at Hamhung, MajGen Smith observed the Marine Corps Birthday in traditional fashion. He read the birthday message from the Marine Corps Manual *and then cut the somewhat meager cake with a Korean sword. As tradition prescribes, the first slice went to the oldest Marine present, BGen Craig.*

which I protested to General Almond in connection with direct orders given to my regiments." To Smith's further annoyance, he was ordered to provide a rifle company to guard X Corps command post at Hamhung. The order was passed to the 3d Battalion, 5th Marines, for execution. Taplett received it with "surprise and disgust," not understanding why a headquarters with about 2,000 troops needed extra security so far behind the lines of advance. He detailed Item Com-pany under Captain Harold G. Schrier to do the job—the same Schrier who as a lieutenant had taken his platoon up Mount Suribachi on Iwo Jima to raise the first flag.

That evening there was a Marine Corps Birthday party in General Smith's mess attended by his staff. Punch and cake were served. Smith entered in his log: "I read the paragraphs from the Marine Corps Manual and then cut the cake with a Korean sword."

The weather had turned terrifically cold up on the plateau, well below zero at night. Platoon warming tents were set up in Koto-ri, a hapless little hamlet. As the official history observed, the cold seemed "to numb the spirit as well as the flesh." On the 11th, Company C, 1st Battalion, 7th Marines, had a fight in which it lost four Marines killed, four wounded, and claimed 40 enemy casualties. Otherwise the enemy seemed to have vanished.

5th Marines Operations, 9-13 November

Murray received orders on 9 November to concentrate his regiment on the MSR leading to Chosin Reservoir. Newton's 1st Battalion, coming out of Chigyong on 10 November, was to move to Majon-dong. A patrol sent forward from Newton's battalion was ambushed and had to be rescued with a battalion-sized attack before the battalion could get to the village. On the 13th, another patrol from the 1st Battalion, 5th Marines,

Col "Chesty" Puller cuts the Marine Corps Birthday cake on 10 November at his 1st Marines regimental headquarters outside Wonsan where the weather was still pleasant. Far to the north, on the Chosin plateau, the 7th Marines was already encountering sub-zero temperatures.

Photo by Cpl W. T. Wolfe, National Archives Photo (USMC) 127-N-A4571

Department of Defense Photo (USMC) A4628

On 12 November the villagers at Koto-ri were informed that they have been "liberated" and were now free to elect their own village officials. The large number of Korean civilians, who would later crowd into the Marines' defensive perimeters, would become a huge problem.

ran into a company-sized group of Chinese that killed seven Marines and wounded three more before withdrawing.

Roise's 2d Battalion came out of Sinhung Valley on 13 November with orders to relieve the 7th Marines of the responsibility of defending Koto-ri. Along the way Roise's Marines picked up one Chinese and 12 North Korean prisoners. An airstrip capable of handling light aircraft was opened at Koto-ri that same day. Taplett's 3d Battalion, 5th Marines, was now at Chinhung-ni. Taplett, 32, from South Dakota, had spent most of World War II on sea duty with the cruiser *Salt Lake City* (CL 25). The 3d Battalion was his first infantry command. He was not always an easy personality, but his performance at Pusan, Inchon (where he had led this battalion ashore in the successful seizure of Wolmi-do in the opening phase of the landing), and Seoul had been outstanding. Still stiffer fights were ahead of him.

Smith, making his own road reconnaissance of Funchilin Pass, took a helicopter as far as Chinhung-ni. Helicopters at that time, because of the cold and altitude, were not going farther north; there being problems with gear boxes freezing up. Smith borrowed a jeep from Taplett and drove on up to Koto-ri.

MacArthur Reassesses Situation

By now MacArthur had to accept that the Chinese were in Korea in strength, perhaps as many as 100,000 of them, but he was still of the opinion that China would not make a full-scale intervention.

Almond had moved his headquarters on 11 November from Wonsan to Hamhung with plans to move his command post farther north to Hagaru-ri. Almond must have reflected that 11 November was Armistice Day from the First World War. Many of the senior leaders in Korea had fought in that war, including MacArthur as a brigadier general and Almond as a major. Almond had served with distinction in the U.S. 4th Infantry Division as commander of the 12th Machine Gun Battalion. Armistice

(Continued on page 242)

The wreath with "Merry Christmas," perhaps some Marine's idea of humor, is misleading. The photo was probably taken at Chinhung-ni in mid-November. At left is LtCol Robert D. Taplett, Commanding Officer, 3d Battalion, 5th Marines, and at right is his executive officer, Maj John J. Canny, who would die at Yudam-ni two weeks later.

Photo by Sgt Frank C. Kerr, National Archives Photo (USMC) 127-N-A7349

Coping with the Cold

How cold was it at the reservoir? Most Marines "knew" that the temperature went down to about 25 degrees below zero at night, but how many Marines had a thermometer in their pack?

The cold was no great surprise, unless, perhaps, you were like one Marine from Samoa who had never seen snow before. The division staff knew by late October or early November that Hagaru-ri had the reputation of being the coldest place in North Korea, with a recorded temperature of 35 degrees below zero. The climate is roughly like that of Minnesota or North Dakota. The winter of 1950 was a cold one, but not unusually so. The powers that be had adequate warning that it was coming and considerable preparations had been made.

Those at the top, and some at other levels in the 1st Marine Division, had had some experience with cold weather operations, if not by participation, at least by observation and a bit of training.

The division's commanding general, Major General Oliver P. Smith, had gone with the 1st Provisional Marine Brigade to Iceland in August 1941 to relieve the British garrison, as a major in command of the 1st Battalion, 6th Marines. He remembered Iceland as a "bleak and rugged island—mountains, cliffs, no trees—not a tree" and most of all the violent, never ceasing wind.

National Archives Photo (USN) 80-G-424584

These Marines, looking very fresh, pause enroute to Yudam-ni on 27 November to heat their C-rations over an open fire. They are wearing the newly issued shoe-pacs with boot socks folded neatly over the tops. At the far left the Marine appears to be wearing old-fashioned galoshes over field shoes or boots, a better combination against the cold than the shoe-pacs.

Cold was all-pervasive, even at Hamhung, which with its near sea-level elevation was much milder than up on the Changjin plateau. Here Marines at Hamhung, probably members of a combat service unit, cook bacon and beans on top of a stove made from a gasoline drum.

Department of Defense Photo (USMC) A4617

There were others, besides Smith, in the division who had also been in the Iceland expedition. One of them was Lieutenant Colonel Raymond L. Murray, commander of the 5th Marines. In Iceland he had been a captain and commander of a machine gun company. He had also served in Peiping in North China before World War II. He did not find Iceland as rugged as Korea: "It was not terribly cold. I don't think it ever got much below 10 above zero."

The Marines suddenly on their way to Iceland did not, at first, have any specialized winter clothing. They wore their wool kersey winter service uniforms including their woolen overcoats, supplemented by some items bought on the open market, notably some short, sheepskin-lined, canvas coats purchased from Sears Roebuck and carried as organizational property. Another much-favored addition were pile-lined hats with ear flaps, such as Marines had worn in North China.

The Marines in Iceland did not live in tents or in

the open. They were billeted in Nissen huts, "an elongated igloo covered with corrugated iron roofing and lined with beaver board," the flimsy British equivalent of the more substantial American Quonset huts. Marines piled sod on the sides of the Nissen huts to improve insulation. Each battalion had a different camp in a different part of the island.

In many ways the deployment, as an opportunity for winter training, was a disappointment. Finnish success with ski troops in the Winter War with the Soviet Union in 1939 had been much publicized (and romanticized). But it did not get as cold in Iceland as it was supposed to get and there was not much snow, seldom as much as a foot.

Marine experiments with skis and more work-a-day snowshoes did not come to much. Nine years later, the Marines at Chosin Reservoir did not have skis or snowshoes and it was just as well. They would not have been useful.

The brigade came back in February and March 1942 wearing the British Polar Bear shoulder patch—and were ordered to take it off. Most of the Marines would soon be on their way to Guadalcanal, and beyond that to Tarawa, and would earn another shoulder patch, either that of the 1st Marine Division or 2d Marine Division.

A larger percentage of Marines in the division than those few who had been in Iceland were those who had served in North China after the end of World War II, a now almost forgotten episode. It began with the 55,000-man deployment of the III Amphibious Corps at the end of September 1945 that included both the 1st and 6th Marine Division and the 1st Marine Aircraft Wing, and ended with the withdrawal of the last battalion in 1949, just a year before the Marines went to Korea. Like Iceland, it was largely a garrison experience, but the China Marines learned about the sub-zero temperatures and the arctic winds that came out of Manchuria and across the Gobi Desert. Marines guarding supply points or critical bridges or riding the coal trains knew how cold it could get.

The clothing they wore, including the Navy parka, was not much different than that which would be worn in Korea. Officially designated as the Marine Corps' 1943 cold weather uniform, it was predicated largely on the Iceland experience and consisted primarily, except for the parka, of U.S. Army components.

Other, older Marines in the division, including the chief of staff, Colonel Gregon A. Williams, the G-2, Colonel Bankson T. Holcomb, Jr., and the commanding officer of the 1st Marines, Colonel Lewis "Chesty" Puller, had had substantial service in pre-World War II China, including a chance to observe operations by Chinese Communist forces in the cold. They knew about the padded Chinese winter uniforms. Some, including Chesty Puller who was much better read and more of a student of military history than his flamboyant reputation would suggest, had studied Japanese winter operations in northern Korea and Manchuria in the Russo-Japanese War of 1904-05.

The Quartermaster General of the Marine Corps, Major General William P. T. Hill, himself an old China hand and an explorer of the Gobi Desert, began shipping our cold-weather clothing, including Navy parkas, to Korea in October 1950. Beginning in November, the battalion-sized replacement drafts being sent to restore combat losses received rudimentary cold weather training, at least in the wearing of cold-weather clothing.

The Marines, and, for that matter, the U.S. Army, used the "layer principle" for winter clothing, which simply meant that the Marine or soldier piled on as many layers of clothing as he could find. From the skin out he might have on cotton underpants and shirt or "skivvies," winter underwear or "long johns," mustard-colored flannel shirt, utility trousers or green kersey service trousers if he had them, sweater, green sateen winter trousers, alpaca vest, utility coat, a woolen muffler, and perhaps an M1943 field jacket, all crammed under a long, hooded, pile-lined Navy parka. The parka was warm but heavy and clumsy. Some Marines managed to find the shorter anorak-type parka worn by the Army and liked it better. Also popular, when they could be found, were the Army's "trooper" style pile-lined winter hats with earflaps. Several styles of gloves were issued. The most common had a leather and fabric outer shell and an inner mitten of knitted wool.

On their feet, Marines, unless they could find a substitute, wore "shoe-pacs"—waterproof rubber bottoms with laced leather uppers. They were issued with two sets each of felt innersoles and heavy woolen boot socks. The Marines were told to keep one set of the socks and innersoles inside their clothing next to their body and to change them frequently. These instructions were good in theory but difficult to follow in practice. Excessive perspiration, generated by marching, soaked the

National Archives Photo (USMC) 127-N-A5351

When possible, as here at Koto-ri, sleeping holes were dug behind fighting holes and frequently covered with ponchos or shelter halves. In this case, the occupants have managed to incorporate a stove. Most often the only relief from the cold for the infantry was in the form of warming tents set up to the rear of their position.

innersoles and socks. When the Marine halted, the felt innersoles and stockings quickly froze and so did the wearer's feet.

The shoe-pacs were hated, but the sleeping bags that the Marines had carried ashore at Inchon, now had a heavier lining and were much loved and, indeed, were indispensable. Marines found the bags, which could be rolled and tied to the bottom of their haversacks, good for sleeping, for warming feet, and for keeping casualties from freezing to death. There was a problem, though, of Marines standing watch in foxholes being allowed to pull their sleeping bags up to their knees or waists, and then, giving into temptation, slipping further into the bag and falling asleep. A Chinese soldier suddenly upon him killed more than one Marine, caught in his sleeping bag.

In fighting the cold, the Marines learned or relearned certain principles including the importance of keeping moving to generate body heat. The drawback to this was, of course, the sweat-soaked shoe-pacs that invited frostbite. The digging in of a foxhole, which could require six to eight hours of effort, often at the end of a long march, also generated heat, sometimes presenting the paradoxical sight of a Marine, stripped almost to the waist, hacking away at the frozen earth. In last analysis, the imposition of cold weather discipline depended upon the small-unit leadership of lieutenants, sergeants, and corporals. All things considered, they did amazingly well.

The Marines were still using their World War II pack; a well designed but complicated piece of equipment with a haversack, knapsack, a bedding roll, and many straps and buckles. Ordinarily, a Marine in combat carried nothing but his haversack and sleeping bag, and, of course, his rifle belt with its load of canteens, bayonet, first aid packet, ammunition, and possibly a few grenades. Most Marines preferred a pack board whenever they could get one.

A Marine also had to find a place for his daily C-ration when it was issued. It came in a clumsy cardboard box about the size of a shoe box, six cylindrical cans in all, three "heavies" and three "lights," plus an assortment of packets that included a day's supply of toilet paper and a neat little box with four cigarettes. The "heavies" were the meat components, much improved and with a much wider variety of items than the disliked World War II C-ration. The Army's Quartermaster Corps had worked hard on the improvements, basing them on regional favorites. Among the offerings were hamburgers (highly prized), chicken with vegetables, ham and lima beans, meat and beans, and sausage patties (the least favorite). The "lights" included at least one half-sized can of some kind of fruit, easily the best-liked element in the ration and one or more "bread units" which were biscuits of one sort or another, descended from Civil War hardtack, and something that passed for cake. Also to be found were different sorts of candy (disks of chocolate were preferred), salt, pepper, and packets of soluble coffee and cocoa. Most often a Marine took out what he liked or could trade and threw the rest away. What he retained he would fit into various pockets. He wondered why the ration could not be packed in flat cans that he could pocket more easily. His largest problem, though, was heating the meat component. Best method was to heat it in a bucket or GI can of boiling water, but these were seldom available. Cooking fires made with available wood usually did more burning than cooking. Unused mortar increments and bits of C-3 plastic explosive, when they could be found, burned with a quick hot heat. Dirt in a larger can, doused with gasoline, gave an improvised stove. But such open fires did not do well, tending to scorch the meat closest to the can and leaving the interior still frozen. Jeep and truck drivers could wire a can to their engine and when their run was finished, have a hot meal.

C-ration meat components would begin to freeze as soon as their cans were removed from the heat. Drinking coffee from an aluminum mess cup could be a dangerous process, the drinker's lip or tongue freezing to the cup. On the march it was often impossible to heat the meat component. Consequently the bread unit and fruit component were the first to be consumed.

Marines soon learned that keeping a thin coat of oil on their weapons, as taught to them emphatically by their drill instructors at boot camp, was not a good idea in sub-zero temperatures. Even a thin coat of oil tended to congeal and freeze the weapon's action. The word went out to wipe all weapons dry of oil. There was some argument over this. Some Marines thought that an infinitesimally thin coat of oil was best. There were arguments, pro and con, on the advisability of keeping personal weapons in sleeping bags or taking them into warming tents, or leaving them out in the cold.

By and large the weapons of the Marines worked well. A notable exception was the caliber .30 M1 and M2 carbine. Already suspect in World War II, it proved to be a miserable failure in sub-zero weather. Its weak action failed to feed rounds into the chamber, the bolt failed to close, and the piece often failed to fire. The release for its box magazine was a fraction of an inch from the safety. Mittened or cold-stiffened fingers sometimes pressed both, dropping the magazine into the snow. Even when a carbine did fire, the round had no stopping power. Most Marines carrying carbines replaced them as quickly as they could (and most often informally) with the prized M1 "Garand" rifle.

The Browning automatic rifle, M1918A2, continued to be a favorite Marine weapon. It functioned in proportion to the care it was given. Ice tended to form in the buffer group and inside the receiver.

As with all weapons with a recoil mechanism, machine guns, in general, were sluggish in their rate of fire. The old reliable Browning water-cooled M1917A1 fired well as long as there was antifreeze (not always easy to get) in the water jacket. Without liquid, the barrels quickly overheated. The barrels of the M1919A4 light machine gun tended to burn out and there were not enough spares. The 60mm and 81mm mortars fired reliably although there was considerable breakage of base plates and optical sights. It was remarked that the 81mm mortar shells looping across the sky left fiery tails more like rockets.

As to the cold, some units did claim nighttime temperatures of 35 degrees and even 40 degrees below zero. Best-documented temperatures, though, are the records kept by the battalions of the 11th Marines, the artillery regiment, that had to factor in the temperature as an element of gunnery. These battalions routinely recorded temperatures of

Department of Defense Photo (USMC) A5356

As shown in this photo of Marines marching out of Koto-ri on 8 December, each Marine carried what he considered necessary to live and fight, a considerable load of upwards of 60 pounds. Some got along with just their sleeping bag slung below their haversack. More carried a horseshoe-shaped bedding roll that could contain as much as a sleeping bag, a blanket, a poncho, and a shelter half.

20 and 25 below zero. Snow showers were frequent but not much snow accumulated. The winds of 35 and 40 miles per hour tended to blow the rock and frozen earth free of the thin snow. When morning came there would be ice crystals in the air, glinting in the sun like "diamond dust."

Water in five-gallon "Jerry" cans and individual canteens turned into blocks of ice. Some Marines carried a canteen inside their clothing to keep it thawed. Since World War II and the thirst of the Pacific War it had been the Marine Corps habit of having each man carry two canteens. This continued in the Korean War. Some Marine officers and senior noncommissioned officers carried whiskey in their left or "port side" canteen, which they doled out to their subordinates on a most-needed basis. The surgeons also had a carefully controlled supply of two-ounce bottles of medicinal brandy. Those lucky enough to get a bottle might use it to thaw out a C-ration can of fruit and then comment wryly on the luxury of "dining on brandied peaches."

Immersion heaters seldom provided enough warmth to thaw the contents of a water trailer. All valves and piping froze solid. Fires built beneath the trailers were a sometime effective expedient. Some men ate snow. The favorite beverages, when the water for them could be heated, were the soluble coffee and cocoa to be found in the C-ration, or better yet, the more generous allowance in larger rations.

A-rations, the full garrison ration with fresh or frozen meat, fruits, and vegetables, was, of course, unavailable except in an extraordinary set of circumstances such as the celebrated Thanksgiving dinner. B-rations, where canned items replaced the fresh or frozen items, were available but hard to use. Indeed, most of them were wasted, as they required a field kitchen for preparation. Efforts were made in the defensive perimeters to set up consolidated field messes serving hot chow, but this seldom benefited the men actually serving in the frontline. An exception were the flapjacks or pancakes made around the clock by a battalion mess at Hagaru-ri and served to thousands of Marines and soldiers. Artillery batteries sometimes managed to set up their own small messes. Captain Andrew Strohmenger's battery, also at Hagaru-ri, was known for its doughnuts.

Big square cans of ground coffee were a component of the B-ration. These, where space in a jeep trailer could be found, would be kept hoarded until circumstances permitted the boiling up of a batch of real coffee in a can or pail. Oatmeal, also to be found in the B-ration with the cooperation of a friendly mess sergeant, boiled in similar manner and flavored with sugar and powdered milk, was another favorite that riflemen, unfortunately, seldom enjoyed. Canned peanut butter, passed from Marine to Marine and dug out of the can by grimy fingers, was popular and more portable.

As a variant to the C-rations, there were sometimes the larger "five-in-ones" and "ten-in-ones,"

As the march continued south from Koto-ri, the Marines took an increasing number of Chinese prisoners. The Chinese, who had padded uniforms, but little protection for their hands and feet, and no tentage, suffered much more from the cold than did the Americans.

National Archives Photo (USMC) 127-N-A5387

much the same in content but with a more varied menu and intended for group consumption by a fire team or squad. For some reason these always seemed to be more available at the higher echelons and seldom at the rifle company level.

Post exchange supplies—cigarettes, candy, writing paper, and such—nominally there to be sold to the Marines, were given to them at no cost in the forward areas. Not many letters were being written, but the candy was a great favorite, particularly the chocolates and hard candies that gave quick energy. Brand-name choices were Tootsie Rolls and Charms.

An enormous advantage that the Marines had over the Chinese was the availability of tentage. The standard tent was the same as used in World War II, four-sided or pyramidal in shape, 16 feet on a side, and with a center pole. A practiced crew could erect one in 15 or 20 minutes even in the cold. Heat was provided by an M1941 stove or space heater the size and shape of a quarter keg of beer. Diesel oil was the preferred fuel, but it thickened in the cold and was frequently—if dangerously—thinned with gasoline to make it flow through the stove's carburetor. The stove stood at the base of the center pole and was good for many things besides

Amidst a snowstorm, a 60mm mortar squad rests by the side of the road south of Koto-ri on 8 December. In general, the Marines' winter clothing was cumbersome but effective—except for the shoe-pac. The two Marines in the foreground manage a grin for the cameraman. Note the mittens worn by these Marines.

National Archives Photo (USMC) 127-N-A5359

least the center portion of the tent; the sidewalls of which were usually, despite the stove, rimed with frost.

General practice was one warming tent per platoon, even rifle platoons. This was possible within the perimeters. Marines were cycled through the tents in relays as frequently as the situation would permit, usually not more than six at a time nor for longer than 20 minutes. They were not a place to sleep. Exact practices varied with location and unit. The warming tents had odd psychological effects. The canvas sidewalls seemed to shut off the war, offering a non-existent protection. Too many Marines could not be clustered together at one time in a tent that might be hit by a mortar shell or machine gun fire. One common practice was to have a communal pot—one company headquarters, as its most prized possession, had a stainless steel pail it used as such, rescued months before from a hospital in Seoul—filled with stew or "slum" constantly simmering on the stove. A Marine, entering the tent, would take his share of the heated slum and then replace it with the contents of one of his C-ration cans. A favorite condiment to make the stew palatable was Tabasco red-pepper sauce, a bottle of which always seemed to materialize. Short sections of wood were often nailed as cross trees to the tent pole as a drying place for sweat-soaked socks and felt shoe-pac liners.

On the march there was some attempt, not very successful, at having warming tents as way stations. Within the perimeters other tents, protected with sandbags, were designated as command posts, usual for regiment and battalion and sometimes, but not often, at the company level. Each perimeter had a field hospital of sorts, using a convenient schoolhouse or some such building. A battalion surgeon might have a cluster of tents, and there was such a thing as a hospital tent, considerably larger than the pyramidal tent. Company-level corpsmen often had a pyramidal tent to use as a sick bay for a few sick, lightly wounded, or exhausted Marines, and as a place to stash their stretchers and medical supplies. Life-saving plasma needed warmth in order to flow. Corpsmen working in the field during a firefight commonly carried morphine Syrettes in their mouths to keep them warm enough for injection.

Elimination of body waste was an unending problem. Within the defensive perimeters at Yudam-ni, Hagaru-ri, Koto-ri, and Chinhung-ni there were certain niceties of expeditionary plumbing available to the headquarters, artillery, and service units if not to the infantry. Packing tubes from mortar and artillery shells provided al fresco urinals. "Four-holers," collapsible plywood "heads" or "shitters," reportedly a Marine Corps invention dating back to the Banana War days, were set up in warming tents. These conveniences, almost never at hand for the rifle units, were not available to anyone on the march out. A much-repeated dark joke involved the problem of finding one's cold-shriveled penis through the many layers of clothing. Urine froze immediately on hitting the cold ground. Defecation was such a difficult procedure that some Marines simply stopped defecating. Later battalion surgeons and company hospital corpsmen would have to contend with impacted colons.

By the time the Marines, after their rehabilitation at Masan, began to move north at the beginning of 1951, some things had gotten better. A small mountain-type gasoline camp stove, about the size and shape of a quart oilcan, was issued on the basis of one stove to every four Marines. It largely solved the task of heating C-rations and at the same time produced boiling water for soluble coffee or cocoa. Inflatable rubber mattresses, to be used as insulation under the much-treasured sleeping bags, also began to appear. They worked best on a canvas cot, but riflemen seldom had the luxury of a cot even when in reserve. Not until the next year, however, would a thermal "Mickey Mouse" boot replace the hated shoe-pac

Meanwhile, in the United States, the Marine Corps sought a cold-weather training site in California. Big Bear was tried, but serious training did not mix with a ski resort. General Smith, after arriving at Camp Pendleton in May 1951 to be the base commander, took an active personal interest in finding a suitable location. Reconnaissance parties were sent out and by late summer a site was found 450 miles north of Camp Pendleton in the Toiyabe National Forest in the High Sierras. With a valley floor at 6,800 feet, elevations went up to more than 11,000 feet. Weather records promised winter temperatures of 20 below zero and 20 feet of snow. Marines called it "Pickle" Meadow, but it was really Pickel Meadow, named for Frank Pickel, a trapper who had built a cabin there in the 1860s. By fall 1951 all replacement drafts and other units headed for Korea would have a week's in-the-field training at Pickel Meadow.

(Continued from page 233)
Day 1950 was marked in X Corps by the landing at Wonsan of the 15th Infantry, largely schools troops from Fort Benning. The regiment, under Colonel Dennis M. "Dinty" Moore, was to relieve Puller's 1st Marines in and around Wonsan. Almond was not pleased to learn that the 3d Battalion, 15th Infantry, was a "Negro" unit and therefore, in his mind, not completely trustworthy. The 3d Division's third and last regiment, the 7th Infantry, commanded by Colonel John S. Guthrie, came from Fort Devens, Massachusetts, by way of Japan and would disembark at Wonsan on 17 November.

Almond celebrated Armistice Day with an order at midnight calling for an advance to the border, the ROK I Corps on the right, the 7th Infantry Division in the center, and the 1st Marine Division on the left. The Marines were allotted a 40-mile stretch along the Yalu as their ultimate objective. That same day MacArthur indirectly ordered Almond, by way of a personal letter from his G-3, Major General Edwin K. "Pinky" Wright, to do everything he could to assist the Eighth Army in its drive to the Yalu.

Almond had his staff prepare an analysis of the Eighth Army's situation. It credited Walker with having 120,000 troops with which to oppose 100,000 of the enemy and having the advantages of air supremacy and superior artillery support. Almond's study concluded that the severing of the enemy's MSR at Mupyong-ni by X Corps would greatly assist Eighth Army's advance. Almond, in his reply to Wright on 14 November, proposed that he attack north and then west to link-up with the Eighth Army.

Members of the 7th Marines "answer up" at a mail call at Koto-ri on 15 November. A large amount of mail had accumulated, some of it intended for Christmas. Airmail arrived in a prompt five or six days; packages could take five to six weeks or longer. The tall Marine with a letter in his hand is carrying a carbine, a weapon that would prove worthless in the cold weather ahead.

Photo by Cpl L. B. Snyder, National Archives Photo (USMC) 127-N-A4632

As the Chinese Saw It

Peng Dehuai's chief of staff, Xie Fang, at about this time made his own assessment of the situation:

> Our 9th Army Group main forces have successively entered Korea from J'ian and Linjiang to assume eastern front operations. . . . We have over 150,000 men on the eastern front, the enemy over 90,000, giving us a 1.66 advantage over him. We have 250,000 men on the western front, the enemy 130,000, giving us a 1.75 advantage over him. Our forces are superior on the eastern and western fronts.

On 16 November, Xie Fang reported: "Our forces on the eastern front abandoned Hwangch'o [Funchilin] Pass on the 7th. On the 10th. . . the enemy on the eastern front continued advancing northward along three separate routes: From Hwangch'o Pass, P'unsan [Pungsan], and Myongchon . . . still far from our pre-selected killing zones."

Monday, 13 November

Meanwhile, in a division order dated 13 November, Smith directed RCT-1 to take Huksu-ri, RCT-7 to seize Hagaru-ri and on order advance on Yudam-ni, and RCT-5 to protect the MSR from positions at Majon-dong, Chinhung-ni, and Koto-ri, and to be prepared to pass through RCT-7 at Hagaru-ri and advance to Changjin 40 miles to the north.

The road leading north from Koto-ri to Hagaru-ri followed a valley formed by the Changjin River. As Litzenberg's Marines moved on toward Hagaru-ri, 11 miles north of Koto-ri and at the southern tip of the Chosin Reservoir, they could see parties of Chinese in the distance.

On 15 November Rear Admiral Albert K. Morehouse, chief of staff of U.S. Naval Forces, Far East, visited Smith. Smith, feeling he was speaking within the naval family, outlined for Morehouse, to be passed on to Vice Admiral C. Turner Joy, the Commander, Naval Forces, Far East, his concern over what he considered Almond's unrealistic planning and tendency to ignore enemy capabilities. Smith may or may not have shown Morehouse a letter he had just drafted to General Cates.

Photo by Cpl Peter W. McDonald, National Archives Photo (USMC) 127-N-A4875

Marines found the 75mm recoilless rifle, shown here in action near Koto-ri in mid-November, of increasing use in the mountains. It gave them a direct-fire weapon of great accuracy and lethality. But the shells were heavy and difficult to lug up into the hills in any great number.

Alarmed at the prospect of attacking simultaneously in two different directions, Smith had stepped out of the chain-of-command to write a personal letter to the Commandant of the Marine Corps. In it he said:

> Someone in high authority will have to make up his mind as to what is our goal. My mission is still to advance to the border. The Eighth Army, 80 miles to the southwest, will not attack until the 20th. Manifestly, we should not push on without regard to the Eighth Army. We would simply get further out on a limb. If the Eighth Army push does not go, then the decision will have to be made as to what to do next. I believe a winter campaign in the mountains of North Korea is too much to ask of the American soldier or marine, and I doubt the feasibility of supplying troops in this area during the winter or providing for the evacuation of sick and wounded.

In conclusion Smith underscored his concern over "the prospects of stringing out a Marine division along a single mountain road for 120 air miles from Hamhung to the border."

Asked years later to comment on this extraordinary action by Smith, Almond said tartly: "My general comment is that General Smith, ever since the beginning of the Inchon landing and the preparation phase, was overly cautious of executing any order that he ever received."

In 1952, General Shepherd, by then Commandant, would report to the Secretary of the Navy: "By orders of higher authority the division was placed in a situation, which, when the Chinese struck in force on 28 November 1950, resulted in the division being in effect deployed in column for a distance of 35 miles within enemy territory. The wide separation of elements of the Tenth Corps of which the First Marine Division was a part, and the gap existing between the Tenth Corps and the Eighth Army had permitted the Chinese to flow around the First Marine Division preparatory to an all-out attack."

MacArthur, responding to Almond's 15 November proposal, asked Almond for an alternate plan giving priority to taking off the pressure confronting the Eighth Army. Accordingly, Almond now visualized an attack by the 1st Marine Division on the Hagaru-ri—Mupyong-ni axis with a regimental combat team from the 7th Division protecting the division's right flank by taking Changjin. This became the operative plan. Almond recognized that extreme minimum temperatures of from 30 to 40 degrees below zero would severely restrict both friendly and enemy operations.

Photo by TSgt J. W. Helms, Jr., Department of Defense Photo (USMC) A155763

A Marine squad trudges through the snow-encrusted streets of Hagaru-ri. The 7th Marines occupied the town on 15 November. By then the weather had turned cold. Weather records indicated that Hagaru-ri could be the coldest spot in North Korea.

7th Marines Reach Hagaru-ri

While the commanders exchanged proposals and plans, the 7th Marines occupied Hagaru-ri on 15 November. The nighttime temperature had dropped to four degrees below zero. Hagaru-ri was a medium-sized town, fairly well flattened by bombing. Just north of Hagaru-ri in the hamlet of Sasu-ri there was a sawmill and a great deal of fresh-cut lumber. Once tents began to spring up, the town reminded at least one Marine officer of an Alaskan gold camp with its mud-and-snow streets, its tents, and rough construction with raw lumber. General Craig visited Hagaru-ri and recommended it to Smith as a forward base.

By then RCT-5 had its 2d Battalion at Koto-ri, its 3d Battalion at Chinhung-ni—along with much of the remainder of the division—and its 1st Battalion at Majon-dong. As Murray, the regimental commander, remembered:

> We'd been highly successful in the south, and we had a lot of this carry over as we went north. There wasn't anybody any better than we were, that was the general feeling in the regiment. . . the hills seemed to be a lot steeper than they were in the south. . . . And in some cases, on the road between, I guess it was just below Hagaru-ri a ways, there was a power plant built right into the side of the mountain, and the road ran over a part of this thing. Very easy to blow it up, which was done, done twice as a matter of fact by the Chinese later on.

Smith again visited the Chosin plateau on 16 November, this time driving up in a heated station wagon. At Chinhung-ni he met, by coincidence rather than design, Major General Field Harris, 55, the commanding general of the 1st Marine Aircraft Wing. Harris had flown as far as Chinhung-ni in a helicopter and had planned to go the rest of the way by open jeep. Smith offered him a ride in his station wagon. They drove comfortably to Hagaru-ri with Smith in a rare burst of jocularity promising

A tent camp sprang up at Hagaru-ri. On 18 November there was time for a brief awards ceremony. Col Homer Litzenberg, right, and LtCol Raymond G. Davis, left, Commanding Officer, 1st Battalion, congratulates Cpl Earle R. Seifert, who received a Bronze Star, and SSgt Earle E. Payne, a Navy Commendation Ribbon, for earlier heroism.

Photo by Cpl L. B. Snyder, National Archives Photo (USMC) 127-N- A4641

National Archives Photo (USN) 80-G-421479

VAdm C. Turner Joy, left, is greeted at Wonsan on 19 October by MajGen Field Harris, commanding general of the 1st Marine Aircraft Wing. Joy, as Commander, Naval Forces, Far East, was MacArthur's naval component commander. No direct command line linked Harris and the wing to O. P. Smith and the 1st Marine Division.

Harris a station wagon of his own in exchange for continued close air support.

Almond had asked Field Harris to reconnoiter Hagaru-ri for a site for an airstrip long enough to handle two-engine transports. Smith and Harris walked the ground and found a stretch just south of the town that seemed suitable. "There is plenty of room, but the soil consists of a thick, black loam," Smith entered in his log. "If the ground freezes it will probably be all right for a strip."

Regiments Get New Orders

On 17 November Smith modified his orders to his regimental combat teams. RCT-7 was to protect the left flank of the division between Hagaru-ni and Yudam-ni. RCT-5 was to pass a battalion through RCT-7 at Hagaru-ri and move up the east side of the reservoir and seize Sinhung-ni, about seven miles northeast of Hagaru-ri. (Sinhung-ni, just east of Chosin Reservoir should not be confused with Sinhung in Sinhung Valley previously visited by the 5th Marines.)

Murray had been told to nominate a battalion commander for return to the United States. He picked George Newton, commander of the 1st Battalion. Murray said of Newton: "He was a very competent battalion commander, but he was, I felt, almost killing himself trying to be a good battalion commander. He seemed to stay awake most of the time."

"George left [on 17 November] before we went all the way up," said Murray. "Anyway, George Newton was relieved by a pretty good leader [Lieutenant Colonel John W. Stevens II]. But I did have good battalion commanders. We had an excellent staff. The main thing, as I say, is that we had been

An M-4A3 Sherman tank from the 1st Tank Battalion travels along a well-graded but narrow road coming up Funchilin Pass on 19 November. Tanks gave the Marines enormous firepower and were useful in crushing enemy roadblocks, but weather and terrain tended to keep them road-bound.

Photo by TSgt J. W. Helms, Jr., National Archives Photo (USMC) 127-N-A5343

National Archives Photo (USN) 80-G-423190

MajGen O. P. Smith and RAdm James H. Doyle, shown here at the time of the Inchon landing, were long-time friends. Both were highly experienced in amphibious warfare. They formed an effective and compatible partnership not only at Inchon, but also in both the Wonsan landing and the ultimate evacuation from Hungnam.

successful in the south, and all that was needed was to keep this going." Stevens was a known quantity. He had been the executive officer of the 2d Battalion, 5th Marines, from Pusan on through Inchon and Seoul.

That evening Smith dined on board the amphibious command ship *Mount McKinley* (AGC 7) with Rear Admiral James H. "Jimmy" Doyle. Describing Doyle as "a typical Irishman," Colonel Bowser, Smith's G-3, said: "He is a real fighter when it comes to the clutches. A fun guy to know—always a laugh or a joke." Smith and Doyle, alone in the admiral's cabin, in Smith's words, "let our hair down."

Vice Admiral Arthur D. Struble, commander of the Seventh Fleet, had been superimposed over Doyle's Task Force 90 during the Inchon and Wonsan landings. Doyle disliked Struble and, dubious of his competence, was determined to keep the Seventh Fleet out of direct control of future amphibious operations. After the Wonsan landing, Doyle had complained to his old friend Vice Admiral Joy, commander of Naval Forces, Far East, that he could not and would not come under Struble again. He was successful in his arguments. As commander, Task Force 90, at Hungnam he would report direct to Admiral Joy.

By now engineers had improved the MSR to a point where armor could be sent forward to join Litzenberg. A tank platoon reached Hagaru-ri on 18 November. That same day Smith visited Puller at his command post just west of Chigyong. Smith noted that there was snow on the mountains but that the road was still open. He was resisting an order from Almond to send a battalion to Huksu-ri, about 20 miles to the northwest, to occupy a blocking position. "There is no truck road to take," said Smith in his log. "I do not intend to put Puller out on a limb where he cannot be supplied. Also I would like to close him up behind the regiments moving toward the Chosin Reservoir. The 26th ROK Regiment is attacking toward Huksu-ri. Possibly this will relieve me of concern regarding that place."

Construction of the airstrip at Hagaru-ri began. Smith asked for X Corps engineers, but could get none. The job was given to Lieutenant Colonel John Partridge's 1st Engineer Battalion. Wind-blown Hagaru-ri was at an elevation of about 4,000 feet. For that altitude the engineer manuals prescribed a minimum runway of 3,900 feet for C-47 transport operations. The engineers crossed their fingers and hoped that a strip as short as 3,000 feet might do. Once started, construction of the airstrip proceeded 24 hours a day, with work at night under floodlights.

Marine observation squadron, VMO-6, although part of the 1st Marine Aircraft Wing, was under Smith's operational control. Smith regarded the squadron as his own private air force. On the 19th, he visited the squadron's commander, Major Vincent J. Gottschalk, at Yonpo airfield to discuss the problems of operating helicopters and light aircraft in the cold at high altitudes. Gottschalk, 31, promised to provide solutions. He had come into the Corps in 1941 after graduating from the University of Michigan. For much of World War II he had served as Marine detachment commander in the light aircraft carrier *Langley* (CVL 27)—after the war came two years of flight training.

Early in November, Admiral Joy had asked Smith if he could use the Royal Marines' 41 Independent Commando—14 officers and 221 enlisted men, commanded by Lieutenant Colonel Douglas B. Drysdale. Smith replied he would be glad to get these fine troops, foreseeing 41 Commando operating with the division Reconnaissance Company in screening the flanks of the Marine advance. The British Marines arrived at Hungnam on 20 November, the same day that Almond passed on instructions from higher headquarters that "damage, destruction or disruption of service of power plants will be avoided." In the larger scheme of things, the intention was to leave the hydroelectric generators intact. Marines would wonder why.

On 21 November the division's southern boundary was adjusted to give the responsibility for Huksu-ri to the 3d Infantry Division. Puller's regiment was now available to fill in behind Murray and Litzenberg.

Photo by Sgt John Babyak, National Archives Photo (USMC) 127-N-A4683

The well-meaning, but bumbling Secretary of the Navy Francis P. Matthews, greeted here by BGen Craig, visited the division's rear area at Hambung on 22 November. At the division hospital he was surprised to learn that Navy personnel met the Marine Corps' medical needs. Another politician accompanied him, Senator Claude Pepper of Florida.

Secretary of the Navy Visits

Wednesday morning, 22 November, found O. P. Smith and Field Harris on the Yonpo airfield awaiting the arrival of the Secretary of the Navy, Francis P. Matthews. Behind his back, Matthews was known as "Rowboat" because of his lack of knowledge of naval matters. Accompanying the secretary was Admiral Joy and Senator Claude Pepper of Florida. Arriving at the airfield at the same time was President Syngman Rhee. Matthews had wanted to call on Rhee in Seoul but could not get clearance from the Secretary of State, Dean Acheson, to do so. This left Smith and Harris with the ticklish problem of keeping the two high-level parties apart. They whisked Matthews away from the field before he could learn of Rhee's presence, taking him to the division hospital. There were very few wounded Marines to visit, but Matthews found the Chinese and North Korean prisoner of war patients of great interest. He seemed to have difficulty understanding why Navy personnel were running a Marine hospital. It was a picture-taking opportunity for the secretary followed by another picture-taking opportunity at the division cemetery. Next the division staff gave the visitors a briefing fol-

LtCol Douglas B. Drysdale, RM, and his 41 Independent Commando, Royal Marines, were billeted briefly with the 1st Engineer Battalion in Hambung before moving up to the Chosin Reservoir. Drysdale's command, largely made up of volunteers, had assembled in Japan where it was re-equipped with American infantry weapons.

Photo by TSgt J. W. Helms, Jr., National Archives Photo (USMC) 127-N-A5322

Photo by Sgt Peter Ruplenas, National Archives Photo (USA) 111-SC363307

It was a proud moment for MajGen Almond, center, when, on 21 November, soldiers of the 7th Infantry Division reached the Yalu River. Savoring the moment with him, from left to right: BGen Homer Kiefer, division artillery commander; BGen Henry I. Hodes, assistant division commander; MajGen David G. Barr, division commander; and Col Herbert B. Powell, Commanding Officer, RCT-17.

lowed by lunch in the commanding general's mess. The cooks had embellished the standard ration with biscuits and cookies. Secretary Matthews and Senator Pepper seemed to enjoy these immensely. Secretary Matthews was then escorted back to Yonpo airfield. Smith again managed to keep him unaware of President Rhee who was departing at the same time. Senator Pepper stayed on for the rest of the day. He wanted to visit with some Marines from Florida. Smith found 15 of these. He had not been able to find any Marines from Nebraska for Secretary Matthews.

Thanksgiving, 23 November

Thanksgiving fell on Thursday, 23 November. The holiday menu, accomplished by strenuous effort on the part of many hands, included shrimp cocktail, stuffed olives, roast young tom turkey with cranberry sauce, candied sweet potatoes, fruit salad, fruit cake, mincemeat pie, and coffee. Even the Marine infantry units got at least the turkey.

Admiral Doyle sent in a cooked turkey for General Smith's mess, but Smith himself had been invited to dinner by Almond. As Smith said in his log: "The dinner was complete with cocktails served from a cocktail bar, tablecloths, napkins, Japanese chinaware, regular silverware, place cards, etc. Admiral Struble and Generals Biederlinden (G-1 of GHQ), Harris, Barr, and Ruffner were also present."

Two days before Thanksgiving, elements of the 7th Division's 17th Regiment had reached the Yalu without encountering a single Chinese soldier. Years later General Almond still savored that moment of triumph:

> And on the 21st of November the leading battalion of the 17th Infantry reached the Yalu River and I was present when they did so. . . . I accompanied General Barr, the division commander; General Hodes, the assistant division commander; and General Kieffer, the artilleryman; with the regimental commander, Colonel Powell. We all walked behind the lead company down the road to the river bank. This was the first element of the American forces to reach the Korean-Manchurian border, although earlier elements of the 6th ROK Division with I American Corps on the west flank, Eighth Army front, attempted to get to the river but did not succeed in remaining there.

Almond and his commanders paused on the banks of the Yalu for a ritual urination into the waters of the river. Meanwhile, Colonel Charles E. Beauchamp's 32d Infantry was advancing to the northwest of Powell's 17th Infantry with orders to reach Singalpajin, originally a Marine Corps objective, on the Yalu. A 34-man patrol under Second Lieutenant Robert C. Kingston (a future four-star general) was sent out from the 3d Battalion, 32d Infantry. The patrol reached Samsu, 23 miles south of the Yalu, where it held on for three days, and then, reinforced by tanks, artillery, engineers, and more infantry, plunged forward, still commanded by the 22-year-old second lieutenant. Now designated "Task Force Kingston," it arrived at Singalpajin on 28 November, fought a house-to-house fight with North Koreans, and then took its turn at urinating in the Yalu. The second and last American unit to reach the Chinese border, Task Force Kingston, for all of its adventures, suffered only one casualty: a soldier reportedly killed by a Siberian tiger.

While soldiers and Marines were eating their Thanksgiving turkey, Smith again modified his orders for the 1st Marine Division's advance. RCT-7 was to move on to

Photo by TSgt J. W. Helms, Department of Defense Photo (USMC) A5289

In the 1st Marine Division rear at Hamhung, where the weather was still benign, members of Maj Robert L. Schreier' 1st Signal Battalion queue up for Thanksgiving dinner. This battalion, responsible for both the wire and radio communications of the division, also included ANGLICO—the Air and Naval Gunfire Liaison Company that provided forward observer teams to Army and ROK elements of X Corps.

Yudam-ni. RCT-5 was to continue up the eastern side of the reservoir. RCT-1 was to protect the MSR from positions at Hagaru-ri, Koto-ri, and Chinhung-ni. As Smith said in his log:

> I did not want to push Murray too far or get Litzenberg out on a limb at Yudam-ni until I could close up Puller in rear of them. . . . I had hoped there might be some change in the orders on the conservative side. This change did not materialize and I had to direct Litzenberg to move on to Yudam-ni.

Most of the 7th Marines had their Thanksgiving dinner at Hagaru-ri. The 2d Battalion, 7th Marines, had set up its field mess in the shadow of what would come to be called "East Hill." Private First Class Alfred P. Bradshaw, a reservist who had recently joined Captain Hull's Dog Company, had lost his mess gear. The mess kits consisted of two flat aluminum pans clamped together, not much changed in pattern since the Civil War. Marines in rifle companies seldom had need for mess gear; they subsisted almost entirely on C-rations, thankfully much improved since World War II. Bradshaw, standing in the chow line, had sought to improvise a plate out of a piece of cardboard. Hull saw Bradshaw's plight and gave him one of his pans. Bradshaw would remember that.

The road from Hagaru-ri to Yudam-ni climbed up through Toktong Pass, four miles to the northwest and about 4,000 feet in elevation, and then descended into a narrow valley before reaching Yudam-ni. Smith personally gave Litzenberg orders to drop off a company at Toktong Pass.

On the day following Thanksgiving, 24 November, MacArthur came to Korea to see the jump-off of the Eighth Army on the offensive that was to end the war. He announced to the press that the war would be won in two weeks and that the Eighth Army would spend Christmas in Japan. To complete Walker's victory, MacArthur ordered Almond to execute the already planned attack to the west so as to squeeze the Chinese

Elements of both the 5th and 7th Marines spent Thanksgiving within the perimeter of the burgeoning combat base at Hagaru-ri. Every effort was made to reach every Marine in the division with a traditional holiday dinner. Here a 5th Marines cook ladles pumpkin pie mix into a piecrust spread out in a square pan.

National Archives Photo (USMC) 127-N-A4726

Department of Defense Photo (USMC) A6791

Thanksgiving was a last lull before the Chinese storm broke. At Hagaru-ri, Reverend Lee In Sup, a Presbyterian pastor, and his wife joined the 5th Marines for Thanksgiving services. Lee thanked LtCol Murray, commanding officer of the 5th Marines, for the liberation "of our country and our church." Beaming broadly in the background is the regimental chaplain, LtCdr Orlando Ingvoldstad, Jr.

between the Eighth Army and the still-independent X Corps. Lieutenant Colonel John H. Chiles, USA, Almond's G-3, had carried the final draft of X Corps operations order to Tokyo on Thanksgiving Day. MacArthur approved the plan on Friday.

On Saturday morning, 25 November, O. P. Smith attended a briefing at X Corps headquarters outlining X Corps Operation Order Number 7. He learned that his division was to be the northern arm of a giant pincer. The other arm of the pincer would be the Eighth Army. He was to sever the enemy's lines of communication at Mupyong-ni and then advance to the Yalu. He was to launch his attack on Monday, 27 November. Concurrently, the 7th Division would continue its advance northward to the Yalu. Almost 100 miles separated the two divisions. Strength returns for that day showed the 1st Marine Division as having 25,323 Americans with 110 South Koreans attached, but of that number only about 15,000 were up at the reservoir. Indeed, some units of the division were as far to the rear as Japan. A goodly number of hospitalized Marine patients were also carried in the total. The 7th Division strength on the same day was 16,001 men of whom 6,794 were South Korean KATUSA soldiers.

Smith estimated the road distance from Yudam-ni west to Mupyong-ni, over another mountain pass and then through a narrow valley, as being 55 miles. The division was then to advance northward to the Yalu. Almond's three columns—the 1st Marine Division, the 7th Infantry Division, and, nominally under his control, the ROK I Corps—were diverging like the ribs of an opened fan. The 7th Infantry Division was to complete its advance to the Yalu. The ROK corps was to advance to the Chinese border from the Hapsu and Chongjin areas. To the rear the newly arrived 3d Infantry Division, under General Soule, was given a multiplicity of missions: gain contact with the right flank of the Eighth Army; protect the left flank of X Corps; support the 1st Marine Division on order; protect the harbor and airfield at Wonsan; and destroy guerrillas in its zone of action. The 3d Division was also to have had the task of

By the third week in November a tent camp, mostly for combat service units had sprung up at Hagaru-ri. One observer said that the badly battered town reminded him of an Alaska gold-rush camp. In the foreground a bit of the narrow-gauge railroad track that once served Hagaru-ri can be seen.

Photo by Sgt Frank C. Kerr, National Archives Photo (USMC) 127-N-A4971

National Archives Photo (USA) 111-SC352933

Gen Douglas MacArthur came to Korea on 24 November to see the jump-off of the Eighth Army in the offensive that was to end the war. Two days later the Eighth Army was in full retreat. LtGen Walton H. "Johnnie" Walker, commanding general of the Eighth Army, seated behind MacArthur, would die in a traffic accident one month later.

lage.

Smith's rough plan was to have the 5th Marines pass through the 7th Marines at Yudam-ni and then attack to the west. The 1st Marines, in reserve, was to occupy positions along the MSR at Chinhung-ni, Koto-ri, and Hagaru-ri. Supporting this plan, Almond decided that a regimental-sized force from Barr's 7th Division should relieve Murray's 5th Marines on the east side of the reservoir so that the 5th Marines could join the 7th Marines at Yudam-ni. He ordered Barr to send a regimental combat team for this purpose by 27 November.

Barr, acting on local intelligence that the Chinese in massive numbers had crossed the Yalu at Linchiang and were moving into the gap between his division and defending the area south of Hagaru-ri, but, with its other missions, the best it could promise to do was take over the security of the MSR from Sudong back to Hamhung. It bothered Smith that the 3d Infantry Division had not yet closed behind him and that he would have to leave Puller's 1st Marines strung out along the MSR to keep it open from Hagaru-ri south to Chinhung-ni.

LtCol Don C. Faith, Jr., USA, right, commanding officer of the 1st Battalion, 32d Infantry, 7th Division, pictured with the regiment's commanding officer, Col Allan D. MacLean, in Japan in the spring of 1950. Except for limited experience during the battle for Seoul, Faith had not commanded a unit in combat prior to the Chosin Reservoir.

Courtesy of Col Erwin B. Bigger, USA (Ret)

Advances on Both Sides of the Reservoir

Davis with his 1st Battalion, 7th Marines, had led off the advance to Yudam-ni on Thanksgiving Day. He ran into a defense of Toktong Pass by an estimated 150-200 Chinese, but scattered it with the aid of air and artillery. The battalion paused to celebrate Thanksgiving a day late, and then moved on into Yudam-ni on the 25th against negligible resistance. The 3d Battalion, 7th Marines, and Litzenberg's regimental headquarters followed Davis' battalion into the forlorn vil-

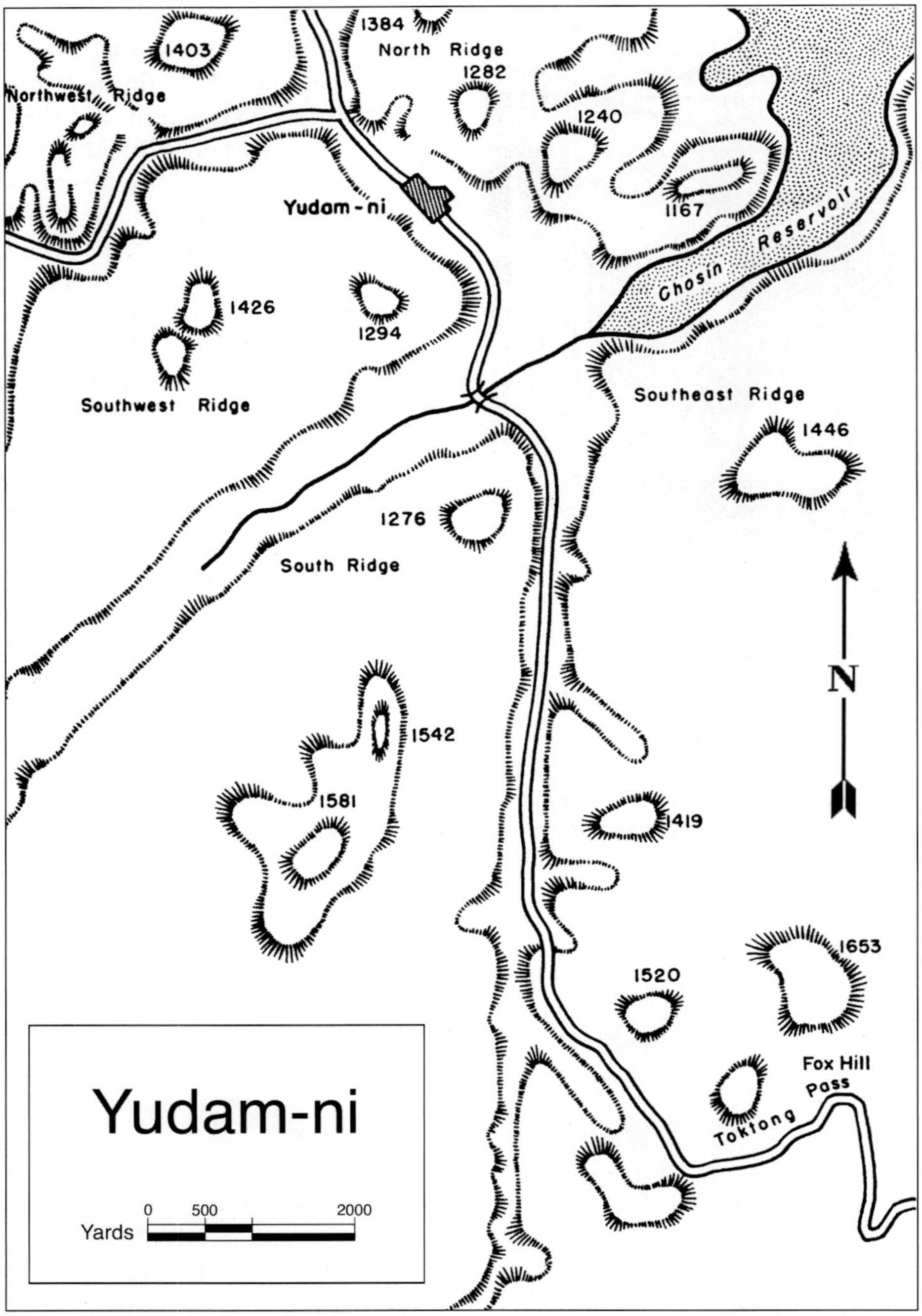

the Marines, had already begun pulling together his scattered battalions.

RCT-31, as assembled by Barr and commanded by Colonel Allan D. MacLean, consisted of the 31st Infantry's Headquarters and Service Company, the regiment's 2d and 3d Battalions, the 31st Tank Company, the 57th Field Artillery Battalion, Battery D of the self-propelled 15th Antiaircraft Artillery Automatic Weapons Battalion, and the 1st Battalion, 32d Infantry, commanded by Lieutenant Colonel Don C. Faith, Jr.

Don Faith, 32, six-feet-tall, handsome, and charismatic, was something of an Army golden boy. The son of an Army brigadier general, he had enlisted in 1941 and won his commission as a second lieutenant the following year. For three years of World War II, he served first in the 82d Airborne Division and then in the XVIII Airborne Corps as an aide to Major General Matthew B. Ridgway with whom he landed at Sicily and jumped into Normandy and Holland. Faith had worked for Barr in China. He had commanded the 1st Battalion, 32d Infantry, for more than a year. In this new war he had been recommended for a Distinguished Service Cross for his performance between Inchon and Seoul.

Barr chose to pull Faith's battalion from the 32d Infantry and assign it to RCT-31 because it, in bivouac northeast of Hamhung, was the Army battalion closest to the reservoir. Faith on 24 November had a strength of 715 Americans and about 300 South Koreans.

Most of Faith's officers were well trained and combat experienced. Some had served in Europe during World War II, some in the Pacific. There was also a layer of battle-hardened senior noncommissioned officers. The mix of Americans and South Koreans in the rank-and-file, however, was a problem, both in language and lack of training. The battalion, in its equipment and preparation for a winter campaign, was about on a par with the Marine battalions, in some ways better and in some ways not as good. "They were short of chains for their trucks. The only tentage they had were tent flies for their kitchens," said one observer. During the previous winter the 31st Infantry, stationed at Camp Crawford, Hokkaido, had received cold-weather training. Most of the men were issued Army winter parkas, shorter and less clumsy than the Navy parkas worn by the Marines. They had sweaters and pile liners of various sorts and shoe-pacs which were really rubber-and-leather hunter's boots. Believed by the troops to have been provided by L. L. Bean, these

National Archives Photo (USA) 111-SC352537

Ill-fated RCT-31, under Col Allan D. MacLean, USA, right, followed RCT-17 ashore at Iwon. Here, on 12 November with the weather still mild, MajGen Almond gives MacLean some words of advice after presenting him with a Silver Star. MajGen Barr, MacLean's division commander, stands stolidly in the center. Two weeks later MacLean would be dead.

boots had been suitable in World War II in the wet cold of northern France and Germany. But, as the Marines were also learning, they were worse than no good in sub-zero temperatures. Faith himself did not like the shoe-pacs and wore galoshes over his leather combat boots. So did many other soldiers and Marines if they could get them.

On Saturday morning, 25 November, Faith and his battalion, the lead element of RCT-31, started up the icy road to the reservoir. At Hagaru-ri they took the right-hand fork in the road. Some miles up the eastern side of the reservoir Faith met Murray, the commander of the 5th Marines. Murray outlined for him the disposition of his three battalions, all of which were now east of the reservoir. Taplett's 3d Battalion, in the lead, had a good defensive position about four road miles north of the Pungnyuri-gang inlet. Earlier that day, a patrol from Taplett's battalion had almost reached the northern end of the reservoir before brushing up against a small party of Chinese. Murray designated an assembly area for Faith's battalion near the village of Twiggae. Faith set up his command post in a hut on a lower slope of Hill 1221.

With the relief of RCT-5 by Faith's battalion, Marine operations east of the reservoir would end. There was no sign of large-scale enemy activity. The soldiers were to stay under the operational control of the 1st Marine Division until the arrival of Colonel MacLean, the commanding officer, 31st Infantry. Faith's command relationship to the 1st Marine Division was not clear. He asked Murray for instructions. Murray, who did not consider Faith to be under his command, said that he had none, but he did caution Faith not to move farther north without orders from the 7th Division. Once Murray departed, the only radio link between Faith and the 1st Marine Division would be that provided by his attached tactical air control party, led by Marine Captain Edward P. Stamford. He and his four-man team had been with the battalion since Seoul.

Just before noon on the 26th, Brigadier General Henry I. Hodes, the 7th's assistant division commander, visited Faith at Hill 1221. Hodes, 51 years old, a West Pointer who had commanded the 112th Infantry in Europe in World War II, and a future four-star general, told Faith that MacLean and the rest of the 31st RCT would soon be arriving.

Smith Visits Yudam-ni

On Sunday morning, 26 November, Smith visited Yudam-ni. During the night he had been informed that the ROK II Corps, on the right flank of the Eighth Army, had been thrown back in the vicinity of Tokchon, about 70 air miles southwest of Yudam-ni. But as yet Smith had no notion of the extent of the disaster that had befallen the Eighth Army. Both Walker's G-2 and GHQ in Tokyo had badly underestimated the

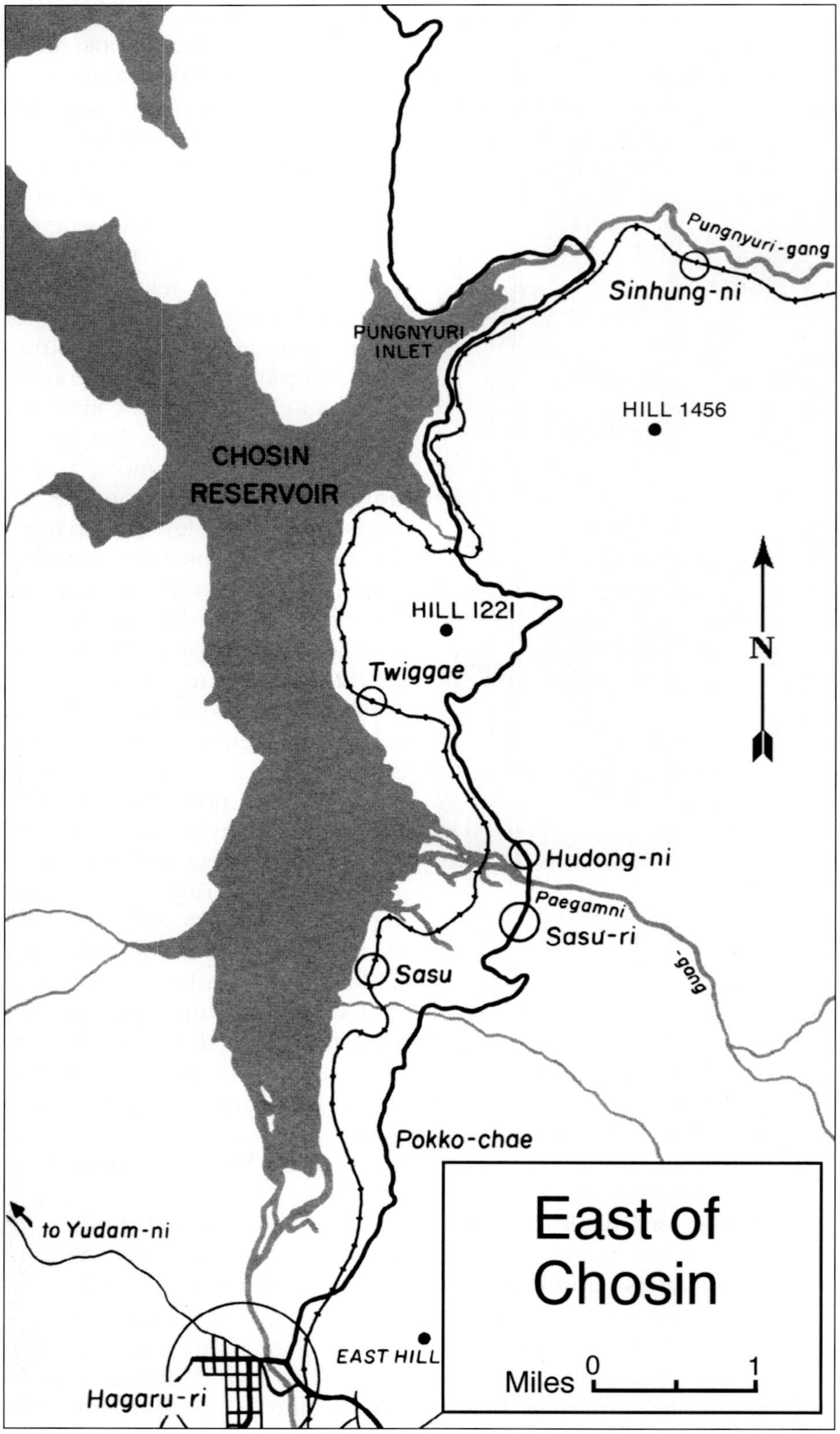

strength of the Chinese. One day into the offensive that MacArthur had blithely informed the press would end the war, the Chinese *Thirteenth Army Group* with 18 divisions counterattacked Walker.

From his helicopter on the way to Yudam-ni Smith could see no signs of enemy activity. As he entered in his log:

> I landed at what I thought was the CP of the 7th, but it proved to be the CP of the 1st Battalion, 7th. I had a visit with LtCol Davis, the Commanding Officer, and got directions from him as to the location of the CP of the 7th, which was about 5000 yards south, up the road to Hagaru-ri. In making the landing at the regimental CP I discovered some of the limitations of helicopters. We first attempted to land on a gentle slope near the CP. As the pilot put his wheels down we slipped backwards on the ice and snow. After 4 or 5 tries we went down to the floor of the valley to land. The elevation here was about 4000 feet. At this altitude the helicopter does not have much hovering capability. There was no air stirring in the bottom of the valley and for the last 10 feet we simply dropped. We hit with quite a bump but no damage was done. Had there been a breeze it might have assisted us in hovering. Litzenberg's role now is to hold the Yudam-ni area while Murray passes through him to continue the advance to the westward. Litzenberg indicated he would like to keep on going.

Yudam-ni lay in the center of a broad valley surrounded by five great ridgelines. Moving counterclockwise from the north, the ridges were given the prosaic but useful designations North, Northwest, Southwest, South, and Southeast. The 7th Marines held a perimeter that commanded four of the five ridges—all but the Northwest Ridge. Yudam-ni itself was a miserable collection of mud-and-thatch houses, battered by air attacks and now abandoned by their owners. The road that was the lifeline of the 1st Marine Division forked at Yudam-ni. One fork continued to the north. The other opened to the west, going as far as Mupyong-ni, before turning north and continuing to Kanggye.

On 26 November the 7th Marines reported the capture of three Chinese soldiers from the *60th CCF Division* and learned from them that the *58th, 59th,* and *60th CCF Divisions*, making up the *20th CCF Army*, were in the vicinity of Yudam-ni

1st Marines Button Up Division Rear

RCT-1 had to wait several days for rail transport to take them the 70 miles north from Wonsan to Chigyong. The regiment's 1st Battalion relieved the 3d Battalion, 5th Marines, at Chinhung-ni on Thanksgiving. Two days later the regiment's 2d Battalion, along with Puller's regimental headquarters, took over Koto-ri from the 2d Battalion, 5th Marines. Smith now had his regiments fairly close together, but further movement was hindered by a shortage of motor transport.

Two-thirds of the 3d Battalion, 1st Marines, arrived at Hagaru-ri during the early evening of Monday, 26 November. The battalion commander, Lieutenant Colonel Thomas L. Ridge, 35 and University of Illinois 1938, had been a naval attaché in Brazil for much of World War II but had reached the Pacific as an intelligence officer in time for Iwo Jima and Okinawa, where he was twice wounded. The motor march to Hagaru-ri was uneventful except for snarls in traffic. Because of the shortage of trucks, Captain Carl L. Sitter's George Company, reinforced with a provisional platoon from Weapons Company, had to be left behind at Chigyong.

Relief of Lieutenant Colonel Randolph S. D. Lockwood's 2d Battalion, 7th Marines, had to wait until morning. Lockwood, 37, U.S. Naval Academy 1936 and Harvard 1940, had just taken over the battalion from Major "Buzz" Sawyer on 9 November. Lockwood had spent most of World War II as a staff officer in Hawaii. The combat-experienced Sawyer stayed on as battalion executive officer.

On 25 November, the 2d Battalion, 1st Marines, along with Col Puller's regimental headquarters, relieved the 2d Battalion, 5th Marines, at Koto-ri. Next day, this heavy machine gun squad, with its water-cooled Browning M1917A1, follows behind two well-deployed rifle platoons making a reconnaissance in force toward the first range of hills.

Photo by Cpl W. T. Wolfe, National Archives Photo (USMC) 127-N-A4866

The new-arrivals at Hagaru-ri watched the engineers hack away at the frozen earth in their effort to build an airstrip capable of handling Air Force C-47 and Marine R-4D transports. The 1st Medical Battalion under Commander Howard A. Johnson set up a clearing station close to the strip for the expected flow of casualties. Extra surgical teams were flown into Hagaru-ri. The hospital ship *Consolation* (AH 15) moved up to Hungnam from Wonsan. The 1st Marine Division 400-bed hospital at Hungnam had an annex of 150 more beds at Hamhung.

Smith informed Fleet Marine Force, Pacific, and Headquarters Marine Corps, that, unless he received word to the contrary, he was sending his assistant division commander, Brigadier General Craig, home on emergency leave. Craig had received the bad news that his father had suffered a cerebral thrombosis and that the prognosis was unfavorable. Craig left for the States on Monday morning, 27 November.

The Chinese

The Marines were gradually learning about the new enemy. The term used for them by the U.S. and other English-speaking forces was "CCF" or "Chinese Communist Forces." Marines would learn that a CCF division, with its three infantry regiments and an artillery battalion (more theoretical than real in 1950), numbered about 8,000 men. A CCF regiment would average about 2,200 men, organized into three infantry battalions, sometimes with an artillery battery, more often with a mortar company, and several meager support companies. In the forward areas the Chinese had little or no motor transport. Things were pulled in carts by man or beast or carried on the backs of

Marine Corps Historical Center Photo Collection

Contrary to popular belief, the Chinese did not attack in "human waves," but in compact combat groups of 50 to 100 men. Here one such group makes its way up a snow-clad hill. The 1st Marine Division came into contact with the major portion of the Chinese Ninth Army Group, *which, with 12 divisions, totaled about 150,000 men.*

men, either Chinese soldiers or impressed Korean porters. The CCF infantry battalion, on paper at least, looked much like the Marines' own battalions: three rifle companies and a machine gun or heavy weapons company. The rifle companies similarly had three rifle platoons and a 60mm mortar section or platoon. The individual Chinese soldier was physically tough, uncomplaining, and used to long marches with few if any creature comforts. Politically he had been thoroughly indoctrinated, but once taken prisoner that indoctrination would tend to crack.

Collectively, his armament was a mixed bag of weapons gained from the surrender of the Japanese, the collapse of the Chinese Nationalist government—and its mixture of American, British, German, Czech, and other weapons—and the more recent issue of Russian weapons by the Soviet Union. But the Chinese army, at least in this stage of the war, was never equipped as uniformly or as well as the North Korean army had been. For the most part, the Chinese soldier wore a two-piece padded uniform with a cap to match, fairly adequate of themselves against the cold, but paired off with canvas "sneakers." They seldom had gloves or mittens and depended upon tucking their hands into the sleeves of their coats to keep them warm. Signal communications were primitive in the extreme. Commonly the Chinese used the SCR-300, captured from the Chinese Nationalists, as their backpacked radio, the same radio used by the Marine infantry. Radio nets almost never went below the regimental level. Telephone wire was seldom strung beneath the battalion level. Below the battalion, communications was by runner supplemented with bugles, whistles, flares, and flashlights.

Lacking adequate communications at the front, Chinese attack patterns tended to be rigid and repetitive. Once committed, a Chinese battalion would usually stay in contact until completely shredded by casualties or until all its ammunition was used up. There was little or no battlefield resupply.

Lin Piao had been concerned over the capability of the poorly equipped Chinese to fight the Americans, but Peng Dehuai hammered home to his senior subordinate officers his belief that Americans were afraid of close combat, a tactic in which the Chinese Communist troops excelled. Peng himself was a specialist in what the Chinese called a "short attack," hammering away at enemy defenses with successive compact combat groups, usually not more than a company in size, until a breakthrough or puncture was achieved, a tactic not unlike that used by German storm troops in the last years of World War I.

U.S. Marines' and soldiers' imaginations sometimes magnified what they saw and heard while under attack. The Western press was soon filled with fantasies of "human sea attacks" by "hordes" of Chinese. Chinese propaganda photographs and films showing wave after wave of Chinese advancing in line across the snow with bravely flying red banners reinforced these exaggerations. The truth was quite different. Hearing or reading such reports, the Marine infantry, those who were really there, would later ask derisively: "How many hordes are there in a Chinese platoon?"

RCT-31 East of the Reservoir

In mid-afternoon on 26 November, Colonel MacLean and his command group arrived at Faith's position on Hill 1221. Faith, ignoring Murray's caution, received MacLean's permission to move his battalion forward the next morning to the position vacated by Taplett's battalion.

MacLean set up his regimental command post in a schoolhouse in Hudong-ni, a village about a mile south of Hill 1221. A big, robust, aggressive man, MacLean was 43, a graduate of West Point, Class of 1930, and a veteran of the European theater. Barr had given him command of the 31st Infantry about two months earlier, replacing a commander who had not done well in the Inchon to Seoul drive. Before that MacLean had been in the G-3 Section of the Eighth Army. Previously, in Japan, he had commanded the 32d Infantry and he knew Faith well.

5th Marines' 27 November Attack

Of the 1st Marine Division's planned attack to the west, Ray Murray later said: "It was unbelievable. The more you think about it, the more unreal it becomes. Well, anyhow, those were the orders and that's what we started to do."

All elements of Murray's RCT-5 were to be relieved by Monday noon, 27 November, so as to take positions at Yudam-ni preparatory to passing through RCT-7 to lead the advance to Mupyong-ni. First objective for the regiment, once it was altogether, was to be the road junction at Yongnim-dong, 27 road miles to the west.

By nightfall on 26 November, Roise's 2d Battalion, 5th Marines, was in its attack position at Yudam-ni. His company commanders gathered in his blackout tent at 2200 to receive the attack order. Two Corsairs from VMF-312 for close support and a "Grasshopper" from VMO-6 for aerial reconnaissance were promised. The 7th Marines would support Roise's attack with patrols and a secondary attack to the southwest. The temperature at Yudam-ni during the night went down to zero degrees Fahrenheit.

In the morning Fox Company, under Captain Uel D. Peters, led off the 5th Marines' attack with an advance up the road leading westward. Peters' first objective was a spur about 500 yards beyond the 7th Marines perimeter. Almost immediately his Marines were engaged by long-range small arms fire. The VMO-6 spotter plane, overhead as promised, reported Chinese positions all across the front. At 1115, Corsairs from VMF-312 dumped rockets and bombs on the Chinese emplacements in front of Fox Company. As Peters began his assault, Chinese soldiers could be seen fleeing to the west. Three prisoners were taken.

Dog Company, under Captain Samuel S. Smith, had followed behind Peters and at about noon joined in the fight. Altogether Roise's battalion advanced about a mile. At 1430, Roise ordered Peters and Smith to break off the attack and set up night defensive positions.

The 3d Battalion, 7th Marines, in its attack to the southwest, had advanced about the same distance, about a mile, before running into stiffening opposition. The battalion had a new commander: Lieutenant Colonel William F. Harris, 32, Naval Academy 1936, who had taken over from Major Roach on 11 November. He was the son of Major General Field Harris. As a captain he had been serving with the 4th Marines when it was surrendered to the Japanese on Corregidor in the Philippines. He had spent the war as a prisoner of war and was one of four former prisoners to witness the Japanese surrender on board the battleship *Missouri* (BB 63). A big man with an easy manner he was immediately liked by the Marines in his battalion.

At noon Taplett's 3d Battalion, 5th Marines, arrived at Yudam-ni, after a hard five-hour motor march from the east side of the reservoir, and was assigned an assembly area

Marines, probably members of the 5th Marines, take a roadside break while on the march from Hagaru-ri to Yudam-ni on 27 November. This photo shows very clearly the nature and condition of the MSR or "main supply route" that was the division's lifeline from Yudam-ni back to Hungnam.

National Archives Photo (USN) 80-G-424585

west of the village where the road forked to the north and west. Taplett understood that his battalion was to follow Roise's 2d Battalion when the attack was resumed in the morning.

The 1st Battalion, 5th Marines, now under Lieutenant Colonel John Stevens, did not arrive until dusk and was given an assembly area east of the village. Meanwhile, the 2d Battalion, 7th Marines, was completing its motor march, company by company, but without Randolph Lockwood, the battalion's new commander, who stayed behind in Hagaru-ri.

Units of the 5th and 7th Marines were now thoroughly intermixed and would become more so, but there was no specific jointure of command. Brigadier General Craig, the assistant division commander, might have been given command of the two regiments combined into a task force, but he was home on emergency leave. Colonel Litzenberg was much senior to Lieutenant Colonel Murray, and perhaps Smith thought that was all the overall command authority needed. Litzenberg had positioned his command post for the 7th Marines in the center of Yudam-ni. Murray's command post for the 5th Marines was some distance away in the northwest corner of the village.

During the day Almond, accompanied by an aide and an assistant operations officer, drove by jeep to Yudam-ni from his command post at Hamhung. Arriving at the 7th Marines command post unexpectedly, he found Litzenberg absent but his executive officer, Lieutenant Colonel Frederick R. Dowsett, 39, present. Tall, lanky, Dowsett briefed him on the enemy situation and the disposition of the regiment. Almond passed out three Silver Stars, one to an officer and two to enlisted Marines, and then late in the afternoon began his return to Hamhung. The MSR was jammed with traffic going in both directions. In his opinion, the traffic was poorly controlled. The drive took nearly five hours. That night he reported to GHQ in Tokyo that the strength of the enemy was considerable and that the disposition of the Marines needed to be reexamined.

Hagaru-ri, 27 November

At Hagaru-ri, Ridge's 3d Battalion, 1st Marines, completed the relief of Lockwood's 2d Battalion, 7th Marines, on the morning of 27 November. Companies D and E of Lockwood's battalion had already arrived at Yudam-ni. While they waited for their own battalion commander, Litzenberg attached the two companies to Davis' 1st Battalion.

Lockwood, in accordance with Smith's directive to Litzenberg, now led forward his remaining rifle company, Fox Company, to occupy Toktong Pass. He gave Captain William E. Barber orders to move off the road, beginning four miles north of Hagaru, with the mission of keeping open three miles of the MSR. Lockwood then returned to Hagaru-ri where his Headquarters Company and the remainder of his Weapons Company were awaiting trucks to take them on to Yudam-ni. The trucks never came. Lockwood himself, and the remainder of his battalion, would never get to Yudam-ni.

When Captain Barber took command of Fox Company on 7 November, he made a little speech, telling his company that he was "an infantryman and a hell of a good one at that." Born in Kentucky in 1919, he had enlisted in the Marine Corps in 1940. He went through parachute training, doing so well that he stayed on as an instructor. He was commissioned in 1943, and as a platoon leader at Iwo Jima with the 26th Marines, he was wounded and evacuated. Refusing to stay hospitalized, he came back to take command of a company. For this he received a Silver Star and his first Purple Heart.

Ridge, faced with the mission of defending Hagaru with two-thirds of a battalion, sent Major Joseph D. Trompeter, his S-3, and Major Edwin H. Simmons, his Weapons Company commander and supporting arms coordinator, on a walking reconnaissance. Trompeter and Simmons found that to enclose all of Hagaru-ri would require a perimeter of four miles, an impossible task for a single infantry battalion at two-thirds strength. Ridge estimated that one to two regiments would be required for a thorough defense.

"Under the circumstances and considering the mission assigned to the 1st Marine Division," General Smith would later comment, "an infantry component of one battalion was all that could be spared for the defense of Hagaru," adding with the benefit of hindsight, "This battalion was very adequately supported by air, and had sufficient artillery and tanks for its purposes."

Captain Benjamin S. Read's How Battery, 3d Battalion, 11th Marines, which had been shooting for the 2d Battalion, 7th Marines, was already in place in the northeast corner of the sketchy perimeter. Now it would have to divide its fire missions between the defense of Fox Company in Toktong Pass and the Hagaru-ri perimeter and at the same time provide its own defense for its segment of the perimeter. "Our lives centered on our 105mm howitzers,

Photo by Cpl L. B. Snyder, National Archives Photo (USMC) 127-N-A4644

Capt William E. Barber took command of Fox Company, 2d Battalion, 7th Marines, on 7 November. A seasoned combat leader, he had commanded a platoon and company at Iwo Jima. He demanded a lot from his Marines and he got it. Shortly before leaving Hagaru-ri to take position in Toktong Pass, he held a rifle inspection for his company.

and our mission was to support the infantry," said Captain Read crisply a short time later.

Captain Andrew J. Strohmenger's Dog Battery, 2d Battalion, 11th Marines, had arrived at Hagaru-ri with Ridge's 3d Battalion, 1st Marines. The battalion and battery had worked together before, notably at Majonni, and were old friends. Strohmenger's battery went into position on the flats just southeast of the village.

The extreme cold affected the recoil systems of the howitzers and the reach of their shells. The guns were slow in coming back into battery and the extreme range was cut down from 12,200 yards to something like 9,000-9,500 yards.

Not being able to be strong everywhere, Ridge decided to concentrate his two rifle companies, How and Item, in a salient southwest of the not yet operational, but all-important, airstrip. The other greatest threat to Hagaru-ri was the hill mass just east of the town that would come to be called "East Hill."

Beyond the airstrip, First Lieutenant Joseph R. "Bull" Fisher's Item Company improved the positions vacated by Barber's Fox Company by blasting deeper foxholes with "Composition C" plastic explosive. On Fisher's left flank, Captain Clarence E. Corley's How Company extended the line until it tied in with the right flank of Strohmenger's Dog Battery, 11th Marines. The frozen marsh in front of Dog Battery was covered with fire but left unmanned. The perimeter picked up again with a roadblock held by a portion of Weapons Company across the road running south to Koto-ri. East Hill remained unoccupied. Ridge planned to put George Company on the hill when it arrived from the south. Service Battalion held the roadblock on the road that led northeast of the reservoir. Somewhere out there on the east side of the reservoir was the Army column that would come to be called "Task Force Faith," named for its doomed commander. The rest of the perimeter was patched together with bits and pieces of the Service Battalion, the division's

A 105mm fires a mission from its position close to the airstrip at Hagaru-ri. The two batteries of artillery—Capt Benjamin S. Read's Battery H, 3d Battalion, and Capt Andrew J. Strohmenger's Battery D, 2d Battalion, both of the 11th Marines—were essential to the defense of the Hagaru-ri perimeter.

Photo by TSgt V. Jobs, National Archives Photo (USMC) 127-N-A130286

Headquarters Battalion, and odds and ends left behind by the 7th Marines, until it closed again on Item Company's right flank. At the northern-most edge of the perimeter, Read's How Battery, 11th Marines, like Strohmenger's battery, was used as a frontline unit.

Lockwood received orders from Litzenberg to move to Toktong Pass to assist Fox Company. He borrowed a platoon from Ridge's battalion as an escort, but the effort went nowhere. Tank-infantry patrols sent out to the north toward Yudam-ni and to the south toward Koto-ri were pushed back in by mid-afternoon.

East of the reservoir, Monday morning, 27 November, Colonel MacLean, commanding RCT-31, went forward, accompanied by Lieutenant Colonel Faith, and together they inspected the lines vacated by the Marines. MacLean then selected a forward command post site south of Faith's intended new position.

Chinese Order of Battle

It was not yet known with certainty, but the scattered Chinese elements encountered earlier by Murray's 5th Marines were from the *80th Division* of the *27th Army, Ninth Army Group.* Commanded by Sung Shih-lun, the *Ninth Army Group,* with a total of 11 and possibly 12 divisions, consisted of three "armies," the *20th, 26th,* and *27th,* each roughly equivalent to a U.S. corps in frontline infantry strength. Sung Shih-lun was the equivalent of a lieutenant general, but the Chinese Communist Forces had not yet adopted Western military grades. Rank was indicated by billet held. Sung, like Peng, was his own political commissar. The *Ninth Army Group* had been poised to invade Taiwan after having captured Shanghai from the Nationalists. At Mao's direction Peng had brought Sung up from the Shanghai area and had sent him into Korea with specific orders to destroy X Corps. Peng's headquarters, it will be remembered, estimated that Sung could bring 150,000 troops against 90,000 men, a close guess at the strength of X Corps, giving him a 1.7 to 1 advantage.

Chinese Order of Battle

Ninth CCF Army Group

20th CCF Army	*26th CCF Army*	*27th CCF Army*
58th CCF Division	*76th CCF Division*	*79th CCF Division*
59th CCF Division	*77th CCF Division*	*80th CCF Division*
60th CCF Division	*78th CCF Division*	*81st CCF Division*
89th CCF Division	*88th CCF Division*	*90th CCF Division*

Source: Montross & Canzona, *The Chosin Reservoir Campaign* (1957). The fourth division in each army was not organic. The *88th, 89th, and 90th CCF Divisions* were attached from the *30th CCF Army.* Some other, later, authorities, Chinese as well as American, show the *90th CCF Division* as the *94th CCF Division* from the *32d CCF Army.*

Mao, in a telegram sent to Peng on 12 November, said: "It is said that the American Marine First Division has the highest combat effectiveness in the American armed forces." Sung would make the destruction of the 1st Marine Division, as the strongest of the American divisions, his main effort.

Sung's information as to the location of Marine Corps units was excellent. His plan, as later pieced together by U.S. intelligence, was as follows: The *27th Army*—except for the *80th Division,* which was to come down the east side of the reservoir—was charged with at-tacking the two Marine regiments at Yudam-ni. The *20th Army* was to cut the MSR or main supply route south of Yudam-ni, including attacks against Hagaru-ri and Koto-ri. The *26th,* initially in reserve, would not come into the fight until somewhat later. Sung was to launch his attack the night of 25 November, simultaneous with the assault to the west against the Eighth Army, but he was not quite ready and he secured Peng's approval to delay his attack for two days.

Early on the afternoon of 27 November, Faith completed the move of his battalion into the positions vacated by Taplett's 3d Battalion, 5th Marines. It was a typical Marine Corps perimeter, horseshoe shaped and occupying the high ground. Each of the exposed sides was occupied by one of Faith's rifle companies, the battalion command post was in the center, and the open side to the rear was covered by elements of his Headquarters and Service Company and Weapons Company. Lacking the strength in men and weapons of a Marine battalion, Faith could not fill all the foxholes.

MacLean, who had returned to Hudong-ni, was told that several hundred Chinese had been sighted east of the Pungnyuri-gang inlet. He sent out his Intelligence and Reconnaissance Platoon to investigate. The platoon roared out of the compound in its machine gun mounted jeeps and was never seen again.

The 3d Battalion, 31st Infantry, commanded by Lieutenant Colonel William R. Reilly, arrived that

National Archives Photo (USMC) 127-N-A4886

LtCol Randolph S. D. Lockwood, Commanding Officer, 2d Battalion, 7th Marines, had allowed his headquarters to become separated from his rifle companies. On 27 November, on orders from Col Litzenberg, he attempted to reach Fox Company in Toktong Pass with elements of his H&S Company and Weapons Company shown here. The effort failed.

afternoon, followed by the 57th Field Artillery Battalion, commanded by Lieutenant Colonel Raymond Embree. MacLean put Embree's battalion—which was minus its Battery C—into a bivouac area near the hamlet of Sinhung-ni, just south of Pungnyuri-gang inlet. The two firing batteries were positioned on the south side of the inlet on low ground surrounded on three sides by ridges. Embree placed his artillery headquarters a mile or so farther south on the slope of Hill 1456. Battery D of the 15th Antiaircraft Automatic Weapons Battalion, with four full-track M19 weapons carriers mounting dual 40mm guns and four half-tracked M16 carriers bearing quad .50-caliber machine guns, was set up close to Embree's headquarters.

The 31st Heavy Mortar Company, with its 4.2-inch mortars, moved into a position close to MacLean's forward command post and about halfway between Faith's battalion and Reilly's battalion. Meanwhile, the 31st Tank Company, with 20 M-4A4 Sherman tanks and two 105mm howitzer tanks, had reached Hudong-ni.

Thus, on the evening of 27 November, elements of MacLean's RCT-31 were stretched out on the road for 10 miles in seven different positions. By nightfall, or shortly thereafter, Faith, on the northern end with his 1st Battalion, 32d Infantry, had registered his artillery and mortar defensive fires. At about this time he received orders from MacLean to attack the next morning toward Kalchon-ni. MacLean himself spent the night at Faith's headquarters.

Sung Shi-lun, it will be remembered, had allocated his *80th Division* to the attack east of the reservoir. Shortly before midnight a firefight developed on Company A's front on the forward edge of Faith's position. The company commander was killed. Stamford, the Marine captain, took temporary command. The Chinese attack spread until it encompassed the rest of the battalion perimeter.

South of the inlet, the two firing batteries of Embree's 57th Field Artillery Battalion and Reilly's 3d Battalion, 31st Infantry Regiment, came under heavy attack from the east. The Chinese overran the 3d Battalion's command post and both artillery batteries. Reilly was severely wounded. Farther south, mortar shells began falling on Embree's artillery headquarters. Embree, in turn, was badly wounded.

Yudam-ni, 27 November

As darkness fell on the 27th at Yudam-ni, Captain Wilcox's Company B, 7th Marines, which had been patrolling South Ridge, came under heavy attack. Lieutenant Colonel Davis, commanding the 1st Battalion, received permission from Litzenberg to take a company to extricate Wilcox. Davis led Charlie Company, less one of its rifle platoons and commanded by Captain

John F. Morris, down the MSR to positions across the road from Hill 1419. Baker Company pulled itself loose from its engagement and Davis took it back into Yudam-ni, leaving Morris' Charlie Company to occupy Hill 1419—about two miles south of the incomplete perimeter. With less than a full company, Morris organized a crescent-shaped defense on an eastern spur of Hill 1419, well below the crest.

Unknown to Litzenberg and Murray as yet was that almost surrounding them at Yudam-ni were the *79th* and *81st CCF Divisions.* Furthermore, the *59th CCF Division* had begun a wide enveloping movement past South Ridge and on south to cut the MSR at Toktong Pass, held only by Fox Company, 7th Marines.

Artillery support at Yudam-ni was provided initially by Major Parry's 3d Battalion, 11th Marines—three batteries of 105mm howitzers, 18 tubes in all, enough to support a regiment in a narrow zone of attack, but not enough to provide adequate 360-degree support for a sprawling two-regiment defensive sector. Fortunately, among the Marine forces converging on Yudam-ni, during that busy 27th of November, was the 4th Battalion, 11th Marines, commanded by Major William McReynolds, with three batteries of its heavier 155mm howitzers—18 more tubes. All would be in action before midnight.

Toward the end of the day on 27 November, two Marines at Yudam-ni help a wounded comrade reach an aid station. Heavy action at Yudam-ni had begun that morning when elements of both the 5th and 7th Marines made an attack westward, were halted, and fell back to defensive positions.

National Archives Photo (USMC) 127-N-A4904

That night the temperature dropped to 20 degrees below zero. Northwest Ridge, the last ridge to be occupied, now had a Marine presence, a frontline of foxholes chipped out of the frozen ground and occupied by tired and cold-benumbed Marines. How Company, 7th Marines, commanded by Captain Leroy M. Cooke, held Hill 1403, the high point on Northwest Ridge. On How Company's left flank were Easy and Fox Companies of Roise's 2d Battalion, 5th Marines, occupying the rest of the ridge until it dropped down to the defile through which passed the road to the west. Roise had his command post behind the juncture of these two companies. A roadblock manned largely by Weapons Company covered the road westward. On the other side of the road, Dog Company curled back toward Southwest Ridge.

Taplett, uneasy with the situation, turned the assembly area assigned his 3d Battalion, 5th Marines, into its own defensive perimeter. His command post was in a draw behind Hill 1282. He sent a platoon from Item Company to outpost a spur of Hill 1384 about 500 yards forward of his command post. The outpost began receiving harassing fire at 2045.

The *89th CCF Division's* attack against Northwest Ridge, with two regiments, the *266th* and the *267th,* began at about 2200. The Chinese suddenly hit all along the line with sub-machine guns and grenades supported by machine gun fire and an intense mortar bar-

Sketch by Sgt Ralph Schofield, USMCR

This BAR-man, as sketched by combat artist Sgt Schofield, sights in his Browning automatic rifle. It still has its bipod, making it an efficient substitute for a light machine gun. Some BAR-men threw away their bipods to lighten their load, reducing the BAR to an assault rifle.

rage. This attack apparently aimed at fixing the Marines in position while a dense column of Chinese assaulted the line on a narrow front against the boundary between Easy and Fox Companies. This assault penetrated the Marines' position, overrunning Fox Company's right flank platoon. Captain Samuel Jaskilka (a future four-star Assistant Com-mandant), commanding Easy Company, turned back his left flank to cover the penetration. Roise pounded the Chinese salient with his 81mm mortars and sent up a platoon from Dog Company to reinforce Fox Company's ruptured right flank. A great number of Chinese were killed and by dawn the break in the line had been repaired.

Things went less well in the fight that had begun for possession of Hill 1403. Captain Cooke, How Company's commander, had deployed his three rifle platoons in a semi-circle on the forward edge of the crest of the hill. His right flank crumbled under the weight of an assault by the *266th CCF Regiment*. Cooke himself bravely led a counterattack to restore his line and was cut down by Chinese machine gun fire. Second Lieutenant James M. Mitchell took temporary charge of the company until First Lieutenant Howard H. Harris, who had earlier commanded the company, could get there from battalion. Harris (no relation to Lieutenant Colonel William Harris, commanding the 3d Battalion) arrived at about midnight and found only one How Company officer, Second Lieuten-ant Minard P. Newton, Jr., still on his feet. The Chinese again assaulted Hill 1403 at about 0300. After an hour of pounding, Lieutenant Colonel Harris ordered the battered How Company to withdraw to the rear of the 2d Battalion, 5th Marines, leaving Hill 1403 in Chinese hands.

Battle for North Ridge

Concurrently with the assault of Northwest Ridge, the *79th CCF Division,* with three regiments, had moved against North Ridge, held by two widely separated companies of the 2d Battalion, 7th Marines—Dog Company on Hill 1240 and Easy Company on Hill 1282. Separating the two hilltops was a long saddle. (The battalion's third rifle company, Fox Company, it will be remembered, had been dropped off at Toktong Pass and Lockwood and his headquarters were still at Hagaru-ri.)

The *235th CCF Regiment* attacked in a column of battalions against Hill 1282 at about midnight. Easy Company, under Captain Walter Phillips, held its ground against the first attack. Simultaneously, the *236th CCF Regiment,* following behind the *235th,* was feeling out Dog Company's position on Hill 1240.

Anticipating an attack against North Ridge, Murray had moved Stevens' 1st Battalion out of its assembly area northward to the reverse slope of Hill 1282. First elements of Able Company reached a spur of Hill 1282 barely in time to reinforce Easy Company, 7th Marines, which was being pummeled by the *1st Battalion, 235th CCF Regiment.* Easy Company's commander, Captain Phillips was killed and would receive a posthumous Navy Cross. His executive officer, First Lieutenant Raymond O. Ball, took over command, was several times wounded, and died in the battalion aid station. Command devolved upon the senior platoon leader, First Lieutenant Robert E. Snyder. Easy Company had been reduced to the size of a single platoon, and by daylight the Chinese had taken the crest of Hill 1282.

The crest of Hill 1240 to the east had also fallen. Chinese from the *3d Battalion, 236th CCF Regiment,* had overrun the command post of big, burly Captain Milton Hull, the company commander of Dog Company. At about 0300, Hull, wounded, counterattacked with the few squads at his disposal, won back a foothold, and was wounded again. When dawn came he could count only 16 Marines left with him, and the enemy had him surrounded.

During the night some Chinese had crossed the saddle that separated the Dog and Easy Company positions and had taken the 5th and 7th Marines' command posts under fire. Some time before midnight, a few half-dressed mortar men from How Company, 7th Marines, beaten back from Hill 1403, found their way into Taplett's 3d Battalion, 5th Marines, perimeter. A message from How Company, 7th Marines—that part that still remained on Hill 1403—reached Taplett, warning him that the Chinese were flanking his position. At about 0145, Taplett's outposted platoon on Hill 1384 received increasingly heavy fire. Shortly thereafter a CCF force, estimated at two companies, overran the outpost. Taplett's command post became the bull's eye of the fight. Major John J. Canney, the battalion executive officer and a World War II aviator turned infantryman, was killed.

South of Yudam-ni

At 0230, with the assaults against North and Northwest Ridges at their height, the Chinese also struck Charlie Company, under Captain Morris, on the spur of Hill 1419 two miles south of Yudam-ni. Morris' Marines held on grimly until dawn when artillery fire finally made the Chinese break off their attack. But, with a third of his men casualties, Morris was effectively pinned into position by Chinese fire continuing to rain down from the heights. His Marines could do nothing more than hold their position and hope that help would come from Yudam-ni.

While the *79th* and *89th CCF Divisions* savaged the Marines on Northwest and North Ridges, the *59th CCF Division* completed its wide sweeping movement to the southeast, putting itself in position to cut the 14 miles of vital MSR between Yudam-ni and Hagaru-ri. Until midnight on 27 November truck traffic on the MSR was still active and unimpeded—mostly empty trucks from Lieutenant Colonel Beall's 1st Motor Transport Battalion rattling their way back to Hagaru-ri, having delivered the last serials of the 1st Battalion, 5th Marines, and the 4th Battalion, 11th Marines, to Yudam-ni.

Captain Barber had gone into position at Toktong Pass with a near full-strength Fox Company reinforced with sections of water-cooled Browning machine guns and 81mm mortars from Weapons Company, 2d Battalion—a total of 240 officers and men. Barber chose to organize his defensive perimeter on a hill at the mid-point of the pass. "We arrived in the late afternoon after which we unloaded and were positioned for the night," remembered Corporal Howard W. Koone. "Our position was off to the right of the road up on a saddle-like hill. The ground was like a sheet of concrete and very barren."

Barber's 3d Platoon, under First Lieutenant Robert C. McCarthy, occupied the high ground at the center of the narrow perimeter. At about 0230, McCarthy's two forward squads were overwhelmed by a company-sized attack. Out of 35 men, McCarthy lost 15 killed, 9 wounded, and 3 missing. The eight survivors fell back to the reserve squad on the reverse slope of the hill. Barber's position was almost cut in half, but his two wing platoons managed to hold their ground. Much was owed to the valor of three Marines: Private First Class Robert Benson and Private Hector A. Cafferatta of the 2d Platoon under Second Lieutenant Elmo G. Peterson on the left, and Private First Class Gerald J. Smith of the 1st Platoon under First Lieutenant John M. Dunne on the right. One party of Chinese penetrated as far as the company command post and the 81mm mortar position. Fighting, some of it hand-to-hand, continued until daybreak when the Chinese broke off the assault but continued to keep the position under fire. In all Barber had lost 20 Marines killed and 54 wounded. Fox Company did not know how many Chinese it had killed but guesses went up to 500.

Howard Koone was one of those wounded. He eventually found himself in a Korean hut being used by Fox Company's corpsmen as a sick bay. He was told that helicopters would be coming to evacuate the wounded and that he would be third on the list, but the helicopters never came.

Yudam-ni, 28 November

Dawn on 28 November saw the tactical situation on Northwest Ridge unresolved. Hill 1403 had been lost to the enemy. Elsewhere both Marines and Chinese were clinging to the high ground. Roise's 2d Battalion, 5th Marines, had a firm grip on its portion of the line. As yet there had been no orders to abandon the offensive begun the day before. Roise had received orders from Murray to continue the attack at daybreak. Taplett's 3d Battalion was to come up on his right flank and add its weight to the assault.

Murray met with Litzenberg at the 7th Marines command post at dawn. Both regimental commanders agreed that the situation dictated that they change from the offensive to the defensive. Murray canceled the attacks to be made by Roise and Taplett.

Murray barely knew Litzenberg. In the south, at Seoul, he had seen him once or twice at division headquarters. The only intimate contact he ever had with him would be at Yudam-ni. Never-theless, the loose command relationship seemed to work. "If he had troops on some hills," said Murray, "then I put troops on some other hills, so that we had a good perimeter defense of the area."

Murray remembered that Litzenberg "had a reputation of being sort of a fussbudget, a stickler . . . he seemed to be a studious type of person, knew his business, and as far as I could tell from talking with people in the 7th Marines, it seemed everyone respected him and his abilities. . . . Many people have asked why he didn't just assume command up there. I can't answer that question definitively. After all, there was a division headquarters over the hill from us, and we were still part of that division, so we had a common head. But in any case, we decided to operate very closely together, and we did."

Taplett had begun his counterattack against the spur of Hill 1384 at about 0300 with two platoons of George Company led by Lieutenants John J. "Blackie" Cahill and Dana B. Cashion. Some time after daylight Cahill and Cashion reached the crest of Hill 1384 with their platoons. About this time Taplett received the order canceling the attack. He, in turn, directed Cahill and Cashion to hold where they were until they received further orders. With their presence on top of the hill, the remainder of How Company, 7th Marines—some 80 officers and men—was able to complete its withdrawal from Hill 1403 and pass on into Taplett's perimeter.

John Stevens' 1st Battalion, 5th Marines, spent the morning consolidating its position on Hill 1240. His Charlie Company, under Captain Jack R. Jones, had moved over during the night to backstop Taplett's battalion and was put under the operational control of the 7th Marines. One platoon was dropped off to rejoin its parent

A row of dead Chinese, frozen in grotesque positions, on the high ground overlooking a command post of the 5th Marines at Yudam-ni, mark the line of their farthest advance. The burden of the Chinese attack was borne chiefly by isolated Marine rifle companies holding ridgeline positions.

Photo by Sgt Frank C. Kerr, Department of Defense Photo (USMC) A4839

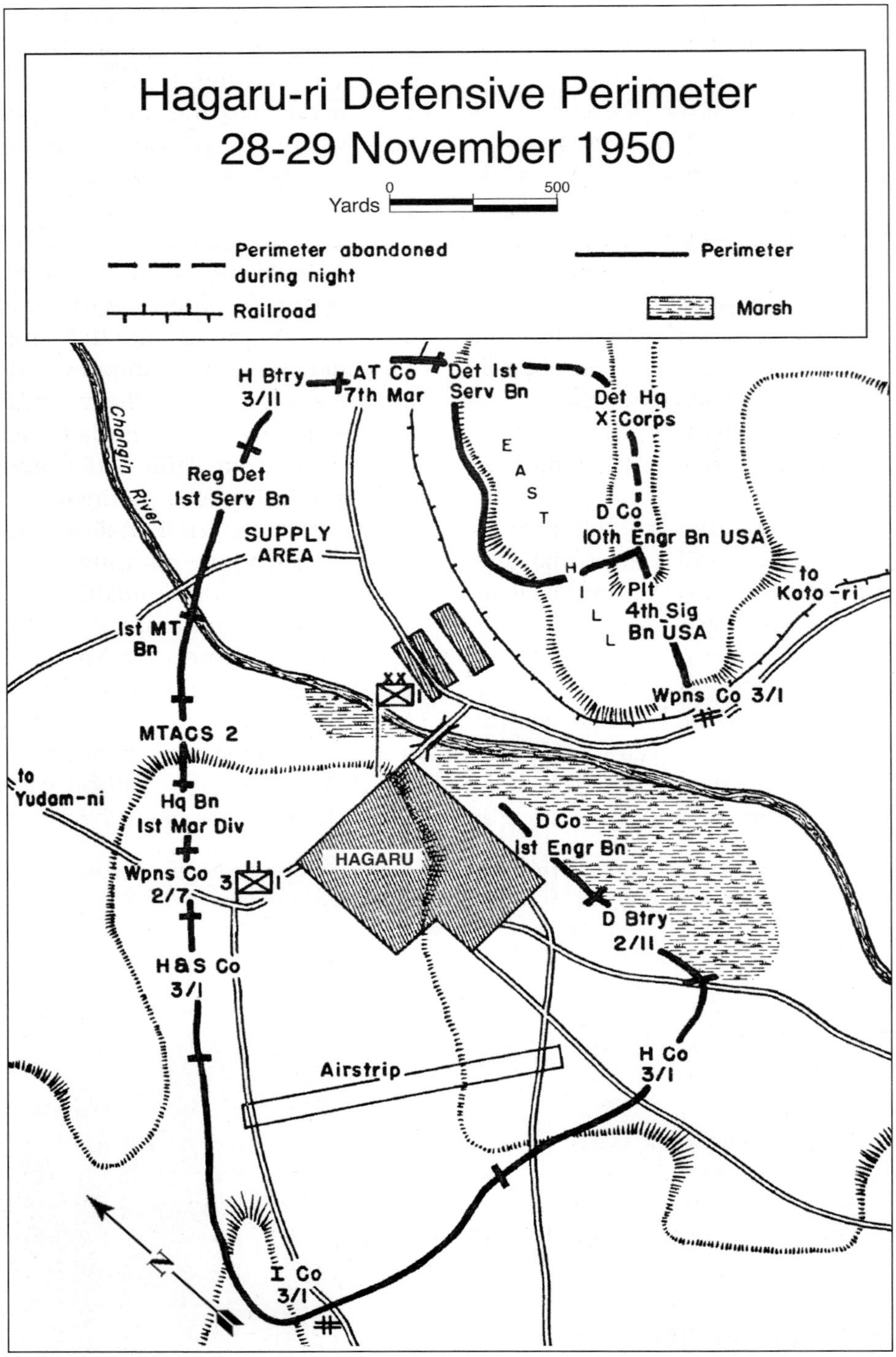

battalion on Hill 1240. The remainder of the company continued on to Hill 1282. A company of Chinese from the *235th CCF Regiment* had lodged itself on the hilltop. Jones led his reduced company in a hand-to-hand assault that won back the hill.

At 1100 Murray ordered Roise to pull his battalion back to Southwest Ridge tying in with Harris' 3d Battalion, 7th Marines, on his left and Taplett's 3d Battalion, 5th Marines, on his right. Early that afternoon Roise brought his battalion back from Northwest Ridge a company at a time. Except for occasional harassing fire, the Chinese did not interfere.

Early that morning, Davis' 1st Battalion, 7th Marines, had set out to the south to relieve both Charlie and Fox Companies from their encirclement on the MSR. Able Company, under First Lieutenant Eugenous M. Hovatter, led off, moving through a gorge separating South from Southeast Ridge. Five hours of fighting found Able Company still a mile short of Charlie Company's position. Baker Company, under Captain Wilcox, joined the attack. Together the two companies reached Charlie Company. Litzenberg, with Charlie Company now relieved and its wounded evacuated, and not wanting to have the 1st Battalion trapped in the gorge, ordered Davis to pull back into the Yudam-ni perimeter. By evening Stevens' 1st Battalion, 5th Marines, had relieved the shattered remnants of Hull's Company D, 7th Marines. The two regiments at Yudam-ni, their perimeter tightened and mended, faced the night of 28 November with considerable confidence. But Barber's Fox Company remained alone in Toktong Pass.

Almond Visits Faith

At Hagaru-ri, a platoon-sized patrol, sent out to the southwest early on the morning of 28 November from Fisher's Item Company, was pushed back into the perimeter. At about the same time as this patrol action, Ridge telephoned Colonel Bowser, the division G-3 and Smith's war chief, recommending that an overall defense commander be designated for Hagaru-ri. He also requested that the arrival of his George Company and the Royal Marine 41 Commando be expedited. Before a decision could be reached, General Smith flew in by helicopter at about 1100 to open his command post at Hagaru-ri. A half-hour later General Almond, along with his junior aide, 26-year-old Captain Alexander M. Haig, Jr., arrived in Almond's L-17 light aircraft, the "Blue Goose."

After meeting with Smith, Almond borrowed a Marine helicopter to take him east of the reservoir to meet with Faith and MacLean. Colonel MacLean, it will be remembered, had spent the night of 27 November at Faith's position. MacLean thought that Faith's battalion had come through the night in fairly good shape. He knew little or nothing about what had happened south of the inlet. At dawn he left to return to his own advance command post. His short jeep trip was not interrupted.

Almond, on arriving at Faith's position in his borrowed Marine Corps helicopter, airily told Faith that there was nothing in front of him except scattered Chinese retreating to the north and that he should try to retake the lost high ground. As further encouragement, Almond informed Faith that he had three Silver Stars to present, one for Faith himself and two more for whomever Faith designated. Faith called forward a wounded platoon leader and a mess sergeant. Almond pinned the three Silver Stars to their parkas, Captain Haig noted their names in his notebook, and the general and his aide got back on board their helicopter. As the helicopter whirled away, Faith and the lieutenant tore the Silver Stars from their parkas and threw them in the snow.

Stopping to see MacLean, Almond advised him that the previously planned attack would be resumed once the 2d Battalion, 31st Infantry, joined the regiment. This battalion and Battery C of the 57th Field Artillery were marooned far south on the clogged MSR.

Photo courtesy of Sgt Norman L. Strickbine, USA

East of the reservoir during 28 November, the Army's 3d Battalion, 31st Infantry, and two firing batteries of the 57th Field Artillery Battalion painfully reorganized after the previous night's devastating attack against their positions just south of Pungnyuri Inlet. Bodies can be seen in the foreground. Sporadic fighting continued north of the inlet.

During the night, the 31st Medical Company, pushing north from Hudong-ni had been ambushed and badly shot up in the vicinity of Hill 1221. Survivors drifting back to the headquarters of RCT-31 at Hudong-ni were the first indication that the road had been cut.

Meanwhile, General Hodes, the assistant commander of the 7th Division, was at Hudong-ni. He directed Captain Richard E. Drake, commander of the 31st Tank Company, to sally forth to the north to see if he could break through to the inlet. Drake moved out with 16 tanks. Hodes rode with Drake as a passenger; he did not take tactical command. Without infantry support, the tanks could not break the Chinese grip on Hill 1221 which effectively blocked the route north. Four tanks were lost. Hodes returned to Hudong-ni in a jeep, intent on getting back to Hagaru-ri for help. He took a tank, at Drake's insistence, for transportation and got back to Hagaru-ri, five miles away, without further incident. He never returned to Hudong-ni.

South of the inlet that day, 28 November, the badly battered 3d Battalion, 31st Infantry, and 57th Field Artillery Battalion painfully reorganized and consolidated their positions. Before nightfall, the Chinese came back into the attack with the M16 and M19 self-propelled guns the focal point of their effort. The automatic 40mm and .50-caliber fire did its lethal work. The perimeter held and many Chinese died.

North of the inlet sporadic fighting had continued. A dominant hill position was lost to the Chinese. Stamford ran close air support strikes with Marine Corsairs with little apparent effect. To the east the battalion could glimpse long columns of Chinese marching south, some of them mounted on

Mongolian ponies, or so it was said. Air strikes were flown against them and claimed good results.

Hagaru-ri, 28 November

Smith had moved into a Japanese-style house, soon overcrowded with the impedimenta of a division command post. On the wall close by Smith's field desk hung a picture of Stalin; Smith let it remain where it was. By nightfall on the 28th Smith had officially sanctioned actions already taken at Yudam-ni. Murray was ordered to halt his attack to the northwest. Litzenberg was told to attack to the south and reopen the MSR to Hagaru-ri. Together, Murray and Litzenberg were to plan for the continued defense of Yudam-ni and the breakout to the south. The joint defense plan worked up by Litzenberg and Murray provided for RCT-5 to take over responsibility for the west and north sectors, RCT-7 for the east, south, and southwest.

"Although the two regimental commanders acted jointly," said Taplett years later, "I harbored the gut feeling that Colonel Litzenberg and not Colonel Murray called the shots simply because of seniority. I confess to having more confidence in Murray."

During the afternoon, Colonel Bowser, the division G-3, telephoned Lieutenant Colonel Ridge confirming his appointment as Hagaru's defense commander. By then Ridge knew that George Company would not be arriving in time to occupy East Hill. George Company under Captain Carl L. Sitter reached Koto-ri that same day. Sitter, 28, had received a field commission in World War II after two years enlisted service. He fought in the Marshalls and at Guam, was twice wounded, and had received a Silver Star. At Koto-ri it soon became obvious that Sitter's company could go no farther without strenuous effort.

Before moving north to the reservoir, the commander and staff of the division's artillery regiment, the 11th Marines, lined up outside their command post at Hambung for a group photograph. Standing, from left to right, are: Maj Donald V. Anderson, LtCol Carl A. Youngdale, the executive officer, Col James H. Brower, the commanding officer, and LtCol James G. Appleyard. Kneeling are Capt William T. Phillips and Maj Floyd M. McCorkle.

Photo by TSgt James W. Helms, Jr., National Archives Photo (USMC) 127-N-A5264

Colonel Brower had arrived at Hagaru-ri with the headquarters of his artillery regiment, the 11th Marines. He set up the fire support control center in juxtaposition with Smith's headquarters. His executive officer, Lieutenant Colonel Carl A. Youngdale, 48, Iowa State University, class of 1936, headed the regiment's fire direction center.

That afternoon, Company D, 10th Combat Engineer Battalion, came in from a tent camp the engineer soldiers had set up just outside the perimeter on the road leading south to Koto-ri. In his expanded role as Hagaru-ri defense commander, Ridge had operational control of the company. He decided to use it to fill the yawning gap on East Hill. He so informed the engineer company commander, Army Captain Philip A. Kulbes. The engineer captain protested, saying that he was at Hagaru to build a new command post for X Corps and that his men—77 Americans and 90 South Koreans—had no training in infantry combat. Aside from individual weapons, the only armament the company possessed was four .50-caliber machine guns, five light .30-caliber machine guns, and six 3.5-inch rocket launchers. Ridge asked Kulbes if he would accept the tactical advice of a Marine officer and Kulbes said he would. Captain John C. Shelnutt, the executive officer of the 3d Battalion's Weapons Company, was assigned as a "liaison" officer. Shelnutt was accompanied by a radioman, Private First Class Bruno Podolak. Major Simmons privately advised Shelnutt that, in face of the Army captain's reluctance, he would have to take de facto command. The Army company procrastinated, taking its time to move its trucks and engineer equipment into a motor park. At dusk the engineers started up the hill. The ascent took them through the roadblock facing south toward Koto-ri.

About 10 Marines under Gunnery Sergeant Bert E. Elliott, the Weapons Company machine gun platoon sergeant, manned the roadblock. Reinforcements for the roadblock came late in the day in the form of a platoon from the Army's 4th Signal Battalion sent to install communications for what was to be General Almond's command post. The Army signal lieutenant, First Lieutenant John A. Colborn, like the Army engineer captain, reported that his men had no infantry training. The 3d Battalion's Weapons Company commander asked him if he would take orders from a Marine gunnery sergeant. The lieutenant eagerly said that he would.

On the north side of East Hill, the commanding officer of the 1st Service Battalion, Lieutenant Colonel Charles L. "Gus" Banks, 36, a World War II Edson's raider, was named a sub-sector commander. He was to coordinate his actions with Lockwood. While Kolbes' engineers climbed the south face of the hill a column of Marines sent by Banks started up the north face. The two columns were supposed to meet on the crest.

At sundown, Simmons pulled his roadblock back about 75 yards to what he thought was a stronger position. With a total of about 40 men Gunnery Sergeant Elliott was able to man the roadblock with his Marines and the knoll on his left flank, which was the first step in the climb up East Hill, with the Army signal platoon. Elliott had been a lieutenant during World War II and he was determined to win back his bars. Tough, battle-wise, and not particularly well-liked, even by his own Marines, he balanced his .45-caliber pistol in the palm of his hand and bluntly advised the soldiers that if they dug in, stayed, and fought, they would be there in the morning; but if they got up to run, he would shoot them himself.

Hagaru-ri Airstrip Defense

While the two columns, Army and Marine, moved toward the crest of East Hill, a major Chinese attack hit the southwest quarter of the perimeter, fortunately striking the strongest segment of the Marine line, that held by Companies H and I of the 3d Battalion, 1st Marines. The two companies, stretched thinly along the far side of the prospective airstrip on which the engineers were laboring under lights, were well dug-in. Holes had been blasted through the top 8 or 10 inches of frozen earth with ration cans filled with C-3 explosive so that foxholes and machine gun emplacements could be dug. The spoil was used to fill sandbags. A meager supply of concertina and other barbed wire was strung out where it would do the most good. Five-gallon cans filled with gasoline were rigged with white phosphorus grenades. Tied to the grenades were strings that could be pulled to explode the grenades and flame the gasoline. An earnest demolitions sergeant explained that these were French devices known as a "foo-gah-say." Three draws led into the Marine position; they had been sown with anti-personnel mines. In all, it would be a tough nut for the Chinese to crack.

A light snow was falling. The two companies were at 100 percent alert. At about 2230, three red flares and three blasts of a whistle signaled that the Chinese were coming. Mortar shells, high explosive mixed with white phosphorus, began crunching down on the frontline positions. Marine supporting arms—some artillery but mostly mortars and machine guns—took the Chinese under fire, but did not stop the enemy from closing to within hand-grenade and burp-gun range. The assault continued for an hour, the Chinese attacking in combat groups of about 50 men each. Most of the Marine line held, but the Chinese succeeded in penetrating the center of How Company's position. The company commander, Captain Clarence Corley, pulled together a scratch squad and tried to plug the gap but was pushed aside. Some few Chinese broke through as far as the airstrip where the engineers killed them.

Ridge dispatched a mixed platoon of Marines and soldiers under First Lieutenant Grady P. Mitchell, Jr., to back up How Company. Mitchell was killed and First Lieutenant Horace L. Johnson, Jr., took over. Johnson deployed his men in a ditch fortuitously behind How Company's ruptured line. The Chinese who had penetrated the position were milling around, seemingly more intent on looting the supply and cook tents than exploiting their success. They were fighting for food, warm clothes, and U.S. ammunition. At least one wounded Marine survived by feigning death when a Chinese soldier stripped him of his parka. Ridge fed in another platoon made up of casuals to build on Johnson's line. By about 0130 the situation appeared to be under control. The engineers relit their floodlights, got back on their dozers, and resumed work on the airstrip.

But bad things were now happening on the other side of the perimeter.

Action on East Hill

The two columns that had been sent up East Hill had failed to reach the crest. Captain Shelnutt, in virtual command of the Army engineers and under heavy fire, reported to the 3d Battalion's Weapons Company commander that Banks' column coming up the other way did not seem to be where it was supposed to be. Shelnutt was told to turn back his left flank and hold for the night. He was promised that artillery fire would fill in the gap.

At about 0115, the Marines and soldiers on the south roadblock were treated to the sight of a company-sized column of Chinese marching up the road toward them. Apparently the pullback of the roadblock earlier in the evening had caused the Chinese to think the position had been abandoned. The column presented the pair of Weapons Company water-cooled machine guns with a perfect enfilade target. Few members of the Chinese column escaped.

At Ridge's command tent heavy small arms fire and grenades could be heard on East Hill itself. The Weapons Company commander reached Shelnutt by radio at about 0200. Podolak, the radio operator, informed him that Shelnutt was dead: "There's nobody up here except me and a couple of doggies." Podolak was sternly enjoined, as a Marine, to take charge. The next time the Weapons Company commander tried to radio him the set was dead.

During the night, stragglers from the Army engineer company, mostly South Koreans, but some Americans, streamed back off the hill and took cover in the ditches and culverts of Hagaru-ri itself. Some few were rallied into a support line, stiffened with a handful of Marines, along the road paralleling the base of the hill. Other soldiers stayed on the hill and fought bravely. Most of these died.

Across the perimeter, Captain Clarence Corley, a spent bullet in his arm, launched a counterattack at about 0430 to restore his main line of resistance. It was successful, but the night had cost How Company 16 men dead and 39 wounded.

Hagaru-ri, 29 November

Ridge's greatest concern now was the situation on East Hill. If the enemy continued to have possession of the crest when daylight came, exposing the defenses of Hagaru-ri to full view, the situation would be critical. At 0530 he decided that he must counterattack. Major Reginald R. Myers, 31, University of Idaho 1941, the battalion executive officer, volunteered to lead a column up the hill. Myers had spent most of World War II on sea duty but joined the 5th Marines in time for Okinawa and North China. There was no tactical unit available to him at Hagaru that could be used. The attack would have to be made by a mixed force of service troops—and some stragglers found skulking in the town—patched together into a provisional company of about 250 men, mostly Marines but including a few soldiers. Myers' improvised company formed up on the road next to the battalion command post and was tolled off into platoons and squads. The first platoon, made up of Marines from the 1st Engineer Battalion and under command of First Lieutenant Robert E. Jochums, was the most homogenous and in the best shape.

Ridge delayed Myers' jump-off until about 0930 by which time the morning mists had cleared and Corsairs for close support could be brought overhead. The south roadblock had held. The soldier signalmen had stayed and fought well, delighting both themselves and the Marines. Myers led his "company" upward through their position. Troubles began almost immediately, if not from Chinese gunfire then from the icy slope. Men stumbled and fell, to be hauled to the rear by others only too willing to carry them to relative safety. Myers' force melted away to about 75 men. Best performance, predictably, was by the platoon led by Jochums. He was wounded in the foot but continued in command. Myers could claim reaching the military crest, but the topographical crest was still firmly in Chinese hands.

A supporting attack was to be made by Company A, 1st Engineer Battalion, under Captain George W. King, coming up the south face and passing through Myers' position. King started up the hill at about noon, his 1st Platoon under First Lieutenant Nicholas A. Canzona in the lead. Orders were changed. King's company was pulled back, marched almost a mile to the north, and then sent up the north face. Like the Myers force, he reached the military crest, but on the north side. His company went into a reverse slope defensive position for the night, separated from Myers by about 500 yards. Ridge had to be satisfied with King and Myers holding these positions. The Chinese continued to hold the topographical crest.

Ridge planned to feed in Sitter's George Company to take the remainder of the hill when it arrived from Koto-ri. From prisoner interrogation Ridge now believed that the *58th CCF Division,* led by the *172d Regiment* and followed by the *173d*, with the *174th* held back in reserve, had attacked Companies H and I. It was not clear what Chinese force was driving west to East Hill or when its deployment to assault Hagaru itself would be completed.

The Corsairs from Frank Cole's VMF-312 flew 31 sorties that day over Hagaru most of them against East Hill. One plane took a bad hit from Chinese small arms. The pilot, First Lieutenant Harry W. Colmery, successfully crash-landed inside the perimeter.

Brigadier General Hodes, the assistant division commander of the 7th Division, had spoken briefly with General Smith upon his arrival from Hudong-ni on the evening of 28 November. At noon on the 29th of November, he met again with Smith, informing him in more detail of the condition of RCT-31 east of the reservoir; that it had taken 400 casualties and was falling back toward Hagaru-ri and probably was unable to fight its way to safety.

LtCol Allan Sutter arrived at Koto-ri with the 2d Battalion, 1st Marines, on 24 November. Col Puller moved his headquarters to Koto-ri on the same day. With the 3d Battalion at Hagaru-ri and the 1st Battalion at Chinhung-ni, Puller would have his infantry battalions at three widely separated combat bases along the MSR.

National Archives Photo (USMC) 127-N-A5546

"The inference was that they should be rescued by a larger force," wrote Smith in his log. "I have nothing now with which to lend a hand except the battalion at Hagaru-ri and it has its hands full. I cannot see why the cutoff battalions cannot at least improve the situation by moving toward us."

Second Night on Fox Hill

Barber was supposed to have brought Fox Company off Toktong Pass and, with the help of Davis' battalion, was to have marched on into Yudam-ni. There was no chance of this. He was already encumbered with 54 wounded. During the morning of 28 November he had the help of a close air strike by Australian F-51 Mustangs. Later he sent out patrols that confirmed that he was completely surrounded. He asked for resupply by air. Marine R-5D four-engine transports, the Marine Corps equivalent of the Air Force C-54, dropped medical supplies and ammunition. Most fell at the base of the hill. Recovering them cost two more Marines wounded.

That night the Chinese came again against Fox Company. Five more Marines were killed, 29 more wounded, among the latter Captain Barber. Hit in the leg, he received first aid and stayed in action. During the day that followed, both Marine and Air Force planes dropped ammunition and other supplies. A Marine helicopter made a precarious delivery of some ammunition and much-needed radio batteries. Lieutenant Peterson, already twice wounded, took a patrol out in front of Fox Company to recover some errant mortar ammunition.

Koto-ri Action, 24-28 November

Lieutenant Colonel Alan Sutter's 2d Battalion, 1st Marines, had arrived at the meager village of Koto-ri on 24 November. Handsome, silver-haired Sutter, 36, Dartmouth 1937, had been a signal officer at Guadalcanal, Guam, and Okinawa. Now at Koto-ri, with his battalion reinforced by the 105mm howitzers of Easy Battery of the 11th Marines, a platoon of 4.2-inch mortars, bits and pieces of the regimental antitank company, and Company D of the 1st Medical Battalion, he set up a conventional perimeter defense.

A patrol from Captain Jack A. Smith's Easy Company brushed up against about 25 Chinese west of the village and brought in two wounded prisoners who said they were part of a Chinese division moving into attack positions. Chesty Puller and his regimental headquarters had joined Sutter at Koto-ri and in the next several days more Marine and Army units jammed their way into the protective envelope of the perimeter. On the morning of 28 November, Smith ordered Puller to send a force up the MSR to meet a tank patrol coming down from Hagaru-ri. Sutter sent out Dog Company under Captain Welby W. Cronk, but it was stopped a mile north of the perimeter by a strong Chinese force entrenched on both sides of the road. Dog Company withdrew under cover of air strikes by the busy Corsairs of VMF-312. The day's fighting cost the Marines four killed and 34 wounded. Three prisoners were taken and they identified their unit as the *179th Regiment, 60th CCF Division.*

Solemn Meeting at GHQ

The 28th of November had been a busy day for General Almond and, when he arrived at his comfortable headquarters at Hamhung that evening, he found urgent orders directing him to report immediately to MacArthur's GHQ in Tokyo. Almond and a small staff left for Tokyo from Yonpo in an Air Force C-54. They arrived at Haneda Airport at 2130 where Almond was told to proceed immediately to General MacArthur's residence at the American Embassy. He learned that MacArthur had called his senior commanders back to GHQ for a secret council of war. General Walker would also be present. The conference lasted two hours. In the west, Eighth Army's "Home-by-Christmas" offensive had gone well for the first two days. Then, on the night of 25 November, Chinese bugles were heard all across the front. By noon on 27 November, Walker had reported to MacArthur that he estimated there were 200,000 Chinese in front of him, that the ROK II Corps had been swept away, and that the U.S. IX Corps was falling back to cover his exposed flank. Walker now informed MacArthur that he thought he could build up a line in the vicinity of Pyongyang. Almond, in a bit of braggadocio, told MacArthur that he expected the 1st Marine and 7th Infantry Division to continue their attack. However, in his own mind Almond had come to realize that the greatest problem facing X Corps was its dispersion over a 400-mile front. He had begun to contemplate concentrating his forces into a perimeter defense around the Hamhung-Hungnam area. MacArthur, after listening to his field commanders, gave his decision: a changeover from the strategic offensive to the defensive. (Some authorities believe MacArthur had already reached this decision before meeting with his senior field commanders.)

Photo by Cpl W. T. Wolfe, Department of Defense Photo (USMC) A4868

A company of Marines, quite possibly Capt Welby W. Cronk's Company D, 2d Battalion, 1st Marines, makes a school-solution advance up a hill outside the Koto-ri perimeter on 28 November. One platoon is deployed as skirmishers followed by the other two platoons in squad column. Cronk took three Chinese prisoners, but the Marines lost four killed and 34 wounded.

Yudam-ni, 29 November

The night of 28-29 November was quiet at Yudam-ni. Division directed that an effort again be made to relieve Fox Company. A composite battalion was pasted together of Able Company from the 5th Marines, Baker and George Companies from the 7th Marines, reinforced with a section of 75mm recoilless rifles and two sections of 81mm mortars. Major Warren Morris, the executive officer of 3d Battalion, 7th Marines, was placed in command. He assembled his force in front of the 1st Battalion, 7th Marines, command post on the morning of 29 November and at 0800 marched out to the south. Three hundred yards outside the perimeter heavy machine gun fire laced into his column. Morris pushed on. Corsairs came overhead to help and dropped messages warning that the Chinese were entrenched along both sides of the road. At 1315, Litzenberg, warned that Morris' column was in danger of being surrounded, ordered him to return to the perimeter. Fox Company would spend another night at Toktong Pass alone. As evening fell, Captain Barber called his platoon leaders together and told them they could expect no immediate relief.

RCT-31 Begins Withdrawal

During the night of 28 November, Faith and MacLean tried to get some sleep in the hut that was now their joint command post. By 0100, 29 November, the Chinese were attacking in strength but were beaten off. An hour later, MacLean ordered Faith to prepare to breakout to the south with the objective of reaching the 3d

Battalion, 31st Infantry. All trucks were to be unloaded of cargo and given over to carrying out the wounded. Equipment and vehicles left behind were to be disabled but not destroyed. Blackout would be observed and there would be no burning of tentage and supplies.

MacLean and Faith began their march-out at about 0600 on Wednesday morning, the 29th. It was strangely quiet as the rifle companies broke contact and came down from the high ground. The truck column, about 60 vehicles in all, formed up and moved south on the road with Marine Corsairs overhead. Leading the way was a command party that included MacLean and Faith. As the party approached the highway bridge over the inlet, it came under fire and split into two parts, MacLean with one, Faith with the other.

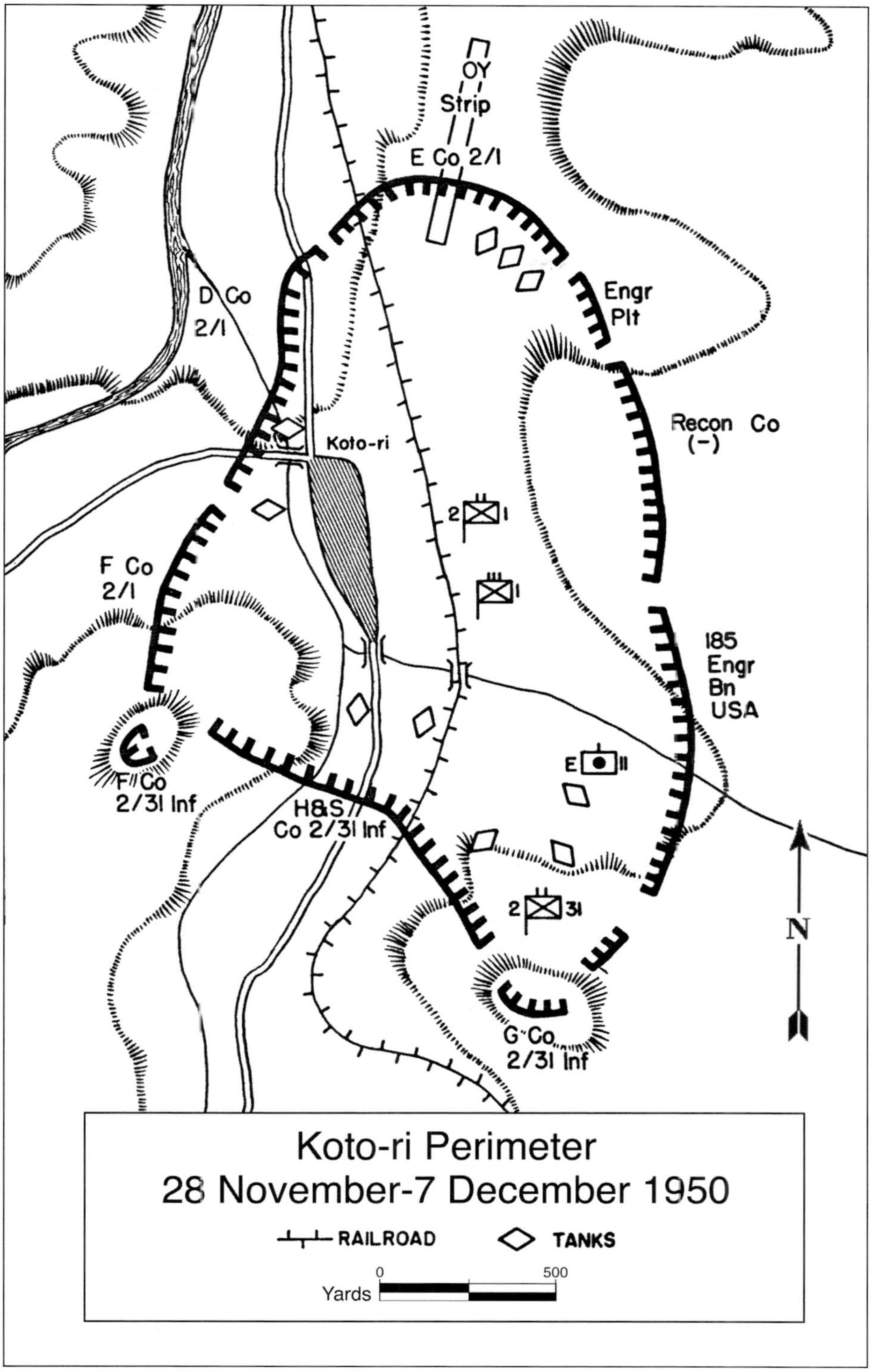

A column of troops was seen coming up the road. "Those are my boys," shouted MacLean and he started on foot across the ice toward them. A crackle of rifle fire was heard. His body was seen to jerk as though hit several times by bullets. He fell on the ice, then got to his feet and staggered on until out of friendly sight. Much later it would be learned that he was taken prisoner, but on the march north died of his wounds. His comrades buried him by the side of the road.

Faith was now the senior surviving officer and the 31st RCT would go into the collective memory of the Korean War as "Task Force Faith," although it would never officially bear that name. The head of Faith's column reached the 3d Battalion's positions by 0900 and by 1300 most elements had closed south of the inlet. Faith formed a new perimeter with the remnants of the two battalions, attempting to incorporate some of the high ground to the south. A helicopter sent in from Hagaru-ri by General Hodes took out the two wounded battalion commanders, Reilly and Embree. Air delivery of ammunition and supplies, called in by Stamford, had mixed results. Much of what was dropped landed outside the new perimeter.

Faith knew nothing of Drake's attempt to reach the inlet with his tanks. Drake tried a second time on 29 November with 11 tanks and a scratch platoon of infantry drawn from the regimental headquarters. After four hours of effort, the tanks fell back once more to Hudong-ni.

Fox's Continued Ordeal

The night of 29-30 November was again relatively quiet at Yudam-ni, but not so at Toktong Pass. At 0200 a voice came out of the dark and in stilted English said: "Fox Company, you are surrounded. I am a lieutenant from the 11th Marines. The Chinese will give you warm clothes and good treatment. Surrender now."

Fox Company threw up some 81mm illumination shells and replied with mortar and machine gun fire. The Chinese were caught in their attack position, perhaps three companies of them. Many died but some got close enough for an exchange of hand grenades. Fox Company, now well dug in, lost only one Marine wounded. At sunrise the protective Corsairs came overhead once again.

Chinhung-ni Action, 26-30 November

Short and feisty Lieutenant Colonel Donald M. "Buck" Schmuck, 35, University of Colorado 1938, had taken over command of the 1st Battalion, 1st Marines, on 8 November. In World War II he had fought as a company commander at Bougainville and Peleliu, and later served at Okinawa. On the night of 26 November the Chinese probed his perimeter at Chinhung-ni at the foot of Funchilin Pass with a series of light attacks. Patrols sent out by Schmuck the next day failed to make contact. That night the Chinese hit his perimeter with another tantalizing, easily repulsed, light attack. Schmuck sent out more patrols during the next two days. What they found or did not find caused him to conclude that a Chinese battalion that attacked him at night and hid in the houses to his west during the day was pestering him.

Photo courtesy of Sgt Norman L. Strickbine, USA

Parachutes laden with ammunition and supplies blossom as they are dropped by an Air Force C-119 of the Far East Air Force's Combat Cargo Command to the entrapped RCT-31 east of the reservoir. Such aerial delivery had mixed results. Much of what was dropped fell outside American lines and into Chinese hands.

A patrol sent out from Captain Wesley Noren's Company B on the 29th more or less confirmed Schmuck's conclusion. Schmuck decided to attack the suspected Chinese position on the following day, using Captain Robert H. Barrow's Company A and a part of Noren's company, reinforced with 81mm and 4.2-inch mortars. Battery F, 11th Marines, under First Lieutenant Howard A. Blancheri, laid down preparatory 105mm howitzer fire, the infantry swept forward, and, in the words of Major William L. Bates, Jr., the battalion's Weapons Company commander, "ran the Chinese right out of the country." The houses that sheltered the Chinese were burned. There was no more trouble at Chinhung-ni.

Task Force Drysdale Formed

On the evening of 28 November three disparate units—41 Commando, Royal Marines; Company G, 3d Battalion, 1st Marines; and Company B, 31st Infantry, 7th Infantry Division—had crowded into the perimeter at Koto-ri after an uneventful motor march up from the south. Puller pasted the three units together into a task force, giving command to Lieutenant Colonel Douglas B. Drysdale of the Royal Marines, with orders to fight his way through to Hagaru-ri the next day.

Drysdale barely had time to uncrate his newly issued American 81mm mortars and Browning machine guns. He moved out at 0945 on 29 November, his truck-borne column followed by a serial of headquarters troops on its way to the new division headquarters at Hagaru-ri. Drysdale's plan was for his Royal Marines to lead off with an assault against the Chinese entrenched on the right of the road just north of Koto-ri. Captain Carl Sitter's George Company—reinforced with a provisional platoon of water-cooled machine guns, rocket launchers, and 81mm mortars—was to follow with an assault against Hill 1236, a mile-and-a-half north of Koto-ri. The soldiers of Baker Company, 31st Infantry, would be in reserve.

The Royal Marines took their objective without much trouble, but Sitter's Marines ran into serious

Photo by PFC C. Wehner, Jr., National Archives Photo (USMC) 127-N-A6593

LtCol Donald M. Schmuck received a Silver Star for his command of 1st Battalion, 1st Marines, at Chinhung-ni from 26 November until 11 December. Making the award sometime in late winter 1951, is Col Francis A. McAlister, who had been the division G-4 but who by this time had succeeded Col Puller in command of the 1st Marines.

resistance before taking Hill 1236. The British and American Marines then moved together about a mile farther up the road where they were stopped by Chinese machine gun and mortar fire coming from Hill 1182. Drysdale received a message from Puller telling him that three platoons of tanks would arrive by 1300 to help. The tanks—two platoons of Pershings from Captain Bruce W. Clarke's Company D, 1st Tank Battalion, and the tank platoon with Shermans from the regiment's Anti-Tank Company—had just arrived at Koto-ri at noon. Drysdale ordered Sitter to break off the action and come back down to the road and await the tanks.

Drysdale found Captain Clarke an "opinionated young man." Drysdale wanted the tanks distributed throughout the length of the column. Clarke insisted that they be kept together at the head of the column to punch a way through. Resignedly, Drysdale resumed his advance at 1350, with 17 tanks leading the way, followed by Sitter's George Company. It was a pulsating advance—short movements followed by pauses while Chinese strong points were reduced with 90mm and machine gun fire. Progress was slow and George Company took heavy losses.

More tanks—Company B, 1st Tank Battalion—arrived at Koto-ri at about 1500. Puller ordered their commander, Captain Bruce F. Williams, to leave one platoon with Sutter's 2d Battalion and to join the rear of the Drysdale column with his remaining two platoons. Meanwhile, Puller had dispatched a platoon from Company E, 2d Battalion, to assist in the evacuation of Drysdale's casualties. The platoon did not get back into the Koto-ri perimeter until

This sketch by Sgt Schofield shows the meeting at Hagaru-ri of two U.S. Marines with two Royal Marines of 41 Independent Commando. The professionalism and sangfroid of the British Marines impressed their American counterparts, who, in turn, impressed the British with their dogged fighting qualities.

Sketch by Sgt Ralph Schofield, USMCR

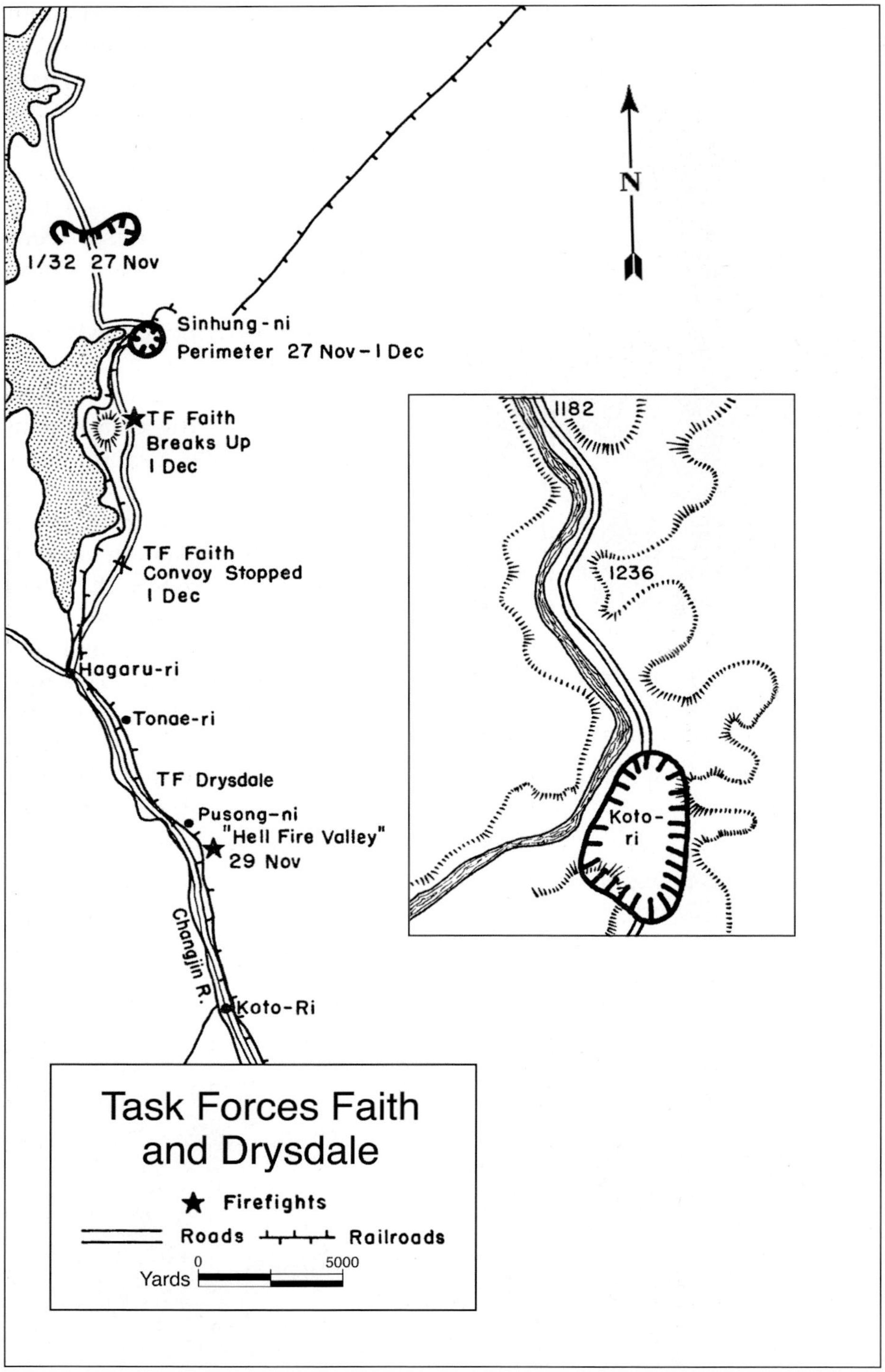

about 1600. Task Force Drysdale came to a halt about four miles north of Koto-ri at about the same time. Shortly thereafter the Chinese began pounding the northern face of the Koto-ri perimeter with mortar fire followed by a company-sized attack that was easily contained by Easy Company.

Clarke and Williams, the tank commanders, advised Drysdale and Sitter that they thought the tanks could get through to Hagaru-ri but were dubious about further movement by trucks. Drysdale put the decision of a further advance up to division. Smith ordered him to continue. The tanks needed to refuel and this took more time. When the column did plunge forward unit integrity was lost and combat troops became intermingled with headquarters elements.

At the midway point to Hagaru-ri there was a valley, about a mile long, high ground on one side and the Changjin River and more hills on the other—Drysdale would name it "Hell Fire Valley." It became the scene of an all-night fight. The column broke in half. George Company, three-quarters of 41 Commando, and a few soldiers, led by tanks from Company D, continued on toward Hagaru-ri. The remainder of 41 Commando; most of Company B, 31st Infantry; and nearly all other headquarters personnel were left on the road which the Chinese closed behind them. The Chinese chopped away at them. The best protection the stalled half of the convoy could find were the shallow ditches on each side of the road. Lieutenant Colonel Arthur A. Chidester, 37, University of Arkansas 1935, the assistant division G-4 and the senior officer in the group, attempted to turn his truncated column around and return to Koto-ri. He was wounded and captured. His place was taken by Major James K. Eagan, soon also wounded and taken prisoner.

The half-column that had been left behind coalesced into one large perimeter and three small ones strung out over a distance of close to a mile. Farthest north, near the hamlet of Pusong-ni, was the largest perimeter, a hodgepodge of about 140 men including Associated Press photographer Frank "Pappy" Noel. Senior officer was Major John N. McLaughlin, 32, Emory University 1941, an assistant division G-3 and a well-decorated veteran who had fought with the 5th Marines at Guadalcanal,

National Archives Photo (USMC) 127-N-A5354

The village of Koto-ri, midway between Hagaru-ri and the Funchilin Pass, was held by the 2d Battalion, 1st Marines, and was the site of Col "Chesty" Puller's regimental command post. Units moving north and south staged through here. Unseen are the fighting holes of the Marine infantry that encircled the camp.

Cape Gloucester, and Peleliu in World War II.

There was some hope that the Company B tanks from Koto-ri would come to the rescue of the four ragged perimeters, but the tanks were stopped by the Chinese at the defile formed by Hills 1236 and 1182, the same hills captured earlier but now reoccupied by the Chinese. The southernmost group on the road worked its way back into the Koto-ri perimeter by 2200 without much trouble. The middle group, mostly headquarters personnel, also made it back by 0230, losing most of its trucks along the way. Its leader, Lieutenant Colonel Harvey S. Walseth, the division G-1, was wounded. (Lieutenant Colonel Bryghte D. Godbold would take his place on the general staff.) By dawn all of the Company B tanks had returned to Koto-ri.

The troops remaining trapped in Hell Fire Valley and still hoping to be rescued by the tanks knew none of this. The Chinese meanwhile seemed more interested in looting the trucks than annihilating the defenders. Major Mc-Laughlin tried sending patrols back to the south to link up with the other perimeters. They were beaten back. He gathered his wounded in a ditch and prayed for daylight and the arrival of Marine Corps aircraft overhead. By 0200 he was out of grenades. A 75mm recoilless rifle, gallantly manned by U.S. soldiers, was knocked out and all its crew killed or wounded. Associated Press photographer Noel and two men attempted to run the gantlet in a jeep and were captured.

The Chinese at about 0430 sent several prisoners into McLaughlin's position bearing a demand that the Americans surrender. McLaughlin and a British Marine went out under a white flag to parley. In a desperate act of bravado McLaughlin pretended that the Chinese wished to surrender to him, but the enemy was neither impressed nor amused. They gave him 10 minutes to capitulate or face an all-out assault. McLaughlin, with only about 40

A gaggle of Marines, looking like giant penguins in their hooded parkas, watch with great interest an air strike against a hill outside Koto-ri on 28 November. The next day these Marines very probably could have been part of Task Force Drysdale, which would depart Koto-ri to fight its way to Hagaru-ri.

Photo by Cpl W. T. Wolfe, Department of Defense Photo (USMC) A4865

able-bodied defenders and almost no ammunition, reluctantly decided to surrender but with the condition that his most serious wounded be evacuated. The Chinese agreed to his terms. The Chinese did not live up to their promise, but they did, however, permit some of the wounded to be placed in houses along the road where they might eventually be found.

While McLaughlin was negotiating his surrender, some few Americans and British Marines and a considerable number of U.S. soldiers managed to slip away from the smaller perimeters to the south. This group, led largely by Major Henry W. "Pop" Seeley, Jr., 33, Amherst College 1939, made its way successfully back to Koto-ri. Seeley had spent four years in the Pacific during World War II and had been well decorated for his service.

Drysdale had continued his start-and-stop progress with Company D's tanks, Company G, and the larger part of 41 Commando, not knowing what had happened to the rear half of his haphazard command. One of the tanks was knocked out by a satchel charge. Drysdale received a grenade fragment in the arm and deferred command of the column, momentarily, to Sitter.

Well after dark the first of Company D's tanks, leading the column, burst through Hagaru-ri's south roadblock, flattening one of Weapons Company's jeeps in the process. Sitter's George Company came into the perimeter, battered but intact. The Royal Marine Commando, in accordance with its training, split into small groups. For most of the night, U.S. Marines on the perimeter were treated to English accents shouting, "Don't shoot, Yanks. We're coming through." Royal Marine troop commander Lieutenant Peter Thomas said later, "I never thought I should be so glad to see an American." At about midnight, Lieutenant Colonel Drysdale, blood dripping down his arm, gave Lieutenant Colonel Ridge a side-winding salute and reported 41 Commando present for duty.

Maj John N. McLaughlin, left, taken prisoner during the disastrous advance of Task Force Drysdale from Koto-ri to Hagaru-ri, is welcomed back from captivity, still in his Chinese cap and coat, on 5 September 1953, by BGen Joseph C. Burger, then the assistant division commander of the 1st Marine Division.

Photo by TSgt Jack E. Ely, Department of Defense Photo (USMC) A1744606.

By best estimates, Task Force Drysdale had begun the day with 922 officers and men. Something like 400 men reached Hagaru-ri, another 300 hundred found their way back to Koto-ri. Killed in action and missing in action were estimated at 162. Another 159 men were identified as wounded. Forty-four Marines, originally listed as missing, were taken prisoner. Of these, just 25 either escaped or survived their captivity. Chidester and Eagan were among those wounded who died in captivity. The column had started with 141 vehicles and 29 tanks. Of these, 75 vehicles and one tank were lost.

George Company Goes Up East Hill

Lieutenant Colonel Ridge's command group had remarkably good intelligence as to the extent of the enemy outside the Hagaru-ri perimeter, the information often brought in by "line-crossers," plainclothes Korean agents who boldly moved in and out of the perimeter. The Chinese *58th Division's* reported intentions to renew its attack against Hagaru-ri seem to have been thwarted by well-placed air attacks during the day and heavy artillery and mortar fires during the night. Ridge's supporting arms coordinator also experimented with night close air support, using converging bands of machine gun tracer fire to point out targets to the Corsair "night hecklers" overhead.

At 0800 on 30 November, the morning after George Company's arrival and a scant night's sleep, Ridge ordered Sitter to pass through Myers' position on East Hill and continue the attack. Drysdale's 41 Commando was held in reserve. This company-sized force of highly trained Royal Marines gave Ridge a small but potent maneuver element, far more promising than the scratch reserve formations he had been forced to use. Drysdale and his officers spent much of the day reconnoitering possible counterattack routes and acquainting themselves with supporting fire plans. "I felt entirely comfortable fighting alongside the Marines," said Drysdale.

Sitter, stoic and unflappable, sent out his 1st and 2d Platoons to pass through Myers' toehold on the hill. They were then to attack on both sides of the ridge. The 3d Platoon and two platoons of Able Company engineers would follow in reserve. Progress was slow and Sitter used his reserve to envelop the Chinese right flank. The attack bogged down and Sitter asked for permission to set up defensive positions on the ground previously held by Myers who had withdrawn his meager force. Corsairs were brought in and worked over the crest of the hill again and again, but George Company could not take the contested ground.

National Archives Photo (USAF) AC40314

Ammunition and supplies for the embattled Marines descend from a C-119 Flying Boxcar of the Far East Air Forces Combat Cargo Command. This was the only photograph the U.S. Air Force photographer was able to take before 23-degree-below-zero weather froze the shutter on his camera.

That same day at Hagaru-ri, 30 November, Colonel Brower, commanding the 11th Marines, came down with a serious liver infection. Command of the regiment passed to his executive officer, Lieutenant Colonel Youngdale.

Disaster Threatens RCT-31

Sung Shi-lun was amazingly well informed as to exactly what his opponents were doing. Chinese reconnaissance was good; and Korean civilians, including line crossers, were at least as useful to the Chinese as they were to the Americans. Moreover, he apparently had a serviceable quantity of signal intelligence from radio intercepts. Stymied by the Marines' stubborn defense at Yudam-ni and Hagaru-ri, he decided to finish off the U.S. Army forces east of the reservoir by adding the weight of the *81st Division* to the *80th Division* already engaged against Task Force Faith.

The curious command relationships at Yudam-ni continued. Without a common commander in place on the ground, the two collocated regiments pursued their separate missions. Smith had issued an order on the afternoon of 29 November directing Murray to assume responsibility for the protection of the Yudam-ni area with his 5th Marines, while Litzenberg was to employ the entire 7th Marines in clearing the MSR to Hagaru-ri "without delay."

Almond Issues New Orders

At noon on 29 November, Almond departed Haneda airfield in Japan on his return flight from his meeting with MacArthur. Enroute to Yonpo he directed his G-3 and other staff members to commence planning the break-off of the offensive and the consolidation of the corps. When Almond arrived at his war room in Hamhung he saw that, in addition to the predicament of the 1st Marine Division and RCT-31 at the Chosin Reservoir, his remaining forces were in considerable disarray. Soule's 3d Infantry Division was headed in two different directions. A CCF column at Sachang far to the southwest of Yudam-ni had already engaged the division's 7th Infantry. The remaining two regiments, the 15th and 65th, were regrouping at Yonghong on the coast preparatory to attacking west, in accordance with orders to relieve pressure on the Eighth Army's dangling flank. In Barr's 7th Infantry Division, MacLean's RCT-31 was already isolated and heavily engaged east of the reservoir. Barr's remaining regiments, the 17th and 32d Infantry, were pulling back to the Pungsan area. By 2100 that evening X Corps Operation Order Number 8, providing for the discontinuance of the attack to the northwest and the withdrawal of forces into the Hamhung-Hungnam perimeter, was ready for Almond's approval.

By that order, Almond placed under Smith's command all Army troops in the Chosin Reservoir area, including Task Force Faith and elements at Hagaru-ri, effective 0800 the next morning. Along with the assignment of these troops came a highly optimistic order from X Corps to Smith to "redeploy one RCT without delay from Yudam-ni area to Hagaru area, gain contact with elements of the 7th Inf Div E of Chosin Reservoir; coordinate all forces in and N of Hagaru in a perimeter defense based on Hagaru; open and secure Hagaru-Koto-ri MSR."

At 0600, 30 November, Litzenberg and Murray issued a joint order for the breakout. (Smith did admit in his log entry for 30 November: "An ADC [assistant division commander; that is, Craig] would have come in handy at this point.") That same morning, Almond gave the senior members of his staff a fuller briefing on the MacArthur decision to go over to the strategic defensive. He made it known that he had also issued orders to the ROK I Corps to pull back. By this time the 3d ROK Division was at Hapsu and the Capital Division above Chongjin. They were now to withdraw to Songjin, a deepwater port about 100 miles northeast of Hungnam.

General Barr, who had established an advance command post at Hungnam, was among those present. After the briefing, Barr—whether at Almond's suggestion or on his own initiative is not clear—flew to Hagaru-ri. There he met with O.P. Smith and Hodes and then borrowed a Marine helicopter to go forward to Faith's position. Smith asked Hodes to draft a message advising Faith that his command was now attached to the 1st Marine Division. Barr at this point was out of the operational chain-of-command to Task Force Faith, but RCT-31 was still, of course, part of the 7th Infantry Division. Barr told Smith that he was recalling Hodes from the Chosin Reservoir area to avoid any misunderstanding as to command arrangements. (Hodes would pay a last visit to Hagaru-ri on 2 December.)

Barr arrived at Faith's command post shortly before noon. He presumably informed Faith of the changed command status, either in substance or by delivering the Hodes dispatch.

On his return to Hagaru-ri, Barr agreed with Smith that Task Force Faith, with Marine and Navy close air support, could extricate itself and get back to Hagaru-ri. Almond arrived at about this time and met with Smith, Barr, and Hodes at Smith's forward command post, a few hundred yards from where Ridge's Marines were contending for possession of East Hill. Almond announced that he had abandoned any idea of consolidating positions in the Chosin Reservoir area. A withdrawal would be made posthaste to Hungnam. Almond authorized Smith to destroy or burn all equipment that would impede his movement. Resupply would be by air. Smith demurred: "I told him that my movements would be governed by my ability to evacuate the wounded, that I would have to fight my way back and could not afford to discard equipment and that, therefore, I intended to bring out the bulk of my equipment."

Almond shrugged. He then directed Smith and Barr to work out a time-phased plan to pull back the three Army battalions of RCT-31 making up "Task Force Faith." Furthermore, if Faith failed to execute his orders, Almond

National Archives Photo (USA) 111-SC353607

Cases of ammunition are sorted out and lashed together at Yonpo airfield on 29 November before loading into C-47 transports of the Far East Air Force's Combat Cargo Command. This ammo was intended for RCT-31, better known as Task Force Faith, east of the reservoir. First airdrops directly from Japan would be to Hagaru-ri on 1 December.

opined that he should be relieved.

Almond later said that in addition to general instructions on withdrawal of forces and a specific plan for the withdrawal of RCT-31, he also asked Smith for an explicit plan for the evacuation of both Army and Marine wounded by way of the airstrip being completed at Hagaru-ri.

Almond told Barr that on his flight up to Hagaru-ri he had passed over a column of trucks halted on the road a few miles south of Koto-ri and recognized it as the 2d Battalion, 31st Infantry, which was working its way up from Chinhung-ni. He suggested to Barr that he relieve the slow-moving battalion commander, West Pointer Lieutenant Colonel Richard F. Reidy. Barr objected, saying he did not know the situation confronting the battalion. He may also have reminded Almond that all movements up to Koto-ri were being coordinated by X Corps. Almond telephoned (by radio link) his chief of staff, the able Major General Clark L. Ruffner, and ordered him to expedite the movement north of the 2d Battalion, 31st Regiment, to join Puller's forces at Koto-ri. Before leaving, Almond told Barr and Smith that Soule's 3d Infantry Division was doing a "magnificent" job of covering the gap with the Eighth Army.

After Almond had departed, Barr and Smith agreed that not much could be done for RCT-31 until the 5th and 7th Marines arrived at Hagaru-ri from Yudam-ni. Smith did order Litzenberg and Murray to expedite their withdrawal, destroying any supplies and equipment that had to be abandoned.

Ruffner sent Captain Joseph L. Gurfein, West Point 1944, to get Reidy, a short, chunky man whose face showed the marks of his boxing days at the Academy, moving. Gurfein found Reidy and his battalion, which had only two rifle companies, stalled about three miles outside Puller's position. (The other rifle company had gone forward earlier, had been with Task Force Drysdale, and was now in place, more or less intact, at Koto-ri.) Reidy's battalion, urged on by Gurfein, made a faltering night attack and eventually pushed its way into Koto-ri. Puller gave the battalion a sector of Koto-ri's defensive perimeter. He also ordered Reidy (who was suffering from a badly infected foot) to take charge of the sizable number of soldiers from various units—including a detachment of the 185th Engineer Battalion—that had now collected at Koto-ri.

With Task Force Faith

During the daylight hours of 30 November, Don Faith worked out counterattack plans to meet a penetration of his perimeter. The Chinese, not waiting for nightfall, began their attack in the afternoon. Task Force Faith's perimeter at Sinhung-ni was now isolated and alone with no friendly forces between it and Hagaru-ri. By midnight the attack against Faith's perimeter had built up to unprecedented intensity. There were penetrations, but Faith sealed these off with local counterattacks. At the aid station, medical supplies were completely exhausted. The dead, frozen stiff, were laid out in rows stacked about four feet high.

Meanwhile, well to Faith's rear, headquarters elements of RCT-31 and the 31st Tank Company at Hudong-ni, with 1st Marine Division approval, had fallen back to Hagaru-ri. Two disabled tanks had to be abandoned along the four-mile route, but otherwise the march, about 325 soldiers altogether, was made without incident. The regimental S-3, Lieutenant Colonel Berry K. Anderson, the senior Army officer present, was in charge. A new 31st Infantry headquarters was being formed at Hamhung, with Colonel John A. Gavin, USA, as its designated commander, but it was not sent forward to Hagaru-ri.

Intimations of an Evacuation

By the end of November it was increasingly obvious that Rear

Admiral James Doyle, who had landed the Marines at Inchon and again at Wonsan, was now going to have to lift them out of Hungnam as part of a massive amphibious withdrawal. Doyle, as Comman-der, Task Force 90, issued plans on 28 November for a redeployment of United Nations forces. Doyle's plans called for the division of his Task Force 90 into two amphibious task groups. Task Group Three, under Rear Admiral Lyman A. Thackrey, would provide for amphibious evacuation on the west coast of Eighth Army units if required. Task Group One, under Doyle's immediate command, would execute the amphibious evacuation of east coast ports, primarily Hungnam. Task Group Three, with two-thirds of the amphibious force, would go to the west coast where the situation, at that moment, seemed more critical. There would not be nearly enough amphibious ships for these tasks; there had to be an enormous gathering of merchant shipping. Vice Admiral Joy's deputy chief of staff, newly promoted Rear Admiral Arleigh Burke, largely ran this effort.

The carrier-based aircraft brought together for the Inchon landing and then the Wonsan landing had by mid-November been largely dispersed. Left with Task Force 77 were the fast carriers *Leyte* (CV 32) and *Philippine Sea* (CV 47). Also still on station was the escort carrier *Badoeng Strait* (CVE 116)—"Bing-Ding" to the Marines and sailors—but her sister ship *Sicily* (CVE 118), having dropped off VMF-214, the "Blacksheep" squadron, at Wonsan, was in port in Japan. Major Robert P. Keller had commanded the squadron until 20 November when he was detached to become the Marine air liaison officer with Eighth Army and Fifth Air Force. In World War, Keller served in the Pacific with Marine Fighter Squadrons 212 and 223 and was credited with at least one aerial victory. Command of VMF-214 was taken over by Major William M. Lundin.

The big carrier *Valley Forge* (CV 45), which on 3 July had been the first carrier to launch combat missions against North Korean invaders, was on her way home for a much-needed refit. Now the emergency caused her to turn about and head for the Sea of Japan. Also on the way was the *Princeton* (CV 37), hurriedly yanked out of mothballs. But until these carriers could arrive, tactical air operations, including all important close air support, would have to be carried out by shore-based Marine squadrons and Navy and Marine squadrons in the *Leyte, Philippine Sea,* and *Badoeng Strait*.

On 1 December the Far East Air Forces relinquished control of all tactical air support of X Corps to the 1st Marine Aircraft Wing, which, in the words of air historian Richard P. Hallion, "performed brilliantly." The *Princeton* arrived on station on 5 December. By then Task Force 77 was giving the Chosin exodus its full attention.

Carrier-based Vought F4U Corsairs and Douglas AD Skyraiders were the workhorses of Task Force 77. A typical ordnance load for the Corsair on a close air support mission was 800 rounds for its 20mm guns, eight 5-inch rockets, and two 150-gallon napalm bombs. With this load the plane had an endurance of two-and-one-half hours. The Navy's Skyraider, much admired by the Marines, packed an ordnance load comparable to a World War II Boeing B-17 heavy bomber, commonly 400 rounds of 20mm, three 150-gallon napalm bombs, and either twelve 5-inch rockets or twelve 250-pound fragmentation bombs. The Sky-raider could stay in the air for four hours with this load.

Bad Night at Hagaru-ri

The Marines at Hagaru-ri had another bad night on 30 November. The Chinese, the *58th Division* now augmented by the *59th*, assaulted Ridge's weary defenders once again in an attack pattern that repeated that of the night of the 28th. One regiment or more came in against the southwest face of the perimeter and unfortunately for them hit Item Company's well-entrenched position. First Lieutenant Joseph Fisher, always optimistic in his counting of enemy casualties, guessed that he killed as many as 500 to 750 of them. His own losses were two Marines killed, 10 wounded.

Another Chinese regiment came across the contested ground on East Hill hitting the reverse slope defenses held chiefly by Sitter's Company G and 1st Engineer Battalion's Companies A and B. General Smith watched the fight from the doorway of his command post, two-thirds of a mile away. Some ground was lost. Ridge sent up a portion of his precious reserve, 41 Commando, to reinforce Company G, and the lost ground was retaken by early morning.

After the night's action, Ridge, the defense force commander, came to see Smith at about 0900. "He was pretty low and almost incoherent," Smith wrote in his log for 1 December. "The main trouble was loss of sleep. He was much concerned about another attack. He felt with the force available to him he could not hold both the airstrip and the ridge [East Hill] east of the bridge. I told him he would have to hold both and would have to do it with what we had."

Casualty Evacuation and Resupply

Friday, 1 December—although no one thought of it that way at the time—was the turning point of the campaign.

Lieutenant Colonel John Partridge's 1st Engineer Battalion had succeeded in hacking out the semblance of an airstrip from the frozen earth in 12 days of around-the-clock dangerous work, the engineers at times laying down their tools to take up rifles and machine guns. Heroic though the engineering effort was, the airstrip was only 40 percent complete. Its rough runway, 50-feet wide and 2,900-feet long, fell considerably short of the length and condition specified by regulations for operation of transport aircraft at those altitudes and temperatures. Smith decided that the urgency of the evacuation problem was such that the uncompleted airfield must be used, ready or not.

Its impossible load of casualties was overwhelming the division field hospital—a collection of tents and Korean houses. Navy Captain Eugene R. Hering, the division surgeon, met with General Smith that morning. Two additional surgical teams had been flown in by helicopter from Hungnam. The two companies, Charlie and Easy, of the 1st Medical Battalion, already had some 600 patients. Hering expected 500 more casualties from Yudam-ni and 400 from the Army battalions east of the reservoir. Grim as his prediction of casualties was, these estimates would prove to be much too low.

Until the airstrip was operational, aerial evacuation of the most serious cases had been limited to those that could be flown out by the nine helicopters and 10 light aircraft of Major Gottschalk's VMO-6, which also had many other missions to perform. From 27 November to 1 December, VMO-6, struggling to fly at the cold, thin altitudes, had lifted out 152 casualties—109 from Yudam-ni, 36 from Hagaru-ri, and 7 from Koto-ri. One of Gottschalk's pilots, First Lieutenant Robert A. Longstaff, was killed on an evacuation flight to Toktong Pass.

At 1430 on Friday afternoon, the first Air Force C-47 transport touched down on the frozen snow-covered runway. A half-hour later the plane, loaded with 24 casualties, bumped its way off the rough strip into the air. Three more planes came in that afternoon, taking out about 60 more casualties. "It takes about a half hour to load a plane with litter patients," Smith noted in his log. "Ambulatory patients go very much faster." The last plane in for the day, arriving heavily loaded with ammunition, collapsed its landing gear and had to be destroyed.

Because of Smith's foresight,

LtCol Murray, exact location and time unknown, but possibly mid-November at Hagaru-ri, briefs Maj Vincent J. Gottschalk, commanding officer of VMO-6, on the locations of units of his regiment. Between 27 November and 1 December, VMO-6's helicopters flew 152 medical evacuation missions in addition to many other reconnaissance and liaison missions.

National Archives Photo (USMC) 127-N-A130916

Photo by Sgt William R. Keating, Department of Defense Photo (USMC) A5683

Casualty evacuation from Hagaru-ri airstrip began on 1 December, although the condition of the field fell far short of what safety regulations required. Here, on a subsequent day, an ambulance discharges its cargo directly into an Air Force C-47 or Marine Corps R-4D transport. Weapons of other, walking wounded, casualties form a pile in the foreground.

the X Corps' deputy chief of staff, visited Smith. McCaffrey, who had been Almond's chief of staff in the 92d Infantry Division in Italy, outlined for Smith the plan for constriction into a Hungnam perimeter and its subsequent de-fense. Soule's 3d Infantry Division was to move elements to the foot of Funchilin Pass and provide a covering force through which the 1st Marine Division would withdraw. The 1st Marine Division would then organize a defensive sector west and southwest of Hungam. The 7th Infantry Division would occupy a sector northeast and north of Hungnam.

The consolidation of X Corps in the Hamhung-Hungnam area included the evacuation of Wonsan to the south. Major General Field Harris, commanding the 1st Marine Aircraft Wing, ordered MAG-12—the Marine aircraft group had three tactical squadrons and a headquarters

Hagaru-ri was already stockpiled with six days of rations and two days of ammunition. The first airdrop from Air Force C-119 "Flying Box Cars" flying from Japan was on that same critical 1st of December. The drops, called "Baldwins," delivered prearranged quantities of ammunition, rations, and medical supplies. Some drops were by parachute, some by free fall. The Combat Cargo Command of the Far East Air Forces at first estimated that it could deliver only 70 tons of supply a day, enough perhaps for a regimental combat team, but not a division. By what became a steady stream of transports landing on the strip and air drops elsewhere the Air Force drove its deliveries up to a 100 tons a day.

Toward the end of the day Lieutenant Colonel William J. McCaffrey, West Point 1939 and

Mountains of supplies were rigged at Yonpo Airfield by Capt Hersel D. C. Blasingame's 1st Air Delivery Platoon for parachute drop, free airdrop, or simply as cargo for the transports flying into Hagaru-ri. The 1st Marine Division needed 100 tons of resupply a day, delivered by one means or another.

Photo by SSgt Ed Barnum, Department of Defense Photo (USMC) A130436

squadron—to move up from Wonsan to Yonpo. The group commander, Colonel Boeker C. Batterton, 46, Naval Academy 1928, had spent most of World War II with a naval aviation mission in Peru. MAG-12 completed the movement of its aircraft in one day—that same busy 1 December. Some planes took off from Wonsan, flew a mission, and landed at Yonpo.

East of Chosin

Total strength of RCT-31 has been calculated at a precise 3,155, but of this number probably not more than 2,500 fell under Don Faith's direct command in "Task Force Faith" itself. On the morning of 1 December, Lieutenant Colonel Faith, on his own initiative, began his breakout from Sinhung-ni to the south. He did not have a solid radio link to the 1st Marine Division, and had nothing more than a chancy relay through Marine Captain Edward P. Stamford's tactical air control net. Faith's own 1st Battalion, 32d Infantry, would lead off, followed by the 57th Field Artillery, with the 3d Battalion, 31st Infantry, bringing up the rear. Trucks would be unloaded so as to carry the wounded. The howitzers were spiked. Most jeeps and all inoperable trucks were to be destroyed, as would be all supplies and equipment. In execution the destruction of the surplus was spotty. About 25 to 30 vehicles—all that were still in operating condition—formed up into column. The overriding mission for Task Force Faith was now to protect the truck convoy with its hundreds of wounded. Much reliance would be placed on the automatic weapons fire of the tracked weapons carriers, down to three in number, two quad .50s and one dual 40mm.

The column began to move out at about 1100. The soldiers could see the Chinese in plain sight on the surrounding high ground. Progress was slow. Mortar rounds continued to fall, causing more casualties, and Chinese infantry began pressing in on the column. The fighting for the first half-mile was particularly intense. Officers and noncommissioned officers suffered disproportionate losses. Control broke down.

Captain Stamford and his tactical air control party tried to keep close air support overhead as continuously as possible. Navy Corsairs from the fast carrier *Leyte* came on station at about 1300. With Stamford calling them in, the Corsairs used napalm and rockets and strafed with 20mm cannon. One napalm drop hit close to Faith's command group causing eight or ten casualties. This ghastly accident was demoralizing, but survivors agreed that without close air support the column would never have cleared the perimeter.

Some of the soldiers on the point began to fall back. Faith drew his Colt .45 pistol and turned them around. Panicky KATUSAs—and some Americans—tried to climb into the trucks with the wounded. Riflemen assigned to move along the high ground on the flanks started to drift back to the road. The head of the column reached a blown bridge just north of Hill 1221 at about 1500. Some of the trucks, trying to cross the frozen stream, broke through the ice and had to be abandoned.

The Chinese held the high ground on both sides of a roadblock that now stood in the way and were in particular strength on Hill 1221. Faith, .45 in hand, gathered together enough men to reduce the roadblock. Other small groups of men clawed their way crossways along the slope of Hill 1221. This fight was almost the last gasp of Task Force Faith. In the words of one major, "[After Hill 1221] there was no organization left."

Even so, by dusk the column was within four-and-one-half road miles of Hagaru-ri when a grenade fragment that penetrated his chest just above his heart killed Faith himself. His men propped up his body, with a blanket wrapped around his shoulders, in the cab of a truck—rather like a dead El Cid riding out to his last battle—hoping that word of his death would not spread through the column causing more demoralization. Just what happened to his body after that is not clear.

As the column struggled on southward the Chinese methodically continued their destruction of the convoy, truck by truck. Individual soldiers and small groups began to break away from the column to attempt to cross the frozen reservoir on foot. Task Force Faith, as such, had ceased to exist.

At Hagaru-ri

During the fighting at Hagaru-ri from 28 November through 1 December, Ridge's 3d Battalion, 1st Marines, had suffered 43 killed, 2 missing, and 270 wounded—a total of 315 battle casualties and a third of its beginning strength. The bits and pieces of Marine and Army units that made up the rest of the Hagaru-ri defense force had casualties perhaps this high if not higher. These casualties, however, did not come even close to those suffered by RCT-31 east of the reservoir.

Throughout the night of 1 December survivors of Task Force Faith drifted into the north side of the perimeter at Hagaru-ri, most of them coming across the ice.

Photo courtesy of Sgt Norman L. Strickbine, USA

Moving toward the perceived safety of Hagaru-ri, soldiers of Task Force Faith march in a well dispersed but terribly exposed single column across the snow-covered, frozen surface of Pungnyuri Inlet. The march across the ice would continue throughout the night of 1 December.

During that last night all semblance of unit integrity dissolved. At about 0230 on the morning of 2 December, Marine Captain Stamford appeared in front of Captain Read's artillery battery position. Stamford had been briefly taken prisoner but escaped. By mid-morning, 670 soldier survivors, many of them wounded or badly frostbitten, had found their way into Hagaru-ri warming tents. They had a terrible tale to tell.

Dr. Hering, the division surgeon, reported to Smith that 919 casualties went out on 1 December, but that among them there was a large number of malingerers. "Unfortunately," Smith entered in his log, "there are a good many Army men, not casualties who got on planes. . . . Men got on stretchers, pulled a blanket over themselves and did a little groaning, posing as casualties. . . .Tomorrow we will get this situation under control and will have MPs at the planes. No man will be able to board a plane without a [medically issued] ticket." Smith, who had ordered Army Lieutenant Colonel Berry K. Anderson to organize physically fit soldier survivors into a provisional battalion, now "talked" to Anderson again and told him to get his soldiers under control.

X Corps had set up a clearing station at Yonpo. Triage determined those casualties who would recover in 30 days or less. They went to the 1st Marine Division Hospital in Hungnam, the Army's 121st Evacuation Hospital in Hamhung, or the hospital ship *Consolation* in Hungnam harbor. Casualties expected to require more than 30 days hospitalization were flown on to Japan.

Lieutenant Colonel Olin L. Beall, 52, a quintessential salty old mustang and a great favorite of General Smith, commanded the 1st Motor Transport Battalion, which held a position on the northeast quadrant of the Hagaru-ri perimeter. Smith, not flamboyant himself, liked colorful leaders such as Beall and Puller. Murray, a very different style of leader, said of Beall: "We all agreed that he would have had to lived a thousand years to have done all the things he claimed to have done. But when people began checking up on some of the things that he had said he had done, by God, he had done them."

Beall had enlisted in the Marine Corps on 5 April 1917, the day before war was declared on

Marine wounded await evacuation at Yudam-ni. A total of 109 went out by helicopter. The rest would go out by ambulance or truck once the 5th and 7th Marines broke their way through to Hagaru-ri. Fixed-wing aerial evacuation from Hagaru-ri began on 1 December.

Photo by Sgt Frank C. Kerr, Department of Defense Photo (USMC) A4858

Photo by Cpl Alex Klein, National Archives Photo (USA) 111-SC353559

PFC Ralph C. Stephens, wounded at Koto-ri, receives whole blood at the 1st Marine Division hospital in Hamhung on 1 December. Administering the blood are HM 2/C Emmett E. Houston and Lt(jg) Vernon L. Summers. Until the airstrips opened at Hagaru-ri and Koto-ri, evacuation of the wounded was virtually impossible.

Germany. He served in Cuba and Haiti, gained a temporary commission, and went to France just as the war ended. He reenlisted in 1920 and served as an officer—as did many other Marine noncommissioned officers—in the *Gendarmerie d'Haiti* chasing bandits. In 1935 he reached the much-respected warrant grade of Marine gunner. By the end of World War II he was a major and a veteran of Okinawa.

On Saturday, 2 December 1950, Beall led a rescue column of jeeps, trucks, and sleds across the ice looking for other Task Force Faith survivors. Marine Corsairs covered his efforts, flying so low that he said, "I could have scratched a match against their bellies." He brought in 319 soldiers, many of them in a state of shock. The Chinese did little to interfere except for long-range rifle fire.

There is no agreement on exact Army casualty figures. Perhaps 1,050 survivors reached Hagaru-ri. Of these only 385 were found to be physically and mentally fit for combat. These soldiers were given Marine weapons and equipment. Not a single vehicle, artillery piece, mortar, or machine gun of Task Force Faith had been saved. When Almond visited Smith that same Saturday, he had, in Smith's words, "very little to say about the tactical situation. He is no longer urging me to destroy equipment."

Coming Out of Yudam-ni

All day long on Thursday, 30 November, at Yudam-ni the Chinese harried the perimeter with long-range small arms fire and minor probing attacks. As a step in the regroupment of their battered regiments, Litzenberg and Murray organized a provisional battalion made up rather strangely of the combined Companies D and E, 7th Marines; and sections of 81mm mortars from the weapons companies of the 2d and 3d Battalions, 7th Marines; and Companies A and G, 5th Marines. Dog-Easy Company, under First Lieutenant Robert T. Bey, was really no more than two under strength platoons—a Dog Company platoon and an Easy Company platoon Litzenberg gave overall command of this odd assortment to Major Maurice Roach, former commander of the 3d Battalion and now the regimental S-3.

A good part of the reason for forming the battalion was to free Lieutenant Colonel Davis of responsibility for Dog-Easy Company and other attachments. The battalion, which was given its own sector in the perimeter, was assigned the radio sign "Damnation" and that became the short-lived battalion's title. Someone tore up a green parachute to make a neckerchief, the practice caught on, and a green neckerchief became the battalion's badge.

The most difficult task in the disengagement probably fell to Roise's 2d Battalion, 5th Marines, which held a long line stretching from Hill 1426 to 1282. Covered by air and artillery, Roise fell back about a mile from Hill 1426 to Hill 1294. This and other movements freed up Harris' 3d Battalion, 7th Marines, to move to a position astride the MSR about 4,000 yards south of Yudam-ni.

Litzenberg and Murray issued their second joint operation order on the morning of 1 December. Essentially it provided that the 7th Marines would move overland and the 5th Marines would move along the axis of the MSR. Both regiments put what were widely regarded as their best battalions out in front. Davis' 1st Battalion, 7th Marines, would take to the hills; Taplett's 3d Battalion, 5th Marines, down to half-strength, would lead the way down the road. They were to converge in the general vicinity of Fox Company on Toktong Pass. Point for the advance along the road would be

Photo by Cpl L. B. Snyder, National Archives Photo (USMC) 127-N-A5667

Maj Maurice E. Roach, the 7th Marines jack-of-all-trades, and for the moment commander of a provisional battalion, pauses for a photo with the regimental commander, Col Litzenberg, before leaving Yudam-ni on 1 December. Roach is wearing the Navy parka that was part of the winter uniform, but Litzenberg finds an Army field jacket sufficient for the daytime cold.

the single Marine tank that had reached Yudam-ni. Staff Sergeant Russell A. Munsell and another crewman were flown up by helicopter from Hagaru-ri to drive it.

Major William McReynolds' 4th Battalion, 11th Marines, was in general support of both regiments but under the command of neither. The arrangement was made to work with Lieutenant Colonel Harvey A. Feehan, commander of the 1st Battalion, 11th Marines, acting as coordinator of supporting artillery fire. McReynolds' battalion was ordered to shoot up most of its 155mm ammunition before pulling out. Excess gunners were then organized into nine rifle platoons. The guns themselves were to bring up the rear of the convoy.

Roach's "Damnation Battalion" was to have followed Davis across country, but Litzenberg reconsidered and broke up the battalion on 1 December returning its parts to their parent organizations except for Dog-Easy Company, which only had about 100 effectives. Litzenberg passed the orphan company to Murray who, in turn, passed it to Taplett, his advance guard commander.

All available Marine aircraft were to be in the air to cover the withdrawal. They were to be joined by carrier aircraft from Task Force 77. On the ground only the drivers and the critically wounded would move by vehicle; the rest would walk. It was decided to leave the dead at Yudam-ni and a field burial was held for 85 Marines.

The grand parade began at 0800 on 1 December. Taplett's 3d Battalion, 5th Marines, came down from its positions north of Yudam-ni, followed an hour-and-a-half later by Stevens' 1st Battalion. Company B of Stevens' battalion under First Lieutenant John R. Hancock made up the rear guard coming out of the town. Meanwhile the 3d Battalion, 7th Marines, under Harris proceeded to clear both sides of the road leading south, Company H going up Hill 1419 east of the road while the rest of the battalion went against Hill 1542 on the west side. Roise's 2d Battalion, 5th Marines, having cleared the town, relieved Davis' 1st Battalion, 7th Marines, on Hill 1276, freeing up Davis to pursue his overland mission. Harris was slow in taking Hill 1542. By mid-afternoon, an impatient Taplett was in position behind Harris, ready to attack south astride the road even with his right flank somewhat exposed.

Company H, 7th Marines, commanded by First Lieutenant

Photo by Sgt Frank C. Kerr, National Archives Photo (USMC) 127-N-A4862

Marines depart Yudam-ni on 1 December. The general plan for the breakout was that the 5th Marines would follow the axis of the MSR and the 7th Marines would move overland. MajGen Smith did not formally designate an overall commander, but LtCol Murray deferred to Col Litzenberg because of seniority.

The 7th Marines command group breaks camp at Yudam-ni on 1 December in preparation for the march back to Hagaru-ri. The jeep in the foreground carries a radio. Note the long antenna. Below zero temperatures caused problems with battery life. Extra five-gallon cans of gasoline are lashed to the front bumper.

Department of Defense Photo (USMC) A5666

Howard H. Harris, met trouble on Hill 1419. Harris' battalion commander, Lieutenant Colonel Harris, was fully occupied with problems on Hill 1542. Litzenberg, realizing that Hill 1419 was too far from Hill 1542 for a mutually supporting attack, detached How Company from the 3d Battalion and assigned it to Davis' 1st Battalion. Davis, Hill 1419 now his responsibility, sent his Able Company to add its weight to How Company's effort.

(Continued on page 292)

OVERLEAF: *"Band of Brothers" by Colonel Charles H. Waterhouse, USMCR (Ret), widely regarded by many Marine veterans of the Korean War as Waterhouse's masterpiece, shows the column of Marines winding its way down Funchilin Pass.* Courtesy of Col Waterhouse and the Chosin Few Association.

National Archives Photo (USMC) 127-N-A5670

The heaviest guns in the defense of Yudam-ni were the 155mm howitzers of the 4th Battalion, 11th Marines, commanded by Maj William McReynolds. Shown here, the big guns are preparing to leave the perimeter, towed by their tractor prime movers. Excess gunners were reorganized into provisional rifle units.

Meanwhile, Lieutenant Harris had been wounded and Second Lieutenant Minard P. Newton, Jr., had taken over the company. Able Company, under Captain David W. Banks, passed through How Company and took the hill at about 1930.

Davis now stripped his battalion down for its cross-country trek. Everything needed for the march would have to be hand carried. He decided to go very light, taking only two 81mm mortars and six heavy machine guns (with double crews) as supporting weapons from his Weapons Company. His vehicles, left behind with his sick, walking wounded, and frostbite cases as drivers, were to join the regimental train on the road. Baker Company, now commanded by First Lieutenant Joseph R. Kurcaba, led off the line of march, followed by Davis and his command group, then Able Company, Charlie Company, battalion headquarters, and How Company, still attached.

It was a very dark night. The guide stars soon disappeared. The snow-covered rock masses all looked alike. The point had to break trail, through snow knee-deep in places. The path, once beaten, became icy and treacherous. Marines stumbled and fell. Radios would not work reliably. Davis, moving ahead and floundering in the snow, lost touch with the forward elements of his battalion for a time. He continued forward until he reached the point. His map, hurriedly read by a flashlight held under a poncho, told him that they were climbing Hill 1520, the slopes of which were held by the Chinese. Baker and Charlie Companies converged on the Chinese who were about a platoon in strength, taking them by surprise. Davis stopped on the eastern slope of Hill 1520 to reorganize. Enemy resistance had slackened to small arms fire from ridges across the valley but Davis' men were numb with cold and exhausted. At 0300 he again halted his advance to give his Marines a rest, sending out small patrols for security. Now, for the first time, he gained radio contact with regiment.

On the MSR with the 5th Marines

Taplett, meanwhile, was marching southward astride the MSR, led

Before leaving Yudam-ni, as a matter of professional pride, the 5th and 7th Marines "policed up the area." Trash, including remnants of rations, was collected into piles and burned. Slim pickings were left for Chinese scavengers who, desperate for food, clothing, and shelter, followed close behind the departing Marines.

Photo by Sgt Frank C. Kerr, Department of Defense Photo (USMC) A4848

National Archives Photo (USMC) 127-N-A5666

In coming back from Yudam-ni, the 1st Marine Division's large number of road-bound wheeled vehicles was both an advantage and a handicap. They carried the wherewithal to live and fight; they also slowed the march and were a temptation for attack by the Chinese. Here an 11th Marines howitzer can be seen in firing position to cover the column's rear.

by the solitary Pershing tank, followed by a platoon from his How Company and a platoon of ever-useful engineers. His radio call sign, "Darkhorse," suited his own dark visage. He advanced for about a mile before being halted by heavy fire coming from both sides of the road. He fanned out How and Item Companies and they cleared the opposition by 1930.

Taplett gave his battalion a brief rest and then resumed the advance. Item Company, led by Captain Harold Schrier, ran into stiff resistance on the reverse slope of still-troublesome Hill 1520 east of the road. Schrier received permission to fall back to his jump-off position so as to better protect the MSR. The Chinese hit with mortars and an infantry attack. Schrier was wounded for a second time and Second Lieutenant Willard S. Peterson took over the company. Taplett moved George Company and his attached engineers into defensive positions behind Item Company. It was an all-night fight. In the morning, 2 December, 342 enemy dead were counted in front of Item Company. Peterson had only 20 Marines still on their feet when George Company passed through his position to continue the attack against Hill 1520. George and How Company were both down to two-platoon strengths.

As a reserve Taplett had Dog-Easy Company, 7th Marines, detached from the now-dissolved "Damnation Battalion." Dog-Easy Company moved onto the road between How and George Companies. By noon George Company, commanded by Captain Chester R. Hermanson, had taken Hill 1520 and Dog-Easy had run into its own fight on the road. Second Lieutenant Edward H. Seeburger, lone surviving officer of Dog Company, was severely wounded while giving a fire command to the solitary tank. He refused evacuation. (Seeburger faced long hospitalization and after a year was physically retired as a first lieutenant. In 1995 he received a belated Navy Cross.)

Corsairs reduced the roadblock that held up Dog-Easy Company. "Darkhorse" trudged on, How and George Companies on both sides of the MSR and Dog-Easy moving down the middle, followed by the engineers and the solitary tank.

After leaving Yudam-ni, a unit of the 7th Marines, possibly a company of LtCol Davis' 1st Battalion, leaves the road to climb into the hills. Davis' objective was to come down on Toktong Pass from higher ground so as to relieve Capt Barber's embattled Fox Company.

Photo by Sgt Frank C. Kerr, National Archives Photo (USMC) 127-N-A4849

With Davis' Battalion in the Hills

East of the MSR at daybreak, Davis reoriented the direction of his march. The 1st Battalion, 7th Marines, passed over the east slope of Hill 1520 and attacked toward Hill 1653, a mountain a mile-and-a-half north of Toktong Pass. Davis' radios could not reach Barber on Fox Hill nor could he talk directly to the Corsairs circulating overhead. Fortunately opposition was light except for Chinese nibbling against the rear of his column where Company H, 3d Battalion, was bringing up the wounded on litters. Davis converged on Hill 1653 with his three organic rifle companies.

At last, radio contact was made with Captain Barber on Fox Hill. Barber jauntily offered to send out a patrol to guide Davis into his position. Davis declined the offer but did welcome the control of VMF-312's Corsairs by Barber's forward air controller. Just before noon, lead elements of Company B reached Barber's beleaguered position.

Company A halted on the north side of Hill 1653 to provide manpower to evacuate casualties. Twenty-two wounded had to be carried by litter to safety. The regimental surgeon, Navy Lieutenant Peter E. Arioli, was killed by a Chinese sniper's bullet while supervising the task. Two Marines, who had cracked mentally and who were restrained in improvised strait jackets, died of exposure before they could be evacuated. Marines of Kurcaba's Company B celebrated their arrival on Fox Hill with a noontime meal of air-dropped rations. They then went on to take the high ground that dominated the loop in the road where the MSR passed through Toktong Pass. First Lieutenant Eugenous M. Hovatter's Company A followed them and the two companies set up a perimeter for the night. Meanwhile the balance of Davis' battalion had joined Barber on Fox Hill. Barber's Company F had suffered 118 casualties—26 killed, 3 missing, and 89 wounded—almost exactly half of his original complement of 240. Six of the seven officers, including Barber himself, were among the wounded.

5th Marines on the Road

At the rear of the column on the MSR, Lieutenant Colonel Roise's 2d Battalion, 5th Marines, the designated rear guard, had troubles of its own on Hill 1276 during the early morning hours of 2 December. Captain Uel D. Peters' Company F was hit hard. Night fighters from VMF(N)-542 came on station and were vectored to the target by white phosphorus rounds delivered by Company F's 60mm mortars. Strafing and rockets from the night fighters dampened the Chinese attack, but the fight continued on into mid-morning with Fox Company trying to regain lost ground. By then it was time for Roise to give up his position on Hill 1276 and continue the march south.

Lieutenant Colonel Jack Stevens' 1st Battalion, 5th Marines, had its fight that night of 1-2 December east of the road, being hit by a Chinese force that apparently had crossed the ice of the reservoir. Stevens guessed the number of Chinese killed at 200, at least 50 of them cut down in front of Charlie Company by machine guns.

Lieutenant Colonel William Harris' 3d Battalion, 7th Marines, meanwhile was continuing to have trouble on Hill 1542. Litzenberg reinforced Harris with a composite unit, called "Jig Company," made up of about 100 cannoneers, headquarters troops, and other individuals. Command of this assortment was given to First Lieutenant Alfred I. Thomas. Chinese records captured later indicated that they thought they had killed 100 Americans in this action; actual Marine losses were something between 30 and 40 killed and wounded.

Yudam-ni to Hagaru-ri, 2-3 December

At the head of the column, Taplett's Darkhorse battalion on the morning of 2 December had to fight for nearly every foot of the way. George Company still had Hill 1520 to cross. Dog-Easy Company was moving along the road itself. South of Hill 1520 at a sharp bend in the road a bridge over a ravine had been blown, and the Chinese covering the break stopped Dog-Easy Company with machine gun fire. Twelve Corsairs came overhead and ripped into the ravine with strafing fire and rockets. Dog-Easy Company, helped by How Company, re-sumed its advance. The attached engineer platoon, now commanded by Technical Sergeant Edwin L. Knox, patched up the bridge so vehicles could pass. The engineers had started out with 48 men; they were now down to 17. Taplett continued his advance through the night until by 0200, 3 December, he was only 1,000 yards short of Fox Hill. Taplett could only guess where Davis might be with the 1st Battalion, 7th Marines.

To the rear, the Chinese pecked away at the Marines withdrawing from Hills 1276 and 1542. Stevens' 1st Battalion, 5th Marines, continued to provide close-in flank protection. Marine air held off much of the harassment, but the column of vehicles on the road moved slowly and the jeep and truck drivers became targets for Chinese snipers.

Photo by Sgt Frank C. Kerr, National Archives Photo (USMC) 127-N-A4840

Dead Marines at a Yudam-ni aid station wait mutely on their stretchers for loading onto the truck that will take them to eventual burial. Once on the march, practice was to carry those who died in the savage fighting in the hills down to the road where they could be picked up by the regiments' Graves Registration detachments.

That night the Chinese got through to Lieutenant Colonel Feehan's 1st Battalion, 11th Marines, and the artillerymen had to repulse them with howitzer fire over open sights.

Six inches of new snow fell during the night. In the morning, 3 December, Taplett combined the remnants of Dog-Easy Company with George Company and returned the command to First Lieutenant Charles D. Mize, who had had George Company until 17 November. From up on Fox Hill, Davis made a converging attack against the Chinese still holding a spur blocking the way to Hagaru-ri. He pushed the Chinese into the guns of Taplett's battalion. An estimated battalion of Chinese was slaughtered. By 1300, Davis' "Ridgerunners" had joined up with Taplett's "Darkhorses."

Davis and Taplett conferred. The senior Davis now took the lead on the MSR with his battalion. The lone tank still provided the point. The truck column reached Toktong Pass. The critically wounded were loaded onto the already over-burdened vehicles. Less severely wounded would have to walk. Stevens' 1st Battalion, 5th Marines, followed Davis' battalion, passing through Taplett's battalion. Taplett stayed in Toktong Pass until after midnight. Coming up from the rear on the MSR was Roise's 2d Battalion, 5th Marines, followed by Harris' 3d Battalion, 7th Marines, now the rear guard.

Sergeant Robert B. Gault, leader of the 7th Marines Graves Registration Section, came out of Yudam-ni in the column on the MSR with his five-man section and a truck with which to pick up Marine dead encountered along the way. As he remembered it a few months later: "That was the time when there was no outfit, you was with nobody, you was a Marine, you were fighting with everybody. There was no more 5th or 7th; you were just one outfit, just fighting to get the hell out of there, if you could."

Column Reaches Hagaru-ri

The six fighter-bomber squadrons of Field Harris' 1st Marine Aircraft Wing flew 145 sorties on Sunday, 3 December, most of them in close support of the 5th and 7th Marines. Under this aerial umbrella, Davis' 1st Battalion, 7th Marines, marched along almost unimpeded. In the early evening, Ridge sent out Drysdale with 41 Commando, supported by tanks from Drake's 31st Tank Company, to open the door to the Hagaru-ri perimeter. At about 1900, a few hundred yards out, Davis formed up his battalion into a route column and they marched into the perimeter, singing *The Marines' Hymn*. Hagaru's defenders greeted the marchers with a tumultuous welcome. A field mess offered an unending supply of hot cakes, syrup, and coffee. Litzenberg's 7th Marines command group arrived shortly after Davis' battalion and was welcomed into the motor transport area by Litzenberg's old friend, Olin Beall.

In Tokyo that Sunday, MacArthur sent a message to the Joint Chiefs of Staff that X Corps was being withdrawn to Hungnam as rapidly as possible. He stated that there was no possibility of uniting it with Eighth Army in a line across the peninsula. Such a line, he said, would have to be 150 miles long and held alone by the seven American divisions, the combat effectiveness of the South Korean army now being negligible.

Sketch by Sgt Ralph Schofield, USMCR

The 5th and 7th Marines on arrival at Hagaru-ri combat base found hot chow waiting for them. The mess tents, operating on a 24-hour basis, provided an almost unvarying but inexhaustible menu of hot cakes, syrup, and coffee. After a few days rest and reorganization, the march to the south resumed.

The Chinese made no serious objection to the last leg of the march from Yudam-ni to Hagaru-ri until about 0200 on Monday morning, 4 December, when the prime movers hauling eight of McReynolds' 155mm howitzers at the rear of the column ran out of diesel fuel. That halted the column and brought on a Chinese attack. Taplett's battalion, unaware of the break, continued to advance. The artillerymen—assisted by bits and pieces of the 1st and 2d Battalions, 5th Marines, who were on the high ground to the flanks—defended themselves until Taplett could face around and come to their rescue.

It was a bad scene. The eight heavy howitzers had been pushed off the road, perhaps prematurely, and would have to be destroyed the next day by air strikes. A half-mile farther down the MSR was a cache of air-delivered diesel fuel that would have fueled the prime movers. By 0830 the road was again open. Chinese losses were guessed at 150 dead.

At 1400 on Monday, the rear guard, still provided by the 3d Battalion, 7th Marines, marched into Hagaru-ri and the four-day, 14-mile, breakout from Yudam-ni was over. The Marines had brought in about 1,500 casualties, some 1,000 of them caused by the Chinese, the rest by the cold. Smith observed in his log: "The men of the regiments are. . . pretty well beaten down. We made room for them in tents where they could get warm. Also they were given hot chow. However, in view of their condition, the day after tomorrow [6 December] appears to be the earliest date we can start out for Koto-ri."

Reorganization at Hagaru-ri

Ridge's Marine defenders of Hagaru-ri breathed much more easily after the arrival in their perimeter of the 5th and 7th RCTs. A sanguine corporal opined to his company commander: "Now that the 5th and 7th Marines are here, we can be resupplied by air, hold until spring, and then attack again to the north."

General Almond flew into Hagaru-ri on Monday afternoon to be briefed on the breakout plan and while there pinned Army Distinguished Service Crosses on the parkas of Smith, Litzenberg, Murray, and Beall. Almond then flew to Koto-ri where he decorated Puller and Reidy (who had been slow in getting his battalion to Koto-ri) with Distinguished Service Crosses. Nine others, including

Gurfein, who had nudged Reidy into moving, received Silver Stars. Reidy was relieved of his command not much later.

For the breakout, Murray's RCT-5, with Ridge's 3d Battalion, 1st Marines, and 41 Commando attached, would briefly take over the defense of Hagaru-ri while Litzenberg's RCT-7, beginning at first light on Wednesday, 6 December, would march to the south. Puller's RCT-1 would continue to hold Koto-ri and Chinhung-ni. All personnel except drivers, radio operators, and casualties were to move on foot. Specially detailed Marines were to provide close-in security to the road-bound vehicles. Any that broke down were to be pushed to the side of the road and destroyed. Troops were to carry two-days of C rations and one unit of fire, which translated for most into full cartridge belts and an extra bandoleer of ammunition for their M-1 rifles. Another unit of fire was to be carried on organic vehicles. The vehicles were divided into two division trains. Lieutenant Colonel Banks, commanding officer of the 1st Service Battalion, was put in command of Train No. 1, subordinate to RCT-7. Train No. 2, subordinate to RCT-5, was given to Lieutenant Colonel Harry T. Milne, the commander of the 1st Tank Battalion. Although Smith had stated that he would come out with all his supplies and equipment, more realistically a destruction plan, decreeing the disposal of any excess supplies and equipment, was put into effect on 4 December. Bonfires were built. Ironically, loose rounds and canned foods in the fires exploded, causing some casualties to Marines who crowded close to the fires for warmth.

Air Force and Marine transports had flown out over 900 casualties on Saturday, 2 December, from Hagaru-ri, and more than 700 the next day. To the south that Sunday, 47 casualties were taken out by light aircraft from the strip at Koto-ri. But casualties kept piling up and by the morning of Tuesday, 5 December, some 1,400 casualties—Army and Marine—still remained at Hagaru-ri. In a magnificent effort, they were all flown out that day. Altogether, in the first five days of December, by best count, 4,312 men—3,150 Marines, 1,137 soldiers, and 25 Royal Marines—were air-evacuated.

Even a four-engine Navy R5D ventured a landing. Takeoff with a load of wounded in an R5D was so hairy that it was not tried again. An R4D—the Marine equivalent of the sturdy C-47—wiped out its landing gear in landing. An Air Force C-47 lost power on take-off and crashed-landed outside the Marine lines. Marines rushed to the rescue.

Members of this patrol, moving along the abandoned narrow gauge railroad track that paralleled the main supply route, help along a wounded or exhausted comrade while the point and rear riflemen provide watchful cover. No wounded Marine need worry about being left behind.

Sketch by Sgt Ralph Schofield, USMCR

National Archives Photo (USMC) 127-N-A4909

Aerial evacuation of wounded and severely frostbitten Marines and soldiers from Hagaru-ri saved many lives. During the first five days of December, 4,312 men—3,150 Marines, 1,137 soldiers, and 25 Royal Marines—were air evacuated by Air Force and Marine transports that also brought in supplies and replacements.

The plane had to be abandoned and destroyed, but there were no personnel casualties during the entire evacuation process.

During those same first five days of December, 537 replacements, the majority of them recovering wounded from hospitals in Japan, arrived by air at Hagaru-ri. Most rejoined their original units. A platoon sergeant in Weapons Company, 3d Battalion, 1st Marines, wounded in the fighting in Seoul, assured his company commander that he was glad to be back.

Visitors who could wangle spaces on board the incoming transports began arriving at Hagaru-ri. Marguerite "Maggie" Higgins of the *New York Herald-Tribune*, well known to the Marines from both the Pusan Perimeter and the Inchon-Seoul campaigns, was among the gaggle of war correspondents that arrived on Tuesday, 5 December, including former Marine combat correspondent Keyes Beach. Higgins an-nounced her intention to march out with the Marines. General Smith disabused her of her intention and ordered that she be out of the perimeter by air by nightfall.

A British reporter made the impolite error of referring to the withdrawal as a "retreat." Smith patiently corrected him, pointing out that when surrounded there was no retreat, only an attack in a new direction. The press improved Smith's remark into: "Retreat, hell, we're just attacking in a new direction." The new television technology was demonstrated by scenes taken of the aerial evacuation of the casualties and an interview with General Smith and Lieutenant Colonel Murray.

Major General William H. Tunner, USAF, commander of the Combat Cargo Combat and greatly admired by the Marines because of the sterling performance of his command, was one of the visitors. Tunner had flown the Hump from Burma into China during World War II and later commanded much of the Berlin Airlift. He solicitously offered to evacuate the rest of the troops now in Hagaru-ri. Smith stiffly told him that no man who was able-bodied would be evacuated. "He seemed somewhat surprised," wrote Smith.

Almond met with Major General Soule, commander of the 3d Infantry Division, that Tuesday, 5 December, and ordered him to form a task force under a general officer "to prepare the route of withdrawal [of the 1st Marine Division] if obstructed by explosives or whatnot, especially at the bridge site." The site in question lay in Funchilin Pass. Almond apparently did not know that the bridge had already been destroyed. The downed span threatened to block the Marines' withdrawal. Soule gave command of what was designated as "Task Force Dog" to his assistant division commander, Brigadier General Armistead D. "Red" Mead, a hard-driving West Pointer who had been G-3 of the Ninth Army in the European Theater in World War II.

Hagaru-ri to Koto-ri, 6 December

At noon on Tuesday, 5 December, Murray relieved Ridge of his responsibility as Hagaru-ri defense commander, and the battalions of the 5th Marines plumped up the thin lines held by the 3d Battalion, 1st Marines. The Chinese did not choose to test the strengthened defenses, but at about 2000 that evening an Air Force B-26 mistakenly dropped a stick of six 500-pound bombs close to Ridge's command tent. His forward air controller could not talk to the Air Force pilot because of crystal differences in their radios, but an obliging Marine night-fighter from Lieutenant Colonel Max J. Volcansek, Jr.'s VMF(N)-542 came overhead and promised to shoot down any Air Force bomber that might return to repeat the outrage.

The well-liked Max Volcansek, 36, born in Minnesota, had come

Photo by Cpl Arthur Curtis, National Archives Photo (USA) 111-SC354492

After her adventures at Chosin, including her aborted efforts to march out with the Marines, Marguerite "Maggie" Higgins, wearing a Navy parka and shoe-pacs, arrived safely at Haneda Air Force Base, Tokyo, on 15 December. Her broadly smiling traveling companion is Army MajGen William F. Marquat of MacArthur's staff.

into the Marine Corps as an aviation cadet in 1936 after graduating from Macalester College. During World War II he had flown Corsairs while commanding VMF-222 in the Pacific and had scored at least one Japanese plane. During the battle for Seoul he had been wounded and shot down but quickly recovered and continued in command of VMF(N)-542.

The plan of attack for Wednesday, 6 December, called for the 5th Marines to clean up East Hill while the 7th Marines moved south along the MSR toward Koto-ri. Close air support for the attack against East Hill was to be on station at 0700. With a touch of condescension, Murray's Marines told Ridge's Marines to stand back and watch for a demonstration of how a hill should be taken.

Smith wanted to march out with his men, but Shepherd ordered him to fly to Koto-ri. Death or wounding of Smith, or worse, his capture by the Chinese, could not be risked. By this time the lurking presence of seven CCF divisions had been identified by prisoner of war interrogations—the *58th, 59th, 60th, 76th, 79th, 80th,* and *89th*. Two more divisions—the *77th* and *78th*—were reported in the area but not yet confirmed.

Later it would be learned that the *26th CCF Army*—consisting of the *76th, 77th,* and *78th Divisions*, reinforced by the *88th Division* from the *30th CCF Army,* had moved down from the north into positions on the east side of the MSR between Hagaru-ri and Koto-ri. They had relieved the *60th Division,* which had moved into positions south of Koto-ri. Elements of the *60th Division* were preparing for the defense of Funchilin Pass including positions on the dominant terrain feature, Hill 1081. Even farther south the *89th Division* was positioning itself to move against the defenders of Chinhung-ni.

Murray has given a characteristically laconic account of the attack by the 5th Marines against East Hill:

> I had been ordered to take a little hill, and I had Hal Roise do that job. When he got over there, he found about 200 Chinese in a mass, and he captured the whole crowd of them. So we had about 200 prisoners we had to take care of. . . . A lot of them were in such bad shape that we left them there, left some medical supplies, and left them there for the Chinese to come along and take care of them after we left.

It was not quite that simple. Heavy air, artillery, and mortar preparation began at 0700 on Thursday, 6 December. Captain Samuel S. Smith's Dog Company jumped off in the assault at 0900, beginning a fight that would go on until daylight the next morning. All

Photo by Sgt Frank C. Kerr, National Archives Photo (USMC) 127-N-A5520

In a ceremony at Masan on 21 December, LtCol Harold S. Roise, the stalwart commander of 2d Battalion, 5th Marines, receives a Silver Star for actions incident to the seizure of Kimpo Airfield after the Inchon landing. Further awards for Roise for heroism at Chosin would come later.

three rifle companies of Roise's 2d Battalion and Charlie Company of the 1st Battalion were drawn into it. Estimates of enemy killed ran as high as 800 to 1,000. East Hill was never completely taken, but the Chinese were pushed back far enough to prevent them from interfering with the exit of the division from Hagaru-ri.

RCT-7 Attacks South

General Smith planned to close his command post at Hagaru-ri on Wednesday morning, 6 December. Before he could leave General Barr, commander of the 7th Infantry Division, who arrived to check on the status of his soldiers, visited him. The survivors of Task Force Faith coupled with units that had been at Hagaru-ri and Hudong-ni added up to a provisional battalion of 490 able-bodied men under command of Army Lieutenant Colonel Anderson. As organized by Anderson, the "battalion" actually was two very small battalions (3d Battalion, 31st Infantry, under Major Carl Witte and 1st Battalion, 32d Infantry, under Major Robert E. Jones) each with three very small rifle companies. Smith attached Anderson's force to RCT-7 and it was sometimes called "31/7."

Litzenberg had about 2,200 men—about half his original strength—for the breakout to Koto-ri. His attack order put Lockwood's 2d Battalion, with tanks, on the MSR as the advance guard; Davis' 1st Battalion on the right of Changjin River and the MSR; Anderson's provisional Army battalion on the left of the road; and Harris' 3d Battalion on the road as the rear guard.

Lockwood, it will be recalled, had stayed at Hagaru-ri with his command group and much of his Weapons Company while Companies D and E went forward to Yudam-ni and Company F held Toktong Pass. At 0630, tanks from Company D, 1st Tank Battalion, led Lockwood's reunited, but pitifully shrunken battalion out of the perimeter through the south roadblock. Almost immediately it ran into trouble from Chinese on the left side of the road. The morning fog burned off and air was called in. A showy air attack was delivered against the tent camp south of the perimeter, abandoned days earlier by the Army engineers and now periodically infested with Chinese seeking warmth and supplies. Lockwood's two rifle companies—Fox Company and Dog-Easy Company—pushed through and the advance resumed at noon. Meanwhile, barely a mile out of Hagaru-ri, Captain John F. Morris' Company C, 1st Battalion, surprised an enemy platoon on the high ground to the southeast of the hamlet of Tonae-ri and killed most of them.

At 1400 Smith received a reassuring message from Litzenberg that the march south was going well. Smith decided that it was time to move his command post to Koto-ri. His aide, Major Martin J. "Stormy" Sexton, World War II raider, asked the commander of Weapons Company, 3d Battalion, 1st Marines, for the loan of a jeep to take his boss to the airstrip. A 10-minute helicopter ride took Smith and Sexton to Koto-ri where Puller was waiting. Smith began planning for the next step in the withdrawal.

Photo by Sgt Frank C. Kerr, Department of Defense Photo (USMC) A5464

Col Litzenberg's 7th Marines led off the march south from Hagaru-ri on 6 December. Here one of his units pauses at the roadblock held by Weapons Company, 3d Battalion, 1st Marines, before exiting the town to watch a drop of napalm by a Marine Corsair against a camp abandoned by Army engineers, now infested with Chinese.

Meanwhile Lockwood's 2d Battalion, 7th Marines, had run into more serious trouble another mile down the road. Davis' 1st Battalion, up in the hills, could see the enemy; Lockwood's battalion, on the road itself, could not. Fox Company, with some help from Dog-Easy Company and the Army mini-battalions under Anderson, pushed through at about 1500. Davis' battalion continued to play company-sized hopscotch from hilltop to hilltop on the right of the road. By dark, lead elements of RCT-7 were about three miles south of Hagaru-ri. Enemy resistance stiffened and air reconnaissance spoke of Chinese columns coming in from the east, but Litzenberg decided to push on. After two more miles of advance, Lockwood's battalion was stopped in what Drysdale had called Hell Fire Valley by what seemed to be a solitary Chinese machine gun firing from the left. An Army tank solved that problem. Another half-mile down the road a blown bridge halted the column. The engineers did their job, the march resumed, but then there was another blown bridge. At dawn on Thursday things got better. Air came overhead, and 2d Battalion, 7th Marines, had no more trouble as it marched the last few miles into Koto-ri. Through all of this Lockwood, sick with severe bronchitis, had sat numbly in his jeep. Early that morning his executive officer, Major Sawyer, had been wounded in the leg by a mortar

Coming out of Hagaru-ri, Col Litzenberg used LtCol Lockwood's 2d Battalion, 7th Marines, reinforced with tanks, as his advance guard. Here a heavy machine gun squad rests by the side of the road while a M-26 Pershing medium tank trundles by. The M-26 mounted a powerful flat-trajectory 90mm gun.

Photo by Sgt Frank C. Kerr, National Archives Photo (USMC) 127-N-A5469

Photo by Sgt Frank C. Kerr, Department of Defense Photo (USMC) A5428

Marching, encumbered with weapons, packs, and winter clothing, on the road from Hagaru-ri to Koto-ri was heavy work, even without Chinese interference. This unit of the 7th Marines, taking advantage of a halt, manages a quick nap. Most of the fighting was on the high ground on both sides of the MSR, but occasionally the Chinese reached the road.

fragment and was out of action. Major James F. Lawrence, Jr., 32, University of North Carolina 1941, the battalion S-3, had become the de facto commander.

Things were going even less well on the left flank and rear of the column. The Army provisional battalion, fragile to begin with, had fought itself out and was replaced by Harris' 3d Battalion, 7th Marines. By 2100 the Chinese had come down to within hand-grenade range of the trucks on the road. Harris deployed his George and Item Companies to push them back. Sometime before dawn, Lieutenant Colonel William Harris, son of Major General Field Harris, disappeared. He was last seen walking down the road with two rifles slung over his shoulder. A search for him found no body and it was presumed he had been taken prisoner. Major Warren Morris, the executive officer of the 1st Battalion, took over command of the 3d Battalion and it reached Koto-ri at about 0700 on Thursday morning.

Chinese prisoners taken along the road from Hagaru-ri to Koto-ri were identified as being from the *76th* and *77th Divisions* of *26th CCF Army*.

For Almond, most of Wednesday, 6 December, was absorbed with a visit by General J. Lawton Collins, the Army chief of staff. Collins and Almond dropped in at the command posts of the Army's 7th and 3d Infantry Divisions, but "weather precluded flying to Koto-ri" for a visit with Smith. Collins left at nightfall for Tokyo. The visit had gone well and Almond noted contentedly in his diary: "Gen. Collins seemed completely satisfied with the operation of X Corps and apparently was much relieved in finding the situation well in hand."

At Koto-ri, 7 December

First Lieutenant Leo R. Ryan, the adjutant of the 2d Battalion, 7th Marines, alarmed by Lieutenant Colonel Lockwood's apathy, pressed the battalion surgeon and assistant surgeon, Lieutenants (jg)

The role of the rifle companies in the breakout from Hagaru-ri on 6 December was to take the high ground on both sides of the road. Much fought over "East Hill" dominated the exit from Hagaru-ri. It was never completely taken, but the Chinese were pushed back far enough to permit the relatively safe passage of the division trains of vehicles.

Photo by Sgt Frank C. Kerr, Department of Defense Photo (USMC) A5465

Photo by Sgt Frank C. Kerr, Department of Defense Photo (USMC) A5466

Soldiers, readily recognizable as such in their short parkas, march in single file on 6 December along the MSR south of Hagaru-ri. RCT-31, badly mauled east of the reservoir and reduced in combat effectives to a small provisional battalion commanded by LtCol Berry K. Anderson, USA, was attached to the 7th Marines for the breakout.

Laverne F. Peiffer and Stanley I. Wolf, to examine him. Neither doctor was a psychiatrist, but they came to the conclusion that Lockwood was suffering from a neurosis that made him unfit for command. This was communicated to Colonel Litzenberg who confirmed Major Lawrence as the acting commander.

In mid-morning, Thursday, 7 December, to ease the passage of the division train, both the 2d and 3d Battalions, 7th Marines, were ordered to face about, move north again, and set up blocking positions on both sides of the road between Koto-ri and Hill 1182. On the way the 2d Battalion picked up 22 Royal Marine survivors who had been laagered up in a Korean house ever since Task Force Drysdale had passed that way. A VMO-6 pilot had spotted them three days earlier by the letters "H-E-L-P" stamped in the snow and had dropped rations and medical supplies.

Elsewhere on Thursday, 7 December, X Corps and Eighth Army had received orders from GHQ Tokyo to plan to withdraw, in successive positions if necessary, to the Pusan area. Eighth Army was to hold on to the Inchon-Seoul area as long as possible. X Corps was to withdraw through Hungnam and eventually to pass to the command of Eighth Army.

Almond visited Smith at Koto-ri and assured him that Soule's 3d Infantry Division would provide maximum protection from Chinhung-ni on into Hamhung. Smith

LtCol William F. Harris, commanding officer of the 3d Battalion, 7th Marines, and son of MajGen Field Harris, disappeared on the march from Hagaru-ri to Koto-ri. He was last seen moving down the road with two rifles slung over his shoulder. His exact fate remains a mystery.

National Archives Photo (USMC) 127-N-A45353

National Archives Photo (USA) 111-SC354254

General J. Lawton Collins, the Army chief of staff, left, visited X Corps on 6 December and warmly praised MajGen Almond, right, for his conduct of the battle. Collins got to the command posts of the 3d and 7th Infantry Divisions, but bad weather kept him from seeing MajGen Smith and the command post of the 1st Marine Division at Koto-ri.

was concerned over the coordination of artillery fire by the 3d Division. Almond promised that it would be under Marine control. He spoke briefly with Puller and Litzenberg and noted that night in his diary, "Morale is high in the Marine Division."

All elements of RCT-7 had closed into Koto-ri by 1700 on the evening of 7 December; but Division Train No. 1, which they were to have shepherded, did not get out of Hagaru-ri until 1600 on the 6th. A little more than a mile out of Hagaru-ri the Chinese came down on the column. They might have thought the train would be easy pickings; if so, they were wrong. They hit Major Francis "Fox" Parry's 3d Battalion, 11th Marines. The artillerymen, fighting as infantry, held them off. Another mile down the road and the process was repeated. This time the gunners got to use their howitzers, firing at pointblank range, and happily, if optimistically, guessed that they had killed or wounded all but about 50 of the estimated 500 to 800 attackers.

As the night wore on there was more fighting along the road. The division headquarters had a stiff scuffle sometime after midnight. The members of the division band were given the opportunity to demonstrate their skills as machine gunners. The Military Police Company was bringing out a bag of 160 able-bodied prisoners of war. The prisoners got caught between Chinese and American fires and most were killed. Night hecklers from David Wolfe's VMF(N)-513 helped and at dawn the omnipresent Corsairs from Frank Cole's VMF-312 came on station and resolved the situation. The column moved through the stark debris—there were still bodies lying about and many broken vehicles—of Hell Fire Valley and by 1000 on 7 December Division Train No. 1, after an all-night march, was in Koto-ri.

Eleven miles away in Hagaru, Division Train No. 2, unable to move onto the road until Division Train No. 1 had cleared, did not get started until well after dark on 6 December. At midnight, the head of the train was still barely out of the town. Lieutenant Colonel Milne, the train commander, asked for infantry help. Taplett's 3d Battalion, 5th Marines, was detailed to the job. Taplett moved forward with two companies. Nothing much happened until dawn on 7 December when the column was able to continue on under air cover.

In Hagaru-ri engineers and ordnance men were busy blowing up everything that could be blown up and burning the rest. Stevens' 1st Battalion, 5th Marines; Ridge's 3d Battalion, 1st Marines; and Drysdale's 41 Commando stood poised to leave but could not get out of town until Thursday morning, 7 December, after some fighting in Hagaru-ri itself, because of the clogged roads. Roise's 2d Battalion, 5th Marines, came off East Hill and fell in behind them as rear guard at about 1000. The Chinese once again seemed more interested in looting what was left of the town than in further fighting. After some light interference on the road, all elements of RCT-5 were safely tucked into the Koto-ri perimeter before midnight on the 7th.

A number of units—including Roise's 2d Battalion, 5th Marines; Ridge's 3d Battalion, 1st Marines; and Drake's 31st Tank Company—assert that they provided the rear point coming out of Hagaru-ri. Able Company engineers, however, busy with last-minute demolitions in the already burning town, probably have the best claim. In round figures, 10,000 Marines and soldiers, shepherding 1,000 vehicles, had marched 11 miles in 38 hours. Marine losses were 103 dead, 7 missing, and 506 wounded.

Marine engineers, arguably the greatest heroes of the campaign, had widened and improved the airstrip at Koto-ri so that it could handle World War II TBMs, no longer used as torpedo bombers, but now stripped-down utility aircraft that could bring in a few passengers—as many as nine—and lift out a corresponding number of wounded. The TBMs, plus the light aircraft and helicopters from VMO-6, took out about 200 casualties on 7 December and 225 more on the 8th. Most of the TBMs were piloted, not by squadron pilots, but by otherwise desk-bound aviators on the wing and group staffs.

National Archives Photo (USMC) 127-N-A5457

Chinese prisoners within the Hagaru-ri perimeter were herded into a stockade guarded by a detachment of Capt John H. Griffin's Military Police Company. When the Marines evacuated the town they left the wounded prisoners of war behind in a compound, telling them that their comrades would soon be down from the hills to help them.

March South from Koto-ri

There would be no rest at Koto-ri. By somebody's count 14,229 men had piled into Koto-ri, including the long-waited Army's 2d Battalion, 31st Infantry, which had arrived far too late to go forward to join its regiment, the shredded RCT-31, east of the reservoir. Reidy's battalion was to continue as part of Puller's RCT-1 in the break out.

Anderson's two-battalion collection of soldiers, quite separate from Reidy's battalion, had suffered additional casualties—both battle and from the cold—coming in from Hagaru-ri. Major Witte, one of the battalion commanders, was among the wounded. Anderson reorganized his shrinking command into two companies: a 31st Company under Captain George A. Rasula, a canny Finnish-American from Minnesota who knew what cold weather was all about, and a 32d Company under Captain Robert J. Kitz, who had been a company commander in Reilly's 3d Battalion, 31st Infantry, in Task Force Faith. Anderson then stepped aside from immediate command, giving the battalion to Major Robert E. Jones who had been Don Faith's S-1 and adjutant. As a paratrooper in World War II, Jones had jumped with the 101st Airborne Division near Eindhoven, Holland. Now, coming out of Koto-ri, his improvised battalion remained part of Litzenberg's RCT-7.

The Marines left Hagaru-ri in flames, wanting to leave no shelter for the Chinese. Veterans still argue as to which unit was the last to leave the town. Marines from Company A, 1st Engineer Battalion, charged with last minute demolitions, probably have the best claim to this honor.

National Archives Photo (USMC) 127-N-A5458

Photo by David Douglas Duncan

As the Marines moved south from Hagaru-ri to Koto-ri they had to pass through "Hell Fire Valley," site of Task Force Drysdale's heaviest losses. The Chinese had made no attempt to salvage the abandoned vehicles, but the Marines did and learned that some needed nothing more than a push to start them.

The march south was to be resumed at first light on Friday, 8 December. It would be a "skin-the-cat" maneuver with the rifle companies leap-frogging along the high ground on each side of the road while the heavily laden vehicles of the division trains made their way toward Funchilin Pass and then down the pass to Chinhung-ni. At the foot of the pass the Marines could expect to find elements of the Army's 3d Infantry Division manning the outer defenses of the Hamhung-Hungnam area. But the road was not yet open. Smith had been warned, as early as 4 December, that the Chinese had blown a critical bridge halfway down the pass. Here water came out from the Changjin Reservoir through a tunnel into four giant pipes called "penstocks." The bridge had crossed over the penstocks at a point where the road clung to an almost sheer cliff. If the division was to get out its tanks, artillery, and vehicles the 24-foot gap would somehow have to be bridged.

Lieutenant Colonel Partridge, the division engineer, had made an aerial reconnaissance on 6 December and determined that the gap could be spanned by four sections of an M-2 steel "Treadway" bridge. He had no such bridge sections, but fortuitously there was a detachment of the Treadway Bridge Company from the Army's 58th Engineer Battalion at Koto-ri with two Brockway trucks that could carry the bridge sections if they could be air-delivered. A section was test-dropped at Yonpo by an Air Force C-119 and got smashed up in the process. Not discouraged, Partridge pressed for an airdrop of eight sections—to give himself a 100 percent insurance factor that at least four sections would land in useable condition. The 2,500-pound bridge sections began their parachute drop at 0930 on 7 December. One fell into the hands of the Chinese. Another was banged up beyond use. But six sections landed intact. Plywood center sections for wheeled traffic were also dropped. Next, the Brockway trucks would have to deliver the sections to the bridge site three-and-a-half miles away, a location likely to be defended fiercely by the Chinese.

Partridge met with Litzenberg and it was decided that the Brockway trucks would move at the front of the 7th Marines' regimental train after RCT-7 jumped-off at 0800 on 8 December. The bridge site was dominated by Hill 1081 so Lieutenant Colonel Schmuck's 1st Battalion, 1st Marines, at Chinhung-ni was

Photo by Sgt William R. Keating, Department of Defense Photo (USMC) A546[illegible]

One of the vehicles salvaged by the Marines in "Hell Fire Valley" was a bullet-ridden Army ambulance. Battered, but still in running condition, the Marines used it to evacuate casualties incurred along the line of march. Once safely in Masan a large number of Army vehicles were returned to the Army.

ordered to advance overland three miles to the north to take the hill. All of this required exquisite timing.

First objective for Litzenberg's 7th Marines coming out of Koto-ri was the high ground on the right of the road for a distance of about a mile-and-a-half. Murray's 5th Marines would then pass through the 7th Marines and take and hold the high ground for the next mile. Puller's 1st Marines was to stay in Koto-ri until the division and regimental trains had cleared and then was to relieve the 5th and 7th Marines on their high ground positions so the trains could pass on to Funchilin Pass. The 5th and 7th Marines, relieved by the 1st, would then move on down the pass toward Hamhung. The 11th Marines artillery would displace from battery firing position to battery firing position but for much of the time would be limbered up and on the road. Heavy reliance for fire support would be placed on the Corsairs and organic mortars. Tanks would follow at the end of the vehicular column so there would be no chance of a crippled tank blocking the road.

Task Force Dog, under Brigadier General Mead and consisting of the 3d Battalion, 7th Infantry, liberally reinforced with tanks and artillery, had started north on 7 December, passed through Su-dong, and by late afternoon had reached Chinhung-ni. Schmuck's 1st Battalion, 1st Marines, after being relieved by Task Force Dog, moved into an assembly area several miles north of Chinhung-ni.

The jump-offs from both Koto-ri and Chinhung-ni on the morning of 8 December were made in a swirling snowstorm. Schmuck's Marines started the six-mile march up the MSR to the line of departure at 0200. His plan was for Captain Robert P. Wray's Company C to take Hill 891, the southwestern nose of Hill 1081, and hold it while his other two rifle companies passed through and continued the attack. Captain Barrow's Company A was to attack east of the road and on up to the summit of Hill 1081. Captain Noren's Company B

Koto-ri as it looked on 8 December, the day that the march to the south continued. Virtually all the combat strength of the 1st Marine Division, plus some Army troops, had concentrated there for the breakout. The next critical terrain feature would be Funchilin Pass where a blown bridge threatened to halt the march. Its repair posed a problem for Marine engineers.

Photo by Sgt William R. Keating, Department of Defense (USMC) A5354

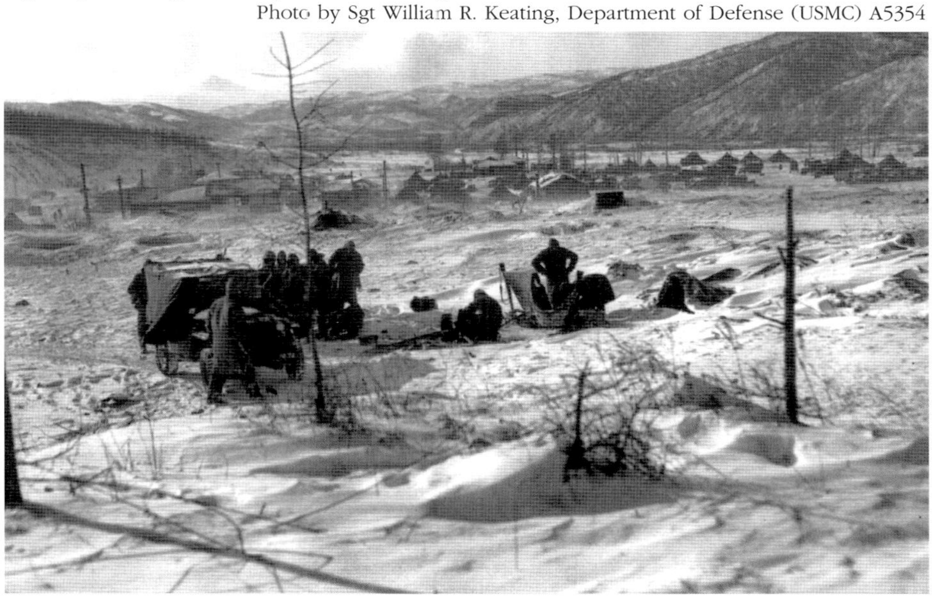

Photo by Sgt William R. Keating, Department of Defense Photo (USMC) A5361

Snow-dusted M-4A3 Sherman tanks of LtCol Harry T. Milne's 1st Tank Battalion await the word at Koto-ri to move out to the south. The tanks had turned in a disappointing performance with Task Force Drysdale. There would be further problems with the tanks in Funchilin Pass.

Immediately outside of Koto-ri, two Chinese soldiers willingly surrender to members of a Marine rifle company early on 9 December. Each leg of the withdrawal, from Yudam-ni to Hagaru-ri, from Hagaru-ri to Koto-ri, and from Koto-ri to Chinhung-ni, showed a marked improvement in Marine tactics to deal with the situation.

Photo by Cpl Peter W. McDonald, National Archives Photo (USMC) 127-N-A5388

Photo by SSgt Ed Barnum, Department of Defense Photo (USMC) A130504

Aerial view taken from one of VMO-6's light observation aircraft, flown by 1stLt John D. Cotton, shows the power station, the pipes or "penstocks" that carried off the water, and the precarious nature of the road occupying a thin shelf cut into the precipitous slope.

would be on the left flank, moving along the slope between Barrow and the road.

Wray had his objective by dawn. On it Schmuck built up a base of fire with his 81mm mortars and an attached platoon of 4.2-inch mortars—the effective, but road-bound "four-deuces." Also effective, but tied to the road, were five Army self-propelled antiaircraft guns—quad .50-calibers and duel 40mms—attached from Company B, 50th Antiaircraft Artillery (Automatic Weapons) Battalion.

Things went like clockwork. Schmuck's main attack jumped off at 1000. Barrow clambered up the hogback ridge that led to the summit of Hill 1081; Noren advanced

A mixed group of Marines and soldiers struggle up an ice-covered slope somewhere south of Koto-ri. The weather and the terrain were at least as much of an enemy as the Chinese. Marines, disdainful of the Army's performance east of the reservoir, learned in the march-out from Hagaru-ri that soldiers, properly led, were not much different from themselves

Photo by Cpl Peter W. McDonald, National Archives Photo (USMC) 127-N-A5389

Sketch by Sgt Ralph Schofield, USMCR

Going downhill was easier for the most part than going up, but wherever it was the march was single file of Marines, or at best a double file. Even stripped down to essentials, the average Marine carried 35 to 40 pounds of weapons, ammunition, rations, and sleeping bag. Anything more, such as toilet articles, shelter half, or poncho, was a luxury.

along the wooded western slope of the hill. Noren met scattered resistance, was stopped momentarily by two enemy machine guns, which he then took out with a tidy schoolbook solution—engaging the enemy with his own machine guns and 60mm mortars while a platoon worked around in a right hook. He then ran into a bunker complex, took it after a savage fight, and found a kettle of rice cooking in the largest bunker. Schmuck moved his headquarters forward and set up his command post in the bunker only to find it louse-ridden. The day cost Noren three killed and six wounded.

Barrow had gone up the ridge against no enemy whatsoever, impeded only by the icy ridgeline, so narrow that he had to march in a dangerous single file. Through a break in the snowstorm, Barrow got a glimpse of a strongly bunkered Chinese position on a knob between his company and the crest of the hill. He elected to do a double envelopment, sending his 2d Platoon around to the left and his 1st Platoon around to the right. He went himself with the 3d Platoon up the center in a frontal attack. It all came together in a smashing assault. Barrow's Marines

Fresh snow fell during the march from Koto-ri to the top of Funchilin Pass. When the column on the road halted, as it frequently did, the Marines tended to bunch up, making themselves inviting targets for Chinese mortar and machine gun fire. March discipline had to be enforced by tough corporals and sergeants more than by orders from the top.

Photo by Sgt Frank C. Kerr, National Archives Photo (USMC) 127-N-A5358

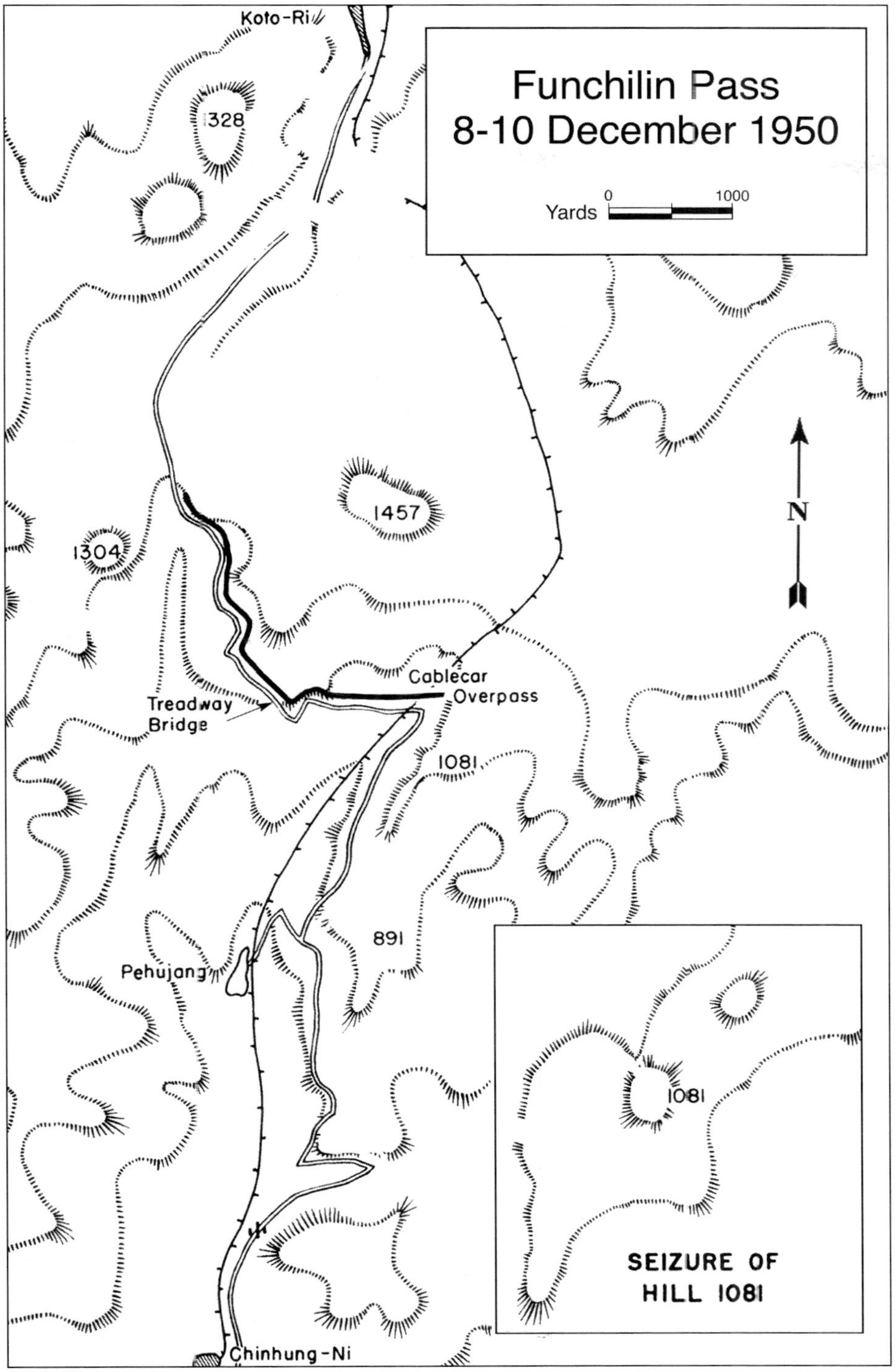

counted more than 60 Chinese dead. They themselves lost 10 killed, 11 wounded. The snow ended and the night was clear. At midnight a Chinese platoon bravely but foolishly tried to evict Barrow's Marines and lost 18 killed. To Barrow's left, all was quiet in front of Noren's position.

To their north, Litzenberg's 7th Marines had come out of Koto-ri on schedule on the morning of 8 December. Counting the Army provisional battalion he had four battalions. Two were to clear each side of the road. One was to advance along the MSR, to be followed by the regimental train and the reserve battalion. Major Morris had been assigned to take Hill 1328 on the right of the road with his 3d Battalion. Going was slow. By mid-morning Litzenberg grew impatient and urged him to commit his reserve company. Morris snapped back: "All three companies are up there—50 men from George Company, 50 men from How, 30 men from Item. That's it." Shortly after noon, Litzenberg committed his regimental reserve, Lawrence's 2d Battalion, to come to the assistance of Morris. By nightfall the two battalions had joined but not much more was accomplished.

Left of the road, the provisional Army battalion, under Major Jones, had jumped off on time and, with the help of two Marine tanks, had moved along against light resistance. In two jumps Jones reached Hill 1457 where his soldiers dug in for the night. Their position was raked by Chinese automatic fire, and in a brief nasty action 12 enemy were killed at a cost of one soldier killed, four wounded.

Litzenberg's executive officer, Lieutenant Colonel Frederick Dowsett, had been shot through the ankle the day before. Litzenberg moved Raymond Davis up to executive officer to replace him and gave the 1st Battalion, still his strongest battalion, to Major Sawyer, whose wound had proved superficial. Sawyer's initial mission was to move a mile down the road and wait for the 3d Battalion to come up on his right flank. The 1st Battalion now had it own fight.

Sawyer's lead platoon came under fire from Hill 1304. Baker Company continued to move against the high ground just left of the road while Able and Charlie Companies moved more deeply to the right against the hill. Baker Company was caught in a crossfire; the company commander, Lieutenant Kurcaba, was killed, two of his platoon leaders were

Photo by Sgt Frank C. Kerr, National Archives Photo (USMC) 127-N-A5357

A section of two 81mm mortars set up in the snow to give fire support for a rifle company in the attack. These high-angle medium mortars, with a shell almost as lethal as a 105mm howitzer round, were considered to be the infantry battalion commander's own "artillery." Mortars were particularly effective in the defense of a perimeter.

wounded. First Lieutenant William W. "Woody" Taylor took over command of the company and had his objective by nightfall. Able and Charlie Companies meanwhile had taken Hill 1304 without much trouble. Sawyer divided his battalion into two perimeters for the night. Vehicular movement along the MSR was halted.

It had been nearly noon on 8 December before Murray's 5th Marines, following behind the 7th Marines, moved out of Koto-ri. Stevens' 1st Battalion was in the lead. Stevens sent out his Baker and Charlie Companies to take Hill 1457. Charlie Company joined up unexpectedly with the Army's provisional battalion and the soldiers and Marines had the Chinese off the high ground by mid-afternoon. Baker and Charlie Companies, combined with the Army troops, formed a perimeter for the night. Able Company had its own perimeter closer to the MSR. Murray moved 41 Commando, in reserve, up behind the 1st Battalion.

Meanwhile, the 2d and 3d Battalions of Puller's 1st Marines held Koto-ri itself. For the defenders the problem was not the scattered small arms fire of the Chinese, but the flood of civilian refugees coming down the road from the north. They could not be admitted into the perimeter because of the probability that the Chinese had infiltrated them. During the bitterly cold night, two babies were born with the help of Navy doctors and corpsmen. In all their misery these thousands of civilians had to wait outside the lines until Koto-ri was vacated. They then followed behind the Marines, as best they could, until the presumed safety of Hamhung-Hungnam might be reached.

During the day Smith, always conscious of his dead, attended a funeral at Koto-ri. What had been an artillery command post, scraped more deeply into the frozen ground by a bulldozer, became a mass grave. A total of 117 bodies, mostly Marines but some soldiers and Royal Marines, were lowered into the hole. A Protestant and a Catholic chaplain officiated. The bulldozer covered the bodies with a mound of dirt.

Sergeant Robert Gault, head of the Graves Registration Section of the 7th Marines, remembered the funeral this way:

> We had a chaplain of each faith, and the fellows had made a big hole and laid the fellows out in rows the best

Task Force Dog included these self-propelled 155mm howitzers shown in firing position covering Funchilin Pass on 10 December near Chinhung-ni. Marine 155mm howitzers at this time were still tractor-drawn and some were lost coming out of Yudam-ni. At a greater distance an Army tank can be glimpsed.

Photo by TSgt James W. Helms, Jr., Department of Defense Photo (USMC) A156320

Photo by Cpl James Lyle, National Archives Photo (USA) 111-SC354456

A squad-sized Army patrol, led by Sgt Grant J. Miller, from the 3d Infantry Division's Task Force Dog, moves up from Chinhung-ni into Funchilin Pass on 9 December. The tank at the side of the road appears to be a Soviet-built T-34 knocked out a month earlier by the 7th Marines.

we could and put ponchos over them. As soon as each chaplain had said his little bit for the fellows, we would cover them up and close them in. Everyone was given—I think under the circumstances—a very fine burial. It wasn't like the one back at Inchon and Hungnam. It wasn't like the one where we had crosses for the boys painted white and all the preliminaries: flowers that we could get for them—we'd go out and pick them. It wasn't like that, no. It was one where we were just out in a field, but it was one with more true heart.

There was more snow during the night, but Saturday, 9 December, dawned bright, clear, and cold. South of Funchilin Pass, Noren moved his Baker Company to the next high ground to his front and Barrow had his Able Company test-fire their weapons before beginning the assault of Hill 1081. Barrow then attacked in a column of platoons behind a thunderous preparation by close air, artillery, and mortars. Even so his lead platoon, under First Lieutenant William A. McClelland, was hard hit as it moved forward by rushes, stopping about 200 yards from the crest. Under cover of air strikes by four Corsairs and his own 60mm mortars, Barrow moved his 2d and 3d Platoons forward and by mid-afternoon his Marines had the hill. The two-day battle cost Barrow almost exactly half his company. He had started up the hill with 223 Marines; he was now down to 111 effectives. But 530 enemy dead were counted and the Marines held the high ground commanding Funchilin Pass.

On the MSR that Saturday, moving south from Koto-ri, the 7th Marines resumed its attack. The rest of Hill 1304 was taken. Captain John Morris with his Company C and a platoon from Company B moved down the road and secured the bridge site. The rest of Company B, following behind, overran an enemy position garrisoned by 50 Chinese so frozen by the cold that they surrendered without resistance.

The old war horse, General Shepherd, arrived from Hawaii the day before on what was his fifth trip to Korea, this time as "Represen-tative of Commander Naval Force, Far East, on matters relating to the Marine Corps and for consultation and advice in connection with the contemplated amphibious operations now being planned." Shepherd may have thought he had more authority than he really had. In his 1967 oral history he said:

> When reports came back that the cold weather had set in and they weren't able to make the Yalu River and things began falling apart, Admiral Radford sent me to Korea—I think [the orders] came from the Chief of Naval Operations on the recommendation of Admiral Joy—that [I] was to take charge of the evacuation of the Marines from Hungnam.

More accurately, the Chief of Naval Operations, Admiral Forrest P. Sherman, had probably been prompted by back-channel messages to Admiral Radford to send Shepherd to the Far East "for the purpose of advising and assisting Commander Naval Forces, Far East [Admiral Joy], with particular emphasis on Marine Corps matters."

Shepherd did recognize that there could be a conflict of command because of Almond's actual command of X Corps. On arriving in Tokyo on 6 December he met with General MacArthur and noted "General MacArthur was unqualified in his admiration and praise for the effective contribution which Marines had made throughout the whole of the Korea fighting. His general demeanor [however] was not one of optimism."

After more conferences and meetings in Tokyo, Shepherd left on 8 December for Hungnam and

Photo by Sgt Frank C. Kerr, Department of Defense Photo (USMC) A5370

Litzenberg's 7th Marines led the way out of Koto-ri at dawn on 8 December. Murray's 5th Marines followed the 7th Marines and in turn was followed by two battalions of Puller's 1st Marines. Infantry units, moving from hilltop to hilltop on both sides of the road covered the movement of the division trains. Some called the maneuver "hop-scotch," others called it "skinning-the-cat."

on arrival went immediately to the *Mount McKinley*. Here on the next day he attended a meeting on outloading and naval support also attended by Admirals Joy, Struble, and Doyle, and General Harris. A press conference followed. Shepherd praised the operations of the X Corps and said that he was there to assist General Almond. He was anxious to get up to the reservoir to see things for himself. He made the trip in a TBM, landing at Koto-ri after circling Hagaru-ri. He then met with Smith for an hour or more. Smith told him that all casualties would be out by the end of the day. Shepherd announced that he intended to march out with the division. Smith dug in his heels and said absolutely not.

Shepherd returned to the airstrip. A number of war correspondents, among them "Maggie" Higgins, Keyes Beach, and the photographer David Douglas Duncan, had wangled their way to Koto-ri. While Shepherd's plane was warming up, Colonel Puller arrived leading Higgins by the hand. Puller said: "General Smith says take this woman out of his hair and see that she goes out on your plane." Shepherd turned to Higgins, whom he had met at Inchon, saying, "Maggie, it's too bad. I wanted to march down too." The plane completed its loading of wounded and taxied to the end of the strip. It was dusk and as the plane took off Shepherd could see machine gun tracer bullets reaching up at the underside of the

As the march went on, cold-benumbed Chinese soldiers surrendered in increasing numbers. This group, probably the remnants of a platoon or perhaps a company, surrendered to Company C, 1st Battalion, 7th Marines, south of Koto-ri on 9 December.

Photo by Sgt Frank C. Kerr, National Archives Photo (USMC) 127-N-A5377

Two Chinese, anxious to surrender, get a quick pat-down search for weapons by members of Company C, 7th Marines, but there was no fight left in them. Once given a cigarette and perhaps a chocolate bar by their captors, they would follow along uncomplainingly into eventual captivity.

National Archives Photo (USMC) 127-N-A5378

Photo by Cpl Peter W. McDonald, National Archives Photo (USMC) 127-N-A5379

First step in evacuation of casualties, alive or dead, during the march-out was to get them down to the road by stretcher. Here four Marines, without helmets or packs, carry a stretcher down from the high ground somewhere south of Koto-ri on 9 December. Fresh snow had fallen the previous night.

plane. Leaning over to Higgins, the irrepressible Shep-herd said to her, "If we get hit, we will die in each other's arms."

From Koto-ri to Chinhung-ni

The column of division vehicles, protected on both sides by the Marine infantry in the hills, crawled along the road south of Koto-ri at a snail's pace and with frequent stops. The Marines, who watched the crawling column from their perches in the hills, wondered profanely why the vehicles had to be piled high with tent frames, wooden doors, and other luxuries of life.

Partridge had held back the Army's Brockway trucks, with their precious cargo of bridge sections, in Koto-ri until first light on 9 December when he considered the MSR secure enough for him to move them forward. He then joined Sawyer's 1st Battalion at the head of the column. Everything worked at the bridge site like a practiced jigsaw puzzle. Army and Marine engineers rebuilt the abutments with sandbags and timbers. A Brockway truck laid the steel treadways and plywood deck panels. At noon, Almond flew overhead in his "Blue Goose" to see for himself that things were going well. Installation was done in three hours and at 1530 Partridge drove his jeep back to the top of the pass to tell Lieutenant Colonel Banks that he could bring Division Train No. 1 down the defile. The first vehicles began to cross the bridge at about 1800. Sawyer's Marines kept the enemy at a distance and captured 60 prisoners in the process. All night long vehicles passed over the bridge.

At 0245 on Sunday morning, 10 December, the head of the column reached Chinhung-ni. Colonel Snedeker, the division's deputy chief of staff, had positioned himself there to direct the further movement of the vehicle serials. The 7th Marines followed Division Train No. 1 down the pass. Up on the plateau Ridge's 3d Battalion, 1st Marines, had come out of Koto-ri and relieved the 3d Battalion, 7th Marines, on Hill 1328 where it had a fight with

Frozen corpses are unloaded from a truck at Koto-ri where they will be buried in a mass grave. A 155mm howitzer can be seen in the background. The dead, 117 of them, mostly U.S. Marines but some soldiers and Royal Marines, were interred in a hole originally bull-dozed into the ground to serve as an artillery fire direction center.

Photo by Sgt Frank C. Kerr, National Archives Photo (USMC) 127-N-A5366

National Archives Photo (USMC) 127-N-A402841

LtGen Shepherd, right, Commanding General, Fleet Marine Force, Pacific, arrived on 8 December on his fifth trip to Korea. Although not in the operational chain-of-command, he had arrived to oversee the evacuation of the Marines. RAdm Doyle, left, and MajGen Harris, center, greet Shepherd. All three look glum.

about 350 resurgent Chinese. At 1030, General Smith closed his command post at Koto-ri and flew to his rear command post at Hungnam.

Puller brought out the remainder of RCT-1 from Koto-ri on the afternoon of the 10th. Milne's tanks, including the tank company from the 31st Infantry, followed behind the elements of RCT-1 on the road. Ridge's 3d Battalion was already deployed on the high ground on both sides of the MSR south of Koto-ri. The plan was for Sutter's 2d Battalion to relieve Stevens' 1st Battalion, 5th Marines, on Hill 1457.

As the last Americans left Koto-ri, the Army's 92d Field Artillery, firing from Chinhung-ni, shelled the town with its long-range 155mm guns. There was confusion at the tail of the column as Korean refugees pressed close. The tankers fired warning shots to make them stay back. Panic developed as the rumor spread that the Marines were shooting the refugees. The tanks passed on down the road, protected on both sides at first by Ridge's Marines in the hills. But Sutter, having begun his climb up Hill 1457 and finding it a long way off and with no enemy in sight, asked Puller's permission, which he received, to return to the road.

Ridge pulled his companies off Hill 1304 and the high ground on the opposite side of the MSR at about 2100. Ridge's battalion was the last major unit to descend the pass, following behind Jones' provisional battalion of soldiers and the detachment of the 185th Engineers. Harry Milne's tanks were behind Ridge with no infantry protection except the lightweight division Reconnaissance Company mounted in jeeps. It was now about midnight.

By then both division trains, all of RCT-7, and most of the 11th Marines had reached Chinhung-ni. The 5th Marines followed the 7th Marines. Beyond Chinhung-ni, guerrillas were reported to be active in the vicinity of Sudong, but the division trains and both the 5th and 7th Marines passed through without interference. Some time after midnight when the vehicles of RCT-1 reached the town sudden swarms of Chinese came out of the houses of the village with burp guns and grenades. Truck drivers and casuals, both Army and Marine, fought a wild, shapeless action. Lieutenant Colonel John U. D. Page, an Army artillery officer, took charge, was killed, and received a posthumous Medal of Honor. Lieutenant Colonel Waldon C. Winston, an Army motor transport officer, took his place. It was dawn before the place was cleaned up. RCT-1 lost nine trucks and a personnel carrier; 8 men killed and 21 wounded.

Meanwhile, Milne's tanks, some 40 of them, descending the narrow, icy-slick road of Funchilin Pass had run into trouble. About a mile short of the Treadway bridge the brakes of the ninth tank from the end of the column froze up. The tanks to its front clanked on, but the immobile ninth tank blocked the eight tanks to the rear. Close behind came the refugees. Left guarding the nine tanks was First Lieutenant Ernest C. Hargett's 28-man reconnaissance platoon. Five Chinese soldiers emerged from the mass of refugees and one, in English, called upon Hargett to surrender. Hargett, covered by a BAR-man, approached the five Chinese cautiously. The English-speaking one stepped aside and the four others produced burp guns and grenades. A grenade wounded Hargett. His BAR-man, Corporal George A. J. Amyotte, cut the five Chinese down, but more Chinese materialized on the road and the steep slope of the hill. Hargett backed away with his platoon. The last tank in the column was lost to the

Department of Defense Photo (USMC) A5376

The Chinese had blown a critical bridge halfway down Funchilin Pass where water flowed downward from Changjin Reservoir and passed through four giant pipes called "penstocks." As early as 4 December, MajGen Smith knew that this 24-foot gap would have to be bridged if his vehicles were to reach Hungnam.

Chinese. Meanwhile, the crew of the tank that blocked the road had succeeded in freeing the frozen brakes and was ready to proceed. But the crews of the remaining seven tanks had departed, leaving the hatches of their tanks open. A member of Hargett's platoon, who had never driven a tank, managed to bring out one tank. The night's adventure cost Hargett two men killed and 12 wounded.

Engineers were waiting at the Treadway bridge, ready to blow it up. They thought the two tanks and Hargett's platoon were the last to come by. They blew the bridge, but one Marine had been left behind. Private First Class Robert D. DeMott from Hargett's platoon had been blown off the road by a Chinese explosive charge. Regaining consciousness, he got back on the road and joined the refugees.

Marines from Litzenberg's regiment, along with some attached soldiers, on 9 December reached the blown bridge in Funchilin Pass, which, unless replaced, would stop any further southward movement of wheeled or tracked vehicles. Plans were already afoot to bridge the gap with a Treadway bridge to be airdropped in sections at Koto-ri.

Photo by Sgt Frank C. Kerr, Department of Defense Photo (USMC) A5375

Photo by Sgt William R. Keating, Department of Defense Photo (USMC) A5408

By evening on 9 December the Treadway bridge was in place and men and vehicles could move unimpeded down the MSR through Funchilin Pass. From here on enemy resistance was limited to small-scale firefights and ambushes. The most sizable resistance would come near Sudong.

The wind that blew from Manchuria and beyond, "down over the Yalu and the mountains all around. . . down into the gorges with their frozen streams amd naked rocks. . . down along the ice-capped road—now shrieking and wild—that wind," said noted photographer David Douglas Duncan, "was like nothing ever known by the trapped Marines, yet they had to march through it."

Photo by David Douglas Duncan

Photo by Sgt Frank C. Kerr, Department of Defense Photo (USMC) A5372

Marines march along a particularly precipitous portion of the road winding down through Funchilin Pass on 9 December. The day before had seen the launching of an exquisitely timed maneuver—the exit of the 5th and 7th Marines from Koto-ri and the simultaneous advance of the 1st Battalion, 1st Marines, from Chinhung-ni to take the high ground controlling the pass.

He heard the detonation that blew the bridge, but figured that he could make his way on foot through the gatehouse above the penstocks. This he did as did many of the following refugees.

Warm Welcome at Hungnam

Donald Schmuck, from his position on Hill 1081, watched the lights of the tanks descending the pass and at 0300 gave orders for Barrow's Company A to begin its withdrawal. At 1300 on 11 December the last units of the division passed through Chinhung-ni. By 1730 they had gone through Majon-dong and by 2100 most had reached their Hamhung-Hungnam assembly areas. They found a tent camp waiting for them. Lieutenant Colonel Erwin F. Wann, Jr.'s 1st Amphibian Tractor

Coming down Funchilin Pass on 10 December, the Marine column was intermixed with many Korean refugees fleeing the Chinese. Numbers were such as to interfere with military traffic. Behind them, Yudam-ni, Hagaru-ri, and Koto-ri were left as deserted ghost towns.

Photo by Sgt William R. Keating, National Archives Photo (USMC) 127-N-A5407

National Archives Photo (USMC) 127-N-A5901

LtCol John H. Partridge, whose engineers performed miracles, particularly in scraping out the airstrip at Hagaru-ri and installing the airdropped Treadway bridge in Funchilin Pass, received a second Bronze Star from MajGen Smith at Masan in early January 1951.

Battalion had done much of the preparation for their arrival. Chow lines were open for the continuous serving of hot meals. Wann, 31 and Naval Academy 1940, had been an amphibian tractor officer at Bougainville, Guam, and Iwo Jima. The weather seemed almost balmy after the unrelieved subzero temperatures of the plateau. Milne's tanks continued on to the LST staging area, arriving just before midnight. From 8 through 11 December, the division had lost 75 men dead, 16 missing, and 256 wounded, for a total of 347 casualties.

As late as Saturday, 9 December, General Smith believed that the 1st Marine Division, once concentrated, would be given a defensive sector to the south and southwest of Hungnam. A day earlier his deputy chief of staff, Colonel Snedeker, who was running his rear headquarters, issued tentative orders for Puller's RCT-1 to organize defensively at Chigyong, with Murray's RCT-5 and Litzenberg's RCT-7 preparing to defend Yonpo airfield.

But on that Saturday, Almond received his formal orders from MacArthur to redeploy X Corps to South Korea and Smith learned that his division would be loading out immediately on arrival. At this point Almond regarded the 1st Marine Division as only marginally combat effective. He considered the 7th Infantry Division, except for its loss of almost a complete regimental combat team, to be in better condition. In best condition, in his opinion, was the 3d Infantry Division, which he visited almost daily.

Almond therefore decided that once the 1st Marine Division passed through the Hamhung-Hungnam perimeter defense it would be relieved from active combat and evacuated. Second priority for evacuation would be given the 7th Infantry Division. Last out would be the 3d Infantry Division.

The Hungnam-Hamhung defensive perimeter, as neatly drawn on the situation maps in Almond's headquarters, consisted of a main line of resistance (MLR) about 20 miles long arcing in a semicircle from north of Hungnam around to include Yonpo. In front of the MLR was a lightly held outpost line of resistance. The northernmost sector, beginning at the coastline, was given to Major General Kim Pak Il's ROK I Corps, which, having arrived uneventfully from Songjin, began moving into line on 8 December. The lift-off from Songjin by LSTs, merchant ships, and the attack transport USS *Noble* (APA 218) had been completed in three days. Counterclockwise, next in line on the perimeter, came Barr's 7th Infantry Division with two sound regiments, followed by Soule's 3d Infantry Division. The southern anchor of the perimeter was held by the 1st Korean Marine Corps Regiment, which had the mission of defend-

LtCol Harry T. Milne, commanding officer of the 1st Tank Battalion, in a post-war photograph. Tanks were vital to the Marine breakout, but their record was marred by poor performance with Task Force Drysdale and, later, in Funchilin Pass.

National Archives Photo (USMC) 127-N-A552011

Photo by Sgt William R. Keating, National Archives Photo (USMC) 127-N-A5404

Once the high ground commanding Funchilin Pass had been taken and the Treadway bridge was in place, the 1st Marine Division could descend almost unimpeded to the sea. All day long on 10 December Marine troops marched down the pass until they reached first Chinhung-ni and then went beyond Sudong-ni where they took trucks to Hamhung.

ing the Yonpo airfield. As the evacuation progressed, the MLR was to shrink back to successive phase lines.

Admiral Doyle, as commander, Task Force 90, assumed control of all naval functions on 10 December. Marine Colonel Edward Forney, a X Corps deputy chief of staff whose principal duties were to advise Almond on the use of Marine and Navy forces, was now designated as the Corps' evacuation control officer. The Army's 2d Engineer Special Brigade would be responsible for operating the dock facilities and traffic control. A group of experienced Japanese dock workers arrived to supplement their efforts,

Years later Almond characterized Forney's performance as follows: "I would say that the success of [the evacuation] was due 98 percent to common sense and judgment and that this common sense and judgment being practiced by all concerned was turned over to General Forney who organized the activities in fine form I mean *Colonel* Forney, he should have been a *General*!"

General Field Harris briefed General Shepherd on 10 December on the status of the 1st Marine Aircraft Wing. Ashore were Marine fighter squadrons -312, -542, -513, and -311. Afloat were VMF-212 in the *Bataan*, VMF-214 in the *Sicily*, and VMF-323 in the *Badoeng Strait*. Shepherd learned that the wing had been offered either the K-10 airfield near Masan or K-1 near Pusan. K-1 was preferable because it was the better field and close to Pusan's port facilities.

The 11th of December was a busy day. X Corps issued its Operation Order 10-50, calling for the immediate embarkation of the 1st Marine Division. The perimeter would shrink progressively as then the 7th and 3d Infantry Divisions, in turn, were withdrawn. As the perimeter contracted, naval gunfire and air support would increase to defend the remaining beachhead. General MacArthur himself arrived that day at Yonpo, met with Almond, and approved the X Corps evacuation plan. He told Almond that he could return to

(Continued on page 326)

A Marine, left, and a Korean soldier, right, check the meager baggage of a Korean family on 10 December at some point east of Chinhung-ni before allowing the family to proceed to Hungnam. Some 91,000 refugees, driven by hunger, the cold, and fear of the Chinese, would be evacuated.

Photo by Cpl James Lyles, National Archives Photo (USA) 111-SC354459

Medals of Honor

The Medal of Honor, the Nation's highest award for valor, has been given to 294 Marines since its inception in 1862. The Korean War saw 42 Marines so honored. Of this number, 14 awards were made for actions incident to the Chosin Reservoir campaign. Seven of these awards were posthumous.

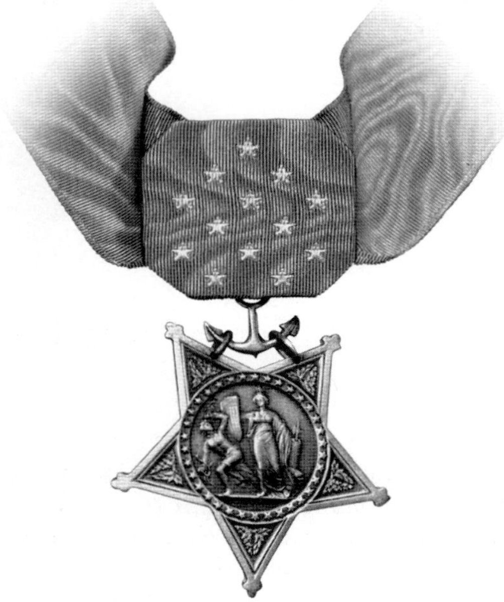

Staff Sergeant Archie Van Winkle

Staff Sergeant Archie Van Winkle, 25, of Juneau, Alaska, and Darrington, Washington, a platoon sergeant in Company B, 1st Battalion, 7th Marines, awarded the Medal of Honor for gallantry and intrepidity in action on 2 November 1950 near Sudong wherein he led a successful attack by his platoon in spite of a bullet that shattered his arm and a grenade that exploded against his chest.

Sergeant James I. Poynter

Sergeant James I. Poynter, 33, of Downey, California, a squad leader with Company A, 1st Battalion, 7th Marines, posthumously awarded the Medal of Honor for his actions on 4 November south of Sudong, where, although already critically wounded, he assaulted three enemy machine gun positions with hand grenades, killing the crews of two and putting the third out of action before falling mortally wounded.

Corporal Lee H. Phillips

Corporal Lee H. Phillips, 20, of Ben Hill, Georgia, a squad leader with Company E, 2d Battalion, 7th Marines, posthumously awarded the Medal of Honor for actions on 4 November 1950 near Sudong where he led his squad in a costly but successful bayonet charge against a numerically superior enemy. Corporal Phillips was subsequently killed in action on 27 November 1950 at Yudam-ni.

Second Lieutenant Robert D. Reem

Second Lieutenant Robert D. Reem, 26, of Elizabethtown, Pennsylvania, a platoon leader in Company H, 3d Battalion, 7th Marines, posthumously awarded the Medal of Honor for actions on 6 November 1950 near Chinhung-ni. Leading his platoon in the assault of a heavily fortified Chinese position, he threw himself upon an enemy grenade, sacrificing his life to save his men.

First Lieutenant Frank N. Mitchell

First Lieutenant Frank N. Mitchell, 29, of Indian Gap, Texas, a member of Company A, 1st Battalion, 7th Marines, posthumously awarded the Medal of Honor for extraordinary heroism in waging a single-handed battle against the enemy on 26 November 1950 near Yudam-ni to cover the withdrawal of wounded Marines, notwithstanding multiple wounds to himself.

Staff Sergeant Robert S. Kennemore

Staff Sergeant Robert S. Kennemore, 30, of Greenville, South Carolina, a machine gun section leader with Company E, 2d Battalion, 7th Marines, awarded the Medal of Honor for extraordinary heroism during the night of 27-28 November 1950 north of Yudam-ni in deliberately covering an enemy grenade whose explosion cost him both of his legs.

Private Hector A. Cafferata, Jr.

Private Hector A. Cafferata, Jr., 21, born in New York City, a rifleman with Company F, 2d Battalion, 7th Marines, awarded the Medal of Honor for his stouthearted defense on 28 November 1950 of his position at Toktong Pass despite his repeated grievous wounds.

Captain William E. Barber

Captain William E. Barber, 31, of Dehart, Kentucky, commanding officer of Company F, 2d Battalion, 7th Marines, awarded the Medal of Honor for his intrepid defense of Toktong Pass from 28 November to 2 December in spite of his own severe wounds.

Private First Class William B. Baugh

Private First Class William B. Baugh, 20, born in McKinney, Kentucky, a member of the Anti-Tank Assault Platoon, Weapons Company, 3d Battalion, 1st Marines, posthumously awarded the Medal of Honor for covering with his body an enemy grenade thrown into the truck in which his squad was moving from Koto-ri to Hagaru-ri on the night of 29 November 1950 as part of Task Force Drysdale.

Major Reginald R. Myers

Major Reginald R. Myers, 31, of Boise, Idaho, executive officer of the 3d Battalion, 1st Marines, awarded the Medal of Honor for leading a hastily organized provisional company of soldiers and Marines in the critical assault of East Hill at Hagaru-ri on 29 November 1950.

Captain Carl L. Sitter

Captain Carl L. Sitter, 28, born in Syracuse, Missouri, commanding officer of Company G, 3d Battalion, 1st Marines, awarded the Medal of Honor for his valiant leadership in bringing his company from Koto-ri to Hagaru-ri as part of Task Force Drysdale on 29 November. He then led it in the continued assault of vital East Hill on 30 November 1950.

Staff Sergeant William G. Windrich

Staff Sergeant William G. Windrich, 29, born in Chicago, Illinois, a platoon sergeant with Company I, 3d Battalion, 5th Marines, posthumously awarded the Medal of Honor for his extraordinary bravery in taking and then holding a critical position near Yudam-ni on 29 November despite two serious wounds which eventually caused his death.

Lieutenant Colonel Raymond G. Davis

Lieutenant Colonel Raymond G. Davis, 35, of Atlanta, Georgia, commanding officer of the 1st Battalion, 7th Marines, awarded the Medal of Honor for his conspicuous gallantry and skill in leading his battalion, from 1 to 4 December 1950, across mountainous and frigid terrain to come to the relief of the beleaguered company holding Toktong Pass.

Sergeant James E. Johnson

Sergeant James E. Johnson, 24, of Washington, D.C. and Pocatello, Idaho, regularly a member of the 11th Marines but serving as a platoon sergeant of a provisional rifle platoon attached to the 3d Battalion, 7th Marines, posthumously awarded the Medal of Honor for continuing to engage the enemy single-handedly in hand-to-hand combat on 2 December south of Yudam-ni after being severely wounded.

Photo by Sgt Frank C. Kerr, National Archives Photo (USMC) 127-N-A5754

LtCol Murray's 5th Marines found a tent camp waiting for them when they arrived at Hungnam after the long march. Much of the camp was the work of the 1st Amphibian Tractor Battalion. Although there was still snow on the ground, the weather at Hungnam was mild in comparison with the sub-zero temperatures of the Changjin plateau.

GHQ and pick up his duties as MacArthur's chief of staff or he could remain in command of X Corps. Almond replied that he wished to stay with X Corps even if it became part of Eighth Army. The 27th of December was set as the day that X Corps would pass to Eighth Army control.

The evacuation of the 1st Marine Division began with the loading-out of the 7th Marines in the MSTS *Daniel I. Sultan*. The 5th Marines would follow on the 12th and the 1st Marines on the 13th. It was anticipated that the ships would have to make a second, even a third turn-around, to lift the entire division. The docks could berth only seven ships at a time. To compensate, there would be some double berthing, but most of the passengers would have to load out in the stream. Approximately 1,400 vehicles had been brought down from the Chosin plateau, about the same number as had gone up, but now some of the complement bore U.S. Army markings. Most of the division's vehicles would go out in LSTs. Green Beaches One and Two could handle 11 LSTs simultaneously. Thankfully there was no great tide to contend with, only one foot as compared to Inchon's 30.

Marine Close Air Support at Its Finest

The Marines' ground control intercept squadron, MGCIS-1, Major Harold E. Allen commanding, shut down at Yonpo on the 11th, passing control of air defense of the perimeter to the *Mount McKinley*. The sky remained empty of enemy aircraft. Overall air control stayed with MTACS-2, the Marines' tactical air control squadron, under Major Christian C. Lee. Each of the infantry battalions had gone into the Chosin campaign with two forward air controllers assigned, most of them reserves, all of them qualified Marine aviators. They brought with them both expertise in close air support and rapport with the fighter-bombers overhead. Although inclined to lament not having a cockpit assignment they realized they were providing an unmatched link to the air. They knew how to talk a pilot onto a target.

Between 1 and 11 December, Marine aviators, ashore and afloat, flew more than 1,300 sorties in support of their comrades on the ground. Of these, 254 were flown from the *Badoeng Strait* and 122 from the late-arriving *Sicily*. (Lundin's Blacksheep squadron, still at Wonsan, reembarked in the

MajGen Almond gave his Marine deputy chief of staff, Col Edward H. Forney, much of the credit for the orderly departure of X Corps from Hungnam. Here, on 14 December, he presents Forney with a Legion of Merit.

National Archives Photo (USA) 111-SC355068

Photo by Sgt Jack T. McKirk, National Archives Photo (USA) 111-SC354464

MajGen Almond, left, always generous with medals and commendations, on 11 December congratulates bareheaded BGen Armistead D. Mead, commander of Task Force Dog, on keeping the MSR open from Chinhung-ni to Hamhung. Few Marines were aware of this Army contribution to their march back from Chosin

Sicily on 7 December.) The rest had been by the shore-based squadrons at Wonsan and Yonpo. The first Marine jet squadron to arrive in Korea, VMF-311, with McDonnell F9F Panther jets, Lieutenant Colonel Neil R. McIntyre commanding, had arrived at Yonpo on 10 December and managed to fly four days of interdiction missions before moving back to Pusan to aid the Fifth Air Force in its support of the Eighth Army.

Flight conditions both ashore at Yonpo and afloat in the carriers were hellish—in the air, poor charts, minimal navigational aids, and capricious radios; at Yonpo, primitive conditions and icy runways; and, afloat, ice-glazed decks and tumultuous seas for the carrier-based aircraft. The *Badoeng Strait* reported scraping off three inches of ice and snow from the flight deck. The *Sicily* at one point had to stop flight operations for VMF-214's Blacksheep in the face of heavy seas and 68-knot winds. Planes were lost. Three night fighters went down. There were other crashes. It was estimated that a pilot who had to ditch at sea in the arctic waters had only 20 minutes before fatal hypothermia. Two VMF-212 pilots from Yonpo, out of gas, managed to save themselves and their planes by landing on the *Badoeng Strait*. By strenuous effort on the part of all hands, aircraft availability at Yonpo hovered around 67 percent and a remarkable 90 percent on board the carriers. About half the missions flown were not for the Marines but for someone else. Statistics kept by the wing reported a total of 3,703 sorties in 1,053 missions controlled by tactical air control parties being flown between 26 October and 11 December. Of these missions, 599 were close support—468 for the 1st Marine Division, 67 for the

Gen MacArthur made one of his quick trips to Korea on 11 December, this time to Yonpo Airfield to meet with MajGen Almond and approve the X Corps evacuation plan. MacArthur is in his trademark peaked cap and Almond is in a bombardier's leather jacket. No one would mistake the accompanying staff officers, with their well-fed jowls, for combat soldiers.

National Archives Photo (USA) 111-SC3544110

Photo by SSgt Ed Barnum, Department of Defense Photo (USMC) A130426

Marine crewmen check the loading of rockets on the rails of an F-4U Corsair flying from Yonpo Airfield early in December. Despite minimum facilities and terrible weather, the hard-working crews maintained an availability rate of 67 percent for aircraft flying from Yonpo.

ROKs, 56 for the 7th Infantry Division, and 8 for the 3d Infantry Division. Eight Marine pilots were killed or died of wounds, three were wounded, and four were missing in action.

Marine transports—twin-engine R4Ds and four-engine R5Ds from VMR-152, commanded by 44-year-old Colonel Deane C. Roberts—supplemented General Tunner's Combat Cargo Command in its aerial resupply and casualty evacuation from Hagaru-ri.

The squadron that the 1st Marine Division considered its own private air force, Major Gottschalk's VMO-6, with 10 light fixed-wing aircraft and nine helicopters, racked up 1,544 flights between 28 October and 15 December. Of these 457 had been reconnaissance, 220 casualty evacuation, and 11 search-and-rescue.

Time To Leave

Wonsan closed as a port on Sunday, 10 December. Outloading for the evacuation, conducted from 2 to 10 December, was under Lieutenant Colonel Henry "Jim" Crowe with muscle provided by his 1st Shore Party Battalion. The attached Company A, 1st Amphibian Truck Battalion, found employment for its DUKWs (amphibian trucks) in shuttling back and forth between docks and ships. In the nine-day period, 3,834 troops (mostly Army), 7,009 Korean civilians, 1,146 vehicles, and 10,013 tons of bulk cargo were evacuated. Defense of the immediate harbor area was shared with two battalions of South Korean Marines and a battalion from the 3d Infantry Division.

General Craig, Smith's sorely missed assistant division commander, returned from emergency leave on the 11th. Marines were left to wonder what his tactical role might have been if he had come back earlier. Smith sent him south to Pusan to arrange for the division's arrival. "I took 35 people of various categories with me and left for Masan," said Craig years later. "[I] conferred with the Army commander there about replacement of enormous losses of equipment of various kinds. He assured me that he would open his storerooms to us and give us anything we required that was in his stock. And this he did."

On Tuesday evening, 12 December, General Almond called his generals together for a conference and a dinner at X Corps headquarters. The division commanders—Smith, Barr, and Soule—listened without comment to a briefing on the evacuation plan. They then learned that the true purpose of the dinner was Almond's 58th birthday. General Ruffner, Almond's chief of staff, eulogized his commander, saying, in effect, that never in the history

Nerve center for the tactical air support of X Corps was the 1st Marine Aircraft Wing's Tactical Air Control Center at Hamhung. After the last of the Marine squadrons departed Yonpo on 14 December, control of air operations passed to Navy air controllers on the command ship Mount McKinley *(AGC 7).*

Photo by SSgt Ed Barnum, National Archives Photo (USMC) 127-N-A130453

National Archives Photo (USN) 80-G-422892

Wings folded, Marine F-4U Corsairs wait on the ice-glazed deck of the escort carrier Badoeng Strait *(CVE 116). As much as three inches of ice had to be scraped from the flight deck. A remarkable aircraft availability rate of 90 percent was maintained on board the carriers. Only half the missions were in support of the Marines. The rest went to the Army and South Koreans.*

of the U.S. Army had a corps in such a short time done so much. General Almond replied and General Shepherd added a few complimentary remarks. Earlier Almond had asked Smith if he thought it feasible to disinter the dead buried at Hungnam. Smith did not think it feasible.

Interdiction fires by Army artillery, deep support by naval gunfire, and air interdiction bombing by Air Force, Navy, and Marine aircraft provided a thunderous background of noise for the loading operations. By 13 December the 5th and 7th Marines were loaded and ready to sail. At 1500, General Smith closed his division command post ashore and moved it to the *Bayfield* (APA 33). Before departing Hungnam, Smith paused at the cemetery to join a memorial service for the dead. Protestant, Catholic, and Jewish chaplains officiated. Volleys were fired and taps sounded. Meanwhile, the 3d and 7th Infantry Divisions had nothing to report except light probing of their lines and minor patrol actions.

The loading of the Marines and attached Army elements was completed on the 14th. That day saw the last of the Marine land-based fighter-bombers depart Yonpo for Japan. Shortly after midnight the air defense section of MTACS-2 passed control of all air to the Navy's Tactical Air Control Squadron One on board the *Mount McKinley*, but, just to be sure, a standby Marine tactical air control center was set up on an LST and maintained until the day before Christmas.

The *Bayfield,* an attack transport and the veteran of many landings, with General Smith embarked, lifted her hook and sailed at 1030 on 15 December. The ship had been experimenting with C-rations, but with the embarkation of the Marines she

In a landing exercise in reverse, Marines in an LCM landing craft head for a transport waiting for them in Hungnam harbor. The docks could only berth seven ships at a time, so most soldier and Marines had to load out in the stream. Collecting enough ships, both U.S. Navy and Merchant Marine, for the evacuation was a monumental effort.

National Archives Photo (USN) 80-G-424506

Department of Defense Photo (USA) SC355243

These members of the 5th Marines move by way of a cargo net from an LCM landing craft into a side hatch of the transport that would take them to the southern tip of the Korean peninsula. Loading out of the regiment was essentially accomplished in one day, 12 December. Destination was the "Bean Patch" at Masan.

returned to a more appetizing Navy diet. A total of 22,215 Marines had boarded an assemblage of 4 transports, 16 landing ships, an assault cargo ship, and 7 merchant ships. General Shepherd, with his Marines safely embarked, left Hungnam the same day for Hawaii by way of Tokyo. Just before leaving Hungnam he attended a ceremony at which General Almond presented a Distinguished Service Cross to General Barr.

A day's steaming on board the jam-packed ships took the Marines to Pusan. They landed at Pusan and motor-marched to the "bean patch" at Masan where a tent camp was being set up. Smith moved into a Japanese-style house. "The toilet works, but the radiators are not yet in operation," he noted.

The Commandant reported to the Secretary of the Navy 4,418 Marine casualties for the period 26 October to 15 December. Of these, 718 were killed or died of wounds, 3,508 wounded, and 192 missing in action. In addition, there were 7,313 non-battle casualties, mostly frostbite. Roughly speaking, these non-battle casualties added up to a third of the strength of the division. (From 26 November until 11 December, Commander Howard A. Johnson's 1st Medical Battalion had treated 7,350 casualties of all categories.) The three infantry regiments had absorbed the lion's share of the casualties and arrived at the Bean Patch at about 50 percent strength. Some rifle companies had as little as 25 or 30 percent of their authorized allowance.

Chinese Casualties

Captured documents and prisoner interrogations confirmed that the Marines had fought at least nine and possibly all 12 CCF divisions. These divisions can be assumed to have each entered combat at an effective strength of about 7,500—perhaps 90,000 men in all. Other estimates of Chinese strength go as high as 100,000 or more. Peng's chief of staff said, it will be remembered, that the *Ninth Army Group* had started across the Yalu with 150,000 troops, but not all of these had come against the 1st Marine Division. The Marines could only guess at the casualties they had inflicted. The estimates came in at 15,000 killed and 7,500 wounded by the ground forces and an additional 10,000 killed and 5,000 wounded by Marine air.

Still waiting in the surrounding hills above Hamhung, Sung Shilun's *Ninth Army Group*—assuming non-combat casualties at least equal to battle casualties—probably had at most no more than 35,000 combat effectives. Almond's X Corps had three times that number. Rank-and-file Marines who grumbled, "Why in the hell are we bugging out? Why don't we stay here until spring and then counterattack?" may have had it right.

Last Days of the Evacuation

The light carrier *Bataan* (CVL 29) joined Task Force 77 on 16 December, too late to help the

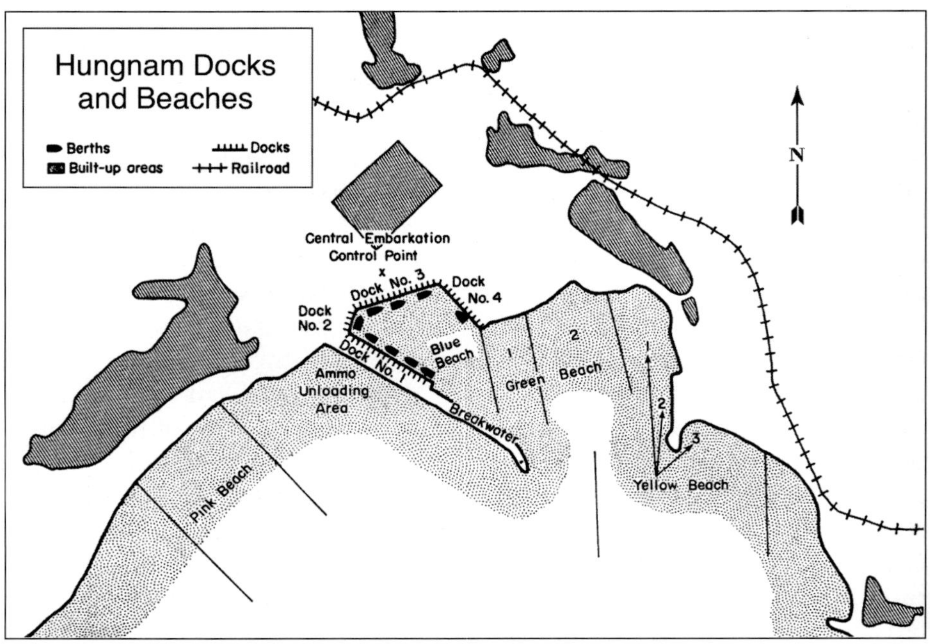

Photo by Cpl W. T. Wolfe, Department of Defense Photo (USMC) A5414

MajGen Smith, a deeply religious man, paid a last visit to the division's Hungnam cemetery before boarding the Bayfield *(APA 33) on 13 December. A memorial service with chaplains of three faiths, Protestant, Catholic, and Jewish, was held. Volleys were fired and taps sounded. During the course of the Chosin Reservoir campaign, 714 Marines were killed or died of wounds.*

Marines, but in time for the last stages of the Hungnam evacuation. Airlift from Yonpo continued until 17 December after which that field was closed and a temporary field, able to handle two-engine transports, opened in the harbor area. The only Marine units still ashore were an ANGLICO (Air Naval Gunfire Liaison Company) group, a reinforced shore party company, and one-and-a-half companies of the 1st Amphibian Tractor Battalion manning 88 amphibian tractors. These Marines had been left behind to assist in the outloading of the remainder of X Corps. General Smith had resisted this detachment, and General Shepherd, before departing, had advised Smith to stress to X Corps the irreplaceable character of the tractors. Admiral Doyle, as a safeguard, had earmarked several LSDs (landing ship, docks) to lift off the tractor companies and their vehicles.

The last of the ROK Army units sailed away on the 18th. General Almond closed his command post ashore on 19 December and joined Admiral Doyle in the *Mount McKinley*. Doyle reminded Almond that, in accordance with amphibious doctrine, all troops still ashore were now under his command as amphibious task force commander. By the 20th all of the 7th Infantry Division was embarked. On the morning of 24 December the 3d Infantry did its amphibious landing in reverse, coming off seven beaches into landing ships in smart style marred only by the premature explosion of an ammunition dump, set off by an Army captain, that killed a Marine lieutenant and a Navy sea-

Department of Defense Photo (USN) 424527

Marines, probably members of the division headquarters, board the Bayfield *by way of the gangway ladder. With MajGen Smith and his command group embarked, the* Bayfield *lifted her hook and sailed before noon on 15 December. A day's steaming would take the ship to Pusan.*

man and wounded 34 others. Three Marine amphibian tractors were lost in the explosion.

Totting up the statistics: 105,000 U.S. and ROK service men, 91,000 Korean refugees, 17,500 vehicles, and 350,000 measurement tons had gone out in 193 shiploads in 109 ships—some ships made two or even three trips.

The carrier *Valley Forge* came on station on 23 December, in time for the final curtain. By mid-afternoon on the 24th, all beaches were clear and the planned pyrotechnic display of demolitions and final naval gunfire began. The whole waterfront seemed to explode as prepared explosive charges went off, sending skyward such ammunition, POL, and other stores as could not be lifted off. On board the *Mount McKinley* the embarked brass enjoyed the show and then the command ship sailed away.

More naval shells were used at Hungnam than at Inchon. Navy records show that during the period 7 to 24 December the expenditure, headed off by 162 sixteen-inch rounds from the battleship *Missouri* (BB 63), included 2,932 eight-inch, 18,637 five-inch, and 71 three-inch shells plus 1,462 five-inch rockets. The Chinese did not choose to test seriously the Hamhung-Hungnam perimeter defenses. Not a man was lost to enemy action.

After the short run south, General Almond went ashore from the *Mount McKinley* at Ulsan at mid-afternoon with Admiral Doyle to inspect unloading areas. Late in the evening they returned in the admiral's barge to the flagship and then went ashore again for Christmas dinner, Doyle explaining to Almond that no alcoholic drinks could be served on board ship.

Chairman Mao Is Pleased

On 17 December the Chinese occupied Hamhung. On the 27th they moved into Hungnam. Chairman Mao sent the *Ninth Army Group* a citation: "You completed a great strategic task under extremely difficult conditions."

But the costs had been high. The assaults against Yudam-ni and Hagaru-ri had almost destroyed the *20th* and *27th CCF Armies*. From Koto-ri on most of the Chinese fight was taken up by the *26th CCF Army*.

Zhang Renchu, commander of the *26th CCF Army* lamented in his report:

> A shortage of transportation and escort personnel makes it impossible to accomplish the mission of supplying the troops. As a result, our soldiers frequently starve. From now on, the organization of our rear ser-

Department of Defense Photo (USA) SC355244

Vehicles of LtCol Youngdale's 11th Marines are swung up on board a merchant ship at a Hungnam dock on 14 December. Some ships had to make two or even three round trips before the evacuation was completed. About 1,400 vehicles had been brought down to Hungnam by the division. Most would go out by LSTs (tank landing ships).

vice units should be improved.

The troops were hungry. They ate cold food, and some had only a few potatoes in two days. They were unable to maintain the physical strength for combat; the wounded personnel could not be evacuated. . . . The fire power of our entire army was basically inadequate. When we used our guns there were no shells and sometimes the shells were duds.

Zhang Yixiang, commander of the *20th CCF Army*, equally bitter, recognized that communications limitations had caused a tactical rigidity:

Our signal communication was not up to standard. For example, it took more than two days to receive instructions from higher level units. Rapid changes of the enemy's situation and the slow motion of our signal communications caused us to lose our opportunities in combat and made the instructions of the high level units ineffective.

We succeeded in the separation and encirclement of the enemy, but we failed to annihilate the enemy one by one. For example, the failure to annihilate the enemy at Yudam-ni made it impossible to annihilate the enemy at Hagaru-ri.

Zhang Yixiang reported 100 deaths from tetanus due to poor medical care. Hundreds more were sick or dead from typhus or malnutrition to say nothing of losses from frostbite. The *26th CCF Army* reported 90 percent of the command suffering from frostbite.

An Army band greets Col Litzenberg's 7th Marines on arrival at Pusan. From here the regiment moved by motor march to Masan where an advance party had the beginnings of a tent camp ready for them. After a pause for Christmas, the rebuilding of the 1st Marine Division began in earnest.

National Archives Photo (USMC) 127-N-A5704

Photo by David Douglas Duncan

The 1st Marine Division was only a fraction of the total evacuation from Hungnam. In all, along with 105,000 U.S. and South Korean servicemen, 91,000 civilian refugees were evacuated. In materiel, 17,500 vehicles and 350,000 tons of all classes of equipment and supplies were taken out in 193 shiploads in 109 ships. "We never, never contemplated a Dunkirk," Admiral C. Turner Joy later said.

MajGen Robert H. "Shorty" Soule, USA, supervises the loading-out of the last elements of his 3d Infantry Division at Hungnam on 24 December. By then, MajGen Almond, the X Corps commander, considered the 3d Infantry Division, which had not been involved in heavy fighting, his most combat-effective division.

National Archives Photo (USA) 111-SC355587

Peng Deqing, commander of the *27th CCF Army,* reported 10,000 non-combat casualties in his four divisions:

> The troops did not have enough food. They did not have enough houses to live in. They could not stand the bitter cold, which was the reason for the excessive non-combat reduction in personnel. The weapons were not used effectively. When the fighters bivouacked in snow-covered ground during combat, their feet, socks, and hands were frozen together in one ice ball. They could not unscrew the caps on the hand grenades. The fuses would not ignite. The hands were not supple. The mortar tubes shrank on account of the cold; 70 percent of the shells failed to detonate. Skin from the hands was stuck on the shells and the mortar tubes.

In best Communist tradition of self-criticism, Peng Deqing deplored his heavy casualties as caused by tactical errors:

> We underestimated the enemy so we distributed the strength, and consequently the higher echelons were over-dispersed while the lower echelon units were over-concentrated. During one movement, the distance between the three leading divisions was very long, while the formations of the battalions, companies, and units of lower levels were too close, and the troops were unable to deploy. Further-

Department of Defense Photo (USN) 424093

North Korean refugees wait apprehensively to board U.S. Navy LST 845. Some 91,000 civilians were evacuated from Hungnam. This does not count the thousands of others who fled Hungnam and other North Korean ports in fishing boats and other coastal vessels. Family separations occurred that in future years would never be mended.

> more, reconnaissance was not conducted strictly; we walked into the enemy fire net and suffered heavy casualties.

Zhang Renchu, commander of the *26th CCF Army* found reason to admire the fire support coordination of the Marines:

> The coordination between the enemy infantry, tanks, artillery, and airplanes is surprisingly close. Besides using heavy weapons for the depth, the enemy carries with him automatic light firearms which, coordinated with rockets, launchers, and recoilless guns are disposed at the front line. The characteristic of their employment is to stay quietly under cover and open fire suddenly when we come to between 70 and 100 meters from them, making it difficult for our troops to deploy and thus inflicting casualties upon us.

In a 17 December message to Peng Dehuai, Mao acknowledged that as many as 40,000 men had perished due to cold weather, lack of supplies, and the fierce fighting. "The Central Committee cherishes the memory of those lost." Peng asked for 60,000 replacements; it would be April before the *Ninth Army Group* again went into combat.

Christmas at Masan

At Masan on Christmas Eve, Olin Beall, the mustang commander of the 1st Motor Transport Battalion, wrote a letter to his old commanding officer, General Holland M. "Howlin' Mad" Smith, now retired and living in La Jolla, California:

An enormous stockpile of equipment and supplies, including rations, fuel, and ammunition had been built up at Hungnam. Much was evacuated, but even more would have to be destroyed. A detachment of Marines was left behind to help in the destruction. A Marine lieutenant was killed on 24 December in a premature explosion, probably the last Marine casualty of the campaign.

Department of Defense Photo (USA) SC 355021

Department of Defense Photo (USN) 424297

The Begor *(APD 127), a high-speed amphibious transport, lends its weight to the pyrotechnic display ashore on 24 December. In a final farewell to the stricken port there was a crescendo of planned demolitions and naval gunfire. RAdm Doyle and MajGen Almond watched the spectacle from the bridge of the* Mount McKinley.

I just thought that you might like to have a few words on first hand information from an ole friend and an ole timer. . . . I've seen some brave men along that road and in these hills, men with feet frozen, men with hands frozen still helping their buddies, men riding trucks with frozen feet but fighting from the trucks. . . .I think the fight of our 5th and 7th Regts, from Yudam-ni in to Hagaru-ri was a thing that will never be equaled. . . . Litzenberg [7th] and Murray [5th] showed real command ability and at no time did any of us doubt their judgment. The night we came out of Koto-ri the temperature was 27 below zero and still we fought. Men froze to their socks, blood froze in wounds almost instantaneously, ones fingers were numb inside heavy mittens. Still men took them off to give to a wounded buddy. . . . We are now in

Department of Defense Photo (USA) 426954

More naval shells were shot at Hungnam than at Inchon. Here the battleship Missouri *(BB 63) bangs away with its 16-inch guns. Altogether the Navy fired more than 162 sixteen-inch, 2,932 eight-inch, and 18,637 five-inch shells plus 1,462 five-inch rockets during the period 7 to 24 December.*

> Masan in South Korea reoutfitting, training and getting some new equipment. I'm very, very proud to be able to say that in all our operation my Bn [1st Motor Transport Battalion] has lost only 27 trucks and every one of these was an actual battle casualty, so I think my boys did pretty good. . . . Oliver P. Smith and Craig make a fine team and we'd stand by them thru hell and high water.

An epidemic of flu and bronchitis swept through the tent camp at Masan. The Marines were treated with an early antibiotic, Aureomycin, in capsules to be swallowed the size of the first joint of a man's finger. The division rebuilt itself rapidly. Replacements—men and materiel—arrived. Some units found themselves with an "overage" of vehicles and weapons that had to be returned to the Army.

A refrigerator ship brought into Masan a planned double ration of Christmas turkey. Through some mix-up a second shipment of turkey and accessories arrived so that there were four days of holiday menu for the Marines. Working parties pretending to be patrols went up into the surrounding hills to cut pine trees to line the company streets of the tent camp. C-ration cans and crinkled tinfoil from cigarette packages made do for ornaments. Choirs were formed to sing Christmas carols. Various delegations of South Koreans, civilian and military, arrived at the camp with gifts and musical shows.

On Christmas Day, General Smith was pleased to note that attendance at church services was excellent. Afterward he held open house at his Japanese-style house for officers of sufficient rank—his special staff, general staff, and more senior unit commanders. First Lieutenant James B. Soper, serving at Sasebo, Japan, had sent the commanding general's mess a case of Old Grand-Dad bourbon. Mixed with powered milk, sugar, and Korean eggs it made a passable eggnog.

The irrepressible LtCol Olin L. Beall in a photo taken at Camp Pendleton in May 1951. Beall's exploits as commanding officer of the 1st Motor Transport Battalion, which lost nothing in his own telling, delighted MajGen Smith, himself a reserved and rather humorless individual.

National Archives Photo (USMC) 127-N-A215229

National Archives Photo (USMC) 127-N-A5848

On Christmas Day, MajGen Smith held open house for his senior officers and immediate staff at his Japanese-style quarters in Masan. Not all can be readily identified, but easily recognizable on the General's right are Cols Puller, McAlister, and Bowser. On Smith's immediate left, with pipe, is BGen Craig. In the middle of the kneeling row is LtCol Beall.

now part of the Eighth Army, which had a new commander. General Walker had been killed when this jeep collided with a South Korean weapons carrier north of Seoul on 23 December. Lieutenant General Matthew B. Ridgway, known to the Marines as a fighting paratrooper in World War II, took his place. General Smith met him for the first time at a conference at X Corps headquarters on 30 December. Ridgway told his listeners that he wanted less looking backward toward the MSR, saying that when parachutists landed their MSR was always cut. Smith, not sure if this was praise or criticism, was nevertheless cautiously impressed by the new commanding general.

By the first of the year the 1st Marine Division would be ready to return to combat. There would be new battles to be fought—and won.

Drysdale's 41 Commando also held an open house. The British embassy in Tokyo had sent over a supply of Scotch whisky and mincemeat pies. Most of the guests were officers of the 1st and 5th Marines.

On 27 December, for the benefit of his log, General Smith added up his division's losses since the Inchon landing on 15 September:

Killed in action	969
Died of wounds	163
Missing in action	199
Wounded in action	5,517
Total	6,848
Non-battle casualties	8,900
Prisoners of war taken	7,916

On the 28th of December the division was placed once again under the operational control of X Corps, still commanded by Almond who would soon be promoted to lieutenant general. X Corps was

Dressed for the occasion of MajGen Smith's Christmas party, LtCol Murray is wearing an Army winter trench coat and Col Puller a brand-new M1943 field jacket. Both wear the highly prized cuffed Army boots, but O. P. Smith, his inseparable pipe clutched in his left hand, is wearing, as always, regulation leggings with his high-top "boondockers."

Department of Defense Photo (USMC) A5850

What Happened to Them?

CHARLES L. "GUS" BANKS, commander of the 1st Service Battalion, received a Navy Cross for his actions at Hagaru-ri. He retired in 1959 with a promotion to brigadier general in recognition of his combat decorations and died in 1988.

BOEKER C. BATTERTON, commanding officer of MAG-12, retired in 1958 with a promotion to brigadier general in recognition of his combat decorations. He died in 1987.

OLIN L. BEALL, commanding officer of the 1st Motor Transport Battalion, retired as a colonel, with both a Navy Cross and a Distinguished Service Cross for his actions at Chosin Reservoir. He died in 1977.

ALPHA L. BOWSER, JR., the division's G-3 or operations officer, retired in 1967 as a lieutenant general and presently lives in Hawaii.

JAMES H. BROWER, commander of the 11th Marines, the artillery regiment, retired as a colonel in 1960 and died in 1984.

J. FRANK COLE, commanding officer of VMF-312, retired as a colonel in 1965 and died in 1969.

HENRY P. "JIM" CROWE, commanding officer of 1st Shore Battalion, retired in 1960 as a colonel, became chief of police in Portsmouth, Virginia, and died in 1991.

RAYMOND G. DAVIS, commander of the 1st Battalion, 7th Marines, went on to command the 3d Marine Division in Vietnam and was a four-star general and Assistant Commandant of the Marine Corps before retiring in 1972. He now lives near Atlanta, Georgia.

FREDERICK R. DOWSETT, the executive officer of the 7th Marines, retired as a colonel and died in 1986.

VINCENT J. GOTTSCHALK, commanding officer of VMO-6, received a Silver Star for his service in Korea. He retired as a colonel in 1968 and died in 2000.

FIELD HARRIS, commanding general of the 1st Marine Aircraft Wing, retired in 1953 and was advanced to lieutenant general because of his combat decorations. He died in 1967 at age 72.

WILLIAM F. HARRIS, commander of the 3d Battalion, 7th Marines, was listed as missing in action. No trace of him was ever found and he was eventually presumed dead. He received a posthumous Navy Cross.

BANKSON T. HOLCOMB, JR., the division's G-2 or intelligence officer, retired as a brigadier general in 1959. An expatriate, he lived for many years in Inverness, Scotland, where he died in 2000 at the age of 92.

MILTON A. HULL, company commander, Company D, 2d Battalion, 7th Marines, twice wounded, received both a Silver Star and Navy Cross for his actions. He retired as a colonel in 1969 and died in 1984.

ROBERT P. KELLER, commander of the "Blacksheep Squadron" and air liaison officer to Fifth Air Force, retired in 1974 as a lieutenant general. He lives in Pensacola, Florida.

RANDOLPH S. D. LOCKWOOD, commander of the 2d Battalion, 7th Marines, was not evacuated after being relieved but continued to move with the 7th Marines. On arrival at Masan he was sent to an Army hospital for psychiatric observation. The Army psychiatrist concluded he had suffered a situational neurosis, which disappeared after the evacuation. Lockwood returned briefly to the 2d Battalion, 7th Marines, but was soon transferred to administrative duties. He retired as a lieutenant colonel in 1960 and resides in Texas.

JAMES F. LAWRENCE, JR., who assumed command of 2d Battalion, 7th Marines, after Lockwood's relief, received a Navy Cross for his actions at the reservoir. After distinguished service as a Marine Corps lawyer, he retired in 1972 as a brigadier general. He lives in northern Virginia.

HOMER L. LITZENBERG, JR., commanding officer of the 7th Marines, rapidly ascended in grade to major general and as such in 1957 served as the senior member of the United Nations component negotiating the peace talks at Panmunjom. He retired in 1959, was elevated to lieutenant general because of his combat decorations, and died in 1963 at age 68.

FRANCIS M. MCALISTER, the division's G-4 or logistics officer, succeeded Puller as the commander, 1st Marines, a position he held until wounded in May 1951. He retired as a major general in 1960 and died in 1965.

JOHN N. MCLAUGHLIN, survived his captivity and went on to become a lieutenant general and chief of staff at Headquarters Marine Corps. He retired in 1977 and lives in Savannah, Georgia.

RAYMOND L. MURRAY, commander of the 5th Marines, rose to the grade of major general before retiring in 1968. He lives in Southern California.

REGINALD R. MYERS, Executive Officer, 3d Battalion,

1st Marines, received a Medal of Honor for his actions on East Hill. He retired as a colonel in 1967 and now lives in Florida.

George R. Newton, commander of the 1st Battalion, 5th Marines, received a Silver Star for his service in Korea and retired as a colonel in 1964. He died in 1993.

Francis F. "Fox" Parry, commander of 3d Battalion, 11th Marines, retired as a colonel in 1967. He published his memoir, *Three War Marine*, in 1987. He lives in Flourtown, Pennsylvania.

John H. Partridge, the division engineer, retired as a colonel in 1965 and died in 1987.

Lewis B. "Chesty" Puller, the 1st Marines' commanding officer, was promoted to brigadier general and became the division's assistant commander in February 1951. He received his fifth Navy Cross for his performance at the Chosin Reservoir and rose to the grade of major general on active service and to lieutenant general on the retired list when he retired in 1955. He died in 1971 at the age of 73.

J. Robert Reinburg, commander of VMF(N)-513, retired as a colonel in 1978 and died in 1997.

Thomas L. Ridge, Commanding Officer, 3d Battalion, 1st Marines, received a Silver Star for his defense of Hagaru. He retired as a colonel in 1964 and died in 1999.

Maurice E. Roach, Litzenberg's jack-of-all-trades, commander of 3d Battalion, 7th Marines, retired as a colonel in 1962 and died in 1988.

Deane C. Roberts, commander of VMR-152, retired as a colonel in 1957 and died in 1985.

Harold S. Roise, commander of 2d Battalion, 5th Marines, received two Navy Crosses for his heroic actions. He retired as a colonel in 1965 and died in 1991.

Webb D. "Buzz" Sawyer, Litzenberg's roving battalion commander, received two Silver Stars for his actions at Chosin Reservoir and a Navy Cross for later heroics during the Chinese spring counteroffensive in April 1951. He retired as a brigadier general in 1968 and died in 1995.

Henry W. "Pop" Seeley, Jr., retired as a colonel in 1963 with his last years of active duty as a highly regarded logistics officer. He lives in Florida.

Donald M. "Buck" Schmuck, Commanding Officer, 1st Battalion, 1st Marines, was later advanced to executive officer of the regiment. He retired in 1959 and because of his combat decorations was advanced in grade to brigadier general. He lives in Wyoming and Hawaii.

Carl L. Sitter, company commander, Company G, 3d Battalion, 1st Marines, recipient of a Medal of Honor, retired as colonel in 1970. He was a long-time resident of Richmond, Virginia, until his death in 2000.

Oliver P. Smith, Commanding General, 1st Marine Division, was promoted to lieutenant general in 1953 and given command of Fleet Marine Force, Atlantic. He retired in 1955 and for his many combat awards was raised in grade to four-star general. He died on Christmas Day, 1977, at his home in Los Altos Hills, California, at age 81.

Edward W. Snedeker, the division's deputy chief of staff, retired as a lieutenant general in 1963. In retirement he was known as a world-class stamp collector. He died in 1995.

Edward P. Stamford, the Marine tactical air controller with Task Force Faith, retired as a major in 1961 and lives in Southern California.

Allan Sutter, commander of the 2d Battalion, 1st Marines, retired as a colonel in 1964 and died in Orange, Virginia, in 1988.

Robert D. Taplett, commander of 3d Battalion, 5th Marines, received a Navy Cross for his heroic actions and retired as a colonel in 1960. He published his memoir, *Dark Horse Six*, in 2003.

Max J. Volcansek, Jr., commander of VMF(N)-542, retired in 1956 and was advanced in grade to brigadier general because of his combat decorations. He died in 1995.

Harvey S. Walseth, the division's G-1 or personnel officer, after recovering from his wounds, returned to the division to serve as deputy chief of staff and commanding officer, rear echelon. He retired in 1960 as a colonel and resides in Santa Barbara, California.

Erwin F. Wann, Jr., commander of the 1st Amphibian Tractor Battalion, retired as a colonel in 1965 and died in 1997.

Gregon A. Williams, the division's chief of staff, retired as a major general in 1954 and died in 1968.

David C. Wolfe, successor to Reinburg as Commanding Officer, VMF(N)-513, served as the head of the U.S. military mission in the Dominican Republic before retiring as a colonel in 1965. He died in 1992.

Carl A. Youngdale, who relieved Brower as Commanding Officer, 11th Marines, went on to command the 1st Marine Division in Vietnam and retired as a major general in 1972. He died in 1993.

About the Author

Edwin Howard Simmons, a retired Marine brigadier general, was, as a major, the commanding officer of Weapons Company, 3d Battalion, 1st Marines, throughout the Chosin Reservoir campaign. His active Marine Corps service spanned 30 years—1942 to 1972—during which, as he likes to boast, successively in World War II, Korea, and Vietnam he had command or acting command in combat of every size unit from platoon to division. A writer and historian all his adult life, he was the Director of Marine Corps History and Museums from 1972 until 1996 and is now the Director Emeritus.

Born in 1921 in Billingsport, New Jersey, the site of a Revolutionary War battle, he received his commission in the Marine Corps in 1942 through the Army ROTC at Lehigh University. He also holds a master's degree from Ohio State University and is a graduate of the National War College. A one-time managing editor of the *Marine Corps Gazette* (1945-1949), he has been widely published, including more than 300 articles and essays. His most recent books are *The United States Marine: A History* (1998), *The Marines* (1998), and a Korean War novel, *Dog Company Six*. He is the author of an earlier pamphlet in this series, *Over the Seawall: U.S. Marines at Inchon*.

He is married, has four grown children, and lives with his wife, Frances, at their residence, "Dunmarchin," two miles up the Potomac from Mount Vernon.

Sources

The official history, *The Chosin Reservoir Campaign* by Lynn Montross and Capt Nicholas A. Canzona, volume three in the five-volume series *U.S. Marine Operations in Korea, 1950-1953*, provided a starting place for this account. However, in the near half-century since this volume was published in 1957, there has been a great deal of new scholarship as well as release of classified records, particularly with respect to Chinese forces. This pamphlet attempts to benefit from these later sources.

With respect to Chinese forces, *The Dragon Strikes* by Maj Patrick C. Roe has been especially useful as have various articles by both Chinese and Western scholars that have appeared in academic journals. *The Changjin Journal*, the electronic newsletter edited by Col George A. Rasula, USA (Ret), has provided thought-provoking detail on the role of U.S. Army forces, particularly RCT-31, at the reservoir. The as-yet uncompleted work on the Hungnam evacuation by Professor Donald Chisholm has yielded new insights on that critical culminating event.

Books, some new, some old, that have been most useful include—listed alphabetically and not necessarily by worth, which varies widely—Roy E. Appleman, *East of Chosin* and *South to the Naktong, North to the Yalu*; Clay Blair, *The Forgotten War;* Malcolm W. Cagle and Frank A. Manson, *The Sea War in Korea*; T. R. Fehrenbach, *This Kind of War*; Andrew Geer, *The New Breed;* D. M. Giangreco, *War in Korea, 1950-1953;* Richard P. Hallion, *The Naval Air War in Korea*; Max Hastings, *The Korean War;* Robert Leckie, *The March to Glory;* Douglas MacArthur, *Reminiscences*; Francis Fox Parry, *Three War Marine;* Russell Spurr, *Enter the Dragon*; Shelby L. Stanton, *America's Tenth Legion;* John Toland, *In Mortal Combat: Korea 1950-1953*; Rudy Tomedi, *No Bugles, No Drums*; and Harry Truman, *Memoirs*.

The official reports that proved most helpful were the Far East Command's *Command Report, December 1950*; the 1st Marine Division's *Historical Diary for November 1950;* the Commander, Task Force 90's *Hungnam Redeployment, 9-25 December 1950*; and the Headquarters, X Corps, *Special Report on Chosin Reservoir, 17 November to 10 December 1950.*

Oral histories, diaries, memoirs (published and unpublished), and personal correspondence were extremely useful, especially those papers originating with Generals Almond, Bowser, Craig, Litzen-berg, Murray, Shepherd, and Smith.

Resort was made to scores of biographical and subject files held by the Reference Section of the Marine Corps Historical Center.

The author also unabashedly put to use his own recollections of events and recycled materials that he had first developed on Chosin Reservoir in various essays, articles, and lectures during the past half-century.

As is invariably the case, the author had the unstinting and enthusiastic support and cooperation of the staff at the Marine Corps Historical Center.

The text has benefited greatly from the critical reviews by the editorial ladder within the Marine Corps Historical Center—Mr. Charles R. "Rich" Smith, Mr. Charles D. Melson, LtCol Jon Hoffman—and externally by Col Joseph Alexander, Col Thomas G. Ferguson, USA (Ret), BGen James F. Lawrence, Col Allan R. Millett, Mr. J. Robert Moskin, Col George A. Rasula, USA (Ret), and Maj Patrick C. Roe. The author, of course, remains responsible for any defects remaining in the book.

A fully annotated draft manuscript is on deposit at the Marine Corps Historical Center. Virtually all of the reference materials published and unpublished, used can be found at the Marine Corps Historical Center in Washington, D.C., or at the Marine Corps Research Center at Quantico, Virginia.

COUNTEROFFENSIVE
U.S. Marines from Pohang to No Name Line

by Lieutenant Colonel Ronald J. Brown, USMCR (Ret)

At Hungnam, the 1st Marine Division, following the withdrawal from the Chosin Reservoir, embarked all of its equipment and personnel in record time and sailed for Pusan. The trip south for the half-starved, half-frozen Marines was uneventful except for the never-closed chow lines, salt-water showers, a complete change of clothes, and a widespread outbreak of colds or mild cases of pneumonia. "For the first time in weeks we felt clean," wrote one Marine, "and our lice were gone forever—washed down a drainhole into the cold Sea of Japan." In addition to a scrub down and new dungarees, there was a good deal of conjecture and discussion on the possible employment of the division; many hoped that instead of landing at Pusan, the convoy would proceed directly to Japan or the United States and relief by the 2d Marine Division. Both officers and enlisted men alike held that it was impossible to visualize the employment of the division in the near future and that rest, reorganization, and rehabilitation was an absolute necessity. Then, too, there were those who had fought around the Pusan Perimeter and were "not too happy or not too eager to see the dreadful country they had fought over." Regardless of the speculation, the convoy steamed on, and on 16 December arrived at Pusan. Although several tank landing ships sailed past Pusan and put in at Masan, a majority of the division's Marines traveled by rail and road from Pusan 40 miles west to their new area outside the small seaport untouched by war.

AT LEFT: *Marine riflemen hug the ground as they advance under fire during Operation Ripper.* National Archives Photo (USMC) 127-N-A6862

In an area previously occupied by the 1st Provisional Marine Brigade, a tent city quickly sprang up—pyramidal tents for all members of the command and squad tents for each battalion. Hospital tents and mess halls were erected and with the help of Korean laborers mess tables and other improvements soon began to appear. A large barracks in the outskirts of Masan served as the administrative headquarters for the regiments, while the division's service and support units occupied areas near the docks and south of town. The men observed the division's first Christmas in Korea with a memorable display of holiday spirit despite a chilling drizzle. A choir from the 5th Marines serenaded the division headquarters with carols, many attended a series of shows put on by troupes of U.S. Army and Korean entertainers, and the U.S. Navy sent Christmas trees and decorations. It was not only a time to be thankful, but also a period of rapid recuperation from fatigue and nervous tension.

As 1950 drew to a close the military situation in Korea was so bleak American policy makers were seriously contemplating the evacuation of U.S. forces from that embattled country, and American military leaders had already formulated secret contingency plans to do so. The Korean Conflict had been raging for six months during which time the fighting seesawed up and down the 600-mile length of the mountainous peninsula with

Gen Douglas MacArthur, America's longest-serving soldier, was Commander in Chief, Far East, and also commanded the multinational United Nations forces in Korea. Although the situation appeared ominous in early 1951, MacArthur later said he never contemplated withdrawal and "made no plans to that effect."
Department of Defense Photo (USA) SC362863

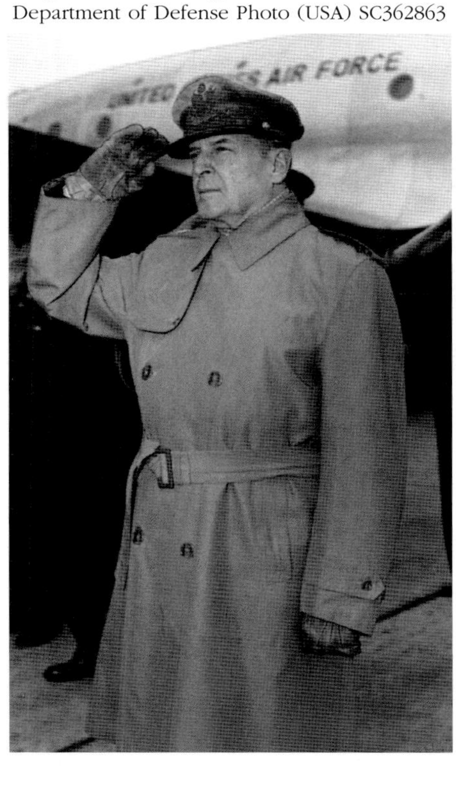

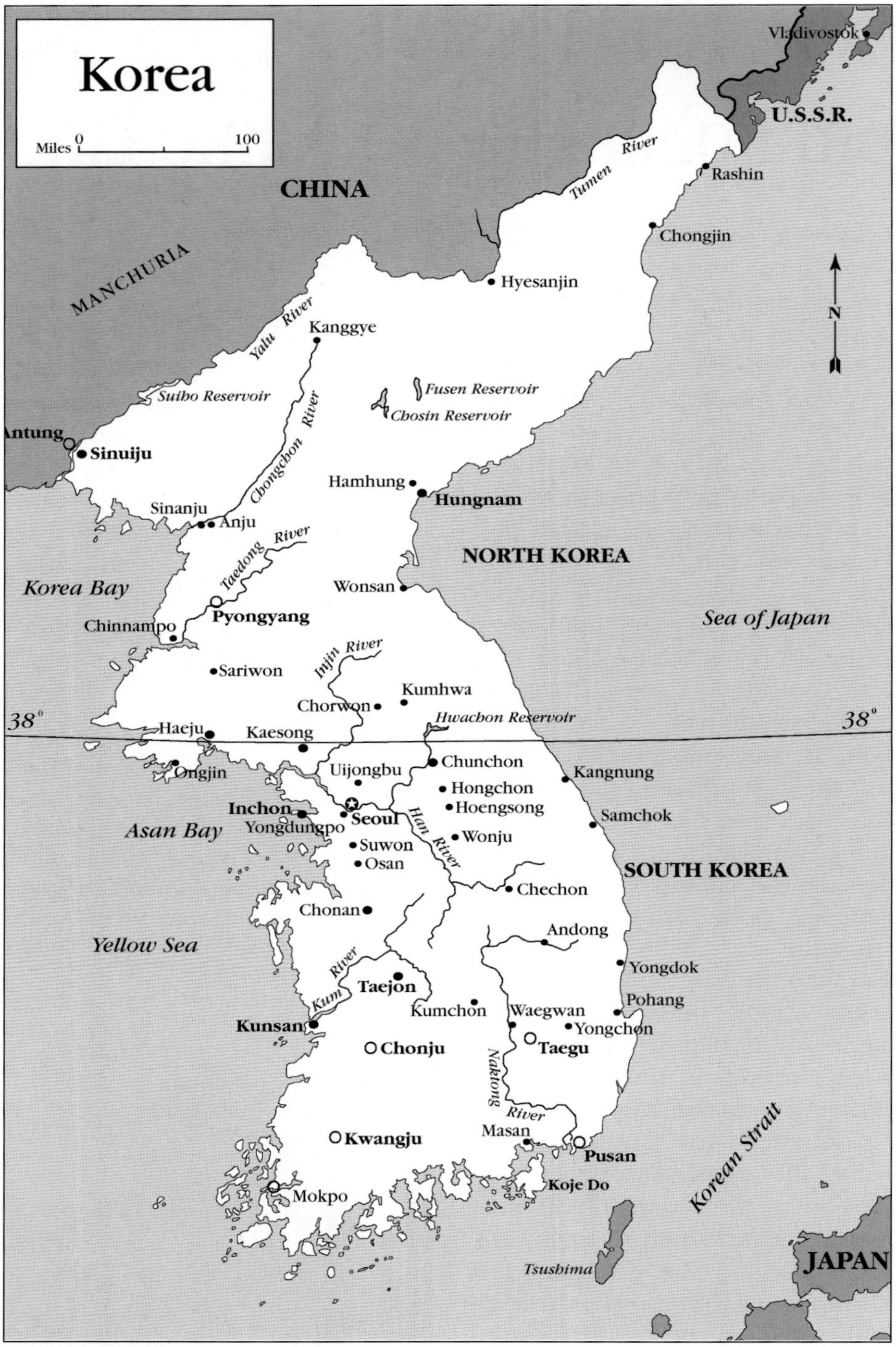

first one side and then the other alternately holding the upper hand. With 1951 only a few days away the Communist forces—consisting of the North Korean People's Army (NKPA) reinforced by "volunteers" from the People's Republic of China, known as the Communist Chinese Forces (CCF)—appeared on the verge of victory. In a series of stunning blows that began the previous November, the United States-led United Nations Command had been pushed back from the Yalu River at the North Korean-Chinese border all the way south of the 38th Parallel, which divided North and South Korea. A momentary lull in the action, however, allowed the energetic new United Nations field commander, Lieutenant General Matthew B. Ridgway, USA, to rally his troops just as the outlook was darkest. This fortuitous event began a dramatic reversal of fortunes, a turnaround so startling that within six months it was the Communists who were on the ropes.

The combined NKPA and CCF armies had more than a half million men inside Korea while the United Nations Command numbered only about two-thirds that many. The U.N. commander was American General of the Army Douglas MacArthur who was concurrently Commander in Chief, Far East. The major Service components of the Far East Command were the Eighth Army, the Fifth Air Force, and elements of the Seventh Fleet. Recently appointed Lieutenant General Ridgway commanded the Eighth Army; Major General Earl E. Partridge, USAF, the Fifth Air Force; and Vice Admiral Arthur D. Struble, USN, the Seventh Fleet. Major General Oliver P. "O. P." Smith's 1st Marine Division and Major General Field Harris' 1st Marine Aircraft Wing were the two major Marine units in Korea. Unlike today's expeditionary force structure, at that time there was no Marine component headquarters so the non-Marine theater commander was the only common superior officer for both the division and aircraft wing in Korea. The nearest senior Marine was Lieutenant General Lemuel C. Shepherd, Jr., Commanding General, Fleet Marine Force, Pacific, in Hawaii, who was responsible for the logistical support of both the division and wing. Despite the fact that no official direct command link existed between Marine air and ground units in Korea, the respective Marine commanders maintained close liaison and carefully coordinated their actions.

Several important new com-

Department of Defense Photo (USMC) A6087

The general staff of the 1st Marine Division assembled for an informal photograph shortly after the New Year. Pictured from left are: Capt Eugene R. Hering, USN, Division Surgeon; Col Alpha L. Bowser, Jr., G-3; Col Bankston T. Holcomb, Jr., G-2; MajGen Oliver P. Smith, Commanding General; Maj Donald W. Sherman, Assistant G-1; BGen Edward A. Craig, Assistant Division Commander; Col Edward W. Snedeker, Chief of Staff; and Col Francis M. McAlister, G-4.

mand relationships developed after the Marines' fighting withdrawal from the Chosin (Changjin) Reservoir. Marine aircraft, which had provided superb close air support for Marine ground units for the previous five months, would no longer be on direct call. Instead, the potent Marine air-ground team was broken up so land-based aircraft of the 1st Marine Aircraft Wing could be incorporated into the Fifth Air Force. The U.N. ground command also underwent some changes. The 1st Marine Division passed from X Corps to Eighth Army control in mid-December 1950, just about a week before the tough and energetic Army paratrooper, General Ridgway, was named Eighth Army commander after his predecessor, Lieutenant General Walton H. Walker, USA, was killed in a traffic accident.

The Masan Bean Patch

After the ordeal at the Chosin Reservoir, the 1st Marine Division moved to Masan in southern Korea where it became part of Eighth Army reserve. Concurrently, the 1st Marine Aircraft Wing was flying from aircraft carriers and airfields in Korea but was about to temporarily deploy to Japan. On the home front, three replacement drafts (the 3d, 4th, and 5th) were either already enroute or were preparing to ship out. Hopefully, their arrival would bring the depleted Marine ranks in Korea back up to strength before the next round of combat began.

The battered 1st Marine Division spent two weeks licking its wounds in a rest area known as the "Bean Patch" about 200 miles south of the main line of resistance. Its three rifle regiments, each of which was led by a future lieutenant general, occupied the agricultural flat lands on the north-

National Archives Photo (USN) 80-G-424655

Marines encamped near Masan in the wake of their return from the Chosin Reservoir. The area had been a rest area when the 5th Marines fought to save the Pusan Perimeter four months earlier.

ern outskirts of Masan, which gave the area its name. Division headquarters and most of the combat support and service elements, including the helicopters and observation aircraft of Major Vincent J. Gottschalk's Marine Observation Squadron 6 (VMO-6), were located nearby. The 1st Marine Division was in very good hands. Devout, pipe-smoking, white-haired O. P. Smith was tall and thin with a scholarly manner, factors that led some observers to remark that he looked more like a preacher than a Marine general. Fortunately, appearances can be deceiving. Smith's performance as a commander thus far in Korea had been outstanding. A respected military analyst studying the Chosin campaign noted that Smith was a careful planner and superb tactician who repeatedly resisted pressure to execute rash orders issued by his corps commander, actions that probably saved the 1st Marine Division from piecemeal destruction.

The 1st Marine Division was also blessed with four of the finest regimental commanders in Korea. The 1st Marines was led by legendary Colonel Lewis B. "Chesty" Puller, whom General Ridgway proudly lauded as "a man of indomitable spirit . . . the officer with the most combat experience in Korea." The 5th Marines commander was lanky Lieutenant Colonel Raymond L. Murray, another seasoned combat veteran. An "Old China Hand" who fought at Guadalcanal, Tarawa, and Saipan during World War II, Murray brought the 5th Marines ashore in August 1950, and ably led his regiment through every Marine engagement in Korea thus far. He would later gain some literary notoriety as the role model for the fictional "High Pockets" Huxley in Leon Uris' best selling novel *Battle Cry*. Colorful, fiery-tempered, hard-driving Colonel Homer L. Litzenberg, Jr., commanded the 7th Marines.

Lieutenant Colonel Carl A. Youngdale led the division's artillery regiment. Youngdale served with the 14th Marines of the 4th Marine Division throughout World War II. In 1950, he came to Korea as the 11th Marines' executive officer and then took over the unit when its commander, Colonel James H. Brower, was evacuated from the Chosin Reservoir in November. Almost two decades later Major General Youngdale would command the 1st Marine Division in Vietnam.

On New Year's Eve the Communists opened their Third Phase Offensive. This massive attack pushed overextended U.N. lines back under heavy pressure, and the United Nations Command was forced to cede the South Korean capital city of Seoul to the enemy for a second time. But this fighting withdrawal was not at all like the helter-skelter retreats following the North Korean invasion of June 1950. This time the Eighth Army fell back in good order to a series of preplanned defensive lines, the last of which would be, if needed, just outside the port of Pusan much farther back than the original Pusan Perimeter. American units traded ground for time while inflicting maximum casualties upon their advancing foe. In short, the U.N. lines were bending but not breaking, and there was no sense of panic. "We came back fast," General Ridgway admitted, "but as a fighting army, not as a running mob. We brought our dead and wounded with us, and our guns, and our will to fight."

Fortunately, the United Nations Command stemmed the oncoming tide so the 1st Marine Division never had to assume the role of rear guard. Instead, the division rested, rehabilitated, restored broken equipment, rearmed, and absorbed almost 3,000 replacements during the last days of 1950, most filling shortages in the infantry and artillery regiments. Daily security patrols were mounted with the purpose of making a reconnaissance of roads and questioning Korean civilians about the nature of guerrilla activity in the area, but no enemy were encountered and there was no evidence of any inclination on the part of the enemy or guerrillas to harass the division.

There were many high-level visitors at Masan during the division's brief stay. General Ridgway dropped in to inspect the Marines and observe field training. "He fully expected to find a division which was so weary and so beaten up, and perhaps even with a defeatist attitude after all he'd read and heard about the Chosin Reservoir," as Colonel Alpha L. Bowser, the division's assistant chief of staff for operations, later noted. But "much to his surprise the thing that impressed him most, everywhere he went in the division, and everybody he talked to, [was] their attitude about what they were doing to get themselves back in shape to get back into the battle." Very satisfied with what he saw, he complimented General Smith for the division's quick recovery. Although it was short of men and equipment, Ridgway still deemed the 1st Marine Division his most effective combat unit. He was, in fact, holding the Marines in reserve in case the pending crisis in the north worsened. Should the U.N. lines break, Ridgway wanted the 1st Marine Division to hold open a corridor to the port of Pusan and then act as a rear guard to cover the U.N. evacuation. Ridgway "gave the impression to us that he was a commander, with plenty on the ball, who had combat experience and the will to fight," recalled Brigadier General Edward A. Craig, the assistant division commander.

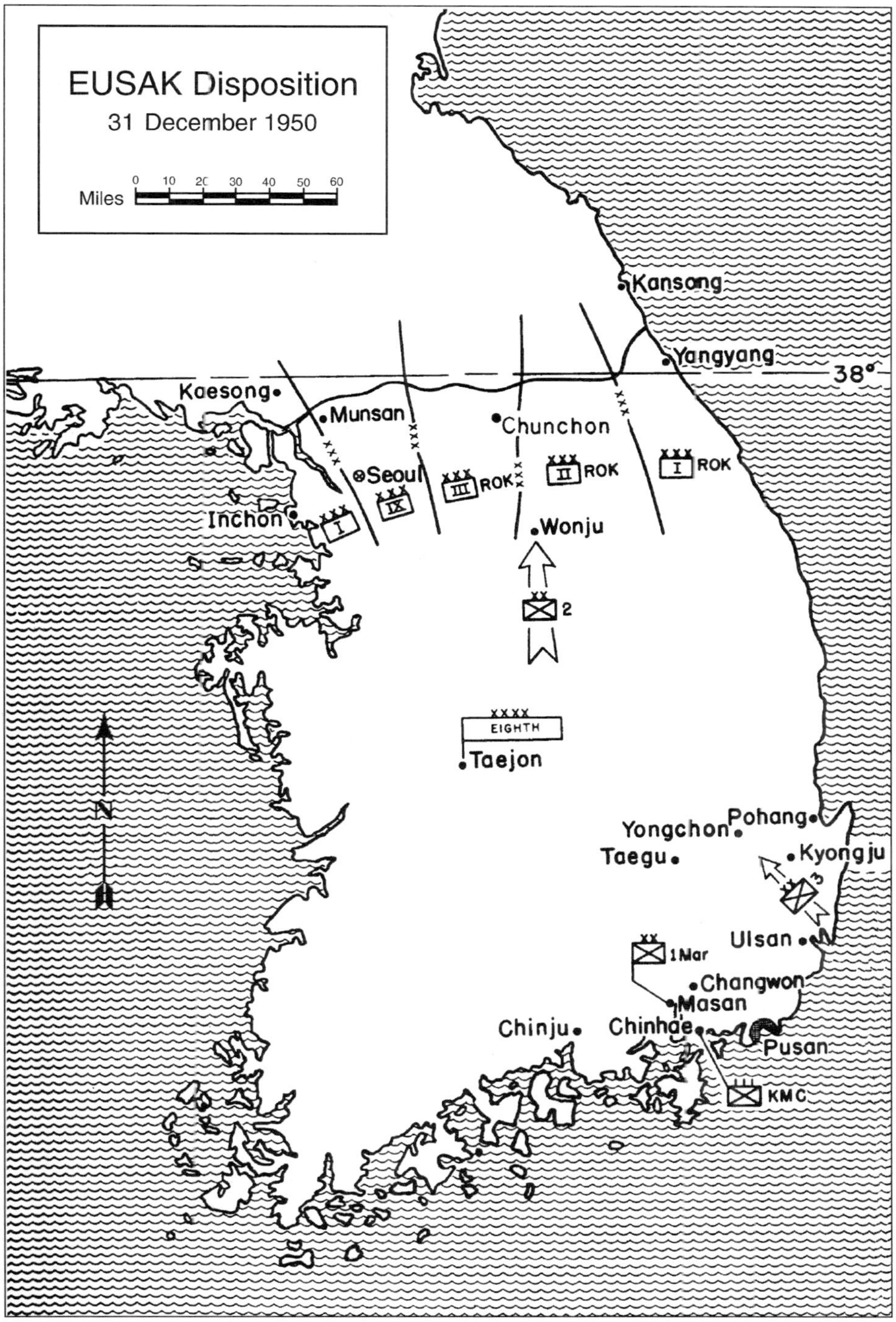

Among other noteworthy visitors was Captain John Ford, USNR, the famous motion picture director, who had been recalled to active duty by Rear Admiral Arthur W. Radford, commanding the Pacific Fleet, who felt the Navy and Marine Corps had not received enough war coverage. Ford gathered background for a documentary he was filming, and some film clips shot at Masan were later used in the feature film "Retreat Hell." Also on hand was military historian Colonel Samuel L. A. Marshall, USAR, who interviewed numerous Marines for a classified report about infantry combat in Korea, portions of which were later published in his book *Battle at Best*. The most welcomed visitors, however, were entertainer Bob Hope and his traveling USO show. The Marines at Masan thoroughly enjoyed a chance to laugh heartily, and many of them stared in awe at the first American women they had seen in months.

Sleep, sports, and good chow

National Archives Photo (USMC) 127-N-A5640

During the brief stay at Masan, Marines rested, gained back some of their lost weight, and found time to engage in an impromptu volleyball game. Despite the efforts of Fleet Marine Force, Pacific, to scrape the bottom of the manpower barrel, the division was still short more than 3,500 officers and men.

were the watchwords at Masan. As Lieutenant Colonel Francis F. Parry, 3d Battalion, 11th Marines' commanding officer, remarked: "We had so much turkey it was coming out of our ears." Impromptu softball, basketball, touch football, and volleyball games became daily rituals, and these were occasionally followed by a well appreciated, albeit limited, beer ration. Weapons familiarization and small unit tactics dominated the training schedule. And as was done prior to the trek down from Hagaru-ri, the division's medical staff examined all personnel, surveying the men for those who, noted Lieutenant Colonel Parry, might have "hidden the fact that they were frostbitten or didn't consider it was worthy of note till we got down to Masan." The serious cases were evacuated. Although the men eventually were allowed go into town, visit the stores, and purchase a lot of useless things, such as artificial flowers and non-regulation fur hats, "there was no liberty," Parry recalled. "A few troops got drunk on native brew and went blind and a few of them caught a venereal disease, but there was no liberty to amount to anything, no recreation that could properly let the troops relax and enjoy themselves for a while, such as could have been obtained in Japan." Despite a fortnight's respite and frantic efforts to bring the 1st Marine Division back up to full strength, General Smith was still short of men, tanks, and communications equipment when the call to return to action finally came.

Notwithstanding the short period of recuperation, fatigue among the officers and men of the division was apparent after more than four months of combat. Concerned, General Smith told his unit commanders that "we had to get our men in hand, do everything we could for them, but not let them begin to feel sorry for themselves." Some of the officers and men, primarily the commanding officers, noted Lieutenant Colonel Parry, "started to lose a little of their zip and hard-charging qualities. Some of the battalion commanders of the 5th who had been through three campaigns were getting to be pretty sick men. They weren't charging up hills the same way they had when they first got there."

In early January, the Communist's strategic goal was to

Veterans of the exhausting Chosin Reservoir campaign used their time at the Masan to hone basic military skills. Here Marines review marksmanship techniques under the watchful eye of a noncommissioned officer.

Department of Defense Photo (USMC) A5628

Lieutenant General Matthew Bunker Ridgway, USA

Lieutenant General Matthew B. Ridgway was called suddenly to Korea to take over the Eighth U.S. Army following the death of its previous commander, Lieutenant General Walton H. Walker. When Ridgway arrived the Eighth Army was in disarray, its morale shattered by heavy losses suffered during the longest withdrawal in American military history. The new Eighth Army commander promptly engineered, to use the words of General of the Army Omar N. Bradley, Chairman of the Joint Chiefs of Staff, "a battlefield turnaround unlike any within American History." Within four months the United Nations Command had regained all lost territory south of the 38th Parallel. As military historian Colonel Harry G. Summers, a veteran of Korea, noted: "Under Ridgway the Eighth Army toughened up and became as good a fighting force as this country has ever fielded." Just as his troops were about to reenter North Korea, General Ridgway was again unexpectedly thrust into higher command when he replaced General Douglas MacArthur as commander of United Nations forces in April 1951.

The son of an artillery colonel and a West Point graduate, Ridgway was an intellectual and diplomat as well as a superb tactical commander. He possessed brains, courage, and decisiveness—traits that served him well in Korea. His peacetime military assignments included overseas stints in the Far East, Latin America, and Europe. In 1942, he was given command of the elite 82d Airborne Division and led the unit in operations against Axis forces in Sicily, Italy, and France. He was "a kick-ass man," one subordinate said, who became known among his men as "Tin-tits" because of the hand grenades so prominently strapped to his chest. Taking command of XVIII Airborne Corps in 1944, Ridgway participated in the Battle of the Bulge and subsequent operations leading to Germany's surrender in 1945. He was serving as the deputy Army chief of staff for plans in Washington, D.C., when his call to Korea came.

Unlike his predecessor, General Ridgway was given a free hand in Korea. When he asked for instructions, General MacArthur simply told him: "The Eighth Army is yours, Matt. Do what you think best." Following an initial tour of the combat area, Ridgway was astonished at the decided lack of morale and purpose, shoddy discipline, and atmosphere of defeat. Problems meant opportunity for the battle-hardened, disciplined paratrooper. First, the men of the Eighth Army needed an adequate answer from their commanding general to the question: "What are we fighting for?" "To me the issues are clear," he wrote:

> It is not a question of this or that Korean town or village. Real estate is, here, incidental. It is not restricted to the issue of freedom for our South Korean Allies, whose fidelity and valor under the severest stresses of battle we recognize; though that freedom is a symbol of the wider issues, and included among them.
>
> The real issues are whether the power of Western civilization, as God has permitted it to flower in our own beloved lands, shall defy and defeat Communism; whether the rule of men who shoot their prisoners, enslave their citizens, and deride the dignity of man, shall displace the rule of those to whom the individual and his individual rights are sacred; whether we are to survive with God's hand to guide and lead us, or to perish in the dead existence of a Godless world
>
> If these be true, and to me they are, beyond any possibility of challenge, then this has long since

National Archives Photo (USA) 111-SC360069

ceased to be a fight for freedom for our Korean Allies alone and for their national survival. It has become, and it continues to be, a fight for our own freedom, for our own survival, in an honorable, independent national existence.

The sacrifices we have made, and those we shall yet support, are not offered vicariously for others, but in our own direct defense.

In the final analysis, the issue now joined right here in Korea is whether Communism or individual freedom shall prevail, and, make no mistake, whether the next flight of fear-driven people we have just witnessed across the HAN, and continue to witness in other areas, shall be checked and defeated overseas or permitted, step by step, to close in on our own homeland and at some future time, however, distant, to engulf our own loved ones in all its misery and despair.

Ridgway not only was determined to recapture moral leadership, but also insisted that the Eighth Army needed to return to infantry combat fundamentals. He sternly ordered his corps commanders to prepare for coordinated offensive action, and he forcefully reminded his division and regimental commanders to get off the roads, to take the high ground, and to use perimeter defenses. He studied previous campaigns and recognized a pattern to Communist operations; they would advance, attack, and then suddenly break contact until resupplied. Ridgway decided the answer was to fall back in an orderly manner trading space to inflict casualties then, once the Communists stopped, to attack and relentlessly pursue them. His plan proved to be very successful. Ridgway's offensive, also known as the "meatgrinder" because of heavy Chinese and North Korean casualties, had by early spring 1951 resulted in the recapture of Seoul and the recovery of all of South Korea by mid-April, when he left to take over as theater commander in Tokyo.

Although Marines admired General Ridgway's offensive spirit and his professionalism, they were disappointed with two of his high-level decisions. First, he pulled the 1st Marine Division away from the sea and began to use that highly trained amphibious unit as just another infantry division; second, he acquiesced to the breakup of the Marine air-ground team by allowing Marine aircraft squadrons to be directly controlled by the Fifth Air Force. Lieutenant General Lemuel C. Shepherd, Jr., the Fleet Marine Force, Pacific, commander, viewed both of these actions as tactical mistakes. Previously, General MacArthur—who had commanded Marine units in the Southwest Pacific during the World War II and fully appreciated their unique capabilities—always kept the very successful Marine air-ground team intact, and he usually tried to keep the Marines near the sea as well. For a variety of reasons, Ridgway did not.

As theater commander, General Ridgway reorganized the U.S. Far East Command to make it a true joint headquarters, never meddled in the tactical handling of forces in Korea, and maintained a good relationship with his superiors in Washington. After leaving the Far East, Ridgway succeeded General Dwight D. Eisenhower as Supreme Commander of Allied Forces in Europe and in 1953 he was named Chief of Staff of the Army. His tenure as Army Chief of Staff was a series of bitter quarrels with what he took to be President Eisenhower's refusal to remember, "most of what counts in battle is the Infantry." A few months short of mandatory retirement, he left the Army in June 1955. He later served as executive director of various business firms until his death in 1993 at the age of 93.

Military historians frequently hold him up as the epitome of the modern "soldier-statesman," but it was the men who served with him in battle who had the most praise. As Major William L. Bates, Jr., commanding officer, Weapons Company, 1st Battalion, 1st Marines, and later operations officer, 1st Battalion, 1st Marines, said of General Ridgway at the time: "He is a real, down-to-earth, honest-to-God soldier. He is a general who can visit a battalion, go into the attack with it, watch it operate, remain several hours, and never try to tell you how to run your own outfit. He is personally courageous and spends much of his time following Patton's suggestion of letting the troops see you at the front. He has sound ideas on the employment of infantry troops, and he knows how to fight, small and large scale. He is, I would guess, the best field commander the Army has had in a long time."

---

divide Korea in half, to separate the U.S. and South Korean forces. The Chinese first carried the main attack aimed at Seoul and Inchon. The NKPA attacked Korean-held Line D from Hoengsong in central Korea not long after the CCF mounted its western offensive. Farther east, North Korean commander General Kim Chaek wanted to drive straight down the center of the peninsula to capture the U.N. staging area at Taegu. His plan was to take Wonju with a frontal attack by Major General Pang Ho San's *V Corps* while outflanking the U.N. lines from the east using Lieutenant General Choe Hyon's *II Corps*; the *III Corps* was his reserve. All went well at first. *II Corps* cracked through the South Korean lines and proceeded down Route 29 peeling off divisions to cut the U.N. line of retreat. The *27th Division* invested Chechon; the *31st Division* attacked Tanyang; the *2d Division* cut Route 29 north of Yongju; and the *10th Division* headed for Andong. If all went according to plan, the United Nations Command would lose control of its main supply route and be denied passage to the port at Pusan. This, combined with the loss of Seoul and Inchon, would effectively end the war.

The Third Phase Offensive presented problems, but General Ridgway was confident his revitalized Eighth Army could handle the situation. Obviously, Ridgway's first priority was to stop the

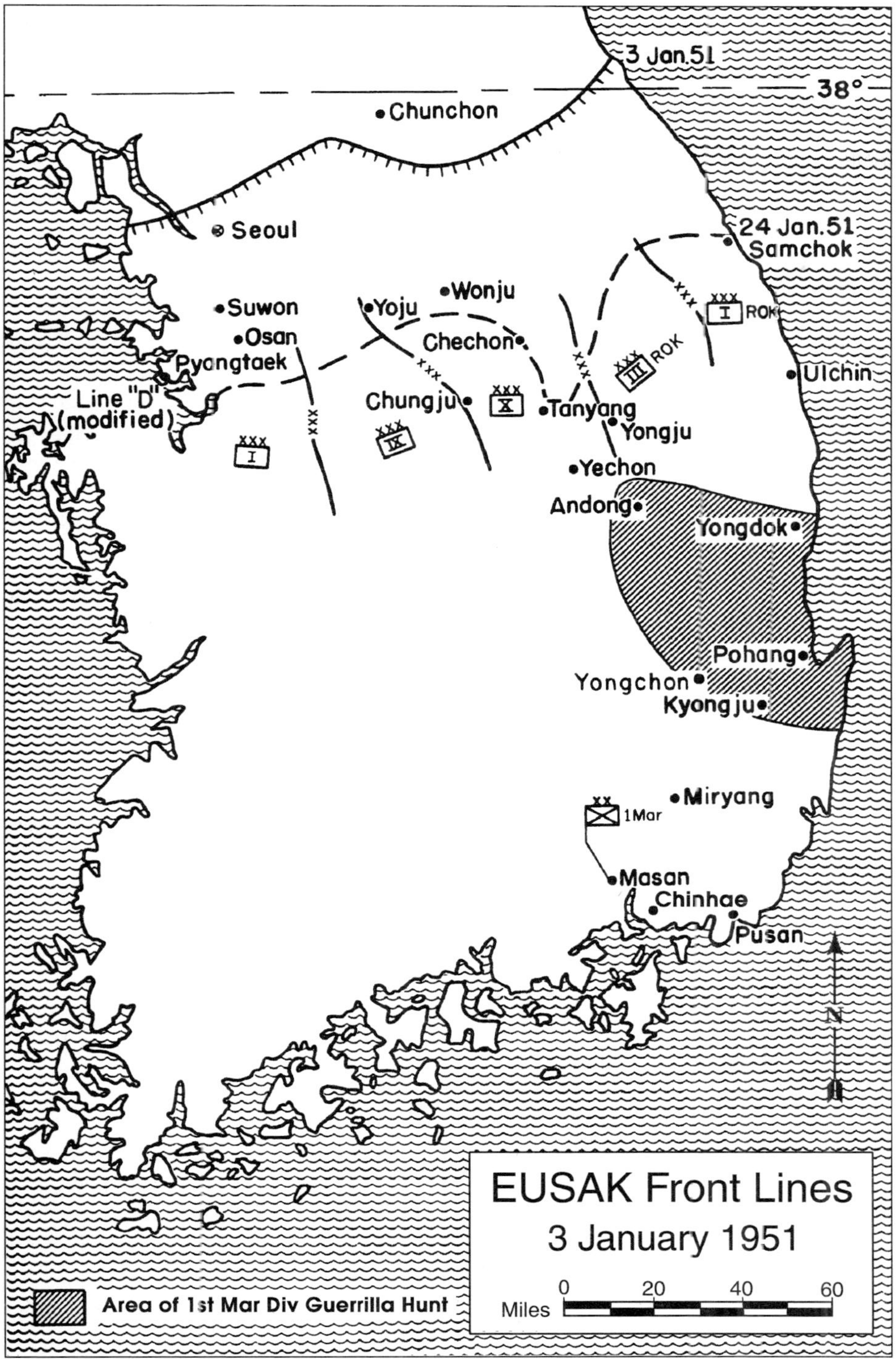

Chinese, so he committed the bulk of his forces near the west coast. He knew that once the Chinese offensive was blunted he could safely shift forces to central Korea. The U.S. 2d Division, the only American unit he had available for the Central Front, was hastily sent forward to defend Wonju. Ridgway's decisive actions set the stage for all Marine combat operations in the spring of 1951.

While the Marines were resting in the Bean Patch, the struggle shifted from Seoul to central Korea. Fighting in knee-deep snow and bitter cold, outnumbered U.N. defenders grudgingly fell back as the enemy poured through a gap in Republic of Korea (ROK) Army lines, endangering Wonju, a vital road and rail junction south of Hoengsong. Responding to this threat, General Ridgway flashed a series of messages to Smith's headquarters. One of these was a warning order for elements of the 1st Marine Division to be ready to move 65 miles northeast to Pohang-dong, a sleepy fishing village about a third of the way up Korea's east coast, in order to protect Eighth Army lines of communication and backstop some shaky Korean divisions. The Pohang area had great strategic importance because it included a significant stretch of the Eighth Army main supply route (National Route 29), housed several key road junctions, included the only protected port on the east coast still in U.N. hands, and was the site of one of the few modern airfields (Yongilman, a former Japanese fighter base labeled "K-3" by the Americans) in eastern Korea. This mission was confirmed on 8 January, but it had by then been modified to include the entire 1st Marine Division which was not assigned to a corps, but would instead be directly under Eighth Army operational control. The division staff cut orders on the 9th, and the Marines began moving out the next day with the maneuver elements going by truck and the support units by air, rail, and ship. The brief Masan interlude was over. The 1st Marine Division was headed back into action.

The Pohang Guerrilla Hunt

Marine activities along the east coast of Korea in late January and early February of 1951 eventually came to be known as the "Pohang Guerrilla Hunt" by the men of the 1st Marine Division. This period began with a week-long movement from Masan to Pohang that started with the departure of the 1st Marine Division vanguard, "Chesty" Puller's 1st Marines organized as a regimental combat team, on 10 January. A motor convoy carried elements of the 1st Marines; the division Reconnaissance Company; the 2d Battalion, 11th Marines; Company C, 1st Engineers; and Company D, 1st Medical Battalion, on a tedious 10-hour journey from Masan to Yongchon. Upon arrival at Uisong the next day, the regimental combat team, later dubbed "Task Force Puller" by General Smith, began patrolling a 30-mile section of road. Two days later, the reinforced 1st Battalion, commanded by Lieutenant Colonel Donald M. Schmuck, a Colorado native and Peleliu veteran, moved 15 miles north to occupy Andong. A key crossroads about 40 miles inland from the sea, it was the site of X Corps rear headquarters as well as two dirt airstrips (one of which was long enough to handle cargo planes, but the other able only to service light observation aircraft and helicopters). As the 1st Marines edged closer to Andong, Puller was convinced, despite General Ridgway's promise to keep the division intact, that the next step would be to attach his unit to X Corps and he would be "off to the races again." Puller, as General Smith later noted, "was apprehensive about being put out on a limb. The basic difficulty was that he had no confidence in the staying power of the Army units deployed north of Andong. Puller felt they might 'bug out' and leave him 'holding the bag.' As far as the Division was concerned, "RCT-1 was strong enough to protect its own withdrawal if it came to that." With the arrival of the division's two other regimental combat teams, soon-to-be colonel, Raymond Murray's 5th Marines patrolled the coast from Pohang to Yongdok and defended the main airfield, while Colonel Homer Litzenberg's 7th Marines occupied

1stMarDiv Historical Diary Photo Supplement, Jan-Feb51

A 75mm recoilless rifle position covers a likely avenue of approach in the vicinity of Andong. Crew-served recoilless rifles were more reliable and had much greater range than the individually carried 3.5-inch rocket launchers.

An aerial photograph of Pohang shows the rugged, irregular hill masses where North Korean guerrillas sought refuge. This village on Korea's east coast was the 1st Marine Division's base of operations in January and February 1951.

National Archives Photo (USA) 111-SC346705

centrally located Topyong-dong. The last Marine units disembarked from tank landing ships at Pohang on 17 January.

Although there was some limited discussion about small-scale amphibious operations by General Ridgway when he visited the 1st Marine Division command post at Pohang, these never came to fruition. Instead, he ordered the Marines to defend an east-west line just north of the Andong-Yongdok Road and to simultaneously protect the north-south-running Eighth Army main supply route. General O. P. Smith faced a dilemma because he was at first uncertain about which of these assignments should receive the highest priority. Should he deploy to guard against an all-out attack on the main line of resistance by Communist regular forces from the north or be prepared for counter-guerrilla operations against small groups of infiltrators? Intelligence reports indicated that the latter was the most likely course of action. Small enemy bands had already proved extremely troublesome by intermittently cutting supply lines and occasionally attacking outposts between Wonju and Taegu so continued guerrilla actions were considered probable. General Smith was well aware that the Marines would not be manning an exposed position. Several South Korean divisions screened the Marine northern flank, the Sea of Japan protected his eastern flank, and hilly terrain made the western approaches inaccessible to armor. Smith, therefore, decided to emphasize mobile security operations and made linear defense a secondary mission.

The enemy threatening Pohang was believed to consist of about 6,000 light infantry troops from Major General Lee Ban Nam's widely respected *10th NKPA Division*. (Post-war analysis revealed that before its destruction by the 1st Marine Division, the *10th Division* inflicted more casualties and captured more equipment than any other North Korean unit.) Although a division in name, the *10th* was short of personnel and lacked artillery, armor, and motor transport. Its only support weapons were a few heavy mortars and some heavy machine guns. These shortfalls limited General Lee's tactical options to hit-and-run raids, roadblocks, and ambushes. The *10th Division* was, therefore, expected to conduct low-intensity operations remaining under cover during the day and attacking only in darkness. General Lee's troops seeped south through a hole in the fluid South Korean lines east of the Hwachon Reservoir in central Korea during the U.N. retreat in late-December 1950, and the division's lead elements were thought to be just arriving in the Pohang area in mid-January.

A Sikorsky HO3S sets down at a landing zone in the Pohang sector of operations. These utility helicopters were invaluable in providing communications in the search for Communist guerrillas.

1stMarDiv Historical Diary Photo Supplement, Jan-Feb51

The 1st Marine Division zone of action was roughly 40 miles square, an area composed of 1,600 square miles of extremely rugged interior terrain enclosed by a semicircular road network joining the coastal villages of Pohang and Yongdok with the inland towns of Andong and Yongchon. Seventy-five miles of the vital Eighth Army main supply route were located inside the Marine zone. That part of the supply route ran north from Kyongju to Yongchon then bent about 25 miles westward until it once again turned north to pass through Andong. A secondary road (Route 48) joined Andong in the northwest corner with centrally located Chinbo and Yongdok on the coast. The valley lowlands were dotted with small villages whose adjoining terraced rice paddies edged roadways and agricultural flat lands. The center of the Marine area of responsibility consisted of snow-capped mountains traversed only by a series of winding trails and narrow pathways that worked their way up and down the steep ridges. The weather was generally cold and often damp with frequent snow flurries, but

National Archives Photo (USMC) 127-N-A6145

In a long, winding, single-file column, a Marine rifle company inches its way down a steep mountain path before assaulting a guerrilla-held village in the valley below. The "guerrilla hunt" was marked by numerous small-scale clashes with enemy discovered by such patrols.

with little accumulation. The occasional high winds and overcast hindered flight operations and limited visibility.

On 16 January, General Smith opened a forward command post at Sinhung, about five miles southeast of Pohang. Division Operations Order 3-51 assigned the Marines three missions. One was to protect the Kyongju-Pohang-Andong portion of the main supply route. A second was to secure the village of Andong and the two nearby airstrips. The third mission was to prevent penetration in force of the Andong-Yongdok defense line. Widely known throughout the Marine Corps as a "by-the-book" man, Smith kept this image intact by mounting a textbook anti-guerrilla campaign. The long-service veterans of the 1st Marine Division were well aware of the travails of guerrilla warfare. A few senior officers and veteran sergeants had fought local insurgents during the so-called "Banana Wars" between the World Wars, some others had fought Chinese guerrillas in North China after World War II, and most field grade officers had closely studied the *Small Wars Manual* at Quantico. These veteran campaigners knew that counter-guerrilla operations were primarily small unit actions that tested individual stamina and required strong leadership at the fire team, squad, and platoon levels. Accordingly, General Smith decentralized operations. He created five defensive areas, formed mechanized task forces to patrol the roads, and saturated the hilly terrain with infantry patrols to keep the enemy constantly on the move. The 1st Marines, at Andong, was assigned Zone A in the northwest; the 5th Marines manned Zone B from Yongchon in the southwest quadrant; the 7th Marines operated out of Topyong-dong in Zone C, a centrally located 20-by-25 mile corridor running north from Pohang; the 11th Marines held a narrow coastal strip north of Pohang known as Zone D; and Lieutenant Colonel Harry T. Milne's 1st Tank Battalion operated in Zone E southeast of Pohang. The light utility aircraft of VMO-6 were in general support.

Anti-guerrilla doctrine called for constant vigilance by static units and aggressive action by mobile forces. A commander's primary concern was force protection, and the best way to accomplish that was to keep the enemy off balance. Guerrillas had to be located, engaged, rendered ineffective, and relentlessly pursued to do this. For large units (regiments or battalions) the favored tactics were "raking" (later known in Vietnam as "search and destroy") operations and encirclements ("cordon and search"). Smaller infantry units relied upon saturation patrols to find, fix, and eliminate the enemy. Most of these so-called "rice paddy patrols" consisted of fire teams and squads operating from platoon or company patrol bases. The 5th Marines was particularly aggressive and once had 29 such patrols in the field at the same time. Ambushes were an effective way to keep the enemy off balance by hindering movement and destroying small units piecemeal. Squad- and platoon-sized ambushes set up nightly along mountain trails or fanned out to cover likely avenues of approach to nearby villages. Motorized road patrols consisted of machine gun-mounted jeeps that roved the main supply route at irregular intervals. Convoys were escorted by gun trucks, tanks, or self-propelled guns.

The anti-guerilla campaign placed a heavy burden on the firing batteries of the 11th Marines. Once the patrols had tracked down groups of enemy troops, the regiment's batteries had to fire on short notice and in any direction. "It was not uncommon to see a battery sited by platoon—two guns to the east, two to the west, and two to the south," noted Lieutenant Colonel Francis Parry. "Two platoons might be laid for low-angle fire and the other for high-angle fire to enable it to reach over and behind a nearby ridge . . . . I doubt if field-artillery batteries anywhere ever surpassed the sophistication and competence . . . demonstrated routinely in January and February of 1951."

Although aggressive, the patrols soon took on an air of routine, according to Private First Class Morgan Brainard of Company A, 1st Marines:

> Each day was much like the one before: we would board trucks in the morning following chow and in full gear minus packs, roll out five miles or more to some predetermined spot, dismount, and undertake a sweep of the nearby hills and valleys, clearing all the villages in our path. And then we would return to camp in late afternoon, wash ourselves in the battalion shower tent (a real luxury), have chow, clean our gear, write letters and engage in bull sessions until it was time to stand watch.

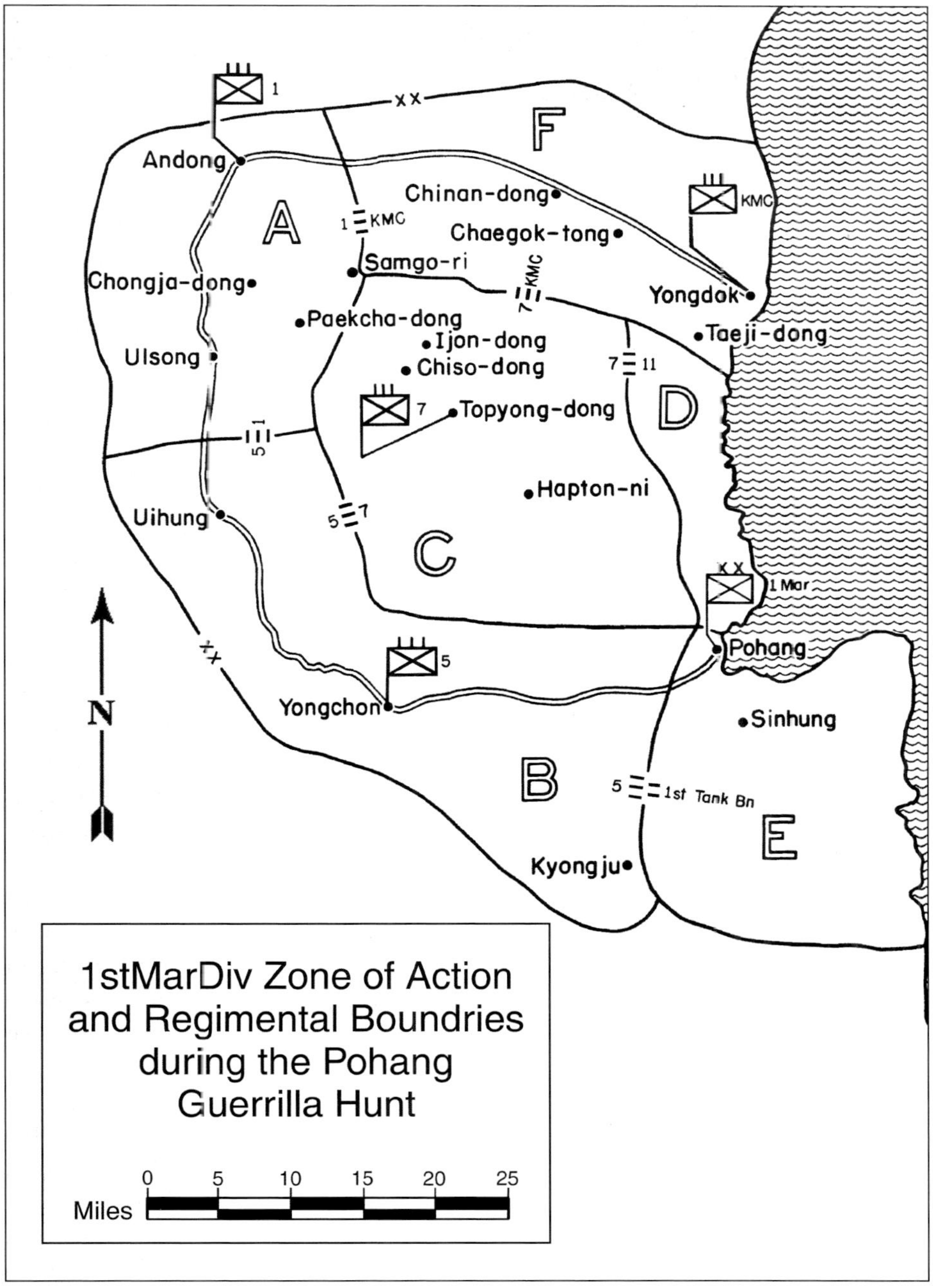

1stMarDiv Zone of Action and Regimental Boundries during the Pohang Guerrilla Hunt

Miles 0 5 10 15 20 25

The constant patrols harried the NKPA and kept it on the run. General Lee's troops were forced to break up into ever-shrinking groups just to survive. Soon, hard-pressed guerrilla bands were reduced to foraging instead of fighting, and the situation was so well in hand that the Marines could be relieved in order to fight elsewhere by mid-February.

The first contact with the enemy in the Pohang zone occurred on the afternoon of 18 January. A patrol from Lieutenant Colonel Thomas L. Ridge's 3d Battalion, 1st Marines, discovered an unknown number of North Koreans east of Andong. The enemy quickly fled, but three of their numbers were captured after a wild chase. These prisoners from the *27th Regiment* confirmed their parent unit was the *10th NKPA Division* and reported that elements of that division's *25th* and *29th Regiments* were also in the area. Four days later a patrol from the 1st Marines discovered an estimated enemy battalion near Mukkye-dong south of Andong just before sunset and promptly got the best of a one-sided exchange of small arms and mortars. Captain Robert P. Wray's Company C suffered no casualties while the NKPA lost about 200 killed or wounded. Unfortunately, nightfall prevented full pursuit. The enemy escaped under cover of darkness by breaking into squad- and platoon-sized exfiltration groups.

On 24 January, Colonel

National Archives Photo (USMC) 127-N-A6080

Col Homer L. Litzenberg set up his command post in a ravine near Topyong-dong. The 7th Marines was assigned to the centrally located Sector D during the Pohang guerrilla hunt.

Litzenberg's 7th Marines began a three-day raking operation to clear the enemy from its zone of action. The *In Min Gun* retaliated by hitting the regimental command post at Topyong-dong and the 1st Battalion three miles to the northwest, but both attacks by the *25th NKPA Regiment* failed. On the 26th, Major Webb D. Sawyer's 1st Battalion isolated an enemy company atop Hill 466 that held the attackers at bay with mortars, small arms, and hand grenades. The Marines answered with their own artillery, mortars, and automatic weapons. The outgunned enemy quickly abandoned the position after suffering an estimated 50 dead and about twice that many wounded. That same afternoon, Lieutenant Colonel Robert L. Bayer's 2d Battalion repulsed a NKPA counterattack and counted 44 enemy dead in the aftermath. During the entire operation, Colonel Litzenberg reported enemy losses at about 250 killed and 500 wounded with a dozen prisoners taken. These one-sided fights left little doubt about who held the upper hand. Consequently, General Lee ordered his troops to cease offensive operations until

A tank-led column from Company C, 1st Tank Battalion, stands by as a patrol from the 5th Marines searches a nearby village for guerrillas. The 90mm gun of the M-26 Pershing tank in the foreground and the 75mm gun of the following M-4 Sherman tank provided the requisite firepower.

National Archives Photo (USMC) 127-N-A6076

they could withdraw into the mountains to regroup. "They appear," noted General Smith, "to be as confused as we are."

The actions at Pohang thus far typified the frustrations of anti-guerrilla warfare. On every occasion the Marines hammered their opponents but were unable to pin down the elusive enemy so decisive action could be affected. "It became a game," Colonel Litzenberg reported. "We would find them about 1400 in the afternoon, get our artillery on them, air on them, and then they would disappear. The next day we would have to find them again." This disconcerting pattern continued throughout January and February 1951, much as it had in the Philippines at the turn of the century and would again in Vietnam little more than a decade later. But as Litzenburg noted, "the operations in this area constituted a very, very successful field exercise from which we derived great benefit." "It was excellent training for the new replacements," echoed General Smith's aide de camp, Major Martin J. Sexton. "It gave them the opportunity of getting a conditioning, and an experience of the hardest type of warfare, mountainous warfare, and fast moving situations. They also had the opportunity to utilize supporting fire of all types, including naval gunfire."

National Archives Photo (USMC) 127-N-A6142

A tired platoon patrol pushes up another hill as it pursues fleeing remnants of the 10th NKPA Division. Such marches provided excellent physical conditioning while at the same time developing unit cohesion and tactical proficiency.

A Marine sentry and his interpreter check passes and obtain information from Korean civilians passing through a roadblock near Andong. Far from supporting the Communists, the inhabitants readily reported North Korean guerrilla movements to the Marines.

National Archives Photo (USMC) 127-N-A6298

A welcome addition to the 1st Marine Division in late January was Colonel Kim Sung Eun's 1st Korean Marine Corps (KMC) Regiment. The Korean Marine regiment brought four rifle battalions (1st, 2d, 3d, and independent 5th). The original Korean Marines had trained under the tutelage of the 5th Marines while enroute to Inchon the previous September. They fought well beside the 1st Marine Division during the liberation of Seoul before being detached for other duties. The Korean Marines were attached administratively to the 1st Marine Division on 21 January, but were not trucked up from Chinhae until about a week later. On 29 January, Marine Lieutenant Colonel Charles W. Harrison, a veteran of pre-war service with the 4th Marines in Shanghai and now the senior Korean Marine advisor, finally reported that the Korean Marine command post was in place at Yongdok. General Smith created a new sector in the northeast to accommodate the new arrivals. This area, Zone F, included

National Archives Photo (USMC) 127-N-A6645

As two Marines guard a captured North Korean soldier, a corpsman administers first aid to him. While the number of counted enemy dead was low, there was little doubt that the total North Korean casualties were crippling.

Yongdok, Chaegok-tong, and Chinandong. The 1st, 2d, and 3d Battalions patrolled subsectors in Zone F while the 5th Battalion worked with the 1st Marines. The U.S. Marines provided combat and logistics support for their South Korean counterparts. The Korean Marines acquitted themselves well at Pohang, just as they had before and would again. In fact, the 1st KMC Regiment would become the 1st Marine Division's fourth rifle regiment for the remainder of the Korean War. The bond between Korean and American Marines was a strong one; so strong that when asked by a reporter about the oriental soldiers nearby, an anonymous U.S. Marine rendered the ultimate compliment when he replied: "They're Marines!"

It soon became obvious that the NKPA had bitten off more than it could handle. Enemy prisoners confirmed signal intercepts and agent reports that the *10th NKPA Division* had been ordered to leave Pohang to rejoin the NKPA *II Corps*. Concurrently, aerial observers noted a general movement to the west out of the 7th Marines' Zone C into Zones A and B (1st and 5th Marines, respectively). The resulting attempts to slip out of the Marine noose resulted in several very one-sided clashes during the first week of February. On the night of 31 January-1 February, a company-sized patrol from the 1st Battalion, 1st Marines, engaged an estimated enemy battalion near Sanghwa-dong. The enemy suffered about 50 casualties and three North Koreans were captured along with several mortars and small arms. A few days later Lieutenant Colonel Allan Sutter's 2d Battalion and Lieutenant Colonel Virgil W. Banning's 3d Battalion pushed fleeing NKPA troops into blocking positions manned by the Korean Marines' 22d Company during a successful "hammer-and-anvil" combined operation. In the 7th Marines zone of action, Lieutenant Colonel Wilbur F. Meyerhoff's 3d Battalion killed about 45 NKPA in a sharp action northwest of Wolmae-dong, and Lieutenant Colonel Bayer's 2d Battalion overcame fierce resistance to take Hill 1123. To the southwest, a trap set by Lieutenant Colonel John W. Stevens II's 1st Battalion, 5th Marines, turned out to be a bust, but Lieutenant Colonel Robert D. Tapplett's 3d Battalion destroyed four roadblocks, killed 30 enemy, and captured three more in the vicinity of Yongchon. Lieutenant Colonel Harold S. Roise's 2d Battalion occupied Hill 930 after ejecting some stubborn defenders. Along the northern coast, Colonel Kim's Korean Marines took Paekcha-dong and forced its defenders to scatter. A unique approach was tried on 4 February when a loudspeaker-equipped Marine Douglas

MajGen Smith pins a single star on newly promoted BGen Lewis B. "Chesty" Puller on 2 February 1951. Not long thereafter Puller, arguably the best-known Marine in modern history, took command of the division when MajGen Smith temporarily took over IX Corps.

National Archives Photo (USMC) 127-N-A6175

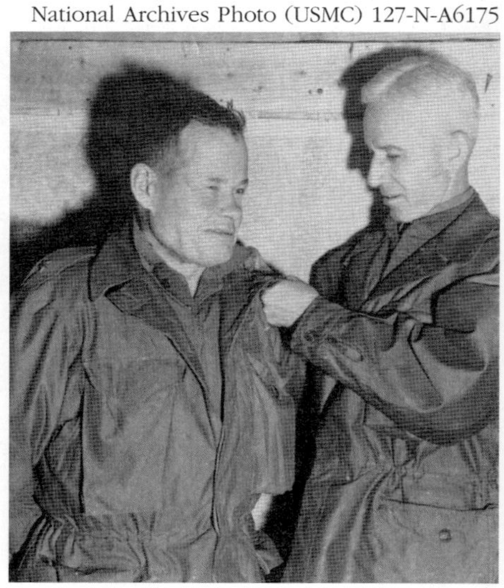

Department of Defense Photo (USMC) A6198

Marines and South Korean laborers bring a Marine casualty down from the scene of a skirmish with North Korean guerrillas. During the month at Pohang, Marine nonbattle casualties outnumbered battle casualties by a ratio of nine to one.

R4D Skytrain transport plane broadcast appeals to surrender. About 150 individuals answered the call, but most of them turned out to be South Korean laborers who had been forced into service by the NKPA. Chance-Vought F4U Corsairs from Marine Fighter Squadron 323 then dropped bombs, rockets, and napalm upon the remaining NKPA. The last major action of the "guerrilla hunt" occurred when two battalions of the 1st Marines, commanded by the division's former logistics officer, Colonel Francis M. "Frank" McAlister, who replaced newly promoted "Chesty" Puller on 25 January, routed an estimated battalion of the *27th NKPA Regiment*, south of Samgo-ri. More than 75 enemy were killed and an unknown number were seriously wounded by the time the North Koreans fled the field of battle on 5 February. Only scattered resistance by diehard individuals or small groups was reported to headquarters from then until the Marines departed Pohang.

Enemy deserters told interrogators that disease and low morale took a heavy toll. They reported an NKPA battalion commander had been shot for desertion and that General Lee was immobilized by severe depression. Other measures of enemy desperation were that women were increasingly being drafted to serve as porters and combat troops were donning captured American clothing to cover their escape. Although the *10th Division* still could muster about 1,000 men, captured dispatches indicated CCF headquarters ordered the remaining NKPA to break out of the Marine encirclement. General Smith's situation report to Eighth Army headquarters on 11 February stated that the enemy had been appreciably reduced and declared "the situation in the Division area is sufficiently in hand to permit the withdrawal of the Division and the assignment of another mission." Armed with this knowledge, intelligence officers at Eighth Army rated the *10th Division* as combat ineffective, and General Ridgway decided the 1st Marine Division could be put to better use elsewhere.

There were several important administrative changes in the 1st Marine Division at Pohang. On 2 February, Brigadier General Puller became the assistant division commander when Major General Edward A. Craig departed for the United States. This was the first in a series of command changes wrought by new rotation policies. In the next three weeks, 12 of 16 maneuver battalions would change hands. Thirty officers and 595 enlisted men, all former members of the 1st Provisional Marine Brigade, were sent back to Pusan to await conveyance to the United States. That much-longed-for transportation arrived when the troop ship USS *General J.C. Breckinridge* (AP 176) delivered 71 officers and 1,717 enlisted men of the 5th Replacement Draft to Pohang on 16 February. These new arrivals were rather hurriedly assimilated and they brought the 1st Marine Division back up to fighting strength just as it shipped out for central Korea. Unfortunately, the combined U.S.-Korean Marine team was broken up again. On 2 February, the independent 5th KMC Battalion was transferred to X Corps Headquarters, and General Smith learned the 1st KMC Regiment would stay behind when the 1st Marine Division moved out.

There were no pitched battles or epic engagements at Pohang, but the Marines had rendered an enemy division ineffective. Marine battle losses during the period 12 January to 15 February numbered 26 dead, 148 wounded, and 10 missing in action. There were also a large number of nonbattle casualties, primarily the result of frostbite or minor injuries, most returned to their units. Enemy casualties and non-combat losses were estimated at more than 3,000 men. The "guerrilla hunt" was also particularly useful for training and physical conditioning. Constant movement over rough terrain ensured all hands were in good shape, rifle squads and mortar sections developed into coherent and tactically proficient units, and most of the 3,387 Marine replacements got at least a brief taste of combat conditions. With respect to operations, the Marines functioned as a truly integrated air-ground team. Although there were few opportunities to use Marine close air support, VMO-6's Consolidated OY "Sentinel" light observation aircraft and Sikorsky HO3S-1 "Dragonfly" helicopters served as airborne scouts and rescue craft while bubble-top Bell HTL helicopters were most often used as aerial ambulances. Indeed, the 1st Marine Division was so well honed after Pohang that five decades later

MajGen Smith distributes clothing donated by the Marine Corps League to Korean children in one of the nearby villages. "No attempt is made to obtain an exact fit," Smith said, "as there is not time."

National Archives Photo (USMC) 127-N-A6437

The Enemy

In the spring of 1951, the forces opposing the United Nations Command consisted of more than a half million men of the North Korean People's Army (NKPA) and the Communist Chinese Forces (CCF) under the leadership of Chinese General Peng Teh-Huai. The CCF played no role during the initial stages of the Korean Conflict, but the Chinese *Fourth Field Army* serving under the questionable rubric "Chinese People's Volunteers" began secretly infiltrating North Korea in the fall of 1950. Such cooperation was nothing new; the Communists in China and North Korea had often worked together in the recent past. North Korea had been a Communist sanctuary during the Chinese Civil War, and North Korean volunteers fought side-by-side with the Chinese Communists since the mid-1930s.

The NKPA

The North Korean People's Army, more formally the *In Min Gun*, entered the Korean Conflict as a well-armed and well-trained military organization. The NKPA was modeled after the Soviet Red Army and was primarily armed with Soviet-made weapons. Specially selected veteran officers attended Soviet military schools in 1948 then became the cadre around which the NKPA was built. A few senior leaders and many enlisted men were veterans of the Korean Volunteer Army (KVA), which fought side-by-side with Mao Tse Tung's Communist guerrillas who successively defeated the Japanese during World War II and Chang Kai Shek's Nationalists during the Chinese Civil War. The Korean volunteers returned to North Korea in 1949 and were promptly integrated into the NKPA.

In the summer of 1950, the *In Min Gun* rolled over the surprised and outnumbered South Koreans. But, just as the victorious NKPA prepared for its final thrust to oust the United Nations from the Pusan Perimeter, General Douglas MacArthur conducted one of the most successful amphibious operations in military history when X Corps, spearheaded by the 1st Marine Division, landed at Inchon and then quickly recaptured Seoul. Outflanked and cut off from its supply bases, the NKPA was quickly routed and its remnants fled to the dubious safety of North Korea with the U.N. in hot pursuit. The sudden intervention of the CCF around Thanksgiving stopped MacArthur's northern advance, and by Christmas the United Nations Command was in full retreat. At that time the disorganized and demoralized NKPA underwent a complete make over. The NKPA was placed under Chinese command and was reorganized into light infantry units similar to those of the CCF.

During the spring of 1951, the Marines faced the NKPA *V and II Corps*. These units were armed with heavy mortars and machine guns, but only occasionally received adequate artillery support. The *10th NKPA Division* was the guerrilla force the Marines encountered at Pohang. The NKPA *V Corps* screened CCF movements and acted as the rear guard battling the Marines on the Central Front.

DIVIC (USMC) HDSN9903152

Communist Chinese Forces

The Communist Chinese *Peoples Liberation Army* (PLA) was a massive, mostly illiterate, peasant army that had been fighting for almost two decades without a break in 1951. Its sheer size and vast combat experience made it a formidable opponent. The PLA was, however, basically a light infantry force that possessed few tanks and its artillery arm was vastly undergunned by western standards. In November 1950, Mao Tse Tung sent more than 500,000 men into Manchuria and North Korea. The men entering Korea called themselves "Volunteers," but were labeled "Communist Chinese Forces" by the United Nations.

Initially, the CCF was actually the Chinese *Fourth Field Army* in Korea. This organization was divided into group armies, armies, and divisions. The 10,000-man Chinese divisions included only about two-thirds as many troops as an American infantry division but, ironically, mustered a much larger number of "trigger pullers" because the spartan CCF had so few support personnel. A CCF division was lucky if it had more than a single artillery battalion armed with 120mm mortars or 76mm antitank guns. The lack of fire support, motor transport, and modern communications dictated CCF tactics, which primarily consisted of night infiltration or massive frontal or "human wave" assaults conducted under cover of darkness. The Marines encountered the CCF *39th, 40th,* and *66th Armies* during the fighting around Hwachon in the spring of 1951.

Marine historian Edwin H. Simmons reminisced about that time, stating: "The 1st Marine Division in Korea was the finest fighting outfit I ever served with"; no small praise from a combat veteran of three wars.

Back to the Attack

While the 1st Marine Division was busy rebuilding at Masan and chasing guerrillas at Pohang, the vastly outnumbered Eighth Army continued to fall back in what eventually became the longest retreat in American military history. But, as soon as the CCF Third Phase Offensive ran out of gas, General Ridgway resumed offensive operations. In mid-January, he initiated the first in a series of drumbeat attacks that eventually carried U.N. forces back above the 38th Parallel. Unlike the reckless rush to the Yalu the previous year, however, this time Eighth Army relied upon cautious advances, which were both limited in scope and closely controlled by higher headquarters, carefully coordinated actions intended to punish the enemy as well as to gain ground. In general, Ridgway eschewed flanking movements and objectives deep in the enemy rear. Instead, numerous phase lines strictly controlled U.N. activities and attacking units kept pace with those on each flank. The Marines—except for pilots flying close air support missions—missed the first three offensive operations (Wolfhound, Roundup, and Thunderbolt), but the 1st Marine Division was destined to play key roles in Operations Killer, Ripper, Rugged, and Dauntless.

Each of Ridgway's successive operations was more ambitious than the previous one. By mid-February, the Eighth Army had gathered momentum and was on the move all across Korea. At that time the U.N. front was held from left to right by the U.S. I Corps, IX Corps, X Corps, and units of the South Korean army. The United Nations Command was in the process of rolling back the Communists in western Korea when General Ridgway met with General Smith at Suwon in late January to discuss the 1st Marine Division's next mission. Ridgway wanted to send the Marines to central Korea, but Smith lobbied hard to have the 1st Marine Division placed on the far right flank in order to stay near the coast. Smith noted that his division was the only Eighth Army unit trained for amphibious operations and added that a position near the sea would allow the Marines to make maximum use of naval gunfire and carrier-based air, supporting arms with which they were intimately familiar and well-practiced in using. Such a disposition would also allow the Marines to use Pohang as the principal port of entry, a factor that would ease the logistical burden by shortening supply lines. Amplifying, Smith pointed out that "the 1st Marine Division with a strength of approximately 24,000 was larger that any of the Army infantry divisions or ROK divisions at this time and in that there were single . . . supply routes, for the corps and the divisions and it would be less of a strain upon transportation and less of a logistical problem to supply a smaller army division or ROK division inboard, well inland in Korea, than it would be in the case of the Marine division." This was a critical consideration because it would reduce overland transportation problems. Marine trucks were both few in number and in poor shape after hard use at the Chosin Reservoir. Smith's logic won over the Eighth Army commander, and after that meeting Ridgway directed his staff to prepare plans for the Marines to remain on the east coast.

Unfortunately, these plans were overcome by events before they could be put into effect. The catalyst for the movement of the 1st Marine Division into central Korea was the third battle for Wonju, a vital communications and road link whose loss might well force the evacuation of Korea by U.N. forces. Wonju was put at great risk when Hoengsong, located about 10 miles north on Route 29, was lost. This near disaster occurred when the Communists launched their Fourth Phase Offensive in which the CCF *40th* and *66th Armies* and the NKPA *V Corps* initiated a series of devastating attacks out of the swirling snow beginning on the night of 11-12 February. The U.S. X Corps suffered a serious setback when three Republic of Korea Army divisions disintegrated and combat support elements of the U.S. 2d Division were cut off and then annihilated in "Massacre Valley" just north of Hoengsong. The 23d Infantry was cut up, the artillery overrun, and "only 800 had come in so far and only one in twenty had weapons," General Puller told Smith. With the key city of Wonju threatened and X Corps reeling back, General Ridgway had no choice but to commit what he called "the most powerful division in Korea" to "where a great threat existed to that portion of the Eighth Army's lines." On 12 February 1951, General Smith received a warning order to prepare the 1st Marine Division to move to Chungju in south-central Korea "on 24 hours' notice at any time after 0700, 14 February." As Major Martin Sexton later commented: "The 1st Marine Division was deployed right in the center of Korea and its amphibious

National Archives Photo (USMC) 127-N-A7274

A Marine light machine gun team boards a train for the trip from Pohang to Chungju in central Korea. The 1st Marine Division was to be positioned astride what LtGen Matthew Ridgway considered to be the logical route for the expected enemy counterthrust.

capabilities destroyed at the same time."

The Korean peninsula can be roughly described as a 600-by-150 mile parallelogram that descends ever downward from the Manchurian border and also slants down from the hilly eastern one-third that abuts the Sea of Japan until it gradually levels off along Yellow Sea to the west. The peninsula can be easily divided into several unique geographic areas: two horizontal sections, one in the north and one in the south, comprise the basic economic sectors; three vertical sections—the east, central, and west corridors—each comprise about one-third the width of the peninsula. Most of North Korea is rugged mountain territory whose fast-flowing rivers provide the water and electric power necessary for industrial development. South Korea, on the other hand, includes most of the agricultural land. Trans-peninsular communications, particularly roads and railways, are hampered by geography. The craggy Taebaek Mountain range roughly parallels the east coast where its irregular cliffs severely limit the number of suitable landing areas and provide no spacious flat lands to support agricultural or urban development. There are few east-west overland links, and the only north-south route in the east runs tenuously along the narrow coast. Numerous fingers of the Taebaek intermittently reach west across the central corridor creating a washboard of alternating river valleys and spiny ridgelines. The flattest expanses are located along the west coast, an area that includes both Korea's major port Inchon and its largest city, Seoul.

General Smith's new orders focused Marine attention upon the central corridor, and all major combat actions during the spring of 1951 would take place in that zone. The dominating terrain feature was the Hwachon Reservoir, a 12-mile basin that blocked the southern flowing Pukhan River using a sizable dam. The reservoir was located just about at the peninsula's dead center. It provid-

Marines move up to the front during Operation Killer, the fourth of the so-called Ridgway Offensives. During that action, the Marines advanced up the Som River Valley from Wonju to Hoengsong against light to moderate resistance.

National Archives Photo (USMC) 127-N-A6851

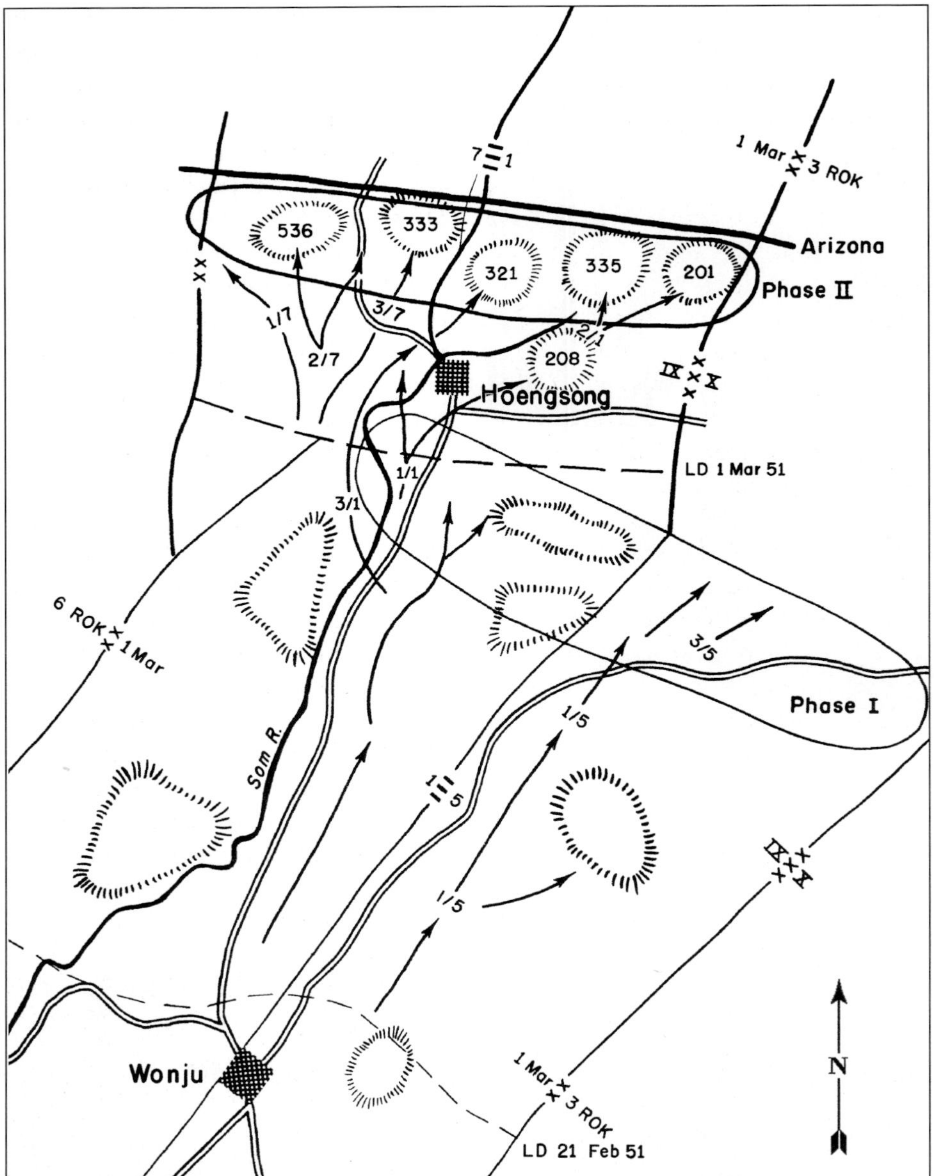

1st MarDiv Zone of Action
Operation Killer
Showing Phase I & II Objectives

ed pre-war Seoul with most of its water and electricity, but that was no longer true. The Hwachon Reservoir did, however, have significant tactical value. Just north of the 38th Parallel and at the southern edge of a mountainous shelf, it marked both the political and geographic divisions of North and South Korea. This barrier effectively channeled all movement to either the east (Yanggu) or west (Hwachon), and the side holding the dam could threaten to flood the low-lying Chunchon and Soyang Valleys at will.

Korea's central corridor also included all of the major communications links between both the east and the west and the north and south. A string of road junctions spiraled south along Routes 17 and 29 from Hwachon at roughly 15-mile intervals. These included Chunchon, Hongchon, Hoengsong, and Wonju—each of which would become a major objective during the Ridgway offensives on the Central Front.

Operation Killer

Beginning on 16 February, the Marines mounted out from Pohang by regimental combat teams for Chungju. Fortunately, by that time the CCF and NKPA were being pounded by air and artillery until their attacks ran out of steam north of Wonju. Thus, when the Marines finally arrived at Chungju, they could be used to spearhead a U.N. counteroffensive, a closely coordinated pincer attack by the U.S. IX and X Corps intended to trap the NKPA *III* and *V Corps* called Operation Killer. Eighth Army released the 1st Marine Division from direct control when it joined Major General Bryant E. Moore's IX Corps for Operation Killer, a two-phase drive up the Wonju basin to retake and secure Hoengsong. An Army officer, General Moore had served side-by-side with the Marines at Guadalcanal in 1942. He ordered General Smith to seize the high ground south of Hoengsong hoping to cut off enemy forces to the south by denying them use of their main egress routes. Although Smith lost tactical control of the 1st KMC Regiment when the Marines departed Pohang, U.S. Army artillery and transportation units reinforced the division. Particularly welcome additions were the much-needed vehicles of the U.S. Army's 74th Truck Company, and the "Red Legs" of the 92d Armored Field Artillery, commanded by Lieutenant Colonel Leon F. Levoie, Jr.,

USA. This was a first-rate Army self-propelled howitzer unit that had rendered outstanding support at the Chosin Reservoir. The artillerymen of the 92d were, as Major Martin J. Sexton noted, "trained basically as Marines are in that they were essentially riflemen too, if not first."

On 19 February, Smith and Puller attended a commander's conference. There, they learned the 1st Marine Division was to be the focus of the main effort for Operation Killer and would advance with the 6th ROK Division on the left and X Corps to the right. General Ridgway's orders were to "seek out the enemy and inflict the greatest possible damage." In Marine terms, Operation Killer was going to be "buttoned up"; all U.N. forces were to keep close lateral contact, to maintain tactical integrity at all times, and to strictly adhere to the timetable. Units would not bypass enemy positions and had to stop at each phase line even if there was no enemy resistance. Regrettably, the conference closed on a less than happy note for the Marines. Generals Smith and Puller were taken back to learn Operation Killer was to kick off in less than 48 hours, too short a time to move the entire division to the line of departure. The Marines were further dismayed when they were denied the use of a dedicated Marine fighter squadron. Their arguments to hold up the attack until the entire division could be assembled were dismissed by Ridgway who also refused to intervene to assure the Marines adequate close air support. In spite of Marine objections, H-hour was set for 1000 on 21 February.

General Smith elected to use two regiments (the 1st and 5th Marines) in the attack and keep one, the 7th Marines, in reserve. The line of departure was located just north of Wonju. The area in the Marine zone was uninviting, to say the least. In the words of official Marine Corps historian Lynn Montross: "There were too many crags [and] too few roads." Rocky, barren, snow-covered ridges boxed in the narrow Som Valley whose lowlands were awash with runoff from melting snow and flooded by overflowing streams. The weather was terrible, "a mixture of thawing snow, rain, mud, and slush," according to 3d Battalion, 5th Marines' commanding officer Lieutenant Colonel Joseph L. Stewart. The axis of advance was generally northwest along Route 29, which was sarcastically known as the "Hoengsong-Wonju Highway" (it was actually a primitive one-lane packed-dirt trail totally unsuited to support vehicular traffic) that generally paralleled

LtGen Ridgway ordered Eighth Army troops to "get off the roads and seize the high ground." Here a Marine patrol makes its way up a difficult trail during Operation Killer.

National Archives Photo (USMC) 127-N-A6952

1stMarDiv Historical Diary Photo Supplement, Jan-Feb51

A Marine Corsair circles after delivering napalm upon a suspected enemy position near Wonju in central Korea. The controversial breakup of the Marine air-ground team by Fifth Air Force in 1951 remained a touchy and unresolved issue throughout the rest of the Korean Conflict.

the Som River. The final objective was an east-west running ridgeline south of the ruins that had once been Hoengsong. The enemy defending this area was identified as the *196th Division* from Chinese General Show Shiu Kwai's *66th Army*.

Unfortunately, a series of events beyond General Smith's control hampered the start of operations. Transportation shortfalls meant that the 7th Marines would not be immediately available, so the 2d Battalion, 5th Marines, commanded by Lieutenant Colonel Glen E. Martin, a reserve officer who had been awarded a Navy Cross and served as a platoon and company commander during World War II, was designated the 1st Marine Division reserve for the first phase of Operation Killer. Snarled traffic, sticky mud, General MacArthur's visit to the 187th Airborne's zone, and the lack of trucks conspired to postpone the planned jump-off time. But, even with the logistical problems, the assault units of the 5th Marines—just like their World War I predecessors at Soissons—had to double time to get to the line of departure in time. Luckily, there was little enemy resistance. Under the watchful eyes of Generals MacArthur, Ridgway, and Moore, the Marines advanced rapidly in a torrential rain opposed only by some ineffectual long distance small arms fire. Colonel Francis McAlister's 1st Marines moved up the muddy road in a column of battalions (Lieutenant Colonel Donald M. Schmuck's 1st Battalion, Major Clarence J. Mabry's 2d Battalion, and Lieutenant Colonel Virgil W. Banning's 3d Battalion, respectively). Colonel Raymond Murray's 5th Marines had a harder row to hoe advancing north (actually climbing up and sliding down the snow-covered terrain) across a series of steep ridges and narrow valleys. Lieutenant Colonel John L. Hopkins' 1st Battalion, with Lieutenant Colonel Joseph L. Stewart's 3d Battalion in trail, maintained the high ground by hugging the regimental left boundary. The Marines made it almost four miles the first day and then covered half as much ground the following day. "Unlike the Inchon-Seoul Campaign," recalled Private

Marines advance across a fog-filled valley in the Wonju-Hoengsong sector supported by machine gun fire. Elaborate weapons positions, common in the latter stages of the war, were unusual during the seesaw fighting in the spring of 1951.

National Archives Photo (USMC) 127-N-A6843

First Class Morgan Brainard, "we were not moving out with what we infantrymen could recognize as a set goal, other than to kill gooks, and to move the lines steadily back north."

The first real resistance occurred when the 1st Marines, moving with the 1st and 2d Battalions abreast, and the 5th Marines in column, neared Objectives A (Hill 537 and Ridge 400) and B (Hill 533), overlooking Hoengsong. The 1st Marines was stopped by small arms and heavy automatic weapons fire from Hill 166 on the left and dug in for the night. Supported by air and artillery, the 1st Marines secured the heights at 1015 on 23 February. That afternoon, the 1st and 2d Battalions conducted a successful flanking attack to take the final hill line and were overlooking Hoengsong as darkness fell. The 5th Marines waited in vain that same morning for an air strike before mounting a two-battalion assault to clear a pair of hills on the left. The next morning, a mechanized patrol from the 1st Marines passed through Hoeng-song on its way to rescue several survivors still holding out in Massacre Valley where U.S. Army artillery units had been overrun almost two weeks earlier. The Marines found a gruesome sight. Burned out vehicles, abandoned howitzers, and more than 200 unburied dead lay strewn across the valley floor. This movement also stirred up a hornet's nest. Enemy mortars and artillery ranged the ridgeline held by the 1st Marines. The major combat action of the day then occurred when Marine cannoneers of Major Francis R. Schlesinger's 2d Battalion, 11th Marines, bested their CCF counterparts in the ensuing late afternoon artillery duel. By dusk on 24 February, all Marine objectives for the first phase of Operation Killer had been secured. The Marines had suffered 23 killed in action and 182 wounded thus far.

The follow-on advance had to wait almost a week. Operations were placed on temporary hold for several reasons. One major reason was that General Moore, the IX Corps commander, died following a helicopter crash. Additionally, the low ground around Hoeng-song had become a rain-soaked, muddy, impassible bog. Bad roads and poor weather stopped the Marines in their tracks because the assault units needed ammunition and food before they could renew the attack. Additionally, the Marine division was still fragmented because the 7th Marines was stranded at Wonju where a severe gasoline shortage idled most trucks. This shortfall was compounded by the poor trafficability of the road net, which had become a gooey morass due to the incessant rain. Army and Marine engineers labored night and day to shore up the deteriorating roads and bridges, but the supply situation became so critical that airdrops—an inefficient method heretofore used for emergencies only—had to become a logistics mainstay. Thirty-five airdrops were required to resupply the assault elements of the 1st Marine Division. Marine transport planes augmented the U.S. Air Force Combat Cargo Command participated in such drops all across the U.N. front. Also used in the resupply effort were 1,200 cargo handlers of the South Korean Civil Transport Corps. These hard-working indigenous laborers toiled under the direction of division Civil Affairs Officer First Lieutenant Oliver E. Dial. These hardy individuals each carried up to 50 pounds of supplies on A-frame backpacks to the forward most Marine units.

On 24 February, Major General Smith became the third Marine to assume command of a major U.S. Army formation (Brigadier General John A. Lejeune had commanded the U.S. 2d Division in France in 1918 and Major General Roy S. Geiger commanded the U.S. Tenth

Marine riflemen move across a frozen rice paddy during the drive back to the 38th Parallel. Although there were a few sharp actions, Operations Killer and later Ripper were remembered as "a long walk" by Capt Gerald P. Averill in his memoir Mustang.

National Archives Photo (USMC) 127-N-A6869

Department of Defense Photo (USMC) A6946

An automatic rifleman supported by another Marine with a carbine fires into a Communist-held position. Because of its great firepower, the Browning Automatic Rifle was the most vital weapon of a Marine fire team.

Army at Okinawa in 1945) when he took over IX Corps after General Moore's fatal heart attack. Smith's reception at Yoju was subdued, according to his aide, Major Sexton:

> It was a very modest and unassuming entrance that was made when General Smith stepped out of the helicopter and was met by General [George B.] Peploe, who was the chief of staff of IX Corps. A brief introduction to the staff—the corps staff officers—followed, and I would say that it was approximately an hour, possibly an hour and a half after General Smith's arrival, that the first decision which he was required to make arose. It involved a seemingly complicated scheme of maneuver wherein IX Corps would execute a flanking maneuver to envelop a sizable North Korean force which lay in front of the left portion of X Corps' zone of action. After deliberation with the G-3 and the chief of staff, the decision was forthcoming. It was simply "No, thank you." At which time, the G-3 excitedly called his . . . G-3 of X Corps, repeated these words and happily hung up the phone. As there were at this time smiles all around the staff, it was my impression that the general had been accepted rapidly.

The following day, Smith conferred with General Ridgway regarding future operations. Although Ridgway warmed Smith with glowing words of encouragement, he concluded by saying "he didn't know what the War Department would do." Smith knew. Despite the recommendations of U.S. Army Major General Frank E. Lowe, sent to Korea by President Harry S. Truman to evaluate American units, that Smith be elevated to corps command, it was obvious no Marine general was going to be allowed to do so on a permanent basis; accordingly, Smith's tenure lasted only until a more senior U.S. Army general arrived in Korea. He also asked for recommendations as to the future employment of the 1st Marine Division, to which General Smith replied that he knew "of no better use for the Division than to continue north on the Hoensong-Hongchon axis" as the main threat would come from that direction. In addition, Ridgway announced that Operation Killer would not resume until 1 March and that he wanted a change in zones to reorient the division more to the north. The IX and X Corps boundary was shifted west in accord with the Eighth Army commander's wishes. To do this Brigadier General Puller, who was filling in as the 1st Marine Division commander, rearranged the Marine dispositions. He pulled the 5th Marines out of the line to become the division reserve and moved the 7th Marines up into the line on the left to replace a South Korean unit that had been holding that position.

Arguably, the 1st Marine Division had the most difficult assignment of any unit in the Eighth Army. It had to cross a muddy triangular open area and then eject a dug-in enemy from a ridgeline located about a mile-and-a-half north of Hoengsong. Phase Line Arizona, as the final objective was known, consisted of five distinct hill masses (Hills 536, 333, 321, 335, and 201). The 1st Marines' commander, Colonel McAlister, assigned two intermediate objectives (Hills 303 and 208) as well. The nature of the terrain, which required a river crossing

1stMarDiv Historical Diary Photo Supplement, Mar51

A Marine tank pushes through the wreckage from ambushed U.S. Army units, which clogged the road to Hoengsong. Bad weather, poor roads, and supply difficulties held up the Marines more than enemy resistance during the spring of 1951.

and prohibited extensive vehicle movement, dictated a complex scheme of maneuver. The 7th Marines on the left would have to seize the hills in its zone to eliminate flanking fires before a battalion of the 1st Marines advancing through the 7th Marines' zone could assault its assigned objectives on the right. The CCF rearguard, consisting of elements of the *196th* and *197th Divisions*, was situated inside a sophisticated reverse-slope defense system anchored by log bunkers and zigzag trenches immune to direct fire. Where possible, the Marines would use fire by tanks and self-propelled guns to reduce point targets, but emplacements on reverse slopes would have to be hit by unobserved close air support or high angle artillery and mortar fires. This meant if the Chinese defended in place the Marines would have to reduce the reverse slope defenses using close combat. The most important terrain obstacle was the chilly, chest-deep, fast-flowing Som River. During the plenary conference Colonel McAlister was informed that no engineer support would be available and was further told the river was not fordable. Major Edwin H. Simmons, commanding 3d Battalion's Weapons Company, offered a solution. He recommended building a "Swiss bent bridge" composed of "A" shaped timber platforms with planking held in place by communications wire. This field expedient did the trick, and the 3d Battalion safely crossed the Som the night before the attack began.

The battle for Hoengsong was a classic four-day slugging match in which the Marines slowly advanced against dug-in enemy troops under the cover of a wide array of supporting arms. On the first day (1 March), the 11th Marines fired 54 artillery missions, Marine Grumman F9F Panther jets and Corsairs flew 30 sorties, and Marine tanks lined up like a row of battleships using their 90mm guns to clear the way. Colonel Litzenberg's 7th Marines moved out with Major James I. Glendinning's 2d Battalion on the left and Major Maurice E. Roach, Jr.'s 3d Battalion on the right headed north toward Hills 536 and 385 respectively. Lieutenant Colonel Banning's 3d Battalion, 1st Marines, which was in the 7th Marines' zone, moved in echelon with Hill 303 in the 1st Marines' zone on the right as its final goal; concurrently, the other two battalions of the 1st Marines held fast and furnished fire support. Three artillery battalions (3d and 4th Battalions, 11th Marines, and the U.S. Army's 92d Armored Field Artillery Battalion) were on call to support the 7th Marines. Major Webb D. Sawyer's 1st Battalion patrolled the left flank and main-

The shattered remains of Hoengsong as they appeared after the four-day battle. During the battle, the 11th Marines fired more than 50 artillery missions while 1st Marine Aircraft Wing Corsairs and Panther jets flew 30 sorties in support.

1stMarDiv Historical Diary Photo Supplement, Mar51

National Archives Photo (USA) 111-SC361108

MajGen William M. Hoge, USA, seated in front, was sent out to relieve MajGen Oliver P. Smith, who had been given temporary command of IX Corps.

tained contact with the 6th ROK Division. The assault battalions advanced about a half-mile before they came under heavy mortar and automatic weapons fire, then were completely halted by a minefield. With the main attack stalled and darkness closing in, it was decided to wait until the following day to finish the job.

At 0800 on 2 March, both regiments jumped off. In the 1st Marines' zone, Banning's 1st Battalion went after Hill 303 on the left using a flanking attack while the 2d Battalion—well supported by rockets, artillery, and tanks—passed through the debris of Hoengsong, then took Hill 208 on the right with a frontal assault. Both attacks went smoothly, and the 1st Marines secured its intermediate objectives by midday. The 7th Marines, assisted by air strikes and 1,600 rounds of artillery, gained about a half-mile on the left. The toughest fighting occurred in the 2d Battalion's zone where the Marines had to crawl forward over rocky terrain. Unfortunately, Phase Line Arizona remained out of reach when darkness fell.

The attacks on 3 March to secure the heights north of "Massacre Valley" featured the bloodiest single day of the operation. The 1st Marines secured Hills 321, 335, and 201 after some very tough hand-to-hand fighting which required the intervention of Captain Thomas J. Bohannon's Company A. The 7th Marines also continued the attack against a determined foe. Major Sawyer's 1st Battalion was called up from reserve to take and secure Hill 536 on the extreme left. The 3d Battalion then attacked Hill 333 with fire support from the stationary 2d Battalion. It was slow going for both assault units, and neither was able to secure its objectives before nightfall despite suffering 14 killed and 104 wounded since daybreak. These same two assault battalions determinedly "went over the top" amid snow flurries the next morning (4 March) only to discover most of the enemy had quietly slipped away during the night. Combat clearing duties ended at dark with the Marines firmly in possession of Phase Line Arizona.

The Marines had suffered almost 400 casualties (48 dead, 345 wounded, and 2 missing), while eliminating an estimated 2,000 enemy in two weeks of combat near Hoengsong. The bottom line, however, was that Operation Killer closed on an anticlimactic note. The Marines drove the enemy out of Hoengsong, but General Ridgway was dissatisfied with the punishment meted out. Although all terrain objectives had been taken, the enemy had deftly avoided a costly set piece battle and slipped out of the United Nations Command's trap. As a result, the Eighth Army commander ordered a new attack, Operation Ripper, to begin immediately.

There were several administrative changes during the brief respite between the end of Operation Killer and the onset of Operation Ripper. On 4 March, the 6th Replacement Draft (29 officers and 1,785 enlisted men) arrived, bringing with them 63 postal pouches—the first mail the Marines received since leaving Pohang. Concurrently, Lieutenant Colonel Erwin F. Wann, Jr.'s 1st Amphibian Tractor Battalion and Lieutenant Colonel Francis H. Cooper's 1st Armored Amphibian Battalion were detached to support the U.S. Army crossing of the Han River and were thereafter sent to Inchon to conduct amphibious training. The amphibian tractors would not rejoin the 1st Marine Division until it moved to western Korea the following year. On the plus side, a 250-man South Korean National Police company joined the Marines. These "Wharrangs" primarily served as scouts and interpreters, but were occasionally used as auxiliary combat troops as well. On 5 March, the day after Operation Killer ended, General Smith returned to the 1st Marine Division upon the arrival at IX Corps headquarters of his replacement, Major General William H. Hoge, USA, who had quickly flown out from Trieste, Italy. An engineer by training, Hoge supervised the Alaska-Canadian Highway effort and commanded Combat Command B, 9th Armored Division, during World War II, the lead elements of which seized the only major bridge over the Rhine River at Remagen.

Operation Ripper

Operation Ripper was the fifth consecutive limited U.N. offensive. It would follow the same basic design as the previous attacks. As before, the real goals were to inflict maximum punishment and keep the Communists off balance, but this time General Ridgway added a major territorial goal as well. He wanted to outflank the enemy near Seoul and force them to withdraw north of the Imjin River, a movement that would carry the United Nations Command almost back to the 38th Parallel. The plan was for IX and X Corps to drive north with I Corps holding in the west and the South Korean Army maintaining its positions along the east flank. The Central Front would be Eighth Army's primary arena. Hoge's plan was to drive north with the towns of Hongchon and Chunchon as major objectives in the IX Corps zone. Intermediate objectives included Phase Lines Albany, Buffalo, and Cairo; the final objective was Line Idaho. General Hoge inserted an intermediate phase line, Baker, between Eighth Army-designated Lines Albany and Buffalo. The enemy in zone continued to be the CCF *66th Army*, but intelligence officers were uncertain as to whether the enemy would continue to retreat or would finally stand and fight. Operation Ripper would, therefore, once again be a cautious advance, another limited, strictly controlled, "buttoned up" operation.

As in just-ended Operation Killer, the Marines would again be the focus of the IX Corps' main effort. A pair of U.S. Army units, the 1st Cavalry Division on the left and the 2d Infantry Division on the right, would guard the Marine flanks. Hongchon, an important communications hub located in the shadow of towering Oum Mountain, about five miles north of the line of departure, was the initial Marine objective. The coarse terrain included formidable Hill 930 and consisted of thickly wooded hills and swift-flowing streams. There were so few roads and trails that gravel-bottom streambeds were often pressed into use as roadways. The only thoroughfare (single-lane National Route 29) passed through Kunsamma Pass as it wound its way north to Hongchon from Hoengsong, and it initially served as the regimental boundary line inside the Marine zone. Intermittent snow, cold nights, and rainy days meant that the weather would continue to be a factor with which to be reckoned. The Marines in Korea, just as had Napoleon's army in Russia a century-and-a-half before, would have to deal with "General Mud" as well as enemy soldiers.

The 1st Marine Division mission was to seize all objectives and destroy all enemy south of Line

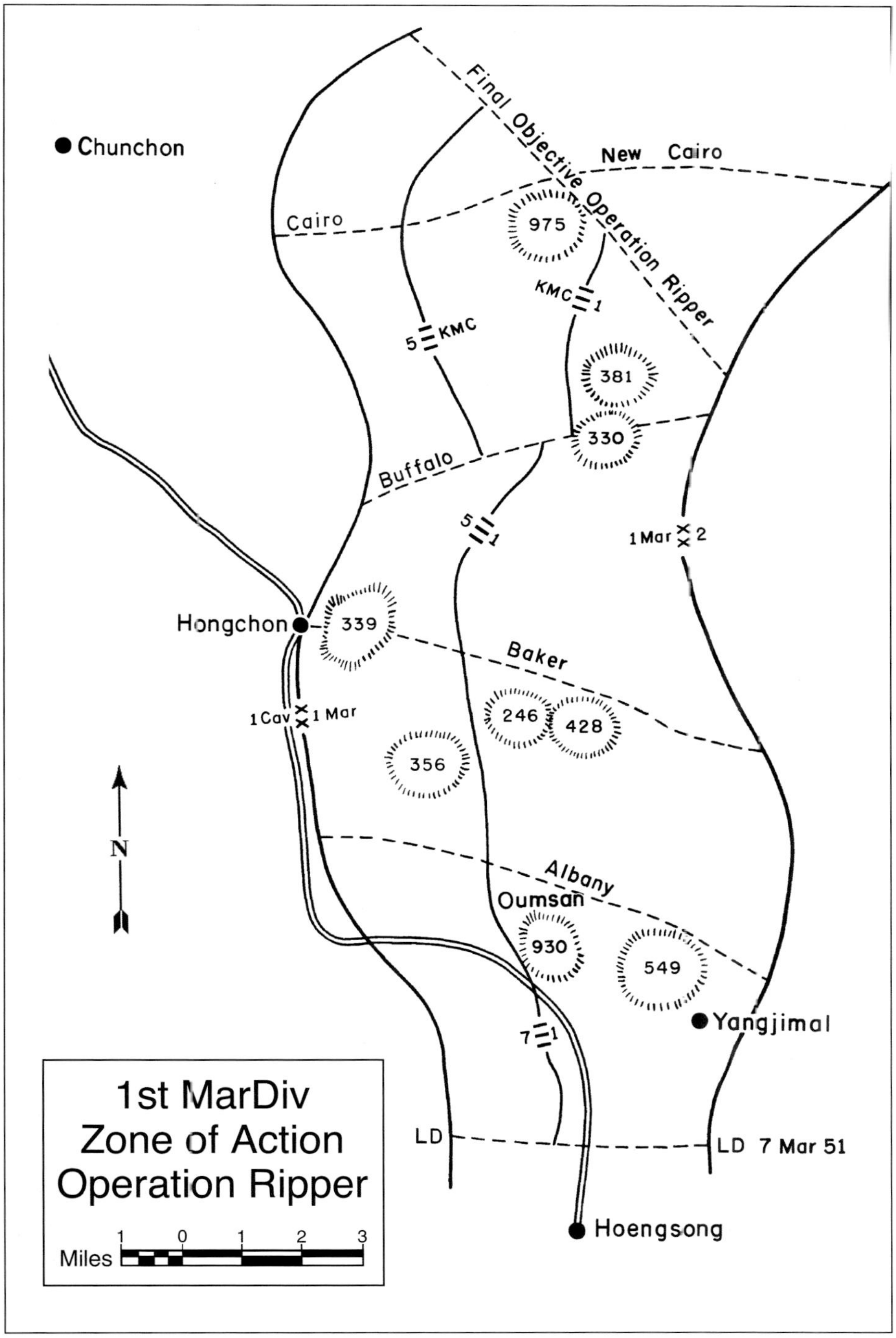

1stMarDiv Historical Diary Photo Supplement, Mar51

Tank and infantry teams move to the flanks of the advancing Marine column north of Hoengsong. The cold weather and sticky mud were as difficult to overcome as was the enemy during Operation Killer.

Albany, then seize Hongchon and destroy all enemy south of Line Buffalo and be prepared to continue the attack to Lines Cairo and Idaho on order, with operations commencing at 0800 on 7 March. One regiment would constitute the corps reserve and would be under the operational control of the IX Corps commander during the latter portion of the operation. The 1st Marine Division would advance up the Hoengsong-Hongchon axis with "two up and one back." General Smith initially placed 1st and 7th Marines in the assault and earmarked the 5th Marines as the reserve. The two assault regiments (7th on the left and 1st on the right) were to advance astride Route 29 with all three battalions on line whenever possible. The difficult supply situation left Colonel Joseph L. Winecoff's 11th Marines short of artillery ammunition, so an emergency agreement between Major General Field Harris, Commanding General, 1st Marine Aircraft Wing, and Major General Earl E. Partridge, Commanding General, Fifth Air Force, temporarily placed a Marine fighter squadron in direct support of the 1st Marine Division.

The operation started as ordered. The Marines advanced in the afternoon snow against only light resistance, primarily small arms and mortar fires. The enemy once again relied upon delaying tactics, opening fire at long range to slow the attackers then withdrawing before close combat could be initiated. Additionally, Marine tankers noted increased use of road mines. During the first three days awful weather and difficult terrain were the main obstacles. The Marine attack was finally stopped in place by orders from above which halted the advance on 9 March until flanking units could catch up. The next two days were devoted to reconnaissance and security patrols as the division marked time. On 11 March, the Marines resumed the advance. This time the enemy put up a stiff fight in the 1st Marines zone, and the 1st Battalion had to use artillery and tank guns to reduce log bunkers atop Hill 549 before that position fell. This single battle cost the Marines more casualties (one killed and nine wounded) than had been inflicted in the previous five days (seven wounded). The first phase of Operation Ripper ended on 13 March when the 1st Marine Division successfully occupied all of its objectives on Line Albany.

General Ridgway decided to change tactics for the next phase of Ripper. This time he opted to maneuver instead of slugging forward on a single line. His plan was a complex one. He decided to try an airborne drop north of Hongchon to be coordinated with a double envelopment by the 1st Marine Division and the 1st Cavalry Division, but this bold strike never came about because the Chinese *39th Army* slipped away before Ridgway's trap could be slammed shut. The reasons for this were divulged in a later intelligence find. A captured CCF directive indicated the enemy had adopted a "roving defensive" whereby units were no longer to hold at all costs, but should defend using movement to entice the United Nations Command to overextend itself as it had the previous November so the CCF could launch a "backhand" counteroffensive to isolate and annihilate the U.N. vanguard. It was a good scheme, but the wily Ridgway did not take the bait. Instead of mounting a headlong rush, his offensives continued to be strictly

Maj Vincent J. Gottschalk, commanding Marine Observation Squadron 6, discusses the tactical situation with Col Richard W. Hayward, commanding officer of the 5th Marines. The high-wing, single-engine OY-1 Sentinel in the background was used primarily as an artillery spotter.

Department of Defense Photo (USMC) A131207

1stMarDiv Historical Diary Photo Supplement, Mar51

North of Hoengsong, Marines move up into the hills. Stamina and physical conditioning were important during the unending series of hill fights that led back to the 38th Parallel.

"buttoned up" affairs.

In the 1st Marine Division zone, General Smith retained the same basic plan of attack. The 1st and 7th Marines would mount the assault, and the 5th Marines, now commanded by Nicaragua-veteran and World War II Marine parachutist Colonel Richard W. Hayward, would be the reserve. As before, the Marine advance on 14 March moved forward against almost no resistance. The 7th Marines did not need to call for artillery or close air support, and the 1st Marines encountered only scattered fire as it moved forward. General Ridgway's hopes of cutting off the enemy at Hongchon were dashed when an intercepted message from the enemy commander reported, "We must move back . . . . Enemy troops approaching fast," before the planned airdrop could be made. True to his word, General Liu Chen's troops were long gone by the time a motorized patrol from Major Sawyer's 1st Battalion, 7th Marines, entered the devastated town of Hongchon. Although the patrol located no enemy, it did discover a large number of "butterfly" bomblets dropped by U.S. Air Force aircraft. These deadly missiles so inundated the area that it took Captain Byron C. Turner's Company D, 1st Engineers, and all available division personnel three days to locate and disarm most of the explosive devices before they could produce casualties. The 7th Marines also found a treasure trove: three ammunition dumps that yielded more than two thousand small arms; a dozen heavy machine guns; a dozen mortars; a dozen recoilless rifles; numerous captured U.S. weapons; assorted demolitions; and four dozen cases of ammunition. This was one of the biggest finds of the war.

When the 7th Marines attacked the high ground north of Hongchon on the 15th, the 2d and 3d Battalions ran into a buzz saw. 120mm mortars and 76mm antitank guns pinned them down as they approached Hill 356. This Chinese fire was unusually accurate and intense, so much so that three 81mm mortars were knocked out. Likewise, the enemy was holding firm at Hills 246 and 248 in the 1st Marines zone. Lieutenant Colonel Robert K. McClelland's 2d Battalion, 1st Marines, performed an extremely complicated maneuver when it moved from the right flank on the east across the entire zone behind the front lines, trucked up to the village of Yangjimal in the 7th Marines zone

A 81mm mortar crew fires in support of an attack. Under the leadership of a sergeant, a mortar squad was composed of seven Marines and was known as the infantry commander's "hip pocket" artillery.

1stMarDiv Historical Diary Photo Supplement, Mar51

on the west, then dismounted for a difficult overland march to join the 3d Battalion, 1st Marines, for the assault on Hill 248. Unfortunately, the ensuing joint attack—including an air strike by Corsairs of Marine Fighter Squadron 214 and plentiful mortar and artillery fire—was not successful. After suffering about 100 casualties, the Marines pulled back to Hill 246 as darkness closed in. Another rifle battalion joined the assault force when Lieutenant Colonel Donald R. Kennedy's 3d Battalion, 5th Marines, was attached to the 1st Marines that night. Fortunately, a morning assault by the 1st Marines the next day found Hill 248 undefended. Back in the 7th Marines' zone the 1st Battalion had to clear Hill 399 with hand grenades and bayonets on its way to Line Baker.

Despite the progress, IX Corps complained of the lack of speed in the advance. General Smith pointed out that the division was "making a conscientious effort to comply with the Army's directive to keep buttoned up and comb the terrain." "This takes time," he said. Smith asked if there was any relaxation of the Army directive. General Hoge's answer was " 'No,' but he still wanted more speed." On 17 March, Hoge ordered the 1st Marine Division to continue the attack to Line Buffalo and beyond. To comply, Smith moved the 5th Marines up on the left and pulled the 7th Marines out of the line. The 5th Marines advanced against scattered resistance and reached Line Buffalo without a major fight. In fact, no Marine in that zone was killed or wounded in action for three straight days. The CCF had pulled back, but left elements of the *12th NKPA Division* behind to delay the Marines. The biggest engagement occurred on 19 March in the 1st Marines' zone. There, the enemy was well dug in on a series of north-south ridges joining Hills 330 and 381. Fortunately, the terrain allowed the tanks of Captain Bruce F. Williams' Company B, 1st Tank Battalion, to support the 2d Battalion attack. After F4U Corsairs from Marine Fighter Squadrons 214 and 323 delivered napalm and high explosive bombs on suspected enemy entrenchments, artillery pounded the objective, then tanks moved up on each side of the ridge keeping pace with the advancing infantry. The powerful 90mm tank guns eliminated enemy bunkers with very accurate direct fire as their machine guns kept enemy heads down. This coordinated direct fire allowed the rifle companies to successfully leapfrog each other over the next couple of days. This formula was so successful that the NKPA finally panicked on 20 March. At that time the enemy fleeing Hill 381 were hammered by supporting arms and infantry fires until they were virtually wiped out. With the end of that action the 1st Marine Division was ready to renew the attack.

A Marine searches an enemy bunker. The Chinese often used small squad-sized bunkers as a limited defense to cover the withdrawal of larger units.
National Archives Photo (USMC) 127-N-7410

The advance to Line Cairo was made with Colonel Kim's 1st KMC Regiment, once again attached to the 1st Marine Division. This allowed General Smith to use three regiments (5th Marines, 1st Korean, and 1st Marines) on line. The 7th Marines was placed in corps reserve. The 5th Marines made it to Line Cairo without serious opposition, but this was not true for the 1st Marines or the Korean Marines. The Koreans relied upon aerial resupply as they moved forward in the undulating and trackless central sector. The biggest fight took place when the Korean regiment, supported by Major Jack C. Newell's 2d Battalion and Lieutenant Colonel William McReynolds' 3d Battalion, 11th Marines, fought a three-day battle to capture Hill 975. The position finally fell to a flanking maneuver. On 22 March, the 1st Marines encountered some fire from Hills 505, 691, and 627 before reaching the Idaho Line where it made contact with the U.S. Army's 38th Infantry Regiment. Elements of the

1st Marines encountered some stiff resistance at Wongo-ri in the Tuchon-Myon hills while patrolling on the 27th. Two days later, the 1st Marines and the Korean Marine regiment extended their lines north to the New Cairo Line without a fight and brought Operation Ripper to a close.

On 31 March, the 1st Marine Division mustered 21,798 men in addition to 3,069 Korean Marines and 234 attached U.S. Army soldiers. Most Marines were at Hongchon, but some service support detachments were located farther back at Masan, Pohang, and Wonju. Unfortunately, when the artillery ammunition crisis abated Marine air once again reverted to Fifth Air Force control, and Marine aviators were no longer in direct support of their comrades on the ground. In two major operations (Killer and Ripper), the 1st Marine Division suffered 958 combat casualties (110 killed and 848 wounded), while inflicting an estimated 7,000 enemy casualties and capturing 150 enemy prisoners. For five weeks the Marines spearheaded each of IX Corps' advances from Wonju to Chunchon and lead the Eighth Army in ground gained during that time.

Although the men at the forward edge of the battlefield did not yet realize it, the nature of the Korean War had changed radically. In fact, strategic discussions now centered on whether to once again invade North Korea or not and, if so, how far that penetration should be. The military situation was so favorable that U.N. diplomats actually began to entertain the notion that the other side might be ready to ask for an armistice if the pressure was kept up. The most controversial element of strategy thus became what to do when U.N. forces reached the 38th Parallel. After much high-level discussion, American national command authorities agreed to go ahead and cross, but they warned General Douglas MacArthur that the conclusion of the next offensive would probably mark the limit of advance. Concurrently, the Joint Chiefs of Staff informed the U.N.

1st Marine Division

The 1st Marine Division was the senior Marine Corps ground combat unit in Korea. By Table of Organization and Equipment it rated 22,343 men divided into combat, combat support, and service support units. The teeth of the division were its three 3,902-man rifle regiments which were subdivided into a headquarters and service company, an anti-tank company, a 4.2-inch mortar company, and three rifle battalions (1,123 men), each composed of a headquarters and service company, three rifle companies, and a weapons company. The firing batteries of an artillery regiment, a tank battalion, and a combat engineer battalion furnished combat support. Service support came from assorted organic battalions and an attached combat service group. A unique attachment in Korea was a composite aircraft squadron that included helicopters and observation aircraft.

For the most part, the Marines were equipped with weapons of World War II vintage: small arms included .45-caliber automatic pistols, .30-caliber Garand semiautomatic rifles, carbines, Thompson submachine guns, and Browning automatic rifles; crew-served weapons included 4.2-inch, 81mm, and 60mm mortars, 155mm medium and 105mm light howitzers, 4.5-inch multiple rocket launchers, and .30- and .50-caliber machine guns. Two notable new weapons were the 3.5-inch rocket launcher and the M-26 Pershing tank.

1st Marine Division (Reinforced)

- Headquarters Battalion
    - Headquarters Company
    - Military Police Company
    - Reconnaissance Company
- 1st Marines
- 5th Marines
- 7th Marines
- 11th Marines
- 1st Amphibian Tractor Battalion
- 1st Armored Amphibian Battalion
- 1st Amphibian Truck Company
- 1st Combat Service Group
- 1st Engineer Battalion
- 1st Medical Battalion
- 1st Motor Transport Battalion
- 7th Motor Transport Battalion
- 1st Ordnance Battalion
- 1st Service Battalion
- 1st Shore Party Battalion
- 1st Signal Battalion
- 1st Tank Battalion
- Battery C, 1st 4.5-inch Rocket Battalion

Attachments

- 1st Joint Assault Signal Company
- 1st Provisional Truck Company
- Det, 1st Signals Operations Battalion
- Marine Observation Squadron 6
- 1st Korean Marine Corps Regiment

commander that, since these actions would terminate maneuver warfare, a diplomatic settlement to end the conflict after all pre-war South Korean territory had been liberated would be pursued. Unfortunately, these high hopes for an early end to the fighting were dashed by General MacArthur's imprudent public ultimatum demanding that the enemy either stop fighting or face annihilation. This presumptuous announcement on 24 March had several far-reaching effects. First, it so offended the Communists that it torpedoed some promising secret negotiations and actually triggered an aggressive battlefield response. Second, this action also sowed the seed that later sprouted into one of the most famous controversies in American military history.

Operations Rugged and Dauntless

All of Operation Ripper's terrain objectives had been taken, but General Ridgway felt not enough punishment had been meted out and this, coupled with a desire to secure a more defensible line, led him to continue offensive operations without a break. This time he envisioned a "double whammy" in the form of Operations Rugged and Dauntless. The goal of Rugged was to carry the Eighth Army back above the 38th Parallel to occupy a trans-peninsular defense line anchored upon the centrally located Hwachon Reservoir. Dauntless, on the other hand, was to be a spoiling attack to threaten the enemy's major staging area located northwest of Hwachon. This was the so-called "Iron Triangle" that included the terminus of several railway lines running down from Manchuria and incorporated the intersection of all major roads in north-central Korea. Its forested flat lands were bounded by protective ridges and included the towns of Chorwan, Pyonggang (not to be confused with the North Korean capital of Pyongyang), and Kumhwa. The geographic shape of the road net connecting these towns gave the Iron Triangle its name.

The battlefield situation was very complicated. Eighth Army intelligence officers were not sure if the enemy was going to defend in place along the former international dividing line or continue to give ground. Large troop movements into the Iron Triangle had been noted, but it was a point of contention as to whether these were part of a Communist "rotation" policy or if they constituted an offensive build up. (Actually, both events were occurring simultaneously; worn out elements of the CCF *Fourth Field Army* were moving back to Manchuria while the fresh CCF *Third Field Army* was entering Korea.) Unsure of enemy intentions, General Ridgway ordered a cautious advance, but warned his corps commanders to be ready to fall back to prepared defensive lines if ordered to do so. Ridgway's primary intent was to seize Line Kansas, a phase line purposely drawn so that it included the best defensive terrain in the vicinity of the 38th Parallel. In IX Corps' zone this line carried eastward from the Imjin River to the western tip of the Hwachon Reservoir and included that body of water's southern shoreline. IX Corps' axis of advance was to be about a dozen miles almost due north from Chunchon astride the Pukhan River using National Route 17 as the main supply route. The terrain in this zone was uneven. It was mostly flat west of the Pukhan River, but high hills on the right dominated the approaches to the Hwachon Reservoir. The enemy was believed to be stay-behind elements of the CCF *39th Army*, but it was uncertain if those forces would flee or fight.

On 29 March, Ridgway issued orders to initiate Operation Rugged. This time the hard-working Marines did not spearhead the attack as they had during Killer and Ripper. The U.S. 1st Cavalry and 6th ROK Divisions would carry that load, while the 1st Marine Division was IX Corps

Tank and infantry teams search out possible enemy positions on either side of the road to Chunchon. Unfortunately, the mountainous terrain of central Korea hampered effective tank support.

1stMarDiv Historical Diary Photo Supplement, Apr51

reserve. General Smith hoped his men could get some well-deserved rest after replacing elements of X Corps at Line Ready near Chunchon. An unusual exception was the 7th Marines, which was actually slated to participate in the drive north.

There was a small modification to the plan almost immediately. Instead of going into reserve, the 1st Marine Division (less the 7th Marines) was ordered to continue the attack. "This arrangement," noted General Smith, "gives me responsibility for 28,000 meters of front and I have for the time being no reserve." The 1st Marines became IX Corps reserve and moved back to Hongchon. The 5th Marines and the Korean Marines continued to move forward toward Line Ready. To do this the 5th Marines had to force a crossing of the Soyang River and seize Hills 734, 578, and 392 against moderate to heavy resistance. Once this was accomplished, elements of the U.S. Army's 7th Infantry Division took over, and the Marines began making their way back to the assembly area near Chunchon on the afternoon of 4 April for what promised to be five days off the firing line, the first real rest for the division since moving up from Pohang in mid-February.

On 1 April, the 7th Marines was placed under the operational control of Major General Charles D. Palmer's 1st Cavalry Division. The plan was for the division to advance about three miles from Line Dover to secure Line Kansas just north of the 38th Parallel. Colonel Litzenberg's regimental combat team, composed of the 7th Marines; 3d Battalion, 11th Marines; Company D, 1st Tanks; Company D, 1st Engineers; and various service detachments, was assigned the left (western) sector for the advance with specific

National Archives Photo (USN) 80-G-442505

Hospital Corpsman Richard De Wert, USNR

Born in 1931 in Taunton, Massachusetts, Richard De Wert enlisted in the Navy in 1948. Following "boot camp" and Hospital Corps training at Great Lakes, Illinois, he was assigned to the Naval Hospital, Portsmouth, Virginia. Attached to the 1st Medical Battalion, 1st Marine Division, in July 1950, he participated in the Inchon, Seoul, and Chosin operations. On 5 April 1951, while serving with the 2d Battalion, 7th Marines, on Hill 439 near Hongchon and the 38th Parallel, he gave his life while administering first aid to an injured Marine. His Medal of Honor award said, in part:

> When a fire team from the point platoon of his company was pinned down by a deadly barrage of hostile automatic weapons fire and suffered many casualties, De Wert rushed to the assistance of one of the more seriously wounded and, despite a painful leg wound sustained while dragging the stricken Marine to safety, steadfastly refused medical treatment for himself and immediately dashed back through the fire-swept area to carry a second wounded man out of the line of fire.
>
> Undaunted by the mounting hail of devastating enemy fire, he bravely moved forward a third time and received another serious wound in the shoulder, after discovering that a wounded Marine had already died. Still persistent in his refusal to submit to first aid, he resolutely answered the call of a fourth stricken comrade and, while rendering medical assistance, was himself mortally wounded by a burst of enemy fire.

The Secretary of the Navy on 27 May 1952 presented Corpsman De Wert's Medal of Honor to his mother, Mrs. Evelyn H. De Wert. The guided missile frigate USS *De Wert* (FFG 45), bears his name.

— Captain John C. Chapin, USMCR (Ret)

instructions to keep the main supply route clear, protect the ferry site, and maintain liaison with the 6th ROK Division.

Colonel Litzenberg closed his command post at Hongchon and moved it to the assembly area near Chunchon. By 1000 on 2 April, the lead element of the 7th Marines, Lieutenant Colonel Wilbur F. Meyerhoff's 2d Battalion, had crossed the line of departure and was moving up Route 17 with the other two battalions in trail. The attack proceeded against very light opposition, and no Marine casualties were reported. The major holdup was the time it took to ferry the Marines across the Pukhan River. Army amphibian trucks took the men across. Most large vehicles were able to ford the river, and smaller ones used rafts operated by an Army assault boat detachment. By the end of the day all objectives had been secured. The next day's mission was to take an intermediate objective, Phase Line Troy. Again, all assigned objectives were reached, without enemy interference, by darkness on 3 April. The main stumbling blocks were torturous terrain, craters and debris blocking the road, and land mines.

The 6th ROK Division on the left moved up against virtually no opposition until it reached Line Kansas. Unfortunately, things did not go so smoothly in the 1st Cavalry zone of action where the enemy increased the pressure near the 38th Parallel and stubbornly held out in the hills south of Hwachon. While the 7th Marines had thus far encountered few enemy, the 7th and 8th Cavalry Regiments became entangled in a fierce slugging match and fell far

Truman Fires MacArthur

Fighting men in Korea, as were people all across the United States, were shocked to learn American President Harry S. Truman had relieved General of the Army Douglas MacArthur of his commands (Commander in Chief, Far East; United Nations Command; and Supreme Commander, Allied Powers) on 11 April 1951. This removal of America's longest-serving warrior turned out to be one of the most controversial military decisions in American history. The President was well within his constitutional authority to depose a field commander, but Truman's action initiated an acrimonious debate about both U.S. political leadership and America's proper role in world affairs that deeply divided the country.

This incident was the result of long-standing policy disagreements about grand strategy and the ultimate purpose of military action. General MacArthur sought an absolute commitment to victory and felt anything less than surrender by the Communists was an unacceptable outcome of the conflict. President Truman, on the other hand, believed that Korea was only one theater in the Cold War and insisted the Communists would be deterred in other arenas if the viability of South Korea could be maintained. In short, the general wanted a military victory akin to those that ended the two World Wars, but the leader of the free world viewed Korea as a limited military action intended to achieve very specific diplomatic aims without embroiling the world in global warfare. President Truman framed the essence of this dispute when he wrote about Korea: "General MacArthur was willing to risk general war; I was not."

The roots of the dispute began almost as soon as the United States became enmeshed in Korea. MacArthur bristled over what he considered political meddling in military affairs in August 1950, and then more frequently and more vociferously aired his views in public as time wore on. The U.N. commander felt he was being saddled by unrealistic restrictions and chafed at not being given the resources he needed to successfully fight the war. Controversial statements about these issues caused Truman to personally confront MacArthur at Wake Island in October 1950 and then led the President to issue several "gag" orders over the next few months.

National Archives Photo (USA) 111-SC353136

Gen Douglas MacArthur greets President Harry S. Truman on his arrival at Wake Island for their October 1950 conference. Within five months the President would be forced to relieve his Far East commander.

One problem was that MacArthur's track record with respect to Korea was enigmatic, one marked by brilliant successes offset by seriously flawed diplomatic and military judgements. The general predicted, despite intelligence reports to the contrary, that China would not intervene in Korea; then, after his U.N. forces were roughly pushed out of North Korea, demanded measures well outside of the U.N. mandate (i.e. bombing Red China, blockading the Chinese coast, and intervention by Nationalist Chinese forces). His recommendations were immediately rejected by all allied nations even though MacArthur proclaimed failure to adopt his plan would mean the annihilation of the United Nations Command. MacArthur suffered a loss of face when his dismal forecast did not come true, but instead United Nations forces rolled back the enemy and regained the 38th Parallel without drastic measures in the spring of 1951.

The most egregious of MacArthur's forays into the diplomatic arena came when he purposely torpedoed secret peace feelers in late March 1951 by publicly taunting the enemy commander and threatening to widen the war. The near simultaneous publication of an earlier letter to Republican House Leader Joseph Martin, which many viewed as a thinly veiled attack on the Truman Administration that closed with the inflammatory statement, "There is no substitute for victory," finally brought the Truman-MacArthur controversy to a head. Thus, at half-past midnight on 11 April 1951, President Truman issued orders to recall General MacArthur.

This unexpected and seemingly rash act, spurred by the insensitive manner in which the relief was handled, created a firestorm on the home front. MacArthur returned from Korea a hero. He was welcomed across the country by an adoring public before he culminated his 52-year military career with a moving and an eloquent speech to Congress. MacArthur's popularity was at an all-time high as he enjoyed his final triumph—a gala ticker tape parade through New York City—before, like the old soldier in his speech, he "faded away" by dropping out of the public eye. On the other hand, Truman's action was so controversial that his popularity dropped to an all-time low. The President's opponents flamed the fires of public dissatisfaction with the war when they demanded public hearings. These were held, but did not turn out as expected. In the end, the Senate reaffirmed the President's right to dismiss a subordinate and surprisingly vindicated Truman's decision after equally venerated General of the Army Omar N. Bradley, Chairman of the Joint Chiefs of Staff, asserted MacArthur's plan would have resulted in "the wrong war, at the wrong time, at the wrong place, and with the wrong enemy."

---

behind schedule as they battled their way north. Still, a milestone was achieved on 4 April when a Marine patrol from First Lieutenant Orville W. Brauss' Company B became one of the first Eighth Army units to recross the 38th Parallel. The 11th Marines fired 17 missions hitting some enemy troops in the open and peppering suspected emplacements with excellent results, and a four-plane flight scattered an enemy column.

The next day, the Marines became heavily engaged and had to fight their way forward for the next 48 hours. On 5 April, the 1st Battalion met very stiff resistance. Automatic weapons and mortar fire pinned down two companies. The 2d Battalion likewise met stubborn resistance and had to call for tank support to overrun its objectives. Navy Corpsman Richard D. De Wert, serving with Company D, was mortally wounded after fearlessly exposing himself to enemy fire four times and being hit twice as he dragged injured men to safety at Mapyong-ni. De Wert was awarded a posthumous Medal of Honor for his actions. Ten very accurate close air strikes coordinated with artillery fire enabled the Marines to push forward late in the day. Good coordination between the assault and support companies (Captain Jerome D. Gordon's Company D, Captain Merlin T. Matthews' Company E, Captain Raymond N. Bowman's Company F, and Captain William C. Airheart's Company G) provided textbook examples of infantry fire and maneuver. The next day enemy opposition was less formidable, but First Lieutenant Victor Stoyanow's Company I took a beating when it became pinned by automatic weapons fire in some low ground which was also zeroed in on by enemy mortars. There was no air support available because of weather (low-lying cloud cover, high winds, and heavy rain), but artillery counter-battery fire, as well as American tank and mortar fires, eventually silenced the enemy guns. The enemy suffered about 150 casualties trying to hold out. The Marines lost five killed and 22 wounded.

On the afternoon of 6 April, the 7th Marines finally reached the Kansas Line after some tough fighting. Twenty Marines were wounded during the day, most by enemy 76mm fire but some to small arms and mortars. With the Kansas Line reached, the men of Colonel Litzenberg's regiment patiently waited for the 1st Marine Division to relieve the 1st Cavalry. General Smith received orders to do so on 8 April, and the relief was tentatively slated for the 10th. General Ridgway also told General Smith that the 1st Marine Division (less the 1st Marines in corps reserve) would then attack north to seize the northwest end of the Hwachon Reservoir. The situation did not look promising. The 1st Cavalry Division had been stopped by

Department of Defense Photo (USMC) A131342

Generals meet along the central Korean fighting front. Pictured from left are MajGen Oliver P. Smith, USMC, LtGen James A. Van Fleet, USA, and MajGen William M. Hoge, USA. LtGen Van Fleet replaced Gen Ridgway as commander of the U.S. Eighth Army.

uncharacteristically fierce resistance and was stalled almost three miles from its final objective, and the Chinese still controlled the Hwachon Dam.

This situation became serious when the enemy opened some of the dam's sluice gates on 8 April sending a massive wall of water around the river bend and onto the Chunchon flood plain. Luckily, the low level of water within the reservoir and the fact that not all the gates were opened kept the damage to a minimum. Only one bridge was knocked out, although several other pontoon bridges had to be disconnected until the rising water subsided. In the end, this man-made flash flood only raised the river level about a foot downstream. Still, the pent-up waters of the reservoir represented a potential threat to future operations. Accordingly, seizure of the dam itself or destruction of the gate machinery became a high priority. Unfortunately, several Army ground attacks and a night raid failed to achieve that goal. The latter was a water-borne raid by Army rangers paddling rubber assault boats, not an amphibious assault as is sometimes claimed; and, contrary to some sources, no Marine units were involved in either the planning or execution. Failure to take or knock out the Hwachon Dam meant its capture unexpectedly became the next major Marine task for Operation Dauntless.

The Marines began arriving at the Kansas Line as scheduled on 10 April, but not all units were in place until two days later when Korean Marines relieved the last elements of the 1st Cavalry Division. Seizing the Hwachon Dam as well as securing the main supply route leading north to Kumhwa and reaching the Wyoming Line were now the objectives for an expanded Operation Dauntless. With this in mind, General Smith assigned his division an intermediate phase line. The Quantico Line included the heights overlooking the Hwachon Dam and the hills north of the village of Hwachon, while the exact positions held on the Marine left flank were to be tied to the advances made by the 6th ROK Division. This was the plan when the 1st Marine Division deployed along the line of departure. Then, Operation Dauntless was suddenly postponed.

Although a time of general tranquility on the Central Front, the break between operations was one of international tumult. Its root cause was President Harry S. Truman's decision to relieve General MacArthur of command. This unexpected announcement was greeted for the most part by stunned silence in Korea, but created a considerable stateside uproar known as "the great debate."

General Ridgway was named the new commander of United Nations forces and was in turn replaced as Eighth Army commander by Lieutenant General James A. Van Fleet, USA. A former football coach, 59-year-old "Big Jim" Van Fleet was an aggressive leader who favored expending fire and steel instead of men. A veteran of both World Wars and the general officer who had seen the most frontline combat in the European theater during World War II, Van Fleet had recently served with the Joint Military Aid Group that saved Greece from Communist insurgents. Like-minded Generals Van Fleet and Ridgway made a good team. This was fortunate because Ridgway had planned Operation Dauntless, but Van Fleet was going to have to carry it out. Obviously,

National Archives Photo (USMC) 127-N-A7862

After 10 days of patrolling and preparation of defensive works on Line Kansas, the 5th Marines resumed the advance toward Line Quantico.

the new commander needed a few days to "snap in" before leading a new offensive.

Van Fleet was greeted with some ominous news. In the wake of Operation Ripper intelligence officers began to grasp that another Communist offensive was near, the fifth such major effort since the CCF intervened in Korea the previous fall. Prisoners of war reported the attack could begin within one week, and captured documents claimed the ultimate goal was to eject U.N. forces from Korea after the Communists celebrated May Day in Seoul. To this end more than 700,000 CCF and NKPA troops had been amassed. The enemy's main force, 36 Chinese divisions, gathered inside the Iron Triangle. About half of the NKPA divisions were also poised to strike in the east. Although the time and place of the expected offensive had been generally deduced, an unforeseen development—a deep penetration of South Korean lines far from the enemy's planned main effort—unexpectedly placed the Marines of the 1st Division in the center of the action, and the period from late April until mid-May featured a series of desperate fights and some intricate maneuvers that kept the enemy at bay until the Chinese Spring Offensive lost its momentum.

CCF Spring Offensive

Spring finally arrived in mid-April. The days were generally warm and sunny with the temperature reaching into the mid-60s. The nights were mostly clear and cool, but there was no longer the need for heavy winter clothing or arctic sleeping bags. All of the snow had melted, and patches of flowers were sprouting up among the scrub pines. And, although there were still a few April showers, the heavy rains let up and the mud was finally drying out.

Thanks to the high-level turmoil caused by the sudden change of command, the 1st Marine Division spent 10 quiet days on the Kansas Line before beginning Operation Dauntless on 21 April. The IX Corps objective was the Wyoming Line, but the Marines were also given an intermediate objective labeled Quantico Line, which included the Hwachon Dam and the meandering Pukhan River as well as Route 17 and a line of hills north and west of the village of Hwachon.

At 0700 on the 21st, the 1st Marine Division resumed the attack with the 7th Marines on the left, the 5th Marines in the center, the Korean Marines on the right, and the 1st Marines in reserve. The 5,000- to 9,000-yard advance, in the words of one regimental commander, was "made into a vacuum." Strangely, there was almost no sign of the enemy other than a few pieces of lost equipment and the ashes of a few cooking fires—the flotsam and jetsam left behind when any large body of troops moves out in hurry. Korean Marines made the only significant contact by killing one straggler and capturing another. About the only reminder that an unseen enemy was lurking nearby was a green haze of deliberately set fires that hugged the damp earth.

The lack of enemy activity was welcome, but it was also baffling. The front was eerily quiet, too quiet for many wary veteran Marines who felt something big was about to break. Lieutenant Colonel John L. Hopkins, commanding officer of 1st Battalion, 5th Marines, thought this "strange atmosphere of silence . . . was much like the stillness which had preceded the first CCF attack on Yudam-ni on 27 November." This nearly universal feeling of unease along the front lines was supported by several ominous signs. Aerial observers suspected the enemy was up to no good, but could not be specific because the area was shrouded by smoke that masked troop movements. There were unconfirmed reports of several thousand troops on the move, but the Marines spotted no actual enemy. Enemy prisoners of war taken in other sectors indicated

1stMarDiv Historical Diary Photo Supplement, Apr51

By 22 April the Marines had seized and held Line Quantico. To the north Chinese Communist Forces were poised to attack and the IX Corps zone was to be the target area for the attempted breakthrough. Heavy machine gun teams were the backbone of the defense.

that at least four CCF armies were poised to take on IX Corps, and they named 22 April as the date of the attack. A particularly disturbing bit of information was that the 6th ROK Division on the Marine left had opened a 2,500-yard gap, and all physical contact with that unit had been lost. Numerous patrols failed to find the elusive South Koreans. Consequently, on the eve of what appeared to be a major enemy effort the Marine western flank was dangling.

At 0830 on the morning of 22 April, preceded by low-flying observation aircraft and jeep-mounted ground reconnaissance units, the 1st Marine Division proceeded up the Chunchon Corridor west of the Hwachon Reservoir. Unlike the day before, however, this time the enemy harassed the Marine advance with small arms, automatic weapons, and mortar fire. Captain Robert L. Autry's Reconnaissance Company, aided by a tank detachment, entered Hwachon village under intermittent fire. They found more than a dozen badly wounded men left behind by the Communists and spotted several dozen more fleeing north. The 7th Marines, commanded by pre-war China veteran and World War II artilleryman Colonel Herman Nickerson, Jr., who had relieved the ailing Colonel Homer Litzenberg, advanced several miles on the left flank with only one man wounded. Air strikes hit suspected enemy assembly areas and possible field fortifications. The 5th Marines moved up Route 17 and occupied the hills on either side of a slender valley encompassing the village of Hwachon against moderate to heavy fire. Korean Marines seized the Hwachon Dam and the heights protecting it, but were then pinned down for a while by accurate enemy indirect fire. Total losses when the Marines reached Quantico Line were five men (two America and three Korean) killed and two dozen wounded (20 U.S. and four Korean).

At the end of the day, the 1st Marine Division was at the Quantico Line arrayed on a nearly straight line north and west of the Hwachon Reservoir with the 7th Marines, 5th Marines, and the Korean Marines from left to right. Two tank companies (B and C, in support of the 5th and 7th Marines, respectively) were forward deployed. The 11th Marines, reinforced by corps artillery (including the 8-inch guns of the 213th and 17th Field Artillery Battalions and the 155mm howitzers of the 92d Armored Field Artillery) was set up in the flat land just behind the front line troops. The Army guns were positioned near the west flank so they could reinforce either the ROKs or U.S. Marines as needed. Artillery ammunition trucks and prime movers jammed the narrow road making resupply and overland travel difficult. The 1st Marines was in reserve several miles away across the Pukhan River at Chunchon.

Enemy resistance seemed to be stiffening, but there was no reason for alarm as the Marines settled in on the night of 22-23 April. The evening promised to be crisp and clear with a full moon. At about 1800, General Smith issued instructions for the 1st Marine Division to continue its advance to seize the Wyoming Line at 0700 the next morning. These orders, however, were overcome by events three-and-a-half hours later. Although unrealized at the time, nearly 350,000 enemy troops were pushing silently forward between Munsan-ni in the west and the Hwachon Reservoir in the east. The CCF Fifth Phase Offensive was underway just as the enemy prisoners had predicted. Furious mortar and artillery barrages struck United Nations lines all across the

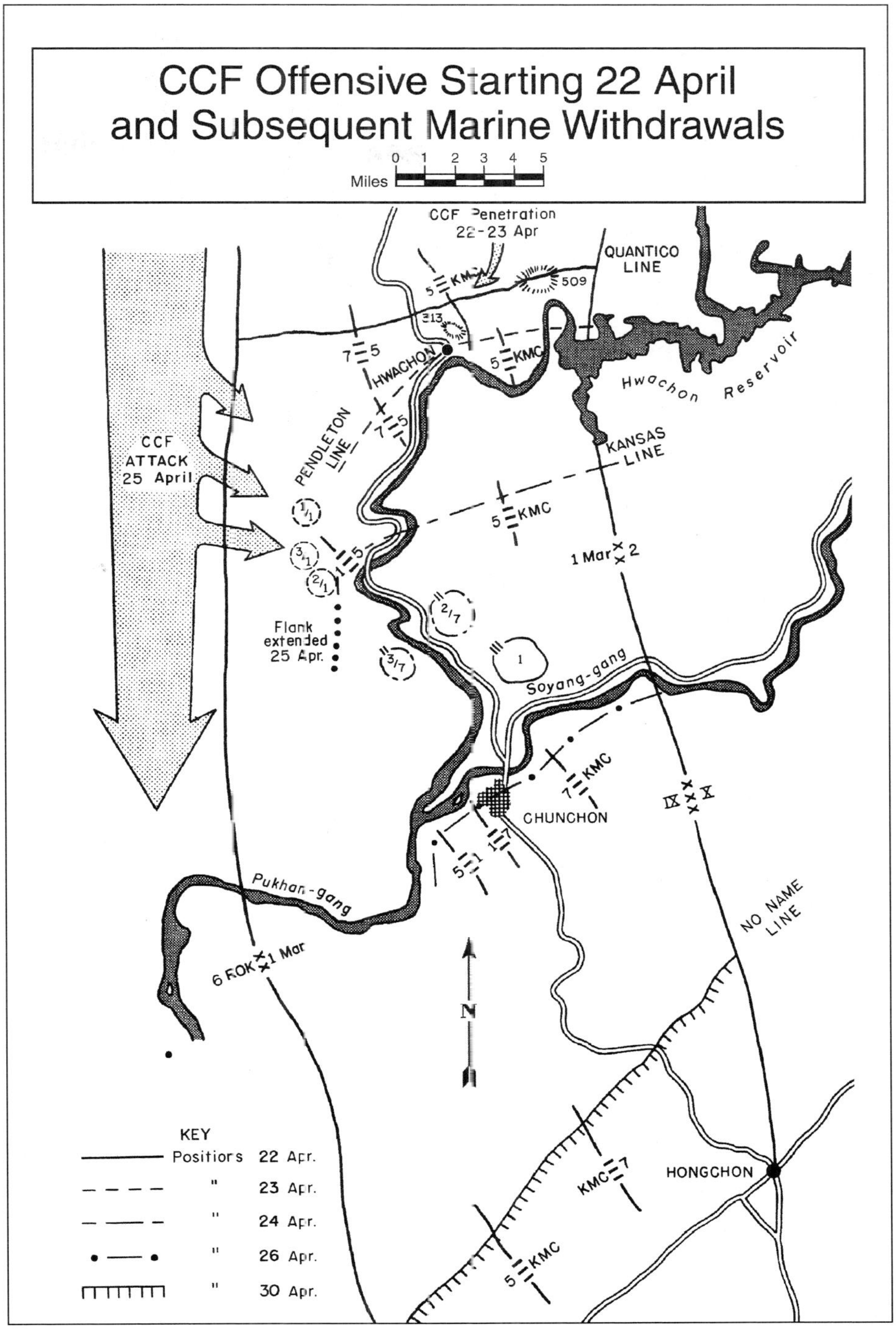

front before midnight. The first blows on IX Corps' front were not directed at the Marines, but at the shaky 6th ROK Division on their left, which was hit full force. That hapless unit simply evaporated as its frightened soldiers fled the field of battle. Facing only token resistance, the CCF *40th Army* was on its way south in full gear by midnight. Soon, a 10-mile penetration was created and the 1st Marine Division was in serious jeopardy. Some of the toughest fighting of the Korean War marked the next 60 hours, and the magnificent defenses of Horseshoe Ridge and Hill 902 were reminiscent of similar heroic Marine stands at Les Mares Farm in World War I and Guadalcanal's Bloody Ridge during World War II.

The dull mid-watch routine at the 1st Marine Division command post was interrupted when the duty officer was informed at about 2130 that the Chinese had penetrated South Korean defenses and were headed toward Marine lines. Not long after the message arrived the vanguard of a long line of demoralized South Korean army soldiers began filing in. By midnight, the Reconnaissance Company and Captain Donald D. Pomerleau's Military Police Company were rounding up stragglers and placing them under guard at the ferry site just south of the 5th Marines' command post. These dejected remnants of the 6th ROK Division reported their unit was in full retreat and further noted that thousands of enemy troops were rapidly moving south. Despite attempts to reconstitute the division as a fighting force, the 1st Division's liaison officer called and said: "to all intents and purposes, the 6th ROK Division had ceased to exist." This was alarming news because the Marine left flank was wide open, and the division's main supply route and all crossing points of the Pukhan River were at great risk.

The first U.S. troops to confirm the disaster on the left were cannoneers from Army artillery units that earlier had been sent west to shore up the South Koreans. Elements of the battered 987th Armored Field Artillery came pouring back into the American lines after being ambushed. The ill-fated artillery unit had lost about half of its 105mm howitzers to the ambush, and the 2d Rocket Artillery Battery lost all of its weapons when its defensive position was overrun. As Lieutenant Colonel Leon F. Lavoie, commanding officer of the 92d Armored Field Artillery Battalion acidly observed: "there had been more

Private First Class Herbert A. Littleton

On the night of 22 April 1951, radio operator Herbert A. Littleton serving with an artillery forward observer team of Company C, 1st Battalion, 7th Marines, sacrificed his life to save the lives of his team members.

Born in 1930, in Arkansas, he attended high school in Sturgis, South Dakota, where he played football and basketball and then worked for Electrical Application Corporation in Rapid City. Shortly after his eighteenth birthday, he enlisted in the Marine Corps, received recruit training at Marine Corps Recruit Depot, San Diego, and additional training at Camp Pendleton before being sent to Korea with the 3d Replacement Draft in December 1950. His Medal of Honor citation read, in part:

> Standing watch when a well-concealed and numerically superior enemy force launched a violent night attack from nearby positions against his company, Private First Class Littleton quickly alerted the forward observation team and immediately moved into an advantageous position to assist in calling down artillery fire on the hostile force.
>
> When an enemy hand grenade was thrown into his vantage point, shortly after the arrival of the remainder of the team, he unhesitatingly hurled himself on the deadly missile, absorbing its full, shattering impact in his own body. By his prompt action and heroic spirit of self-sacrifice, he saved the other members of his team from serious injury or death and enabled them to carry on the vital mission that culminated in the repulse of the hostile attack.

Departmentt of Defense Photo (USMC) A46967

Private First Class Littleton's heroic actions were later memorialized at Camp Pendleton by a marksmanship trophy, a baseball field, and a street, all named in his honor.

— Captain John C. Chapin, USMCR (Ret)

---

artillery lost in Korea up to that point than there was lost in the whole of the European theater in the last war by American forces."

By 2224, the impact of the disaster on the left was apparent, so all plans to attack the next day were abruptly canceled. Units along the forward edge of the battlefield were placed on full alert with orders to button up tight. Commanders hurriedly sent out combat patrols to locate the enemy and to try fix his line of march, while the Marines at the main line of resistance dug in deep and nervously checked their weapons. In addition, Smith ordered Colonel McAlister to send Lieutenant Colonel Robley E. West's 1st Battalion, 1st Marines, up from Chunchon to tie in with the artillery and tanks located in the valley on the far west flank. West's battalion was soon on trucks headed for its new position, but the convoy could only creep along over roads choked with panic-stricken South Korean soldiers escaping the battle zone. Captain John F. Coffey's Company B led the way. At about 0130, while still 1,000 yards short of its assigned position, the long column of vehicles stopped at the tight perimeter formed by the 92d Armored Field Artillery Battalion, which a short time before had established a road block, collected more than 1,800 South Koreans, and attempted by machine gun and bayonet with little success to deploy them to slow the Chinese advance. Moving west, Coffey's company assisted in the extricating the 987th Artillery's 105mm howitzers that were stuck in the mud. After as many guns as possible were freed, Coffey returned to friendly lines where the 1st Battalion was manning a wooded semi-circular ridge with Captain Thomas J. Bohannon's Company A on the right, Captain Robert P. Wray's Company C in the center, and the 81mm mortars of First Lieutenant Wesley C. Noren's Weapons Company on relatively level ground in the immediate rear. Company B was promptly assigned the battalion left flank.

The enemy began probing Marine lines around 2300 on 22 April and then mounted an all-out assault to turn the Marine flanks about three hours later. The 7th Marines on the left was the hardest hit U.S. unit. Enemy mortar, automatic weapons, and small arms fire began at about 0200 on the 23d. This reconnaissance by fire was followed by a very determined ground assault an hour later. Shrieking whistles, clanging cymbals, and blasting bugles signaled the onslaught. Up and down the line grizzled veterans of the Chosin Reservoir walked the lines to settle down young Marines who had not yet experienced a terrifying "human wave" ground assault. Noncommissioned officers force-

1stMarDiv Historical Diary Photo Supplement, Apr51

Near the Quantico Line, a Marine 75mm recoilless rifle crew opposes the Chinese Communist attack on 23 April. Recoilless rifles provided long-range pinpoint accuracy, but were light enough to be carried up and down Korea's mountainous terrain.

fully and profanely reminded their charges not to use grenades until the enemy was close at hand, and more than one of them tried to calm the new men by remarking about the frightening cacophony: "Those guys sure could use some music lessons!"

At least 2,000 enemy troops hit Major Webb D. Sawyer's outmanned 1st Battalion, 7th Marines, full force. That attack by the CCF *358th Regiment, 120th Division*, primarily directed at Captain Eugene H. Haffey's Company C and Captain Nathan R. Smith's Company A, was repulsed by hand-to-hand fighting that lasted almost until dawn. Private First Class Herbert A. Littleton, a radioman with the forward observer team attached to Company C, was standing the mid-watch when the enemy appeared. He sounded the alarm then moved to an exposed position from which he adjusted supporting arms fires despite fierce incoming machine gun fire and showers of enemy grenades. Forced back into a bunker by enemy fire, Littleton threw himself upon a grenade to save his comrades in that crowded space at the cost of his own life. He was awarded a posthumous Medal of Honor for his selfless actions that night. Heavy fighting—much of it grenade duels and close quarters combat—lasted several hours. Enemy mortar fire and small arms continued throughout the night and into the next day. As always, supporting arms were a critical Marine advantage. The 11th Marines ringed the endangered position with a wall of steel, and Marine tanks successfully guarded the lowland approaches.

In the division's center, Chinese infiltrators silently slipped through the 5th Marines' outpost line to occupy Hill 313. A futile counterattack was quickly launched, but despite tremendous heroism (three Marines received the Navy Cross for their actions) the assault platoon was held in check and suffered heavy casualties. It was not until the next morning that elements of the 1st and 2d Battalions, 5th Marines, retook the hill. At around 0300, Korean Marines on the right came under heavy attack in the vicinity of Hill 509. The stalwart Koreans threw back successive enemy attacks throughout the night and had ejected the enemy by the next morning. Particularly hard hit was a single rifle company of the 1st KMC Battalion holding the left flank. The 150-man company was reduced to only about 40 men ready for duty by daylight.

The timely arrival of the reserve 1st Battalion, 1st Marines, eased the pressure on Major Sawyer's 1st Battalion, 7th Marines, and solidified the division west flank, but fierce fighting continued into the next morning. The Chinese, well aware that they would be pounded from the air during the day, hurriedly retreated as the first rays of light began to creep over the horizon. Eight Marine Corsairs swooped over the battlefield guided onto their targets by aerial observers flying vulnerable OY observation aircraft; Marine Fighter Squadron 323 flew in support of the 5th Marines while Marine Fighter Squadron 214 worked over the Chinese in the 7th Marines' zone. The retreating enemy was slaughtered by this blitz from above. Enemy casualties by all arms were estimated to be well above 2,000 men. By noon on the 23d, it was obvious the Marines had won the first round, but it was also obvious that the fight was far from over. For his skillful and aggressive leadership in securing the division's vital flank, Major Sawyer, the recipient of two Silver Stars for the Chosin Reservoir campaign, was awarded a Navy Cross.

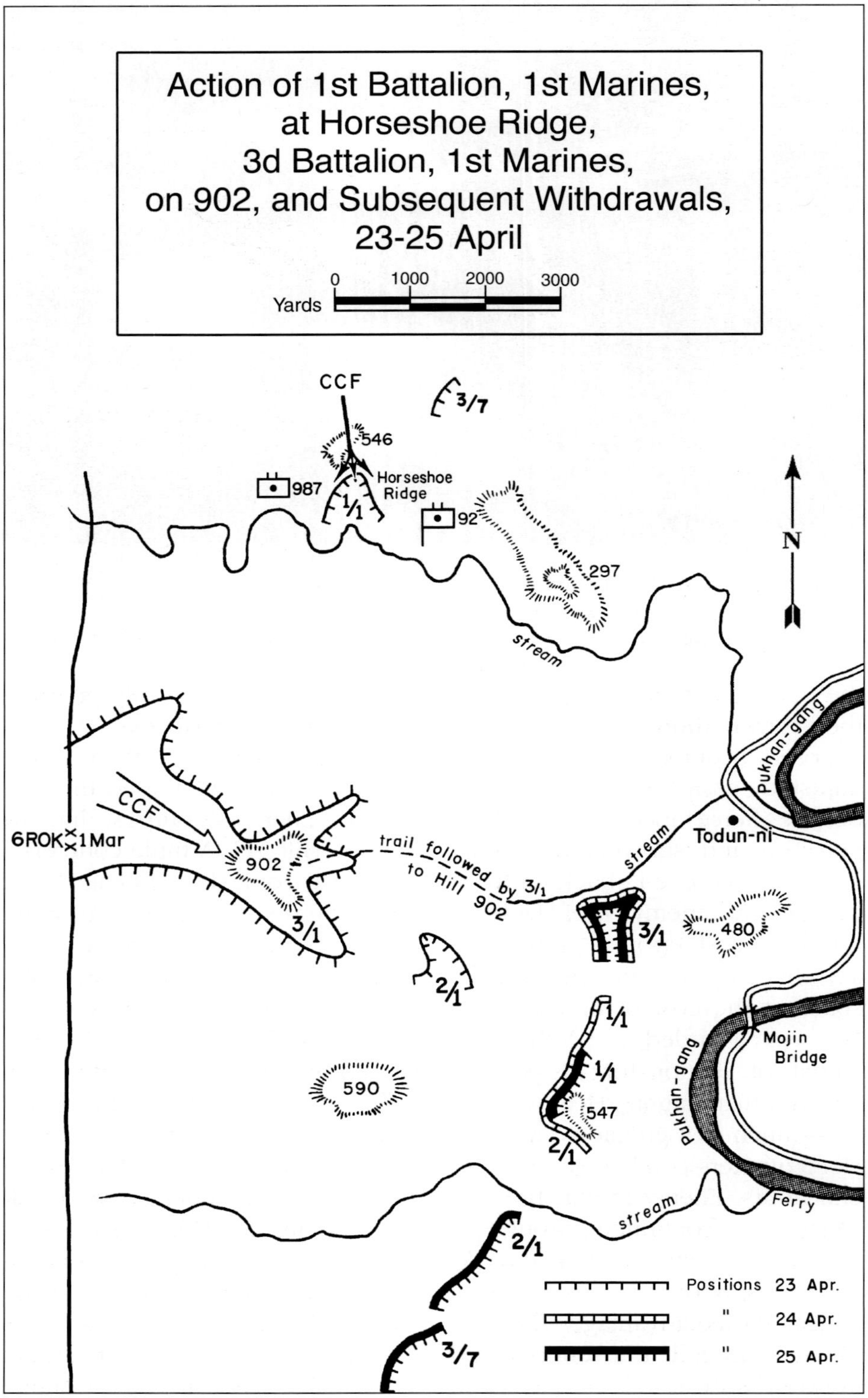

Although the Marines held fast and remained a breakwater that stemmed the onrushing Red tide, the Chinese were still pouring through South Korean army lines. "The position of the 1st Marine Division was beginning to appear to some persons," noted Major Martin Sexton, "very similar to the situation at the Chosin Reservoir." On the Marine left a deep envelopment threatened. As a result General Hoge ordered the 1st Marine Division to fall back. Consequently, General Smith passed the word for his units to retire to new defensive positions on the Pendleton Line at 0935. This would be no small feat. The enemy threat was so great that Smith was forced to place the entire 1st Marine Division on the high ground north of the Chunchon Corridor to protect the vital Mujon Bridge and several ferry crossings. This was a bold move because the Marines would have an unfordable river at their back and there was no division reserve in place. It also required a complex set of maneu-

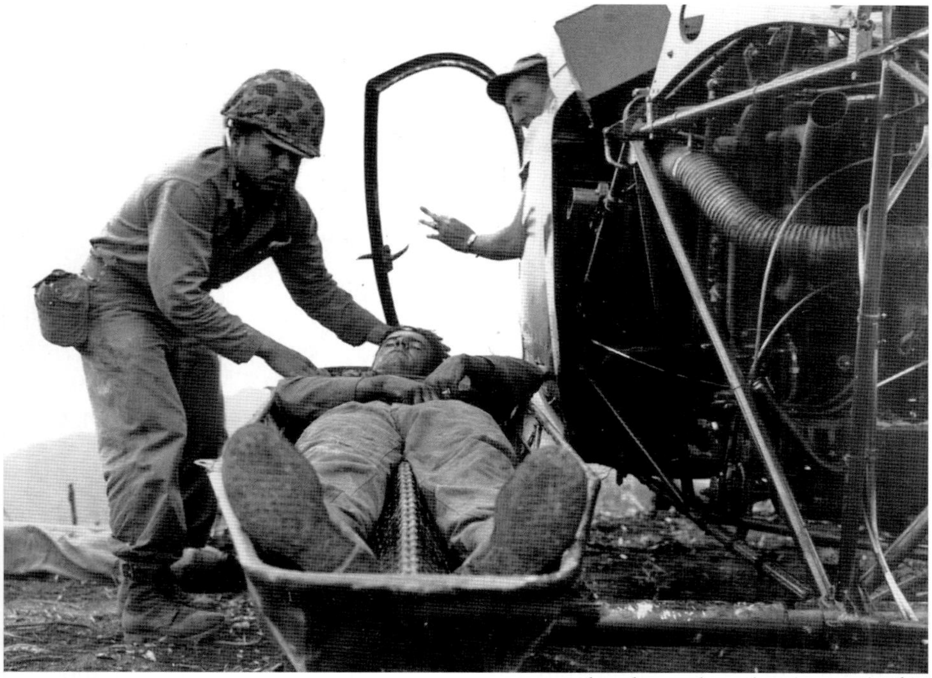

National Archives Photo (USN) 80-G-429642

A Bell HTL-4 light helicopter waits for a badly wounded Marine to be loaded on board for a trip to a rear area hospital. These bubble-top aircraft saved lives by cutting the amount of time it took seriously wounded men to get to medical attention.

vers whereby the Marines would have to defend the Pukhan River line, and at the same time move back to Chunchon. General Smith would have to carefully coordinate his supporting arms as well as effect a passage of lines under fire. Air and artillery would keep the enemy at bay while armor and the division's heavy weapons protected the avenues of approach and the river crossings. Smith's plan was to give ground rapidly in the north while slowly pulling back in the south, letting his westernmost units alternately pass through a series of blocking positions. Engineers would finally blow the bridges once the rear guard made it over the river. A key element was the Marine aviators whose fighter-bombers would be guided onto their targets by airborne spotters to delay enemy pursuit. All hands were called upon to contribute during this fighting withdrawal. Cooks, bakers, and typists—even a downed pilot—were soon shouldering M1 rifles or carrying stretchers under fire. Just as at the Chosin Reservoir, the creed that "every Marine is a rifleman regardless of his military occupational specialty" saved the day.

General Smith wanted to form a semi-circular defense line that arched southwest atop key ground from the tip of the Hwachon Reservoir west for a few miles then bending back along the high ground abutting the Pukhan River and over looking the Chunchon Valley. To do this he immediately ordered the rest of the 1st Marines forward from Chunchon to hold the hills in the southwest while in the north he instituted a "swinging gate" maneuver whereby the Korean Marines anchored the far right, the 5th gave ground in the center, and the 7th Marines pulled back in echelon to link up with the 1st Marines

Fighting continued throughout the day. In the west, the 7th Marines had its own 3d Battalion and the attached 1st Battalion, 1st Marines, to cover the retrograde. The hard-hit 1st Battalion pulled back covered by fires from the 2d Battalion. Major Maurice E. Roach Jr.'s 3d Battalion, 7th Marines, seized some fiercely held high ground while Lieutenant Colonel Robley E. West's 1st Battalion, 1st Marines, fought off repeated enemy probes that lasted until nightfall. Units of the 1st Marines held the southernmost positions. The remaining two reserve battalions had moved out of their assembly areas that morning, crossed the Pukhan River, then occupied a pair of hills protecting the main supply route and several crossing points. Actually, the arrival of the 2d and 3d Battalions, 1st Marines, was a close run thing. The Marines had to virtually race up the hills to beat the Chinese who were also on the way to take what was obviously the most important terrain feature in the area. Hill 902 (actually a 4,000-foot mountain top) dominated the road to Chunchon and protected the concrete Mojin Bridge as well as two ferry sites. Its defense became the focal point of the Marine retrograde. In the center, Colonel Richard W. Hayward's 5th Marines moved back under scattered small arms and mortar fire, but encountered no enemy ground units. On the division right, Korean Marines pulled back and then dug in just before being ranged by enemy mortar and artillery fire. Unfortunately, the 1st Marine Division's line was fragmented, not continuous, with units of the 1st and 7th Marines holding widely separated battalion-sized perimeters located atop key terrain. The 11th Marines, reinforced by several Army artillery battalions, was busy registering defensive fires as night fell on 23 April.

That day also marked the first mass helicopter medical evacuation in history. All of VMO-6's Bell HTL-4 "bubble top" helicopters (able to carry two litter cases and one man in the observer seat) were airborne at first light. Fifty critically wounded men were flown out by these Marine "egg-beaters" between 0600 and 1930. A total of 21 sorties (22.6 flight hours) were made from Chunchon to the front lines then back to the 1st Medical Battalion collecting and clearing station. Every flight encountered some type of enemy fire, but there were no losses of aircraft or personnel. Captain Dwain L. Redalin logged 9.7 flight hours while carrying 18 wounded men to safety. First Lieutenant George A. Eaton accounted for 16 more evacuations. The final flight had to be guided in with hand-held lights because the airfield had been officially blacked out. Ground personnel and flying officers alike were formed into provisional platoons and assigned defense sectors in case the enemy broke though, and all excess material and equipment was loaded on trucks for movement back to Hongchon that night.

On the night of 23-24 April, the 1st Marines caught the brunt of the CCF *120th Division* attack. In the north, the 1st Battalion, 1st Marines, still under the operational control of Colonel Nickerson's 7th Marines, was dug in on Horseshoe Ridge. This was a key position which, if lost, would split the 1st Marine Division wide open and allow the enemy to defeat it in detail. Farther south, the 2d and 3d Battalions, 1st Marines, manned separate perimeters on Hill 902 overlooking the flat lands of the Chunchon Corridor. These positions constituted the last line of defense, and if they were lost the division would be surrounded and cut off. In short, the situation that night was as desperate as any in

Technical Sergeant Harold E. Wilson

Department of Defense Photo (USMC) A46634

Born in 1921, in Birmingham, Alabama, Harold E. Wilson enlisted in the Marine Corps Reserve and was assigned to active duty in April 1942. During World War II, he served 27 months overseas stationed on Midway Island. In addition to his Pacific service, he was stationed at Parris Island, South Carolina; Camp Lejeune, North Carolina; and Portsmouth, Virginia. Sergeant Wilson was honorably discharged in 1945.

Recalled to active duty in August 1950, he was assigned to Company G, 3d Battalion, 1st Marines, and participated in the Wonsan landing and was wounded during the Chosin Reservoir Campaign. In March 1951, he was awarded a Bronze Star Medal for "fearless and untiring leadership" of his platoon. While serving as a platoon sergeant, his bravery on the night of 23-24 April 1951 brought an award of the Medal of Honor, with a citation that read, in part:

> Wilson braved intense fire to assist the survivors back into the line and to direct the treatment of casualties. Although twice wounded by gunfire, in the right arm and the left leg, he refused medical aid for himself and continued to move about among his men, shouting words of encouragement. After receiving further wounds in the head and shoulder as the attack increased in intensity, he again insisted upon remaining with his unit. Unable to use either arm to fire, and with mounting casualties among our forces, he resupplied his men with rifles and ammunition taken from the wounded.
>
> After placing the reinforcements in strategic positions in the line, [he] directed effective fire until blown off his feet by the bursting of a hostile mortar round in his face. Dazed and suffering from concussion, he still refused medical aid and, despite weakness from loss of blood, moved from foxhole to foxhole, directing fire, resupplying ammunition, rendering first aid and encouraging his men.

Following the April 1951 action, Wilson was evacuated to the Yokosuka Naval Hospital in Japan and five months later returned to the United States. He was awarded a meritorious promotion to master sergeant in 1951 and commissioned as warrant officer in 1952. After a number of assignments, he assumed the post of Adjutant, Marine Corps Engineer Schools, Camp Lejeune, in December 1962, and a year later, was assigned to Force Troops, Fleet Marine Force, Atlantic, serving as adjutant and personnel officer of the 2d Tank Battalion.

During the Vietnam War, Chief Warrant Officer Wilson served with Marine Aircraft Group 13 prior to being assigned as the 6th Marine Corps District's personnel officer in November 1968. He retired from the Marine Corps in 1972 and died in Lexington, South Carolina, on 29 March 1998.

— Captain John C. Chapin, USMCR (Ret)

the history of the Marine Corps.

The Marines were hit by artillery, mortar, small arms, and automatic weapons fire all through the night. The 1st and 7th Marines on the left flank were probed as Chinese forces searched for crew-served weapons positions and weak spots in the line. The four-hour fight for Horseshoe Ridge began at about 2000. There, the men of the 1st Battalion, 1st Marines, managed to blunt an attack by the CCF *358th Regiment* in savage hand-to-hand fighting. Farther north, the 3d Battalion, 7th Marines, repelled enemy probes all night long. As part of that action, the "Redlegs" of the Army's 92d Armored Field Artillery Battalion acquitted themselves well by repelling a dawn ground attack using machine guns and direct fire artillery to eliminate several hundred enemy troops, while continuing to deliver fire for the hard-pressed Marines on Horseshoe Ridge. Marine M-26 Pershings from Lieutenant Colonel Holly H. Evans' 1st Tank Battalion eventually joined the hard-fighting cannoneers, scattering the enemy with deadly flat-trajectory fire. Enemy stragglers were cleared out by joint Army-Marine patrols before the Army artillerymen displaced to new positions.

The enemy's main thrust that night, however, was directed farther south where the CCF tried to turn the open Marine flank but instead ran headlong into Lieutenant Colonel Virgil W. Banning's 3d Battalion, 1st Marines, atop Hill 902. A series of full-scale assaults began at about midnight. The CCF *359th* and *360th Regiments* repeatedly crashed into the 3d Battalion's exposed perimeter, but all efforts to eject the determined defenders were unsuccessful. After enemy mortars pounded Banning's Marines for several hours a "human wave" ground assault almost cracked First Lieutenant Horace L. Johnson's Company G. That this did not happen was a tribute to the actions of Technical Sergeant Harold E. "Speed" Wilson. Despite being wounded on four separate occasions, he refused evacuation and remained in command of his platoon. Unable to man a weapon because of painful shoulder wounds, Wilson repeatedly exposed himself to enemy fire while distributing ammunition and directing tactical movements even though he was hit several more times. Wilson was later awarded the Medal of Honor for his stirring leadership that night. The Marines took heavy casualties during fierce hand-to-hand fighting, but the Chinese were unable to dislodge them. At 0930 on 24 April, the battered Marines were almost out of ammunition and their ranks had been severely thinned, but they were still standing tall. The Chinese plan to trap and annihilate the 1st Marine Division had been a costly failure.

A Marine 105mm howitzer battery near Sapyong-ni fires on suspected enemy positions. The guns of the 11th Marines rendered outstanding fire support regardless of time of day or weather limitations.

1stMarDiv Historical Diary Photo Supplement, Apr51

General Hoge ordered the Marines to pull back to the Kansas Line as part of a general realignment of IX Corps. This would not be an easy maneuver because it would require disengaging under fire and making several river crossings. To do this, General Smith had to restore tactical unity prior to movement. The 1st Marines was reunited on the morning of the 24th when 1st Battalion, 1st Marines, which had been hotly engaged while attached to the 7th Marines for the past few days, rejoined the regiment. Concurrently, the 3d Battalion, 1st Marines, conducted a fighting withdrawal protected by Marine, Navy, and Air Force air strikes and artillery fire by Marine and Army units. The battered 3d Battalion passed through the 2d Battalion and then both units fought their way back to the high ground covering the river crossing. The regiment was under continuous fire during the entire movement and suffered numerous casualties enroute. At the same time, Major Roach's 3d Battalion, 7th Marines, set up farther south on Hill 696 to defend the Chunchon-Kapyong road as well as the southern ferry sites. This key position, the southernmost high ground, dominated the Chunchon Corridor and the Pukhan River and would be one of the last positions vacated. On the right, the 5th Marines and the Korean Marine battalion pulled back harassed by only scattered resistance. The resultant shortening of the division front allowed Smith to pull the 7th Marines out of the lines and use it as the division reserve. By the evening of 24 April, the 1st Marine Division's lines resembled a fishhook with the Korean Marines at the eye in the north, the 5th Marines forming the shank, and the 1st Marines at the curved barb in the south. The 7th Marines, less the 3d Battalion, was charged with rear area security and its 1st and 2d Battalions were positioned to protect river crossings along the route to Chunchon as well as the town itself.

The 24th of April was another busy day for Marine aviators as well. First Lieutenant John L. Scott evacuated 18 wounded in his HTL-4 to become the high-rescue-man that day. Another HTL-4, piloted by First Lieutenant Robert E. Mathewson, was brought down near Horseshoe Ridge by enemy fire. Mathewson escaped unhurt, but had to wave off a rescue attempt by First Lieutenant Harold G. McRay because enemy fire was so intense. The downed pilot was promptly given a rifle and joined his fellow Marines as they broke out of the Chinese encirclement. Over the battlefield an OY observation plane flown by Technical Sergeant Robert J. Monteith, struck a Corsair in midair and crashed. He and his artillery spotter, First Lieutenant Roscoe F. Cooke, Jr., were both killed when their plane spun out of control, hit the ground, and burned.

Department of Defense Photo (USMC) A8030

MajGen Oliver P. Smith bids farewell to division staff officers before turning command over to MajGen Gerald C. Thomas, center, on 25 April. BGen Lewis B. Puller, the assistant division commander, would follow Smith a month later.

The 1st Marines again bore the brunt of Chinese probes on the night of 24-25 April, but accurate close-in fires by 105mm and 155mm howitzers kept potential attackers at a distance. The 2d Battalion repelled an enemy company in the only major action of the evening. But the Chinese were still lurking in the west as became evident when patrols departing friendly lines in that area quickly struck an enemy hornet's nest the following morning. One such patrol was pinned down less than 200 yards from friendly lines. Another platoon suffered 18 casualties and had to be extricated from an ambush by tanks. On the other hand, 5th Marines and Korean Marine scouts ventured a mile to the north without contact. Air and artillery plastered the western flank, but enemy machine gun, mortar, and artillery fire continued to hit Marine positions. In the 1st Marines' zone Chinese gunners found the 3d Battalion command post, wounding Colonel McAlister; Lieutenant Colonel Banning; Major Reginald R. Meyers, the executive officer; and Major Joseph D. Trompeter, the operations officer. Banning and Meyers had to be evacuated, and Major Trompeter took over the battalion. Colonel McAlister refused evacuation and remained in command of the regiment.

It was obvious the Chinese were biding their time until they could

National Archives Photo (USMC) 127-N-A8362

Marine infantry and vehicles start the long haul back to Chunchon, where they would defend along the south bank of the Soyang River until service units could move their large supply dumps.

gather enough strength for another try at the Marine lines. There was continual pressure, but the 11th Marines artillery harassment and interdiction fires, direct fire by Marine tanks, and an exemplary air umbrella prevented a major assault. Enemy action was limited to only a few weak probes and a handful of mortar rounds as the Marines moved back. The 1st Marine Division reached the modified Kansas Line in good order. Despite suffering more than 300 casualties in the last 48 hours, the Marines handled everything the enemy threw at them and still held a firm grip on the IX Corp right flank when the Chinese Fifth Phase, First Impulse Offensive ground to a halt.

During this very brief break in the action a new division commander took over. Major General Gerald C. Thomas became the 1st Marine Division's commanding general at a small ceremony attended by the few available staff members on the afternoon of 25 April. Thomas had been awarded a battlefield commission in recognition of his outstanding combat performance during World War I, then pulled sea duty and fought in the Banana Wars between the World Wars. His experience as a highly respected staff officer in the Pacific during World War II prepared him to handle a division, and his postwar duties at Headquarters Marine Corps and Quantico gave him a good look at the "big picture" as well. Despite the hurried nature of the command change and the fact that it occurred in the midst of a complex combat action, the transition was a smooth one that did not hinder operations.

The first order General Thomas received was one no aggressive commander relishes. He was told to pull the 1st Marine Division back to a new position where Korean laborers were toiling night and day to construct a defensive bulwark. The Marine movement was no isolated withdrawal. All across the front, the United Nations Command was breaking contact in order to man a new main line of resistance known as the No Name Line. This unpressured retrograde marked a radical change in U.N. tactics. As will be recalled, upon taking charge of Eighth Army General Ridgway adopted mobile defensive tactics to deal with enemy attacks. Instead of "hold your ground at all cost," he instituted a "roll-with-the-punches" scheme whereby U.N. units traded ground to inflict punishment. To do this Ridgway insisted that his troops always maintain contact with both the enemy and adjoining friendly forces during retrograde movements. This time, however, General Van Fleet decided to completely break contact. He opted to pull back as much as 20 miles in places. There, from carefully selected positions, his troops could trap exposed attackers in preplanned artillery kill zones at the same time air power pummeled ever-lengthening enemy supply routes. In hindsight, this sound combined-arms approach fully utilized United Nations Command strengths while exploiting enemy weaknesses, but at the time it befuddled many Marines to have to abandon hard-earned ground when there seemed to be no serious enemy threat. Such was the case when the 1st Marine Division was told to fall back to a section of the No Name Line located near Hongchon far to the south.

This movement would be done in two stages. The first leg of the journey was back to Chunchon where the rifle units would cover the support units as they pulled out. When that was accomplished the combat units would continue on to the No Name Line. Luckily, there was no significant enemy interference with either move. The initial departure began at 1130 on 26 April. The 5th Marines and Korean Marines retired first, followed by 1st Marines, with 3d Battalion, 7th Marines, attached. A curtain of close air support supplemented by rocket and artillery fires shrouded these movements. All units, except the rear guard, were safely across the meandering Pukhan River before dark. The last remaining bridge across the chest-deep river was blown up at 1900, forcing 2d Battalion, 1st Marines, and 3d Battalion, 7th Marines, to wade across the chilly barrier in the middle of the night. The movement back to Chunchon was completed by noon, and the Marines took up defensive positions along the southern banks of the Soyang River on the afternoon of the 27th without incident. The only enemy encountered during the pull-back was one bewildered Chinese straggler who had inadvertently fallen in with the Marine column in the darkness. Needless to say, he was more than somewhat surprised to discover himself in the midst of several thousand Americans when daylight came.

On 28 April, the second phase of the withdrawal began. The Marine retrograde was again unpressured, but it took three days to finish the move south due to serious transportation problems. Finally, on 30 April, the Marines settled in at the No Name Line with the 5th Marines on the left, the 1st Korean Marine Regiment in the center, the 1st Marines on the right, and the 7th Marines in reserve.

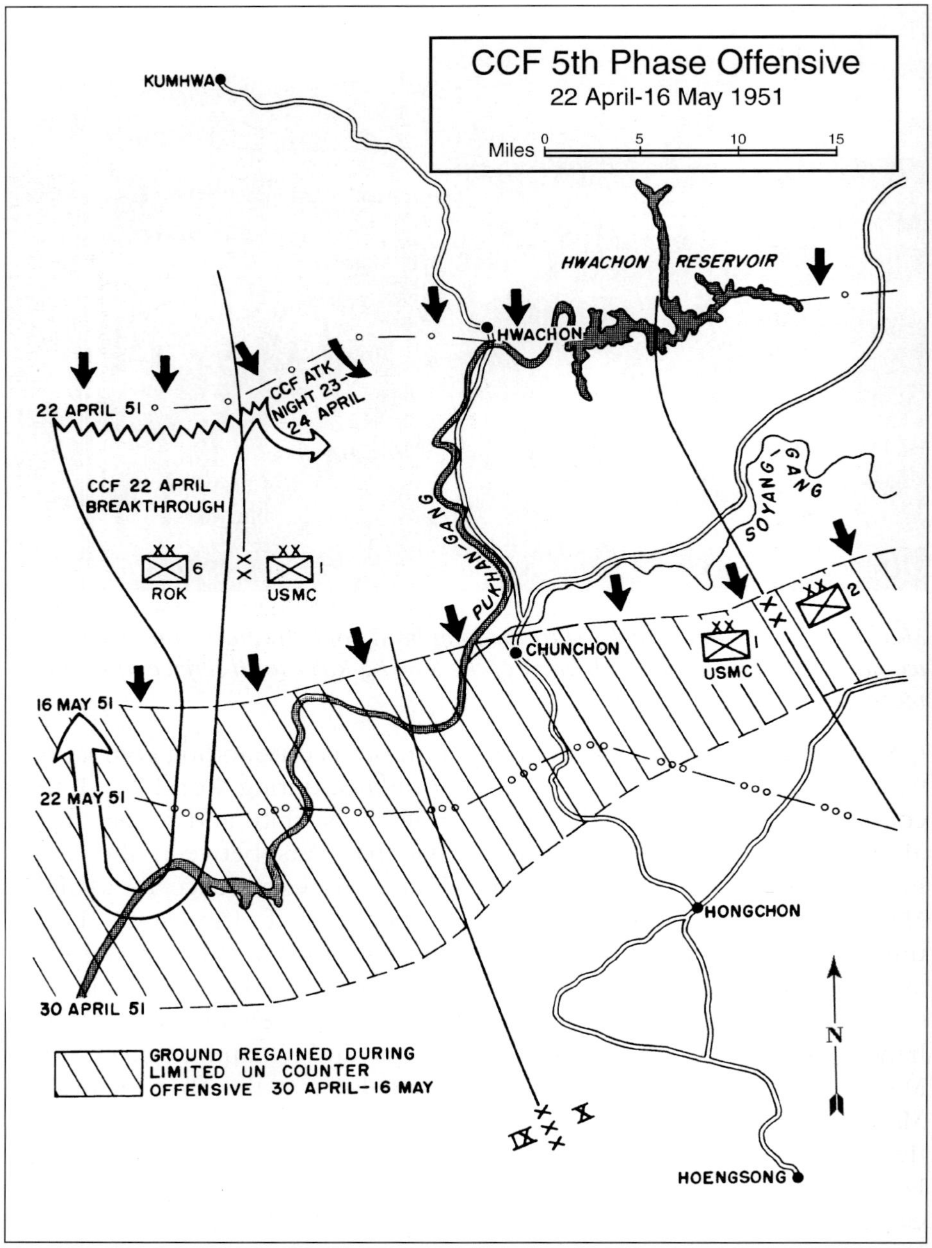

CCF 5th Phase Offensive
22 April-16 May 1951

The month of April cost the Marines 933 casualties (93 killed, 830 wounded, and 10 missing), most lost during the First Impulse of the Chinese Fifth Phase Offensive. The enemy enjoyed some local successes, but overall their attacks fell far short of expectations. The U.N. counteroffensive had been stopped in its tracks, but what little ground the enemy gained had been purchased at a fearful cost; the CCF lost an estimated 70,000 men. The headlong U.N. retreat the Chinese expected did not materialize. This time there was no "bug out," to use a popular phrase of the day. Instead, most breaks in the line were quickly sealed, and the United Nations Command was holding firm at the No Name Line. By the last day of April, it was apparent to both sides that the Communists would not be parading through the streets of Seoul on May Day as their leaders had promised.

The first days of May were so quiet that no Marine patrols made contact. This temporary lull, however, was about to end because a Second Impulse Offensive was aimed at eastern Korea. To meet this threat, General Van Fleet redeployed his command. As part of this reorganization the 1st Marine Division was taken from IX Corps and was once again assigned to Major General Edward M. Almond's X Corps (it will be recalled that the Marines landed at Inchon, liberated Seoul, and fought their way out of the Chosin Reservoir as part of X Corps). This was easy to do because the 1st

Department of Defense Photo (USMC) A155669

A tank-infantry patrol from 3d Battalion, 7th Marines, moves through the deserted city of Chunchon. The ebb and flow of Korean fighting ended when the U.N. lines stabilized after the Marines reached the Punchbowl in June 1951.

Marine Division was located on the IX and X Corps boundary. That imaginary line was simply shifted about 12 miles west, and only one battalion of the 5th Marines had to actually move. Other than that the only action required was to redraw the grease pencil lines on tactical maps.

The next two weeks were devoted primarily to improving defensive positions, but some tactical issues came to the fore. General Thomas was particularly disturbed by two Eighth Army orders. First, the 1st Marine Division was told to establish an "outpost line of resistance" to maintain contact with the enemy, provide early warning of a major attack, and delay the enemy advance as long as possible. Second, the 11th Marines was ordered to shoot a unit of fire each day whether there were observed targets or not. Thomas felt he could adequately cover his zone of action using aerial observation and long-range reconnaissance patrols, so he protested the placement of an entire battalion outside of 105mm artillery range. When told that the post must be manned, Thomas requested that an entire regiment be located at the exposed position. When this request was granted, he sent Colonel Nickerson's 7th Marines onto some high ground over looking the Chunchon Valley with orders to keep the road open and be prepared to fight its way out if the Chinese came down in force. Thomas also protested that shooting a unit of fire each day was a wasteful practice, one that would surely cause an ammunition shortage sooner or later. He was overruled in this case.

The expected Second Impulse of the Fifth Phase Offensive fell upon units of the Republic of

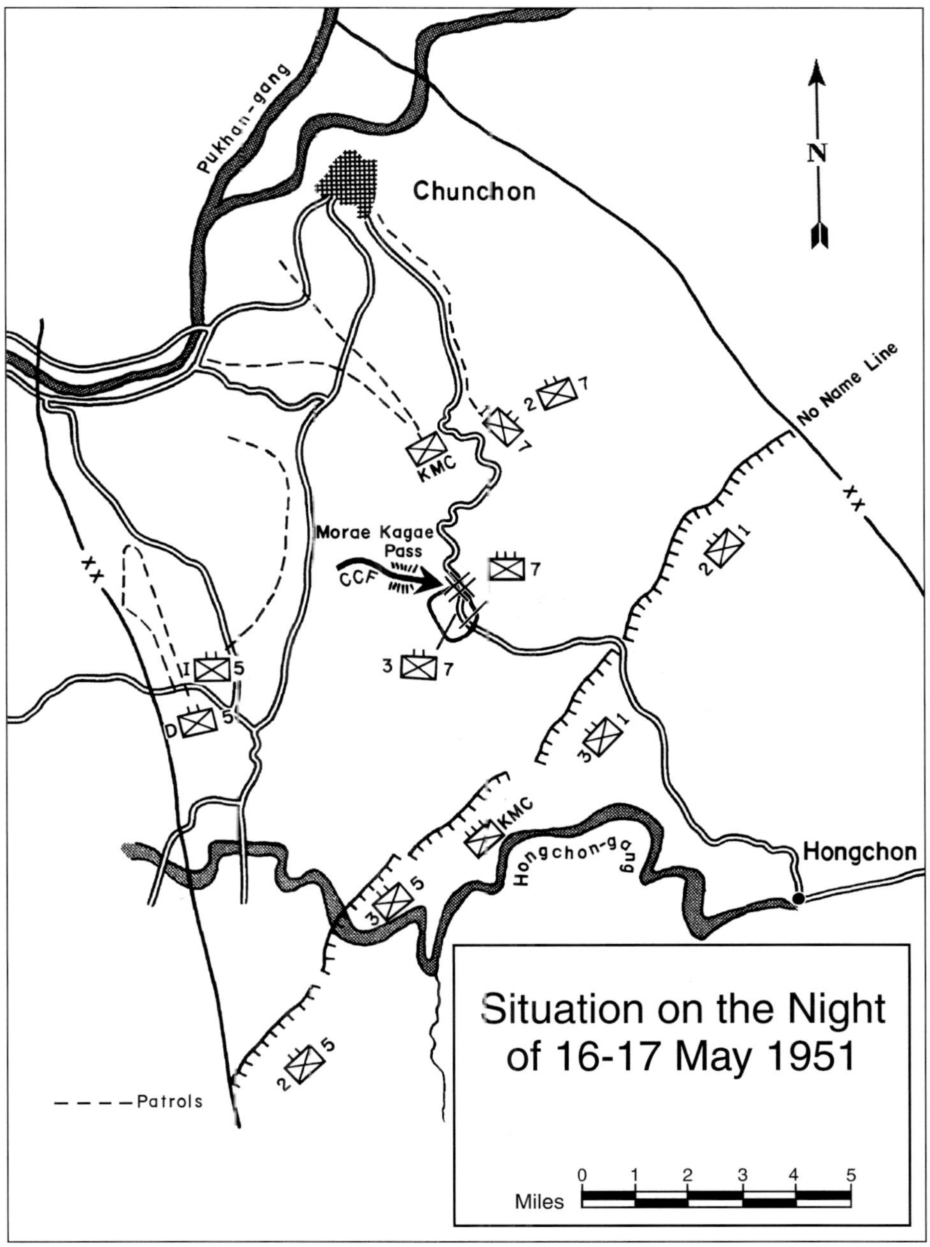

Korea Army in the east on 16 May, and soon a 30-mile penetration threatened the U.S. 2d Infantry Division on the Marine right. That night Chinese forces entered the Marine zone in regimental strength where the 5th Marines and the Korean Marines had several company-sized patrol bases well north of the main line of resistance in the left and center sections respectively. To the right, Colonel Nickerson's 7th Marines had Lieutenant Colonel John T. Rooney's 1st Battalion patrolling the Chunchon Road, 2d Battalion (now commanded by Lieutenant Colonel Wilbur F. Meyerhoff, formerly the 3d Battalion, 7th Marines, commanding officer) manning the outpost, and Lieutenant Colonel Bernard T. Kelly's 3d Battalion, 7th Marines, holding Morae Kogae Pass—-a vital link on the road leading from the forward edge of the battle area back to the main front line. Well aware that whoever controlled the pass controlled the road, the Chinese made Morae Kogae a key objective. Under cover of darkness, they carefully slipped in behind the Korean Marines and headed straight for the pass, which they apparently thought was unguarded. The assault force unexpectedly bumped into the northern sector of the 7th Marines perimeter at about 0300 and a furious fight broke out. Within minutes the 11th Marines built up a wall of fire at the same time the infantrymen initiated their final protective fires. Burning tracer rounds crisscrossed all avenues of approach and exploding shells flashed in the night as Marine artillery pinned the enemy in place from the rear while Marine riflemen knocked them down from the front. In spite of the curtain of steel surrounding the Marine positions, the quilt-coated enemy closed the position. Amid the fierce hand-to-hand fighting First Lieutenant Victor Stoyanow led a counterattack to throw the enemy back out of Company I's lines. The critical battle for the pass did not end until daybreak when the Chinese vainly tried to pull back but were instead caught in the open by Marine artillery, mortars, and some belated air strikes. The Chinese lost an estimated 530 men. By actual count, they left behind 112 dead, 82 prisoners, and a wealth of abandoned weapons that included recoilless rifles, mortars, machine guns, and even a 76mm antitank gun. Marine losses in this one-sided battle were seven dead and 19 wounded.

Col Francis M. McAlister, left, extends congratulations to Col Wilburt S. Brown, as the latter assumed command of the 1st Marine Regiment.

Department of Defense Photo (USMC) A8654

The following day, 18 May, the 1st Marine Division performed a very tricky maneuver to readjust defensive dispositions that allowed the U.S. 2d Infantry Division to move east to reinforce its right flank which was bearing the brunt of the new Chinese offensive. The 7th Marines pulled back to the No Name Line to relieve the 1st Marines which then sidestepped east to take over an area previously held by the U.S. Army's 9th Infantry Regiment and the 5th Marines swung over from the far left flank to relieve the 38th Infantry Regiment on the extreme right.

By noon on the 19th, all four regiments (1st Korean Marine, 7th Marines, 1st Marines, and 5th Marines) were aligned from left to right on the modified No Name Line as the enemy's offensive lost its momentum. That same day, Colonel Wilburt S. Brown, an experienced artilleryman known throughout the Marine Corps as "Big Foot" because of his large feet, took over the 1st Marines. There was also a change at division headquarters. Brigadier General William J. Whaling—-an avid sportsman and Olympic marksman who commanded regiments at Guadalcanal, New Britain, and Okinawa during World War

National Archives Photo (USMC) 127-N-A133537

A Grumman F7F Tigercat armed with napalm flies over North Korea seeking a suitable target. The twin-engine, single-seat, carrier-borne Tigercats were primarily used as night fighters, but sometimes conducted bombing and aerial reconnaissance missions.

II—became the assistant division commander on 20 May.

The final action of the Chinese Spring Offensive occurred at about 0445 on 20 May when Major Morse L. Holladay's 3d Battalion, 5th Marines, caught elements of the CCF *44th Division* in the open. The Marines on the firing line opened up with everything they had as Major Holladay directed rockets, artillery, and air support during a five-hour battle that cost the enemy 152 dead and 15 prisoners. This action marked the end of Marshal Peng's attempts to drive the 1st Marine Division into the sea. The enemy, short of men and supplies after the previous month's heavy combat, had finally run out of steam and was now vulnerable.

With the Chinese Fifth Phase Offensive successfully blunted, General Van Fleet was ready to shift back into an offensive mode to exploit what was clearly a devastating Communist defeat. The United Nations Command had come through the last month with relatively light casualties and for the most part had only ceded territory on its own terms. Many Marine veterans of both campaigns, however, later recalled that the hard fighting to hold the Pendleton Line was as desperate as any they encountered at the Chosin Reservoir. The 1st Marine Division not only weathered the storm, it had given the enemy a bloody nose on several occasions and performed many complex maneuvers well. Reiterating his experiences in Korea, General Smith said that blunting the Chinese counterattacks in April "was the most professional job performed by the Division while it was under my command." Likewise, by the time the CCF Spring Offensive ended General Thomas remarked that he commanded "the finest division in Marine Corps history."

Marine Air Support

Major General Field Harris' 1st Marine Aircraft Wing comprised of two aircraft groups, Colonel Boeker C. Batterton's Marine Aircraft Group 12 (MAG-12) and Lieutenant Colonel Radford C. West's Marine Aircraft Group 33 (MAG-33), and flew more than a dozen different aircraft types. Lieutenant Colonel "J" Frank Cole's Marine Fighter Squadron 312 (VMF-312), Lieutenant Colonel Richard W. Wyczawski's VMF-212, Major William M. Lundin's VMF-214, and Major Arnold A. Lund's VMF-323 all flew "old reliable and rugged" propeller-driven Chance-Vought F4U-4 Corsair fighter bombers. Lieutenant Colonel Neil R. MacIntyre commanded the "hottest" squadron, VMF-311, which flew Grumman F9F-2B

A VMF-323 "Death Rattler" F4U armed with 5-inch rockets and napalm readies for take off from the Badoeng Strait *(CVE 116). At least one Marine squadron was on board an aircraft carrier at all times during the spring of 1951, as this duty rotated among the Corsair squadrons.*

Photo Courtesy of LtCol Leo J. Ihli, USMC

1st Marine Aircraft Wing, 1951

Marine land-based tactical and support aircraft, except for the observation planes and helicopters attached to the 1st Marine Division, comprised the 1st Marine Aircraft Wing in Korea. The wing had two aircraft groups (MAGs -12 and -33) that flew more than a dozen different aircraft types in 1951. Its most famous airplanes were World War II vintage F4U Corsairs and brand new F9F Panther jets, but also included in the combat aircraft mix were F7F Tigercat and F4U-5N Corsair all-weather fighters. Most Marine land-based aircraft were under the operational control of the U.S. Fifth Air Force, and the Joint Operations Center coordinated most air operations. Marine carrier-based aircraft, on the other hand, were under the operational control of the U.S. Navy task forces to which their respective carriers were assigned. A few utility aircraft (SNBs and TBMs) were assigned to headquarters squadrons. The aircraft of VMO-6 (OY "Sentinels," as well as HO3S and HTL helicopters) flew in direct support of the 1st Marine Division. Marine R4Q Packets and parachute riggers of the 1st Air Delivery Platoon supported the U.S. Air Force Combat Cargo Command. Marine transport planes (R4D Skytrains and R5D Skymasters) flew in support of the Naval Air Transport Service and the Combat Cargo Command.

1st Marine Aircraft Wing
- Marine Aircraft Group 33
- Marine Aircraft Squadron 12
- Marine Wing Service Squadron 1
- Marine Ground Control Intercept Squadron 1
- Marine Fighter Squadron 212
- Marine Night-Fighter Squadron 513
- 1st 90mm AAA Gun Battalion
- Marine Night-Fighter Squadron 542
- Marine Fighter Squadron 323
- Marine Air Control Group 2
- Marine Tactical Air Control Squadron 2
- Marine Ground Control Intercept Squadron 3
- Marine Fighter Squadron 214
- Marine Fighter Squadron 312
- Marine Fighter Squadron 311
- Detachment, Marine Transport Squadron 152

Supporting Naval Air Transport Service
- Marine Transport Squadron 242
- Marine Transport Squadron 152
- Marine Transport Squadron 352

Attached to 1st Marine Division
- Marine Observation Squadron 6

Panther jets. Lieutenant Colonel David C. Wolfe led Marine Night (All-Weather) Fighter Squadron 513 (VMF[N]-513) mounted in F4U-5N Corsair night fighters. The other night fighter squadron, Lieutenant Colonel Max J. Volcansek, Jr.'s VMF(N)-542, flew twin engine Grumman F7F-3N Tigercats. Wing headquarters had specially configured General Motors (TBM) Avenger single-engine torpedo bomber radio relay planes, F7F-3P and F4U-5P photo reconnaissance planes, Douglas twin-engine R4D Skytrain and SNB light utility transports. Major Vincent J. Gottschalk's Marine Observation Squadron 6 (VMO-6), flying Consolidated OY Sentinel light observation planes and Sikorsky HO3S and Bell HTL helicopters, was attached to the 1st Marine Division and did not come under the operational control of the 1st Marine Aircraft Wing. Other Marine aircraft serving the Korean theater but not part of the wing included Marine transport planes such as four-engine Douglas R5D Skymasters and twin-boom Fairchild R4Q Packets.

Normal operational relationships were disrupted by the CCF Winter Offensive, which forced retreating U.N. forces to close air bases at Yonpo, Wonsan, Seoul, Kimpo, and Suwon as they pulled back. The few airfields still in U.N. hands in early January 1951 could not handle all United Nations Command aircraft, and the resulting ramp space shortfall scattered Marine air assets throughout Korea and Japan. This unanticipated diaspora placed Marine squadrons under several different control agencies. The "Checkerboard" Corsairs of VMF-312 were in Japan at Itami Air Base on the island of Honshu along with the wing rear support units. The other three Corsair squadrons were carrier-borne. The "Devil Cats" of VMF-212 were on the light carrier USS *Bataan* (CVL 29) under the operational control of combined Task Group 96.8 operating in the Yellow Sea near Inchon, while VMF-214's "Black Sheep" were on the USS *Sicily* (CVE 118) and the "Death Rattlers" of VMF-323 were flying off the USS *Badoeng Strait* (CVE 116) under the operational control of U.S. Navy Task Force 77 in the Sea of Japan. The only land-based fighter squadron still in Korea was the "Panther Pack" of VMF-311 operating from airfield K-9 at Pusan. Unfortunately, the Panther jets were temporarily out of service due to mechanical and electronic teething problems serious enough to ground the entire squadron until it could be pulled back to Japan for maintenance. The two Marine night fighter squadrons, the "Flying Nightmares" of VMF(N)-513 and the "Tigers" of VMF(N)-542, were in Japan under the direct control of U.S. Fifth Air Force flying air defense missions as part of the 314th Air Division. VMO-6 was attached to and collocated with the 1st Marine Division at Masan. Two Marine transport squadrons supported the Naval Air Transport Service. Colonel William B. Steiner's Marine Transport Squadron 352 (VMR-352) shuttled between California and Hawaii, while Colonel Deane C. Roberts' VMR-152 flew two legs, one from Hawaii to Japan and the other from Japan to Korea.

Department of Defense Photo (USMC) A132120

A Vought F4U Corsair from VMF-214 is guided into position for take-off on its way to a close air support mission. The bent-wing, single-seat, propeller-driven "Dash Fours" featured six 50-caliber machine guns.

All four Marine fighter-bomber squadrons flew daily sorties during the first week of January. Their missions included close air support for the Eighth Army, combat air patrols, armed reconnaissance, coastal surveillance, and interdiction bombing. By mid-month the wing administrative and service units, the Corsairs of VMFs-214 and -323, and VMF-311's jets were temporarily ensconced at Itami until facilities at Bofu on Honshu and K-1 (Pusan West) in Korea were activated. Wing headquarters stayed at Itami, MAG-33 was slated to move to Bofu once the airfield was operational, and MAG-12 was temporarily assigned to K-9 (Pusan East) until all of its squadrons returned to Korea.

This Japanese interlude was a period of transition for Marine aviation. The 1st Marine Aircraft Wing was reorganized, some command changes occurred, and several moves were accomplished. As part of the wing reorganization, squadrons were realigned among the air groups. The 1st Marine Aircraft Wing had to be realigned because its elements were going to be split up, some operating from air bases in Japan while others would be stationed in Korea, and one squadron would be afloat. Lieutenant Colonel Paul J. Fontana replaced Lieutenant Colonel Radford C. West as MAG-33's commanding officer. New squadron commanders included Major Donald P. Frame (VMF-312), Major Stanley S. Nicolay (VMF-323), Major James A. Feeley, Jr. (VMF-214), and Lieutenant Colonel Claude H. Welch (VMF-212). Lieutenant Colonel James R. Anderson took over both night fighter squadrons (VMF[N]-513 and VMF[N]-542) in February, a unique arrangement that lasted until VMF(N)-542 returned to the United States in mid-March. The squadrons slated to move to Bofu were assigned to MAG-33 and the squadrons returning to Korea were assigned to MAG-12. In addition, the night fighter squadrons returned to Marine control.

This temporary turmoil was a source of irritation, but it was far less ominous than an emerging doctrinal issue. The 1st Marine Division and the 1st Marine Aircraft Wing were separated for the first time since they arrived in the Far

Calling "Devastate Baker." A Marine pilot serving with a ground unit directs a close air support mission. The assignment of Marine aviators to ground units ensured proper ground-to-air liaison.

National Archives Photo (USMC) 127-N-A9458

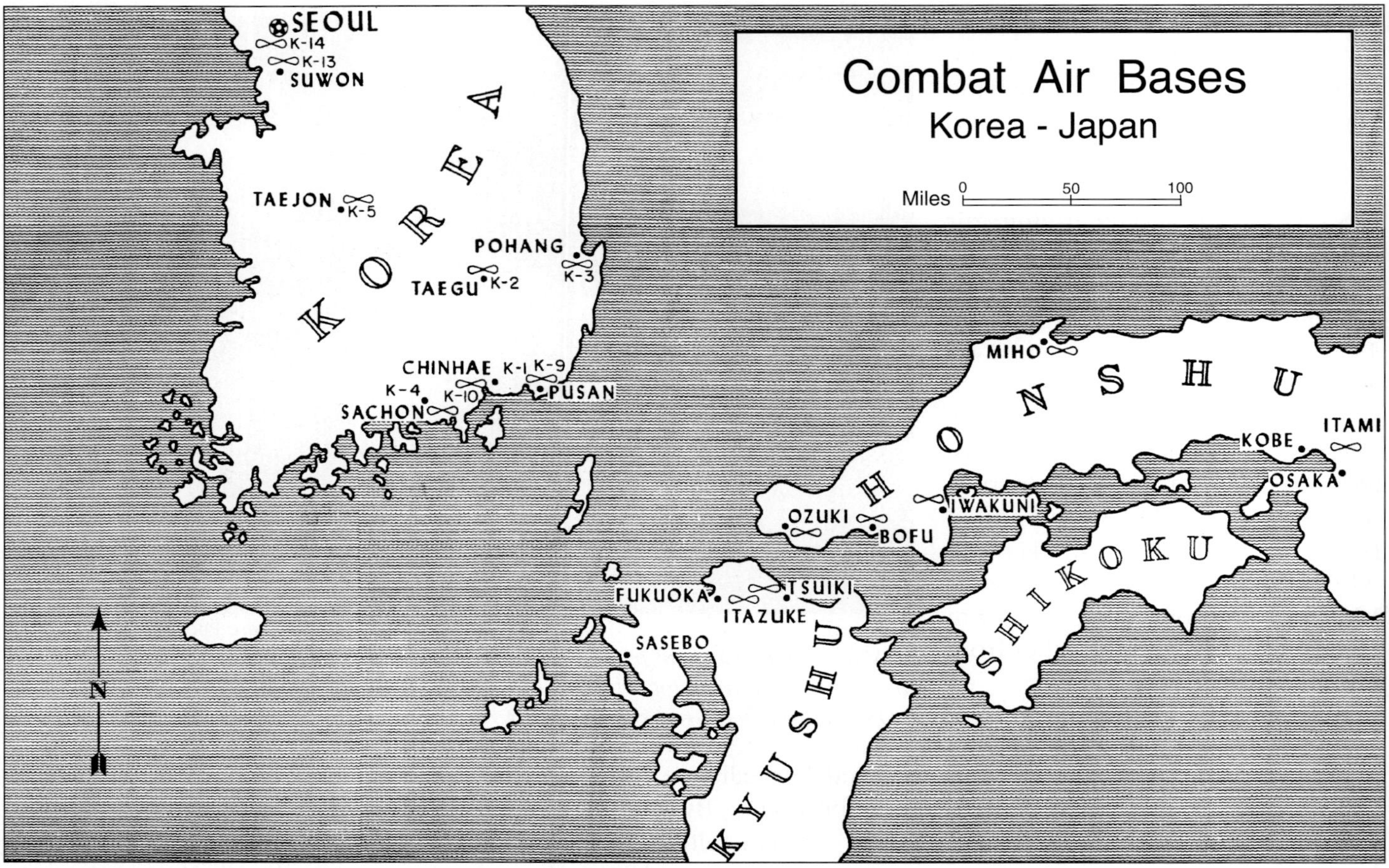

East. Marine land-based aircraft had been under the titular control of the Fifth Air Force for months, but a verbal agreement between Marine General Harris and U.S. Air Force General Earl E. Partridge allowed the wing to regularly support ground Marines. As the wing pulled back to Japan, however, Harris' de facto control of Marine air was lost and this agreement went by the wayside. Thereafter, all land-based wing aircraft would be under the operational control of the Fifth Air Force, and all missions would be assigned by the Fifth Air Force-Eighth Army joint operations center. Leery veteran Marine aviators foresaw procedural and allocation problems and, needless to say, there was great trepidation by all

A Sikorsky HO3S helicopter sits on a mountaintop landing zone while Navy Corpsmen prepare three wounded Marines for evacuation. In addition to standard command, liaison, and observation duties, these helicopters also often flew search and rescue missions behind enemy lines.

National Archives Photo (USN) 80-G-439571

Marines about the breakup of the combat-proven Marine air-ground team. These concerns were acknowledged, but General Partridge insisted that a vastly increased enemy air threat and plans to initiate a deep air interdiction campaign demanded new air control measures. Unfortunately, Marine reservations about this system were soon justified by events on the battlefield. After the joint operations center took over, Marine air and ground commanders chafed at what they considered inordinate delays and inappropriate use of aircraft. The problems were so serious that every commander of the 1st Marine Division (Generals Smith, Puller, and Thomas) filed formal complaints about the quality, quantity, and timeliness of close air support.

Late January and early February 1951 were devoted to maintenance, training, and movement back to Korea. General Harris opened his command post at Itami and MAG-33 completed its temporary move to Bofu during the third week of January. The only Marine combat sorties during the 1st Marine Aircraft Wing stand-down from 16 to 23 January were conducted by VMF-212 on board the *Bataan*. When the land-based Corsair squadrons returned to action, most sorties were flown in support of Eighth Army units conducting Operations Thunderbolt and Roundup in western Korea. This was because the 1st Marine Division needed few air strikes during the "guerrilla hunt" at Pohang, but on 26 January MAG-12 aircraft flying from K-9 (Pusan East) did manage to conduct close air support strikes for the division for the first time since the Chosin campaign.

The next month saw the return of the wing to Korea. In mid-February, K-1 at Pusan became the new home of MAG-12, and MAG-33 moved from Japan to K-3 at Pohang. The night fighters of VMF(N)-513 and -542 moved to K-1 and K-3 respectively. Major Donald S. Bush's task-organized "Marine Photographic Unit" operated its reconnaissance planes from K-1 under the auspices of the Air Force's 543d Tactical Support Group. Thus, all Marine tactical squadrons were back in Korea in time for the upcoming U.N. spring offensives.

The 1st Marine Aircraft Wing flew most of its sorties in support of Eighth Army units during Operation Killer, but Operation Ripper found the Marine air-ground team once more in action as wing aircraft cleared the way for the 1st Marine Division's rapid advance from Hoengsong to Hongchon. Responding to intense criticism from ground commanders, General Partridge reluctantly granted General Harris at least 40 sorties per day in support of the gravel-crunching Marine infantry. In the way of organizational changes, VMF-312 became the carrier squadron when it replaced VMF-212 on board the *Bataan*, VMR-152 established a five-plane forward echelon at Itami, and an additional Marine Air Control Squadron (MACG-2) was sent to Korea. The efficient performance of Lieutenant Colonel John F. Kinney's refurbished Panther jets of VMF-311 for armed reconnaissance and close air support was a pleasant surprise after their inauspicious introduction to combat.

There were several important command changes in April and May. Lieutenant Colonel Fontana departed MAG-33 on 31 March and Lieutenant Colonel Richard A. Beard, Jr., became acting commander until Colonel Guy M. Morrow arrived on 9 April. When Major Donald P. Frame was killed in action on 3 April, the "Checkerboard" executive officer, Major Frank H. Presley, assumed command of VMF-312. Major David W. McFarland took over VMO-6 on 5 April. On 3 May, Major Charles M. Kunz replaced Major Donald L. Clark who had commanded VMF-323 since 1 March. On 16 May, Lieutenant Colonel James W. Poindexter took the reins of VMF-214 from Major Edward Ochoa and Colonel Stanley W. Trachta assumed command of MAG-12. On the 28th, Brigadier General Thomas J. Cushman became commanding general of the wing when General Harris rotated back to the United States. Cushman was a veteran aviator who had commanded the 4th Marine Base Defense Wing in the Central Pacific during World War II

Panther jets of VMF-311 are gassed up at K-3 (Pohang). Refueling operations were a slow and laborious process. Fuel had to be transferred ashore in landing ships, hand pumped into fuel trucks, and then hauled out to the airfield.

National Archives Photo (USMC) 127-N-A130478

and brought MAG-33 to Korea in August 1950.

Marine air was used all along the U.N. front during the CCF Spring Offensive, and close air support played an important, if not decisive, role during that hectic time. Fifth Air Force regularly used Marine planes not earmarked to support the Marine division for armed reconnaissance and battlefield interdiction beginning in late April. On 20 April, a pair of VMF-312 pilots flying off the *Bataan*, Captain Phillip C. Delong and First Lieutenant Harold D. Daigh, encountered four North Korean Yakovlev YAK-9 fighters over central Korea. Delong, a double ace with 11 kills during World War II, shot down two of them. Daigh knocked one YAK out of the sky and left the other one trailing smoke as it fled north. These were the first Marine aerial victories in Korea, and they were among the very few kills scored by Marines not on exchange duty with the U.S. Air Force or flying a night intercept mission. Seventy-five Marine aircraft, Panthers and Corsairs, participated in the largest air raid to date as part of a 300-plane sweep that hit Communist airfields at Sinuiju just south of the Yalu River on 9 May.

One reason for pulling the 1st Marine Aircraft Wing away from the 1st Marine Division was that the Fifth Air Force instituted an all-out effort to halt enemy traffic south with a deep interdiction campaign codenamed Operation Strangle. The goals of the campaign were to cut enemy supply routes, which were channelized by the mountainous terrain, and to destroy supply columns halted by swollen streams. Bomb damage assessments credited the wing with the destruction of more than 300 enemy troops, more than 200 trucks, about 80 boxcars, and 6

1stMarDiv Historical Diary Photo Supplement, Jan-Feb51

A Marine 105mm howitzer sets up for a fire mission in the Andong area. The trusty "105" was the backbone of the 11th Marines in Korea, just as it had been during much of World War II.

locomotives. The price of this success was, however, high; the Marines lost a plane a day during the first week. Much to the dismay of ground and aviation Marines alike, close air support became a secondary mission. This change in priority abruptly cut the number of sorties allocated to ground units almost in half. In addition, cumbersome joint operations center request procedures often delayed air strikes for excessively long periods of time. Generals Puller and Thomas successively complained directly to the Fifth Air Force commander, and Lieutenant General Lemuel C. Shepherd, Jr., Commanding General, Fleet Marine Force, Pacific, took the issue up with the theater commander, all to no avail. The controversial joint operations center control and allocation procedures remained in force. The official Marine Corps history describes the questionable success of the Operation Strangle deep interdiction campaign: "There can be little doubt [Operation Strangle] added enormously to the Communists' logistical problem. It is equally certain that . . . . their combat units were never at a decisive handicap for lack of ammunition and other supplies [so] air interdiction alone was not enough to knock a determined adversary out of the war."

Despite these problems, many innovations were instituted in Korea. In addition to well-practiced daylight air-ground combat procedures, new techniques improved nighttime close air support. Marine R4D transport planes were put to use dropping flares that illuminated the battlefield and allowed VMF(N)-513 to deliver accurate night close air support. This experiment was so successful that the U.S. Navy provided the wing with four-engine, long-range PB4Y Privateer bombers, nicknamed "Lamplighters," whose bigger payloads and longer linger time were put to good use.

In characterizing Marine air support from January to May 1951, Marine aviators provided crucial support to their ground brethren throughout. Venerable per-formers—both aircraft and personnel—from World War II once again proved their mettle, and new types of aircraft and pilots were introduced to combat. The ground Marines were well served by the attached observation squadron,

1stMarDiv Historical Diary Photo Supplement, Jan-Feb51

A tank commander carefully scans the hills near Pohang for signs of the enemy. Pershing tanks like the one shown were the forerunners of the "Patton" tanks that served as the Marines' main battle tanks in Korea, Vietnam, and the Persian Gulf.

which directed artillery fire and close air support, evacuated wounded, and brought in emergency supplies. Transports delivered badly needed replacements and carried returning veterans safely home as well as dropping vital supplies by parachute to forward units. Aerial reconnaissance kept ground commanders informed of enemy movements and locations. The pilots of the 1st Marine Aircraft Wing relentlessly attacked the enemy at every possible opportunity, and Marine close air support was the envy of every United Nations Command commander. The appearance of Marine air on the scene almost always forced the enemy to rush for cover, and occasionally caused him to surrender or abandon key positions. It was with great reluctance that Marine fliers were diverted from their close air support mission, and all Marines became extremely frustrated when that vital support was gradually diminished due to circumstances beyond their control.

Combat and Service Support

The 11th Marines was the 1st Marine Division artillery regiment. Commanded by Lieutenant Colonel Carl A. Youngdale and then Colonel Joseph L. Winecoff, the regiment mustered 54 M2A1 105mm towed howitzers (18 each in the 1st, 2d, and 3d Battalions) while the 4th Battalion had 18 M2 155mm towed howitzers. The 105mm units were most often used for direct support with one artillery battalion assigned to fire exclusively for a particular rifle regiment, and the 155mm were most often in general support so they could use their longer range and heavier firepower to the best advantage. This was not always true, because the U.S. Marines provided artillery support for the 1st KMC Regiment as well as its organic units, and when all four rifle regiments were on the main line of resistance every artillery battalion had to be used for direct support. The nature of the fighting in Korea dictated that additional firepower was needed so the 11th Marines had Battery C, 1st 4.5-inch Rocket Battalion, permanently attached. Marine units were often supported by U.S. Army artillery as well. It was common for the corps commander to furnish at least one self-propelled howitzer battalion and a battery of 8-inch heavy guns to the 1st Marine Division for additional firepower. Army artillery units working with the 11th Marines at various times included the 17th Field Artillery, the 92d Armored Field Artillery Battalion, the 96th Armored Field Artillery Battalion, and the 987th Armored Field Artillery Battalion. Offensive artillery missions included supporting maneuver units, neutralizing enemy fire, and isolating the battlefield. On defense, artillery fire was used effectively against CCF mass infantry assaults. Forward

Marine engineers construct a bridge near the Kansas Line. Road construction and bridge building took the lion's share of the 1st Engineer Battalion effort in the spring of 1951.

1stMarDiv Historical Diary Photo Supplement, Apr51

1stMarDiv Historical Diary Photo Supplement, Apr51

Marine combat engineers remove a land mine on a road near Hongchon. Trained to install and maintain friendly minefields, the engineers were the first called to remove enemy mines.

observation teams at the leading edge of the battlefield controlled most artillery fires, but airborne spotters flying in light observation planes also sometimes directed them. The main problems encountered by the cannoneers of the 11th Marines were transporting heavy guns over poor roads and intermittent ammunition shortages. Generals Ridgway and Van Fleet preferred to "use steel instead of men" and artillery was the favored combat arm under both men. Ammunition expenditure was much heavier in Korea than during World War II, and shooting several units of fire on a single mission was referred to as a "Van Fleet load" by Marine artillerymen. Unfortunately, this practice sometimes drained carefully hoarded ammunition caches that were not easy to replenish, so orders to deliver specific amounts of unobserved ("harassment and interdiction") fire became a bone of contention between the Eighth Army and the Marines.

The 1st Tank Battalion, under the command of Lieutenant Colonel Harry T. Milne, also provided excellent combat support. The battalion was divided into four companies (A, B, C, and D) each with 17 medium tanks. These companies were usually placed in direct support of a specific rifle regiment. It was not uncommon for five-vehicle tank platoons to accompany combat patrols. When the regiment they supported was in reserve, the tankers tried to use that time for maintenance and rest. The super-accurate 90mm guns of the M-26 Pershing tanks were particularly well suited for long range "bunker busting" and were occasionally used to supplement artillery fires (much to the chagrin of the tankers who felt this practice was a deplorable misuse of their point target guns). Tanks were also sometimes pressed into service as armored ambulances In addition to the modern M-26 Pershing main battle tanks, there were also a dozen or so World War II-vintage M-4A3 Sherman bulldozer tanks with 105mm short-barrel guns and front-mounted plows used for mine clearing, hasty engineering, and tank recovery as well as fire support. Although Korea's mountainous terrain was generally unsuited for armor operations, frequent use was made of separate axis attacks whereby the road-bound tanks in the valleys supported infantry units as they worked their way along ridgelines. During the CCF Spring Offensive tanks were used to protect lines of communication and river crossings or cover nearby flatlands with their machine guns and main guns.

The 1st Engineer Battalion, commanded by Lieutenant Colonel John H. Partridge, provided services including rebuilding airstrips, constructing and repairing roads and bridges, emplacing and clearing mines, demolitions, manning water points, and preparing field fortifications. Although the 1st Engineers did all of these things, Lieutenant Colonel Partridge's number one priority throughout the spring of 1951 was keeping the main supply route open. The 1st Engineer Battalion spent most of its time and energy constructing, improving, and maintaining the supply route. Korea's primitive roadways were neither designed nor built to meet the demands of a major modern military force. There were few hard-surface roads, and there was no true road network. Most roads were little more than narrow dirt pathways that simply ran between local villages by the most direct route. Almost all roadways were poorly drained, inadequately bridged, and unpaved.

Snow and ice hampered movement in cold weather, the dry season choked the roads with dust, and spring thaws and summer rains often turned them into impassable bogs. Unfortunately, the need for constant road maintenance sometimes required foregoing other vital engineer functions, which were then left to the combat units.

Logistics—the acquisition and distribution of the means to wage war—encompassed the supply, maintenance, medical, transportation, and administrative services necessary to support combat operations. Although the efforts of the men who furnish the beans, bullets, and bandages are often overlooked, logistics are no less important than tactics in determining the outcome of a battle because—according to an old military adage— "logistics set operational limits." This was particularly true in Korea where Marine logisticians faced a wide array of challenges. Most short-term problems were the result of Korea's poorly developed infrastructure, rugged terrain, inhospitable weather, the rapidly changing tactical situation (which saw the entire 1st Marine Division go from offense to defense within a matter of hours on several occasions), and the wide physical separation of Marine air and ground elements. Unfortunately, some nagging problems also stemmed from doctrinal shortcomings. In 1951, U.S. joint operations did not feature the smooth multi-Service integration common among today's branches of the Armed Forces. The Marine air-ground task force concept was not developed, hence, there was no single Marine component commander in Korea so the Marine air and ground combat elements had no common superior below the theater commander. For the most part Marine ground and aviation units remained separate logistical entities operating without central direction because no equivalent of Vietnam's Force Logistics Command or modern force service support groups emerged in Korea. Luckily, Lieutenant General Lemuel Shepherd, the commanding general of Fleet Marine Force, Pacific, was an energetic leader who took an active role. His forceful suggestions and direct intervention unclogged many bottlenecks and kept the personnel and supply pipelines flowing smoothly.

Difficult terrain, bad weather, and the inadequate road and rail networks were physical obstacles not easily overcome, but doctrinal issues and equipment shortages also created logistics problems. The 1st Marine Division, specifically structured for amphibious warfare, was neither organized nor equipped for sustained inland operations like those on Korea's Central Front. Unfortunately, this simple fact was either misunderstood or ignored by the high command. Repeated requests to keep the Marines close to the coast in order to minimize logistical concerns fell upon deaf ears at Eighth Army and United Nations Command headquarters. Service support challenges were further complicated by the physical separation of the 1st Marine Division and the 1st Marine Aircraft Wing. Additionally, during the spring of 1951 the 1st Marine Division provided much of the logistical support for the Korean Marines.

Logistical support in Korea was a massive multi-Service operation; it was a complicated logistical maze, one not easily traversed by the uninitiated, that existed because Marine units had to draw upon the resources of all four Services as well as indigenous labor. At the lowest level the Marines relied upon their own robust organic service and support units. The 1st Combat Service Group functioned as an intermediate clearing house and established liaison with the other Services. The Marines drew upon Eighth Army for theater-level support and further relied upon Navy and Marine service support from Pacific Command. The Marines also obtained support from the Republic of Korea.

The first option when answering logistics challenges, of course, was to make the most effective possible use of organic assets. 1st Marine Division logistics units included Commander Howard A. Johnson's, and after 23 January, Commander Clifford A. Stevenson's 1st Medical Battalion; Lieutenant Colonel Olin L. Beall's (Lieutenant Colonel John R. Barreiro, Jr., commanded after 16 March) 1st Motor Transport Battalion; Lieutenant Colonel Carl J. Cagle's 7th Motor Transport Battalion; Major Lloyd O. Williams' 1st Ordnance Battalion; the 1st Service Battalion (commanded successively by Lieutenant Colonel Charles L. Banks, Colonel Gould P. Groves, Lieutenant Colonel Horace E. Knapp, and Lieutenant Colonel Woodrow M. Kessler); and 1st Shore Party Battalion (Lieutenant Colonel Henry P. "Jim" Crowe until 10 May and thereafter commanded by Lieutenant Colonel Horace H. Figuers). The 1st Marine Division was specifically tailored for amphibious operations, but in Korea the specific needs of the moment very often superseded doctrine. Amphibious combat support units, such as the 1st Amphibian Tractor Battalion and the 1st Armored Amphibian Battalion, could not be fully utilized by the 1st Marine Division when it operated far from the coast, so one amphibian tractor company provided ship-to-shore

Saving Lives

No cry for help on the battlefields of Korea carried more urgency than the plea "Corpsman up!" This chilling entreaty invariably meant that a Marine was seriously wounded. Within moments, a medical corpsman would come scurrying forward through a hail of fire to lend life-saving assistance, often conducted in full view of the enemy and done at great peril to the caregiver.

The U.S. Navy provided medical (doctors, nurses, and corpsmen) and morale (chaplains) personnel to the United States Marine Corps. The chaplains were known by a variety of names that indicated their particular status or religious affiliation; "Father," Rabbi," "Reverend," and "Padre" were among the most common nicknames. On the other hand, Navy medical personnel—from the lowest ranking hospital apprentice all the way up to the chief surgeon of the Medical Corps—were simply known as "Doc" to the Marines they served.

Most medical personnel assigned to the 1st Marine Division in Korea came from the 1st Medical Battalion, which was successively commanded by Navy Commanders Howard A. Johnson and Clifford A. Stevenson. That parent unit was divided into a Headquarters and Service Company and five medical companies—two hospital companies and three collecting and clearing companies. Headquarters and Service Company (Commander William S. Francis and Lieutenant Commander Gustare T. Anderson, successively) provided administrative and support personnel and functions. Hospital Companies A (Commanders Buron E. Bassham, Philip L. Nova, and James A. Addison, respectively) and B (Lieutenant Commanders James A. Kaufman) were staffed and equipped to operate one 200-bed hospital each. The three collecting and clearing companies were: Company C (Commanders Harold A. Streit and Lewis E. Rector), Company D (Lieutenant Commanders Gustare T. Anderson and Daniel M. Pino), and Company E (Lieutenant Commanders Charles K. Holloway and John H. Cheffey). Generally speaking, Company C worked in direct support of the 5th Marines, Company D in support of the 1st Marines, and Company E in support of the 7th Marines during the spring of 1951.

The lowest rung on the medical evacuation chain was the individual hospital corpsman. Generally, two junior ratings of the 40 corpsmen assigned to each infantry battalion accompanied each rifle platoon into action. The primary jobs of these men, most of whom had only six

National Archives Photo (USMC) 127-N-A155354

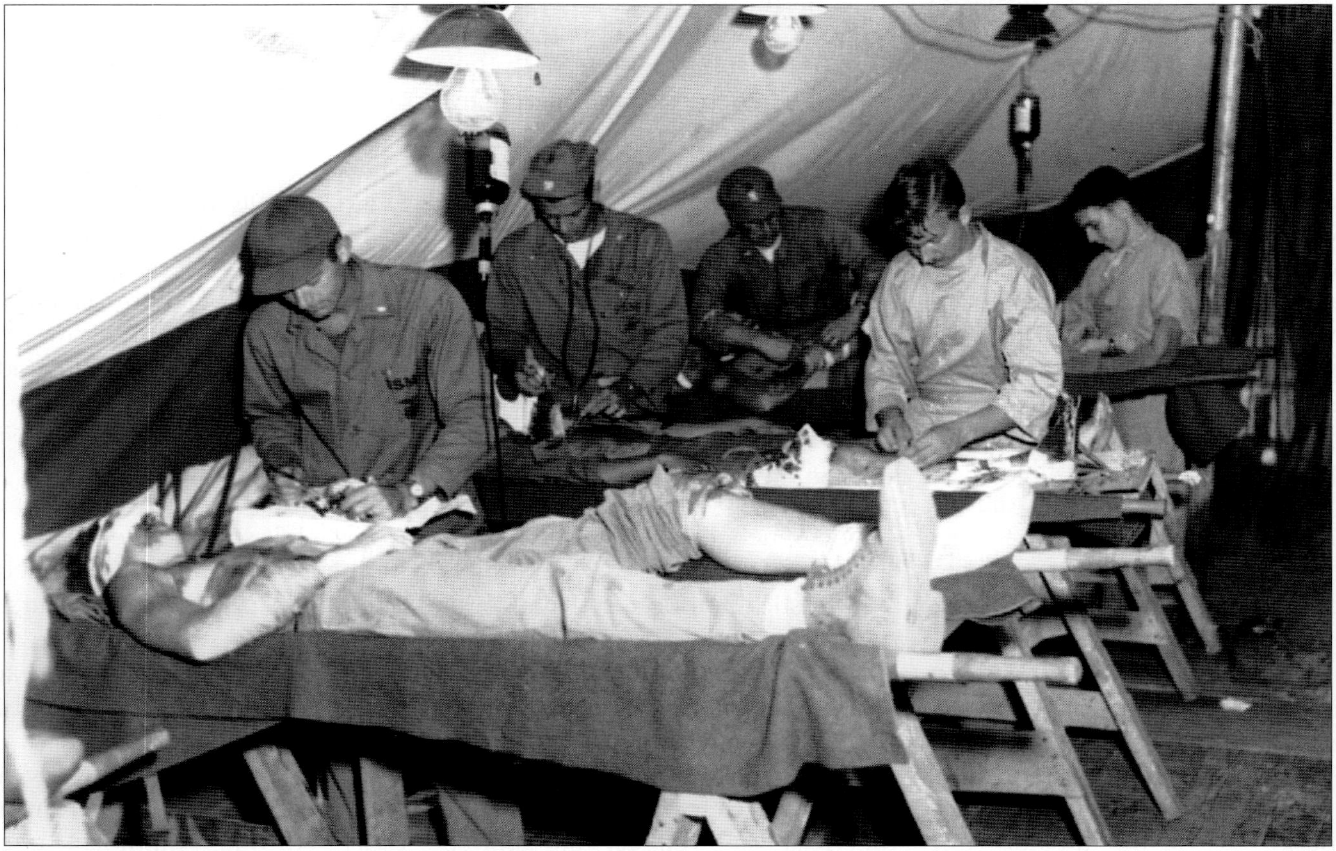

weeks of advanced medical training under their belts, were to stabilize wounded men and to supervise the initial evacuation process. Under fire on the battlefield they would conduct a hasty exam and apply necessary first aid measures (start the breathing, stop the bleeding, stabilize or bandage the crucial area, and treat for shock). Once this was done, the corpsman would arrange for evacuation. Usually, this meant four Marines or Korean litter bearers would carry the wounded man to the nearest collection point (usually the company command post) for transportation to the battalion aid station. The 28 chaplains assigned to the 1st Marine Division often played a critical role in this stage as well. They frequently lent a hand as stretcher-bearers or administered first aid in addition to performing last rites or building up the sagging spirits of the wounded.

Two Navy doctors, usually lieutenants, manned the battalion aid station (called the BAS), along with 10 or so enlisted corpsmen headed by a chief pharmacists mate. Incoming casualties were quickly inspected by an experienced corpsman so they could be categorized for treatment precedence ("triage"). The BAS facility was simple: usually an open air or tent operating arena, where rudimentary "meatball" surgery was performed while the patient's stretcher was placed upon a pair of sawhorses. This procedure saved time and minimized the amount of uncomfortable shifting. The battalion medics applied either life-saving surgery or gave just enough treatment to get the casualty ready for further evacuation.

The collecting and clearing companies then evacuated patients from the BAS to one of the 60-bed mobile field hospitals (in Army parlance, a MASH; to the naval services, depending upon which letter company was used, the nomenclature was something like "Charlie Med"). Here the facilities and care were more advanced. Surgical teams treated non-evacuables requiring resuscitation or immediate surgery then sent them on their way to semi-permanent division hospitals, which provided definitive care and short stay hospitalization. Extreme cases that were stable but could not return to duty in the near future were sent on to theater-level hospitals from whence they usually were returned to the United States.

Two intermediate steps in the evacuation process came into their own during the Korean War, use of hospital ships and aerial evacuation. Prior to the Second World War, hospital ships were used only to transport badly wounded men home. During World War II, however, hospital ships could often be found waiting off the landing beaches to provide a safe haven for treating casualties incurred during the opening rounds of amphibious operations. In Korea it was common practice to keep at least one hospital ship nearby at all times. These Haven- and Comfort-class vessels mustered about

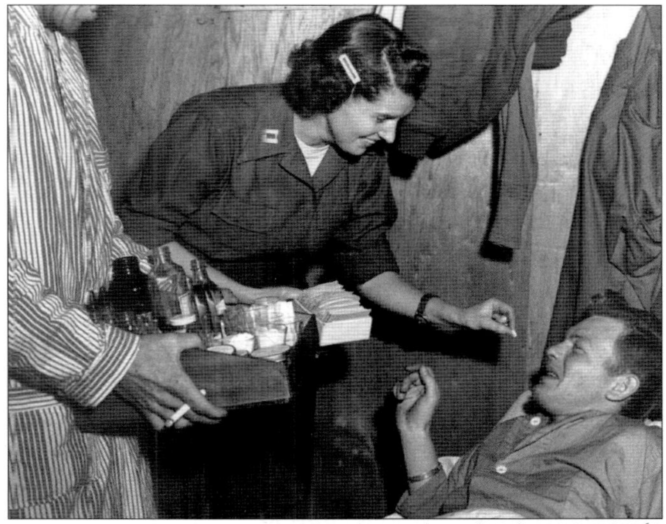

National Archives Photo (USA) 111-SC358063

150 officers and more than 1,000 enlisted men to man the operating rooms and healing wards which could accommodate several hundred critical short-term patients at one time. This practice, combined with the increasing use of helicopters for medical evacuations, ensured rapid advanced medical treatment was available. Several Haven- and Comfort-class hospital ships rotated station watches during the spring of 1951, and the USS *Consolation* (AH 15) was fitted with a helicopter landing pad—an adaptation that soon thereafter became standard practice.

Many view the advent of rotary-wing aircraft as the most important aviation innovation during the Korean Conflict. Inevitably, the nimble helicopters soon became an important means of medical evacuation because they could fly directly to the forward areas, pick up wounded men from previously inaccessible locations, then deliver them to an advanced care facility within a matter of minutes rather than hours or days. Helicopters could land atop the mountains and ridges that dotted Korea eliminating the rough handling and long movements necessary for overland evacuation. Unfortunately, the Sikorsky HO3S-1 could carry only one stretcher case at a time (and the patient's lower extremities would have to extend out the rear hatch), limiting their utility as an evacuation machine. By the spring of 1951, the bubble-topped Bell HTL, which mounted a pair of stretchers on each side and could carry a sitting evacuee as well, augmented these older machines. Eventually, even more capable evacuation helicopters (Sikorsky HO5S and HRS) made their way to Korea. Fixed-wing observation aircraft were sometimes pressed into service for emergency evacuations as well. Twin- and four-engine fixed-wing transport planes were used to deliver men to in-country theater-level facilities, hospitals in Japan, or to take the badly wounded back to the States.

transportation at Pohang while the remaining tracked landing vehicles were used by Eighth Army for non-Marine support. The 1st Engineer Battalion often used Shore Party motor transport and engineer assets. In addition, U.S. Army transportation units or trucks on temporary loan from other Marine units often reinforced the motor transport battalions. Navy Seabee Construction Battalions regularly furnished construction engineer support, Army engineer assets were often temporarily attached to Marine units, the U.S. Air Force provided equipment and materials for air base construction and maintenance, and the Korean Service Corps furnished laborers.

Colonel John N. Cook, Jr.'s 1st Combat Service Group at Masan furnished Marine general logistics support. The 1,400-man group was composed of headquarters, maintenance, supply, support, and truck companies. It furnished most service support functions: advanced maintenance and repair, central storage, general administration, and laundry services. Colonel Cook coordinated inter-Service logistics efforts, requisitioned supplies and equipment from higher echelons, controlled and maintained rear area depots, stored spare parts and high demand items, and distributed these to the division and the wing. The group also mustered special support units including a bath and fumigation platoon and an air delivery platoon. Although it provided support to the wing, the 1st Combat Service Group was actually attached to the 1st Marine Division. Group detachments were located in Japan, Pusan, Pohang, and operated forward area supply terminals at Wonju, Hoengson, and Chunchon.

The Military Sea Transportation Service, Military Air Transportation Service, and Naval Air Transport Service furnished inter-theater lift of supplies, personnel, and equipment. Army Brigadier General Crump Garvin's 2d Logistical Command replenished common use items for all Services in Korea. Fleet Marine Force, Pacific's Service Command furnished unique Marine equipment and supplies. The situation was more complex with regard to aviation. The 2d Logistical Command provided a few aviation-related items but for the most part did not stock technical equipment such as aircraft parts, special maintenance tools, or aircraft ordnance. The U.S. Navy Pacific Service Command handled most of these, although Marine-specific items came from Fleet Marine Force, Pacific. Emergency resupply procedures allowed critical items to be flown to Korea from the United States.

1stMarDiv Historical Diary Photo Supplement, Apr51

A column of Korean Civil Transport Corps bearers brings supplies from a rear area to the main battle line. Rugged terrain and lack of roads often dictated that man-packing supplies was the only way they would reach the front lines.

Marine logistics problems mirrored the tactical situation. In January 1951, the major challenge was filling critical personnel and equipment shortfalls in the wake of the costly Chosin Reservoir campaign. After that the major logistics challenge became sustaining units almost constantly on the move.

Major equipment shortages occurred in communications and transportation. The Marines had only half their authorized radios and only 58 (of 1,162) EE8 telephones. The division was also short 58 jeeps and 33 two-and-a-half ton trucks. Not revealed in these statistics is the poor condition of the trucks that survived the Chosin campaign. Most were in terrible shape and badly needed advanced maintenance and new tires. The only significant combat arms shortfall was tanks; the 1st Tank Battalion had only 78 of its 97 authorized M-26 and M-4 tanks.

After the Marines left Masan in mid-January, resupply became the overriding logistics concern. The supply pipeline ran from the United States to Japan then on to Korea. Cargo and transport ships and long-range airplanes carried men, supplies, and equipment from the United States to depots and processing centers in Japan. The 1st Combat Service Group maintained an administrative processing center and a supply receiving area at Kobe, Japan. Unfortunately, there was a poor supply flow from Japan to Korea, partially due to labor and transportation shortages and partially due to red tape. The Marines in Korea had few rear area storage facilities and inadequate transportation assets There was only one true deep-water seaport in all of Korea, Pusan, and it was located at the peninsula's southernmost tip, which was serviced by a very limited road and rail network. This created a tremendous supply bottleneck. The Marines were able make some use of Pohang as a port of entry, but unloading there was a cumbersome and time-consuming process. The U.S. Army 55th Quartermaster Depot which handled joint-Service requests did not back-order most types of supplies, hence, requests were routinely denied if a particular item was not on hand. Eventually, 1st Service Battalion assigned a Marine liaison team to smooth out this problem. Regardless, there was a constant shortage of expendable items, such as steel wool or stationery supplies, and individual requests sometimes required a four-week lead-time before issue. The 1st Combat Service Group ran railheads at Masan and Dalchon, the 1st Shore Party Battalion handled incoming supplies at Pohang and ran the railhead at Yodo-nae, and the division established truckheads as far forward as possible.

The poor roads, inadequate railroad system, and fluid nature of the fighting made resupply of forward units a never-ending headache. Trains, trucks, and airplanes carried in-country supplies from rear areas to forward supply points. From there, however, it was division's job to get those supplies to its troops in the field. Unfortunately, there was no rail line north of Wonju, and there were often too few trucks to move the supplies that did arrive in a timely manner. The closing of forward supply points and ammunition storage areas during the CCF Spring Offensive also created problems. The closures created temporary ammunition shortages and stopped the flow of "A" and "B" rations so the troops had to rely upon less tasty and less filling "C" and "K" field rations. The only solution to this problem was to airdrop supplies and ammunition. Poor flying weather and limited airfield facilities made air transportation an iffy proposition, and airdrops were inefficient in terms of equipment, manpower, and loss rates, but there was simply no other choice. The multi-Service Combat Cargo Command accomplished airdrops. Marine transport planes joined those of the Air Force and the Navy to deliver supplies all across the front. The most unique air delivery was a single size 16 EEE combat shoe dropped over the 1st Marines headquarters from an OY light observation aircraft. This jocular package was addressed to Colonel Wilburt S. "Big Foot" Brown and included a note that the pilot did not have sufficient space in his small plane to carry two such gigantic "Boondockers" at the same time. This joke, however, must have tried Brown's patience because pilots in Nicaragua had first used it three decades earlier.

One of the most difficult logistics challenges was overland transportation. Gasoline and tire shortages often idled much-needed trucks, jeeps, and weapons carriers. The Marines were also constantly hampered by lack of vehicles; for example, in April the division was short 1 tracked landing vehicle, 13 tanks, 18 jeeps, and 59 trucks. Although the 1st Marine Division had been augmented with an extra motor transport battalion, there were still insufficient trucks to move men and supplies in a timely manner. Heavy demands, combat losses, accidents, and hard use all contributed to the problem. Pooling Marine resources and borrowing U.S. Army trucks sometimes addressed this concern, but

Marines line the rail as the attack transport General J. C. Breckinridge *(AP 176) docks in San Francisco. These veterans, some with more than six months of combat, were among the first Marines to return from Korea.*

National Archives Photo (USN) 80-G-428299

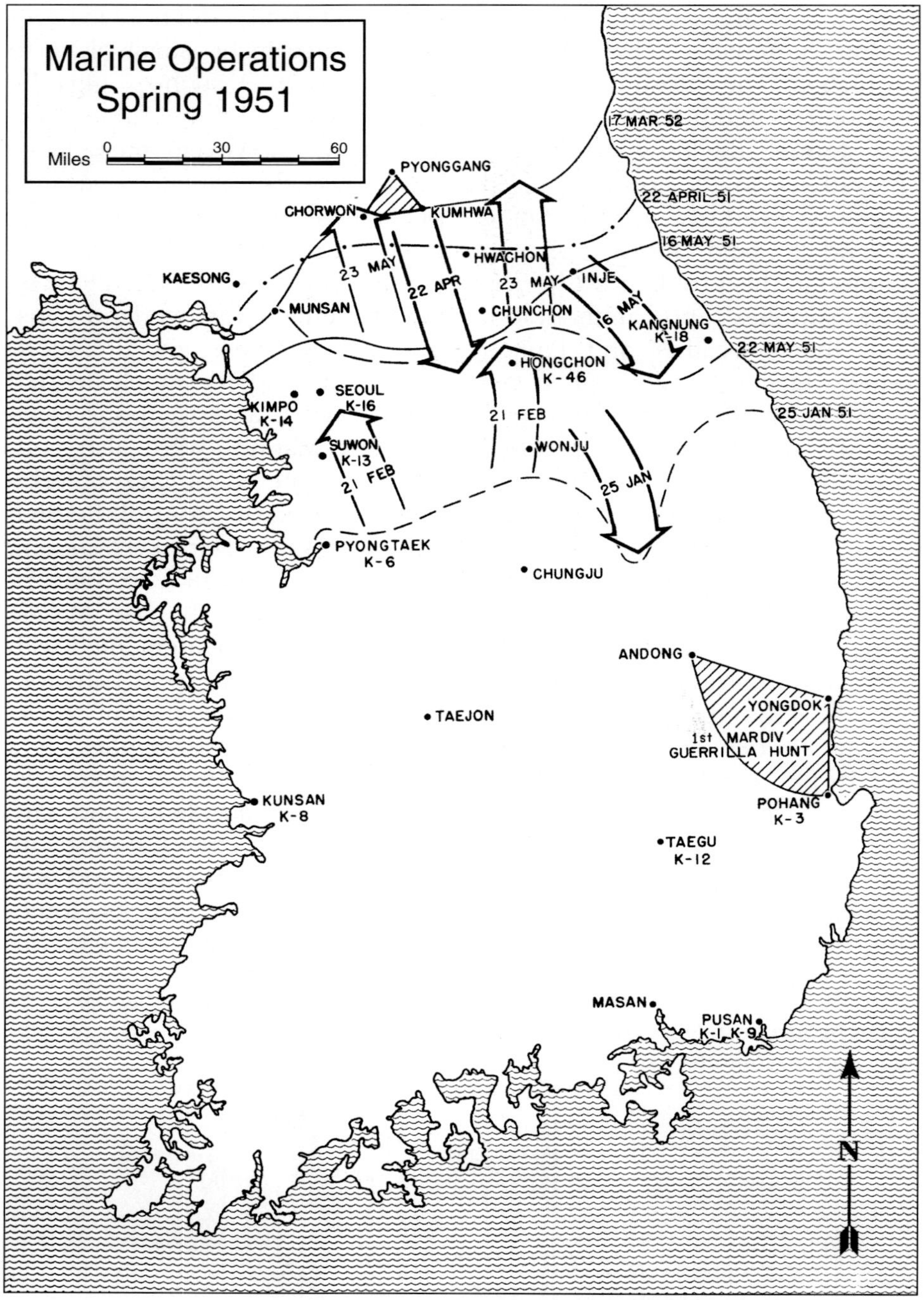

even additional vehicles could do nothing to alleviate the major transportation obstacle, the inadequate Korean transportation infrastructure.

Food, clothing, ammunition, and other necessities slowly made their way forward to regimental and battalion supply dumps in trucks, jeeps, and weapons carriers, but then most often had to be hand carried to the front lines. This was a labor-intensive process that few combat units could spare men for. The South Korean government, at the request of Eighth Army, organized a pair of quasi-military organizations—the Korean Service Corps and the Korean Transportation Corps—to fill this need. Members of the Korean National Guard and volunteers from refugee camps manned these organizations. The Korean Service Corps included "Yoboe" construction gangs, and the Transportation Corps comprised "cargodore" companies consisting of about 300 "Chiggy Bear" porters. The Korean government provided almost 300 laborers to the 1st Marine Division. Yoboes were used for roadwork and manual labor by combat and service support units. The Chiggy Bears were parceled out to each rifle regiment where they labored under the supervision of a senior Marine noncommissioned officer or junior lieutenant. Organized as a unit under a headman and a straw boss, these never-ending columns of porters, called "Mule Trains" after a popular song of the day, kept frontline Marines supplied under the most trying circumstances. There are no specific figures as to how many of these loyal workers were killed or wounded in action, but those numbers were undoubtedly high. Although sacrifices of the Chiggy Bears may have gone unrecorded, their tireless efforts were certainly not unappreciated by the cold, thirsty, hungry Marines at the front.

The 1st Marine Aircraft Wing's major engineering headaches were airfield renovation and upkeep. This was particularly difficult because the wing was almost constantly on the move. The wide dispersal of Marine air units located at air bases in Japan (Itami, Itazuke, and Bofu) and Korea—Pohang (K-3), Pusan (K-1 and K-9), Hoegsong (K-46), and Seoul (K-16). Marine Wing Service Squadron 1 (successively commanded by Chief Warrant Officer Aubrey D. Taylor, Lieutenant Colonel James C. Linsay, and Colonel Roger T. Carleson) was the unit charged to provide such support, but the overworked Marines often had to ask for help from Navy Seabee construction units as well as Army and Air Force engineers. When such support was not immediately forthcoming, as it often was not, Marine technicians had to be pulled away from other jobs to pick up shovels. Fuel handling was also a problem. For example, Marine Aircraft Group 33 at K-3 (Pohang) had to rely upon tracked landing vehicles to haul fuel drums ashore, which then had to be hand pumped into 1,200-gallon fuel trucks. This slow, inefficient, labor-intensive process siphoned off men whose skills could have been put to better use. Additionally, vehicles designed to handle World War II ordnance were ill-suited to service modern aircraft. The primitive conditions in Korea also took a toll on wing motor transport. These problems required constant attention throughout the spring of 1951.

That operations only intermittently suffered for lack of service support is a tribute to Marine service and support personnel. The Marines faced seemingly insurmountable logistics challenges between January and May 1951, yet—despite a few hiccups—the only serious long-term supply shortfall was the lack of artillery ammunition caused by Eighth Army policies dictated from above over the strenuous objections of Marine commanders. That this was the case is a testament to the hard working, but too often unsung, Marines of the combat service support units.

National Archives Photo (USMC) 127-N-A157778

Marine replacements come ashore from a Navy landing ship. Three replacement drafts were rushed to Korea after the Chosin campaign, and about one replacement draft with about 100 officers and more than 1,000 enlisted men arrived each month thereafter.

The 1st Marine Division received two replacement drafts in December 1950, but was still short almost 3,000 men on New Year's Day. The initial personnel deficit was partially alleviated by the return to duty of 945 men, most of whom had been frostbite evacuations, and the arrival of 700 veteran Marines pulled from posts and stations in the Far East. Two replacement drafts were also formed at Camp Pendleton. The largest part of these drafts consisted of recalled reservists, but there were also some veteran regular Marines included. Freshly minted Marines from the recruit depots and "shiny-bar" second lieutenants just arrived from officer training filled out replacement rosters. Two hundred and thirty men with critical military occupational specialties were flown directly to the combat zone. The 4th Replacement Draft sailed for Korea on board the fast transport USNS *General William O. Darby* (AP 127) and was due in mid-January. The just-forming 5th Replacement Draft was assigned to the USS *General J. C. Breckinridge* (AP 176) and was slated to arrive in mid-February. Replacement drafts containing about 1,700 men each continued arriving on a monthly basis from then on. This personnel replacement system was adequate, but it was not perfect. The adoption of a combat rotation system primarily based upon time served meant that the most experienced Marines were constantly leaving Korea and their places taken by inexperienced replacements. The introduction of new men, as individuals rather than units, created cohesion problems in small units. Personnel shortages after major engagements remained a nagging problem throughout the spring of 1951.

Extraordinary Heroism

The period from January to May 1951 encompassed three designated U.N. campaigns: Chinese Intervention from 3 November 1950 to 24 January 1951; the First U.N. Counteroffensive from 25 January to 21 April; and the CCF Spring Offensive from 22 April to 8 July 1951. It is ironic that the spring of 1951 is one of the most overlooked periods in American military history because that period featured some of the most intense and hard-fought Marine actions of the Korean Conflict. The anonymous battles of that time were as desperate and bloody as those at the Pusan Perimeter, the Inchon landing, and the liberation of Seoul, yet they remain almost unknown except to those who fought there. Too often relegated to the dustbin of history is the fact that some Marine units suffered more casualties during the drive to the Punchbowl than they had during the legendary fighting at the Chosin Reservoir. Indeed, the events of that time might well be called the "Forgotten Campaigns" of what is now often termed the "Forgotten War." What should be remembered is the key role played by the Marines, both on the ground and in the air. The 1st Marine Division rendered ineffective one NKPA division at Pohang, spearheaded the United Nations' recapture of the Hwachon Reservoir during Operations Killer and Ripper, and stabilized the center of the U.N. line in the midst of the CCF Spring Offensive. The versatile 1st Marine Aircraft Wing flew a wide variety of missions; helicopters proved their utility in combat and Marine close air support was unsurpassed in efficiency. These accomplishments did not go unrecognized at the time; both the 1st Marine Division and the 1st Marine Aircraft Wing were awarded Presidential Unit Citations for their actions in the spring of 1951.

Luckily, the military lessons of the day were not forgotten. The Marines in Korea fought well, but they were not employed in accord with their envisioned inter-Service role. They, even more than their antecedents in World War I, became an integral part of a United States field army fighting far from the sea for an extended period. Instead of acting as a semi-independent combined arms team, as had been the case in 1950, in the spring of 1951 the 1st Marine Division was stripped of its direct air support and became just one more Eighth Army ground maneuver unit. Thus, contrary to the wishes of Marine commanders, the 1st Marine Division was used as a "second land army." The forced separation of the 1st Marine Division and the 1st Marine Aircraft Wing and the lack of an in-theater Marine commander prompted the later creation of permanent Marine air-ground task forces.

Another factor that affected the future of the Marine Corps was the performance of the Marine Corps Reserve. Without the Reserve, it is doubtful that the Marines would have been able to deploy an entire division and aircraft wing to Korea. The character of the 1st Marine Division underwent a drastic change in the spring of 1951. When the 1st Provisional Marine Brigade arrived in Korea in August 1950 it was virtually an all-regular formation, by the time of the Chosin Reservoir campaign in November about one-third of the Marines were reservists, but by the end of May 1951 almost two-thirds of the U.S. Marines in Korea were reservists. There were very few regular officers below the rank of captain and almost no regular enlisted men other than staff noncommissioned officers by the time the 1st Marine Division reached the No Name Line. Similar figures also apply to the 1st Marine Aircraft Wing. This proved that the Marine Corps could count on its Reserve when the chips were down. This lesson was validated in the Persian Gulf some 40 years later when Marine reservists once again answered the call to the colors during the Gulf War and acquitted themselves well.

The period January to May 1951 was one of transition and tumult during which United Nations forces traveled from the brink of defeat to the edge of victory several times as fierce fighting ebbed and flowed across Korea's midlands. The enemy still remained a potent and dangerous foe after the spring of 1951, but the United Nations Command had become a seasoned force that was not about to be ejected from the peninsula. All talk of evacuating Korea due to enemy pressure was silenced by the recent stellar performance on the battlefield. This favorable reversal of fortunes in Korea between January and May has been characterized by the eminent military historian Colonel Harry G. Summers, Jr., USA, as "the single greatest feat of arms in American military history," and the Marines played a key role in that amazing reversal of fortune.

The impact of that stunning turnaround was, however, not realized on the home front. By mid-1951 many Americans were dissatisfied with "Truman's Police Action," and there was deeply felt sentiment across the country for an end to the fighting. The resulting political pressure led to a fundamental change in American foreign policy. A Joint Chiefs of Staff directive stated that the military objective was no longer to unify Korea, but "to repel aggression against South Korea." In fact, both sides unofficially accepted a mutual cessation of major offensive actions after the U.N. regained the modified Kansas Line in June. The Korean War then passed its first anniversary without fanfare or celebration, and not long after peace talks began. The United Nations Command briefly mounted a limited offensive after the talks broke down, but the Korean Conflict thereafter became a bloody stalemate marked by two more years of contentious negotiations and inconclusive fighting.

About the Author

Lieutenant Colonel Ronald J. Brown, USMCR (Ret), is a freelance writer and scoring director for Measurement Incorporated, an educational testing firm. The author of two monographs in the Persian Gulf series and two official unit histories, he was also a contributing author for the best-selling book *The Marines*, and has been a frequent contributor to professional journals. He is working on a second Korean commemorative pamphlet on Marine helicopter operations. Lieutenant Colonel Brown served as an active duty infantry officer from 1968 to 1971 and saw combat in Vietnam. He joined MTU DC-7 at its inception in 1976 and served continuously with that unit until his retirement. He went to Korea during Exercise Team Spirit-84. Six years later he was activated during the Persian Gulf War and was assigned to I Marine Expeditionary Force. After Operation Desert Storm, he became the Marine component historian for Combined Task Force Provide Comfort in northern Iraq. Lieutenant Colonel Brown, then commanding MTU DC-7, retired in 1996. In civilian life, Ronald Brown was a high school history teacher for three decades and is a nominee for the Michigan High School Football Coaches Hall of Fame.

Sources

The basic source for this pamphlet was the fourth volume in the series *U.S. Marine Operations Korea, 1950-1953: The East-Central Front* (Washington, D.C.: Historical Branch, G-3 Division, HQMC, 1962), written by Lynn Montross, Maj Hubbard D. Koukka, and Maj Norman W. Hicks. Marine-specific books consulted were: Robert D. Heinl, *Soldiers of the Sea* (Annapolis: U.S. Naval Institute Press, 1962); Allan R. Millett, *Semper Fidelis* (New York: McMillan, 1980); J. Robert Moskin, *The U.S. Marine Corps Story* (New York: McGraw-Hill, 1977); LtCol Gary W. Parker and Maj Frank M. Batha, Jr., *A History of Marine Observation Squadron Six* (History and Museums Division, HQMC, 1982); Col Gerald R Pitzel, *A History of Marine Fighter Attack Squadron 323* (History and Museums Division, HQMC, 1987); Maj William J. Sambito, *A History of Marine Attack Squadron 311* (History and Museums Division, HQMC, 1978); Col Francis F. Parry, *Three Marine War: The Pacific, Korea, Vietnam* (Pacifica: Pacifica Press, 1987); and BGen Edwin H. Simmons, *The United States Marines: The First Two Hundred Years* (Annapolis: Naval Institute Press, 1998).

Overviews of the Korean Conflict included: Roy E. Appleman, Ridgway *Duels For Korea* (College Station: Texas A&M Press, 1990); Clay Blair, *The Forgotten War: America in Korea, 1950-1953* (New York: Times Books, 1987); Russell A. Gugeler, *Combat Actions in Korea* (Washington, D.C.: Office Chief of Military History, 1970); Robert Leckie, *Conflict, The History of the Korean War, 1950-53* (New York: Putnam's, 1962); Billy C. Mossman, *Ebb and Flow—U.S. Army in the Korean War* (Washington, D.C.: Government Printing Office, 1990); and Matthew B. Ridgway, *The Korean War* (Garden City: Doubleday, 1967). Individual perspectives included Burke Davis, *Marine! The Life of Chesty Puller* (New York: Little-Brown, 1962); Paul N. McCloskey, Jr., *The Taking of Hill 610* (Woodside: Eaglet Books, 1992); LtGen Matthew B. Ridgeway, *Soldier: The Memoirs of Matthew B. Ridgway* (New York: Harper, 1956); Allan R. Millett, *In Many A Strife: Gerald C Thomas and the U.S. Marine Corps, 1917-1956* (Annapolis: U.S. Naval Institute Press, 1993); and Morgan Brainard, *Men in Low Cut Shoes: The Story of a Marine Rifle Company* (New York: Tood & Honeywell, 1986).

Primary documents and military periodicals held by the History and Museums Division in Washington, D.C., include unit diaries, after action and special action reports, biographical files, subject files, comment files, personal diaries, and articles in the *Marine Corps Gazette* and the *U.S. Naval Institute Proceedings*. Among the oral interviews consulted were those of Gen Oliver P. Smith, LtGen Alpha L. Bowser, Maj Martin J. Sexton, LtCol John L Hopkins, Col Homer L. Litzenberg, Jr., LtCol Francis F. Parry, Maj William L. Bates, Jr., and MajGen Edward A. Craig. The author also used personal files compiled during Exercise Team Spirit-84, and wishes to acknowledge the recollections of retired BGen Edwin H. Simmons, USMC (Ret), 1stLt Robert Harding, USAR, and SSgt Edward Huffman, USMCR, all of whom served with the 1st Marine Division in Korea in 1951.

DRIVE NORTH
U.S. Marines at the Punchbowl

by Colonel Allan R. Millett, USMCR (Ret)

The rumble of American field artillery through the morning mists in the valley of the Soyang River gave a sense of urgency to the change-of-command ceremony inside the headquarters tent of the 1st Marine Division. Four days of hard fighting in the withdrawal from the Hwachon Reservoir had brought the division safely to the river on 25 April 1951. The trek away from the Chinese *39th* and *40th Armies* had not yet, however, brought the division to the No Name Line, the final defensive position 15 miles south of the river designated by Lieutenant General James A. Van Fleet, commanding the U.S. Eighth Army. In a simple rite that included only the reading of the change-of-command orders and the passing of the division colors, Major General Gerald C. Thomas relieved Major General Oliver P. "O. P." Smith and took command of a division locked in a battle to stop the Chinese Fifth Offensive.

The ceremony dramatized the uncertainty of the Marines in the second year of the Korean War. Understandably, General Smith did not want to turn over command in the middle of a battle. On the other hand, General Van Fleet wanted Thomas to take command of the division as soon as possible, something Thomas had not planned to do since his formal orders from the Commandant, General Clifton B. Cates, designated 1 May 1951 as turn-over day. Thomas had planned to spend the intervening week on a familiarization tour of Korea and the major elements of the Eighth Army. He had thought his call on Van Fleet the day before had been simply a courtesy visit, but instead he found himself caught in a delicate matter of command relations.

General Thomas arrived in Korea to face an entirely new war. The October 1950 dream of unifying Korea under the sponsorship of the United Nations (U.N.) had swirled away with the Chinese winter intervention. The war still hung in the balance as the United Nations Command attempted to drive the Communist invaders out of the Republic of Korea (ROK) for the second time in less than a year. The U.S. Eighth Army and its Korean counterpart, the *Hanguk Gun* (South Korean Armed Forces) had rallied in January and February 1951, under the forceful leadership of Lieutenant General Matthew B. Ridgway, USA. United Nations

AT LEFT: *Marines quickly demolish enemy bunkers with grenades and planted charges before moving north.*
Department of Defense Photo (USMC) A8504

MajGen Gerald C. Thomas, right, meets with MajGen Oliver P. Smith prior to the spartan change-of-command ceremony witnessed by a handful of participants drawn from the 1st Marine Division's staff.
Gen Oliver P. Smith Collection, Marine Corps Research Center

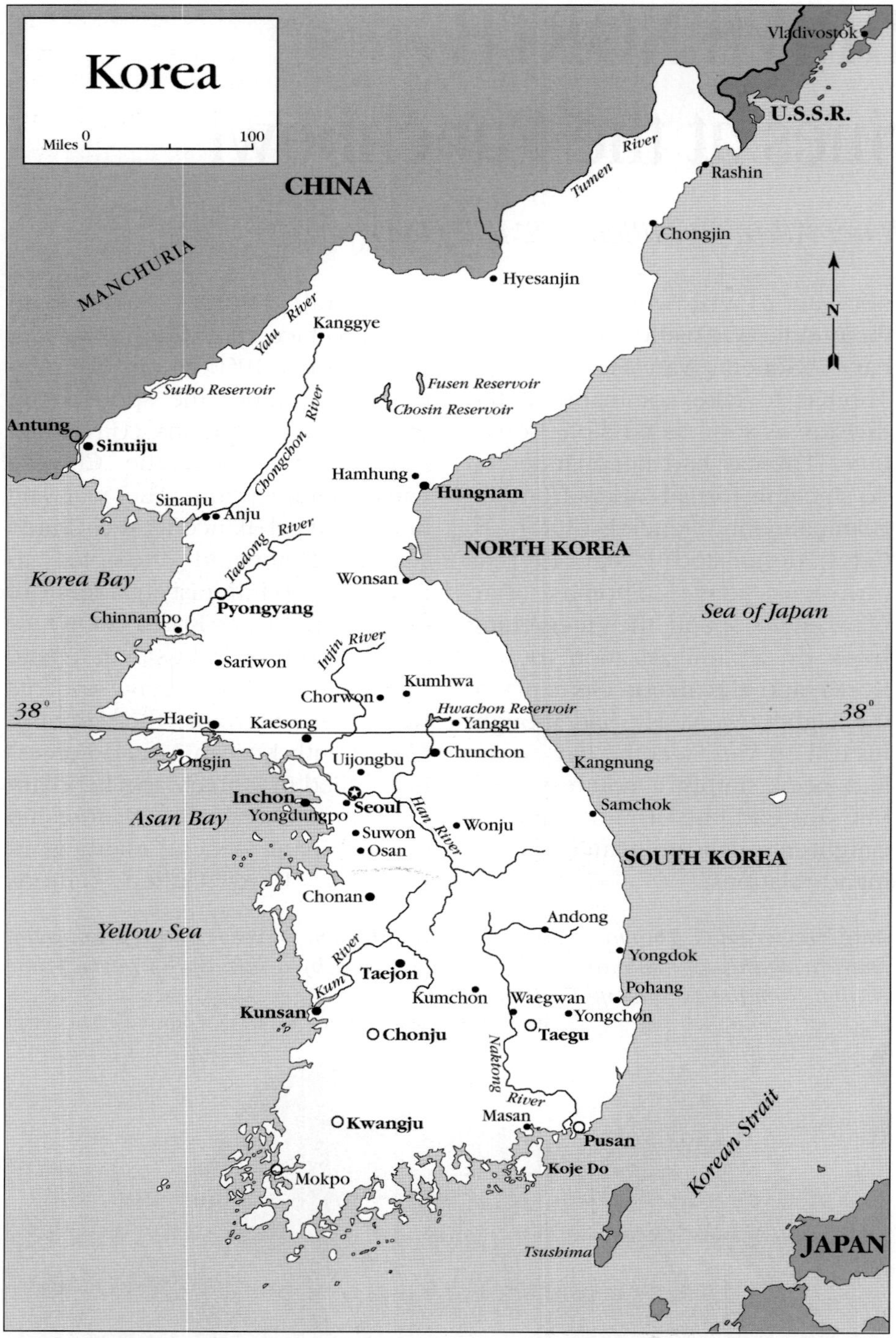

Command had then driven back the Chinese Communist Forces (CCF) and the North Korean People's Army (NKPA). The allies had advanced well north of the 38th Parallel in central and eastern Korea. Goaded by Mao Zedong, General Peng Dehuai ordered his joint expeditionary force of 693,000 Chinese and North Korean soldiers to mount one more grand offensive. Eleven Chinese armies and two North Korean corps (40 divisions) would smash south just west of Hwachon Reservoir in the sectors held by the U.S. I and IX Corps. At a minimum the Communist forces, about half of Peng's total army, would drive United Nations forces below the 38th Parallel. The maximum objective would be to threaten the Han River valley and the corridors to Seoul while at the same time recapturing the territory south of the Soyang River, which opened an alternative corridor south to Hongchon.

When General Thomas called on General Van Fleet on 24 April, the Eighth Army commander, a combative 59-year-old Floridian with a World War II record of successful command from regiment to corps in Europe, felt confident that his forces had blunted the four-day-old Communist offensive. However, he had an organizational problem, which was that the 1st Marine Division should be shifted back to X Corps and redeployed to the No Name Line under the command of Lieutenant General Edward M. Almond, USA, the division's corps commander throughout 1950. The relationship between O. P. Smith and Almond, however, had become so venomous that Ridgway assigned the Marine division to IX Corps in January 1951 and promised Smith that he would not have to cope with Almond, whose style and substance of command angered Smith and his staff. Van Fleet had honored Ridgway's commitment, but the operational situation dictated that the Almond-Smith feud could not take precedence.

Van Fleet explained the plan to shift the 1st Marine Division back to X Corps to Thomas without going into the Almond-Smith problem. Van Fleet did not give Thomas a direct order to proceed immediately to the 1st Marine Division headquarters near Chunchon. Thomas believed, however, that Van Fleet had sound reasons to want a change of command now, so he caught a light plane furnished by the 1st Marine Aircraft Wing and flew to the primitive airstrip that served the division. Escorted by the new assistant division commander, Brigadier

General Lewis B. "Chesty" Puller, Thomas went directly to Smith's van and told him of Van Fleet's request and future plans. Smith refused to relinquish command. Without mounting an argument, Thomas left the van and went to the operations center to confer with Colonel Edward W. Snedeker, the chief of staff, and Colonel Alpha L. Bowser, the G-3, both of whom sympathized with Thomas but thought Smith should remain in command. Thomas thought the division was well positioned to refuse the open left flank of X Corps, but he also felt the tension in the command post.

Thomas decided that the issue of command could not be postponed—and now at least Smith knew he faced the prospect of again serving under Almond. Thomas returned to Smith's van within the hour and stated simply: "O. P., the table of organization calls for only one major general in

Major General Gerald C. Thomas

National Archives Photo (USMC) 127-N-A132593

Gerald Carthrae Thomas spent a lifetime dealing with challenging command relationships and operational problems inside and outside the Marine Corps. Born on 29 October 1894 on a farm near Slater, a western Missouri railroad town, Thomas grew up as a working boy in a working family. He was also a good student and versatile high school athlete. Living in Bloomington, Illinois, he attended Illinois Wesleyan University (1915-1917) before enlisting in the Marine Corps in May 1917 to fight the Germans. Thomas, age 22, mustered in at five feet, nine inches and 160 pounds, strong of wind and limb from athletics and labor. Dark hair and heavy eyebrows set off his piercing blue eyes and strong jaw. He would need every bit of his emotional balance and physical stamina—lifelong traits—for the Marine Corps placed him in the Germans' sights for most of 1918. As a sergeant and lieutenant in the 1st Battalion, 6th Marines, Thomas learned about war at Belleau Wood, Soissons, and the Meuse-Argonne. When the Silver Star and Purple Heart medals were authorized in 1932, Captain Thomas, professional officer of Marines, pinned on one award for gallantry and another for being gassed.

In the interwar years, Thomas had already fought against Haitian guerrillas, served a second tour in Haiti as a staff officer, and commanded a Marine detachment on a Navy gunboat in the Caribbean and Central America. He also lost one wife to disease, married again (Lottie Capers Johnson of Charleston, South Carolina) and started a family of two sons and two daughters. The Marine Corps recognized his potential value in wartime by sending him to five different Army schools (including the prestigious U.S. Army Command and General Staff College, Fort Leavenworth) and assigning him twice as an instructor in Marine Corps schools.

Between 1940 and 1950, Thomas proved that the Marine Corps had not wasted a minute or a dollar on his professional education. In a decade that saw him advance in rank from major to major general, Thomas prepared the Fleet Marine Force for war as an instructor at Quantico, military observer abroad, and a staff officer in the 1st Marine Division as the division conducted its last pre-Pearl Harbor amphibious exercises. When the division deployed to the South Pacific, Lieutenant Colonel Thomas went to war as Major General Alexander A. Vandegrift's operations officer (G-3), an assignment that made him one of the architects of victory on Guadalcanal. A trusted intimate of Vandegrift's,

Thomas served as the general's chief of staff during the final months of the Guadalcanal campaign and then played the same role in the I Marine Amphibious Corps' landing on Bougainville. He returned to Washington with Vandegrift when the general became Commandant in 1944. As a brigadier general, his second "spot" promotion in a row, Thomas fought the battles of demobilization and postwar defense reorganization, 1944-1947, as the Director of Plans and Policies on the Headquarters staff and played a critical role in winning legislative protection for the Fleet Marine Force in the National Security Act, 1947. He then spent two years as commanding general, Fleet Marine Force, Western Pacific, a brigade-sized force that garrisoned the Shantung peninsula and the city of Tsingtao until the Chinese Communist military victories in North China in 1948 made the American enclave irrelevant. Thomas successfully withdrew his force without incident in February 1949 and returned to educational and developmental billets at Marine Corps Base, Quantico.

Thomas' rich and exciting career had not, however, been without professional risks and cost. His aggressive personality, the force with which he defended his convictions, and his unwillingness to tolerate leadership lapses that endangered Marines had made him anathema to some of his peers, two of whom stopped his first promotion to major general. Others thought him too demanding a colleague. In 1951 only two opinions counted, those of Commandant Clifton B. Cates and Lieutenant General Lemuel C. Shepherd, Jr., Commanding General, Fleet Marine Force, Pacific, and Cates' likely successor as Commandant. Although neither Cates nor Shepherd were part of Vandegrift's "Guadalcanal gang," they knew Thomas well and recognized his special qualifications to go to Korea. In addition to his recent service as Commanding General, Fleet Marine Force, Western Pacific, Thomas had done a pre-war China tour in the Peiping legation guard. And despite his dogged defense of the Marine Corps in the Battle of the Potomac, 1945-1947, he got along well with the U.S. Army. As head of research and development, Thomas also understood the 1st Division's importance as test bed for future techniques like vertical envelopment.

After his successful command of the 1st Marine Division in Korea, April 1951-February 1952, Thomas returned to Headquarters Marine Corps as a lieutenant general and Assistant Commandant/Chief of Staff for General Shepherd. For the next two years, Thomas focused on reorganizing the Headquarters staff on functional general staff lines, on improving Marine Corps relations and representation within the Department of the Navy, and planning the postwar Fleet Marine Force of three divisions and three aircraft wings, a force more than twice as large as the Fleet Marine Force in June 1950.

For his "twilight cruise" Thomas became Commandant, Marine Corps Schools, Quantico (1954-1955), his favorite post and role as officer-instructor. Upon retirement he remained in government service as the first executive director, Net Evaluation Committee, National Security Council staff from 1956 to 1958. He then entered private business in real estate and insurance in the Washington, D.C. area. He regularly attended 1st Marine Division Association functions and events related to Marine Corps history; his four sons and sons-in-laws all served as Marines, two retiring as colonels. General Thomas died on 7 April 1984 at the age of 89.

---

a division. Either you turn over to me, or I'm going to leave." Smith did not respond, and Thomas again left the van. After several minutes of more tension, Smith emerged from his van and told Thomas that the change-of-command ceremony would be held at 0800 the next morning. The 1st Marine Division had a new commanding general as it entered a new era in its service in Korea.

The New Division

Although the last veterans of the campaigns of 1950 did not leave Korea until the autumn of 1951, the 1st Marine Division had started a process of transformation in April 1951 that did not depend solely on Communist bullets. Headquarters Marine Corps now sent out replacement drafts not just to fill holes in the ranks from casualties, but also to allow the surviving veterans of longest service to return to new assignments in the United States or for release from active duty. The 9th Replacement Draft reached Korea in early June, bringing 2,608 Marine officers and enlisted men to the division and 55 officers and 334 men to the 1st Marine Aircraft Wing. New naval personnel for both Marine organizations totaled six officers and 66 sailors, mostly medical personnel. The incoming Marines had a departing counterpart, the 3d Rotation Draft, composed of 62 Marine officers, 1,196 enlisted men, and 73 sailors; the draft included 103 convalescing wounded. The 10th Replacement Draft arrived late in June, adding 74 more officers and 1,946 men to the division and 12 officers and 335 men to the aircraft wing. One naval officer and 107 sailors joined the division and wing.

Nevertheless, Thomas thought that the manpower planners had cut their estimates too close and requested that subsequent drafts be increased by a 1,000 officers and men. Despite the personnel demands of forming the new 3d Marine Brigade at Camp Pendleton, Fleet Marine Force,

Department of Defense Photo (USMC) A156022

The command team of the 1st Marine Division stands outside a briefing tent at the division headquarters. Pictured from left are BGen William J. Whaling, assistant division commander, MajGen Gerald C. Thomas, division commander, and Col Victor H. Krulak, division chief of staff.

Pacific, honored Thomas' request. The 11th Replacement Draft (14 July 1951) brought 3,436 Marines and 230 naval personnel to the division and 344 Marines to the aircraft wing, accompanied by 22 sailors. Nevertheless, the division remained short of majors, company grade artillery officers, and officers and enlisted men in almost every technical specialization, especially communications and logistics.

General Thomas had no complaint about the quality of the Marines he had inherited from O. P. Smith or those sent to him by Lieutenant General Lemuel C. Shepherd, Jr., Commanding General, Fleet Marine Force, Pacific. The senior officers and company commanders were proven World War II veterans, and the lieutenants were an elite of Naval Academy graduates, NROTC graduates, and officer candidate school products that more than matched the company grade officers of World War II. The enlisted Marines were a solid mix of career noncommissioned officers and eager enlistees. Thomas recognized that the division he now commanded was "in splendid shape" and prepared to fight and win in terrain and weather "never designed for polite warfare." He wrote retired Major General Merritt A. Edson that the 1st Marine Division was "the best damn division that ever wore an American uniform."

Thomas went ahead with plans, coordinated with Shepherd, to form his own team as the division staff and to appoint new regimental commanders. Thomas arranged for Brigadier General William J. Whaling, an old friend who had been Thomas' alter ego on Guadalcanal, to become the assistant division commander on 20 May. Whaling became his eyes and ears on tactical issues with his superb knowledge of men, weapons, and fieldcraft. Colonel Snedeker remained chief of staff until he gave way on 23 May to Colonel Francis M. McAlister, whose command of the 1st Marines was cut short by wounds on the 18th. Since he had come to Korea in 1950 as the division G-4, McAlister rotated home, to be replaced temporarily by Colonel Richard G. Weede, who had taken Colonel Bowser's place as G-3 on 8 May. Shepherd and Thomas had someone else in mind for the division chief of staff's post, Colonel Victor H. Krulak, the G-3 of Fleet Marine Force, Pacific, and a trusted colleague of both generals through World War II and the postwar years. Krulak became division chief of staff on 29 June with a special charge to begin experiments with the Marine Corps' one operational helicopter squadron.

The rest of the division staff brought enough character and expertise to their jobs to please Thomas. The G-1, Colonel Wesley M. Platt, had spent World War II as a Japanese prisoner of war; his leadership among the prisoners had won him the admiration of his peers and great influence on the staff. Thomas' two G-2s, Lieutenant Colonels Joseph P. Sayers and James H. Tinsley, did a workmanlike job. Like Weede, Colonel Bruce T. Hemphill, and Lieutenant Colonel Gordon D. Gayle served as Thomas' G-3s under the close and critical scrutiny of Colonels Krulak and Weede, who also served as division chief of staff. Colonels Frank P. Hager and Custis Burton, Jr., performed the thankless task of G-4 until they rotated to the command of the 5th and 11th Marines, respectively, although Burton later returned as chief of staff in February 1952 to replace Weede.

The commanders of the infantry regiments were all tested veterans of the Fleet Marine Force, and their styles varied more than their competence. After Francis McAlister fell in a precision Chinese mortar barrage on his command group, Thomas assigned his regiment to the legendary Colonel Wilburt S. "Big Foot" Brown, an artillery officer sent out to command the 11th Marines. The irrepressible "Big Foot" Brown (whose homeric 14F sized-feet required special supply arrangements, including the airdrop of field brogans into the wilds of Nicaragua) took command only to issue an order to withdraw. As the 1st Marines trooped by his jeep on the way south to the No Name Line, the files of men broke into chicken-like cackles, showing that their "red leg" colonel looked "yellow" to them. Colonel Brown soon showed that their judgment was a short round by a mile. When the veteran of World War I, Nicaragua, and World War II surrendered command to another World War I veteran, Colonel Thomas A. Wornham, Brown had won the affection of the 1st Marines, "Chesty Puller's Own," a very tough bunch of Marines to impress.

The other two infantry regiments went to colonels of high ability. Colonel Richard W. Hayward brought intelligence and personal elegance (too much some of his troops thought) to the 5th Marines, succeeded by Weede on 7 August whose energy and force exceeded Hayward's. Almond liked them both, a dubious recommendation. Weede then turned over command to Frank Hager on 19 November. The 7th Marines bid farewell to Colonel Homer L. Litzenberg, Jr., on 15 April and welcomed Colonel Herman Nickerson, Jr., no stranger to the Korean War since he had been in the combat zone since the Inchon landing as the senior Marine liaison officer from Fleet Marine Force, Pacific to Far East Command and Eighth Army. No less professional than the other regimental commanders, Nickerson brought a driving, no-nonsense command style to the 7th Marines that made the regiment, in Thomas' opinion, the best in the division. Nickerson appreciated the contributions of his two executive officers, the incomparable Lieutenant Colonel Raymond G. Davis, Jr., and Lieutenant Colonel John J. Wermuth. Promoted to colonel, Wermuth assumed regimental command on 20 September when Nickerson's extended overseas tour ended.

The high level of competence at the regimental level did not drop off in the division's separate battalions. With an officer corps created by service in six divisions in the Pacific War, the Marine Corps

Courtesy of the Naval Institute Press

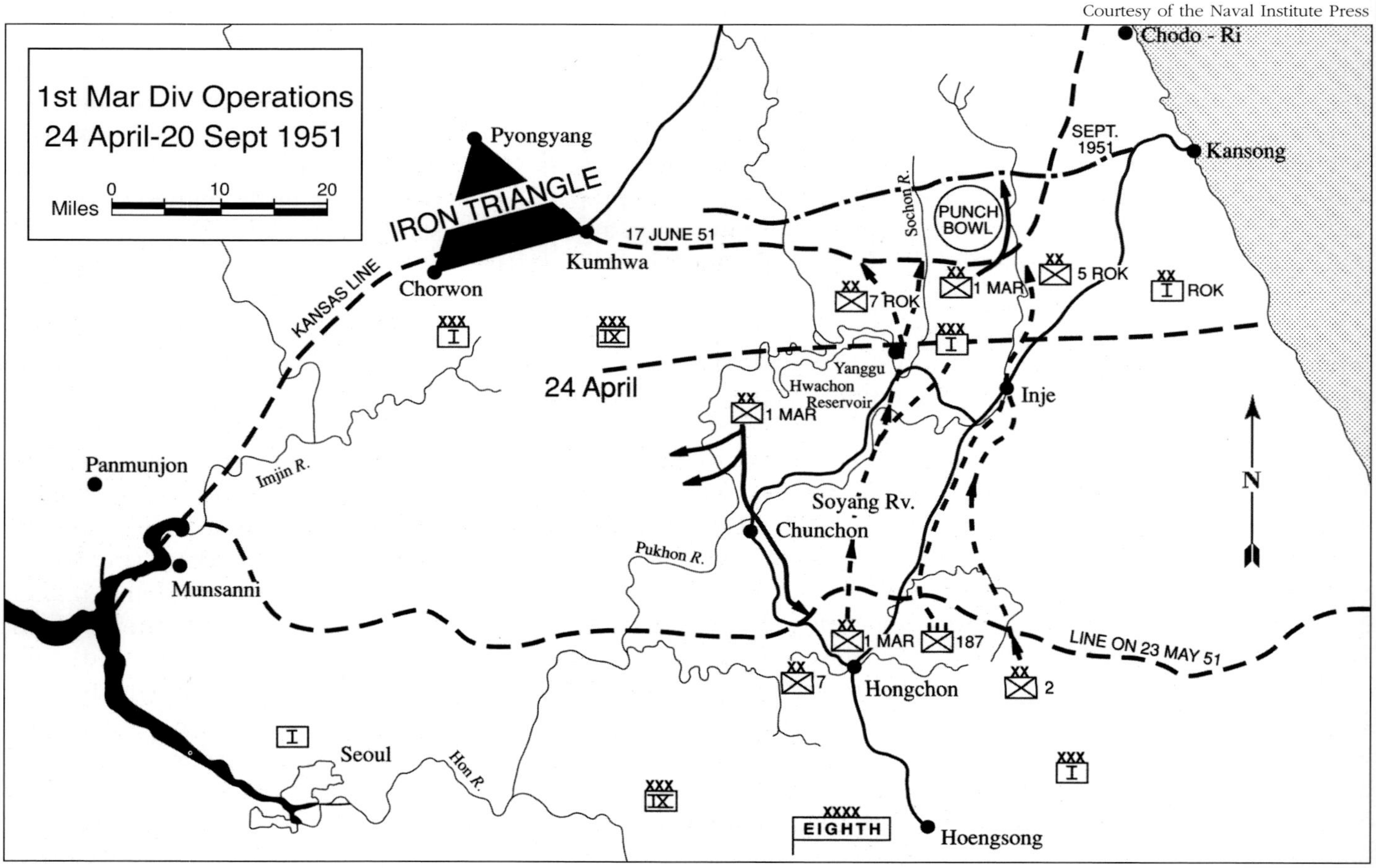

Clarence Jackson Davis: Every Marine

In the soft spring of his senior year (June 1950) at Hillsboro High School, Nashville, Tennessee, Clarence Jackson Davis, called "Jack" by his family and friends, discovered several reasons to join the Marine Corps Reserve. Going to war was not one of them. Jack Davis planned instead to go to Vanderbilt University, where his older brother Vince was already a sophomore and a keen midshipman in the Naval ROTC unit. Jack admired Vince, but he did not fancy himself a naval aviator like his older brother. On the other hand, the local Marine Corps Reserve infantry company had some openings, and he and some high school football and baseball teammates liked the idea—advanced by some sweet-talking Marine sergeants—of keeping their baseball team together under the sponsorship of the Marine Corps. Marine training seemed little more than another athletic challenge; the recruiters mentioned that weekly drill often included a basketball game. The new recruits had no active duty requirement and the two weeks summer training sounded like a Boy Scout camp with guns. Besides, the recruiters insisted, participating in reserve training made a young man draft-proof from the U.S. Army.

Jack also saw his enlistment as a potential way to help pay for his college education and, if all went well, become a Marine officer. More farsighted than many of his friends and counseled continuously by Vince, Jack had already talked with the Marine major on the Vanderbilt NROTC staff, who advised him that enlisted service would strengthen his chances for selection for the next summer's Platoon Leaders Class. Serving as an enlisted officer-candidate in the Marine Corps Reserve seemed a less-demanding way of helping pay for his education than attempting to win a football grant-in-aid playing for the hapless Commodores. The Davis brothers calculated that their military commitments would allow them to attend school without facing a demanding working schedule, a financial relief they could stretch by living at home to study and avoiding the temptations of campus social life.

Jack enlisted in the Marine Corps Reserve at the age of 17 in March 1950. If not quite a youthful lark, his decision did not seem very momentous, but a combination of good planning, reasonable sense, and anticipated adventure. He would try the life of a Marine, and he would be paid to camp out and play sports for the Marine Corps. His life after high school, however, "did not work out as planned." One night in June 1950, after graduation, Jack watched a newsreel at a local movie theater and learned about some distant war in Korea. His first reaction: "I was thrilled I was not there."

Lieutenant General Clifton B. Cates, Commandant of the Marine Corps and a fellow Tennessean, made sure that Jack Davis learned the true meaning of being a volunteer Marine from the Volunteer State. After the Fourth of July the ground units of the Marine Corps Reserve received a warning order that they would soon be mobilized, and on 20 July the Commandant made it official: Marine reservists in ground units would be called to active duty "for the duration." Certainly most (if not all) of them would go to Korea as part of the 1st Marine Division. After the confusion of in-processing, medical examinations, and additional issues of 782 gear, the Nashville Marines of Company C, 14th Battalion, Marine Corps Reserve, marched off to war from their reserve center at Shelby Park down Broad Street until they reached the 11th Street railroad siding off old Union Depot. Curious spectators watched the young men march off, their parade dominated by the blaring of high school bands at the front and rear of the column. With M-1s and dressed in green utilities the Marines did not look like their predecessors of the 11th Tennessee Volunteer Infantry, Confederate States of America, but the spirit of those young Nashvillians, equally perplexed and determined in 1861, stiffened the backs of their 1950 successors.

Whatever his expectations, Private Clarence Jackson Davis, U.S. Marine Corps Reserve, did not go off to war untrained—as did his Confederate ancestors and goodly part of the U.S. Eighth Army in 1950. At Camp Pendleton, his "station of initial assignment," Jack's Company C received the triage of personnel mobilization: true veterans of active duty were culled out for immediate assignment to the Fleet Marine Force, probably directly to the 1st Marine Division; Marine reservists whose drills and summer camp more or less approximated boot camp went on to eight more weeks of predeployment field training and physical conditioning; and the untrained true "boots" like Jack Davis went south to the Marine Corps Recruit Depot, San Diego, to begin their life as real Marines.

The temporary mission of the Marine Corps Recruit

Depot was not turning young men into Marines but into day laborers and stevedores. All hands spent much of every day mobilizing the depot for the expected waves of new recruits; the reservists set up bunks, hauled mattresses out of storage, and carried footlockers into the reopened barracks. Every night for two weeks the reservists went to the North Island docks to load ammunition and mount-out boxes for the 1st Marine Division. Not until the division cleared the harbor for Japan did the reservists start their formal boot camp schedule, which now seemed like welcome relief from the role of slave laborers. Jack Davis found boot camp no special challenge.

The follow-on field training, eight weeks and mandatory for every Marine regardless of assignment, proved more and less fun. Jack enjoyed the long days on the ranges of Camp Elliott. He qualified with ease with the M-1 and fired the entire range of individual and crew-served weapons found in a Marine infantry battalion. Jack liked them all except the M-1 carbine, which riflemen did not carry anyway. The last phase of the training focused on cold weather, mountain training at Pickel Meadow, which was neither meadow-like nor cold. Temperatures that December reached the 70s, and the Marines trudged around in the mud and slush in their layers of cold-weather clothing issue, all of them perilously close to heat exhaustion and dehydration. Apparently the Marine Corps had found a way to train the replacements for Korea's winters and summers at the same time.

In January 1951, Jack Davis and his comrades of the 6th Replacement Draft boarded the Army transport *Randolph* for the trip to Korea. The *Randolph* had come directly from Pusan where it had disembarked a part of the 1st Marine Division, recently evacuated from Hungnam. When Jack and his messmates reached their troop compartment, they found the canvas bunks decorated with messages from the survivors of the Chosin Reservoir campaign. Jack did not find the words of wisdom very comforting. A few were clever and humorous, but most of the collected battlefield folk wisdom struck Jack as sad, depressing, bewildered, stunned, and even suicidal. As the messages from the veterans attested, the war had almost been lost, had almost been won, and then almost lost again. As General Douglas MacArthur had just reported to the Joint Chiefs of Staff, Jack Davis and his comrades now faced the Chinese army and an entirely new war.

---

could usually find the right combination of leadership and technical skill to give the separate battalions a strong commander. Among the commanders General Thomas inherited in these battalions were Marines whose accomplishments had made them legends: Lieutenant Colonel John H. Partridge of the 1st Engineers, who had opened the route out of the Chosin Reservoir; Major Lloyd O. Williams, the master marksman, who commanded the 1st Ordnance Battalion; and Lieutenant Colonel Henry J. "Jim" Crowe of the 1st Shore Party Battalion, a heroic battalion commander at Tarawa and Saipan as well as another team shooter.

As the commander of a great division, whose management he shared with an able staff, General Thomas could focus on his relations with Generals Van Fleet and Almond. Although Van Fleet took a practical, unpretentious approach to commanding Eighth Army, General Almond had become not one whit more subdued by the twin blows of surrendering the independent status of X Corps and the abrupt removal of his patron, General of the Army Douglas MacArthur. Almond retained the imperiousness and elegant field life style that characterized many Army generals of his generation and, especially, his two models, MacArthur and General Mark W. Clark, his army commander in World War II. Surrounded by staff officers from central casting—albeit very talented—Almond favored high-fashion field uniforms, opulent vans and messes, and imperial gestures worthy of Napoleon himself, including the haphazard awarding of medals. Almond's airy disregard for time-space factors and enemy capabilities, as well as his habit of ignoring the chain-of-command, had driven General Smith into tight-lipped rebellion. Thomas had dealt with some difficult Marine generals, but Almond would be a challenge.

Thomas first made sure that no one would mistake him for the Almond model of a modern major general. He truly preferred the look of the Old Corps of World War II, not a U.S. Army that had remade itself in the image of its flamboyant armor and airborne generals like Walton H. Walker and Ridgway. Thomas wore a uniform that was strictly issue from his battered utility cap and standard helmet to his canvas leggings and worn, brown field shoes. Instead of the "generals" version of the Colt .45-caliber automatic worn the Army-way with a fancy leather belt, he carried an issue pistol in a black shoulder-holster. None of his regular field wear—jacket, sweaters, shirts, and trousers—would be mistaken for tailors' work. His only personal affectation (a very useful one at that) was his old Haitian *coco-macaque* walking stick, whose only local counterpart was carried by General Shepherd. On special occasions General Thomas and his regimental commanders might sport white scarves and the division staff red scarves, but the idea was Van Fleet's, who thought the scarves would show the troops that senior officers were not allergic to frontline visits.

Still a part of IX Corps—one wonders now about the urgency in the change of command—the 1st Marine Division disengaged from the Chinese *39th* and *40th Armies* and fell back unmolested 15 miles to the No Name Line, a belt of prepared positions dug by Korean laborers and Army engineers. The 1st Marines, reinforced by a battalion of the 7th Marines, protected the bridges and passes while the rest of the division withdrew in good order over the Pukhan and Soyang Rivers, both in flood from rain and melting snow. By 29 April the division had put the rivers at its back and filed into the No Name Line positions with the 5th Marines, the 1st Regiment of the Korean Marine Corps (KMC), and the 7th Marines on line from west to east. The 1st Marines went into division reserve. Colonel Bowser thought that the retrograde movement—which he and Thomas did not think necessary—proved that the division had lost nothing of its 1950 ability to march and fight superior numbers of Chinese troops without prohibitive losses. Meanwhile General Van Fleet met with his corps commanders on 30 April to discuss Eighth Army's next move: an active defense of the No Name Line and maximum readiness to meet another Chinese-North Korean offensive, predicted for mid-May by Van Fleet's intelligence staff. In the reorganization of the front, the 1st Marine Division would rejoin X Corps, effective 1 May.

General Thomas first had to fight off General Almond before he could focus on killing Chinese. The two generals met every day for three days (1-3 May), and Thomas emerged victorious in establishing new ground rules for the Marines' dealing with X Corps. Thomas had already told his staff that it would take a hard-line with "suggestions" from any corps staff member; one of his assistant operations officers tested the guidance by telling his Army counterparts from Almond's headquarters that he could "go to hell" for giving orders in Almond's name.

The 1st Marines regimental headquarters occupies a tent-camp along the road to Hongchon, just south of the No Name Line. United Nations Command air superiority allowed such administrative arrangements.

Department of Defense Photo (USMC) A8728

Thomas took a disgruntled Almond head on. He could be charming in his own way—he pointed out his own Virginian and Confederate roots to Virginia Military Institute "Old Grad" Almond—but he insisted that Almond stop bypassing the chain-of-command or allowing his staff to run roughshod over the proper channels in the 1st Marine Division. Almond insisted he was an active corps commander. (Meddlesome was the word the Marines chose.) Thomas told him that he was an active division commander and that he intended to make as many visits to regiments and battalions as Almond made. Thomas added that he would "execute any order proper for a soldier to receive." Almond pressed Thomas to go on, but Thomas now remained silent as if expecting a further elaboration of policy. Almond said he would make many visits. "Is that all right with you?" Almond continued. More silence. Almond went on: "But I can assure you that I will never issue an order affecting one of your units except through you." Now Thomas responded: "On that basis . . . you are always welcome." The two generals remained true to their word.

Thomas tested the era of good feeling with X Corps almost immediately and to positive effect. Van Fleet had the notion that each division should establish a battalion-sized outpost from which it could patrol northwards to make contact with the Chinese. For the 1st Marine Division the best place to establish such a base—which Thomas and Bowser thought was a miserable idea—was south of Chunchon but north of the critical

Morae Kagae Pass, the only route of escape to the No Name Line. The position would be outside the artillery fan of the 11th Marines, and close air support alone (now complicated by Air Force scheduling practices) would be no substitute. Thomas argued with Van Fleet and Almond that he would perform the mission, but that he should dictate the size of the force and its rules of engagement—and disengagement. When Thomas put his "patrol base" in place on 5-7 May, he sent the entire 7th Marines (artillery and tank reinforced) north toward Chunchon, and he added the 1st KMC Regiment to Nickerson's task force. In addition, he had the 5th Marines put a screening company in front of each of its frontline battalions, but kept the companies well within artillery support.

National Archives Photo (USMC) 127-N-A8898

A Marine patrol secures a hill and moves forward. Its mission was to maintain contact with the enemy, warn of an impending attack, and delay its progress as much as possible.

Thomas continued to press X Corps for more artillery since Van Fleet's intelligence staff insisted that the next Chinese offensive might focus on the 1st Marine Division. Thomas' own ground and aerial patrols found ample evidence of Chinese troop movements between the Pukhan River and the No Name Line. The commanding general had also heard Van Fleet insist that no Eighth Army unit, a company or larger, should be isolated and cut-off; Van Fleet told his generals that night withdrawals and counterattacks should be abandoned as operational options. He also insisted that every division artillery groupment (the 11th Marines for Thomas) should use its daily allowance of shells (the "Van Fleet unit of fire" or five times the normal allotment of shells) to fire upon suspected enemy concentrations and transportation routes. Thomas persuaded Almond that the 1st Marine Division could not meet Van Fleet's expectations without some Army help, and Almond committed two X Corps general support artillery battalions to reinforce the fires of the 11th Marines. Thomas also negotiated a shortening of his frontage since he had to put two battalions of the 1st Marines into the line to replace the 7th Marines, which left only one infantry battalion as division reserve. Even though he had come to conclude that the Chinese were massing to the east instead of to his front, Thomas had no intention of allowing any part of the 7th Marines to be cut off between the Morae Kagae Pass and the No Name Line. He approved a Nickerson-Davis plan to garrison the pass with a reinforced battalion (less one rifle company) and simply announced the change to Almond, who did not object to the fait accompli. Thomas also planned to extract the 7th Marines from its advanced position as soon as he though he could justify such an action to Almond. He anticipated that trouble would develop along the boundary of the 1st Marines and the U.S. 2d Infantry Division, not to his front. The 7th Marines would be his new division reserve, ready to attack to the northeast. The plan proved to be prescient.

Offensive and Counteroffensive

Changing their operational style of nighttime infiltration attacks, characterized by surprise and the limited use of artillery, the Chinese *Ninth* and *Third Army Groups*, augmented by the North Korean *II* and *V Corps*, opened the Fifth Offensive (Second Phase). On the morning of 16 May 1951, the offensive began with a Soviet-style preparatory artillery bombardment. Frustrated in his April offensive, Peng Dehuai decided that the limited road network and sharp, rugged mountains of eastern Korea offered a better area of operations for a renewed offensive. Van Fleet and his corps commanders would find it more difficult to shift reinforcements against the shoulders

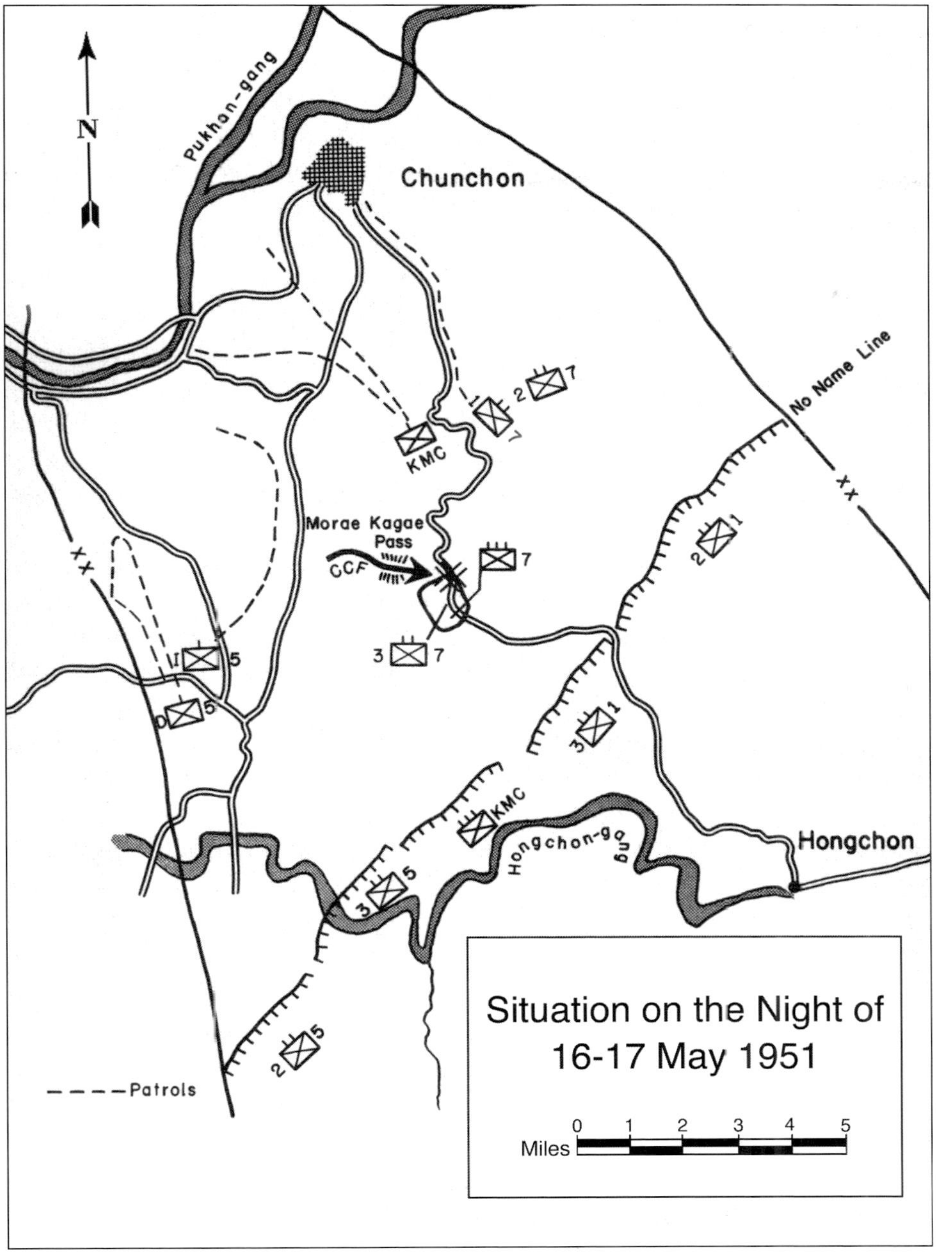

Situation on the Night of 16-17 May 1951

of any breakthrough, and the steep mountains made it difficult to mass United Nations artillery fire. The broken, forested terrain would provide welcome cover and concealment from United Nations Command air strikes. The weight of the Chinese offensive (27 divisions with three artillery divisions in support) fell on (from west to east) the U.S. 2d Infantry Division, the 5th ROK Division, and the 7th ROK Division of X Corps with additional attacks upon the neighboring 9th ROK Division of the Republic's III Corps. The Chinese did not ignore the western-most division of X Corps, the 1st Marine Division, which would be pinned in its part of the No Name Line by attacks from the Chinese *60th Army*. The minimal operational goal was to destroy one or more U.N. divisions; a major victory would be the fragmentation of either X Corps or the ROK III Corps and a return to a campaign of movement that would dislodge the Eighth Army from the Taebaek Mountains to the Han River valley.

Eventually described by X Corps as the battle of the Soyang River, 15-21 May 1951, the Chinese offensive overran various parts of the frontline positions and the patrol bases of the hard-luck 2d Infantry Division and the three ROK divisions to its right. Despite some dogged defensive action by American and South Korean soldiers, the Chinese advanced 30 miles, forcing the three ROK divisions to the south and threatening to roll-up the right flank of the 2d Division, which lost the better part of the 38th Infantry and its attached Dutch battalion in slowing the Chinese attack. General Almond decided he needed to insure that the western side of the Chinese salient was secure first; he requested reinforcements from Van Fleet, who sent the 187th Airborne Infantry Regiment and U.S. 3d Infantry Division to blocking positions behind X Corps. In the meantime, Almond wanted the 2d Infantry Division to refuse its right flank. Such a redeployment required the 1st Marine Division to extend its sector of the No Name Line to the east and to do so while in contact with the enemy.

On the first day of the Chinese offensive, General Thomas visited Almond's command post at Hoengsong and saw the crisis build in X Corps' eastern sectors. Thomas and Almond discussed what situations the 1st Marine Division might face, but Thomas would make no commitments until he was sure he could withdraw the 7th Marines (Reinforced) from the ill-conceived "patrol base" north of the No Name Line. Closer to the anticipated Chinese attack, Colonel Nickerson reinforced the outpost at Morae Kagae Pass, bringing the defenders to battalion strength and including the regimental headquarters and a tank platoon. Having just joined the 7th Marines—his

Department of Defense Photo (USMC) A155692

Marines of the 3d Battalion, 7th Marines, gather the bodies of the Chinese 179th Division, *which attacked the regiment's patrol base at Morae Kagae Pass on the night of 16 May. As part of the Chinese Fifth Offensive (Second Phase), the attack did not pin the 1st Marine Division to the No Name Line, which allowed its redeployment to the east to aid the U.S. 2d Infantry Division.*

regiment in World War II—Second Lieutenant Earl F. Roth wondered who had placed the regiment so far from the rest of the division. He reached the Morae Kagae Pass and the 7th Marines rear defenses only after a long and lonely jeep ride across an empty countryside, but he felt eyes watching him from every hill. When he later saw the piles of Chinese bodies at the pass, he remembered similar scenes from Peleliu. On the evening of 16 May, a Chinese regiment attacked the pass in force and lost 112 dead and 82 captured before breaking off the action. Nickerson's force lost two tanks, seven dead, and 19 wounded. The attack gave Thomas plenty of reason to pull back Nickerson's entire regiment, ordered that night with Almond's approval. Colonel Frank T. Mildren, X Corps' operations officer, correctly assumed that the Chinese wanted no part of the 1st Marine Division: "The Marines [are] just wrapped up in their usual ball." Mildren's assessment did not accurately picture the 1st Marine Division's skillful redeployment to

Gen Edward M. Almond Collection, U.S. Army Military History Institute

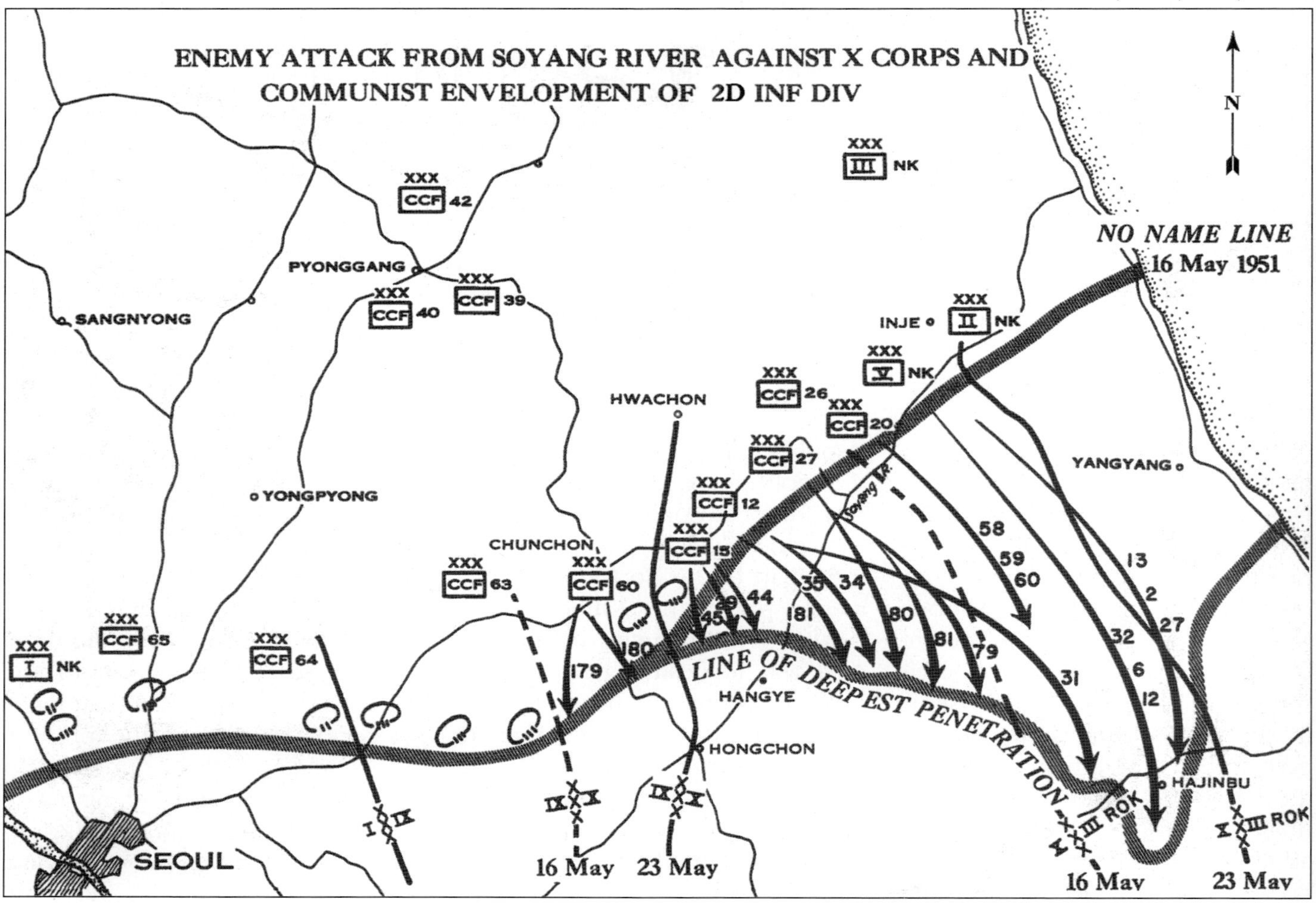

release the U.S. 9th Infantry Regiment for a new mission, saving the rest of its parent division.

After artillery and air strikes insured that the Chinese *60th Army* marched east to the sound of somebody else's guns, Thomas ordered the 1st Marines to shift right and take the 9th Infantry's positions while the 7th Marines marched back to the No Name Line and took over the 1st Marines sector. In the meantime, two battalions of the 5th Marines moved eastwards behind the No Name Line to refuse the division right flank north of the crucial road junction of Hongchon. The 7th Marines and the 1st KMC Regiment slipped to the left to take over part of the 5th Marines' former sector. General Thomas reported at 1730 on 18 May to Almond that the realignment had been accomplished, but that he also wanted more corps artillery ready to fire defensive fires along his thinly-manned front. He requested and received more aerial reconnaissance from the Cessna light patrol aircraft (L-19 "Bird Dogs") assigned to X Corps. Thomas had already improved his defensive posture by placing the 1st Marines in positions almost four miles south of the original No Name Line. The only contact occurred on 20 May when elements of the Chinese *44th Division* marched unawares into the defenses of the 3d Battalion, 5th Marines, and left behind almost 170 dead and prisoners when the Marines shattered the lead regiment with their battalion weapons, artillery, and air strikes. The 1st Marine Division awaited more orders. It did not expect Almond to remain on the defensive since X Corps now had fresh troops and the two Chinese army groups had placed themselves inside a vulnerable salient.

Department of Defense Photo (USMC) A8875

MajGen Gerald C. Thomas checks the frontline situation map with the 5th Marines' commanding officer, Col Richard W. Hayward, center, and the regiment's operations officer, Maj Robert E. Baldwin.

Although he had won the respect of Almond and his staff in his first month of division command, General Thomas had no intention of becoming a compliant subordinate commander when he thought Army generals paid too little attention to tactical realities. Thomas and Almond conferred twice on 19 May and again on 20 May at the 1st Marine Division command post. The issue was a counteroffensive order from Van Fleet to I and IX Corps, a movement that began on 20 May for the 7th Infantry Division, the IX Corps element on Thomas' left flank. Almond wanted the 1st KMC Regiment to advance beyond the No Name Line to conform to IX Corps' advance, but Thomas "expressed reluctance" to send a regiment on an axis of advance that took it away from the rest of the division and opened a gap in the division's defensive alignment. Thomas won a concession from Almond immediately: he could make his own arrangements to secure X Corps' left flank and coordinate the movement directly with IX Corps.

As Peng Dehuai acknowledged, the collapse of the Fifth Offensive (Second Phase) gave United Nations Command an unprecedented opportunity to mount a counteroffensive of potential strategic consequences. Even though his army group commanders protested his withdrawal orders, Peng called off the offensive on the afternoon of 21 May and issued orders that the eastern armies should withdraw during the night of 23-24 May to a defensive line that would run from the Imjin River to Hwachon to Kansong, roughly the line occupied by United Nations Command when the Fifth Offensive began in April. Five Chinese armies and three North Korean corps would defend the line.

Prodded by General Ridgway, who flew to Korea to inject some of his special bellicosity into a flagging Eighth Army, Van Fleet had stolen half a march on his Chinese counterpart by ordering I and IX Corps to start a drive to the Topeka Line, a phase line on the ground about halfway to the contemplated

Department of Defense Photo (USMC) A8867

A .30-caliber machine gun team and a Marine with a Browning Automatic Rifle occupy recently abandoned enemy foxholes, using them for cover while pursuing Chinese and North Korean forces.

Chinese defenses. Van Fleet and the corps commanders of I and IX Corps, however, could not create much urgency in their divisions. Neither Ridgway nor Van Fleet thought I and IX Corps had seized the moment. They were thus pleasantly surprised when Almond, who seized moments whether they were there or not, proposed that he could shift to the offensive as soon as noon on 23 May as long as he retained control of the 187th Airborne and the 3d Infantry Division and gained the use of the brand-new 8th ROK Division as well. Instead of driving almost directly north like I and IX Corps, however, Almond planned to use his South Korean divisions to keep the Chinese and North Koreans engaged at the forward edges of the salient. His American divisions would cut across the base of the salient from southwest to northeast, roughly on an axis that followed Route 24 through Chaun-ni—Inje—Kansong where X Corps would link up with ROK I Corps. The counteroffensive, supported by massive aerial bombardment and Van Fleet-directed artillery barrages of World War I profligacy, would bag the survivors of the Chinese *Third* and *Ninth Army Groups*. Van Fleet approved Almond's plan, and X Corps issued its attack order on 21 May.

For the 1st Marine Division the development of two Eighth Army counteroffensives with different

Elements of the 2d and 3d Battalions, 5th Marines, hit the dirt after taking heavy enemy mortar and machine gun fire from Chinese forces occupying Hill 1051. Air and artillery forced the enemy to retire northward and the regiment secured the commanding high ground.

National Archives Photo (USA) 111-SC368657

axis of advance provided General Thomas and his staff with new challenges. A shift of corps boundaries as far east as a line Hongchon-Hwachon Reservoir helped some, but not much. As he himself later admitted, Almond had once again promised too much, too soon in the way of decisive action. For once he had not underestimated the enemy; the Chinese army groups in his zone of action were indeed wounded, but not as seriously as Eighth Army estimated. (United Nations Command estimated total Chinese casualties for the Fifth Offensive at 180,000, but the Chinese put their own losses at half this total.) The difficulty was the time and effort necessary to get the offensive moving with task forces drawn from the 2d Infantry Division, the 3d Infantry Division, the 187th Airborne, and the divisional and corps tank battalions. The result was that the attacks at the tip of the salient jumped off on time (mid-23 May), but the big drive across the base of the salient did not begin until 24 May and the serious, organized advance up Route 24 did not begin until the next day. In the meantime the Chinese, attacked 12 hours before they began their own withdrawal, fought back sluggishly as they moved up their withdrawal schedule, a euphemism for—in some cases—a Chinese 'bug out."

The result of the gelatinous attack by Major General Clark L. Ruffner's 2d Division and its attached task forces was that the 1st Marine Division advance, also dutifully begun on 23 May, had to conform to the Army units on its right. The Marine advance of 24-31 May developed into a two-axis attack with the 1st Marines and the 1st KMC Regiment moving through the hills south of Soyang, crossing the river on 28 May, and reaching the heights above the Hwachon Reservoir on 31 May. The 5th and 7th Marines started the march north in a column of regiments, but the 7th Marines pulled ahead while the 5th Marines took the commanding heights of Kari-san (Hill 1051). The 7th Marines then turned northeast away from Route 24 to take the shortest route to the town of Yanggu, just east of the eastern end of the Hwachon Reservoir. The 7th Marines assaulted and captured the Yanggu heights, but watched the Chinese flee through the open zone of the tardy U.S. 2d Infantry Division. The

Gen Edward M. Almond Collection, U.S. Army Military History Institute

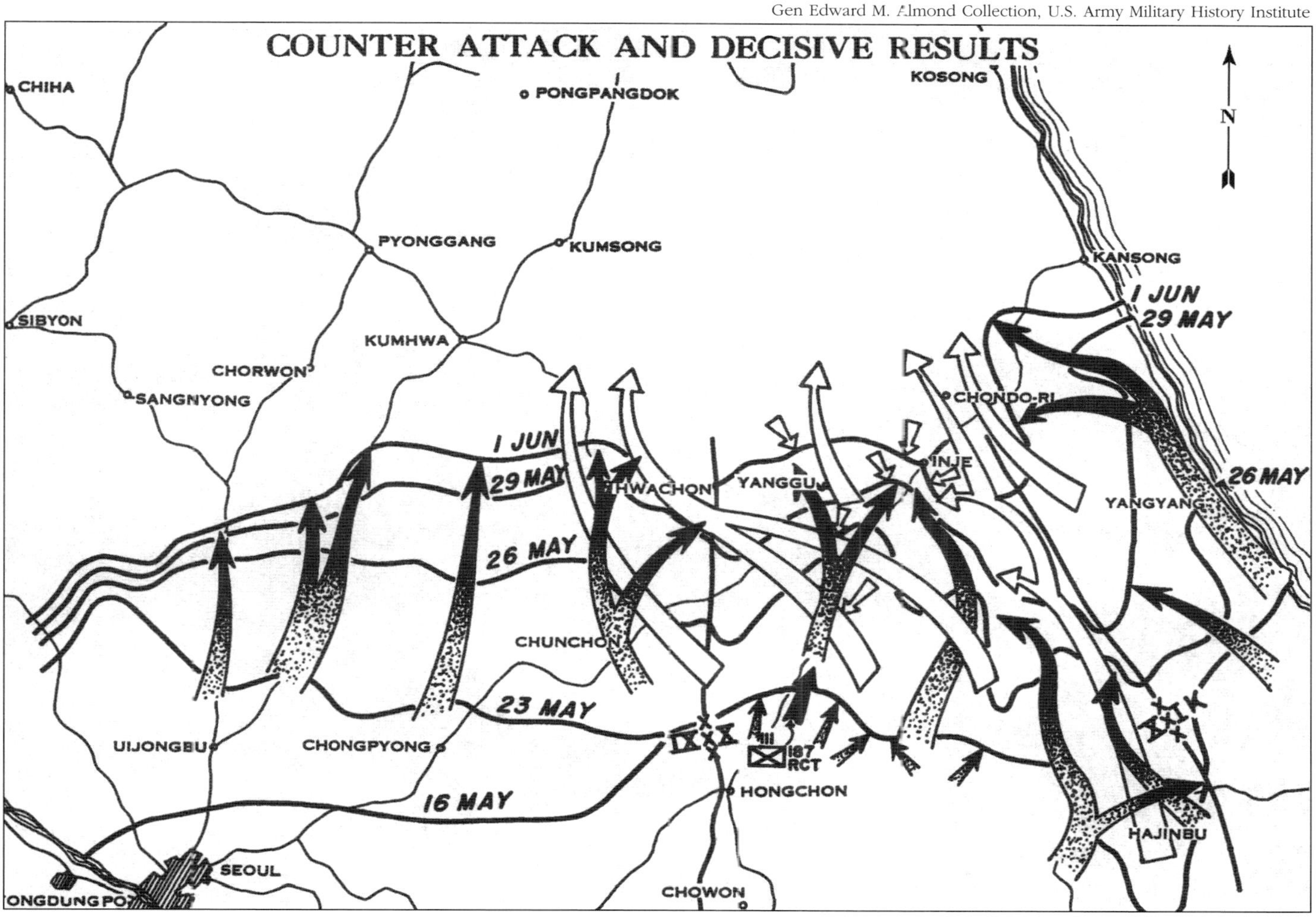

National Archives Photo (USN) 80-G-429689

After securing Kari-san (Hill 1051), Marines search two Chinese prisoners of war for weapons and documents.

5th Marines shifted right to the hills east of the road to Yanggu and drew abreast of the 7th Marines on 29-30 May. The next day all of Thomas' four regiments occupied their portion of Line Topeka.

For the rifle companies at the head of each pursuing battalion, the war did not look much like the reassuring blue arrows on an acetate-covered 1:25,000 map. The last two weeks of May 1951 proved to be hot and very dry during the day, but cold and wet at night as unusual spring rains kept the hills slick and the valleys a slough. Water to drink, however, proved harder to find than water for discomfort. Few Marines were willing to chance the ground water or local streams, but potable water seemed to take second place to ammunition in the columns of Korean bearers. In an era when "water discipline" made "exces-

Hugging the crest of a ridgeline, the 7th Marines prepare to "pour hot lead" into enemy positions as a prelude to a general assault by other Marine units.

National Archives Photo (USMC) 127-N-A8888

National Archives Photo (USMC) 127-N-A155279

Marines of Battery L, 4th Battalion, 11th Marines, prepare their 155mm howitzer for a fire mission in support of Marine units around Yanggu. Observed, adjusted artillery fire provided the Marines with essential support against North Korean defensive positions.

sive" drinking a sin in the Marine Corps, dehydration stalked the struggling columns of laden troops. The columns not only fought groups of Chinese, but marched through the Eighth Army's dying fields of February and May, passing the bodies of soldiers from the 2d Infantry Division.

Despite the profligate use of artillery and air strikes, the Marine rifle companies found their share of close combat in the last week of May. Moving along a steep hillside only by hanging from the trunks of shattered trees, Second Lieutenant Earl Roth's platoon saw enemy mortar rounds fly by them and explode in the gully below. Roth suppressed a strong urge to reach out and catch a mortar round as it passed by, a vestige of his football playing days at the University of Maryland. Although the firefights seldom involved even a whole company, they were a world of war for the engaged Marines. One platoon of Company C, 1st Battalion, 5th Marines, stormed a Chinese ambush position only to have the defenders charge right back at them in the most intimate of meeting engagements, a brawl won by the Marines with grenades, clubbed rifles, bayonets, and fists. Urged on by the company's Beowulf, Second Lieutenant Paul N. "Pete" McCloskey, the Marines left few survivors, but their post-fury victory celebration was cut short by a deluge of 120mm mortar rounds pre-registered on top of the position. The company lost is commanding officer and other Marines in the swift reversal of fortune.

The six days of offensive action in the last week of May 1951 demonstrated to friend and foe alike that the 1st Marine Division remained a fearsome killing machine. Using artillery and tank fire, supplemented with battalion mortars and machine guns, the infantry regiments methodically took their objectives with minimal casualties and no operational crises. The Marines continued to run into scattered battalion-sized remnants of Chinese divisions, none willing to hold any position against the deluge of fire poured upon them. On 28 May, however, the Marines started to discover organized, company-sized defensive positions manned by North Koreans and ringed with mines. By 31 May, the day of the division's heaviest casualties for the week (126 killed and wounded), the Chinese had disappeared from the battlefield. During the week the division intelligence staff estimated that the division had inflicted 10,000 casualties; what it knew for certain was that the regiments had counted 1,870 enemy bodies and taken 593 prisoners. The 1st Marine Division's losses for the entire month of May were 83 killed in action or died of wounds and 731 wounded. The "exchange ratio" against an enemy still considered dangerous and willing to fight was about as good as could be expected.

The week of divisional attack brought its share of surprises. The enemy provided some of them. The Chinese, aided by the slow advance of the 2d Infantry Division, refused to wait for their entrappers and poured out of the salient after the first attacks of 23 May. Chinese soldiers from five different divisions of the *Third Army Group* crossed the path of the Marines on their way to rally points at Yanggu and Hwachon; the chaotic pattern of the Chinese withdrawal meant that enemy bands might appear at any time from the east and south, which lead Almond and Thomas to confer daily on flank security issues. When the Marines met the better-armed and trained infantry of the North Korean *12th Division*, they also came under fire from Soviet-made artillery and mortars. The Chinese withdrawal, however, gave Marine artillery a field day; between 10 May and 7 June the 1st

Private First Class Whitt L. Moreland

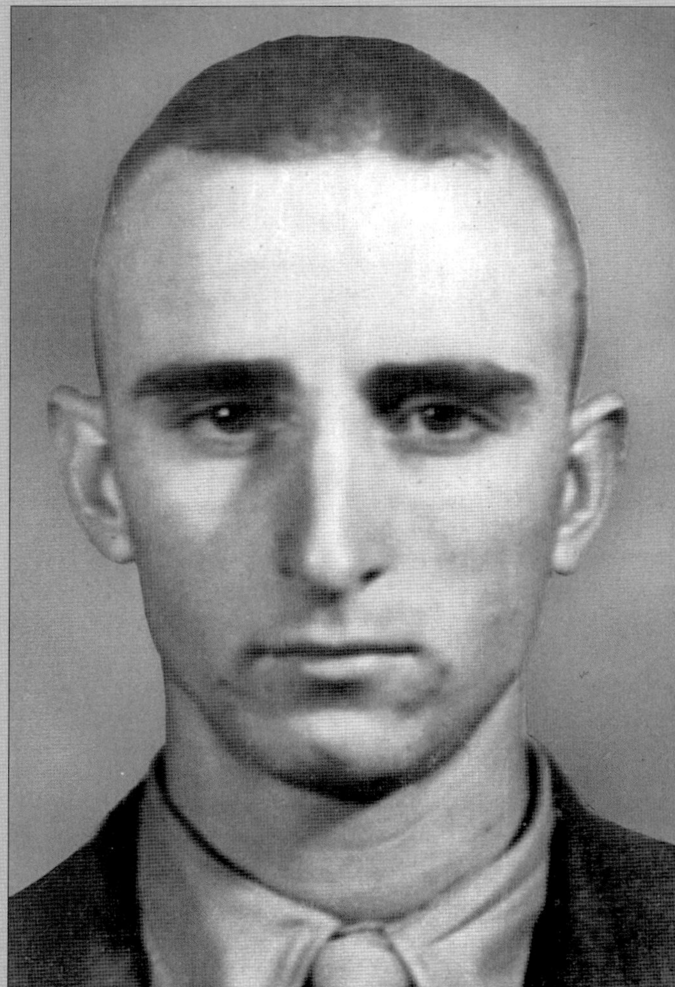

Department of Defense Photo (USMC) A46966

Born in 1930 in Waco, Texas, he enlisted in the Marine Corps in 1948, following graduation from Junction City High School. After serving out his active duty, he reverted to Reserve status. In November 1950, he was recalled to active duty and sent to Korea. While serving as an intelligence scout while attached to Company C, 1st Battalion, 5th Marines, he was killed at Kwagchi-dong on 29 May 1951. The citation of his posthumous Medal of Honor award reads, in part:

> Voluntarily accompanying a rifle platoon in a daring assault against a strongly defended enemy hill position, Private First Class Moreland delivered accurate rifle fire on the hostile emplacement and thereby aided materially in seizing the objective. After the position had been secured, he unhesitatingly led a party forward to neutralize an enemy bunker which he had observed some 400 meters beyond and, moving boldly through a fire swept area, almost reached the hostile emplacement when the enemy launched a volley of hand grenades on his group. Quick to act despite the personal danger involved, he kicked several of the grenades off the ridgeline where they exploded harmlessly and, while attempting to kick away another, slipped and fell near the deadly missile. Aware that the sputtering grenade would explode before he could regain his feet and dispose of it, he shouted a warning to his comrades, covered the missile with his body and absorbed the full blast of the explosion, but in saving his companions from possible injury or death, was mortally wounded.—Captain John C. Chapin, USMCR (Ret)

Marine Division artillery fired 13,157 tons of shells, second only to the 2d Infantry Division (15,307 tons). The corps artillery group kept pace, especially since its fires supported the South Korean divisions. All X Corps divisions surrendered their trucks to keep X Corps guns supplied with shells. By the end of May ammunition shortages had become an operational concern. The artillery expenditures and the stiffening Communist defenses suggested that the "happy time" of X Corps exploitation operations had come to an end.

General Thomas had every reason to be proud of his division, for Generals Ridgway, Van Fleet, and Almond all visited his command post and praised the division's performance. General Shepherd and his senior staff visited the division on 28-29 May, and Shepherd added his congratulations not only for the operational successes, but also for the good relations with the Army. And Almond went out of his way to tell the other generals how much he valued Thomas' wise counsel. (Thomas was not so sure that Almond listened to anyone, but at least the corps commander now observed the chain-of-command.) Finding another way to celebrate a victory, the commander of the 3d Battalion, 1st Marines, sent one of his lieutenants out on a desperate mission: find ice somewhere around Yanggu to cool the battalion's beer ration. Second Lieutenant Harold Arutunian's patrol returned with ice—stolen from the body bags of an Army graves registration unit. For at least one week the 1st Marine Division had fought by its book, and it suffered negligible casualties by pounding every objective with preparatory air strikes and artillery concentrations. For once Almond did not exaggerate when, on 31 May, he characterized the Marines as "fatigued, but spirits high."

North to the Kansas Line

Perched in their most recent foxholes above the Hwachon Reservoir and the blackened ruins of Yanggu—so flattened and incinerated that only the charred bank vault gave the town a skyline—the forward infantry battalions of the 1st Marine Division could see only more sharp hills to the north, rising ever higher into the smoky dusk of the last day of May. They did not know that conferences elsewhere were already deciding their fate in the month ahead.

The Chinese Fifth Offensive and its crushing defeat had opened the way for a second "entirely new war," but not one that made any of the belligerents very happy. The Communist coalition shared a common problem with United Nations Command: was there any operational option that offered advantage worthy of the risks of strategic escalation? What if the Soviet air forces, for example, mounted attacks on the American airbases in Korea? What if the Soviet navy mounted submarine or maritime aviation attacks upon the U.N. naval forces that roamed the east and west seas with impunity?

Relatively certain that Joseph Stalin would not authorize any attacks that might bring American retaliation on Soviet bases in the Far East, Mao Zedong sought some employment of the Chinese Communist Forces that would eventually destroy the will of the United Nations and the Republic of Korea to continue the war. On 27 May 1951, Mao Zedong opened discussions on strategy with his principal commanders in Korea. Within a week Mao conferred with eight senior officers of the CCF, especially First Deputy Commander Deng Hua and Chief of Staff Xie Fang. Mao told his field commanders that the CCF would con-

National Archives Photo (USA) 111-SC382822

The Joint Chiefs of Staff directed the Commander in Chief, U.N. Command, Gen Matthew B. Ridgway, USA, to continue the offensive but only by advancing to the Wyoming-Kansas Line, a phase line in the mountains north of the 38th Parallel. The underlying objective of these operations was designed to support a negotiated end to hostilities.

duct *niupitang* attritional warfare of position until United Nations Command casualties reached unbearable proportions.

Mao's use of the word *niupitang* could not have been more apt since *niupitang* was a delicious but very sticky candy from his native Hunan Province, an irresistible sweet that took a very long time to eat and usually made a mess. The *niupitang* strategy would work well with a policy of *biantan bianda* or simultaneous negotiating and fighting. Within two months Mao replaced three of the four army group commanders, retaining only Yang Dezhi, a modern commander and a protégé of Deng and Xie, and promoting him to second deputy commander and de facto director of operations for the Communist field forces. Peng Dehuai remained the titular commander of the CCF, but Deng Hua, Xie Fang, and Yang Dezhi directed the new strategy, "On the Protracted War in Korea," announced in July 1951.

The other Communist co-belligerents reacted to *niupitang* in much different ways, but neither the Soviets nor the North Koreans had much leverage on Mao Zedong. If they wanted the war to continue—and they did—they depended upon the Chinese army to bear the brunt of the fighting. Now that the war had not produced a great Communist victory, Stalin (beset with political problems at home) saw no reason to go beyond his commitment of Soviet air defense forces to "MiG Alley" along the Korean-Manchurian border and to rearm the Chinese army. The Soviets, in fact, saw truce negotiations as a way to increase their influence in the United Nations as well as to buy time to rebuild and rearm the Chinese forces. The North Koreans—represented by the pestiferous Kim Il Sung—wanted only more war and no talks, unless a truce brought an end to American air strikes. Kim and his inner circle agreed, however, that the 38th Parallel should be restored as an international border and that all foreign troops (including the Chinese) should leave Korea—after the South Korean army had been fatally weakened and the North Korean People's Army restored to fighting trim and much-enlarged. Kim ordered his generals to fight to the death for every rocky foot of North Korean soil.

The process of political-strategic reassessment, which had begun with the Chinese intervention in November 1950, blossomed in May 1951 like the cherry-blossoms in Washington, D.C. and Korean coastal resort town of Chinhae. Hints of peace negotiations sprout-

Close Air Support Controversy

By the spring of 1951, the question of close air support for United Nations Command ground forces had become a serious inter-service controversy that pitted the Marine Corps and some of the senior commanders of the Eighth Army against the United States Air Force and General Matthew B. Ridgway, the United Nations and American theater commander. To some degree the controversy involved the employment of the 1st Marine Aircraft Wing and several lesser and often false issues, e.g. jets versus propeller aircraft, but the heart of the problem was simply that the Air Force did not want to perform the mission. It regarded close air support as a wasteful and dangerous misuse of offensive tactical air power. Marine Corps aviation and Navy carrier-based aviation regarded close air support as an essential contribution to the ground campaign. The victim in all this inter-service wrangling was the Eighth Army and the 1st Marine Division.

From the Air Force perspective, the close air support mission belonged at the bottom of its offensive air missions, although the leaders of the Army Air Forces as early as 1943 insisted that air power was the equal of ground combat power in the conduct of war. The same senior officers donned new uniforms in 1947, but did not drop their old ideas about close air support, despite the relatively effective use of ground-directed air strikes against the German army in 1944-1945. The Air Force position was rooted in negative experiences: the bombing and strafing of friendly troops; the extraordinary losses to ground fire in making front-line, low-level bombing runs; and the conviction that Army ground commanders knew nothing of fighter-bomber capabilities and would scream for close air support when artillery was a more rapid and appropriate response to their indirect fire support requirements. The guidance in effect for Air Force-Army close air support operations in 1950 was the "Joint Training Directive for Air-Ground Operations," an agreement only between Tactical Air Command and Army Field Forces, not the Service headquarters.

In theory and application in Korea in 1950 the doctrine of the "Joint Training Directive," which the Air Force embraced as authoritative, made close air support difficult for a ground command to obtain. Basically, the Air-Ground Operations System (AGOS) required that a ground commander request air support prior to an operation and be very specific about his needs. Requests had to be processed through an Army operations officer (G-3 Air) from regiment through field army and reviewed by an Air Force officer at each echelon of command (the air liaison officer) until the request reached the Joint Operations Center (JOC), run by an Air Force general, which would allocate the available air strikes. The request system insured that close air support strikes were not likely to be tactically relevant, but the air direction system the Air Force preferred also added to the problem. The definition of close air support was that air strikes should be coordinated with the fire and maneuver of the ground forces through the positive direction by a forward air controller (FAC) who was fully knowledgeable about the ground combat situation. There was no fundamental disagreement that a Tactical Air Control Party (TACP) with reliable air-ground communications (vehicle- or ground-mounted) should be available so the FAC could direct air strikes by sight, just like an artillery forward observer. The Air Force, however, did not want to use its own personnel for such missions, and it did not trust the Army to provide a competent FAC. The Air Force might provide an Air Liaison Party down to the regimental level to do air strike planning, but it was not going to send Air Force officers (presumably pilots) out to the front to direct air strikes. In some fairness, the Fifth Air Force did provide such Tactical Air Control Parties to the Eighth Army in 1950, and they were shot to pieces—radio-jeeps and people alike.

The Fifth Air Force in 1950 created an air strike direction system that depended on airborne air controllers, basically the World War II system. During the course of the fighting in 1950 the Fifth Air Force and Eighth Army committed people and equipment to form the 6147th Tactical Control Squadron, later expanded to wing status. The "Mosquitoes," as the forward air controllers (airborne) came to be known, did yeoman work throughout the war, directing air strikes from their two-seat, propeller-driven North American AT-6 "Texan" aircraft, a World War II pilot trainer. The "Mosquitoes" lacked nothing in courage and skill, but they were still hostages to the JOC system. Either the air strikes had to be preplanned or they had to be requested as a matter of dire

National Archives Photo (USMC) 127-N-A131261

Department of Defense Photo (USMC) A130146

emergency or diverted from other missions.

The Navy-Marine Corps system, developed for amphibious operations in World War II, offered a different approach. The Air Force tried to brand the system as driven by amphibious operations, which it was to some degree, but the system had proved itself in land campaigns on Saipan, Guam, Peleliu, Iwo Jima, the Philippines, and Okinawa. The Navy and the Marine Corps brought the same system to Korea, and it worked. It worked so well that Army generals, especially Major General Edward M. Almond, embraced it without reservation. His successor as commander of X Corps, Major General Clovis E. Byers, also became a convert, and it cost him his command. Other Army commanders at the division level envied the system and wondered why they could not receive adequate support, but they were too intimidated by Lieutenant General Matthew B. Ridgway to push the issue.

The Air Force consistently misrepresented the essence of the Navy-Marine Corps system. The naval services never challenged the important of air superiority or interdiction operations. The naval services simply argued—and placed in their own doctrine—that if close air support missions were to be flown at all, they should be rapid, responsive, appropriate, and effective. The 1st Marine Aircraft Wing might be best trained to perform such missions, but it was the system that counted, not the uniforms of the pilots or the type of planes they flew. The senior Marine ground commander did not command aviation units, as the Air Force charged. Either X Corps or 1st Marine Division did not command the 1st Marine Aircraft Wing.

The Navy-Marine Corps system accommodated planned requests, but its strength was its tactical flexibility. Each infantry battalion in the 1st Marine Division had a Tactical Air Control Party of two elements. One group served as the Air Liaison Party, part of the battalion operations staff. The other group was the Forward Air Control Party, an officer and communicators who could process requests for air support and direct air strikes from the ground, usually well forward with an infantry company. In practical terms, this system meant that each Marine infantry battalion had two Marine officers (naval aviators) as part of the battalion staff to insure that air strikes hit the enemy and did so soon enough to affect the tactical situation. The system worked, and the Marine Corps saw no reason to abandon it

As X Corps commander, General Almond liked the Navy-Marine Corps system, which he saw at close quarters during the Inchon-Seoul campaign and again during the withdrawal to the Hungnam enclave. In fact, he ordered his Army divisions to form their own TACPs or he arranged for the 1st Marine Aircraft Wing to send

TACPs to Army units (American and Korean) within his corps. The ability of the TACPs to direct strikes naturally drew most of the sorties flown in December 1950 by the Marine squadrons and the naval aviators flying from the decks of Task Force 77.

The operational conditions and requirements of 1950 made it appear that the 1st Marine Aircraft Wing had come to Korea to be General Almond's corps aviation component. In fact, X Corps functioned much like a modern Marine air-ground task force, even if Almond had no direct authority over any of his supporting tactical aviation squadrons. Fifth Air Force, however, thought this ad hoc arrangement should not continue. In early 1951, General Ridgway and Lieutenant General George E. Stratemeyer, Commander General, Far East Air Forces, insured that General Douglas MacArthur placed X Corps in the Eighth Army and the 1st Marine Aircraft Wing under the operational control of the Fifth Air Force and the Joint Operations Center. The Marines could perform their close air support magic for all of Eighth Army, not just X Corps. Ridgway, however, demanded that Fifth Air Force study the whole close air support question and find ways to make the JOC system more responsive to unplanned ground requests for air strikes.

While Fifth Air Force and Eighth Army both conducted reviews of the Air-Ground Operations System, the war went on. The 1st Marine Division returned to the fray in February 1951 without its usual customary air support, either in quality or quantity. Marine fighter-bomber squadrons (F4U Corsairs or F9F Panthers) flew missions for all of Eighth Army with results that depended entirely upon the ability of either the airborne "Mosquitoes" or ground spotters (if any) to identify the targets and communicate with the aircraft. In the meantime, Task Force 77 sailed north to attack Communist railroads and highways ("the bridges of Toko-ri"), and Air Force fighter-bombers of varying nationalities (predominately American or Australian) showed up to conduct missions for the 1st Marine Division with mixed results. Major General Oliver P. Smith asked Ridgway to use his influence with Fifth Air Force to give Smith operational control of just one Corsair squadron. Ridgway refused to raise the issue and breach the "single management" doctrine. "Smith, I'm sorry, but I don't command the Air Force!"

Even though Eighth Army and Fifth Air Force made serious efforts to establish all the personnel and communications elements of the AGOS request and direction organization, the Air Force's lack of interest and ability in close air support still discouraged ground commanders from making pre-attack requests. The system virtually guaranteed that emergency requests would be answered late, if at all. With their own TACPs at the battalion level, the Marines could and did short circuit the system by making emergency requests to an airborne Mosquito, who would then divert either outgoing or returning interdiction strikes to the Marines and release direction of the strikes to the forward air controllers. If the attacking aircraft happened to be flown by trained Marines, so much the better. The 1st Marine Division FACs, however, reported that in April 1951 the JOC had answered 95 percent of their requests, but only 40 percent of the missions were flown by adequate numbers of aircraft, properly armed, and arrived in time to make some difference in the battle. In the meantime losses of aircraft and pilots soared in the 1st Marine Aircraft Wing, in part because non-Marine controllers provided poor information about the terrain and enemy situation. In April 1951, the Marines lost 16 aircraft and 10 pilots (one captured, nine killed) to enemy ground fire.

Although X Corps received ample close air support during the Fifth Offensive (Second Phase), Almond still criticized the AGOS practices. Major General Gerald C. Thomas entered the fray when he learned that his division had received only two-thirds of its requested air strikes in late May. Only about half of the delivered sorties were effective, and almost all were over an hour or more late. The only concession Almond and Thomas received was the stationing of one mixed Corsair squadron from Marine Aircraft Group 12 at K-46 a primitive strip near Hoengsong, but the JOC (Kimpo Airfield) still had to approve the missions. With the AGOS still in place—albeit somewhat more efficient and flexible—the war against the *niupitang* Chinese and North Korean defenders would go on—and the 1st Marine Division would indeed get stuck.

---

ed everywhere—most planted by shadowy Soviet sowers in the worlds' capitals and at the United Nations. The Joint Chiefs of Staff kept General Matthew B. Ridgway informed on the flood of speculation and hope. Until the last week of May, however, Ridgway had no reason to link his sense of the strategic shifts underway with the continuing operations in Korea. Then Van Fleet, pressed by Almond, proposed a significant change in the exploitation campaign that followed the defeat of the Fifth Offensive. When the bulk of the Chinese forces had already escaped the bag between X Corps and ROK I Corps, Van Fleet proposed a series of amphibious envelopments up the east coast that would conclude with the creation of an enclave at Tongchon, still short of Wonsan, but well north of the "Iron Triangle," the central Korean network of transportation connections and mountain corridors bounded by Chorwon-Pyonggang-Kumhwa from west to east. Ridgway and Van Fleet agreed with Almond that control of the "Iron Triangle," even from Tongchon, would give either

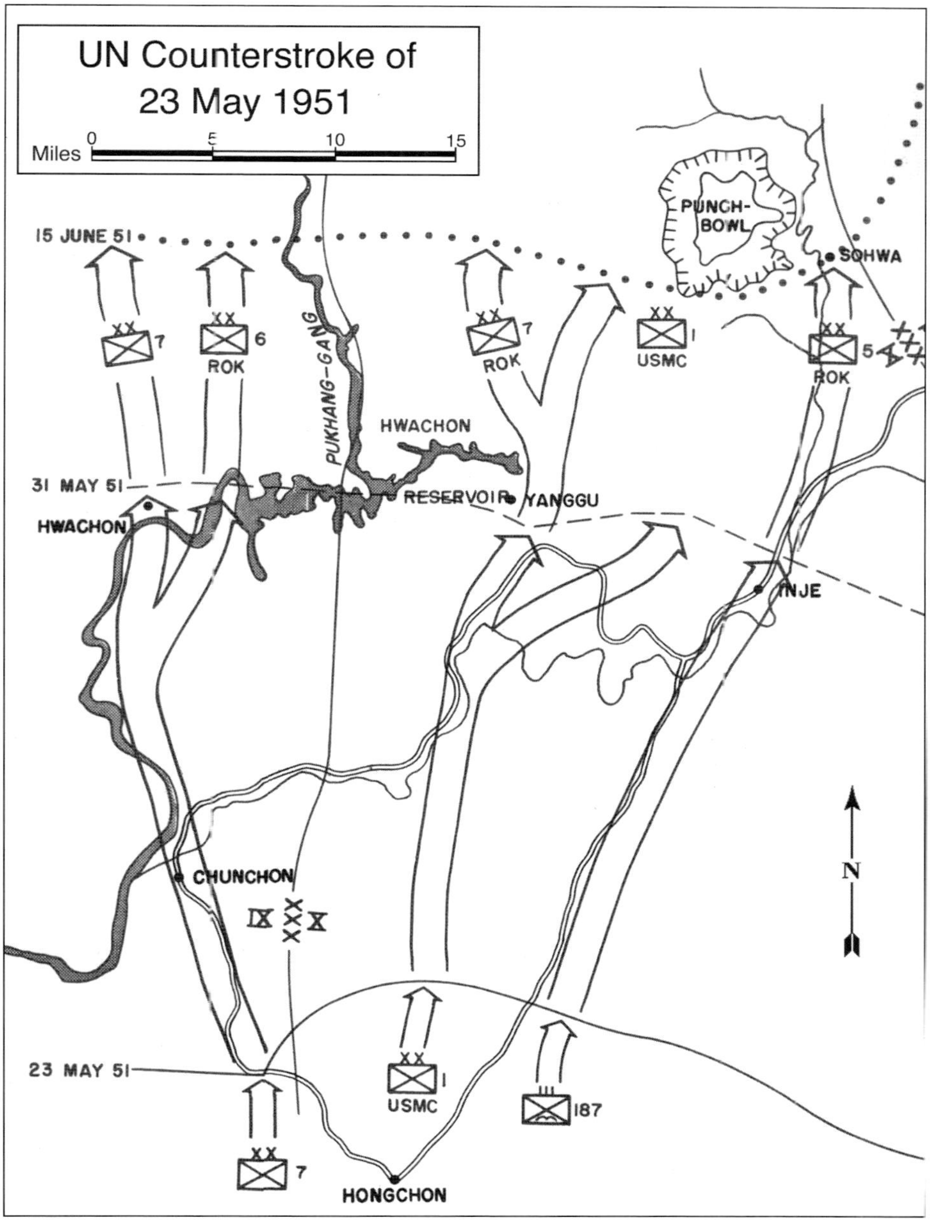

side an advantage in ending or continuing the war.

General Ridgway, however, did not agree that Van Fleet's proposed Operation Overwhelming could be mounted because of resistance in Washington and sheer operational feasibility. Even a modest shore-to-shore movement would require disengaging the 1st Marine Division and (probably) the 3d Infantry Division and transporting them to a port for embarkation. From Ridgway's perspective, time was of the essence, and the requirements of Overwhelming were too overwhelming with truce talks in the wind. Ridgway's greatest fear was that someone would give away the territorial gains already made in May and the additional ground he wanted to control in June after Operation Piledriver, a straight-ahead push by all four of Van Fleet's corps. The Eighth Army's goal would be the seizure and defense of a cross-peninsula line (Wyoming-Kansas) that would retake Kaesong, hold the mountain ranges and passes northwest of the Imjin River, secure at least part of the "Iron Triangle," and hold the mountains north of the Hwachon Reservoir all the way to the coast at Tongchon. Anticipating that a ceasefire would entail the creation of some sort of territorial buffer zone, Ridgway wanted to reach a line (Kansas) well north of the Wyoming Line, his non-negotiable position for ensuring the ground defense of the expanded Republic of Korea.

Two other considerations shaped Ridgway's thinking about the conduct of the war. Some of the general's critics and champions later suggested that he had become too interested in his personal goal of becoming Army chief of staff or faint-hearted at the prospect of excessive American casualties in Korea. Ridgway's ambition was well-known to his Army peers, but he realized that trying to please Washington was a fool's errand. Nor had Ridgway, notoriously ruthless in relieving non-fighters, suddenly become casualty-shy. He simply saw no purpose in risking lives in adventures that probably would not produce the promised results. Moreover, Ridgway had become convinced that air power could give him an offensive option to punish the Communist armies beyond bearing, his own high explosive, high altitude version of *niupitang*. Recent changes in the Air Force high command in the war zone placed very aggressive and persuasive air generals in Ridgway's inner council. General Otto P. Weyland, the Far East Air Forces director of operations since early in the war, became the commander, and Lieutenant General Frank F. Everest assumed command of the Fifth Air Force. Both Weyland and Everest, tactical aviation commanders in World War II, championed aerial interdiction as the most decisive way to use air power in a war like the Korean conflict. Both also insisted that the senior theater Air Force officer should have operational control of all aviation units with combat capability, including the carrier air groups of Task Force 77 and the 1st Marine Aircraft Wing. Ridgway's coolness to any amphibious operation and his warmth toward the Weyland-Everest interdiction campaign, Operation Strangle, would have critical effect on both the 1st Marine Division and the 1st Marine Aircraft Wing for the rest of the war.

For the 1st Marine Division the high-level discussions on the relative weight of the ground and air wars on bringing the Communists to terms had no immediate effect since X Corps' mission remained unchanged: seize the Kansas Line. Nevertheless, the Marines needed at least a brief pause, which Almond would not grant the division on 31 May. He ordered the attacks to the north to continue, and on 1 June the 5th and 7th Marines dutifully pushed on—and went nowhere. With the 11th Marines short of artillery shells, air arrangements uncertain, and the 7th Marines in need of a break, Thomas did not push Hayward and Nickerson until they all had a chance to think about the new attacks.

A rifle platoon of the 5th Marines does some "ridge-running" as it moves to an assault position in the broken terrain south of the Punchbowl.

Department of Defense Photo (USMC) A8868

The terrain alone appeared formidable. A long, high ridge of the Taebaek Mountain chain dominated the 1st Marine Division's zone of action. The ridge was known as Taeu-san and Taeam-san for its two highest peaks, 1,179 meters for the northern most Taeu-san and 1,316 meters for the southern Taeam-san. Taeu-san/Taeam-san were bordered on the west by the Sochon River, which ran into the Hwachon Reservoir just past Yanggu. The Marines also inherited the southern part of another parallel ridge to the west, but dominated by the Taeu-san/Taeaum-san hill mass to the east, which meant that any force attacking directly north of Yanggu would receive fire from its right flank. The terrain situation to the east was even more daunting. The division's eastern boundary ran generally along the Soyang; the distance between the two rivers was 15 miles, more or less, and the entire zone stretched another four miles to the west. The Taeu-san/Taeam-san ridge, however, did not uniformly run northwards. The whole ridge complex had once been a volcano, and the crater created a depression in the mountain, the "Punchbowl," open at its eastern edge where the Soyang River had eroded a hole in the crater wall. The southern lip of the crater, remained, however, as a formidable extension at a right angle east of the main ridgeline, which provided a transverse position for fire directly along all the lower ridges to the south. In a sense the whole Taeu-san/Taeam-san complex looked like a giant leaf with its thin tip to the south and its thicker (higher) base to the north; many veins (ridges) ran west and east from the central spine, some creating separate compartments to cross, others echeloned southwest or southeast and running uphill to the central stem, dominated by a series of separate peaks. The terrain is a defensive commander's dream.

The 1st Division attack on the Taeu-san/Taeam-san massif and the ridge adjoining it to the west began on 2 June and ended almost three weeks later with all four infantry regiments very bloodied, but unbowed and with three of them on or beyond the original Kansas Line. The advance uphill for about eight miles took the measure of the entire division as had no fight since the Chosin Reservoir campaign. For the 1st Marines, its losses exceeded those of December 1950, and the entire division suffered 183 dead and 1,973 wounded. Both Generals Van Fleet and Almond questioned General Thomas about his division's losses. Especially aggravated about the poor quality of his close air support and the Eighth Army's timorous treatment of the Fifth Air Force, Thomas felt no need to apologize to Van Fleet for completing his mission. "Well, General, you told us to take the Kansas Line, and we took it for you. I'm sure we paid for what we got, but we got what we paid for." Thomas wrote his family that his Marines were the best he had ever seen, and "Big Foot" Brown told his friends that the feats of his regiment had to be seen to be believed. Thomas fully appreciated the North Koreans' tenacity: "They fight like Japs!"

The battle began in earnest on 2 June with the 1st Marines and 5th Marines attacking abreast, each with two battalions, with the 7th Marines and 1st KMC Regiment in reserve. The 1st Marines took one intermediate objective (a small hill called X-Ray) and entered the lower ridges of the hill mass north of the Hwachon Reservoir and west of the Sochon River. The fight was an uphill slog all the way. General Thomas learned that the press identified the engaged Americans as "GIs." He wrote home: "That is us, and we are not GIs." Expert at the coordination of supporting arms, Colonel Brown used artillery to the limit of its effectiveness, but each objective ultimately had to be taken by Marine infantry, savaged with grenades and mortar shells as they literally crawled uphill. Brown had to pay special attention to his left flank, his boundary with the 7th ROK Division, and he often had to deploy one battalion against flanking attacks while the other two continued their forward crawl. As Brown recalled: "it was the toughest fighting I have ever seen."

Over the same period (2-10 June), the 5th Marines faced an even greater ordeal, especially its 1st Battalion, whose zone included seven ascending peaks before it could reach the crest at Hill 1316 (Taeam-san). The 2d Battalion's zone was somewhat less demanding, and Lieutenant Colonel Glen E. Martin more deft in paving the way with air strikes and artillery, and the weight of the North Korean defense faced the 1st Battalion anyway. It took two long days for the 1st Battalion to capture Hills 610, 680, and 692, a distance of about 2,000 yards. In addition to the stubborn defense by the North Koreans, the three rifle companies survived one "friendly" artillery barrage and one errant air strike as well as tank fire from the valley below to the west that, while welcome in bunker-busting, did not seem especially concerned about the position of friendly troops. At one point an inexperienced company commander allowed his men to be trapped in a North Korean mortar barrage, and another company, run off its objective by Corsair-dropped napalm, found itself the target of Communist artillery. McCloskey's platoon in Company C started the two-day ordeal with sergeants as squad leaders and ended it with a corporal and two private first classes in command; almost every platoon commander suffered at least minor wounds. In the meantime, the 2d Battalion had advanced almost 5,000 yards along the eastern edge of the ridge, but its movement did not put it on terrain that menaced the North Koreans on Taeam-san.

National Archives Photo (USMC) 127-N-A9304

The look on the Marine's face tells the whole story. Having just engaged in a savage firefight with enemy forces, these Marines continue their grueling upward climb.

At this point, General Thomas decided he needed to bring his two uncommitted regiments into the battle since the burden of close combat in May-June 1951 had fallen disproportionately upon the 5th Marines. (Nine members of the regiment received Navy Crosses for heroism, the 7th Marines four, and the 1st Marine two.) The 7th Marines, after all of two days rest, went into the attack on the right of the 1st Marines, which allowed Brown to slide left to guard his loose connection with the 7th ROK Division. Nickerson's regiment also inherited the highest and most heavily defended ridgelines that ran eastwards to Taeu-san (Hill 1179) and the western rim of the Punchbowl. Thomas put the 1st

Private First Class Jack Davis: Combat Marine

After a short stop at Kobe, Japan, the Marines of the 6th Replacement Draft joined the 1st Marine Division in late January 1951. In the process of retraining and reorganizing, the division was conducting counter-guerrilla operations around Pohang, an east coast port within the Republic of Korea. Jack Davis (pictured on the left in the first row) was assigned to Company G, 3d Battalion, 1st Marines, somewhere near Andung, northwest of Pohang. Known within the regiment as "Bloody George" Company, Jack's new comrades were little more experienced than he was with the exception of a handful of officers, noncommissioned officers, and privates. The original Company G had landed across Blue Beach at Inchon and fought in the liberation of Yongdung-po and Seoul. It earned its nickname during the Chosin Reservoir campaign. The company first met the Chinese at Majon-ni and lost nine dead and 15 wounded (including attachments) when its truck convoy was ambushed. Filled with wide-eyed reservists from the 1st and 2d Replacement Drafts, the company started north toward the Yalu with a full complement of seven officers and 224 enlisted men. It also had a new company commander, Captain Carl L. Sitter, a World War II combat veteran. He had replaced the first company commander, reassigned to Quantico, Virginia, as an instructor at The Basic School.

In just about one month, the second Company G practically disappeared, lost to battlefield deaths, wounds, and frostbite. The company fought its way into Hagaru-ri on 29 November as the spearhead of Task Force Drysdale, taking 48 casualties from the gauntlet of fire the Chinese created for the convoy of tanks and vehicles. At Hagaru-ri the company tried to retake East Hill, but faced too many Chinese with too few Marines. The dwindling ranks of Company G, nevertheless, held the shortened perimeter and took 60 more casualties.

Captain Sitter received the Medal of Honor, and 10 other company Marines were awarded decorations for valor, including a Silver Star for the first sergeant, Master Sergeant Rocco A. Zullo. Within 10 days of battle, the company lost all but 87 officers and men, and fully a third of these "originals" had been wounded and returned to duty. Jack Davis had not yet experienced the physical and emotional ravages of combat when he joined Company G, but he could appreciate having even a handful of veterans around to stiffen the third Company G.

Assigned to the 3d Platoon as a BAR-man, Jack soon learned that Sergeant Robert W. "Blackie" Jones, new to the company but a World War II veteran, had strong opinions about weapons. Sergeant Jones liked the Browning Automatic Rifle, and he had a way of finding additional BARs for his squad. Sometime in February, between Operations Ripper and Killer, as the Eighth Army ground its way back toward the 38th Parallel, Colonel Lewis B. "Chesty" Puller inspected Company G. Puller found Jones' squad armed in an unusual manner. "How many BARs are there in a Marine rifle squad?" he asked the squad sergeant. "Three, sir!" Jones smartly responded. "How many BARs do you have in your first squad, sergeant?" Puller continued. "Six, sir!" Puller then asked: "How did you get these weapons?" Jones responded with even more snap in his voice: "We liberated them from the Army, sir!" Puller grunted his approval and went on without further comment. Jack, standing next to Johnson in the ranks, almost laughed at the spirited exchange.

Sergeant Jones also demonstrated quick thinking under fire. Sitting on a rice paddy dike somewhere between Wonju and Hoengsong in late February, Jack's squad watched spouts of cold, muddy water rise from the paddy less than a foot beyond their outstretched boots. "Blackie" did a back-gainer off the dike to a lower-level paddy and screamed at his men to take cover. The Chinese burp-gunner faded into the woods without molestation. Jack also learned the value of water from Sergeant Jones and soon carried two canteens, the only man in the squad to do so. He checked his water sources carefully, especially after he found a rotting horse upstream in one clear, bubbling brook, and used halazone tablets liberally. His health remained good despite his constant fatigue and unrelenting diet of C-rations. Nevertheless, he lost weight and seemed to shrink within his parka and field equipment.

The new Company G received another opportunity to add to its "bloody" reputation in the first days of the Chinese Fifth Offensive, April 1951. Upon the collapse of the 6th ROK Division on the division's left flank, Major General Oliver P. Smith sent the 1st Marines west of the Pukhan River to seize the critical hills that dominated the river valley and the only road by which he could extract the 5th and 7th Marines from the Hwachon Reservoir sector to the north. Hill 902, a 3,000-foot tower, became the 3d Battalion's objective; the hill dominated a road into the Pukhan River valley and a critical bridge on the road south to Chunchon and the No Name Line where the division was to establish a new defensible position. The battalion beat the Chinese to the peak (aided by Marine trucks) and moved down three parallel western ridges where Lieutenant Colonel Virgil W. Banning, the battalion commander, expected to meet the Chinese on their belated climb toward the peak of Hill 902. Banning placed one rifle company on each of the three entrant ridges and arranged his supporting arms into the evening of 23 April. Around 2000 the Chinese attacked, first striking Company G, the most advanced (by plan) and defender of the center ridgeline.

Huddled behind barriers of rocks—no foxholes could be dug here—Company G threw back a Chinese regiment with the assistance of Companies H and I, which fired across its flanks and sent reinforcements. Marine and Army howitzers and mortars showered the attackers with shellfire. Rallied by Technical Sergeant Harold E. Wilson, an Alabama reservist, the center platoon of Company G barely held. Jack Davis' platoon held its hillside position along the southern slope of George Ridge; the platoon suffered two or three killed and several more wounded, popping up to spray the Chinese in between the artillery barrages. Jack made it through the night unscathed, but the next morning, as the battalion backed away from the Chinese under an umbrella of close air support strikes, Jack fell victim to some unfriendly "friendly fire." As he and some other Marines struggled down the steep eastern slope of Hill 902 with stretchers loaded with dead and wounded Marines, two Marine F4U Corsairs strafed the column, showered the cowering infantrymen with ricocheting bullets and rock fragments. Although he took no life-threatening hits, Jack found himself a bleeding, lacerated, and thoroughly enraged member of the "Society of Walking Wounded" and headed for the battalion aid station for treatment. Rested, fed, and patched up, he returned to the company on the No Name Line.

Although he avoided telling his parents about his combat experience, Jack wrote his brother Vince that he should forget about leaving Vanderbilt and joining the Marine Corps to fight in Korea. If Vince became an officer, his chances of surviving would not be as good as an enlisted man's since the Marines expected all officers to lead from the front. Artillery officers—to which Vince aspired—had no greater chances of survival since they all had to serve as forward observers with rifle companies before assigned to the comparative safety of a firing battery. "Stick to the Navy for my sake as well as the folks. I'll do both our shares of the dodging." Jack's war had just begun.

KMC Regiment into the 5th Marines hard-earned foothold below Hills 1122, 1216, and 1316 (Taeam-san). Hayward's regiment (with the exhausted 1st Battalion in reserve) moved into an expanded sector east of the Taeu-san/Taeam-san massif and started to work its way north toward the southern lip of the Punchbowl. Thomas did not pressure Hayward to move aggressively since such an advance would have put the 5th Marines in a salient below an L-shaped hill mass still occupied by much of the *12th NKPA Division*. Before the 5th Marines could press forward to its share of the Kansas Line, the South Korean Marines would have to take Taeam-san.

National Archives Photo (USMC) 127-N-A155066

The natural beauty of this quiet scene in North Korea means little to these Marines as they rest during a lull in the struggle for the Punchbowl. In the wide, calm valley before them, each green field may hide a Communist gun position, each tree an enemy sniper.

For five days (5-10 June), the 1st KMC Regiment repeatedly assaulted the Hill 1122-1218-1316 complex but, despite maneuvering to the right and left of the peaks, the Korean Marines made no progress and lost over 500 men without taking even one objective. Anytime the Marines gained a foothold, a North Korean counterattack threw the Marines back. Neither side took prisoners; one South Korean assault discovered 10 bound ROK Marines executed with neat headshots. In desperation, Colonel Kim Suk Bum, the Korean Marine regimental commander, decided to abandon the American way-of-war and ordered a three-battalion unsupported night attack on Hill 1122, the most exposed North Korean position. Advancing by slow infiltration, the South Korean Marines fell on the Communists at 0200 with complete surprise and ran the defenders off to Hill 1216. With a solid hold on at least a part of the crest, the Korean Marine regiment held its ground while its American advisers called in artillery and air strikes on Hills 1216 and 1316. The North Koreans soon fell back to the north to Taeu-san to avoid being cut off by the American Marines now advancing steadily on both their flanks.

Service with the 1st KMC Regiment came as a surprise to some Marine officers. Assigned against his wishes to the 1st Shore Party Battalion, Second Lieutenant David J. Hytrek, a former private first class in the 5th Marines in 1950, wanted an infantry assignment to avenge the deaths of his comrades who had already fallen in Korea. Instead a crusty master sergeant serving as a personnel officer assigned many of the former enlisted men of the 7th Basic Class to combat service support battalions. "Let the college boys get killed in this war," he growled. Hytrek, however, had barely arrived at his new unit when he received orders to report to the Korean Marines as a liaison officer. General Thomas wanted experienced lieutenants sent to assist the Koreans, so David Hytrek found plenty of war in the battles fought by the Korean Marines around the Punchbowl.

To the west the 1st and 7th Marines fought from one hill to the next hill with consistent but costly success. The 1st Marines reached a line of hills identified as the Brown Line, a more defensible position than the original Kansas Line, which ran through the Sochon River valley to the regiment's rear. The 1st Marines started the regimental advance on 6 June and completed its mission on 14 June. The experience of the 2d Battalion represents the regimental ordeal. After two days of modest advances, the battalion, with the 1st Battalion on its left flank, ran into a very stubborn and skilled North Korean force on Hill 676. The attack stalled, in part because a heavy mortar concentration fell short and inflicted 40 casualties, including the battalion commander, Lieutenant Colonel Robert K. McClelland. On 10 June, the battalion sent two companies against the eastern face of the hill since it could then take advantage of supporting tank fire from the valley below. Many of the North Korean bunkers, however, were sited to protect them from tank guns, 75mm recoilless rifles, and 3.5-inch rocket launchers. Air strikes would have eliminated them, but repeated requests for close air support went unanswered until 2000 when one four-plane strike broke the North Korean defense. All day long, Marine squads inched upwards through the bunker complex, eventually destroying the bunkers with grenades and satchel charges. In one case a lone Marine jumped into a bunker, killed three Koreans with his rifle and strangled the fourth with his bare hands. Throughout the day "chiggy bearers" struggled forward through constant shelling with ammunition and water and stumbled backwards with loaded stretchers. In two days the battalion took more than 300 casualties and lost more than 200 members of its loyal force of Korean porters. The Marines found more than 100 North Korean bodies in the bunkers, including the NKPA battalion commander. The battalion went into reserve on 12 June when the 3d Battalion replaced it.

Wedged into a narrow but difficult sector between the 1st Marines and the 1st KMC Regiment, the 7th Marines fought for 10 days (9-19 June) to establish the regiment (two battalions abreast) along the critical hill complex to the west and the Taeu-san/Taeam-san peaks to the east above the Punchbowl. Colonel Nickerson used his supporting tank company to good effect, but Communist mines in the Sochon River valley put more than half of the company (10 of 17 tanks) eventually out of action despite heroic and costly efforts by Marine engineers to sweep the ground. Nickerson's use of supporting arms mirrored Brown's—long on artillery and short of crucial close air support. If the 7th Marines rifle companies took their assigned hills with slightly less cost than the 1st Marines, they had to defend them against even more stubborn nightly counterattacks. The NKPA battalion commander in this sector

Corporal Charles G. Abrell

Born in Terre Haute, Indiana, in 1931, he attended public schools in Las Vegas, Nevada, before enlisting in the Marine Corps in 1948 at the age of 17. Following recruit training at Parris Island, South Carolina, and a short assignment on board the USS *Noble*, he was sent to Korea in 1950 where he took part in five successive operations: Inchon, Seoul, Chosin, and two against the Chinese Communists. For his bold actions on 7 November 1950, he was awarded the Commendation ribbon with Combat "V."

Department of Defense Photo (USMC) A46965

As a fireteam-leader with Company E, 2d Battalion, 1st Marines, he gave his life on 10 June 1951 at Hill 676 near Hangnyong. His Medal of Honor citation reads, in part:

> While advancing with his platoon in an attack against well-concealed and heavily-fortified enemy hill positions, Corporal Abrell voluntarily rushed forward through the assaulting squad which was pinned down by a hail of intense and accurate automatic-weapons fire from a hostile bunker situated on commanding ground. Although previously wounded by enemy hand-grenade fragments, he proceeded to carry out a bold, single-handed attack against the bunker, exhorting his comrades to follow him. Sustaining two additional wounds as he stormed toward the emplacement, he resolutely pulled the pin from a grenade clutched in his hand and hurled himself bodily into the bunker with the live missile still in his grasp. [He was] fatally wounded in the resulting explosion which killed the entire enemy gun crew within the stronghold.—Captain John C. Chapin, USMCR (Ret)

employed reverse slope defenses, which swept each topographical crest with fire and put the North Korean soldiers close enough for sudden assaults. One 7th Marines company had to throw back five such attacks in one night before it could call its hill secure.

On the eastern side of Taeusan/Taeam-san ridge, the 5th Marines advanced through the ridges that ran down to the Soyang River valley. Alternating in the attack, Hayward's three battalions had to cross five different east-west transverse spur ridges before they reached the last (and highest) ridgeline above the Punchbowl, some 8,000 yards from the regiment's original line of departure on 6 June. As the regiment pushed north, the North Korean defenders took their toll, although somewhat less than the regiments to the west. Again, supporting arms and close air strikes that arrived broke the defensive positions until the regiment, lead by the 1st Battalion, reached the last objective, the Hill 907-Hill 920 ridgeline. No longer able to fall back to another defensive position, the remaining soldiers of the defending North Korean regiment went into their bunkers with no intention of conceding Hill 907 to the oncoming Marines.

The final assault on Hill 907, the regimental objective of the 5th Marines, caught the desperate character of the mountain war in Korea in June 1951. The long, narrow ridge that led to Hill 907 allowed no more than a reinforced platoon to deploy against the line of North Korean bunkers that stretched to the peak. So the 1st Battalion, 5th Marines, had the objective, which it assigned to Company B, which passed the mission (at Lieutenant Colonel John L. Hopkins' direction) to Second Lieutenant Charles G. Cooper's 3d Platoon. As the Marines worked methodically through the bunker system, supported by mortar and machine gun fire, their casualties mounted. Cooper called in more artillery and air strikes, but enemy fire from his front and from two flanking ridgelines cut his ranks down to squad size. He lost two radio operators and then had the disconcerting experience of listening to Hopkins, who had turned ferocious on the eve of his change of command, screaming obscene challenges to the North Koreans over the battalion tactical net, presumably to confuse the listening enemy. Cooper managed to arrange for one more air strike, Air Force jets armed with napalm. Marking the target with white smoke, Cooper ordered an advance through the swirling mess, only to find the North Koreans attacking him. Knocked down by a ricocheting bullet in the back, Cooper lost his carbine to another bullet and ended the fight with a Ka-Bar and a hole in his left side that filled with blood and a damaged kidney. Just as the surviving Koreans reached the "Last Stand of the 3d Platoon," the Air Force jets—which had flown one dummy run to get the route right—returned and dumped their napalm tanks in the middle of the melee, only 30 yards from Cooper's position. The Marines almost suffocated, and most of them suffered burns, but the North Koreans disappeared, incinerated in the flames. The Marine attack ended 100 yards short of the summit, but the next day the 3d Battalion occupied Hill 907, abandoned by the North Koreans after the division headquarters they were protecting had displaced.

Department of Defense Photo (USMC) A8745

Crewmen of Battery C, 1st 4.5 Inch Rocket Battalion, reload their multiple-rocket launcher for another devastating ripple against North Korean troops. The battery of six launchers could fire 144 rounds on target in less than a minute.

On 18 and 20 June, General Almond and General Thomas visited the high ground now held by the 1st Marine Division, and the X Corps commander agreed that patrols in the mountains ahead would be all the offensive action required of the Marines. In the meantime, the defensive positions of the Kansas Line should be developed into complexes of trench lines, barbed wire, bunkers, and minefields, and before the summer monsoon made the supply effort even more difficult than it already was. Thomas could tell that major changes in the war might be underway since he had to entertain an endless stream of visiting military officers of all the Services, most of whom simply wanted to see the Punchbowl from the 5th Marines' observation post. Only admirals bearing gifts of good bourbon were truly welcome. General Thomas knew his division needed rebuilding with replacements and some rest. In the meantime, he had some unfinished business with Eighth Army over the issue of close air support.

A Summer of Discontent

When the 1st Marine Division settled down to a life of night patrols and the daytime construction of trenches and bunkers, two different changes of climate enveloped the men spread along the mountain ridges of the Kansas Line. The changes started a summer of discontent, a season of discomfort and uncertainty that did not reach the level of demoralization, but nevertheless took its toll on the morale of the Marines. The first change in the weather was predictable, the arrival of the summer monsoon, which advances northward from the island of Cheju-do until it reaches central Korea in late June and blankets the hills with daily showers and occasional downpours that seem to wash half of Korea into the west sea. The summer rains of 1950 had been light, a welcome blessing for American airmen. Even though it arrived weeks behind schedule, the next monsoon reversed the trend. The rains of 1951 gave Korea its normal ration of water. Twenty-six inches fell in July, and August brought about 20 inches more rainfall before the deluge stopped in September. The omnipresent mud and cascading streams made the patrols and construction an ordeal, even without an active enemy.

Marines wait for an air strike by Marine aircraft before moving on an enemy position.

National Archives Photo (USMC) 127-N-A9308

The other atmospheric change began with the preliminary truce negotiations on 7 July between the military delegations of China and North Korea on one side and a group of American officers on the other. When the negotiators at Kaesong—a neutral enclave within Communist lines—finally came to an incomplete agreement on an agenda, the one that most affected the Marines was the question of a ceasefire boundary between the two armies. The Communists wanted a return to the 38th Parallel. The United Nations demanded a line based on the forward edge of the battlefield if and when an armistice went into effect. Presumably the two forces would fall back by some agreed distance, and the intervening No Man's Land would become a demilitarized zone. To those with no sense of military geography, one hill seemed no different from another, but the relationship of dominant peaks, road networks, river valleys, and intersecting corridors in the mountains made the control of terrain an important issue, not just a matter of "face." From the frontline foxholes, however, the gloomy mountains all looked alike and simply reinforced the sense that no disputed peak could be worth dying for. Conditioned by World War II to think of victory in terms of geographic advances, the combat troops of the Eighth Army felt their martial ardor wash away with the rain.

No stranger to the challenges of command created by poor weather and endless action—Guadalcanal had provided both—Gerald C. Thomas pressed his regimental

National Archives Photo (USN) 80-G-432028

United Nations delegates to the Kaesong ceasefire talks pose with Gen Matthew B. Ridgway at Munsan-ni. Pictured from left are RAdm Arleigh A. Burke, USN, MajGen Laurence C. Craigie, USAF, MajGen Paik Sun Yup, ROKA, VAdm C. Turner Joy, USN, Gen Ridgway, and MajGen Henry I. Hodes, USA.

Pictured from left are Chinese and North Korean negotiators, MajGen Hsieh Fang and LtGen Teng Hua of the Chinese People's Army, and Gen Nam Il, MajGen Lee Sang Cho, and Gen Chang Pyong San of the North Korean People's Army.

National Archives Photo (USN) 80-G-431929

National Archives Photo (USA) 111-SC378978

Riflemen of the 5th Marines are issued a portion of the initial 40 armored vests developed by the Naval Medical Field Research Laboratory, Camp Lejeune, North Carolina, for field-testing in Korea. The new vest weighed eight-and-one-half pounds and combined curved, overlapping doron plates with flexible pads of basket-weave nylon. The garment was said to stop a .45-caliber pistol or Thompson submachine gun bullet, all fragments of a hand grenade at three feet, 75 percent of 81mm mortar fragments at 10 feet, and the full thrust of an American bayonet.

commanders to do the digging and patrolling Almond ordered. Sporadic shelling by the Communists provided extra incentives, and the Marines still took casualties, 39 in the last week of June. Thomas fought a successful rearguard action against Eighth Army and X Corps to hold pointless casualties down. On 22 June, Almond ordered Thomas to execute an Eighth Army plan to push forces northwards to the Badger Line, between a mile-and-a-half to two-and-a-half miles in front of the Kansas Line defenses. (Later in the war the Badger Line would be called the Combat Outpost Line.) Each frontline regiment was supposed to occupy a combat outpost of battalion strength; Thomas got Almond on 26 June to agree that one outpost was sufficient for the entire 1st Marine Division front, given the nature of the terrain. The 1st Marines sent its 3d Battalion forward to Hill 761 and received a 7th Marines battalion to plug the gap. Like Thomas, "Big Foot" Brown thought the patrol base concept dangerous and pointless; both sides had maintained very close contact with shelling and patrols and needed no additional action. The North Koreans immediately shelled the patrol base with such enthusiasm that Thomas and Brown withdrew the battalion and then told Almond that they would meet X Corps reconnaissance requirements in other ways.

Aware that Almond would soon leave command of X Corps, Thomas had one overriding reason to remain on friendly terms with his difficult corps commander: the close air support controversy. With a pause in the action, Almond marshaled an array of studies for Eighth Army that demonstrated that the Fifth Air Force's close control of each day's quota of close air support sorties limited the ground advances and caused avoidable casualties. Thomas consistently raised the issue with high-ranking military visitors to his headquarters, including Van Fleet, who dropped in on 8 July to give Thomas, Nickerson, Hayward, and five other Marines the Distinguished Service Cross. Thomas persuaded Major General Frank F. Everest to approve the movement of Marine Aircraft Group 12, the premier close air support group of Marine Corsairs, from Hoengsong to the east sea fishing town of Kangnung. The move to Airfield K-18 put the Marine fighter-bombers closer to their supply sources and only 40 miles from the front. Sheer proximity offered new opportunities to circumvent the Joint Operations Center request system, including Everest's promise to allocate 40 sorties a day for offensive operations. Closer division-wing relations seemed at least temporarily acceptable to Eighth Army and Fifth Air Force because Van Fleet had his planners hard at work on another version of Operation Overwhelming, the amphibious landing up the east coast that would involve the 1st Marine Division.

General Almond, however, did not relent in his demands for more fighting of dubious value. His aggressiveness brought General Thomas' only embarrassment as a division commander, the Taeu-san Affair, an abortive operation that remained unnoticed because the victims were the valiant men of the Korean Marine Corps' 1st Regiment. Almond had convinced himself that the North Koreans (despite the Hill 761 experience) would not fight for the lines they currently held. Therefore, Almond ordered the 1st Marine Division to capture the peak of Taeu-san (Hill 1179) and develop it into a regi-

Private First Class Jack Davis: Seasoned Infantryman

With a chastened Jack Davis back in its ranks, the 3d Battalion, 1st Marines, held its part of the No Name Line and watched the last Chinese offensive of May 1951 slide past the 7th Marines to its front and fall upon the left flank of the 2d Infantry Division. With the 1st Marines on the corps and division left flank, the Marines went on the attack on 23 May and a week later reached the high ground overlooking the reservoir. Squeezed out of the advance, the 1st Marines reverted to division reserve.

With the 5th and 7th Marines struggling to penetrate the Communist hilltop positions north of Yanggu, the 1st Marines soon joined the slugfest. Somewhere in the barren ridges Jack Davis' platoon found itself in a grenade-throwing match with the stubborn—and uphill—Chinese defenders. Chinese mortar shells fell among the attacking Marines and took their toll, mostly in wounded. Jack saw five Marines from his squad go down in one shower of grenade and mortar fragments. Amazed by his own apparent invulnerability, Jack attacked a Chinese position with his rifle and grenades after crawling to a protected firing position. His attack and a flank assault by his buddy Frank Brown (carrying a BAR) wiped out the Chinese bunker and spider-traps. More American grenades completed the task. Jack thought he might have killed three Chinese, his only victims of the war.

During the fight, Jack received his second wound of the war, a grenade fragment that tore open his upper left arm and made him a one-armed Marine. While a corpsman bandaged Jack, his platoon commander asked him if he would take charge of three other walking wounded and lead them down the mountain to the battalion collecting and clearing station. Jack agreed, and off he went—slowly—trailed by his more seriously wounded comrades, one of whom had both eyes bandaged. As night fell, Jack's forlorn band had reached the foot of the mountain, but had strayed through a "no man's line" into the lines of the 5th Marines. Jack had no idea what the challenge and password was, so he simply screamed: "Wounded Marines! Wounded Marines!" Persuaded that no Chinese could scream with a Tennessee accent, the Marines brought in the wounded and sent them off to safety by jeep. Jack had a second Purple Heart, but his second wound was not severe enough for the Navy doctors to invoke a welcome Marine Corps policy: two wounds serious enough to require hospitalization bought a Marine a trip home.

Jack's arm healed more rapidly than his spirit. After almost six months in a rifle company with no real escape from the most primitive and exhausting field living conditions as well as combat, Jack Davis felt himself weakening in the psychological sense. He wrote Vince that he was not sure whether he could take the constant mortar and artillery fire. "You can't imagine what it does to a man's insides to see a big, six-foot man crying and shaking with fear, just because his mind has had all the killing and bloodshed it can take. When this happens to man, it also [is] because he is scared to death and wants to run but his loyalty won't let [him] and that if he did run, there's no place to go. Sometime they get evacuated and sometimes they don't . . . if they do come back in a couple of weeks . . . as soon as the first mortar or artillery shell comes screaming over and explodes nearby they are worse than ever." Jack was proud that he had not yet broken down, but he had some doubts about his ability to carry on. "I was a dope fiend about the last month I was in the hills." He took a quarter grain of Phenobarbital, dispensed by a corpsman, so that he would quit shaking from cold and fear while he stood watch at night. He could not eat or sleep without drugs. Unfortunately, the barbiturates gave him a "don't give a shit attitude" that worried him.

Upon his return to Company G, Jack requested an interview with the new company commander, Captain Varge G. Frisbie, and asked if he could get some credit for his two Purple Hearts and be transferred somewhere out of the battalion. Frisbie promised to take the matter up with the battalion personnel officer, and within days Private First Class Jack Davis had orders to report to the Service Battery, 3d Battalion, 11th Marines, for retraining as an artilleryman—and a survivor.

mental patrol base on which to anchor the Badger Line. Thomas objected to the mission, pointing out that all the evidence suggested that Taeu-san anchored the main defensive position of the entire North Korean *V Corps*. Unmoved, Almond ordered the attack to be made, and Thomas assigned the mission to the 1st KMC Regiment, whose lines were closest to Taeu-san and who had shown some aptitude for mountain warfare. One suspects that Thomas saw no reason to squander one of his own Marine regiments on a forlorn hope. Colonel Kim Dae Sik accepted his assignment without a murmur, and the 1st Marine Division provided all the fire support it could possibly mount on behalf of the 1st KMC Regiment. *Han Pon Haepyong Un Yongwon Han Haepyong!* (Once a Marine, Always a Marine!)

For five days (8-12 July), the Korean Marines—one battalion at a time—tried to take and hold Taeu-san but managed only to hang on to Hill 1001, a hillock only halfway to Taeu-san. Successive assaults on Hill 1100 produced dead Korean Marines, but no permanent foothold on the Taeu-san main ridge. All combinations of shelling, air strikes, and infantry attacks did not break the North Korean defenses. Colonel Gould P. Groves, senior adviser to the 1st KMC Regiment, demanded that the fruitless attacks cease before the regiment became permanently ruined by the loss of its key leaders; one KMC battalion lost all its company grade officers and all but five of its sergeants. Thomas insisted to X Corps that Taeu-san would take an entire American regiment to capture (as indeed it later did) and that the security mission could be performed without the Badger Line. Almond insisted, however, that the Koreans hold on to the outpost on Hill 1001 even if the 1st KMC Regiment returned to the Kansas Line, which it did on 12 July. Of the 77 Marines killed or missing and 360 wounded in July, 55 of the dead or missing and 202 of the wounded were South Koreans.

National Archives Photo (USA) 111-SC382938

MajGen Gerald C. Thomas joins MajGen Clovis E. Byers, left, Gen Edward M. Almond's replacement as commander of X Corps, on board a helicopter at Kwandae-ri, the Corps' airstrip. Thomas and Byers developed a strong working relationship that profited the 1st Marine Division.

At the 1st Division headquarters the bad taste of the Taeu-san Affair faded with two bits of welcome news: Almond was finally leaving X Corps for a new posting in the United States and the division had been ordered to turn over its sector to the U.S. 2d Infantry Division and withdraw to corps reserve. Almond flew off to Seoul after giving Thomas a Distinguished Service Medal. He left X Corps in the capable hands of Major General Clovis E. Byers, a 52-year-old Ohioan and Military Academy graduate (class of 1920) with an impeccable professional reputation and companionable personality. Thomas, who knew Byers, could not have been happier. In World War II, Byers had served with distinction in the Southwest Pacific theater as commanding general, 32d Infantry Division, chief of staff of I Corps, and chief of staff of Eighth Army. He had then commanded the 82d Airborne Division, the Army's only combat-ready contingency force, before becoming the G-1 (Personnel) of the Army Staff. Byers, however, had one glaring weakness. In a faction-ridden Army, he was a protégé of Lieutenant General Robert L. Eichelberger, just retired, and not a member of the European clique of Generals Eisenhower, Bradley, Collins, Ridgway, and Van Fleet.

After the various elements of the 1st Marine Division reached their reserve areas, Thomas ordered a demanding training program of live-fire exercises, designed by his new chief of staff, Colonel Victor H. Krulak, and the G-3, Colonel Richard W. Hayward, former commander of the 5th Marines. Thomas prowled the regimental training areas by helicopter and jeep: the 5th Marines near Inje, the 7th Marines near Yanggu, and the 1st Marines near Hongchon. The pattern of deployment (with the battalions of the 11th Marines positioned to either fire for the 2d Infantry Division or train with the Marine infantry regiments) reflected Byers' concern about a sudden attack on the 2d Infantry Division or the 5th ROK Division. Byers also felt some anxiety about his eastern flank with the South Korean I Corps. Eighth Army's nervousness exceeded Byers', and Van Fleet ordered X Corps to form a task force built around the 1st Marines (Task Force Able) to be prepared to move east for a preemptive offensive. Thomas liked none of this business and said so to Byers, who supported Thomas' insistence that Army ad hocery would give way to Marine command if a real crisis arose. There was none, but Thomas and Byers cemented their sound working relationship. As Byers wrote another Army general: "the 1st Marine Division under the command of Major General Thomas, with Brigadier General Whaling as Assistant Division Commander and Col. Krulak as Chief of Staff, has become a vastly different outfit from that which it was under its former commander. They cooperate with the other divisions of the Corps smoothly and willingly."

Combat-ready division replacements disembark from a U.S. Navy landing ship. In the movement of Marines the Corps functioned as a single great unit, even though an ocean separated the vanguard in Korea from rear echelons in the United States.

Department of Defense Photo (USMC) A157123

Byers showed his appreciation in tangible ways. His staff ensured that the equipment rehabilitation of the division went forward without friction. X Corps engineers and artillery helped the Marines turn swamps into muddy camps with a few amenities like shower and mess tents with floors and drainage. Army and Marine technical experts worked together to train novice personnel and put everything from ordnance, tanks, radios, watches, motor vehicles, to engineering equipment in working order. The military policemen of both Services cooperated in trying to control the flood of Koreans sweeping toward the Marine tent camps to sell carnal and alcoholic pleasures. In turn, Thomas ordered 12 special Marine training teams from his infantry regiments to work with the 1st KMC Regiment to improve the regiment's use of supporting arms. All units conducted at least a third of their training at night. Night patrols went to work with rounds in the chamber and engaged guerrillas along the rear area roads. Marines worked with Korean security forces and laborers constructing additional defensive positions to protect both I Corps flanks.

Although only Thomas and his immediate staff knew about the continuing exchanges between Ridgway and Van Fleet over future operations, the focus and pace of the 1st Marine Division training program suggested that the division might provide the spearhead of a new Eighth Army offensive. Van Fleet urged Operation Overwhelming upon Ridgway, but only if Eighth Army received American reinforcements. The new Ridgway, a paragon of caution, did not embrace the plan. Thomas and Krulak anticipated a landing until they received a clear signal that there would be no Inchon in their future when Van Fleet on 3 August ruled that the 1st Amphibian Tractor Battalion and 1st Armored Amphibian Battalion would not be returned to the division's operational control.

The disappointment did not change the urgency of bringing the 1st Marine Division to a new peak of strength in numbers and effectiveness. Two new replacement drafts (the 11th and 12th) would arrive in August and early September with more than 4,000 officers and men, more than replacing casualties and a small rotation draft. By mid-August the division had the responsibility of caring for 32,000 American and Korean personnel, making the division almost a small Army corps. The division's combat power—enhanced by its ability to use close air support as available—made it difficult for Van Fleet to move it from the eastern part of the front where the only other American division was the hard-used 2d Infantry Division.

From General Byers' perspective X Corps and the neighboring ROK I Corps occupied a vulnerable part of the Kansas Line, more vulnerable than the western sectors and the "Iron Triangle" where U.S. I and IX Corps faced the bulk of the recovering Chinese expeditionary force. Byers' G-2 made a special study of the activities of the North Korean *III Corps* (three divisions of 8,500 each) and concluded that the North Koreans had the capability to mount a serious offensive on any X Corps division sector along the Kansas Line. Well ahead of the Chinese rearmament programs, *III Corps* had accepted a full set of new Soviet weapons and showed every intention of using them again in the attack. Over the latter part of August the intelligence analysts saw the usual omens of an attack: increased patrolling and counter-patrolling ambushes, increased desertions, a reduced flow of refugees, tank sightings, the mass distribution of ammunition and rations, a decline in vehicle movement, and the imposition of radio silence. In the meantime, at Van Fleet's insistence, Byers had ordered the 2d Infantry Division into action west of the Punchbowl, and the division had exhausted itself again fighting the North Koreans and the rain over terrain only too familiar to the Marines. Only the names of the hills ("Bloody Ridge" and "J Ridge") and the Service of the bodies changed. To the west Major General Paik Sun Yup's ROK I Corps made no significant progress against three NKPA divisions, all entrenched and very combative in the hills east of the Punchbowl. The campaign wrecked the North Korean *II Corps*, but *III Corps* remained ready to enter the fray, perhaps in a major counteroffensive. The only fresh force in the eastern sector was the 1st Marine Division.

Once More into the Breach

With neither a ceasefire nor great offensive in prospect, General Van Fleet ordered his corps commanders to plan operations that would improve their control of the critical terrain in their sectors. They should prepare either for some later offensive (should the truce talks remain in recess) or to defend South Korea for the indefinite future. With the defenses of the Kansas Line largely completed, Van Fleet on 30 July decided to convert the combat outpost line (the Wyoming Line) into an advanced main line of resistance where the terrain allowed. The distance between Kansas Line and the Wyoming Line varied between two miles and 10 miles. In X Corps' sector Van Fleet thought that the trace of the front in July did not allow Byers to dominate the Punchbowl and the Sochon and Soyang River valleys. Van Fleet wanted X Corps to shift the focus of its attacks to the high ground (including Taeu-san) west of the Punchbowl, but the heavy rains of early August made it impossible for Byers to begin the attacks of the U.S. 2d Infantry Division and the 7th ROK Division (Brigadier General Kim Yong Bae). The 8th ROK Division (Brigadier General Choi Yong Hee) would attack the dominant hills east of the Punchbowl.

Having designated an intermediate phase line (Hays) between the Kansas and Wyoming Lines, Byers quickly learned that the terrain, the weather, and the North Koreans would prevent any easy victories. The battles west of the Punchbowl produced such disappointing results and bad blood between the American and Korean commanders that Byers narrowed the division sectors and committed the 5th ROK Division (Brigadier General

Min Ki Shik) west of the 2d Infantry Division, which meant that X Corps had three committed divisions west of the Punchbowl and only the 8th ROK Division in the Soyang River valley and the dominant hills on either side of the valley. On 23 August, Byers warned Van Fleet that he might have to relieve the 2d Infantry Division with the 1st Marine Division, which was "very anxious to take action," but Van Fleet still had an amphibious role in mind for the Marines, and he vetoed the idea. Van Fleet thus spared the 1st Marine Division the mission of capturing "Heartbreak Ridge." Only the 8th ROK Division had done better than anticipated, capturing some of the high ground east of the Punchbowl, but the South Korean divisions on its eastern flank had not kept pace, thus giving Byers some concern about his corps boundary.

Meeting on both 25 and 26 August, Van Fleet and Byers concluded that they could no longer hold the 1st Marine Division in reserve since all the rest of X Corps divisions had bogged down, and the corps could not change the tactical balance with artillery and close air support alone. Ammunition shortages, caused principally by transportation problems, had already affected operations. Troop movements, for example, on 28-30 August prevented the stockpiling of 1,800 tons of munitions. The Fifth Air Force, anticipating a break in the weather that would allow a surge in the interdiction bombing campaign, announced on the 23d that the Eighth Army would have to manage with less close air support through the end of the month. On 26 August, Byers called General Thomas and told him to move at least part of his division to the front east of the Punchbowl where the Marines would take up the missions of the 8th ROK Division. Thomas had four days warning since Byers alerted him to a possible move on 23 August. With the plans already in place, Thomas ordered the 7th Marines to start for the front that night, followed by the 1st KMC Regiment. The 5th Marines would move last, and the 1st Marines not at all since the regiment would be the only corps reserve.

Thomas knew that the division

Rain and mud fail to halt the mortarmen of the 5th Marines' 4.2 Inch Mortar Company as they fire their heavy mortars at enemy-held positions.

National Archives Photo (USA) 111-SC380808

would receive an offensive mission: capture a ridgeline, an eastern extension of the hill mass that formed the northern rim of the Punchbowl. A corps objective designated Yoke, the ridge had four dominant west-to-east peaks (Hills 930, 1026, 924, and 702) and another north-south extension that began at Hill 702 and ran south through Hills 680, 755, and 793, thus forming a large L just west of the Soyang River. The river itself curled westwards, bounding Yoke Ridge on the north. Since the North Koreans showed no sign of reduced morale and fighting tenacity—they, in fact, had mounted many aggressive counterattacks west of the Punchbowl—the assignment had nothing easy about it. The rains and planning changes made 27-31 August some of the most discouraging days Thomas and his Marines had faced together.

From the division commander's perspective, the mudslides and floods that slowed his truck convoys were bad enough, but the operational confusion within X Corps, fed by tactical errors and bad blood between the 2d Infantry Division and 8th ROK Division, made the changes of orders reach epidemic proportions. Before it could displace, the 5th Marines detached a battalion to the operational control of the 2d Infantry Division to defend the Kansas Line while the 23d Infantry slipped to the west. The 1st KMC Regiment also picked up part of the Kansas Line defense, which meant that only the 7th Marines, struggling to cross the swollen Soyang River by wading or by a shuttle of DUKWs (amphibian trucks) could man the sketchy positions on the edge of Yoke Ridge held by dispirited soldiers of the 8th ROK Division. Confusion reigned, and the rain fell. Warning orders flooded the

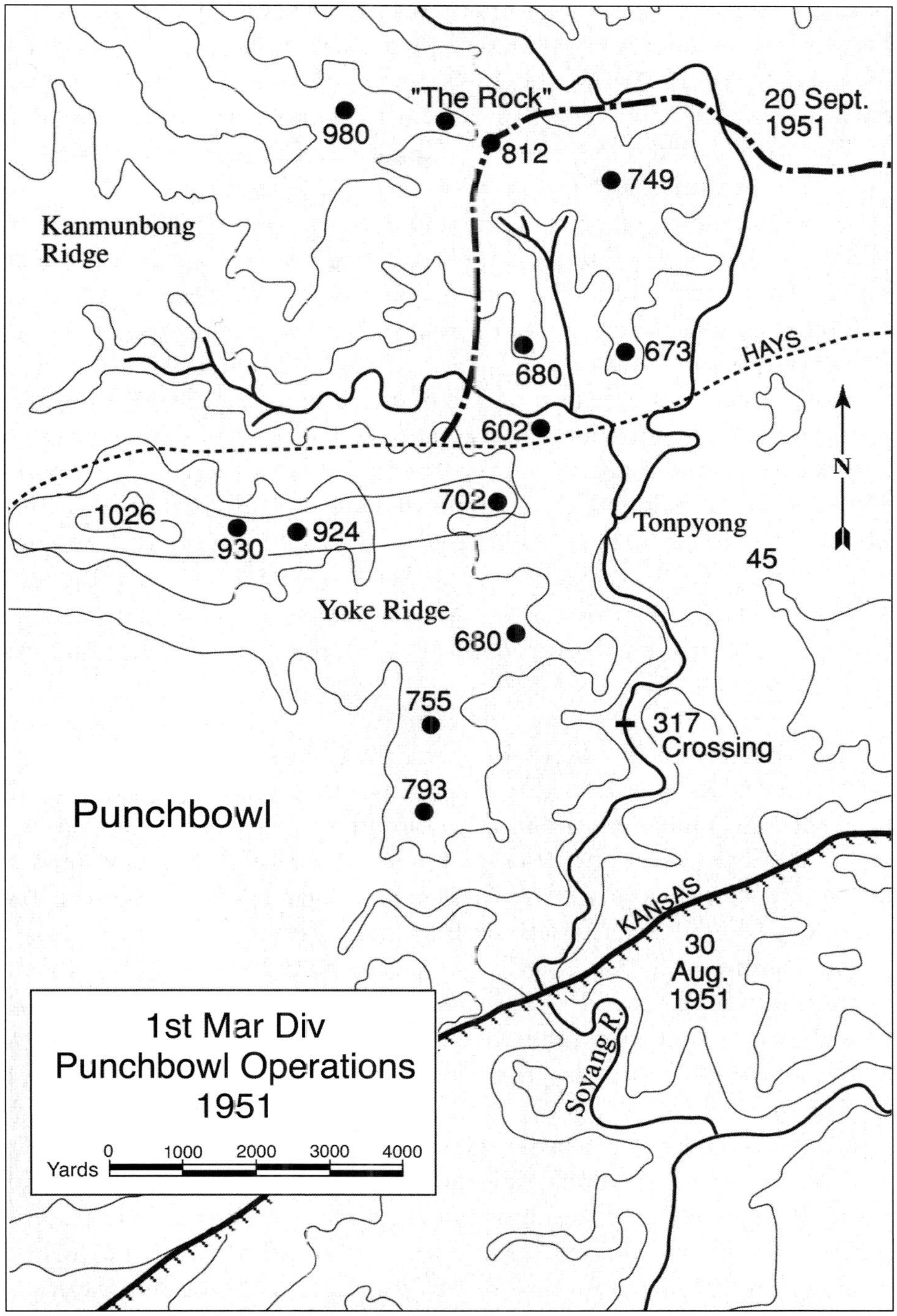

Courtesy of the Naval Institute Press

airwaves, and commanders and staff officers scurried by helicopter and jeep from headquarters to headquarters. General Byers, for example, made 12 commands calls in one week (25-31 August) and received General Van Fleet three times. General Thomas and his staff made the best of a bad situation, pushing the 7th Marines and 1st KMC Regiment into their forward positions. He tried to prevent the diversion of the 5th Marines to the 2d Infantry Division and kept the 1st Marines ready for such time, as Byers would release the regiment from corps control. In the meantime, the 11th Marines fired missions all along the corps front, scattered about the valleys in a desperate attempt to stay close to its ammunition supply and to avoid having its fires masked by the hills to its front.

During a Van Fleet-Byers conference on 29 August, the army and corps commanders agreed that they could not wait for more success west of the Punchbowl before ordering the 1st Marine Division into action. Byers passed the news to Thomas the next day: attack Yoke Ridge on 31 August. Two factors related to the enemy situation helped shape Thomas' plan. Patrols by the division's Reconnaissance Company and the 5th Marines discovered enemy patrols active on either side of the Kansas Line, but no more than a nuisance. On the other hand, North Korean prisoners taken by the 8th ROK Division and the Marines reported large troop movements to the north and much talk about another Communist offensive while the weather limited United Nations Command air support. Visual sightings and other intelligence sources

Corporal Jack Davis: Truck Driver and Short Timer

Jack Davis, an old man at 19, found a new home in Service Battery, 3d Battalion, 11th Marines. His principal responsibility was driving a dump truck and working as a laborer on the battalion's gun positions and other construction projects. As the weather cooled in the fall of 1951, the 1st Marine Division resumed its attacks on the high ridges northeast of the Punchbowl. Its opponents were troops of the re-born Korean People's Army *2d Division*. Another enemy was a monsoon season that lasted through the entire month of August, washing away roads and bridges and making life generally miserable for all hands. Jack Davis found his dump truck in high demand. In addition to the usual construction materials, Jack hauled cut wood for the battalion's stoves. He became an expert at fitting out bunkers with furniture made from used shell boxes and other handy materials; he and his fellow engineers used layers of sandbags, logs, and loose dirt to build sleeping bunkers that could withstand a direct shell hit. Jack estimated that they made seven-foot thick ceilings to provide overhead protection.

Even if the pace of the combat froze along with the weather and Panmunjom peace talks, danger still waited for the unwary and unlucky. Employing their new Russian field artillery, the Chinese and North Koreans started to fire short counterbattery barrages late every

second or third afternoon. Even the bunkers to which the Marines fled to avoid the shelling could be death traps of their own; weakened by the rains and shellings and too heavy for their supporting walls, bunker roofs habitually collapsed. One such roof fell on top of Jack, bruising his body and pride and burning parts of his body when a stove overturned and ignited the bunker's interior. Jack took his third trip to sickbay with cracked ribs.

There were few diversions north of the Soyang River. Jack grew his third mustache, not as long and menacing as his "infantry mustache" of the summer. One day he received a call to report to battery headquarters, only to learn that the battery commander and first sergeant had arranged a little ceremony to award Jack his first and second Purple Hearts (a medal with gold star affixed). Jack had no idea what to do with the medal and presentation box until the first sergeant suggested he send it home. The final package featured paper torn from boxes in the mess tent, secured with communications wire. The Davis family received the box and properly concluded that Jack had not been entirely honest in his summer letters.

Although his anxiety about dying eased some, Jack's fears about living grew as his tour in Korea shortened. Under the rotation policy adopted in 1951, he could expect to rotate home sometime in early 1952, and the Marine Corps, having little need for short-timer reservists at the end of a two-year contract, promised to release him go to college short of his two-year obligation. Jack thought about getting his personal life in some order. He wrote a "Dear Jane" letter to a girl friend whose religiosity and immaturity now struck Jack as intolerable. He warned Vince that no one in Tennessee should discuss his love life. Jack also continued to send money home for his college savings account. His sense of duty received a jump-start with his promotion to corporal in November. His greatest leadership accomplishment to date was organizing the theft of an Army jeep that the battery sorely needed. He did his work, and he stayed out of trouble as he watched veterans of earlier replacement drafts turn in their equipment and head for processing for a flight or transport berth back to California.

By Thanksgiving the 3d Battalion had endured two snowfalls, general freezing, and the news that it was not on the itinerary for Bob Hope's Christmas show. Jack bought a contraband bottle of Canadian Club to hoard until Christmas. He liked the brand new thermal boots issued to the battalion—until he had to change his sweat-soaked socks in the cold. After a muted celebration of Christmas, Jack started watching the organization of each rotation group. He wrote Vince that he now stood 29th on the list and that 37 men had started home in December. Jack reported that he was "kinda nervous about coming home. I'm still not doing much work per usual.' He worried about his future relations with his parents, whom he remembered as full of sermons about all the things he should not do and think. "If they start a bunch of harping and bullshit, I ship into the regular Marines because I really like this outfit." He admitted to Vince, however, the he would really have to be aggravated with civilian life to re-up for a second tour. He certainly was not going to miss his ride back to the United States. On 18 February 1952, Jack Davis left Korea for home.

---

confirmed that fresh enemy troops were going into position on Yoke Ridge. The 1st Marine Division attack of 31 August was designed to squeeze out the Koreans on the eastern part of Yoke Ridge and to prevent the objective area from being reinforced from the north while the battle raged. Two Korean Marine battalions advancing in column from their position on Hill 755 would attack north to take Hills 1026 and 924 while two battalions on the 7th Marines would attack westwards from the Soyang River valley with two battalions abreast. They would seize the ground east and north through Hill 702 to Hill 602, another lower ridge that ended at the river as it changed its direction from east-west to north-south. Catching the North Korean *2d Division* in the process of moving into the bunkers of the North Korean *1st Division* on the morning of 31 August, the initial Korean Marine and 7th Marines attacks still faced extensive minefields and mortar barrages as the troops worked their way uphill. Marine artillery fire damped some of the enemy fire. The two 7th Marines battalions took their objectives, but the 1st KMC Regiment advanced no farther than the base of Hill 924, the most heavily-defended position encountered on eastern Yoke Ridge. Almost all the division's casualties for August (three killed and 57 wounded) fell on the first day of the Battle for Yoke Ridge.

Second Lieutenant Frederick F. Brower moved into his first big fight at the head of the 1st Platoon, Company H, 3d Battalion, 7th Marines. Occupying Hill 680 on 30 August, the company had endured a heavy mortar barrage and learned that North Korean regulars had replaced the scattered Chinese the company had chased north of the Punchbowl. The next day the company attacked Hill 702, Yoke Ridge, against "light resistance." Brower had commanded his platoon for three months, but he and his Marines had not yet closed with the enemy since they always seemed to be patrolling the division's western-most flank, keeping an eye on the neighboring South Korean division. As the skirmish line approached Hill 702, the North Koreans greeted it with a barrage of mortar fire. Only minutes into the battle, Brower crumpled with multiple wounds in his left leg, and he looked with dismay at his bloody and misshapen left knee. Pistol marksman, model Marine platoon commander, dedicated to a career in the Marine Corps, Brower ended his first battle on a stretcher carried by nervous Korean "chiggy bearers." Although he eventually served his full Korean tour as a semi-cripple, his career in a rifle company ended on 31 August 1951, and his damaged knee forced him into disability retirement in 1955. It had been a short but final war for Second Lieutenant Brower.

The fight gave few hints of the ordeals ahead. On 1 September, General Shepherd visited General Thomas and found no cause for alarm. Thomas felt confident that the attacks that day would take care of the Yoke Ridge problem. After seeing Byers, they agreed that X Corps had problems west of the Punchbowl where the 2d Infantry Division still had not secured all of "Bloody Ridge" despite the loss of 2,772 American and attached Korean soldiers since 18 August. For the Marines, however, the attacks of 2 September took only Hill 924 (but not Hill 1026) and consolidated the 7th Marines defenses on Hill 602. Throughout the day and the next, the North Koreans bombarded Yoke Ridge and mounted counterattacks of up to battalion-size. The 1st Marine Division's modest successes came in no small part from the artillery fire from two 11th Marines battalions and three Army corps artillery battalions, which fired 8,400 rounds on 1-2 September, an amount of fire that exceeded the "Van Fleet Day of Fire" for the five battalions (6,000 rounds). The battle drew in the remaining battalions of the 1st KMC Regiment and the 7th Marines. With American Marines holding the northern edge of Yoke Ridge, the South Korean Marines finally took Hills 924 and 1026, which completed the mission. It did not end enemy counterattacks and shelling, but the two regiments held the objective. The 7th Marines suffered five dead and 75 wounded, the Korean Marines 70 dead and missing and 274 wounded. The North Koreans left behind almost 600 bodies to be counted and 40 prisoners. None of the allies thought the victory had been easy.

The capture of Yoke Ridge might have been less costly if the Marines had received more effective close air support. General Shepherd made it one of the highest priority issues when he visited the war zone from 27 August to 12 September. The Commanding General, Fleet Marine Force, Pacific (and likely Commandant) met with the Major General Christian F. Schilt, commander, and Brigadier General William O. Brice, deputy commander of the 1st Marine Aircraft Wing. "A discussion of the close air support problem revealed that unsatisfactory conditions still prevail in regard to close air support for the 1st Marine Division." Shepherd then complained about the poor air support to Van Fleet and Everest even before he consulted with Byers and Thomas on 1 September. Shepherd recruited Vice Admiral C. Turner Joy, Commander, Naval Forces, Far East, to join a coalition of senior officers who would force the issue with Ridgway.

During his Korean inspection trip, LtGen Lemuel C. Shepherd, Jr., Commanding General, Fleet Marine Force, Pacific, left, discussed not only close air support, but also the performance of the Sikorsky HRS-1 helicopter. Pictured to his left are MajGen Christian F. Schilt, Commanding General, 1st Marine Aircraft Wing, LtCol George W. Herring, commander officer of Marine Helicopter Transport Squadron 161, and newly-promoted MajGen William O. Brice, the wing's deputy commander.

Department of Defense (Photo) A131870

The battle of Yoke Ridge provided ample evidence that the Fifth Air Force would not modify the request system and that the real purpose of the Joint Operations Center was to prevent the diversion of fighter-bombers from Operation Strangle, the campaign against the Communist lines of communication. The 1st Marine Division had requested 26 aircraft to support the attack of 1 September. Despite the fact that the requests had been made 40 hours before the mission, only 12 aircraft were assigned. Requests made by the forward and other air controllers in the heat of combat took more than an hour to produce aircraft on station. One 7th Marines request for air strikes against a heavy North Korean counterattack had been canceled by X Corps' G-2 because he did not believe the counterattack was real. Despite the mounting evidence—and much of it came from the 2d Infantry Division's ordeals to the west—the Fifth Air Force made no concessions. The 1st Marine Division's fire support coordinator made the point in his briefing for Shepherd: "Close air support furnished by the Fifth Air Force JOC was inadequate and often not opportune."

Shell shortages, complex planning by both Eighth Army and X Corps headquarters, and the determination of the North Koreans brought a pause of six days to 1st Marine Division operations. The likely artillery shell expenditures of any future offensive—combined with road conditions between Hongchon and the front—would make an immediate offensive beyond Yoke Ridge difficult. The division goal was to stockpile 10 days of fire in artillery shells ("Van Fleet days") at ammunition supply post-60B, the ammunition dump and distribution point run by X Corps and division ordnance men located 48 miles from Hongchon and five miles from the gun line. Until the roads dried and engineers repaired the washouts and strengthened the roadbed, the round trip to ASP-60B took 25 hours. Some trucks still had to be diverted to lift troops to and from the front. In fact, the estimates for shells fell short of the actual expenditures, 24,000 tons (874,000 rounds) for X Corps in September 1951.

The body of a Communist soldier lies atop a bunker captured by elements of the 7th Marines during the assault against Hill 673.

National Archives Photo (USA) 111-SC380918

Intelligence officers believed that X Corps would need every shell it could find. The combat around the Punchbowl revealed a system of defensive fortifications that had been built before 1951 and strengthened since April. Much of the NKPA *I Corps* had been withdrawn, but its replacement—the NKPA *III Corps*—was one of the largest (30,000 soldiers) and best-trained in the North Korean army. Unlike the Chinese, the North Koreans had plenty of artillery, too, out-numbering Marine artillery pieces in the Punchbowl sector. In the Marine division's zone of action the NKPA *1st Division* appeared to be assigned the bunker defense role while the NKPA *45th Division* mounted counterattacks.

General Van Fleet did not win approval of his amphibious hook north to Tongchon, but his planners produced some more modest variants that might have put all or part of the 1st Marine Division within the ROK I Corps area and closer to the air and naval gunfire support that Task Force 77 could provide. An offensive westward from the coast might bring the Marines and the ROK I Corps in behind the fortified belt so well-manned by the North Koreans. For almost 10 days, Van Fleet and Byers examined their contingency plans and ruled them out as too risky and subject at any moment to another Ridgway veto. The result of the operational paralysis was that General Thomas learned on 8-9 September that he would repossess the 1st Marines from corps reserve, which would release the

The "Chiggy Bearers"

They could be found trudging along after every Marine rifle company in Korea's mountains in the summer of 1951. Small men, powered by muscular but thin legs, bent under the loads of their A-frames or *chigae*, struggling along with ammunition, rations, and water, they were the "chiggy bearers." The 1st Marine Division depended upon them to close the gap between the supply points served by trucks and the Marine companies engaged in battle. The "chiggy bearers" made it possible for the Marines to search out and destroy the enemy.

Organized by the U. S. Eighth Army in 1950 and originally called the Civil Transportation Corps, this army of Korean laborers provided the United Nations forces with construction workers and pack bearers. For carrying supplies, the Koreans relied upon their traditional wooden A-frame packboard or *chigae*. Although renamed the Korean Service Corps (KSC) in 1951, the bearer corps remained the *chigaebudae* (A-Frame Army) or "chiggy bearers" to the Marines.

The "chiggy bearers" had either been drafted into their country's service or had volunteered. Members of the KSC had to be medically unfit for duty in the South Korean army or be over age 38. Marines often characterized the "chiggy bearers" as "elderly," but, in fact, the KSC included men and boys who had convinced someone that they were unfit for frontline service in the South Korean army. The South Korean government had almost absolute power to commandeer people and things for the war effort, but in reality the KSC competed with other American-financed Korean service agencies for personnel and could count only on unskilled workers (often displaced farmers and farm laborers) for the bulk of its manpower.

In many ways the lot of the average "chiggy bearer" was not a happy one, however essential. His contract said that he would carry up to a 50-pound load for as many as 10 miles each day, but the bearers often carried heavier loads for longer distances, especially if measured from valley floor to hilltop. The lines of bearers, shepherded by Korean soldiers assigned as KSC cadremen, often came under artillery and mortar fire. American divisions did not keep track of KSC casualties. Any man could be pressed into service as a bearer for six months, and the living and medical conditions for the bearers were no better than most refugee camps.

the end of the war an estimated 300,000 Koreans had served a tour as a "chiggy bear," and at the height of the fluid war of 1951 the South Korean government impressed an average of 3,000 men and boys a week into the KSC. A postwar accounting of KSC personnel listed 2,064 porters killed in action, 2,448 missing in action, and 4,282 wounded in action.

If a KSC "regiment"—with one assigned to each American division—had efficient and honest officers, the KSC bearer did not fare badly—provided he lived to collect his pay. Clothing and food were not a problem, which could not be said for his countrymen; the "chiggy bearer" ration was supposed to provide 3,500 calories a day and included a ration of 10 cigarettes. After some strident protests in 1951, KSC pay scales moved from those set for the South Korean army toward those paid other Koreans working as civilians for the United Nations Command.

One American army logistician calculated that an American infantry company required just about as many bearers as its own strength, around 150-200. If so, the 1st Marine Division had a "chiggy bearer" shortage since it had only 1,922 KSC members in support in May 1951. The bearer "gap," however, applied to all of United Nations Command. By war's end the KSC had a paper strength of 133,000, but its "A-frame strength" was about 100,000 or roughly one bearer for every six American and allied soldiers in Korea. Like everyone else on the United Nations side of the war, the "chiggy bearers" carried more than their prescribed load.

Department of Defense Photo (USMC) A8434

5th Marines from the Kansas Line for additional offensive operations east of the Punchbowl. Except for the 8th ROK Division on Thomas' right flank, the rest of X Corps would seize another hill mass soon called "Heartbreak Ridge." Byers expected the Marines to resume the attack on 11 September.

With only 48 hours to mount an attack, Thomas had little alternative but to look again to the 7th Marines to lead the advance on Kanmubong Ridge, the hill mass directly north of Yoke Ridge and the division's next objective. The concept of the operation envisioned a two-phase operation that would begin with the 7th Marines seizing the two most dominant peaks at the eastern edge of the ridge, Hills 673 and 749. To eliminate a transverse ridge spur (Hill 680), a secondary attack would strike directly north from the Hays Line on Yoke Ridge. This mission went to 3d Battalion, 7th Marines, commanded by Lieutenant Colonel Bernard T. Kelly, with the main attack to 1st Battalion, 7th Marines, under Lieutenant Colonel James G. Kelly, relatively untouched by the fight for Yoke Ridge. When the 7th Marines had secured the Hill 673-Hill 749 area, the 1st Marines would come forward and continue the attack up Kanmubong's long axis, "ridge-running," to capture a series of peaks designated (east to west) Hills 812, 980, 1052, and 1030. The scheme of maneuver would allow tanks to fire across the front of the advancing troops and artillery fire (even naval gunfire) to converge in concentrations from the firing positions to the south and southeast. The advances had to be supported by hundreds of "chiggy bearers" since there were no roads of any kind to bring the ammunition, food, and water forward in any other way.

Fighting from cleverly-concealed and strongly-built bunkers and trench systems, the North Koreans made the 7th Marines (all three battalions) pay dearly in three days of fighting, 34 dead and 321 wounded. The assault companies that crossed the line of departure in the morning fog of 11 September did not expect a walk-over. Despite the hour of intense artillery preparation, the North Korean defenders fought with unflagging tenacity until killed. Each bunker system came ringed with mines and booby-traps, and Korean mortar shells and grenades showered crippling fragments across every contested position. Long-range heavy machine gun fire from higher up Kanmubong Ridge took its toll among the Marine assault units that struggled forward with flamethrowers and satchel charges.

Once again dark memories of Iwo Jima and Okinawa came to the veterans. More heirs of the Japanese military tradition than the Soviet, the North Koreans showed no hesitation in launching counterattacks large and small and at unexpected times and from unexpected directions. Although the enemy did not overrun any Marine positions, only quick shooting and quick thinking broke the backs of the attacks with bullets and artillery shells. Although the 3d Battalion took its objective with no assistance, Colonel Nickerson had to commit his 2d Battalion to aid the 1st Battalion on 12 September. Only a converging two-battalion attack—the companies in column—finally seized Hill 673, and the subsequent 2d Battalion attack on Hill 749 fought itself out far short of the crest. In all the fighting tank fire proved decisive when the bunkers could be identified and fired upon, line-of-sight. Many bunkers, however, could have been reached by close air support, conspicuously absent. The key ground maneuver came from a company of the 1st Battalion that made an undetected night march to reach a poorly-defended entrant to Hill 673, then assaulting through a breach in the North Korean defenses. Nevertheless, the 2d Battalion's attack on Hill 749 stalled with the three rifle companies reduced, scattered, and battling back small counterattacks in the dark before a battalion of the 1st Marines replaced them on 13 September. So hard-pressed and scattered were the Marines of 2d Battalion, 7th Marines, that the battalion misreported its location and gave Nickerson the impression that his regiment had taken Hill 749, which it had not. Moreover, the approaches to the hill were still held by some very combative North Koreans. Assuming operational control of the 2d Battalion, the 1st Marines, under Colonel Thomas A. Wornham, picked up the responsibility for occupying Hill 749. Only a helicopter reconnaissance proved that Hill 749 would have to be taken first.

The logistical burden of supporting five committed infantry battalions (the situation on 13 September) proved too much for the "chiggy bearers" of the Korean Service Corps 103d Division, but the Marines now had an alternative for the emergency resupply of ammunition and medical goods and the evacuation of the seriously-wounded: the Marine Corps helicopter. Although the light helicopters of Marine Observation Squadron 6 (VMO-6) had been a fixture in operations since August 1950, the battle for Kanmubong Ridge opened a new era in Marine Corps history, the combat employment of helicopters as an integral part of Marine air-ground operations. General Thomas and Colonel Krulak had both played key roles in developing the concept of vertical envelopment and fighting for funds to procure and test helicopters in HMX-1, the experimental helicopter squadron created at Marine Corps Air Station, Quantico, Virginia. HMX-1 gave birth in January 1951 to Marine Transport Helicopter Squadron 161 (HMR-161), commanded by a helicopter pioneer, Lieutenant Colonel George W. Herring. Herring brought HMR-161 to Korea in August 1951 ready to make its combat debut under the sharp eye of Krulak, who had made vertical envelopment his latest magnificent obsession. Herring's squadron of 300 Marines and 15 Sikorsky HRS-1 transport helicopters arrived at the airstrip (X-83) near the division command post at Sohwa-ri and moved in with VMO-6. Anticipating some employment in the weeks ahead, Krulak and Herring prepared the squadron for operations in combat landing zones and declared it ready for commitment on 12 September. Thomas told HMR-161 to carry supplies to the embattled Marines near Hill 793.

Operation Windmill I on 13 September lasted only about three hours, but its impact stretched into the future by years. In the short term it made sure that 2d Battalion, 1st Marines, commanded by Lieutenant Colonel Franklin B. Nihart, faced another day's battle with plenty of ammunition, water, and rations and without the burden of casualties. The first lift brought in a helicopter support team from the 2d Battalion to run the landing zone, and the remaining 27 flights delivered nine tons of cargo and evacuated 74 casualties. Not one helicopter was lost to ground fire or accident. A similar resupply mission would have required almost 400 Korean bearers and a full day to accomplish. Unlike an earlier parachute resupply mission to the Korean Marines,

Second Lieutenant George H. Ramer

Born in 1927 at Meyersdale, Pennsylvania, he enlisted in the Navy in 1944. After the war, he entered Bucknell University, from which he graduated in 1950 with a degree in Political Science and History. While attending Bucknell, he enrolled in the Marine Corps Reserve Platoon Leader's program and was commissioned in the Marine Corps Reserve. He taught high school civics and history in Lewisburg, Pennsylvania, before being called to active duty in January 1951 at his own request.

As a platoon leader with Company I, 3d Battalion, 7th Marines, in Korea, his bravery in covering the withdrawal of his platoon on Kanmubong Ridge on 12 September 1951 was recognized by the posthumous award of the Medal of Honor. His citation reads, in part:

> Second Lieutenant Ramer fearlessly led his men up the steep slopes and, although he and the majority of his unit were wounded during the ascent, boldly continued to spearhead the assault. . . . he staunchly carried the attack to the top, personally annihilated one enemy bunker with grenade and carbine fire and captured the objective with his remaining eight men.
>
> Unable to hold the position against an immediate, overwhelming hostile counterattack, he ordered his group to withdraw and single-handedly fought the enemy to furnish cover for his men and for the evacuation of three fatally wounded Marines. Severely wounded a second time, Second Lieutenant Ramer . . . courageously manned his post until the hostile troops overran his position and he fell mortally wounded.

Department of Defense Photo (USMC) A48025

In 1963, a facility for physical conditioning at Marine Corps Base Quantico, Virginia, was named in his memory.—Captain John C. Chapin, USMCR (Ret)

Whirlybirds

When Marine Transport Helicopter Squadron (HMR) 161 deployed to Korea, the squadron took with it an aircraft that pushed the technical state-of-the-art in helicopter design into a new frontier. Designated the HRS-1, the Sikorsky-designed and built helicopter had endured the inevitable ups and downs that characterized the introduction of any pioneering aircraft. Without government contracts, the Sikorsky Aircraft Division of the Vought-Sikorsky Corporation, Stratford, Connecticut, produced an aircraft designated the S-55, first flown in 1949. Initially marketing the aircraft as a commercial utility helicopter, Igor Sikorsky hoped the S-55 could compete with the Piasecki H-21 (or PD-22), which had been adopted by the U.S. Air Force for its air rescue service. The Navy, however, was in the hunt for a general-purpose helicopter that could be adopted for shipboard use. Naval aviators liked the S-55 because of its economical design, modest size, and serviceability.

Redesignated the HO4S-1 in its naval model, the S-55 represented at least two major engineering advances: the addition of a tail rotor for greater stability in flight and a front-mounted Pratt & Whitney R-1340-57 engine that could generate a respectable 600 horsepower. The engine placement helped solve a nagging problem of weight-distribution and flight characteristics. Prior helicopter models placed the engine directly under the rotor-blades, a design that gravely limited any so-designed helicopter to very light loads and insured flight instability. The front-mounted engine dramatically increased the helicopter's carrying capacity and simplified maintenance since the HRS-1 had clam-shaped nose doors that provided easy access to the engine for the ground crew mechanics. The new design also improved vertical flight stability.

In the earliest stage of evaluation, 1948-1949, Navy and Marine Corps officers, encouraged by Sikorsky, saw capabilities the helicopter did not yet have, even under optimum weather and altitude conditions. The original requirement the naval aviators placed on the helicopter was a 10-man load (225 pounds per Marine) to be carried 150 miles. The requirements shrank, as it became more and more obvious that the HRS-1 was not going to be a two-ton-plus lifter. All the helicopter's other characteristics, however, made it the aircraft of choice for the Bureau of Aeronautics, and the Marine Corps joined the program in August 1950, with an initial order of 40 aircraft.

The HRS-1s that went to Korea came into service with a gross weight rating (7,000 pounds at sea level) about 1,000 pounds slighter than originally designed with a payload reduced to 1,420 pounds under optimal flight conditions. Its troop load dropped from 10 to four to six. The helicopter's maximum speed remained at 90 knots, but its range had dropped by half to 70-mile round trips. Nevertheless, the HRS-1 was not a "whirlybird" of disappointment, but promise.

National Archives Photo (USN) 80-G-433347

none of the cargo drifted off to places and users unknown. The use of externally-slung, quick-release loads in cargo nets made easy. For the corpsmen and wounded Marines, helicopter evacuation meant that a hard-hit casualty could be transported to a medical clearing station ("battalion med") in 30 minutes, not doomed to a day-long stretcher ride. Even without accumulated statistics, medical personnel could already tell that medical evacuation helicopters would save lives and boost morale.

The plan for the 1st Marines to attack up Kanmubong Ridge continued to unravel despite the helicopter resupply and the commitment of two battalions, the 2d Battalion to take Hill 749 and the 3d Battalion to seize the ridgeline across the Soyang on Nihart's right flank. Nihart's battalion finally cleared Hill 749 after sharp fighting one company at a time with only a platoon in battle by the evening of 14 September. Before Nihart could mount another attack the next day, the North Koreans deluged his Marines with heavy artillery and mortar fire, pinning them to their Hill 749 positions. The North Korean regiment with accompanying artillery tried to throw the 2d Battalion off Hill 749 for four hours during the night of 15-16 September and left almost 200 bodies and many blood trails behind when it withdrew, but the battle cost the 2d Battalion almost 200 casualties and limited it as an offensive threat. Two Korean deserters reported that their regiment had 1,200 casualties.

A wounded squad leader of the 5th Marines ensures that a North Korean emplacement no longer threatens the advance along Kanmubong Ridge. The September attacks into the ridge mass north of the Punchbowl produced the most intense combat since the Chinese Fifth Offensive of April and May.

National Archives Photo (USMC) 127-N-A156867

Wornham now had to commit his reserve 1st Battalion to ensure that the complete Hill 673-Hill 749 complex was secure, leaving Thomas only one unbloodied regiment (the 5th Marines) to assault the heights of Kanmubong Ridge. At the cost of more than 800 casualties in the 7th and 1st Marines, the 1st Marine Division had only seized the ground identified five days before as the departure point for the more demanding advance up the spine of the ridge. Now it was the turn of Colonel Richard G. Weede's 5th Marines to continue the attack.

The battle of Kanmubong Ridge continued for four more days (16-20 September) and ended with the 5th Marines reduced by some 250 casualties and only Hill 812 securely under Marine control. The commanders of Weede's two assault battalions believed they could also have taken Hill 980, but it would have been difficult to hold with the peak (Hill 1052) still under North Korean control. The problems of Communist enfilade fire from the north simply got worse as the Marines worked their way to the west along the ridge. The Lieutenant Colonel Donald R. Kennedy's 3d Battalion in the left zone of action had its flank protected by Yoke Ridge and by tank fire, but the 2d Battalion working along the opposite slope enjoyed no advantages in cover and friendly fire, except close air support—which did not arrive. Staggered by its mounting casualties, the 2d Battalion stormed Hill 812 on the evening of 17 September. Without physical contact, the two battalions went into perimeter defenses, expecting North Korean counterattacks from the heights to their front or, in the case of the 2d Battalion, from the broken ground to the

Sergeant Frederick W. Mausert III

Department of Defense Photo (USMC) A46970

Born in 1930 in Cambridge, New York, he enlisted in the Marine Corps in 1948. Following recruit training at Parris Island, South Carolina, he was stationed at Cherry Point and Camp Lejeune, North Carolina, before going to Korea, where he participated in campaigns in South and Central Korea. Serving as a squad leader with Company B, 1st Battalion, 7th Marines, he was wounded on 10 September 1951. Two days later at Songnap-yong (Punchbowl), he was killed in a courageous action for which he was awarded the Meal of Honor. His citation reads, in part:

> Sergeant Mausert unhesitatingly left his covered position and ran through a heavily mined and fire-swept area to bring back two critically wounded men to the comparative safety of the lines. Staunchly refusing evacuation despite a painful head wound . . . [he] led his men in a furious bayonet charge against the first of a literally impregnable series of bunkers. Stunned and knocked to the ground when another bullet struck his helmet, he regained his feet and resumed his drive, personally silencing the machine-gun and leading his men in elimination several other emplacements in the area. Promptly reorganizing his unit for a renewed fight to the final objective on top of the ridge, Sergeant Mausert . . . still refused aid and continued spearheading the assault to the topmost machine-gun nest and bunkers, the last bulwark of the fanatic aggressors. Leaping into the wall of fire, he destroyed another machine-gun with grenades before he was mortally wounded by bursting grenades and machine-gun fire.—Captain John C. Chapin, USMCR (Ret)

north and east. Acutely aware of his danger and reduced supply circumstances, the 2d Battalion commander, Lieutenant Colonel Houston "Tex" Stiff, asked for helicopter resupply. In Windmill II, the helicopters of HMR-161 delivered six tons of scarce ammunition and engineering supplies in less than an hour on the afternoon of 19 September.

Just to hold his line across the ridge line, Colonel Weede had to bring up his uncommitted 1st Battalion, which fell in on Stiff's right and rear and allowed the hard-pressed 2d Battalion to consolidate its hold on Hill 812 and to find the 3d Battalion on its left. Even with the 5th Marines' lines more or less connected, the North Koreans made life miserable for the troops by sniping with long-range, high velocity antitank artillery guns and by attacking any patrols or outposts pushed forward of the main line of resistance. Two companies of the 2d Battalion became embattled for two days over control of "The Rock," a granite knob about 700 yards west of Hill 812. In a close-quarters melee of almost 24 hours, the Marines finally chased off the last of the Korean raiders. The Marine victors found 60 dead North Koreans scattered among the shattered rocks, but the victory cost the 5th Marines five dead and almost 50 wounded. Major Gerald P. Averill, the battalion operations officer, watched Marines shoot fleeing Koreans from the off-hand position while one Marine took photographs of the Korean corpses.

The battle for "The Rock" seemed almost symbolic since it had been a no-quarters fight for a piece of ground of little tactical significance. It also was the last part of the battle for Kanmubong Ridge, for General Van Fleet on 20 September ordered Byers to stop the offensive. The simultaneous battle of Heartbreak Ridge to the west of the Punchbowl had produced few results except soaring casualties in the 2d Infantry Division, and Van Fleet wanted all of X Corps fire support committed to that struggle. In addition, he approved of Byers' plan to shift the 8th ROK Division since the division—one of the better units in the South Korean army—had taken its objectives east of the Soyang River, though at prohibitive cost. Only in disgruntled retrospect did the Marines realize that they had fought in their last division offensive in Korea. The relief of the 8th ROK Division simply meant that the South Koreans would shift from the eastern to the western flank of X Corps.

The change meant that the 1st Marine Division assumed about five miles more of front in a sector already 15 miles in length. With the 1st KMC Regiment still holding the northern lip of the Punchbowl and the 5th Marines defending part of Kanmubong Ridge and the Soyang River valley, the 1st Marines assumed the mission of

National Archives Photo (USMC) 127-N-A157221

A rifleman of the 3d Battalion, 7th Marines, disembarks from a Sikorsky HRS-1 helicopter of Marine Helicopter Transport Squadron 161 during Operation Bumblebee. The operation allowed two Marine infantry battalions to exchange positions on Hill 702 by airlift, not overland march.

occupying the sector east of the river since the 7th Marines were the corps reserve. To take possession of its sector the 1st Marines had to control Hill 854 and Hill 884. Lieutenant Colonel Foster C. LaHue's 3d Battalion found to its dismay that the 21st ROK Regiment occupied the summit of Hill 854, but that the North Koreans still held almost all the northern face. For two days the battalion attacked and ran the survivors off, but the Marines lost 11 men to uncharted South Korean minefields and 50 more casualties in the fighting. LaHue's requests for essential air strikes were answered too late or not at all, again bringing the close air support issue to a boiling point with Thomas and Krulak.

Marine helicopters, however, provided one of the bright spots in the sector extension. To buy some time for another 1st Marines battalion to move to Hill 884 and to explore the possible routes to the hill—and any enemy ambushes or friendly minefields—Thomas ordered the division's Reconnaissance Company to move by helicopter to the summit of Hill 884 and to establish a patrol base there as well as replace the South Koreans. On 21 September, HMR-161 carried the first fully operational combat unit into a potential battle. Despite poor landing sites and marginal weather, the helicopters delivered 224 Marines and almost nine tons on supplies and equipment in four hours. The troops disembarked by "hot rope," a rappelling technique that does not require a snap-ring hook-up; the Marines and accompanying load slung from each aircraft could be delivered in 90 seconds after an eight minute flight from X-83, 15 miles away. General Thomas and Colonel Krulak complimented all hands with glowing messages. Van Fleet and Byers sent their own congratulations, fully realizing the potential of helicopter operations. The next operations, Blackbird on 27 September and Bumblebee on 11 October, produced mixed results, but Blackbird proved that HMR-161 could do limited night-work, and Bumblebee demonstrated that HMR-161 could move an entire battalion into a non-hostile landing zone. A helicopter-mobile briefing became a standard stop, dictated by Van Fleet, for VIPs, which included the Chairman of the Joint Chiefs of Staff, General Omar N. Bradley, USA. After his 1 October briefing, Bradley, no fan of the Marine Corps, admitted to his staff and accompanying journalists that the Marines might have discovered an operational technique that might change the conduct of land warfare. The September fighting that might not have been Iwo Jima II for the 1st Marine Division, but Bradley's faint praise gave a little more meaning to the battle for Yoke Ridge and Kanmubong Ridge. The surviving Marines felt that strange mixture of grief, guilt, relief, and satisfaction of veterans. They had upheld the reputation of the 1st Marine Division and X Corps for never shirking the most dangerous and onerous missions.

With the 1st Marine Division in place in its part of the Hays-Kansas Line, the division could assess its latest month of Korean combat. First, the North Korean army had proved more skilled and determined than the Chinese, but not immortal. The division intelligence section estimated that the Marines had inflicted about 10,000 casualties on the enemy. The number of enemy bodies actually counted numbered 2,799, and the Marines had taken 557 prisoners. Measured against its most taxing battles in World War II, Peleliu and Okinawa, the 1st Marine Division losses, compared with the casualties inflicted, appeared acceptable: 227 killed in action, and 2,125 wounded in action for a total of 2,452 casualties. Almost all the casualties occurred in the four infantry regiments and their attachments. The single most costly 24-hour period (39 killed and 463 wounded) was 13-14 September in the first phase of the attack on Kanmubong Ridge, which involved two battalions each of both the 1st and 7th Marines.

What gave the battles for Yoke Ridge and Kanmubong Ridge a special quality was the discouraging impact of the geography. If one stands along the Demilitarized Zone today—as the author did in 1994 and 1998—in the sectors in which the Marines fought around the Punchbowl, the mountain ranges stretch off without visible end into North Korea. It is difficult not to feel that there must be a better way to conduct war than to mount one attack after another against those forbidding (and still fortified) mountains. Surely the same thoughts came to the Marines of 1951 as they felt the first chill winds of winter on the Hays-Kansas Line.

A Long Winter and a Longer War

While battles still raged to the western zones of I and IX Corps and most of X Corps focused on the capture of Heartbreak Ridge, the 1st Marine Division secured its own portion of the new Minnesota Line. For the Marines the front became the Kansas Line bent, twisted, and renamed to include the terrain captured on the Yoke and Kanmubong Ridges. The signs of approaching winter were many. The distribution of cold weather clothing and equipment went on throughout the division, and the Marines' bunkers started to include stoves and makeshift furniture

Private First Class Edward Gomez

Department of Defense Photo (USMC) A46968

Born in 1932 in Omaha, Nebraska, he attended Omaha High School before enlisting in the Marine Corps Reserve in 1949. In Korea, he participated in three operations and was wounded in June 1951. With a strong premonition of death, he wrote his mother in September: "I am writing this on the possibility that I may die in this next assault . . . . I am not sorry I died, because I died fighting for my country and that's the Number One thing in everyone's life, to keep his home and country from being won over by such things as communism . . . .Tell Dad I died like the man he wanted me to be."

On 14 September, he was killed on Kanmubong Ridge, while serving as an ammunition bearer with Company E, 2d Battalion, 1st Marines, and saving the lives of four of his squad members. His Medal of Honor citation reads, in part:

> Boldly advancing with his squad in support of a group of riflemen assaulting a series of strongly fortified and bitterly defended hostile positions on Hill 749. Private First Class Gomez consistently exposed himself to the withering barrage to keep his machine-gun supplied with ammunition during the drive forward to seize the objective. As his squad deployed to meet an imminent counterattack, he voluntarily moved down an abandoned trench to search for a new location for the gun and, when a hostile grenade landed between himself and his weapon, shouted a warning to those around him as he grasped the activated charge in his hand. Determined to save his comrades, he unhesitatingly chose to sacrifice himself and, diving into the ditch with the deadly missile, absorbed the shattering violence of the explosion in his own body.

After the war, a plaque was dedicated in his honor at the Omaha Boys Club.

—Captain John C. Chapin, USMCR (Ret)

Corporal Joseph Vittori

Born in Beverly, Massachusetts, in 1929, he attended high school and worked on his father's farm before enlisting for three years in the Marine Corps in 1946. After being discharged, he joined the Marine Corps Reserve in 1950 for an indefinite tour of active duty. He trained at Camp Lejeune, North Carolina, until January 1951, when he joined Company F, 2d Battalion, 1st Marines, in Korea. Having been wounded in June near Yanggu, he was killed in the fight for Hill 749 in the Punchbowl on 15 September 1951 and became the second Marine of 2d Battalion, 1st Marines, within a 48-hour period to receive the Medal of Honor. His citation reads, in part:

> Corporal Vittori boldly rushed through the withdrawing troops with two other volunteers from his reserve platoon and plunged directly into the midst of the enemy. Overwhelming them in a fierce hand-to-hand struggle, he enabled his company to consolidate its positions . . . he assumed position under the devastating barrage and, fighting a single-hand battle, leaped from one flank to the other, covering each foxhole in turn as casualties continued to mount, manning a machine-gun when the gunner was struck down . . . With the situation becoming extremely critical . . . and foxholes left practically void by dead and wounded for a distance of 100 yards, Corporal Vittori continued his valiant stand, refusing to give ground as the enemy penetrated to within feet of his position. . . . Mortally wounded by enemy machine-gun and rifle bullets while persisting in his magnificent defense of the sector, where approximately 200 enemy dead were found the following morning, Corporal Vittori . . . undoubtedly prevented the entire battalion position from collapsing.

Department of Defense Photo (USMC) A46971

In 1986 there was a parade and memorial service in his honor, with a park named after him in his hometown of Beverly, Massachusetts.

Corporal Jack A. Davenport

An ardent athlete and a Golden Gloves champion, he was born in 1931 in Kansas City, Missouri, and enlisted in the Marine Corps in July 1950. Sent to Korea that December, he took part in four successive operations. Then, as a squad leader with Company G, 3d Battalion, 5th Marines, he died in a valorous action at the Punchbowl on 21 September 1951. His citation for the Medal of Honor read, in part:

Department of Defense Photo (USMC) A48023

> While expertly directing the defense of his position during a probing attack by hostile forces attempting to infiltrate the area, Corporal Davenport, acting quickly when an enemy grenade fell into the foxhole which he was occupying with another Marine, skillfully located the deadly projectile in the dark and, undeterred by the personal risk involved, heroically threw himself over the live missile, thereby saving his companion from serious injury or possible death. His cool and resourceful leadership were contributing factors in the successful repulse of the enemy attack.

The man in that same foxhole was Private First Class Walter L. Barfoot, and, due to Davenport's heroic self-sacrifice, he survived the war. Later, a gymnasium at Camp Pendleton, California, was named in honor of Corporal Davenport.

—Captain John C. Chapin, USMCR (Ret)

made from ammunition and equipment boxes. American fighting men make their lives as comfortable as possible; the Marines had no desire for a Valley Forge in the Taebaek Mountains. They also followed their instructions to make the Minnesota Line defensible without unnecessary casualties or the commitment of reserve forces. For the first time division staff officers reported how many sandbags the troops filled and placed and how many yards of barbed wire they strung in front of their positions.

Even without some progress in the armistice negotiations—and the plenary sessions did not resume until 25 October at a village along the Kaesong-Masan road called Panmunjom—the onset of winter alone would have given urgency to the 1st Marine Division's energetic development of its defensive area, some 14 miles across and 30 miles deep. The division could hardly afford to take its defensive mission lightly. On its left the 11th ROK Division showed some reluctance to either man the boundary or patrol it very carefully, which concerned the division staff. The U.S. 7th Infantry Division in the ridges west of the Punchbowl did better. The only advantage to the west was that the terrain and corridors offered less opportunity for the North Koreans. The division had few troops to spare since Byers decreed that one of the American Marine regiments would be the corps reserve and occupy positions 17 miles to the rear, but at least the reserve (initially the 7th Marines) could conduct rear area security patrols. After trying several combinations, Thomas committed six battalions on the main line of resistance and held three battalions in either regimental and/or division reserve. The division reserve battalion had the mission of patrolling the Kansas Line and protecting the main supply routes.

MajGen Gerald C. Thomas talks with MajGen Clovis E. Byers, USA, X Corps commander, center, and Gen Omar N. Bradley, USA, during the Chairman of the Joint Chiefs of Staff's visit to Marine frontlines. Among the tactics and problems discussed, Gen Bradley made note of the Marines' unique use of the helicopter in battle.

National Archives Photo (USMC) 127-N-A132072

The principal objective of the operations along the Minnesota Line was to drive the North Koreans away from their observation posts, combat outposts, and forward slope bunker defenses, and the Marines made major advances in this terrain cleansing in October and November 1951. The 11th Marines, occasionally reinforced by corps artillery, provided the umbrella of counterbattery fire that kept the Communist artillery in its caves. Their fire support burden eased with the arrival in January 1952 of an artillery battalion (four firing batteries) of Korean Marines; the U.S. Marine artillery advisors who had trained the battalion accompanied it to the front. The war on the bunkers, however, required much more than saturation shelling, given the strength of the Korean positions. The Marines went after the bunkers with 90mm tank guns, 75mm recoilless rifles, rocket launchers, and flamethrowers, supplemented by snipers with .50-caliber machine guns and scoped rifles. Some of the operations did not require close combat, but in other cases only heavy combat patrols, sometimes with tanks, would suffice. Such actions—night or day— cost some American lives. The Marines killed many more Koreans.

Some of the raids took the Marines time and again back to terrain they had learned to hate in September. Hill 1052 on Kanmubong (a North Korean strong point and observation post) became a favorite target, and "The

The Year of the Boot

In the autumn of 1951, the 1st Marine Division received a new piece of cold weather clothing: the boot, combat, rubber, insulated or Insulated Rubber Boot. No one called it anything else but "Mickey Mouse Boots" since their outsized shape and black color gave the wearer some podiatric similarity to Hollywood's famous rodent. Other names for the boots were less complimentary, but compared with the "shoe-pacs" they replaced, the Mickey Mouse boots quickly proved their value in preventing frozen feet.

The U.S. Army had conducted experiments with a cold weather boot during and after World War II, but by 1949 it had abandoned the effort since all the experimental prototypes did not meet Army standards for long-distance marching. Less concerned about the marching requirement, the Navy and Marine Corps conducted their own boot tests, 1948-1951, and concluded that one boot had merit. The field tests included wear in all sorts of cold weather and terrain conditions, and the Marines hiked in the boot and found it at least acceptable as winter footwear since no one marched very fast or far in inclimate conditions anyway. The Mickey Mouse boots arrived in Korea in August 1951.

The design of the insulated rubber boot was based on the concept that body-heat from the feet could be stored as a vapor barrier between two layers of felt-lined rubber. The airtight boot allowed the wearer to keep his feet warm with captive air, created by the wearer's own movement. The vapor barrier principle and the boot's all-rubber construction meant that cold and moisture from outside the boot would be defeated before they reached a Marine's precious feet. Only a boot puncture by shrapnel or some sharp object could ruin the boot's airtight integrity, and the boot, like early automobile tires, came with a patching kit.

Mickey Mouse boots, however, could turn the unwary and careless Marine into a frostbite casualty. The boots trapped more than heat. They also trapped sweat, and even if the feet remained warm, the moisture—with its ability to transfer heat four times more rapidly than dry air—accumulated, too. If a Marine did not stay on the move, the feet cooled, and the more sweat-soaked one's socks, the faster one's feet froze. One hour of inactivity could bring on an attack of frostbite. The standing operating procedure, therefore, for Mickey Mouse bootwear included a provision that each Marine had to dry his feet and change to dry socks at least once a day and preferably more often.

The next worse thing to having frozen feet, however, was preventing frozen feet. Changing socks and drying feet in the open air of a Korean winter tested the staunchest Marines. Units tried to establish a warming tent of some sort where the machocistic ritual could be performed with a hint of comfort and adequate time. Fortunately, the static winter war of 1951-1952 allowed such luxuries and cases of frozen feet in the 1st Marine Division dropped dramatically. The Mickey Mouse boots had come to stay.

Department of Defense Photo (USMC) A167955

Rock" received a new name that suggested the permanency of its final residents, "Luke the Gook's Castle." Marine patrols prowled the unoccupied terrain at night to discourage infiltrating Koreans. The frontline battalions had no monopoly on armed nighttime strolls. The rear areas of the 1st Marine Division (like those throughout Korea) were not safe from guerrillas, who preyed on road transport. The greater threat was teams of Communist artillery observers who infiltrated the 11th Marines positions and called in counterbattery fire of considerable accuracy, if not heavy weight of shell. Marine gun batteries lost both men and field pieces in such shoots. The rear area patrols used one capability to advantage: helicopter transportation. With every mission, the work of HMR-161 became more routine, and only the lift of entire battalions to and from the main line of resistance now justified codenames and special publicity. The squadron also received an aggressive new commander, Colonel Keith B. McCutcheon, whose work in aerial innovation had made him a legendary figure in the Corps.

None of the virtuoso campaign against the North Korean bunker system could reach the growing Communist in-depth system of fortifications, all duly observed and photographed by aerial observers.

National Archives Photo (USMC) 127-N-A157015

Marine foot patrols ranged farther into enemy territory during October, while tank-infantry raids in company strength, supported by air and artillery, were launched at every opportunity.

Naval gunfire—8- and 16-inch shells from cruisers and the battleship *New Jersey* (BB 62)—contributed to the bunker-busting, but not enough. The missing ingredient was close air support. Two Marine aircraft groups operated fighter-bombers within an hour's flight from the front, but they seldom came when called, and their Air Force and Navy comrades seemed even less available as they flew off to bomb railroads, tunnels, and roads off a target list dictated by Fifth Air Force.

General Thomas had grown ever unhappier with the lack of close air support. When he learned that Van Fleet had told Byers on 28 September that X Corps requested too much air, he ordered his staff to do a study on the lack of close air support in the September battles. His anger grew with his division's casualties. As a veteran of World War I, Thomas had his heart hardened early, but he never measured success by counting his own losses. He knew that the fights for Yoke Ridge and Kanmubong Ridge would have been much easier with Marine air on call. His own anger was fueled by deaths that touched him personally. One was the loss of his G-1, Colonel Wesley M. "Cutie" Platt, on 27 September. The most senior Marine officer to die in Korea, Platt had earned a special place in the Corps' history as one of the heroes of the defense of Wake Island. Now a shell ended a distinguished career and a special person. Thomas also knew that his division included as many as 20 sons of Marine generals and colonels, all eager to prove their own mettle. He wrote his wife that he worried about these "juniors" constantly, but could hardly ruin their careers and lives by protecting them. When one of the "juniors," First Lieutenant John C. Breckinridge died in combat, Thomas immediately pulled First Lieutenant James Breckinridge out of his infantry battalion and did so without regret since he wanted the family line of Major General James C. Breckinridge to survive Korea. The September battles had also turned Thomas into a critic of the Truman Administration, and he wrote his brother that the concept of "limited war" was ridiculous. China should be ruined as a Communist revolutionary power, and it would be easier to do it now, not later. The administration's concept of limited war offended him because "the wounds and worse acquired by Americans here had a real one hundred percent appearance . . . our guys are off base in D.C. and plenty."

Against this emotional background, Thomas again challenged the U.S. Air Force. Thomas fired the first shot in the new war with a letter asserting that his September casualties could have been reduced with timely close air support. He forwarded a study done by his staff that showed how much air he had requested and how little he had received. The 1st Marine Division had made 271 requests and had only 187 granted. More

Corporal Jack Davis: Veteran

After an ocean voyage to California, three or four days of out-processing, and an air flight to Tennessee, Corporal Jack Davis, U.S. Marine Corps Reserve, returned to Nashville in March 1952. He had been gone almost two years, and he was not quite 20 years old. In his well-tailored green service uniform, bearing ribbons for his two Purple Hearts, a Presidential Unit Citation, and a row of Korean service ribbons, Jack marched across the tarmac towards his waiting parents and older brother. They did not recognize him. He had been the only Marine in uniform on the plane. He also had lost 30 pounds and still sported "a Joe Stalin" mustache. Jack walked past his mother and father, then turned back to them: "You don't know your own son?" As Jack recalled, "mild bedlam ensued for a few moments."

After three or four days at home, however, Jack realized that he was having serious difficulty adjusting to life outside a combat zone. His brother went off to classes at Vanderbilt, and his parents both had full-time jobs. He even felt strange driving a new Nash Rambler after months of driving a Marine truck on Korean country roads. After about a week, he went out and found a job driving a big truck for a local construction company, and for the first time he felt mildly comfortable, earning 75 cents an hour, and enjoying the company of hard men doing hard work. He did not miss being shelled. His parents found his job puzzling and worried about his sanity, but his new job started Jack on the road to normalcy. "The Marine Corps does a marvelous job making civilians into warriors, but then to turn those persons back to civilian life approximately thirty days after leaving a combat zone, with no decompression period stateside, acclimatizing back to a quasi-polite environment, leaves a bit to be desired."

With his usual enthusiasm, Vince Davis tried to help his brother reenter Nashville—actually Vanderbilt—social life, but the well-intentioned effort on Jack's behalf did not go well. Jack found little in common with Vince's friends in Vanderbilt's senior class, and he concluded they would find his service stories meaningless. Besides, Jack hardly trusted his ability to carry on a social conversation without lapsing into standard Marine language, which would have set all of Davidson County on fire. "I was so unsure of my polite verbal utterances that I was barely better than a mute. And for a while I preferred not to go out in polite society until I had reacquired some social skills."

In the autumn of 1952, Jack Davis entered Vanderbilt University, his education partially funded by a disability pension and the GI Bill. When he graduated with a degree in geology in 1956, he also had a new wife, Joan Fortune of Lafayette, Georgia. Even with a Master's of Science in geology (1959) Jack found the life of a petroleum geologist too uncertain for a family man (two sons), and he shifted into sales management in the oil and pharmaceutical industries until his semi-retirement and return to Tennessee in 1971. A small advertising business he and some local friends created keep him busy enough—along with church and civic activities in his new hometown of Winchester, Tennessee—until the partners sold the business to a larger Texas-based company in 1998. Jack and Joan now have more time to enjoy their hobbies and extended family.

Jack has never returned to Korea, but he is proud to have been the teenage rifleman of 1951 who found himself in Korea with "Bloody George" Company, 3d Battalion, 1st Marines.

---

serious to operations, only 32 had arrived within 30 minutes, the Marine Corps standard. During various conferences in October, Thomas had an opportunity to discuss the issues with General J. Lawton Collins, USA, the Army Chief of Staff, and General Ridgway. Although Byers appreciated his aggressiveness, Van Fleet and Everest did not. Thomas thought Collins and Ridgway liked his letter, "a stick to beat the Air Force with." Ridgway said he wanted Van Fleet and Everest to look at the Joint Operations Center system and see if it could be adjusted, at least to give the 1st Marine Division 40 sorties a day. Van Fleet, however, argued that X Corps got too much air support already. Thomas and Byers decided to push the issue; Byers' outrage was fueled by another problem, the unwillingness of the Army to send its best officers and noncommissioned officers to combat assignments in Korea. Byers also bridled at Van Fleet's suggestion that his faltering generalship

Birthday with a heavy strike north of Kanmubong Ridge. The X Corps study, directed by Lieutenant Colonel Ellis W. Williamson, a pioneer in Army aviation and air mobility operations, produced similar results. For 1-20 November, the entire corps had made 224 requests and had 145 requests filled in some way. Forty-six requests took more than two hours to fill, and Williamson, X Corps G-3, judged that 42 of these strikes came too late to have the anticipated results.

The X Corps' analyses, like studies submitted earlier in the year by General Almond, changed nothing. Van Fleet and Ridgway saw no immediate advantage in pressing the issue, whatever their personal views. The major influence on their commitment to the Joint Operations Center-status quo was the news from Panmunjom. The negotiators had accepted the United Nations position that the armistice line should be the line of troops when the armistice was signed, not the 38th Parallel. Each belligerent coalition would withdraw four kilometers from the point of contact and thus establish a neutral zone between the armies. Van Fleet had anticipated the agreement on 14 November and ordered that no operations by a battalion or larger formation could be mounted without the approval of the corps commander. When the negotiators signed an agreement on 25 November on the line-of-contact solution, Eighth Army interpreted the agreement as an omen of an early ceasefire. The word went forth throughout the front to hold down casualties, conserve ammunition, defend the current positions, and even to limit patrols to those areas where earlier patrols had made contact with the enemy. These rules of engagement would be in effect for at least

National Archives Photo (USA) 111-SC382827

Gen Matthew B. Ridgway, left to right, Gen J. Lawton Collins, U.S. Army Chief of Staff, Gen James A. Van Fleet, U.S. Eighth Army commander, and MajGen Clovis E. Byers, commanding general of U.S. X Corps, study the situation map during Gen Collins' tour of the frontlines in Korea. Within a month of Collins' visit, Ridgway relieved a stunned Byers for what some believed was his support for the Marine position on close air support.

explained the losses of the 2d Infantry Division on Heartbreak Ridge. Byers put his staff to work on studies like those underway in the 1st Marine Division. Byers and Thomas also raised the related issue of dictated artillery shell expenditures, which they claimed produced predictable shortages when real fighting occurred. Van Fleet was not happy with the two senior generals of X Corps.

Byers and Thomas mustered more evidence in November that the Joint Operations Center had willfully prevented X Corps from receiving effective close air support. The 1st Marine Division claimed that it had made 188 requests for air support and received only 53 strikes in response. In the case of 86 requests, Fifth Air Force said it had no aircraft available, and poor weather had affected most of the other requests. The 1st Marine Division's statistics provided an even more damning picture. During the 30 days of November, the division had requested air support on 26 days and received no response at all on 12 of those days. On the days that the Joint Operations Center responded, it approved only 52 of 125 close air support requests. In terms of aircraft and ordnance, only four missions were flown as requested, and only one arrived in less than an hour from the original request. The only mission that went as planned occurred on 10 November when 89 aircraft from the 1st Marine Aircraft Wing celebrated the Marine Corps'

30 days. The instructions had "a considerably inhibitory effect on the operations of the division." They also meant that a change in the system of air support had been overtaken by events.

An examination of the fighting by the 1st Marine Division in October and November 1951 suggests how large an opportunity cost the ground forces paid for the Air Force interdiction campaign. The aggressive ground operations and use of artillery and naval gunfire demonstrated that even without close air support, the 1st Marine Division (and probably most of the U.S. Army divisions in Korea) could still inflict substantial casualties on the enemy. The Chinese might have unlimited manpower to throw into the battle, but the North Koreans did not. In two months of operations that can only be characterized as "defensive," the 1st Marine Division killed 1,117 North Koreans and captured 575 more at a cost of 87 dead or missing Marines (both American and South Korean) and 573 wounded. The 1st Marine Division order-of-battle analysts estimated that the division and air strikes had caused as many as 12,000 more casualties in the three North Korean divisions that faced the Marines. Even if wildly optimistic, the estimates were probably not completely illusory. If, for example, the Marines had killed or wounded only one-third of the enemy estimated, they still would have accounted for more than 6,000 enemy casualties at a cost of 660 losses of their own. An exchange ratio of 10:1—given the rules of engagement—is an operational achievement in any war, but the Eighth Army missed the lesson.

As the pace of the war congealed with the coming of winter and the hope of an armistice, the 1st Marine Division passed through another set of organizational milestones. It celebrated the Marine Corps' 176th birthday with the traditional ceremonies of reading John A. Lejeune's birthday message and cake-cutting; General Thomas added a special wrinkle, a one-round salute at noon 10 November from every weapon in position, which the commanding general found "very satisfying." Thomas then hosted Byers and X Corps staff for lunch.

A rifle platoon celebrates the traditional Marine Corps Birthday, 10 November, with a no-frills cake cutting on the reverse slope of battle-scared Kanmubong Ridge.

Department of Defense Photo (USMC) A159434

Thomas also drew satisfaction from an extensive study done by X Corps for the Department of Defense. Sensitive to Army carping about Marine tactics and casualties, Thomas could point to irrefutable statistics: in both raw number and percentages the 1st Marine Division in 1951 had one of the three lowest loss rates in the Eighth Army. Within X Corps its losses in combat were half those of the U.S. 2d Infantry Division in percentage terms and 50 percent lower in raw numbers. Between 1 June and 15 October, the Army division suffered 6,247 casualties, the Marine division, less the 1st KMC Regiment, 4,241. During the first 10 months of 1951, the 1st Marine Division had rotated 11,637 Marines out of Korea and received 13,097 replacements, which did not quite cover the losses from all sources. In the autumn of 1951 there were still almost 5,000 Marines in the division who had joined the division in 1950, but by Christmas these veterans (the vast majority technical specialists) had all gone home. By the end of December, one-third of the division's Marines had come to Korea since early September, but the division showed no signed of reduced effectiveness. The 14th Replacement Draft (2,756 officers and men) closed the gap, but the 11th ("Home for Christmas") Rotation Draft opened it again.

From the perspective of Headquarters Marine Corps and the Commanding General, Fleet Marine Forces, Pacific, the 1st Marine Division had done a splendid job in 1951 as a fighting organization and as a source of favorable publicity. Both Generals Cates and Shepherd visited Thomas in the last two months of 1951 and congratulated him on his successful command. By November, Shepherd knew that he would replace Cates as Commandant and he told Thomas and Krulak that he wanted both of them back in Washington to run the staff of Headquarters Marine Corps. Thomas' successor pleased the incumbent division commander: Major General John Taylor Selden, an accomplished officer who had commanded the 5th Marines on Cape Gloucester and served as the division chief of staff on Peleliu. A Virginian of "First Family" roots as well as another "mustang" of the World War I era, Selden had a deserved reputation for getting along well with the Army without compromising Marine Corps interests.

The change of command for X Corps showed none of the good feeling that accompanied Thomas' departure in early January 1952. On his latest visit to Korea in November, J. Lawton Collins asked Van Fleet how Byers was doing as a corps commander and whether he met the World War II standards of the Army in Europe. Van Fleet responded that Byers did not match the best corps commanders of Eisenhower's army, which meant that Byers was not Collins, Ridgway, or Van Fleet. Without warning, Byers learned from Van Fleet on 24 November at a ceremony presenting a Presidential Unit Citation to the 2d Infantry Division that Byers would be reassigned as commanding general, XVI Corps, the theater reserve force just constituted in Japan. Without ceremony or any chance to visit his subordinate commanders, Byers flew to Japan as

National Archives Photo (USMC) 127-N-A159279

With a resumption of the armistice talks, a lull set in along the 1st Marine Division's front in December. While patrols were sent out to maintain pressure on the enemy, work continued on winterizing bunkers and improving defensive installations, such as this machine gun position.

ordered four days later, his distinguished career in eclipse. His replacement, Major General Williston B. Palmer, was a Europeanist and a Collins intimate. General Thomas remained convinced that Byers had been too friendly with the Marines for Collins' and Ridgway's taste and too assertive in demanding changes in the close air support system.

The best way to deal with the other armed forces, whether the battlefield was in Washington or the entire Pacific theater, had always been obvious to Cates, Shepherd, Thomas, and Shepherd's successor in Hawaii, Lieutenant General Franklin A. Hart. The answer was to attack and withdraw at the same time. Hart's mission was to support the 1st Marine Division and 1st Marine Aircraft Wing, but he had a second responsibility, which was to prepare a new 3d Marine Brigade and supporting aviation elements for possible use (including amphibious operations) in the Pacific. Pending legislation in Congress suggested that in 1952 the Marine Corps would add a third division and aircraft wing to the Fleet Marine Force. The Marine Corps also was investigating arming itself with tactical nuclear weapons. With Colonels Wornham, Hayward, and Nickerson on Hart's staff, the interests of the 1st Marine Division would not be ignored, but as a Service the Marine Corps was not inclined to pursue the close air support issue when the interservice relations landscape looked good for the moment and the future of operations in Korea so uncertain. Another factor was simply that Fleet Marine Force, Pacific, and its Navy superiors, including Admiral Arthur H. Radford, Commander in Chief, Pacific, proposed that the 1st Marine Division and 1st Marine Aircraft Wing be withdrawn from Korea and placed in strategic reserve. When General Hart made his first command tour of Korea and Japan, he found the senior commanders in Korea and Japan convinced that Eighth Army would waste its Marines. The division and aircraft wing should be placed in Japan and brought to a peak state of readiness for battles that mattered.

Nothing that occurred in December 1951 along the Minnesota Line gave any clues that the war would either end or be fought with any rational purpose, as seen from the 1st Marine Division. The division's defensive posture faced no serious menace from the North Koreans, although the Communist troops became progressively more aggressive with their patrolling and use of artillery fire. A summary of one day's operations (8 December 1951) catches the winter war along the Minnesota Line. A 5th Marines patrol exchanged gunfire with a Korean patrol without known results; the regiment called in 117 rounds of naval gunfire on a bunker system and did acceptable damage to seven bunkers. The 7th Marines sent out a patrol to retrieve the body of a dead Marine and engaged in a firefight with an enemy patrol. The Korean Marines sent out two patrols, which were fired upon by machine guns in hidden bunkers, but took no casualties. The 11th Marines fired 14 observed missions on bunkers or in patrol-support. The following days were no different. On Christmas Day, the North Koreans tried to disrupt the division's hot holiday meal and a visit by Cardinal Francis J. Spellman but only drew smothering artillery and naval gunfire on the NKPA combat patrols. The next day a heavy snow slowed the action even more, and on 27 December Eighth Army announced that even though no truce agreement had been signed, the restrictive rules of engagement remained in effect.

Cpl Kevin J. Griffin receives a blessing from Cardinal Francis Spellman during the cardinal's Christmas visit to the 1st Marine Division.

National Archives Photo (USN) 80-G-436954

The dawn of a new year did not change the pattern of warfare for the 1st Marine Division. Such novelties as psychological warfare units—the masters of the surrender leaflet and the insinuating broadcast—became regular fixtures at the front, first a source of amusement and later an object of contempt. Army searchlight batteries added little light to the operations. An epidemic of boredom and carelessness spread throughout Eighth Army. To give at least a hint of battle, Van Fleet's staff dreamed up Operation Clam-Up, 9-15 February 1952, as a way to draw the Chinese and North Koreans into an aboveground killing zone through deception. The basic concept was that the frontline battalions would either go underground or appear to withdraw from the front; allied artillery would reduce their firing to almost nothing; and the usual patrols would not go out. Presumably, the collective impression would be that the United Nations forces had fallen back to the Kansas Line. Van Fleet's grand deception did not fool the Communists—at least not much. In the 1st Marine Division sector—especially on Yoke and Kanmubong Ridges—the North Koreans sent out only patrols, which set off a series of small battles. The North Koreans also deluged the main line of resistance with artillery fire, a sure sign that they had not been fooled. For all the sound and fury the casualties showed how insignificant Clam-Up had been. For February the Marines lost 23 killed and 102 missing; they killed by count 174 Koreans and took 63 prisoners. Whether they had actually inflicted an additional estimated 1,000 more casualties was guess work.

Department of Defense Photo (USMC) A159378

Oblivious to the wet snow and North Korean harassing fires, Marines line up for a Christmas dinner with all the trimmings.

In other aspects of the division's operations, the order-of-the-day became doing less with less. The helicopters of HMR-161 developed stress fractures in their tail assemblies, so Colonel McCutcheon grounded his squadron until the defects could be corrected. The lack of helicopters slowed the modest counter-guerrilla campaign in the rear areas. Certainly the North Korean lines had become impenetrable. The G-2 estimated that 21 infantry battalions and nine artillery battalions, all embedded in the ridges to the north of the main line of resistance, faced the division. Although the Marines did not yet know what great plans Eighth Army held for them, their campaign in east-central Korean was ending not with a bang, but a shrug.

In Retrospect

From its initiation in battle as part of the U.S. Eighth Army in January 1951 until its eventual movement to an entirely new zone of action in western Korea in early 1952, the 1st Marine Division fought with as much distinction as its 1950 edition, the division that landed at Inchon, liberated Seoul, chased the North Korean army away from Korea's northeast coast, and blunted the first appearance of the Chinese army in the battle of Sudong. The advance to the Chosin (Changjin) Reservoir and "the attack in another direction" to Hungnam added more honors to the 1st Marine Division and created a tradition of valor and professionalism that shares pride of place in the memory of Marines with Tarawa and Iwo Jima.

Yet the 1st Marine Division in 1951 added a new and equally useful tradition of valor: that a Marine division in war of diminishing rewards, fought under unpleasant physical conditions and uneven Army leadership, could maintain the Corps' highest standards even without the constant stroking of admiring reporters and camera crews. To be sure, the division suffered more casualties in all

National Archives Photo (USMC) 127-N-A159018

Marine TSgt John Pierce gives some North Korean soldiers hiding in a bunker a warm reception as he throws a white phosphorus grenade into the bunker's entrance.

categories per month of combat in 1950 (1,557) than it did in 1951 (747) or 1952 (712), but there is no convincing evidence that the division inflicted more casualties upon the enemy in 1950 (per month of combat) than it did in 1951, only that its battles had been more dramatic and photogenic.

In an official sense, the 1st Marine Division in 1951 received the same recognition as its 1950-predecessor, the award of a Presidential Unit Citation "for extraordinary heroism in action against enemy aggressor forces in Korea." There are, however, no battles in the citation, no identified geographic locations like Hill 902 or Taeu-san and Taeam-san Mountains or Yoke Ridge or Kanmubong Ridge. The only geographic location mentioned is the Punchbowl and some vague terrain "north of the Hwachon Reservoir." The citation gives only three sets of dates: 21-26 April, 16 May-30 June, and 11-25 September 1951. There is no book like Andrew Geer's *The New Breed* to honor the 1951 Marines, no collection of memorable photographs by David Douglas Duncan to freeze the fatigue and horror of war on the faces of young men turned old in a matter of days in December 1950. The same faces could have been found on Hill 924 or Hill 812 if anyone had looked.

In addition to adding to the heritage of heroism in the second year of the Korean War, the 1st Marine Division and the 1st Marine Aircraft Wing made history for its introduction of the transport helicopter to American ground operations. In January 1951, the Landing Force Tactics and Techniques Board, Marine Landing Force Development Center, approved the first doctrinal study of vertical envelopment in *Employment of Assault Transport Helicopters*. When the first operational squadron, HMR-161, was formed, its original name was "Marine Assault Helicopter Squadron 161" until Marine aviation bureaucrats protested that they and the Navy thought the more comprehensive designator of "transport" was more appropriate. The name mattered less than the mission. When the squadron began operations with Windmill I (13 September 1951) until the tail section fractures grounded the helicopters on 28 February 1952, HMR-161 conducted six major operations and many hundreds of other less dramatic flights with troops, weapons and supplies. A concept developed for amphibious assaults received its first test among the mountains of Korea, an irony that bothered no one among the community of Marine helicopter pioneers. The future arrived to the sound of flailing rotors and storms of ground-effect dirt on a bit of ground on the lower slopes of Hill 749, Kanmubong Ridge. Today the site is somewhere within the Demilitarized Zone.

What make the fighting qualities of the 1st Marine Division, aided without stint by VMO-6 and HMR-161, even more remarkable was the growing difficulty in obtaining close air support and the suspicion after July 1951 that there would be a substitute for victory. As the 1st Marine Division proved in September 1951, the young riflemen and platoon commanders might not be ready to kill and die for a stalemate, but they were more than willing to kill and die for each other, and that was what was important to them then. And it still is.

An alert Marine rifleman, framed in the doorway, provides cover while another Marine searches through an abandoned Korean farmhouse. Guerrilla and infiltrator attacks forced all Marine units to mount security patrols and to defend their positions.

National Archives Photo (USMC) 127-N-A159239

Essay on Sources

The archival sources on the 1st Marine Division, X Corps, and the U.S. Eighth Army for the campaign of 1951-1952 are voluminous, but the place to start is the monthly organizational historical reports, usually containing annexes of other reports and studies, submitted to the service headquarters for permanent retention and reference use. For the 1st Marine Division, I used the monthly historical reports for April 1951 through February 1952, supplemented by similar monthly historical reports made by Headquarters, Fleet Marine Force, Pacific (FMFPac), for the same period. Both 1st Marine Division and FMFPac reports included invaluable appendices. Of special use for this study were two Type C Special Reports:"Employment of Assault Helicopters," 4 October and 15 Novem-ber 1951; and "1st KMC Regiment and Its Relationship to the 1st Marine Division, September 1950-May 1952," 13 June 1952. The FMFPac historical reports include memoranda for the record of the weekly staff conferences and the travel reports for Generals Lemuel C. Shepherd, Jr., and Franklin H. Hart for their visits to Korea and Japan. The original reports are in the Records of Marine Corps Field Commands, Record Group 127, in the National Archives, but copies may often be found at the Marine Corps Historical Center, Washington, D.C., and the Marine Corps Research Center, Quantico, Virginia.

The X Corps Command Reports for April 1951- February 1952 may be found in Command Reports, 1949-1954, Records of U.S. Army Field Commands, Record Group 407, but these records— which include such things as the "Commanding General's Diary," which is schedule and commentary as maintained by his aides— can also be found in duplicate form in key Army educational and research repositories like the U.S. Army Center of Military History, Ft. Leslie J. McNair, Washington, D.C. and the U.S. Army Military History Institute, Carlisle Barracks, Pennsylvania. Like their Marine Corps counterparts, the monthly "Com-mand Reports" include special studies, which for X Corps included analyses of close air support and personnel matters.

The senior Army and Marine commanders in Korea, 1951-1952, maintained extensive personal files that are open to researchers, and I used these invaluable sources extensively. General Matthew B. Ridgway, Commander, United Nations Command, and Commanding General, Far East Command during this study, kept extensive correspondence and memoranda files, Matthew B. Ridgway Papers, U.S. Army Military History Institute (MHI). His successor as Commanding General, U.S. Eighth Army, General James A. Van Fleet, maintained a personal journal and conducted an extensive correspondence with his military and civilian contemporaries, all preserved in the James A. Van Fleet Papers, George C. Marshall Library, Lexington, Virginia. Van Fleet, like Ridgway, kept essential data, studies, maps, memoranda of staff meetings, and orders/instructions, but unlike Ridgway, he never wrote a book about the war. His papers are especially important on interservice and intercoalition command relations.

Lieutenant Generals Edward M. Almond and Clovis E. Byers saved extensive files for their periods of command of the U.S. X Corps, which included all but two months of the period covered in this study. Especially important is Headquarters, X Corps, "Battle of the Soyang River," June 1951, a special report with extensive intelligence studies and fire support studies. Almond's papers are essential sources on the conduct of the Korean War. Almond did an especially good job at creating subject files, two of the most important containing material on close air support and artillery employment. Like Van Fleet, Almond kept extensive personal notes. After his retirement Almond became a subject of extensive interviewing, the most exhaustive by Lieutenant Colonel Thomas Ferguson, USA (Almond's son-in-law), Professor D. Clayton James, and John Toland, and these transcripts are attached to the Almond Papers and the Douglas MacArthur Papers, the General Douglas MacArthur Library and Memorial, Norfolk, Virginia. Throughout his retirement Almond continued to collect documents and add them to his collection. A complementary view of X Corps may be found in the oral memoir of General Frank T. Mildren, the corps G-3 in 1951, Senior Officers Oral History Project, 1980, MHI. The Clovis E. Byers Papers are held at the library of the Hoover Institution on War, Revolution and Peace, Stanford, California, and include his copies of the corps reports and studies and his "Commander's Diary." Byers also corresponded with many of his Army contemporaries and friends in other services. He kept a personal diary that includes the touching story of his "reassignment." Byers' comments on the South Korean divisions in his corps are important in giving a full picture of corps operations.

General Gerald C. Thomas did not maintain the vast correspondence files or intimate diaries of his Army contemporaries, but he wrote his wife, his brother, and his oldest son and son-in-law, both Marine officers and both 1952 members of the 1st Marine Division. The Thomas family allowed me to read this correspondence while I did research on *In Many a Strife: General Gerald C. Thomas and the U.S. Marine Corps, 1917-1956* (Annapolis: Naval Institute Press, 1993). General Thomas also left multiple, extensive oral histories, 1966-1979, and assorted files, now part of the Oral History Collection and Personal Papers Collection at the Historical Center and Research Center. Other personal collections and oral histories from important sources are the late General Lemuel C. Shepherd, Jr., USMC; Lieutenant General Herman Nickerson, USMC (Ret); Lieutenant

General Victor H. Krulak, USMC (Ret); the late General Keith B. McCutcheon, USMC; the late Major General Wilburt S. Brown, USMC; and the late Major General Thomas A. Wornham. During my research on General Thomas, Generals Krulak and Nickerson furnished me with personal papers and did so again on this project. A more extensive discussion of Marine Corps sources may be found in the Thomas biography and my *Semper Fidelis: The History of the United States Marine Corps*, rev. ed. (New York: The Free Press, 1991).

The official documents and perspectives of the senior officers of the X Corps and 1st Marine Division do not provide a complete picture of the campaign of 1951 in human terms. With his complete cooperation, I used the story of Corporal Clarence Jackson Davis, USMCR, as a way to see the fighting from the perspective of the enlisted combat Marine. I focused on the experiences of a special group of Marine officers, the 7th Basic Class, those Marine lieutenants commissioned in the spring of 1950 who became the platoon commanders of 1st Marine Division in 1951. Their contribution began with an interview with Captain Frederick F. Brower, USMC (Ret) in 1998 and went on to access to Lieutenant General Charles G. Cooper, USMC (Ret) "Blood and Tears," an unpublished memoir; Mr. John E. Nolan, "Korea Comments," 11 December 1999; and interviews at the 50th Reunion of the 7th Basic Class (4-7 May 2000) with Colonel Earl T. Roth, USMC (Ret), Mr. Harold Arutinian, and Colonel David J. Hytrek, USMC (Ret).

For sardonic views of the campaign of 1951, see Paul N. McCloskey, Jr., *The Taking of Hill 610* (Eaglet Books, 1992); Lieutenant Colonel Gerald P. Averill, USMC (Ret), *Mustang: A Combat Marine* (Presidio Press, 1987); [Private First Class] Burton F. Anderson, *We Claim the Title* (Tracy Publishing, 1994); and [Sergeant] A Andy Andow, *Letters to Big Jim Regarding Narrul Purigo, Cashinum Iman* (Vantage, 1994).

The official histories of the 1951 campaign for the Marine Corps and Army are much used and often-cited, but should not be used as scripture: Lynn Montross, Major Hubard D. Kuokka, USMC, and Major Norman Hicks, USMC, *The East-Central Front*, Vol. IV, *U.S. Marine Operations in Korea, 1950-1953* (Historical Branch, G-3, Headquarters U.S. Marine Corps, 1962) and Billy C. Mossman, *Ebb and Flow: November 1950-July 1951* in *U.S. Army in the Korean War*, five volumes to date (Office of the Chief of Military History, U.S. Army, 1990) and Walter G. Hermes, *Truce Tent and Fighting Front* (1966), another volume in the same series. The Air Force official history is Robert F. Futrell, *The United States Air Force in Korea, 1950-1953* (rev. ed., Office of Air Force History, 1983). The documentation for the close air support controversy may be found collected in Subject File K239-04291-1, "Close Air Support," Research Archives, Air Power Historical Research Center, Air Force University, Maxwell Air Force Base, Alabama. Allan R. Millett, "Korea, 1950-1953," in Benjamin F. Cooling, ed., *Case Studies in the Development of Close Air Support* (Office of Air Force History, 1990) covers the issues and the source material in detail. Lynn Montross, *Cavalry of the Sky: The Story of U.S. MarineCombat Helicopters* (New York: Harper & Bros., 1954) is a popular account of HMR-161's Korean War service. A more conventional official account is Lieutenant Colonel Eugene W. Rawlins, USMC, *Marines and Helicopters, 1946-1962* (History and Museums Division, HQMC, 1976).

I visited most of the battle sites described in this study in 1994 and 1998, and I have profited from the advice of Brigadier General Edwin H. Simmons, USMC (Ret) and Colonel Franklin B. Nihart, USMC (Ret), both veterans of the campaign in infantry battalions. Gunnery Sergeant Leo J. Daugherty III, USMCR, provided valuable research assistance.

About the Author

The Raymond E. Mason, Jr., Professor of Military History, Ohio State University, Allan R. Millett is a specialist in the history of American military policy and institutions. He is the author of four books: *The Politics of Interven-tion: The Military Occupation of Cuba, 1906-1909* (1968); *The General: Robert L. Bullard and Officership in the United States Army, 1881-1925* (1975); *Semper Fidelis: The History of the United States Marine Corps* (1980, revised edition, 1991); and *In Many a Strife: General Gerald C. Thomas and the U.S. Marine Corps, 1917-1956* (1993). His most recent book, co-authored with Williamson Murray, is *A War to be Won: Fighting the Second World War* (2000). He also co-authored and co-edited several other works on military affairs and has contributed original essays to 25 books and numerous journals on American historiography, foreign and defense policy, and military history. A noted lecturer and officeholder in many prestigious military history societies, Dr. Millett is now president of the U.S. Commission on Military History.

A graduate of DePauw University and Ohio State University, Dr. Millett served on both active and reserve duty, retiring in 1990 with a rank of colonel in the U.S. Marine Corps Reserve.

STALEMATE
U.S. Marines from Bunker Hill to the Hook

by Bernard C. Nalty

On a typical night during 1952, a Marine patrol set out from the very center of a company position on the Jamestown Line in west-central Korea. The group was following the trace of an abandoned trench-line when a Chinese machine gun cut loose, killing the leader, wounding some of his men, and forcing the patrol to return without completing its mission of setting an ambush.

Shortly afterward, about two hours before midnight, Second Lieutenant William A. Watson, who had recently joined the 1st Marine Division, received orders to move out with a squad from his platoon and set up the ambush, finishing what the ill-fated patrol had begun. The powerful searchlight aimed skyward to warn airmen of the location of Panmunjom, where the United Nations forces were conducting truce talks with the North Korean and Chinese, reflected from the clouds creating the impression that Watson's patrol was "walking in bright moonlight."

The lieutenant and his men moved between the spine of a ridgeline and the trench they were following, watching carefully for signs of a Chinese ambush and maintaining enough space between Marines to minimize the effect of a sudden burst of fire. "Creep, sit, wait," Watson told his men. "Move on my order. A few feet and be still." The Marines were confident that their cautious advance, the 50 or so yards separating their route from the nearest concealment the enemy could use, the artificial moonlight, and the trench itself, which provided ready cover in case of an attack, would combine to prevent the Chinese from surprising them.

AT LEFT: *The 1st Marine Division engaged in static warfare during 1952 from typical segments of trench-line on the Jamestown Line.* Department of Defense Photo (USMC) A167091

While two Marines provide protection by watching for enemy snipers, two other members of a patrol probe for mines. The Marines in the foreground wear armored vests. By November 1952 delivery of the new vests to the division was completed, including more than 400 sets of lower torso armor.
National Archives Photo (USMC) 127-N-A160817

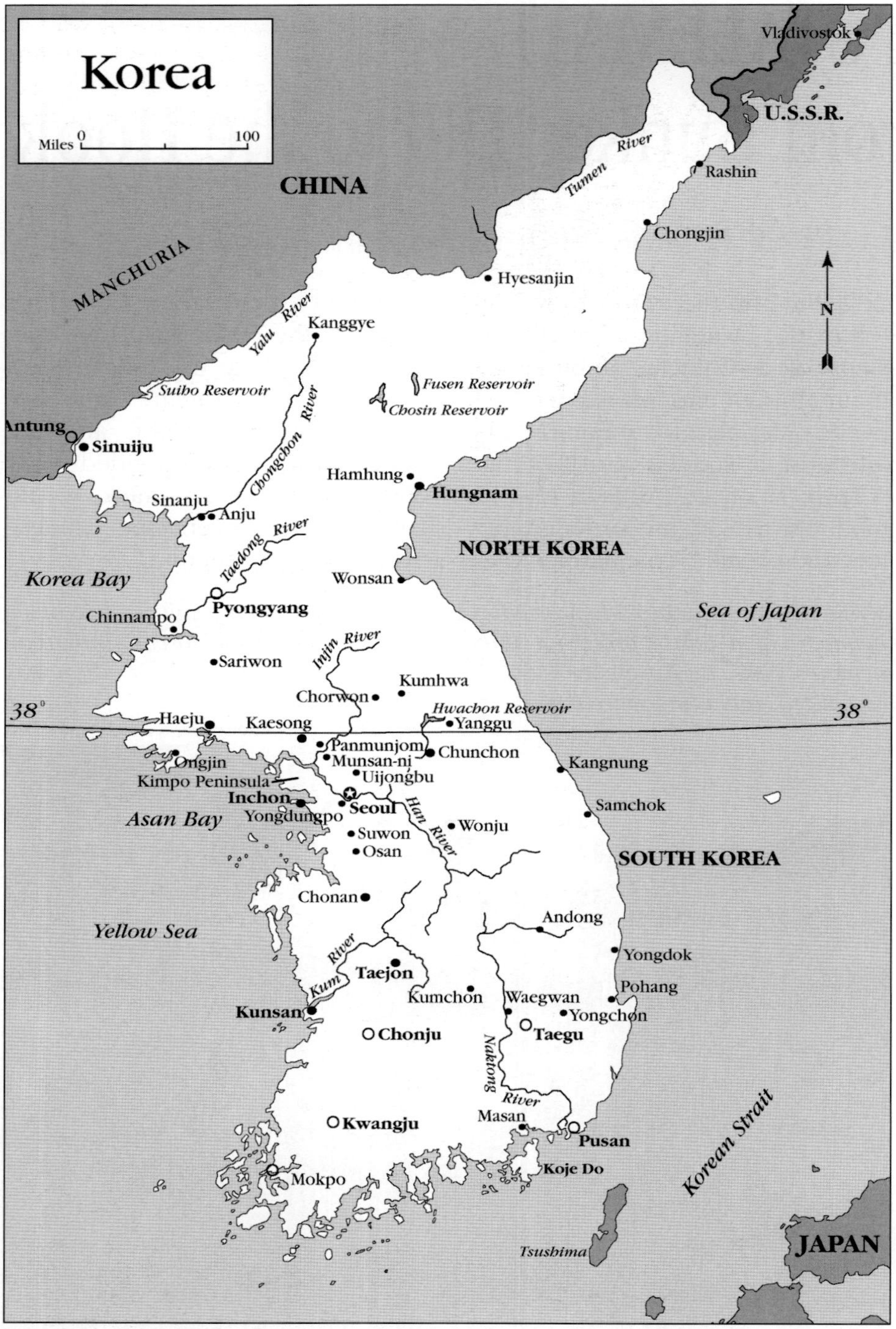

The patrol drew no fire as it made its way to the objective, where the trench the two patrols had followed intersected with another shallower trench. Watson deployed the fire teams in a perimeter. The Marines strained their eyes and ears to detect movement over sandy soil that gleamed almost white in the cloud-reflected light. Nothing moved; Chinese mortars and machine guns remained silent.

At 0300 Watson's patrol started back, the fire team that had led the way out was now at the rear. The return, as cautious and methodical as the advance, took roughly two hours. When the lieutenant at last came through the wire, he realized he was soaking wet from perspiration, more from tension, he believed, than from exertion.

Fighting took place by day as well as by night, but an early morning attack often depended on preparations made under cover of darkness. For example, before Lieutenant Watson's platoon took part in an early morning attack on a Chinese outpost, Marine engineers moved out shortly after midnight to mark a path through the minefields protecting the Jamestown Line. This work took them past marshy ground inhabited by frogs that fell silent at the approach of the Marines, only to resume their croaking at about 0300 when the passage had been marked and the engineers returned to the main line of resistance. After daybreak, Watson's platoon advanced, staying between the lines of white-tape Xs that marked the presence of mines.

New Mission

The night patrol by Watson's Marines was one in a succession of probes and patrols—interspersed with attacks and counterattacks—that occurred during 1952 after the 1st Marine Division moved onto the Jamestown Line. The move there in March 1952 confirmed a shift to position warfare. Instead of making amphibious landings as at Inchon or Wonsan or seizing ground either to break out of encirclement or to advance, the division had the mission of defending its portion of the Jamestown Line and preparing to counterattack as ordered to contain or eliminate any Chinese penetration.

The enemy maintained pressure on the United Nations forces. He probed the line of combat outposts, which provided warning of attacks and disrupted or delayed them until the troops posted there could withdraw, and also tested at times the defenses of the main line

Department of Defense Photo (USMC) A160928

A landing ship disgorges a Marine tank at Inchon during Operation Mixmaster, the deployment of the 1st Marine Division from east-central to western Korea.

of resistance. Because of the threat of a major Chinese offensive, the division assumed responsibility for two other lines, Wyoming and Kansas, which might serve as fallback positions if Jamestown should fail. More important than keeping the Wyoming and Kansas lines ready to be manned, was the division's mission, assigned on April 19, of standing by to rescue the United Nations truce negotiators, should the enemy try to trap them at Panmunjom.

Operation Mixmaster, the transfer of Major General John T. Selden's 1st Marine Division from X Corps positions in the vicinity of the Punchbowl in eastern Korea to the Jamestown Line north of the Imjin River under I Corps control, began on St. Patrick's Day, 17 March 1952. The division's major infantry units—the 1st, 5th, and 7th Marines, and the 1st Korean Marine Corps Regiment—the organic artillery of the 11th Marines, and the service and other support units moved over steep and twisting roads, with almost 6,000 truckloads required for the deployment. The heaviest equipment, totaling an estimated 50,000 tons, traveled on 63 flatbed trailers and 83 railroad cars, along with 14 Landing Ships Dock and Landing Ships Tank that sailed from Sokoho-ri and unloaded at Inchon. Two transport aircraft also figured in the move. By the time the division took over its segment of the Jamestown Line on 25 March, completing the relief of a Korean division, the officers who directed the move realized all too well how much excess equipment the unit had accumulated during the period of comparative stability that followed the capture of the Punchbowl in the summer of 1951.

Area of Operations

The segment of the Jamestown Line assigned to the 1st Marine Division extended southwest from the Samichon River and the left flank of the British 1st Commonwealth Division, crossed the 38th Parallel (the original demarcation between North and South Korea), shifted to the south bank of the Imjin in the vicinity of Munsan-ni,

The 5th Marines with reinforcing artillery, slowed by muddy roads, moves into its sector as the division occupied new positions along the Jamestown Line northeast of Seoul.

National Archives Photo (USMC) 127-N-A160346

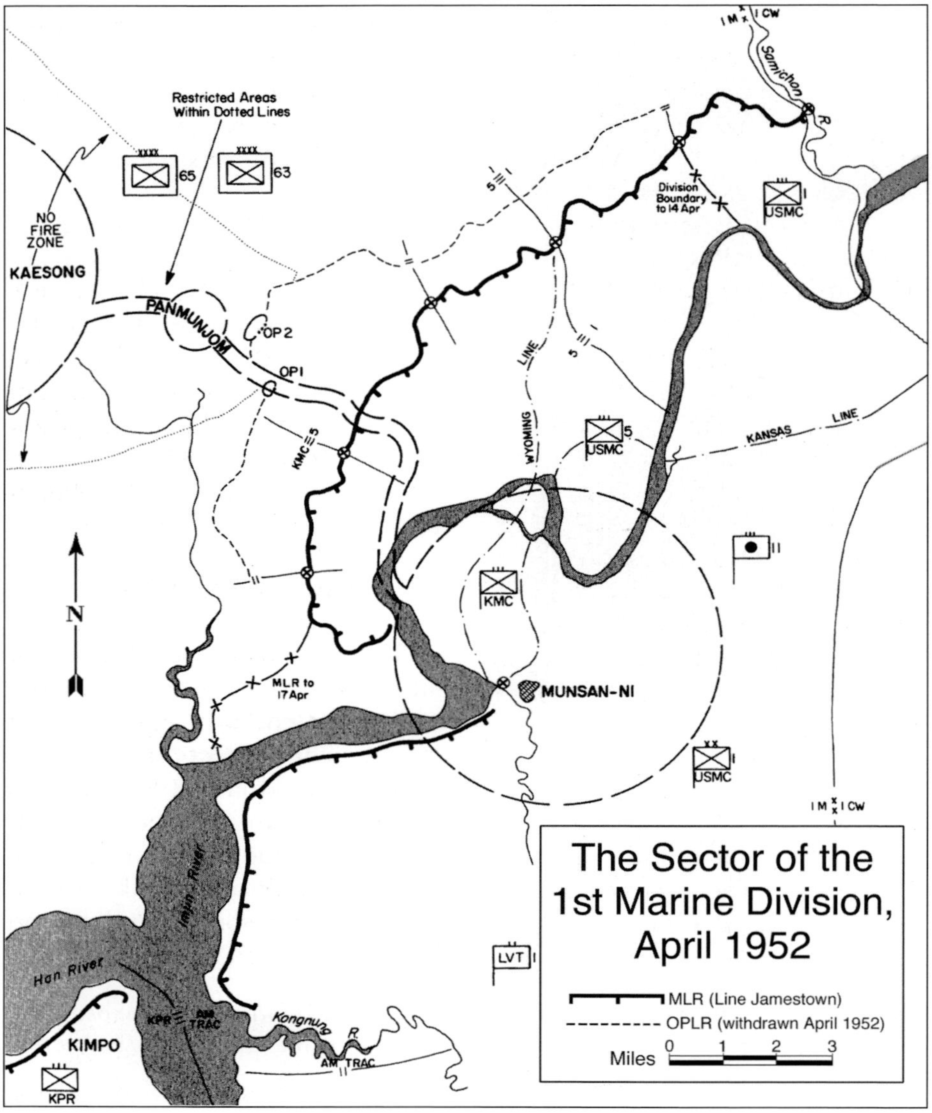

The Sector of the 1st Marine Division, April 1952

continued to the conflux of the Imjin and Han, and then followed the south bank of the Han past the Kimpo Peninsula. Initially, the 1st Marines, under Colonel Sidney S. Wade, held the right of the main line of resistance, the regiment's right flank on the heights beyond the Imjin River, some 1,100 yards north of the 38th Parallel. The 5th Marines, commanded by Colonel Thomas A. Culhane, Jr., held the center of the new line, with a regiment of Korean Marines on the left. Colonel Russell E. Honsowetz' 7th Marines served as division reserve. An adjustment in April resulted in the insertion of the 1st Amphibian Tractor Battalion on the left of the Korean Marines.

The Kimpo Peninsula, bounded by the Han and Yom Rivers, complicated the defense of the 1st Marine Division's segment of the Jamestown Line, even though an attack there would require the Chinese to cross the broad and sometimes raging Han. Defending the peninsula became the mission of the Kimpo Provisional Regiment, led by Colonel Edward M. Staab, Jr., an improvised force made up of American and South Korean soldiers and Marines from a variety of combat and service units, with the 1st Armored Amphibian Battalion providing artillery support (thirty-six 75mm guns) and a battalion of the division reserve, at this time the 7th Marines, serving as a maneuver force.

The 1st Marine Division—including the Kimpo Provisional Regiment, the amphibian tractor battalion, the Korean Marines, and the two Marine regiments on line—defended some 60,000 yards, two to four times that normally assigned to a similarly reinforced division. Within the division, a battalion, one third of the infantry strength of a regiment, held a frontage of from 3,500 to 5,000 yards, while a rifle company, one-third the infantry strength of a battalion, could man a sector as wide as 1,700 yards. A line of outposts of varying strength located on hills as far as 2,500 yards in front of the main line of resistance, improved the security of the Jamestown positions, but forced the Marines to spread themselves even thinner along the front. To defend the division's broad segment of the Jamestown Line, General Selden commanded a total of 1,364 Marine officers, 24,846 enlisted Marines, 1,100 naval officers and sailors—mostly doctors, dentists, and medical corpsmen—and 4,400 Korean Marines.

The Imjin River, flowing southwest from the division's right flank, lay behind the main line of resistance until the defenses crossed the river west of Munsan-ni. Since only three bridges—all of them vulnerable to damage from floods—spanned the Imjin, the stream, when in flood, posed a formidable obstacle to the movement of supplies and reinforcements. A single rail line to Munsan-ni served the region and the existing road net required extensive improvement to support military traffic. The terrain varied from mountainous, with sharp-backed ridges delineating narrow valleys, to rice paddies and mud flats along the major rivers. West-central Korea promised to be a difficult place for the reinforced but widely spread 1st Marine Division to conduct sustained military operations.

General Selden's Marines took over their portion of the Jamestown Line from South Korean soldiers manning an area that had become something of a backwater, perhaps because of its proximity to Kaesong, where truce talks had begun, and Panmunjom where they were continuing. "It was quite apparent," Seldon noted, "that the relieved ROK [Republic of Korea] Division had not been conducting an aggressive defense." As a result, the Marines inherited bunkers built to protect more against the elements than against enemy mortars and artillery. Korean noncombatants, taking advantage of the lull, had resumed farming in the area, moving about and creating concealment for possible Chinese infiltration.

To oppose the Marines on the Jamestown Line, the Chinese Communist Forces (CCF) had the *65th* and *63d Armies*, totaling 49,800 troops. Probing the Marine outposts and the main line of resistance were an estimated 15 infantry battalions, equipped with small arms, automatic weapons, and mortars, and supported by 10 battalions of artillery, totaling 106 guns ranging from 75mm to 155mm. Unlike the defenses the Marines had inherited, the solidly built Chinese bunkers were protected by barbed wire, minefields, and other obstacles, and organized to provide defense in depth. A variety of automatic weapons, including 37mm guns, provided antiaircraft protection.

LtGen John W. "Iron Mike" O'Daniel, USA, I Corps commander, right, joins MajGen John T. Selden, center, commanding general of the 1st Marine Division, and Col Thomas A. Culhane, left, commanding officer of the 5th Marines, on an inspection of the regiment's sector of the main line of resistance

National Archives Photo (USMC) 127-N-A160325

1st Marine Aircraft Wing

Under the command of Major General Christian F. Schilt, who had earned the Medal of Honor during the Nicaraguan campaign for a daring rescue in January 1928, the 1st Marine Aircraft Wing consisted of land- and carrier-based fixed-wing aircraft plus helicopters. The land-based fighter and attack squadrons, whether flying jets or propeller aircraft, came under the operational control of the Fifth Air Force, which in early 1952 was attempting to direct Marine Corps and Air Force activity from a Joint Operations Center at Seoul. With the exception of Marine Aircraft Group 12 (MAG-12), the components of the wing based in eastern South Korea remained there when the division moved westward. MAG-12's night fighter squadron, VMF (N)-513, shifted to the airfield at Kunsan, and the rest of the group, including two fighter outfits, began flying from Pyongtaek, also in April. Unlike the land-based fighter-bombers and attack aircraft—and the new jet-equipped photographic squadron, VMJ-1—the wing's helicopters, light observation planes, and carrier-based fighter-bombers directly supported the 1st Marine Division.

The inventory of Marine rotary-wing aircraft included Bell HTL-4 and Sikorsky HO3S-1 light helicopters and the larger Sikorsky HRS-1. The fixed-wing, piston-engine aircraft ranged in size from the unarmed, lightweight Cessna OE-1 to the Douglas AD-2 Skyraider the most powerful, heaviest, and deadliest single-engine attack plane of the era. Marines

Armistice Talks

During the summer of 1951, the succession of offensives and counteroffensives ended with the establishment of a line that stretched across the Korean peninsula generally along the 38th Parallel. The Chinese had suffered grievous losses after intervening in late 1950. Although they drove the United Nations forces out of North Korea, they failed to hold a bridgehead in the South that for a time included the capital city, Seoul. As the names of two United Nations counterattacks, Operations Killer and Ripper, indicated, the United States and its allies sought to inflict casualties rather than recapture ground. This strategy magnified the effect of the enemy's earlier losses and succeeded so well that Communist Chinese Forces (CCF)—and Chinese society, as well—needed a respite from the cumulative attrition of late 1950 and early 1951.

An armistice also seemed attractive to the United States for reasons of both strategy and domestic politics. The lengthening list of American casualties, and the continuation into a second and third year of a war described in November 1950 as on the verge of being won, undermined public support for the conflict, derisively described as Mr. Truman's war, as though the President had somehow started the fighting. In terms of strategy, Europe, where the Soviet Union and its satellites seemed ready to test the new North Atlantic Treaty Organization (NATO), seemed more important than the Far East. Indeed, a ceasefire that would free American forces from their commitment in Korea, enabling them to strengthen NATO, should work to the long-term strategic advantage of the United States.

As a result, when the Soviet delegate to the United Nations, Jacob Malik, suggested discussing the possibility of negotiating an armistice in Korea, the United States and its allies agreed. The preliminary discussions began on 8 July 1951 at Kaesong, south of the 38th Parallel and some 35 miles northeast of Seoul. The Chinese and North Koreans showed little enthusiasm for negotiations until the United Nations, in July and August, mounted a limited offensive that resulted in the capture of the Punchbowl. On 25 October negotiations resumed at Panmunjom, a village just south of the 38th Parallel, which became a demilitarized island in a sea of fighting and was linked by a road to South Korean territory.

By the end of November, the negotiators had agreed that the battle line, rather than the 38th Parallel, would serve temporarily as the line of demarcation between the two Koreas, a boundary that became permanent, essentially by default as other issues took precedence in the negotiations. Military operations slowed, as did the pace of the talks, which, by the time the Marines entered the Jamestown Line, had encountered several obstacles, the most serious dealing with the repatriation of prisoners of war.

1stMarDiv Historical Diary Photo Supplement, Oct52

National Archives Photo (USMC) 127-N-A133271

BGen Clayton C. Jerome, right, the new commanding general of the 1st Marine Aircraft Wing, tours the front in a transport helicopter piloted by Col Keith B. McCutcheon, commanding officer of Marine Helicopter Transport Squadron 161. BGen Jerome replaced MajGen Christian F. Schilt in command of the wing on 12 April 1952.

also flew the Vought F4U-4 Corsair, a piston-powered fighter-bomber, a dozen of which operated from the escort carriers *Bataan* (CVL 29) and *Bairoko* (AKV 15), and later the light carrier *Badoeng Strait* (CVE 116). A more heavily armored version of the Corsair, the AU, served as an attack aircraft. The Marine Corps jets were the Grumman F9F Panther fighter-bomber, the McDonnell F2H-2P photo plane, and the Douglas F3D Skyknight, a two-seat night fighter. The Skyknight by year's end became the principal Marine night fighter, replacing the piston-engine Grum-man F7F Tigercat, which continued until the spring of 1953 to fly interdiction and close air support during darkness.

Like their fellow Marines on the ground, the airmen operated under restrictions peculiar to a limited war. Air strikes were prohibited in the vicinity of Panmunjom to avoid jeopardizing the truce talks. Moreover, to ease the task of the Joint Operations Center in exercising centralized control over tactical aviation, the number of close-support sorties flown over the battlefront could not exceed 96 each day. In general, the allocation of air power proved flexible enough to satisfy General Schilt. Although conceding that Marines on the ground "did not always get all that they wanted" because the wing was "sometimes . . . tied up with the Air Force," Schilt found that "if there was anything we particularly wanted to do and thought it necessary to support our ground forces, we'd go over and talk to them [representatives of the Fifth Air Force] and they'd go along with us."

Besides affecting aerial operations, the neutral zone around Panmunjom influenced the mission of the Marine division. On 19 April, General Selden, reacting to orders from higher headquarters, directed the regiment with the best access to Panmunjom to draft a plan to rescue the United Nations Truce Team if it should be trapped there. The regiment that fit this description, initially the 5th Marines, organized a tank-infantry team from within its reserve battalion. Supported by tanks and fire from mortars and artillery, a covering force would advance along the demilitarized corridor leading to the negotiation site and seize the dominant ground beyond Panmunjom so that a second group could move in and pick up the negotiators. A third contingent would escort the pick-up force as it brought the truce team to a safe area behind the Jamestown Line.

Artillery and Air

During the spring of 1952, the fighting along the Jamestown Line gradually intensified, requiring the support of artillery and aircraft. The 105mm and 155mm howitzers of the 11th Marines joined tanks and other weapons in battering Chinese

Battalion Tactical Air Control Party

On the Jamestown Line, the Tactical Air Control Party assigned to each battalion on the main line of resistance linked the ground forces with the aircraft supporting them. At this stage of the war, these parties consisted of two Marine Corps officers, both of them naval aviators, and eight enlisted men who handled radio and wire communications and drove the truck assigned to the group. Earlier in the fighting, the truck provided necessary mobility, but the controllers could now operate from command bunkers using radios located there instead of relying on the temperamental set mounted in the vehicle.

While the battlefront remained fluid, one officer served as a forward air controller with each of the two infantry companies, the third company normally being in reserve. After the move to the Jamestown Line, one of the officers took his place at a forward command post, usually a bunker more solidly built than those that sheltered the infantrymen, calling strikes in support of the Marines manning the defenses. The other remained at the battalion commander's supporting arms center, serving as air liaison officer. Every week or two, the airmen changed places.

positions. The artillerymen experimented successfully with variable-time fuses, actuated by radio waves. When fitted to a standard high-explosive shell, the fuse achieved airbursts at a height of about 20 meters above Marine defensive positions, which had overhead cover. Logs, sandbags, and earth protected the Marines, while a deadly hail of shell fragments scourged the attackers. Concentrations of variable-time fire, delivered in conjunction with so-called "box-me-in" barrages that placed a curtain of fire around friendly forces, became standard tactics. On 18 May 1952, for example, Chinese troops cut off a Marine platoon led by Second Lieutenant Theodore H. Watson, as it withdrew from the outpost line. Watson shepherded his men into two abandoned bunkers and called for airbursts overhead, which helped scatter the enemy.

Marine aviation also supported operations along the Jamestown Line. In May 1952, the Fifth Air Force granted the Marines an additional dozen sorties per day to train controllers, ground commanders, and pilots in the techniques of close air support. Although the number of these daily training sources increased to 20, the program lasted only until 3 August, largely because of Army complaints that General Selden's division was getting a lion's share of close air support in the theater.

Stabilization of the battle line enhanced the value of ground-based radar in nighttime close air support. The Air Force had begun using an improvised system in January 1951, and September of that year marked the introduction into combat of the Marine-developed MPQ-14 radar. Despite nagging technical problems, the Marine radar and its operators became increasingly precise until, by mid-1952, the Fifth Air Force granted permission to use the MPQ-14, supplemented by a tactical air controller with the troops on the ground, to direct close air support.

One supporting arm, artillery, sometimes came to the aid of another, Marine Corps aviation. Even before the 1st Marine

Offshore Islands

Even before the Marines occupied positions on the Jamestown Line, they were involved in the defense of several offshore islands on the east and west coasts of Korea. The 1st Marine Division provided officers and enlisted men to direct the Korean Marines actually manning the defenses, but in January Fleet Marine Force, Pacific, took over, usually employing Marines who volunteered for this duty after recovering from wounds suffered while serving with the division. The West Coast Island Defense Unit bore responsibility for six islands astride the 38th Parallel, while the East Coast Defense Unit guarded nine islands, seven clustered around Wonsan, within artillery range of the mainland, and the other two near Songjin.

The eastern islands proved tempting to the Chinese, who occasionally took them under fire, and attacked the two Yang-do Islands, 160 miles northeast of Wonsan, where the South Korean defenders, backed by United Nations ships, beat off the invaders. In March, the United Nations Command sought to strengthen its hold on the western islands by seizing Ho-do, an unoccupied islet north of the 38th Parallel, even though it lay within range of enemy snipers on the mainland and might be attacked in winter by advancing across the ice. The occupation nevertheless went ahead, but a Chinese amphibious attack on the night of 25-26 March overwhelmed the South Korean platoon that held Ho-do. Only six members of the platoon survived, and the United Nations Command made no attempt to recapture the tiny island. The Chinese refrained from further amphibious activity, but instead struck back with a rare aerial attack, bombing Cho-do, one of the western islands, though ineffectually.

Department of Defense Photo (USMC) A161586

Marine 105mm howitzer crews of the 11th Marines prepare to fire in support of the Jamestown Line by "clobbering" Chinese command posts, bivouac areas, artillery and mortar positions, and observation outposts.

Division deployed to the Jamestown Line, the 11th Marines was firing flak suppression missions in support of close air strikes. The batteries tried to neutralize or destroy known antiaircraft positions, some of them discovered when aircraft began an attack only to break it off deliberately after forcing the Chinese guns to cut loose and reveal their locations.

Despite the doctrinal emphasis on close air support, in the summer of 1952 Marine pilots were attacking targets far beyond the battle line as a part of the Fifth Air Force's Operation Pressure, designed to destroy important North Korean industrial facilities. During one such mission, Colonel Robert E. Galer, who commanded MAG-12 in Korea and had earned the Medal of Honor at Guadalcanal in World War II, led 31 attack aircraft against targets in the mountains southwest of Wonsan. His Vought AU Corsair sustained damage from antiaircraft fire that forced him to parachute. One foot became wedged in the cockpit, but he managed to kick free of the doomed airplane, which almost ran him down in its gyrations. He succeeded, however, in opening his chute and drifted to earth within 10 feet of his crashed aircraft. He got away from the wreckage, which was sure to attract the enemy, found concealment, and with his survival radio contacted a rescue force orbiting overhead. As a helicopter darted in his direction at treetop height, he ignited a smoke grenade to mark his position and enable the rescue craft to pick him up. The flight to a ship off the coast proved more dangerous than the actual pick up, for enroute to safety antiaircraft shells exploded so close that the concussion spun the helicopter around, fuel ran low, and patches of fog concealed landmarks making navigation difficult.

Ground Fighting Intensifies

The Marines and the Chinese soon began clashing over the high ground between the frontlines that could accommodate combat

Gen Holland M. Smith, a leader of the amphibious war against Japan and whose Marines fought their way from Tarawa to Okinawa, visits the Jamestown Line in Korea. From the left are: Col Russell E. Honsowetz, commander of the 7th Marines; Col Frederick P. Henderson, commander of the 11th Marines; Gen Smith; and MajGen John T. Selden, commanding general of the 1st Marine Division.

Department of Defense Photo (USMC) A161123

outposts or observation posts. A series of objectives that the Marines designated by letters of the alphabet became a bone of contention early in May. These were Objective S, a small outcropping northwest of the main line of resistance, and V, X, Y, and Z, three separate peaks on a ridge extending northeastward from S and forming an angle of roughly 45 degrees with the main line of resistance. As part of the continued probing that occurred almost every night, First Lieutenant Ernest S. Lee, commander of Company A, 1st Battalion, 5th Marines, led his unit, reinforced with light and heavy .30-caliber machine guns, to occupy the high ground south of Objective Y, arriving there before sunrise on 4 May. The Chinese immediately opened fire with mortars, but an aerial observer spotted a half-dozen of the weapons and called in Marine F4U-4Bs that destroyed them. The enemy then attacked unsuccessfully, but since more powerful attacks seemed certain, the reinforced platoon pulled back.

Twice during the withdrawal, Chinese troops tried to ambush the patrol, which used its own weapons to beat off the first attempt and called down artillery fire to help frustrate the second. Forced from their route by the second ambush, Marines carrying the patrol's casualties, one dead and four wounded, entered an unmarked and uncharted minefield left behind by United Nations troops; two stretcher bearers were killed and three others wounded by the mines, which later were cleared.

Colonel Thomas A. Culhane, in command of the 5th Marines, directed the 1st Battalion, under Lieutenant Colonel Franklin B. Nihart, to drive the enemy from the vicinity of Objective Y, in the process taking prisoners and inflicting casualties, before seizing Objective Z. Nihart decided to capture Objectives S, V, and X before attacking Objective Y; if all went well, he could then move against Z.

Marines on patrol forward of the main line check out a cache of enemy ammunition found in an abandoned farmhouse. In addition to denying the enemy use of critical terrain, inflicting casualties and capturing prisoners were added tasks assigned to daily patrols.

National Archives Photo (USN) 80-G-442340

Nihart used his battalion's Company C to feint toward Objective T, located between the ridge and the Marine division's main line of resistance, in an attempt to neutralize the Chinese there and prevent them from interfering with the attack, which began when Company A, the 1st Platoon leading the way, quickly overran Objective S. Fire from the Marine division's rocket battery shook the defenders of Objective V, enabling the attackers to capture it. Both Marine and Chinese artillery stepped up their firing as the Nihart's men reorganized to advance on Objective X. In preparation for that move, friendly fire from artillery, mortars, tanks, and even machine guns scourged the knob raising a cloud of dust that enveloped it and blinded the attacking Marines, who encountered increasingly savage fire as they climbed the slope.

At this point, the Chinese counterattacked. Although the Marines beat back this thrust, other probes followed, as infiltrators tried to isolate the 1st Platoon from the rest of Company A. To maintain the integrity of his unit, the company commander, First Lieutenant Ernest S. Lee, pulled back the endangered platoon, while Chinese artillery rained fire on Objective X, some 400 rounds exploding in five minutes. The deadly fire forced Company A to abandon the toehold on X and then fall back to the main line of resistance under the cover of fire from the division's tanks. The Marines, however, set up a part-time outpost on

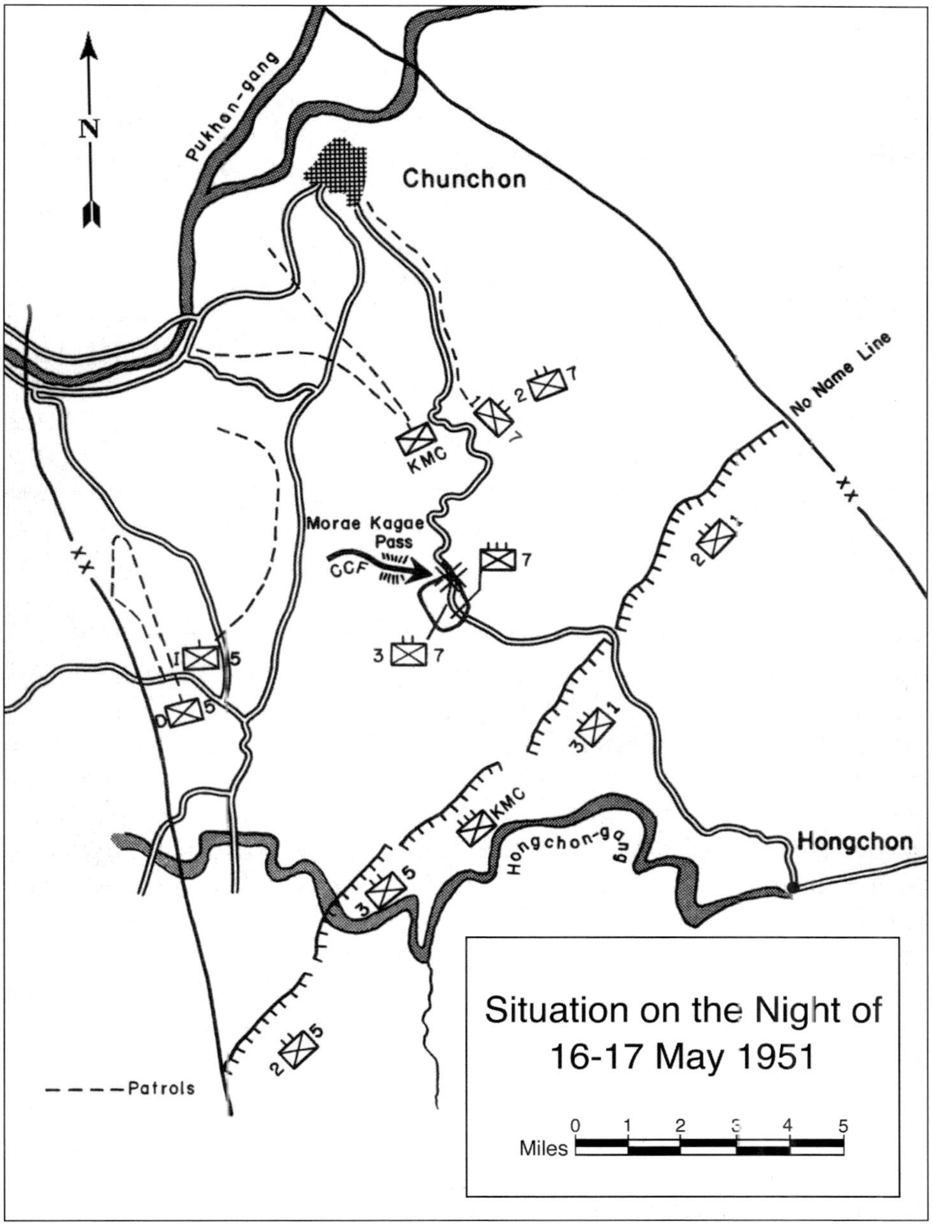

Objective Y, at first manning it mostly during daylight. In the bloodiest single day of fighting since the capture of the Punchbowl, the Marines suffered seven killed and 66 wounded, perhaps one-fourth the number of the Chinese casualties.

The fighting now shifted eastward. After relieving the 5th Marines, the 7th Marines, commanded by Colonel Russell E. Honsowetz, attacked Hill 104 and the adjacent ridgeline, located on the regimental right. Advancing during darkness on the early morning of 28 May, Companies A and C of Lieutenant Colonel George W. E. Daughtry's 1st Battalion, seized their objectives but could not hold them against fierce Chinese reaction and fell back to the Jamestown Line. The fighting proved costlier than the struggle for Objectives S, T, V, W, and X, with seven Marines killed and 107 wounded. Two of those killed in action were honored posthumously with the Medal of Honor: Corporal David B. Champagne for throwing himself on a grenade to save the lives of other Marines; and Private First Class John D. Kelly for sacrificing his life while gallantly attacking enemy positions.

Despite the developing stalemate, the Marine division continued probing, sending out patrols as large as a company to raid Chinese positions, killing or wounding the defenders and keeping the enemy off balance. Both American and South Korean Marines conducted these actions, and the Chinese retaliated in kind, as on the night of 24 June, when they cut off the elements of the 5th Marines manning an outpost on Objective Y, now redesignated Hill 159. Hostile mortar and artillery fire prevented the Marines from withdrawing over the trails leading back to the Jamestown Line, but they were able to take cover in their bunkers while fire from the 11th Marines helped frustrate the attack. The Marines could not hold the hill against a determined enemy, and by the end of the month, a Chinese battalion occupied it.

The 3d Battalion, 7th Marines, used its Company G to attack Hill 159, occupying an assault position on the night of 2 July and attacking at dawn of the following morning. The first phase went smoothly, and the assault began at 0630. Deadly fire from the battalion holding Hill 159 stalled the attack until the leader of a Marine machine gun squad, Staff Sergeant William E. Shuck, Jr., took over a rifle squad whose leader had been wounded. Shuck maneuvered the combined squads up the hill and clung to the exposed position until ordered to withdraw. While pulling his Marines back, the sergeant suffered a third and fatal wound. Shuck's daring and initiative earned him a posthumous Medal of Honor, but the hill remained in Chinese hands, even though the defenders may have suffered 200

National Archives Photo (USMC) 127-N-A162789

On the forward slope of Outpost Yoke (Hill 159) exhausted members of the 34-man 5th Marines outpost relax on the morning of 25 June. The night before they withstood an assault on the position by an estimated enemy battalion, killing or wounding more than 100 Chinese soldiers.

casualties compared to four Marines killed and 40 wounded.

On the right of the division's line, the portion now held by the 5th Marines, Company A of the regiment's 1st Battalion overran two unoccupied outposts on the night of 2-3 July before receiving orders to return to the main line of resistance. A patrol from the regiment's 2d Battalion ambushed a Chinese patrol shortly before midnight on 2 July, suffering no casualties while killing six of the enemy and wounding eight. Another patrol from the same battalion set out shortly after dawn on 3 July and engaged in an hour-long firefight that killed or wounded an unknown number of Chinese at the cost of one Marine killed and 11 wounded.

Within the next few days, two ambitious operations would involve the 1st Marine Division. The first was Operation Firecracker, a fire mission planned for 4 July when I Corps would mass artillery fire on targets all along the battle line, timing the shoot so that all the shells would detonate within one minute, a technique known as time on target. The 11th Marines opened fire with its howitzers, and the 4.5-inch rocket battery joined in as did corps artillery, so that 3,202 shells detonated almost simultaneously on Chinese positions in front of the Marine division.

Besides thus helping celebrate Independence Day, the Marines took part, over General Selden's objections, in large-scale raids, directed by Major General Paul W. Kendall, USA, I Corps commander, to gather additional intelligence on Chinese defenses. The division's commanding general believed that his Marines were spread so thin that he could not pull together a force strong enough to conduct such a raid without jeopardizing the overall security of the Jamestown Line. Selden suggested that smaller patrols could obtain the necessary information with less risk. The Marine general also pointed out that 2,651 officers and enlisted men were in the process of returning to the United States and that their replacements would not be in place until 11 July. Although the British commander of the adjacent 1st Commonwealth Division, Brigadier C. N. Barclay, agreed that the more ambitious raids might well prove too costly for the results achieved, Selden's

One enemy soldier reached the Marine entrenchment at Yoke before being killed. He was armed with nothing but stick hand grenades carried in a belt under his arm and a gas mask, the first known instance of the enemy being equipped with masks in the division's sector of the line.

Department of Defense Photo (USMC) A162793

Corporal Duane E. Dewey

Department of Defense Photo (USMC) A48747

Born in 1931 in Grand Rapids, Michigan, he enlisted in the Marine Corps Reserves in 1951. In Korea, he served as a machine gun squad leader with Company E, 2d Battalion, 5th Marines, and was critically wounded near Panmunjom on 16 April 1952. His Medal of Honor citation reads, in part:

> When an enemy grenade landed close to this position, while he and his assistant gunner were receiving medical attention for their wounds during a fierce night attack by numerically superior hostile forces, Corporal Dewey, although suffering intense pain, immediately pulled the corpsman to the ground and, shouting a warning to the other Marines around him, bravely smothered the deadly missile with his body, personally absorbing the full force of the explosion to save his comrades from possible injury or death.

The survivors of his heroic self-sacrifice never forgot his remarkable shout, as he threw himself on the grenade, "Doc, I got it in my hip pocket!" After presenting the Medal on 12 March 1953, President Dwight D. Eisenhower told him: "You must have a body of steel."

Corporal David B. Champagne

Department of Defense Photo (USMC) A240117

Born in Wakefield, Rhode Island, in 1932, Corporal Champagne enlisted in the Marine Corps in 1951. Serving as a fire team leader with Company A, 1st Battalion, 7th Marines, he was killed on 28 May 1952. His Medal of Honor citation reads, in part:

> Corporal Champagne skillfully led his fire team through a veritable hail of intense enemy machine-gun, small-arms and grenade fire, overrunning trenches and a series of almost impregnable bunker positions before reaching the crest of the hill and placing his men in defensive positions. Suffering a painful leg wound while assisting in repelling the ensuing hostile counterattack, which was launched under cover of a murderous hail of mortar and artillery fire, he steadfastly refused evacuation and fearlessly continued to control his fire team. When the enemy counterattack increased in intensity, and a hostile grenade landed in the midst of the fire team, Corporal Champagne unhesitatingly seized the deadly missile and hurled it in the direction of the approaching enemy. As the grenade left his hand, it exploded, blowing off his hand and throwing him out of the trench. [He was] mortally wounded by enemy mortar fire while in this exposed position.

Corporal Champagne's Medal of Honor was presented to his younger brother during ceremonies held in July 1953 at the Old Mountain Baseball Field in Wakefield.

Private First Class John D. Kelly

Department of Defense Photo (USMC) A403015

A 23-year-old native of Youngstown, Ohio, he enlisted in the Marine Corps in 1951. As a radio operator in Company C, 1st Battalion, 7th Marines, he volunteered to join an assault and was killed on 28 May 1952. His Medal of Honor citation reads, in part:

> Fearlessly charging forward in the face of a murderous hail of machine-gun fire and hand grenades, he initiated a daring attack against a hostile strongpoint and personally neutralized the position, killing two of the enemy. Unyielding in the face of heavy odds, he continued forward and single-handedly assaulted a machine-gun bunker. Although painfully wounded, he bravely charged the bunker and destroyed it, killing three of the enemy. Courageously continuing his one-man assault, he again stormed forward in a valiant attempt to wipe out a third bunker and boldly delivered point-blank fire into the aperture of the hostile emplacement. —Captain John C. Chapin, USMCR (Ret)

National Archives Photo (USMC) 127-N-A161138

Marines hug the trench as a Communist mortar lands nearby. Marine Corsairs were often called upon to destroy troublesome enemy mortar positions in support of the division outpost line.

arguments for waiting until his division returned to full numerical strength and in the meantime dispatching smaller patrols did not prevail.

A tank-infantry team made the Marine division's contribution to large-scale patrolling with Buckshot 2B, an operation launched on 6 July. At 2200, two companies of Lieutenant Colonel Daughtry's 1st Battalion, 7th Marines, supported by elements of the 1st Tank Battalion, advanced against Hill 159. The assault force braved deadly fire to gain a lodgment on the hill. Because they were in danger of encirclement, the Marines had to pull back before daylight. General Selden had been correct; the intelligence gained did not justify the effort and the casualties—12 dead, 85 wounded, and five missing. Until the incorporation of

The war on the Jamestown Line became a battle for the combat outposts that provided security for the main line of resistance. These Marines are preparing to join in the fighting on the outpost line.

Department of Defense Photo (USMC) A163311

Department of Defense Photo (USMC) A164200

Marine observers direct an air strike on Hill 122, later called Bunker Hill, a Communist position critical to the fighting.

replacements had restored the strength of the division, emphasis shifted to smaller patrols with less ambitious objectives than raiding a stoutly defended hill.

Siberia

The bunker symbolized the fighting along the Jamestown Line and its combat outposts like Siberia. To build bunkers for future fighting, Marine engineers and truck drivers, and some 500 members of the Korean Service Corps, cut trees, shaped timbers, and hauled the rough-hewn beams some 50 miles to the sector held by the 1st Marine Division. When some 35,000 timbers proved insufficient, the Eighth Army made up the difference, and work went ahead on the Jamestown Line, its combat outposts, and the two back-up lines, Wyoming and Kansas. Although a company of Marine engineers, assisted as necessary by members of the 1st Shore Party Battalion, provided supervision, infantrymen did most of the work, following plans prepared by the Army for the assembly of the ready-cut timbers. The Marines set up each standard bunker in a hole 12-feet square and seven-foot deep, excavated using shovels, without the aid of earth-moving machinery. Once the timbers were in place, some of them shaped from tree trunks eight inches in diameter, and the basic structure finished, the Marines covered the roof, some four feet of timbers, with another three or four feet of earth, rock, and sandbags. If carefully built, the structure could withstand a direct hit from a 105mm shell, besides affording protection against shrapnel from time-fused shells exploding overhead. The living bunker provided sleeping quarters and the fighting bunker featured firing ports for machine guns and rifles.

Bunker construction failed, however, to keep pace with plans or achieve the desired degree of protection. Fatigue contributed to the shortcomings, since the infantrymen who by day dug holes and manhandled timbers into

PFC James McIntosh of Company H, 3d Battalion, 7th Marines, aims a .50-caliber machine gun with mounted scope at Communist positions from Hill 229. The 750-foot-high Paekhak Hill, a mile east of the road leading to Panmunjom and Kaesong, was the goal of Communist forces who hoped to acquire the dominant terrain necessary for controlling access to Seoul.

National Archives Photo (USA) 111-SC411556

Rotation

Even as the 1st Marine Division became more heavily engaged along the Jamestown Line, replacements had to be absorbed, not only for the growing number of killed and wounded, but also for those whose tours of duty in Korea were ending. In the spring of 1952, for example, the division transferred elsewhere 433 officers and 6,280 enlisted Marines, while adding 506 officers and 7,359 men. The greater number of replacements kept the division slightly above authorized strength.

At this time, a normal tour of duty in Korea encompassed about 10 and one-half months. Infantry lieutenants and captains arrived in such large numbers, however, that a six-month tour became common for these officers, although those in other grades and specialties might continue to serve from nine to twelve months. The turnover among officers, plus reassignments within the division, had mixed results. Although changing assignments every three to five months reduced the effectiveness of the division, the policy broadened the experience of officers, individually and as a group. In the summer of 1952, however, the division chose efficiency over experience and reduced the frequency of reassignments among its officers.

Replacement drafts did not always fill existing vacancies. Indeed, for a time in 1952 the 11th Marines had to retrain infantry officers for artillery duty. Moreover, skilled drivers and gunners for the M-46 tank proved scarce until the training programs at Camp Pendleton, California, could be expanded.

Similar problems affected the 1st Marine Aircraft Wing where tours of duty averaged six to nine months for pilots and 10 to 12 months for non-fliers. As in the division, rotation between Korea and the United States and reassignment within the wing affected efficiency. The turnover in pilots got the blame for a series of accidents on the escort carrier *Bataan*, even though the new arrivals had requalified to fly from a carrier. Moreover, the wing's Marine Air Control Group 2 operated a formal course to train forward air controllers, and recently arrived pilots with rusty skills underwent informal refresher training. A scarcity of aircraft mechanics and electronics technicians persisted.

Department of Defense Photo (USMC) A170762

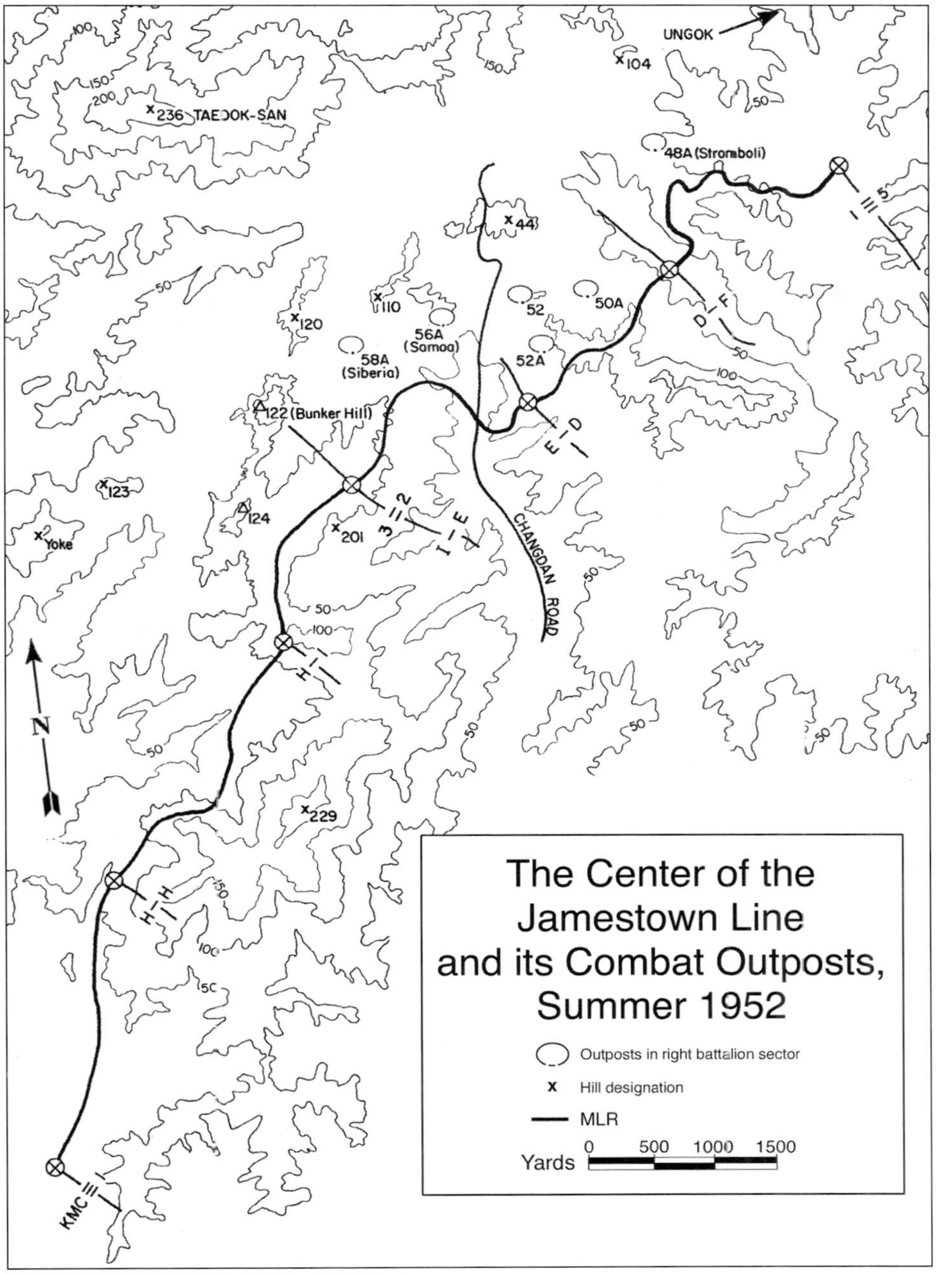

The Center of the Jamestown Line and its Combat Outposts, Summer 1952

place had to guard against attack at any time and patrol aggressively by night. Another explanation of the lagging program of bunker construction blamed the training received by the Marines, who learned to emphasize the attack at the expense, perhaps, of defensive preparations. Whatever the reasons, Marine bunkers, as well as those manned by American soldiers, did not measure up to the standards of the Chinese, who provided as much as 35 feet of overhead cover for frontline positions, which usually were linked by tunnels rather than trenches.

The fighting along the Jamestown Line grew even deadlier. Shortly after midnight on 9 August, the Chinese seized Siberia (Hill 58A), the site of a squad-size outpost, and also probed the positions of the 1st Marines. Siberia lay midway between the Marine main line of resistance and the line of Chinese outposts. The enemy's possession of Siberia would provide observation posts to adjust artillery and mortar fire against the nearest segment of the Jamestown Line. As a result, Colonel Walter F. Layer's 1st Marines, on the right of the division's line, counterattacked at once, using the same unit, Company E, 2d Battalion, that had dispatched the squad driven from Siberia. Chinese artillery and mortar fire, directed from Hills 110 and 120, stopped the counterattack short of its objective.

The Marines called for air strikes and additional artillery fire before renewing the counterattack on Siberia. At 0650, four Grumman F9F jets from the 1st Marine Aircraft Wing struck, dropping napalm and 500-pound bombs. Shortly before 1000, Air Force F-80 jets dropped 1,000-pounders, and a platoon from Company A, 1st Battalion, the regimental reserve of the 1st Marines, immediately stormed the hill, with the support of a platoon from the 2d Battalion's Company E. The Chinese again cut loose with mortars and artillery but could not stop the assault, which seized the crest. The supporting platoon from Company E joined in organizing the defense of the recaptured outpost, which came under a deadly torrent of accurate fire that forced the Marines to seek the protection of the reverse slope, nearer their main line of resistance, where they held out until mid-afternoon before falling back. The enemy's artillery and mortars had fired an estimated 5,000 rounds, wounding or killing perhaps three-fourths of the Marines who had attacked Siberia on the morning of 9 August.

While Companies E and A reorganized, the task of recapturing Siberia fell to Company C, commanded by Captain Casimir C. Ksycewski, who attacked with two platoons starting uphill at 0116. A firefight erupted, lasting four

Department of Defense Photo (USMC) A164331

Col Walter F. Layer, a veteran of the battles for Saipan, Tinian, and Okinawa, assumed command of the 1st Marines in July. He would later serve as the senior advisor to the Korean Marine Corps.

hours, but the Marines gained the crest and held it until dawn, when driven from Siberia by a Chinese counterattack.

The losses suffered by the 1st Marines, 17 killed and 243 wounded within 30 hours, convinced Colonel Layer that his regiment could not hold Siberia if Hill 122, nicknamed Bunker Hill, remained in Chinese hands. He and his staff planned a sudden thrust at Bunker Hill, possession of which would enable his command to dominate Siberia and observe movement beyond the Chinese outpost line.

Fight for Bunker Hill

To disguise the true objective, Lieutenant Colonel Roy J. Batterton attacked Siberia at dusk on 11 August with one company from his 2d Battalion. The 1st Tank Battalion supported the maneuver with four M-46 tanks (M-26 tank with a new engine and transmission), each mounting a 90mm gun and an 18-inch searchlight fitted with shutter to highlight a target in a brief burst of illumination, and four M-4A3E8 tanks, each carrying both a flamethrower and a 105mm howitzer. While the 90mm weapons hammered Hill 110, the flame-throwing tanks climbed Siberia, using bursts of flame to light their way while demoralizing the defenders, and gained the crest before doubling back toward Marine lines. As the flame-throwing M-4s withdrew, the M-46s opened fire on both Siberia and Hill 110, illuminating targets with five-second bursts of light from their shuttered searchlights, and Captain George W. Campbell's Company D overran Siberia, holding the crest until midnight when the diversionary attack ended.

The Bunker Hill assault force, Company B, 1st Battalion, 1st Marines—commanded by Captain Sereno S. Scranton, Jr., and under the operational control of Batterton's 2d Battalion—reached the crest by 2230 and began driving the enemy from the slope nearest the division's main line of resistance. The defenders recovered from their initial surprise, but the bypassed pockets of Chinese soldiers, though they tried to resist, could not check the Marine advance. In the wake of the assault force, other Marines and members

Marines crouch in a trench during the fighting for Siberia and other nearby hills. The struggle was fierce; some Chinese refused to yield and fought to their death. Most briefly held their defensive positions before retiring.

Department of Defense Photo (USMC) A165154

Department of Defense Photo (USMC) A165106

Navy Corpsmen administer blood plasma to a Marine wounded in the fight for Siberia. Intense enemy mortar and artillery fire during the seesaw battle caused most of the casualties.

of the Korean Service Corps manhandled sandbags, wire, and shovels up the hill to help Company B organize the defenses of the objective against the counterattack that was certain to come.

Chinese mortars and artillery harassed the Marines on Bunker Hill until dawn on 12 August, but the counterattack did not come until mid-afternoon, after Company B passed under the operational control of 3d Battalion, 1st Marines. The defense of Bunker Hill became the responsibility of the battalion commander, Lieutenant Colonel Gerard T. Armitage, whose Marines faced a demanding test. The volume and accuracy of the shelling increased at about 1500, a barrage that lasted an hour and forced the Marines to seek the protection from direct fire afforded by the reverse slope. Company I, 3d Battalion, 1st Marines, commanded by Captain Howard J. Connolly, reinforced Scranton's embattled Company B in time to help break up an attack by some 350 Chinese and hold the southern slope of Bunker Hill.

While the battle raged on Bunker Hill, General Selden moved his reserves closer to the fighting. Company I, 3d Battalion, 7th Marines, took the place of Connolly's company on the main line of resistance, and by the end of the day, all of the 3d Battalion, 7th Marines, had come under the operational control of the 3d Battalion, 1st Marines. Selden attached the 2d Battalion, 7th Marines, to Layer's command to strengthen the reserve of the 1st Marines. Meanwhile, Layer moved two provisional platoons from his reserve, the 1st Battalion, to reinforce the 3d Battalion, and the 3d Battalion's reconnaissance platoon established an outpost on Hill 124, linking Bunker Hill with the main line of resistance. This shuffling of units proved necessary because the 1st Marine Division was so thinly spread over an extended front. During the realignment, supporting weapons, ranging from machine guns through mortars and artillery to rocket batteries, prepared to box in the Marines holding the near slope of Bunker Hill, hammer the Chinese at the crest

Marines take a much-needed break on the reverse slope of the main line of resistance during a lull in the fight for Siberia.

Department of Defense Photo (USMC) A165132

Department of Defense Photo (USMC) A164530

On the night of 11-12 August, the 1st Tank Battalion made effective use of the 18-inch searchlight mounted on its M-46 tanks by temporarily blinding the enemy with dazzling beams of light. Behind the "cloak of darkness," elements of the 2d Battalion, 1st Marines, were able to overrun Siberia.

and beyond, protect the flanks, and harass movement on the routes enemy reinforcements would have to use.

As daylight faded into dusk on 12 August, the Marines defending the reverse slope of Bunker Hill struggled to improve their hurriedly prepared fortifications, for the Chinese preferred to counterattack under cover of darkness. Fortunately, the comparatively gentle incline of the reverse slope of the ridge that culminated in Bunker Hill reduced the amount of dead space that could not be covered by grazing fire from the Marine position. Moreover, weapons on the Jamestown Line could fire directly onto the crest, when the expected attack began. By 2000, all the supporting weapons had registered to help the two companies hold the position.

Just as the Marines had attacked Siberia on the evening of 11 August to divert attention from Bunker Hill, the Chinese sought to conceal the timing of their inevitable counterthrust. Shortly before midnight on the night of 12 August, the enemy probed the division's sector at three points. While one Chinese patrol was stumbling into an ambush set by Korean Marines, another harried a Marine outpost east of Bunker Hill. The third and strongest blow, however, landed after midnight at Stromboli, a Marine outpost on Hill 48A at the far right of the sector held by Layer's regiment, near the boundary with the 5th Marines.

In conjunction with the attack on Stromboli, launched in the early hours of August 13, the Chinese hit Company F on the right of the line held by the 1st Marines. The Chinese failed to crack the Jamestown defenses, but they inflicted so many casualties at Stromboli that reinforcements had to be sent. The reinforcing unit, a squad from Company F, came under mortar and machine gun fire from the Chinese probing Company F's defenses and had to return to the main line of resistance. Pressure against Stromboli and its defenders continued until the commander of Company F, Captain Clarence G. Moody, Jr., sent a stronger force that fought its way to the outpost, breaking the Chinese encirclement. The 5th Marines moved one company into a blocking position behind the Jamestown Line near Stromboli in case the fighting again flared at that outpost.

Some 4,500 yards to the southwest, the Chinese attempted to seize Bunker Hill. At about 0100 on the morning of 13 August, savage Chinese artillery and mortar fire persuaded Captain Connolly of Company I to request box-me-in fires, which the 11th Marines provided immediately. Enemy infantry, supported by machine gun fire, advanced behind bursting shells, but the Marines fought back with every weapon they could bring to bear—artillery, mortars, tank guns, rockets, rifles, and automatic weapons. After almost four hours, the violence abated as the enemy relaxed his pressure on Bunker Hill.

Company G, 3d Battalion, 7th Marines, under Captain William M. Vanzuyen, joined Connolly's men before the Chinese broke off the action and withdrew behind a screen of artillery and mortar fire. Except for a determined few, whom the Marines killed, the enemy abandoned Bunker Hill. Colonel Layer took advantage of the lull to send Company H, 3d Battalion, 7th Marines, to relieve the Marines holding the hill. He afterward withdrew all the other elements of the 7th Marines that had reinforced his regiment, but not until a patrol from Company I had reconnoitered the far slope of the hill.

In keeping with their usual tactics, the Chinese tried to divert attention from Bunker Hill before

Staff Sergeant William E. Shuck, Jr.

Department of Defense Photo (USMC) A403016

Born in 1926 in Cumberland, Maryland, he enlisted in the Marine Corps in 1947. Serving as a machine gun squad leader with Company G, 3d Battalion, 7th Marines, he was killed near Panmunjom on 3 July 1952. His Medal of Honor award bore a citation which reads, in part:

> When his platoon was subjected to a devastating barrage of enemy small-arms, grenade, artillery, and mortar fire during an assault against strongly fortified hill positions well forward to the main line of resistance, Staff Sergeant Shuck, although painfully wounded, refused medical attention and continued to lead his machine-gun squad in the attack. Unhesitatingly assuming command of a rifle squad when the leader became a casualty, he skillfully organized the two squads into an attacking force and led two more daring assaults upon the hostile positions. Wounded a second time, he steadfastly refused evacuation and remained in the foremost position under heavy fire until assured that all dead and wounded were evacuated. [He was] mortally wounded by an enemy sniper bullet while voluntarily assisting in the removal of the last casualty.

After the war, enlisted quarters at Marine Corps Base, Quantico, Virginia, were named in his honor.

Hospital Corpsman John E. Kilmer

National Archives (USN) 80-G-708891

A native of Highland Park, Illinois, 22-year-old Kilmer enlisted in the Navy from Texas in 1947. He was assigned to duty with 3d Battalion, 7th Marines, in Korea and was killed on 13 August 1952. His Medal of Honor citation reads, in part:

> With his company engaged in defending a vitally important hill position, well forward of the main line of resistance, during an assault by large concentrations of hostile troops, HC Kilmer repeatedly braved intense enemy mortar, artillery, and sniper fire to move from one position to another, administering aid to the wounded and expediting their evacuation. Painfully wounded himself when struck by mortar fragments, while moving to the aid of a casualty, he persisted in his efforts and inched his way to the side of the stricken Marine through a hail of enemy shells falling around him. Undaunted by the devastating hostile fire, he skillfully administered first aid to his comrade and, as another mounting barrage of enemy fire shattered the immediate area, unhesitatingly shielded the wounded man with his body.

Private First Class Robert E. Simanek

Department of Defense Photo (USMC) A49769

Born in Detroit, Michigan, in 1930, he was inducted into the Marine Corps in 1951. For his bravery in Korea on 17 August 1952, while serving with Company F, 2d Battalion, 5th Marines, he was awarded a Medal of Honor with a citation which reads, in part:

> While accompanying a patrol en route to occupy a combat outpost forward of friendly lines, Private Class Simanek exhibited a high degree of courage and resolute spirit of self-sacrifice in protecting the lives of his fellow Marines. With his unit ambushed by an intense concentration of enemy mortar and small-arms fire, and suffering heavy casualties, he was forced to seek cover with the remaining members of the patrol in the near-by trench line. Determined to save his comrades when a hostile grenade was hurled into their midst, he unhesitatingly threw himself on the deadly missile, absorbing the shattering violence of the exploding charge in his own body and shielding his fellow Marines from serious injury or death.

He miraculously survived the explosion and was retired on disability in 1953. —Captain John C. Chapin, USMCR (Ret) attacking again. Mortars and artillery shelled Combat Outpost 2, overlooking the Panmunjom corridor on the left of the sector held by the 3d Battalion, 1st Marines, and also harassed the main line of resistance nearby. The main Chinese thrust, directed as expected against Bunker Hill, began at about 2100 on the night of August 13. While shells still exploded on Combat Outpost 2, the enemy intensified his bombardment of Bunker Hill, which had been under sporadic fire throughout the afternoon. Chinese troops hit Company H, commanded by Captain John G. Demas, attacking simultaneously near the center of the position and on the right flank. (His was the only element of the 3d Battalion, 7th Marines, not yet pulled back to the Jamestown Line.) High explosive shells boxed in the Marines, and illuminating rounds helped them isolate and kill the few Chinese who had penetrated the position.

Department of Defense Photo (USMC) A161025

Two Marine machine gunners wait for another Chinese onslaught after beating back an attack that seemed to last for hours.

The Chinese battalion that attacked Bunker Hill on the night of August 13 again tested the Marine defenses at 0225 on the following morning. Before this unit's second attack, a Chinese machine gun on Siberia began firing onto Bunker Hill. Marine M-46s stabbed Siberia with brief shafts of illumination from their searchlights and silenced the weapon with 90mm fire, thus revealing the position of the tanks and enabling Chinese artillery fire to wound a crewman of one of them. The enemy may have initiated this flurry of action, which lasted only about four minutes, to protect the recovery of his soldiers wounded or killed in the earlier fighting rather than to challenge the hill's defenses.

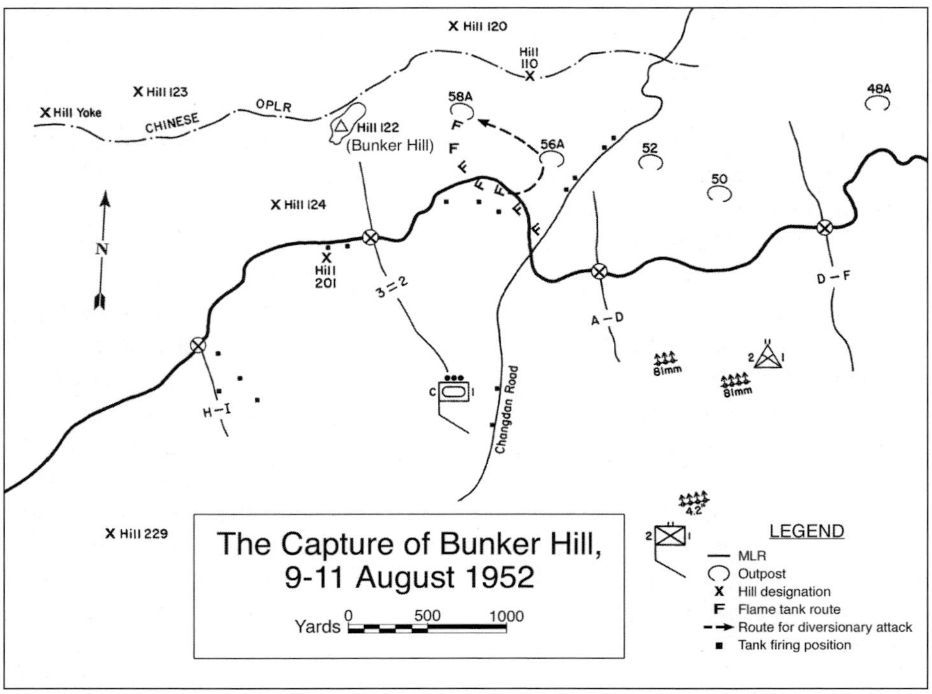

The 1st Marines responded to the fighting of 13 and 14 August by reinforcing both Bunker Hill and the nearest segment of the Jamestown Line, the so-called Siberia Sector, in anticipation of further Chinese attacks. As part of the preparation, Captain Demas, whose Company H, 7th Marines, still held Bunker Hill, patrolled the slopes where the enemy had launched several attacks but found no Chinese, a situation that rapidly changed. At 0118 on 15 August, a deluge of hostile artillery began pummeling the Marine position, while Chinese infantry jabbed at the defenses. Once again, Marine-

Department of Defense Photo (USMC) A164539

Tired Marines of Company F, 2d Battalion, 1st Marines, unwind during a pause in the fighting on Bunker Hill. Tank, artillery, air, and ground Marines participating in the battle gave up one outpost but took another, one that added strength not only to the outpost defense but also to the main line.

supporting weapons laid down final protective fires that prevented this latest attack from gaining momentum. Fate then intervened when the crew of an M-46 tank triggered the shutter of its searchlight and illuminated a force of Chinese massing in a draw. Before these soldiers could launch their assault, fire from tanks, artillery, and mortars tore into the group, killing or wounding many and scattering the survivors.

Even though Marine-supporting weapons had deflected this planned thrust, the Chinese regrouped, called in additional supporting fire, and plunged ahead. The bombardment by mortars and artillery attained a volume of 100 rounds per minute before ending at about 0400, when the enemy realized that he could not overwhelm Bunker Hill and called off his attack, at least for the present. When the threat abated, Demas withdrew his company to the main line of resistance, leaving the defense of Bunker Hill to Captain Scranton's Company B.

The quiet lasted only until late afternoon. At 1640, the Chinese attacked during a thunderstorm, avoiding the use of mortars or artillery, presumably to achieve surprise, but once again the attack failed. The Chinese refused, however, to abandon their attempts to seize Bunker Hill. At 0040, 16 August, a Chinese battalion attacked behind mortar and artillery fire, penetrating to the crest of the hill. Captain Scranton, whose Company B held the hill, called for reinforcements, and a platoon from Company I, 3d Battalion, 7th Marines, arrived as this assault was ending. The enemy again probed the hill with fire but did not press the attack. Before Company C, 1st Battalion, 1st Marines, relieved Scranton's unit, it came under artillery fire three more times.

The succession of Marine companies that took over Bunker Hill had to repel seven attacks before the end of August, but only one, on the night of 25-26 August, threatened to overrun the outpost. The struggle for Bunker Hill cost the Marines 48 killed, 313 seriously wounded, and hundreds of others who suffered minor wounds. The number of known Chinese dead exceeded 400 and total casualties may have numbered 3,200. The month ended with Bunker Hill in Marine hands.

The capture of Bunker Hill by Colonel Layer's Marines and its subsequent defense relied on the deadly use of supporting arms, ranging from the tanks that had spearheaded the diversion against Siberia, through mortars and artillery, to aircraft. The searchlight-carrying M-46s, for example, helped illuminate Chinese troops massing to counterattack the hill and opened fire on them. Marine and Air Force fighter-bombers hit the enemy on Siberia, helped seize and hold Bunker Hill, and hammered the approaches to Stromboli, where F4Us dropped 1,000-pound bombs and napalm. Marine MPQ-14 radar directed nighttime strikes against Chinese artillery positions. In addition, Marine artillery played a critical role, especially the box-me-in fires planned for emergencies. The 11th Marines fired 10,652 shells in the 24 hours ending at 1800 on 13 August, a volume not exceeded until the final months of the war in 1953. Although supplies of explosive shells proved adequate at this time, illuminating rounds tended to be in short supply.

The Marines of 1952 were fighting a war of unceasing attrition far different from the succession of bloody campaigns, interspersed with time for incorporating replacements into units withdrawn from combat, that had characterized the war against Japan. Men and equipment had to be absorbed during sustained fighting to replace not only combat losses, but also administrative attrition

Casualties and Courageous Hospital Corpsmen

The fighting at Bunker Hill would have been even deadlier for the Marines had it not been for their protective equipment and the excellent medical treatment available from battlefield first-aid stations to hospital ships off the coast. According to regimental surgeons, 17 wounded Marines would certainly have died had they not been wearing the new armored vest. Moreover, the steel helmet used in World War II had again proved its worth in deflecting or stopping shell fragments or nearly spent bullets. The heavy vest, though it undeniably saved lives, proved an enervating burden in the heat and humidity of the Korean summer and contributed to dehydration and heat exhaustion, but the advantage of increased safety—along with improved morale—more than outweighed the disadvantages.

The uniformly sloping terrain of Bunker Hill caused the Hospital Corpsmen assigned to the Marines to set up their forward aid stations on the reverse slope, overlooked by the Jamestown Line, to obtain protection from flat trajectory fire and direct observation by the enemy. Wounded Marines being evacuated from the aid stations for further treatment ran the risk of mortar and artillery fire. A shortage of tracked armored personnel carriers, which would have provided a measure of protection from hostile fire and a less jolting ride than ordinary trucks, increased both the danger and the discomfort. Moreover, the stretcher rapidly became a precious item, since so many of the limited supply were being used for the time-consuming trip to surgical facilities some distance to the rear.

The 1st Medical Battalion had three collecting and clearing companies, two of which supported units committed on the division's main line of resistance while the third supported units on the Kimpo Peninsula. Located about six miles behind the line was the division hospital, a 200-bed facility staffed and equipped to provide definitive care for all types of cases. Serious cases or cases requiring specialized care were evacuated by helicopter, ambulance, hospital train, rail bus, or plane to a hospital ship at Inchon, Army Mobile Surgical Hospitals, or to the Naval Hospital, Yokosuka, Japan.

The Surgeon General of the Navy, Rear Admiral Herbert L. Pugh, evaluated the medical support that he observed during the Bunker Hill fighting. He declared that medical treatment had been successful for a variety of reasons: the armored vest and steel helmet, of course; skilled Navy surgeons with access to a reliable supply of blood for transfusions; evacuation by helicopter to surgical facilities ashore or, if necessary, to hospital ships; and the courageous effort of Navy Hospital Corpsmen serving alongside the Marines.

Courtesy of Frank D. Praytor, USMC

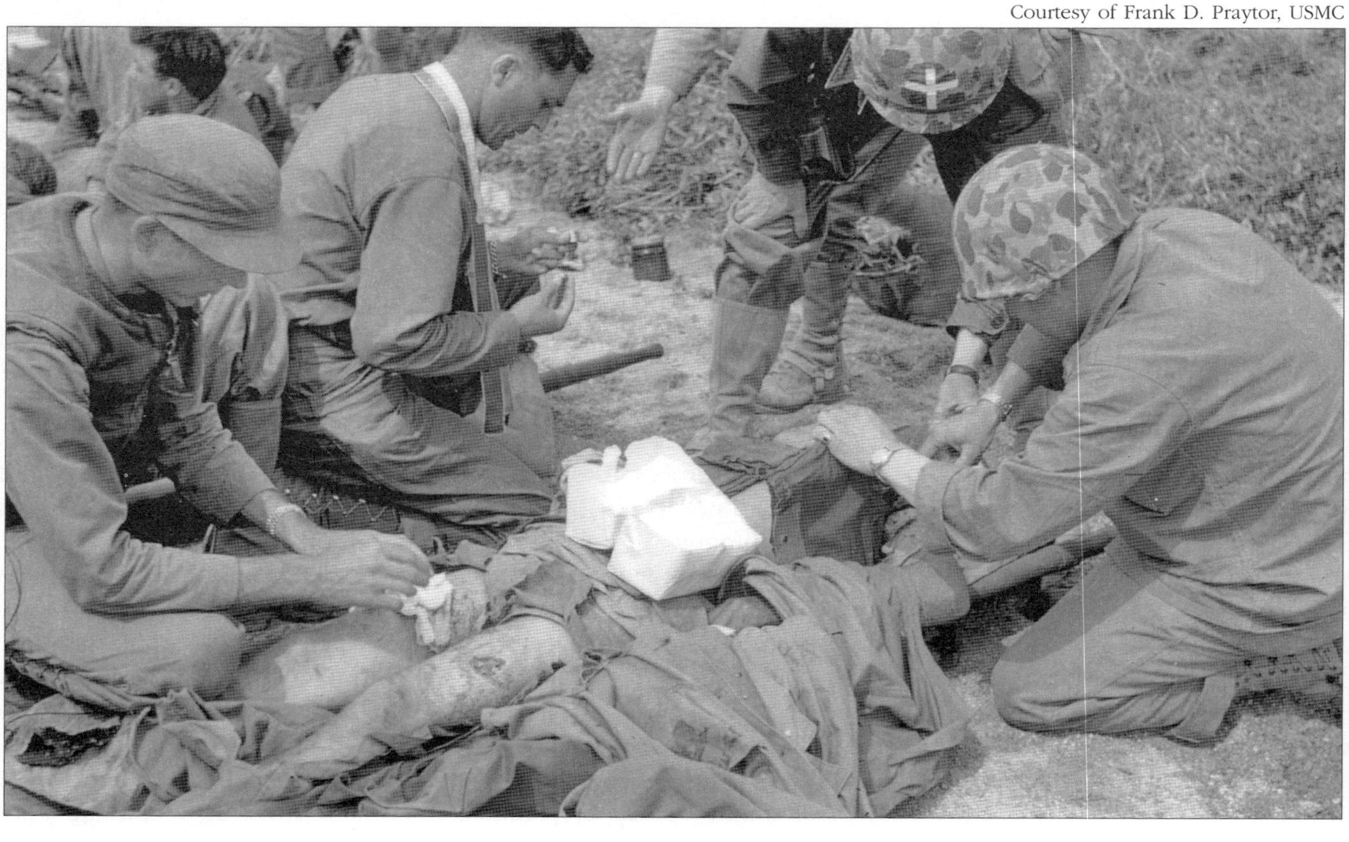

caused by a fixed period of enlistment and a definite tour of duty in Korea. To help man his combat battalions, General Selden on 12 August directed rear-area units to provide replacements for the 1st Marines; within two days, some 200 men, most of them volunteers, joined the regiment's infantry units. The Commandant of the Marine Corps, General Lemuel C. Shepherd, Jr., approved flying 500 infantry replacements to Korea. Lieutenant General Franklin A. Hart, in command of the Fleet Marine Force, Pacific, requested a delay to make sure he could assemble the necessary air transport; he could, and on 21 August the replacements arrived in Korea. The Commandant also approved

Change of Command

National Archives Photo (USMC) 127-N-A164605

On 29 August 1952, Major General John T. Selden, shown on the left in the accompanying photograph, relinquished command of the 1st Marine Division. His tenure as commanding general, which began in January 1952, earned him the U.S. Army's Distinguished Service Medal for conducting "successful operations against the enemy." During World War II, he had commanded the 5th Marines on New Britain and served as Chief of Staff, 1st Marine Division, in the conquest of Peleliu in the Palau Islands. After leaving Korea, he served on the staff of the Commander in Chief, U. S. Forces, Europe, until 1953 when he assumed command of Marine Corps Base, Camp Pendleton, California, from which post he retired in 1955. General Selden died in 1964 and was buried with full military honors in Arlington National Cemetery.

Born at Richmond, Virginia, in 1893, Selden enlisted in the Marine Corps in 1915 and was commissioned a second lieutenant in July 1918. Instead of being sent to France, he received orders to convoy duty in USS *Huntington*. Between the World Wars, he served in Haiti and China and at various posts in the United States. He attended the Senior Officers' Course at Marine Barracks, Quantico, Virginia, joined the 1st Brigade, which formed the nucleus of the 1st Marine Division, and participated in maneuvers at Guantanamo Bay, Cuba. When war came, he served in a variety of assignments in the Pacific before joining the 1st Marine Division in time for the New Britain campaign.

Command of the 1st Marine Division in Korea passed to Major General Edwin A. Pollock, a native of Georgia who graduated from The Citadel, South Carolina's military college, in 1921. He promptly resigned his commission in the Army Reserve and on 1 July accepted an appointment as a second lieutenant in the Marine Corps Reserve. His early career included service in the Dominican Republic and Nicaragua, on board ship, and with various installations and units in the United States.

Promoted to lieutenant colonel in April 1942, he commanded the 2d Battalion, 1st Marines, at Guadalcanal, where on the night of 20-21 August, he earned the Navy Cross for his inspired leadership in repulsing a Japanese attack at the Tenaru River. He became a colonel in November 1943 and afterward saw action on New Britain and at Iwo Jima.

Following the war he served at Marine Corps Schools and Marine Barracks, Quantico, Virginia, and at Headquarters Marine Corps. As a major general, he took over the 2d Marine Division at Camp Lejeune, North Carolina, before being reassigned to Korea, where he succeeded General Selden in command of the 1st Marine Division. General Pollock led the division in Korea from August 1952 until June 1953, in the final weeks of the war.

After Korea, General Pollock returned to Quantico as director of the Marine Corps Education Center, and then was assigned as commander Marine Corps Recruit Depot, Parris Island, South Carolina, before coming back to Quantico as Commandant of Marine Corps Schools. In 1956, he was appointed Commanding General, Fleet Marine Force, Pacific, followed by a tour as Commanding General, Fleet Marine Force, Atlantic, from which post he retired in 1959. General Pollock served in several capacities on The Citadel's Board of Visitors before his death in 1982.

Selden's request to increase by 500 men each of the next two scheduled replacement drafts.

Fighting Elsewhere on the Outpost Line

As the month of August wore on, any lull in the action around Bunker Hill usually coincided with a surge in the fighting elsewhere, usually on the right of the main line of resistance, the segment held by the 5th Marines, commanded after 16 August by Colonel Eustace R. Smoak. On 6 August, while Colonel Culhane still commanded the regiment, the Chinese began chipping away at the outpost line in front of the 2d Battalion, 5th Marines, which consisted of Outposts Elmer, Hilda, and Irene.

Because the battalion manned the outposts only in daylight, the Chinese simply occupied Elmer, farthest to the southwest, after dusk on 6 August and employed artillery fire to seal off the approaches and prevent the Marines from returning after daybreak. The Chinese took over Outpost Hilda on the night of 11 August, driving back the Marines sent to reoccupy it the following morning. The same basic tactics enabled the enemy to take over Outpost Irene on the 17th. During an unsuccessful attempt to regain the third of the outposts, Private First Class Robert E. Simanek saved the lives of other Marines by diving onto a Chinese hand grenade, absorbing the explosion with his body, suffering severe though not fatal wounds, and earning the Medal of Honor.

Heavy rains comparable to the downpour of late July turned roads into swamps throughout the Marine sector, swept away a bridge over the Imjin River, and forced the closing of a ferry. Bunkers remained largely unaffected by flooding, but the deluge interfered with both air support and combat on the ground. Since the rain fell alike on the U.N. and Chinese forces, activity halted temporarily when nine inches fell between 23 and 25 August.

Bunker Hill and Outpost Bruce

The Chinese greeted the new division commander, Major General Edwin A. Pollock, by exerting new pressure against the Bunker Hill complex, now held by Captain Stanley T. Moak's Company E, 2d Battalion, 1st Marines, attached temporarily to the regiment's 3d Battalion. On the night of 4 September, Chinese gunners began shelling the outpost and probing its right flank, but small arms fire forced the enemy to pull back.

The resulting lull lasted only until 0100 on 5 September, when Chinese mortars and artillery resumed firing, concentrating on Hill 122, Bunker Hill. Apparently confident that the barrage had neutralized the defenses, the attackers ignored cover and concealment and moved boldly into an unexpected hail of fire that drove them back. After regrouping, the enemy attacked once again, this time making use of every irregularity in the ground and employing the entire spectrum of weapons from hand grenades to artillery. This latest effort went badly awry when a force trying to outflank Bunker Hill lost its way and drew fire from Marines on the main line of resistance. The attackers tried to correct their mistake only to come under fire from their fellow Chinese who had penetrated the extreme right of Bunker Hill's defenses and may have mistaken their comrades for counterattacking Marines. Amid the confusion, Moak's company surged forward and drove the enemy from the outpost. The Marines of Company E suffered 12 killed and 40 wounded in routing a Chinese battalion while killing an estimated 335 of the enemy.

Yet another diversionary attack on Outpost Stromboli coincided with the thrust against Bunker Hill. The Marines defending Stromboli sustained no casualties in breaking up an attack by an enemy platoon supported by machine guns.

The 2d and 3d Battalions of Colonel Smoak's 5th Marines

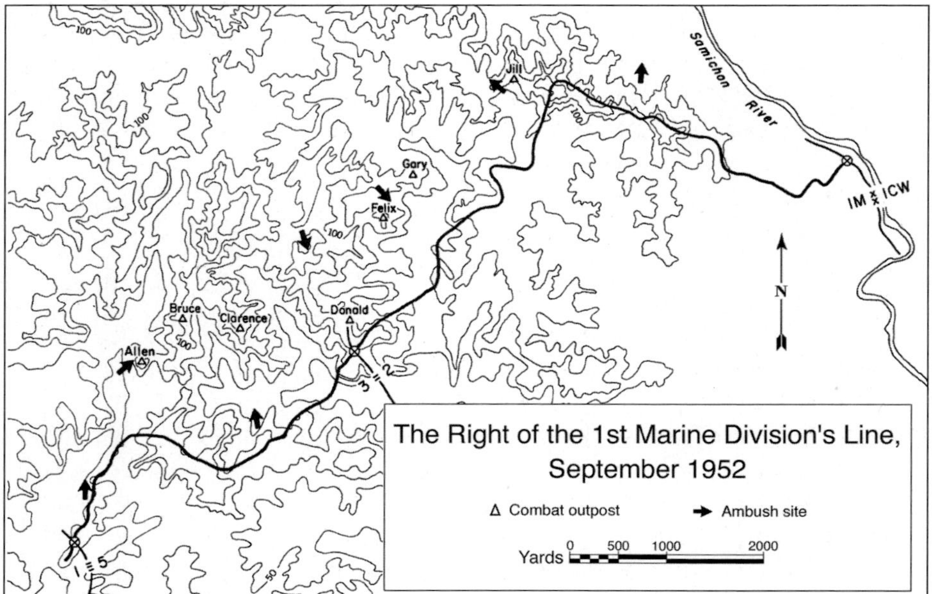

The Right of the 1st Marine Division's Line, September 1952

Private First Class Alford L. McLaughlin

Department of Defense Photo (USMC) A49766-A

A 24-year-old native of Leeds, Alabama, he listed in the Marine Corps in 1945. Awarded a Purple Heart Medal in August 1952, his heroism again was recognized during the struggle on Bunker Hill on the night of 4-5 September 1952, while serving as a machine gunner with Company I, 3d Battalion, 5th Marines. His citation reads, in part:

> When hostile forces attacked in battalion strength during the night, he maintained a constant flow of devastating fire upon the enemy, alternating employing two machine guns, a carbine and hand grenades. Although painfully wounded, he bravely fired the machine guns from the hip until his hands became blistered by the extreme heat from the weapons and, placing the guns on the ground to allow them to cool, continued to defend the position with his carbine and grenades. Standing up in full view, he shouted words of encouragement to his comrades above the din of battle and, throughout a series of fanatical enemy attacks, sprayed the surrounding area with deadly fire accounting for an estimated one hundred and fifty enemy dead and fifty wounded.

Remarkably, he survived this battle, made a continuing career in the Marine Corps, and eventually retired in 1977 as a master sergeant.

Hospital Corpsman Third Class Edward C. Benfold

Navy Medicine Photo Collection

Born in Staten Island, New York, in 1931, he enlisted in the Navy and was assigned in 1951 to the Marine Corps where he bore the designation of "medical field technician." In Korea, while serving with 3d Battalion, 5th Marines, he was killed on 5 September 1952. His Medal of Honor citation reads, in part:

> Benfold resolutely moved from position to position in the face of intense hostile fire, treating the wounded and lending words of encouragement. Leaving the protection of his sheltered position to treat the wounded, when the platoon area in which he was working was attacked from both the front and rear, he moved forward to an exposed ridge line where he observed two Marines in a large crater. As he approached the two men to determine their condition, an enemy soldier threw two grenades into the crater while two other enemy charged the position. Picking up a grenade in each hand, HC3c Benfold leaped out of the crater and hurled himself against the onrushing hostile soldiers, pushing the grenades against their chests and killing both the attackers.

In 1994, the Navy named a guided missile destroyer, USS *Benfold* (DDG 65), in his honor.

—Captain John C. Chapin, USMCR (Ret)

fought to defend their outpost line—from Allen in the west, through Bruce, Clarence, Donald, Felix, and Gary, to Jill in the east—against a succession of attacks that began in the early hours of 5 September. At Outpost Bruce—manned by Company I, 3d Battalion, 5th Marines, under Captain Edward Y. Holt, Jr.—a company of attacking infantry followed up a savage barrage. Private First Class Alford L. McLaughlin killed or wounded an estimated 200 Chinese, victims of the machine guns, carbine, and grenades that he used at various times during the fight, and survived to receive the Medal of Honor. Private First Class Fernando L. Garcia, also earned the nation's highest award for heroism; though already wounded, he threw himself of a Chinese grenade, sacrificing his life to save his platoon sergeant. Hospitalman Third Class Edward C. Benfold saw two wounded Marines in a shell hole on Outpost Bruce; as he prepared to attend to them, a pair of grenades thrown by two onrushing Chinese soldiers fell inside the crater. Benfold picked up one grenade in each hand, scrambled from the hole, and pressed a grenade against each of the two soldiers. The explosions killed both the Chinese, as Benfold sacrificed himself to save the two

Private First Class Fernando L. Garcia

Department of Defense Photo (USMC) A403017

Born in 1929 in Utuado, Puerto Rico, and inducted into the Marine Corps in 1951, he was killed at Outpost Bruce in the Bunker Hill area on 5 September 1952 while serving as a member of Company I, 3d Battalion, 5th Marines. The first Puerto Rican to be awarded a Medal of Honor, his citation reads, in part:

> While participating in the defense of a combat outpost located more than one mile forward of the main line of resistance, during a savage night attack by a fanatical enemy force employing grenades, mortars, and artillery, Private First Class Garcia, although suffering painful wounds, moved through the intense hail of hostile fire to a supply point to secure more hand grenades. Quick to act when a hostile grenade landed nearby, endangering the life of another Marine, as well as his own, he unhesitatingly chose to sacrifice himself and immediately threw his body upon the deadly missile, receiving the full impact of the explosion.

The enemy then overran his position and his body was not recovered. In his honor, a camp at Vieques in Puerto Rico and a destroyer escort, USS *Garcia* (DE 1040), were named after him.

—Captain John C. Chapin, USMCR (Ret)

wounded Marines, earning a posthumous Medal of Honor.

When dawn broke on 5 September, Holt's Company I still clung to Outpost Bruce, even though only two bunkers, both on the slope nearest the Jamestown Line, survived destruction by mortar and artillery shells. The commander of the 3d Battalion, Lieutenant Colonel Oscar T. Jensen, Jr., rushed reinforcements, construction and other supplies, and ammunition to the battered outpost despite harassing fire directed at the trails leading there.

The 1st Marine Division commander, MajGen Edwin A. Pollock, left, welcomes the United Nations commander, Gen Mark W. Clark, USA, to the division's area. Gen Clark had succeeded Gen Matthew B. Ridgway, who took over as Supreme Allied Commander, Europe, from General of the Army Dwight D. Eisenhower.

National Archives Photo (USMC) 127-N-A167677

Marine and Air Force pilots tried to suppress the hostile gunners with 10 air strikes that dropped napalm as well as high explosive.

On the morning of 6 September, the defenders of Outpost Bruce beat off another attack, finally calling for box-me-in fires that temporarily put an end to infantry assaults. At dusk, however, the Chinese again bombarded the outpost, this time for an hour, before attacking with infantry while directing long-range fire at neighboring Outpost Allen to the southwest. The Marines defending Outpost Bruce survived to undergo further attack on the early morning of 7 September. Two Chinese companies tried to envelop the hilltop, using demolitions in an attempt to destroy any bunkers not yet shattered by the latest shelling.

For a total of 51 hours, the enemy besieged Outpost Bruce before breaking off the action by sunrise on the 7th. At Bruce, the site of the deadliest fighting in this sector, the Marines suffered 19 killed and 38 wounded, 20 more than the combined casualties sustained defending all the other outposts manned by the 2d and 3d

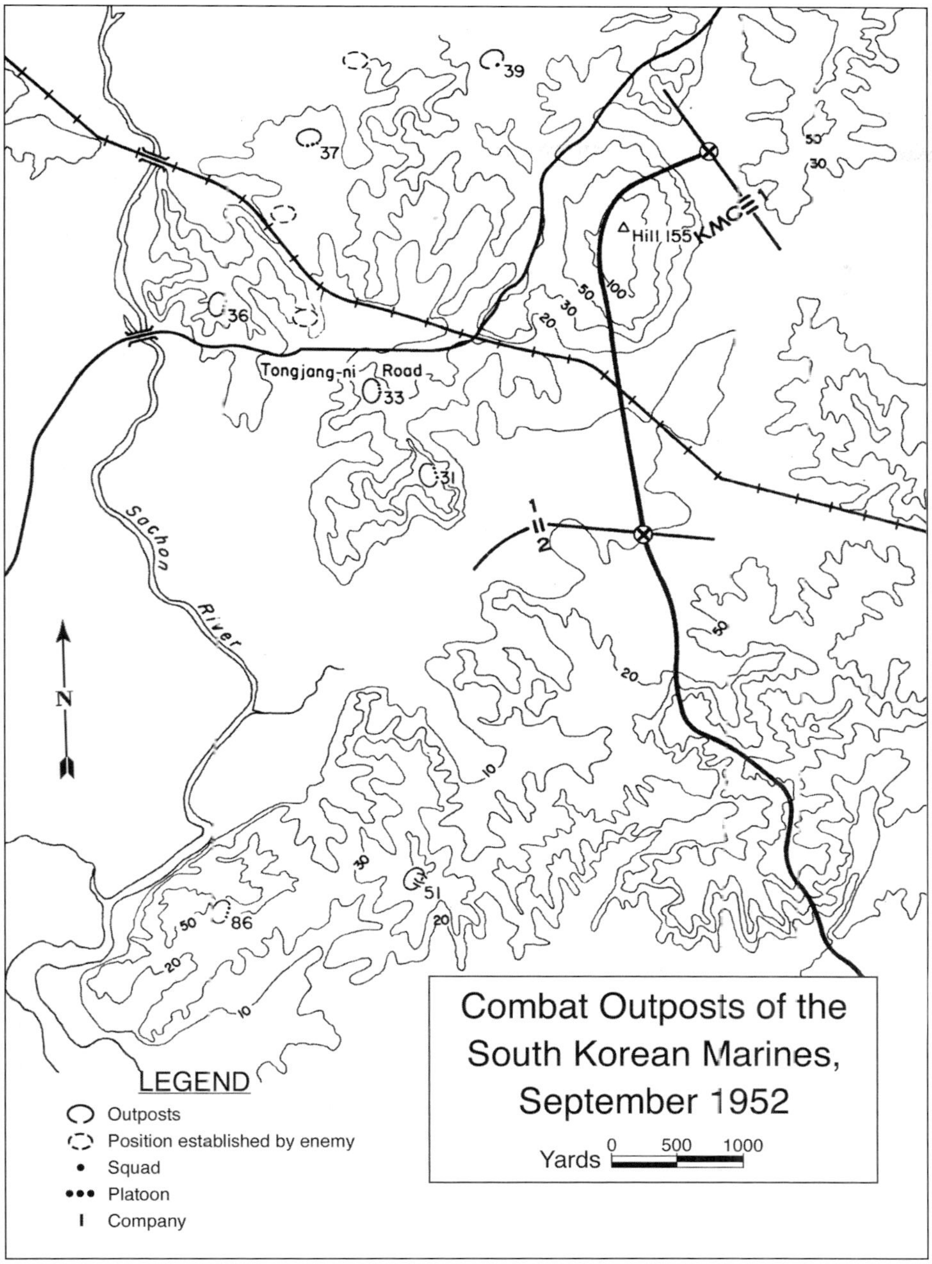

Combat Outposts of the South Korean Marines, September 1952

Battalions, 5th Marines. Chinese killed and wounded at Outpost Bruce may have totaled 400.

Pressure on Korean Marines

At the left of the line held by the 1st Marine Division, the frontage of the Korean Marine Corps veered southward, roughly paralleling the Sachon River as it flowed toward the Imjin. The Korean Marines had established a series of combat outposts on the broken ground between the Sachon and the Jamestown Line. On the evening of 5 September, while the outposts manned by the 5th Marines were undergoing attack, Chinese artillery and mortars began pounding Outpost 37, a bombardment that soon included Outpost 36 to the southwest and an observation post on Hill 155 on the main line of resistance, roughly 1,000 yards from the boundary with the 1st Marines.

Chinese infantry, who had crossed the Sachon near Outposts 36 and 37, attacked both of them. Two assaults on Outpost 37, which began at 1910, may have been a diversion, for at the same time the enemy launched the first of three attacks on Outpost 36. The final effort, supported by fire from tanks and artillery, overran the hill, but the Chinese had lost too many men and too much equipment—at least 33 bodies littered the hill along with a hundred abandoned grenades and numerous automatic weapons—forcing the enemy to withdraw the survivors. The Korean Marines suffered 16 killed or wounded from the platoon at Outpost 36 and another four casualties in the platoon defending Outpost 37.

Clashes involving the outposts of the Korean Marine Corps continued. The South Koreans dispatched tank-infantry probes and set up ambushes. At first, the Chinese responded to this activity with mortars and artillery, firing a daily average of 339 rounds between 12 and 19 September, about one-third of the total directed against Outpost 36. An attack by infantry followed.

Before sunrise on the 19th, Chinese troops again infiltrated across the Sachon, as they had two weeks earlier. They hid in caves and ravines until evening, when they advanced on Outposts 36, their principal objective, and 37, along with Outposts 33 and 31 to the south. A savage barrage of more than 400 rounds tore up the defenses rebuilt after the 5 September attack and enabled the Chinese to overwhelm Outpost 36. Shortly after midnight, a South Korean counterattack seized a lodgment on the hill, but the Chinese retaliated immediately, driving off, killing, or capturing those who had regained the outpost.

With the coming of daylight, Marine aircraft joined in battering

National Archives Photo (USMC) 127-N-A165912

MajGen Edwin A. Pollock, center, accompanied by Col Kim Suk Bum, commanding officer of the 1st Korean Marine Corps Regiment, talks with members of the 1st Air and Naval Gunfire Liaison Company concerning Marine air support for Korean Marine operations.

enemy-held Outpost 36. Air strikes succeeded in hitting the far side of the hill, destroying mortars and killing troops massing in defilade to exploit the earlier success. Two platoons of Korean Marines—supported by fire from tanks, mortars, and artillery—followed up the deadly air attacks and regained Outpost 36, killing or wounding an estimated 150 of the enemy.

Further Action Along the Line

In late September, while fighting raged around Outpost 36 in the sector held by the Korean Marines, the Chinese attacked the combat outposts manned by Colonel Layer's 1st Marines, especially Hill 122 (Bunker Hill) and Hill 124 at the southwestern tip of the same ridge line. The enemy struck first at Hill 124, attacking by flare-light from four directions but failing to dislodge the squad dug in there, even though most of the Marines suffered at least minor wounds.

The entire ridge from Hill 124 to Hill 122 remained under recurring attack for the remainder of September, especially Bunker Hill itself, where the enemy clung to advance positions as close as 30 yards to Marine trenches. The Chinese frequently probed Bunker Hill's defenses by night, and the Marines took advantage of darkness to raid enemy positions, using portable flamethrowers and demolitions to destroy bunkers while fire from tanks and artillery discouraged counterattacks.

Tests earlier in the year had proved the theory that transport helicopters could resupply a battalion manning the main line of resistance. The next step was to determine if rotary wing aircraft could accomplish the logistical support of an entire frontline regiment. For the five-day period, 22-26 September, Marine Helicopter Transport Squadron 161 successfully supplied the 7th Marines with ammunition, gasoline, rations, and made a daily mail run. All but valuable cargo, such as mail, was carried externally in slings or wire baskets.

As October began, the Chinese saluted the new month with heavier shelling, the prelude to a series of attacks on outposts all across the division front from the Korean Marine Corps on the left, past the 1st Marines and Bunker Hill, to the far right, where the 7th Marines, commanded by Colonel Thomas C. Moore, Jr., had taken over from the 5th Marines, now in reserve. To make communications more secure, Colonel Moore's regiment redesignated Outposts Allen, Bruce, Clarence, Donald, Gary, and Jill, replacing proper names, in

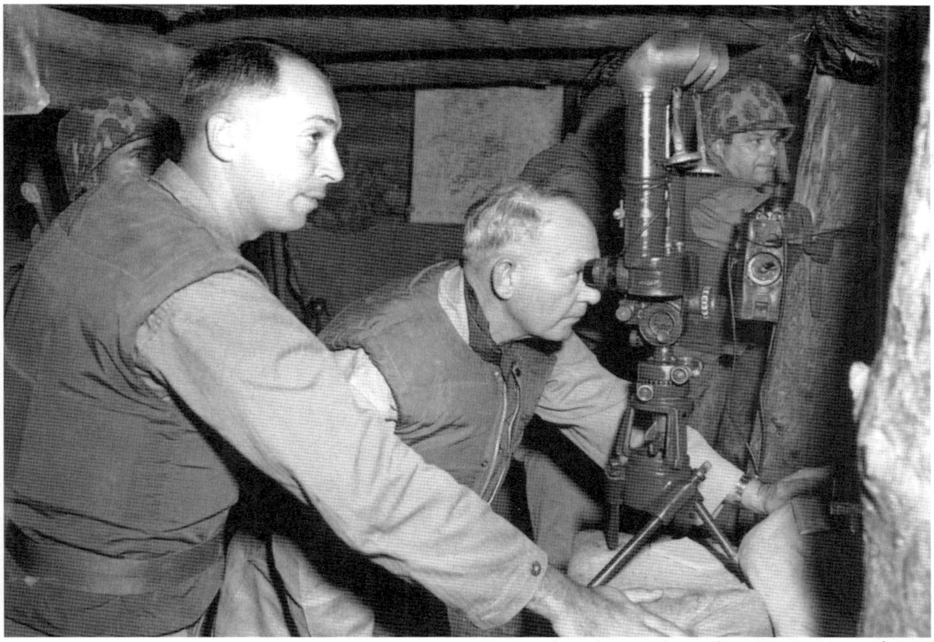

National Archives Photo (USMC) 127-N-A165528

Gen Lemuel C. Shepherd, Jr., center, Commandant of the Marine Corps, looks at enemy positions while touring frontline Marine positions. LtCol Anthony J. Caputo, left, commanding officer of 2d Battalion, 7th Marines, briefs the Commandant on the situation.

alphabetical order, with the randomly arranged names of cities: Carson, Reno, Vegas, Berlin, Detroit, Frisco, and Seattle.

On the left, Chinese loudspeakers announced on the night of 1 October that artillery would level the outpost on Hill 86, overlooking the Sachon River, and warned the defenders to flee. When the South Korean Marines remained in place, a comparatively light barrage of 145 rounds exploded on the hill and its approaches, too few to dislodge the defenders. On the following evening at 1830, the Chinese resumed firing, this time from the high ground beyond Hill 36, and extended the bombardment to all the outposts within range. The Korean Marines dispatched a platoon of tanks to silence the enemy's direct-fire weapons, but the unit returned without locating the source of the enemy fire. After the tanks pulled back, Chinese artillery fire intensified, battering all the outposts until one red and one green flare burst in the night sky. At that signal, the guns fell silent, and infantry attacked Outposts 36, 37, and 86.

Outpost 37, the northernmost of the three, resisted gallantly, forcing the Chinese to double the size of the assault force in order to overrun the position. On 3 October, the Korean Marines launched two counterattacks. The second of these recaptured the crest and held it until Chinese artillery and mortar fire forced a withdrawal. By late afternoon on 5 October, the Korean Marines twice regained the outpost, only to be hurled back each time, and finally had to call off the counterattacks, leaving the hill in enemy hands.

To the south, the Chinese also stormed Outpost 36 on 2 October. The defenders hurled back two nighttime attacks, but the cumulative casualties and damage to the fortifications forced the Korean Marines to withdraw. The enemy immediately occupied the hill and held it.

The most vulnerable of the three, Outpost 86, lay farthest from the main line of resistance and closest to the Sachon River. On the

During Operation Haylift in late September, Sikorsky HRS-2 helicopters of HMR-161 lift cargo from the 1st Air Delivery Platoon area to the sector occupied by the 7th Marines. The cargo net slung under the helicopter greatly increased its lift capability.

National Archives Photo (USMC) 127-N-A166041

Department of Defense Photo (USMC) A166198

Replacements step ashore at "Charlie Pier." Brought from transports by landing craft, the Marines would broad waiting trains for the four-and-one-half-hour trip to the division's lines.

night of 2 October, a Chinese assault overran the outpost, forcing the Korean Marines to find cover at the foot of the hill and regroup. At mid-morning on the 3d, artillery barrages and air strikes pounded the enemy at Outpost 86, scattering the Chinese there and enabling a South Korean counterattack to regain the objective. The Korean Marines dug in, but on 6 October the Chinese again prevailed, holding the outpost until a South Korean counterattack in the early hours of the 7th forced them down the slope. At about dawn on that same morning, the enemy mounted yet another counterattack, advancing behind a deadly artillery barrage, seizing and holding Outpost 86, and ending this flurry of action on the outpost line in front of the Korean Marine Corps.

Focus on 7th Marines

Although October passed rather quietly for the 1st Marines—except for recurring probes, patrols, and ambushes in the vicinity of Bunker Hill—violent clashes erupted along the line held by the 7th Marines. The enemy began stubborn efforts, which persisted into 1953, to gain control of some or all of the nine combat outposts that Colonel Moore's regiment manned on the high ground to its front.

The 7th Marines took over seven outposts when it relieved the 5th Marines, renaming them, from left to right, Carson, Reno, Vegas, Berlin, Detroit, Frisco, and Seattle. At the point—later known as the Hook—where the frontline veered southward toward the boundary with the British Commonwealth Division—the 7th Marines set up Outpost Warsaw. A second new outpost, Verdun, guarded the boundary between the Marine and Commonwealth divisions. An average of 450 yards separated four of these outposts—Detroit, Frisco, Seattle, and Warsaw—which occupied hills lower than those on the left of the regimental line and therefore were more easily isolated and attacked.

As they did so often, the Chinese began with a diversionary thrust, jabbing at Detroit before throwing knockout blows—artillery and mortar fire preceding an infantry attack—at Seattle and Warsaw. An enemy company overwhelmed the reinforced platoon on Warsaw on 2 October, but the Marines fought stubbornly before falling back. Private Jack W. Kelso picked up a grenade thrown into a

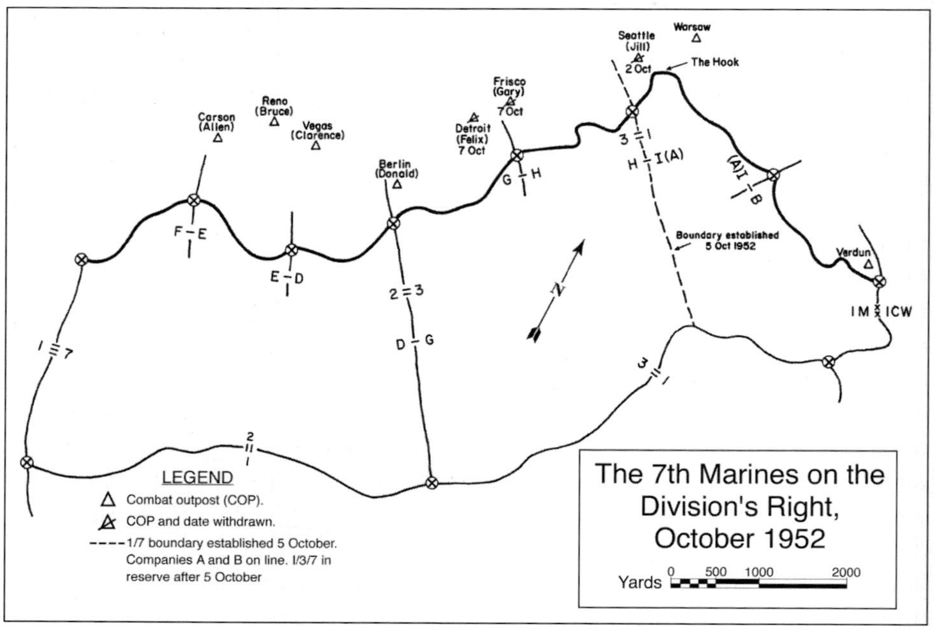

Private Jack W. Kelso

Department of Defense Photo (USMC) A164983

Born in 1934 in Madera, California, he enlisted in the Marine Corps in 1951. In Korea, he was awarded a Silver Star medal for heroism in August 1952. As a rifleman with Company I, 3d Battalion, 7th Marines, he was killed on 2 October 1952 near Sokchon. His Medal of Honor citation reads, in part:

> When both the platoon commander and the platoon sergeant became casualties during the defense of a vital outpost against a numerically superior enemy force attacking at night under cover of intense small-arms, grenade and mortar fire, Private Kelso bravely exposed himself to the hail of enemy fire in a determined effort to reorganize the unit and to repel the onrushing attackers. Forced to seek cover, along with four other Marines, in a near-by bunker which immediately came under attack, he unhesitatingly picked up an enemy grenade which landed in the shelter, rushed out into the open and hurled it back at the enemy.
>
> Although painfully wounded when the grenade exploded as it left his hand, and again forced to seek the protection of the bunker when the hostile fire became more intensified, Private Kelso refused to remain in his position of comparative safety and moved out into the fire-swept area to return the enemy fire, thereby permitting the pinned-down Marines in the bunker to escape.

Staff Sergeant Lewis G. Watkins

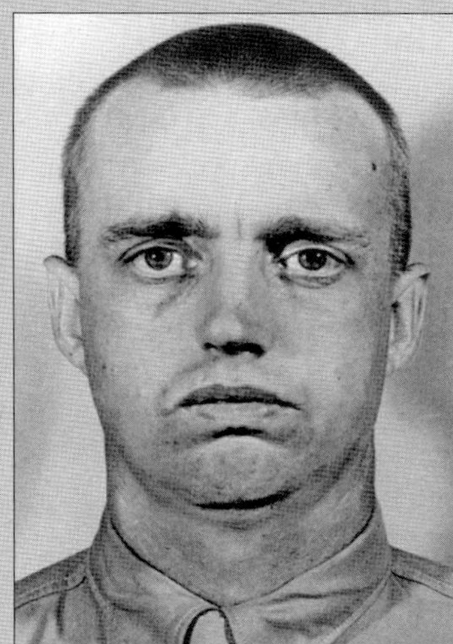

Department of Defense Photo (USMC) A403014

Born in 1925 in Seneca, South Carolina, he enlisted in the Marine Corps in 1950. While serving as a platoon guide for Company I, 3d Battalion, 7th Marines, he gave his life on 7 October 1952. His Medal of Honor citation reads, in part:

> With his platoon assigned the mission or re-taking an outpost which had been overrun by the enemy earlier in the night, Staff Sergeant Watkins skillfully led his unit in the assault up the designated hill. Although painfully wounded when a well-entrenched hostile force at the crest of the hill engaged the platoon with intense small-arms and grenade fire, gallantly continued to lead his men. Obtaining an automatic rifle from one of the wounded men, he assisted in pinning down an enemy machine gun holding up the assault.
>
> When an enemy grenade landed among Staff Sergeant Watkins and several other Marines while they were moving forward through a trench on the hill crest, he immediately pushed his companions aside, placed himself in position to shield them and picked up the deadly missile in an attempt to throw it outside the trench. Mortally wounded when the grenade exploded in his hand, Staff Sergeant Watkins, by his great personal valor in the face of almost certain death, saved the lives of several of his comrades.

—Captain John C. Chapin, USMCR (Ret)

bunker that he and four other Marines from Company I, 3d Battalion, 7th Marines, were manning. Kelso threw the live grenade at the advancing Chinese, but it exploded immediately after leaving his hand. Although badly wounded, Kelso tried to cover the withdrawal of the other four, firing at the attackers until he suffered fatal wounds. Kelso earned a posthumous Medal of Honor, but heroism alone could not prevail; numerically superior forces captured both Warsaw and Seattle.

The Marines counterattacked immediately. Captain John H. Thomas, in command of Company I, sent one platoon against Warsaw, but the enemy had temporarily withdrawn. The lull continued until 0145 on the morning of 4 October, when a Chinese platoon attacked only to be beaten back by the Marines holding Warsaw.

Meanwhile, Captain Thomas mounted a counterattack against Seattle early on the morning of 3 October, sending out two squads from the company's position on the main line of resistance. Despite Chinese artillery fire, the Marines reached the objective, but Seattle proved too strongly held and Thomas broke off the counterattack. As dusk settled over the battleground, Marine aircraft and artillery put down a smoke screen behind which the counterattack resumed, but the Chinese succeeded in containing the two squads short of the crest. To regain momentum, another squad—this one from Company A, 1st Battalion, 7th Marines, which had come under the operational control of the 3d Battalion—reinforced the other two, but the Chinese clinging to Seattle inflicted casualties that sapped the strength of the

Second Lieutenant Sherrod E. Skinner, Jr.

Department of Defense Photo (USMC) A301998

Born in 1929 in Hartford, Connecticut, he was appointed a second lieutenant in the Marine Corps Reserve 1952 and then ordered to active duty. He died heroically at "The Hook" on 26 October 1952 while serving as an artillery forward observer of Battery F, 2d Battalion, 11th Marines. The citation for his Medal of Honor award reads, in part:

> Skinner, in a determined effort to hold his position, immediately organized and directed the surviving personnel in the defense of the outpost, continuing to call down fire on the enemy by means of radio alone until this equipment became damaged beyond repair. Undaunted by the intense hostile barrage and the rapidly closing attackers, he twice left the protection of his bunker in order to direct accurate machine-gun fire and to replenish the depleted supply of ammunition and grenades. Although painfully wounded on each occasion, he steadfastly refused medical aid until the rest of the men received treatment.
>
> As the ground attack reached its climax, he gallantly directed the final defense until the meager supply of ammunition was exhausted and the position overrun. During the three hours that the outpost was occupied by the enemy, several grenades were thrown into the bunker which served as protection for Second Lieutenant Skinner and his remaining comrades. Realizing that there was no chance for other than passive resistance, he directed his men to feign death even though the hostile troops entered the bunker and searched their persons. Later, when an enemy grenade was thrown between him and two other survivors, he immediately threw himself on the deadly missile in an effort to protect the others, absorbing the full force of the explosion and sacrificing his life for his comrades.

In 1991, Skinner Hall at Quantico, Virginia, was dedicated in his honor.

Second Lieutenant George H. O'Brien, Jr.

Department of Defense Photo (USMC) A49768

Born in Fort Worth, Texas, in 1926, he enlisted in the Marine Corps Reserve in 1949. Ordered to active duty in 1951, he entered Officer Candidate School and was commissioned in 1952. As a rifle platoon leader with Company H, 3d Battalion, 7th Marines, in the battle for the Hook on 27 October 1952, he was awarded a Medal of Honor with a citation, which reads, in part:

> O'Brien leaped from his trench when the attack signal was given and, shouting for his men to follow raced across an exposed saddle and up the enemy-held hill through a virtual hail of deadly small-arms, artillery and mortar fire. Although shot through the arm and thrown to the ground by hostile automatic-weapons fire as he neared the well-entrenched enemy position, he bravely regained his feet, waved his men onward and continued to spearhead the assault, pausing only long enough to go to the aid of a wounded Marine. Encountering the enemy at close range, he proceeded to hurl hand grenades into the bunkers and, utilizing his carbine to best advantage in savage hand-to-hand combat, succeeded in killing at least three of the enemy.
>
> Struck down by the concussion of grenades on three occasions during the subsequent action, he steadfastly refused to be evacuated for medical treatment and continued to lead his platoon in the assault for a period of nearly four hours, repeatedly encouraging his men and maintaining superb direction of the unit.

He received a second Purple Heart Medal in January 1953, and after the war, he joined the Reserves and was promoted to major.

—Captain John C. Chapin, USMCR (Ret)

Department of Defense Photo (USMC) A165717

Riflemen of the 1st Battalion, 7th Marines, take a break from frontline duty long enough to buy "luxuries" from the regimental Post Exchange truck. The truck toured the frontline twice a month offering Marines a chance to buy such items candy, cameras, and toilet articles.

counterattack. While the Marines regrouped for another assault, the 11th Marines pounded the outpost with artillery fire. At 2225 on 3 October, the Marines again stormed the objective, but Chinese artillery prevailed, and Seattle remained in Chinese hands.

The loss of Outpost Seattle, the recapture of Warsaw, and a successful defense of Frisco against a Chinese probe on the night of 5 October did not mark the end of the effort to seize the outposts manned by the 7th Marines, but only a pause. The regiment's casualties—13 killed and 88 wounded by 3 October when the Marines suspended the attempt to retake Seattle—caused the 7th Marines to shuffle units. The 3d Battalion, commanded until 13 October by Lieutenant Colonel Gerald F. Russell, had suffered most of the casualties. As a result, while Russell's battalion reduced its frontage, the 1st Battalion, under Lieutenant Colonel Leo J. Dulacki, moved from the regimental reserve to take over the right-hand portion of the Jamestown Line. Dulacki's Marines manned the main line of resistance from roughly 500 yards southwest of the Hook to the boundary shared with the Commonwealth Division, including Outposts Warsaw and Verdun. Colonel Moore thus placed all three battalions on line, Lieutenant Colonel Anthony Caputo's 2d Battalion on the left, Russell's 3d Battalion in the center, and now Dulacki's 1st Battalion on the right.

The 7th Marines completed its realignment just in time to meet a series of carefully planned and aggressively executed Chinese attacks delivered on 6 and 7 October against five combat outposts and two points on the main line of resistance. The Marines struck first, however, when a reinforced platoon from Company C, 1st Battalion, attacked toward Outpost Seattle at 0600 on 6 October. Mortar and artillery fire forced the platoon to take cover and regroup, even as the Chinese were reinforcing the outpost they now held. The attack resumed at 0900. Despite infantry reinforcements, air strikes, and artillery, the Marines could not crack Seattle's defenses and broke off the attack at about 1100, after losing 12 killed and 44 wounded. The attackers estimated that they had killed or wounded 71 Chinese.

On the evening of the 6th, the Chinese took the initiative, by midnight firing some 4,400 artillery and mortar rounds against the outpost line and two points on the main line of resistance. On the left of Colonel Moore's line, the enemy probed Outposts Carson and Reno, and on the right he stormed Warsaw, forcing the defenders to call for box-me-in fire that severed the telephone wire linking the outpost with the Jamestown Line. The first message from Warsaw when contact was restored requested more artillery fire, which by 2055 helped break the back of the Chinese assault, forcing the enemy to fall back.

The most determined attacks on the night of 6 October and early hours of the 7th hit Outposts Detroit and Frisco in the center of the regimental front. To divert attention from these objectives, each one manned by two squads, the Chinese probed two points on the main line of resistance that had already been subjected to artillery and mortar bombardment. At 1940 on the night of the 6th, an attacking company that had gained a foothold in the main trench on Detroit fell back after deadly fire stopped the enemy short of the bunkers. Two hours later, the Chinese again seized a segment of trench on Detroit and tried to exploit the lodgment. The Marines reacted by calling on the 11th Marines to box in the outpost. Communications failed for a time, but at about 2115 the defenders of Detroit requested variable-time fire for airbursts over the bunkers, which would protect the Marines while the enemy outside remained exposed to a hail of shell fragments. The artillerymen fired as the Marines on Detroit asked, but the outpost again lost contact with headquarters of the 3d Battalion.

Two squads set out from the Jamestown Line to reinforce the Marines from Company G, 3d Battalion, 7th Marines, manning Outpost Detroit. The Chinese frustrated this attempt with artillery fire but in the meantime again abandoned their foothold on Detroit, probably because of the shower of fragments from Marine shells bursting above them. The respite proved short-lived, however, for the enemy renewed the attack shortly after midnight and extended his control of the hill despite further scourging from the 11th Marines. A six-man Marine patrol reached the outpost, returning at 0355 to report that the Chinese now held the trenchline and bunkers; only two of the Marines who had manned Detroit escaped death or capture. The attempt to break through to the outpost ended, and by 0630, the Marines engaged in the effort had returned to the main line of resistance.

Meanwhile, at about 2000, a Chinese company hit the two squads from Company H that held Outpost Frisco, north of Detroit. The assault troops worked their way into the trenches, but airbursts from Marine artillery reinforced the small arms fire of the defenders in driving the enemy back. The Chinese renewed the attack just after midnight, and two Marine squads advanced from the main line of resistance to reinforce Frisco, only to be stopped short of their goal by fire from artillery and mortars. Companies H and I of Russell's 3d Battalion made further attempts to reach Frisco during the early morning, but not until 0510 did a reinforced platoon from Company I arrive and take control.

During the battles of October 1952, a number of Marine bunkers on the Jamestown Line were severely damaged by Chinese mortar and artillery fire.

Department of Defense Photo (USMC) A167465

During the final attack, Staff Sergeant Lewis G. Watkins, despite earlier wounds, took an automatic rifle from a more badly injured Marine and opened fire to keep the platoon moving forward. When a Chinese grenade landed near him, he seized it, but it exploded before he could throw it away, fatally wounding Watkins, whose leadership and self-sacrifice earned him a posthumous Medal of Honor. A second platoon from Company I joined the other unit atop the hill, and at 0715 Frisco was declared secure.

To keep Frisco firmly in Marine hands, however, would have invited attrition and ultimately required more men than General Pollock and Colonel Moore could spare from the main line of resistance. Consequently, the 7th Marines abandoned the outpost. The regiment had yielded three outposts—Detroit, Frisco, and Seattle—but forced the Chinese to pay a high price, estimated to include 200 killed. The losses suffered by the 7th Marines totaled 10 killed, 22 missing, and 128 wounded, 105 of them seriously enough to require evacuation.

Since the Chinese also wrested Outposts 37, 36, and 86 from the Korean Marine Corps, the 1st Marine Division had lost six combat outposts of varying tactical importance. The lost outposts and those that remained in Marine hands had no value except to the extent their possession affected the security of the Jamestown Line. As a result, the Marines remained wary of mounting major counterattacks; to provide continued protection for the main line of resistance, General Pollock would rely on nighttime patrols and listening posts to supplement the remaining outposts and replace the captured ones.

Department of Defense Photo (USMC) A166429

The Marine in the foreground reads while others talk or sleep just before jumping off into "no-man's-land" on the Hook.

Unlike the 1st Marine Aircraft Wing, which experienced shortages of spare parts, especially for its newer planes and helicopters, the 1st Marine Division emerged from the outpost fighting with few supply problems except for communications gear. The Army helped with replenishment by releasing essential spare parts from stocks in Japan, and new radio, telephone, and teletype equipment also arrived from the United States. The Army, moreover, tapped its stocks for new trucks to replace the division's worn-out vehicles. To operate the logistics network more efficiently, the division placed the re-equipped truck units in direct support of the infantry regiments instead of keeping them in a centralized motor pool. The change reduced both the total mileage driven and vulnerability to artillery fire. In another attempt to improve logistics, the Marines increased from 500 to 800 the number of Korean laborers serving with each of the front-line regiments.

Chinese Attack the Hook

The fighting for the outposts that raged early in October died down, although the Chinese jabbed from time to time at Bunker Hill and continued their nighttime patrolling elsewhere along the Jamestown Line. During the lull, Colonel Smoak's 5th Marines took over the center of the division's line from the 1st Marines, which went into reserve, improving the fall-back defense lines, undergoing training, and patrolling to maintain security in the rear areas. As the division reserve, the 1st Marines prepared to counterattack if the enemy should penetrate the main line of resistance. Indeed, the regiment had to be ready to help block a Chinese breakthrough anywhere in I Corps, which held the western third of the United Nations line.

With the 1st Marines now in reserve, Colonel Moore's 7th Marines manned the right of the division's line placing all three battalions on line and keeping only one company from the 3d Battalion in reserve. To replace this company, Lieutenant Colonel Charles D. Barrett, the 3d Battalion's new commander, organized a platoon of cooks, drivers, and other members of the headquarters into an improvised platoon that served as his unit's reserve. Barrett's thinly spread 3d Battalion manned two combat outposts—Berlin and East Berlin, the latter established on 13 October—besides defending the center of the regimental line.

To the left of Barrett's unit, the 2d Battalion, 7th Marines, under Lieutenant Colonel Anthony Caputo defended its portion of the regimental line and maintained three combat outposts, Carson, Reno, and Vegas. Like Barrett's battalion in the center, Caputo's Marines held a sector with few vulnerable salients that the aggressive Chinese might pinch off and capture.

On the far right of the regimental line, Lieutenant Colonel Dulacki's 1st Battalion held the dominant terrain feature in the reg-

Distant Strikes to Close Air Support

During the struggle over the combat outpost line, from late July through early October, the 1st Marine Division benefited from a shift in aerial strategy. Long-range interdiction, an important element of U. S. Air Force doctrine, gave way to a policy of hitting the enemy wherever he might be, whether exercising control from the North Korean capital of Pyongyang or massing to attack Bunker Hill or some other outpost.

United Nations aircraft hit Pyongyang hard in July and again in August. The 11 July attack consisted of four separate raids, the last of them after dark. The operation, named Pressure Pump, blasted headquarters buildings, supply dumps, and the radio station, which was silenced for two days. Of 30 individual targets attacked on the 11th, three were obliterated, 25 damaged in varying degrees and only two survived intact. On 29 August, three daylight raids damaged 34 targets on a list of 45 that included government agencies—among them the resurrected Radio Pyongyang—factories, warehouses, and barracks.

According to Colonel Samuel S. Jack, chief of staff of the 1st Marine Aircraft Wing, a shift of emphasis away from long-range interdiction caused the Far East Air Forces to endorse the wing's using its aircraft primarily to respond to requests for support from the 1st Marine Division. Even as the Air Force became more cooperative, Army commanders complained about the air support they received and began agitating for a more responsive system, like that of the Marine Corps, that would cover the entire battlefront. General Mark W. Clark, the United Nations commander since May 1952, refused, however, to tamper with the existing system that in effect gave the Far East Air Forces, an Air Force headquarters, "coordination control" over the 1st Marine Aircraft Wing and channeled requests through a Joint Operations Center. Instead, he called for increasing the efficiency of the current arrangement to reduce reaction time.

During the defense of Bunker Hill in August, 1,000 sorties, most of them by Marines, bombed and strafed the attacking Chinese. In early October, when the fighting shifted eastward to the outposts manned by the 7th Marines, the wing flew 319 close air support sorties.

Besides coming to the aid of their fellow Marines on the ground, crews of the 1st Marine Aircraft Wing supported the U. S. and South Korean armies. In July, for example, eight heavily-armed and armored AU-1 Corsairs from Marine Attack Squadron 323 attacked Chinese mortar positions, antitank weapons, and troop concentrations opposing the Republic of Korea's I Corps. Rockets, 1,000-pound bombs, napalm, and 20mm fire killed an estimated 500 of the enemy. Similarly, four pilots from Marine Fighter Squadron 311 helped the U.S. Army's 25th Infantry Division by destroying three bunkers and two artillery pieces while collapsing some 50 feet of trench.

National Archives Photo (USMC) 127-N-A133864

Col John P. Condon, left, commanding officer of Marine Aircraft Group 33, shows BGen Clayton C. Jerome, the area strafed in the largest Marine jet strike of the Korean War.

In operations like these, weather could prove as deadly as antiaircraft fire. On 10 September, as 22 Panther jets were returning from a strike near Sariwon, fog settled over their base. Sixteen planes landed safely at an alternate airfield, but the others flew into a mountain as they approached the runway, killing all six pilots.

Marine helicopters continued to fly experimental missions, as when Marine Helicopter Transport Squadron 161 deployed multiple-tube 4.5-inch rocket launchers from one position to another, thus preventing Chinese observers from using, as an aiming point for effective counterbattery fire, the dust cloud that arose when the rockets were launched. The same squadron, using 40 percent of its assigned helicopters for the purpose, delivered enough supplies to sustain the 7th Marines for five days. Problems continued to nag the helicopter program, however. The Sikorsky HO3S-1 light helicopter—used for liaison, observation, and casualty evacuation—had to be grounded in October to await spare parts.

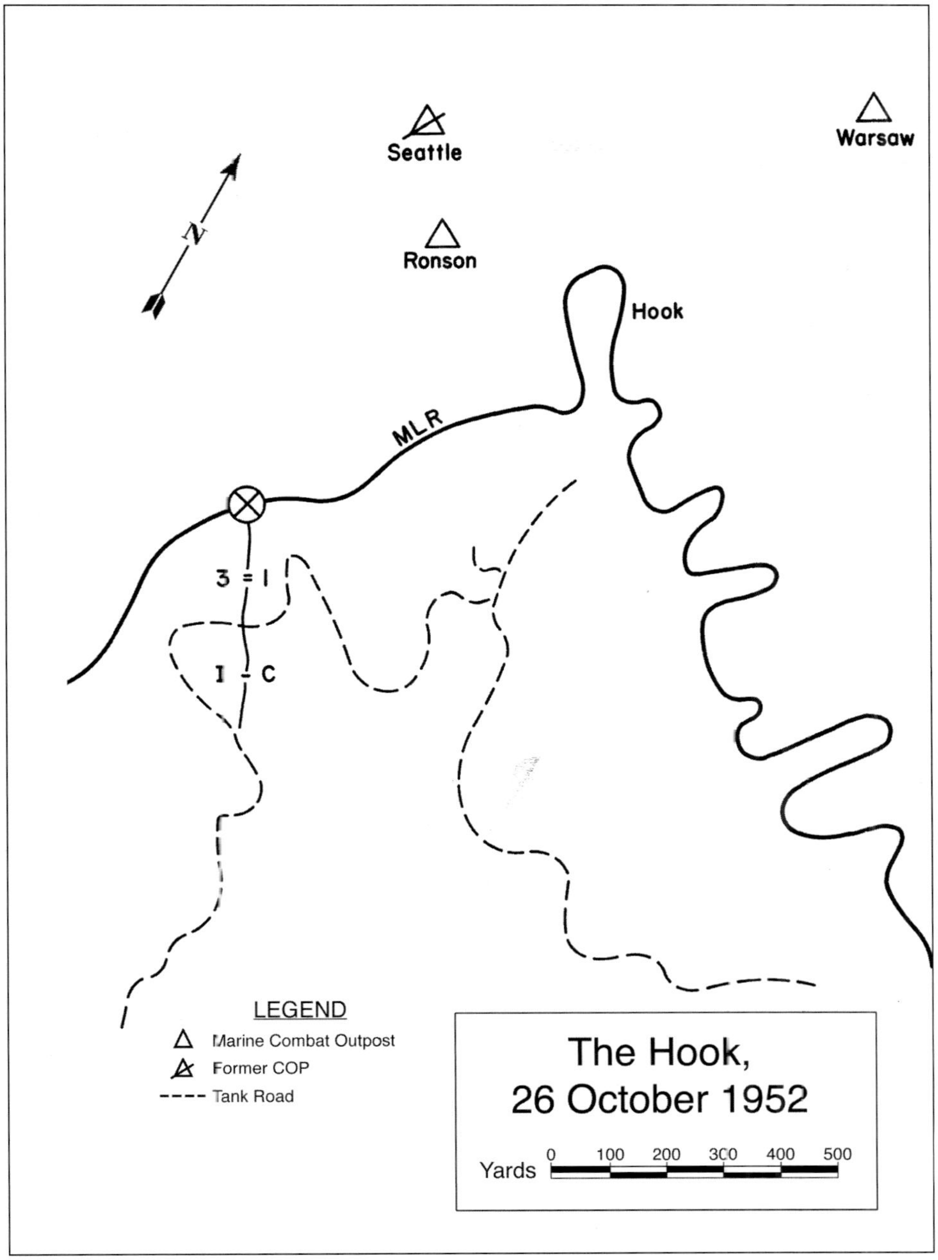

The Hook, 26 October 1952

imental sector, the Hook, where the high ground that defined the Jamestown Line veered sharply to the south. Outposts Seattle and Warsaw had protected the Hook, but only Warsaw remained in Marine hands. To restore the security of the Hook, Dulacki set up a new outpost, Ronson, some 200 yards southeast of enemy-held Seattle and 275 yards west of the Hook. Ronson guarded the western approaches to the Hook, while Warsaw commanded the lowlands east of the Hook and a narrow valley leading eastward toward the Samichon River.

If the Chinese should seize the Hook, disastrous results might follow, for the Hook held the key to controlling the Samichon and Imjin valleys. Its capture could expose the rear areas of the 1st Marine Division and force that unit and the adjacent Commonwealth Division to fall back two miles or more to find defensible terrain from which to protect the northeastern approaches to Seoul. Because of the Hook's importance, Colonel Moore set up his headquarters on Hill 146 near the base of the salient.

Opposite the 7th Marines, the Chinese had massed two infantry regiments, totaling some 7,000 men, supported by 10 battalions of artillery ranging in size from 75mm to 122mm and later to 152mm. The enemy, moreover, had learned during the fighting at Bunker Hill and along the outpost line to make deadlier use of his artillery, massing fires and, when the Marines counterattacked, imitating the box-me-in fires used by the 11th Marines. In preparation for an attack on the Hook, the Chinese massed their artillery batteries within range of the salient, stockpiled ammunition, and dug new trenches that reached like tentacles toward the various elements of the outpost line and afforded cover and concealment for attacking infantry.

Against the formidable concentration of Chinese troops and guns, Colonel Moore's regiment could muster 3,844 Marines, officers and men, supported by 11 Navy medical officers and 133 Hospital Corpsmen, three Army communications specialists, and 746 Korean laborers with their 18 interpreters. As in the earlier fighting, the 7th Marines could call upon the 105mm and 155mm howitzers of the 11th Marines, and other Marine supporting weapons including rocket batteries, tanks, and aircraft. Army artillery and Air Force fighter-bombers reinforced the firepower of the division.

In the words of Staff Sergeant Christopher E. Sarno of the 1st Marine Division's tank battalion: "Korea was an artilleryman's paradise." It seemed to him that the Chinese always fought by night, making effective use of an arsenal of weapons. The burp guns and mines were bad, Sarno said, but the worst was the artillery, which "could blast a man's body to bits so that his remains were picked up in a shovel."

The static battle line obviously placed a premium on artillery, especially the 105mm and 155mm howitzers of the Army and Marine Corps. Indeed, by mid-October 1952 American batteries were firing these shells at a more rapid rate than during the bloody fighting in the early months of 1951, when United Nations forces advanced beyond the 38th Parallel and conducted offensive operations like Killer and Ripper. By the fall of 1952, firepower, especially artillery, dominated the battleground, as probes, patrols, ambushes, and attacks on outposts took the place of major offensives.

Because of the demand of artillery support, the rationing of 105mm and 155mm shells became necessary. During the last 11 days of October, a quota prevailed, at least for purposes of planning. In support of the 7th Marines, each 105mm howitzer might fire a daily average of 20 rounds and each 155mm howitzer 4.3 rounds. The 81mm mortars located in each battalion also suffered from a shortage of shells, and even hand grenades were now scarce. To do the work of mortars and artillery, Colonel Moore employed tripod-mounted, water-cooled .30-caliber machine guns. These weapons employed the techniques of World War I, engaging targets like potential assembly areas with indirect fire based on map data and adjusted by forward observers, as well as aimed direct fire.

Ammunition carriers step aside to let litter barriers pass with a wounded Marine during the battle for the Hook.

Department of Defense Photo (USMC) A166424

The enemy initiated his attack on the Hook by battering the salient and its combat outposts with mortar and artillery fire; an estimated 1,200 rounds exploded among the trenches and bunkers on the Hook and Outposts Ronson and Warsaw between dusk on 24 October and dawn on the 25th. The Marine defenders, aided by detachments from the Korean Service Corps, struggled to keep pace with the destruction, repairing damage as best they could during lulls in the bombardment only to face new damage when the shelling resumed.

While shoring up trenches and bunkers, the 7th Marines fought back. In the hardest hit area, Lieutenant Colonel Dulacki's 1st Battalion returned fire with its own mortars, machine guns, and recoilless rifles, while the regimental mortars and tanks joined in. Despite the shortage of high-explosive shells, the 2d Battalion, 11th Marines, fired some 400 105mm rounds in response to the first day's Chinese bombardment, 575 rounds on the 24 October, and 506 on the 25th, balancing the need to conserve ammunition against the worsening crisis. Air strikes also pounded the enemy massing near the Hook and the two nearby outposts, including attacks by four Panther jets from VMF-311 that dropped high explosives and napalm on Chinese troops massing some 750 yards east of the salient.

During the 24 hours beginning

National Archives Photo (USMC) 127-N-A169052

The crew of Number 5 gun, Battery L, 4th Battalion, 11th Marines, loads its 155mm howitzer after receiving word to prepare for a fire mission.

at 1800 on the 25th, Chinese gunners scourged Colonel Moore's regiment with another 1,600 mortar and artillery shells, most of them exploding on the ground held by Dulacki's battalion. The shelling abated briefly on the 25th but resumed, convincing the division's intelligence officer, Colonel Clarence A. Barninger, that the enemy was planning a major attack to overwhelm the Hook and gain control of the Samichon Valley. Barninger warned the division's commander, General Pollock, well in advance of the actual attack.

The Chinese fire diminished somewhat on the morning of 26 October, but the Hook remained a dangerous place. On that morning, Lieutenant Colonel Dulacki took advantage of a lull to inspect the defenses, only to be knocked down by the concussion from an enemy shell. He escaped with bruises and abrasions and continued his rounds. The intensity of the bombardment increased later in the morning and continued after dark in preparation for attacks on Outposts Ronson and Warsaw and the Hook itself.

Chinese troops stormed Ronson at 1810 on 26 October after three days of preparatory fire that had collapsed trenches, shattered bunkers, and killed and wounded Marines at both outposts and on the Hook. At Ronson, 50 or more of the enemy penetrated the defensive artillery concentrations, overran the position, and killed or captured the members of the reinforced squad manning the outpost.

At about the same time, a Chinese company split into two groups and attacked Outpost Warsaw simultaneously from the east and west. A box-me-in barrage fired by the 11th Marines could not prevent the assault troops for closing in on the defenders, led by Second Lieutenant John L. Babson, Jr. The Marines at Warsaw fought back with grenades, pistols, and rifles, using the latter as clubs when ammunition ran out. Taking cover in the wreckage of the bunkers, they called for variable-time fire directly overhead. Hope lingered that the rain of shell fragments had saved Warsaw, but after four hours of silence from the garrison, Colonel Moore reluctantly concluded that Warsaw was lost, its defenders either dead or captured. Lieutenant Babson was one of those killed.

While the fate of Warsaw still

Chinese shells explode in the valley below as Marines on the Hook move about in their trenches.

Department of Defense Photo (USMC) A166773

remained in doubt and a platoon from Captain Paul B. Byrum's Company C was preparing to reinforce the outpost, a flurry of Chinese shells battered the Hook. Colonel Moore reacted by sending Captain Frederick C. McLaughlin's Company A, which Byrum's unit had just relieved, to help Company C defend the salient. Moore also directed that the 1st Battalion have first call on the regiment's supply of ammunition. In addition, the 1st Marine Division lifted the restrictions on artillery ammunition fired in support of the Hook's defenders.

Under cover of artillery and mortar fire, a Chinese battalion launched a three-pronged attack on Dulacki's 1st Battalion. By 1938 on the 26th, Chinese infantry first threatened the main line of resistance southwest of the Hook itself, to the left of the salient and roughly halfway to the boundary with the 3d Battalion. Within a few minutes, a second attack hit the very nose of the Hook, while a third struck its eastern face. Mingled with the assault troops were laborers carrying construction materials to fortify the Hook after the three prongs of the attack had isolated and overrun it.

The thrust along the ridge that formed the spine of the Hook continued until the Chinese encountered the observation post from which Second Lieutenant Sherrod E. Skinner, Jr., was directing the fire of the 11th Marines. The lieutenant organized the defense of this bunker, running from cover when necessary to replenish the supply of small arms ammunition. He was still calling down artillery fire when the attackers overran the Hook. He then told his men to play dead until other Marines counterattacked. For three hours, they fooled the Chinese who entered the bunker. Finally, an enemy soldier became suspicious and threw a grenade inside. Skinner rolled onto the grenade, absorbing the force of the explosion and saving the lives of two of his men. His sacrifice resulted in the posthumous award of the Medal of Honor.

National Archives Photo (USMC) 127-N-A165743

MajGen Paul A. Kendall, center, I Corps commander, receives a tactical briefing from Col Russell E. Honsowetz, operations officer of the 1st Marine Division. To MajGen Kendall's right is MajGen Edwin A. Pollock, 1st Marine Division commander, and to Col Honsowetz' left Col Thomas C. Moore, commanding officer of the 7th Marines.

To the left of the Hook, the assault force outflanked a platoon of Company C, led by Second Lieutenant John W. Meikle, but he succeeded in pulling back the flanks to form a perimeter. East of the Hook, on the 1st Battalion's right, other elements of Company C formed another perimeter. In the 400 yards separating the two, scattered groups from Byrum's company struggled to close the gap.

Help for Company C began arriving at about 2330, after the Chinese had overrun the Hook, when the first elements of McLaughlin's Company A, sent to reinforce Byrum's Marines, made contact with Meikle's perimeter to the left rear of the captured salient. The arrival of reinforcements enabled the members of Company C, scattered between the two perimeters, to form a blocking position on a ridge running east and west a few hundred yards to the rear of the Hook. At 0300 on the 27th, Colonel Moore committed the regimental reserve, Company H, 3d Battalion. General Pollock ordered the 3d Battalion, 1st Marines, from the division reserve into the sector held by Moore's 7th Marines.

The 3d Battalion, 1st Marines, had the mission of counterattacking the Chinese who had seized the Hook and penetrated the Jamestown Line. Anticipating commitment in this critical sector, the battalion commander, Lieutenant Colonel Sidney J. Altman, had already drawn up a basic plan for such a counterattack and personally reconnoitered the area. Now Altman's Marines prepared to exe-

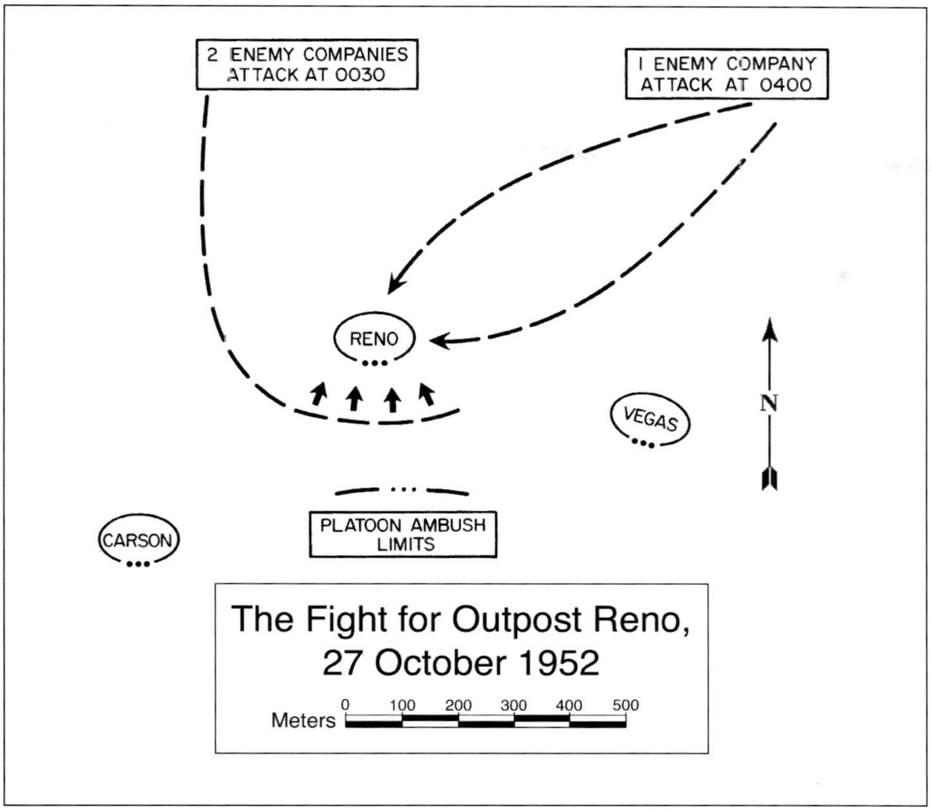

The Fight for Outpost Reno, 27 October 1952

cute that plan. Tank gunners, mortar crews, and artillerymen battered the recently captured Outpost Warsaw, other Chinese troop concentrations, firing batteries, and supply routes. The 1st Marine Aircraft Wing also joined in, as Grumman Tigercat night fighters used the ground-based MPQ radar to hit the main Chinese supply route sustaining the attack, dropping their bombs less than a mile west of the Hook.

Action at Outpost Reno

While the 3d Battalion prepared to counterattack the Hook, the enemy made a new thrust at the 7th Marines. Early on the morning of 27 October, the Chinese attacked Reno, one of the outposts manned by Lieutenant Colonel Caputo's 2d Battalion. Some two miles west of the Hook, the battalion's three outposts formed an arrowhead aimed at Chinese lines, with Reno at the point, Carson on the left, and Vegas on the right.

The pattern of hostile activity opposite the 7th Marines earlier in October persuaded General Pollock's intelligence specialists that the enemy had given first priority to seizing Caputo's three outposts. This estimate of Chinese intentions caused the 11th Marines to plan concentrations in front of the 2d Battalion, while Caputo set up strong ambush positions to protect the threatened outposts.

On the night before the assault on Reno, Captain James R. Flores led a reinforced platoon from Company E into the darkness. The patrol's destination was a camouflaged ambush position about 300 yards south of Reno and halfway between Carson and Vegas. At midnight, noises to the front alerted Flores and his men that the enemy had infiltrated between them and the outpost and were preparing to attack Reno from the rear. The ambush force alerted the Marines defending the outpost of the danger behind them and opened fire when the force, estimated at two Chinese companies, seemed on the verge of attacking. Although raked by fire from the front and rear, the enemy fought back, holding the Marines in check until they could break off the action and make an orderly withdrawal to the main line of resistance.

Quiet enveloped Outpost Reno until 0400 on the 27th, when a Chinese platoon attacked from the northwest, assaulting in two waves. The Marines on Reno beat back this first attack, but a second thrust from the same direction broke through the perimeter. The defenders took cover so that variable-time artillery fire bursting overhead could maul the enemy. The tactics worked to perfection, forcing the Chinese to abandon their lodgment after some 40 minutes of fighting

Counterattacking the Hook

After the ambush by Marines behind Reno, while the outpost's defenders were fighting off the subsequent Chinese assaults, Captain McLaughlin's Company A passed through the line established by Byrum's Marines and began advancing toward the Hook. Enemy mortar and artillery fire stopped McLaughlin's unit short of the objective, forcing him to order his Marines to dig in and hold the ground they had gained.

When McLaughlin's counterattack stalled, Colonel Moore attached the last of the regimental reserve—Company H, 3d Battalion, under Captain Bernard B. Belant—to Lieutenant Colonel Dulacki's 1st Battalion. At 0505, Belant reported to Dulacki, who directed him to renew the counterattack. Within three hours, Company H stood ready to attack the Chinese, whose bridgehead encompassed the Hook itself and a crescent of ridges and draws extending from the spine of that terrain feature and embracing a segment of the main line of resistance about a half-mile wide.

As Belant led his unit forward, it rapidly covered the first 200 yards before Chinese small arms, mortar, and artillery fire shifted to meet the threat. The company commander pressed the attack, however. Second Lieutenant George H. O'Brien, Jr., led his platoon over the ridge to his front, the men zigzagging as they ran toward the Chinese-held trenchline. A bullet struck O'Brien's armored vest, knocking him down, but he scrambled to his feet and continued toward the enemy, pausing only briefly to help a wounded man. Throwing grenades and firing his carbine, he silenced the Chinese weapons in a bunker and led his platoon toward the Hook itself. This headlong assault, which earned O'Brien the Medal of Honor, broke through the Chinese perimeter and approached the enemy-held bunkers on the Hook before being contained by hostile mortar and artillery fire. The remainder of Company H widened the crack that O'Brien's platoon had opened and captured three prisoners as it overran the southeastern portion of the Hook before a fierce shelling forced the advance elements to find cover and yield some of the ground they had taken.

Marines climb an improved trail to bring ammunition to machine gunners on the Hook.
Department of Defense Photo (USMC) A166423

Marine aircraft helped Company H advance onto the Hook, attacking reinforcements moving into battle and the positions from which the Chinese were firing or adjusting fire. Key targets included the former Marine outposts of Irene, Seattle, and Frisco, along with the frequently bombed main supply route and those enemy troops trying to dislodge the Marines who had gained a foothold on the Hook. Fire from Marine tanks and artillery engaged some of the same targets and proved deadly against trenches and bunkers that the Chinese had seized. The howitzers also joined mortars in counterbattery fire.

At midday on 27 October, after General Pollock had released Company I, 3d Battalion, 1st Marines, to Colonel Moore's control, the counterattack to regain the Hook entered its final phase. Company I would drive the enemy onto the Hook, after which Company H, the unit that had penetrated the Chinese perimeter earlier in the day, was to take over the right of the 1st Battalion's line, relieving Company B, which would make a final assault on the Hook and recapture both the Hook and Outpost Warsaw.

Captain Murray V. Harlan, Jr., who commanded Company I, launched his attack early in the afternoon. The 1st Platoon, which led the way, seized the crest of the ridge to its front at 1350 and began advancing toward the Hook behind a barrage laid down by the 11th Marines. The Chinese reacted to the threat from Harlan's company with deadly artillery and mortar concentrations directed against not only the advancing Marines, but also Colonel Moore's command post and the weapons along the Jamestown Line that supported the assault.

Despite severe casualties, Harlan's Marines pushed ahead, at times crawling from one outcropping in the shell-torn earth to another. After pausing to reorganize at about 1635, the company moved, a few men at a time, onto the Hook, regaining the bunker where Lieutenant Skinner had sacrificed his life for his men and forging ahead against die-hard Chinese in collapsed bunkers and trenches. The deadliest fire came, as always, from enemy mortars and artillery shells that plunged steeply onto the Hook before exploding. Company I took such cover as it could find, but terrain afforded concealment and greater protection from flat-trajectory than from high-angle-fire weapons.

By midnight on 28 October, Company B had threaded its way through a maze of shell craters and moved into position to the left of Company I. Shortly afterward Company B began what Dulacki had planned as the final assault on the Hook. Small arms fire and a shower of grenades from the Chinese positions stopped the Marines as they attacked with rifles and grenades of their own. After exchanging fire with the enemy for perhaps 90 minutes the company fell back to obtain cover and called for mortar and artillery fire. The shelling battered not only the strongpoints immediately to the

Department of Defense Photo (USMC) A166425

A Marine casualty arrives at a rear aid station for emergency treatment and evacuation to a hospital. Ammunition carrier in the background starts uphill with belted ammunition.

front of Company B, but also the enemy's supporting weapons and the routes of reinforcement and replenishment that passed through the Chinese-held outposts of Warsaw and Ronson. The Marines renewed the assault at 0340, broke through, and by 0600 overran the Hook. Afterward, elements of Colonel Moore's regiment reoccupied Warsaw and Reno, which the enemy had abandoned.

A dense fog settled over the Hook as Lieutenant Colonel Dulacki's battalion killed or captured Chinese stragglers and reorganized the defenses. The battle for the Hook and its two outposts, along with the diversionary action at Outpost Reno, cost the Chinese 269 dead and wounded, a number verified as best the Marines could, and perhaps another 953 casualties that escaped verification. In preventing the enemy from gaining a permanent tactical advantage at the Hook, the Marines lost 70 killed, 386 wounded, and 39 missing, of whom 27 were prisoners of the Chinese. At Reno, nine Marines were killed and 49 wounded.

During the struggle for the Hook, Chinese mortars and artillery proved as deadly as usual in both defending against Marine attack and battering Marine defenses. The enemy sent additional soldiers immediately behind the assault troops, intending that they exploit any breakthrough by the earlier waves, but the tactics failed to accomplish the intended purpose. Lieutenant Colonel Dulacki believed that the gambit failed because too few junior officers were on hand to commit the reserve at the precise moment to maintain momentum.

Dulacki also found flaws with his Marines. Judging from the frequency of malfunctions, they did not appear to be cleaning and caring for their weapons, as they should. Dulacki also expressed concern that the Marines had become too willing to seek the cover of bunkers during Chinese attacks, gaining protection from shell fragments—though not from satchel charges or direct hits by large-caliber shells—at a sacrifice in fields of fire. He believed that infantrymen should build and fight from individual positions with at least some overhead cover, rather than from large bunkers in which several defenders might be isolated and attacked with explosive charges or trapped if a direct hit collapsed the structure.

Renewed Action Against Korean Marines

From the Hook, the fighting returned to the sector of the South Korean Marines on the left of the 5th Marines. The most vulnerable points along the segment of the Jamestown Line held by the Korean Marines were Combat Outposts 39, 33, 31, and 51. Outposts 39, 33, and 31 were located near the boundary with the 5th Marines and manned by the 5th Battalion, Korean Marine Corps, which had relieved the 3d Battalion on the afternoon of 31 October. The 2d Battalion of the Korean Marine Corps maintained Combat Outpost 51, in front of its lines.

An assault against these four hilltop outposts would represent an extension of the attacks in early October that had overrun three other South Korean outposts—86 to the south, 36, and 37. The dominant terrain feature in this area was Hill 155 on the main line of resistance in the sector of the 5th Battalion, a promontory that overlooked not only the Sachon Valley and the Chinese activity there, but also the Panmunjom corridor and its environs. The enemy made no attempt to disguise his designs on the outposts and ultimately the hill, unleashing a savage bombardment —more than 3,000 rounds during the 48 hours ending at 1800, 31 October—that rocked the South Korean positions, especially Outposts 39 and 33.

The anticipated Chinese attack began at 1830 on the 31st, when the Chinese probed Outposts 39 and 33, apparently in an attempt to exploit any confusion resulting from the 5th Battalion's relief of the 3d. Artillery fire blocked these enemy jabs but did not end the night's fighting. At 2200, an eight-minute bombardment struck the four outposts in preparation for infantry assaults against all of them.

On the right, a Chinese company pressured the platoon of South Korean Marines holding Outpost 31 until 0155, when defensive fire prevailed and the attack ended. At Outpost 33, another enemy company broke through a perimeter manned by only two South Korean squads, which clung to parts of the hill until 0515 when, with the help of artillery fire, they drove off the Chinese.

Two enemy companies attacked Outpost 39, the nearest of the four to Hill 155. The platoon of South Korean Marines deployed there yielded some ground before taking advantage of defensive artillery concentrations to eject the Chinese by 0410. The enemy again probed the outpost two hours later, but soon broke off the action on the right of the sector held by the South Korean Marines.

On the left, four Chinese companies attacked Combat Outpost 51, the most heavily defended of the four, since an entire company had dug in there. It was also the most remote of the outposts, 2,625 yards from the main line of resistance, and this vulnerable location may have persuaded the enemy to scrimp on shelling. Except for 20 rounds of 90mm fire from Soviet-built tanks, the bombardment here was lighter and less effective than at the other three outposts. Three Chinese companies attacked Outpost 51 from the southwest and another from the north. After some initial gains, the attack lost momentum and ended at 0330.

Department of Defense Photo (USMC) A168017

Men of the 5th Marines eat hot chow cooked and brought up to them on the front-lines from the battalion command post to the rear. Marines often ate in relays so as not to hamper operations

Period of Comparative Calm

Despite the clashes involving the Korean Marines, a lull settled over the division following the fierce action at the Hook. Elements of the Commonwealth Division assumed responsibility for the Hook itself on 3 November, but the 11th Marines continued to fire in support of the salient's British defenders. On the night of 18-19 November, for example, Marine artillerymen fired some 2,000 rounds to help break up a Chinese attack.

Meanwhile, the 1st Marines replaced the 7th Marines in manning the right of the division's main line of resistance. On 22 November, the 2d Battalion, 1st Marines, in regimental reserve, provided one company for Operation Wakeup, a raid on Chinese positions opposite Combat Outposts Reno and Vegas. The battalion commander, Lieutenant Colonel Charles E. Warren, assigned the mission to Captain Jay V. Poage's Company D, which attacked just before dawn. The operation succeeded mainly in demonstrating the strength of the Chinese, whose defensive fire stopped the attack short of the objective and frustrated the plan to seize and interrogate

Life in the Bunkers

Upon arriving on the Jamestown Line, the 1st Marine Division made use of log-and-sandbag bunkers, which Second Lieutenant William Watson, described as a simultaneous curse and blessing. Although the structures provided "places of some comfort to which Marines went to get dry and sleep and to escape incoming [fire],' the "sandbag castles" lured men out of their fighting positions. Once inside a bunker, the Marines could be killed or wounded not only by a direct hit by a heavy shell that collapsed the structure, but also by a grenade or explosive charge hurled inside.

Given the vulnerability of bunkers and their effect on aggressiveness, it was no wonder, said Watson, that he and his fellow junior officers had received no formal instruction in bunker placement or construction. "They didn't teach bunker building at Quantico," he recalled. "Who would have dared?" In a Marine Corps trained "to assault and dominate the enemy," anyone foolish enough to advocate the use of defensive bunkers would probably have been shipped at once to Korea where he might well have found himself building the very structures that had resulted in his being sent there.

Life in the bunkers gave rise to unique problems, not the least of which was trash disposal. As the bunkers and trenches proliferated on the main line of resistance and the outpost line, the troops manning them—whether American or South Korean Marines or Chinese soldiers—generated vast amount of refuse. Both sides, reported Second Lieutenant John M. Verdi, a Marine pilot who flew a hundred missions and for a time controlled air strikes from a bunker on the static battle line, did an adequate job of policing the main line of resistance, but the combat outposts posed a more difficult challenge. By day activity in a confined and exposed area could attract hostile fire, and by night policing up might interfere with planned fires, the movement of patrols, or the establishment of listening posts. As a result, trash accumulated around the outposts, especially those of the Marines who not only ate more than their enemy, but consumed food that came in cans or packages that could be more easily discarded than carried away, even with the help of Korean laborers.

Trash attracted rats, nicknamed "bunker bunnies" because of their size. The Marines waged war against them but with partial success at best. Verdi recalled that one Marine used his bayonet to pin a scurrying rat to a sandbag, but the screams of the dying varmint proved so unnerving that the hunter had to borrow a pistol to finish off his prey.

Mines planted by Marines could be as dangerous as those laid by the Chinese, or so Verdi believed. The Marines charted their minefields, selecting a starting point and an azimuth, then planting the mines at specific intervals to form rows a certain distance apart. An error with the compass or a mistake in recording the information would render the chart useless, and the map itself might disappear, destroyed as a result of enemy action or simply lost during the relief of a unit. A return to mobile warfare, or a truce that required removal of the old minefields, would increase the danger.

Department of Defense Photo (USMC) A169578

prisoners. Patrols, ambushes, and less ambitious raids continued for the balance of the year.

On the same day as Operation Wakeup, South Korean police captured two infiltrators in the rear area of the Kimpo Provisional Regiment. Although the Han River interposed a barrier to large-scale attacks, the occasional infiltrator managed to cross and lose themselves among the 80,000 civilians living on the Kimpo Peninsula. A few of these infiltrators struck on 1 December, ambushing a jeep driving on the main supply route. Some two-dozen bullets tore into the vehicle, but only one struck the lone occupant, wounding him in the knee.

Situation at Year's End

During 1952, the Korean fighting assumed a pattern far different from that for which the Marines sent to Korea in previous years had prepared. Mobility gave way to stalemate. The battlefield now resembled the trench warfare of World War I more than the sudden

Department of Defense Photo (USMC) A167563

Two 48-ton Patton tanks wait to receive firing orders before their Marine drivers wheel them up onto earthen ramps behind the main line of resistance where they would fire on enemy positions as artillery.

In December, President-elect Dwight D. Eisenhower made good his campaign promise to "go to Korea," but it was not yet possible to predict the consequences, if any, of his visit. The war remained unpopular with the American people, threats posed by the Soviet Union in Europe seemed more dangerous than the Chinese menace in the Far East, and the negotiations at Panmunjom had stalled over the question whether prisoners of war could refuse repatriation. The only hope, as yet a slim one, for resolving the issue of repatriation lay in the adoption by the United Nations of an Indian proposal to create a special commission to address the issue.

amphibious thrusts and rapid campaigns of World War II in the Pacific. If the amphibian tractor symbolized the role of Marines in the war against Japan, the bunker, built of logs and sandbags on both the main line of resistance and the line of outposts that protected it, represented the war along the Jamestown Line.

The final months of 1952 saw changes of leadership within the 1st Marine Division. On 5 November, Colonel Moore handed over the 7th Marines to Colonel Loren E. Haffner, and Colonel Hewitt D. Adams replaced Colonel Layer in command of the 1st Marines on the 22d. On 10 December, Colonel Lewis W. Walt assumed command of the 5th Marines from Colonel Smoak. Colonel Harvey C. Tschirgi took over the Kimpo Provisional Regiment on 1 December from Colonel Richard H. Crockett, who had replaced Colonel Staab, the unit's original commander, on 31 August. General Pollock remained in command of the division as the new year began and the stalemate continued.

Honors are rendered for President-elect Dwight D. Eisenhower during his visit to the 1st Marine Division. Standing to Eisenhower's left are MajGen Edwin A. Pollock, the division commander, and Gen Mark W. Clark, USA, the United Nations commander.

1stMarDiv Historical Diary Photo Supplement, Dec52

About the Author

Bernard C. Nalty was a civilian member of the Historical Branch, G-3 Division, Headquarters Marine Corps, from October 1956 to September 1961. He collaborated with Henry I. Shaw, Jr., and Edwin T. Turnbladh on *Central Pacific Drive*, a volume of the *History of U. S. Marine Corps Operations in World War II* series. He also completed more than 14 short historical studies, some of which appeared in *Leatherneck* or *Marine Corps Gazette*. He joined the history office of the Joint Chiefs of Staff in 1961, transferred in 1964 to the Air Force history program, and retired in 1994 Mr. Nalty has written or edited a number of publications, including *Blacks in the Military: Essential Documents, Strength for the Fight: A History of Black Americans in the Military, The Vietnam War, Tigers Over Asia, Air Power and the Fight for Khe Sanh*, and *Winged Shield, Winged Sword: A History of the U.S. Air Force*. In addition, he participated in the Marines in World War II commemorative series, writing *Cape Gloucester: The Green Inferno* and *The Right to Fight: African-American Marines in World War II*.

Sources

Clay Blair, Jr., has chosen *The Forgotten War* as the title of his account of the Korean conflict, the history of which tends to be overshadowed by World War II and the Vietnam War. Detailed though it is compared to other such histories, the *Forgotten War: America in Korea, 1950-1953* (New York: *Times* Books, 1987) tends to gloss over the final 18 months of the war, especially the battles fought by the United States Marines. The best account of Marine operations during 1952 and after remains *Operations in West Korea*, volume five of the series *U. S. Marine Corps Operations in Korea, 1950-1953* (Washington, D.C.: Historical Division, HQMC, 1972), by LtCol Pat Meid, USMCR, and Maj James M. Yingling, USMC.

Although the official Marine Corps account is essential, Walter G. Hermes contributes valuable additional information on Marine Corps operations, as well as Army activity and the negotiation of a ceasefire, in *Truce Tent and Fighting Front* (Washington, D.C.: Office of the Chief of Military History, United States Army, 1988), a volume of *United States Army in the Korean War* series.

Personal accounts by Marines are included in *Korean Vignettes: Faces of War* (Portland, OR: Artwork Publications, 1996), a compilation of narratives and photographs by 201 veterans of the Korean War, prepared by Arthur W. Wilson and Norman L. Strickbine.

Other valuable insights and first-hand accounts appear in *The Korean War: Uncertain Victory: The Concluding Volume of an Oral History* (New York: Harcourt Brace Jovanovich, 1988) by Donald Knox, with additional text by Alfred Coppel.

The *Marine Corps Gazette* covered the professional aspects of the war in a number of articles. Es-pecially important are "Outpost Warfare," in the November 1953 issue, and "Back to the Trenches," (March 1955), both by Peter Braestrup, and LtCol Roy A. Batterton's "Random Notes on Korea," (November 1955).

The personal papers collected by the History and Museums Division, Headquarters U. S. Marine Corps proved extremely helpful. The material on file varies tremendously, including journals, photographs, letters, narrative memoirs, and at least one academic paper, a master's thesis on outpost warfare by Maj Norman W. Hicks, a Korean War veteran assigned to help write the history of that conflict. Among the most valuable of these were the submissions by John Minturn Verdi, William A. Watson, and Gen Christian F. Schilt.

With the exception of MajGen John T. Selden, the senior Marine officers serving in Korea at this time participated in the Marine Corps' oral history program. The interviews with Gen Christian F. Schilt and Gen Edwin A. Pollock proved especially valuable.

OUTPOST WAR
U.S. Marines from the Nevada Battles to the Armistice

by Bernard C. Nalty

When 1953 began, the Jamestown Line had become, in the words of Marine Corporal Robert Hall who fought there, "a messy, rambling series of ditches five to seven feet deep" that linked a succession of bunkers constructed of sandbags and timber and used for shelter or fighting. The trenches wandered erratically to prevent Chinese attackers who penetrated the perimeter from delivering deadly enfilade fire along lengthy, straight segments. As for the bunkers themselves, since "piles of trash, ration cans, scrap paper, and protruding stove pipes" revealed their location, the enemy "must have known where every bunker was."

A bunker, therefore, could easily become a death trap. As a result, the Marines had learned to dig and man fighting holes outside the bunkers. Hall described such a hole as "simply a niche in the forward wall of the trench, usually covered with planks and a few sandbags." Within the hole, a crude shelf held hand grenades and a sound-powered telephone linked the hole to the company command post. Along with the fighting holes, Hall and his fellow Marines dug "rabbit holes," emergency shelters near the bottom of the trench wall that provided "protection from the stray Chinese mortar round that sometimes dropped into the trench."

At left: *Spotting targets of opportunity, a Marine crew fires its 75mm recoilless rifle directly against enemy bunkers.* National Archives Photo (USMC) 127-N-A170206

Department of Defense Photo (USMC) A167141

By the end of 1952, the 1st Marine Division defended a static main line of resistance and its outposts, fighting from trenches, covered holes, and bunkers like these manned by Company E, 2d Battalions, 7th Marines.

Some bunkers contained firing ports for .30-caliber or .50-caliber machine guns and accommodations for the crews. Chicken wire strung across the firing ports prevented Chinese assault troops from throwing grenades inside, but fire from the machine guns soon tore away the wire, which could be replaced only at night when darkness provided concealment from Chinese observers.

Other bunkers served as living quarters for five to 10 Marines and might also provide a brief respite for those standing watch in the rain or cold. Because of the emphasis on fighting holes, the living bunkers that Hall remembered had no firing apertures and sometimes a curtain of blanket wool or canvas instead of a door. Candles, shielded so they would not attract Chinese fire, provided light, and kerosene or oil stoves, vented through the roof, supplied heat. The more elaborate living bunkers to the rear of the main line of resistance had electric lights, the power produced by gasoline generators.

By night, during the early

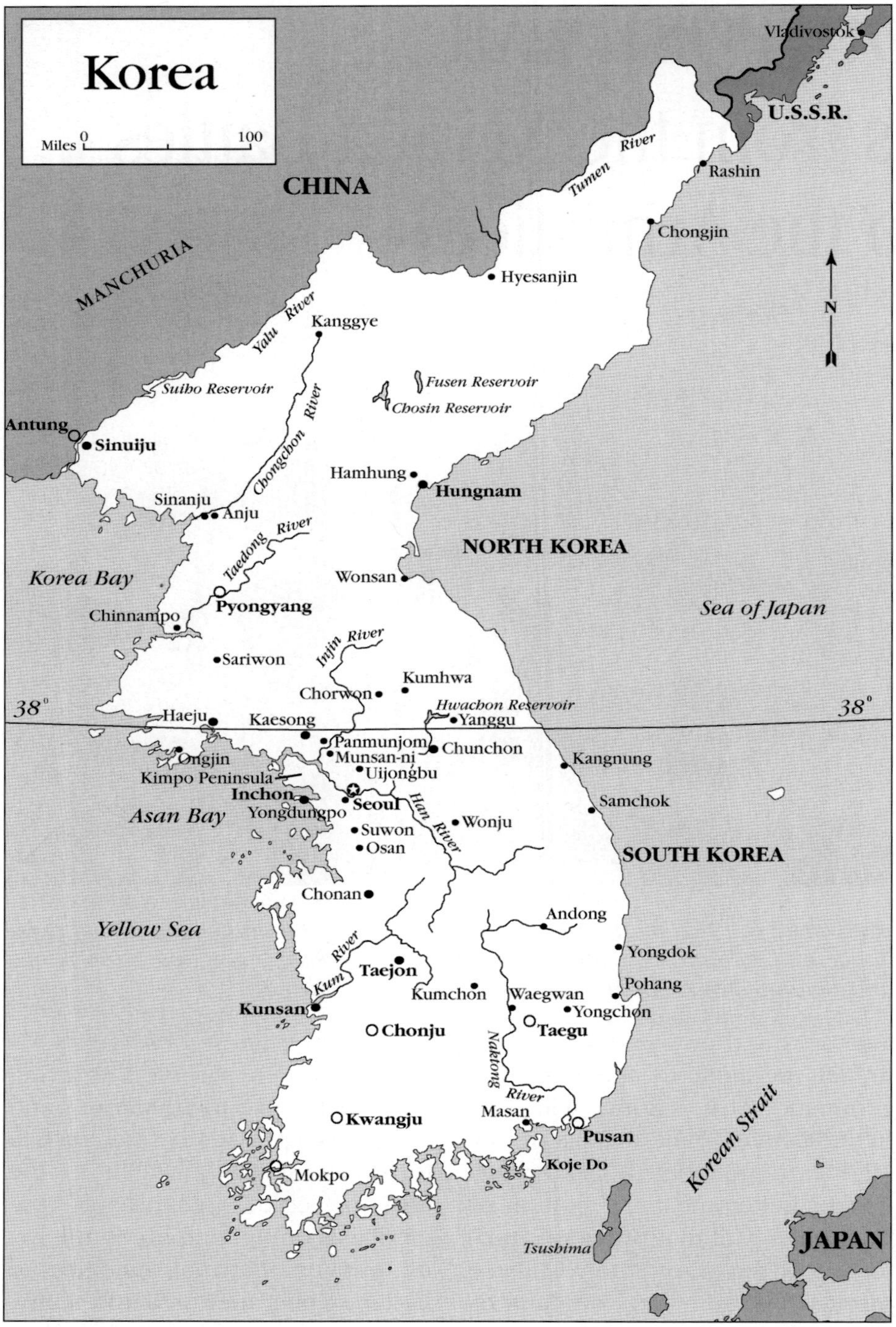

months of 1953, a cold wind usually blew from the north, sometimes bringing with it the sound of Chinese loudspeakers broadcasting English-language appeals to surrender, interspersed with country music. The enemy's propaganda tended to reflect Communist ideology, urging members of the United Nations forces to escape their capitalist masters. The Chinese, however, also tried to take advantage of the fact that the combatants in Korea were discussing a ceasefire even as they fought. Since the summer of 1951, truce talks had taken place at Kaesong and later at Panmunjom, with the United Nations delegation traveling to the site of the talks through a carefully marked demilitarized corridor. When the talks seemed to be making progress, the Chinese used a more subtle approach, trying to persuade members of the United Nations forces not to risk their lives in a war that had almost ended.

Despite cold-weather clothing and insulated boots, the chill of the winter night could numb the senses. As a result, a Marine usually stood nighttime watch in a fighting hole for 30 to 45 minutes before warming himself in a nearby bunker. On some nights, Corporal Hall recalled, an outgoing salvo or ripple of 4.5-inch rockets might swish overhead to explode on some distant hill. "All through the night," he said, "there were sporadic shots, grenades going off, artillery fire"—as outposts came under attack, ambushes were triggered, and patrols drew fire—and "at first light a ripple of random shots would greet the new day," as visibility improved revealing targets previously hidden by darkness.

From time to time, Marines manning the Jamestown Line got a brief respite from the danger, tension, and discomfort. "One of the most pleasurable things" in moving off the line, said Corporal Hall, "was to walk back to battalion for a hot meal and a shower." A Marine just come from the battlefield could sit down to the kind of meal he might have been served at a mess hall in the United States, eat at his own pace instead of the tempo set by the mess sergeant, have a hot shower in tents modified for that purpose, and exchange a filthy uniform for a clean one. A laundered utility jacket, formerly worn by a staff sergeant whose chevrons remained in place, might be issued to a corporal like Hall. After a trip to the shower, he said, "you could never be sure about a person's rank unless you knew him."

Department of Defense Photo (USMC) A170084

Inside view of one of the cramped sleeping caves, which sheltered two to four Marines, on Outpost Carson (Hill 27). A majority of the outpost's strength stood watch and worked on fortifications at night, while a small security team was on duty during the day.

Marines on the battle line, who could not be reached with food prepared at field kitchens and brought forward in insulated containers, relied on C-rations easily transported in cardboard boxes and quickly prepared. The C-ration featured canned foods like sausage patties in gravy, corned beef hash, or beans and frankfurters that could be heated over a Sterno flame, along with candy, a cookie, crackers, instant coffee, cigarettes, canned fruit, toilet paper, and an ingenious can opener. Although the object of many a joke, C-rations were "quite re-markable" in the opinion of one Marine. "Everything was used," he said, "even the oiled cardboard box into which a person could relieve himself."

Marines of Company B, 1st Battalion, 1st Marines, enjoy hot chow on the reverse slope of the main line of resistance. These field mess areas, where Marines got at least one hot meal a day, were set up 50 to 100 yards from the frontlines.

Department of Defense Photo (USMC) A170997

National Archives Photo (USA) 111-SC428270

In February 1953, LtGen Maxwell D. Taylor, USA, left, who led the 101st Airborne Division at Normandy during World War II, replaced Gen James A. Van Fleet, USA, center, in command of the U.S. Eighth Army. Standing to the right is Gen Mark W. Clark, USA, commander of United Nations forces.

Refinements in Position Warfare

Experience on the Jamestown Line in 1952 inspired innovations in static warfare, although not all were immediately applicable to the battlefield. The concept of a main line of resistance and its protecting line of combat outposts persisted into 1953, but the ideal placement and construction of trenches and bunkers changed. Chinese mortar and artillery fire demonstrated that trenches had to be deeper, fighting holes better protected, and bunkers stronger. Moreover, the military crest of a hill or ridgeline—the position on the forward slope with the longest fields of observation and fire—need not be the best location for the defenses protecting vital terrain. The topographic crest—the

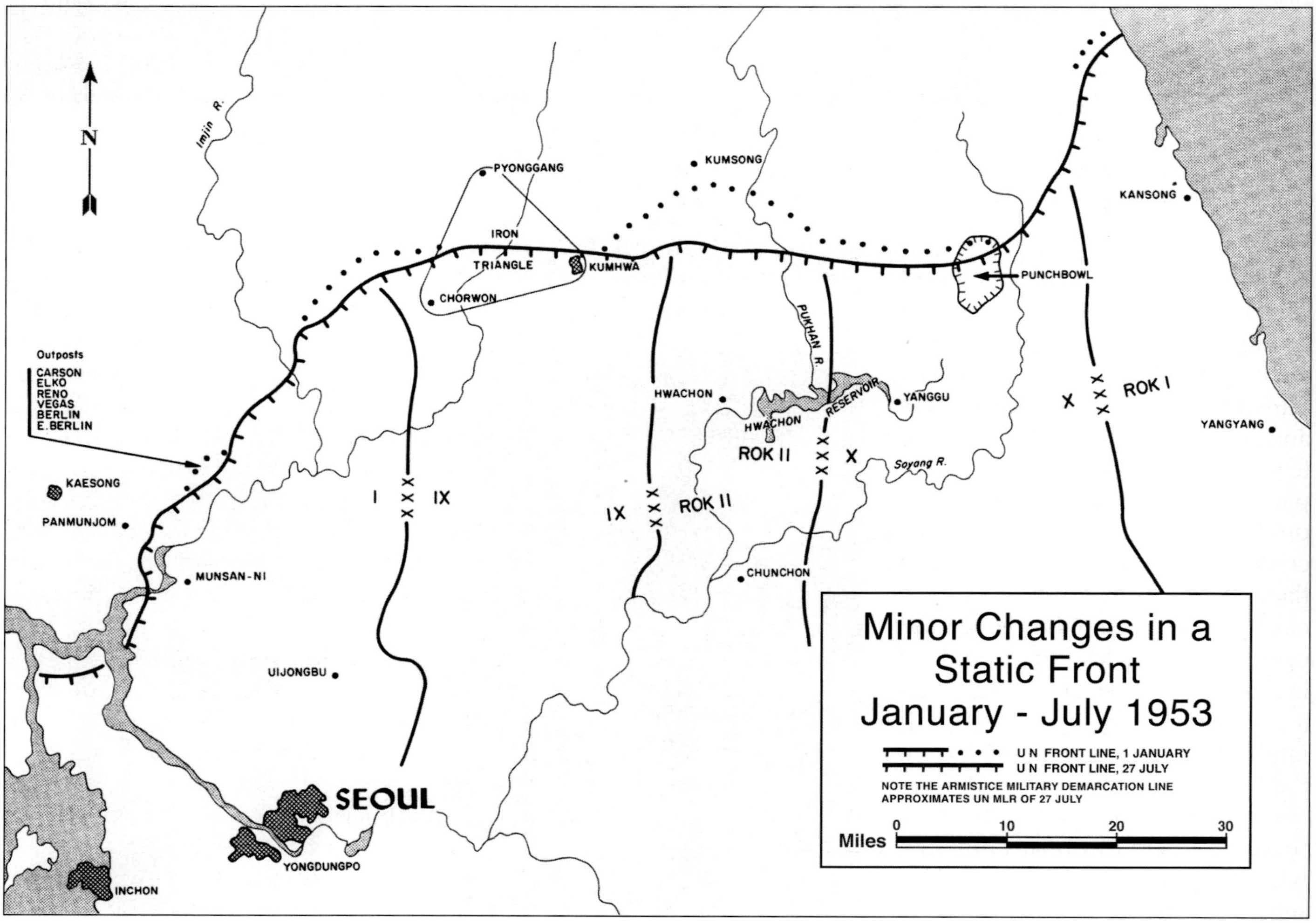

spine of a ridge or top of a hill—might be a better site for the main line of resistance, provided that fighting positions, readily accessible from the main trench, were located a short distance down the forward slope. Formerly, machine guns tended to fire directly into draws that the Chinese might follow to attack an outpost or a portion of the main line of resistance; now the machine gun crews, protected as necessary by riflemen, would dig in near the tips of fingers of high ground extending outward from the main line or its outposts and place interlocking bands of fire across the front. Moreover, units in reserve immediately behind the main line of resistance could move through trenches on the topographical crest without crossing the skyline and risking direct fire from flat trajectory weapons.

Those Marines who could shift from the military crest to the topographical crest would enjoy certain advantages. Plunging fire, directed by mortars or howitzers against trenches on the topographic crest, proved difficult for the Chinese to register or adjust, since observers could not see the explosion of shells that fell beyond Marine positions. In contrast, when firing against the military crest, the enemy could spot shells that detonated both beyond and short of the target and adjust his fire accordingly.

Another defensive innovation proved as effective as it was simple. Instead of two aprons of barbed wire separated by about one yard, the Marines adopted the so-called Canadian system, which featured random strands of wire connecting the parallel aprons in the void between them. Artillery or mortar fire, which might rip apart the old aprons, merely churned and tangled the wire between the aprons, thus making the barrier harder to penetrate.

These changes could easily be incorporated in the fall-back positions, the Wyoming and Kansas Lines, to the rear of the main line of resistance. Major changes to the Jamestown Line itself and its combat outposts proved all but impossible when in contact with an aggressive enemy able to make deadly use of artillery and mortars. The frontline Marines could rarely do more than put out the additional wire, dig deeper, and add timbers and sandbags to the existing defenses.

A fire team from Company I, 3d Battalion, 1st Marines, sets out on a daylight reconnaissance patrol to one of outposts in front of the main line of resistance. Aggressive patrolling and improvement of the secondary defenses of the Wyoming and Kansas Lines occupied much of the reduced wintertime schedule.
Department of Defense Photo (USMC) A168856

Daylight Raids

Besides strengthening the defenses, especially of the Wyoming and Kansas Lines, Marines of the division's reserve regiment continued to undergo training while patrolling the rear area. On the Jamestown Line and its outposts, the lull that settled in after the fight for the Hook ended when one of the regiments on the main line of resistance, the 5th Marines, raided Hills 31 and 31A in the Ungok hill mass. Because the Chinese had fortified this high ground so strongly, the Marines made elaborate preparations, planning feints to divert attention from the main thrust, bringing in pilots from the 1st Marine Aircraft Wing to visit observation posts and become familiar with the battlefield, arranging for air strikes and artillery concentrations, reconnoitering

Department of Defense Photo (USMC) A169911

Marine fliers batter Hill 67 with high explosives and napalm on 3 February in support of the 1st Battalion, 5th Marines' two-reinforced-platoon-raid on the Chinese-held Ungok hill mass to destroy fortifications, inflict casualties, and take prisoners. The air strikes were designed to divert enemy attention from the final objective area.

routes of attack and withdrawal, and clearing mines from them. The complex operation, named Clambake, required a half-dozen rehearsals, the last on 1 February. Unlike Operation Wake-up in November 1952, which had sought to take prisoners and gather intelligence on the Chinese defenses, planners designed Clambake primarily to kill the enemy and destroy his bunkers and trenches.

The raid began at first light on 3 February, when three platoons of tanks roared toward the enemy-held high ground—Hill 104, Kumgok, and Red Hill—a short distance west of the real objective, Ungok. While the armored vehicles cut loose with 90mm guns and flamethrowers, the 1st Battalion, 11th Marines, added to the realism of the feint by shelling the apparent objective.

Taking advantage of the diversion, two reinforced platoons from Company A, 1st Battalion, 5th Marines—armed with bangalore torpedoes to breach barbed wire and flamethrowers, satchel charges, and 3.5-inch rocket launchers to destroy heavier fortifications—stormed Hills 31 and 31A. The tanks taking part in the diversion protected the left flank of the attacking Marines by crossing a frozen paddy to open fire on the trenchline connecting Ungok with the hills to the west. The Chinese blazed away at the Marine tanks that either protected the flank from beyond the rice paddy or accompanied the assault force and succeeded in destroying a flame-throwing M-4 Sherman tank. Supported by air, armor, and artillery, the Marine raiding party prevailed. Clambake captured no prisoners but accomplished its main purpose by collapsing bunkers, trenches, and caves, and killing perhaps 390 Chinese before the attackers withdrew. Marine casualties totaled 14 killed and 91 wounded.

Operation Clambake demonstrated anew the value of planning and rehearsal, the ability of flame-throwing tanks to discourage Chinese tank-killer teams armed with shoulder-fired rocket launchers, and the importance of coordinating air, artillery, and armor in support of an infantry assault. Colonel Lewis W. Walt, commander

Department of Defense Photo (USMC) A169719

After being examined at the company aid station, a Marine of Company A, 1st Battalion, 5th Marines, who was wounded during the battalion's daylight raid on Ungok is put onto a helicopter for evacuation to a hospital ship for further treatment.

of the 5th Marines, believed that Clambake taught his regiment how to forge an effective tank-infantry team, a weapon he soon employed.

On 25 February, Colonel Walt's 5th Marines conducted a raid similar to Operation Clambake, attacking Hill 15, where the Chinese had overrun Outpost Detroit in early October. The commander of the 2d Battalion, Lieutenant Colonel Oscar F. Peatross, chose Company F, under Captain Harold D. Kurth, Jr., to execute the attack. After detailed planning and careful rehearsal, the raid, designated Operation Charlie, kicked off on the morning of the 25th. Tanks and artillery fired as planned, but bad weather interfered with the scheduled air strikes. The two assault platoons reported that the supporting fire had isolated the objective as planned, but they found that "the

Officers of the 5th Marines assemble for a photograph during a lull in the fighting. Pictured in the front row, from left, are: LtCol Oscar F. Peatross, LtCol Jonas M. Platt, Col Lewis W. Walt, LtCol Edwin B. Wheeler, LtCol Robert J. Oddy, and LtCol James H. Finch. Second row, Maj Harry L. Sherwood, Maj Robert L. Willis, Capt Ralph L. Walz, Maj Jack M. Daly, Maj Ross T. Dwyer, Jr., Maj Robert H. Twisdale, Capt Richard G. Gilmore, and Maj William C. Doty. Third row, Capt James R. Schoen, Capt Dean W. Lindley, Capt James E. Hendry, and Capt Arthur J. Davidson.

National Archives Photo (USMC) 127-N-A170216

The Marine Division and Its Weapons

Between 1 January 1953 and the end of the fighting in July, the strength of Marine Corps ground forces hovered between 25,000 and 28,000, fluctuating as casualties occurred, tours of duty or enlistments expired, and replacements arrived. In terms of organization, the 1st Marine Division adhered to a triangular concept, with three organic infantry regiments—the 1st, 5th, and 7th Marines—an artillery regiment, the 11th Marines, and a variety of other combat and support elements under control of the division or its components. The combat elements employed tanks, mortars, and other weapons; the support units provided such specialized activity as transportation, by truck or amphibian tractor, communications, engineering, reconnaissance, and cargo handling.

On the Jamestown Line, a fourth regiment served with the 1st Marine Division, the 1st Korean Marine Regiment, which had been organized, trained, and equipped with the assistance and advice of U. S. Marines. The Korean Marine regiment, with a maximum strength in 1953 of 4,400, had been attached to the 1st Marine Division in time for the Inchon invasion of September 1950, but after the Chosin Reservoir fighting, the Korean Marines passed under the control of the Republic of Korea's Army. During 1951, however, the Korean Marine regiment was again attached to the 1st Marine Division and, along with the 5th Battalion that joined in 1952, remained a part of the American division for the rest of the war. The Korean Marines assigned to the division had their own organic artillery and armor. The 2d Korean Marine Regiment provided troops to man the islands off the east and west coasts.

The basic infantry weapons of the 1st Marine Division had seen action in World War II. Riflemen still used the semi-automatic M1 or, if trained as snipers, the bolt-action M1903 with a telescopic sight. Although designed for World War I, the Browning Automatic Rifle still increased the firepower of the Marines. Officers, or those enlisted men assigned to crew-served weapons like mortars or machine guns, carried the .45-caliber pistol or the lightweight .30-caliber carbine, which came in both automatic and semi-automatic versions.

The standard crew-served infantry weapons had also helped fight World War II. The 60mm, 81mm, and 4.2-inch mortars were at best modifications of older weapons, as were the machine guns, whether heavy .50-caliber weapons, water-cooled .30-caliber guns, or lightweight, air-cooled .30-caliber types. The 3.5-inch rocket launcher, however, had replaced the 2.36-inch Bazooka of World War II, which had failed to effectively penetrate the armor of the Soviet-built tanks used by the North Koreans and Chinese.

Artillery and armor relied heavily on designs used in World War II or intended for that conflict. The 11th Marines, aided by a battery of 4.5-inch rockets on multiple launchers, provided the division's organic fire support with three battalions of 105mm howitzers and one battalion of 155mm howitzers. A battalion of tanks added to the mobile firepower of the division with weapons ranging from rebuilt M-4 Shermans from World War II, some of them mounting a flamethrower as well as a 105mm gun, to new M-46 Pershings with a 90mm gun.

In short, the 1st Marine Division was using the weapons of World War II to fight the kind of trench warfare characteristic of World War I.

majority of enemy installations were relatively undamaged," perhaps because the Chinese bunkers were so solidly built.

Although the raids that culminated in Operation Charlie were the most ambitious attacks during February, Chinese troops and the American and South Korean Marines conducted many smaller probes. On the night of 12-13 February, for example, a Chinese platoon, supported by mortars and artillery, tested the defenses of Outpost Hedy on Hill 124. Two nights later, hostile troops stalked a patrol from the 7th Marines that was attempting to set up an ambush and forced it to turn back. On 19 February, artillery and mortar fire frustrated a Chinese attack on Combat Outpost 33, manned by South Korean Marines.

The Marines, whether American or South Korean, exerted pressure of their own. On the night of 13-14 February, two platoons of South Korean Marines had successfully raided Hill 240, on the west bank of the Sachon River roughly three miles upstream from its conflux with the Imjin. On the morning of the 22d, the 5th Marines raided Hill 35A, some 1,300 yards southwest of the Ungok hill mass, using flamethrowers to deadly effect. On the following night, a reinforced platoon from the 1st Battalion, 7th Marines, supported by four M-46 tanks set out to raid Hill Yoke, southwest of Bunker Hill. Shortly after midnight, however, as the raiding party regrouped for the final assault, the Chinese struck from ambush. Another reinforced platoon went to the aid of the first, and the enemy broke off the action; but not before the raid on Hill Yoke had to be called off, even though the Marines accounted for perhaps five times their own

1stMarDiv Historical Diary Photo Supplement, Feb53

Three Marines from Company F, 2d Battalion, 5th Marines, who participated in the raid on Hill 15 (Detroit), display captured enemy weapons and material. The raid was launched during early daylight hours with a smoke screen in an attempt to gain surprise.

losses of five killed and 22 wounded.

The succession of raids and ambushes continued into March. In a restaging of Operation Clambake, Company B, 1st Battalion, 5th Marines, attacked Hill 31A of the Ungok massif on the 19th. Once again, air strikes, and artillery preparations shattered the pre-dawn calm and forced the defenders to move to positions on the reverse slope until the attacking Marines withdrew. On the same morning, however, the enemy hit Outposts Esther and Hedy and tried unsuccessfully to crack the Jamestown defenses to the rear of Hedy, failing despite a lavish expenditure of mortar and artillery shells.

Improvements in Logistics

During the first three months of 1953, the supply of howitzer

During February and March, as night fighting intensified, Marine M-46 Patton tanks assigned to Company D, 1st Tank Battalion, fire their 90mm guns to harass the Chinese and disrupt their movement.

Department of Defense Photo (USMC) A169267

Department of Defense Photo (USMC) A170050

Cargo handling personnel prepare to hook a load of ammunition to a Sikorsky HRS-2 helicopter of Marine transport squadron 161 during Operation Haylift II, the five-day experiment to resupply two frontline regiments carried out in late February.

ammunition increased, and the restrictions on artillery support, in effect during the autumn, ceased. The availability of hand grenades also improved, but 81mm mortar shells remained in short supply.

Even as the ammunition shortage eased, the Marines had to impose restrictions on the use of gasoline and diesel fuel. In January 1953, consumption of gasoline declined by 17 percent from the previous month, and diesel fuel by seven percent. Stocks were rapidly replenished, however, so that in February consumption returned to normal.

Refinements in equipment also appeared. A thermal boot worn during the Korean winter of 1952-1953 afforded better protection against cold and dampness than the footgear it replaced, but the leather combat boot wore out all too quickly. Redesigned body armor began arriving in November 1952. The new model protected the groin as well as the upper body, greatly improving morale as it reduced casualties still further.

Experiments continued in the use of helicopters as flying pack mules to deliver supplies over broken terrain. In February 1953, Lieutenant Colonel John F. Carey's Marine Helicopter Transport Squadron 161 (HMR-161) carried out Operation Haylift II, resupplying two front-line regiments, the 5th and 7th Marines. This operation, lasting from 23 through 27 February, proved more demanding that Haylift I, conducted in September 1952, which had resupplied only one regiment. On the first day of Haylift II, Carey's squadron had to divert helicopters from the 7th Marines to rush ammunition to the other regiment, and on the final morning fog disrupted the schedule of flights. Nevertheless, Haylift II delivered 1.6 million pounds of cargo; five times the total of the earlier operation.

Fighting Intensifies

When the winter of 1952-1953 ended, the deployment of the 1st Marine Division remained essentially unchanged, although the unit on the right was now the U.S. Army's 2d Infantry Division rather than the British 1st Commonwealth Division. The American Marine regiments held the right of the line—the 5th and 1st Marines occupying the Jamestown positions and the 7th Marines currently in reserve. Beyond the Panmunjom corridor, South Korean Marines defended the portion that extended to the north bank of the Han River. On the south bank of the Han, the division's amphibian tractor battalion and the Kimpo Provisional

Air Support for the Division

Marine Corps aviation continued to play a critical role on the battlefield. Indeed, its value had increased as the Fifth Air Force, which exercised operational control over land-based Marine airmen, shifted emphasis to targets on or near the frontlines and away from industries, transportation links, and command and control facilities, all of them already heavily bombed. Taking advantage of this change, Major General Vernon E. Megee, who in January 1953 replaced Major General Clayton C. Jerome in command of the 1st Marine Aircraft Wing, persuaded Major General Glenn O. Barcus, commander of the Fifth Air Force, to abandon the practice of dealing directly with Marine aircraft groups or even squadrons and work through the wing headquarters. To facilitate planning within the 1st Marine Aircraft Wing, Megee revitalized and enlarged his G-3 section, which, he conceded, had become "somewhat rusty." General Megee also replaced the lone Marine Corps liaison officer at the Joint Operations Center with an element drawn from of the wing's G-3 section that could deal more efficiently with requests for air support.

Barcus endorsed Megee's plan to expand the role of wing headquarters, but the Air Force general retained control over close air support, even though it was a Marine Corps specialty. Policy established jointly by the Army and Air Force, to which the Navy assented, required that the Joint Operations Center, which now had a greater Marine presence, approve requests for this kind of mission. In waging the air war, the Joint Operations Center, paid stricter attention to requests for close air support, tending to screen carefully these urgent strikes while assigning Megee's headquarters greater responsibility for interdiction, armed reconnaissance, and other missions, by day or by night, that had been planned in advance against targets 3,000 yards or more beyond the main line of resistance.

The cautious attitude toward close air support reflected the potential danger to friendly troops inherent in the kind of operations routinely flown by the 1st Marine Aircraft Wing. Indeed, with a mere 14.5 percent of the available tactical aircraft, Marine airmen had undertaken between 30 and 40 percent of all the close air support missions flown for the United Nations forces between January and October 1952. These attacks, some of them within 100 yards of United Nations troops, could accidentally cause friendly casualties. The battlefield itself—a succession of similar hills and ridges, separated by draws and intermittent streams, with few obvious landmarks except for major reservoirs or rivers—contributed to the possibility of error. During the first nine months of 1952, Marine pilots figured in 18 of 63 incidents in which air strikes killed or wounded friendly troops.

News reports appearing in the United States during February 1953 focused on the involvement of Marine airmen in 28.5 percent of the recent accident attacks that killed or wounded friendly troops. Ignoring the dangerous nature of these strikes, which included almost all the targets within 100 yards of friendly forces, the press accused the Marines of carelessness, a charge that had no merit. Given the difficulty in pinpointing targets on the Korean battlefield, effective close air support involved danger to the troops on the ground, especially those manning outposts that were surrounded or under simultaneous attack from various directions. General Megee, when evaluating a January 1953 strafing run by Marine jets that killed one Marine and wounded another, concluded that the incident "resulted from the inescapable operational hazard incident to laying on a real close strike." The same judgment applied to the other similar accidents.

Regiment, the latter an improvised "United Nations" force using armored amphibian tractors as artillery, manned the defenses in addition to controlling civilians within the regimental sector and regulating river traffic.

Except on the far left of the division's line, where the Han River provided a natural barrier, a series of combat outposts contributed to the security of the main line of resistance. From right to left, the principal outposts were East Berlin, Berlin, Vegas, Reno, Carson, and Ava, all manned at the end of March by the 5th Marines. The 1st Marines maintained Corrine, Dagmar, Esther, Ginger, Bunker Hill, Hedy, Ingrid, Kate, and Marilyn. Beyond the Pan-munjom corridor, the South Korean Marines held, from right to left, Outposts 39, 33, 31, and 51.

As it had during the winter now ending, the 11th Marines provided artillery support for the infantry regiments manning the Jamestown Line and its outposts, using the firepower of three battalions of 105mm howitzers, a battalion of 155mm howitzers, and a battery of multiple 4.5-inch rocket launchers. One battalion of 105mm howitzers supported the 5th Marines and another the 1st Marines. The third such battalion provided general support of the division and stood ready to reinforce the fires of the battalion supporting the 5th Marines, which held a critical sector. Both the 155mm howitzers and the rocket launchers rendered general support for the division. The South Korean Marines depended

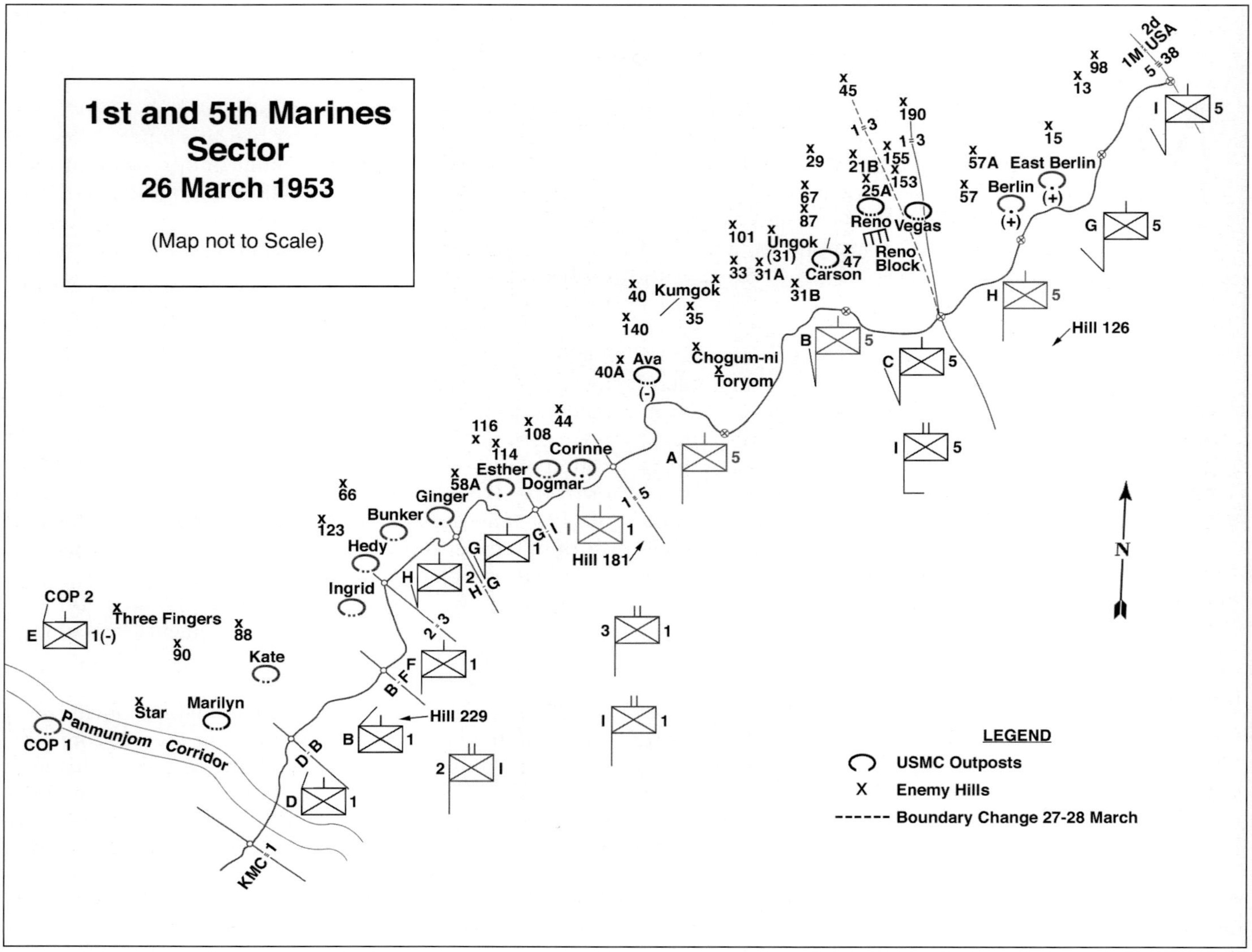

primarily on a battalion of 75mm guns, attached to the 11th Marines. U.S. Army artillery battalions, assigned to I Corps, could reinforce the fires of the 11th Marines anywhere along the line with 155mm howitzers and 8-inch howitzers. To protect the bridges across the Imjin River to the rear of the Jamestown Line against possible aerial attack, the Marines deployed a provisional antiaircraft artillery platoon armed with automatic weapons.

The 1st Marine Division's tank battalion continued to support the defenses of the Jamestown Line, mainly with M-46 tanks mounting 90mm guns, though the older M-4s armed with a 105mm howitzer and a flamethrower were available. The battalion assigned one tank company to support each of the line regiments and designated a third as a forward reserve to reinforce the main line of resistance or spearhead counterattacks. The fourth company became the rear reserve, undergoing unit training and conducting maintenance for the entire battalion.

The 1st Marine Aircraft Wing, now participating more directly in planning air strikes, supported the division with an array of piston-engine and jet types, fixed-wing models and helicopters. Besides conducting strikes, the wing placed its helicopters and light observation planes at the disposal of the division. Helicopters evacuated the wounded and delivered supplies, while the light planes flew reconnaissance and liaison missions, directed air strikes and adjusted artillery fire.

The coming of spring brought rain and warmer temperatures that melted snow, thawed frozen rivers, and caused flooding. Roads became all but impassible, and trenches turned into streams of mud. Water-soaked aging sandbags, which rotted and split, undermined timbers already weakened by sustained Chinese shelling. In March, noncommissioned officers from the division's engineer battalion inspected the Jamestown Line and evaluated the condition of the defenses, deter-

Department of Defense Photo (USMC) A169051

After loading the 155mm howitzer, artillerymen of the 4th Battalion, 11th Marines, make a final check on the sights. The gunner stands by with lanyard in hand, waiting for the command to fire.

mining which bunkers and fighting holes had to be repaired or rebuilt. Marine infantrymen did the actual work under the supervision of the engineers, using materials manhandled into position by the Korean Service Corps. At the outposts or on those portions of the main line of resistance under enemy observation, the work had to be done at night.

Warmer weather heralded the appearance of a beer ration—two cans per day for each Marine not engaged in actual fighting. Although the beer issued to Marines had a lesser alcohol content than that sold to civilians in the United States, it was welcome indeed. Those who drank made friends with those who did not, and a brisk trade in beer ensued.

Spring also brought the certainty that, as soon as weather permitted, the Chinese would renew their attacks on the Jamestown Line and its outposts, duplicating the intensity of the fight for the Hook (now a responsibility of the 2d Infantry Division) that had raged in October 1952. The enemy's capture of key terrain could yield political advantage as well as immediate tactical gain. Chinese success might force the Marines back to the Wyoming or Kansas Lines, both of them fallback positions, or even open the way to Seoul. Smaller gains could combine to exert pressure on the United Nations to accept a truce and, if the ceasefire should fail, leave the Chinese in a stronger

A Marine 4.5-inch rocket crew launches a fire mission against Communist positions. The launcher could discharge 24 rounds in rapid succession. Two helicopters from Marine Helicopter Transport Squadron 161 wait in the background to airlift the crew and their rocket launcher to the rear or a new position.

National Archives Photo (USMC) 127-N-A346664

1stMarDiv Historical Diary Photo Supplement, Mar53

Each spring, protracted periods of rain and the seasonal thaw turn the earth into a quagmire, impeding movement even in trenches. Although road conditions became a serious problem, frontiline units were kept supplied.

position to renew the fighting.

Certain of the Marine combat outposts had already demonstrated their importance, among them Bunker Hill, the site of deadly fighting during 1952. The list of critical outposts also included Reno, Carson, and Vegas, northeast of the Ungok hills, a massif that had been the objective of a number of Marine raids during the winter. The importance of the three outposts named for cities in Nevada, which the 5th Marines manned, derived from their proximity to a potential invasion route leading toward Seoul, a road that passed between Carson and the Ungok hills. In addition, control of the three outposts provided observation of the Chinese main line of resistance and certain areas immediately beyond, while screening portions of the Jamestown Line and some its rear areas from enemy observers.

Forty or fifty Marines, with two Navy hospital corpsmen, manned each of the combat outposts. The Marines, often drawn from various squads in different platoons, formed composite units with numbers and firepower comparable to a reinforced platoon. In addition to their rifles, Browning automatic rifles, carbines, and pistols, the defenders usually were reinforced

The Korean Service Corps

The government of the Republic of Korea drafted men already rejected for service in the army and assigned them to a labor force, the Korean Service Corps, organized into companies, battalions, and regiments that carried supplies, food, ammunition, and building materials to combat units and performed other necessary logistics duties. Although the service troops wore a uniform, they were not issued weapons. On the Jamestown Line, a Korean Service Corps regiment, usually numbering more than 5,000 men, supported the components of the 1st Marine Division. Besides forming human pack trains, these Korean laborers helped evacuate the wounded, buried enemy dead, and retrieved weapons abandoned on the battlefield.

After the Marines moved onto the Jamestown Line in the spring of 1952, some 500 members of the Korean Service Corps helped cut timbers for the construction of bunkers along the frontline and for the two fall-back positions, the Wyoming and Kansas Lines. By July, the Koreans had helped cut and shape some 35,000 lengths of timber. Once the bunkers were completed, the Korean Service Corps carried new timbers, sandbags, barbed wire, and other materials to strengthen them and repair battle damage, along with food and ammunition. When manhandling cargo, each Korean laborer was expected to carry 50 pounds a distance of 10 miles, a burden affixed to an A-frame on the porter's back. Over long distances or rugged terrain, the laborers might adopt a relay system, dividing the journey into manageable segments.

Two members of the Korean Service Corps work under the supervision of an explosive ordnance disposal Marine and taught how to remove explosives from bombs and artillery shells. Once the explosives were removed, the cases were disposed of.

Department of Defense Photo (USMC) A169560

National Archives Photo (USMC) 127-N-A170193

Members of the 1st Squad, 1st Platoon, Company C, 1st Battalion, 5th Marines, wait to be briefed on a night combat patrol involving the setting of an ambush. Since early March, the 1st Battalion had conducted nearly a dozen such patrols to test the enemy in the Carson-Reno-Vegas area.

with two portable flamethrowers and as many as five light machine guns. One or more forward observers adjusted the fire of 60mm and 81mm mortars in defense of each outpost.

The three outposts of Carson, Reno, and Vegas differed from one another according to their location, the terrain to be defended, and the threat they faced. Combat Outpost Carson, on the left, guarded a largely barren hilltop where a cave provided living quarters for the Marines, who manned an oval perimeter protected by barbed wire and including bunkers, tunnels, and a main trench with fighting holes. Except for the slope nearest the Jamestown Line, where a deeper entrenchment was being dug, the main trench on Carson averaged five feet deep by two feet wide. Most of the 28 fighting holes had excellent fields of fire, though the overhead cover on some of them had reduced the opening for observation and firing. During darkness, two listening posts covered the likeliest avenues of enemy attack, from the Ungok hills to the west and Hill 67 to the north.

Reno, in the center, was the most vulnerable of the three. It not only lay closest to Chinese lines, but also occupied a ridge that forced the defenders into a perimeter vaguely resembling the wishbone of a turkey, open end to the north. As at Carson, a cave served as living quarters and might also become a last-ditch redoubt. A tunnel provided access to the cave from the main trench, which varied from five to seven feet deep, but one Marine, Corporal James D. Prewitt, confessed that he hated to go through the entrance. As a boy, he explained: "I had helped dig my brother out of a collapsed play tunnel, and I was left with a real horror of such things."

Two Marines at Outpost Reno make their hourly call to the company command post. Located more than 1,600 yards from the main line of resistance, the outpost was customarily manned by 40 to 43 Marines.

Department of Defense Photo (USMC) A170020

The Marines at Reno built no bunkers, relying exclusively on fighting holes in the trenches and, as a last resort, the cave itself. Outpost Reno had limited fields of fire in the direction of enemy-held Hill 67, also called Arrowhead Hill, but Outpost Carson, on the left, provided fire support in this area. As a result, the approach that seemed to pose the greatest danger to Reno's defenders followed a ridge extending generally southward from Hill 150.

Like the Marines defending the other outposts, those at Reno relied on C-rations and tossed the empty cans into nearby gullies. At night, when the tin cans clattered, the source of the noise might be Chinese moving close to attack behind a sudden barrage or merely rats scavenging for food. Members of the Korean Service Corps kept Reno supplied and performed the unpleasant task, as after the fighting in October 1952, of burying the Chinese dead, a task repeated whenever artillery fire disinterred the corpses. Marines from Reno accompanied the Korean burial details to protect them against ambush and also to keep them at their grisly work. Enemy snipers, as well as mortar and artillery crews, posed a continuing threat, forcing the Marines to remain under cover during daylight, insofar as possible. Indeed, a sniper alerted by the reflection from a forward observer's field glasses, fatally wounded the Marine with a single shot. Because of the danger, tension, and discomfort, the Marines at Reno normally stayed only for a week before being relieved.

To the south of Reno lay Reno Block, an L-shaped trench with a small bunker at the end of the shorter leg and a machine gun position at the point where the

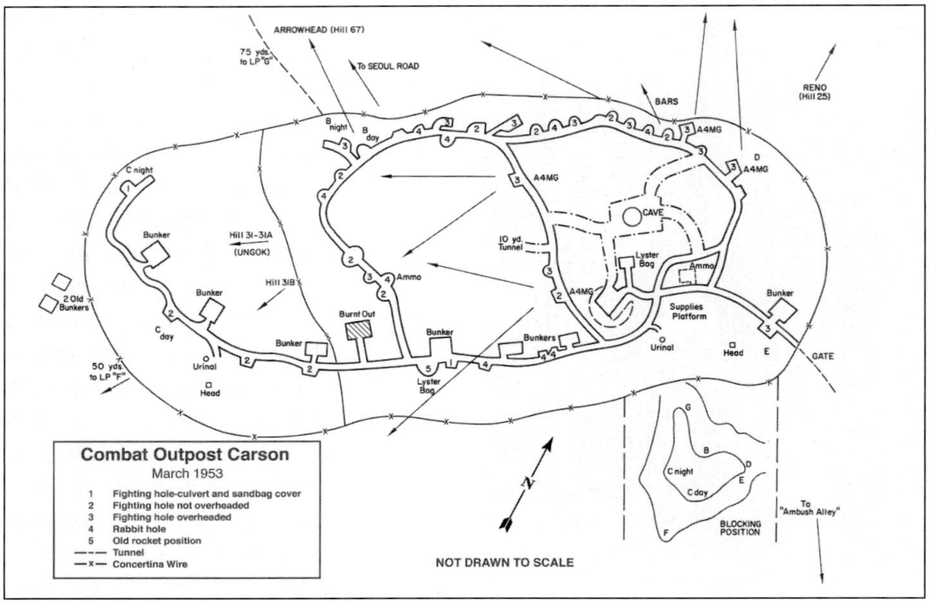

Combat Outpost Carson
March 1953

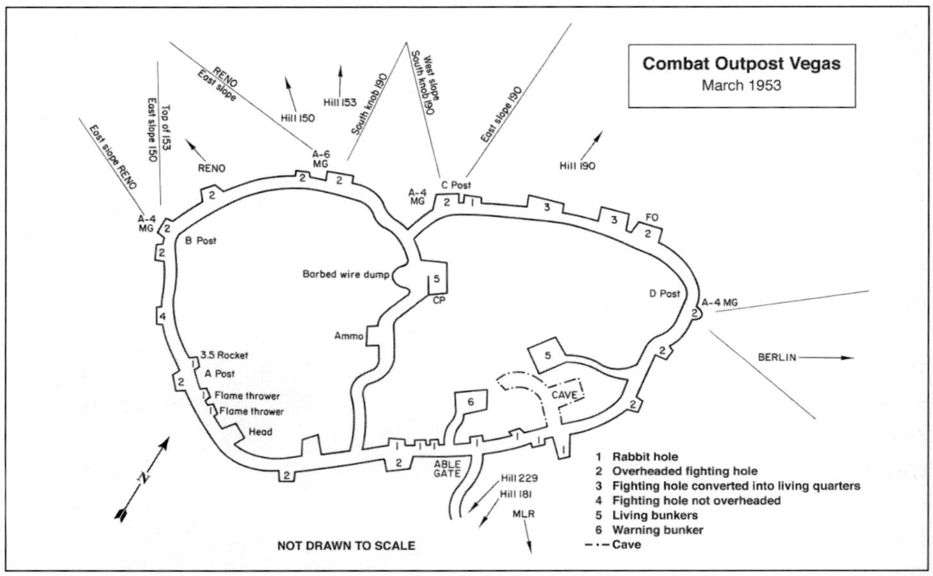

Combat Outpost Vegas
March 1953

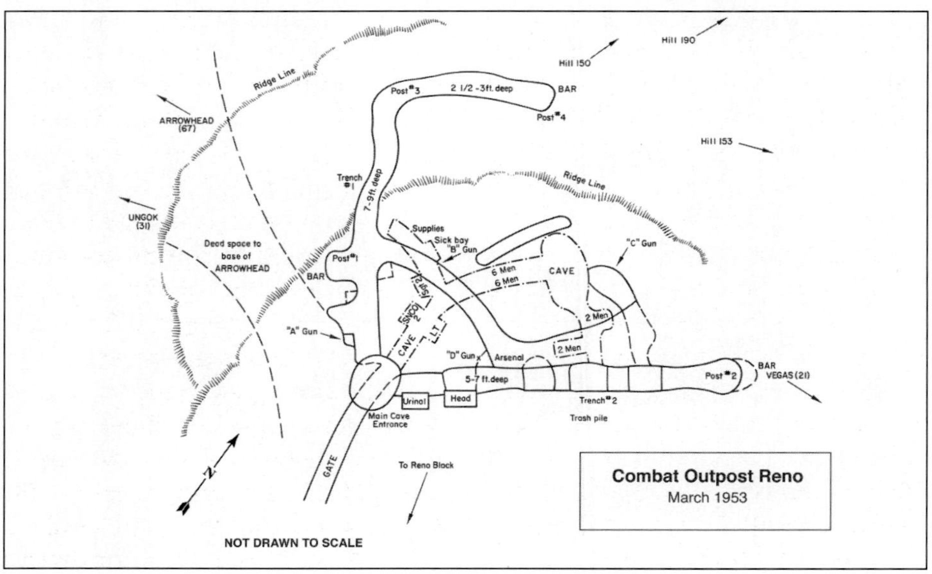

Combat Outpost Reno
March 1953

Department of Defense Photo (USMC) A170011

The trenchline on Outpost Reno; to the left of the "L" where the trench changes direction lies Reno Block, the site of savage fighting on the night of 26-27 March 1953.

legs joined. At night a reinforced squad manned the blocking position, which served as a listening post, helped screen the movement of supplies and reinforcements, and provided a rallying point for relief columns ambushed by Chinese patrols. Perched on a hilltop, Reno Block afforded excellent visibility, but conversely it could easily be seen from Chinese lines. Consequently, as a Marine who served there recalled: "We would light a cigarette under cover of a coat or blanket, then when we took a drag it was with both hands cupped to hide the glow," which could draw sniper fire. Marines manning Reno's east-west trench could fire in support of the blocking position, as could the garrison at Carson.

To the right of Reno loomed Combat Outpost Vegas, which attained a height of 175 meters and, as the tallest of the three, afforded the best fields of observation. Barbed wire and a well-constructed trench encircled the egg-shaped perimeter on Vegas, with its one warming and two living bunkers. Although the fields of fire on Vegas were less than ideal, handicapped in places by a steeply pitched slope too irregular for grazing fire and also by the small firing apertures in some of the covered fighting holes, weapons there could support Reno with long-range fire. By day, Vegas proved a magnet for sniper fire and harassment by mortars and artillery, forcing the Marines to remain under cover. A tour of duty at Vegas usually lasted three days for infantry and no more than five for artillery forward observers

Attack on Carson, Reno, and Vegas

Although the 5th Marines had been active during March—raiding Hill 31A in the Ungok hill mass and patrolling by night, especially in the vicinity of Carson, Reno, and Vegas—the Chinese tended to avoid combat. The lull ended abruptly at 1900 on 26 March, when the Chinese shattered the springtime evening with fire from small arms, machine guns, mortars, and artillery. Almost every Chinese weapon within range raked the left and center of the sector manned by the 1st Battalion, 5th Marines.

Combat Outposts Carson, Reno, and Vegas, each manned by a composite platoon from the 1st Battalion, 5th Marines, underwent a savage bombardment. An estimated 1,200 60mm and 82mm mortar rounds exploded on Carson within roughly 20 minutes, and the shelling continued at the rate of about one round every 40 seconds until 2200. The Chinese gunners also directed counterbattery fire at the howitzer positions of the 11th Marines, sought to interdict movement behind the main line of resistance, and tried to sever the telephone lines and routes of movement between the battalion and the threatened outposts.

The bombardment of Carson, Reno, and Vegas formed one part of a general shelling of outposts all along the Jamestown Line. To the right of the 1st Battalion, 5th Marines, the Chinese lashed out at Berlin and East Berlin, manned by Marines of the regiment's 3d Battalion. To the left of the 5th Marines, artillery fire and sometimes infantry threatened the outposts of the 1st Marines, like Hedy, Bunker Hill, Esther, and Dagmar. Chinese troops also seemed to be positioning themselves to attack in the sector of the Korean Marines.

On the night of 26 March, the general bombardment fell most heavily on Carson, Reno, and Vegas, and just 10 minutes after the shells began exploding there, some 3,500 soldiers from the *358th Regiment, 120th Division*, of the Chinese *46th Army* began converging on the three outposts. Taking advantage of the shelling, two platoons advanced from the Ungok hills to attack Carson, while one Chinese company stormed Reno and another Vegas. Yet another company—this one from Arrowhead Hill and nearby Hill 29—crossed the road to Seoul to attack Reno from the northwest.

5th Marines Historical Diary Photo Supplement, Mar53

Marines on Vegas huddle in the command post bunker as Chinese mortars and artillery pound the outpost. While Marine artillery responded with protective boxing and variable-time fires on the outpost and routes of approach, defending infantry called down 60mm and 81mm mortar barrages.

Meanwhile, Chinese troops from Hill 190 outflanked Reno on its left to hit the outpost from the rear, and others advanced from the high ground north of Vegas to storm the outpost head-on.

The fight for Carson, a part of the enemy's main effort, pitted Chinese numbers, perhaps 20 attackers for every defender, against a determined garrison that could be readily reinforced from the main line of resistance. In the first 35 minutes, the attackers penetrated the outer trenches at Carson, but the Marines fought the Chinese to a standstill in a fierce hand-to-hand struggle. Moving their wounded to the shelter of the centrally located cave, the defenders continued to fire from their fighting holes along the main trench. As squads from Companies C and D of Lieutenant Colonel Jonas M. Platt's 1st Battalion, 5th Marines, were moving out to reinforce Carson, the Chinese relaxed the pressure on that outpost, shifting their attention to Reno and Vegas. The attack began to ebb at about 2135, but mortars and artillery continued to pound the Marines on Carson. The violent though brief assault, followed by sustained shelling, took a psychological as well as physical toll among Carson's surviving defenders. For example, the outpost commander, First Lieutenant Jack F. Ingalls, who survived unwounded, seemed to have aged 10 years, according to a sergeant who knew him.

Reno, where the terrain precluded the establishment of a tight perimeter, proved harder to defend than Carson and, because it was farther from the main line of resistance, more difficult to reinforce. As at Carson, the Chinese gained a hold on the outer works, but at Reno they capitalized on this early success and forced the Marines to fall back to the cave that anchored the position. At 2030, Outpost Reno reported by radio that the enemy controlled everything outside the cave, which was collapsing under the sustained shelling. According to the message, death, wounds, and the lack of oxygen inside the cave left only seven Marines able to fight.

Supporting weapons did their best to save the doomed outpost. Aided by flares from an aircraft and illuminating rounds, machine gunners on the main line of resistance and rocket batteries just to the rear fired into the Chinese swarming over Reno, while variable-time artillery shells burst overhead, showering the attackers with deadly fragments. Two Marine M-46 tanks, on the Jamestown Line just behind Reno, joined in with their 90mm guns. Radio contact with the Marines fighting at the outpost faded, and then failed entirely—never to be restored.

Chinese forces also seemed on the verge of victory at Outpost Vegas. The intensity of the bombardment and overwhelming numbers forced the Marines from their least defensible positions. A breakdown in communication hampered efforts to reinforce Vegas, as Chinese artillery fire tore up telephone wires leading from the outposts to the battalion command post. Radio had to replace wire.

While the defenders of Vegas were undergoing attack, the 5th Marines sought to reinforce Outpost Reno, with the 1st Battalion assuming operational control of those elements of the 2d Battalion involved in the effort. At 2015, advance elements of a platoon from Company F, 2d Battalion, set out from the main line of resistance, fought their way out of an ambush near Hill 47, but were pinned down short of Reno Block, which had yet to be manned that night and had been occupied by the Chinese. The reinforced 3d Platoon, Company

A Platoon at Reno Block

The 3d Platoon, Company C, 1st Battalion—reinforced by an attached machine gun section—manned a portion of the Jamestown Line on the night of 26 March 1953. The men of the platoon, led by Second Lieutenant Warren C. Ruthazer, were standing the usual nighttime alert when they heard Chinese artillery exploding along the combat outpost line. Soon the bombardment began battering the main line of resistance, and Ruthazer summoned his noncommissioned officers to the command post where he told them that a Chinese attack on Outpost Reno had driven the defenders into the cave there and might soon overwhelm them.

Each night, a reinforced rifle squad, occupied Reno Block, a listening post about 100 yards closer to the main line of resistance than Reno itself. On this evening, however, the Chinese bombardment prevented the squad from moving out. As a result, the garrison intended for Reno Block, a reinforced squad from the 1st Platoon of Company C, joined forces with Lieutenant Ruthazer's 3d Platoon.

Shortly after 2000, the reinforced 3d Platoon, and its additional squad, received orders to drive the Chinese from Outpost Reno and bring back the surviving Marines. The men of the rescue force carried shovels and entrenching tools to free Marines trapped in the cave or in collapsed trenches, along with grenades for close-in fighting. Ruthazer's platoon started along a trail that extended some 1,800 yards, served Reno Block, and terminated at Reno. The Marines had to remain close to the trail because of minefields on both flanks, thus becoming more vulnerable to ambush. They avoided bunching up, dropped to the ground when necessary to escape enemy fire, and then jumped to their feet, rushing forward until again forced to hug the muddy earth. One of the many mortar shells that exploded along the trail burst close enough to Sergeant William H. Janzen, the platoon guide, to pelt him with dirt, and another landed directly in front of Private First Class Bobby G. Hatcher as he sprawled for cover alongside the trail, but the round failed to explode. During this ordeal, as he later recalled, Sergeant Janzen kept his sanity, although his face "was buried in the dirt and mud," by concentrating on repeating the Lord's Prayer.

To reach Reno Block, the force had to climb the steep hill on which the trench and bunker lay. The Marines had rigged a strong rope alongside the trail to help the heavily laden troops pull themselves upward. Chinese mortar and artillery fire had cut the rope, however, and the platoon had to claw its way up the slope. Atop the hill, the men entered a trench so shallow that at times the Marines had to crawl toward the blocking position.

The Chinese concentrated on the head of the relief column instead of trying to encircle it. Sergeant Janzen

Department of Defense Photo (USMC) 13964

At a December 1953 battalion review, MajGen Randolph McC. Pate, commander of the 1st Marine Division, presents the Navy Cross, the nation's second highest award for military valor, to Hospital Corpsman Paul N. Polley.

believed that this decision enabled the platoon to cling to the segment of trench, making it the anchor of a ragged perimeter. The arrival of Captain Ralph L. Walz and two platoons from his Company F, 2d Battalion, 5th Marines, tipped the balance in favor of the Marines, at least momentarily. The captain quickly mounted a bayonet charge, described by Janzen as "magnificent, heroic, and ghastly," that overwhelmed the Chinese at Reno Block.

Two corpsmen with Ruthazer's platoon, Hospitalman Third Class Paul N. Polley and Hospitalman Francis C. Hammond, struggled to care for the increasing number of casualties at Reno Block. Temporarily blinded by dirt thrown in his face by an exploding shell, Polley continued as best he could to tend to the wounded by sense of touch. Hammond, though already wounded, voluntarily remained behind with Captain Walz's Marines when the platoon from Company C received orders to withdraw. Hammond, killed when a mortar shell exploded near him, was awarded a posthumous Medal of Honor; Polley lived to receive the Navy Cross.

Casualty evacuation teams began arriving shortly after the 3d Platoon received orders to disengage. The survivors able to move about on their own had gathered at the base of the hill, when someone reported seeing a machine gunner, Private First Class Mario Lombardi, half-buried in a collapsed trench. A final search located Lombardi, whose legs had been broken, and his comrades brought him back. Of the 40 Marines in the reinforced 3d Platoon, fewer than 10 returned unscathed to the main line of resistance.

C, 1st Battalion—led by Second Lieutenant Warren C. Ruthazer—started toward Reno at 2030, together with the squad that had been assigned to Reno Block but had not yet deployed, and reached the enemy-held blocking position despite twice coming under long-range fire and twice being ambushed. Two platoons of Company F, 2d Battalion, followed in the wake of Ruthazer's men, leaving the Jamestown Line at 2227 and advancing toward Reno Block until stopped by fire from the Chinese holding the blocking position. Here these latest reinforcements, under Captain Ralph L. Walz, the commander of Company F, found Ruthazer's Marines, the elements of Company F dispatched earlier, and a platoon from Company D sent to reinforce Outpost Vegas but stopped near Reno Block by Chinese fire. Captain Walz took command of the group and launched an attack that drove the enemy from the blocking position.

The cobbled-together force of Marines clinging to Reno Block, now commanded by Captain Ralph L. Walz of Company F, tried gallantly but unsuccessfully to reach the composite force from the 1st Battalion, 5th Marines, that manned Outpost Reno when the battle began. Chinese attacks on the blocking position continued without respite, forcing the Marines to beat back three separate assaults by midnight and denying them an opportunity to mount a strong attack of their own. While the fighting raged at Reno Block, the Chinese marshaled reinforcements of their own behind Hills 31 and 31D. Fire from Marine artillery and from tanks on the main line of resistance scattered one such group as it massed to join in attacking Reno, but the Chinese prevailed, killing or capturing all the defenders. The survivors consisted of just five Marines—among them the outpost commander, Second Lieutenant Rufus A. Seymour of Company C, 1st Battalion, 5th Marines—and a Navy corpsman. They became

Marines of the 4.2-inch mortar company, 5th Marines, unload ammunition in support of the assault to retake Outpost Vegas. Because of the heavy incoming fire, trucks were unable to get to the company's mortar positions, so the Marines had to haul the ammunition up by carts.

Department of Defense Photo (USMC) A170437

prisoners of war and were ultimately repatriated.

The Chinese capture of Reno freed one of the companies that had helped subdue that outpost. The enemy assigned it the mission of delivering a coup de grace to Reno Block, but Marine artillery and tanks firing from the main line of resistance caught the Chinese as they moved south and frustrated the plan. The blocking position remained in Marine hands, at least temporarily.

During the struggle for Reno and Reno Block, Outpost Vegas continued to hold out, but almost from the time the Chinese attacked, contact with Vegas proved uncertain. To facilitate the restoration of reliable communications by wire, and if necessary with runners, Colonel Walt, the commander of the 5th Marines, shifted operational control of the Marines on Vegas, and those attempting to reinforce them, from the 2d Battalion to the 3d Battalion. Shortly before midnight, contact with Vegas ended. All the Marines there were either killed or captured; the dead included the officer in charge, First Lieutenant Kenneth E. Taft, Jr.

Initial Counterattacks End

At midnight, after some five hours of fighting, the Chinese controlled Vegas and Reno, although Carson remained under Marine control. The Marines trying to break through to Vegas or Reno had thus far got no farther than Reno Block, but they kept trying. At 0144, Captain Walz, in command of Company F and in charge of the composite force at the blocking position, reported that he had only the equivalent of one reinforced platoon to break through to Reno. Within an hour, Walz had launched three attacks, each one stopped by fire from mortars and small arms. For now the outpost would remain in Chinese hands. The attempts to reach Outpost Reno on the night and early morning of 26-27 March resulted in severe Marine casualties that Colonel Walt later estimated as being "as high as 35 percent, with many dead."

National Archives Photo (USMC) 127-N-A170429

A Marine gives a wounded buddy a drink while others prepare him for evacuation. During the initial attempts to regain control of Outposts Reno and Vegas, both Marine and enemy casualties were heavy; an estimated 600 Chinese had been wounded or killed while Marine losses were placed at more than 150.

On the early morning of 27 March, while the attempts to fight through to Outpost Reno were ending in frustration, two platoons organized from Companies D, 2d Battalion, and C, 1st Battalion, of the 5th Marines advanced toward Vegas. They worked their way as close as 400 yards to the entrance to the outpost's trenches before a fresh Chinese assault stopped them. This setback forced the 1st Marine Division to commit a part of its reserve.

The 2d Battalion, 7th Marines, commanded by Lieutenant Colonel Alexander G. Cereghino and functioning as part of the division reserve, placed its Company F under the operational control of the 3d Battalion, 5th Marines, for this new effort to reach Vegas. The platoon leading the way advanced to within 200 yards of the outpost, but could only confirm that the enemy had already seized it. Beginning at about 0300, the Marines who had made this early morning attempt to break through to Vegas withdrew to the main line of resistance, arriving there at 0417. Earlier on the night of the 26th, Colonel Walt had given his 3d Battalion operational control over the attempts to save Vegas; now he ratified the decision by shifting the boundary between the 1st and 3d Battalions some 250 yards to the west.

The eight-hour fight at Vegas and on its approaches cost the Chinese an estimated 600 casualties, four times the total of Marines killed or wounded. Unremitting Chinese mortar and artillery fire claimed most of the Marine victims. A platoon from Company F, 2d Battalion, joined Company C, as did a provisional unit assembled from the 1st Battalion's Headquarters and Service Company. Rather than taking part in an attempt to regain Vegas, these latest reinforcements helped evacuate those Marines wounded in the earlier fighting.

Marines wounded in the vicinity of Outpost Vegas followed one of two routes to a collection point behind the position held by Company H, 3d Battalion, 5th Marines. From here, casualties went either directly to the 3d Battalion aid station or to a camp of the Korean Service Corps enroute to the aid station of the 1st Battalion. The most severely wounded traveled by helicopter directly to a hospital ship, *Haven* (AH 12) or *Consolation* (AH 15), in Inchon Harbor. The others received further treatment from the division's medical battalion before being transferred to better-equipped facilities in the rear.

Chinese Diversionary Attacks

On the night of 26 March, the Chinese diverted attention from Carson, Reno, and Vegas by jabbing at Berlin and East Berlin in the sector of Lieutenant Colonel Robert J. Oddy's 3d Battalion, 5th Marines. Berlin occupied a rounded hilltop roughly the same height as the main line of resistance; a 400-yard ridge linked the two. Supplies or reinforcements bound for Berlin, unlike those destined for Vegas or Reno, were screened by the hill from direct enemy observation. Taking advantage of the concealment, carrying parties of the Korean Service Corps shuttled supplies to the entrance to the outpost. The porters were not allowed to enter the trenches, however, to avoid confusion that might prove advantageous to the enemy.

One reinforced squad from Company G held Berlin, and another East Berlin. Both garrisons beat back the Chinese with the help of barrages that boxed-in the outposts and variable-time fire that rained fragments on the enemy's routes of approach. By midnight, a second squad had reinforced each of the outposts.

The Chinese also probed the outposts held by the 1st Marines on the left of Colonel Walt's regiment. The enemy bombarded four outposts—Hedy, Bunker Hill, Es-ther, and Dagmar—and tested the defenses of all of them. Only at Combat Outpost Dagmar did the attackers break through the wire. The 27 Marines defending the outpost held their own against an approximately equal number of Chinese. Some 300 rounds of mortar and artillery fire helped the defenders, commanded by Second Lieutenant Benjamin H. Murray of Company I, 3d Battalion, to contain the penetration until help arrived. Before midnight on 26 March, Second Lieutenant John J. Peeler, the executive officer of Company I, led a counterattack from the main line of resistance that regained control over all of Outpost Dagmar. The clashes at Dagmar and the other outposts—Hedy, Bunker Hill, and Esther—killed perhaps 10 Chinese and wounded 17 at the cost of four Marines killed and 16 wounded. A feeble jab at Outpost Kate proved an annoyance rather than a threat, and Chinese troops massing in front of the Korean Marines did not attack at this time. Marine artillery again demonstrated its importance during the outpost actions on the night of 26-27 March. The 11th Marines fired more than 10,000 105mm and 155mm rounds at targets from East Berlin and Berlin westward to Outpost Hedy. The bombardment of Vegas and Reno, the outposts

A Marine mans the main trenchline on the forward sloop of Outpost Dagmar. The outpost was the site of a two-squad Chinese diversionary assault with automatic weapons and satchel charges that was beaten back with help from Marines on the main line of resistance.

Department of Defense Photo (USMC) A169923

National Archives Photo (USMC) 127-N-A317670

A graduate of Stanford University and veteran of the British Commando Program, LtCol Alexander D. Cereghino spent most of World War II in command of the 1st Provisional Battalion stationed in Northern Ireland. Following a three-year stint at Headquarters Marine Corps, he was given command of 2d Battalion, 7th Marines, in Korea.

the Chinese had seized, continued into the morning of the 27th in preparation for a more powerful counterattack.

Marine Counterattack of 27 March

Preparations for a major counterattack went ahead on the ground and in the air. At 0345 on the 27th, Lieutenant Colonel Cereghino's 2d Battalion, 7th Marines, an element of the division reserve, came under the operational control of Colonel Walt's 5th Marines. Cereghino's unit had already moved into an assembly area behind the 1st Battalion, 5th Marines, and Company F had taken part in as unsuccessful attempt to break through to Outpost Vegas.

By the time the Marines abandoned the early attempts to fight their way through to Vegas and Reno, observation planes from Marine Observation Squadron 6 (VMO-6) began flying missions to direct friendly artillery fire against Chinese batteries. Aerial observers called down some 60 fire missions against targets that included mortars, artillery, and self-propelled guns. During the morning darkness, Douglas F3D-2 Skyknight night fighters conducted radar bombing against Chinese gunners and troop concentrations As dawn approached, the time chosen for an attempt to recapture Vegas and Reno, air strikes intensified. Grumman F9F Panther jets of Marine Fighter Squadron 115 (VMF-115) arrived overhead at 0650 to help neutralize the Chinese defenders of the two captured outposts. A breakdown of communications forced a postponement, however; H-Hour for the assault on Reno and Vegas slipped until 0900, but persistent problems of coordination caused further delay. Airmen took advantage of the additional time; with two-dozen Marine fighter-bombers joining Air Force jets in delivering strikes.

Marine tanks—which the division commander, Major General Edwin A. Pollock, had ordered to join the rocket battery, mortars, and artillery regiment in supporting the counterattack—made their contribution to the preliminary fires. Company A of the division's tank battalion spotted two groups of Chinese carrying logs for the construction of bunkers on the site of Outpost Reno. The tank company wiped out one group with 90mm fire, but the other presumably reached its destination.

Before the attack finally began, Major General Pollock agreed to a change of plan. Vegas would be the sole objective; while attacking there the Marines would neutralize Reno with fire. Smoke shells bursting on Hills 57A and 90 blinded the Chinese observers there and marked the launching of the attack, which got under way shortly after 1100. Artillery, mortars, tanks, and aircraft hammered Vegas, Reno, and the enemy's firing batteries, including some located by Marine airmen earlier in the day.

As bombs, shells, and rockets exploded on Vegas and the other targets, Company D, 2d Battalion, 5th Marines—an element of Colonel Walt's regimental reserve—advanced from the main line of resistance. The assault company, commanded by Captain John B. Melvin, launched its attack at 1120 but immediately came under fire from Chinese mortars and artillery. Within an hour, the company's first platoon had been reduced to just nine able-bodied Marines, but Melvin's survivors slogged forward through flooded rice paddies and up a rain-soaked slope. Marine casualties and enemy reinforcements slowed Company D, as the men worked their way from one depression or rocky outcropping to another until the counterattack stalled some 200 yards short of the objective. Chinese fire raked the slope where the assault had bogged down, but Marine jet fighters and piston-powered attack planes joined the 11th Marines and the 1st Tank Battalion in trying to silence the hostile weapons.

Shells from Chinese 122mm artillery and 120mm mortars churned the slopes of Vegas; making the detonations from 60mm mortar rounds seem like mere firecrackers. Captain Melvin recalled that the enemy fire "was so intense at times that you couldn't move forward or backward.... You could only hope that the next round wouldn't be on target." Despite the deadly barrage, a handful of

Marines succeeded in entering the outer trench and making a brief lodgment there.

To sustain the counterattack, two forces set out shortly after 1200 from the lines of Company H, 3d Battalion, 5th Marines. Captain Floyd G. Hudson led the way with a provisional unit made up from the 2d Battalion, 5th Marines. Captain Herbert M. Lorence commanded the second, Company F of the same battalion. These reinforcements reached the slope where Melvin's Marines were undergoing their ordeal but could advance no farther. As a result, Company F, 2d Battalion, 7th Marines—available from what had been the division reserve—set out at 1530 under Captain Ralph F. Estey to lend its weight to the counterattack on Outpost Vegas.

Captain Estey's company reached the fire-swept slope leading to the enemy-held outpost and relieved the surviving elements of Melvin's Company D, 5th Marines, which returned to the main line of resistance. While 90mm guns from Captain Clyde W. Hunter's Company A, 1st Tank Battalion, fired from the main line of resistance against a Chinese strongpoint on the crest of Vegas, Marine aircraft maintained a smoke screen that blinded enemy observers. Estey combined his men with those of Company E of the 5th Marines, and formed the equivalent of three platoons. After a savage fight lasting about an hour and a half, the counterattack succeeded by 2000 in overrunning the trenches nearest the Marine main line of resistance. The Chinese who held the rest of Vegas contained the breakthrough short of the summit and forced the Marines to fall back to the base of the hill and dig in. Fire from supporting weapons in the rear, and also from the site of Outpost Reno, helped the enemy prevail.

Pressure on the Marine foothold continued throughout the night. By midnight, Captain Estey's men had beaten off three attacks and clung firmly to the position at the base of Vegas. Marine night fighters and attack planes made nine radar-controlled strikes between darkness on the 27th and 0115 the next morning, dropping some 24 tons of bombs on enemy positions and supply lines. The bombardment of Estey's perimeter gradually abated, as Chinese gunners shifted their fire from the base of Vegas to the main line of resistance.

During the counterattack on Chinese-held Outpost Vegas, Hospitalman Third Class William R. Charette—attached to Company F, 2d Battalion, 7th Marines—repeat-

Grumman F9F Panther jets taxi into position for takeoff against enemy concentrations. During the Marine counterattack on Outpost Vegas, Panther jets from Marine Fighter Squadrons 115 and 311 not only attacked Chinese trenches, bunkers, and mortar emplacements, but also bombed enemy resupply points.

Department of Defense Photo (USMC) A346720

ly risked his life to care for the wounded. While Charette was giving first aid to a Marine, a Chinese grenade landed next to them. Charette pounced on the grenade; his armored vest saved him from serious injury, but the force of the explosion tore away his helmet and his medical kit. Since he could not waste time searching for the kit, he used his clothing to improvise bandages. When he was tending to another wounded Marine, whose armored vest had been blown off by the concussion from an explosion, Charette removed his own vest and placed it over the wounded man. Without either a helmet or an armored vest, he stayed with his platoon throughout the fighting. According to one of the Marines who fought there, the corpsman "was everyplace seemingly at the same time, performing inexhaustibly." In recognition of his bravery and devotion to duty, Charette received the Medal of Honor. Of the corpsmen whose service in Korea earned them the nation's highest award for valor, he alone survived to wear it.

The Marines counterattacking Vegas had seized a strong foothold at the base of the hill where the outpost stood, but the summit and the shell-battered complex of trenches and bunkers remained in Chinese hands. Preparations for another assault up the hill began at 0335 on 28 March when the 105mm and 155mm howitzers of the 11th Marines fired the first of more than 2,300 rounds directed at Chinese weapons positions and assembly areas, as well as the defenses of the summit itself.

After half an hour's bombardment, the men of Captain Estey's Company F, 2d Battalion, 7th Marines, who had entered the fight at mid-afternoon the previous day, worked their way to within grenade range of the Chinese before being driven back by fire from small arms and mortars. While the company regrouped, Marine aircraft joined in battering the enemy. After sunrise, Marine fliers laid a smoke screen to conceal air strikes from Chinese antiaircraft gunners on Hill 67, north of Outpost Carson. During the early morning, Marine aircraft conducted four powerful strikes, the first of a series that lasted throughout the day.

The second infantry assault of the morning, launched at 0600, failed to recapture Vegas, ending when Captain Estey's Marines took cover from enemy fire about 375 yards south of the crest. After further air strikes, Estey's Company F attacked once again and by 1015 penetrated to within 15 yards of the trenches, where it engaged in a firefight that lasted 22 minutes and deprived the assault of its momentum.

When this latest counterattack stalled in front of a machine gun firing from the base of a rock formation, Sergeant Daniel P. Matthews, a squad leader in Company F, saw that the weapon had pinned down a wounded Marine and the corpsman trying to bring him to safety. The sergeant crawled to the formation, scrambled onto the rocks, opened fire with his rifle, and charged the gun. Although wounded almost immediately, Matthews killed two members of the gun crew and silenced the weapon. His bold action enabled the corpsman to save the wounded Marine, but Matthews died of his wounds before members of his squad could reach him. His heroism earned a posthumous Medal of Honor.

Shortly after the one-man assault by Sergeant Matthews, Captain Lorence's Company E, 2d Battalion, 5th Marines, passed through Company F, 2d Battalion, 7th Marines. Captain Estey's company, reduced to just 43 able-bodied men after a half-dozen attempts to recapture Vegas, fell back to regroup at the base of the hill.

While the relief of Estey's company was taking place, aircraft, mortars, and artillery continued to scourge not only outpost Vegas, but also Chinese supply routes, enemy-held Outpost Reno, nearby Hill 25A, and other high ground from which hostile gunners could fire in support of their comrades dug in on Vegas. Taking advantage of the savage bombardment—during a 23-minute period, aerial bombs fell at the rate of more than a ton per minute—the 1st Platoon of Captain Lorence's company launched its assault at 1301. Led by Staff Sergeant John J. Williams, who had taken over when Second Lieutenant Edgar R. Franz was wounded, the platoon killed or drove off the Chinese defending the former site of Outpost Vegas. The platoon took only one prisoner, the second the Marines had captured during the day.

As soon as Captain Lorence's Marines regained the ruins of Outpost Vegas, the Chinese counterattacked, but fire from tanks dug in on the main line of resistance and artillery helped break up the assault. Although the Marines controlled Vegas itself, a few Chinese stubbornly held out at the very summit of the hill. The work of preparing to meet the next counterattack went ahead under the command of Major Benjamin G. Lee, who had earned the Silver Star and Purple Heart as a noncommissioned officer at Guadal-canal during World War II. Although Lee, operations officer of the 2d Battalion, 5th Marines, took command, the resupply of Vegas became the responsibility of the

1stMarDiv Historical Diary Photo Supplement, Mar53

Although Vegas was retaken, Marines still had to keep low when moving around the ruined outpost. Daylight hours were relatively quiet with only an occasional burst of machine gun or sniper fire, while night witnessed renewed Chinese attempts to overrun the outpost.

regiment's 3d Battalion. To defend the newly recaptured outpost, Lee at first could muster only 66 Marines, eight of them members of Company F, 2d Battalion, 7th Marines, who had not fallen back to the base of the hill with the unit's other survivors, and the rest from Captain Lorence's Company E, 2d Battalion, 5th Marines. Fortunately, Captain Thomas P. Connolly's Company E, 2d Battalion, 7th Marines, reached the hill with an additional 150 men and prepared to take over from Lorence's Marines and spearhead a final attack.

Further Action on Outpost Vegas

The Chinese held the initiative, however, and at 1955 on 28 March, as darkness enveloped Vegas, they counterattacked. The preparatory fires included the usual artillery and mortar barrages supplemented by 3.5-inch rockets fired from the Chinese equivalent of the American Bazooka antitank weapon. An enemy battalion advanced from captured Outpost Reno, but fell back after coming under deadly fire from Army and Marine artillery and the 1st Marine Division's 4.5-inch rocket battery.

Outpost Carson came under fire at about the same time from mortars and automatic weapons. Chinese patrols probed the approaches to the outpost, but the garrison, supported by weapons on the main line of resistance, drove off the enemy. The attack, if one was planned, did not take place.

At 2130, Major Lee radioed from Vegas to report that a second Chinese assault was imminent, and within an hour the enemy struck. Box-me-in fires helped Lee's Marines cling to their position, but the Chinese struck again, launching the night's third attack about an hour before midnight. At least 200 Chinese soldiers tried to overwhelm Lee's perimeter but succeeded only in forcing the Marines to yield some non-critical ground. While aircraft dropped flares, howitzer batteries of the 11th Marines dueled with Chinese gunners, firing more than 6,000 rounds by midnight. The attack continued until 0130 on the morning of 29 March, when another savage bombardment by the division's artillery and the fire of Captain Connolly's Company E, 2d Battalion, 7th Marines, forced the enemy to relax his pressure. The bulk of the Chinese assault force, totaling two battalions during the fight, withdrew behind a curtain of fire from weapons emplaced on Reno.

As Connolly's Company E, 7th Marines, and Lorence's Company E, 5th Marines, prepared to eliminate the Chinese die-hards still clinging to the crest of the hill, artillery and 4.5-inch rockets kept on pounding the enemy. More than 4,000 rounds neutralized Reno while others tore into the Chinese-held portion of Vegas. A final Marine assault secured the summit of Vegas at 0450.

Chinese mortar and artillery fire directed at the Marines continued after the recapture of Vegas. Shortly after 0500 a 120mm mortar shell killed both Major Lee and Captain Walz, the commander of Company F, 2d Battalion, 5th Marines, who had led the charge at Reno Block. Also killed in this flurry of shelling was First Lieutenant John S. Gray, a forward observer from the 1st Battalion, 11th Marines.

Major Joseph S. Buntin, the executive officer of the 3d Battalion, 5th Marines, which was responsible for rebuilding the outpost, now took command. Replacements, including corpsmen, arrived that morning, along with weapons, tools, construction supplies, and laborers from the Korean Service Corps. By noon, work was underway on trenches, fighting holes, and bunkers, screened as necessary by smoke missions fired by the 11th Marines, but the muddy, shell-churned earth complicated the efforts of the Marines and the Korean labor troops helping them. As daylight faded, rain and light snow added to the discomfort of the men on the ground and forced the aerial observers, who had been directing the maintenance of smoke screens and other artillery missions, to return to their airstrips.

The Chinese had not yet abandoned their designs on Vegas. At 1805 on 29 March, they advanced from assembly areas in the vicinity of Reno and Hill 153 and hit Vegas on both flanks. This attack triggered the most violent single barrage of the battle for Vegas, as more than 6,400 shells from five artillery battalions, plus two rounds from each launching tube in the division's 4.5-inch rocket battery, exploded among the attacking troops. Army 8-inch howitzers and 4.2-inch mortars also helped break up the assault.

After a brief flurry of activity at about 2045, perhaps an attempt to retrieve men wounded in the earlier attack, the Chinese mounted another major effort early in the morning of 30 March, again striking from Reno and Hill 153. This latest attempt to isolate Outpost Vegas and destroy it also collapsed under an avalanche of artillery and mortar shells.

Sunrise brought clearing skies, enabling Marine AU Corsairs to disrupt Chinese attempts to regroup for further assaults. Throughout the day, Marine airmen attacked troops, fortifications, and firing batteries on Reno and other hills that menaced Vegas. On Vegas, the Marines "were like rabbits digging in," said Corporal George Demars of Company F, 2d Battalion, 5th Marines, who added that the replacements "jumped right in," as members of hastily organized squads worked together, even though "they didn't know half the people on the fire teams." In the afternoon, Company G, 3d Battalion, 5th Marines, relieved Captain Connolly's Company E, 2d Battalion, 7th Marines, and Major George E. Kelly, operations officer of the 2d Battalion, 7th Marines, replaced Major Buntin.

National Archives Photo (USMC) 127-N-A365894

For "outstanding services performed in the line of his professional services," Maj Joseph S. Buntin would later receive the Bronze Star Medal from MajGen Randolph McC. Pate. The award was presented in December 1953.

By daybreak on 31 March, the men on Vegas could relax in the partially reconstructed trench works. The five-day siege involving more than 4,000 ground and air Marines was the bloodiest action Marines on the Western Front had yet engaged in.

National Archives Photo (USMC) 127-N-A170454

At about 1100 on the morning of 30 March, five Chinese soldiers approached Outpost Vegas, giving the impression they intended to surrender. Instead they threw grenades and cut loose with submachine guns in a suicidal gesture of defiance. Marine infantrymen returned the fire, killing three of the intruders outright. Another died of his wounds, and the survivor was taken prisoner.

During the night, Army searchlights provided illumination for Marine artillery, including the half-track-mounted .50-caliber machine guns of the 1st Provisional

Department of Defense Photo (USMC) A170436

Members of the 2d Battalion, 5th Marines, take time out after returning from the fierce five-day battle for Outpost Vegas to eat chow and take a well-deserve breather. With a touch of sarcasm, they called the disputed crest of Vegas "the highest damn beachhead in Korea."

Antiaircraft Artillery Battery (Automatic Weapons), which took a hand in the fighting on the ground. Marine aircrews directed by Marine Corps radar made seven strikes against Chinese positions on the night of the 31st, and aircraft, mortars, and artillery continued to harass the enemy on the following day, as the 5th Marines and 1st Marines continued to relieve the garrisons at their combat outposts.

To prevent the enemy from exploiting his conquest of Outpost Reno, the Marines strengthened Vegas after its recapture, manning it with a company rather than a platoon. The 1st Marine Division also established a new outpost, Elko (named like Reno, Vegas, and Carson for a city in Nevada). Elko stood on Hill 47, southeast of Carson and 765 yards north of the main line of resistance. During the unsuccessful attempt to break through to the Reno garrison, Chinese troops had used Hill 47 to ambush the rescue forces. Marine possession of the hill improved the security of Carson, Vegas, and the main line of resistance.

Outpost Vegas: Summing Up

The five-day battle that ended with the recapture and successful defense of Outpost Vegas cost the

Standing together at I Corps headquarters in early April are, from left, BGen John K. Waters, Chief of Staff, I Corps, MajGen Bak Lim Hang, Commanding General, 1st Republic of Korea Division, MajGen M. A. R. West, Commanding General, 1st Commonwealth Division, MajGen Edwin A. Pollock, Commanding General, 1st Marine Division, LtGen Paul W. Kendall, departing commanding general, I Corps, MajGen Bruce C. Clarke, newly appointed commanding general, I Corps, MajGen James C. Fry, Commanding General, 2d Infantry Division, MajGen Arthur G. Trudeau, Commanding General, 7th U.S. Infantry Division, and Col Pak Ki Sung, Commander, 101st Korean Service Corps.

National Archives Photo (USA) 111-SC429583

1st Marine Division 1,015 casualties or some 63 percent of those suffered by the division, including the Korean Marine regiment, during the entire month of March 1953. Chinese losses were estimated to be 2,221, of whom 536 had been actually counted. Whatever the exact toll, the Marines had crippled the *358th Regiment* of the Chinese army, which faced the task of rebuilding before it could again take the offensive.

During the loss and recapture of Vegas, the 1st Marine Aircraft Wing flew 218 missions in support of the combat outpost line, most of them to help hold positions manned by the 5th Marines. This total represented some 63 percent of the 346 close air support missions flown by the wing during March. Despite rain and snow that sometimes restricted visibility, Marine aircraft dropped some 426 tons of bombs in support of Marines on the ground, provided battlefield illumination, and laid smoke screens. In addition, helicopters evacuated the critically wounded.

Although Marine tanks not only fired some 9,000 rounds from the main line of resistance but also illuminated some targets using their shuttered searchlights, artillery proved the deadliest weapon in support of the infantry. Between 27 and 31 March, the 11th Marines, the 1st Marine Division's rocket battery, and the Army artillery and heavy mortar units reinforcing their fires, delivered almost 105,000 rounds against targets in the vicinity of Vegas, Carson, and Reno. The heaviest artillery action took place during the 24 hours ending at 1600, 29 March, when four Marine howitzer battalions fired 33,041 rounds and two Army battalions another 2,768.

The recapture of Outpost Vegas and the related fighting earned congratulations from the Commandant of the Marine Corps, General Lemuel C. Shepherd, Jr., who praised the "stubborn and heroic defense of Vegas, Reno, and Carson Hills coupled with the superb offensive spirit which characterized the several counterattacks." The sustained and deadly action earned a respite for the 5th Marines, which on 4 and 5 April moved into division reserve, replaced by the 7th Marines after 68 days on line. Meanwhile, on 28 March the Chinese high command, perhaps motivated to some extent by the failure to hold Vegas and crack the Jamestown Line, advised Army General Mark W. Clark, the United Nations commander, of its willingness to discuss an exchange of sick and wounded prisoners of war.

Shuffling Marine Regiments

After the battles for Reno, Carson, and Vegas, the 5th Marines went into division reserve. Shortly afterward, on 14 April, Colonel Harvey C. Tschirgi took command of the regiment, replacing Colonel Lewis Walt, who was reassigned to the division staff. The 7th Marines, under Colonel Glenn C. Funk after he took over from Colonel Loren E. Haffner on 27 March, replaced the 5th Marines on 4-5 April, moving from division reserve to man the portion of the Jamestown Line formerly held by Colonel Walt's regiment. The 1st Marines, commanded by Colonel Hewitt D. Adams from 1 November 1952 until Colonel Wallace M. Nelson took over on May 1, remained in place, holding the segment of the line between the 7th Marines and the Korean Marine regiment. After truce talks resumed at Panmunjom, the 1st Marines under Colonel Adams provided a force of 245 men and five armored vehicles that stood by to rescue the United Nations negotiators if the other side should spring a trap.

As the division reserve, the 5th Marines assumed responsibility for maintaining the fallback positions behind the main line of resistance. One of these, the Kansas Line, had suffered severe structural damage from torrential rains and the spring thaw. Its restoration required the full-time efforts of the regiment's 3d Battalion, which had to cancel its scheduled training. The 2d Battalion, however, participated in a landing exercise at Tokchok-to, an island southwest of Inchon, but high winds and rolling seas cut short the training. While the 5th Marines was in reserve, its staff participated—together with staff officers of the 1st Marines, the Commonwealth Division, and U.S. Army and South Korean forces—in a command post exercise staged by I Corps.

Meanwhile, the Chinese tested the 7th Marines. On 9 April, after mortars and artillery battered the regiment with some 2,000 rounds, the Chinese launched an attack on Carson, followed by a series of probes of that outpost and nearby Elko. At 0345 on the morning of the 9th, some 300 Chinese, advancing in two waves, hit Outpost Carson. After an hour, some of the assault troops reached the trenchline and for another hour exchanged fire at point-blank range with the defenders. A platoon set out at 0530 to reinforce Carson but got no farther than Elko, the newly established outpost some 400 yards southeast of Carson, before fire from mortars and small arms stopped the unit until 90mm tank guns and Marine mortars broke up the ambush. Howitzers and rocket launchers joined in, battering the approaches to Carson until the Chinese fell back.

To strengthen Carson's defenses,

Little Switch

During December 1952, when the truce talks at Panmunjom seemed to have reached a dead end, General Mark W. Clark—who in May of that year had replaced another Army officer, General Matthew B. Ridgway, in command of the United Nations forces—took note of a suggestion by the International League of Red Cross Societies that the combatants in Korea exchange sick and wounded prisoners of war. On 22 February 1953, Clark formally proposed an immediate exchange, but, as he expected, the Chinese and North Koreans did not respond. On 5 March, however, the death of Joseph Stalin deprived the Soviet Union of a forceful dictator and the People's Republic of China of a strong ally.

In this time of transition, as Georgei Malenkov emerged from the shadows to become Stalin's successor, at least temporarily, the Chinese leadership began thinking of Clark's proposal as an anchor to windward in a gale of uncertainty. Indeed, if Malenkov or some other ruler proved less supportive than Stalin, the limited exchange of prisoners might enable China to revive the truce talks and perhaps bring to an end a long and thus far inconclusive war. On 28 March, while the fight for Vegas and the other outposts still raged, General Clark received not only a formal Chinese acceptance of the proposal to exchange prisoners, but also an offer to resume serious armistice negotiations at Panmunjom.

On 6 April, representatives of the two sides began talks at Panmunjom that resulted in an agreement for the exchange, Operation Little Switch, scheduled to begin on the 20th. To prepare for the transfer, 100 Marines from the 1st Engineer Battalion and the 1st Shore Party Battalion built the so-called Freedom Village at Munsan-ni, a tent city complete with roads, a helicopter pad, and facilities for emergency medical treatment, administration, and press coverage.

Frontline Marines watch a U.S. Army ambulance convoy bringing the first freed United Nations prisoners from Panmunjom to Freedom Village. Following the long one-and-a-half hour trip, each Marine prisoner received a medical check and a new utility cap with its Marine Corps emblem.

National Archives Photo (USMC) 127-N-A170964

National Archives Photo (USN) 80-G-480686

Marine Pvt Alberto Pizzaro-Baez talks with MajGen Edwin A. Pollock at Freedom Village following his release. Pvt Pizzaro-Baez was among the 15 Marines and three Navy Corpsmen who had been captured from the 1st Marine Division.

When the exchange began on 20 April, Major General Edwin A. Pollack, the division commander, stood by to greet the Marines among the 149 Americans released through 26 April when Little Switch ended. The first Marine welcomed by Major General Pollack, and by the other dignitaries that included General Clark, was Private Alberto Pizzaro-Baez of Company H, 3d Battalion, 7th Marines, wounded in the leg and captured at Outpost Frisco in October 1952. During his imprisonment, the leg became gangrenous and had to be amputated. The men of the 1st Marine Division returned in Operation Little Switch totaled 15 Marines and three Navy hospital corpsmen. Two of the Marines, Corporal Jimmie E. Lacy and Private First Class George F. Hart, and one of the corpsmen, Hospitalman Thomas "Doc" Waddill, had been wounded and captured when the Chinese overran Outpost Reno on the night of 26 March. The Communist forces released 684 sick and wounded captives from 11 nations, more than half of them South Koreans, in return for 6,670 North Koreans and Chinese. The disparity in numbers may have reflected the desire of the Americans to rid themselves of dedicated and disciplined Communists among the prisoners like those who had seized a prison compound at the island of Koje-do and briefly held the commander hostage. Little Switch did not address, let alone resolve, the question of the forced repatriation of prisoners unwilling to return when a truce finally went into effect.

Company E, 2d Battalion, 7th Marines, moved there from regimental reserve, more than making good the day's losses of 14 killed and 64 wounded. The Marines counted 60 Chinese killed in the fighting on the early morning of 9 April. Another 160 may have been killed or wounded, which would have brought the morning's casualties to more than half the Chinese assault force that triggered the action.

During the first day's fighting on the ground, Marine aircraft appeared overhead after 0715, attacking visually or with the help of ground-based Marine radar. By mid-afternoon, fighter-bombers had dropped more than 67 tons of bombs on Chinese positions north of Carson. Radar-directed strikes took place after dark, and visual strikes resumed after daylight on 10 April.

As aerial action intensified, activity on the ground slowed. On the night of the 9th, three Chinese platoons, possibly searching for the day's casualties, advanced as far as the ruins of a bunker just 50 yards from Outpost Carson. The Chinese activity attracted fire that

Not every veteran of the 1st Marine Division was a human being. The pack horse "Reckless," shown here with her handler, took her name from the recoilless (or rec-less) rifle, like the one next to her, for which she carried ammunition. The horse returned to Camp Pendleton, California, with the division in 1955.

Department of Defense Photo (USMC) A171729

National Archives Photo (USMC) 127-N-A170606

Marine Cpl Dennis W. Burker, using a Browning automatic rifle, fires at enemy positions in front of Company I, 3d Battalion, 7th Marines' place on the main line of resistance. Cpl Burker wears one of the 24,000 armored vests issued the division by 1953.

definitely killed 15, may have killed 15 more, and wounded between seven and 27. Shortly before midnight, some 70 Chinese advanced from the Ungok hills and attacked Carson, only to lose perhaps 20 additional men to Marine mortars, tank guns, and machine guns. A couple of Chinese squads probed Outpost Elko on the night of the 11th, but for the most part, the enemy now contented himself with trying to exploit the renewal of truce talks by showering the Jamestown Line with propaganda leaflets warning the Marines against risking their lives because a ceasefire was at hand. The Chinese reinforced the printed message with loudspeaker broadcasts, and on one occasion they dropped leaflets from an airplane to supplement those scattered from special mortar shells.

Into I Corps Reserve

On 5 May, the Army's 25th Infantry Division, commanded by Major General Samuel T. Williams, and its attached Turkish brigade replaced the 1st Marine Division on what had been called the Jamestown Line. The practice of naming each separate segment of the line, like Jamestown, ended on 28 April, after which the entire front, from coast to coast, was known collectively as the main line of resistance. At the same time, the Eighth U.S. Army in Korea became simply the Eighth U.S. Army.

While the bulk of the Marine division moved 15 miles eastward over muddy roads to occupy the three cantonment areas that comprised Camp Casey, named in memory of U.S. Army Major Hugh B. Casey, the 11th Marines and the division's rocket battery remained attached to I Corps Artillery, in position to provide general support and fire counterbattery missions as necessary. The artillery battalion of the Korean Marines moved into position to reinforce the fires of I Corps Artillery. The Marine 1st Tank Battalion came under the control of the 25th Infantry Division; two companies supported the Turkish brigade, which had no armor of its own, another was assigned to the Army division's 35th Infantry, and the fourth served as a reserve. The Korean Marine tank company

The first contingent of the 3d Turkish Battalion begins the relief of the 3d Battalion, 7th Marines. On 5 May, the U. S. Army's 25th Infantry Division, to which the Turkish troops were attached, took over from the 1st Marine Division, which went into corps reserve.

Department of Defense Photo (USMC) A171351

Women Marines

The role of women in the Marine Corps during the Korean War was the result of a checkered series of events in the preceding years. When World War II ended, there were 820 officers and 17,640 enlisted members of the Marine Corps Women's Reserve. They had served to "free a Marine to fight," as the recruiting slogan proclaimed.

In one tumultuous year of peacetime demobilization, the number of women reservists plummeted to a total of 298 in August 1946. Simultaneously, there was skepticism at the highest levels in Headquarters Marine Corps as to whether there should be any women reservists on active duty, or, in fact, any women at all in a peacetime Marine Corps. With other branches of the Armed Services retaining women, the Marine Corps finally agreed to a minimum step, it would enlist former women reservists in its Reserve and authorize their formation into Volunteer Training Units (January 1947).

The decisive breakthrough occurred on 12 June 1948 with the passage of the Women's Armed Services Integration Act. This led to women in the Organized Reserve and, for the first time, in the Regular Marine Corps. They were now confirmed as a permanent part of a peacetime Corps, with authorization to seek 100 officers, 10 warrant officers, and 1,000 enlisted women during the next two years.

To take charge of this rebirth, a superb leader was essential. There would be endless problems and details in organizing this latest expansion, as well as a crucial need for a firm but diplomatic style at Headquarters Marine Corps. The right person emerged—Katherine A. Towle. She had been a captain in 1943 and had returned to a college campus in 1946. General Clifton B. Cates, Commandant of the Marine Corps, asked her to return to active duty and take charge. On 18 October 1948, she was sworn in as Director of Women Marines with the rank of colonel. It was she who would lead Women Marines throughout the Korean War. Progress came quickly after that—commissioning of women as regulars, with their title changed from women reservists to Women Marines.

The year 1949 saw a variety of activities, which, unknowingly, prepared women reservists and Women Marines for the wartime demands which would erupt the following year. Enlisted training began at Parris Island, South Carolina. Drawing on the Volunteer Training Units, Organized Women's Reserve platoons were activated, with 13 functioning by February 1950. At Quantico a Woman Officer Candidate Course and a Woman Officer Training Class were instituted, and 86, later reduced to 27, appropriate military occupational specialties were targeted as potential for women. By March 1950, there were 28 Regular Women Marine officers, 496 regular enlisted,

Marine Corps Historical Center Photo Collection

The first post-World War II Women Marines arrived at Quantico, Virginia, in 1950 and were assigned to the Administrative Section of the Landing Force Development Center. Although limited to a minute number of occupational fields, Women Marines numbered more than 2,700 by mid-1953.

18 women reserve officers, and 41 women reserve enlisted on active duty.

A bombshell exploded on 25 June 1950 when South Korea unexpectedly was invaded. Now the trials and tribulations that Marine women had experienced in the past years would bear fruit in a time of crisis. With the brutal strength reductions that had been forced upon it in the preceding years, the Marine Corps suddenly needed all the personnel—hopefully trained—that it could scrape together. And here were the women! Their potential went far beyond those on active duty and out into a reservoir of civilians who would soon be joining up.

The Marine Corps immediately called up its 13 women reserve platoons, with 287 enlisted veterans going straight to active duty and 298 non-veterans being sent to Parris Island. This mobilization was characterized by President Harry S. Truman's 19 July call up of all Reserves for active duty, including the Marine Corps' Organized Reserve and Volunteer Reserve.

For women, the Marine Corps had strict standards: an age limit of 18 to 31 (with 20 as the minimum for a regular); single with no dependents; and a high school diploma or its equivalent. Within a year, the eager response had brought the total number of women in the Corps to 2,065.

The flow of recruits led to a battalion of women at Parris Island and a similar company at Camp Lejeune, North Carolina, and the air base at El Toro, California, by October 1950. There they worked five days, spent a half-day on instruction in military subjects, and finished off after their evening meal with close order drill. It was not surprising that the vast majority of their assignments were clerical and administrative, given the culture of the 1940s and 1950s, when many civilians, and not a few Marines, believed that a woman's place was in the home. This also was evidenced when family and friends would often try to dissuade young women from enlisting. And then there were some Marines who made life difficult for the women who did join up.

This narrow limitation of assignments for women resulted in their being untrained for billets that needed them, as well as strange anomalies such as a woman private first class filling a master sergeant's billet.

Colonel Towle, alert as always, wrote a pithy memorandum to the powers at Headquarters Marine Corps in January 1951. She stated unequivocally that Marine Corps policies for women were "unrealistic and short-sighted, as well as uneconomical." Thus, there was an urgent need, she continued for a "systematic, long-range training [plan]."

As the number of women grew, so did the range of their activities. At their assigned bases, they organized athletic teams and had a variety of off-duty interests. There was also a modest expansion of duties, with some women now in billets for motor transport, recruiting, photography, air traffic control, public affairs, and base newspapers. A few even made it to duty in Italy and Germany.

With the increase in active duty billets, women reservists with minor-age children were released while, at the same time, applications for active duty rose.

In January 1952, women reserve platoons were reestablished and by the end of the year totaled 14. In addition to basic military subjects, they received individualized training in one of five occupational specialties: administration, supply, classification, disbursing, and communications. Besides regular "drill nights" these women reservists had a two-week summer training period.

In spite of these expanded roles for women in the Corps, billets for officers were still limited to 10 occupational specialties in April 1952. Nevertheless, by May, there were Women Marine companies at Cherry Point, Camp Pendleton, Fleet Marine Force, Pacific, Norfolk, San Diego, and Quantico, with Kaneohe Bay added in 1953.

The year 1953 saw the establishment at Camp Lejeune of a Staff Noncommissioned Officer Leadership School for women. Military occupational specialties available to officers were expanded in March. Three more women reserve platoons had been formed by April.

There was a change of command on 1 May 1953. Colonel Towle had reached the statutory age limit of 55 for colonels, and she had to retire. She had been a superb leader, and a Letter of Commendation from the Commandant and an award of a Legion of Merit medal recognized her achievements.

Her successor as Director of Women Marines was Lieutenant Colonel Julia E. Hamblet. She was a graduate of the first training class for women officers in 1943, and was only 37 years old when, after a wide variety of active duty, she was promoted to colonel in her new assignment.

When she took office, the Korean War was winding down. June 1953 saw a total of 160 women officers and 2,502 enlisted. Then, on 27 July, the war was over.

As one author summed up these years:

> The Korean War era witnessed a brief, temporary surge of interest in WMs [Women Marines] on the Corps' part, but it did not result in major, long-term changes in either the women's standing within the Corps or in the duties they were assigned. In the wake of the war their numbers began to decrease, the sense of urgency that surrounded their redux subsided, and WM-related issues were shelved indefinitely. The ambivalence the Corps felt about women in the ranks never really disappeared, even when the war was on the WMs were sorely needed.

Besides these institutional evaluations, there was another vital factor to record, the impact that duty in the Corps had on these women. Their later comments were nearly unanimous: "The best years of my life."

—Captain John C. Chapin, USMCR (Ret)

---

remained in its former sector, although now under control of I Corps.

The tanks on the main line of resistance occupied fixed positions, functioning mostly as artillery; they shifted as necessary to alternate sites to prevent Chinese gunners from zeroing in on them. The Army division provided armored personnel carriers to haul food, water, fuel, and ammunition to the tanks. The presence of Marine armor at the front provided a bonus in addition to firepower, for when Chinese artillery ripped up telephone wires leading to the rear, the tank radios ensured reliable communications.

While in reserve, the 1st Marine Division added depth to the I Corps positions by standing ready to counterattack after occupying any of the back-up lines—like the former Wyoming and Kansas—behind the four divisions manning the main line of resistance: the 25th Infantry Division; the Commonwealth Division; the 1st Republic of Korea Division; and the 7th Infantry Division. In addition, the Marines made improvements to Camp Casey, repaired the

Department of Defense (USMC) A171330

Marines of Company H, 3d Battalion, 5th Marines, set up their tents after moving from the frontlines into corps reserve in the central area of the Casey complex. Although the division had been on line for more than 20 months, there was a reluctance to turn over positions.

went a week of training for an amphibious exercise held at Yongjong-ni, on the west coast near Kunsan. Dense fog forced cancellation of a planned rehearsal that might have highlighted the impact of a shallow beach gradient that grounded landing craft far from shore and complicated the landing of vehicles.

While Colonel Tschirgi's regiment was carrying out this exercise, other elements of the division took part in a combined command post exercise and firing exercise designed to help the division prepare for an Eighth Army exercise, then scheduled for the end of May but later canceled. The 7th Marines landed at the Yongjong-ni training area on 5 June, employing a 144-foot pontoon bridge, supplied by the Army, to speed the movement of vehicles over the shallow-sloping beach. However, the possible need for amphibious back-up positions, especially the former Kansas Line which had been severely damaged by rains and the spring thaw, and embarked on a program of instruction that emphasized night combat and began with the individual and small unit. The larger-unit exercises included the use of helicopters in conjunction with a rifle company, proceeded to regimental landing exercises, and culminated in a field exercise. All of this training, which had to be somewhat curtailed, was to take place before the division returned to the main line of resistance in early July.

Although the spring weather that had damaged the old Kansas Line also turned roads to rivers of mud, the division completed its move to Camp Casey and launched the programs of training and rebuilding. The 5th Marines, the reserve regiment when the division moved off line, under-

Members of the 81mm mortar platoon, Weapons Company, 2d Battalion, 7th Marines, receive refresher instruction on the parts and nomenclature of the 81mm mortar during the eight-week reserve period. Lectures such as this were kept to a minimum with at least 50 percent of the tactical training conducted at night.

Department of Defense (USMC) A171518

A New Commanding General

On 15 June 1953, Major General Randolph McCall Pate took command of the 1st Marine Division, replacing Major General Edwin A. Pollack, who became director of the Marine Corps Educational Center, Quantico, Virginia. A native of South Carolina who spent his boyhood in Virginia, General Pate's military career began in 1918, the final year of World War I, when he enlisted in the Army, serving long enough to qualify for the Victory Medal though he did not fight in France. After graduating from the Virginia Military Institute in June 1921, he accepted in September of that year a commission as a second lieutenant in the Marine Corps Reserve. He received a regular commission in May 1922, afterward carrying out a variety of assignments in the Dominican Republic and China, as well as at posts in the continental United States and Hawaii.

During World War II, General Pate served as supply officer on the staff of the 1st Marine Division at Guadalcanal, where he was wounded, and held other wartime staff positions. After Japan's surrender, he became Director of Reserve at Marine Corps headquarters, served on the Navy's General Board, and held the assignment of Chief of Staff, Marine Corps Schools, Quantico, Virginia. While at Marine Corps Schools, he achieved promotion to brigadier general in 1949 and during the following year took over as Director, Marine Corps Educational Center at Quantico.

After an assignment with the Office of Joint Chiefs of Staff as the Joint Staff's Director for Logistics, General Pate returned to Marine Corps headquarters and in August 1952 was promoted to major general, while serving his second tour directing the Marine Corps Reserve. In September of that year, he assumed command of the 2d Marine Division at Camp Lejeune, North Carolina. After taking over the 1st Marine Division in June 1953, he commanded the unit for the remainder of the Korean fighting, finally turning the division over to Major General Robert H. Pepper in May 1954.

Upon leaving Korea, General Pate became Assistant Commandant of the Marine Corps and Chief of Staff, advancing to the rank of lieutenant general. On 1 January 1956, he received a fourth star and became Commandant of the Marine Corps, succeeding General Lemuel C. Shepherd, Jr. General Pate served as Commandant for four years, during which the Marine Corps moved away from the trenches and bunkers and reestablished itself as a highly mobile amphibious force in readiness. Following four years as Commandant, he retired in December 1959 with the rank of general.

Following a brief illness, General Pate died in 1961 and was interred with full military honors in Arlington National Cemetery. Pate's military colleagues described him as: "A man of deep sincerity and untiring integrity, who is thoughtful and considerate of others—a true gentleman of the old school."

National Archives Photo (USMC) 127-N-A172239

MajGen Randolph McC. Pate, left, relieves MajGen Edwin A. Pollock in command of the 1st Marine Division. The departing Pollock was presented the Distinguished Service Medal for his "outstanding success in the defense of Carson, Vegas, and Elko."

shipping to repatriate prisoners of war in the event of a truce caused the cancellation of a landing exercise scheduled for the 1st Marines between 14 and 23 June.

Marine aircraft participated in the training program conducted while the division formed the I Corps reserve. Fighter-bombers covered the landing exercises, for example, and helicopters of HMR-161 landed at Camp Casey for a practice redeployment of two sections of 4.5-inch rocket launchers from the division's rocket battery, along with infantry from the 5th Marines.

The principal mission of the 1st Marine Aircraft Wing, although it took part in the training program, remained the support of United

Department of Defense Photo (USMC) A169791

While the division was in corps reserve in May and June, the 1st 4.5-inch Rocket Battery supported units manning the main line of resistance, as it had in March when the Marines were fighting to hold Reno, Vegas, and Carson.

Nations ground forces by day and by night. The crews of Marine night fighters, Grumman F7F Tigercats and Douglas F3D Skyknights, learned to take deadly advantage of beams from searchlights along the main line of resistance that illuminated terrain features held by the Chinese, bathing in light those on the near slope and creating an artificial horizon for air strikes against those on the shadowed slope farther from United Nations lines. Artillery, firing time-on-target or variable-time concentrations, blanketed known Chinese antiaircraft batteries within 2,500 yards of a target. At first, this kind of bombardment, usually lasting about three minutes, effectively silenced the hostile gunners, but with repetition the technique declined in effectiveness, serving to alert the Chinese that an air strike was imminent. Besides providing close air support for the ground forces, Marine night fighters escorted Air Force B-29s on missions against North Korea and flew long-range interdiction.

On the Main Line of Resistance

While infantry, service, and some support elements of the 1st Marine Division repaired the positions that added depth to the I Corps front and underwent training to the rear, the Chinese exerted renewed pressure on the portion of the main line of resistance that the Marines had formerly held. The enemy struck first at the outposts now held by the Turkish brigade—Berlin and East Berlin, Carson, Elko, and Vegas. Shortly before 0200 on 15 May, advancing behind deadly mortar and artillery fire, one Chinese battalion attacked Berlin and East Berlin, while another hit the Carson-Elko-Vegas complex. The Turks held their ground, thanks in part to accurate fire from Marine tanks on the main line of resistance, which may have inflicted 300 casualties, and from the 11th Marines and the rocket battery, which fired some 5,500 105mm and 155mm how-

Events at Panmunjom

The fighting along the combat outpost line, which caused the commanding general of I Corps to shift the 1st Marines into position for a possible counterattack, coincided with resolution of the lingering issue of forced repatriation. By the end of May 1953, the United States was insisting that there be no forced repatriation, as had happened in Europe after World War II when American troops handed over refugees and prisoners of war to Soviet authorities, the representatives of a wartime ally. The alliance that defeated Hitler's Germany soon collapsed, and the Soviet Union emerged as America's principal antagonist in a Cold War that burst into flame in Korea. China and the Democratic People's Republic of Korea, at first indifferent to the issue, had come to insist on forced repatriation upon realizing how many of their captured soldiers might prefer to remain in South Korea.

In short, both sides faced possible embarrassment. American acceptance of involuntary repatriation would seem a repetition of the transfers of both captives and civilians to Soviet control in the aftermath of World War II. Widespread refusals to return to China or North Korea would shatter the image of a Communist paradise and lend substance to Western propaganda.

On 25 May, Lieutenant General William K. Harrison of the Army, the senior American delegate at the Panmunjom negotiations, persuaded the Communist side to agree to a closed session. The absence of press coverage, Harrison believed, would prevent posturing designed to influence world opinion and instead focus the attention of the delegates on progress toward a settlement. At this meeting, Harrison declared that there could be no forced repatriation and then offered compromises to make this principle palatable to the Chinese and North Koreans.

The United Nations forces would neither force captured Communist troops to return to their homeland nor simply release them. Harrison proposed instead that soldiers who refused repatriation be turned over, as the Communist delegates desired, to a Neutral Nations Repatriation Committee with representatives of five so-called neutral states—Switzerland and Sweden, with ide-

After a high wind blew down the tents in August 1952, the Communist built a more substantial wooden structure for the armistice meetings at Panmunjom. The white tent on the left is the Communist delegates' tent, while the two dark ones on the right are the United Nations delegates' tent and press tent.

National Archives Photo (USN) 80-G-482368

ological ties to the West; Poland and Czechoslovakia, Soviet satellites at the time; and India, which followed an often erratic political course determined largely by its own self-interest. To minimize friction among the neutral nations—Poland and Czechoslovakia against Sweden and Switzerland—Indian troops would assume custody of all newly released prisoners seeking asylum.

The Communist side, troubled by the prospect of mass defections, insisted on an opportunity to persuade its soldiers who refused repatriation to change their minds and return. The United Nations, however, feared that persuasion would give way to coercion, and, as a result, Harrison's compromise specifically forbade the threat or use of force and limited the time allotted for re-indoctrination to 90 days. Those prisoners who persisted in refusing repatriation would then be turned over to the control of the United Nations General Assembly.

The Communist negotiators demanded certain changes. They called for an additional 30 days for re-education and the elimination of the United Nations General Assembly from the repatriation process, since the international organization, as a belligerent, could not serve as a disinterested guardian of the rights of Chinese or North Korean prisoners. On 4 June, however, the Communist parties accepted the principle of voluntary repatriation, although without formally endorsing it, in return for the additional 30 days of persuasion and the substitution of India's Red Cross organization for the General Assembly as custodian of those prisoners who refused to return to their homelands.

---

witzer shells and 4.5-inch rockets during four hours of fighting. In addition, the 1st Marine Aircraft Wing flew 21 strikes against Chinese positions threatening the Turkish troops.

Neither the successful Turkish resistance nor the continuing talks at Panmunjom—where negotiators were addressing both the issue of repatriation and South Korean concerns that a truce could place their country at a fatal disadvantage—dissuaded the Chinese from intensifying their offensive. On 25 May, despite signs of real progress in the truce talks, Chinese gunners resumed pounding the Turkish brigade, and three days later, the enemy launched a series of attacks along much of the combat outpost line held by I Corps. At 1800 on 28 May, the enemy's *120th Division* launched simultaneous attacks against Combat Outpost 2, to the east of the Panmunjom Corridor, on Carson, Elko, and Reno, Berlin and East Berlin, and against the Hook and its two outposts, Ronson and Warsaw, these last three held by the British 1st Commonwealth Division.

The major thrusts by the Chinese division sent a battalion, advancing behind a mortar and artillery barrage, against Outposts Carson and Elko, while another battalion took advantage of a smoke screen to storm Vegas, and a third menaced Berlin and East Berlin. At the same time, still other Chinese troops attacked Combat Outpost 2, on the left of the 25th Infantry Division's sector, and the Hook, along with Outposts Ronson and Warsaw, on the right of the I Corps line.

The 11th Marines, commanded by Colonel James E. Mills—who had taken over from Colonel Harry N. Shea on 22 February 1953—fired some 9,500 rounds during the night of 28 May, and Marine aircraft flew eight strikes against Chinese artillery positions. The Marine division's 4.5-inch rocket battery, now firing from the vicinity of the Hook, supported the Commonwealth Division. Marine tanks, dug-in along the main line of resistance, provided deadly fire from prepared positions, especially in the sector of the Turkish brigade.

The deluge of shells and rockets from Marine weapons helped the 35th Infantry Regiment hold Combat Outpost 2 and the Commonwealth Division to maintain its grip on the Hook, Warsaw, and Ronson. The Chinese, however, succeeded in capturing Carson, although Turkish troops clung stubbornly to Elko and Vegas. The threat to Berlin and East Berlin abated swiftly, suggesting that the enemy launched the probe to divert attention from the adjacent Carson-Elko-Vegas complex.

Marine air, artillery, and armor continued to support I Corps as the fighting entered a second day. Attack planes and fighter-bombers hit Chinese troops, artillery positions, and supply points by day and night. Additional tanks moved into prepared positions on the main line of resistance until 33 of them supported the Turkish brigade. After Chinese counterbattery fire silenced six Turkish howitzers, the 2d Battalion, 11th Marines, took over the direct-fire mission. By dusk on 29 May, the total rounds fired by Marine howitzers and rocket launchers in defense of the outposts exceeded 40,000.

This intense fire, and the tenacity of the Turkish soldiers who suffered 150 killed and 245 wounded, could not save the Carson-Elko-Vegas complex. By mid-day, Lieutenant General Bruce C. Clarke, the corps commander, and General Williams, in command of the 25th Infantry Division, decided to withdraw from these outposts at least temporarily. Carson had fallen to the Chinese on the previous day, and fewer than 40 survivors continued to resist on Vegas. Since

National Archives Photo (USMC) 127-N-A172204

On 29 May, Col Wallace M. Nelson's 1st Marines move up on foot in full gear to set up a hasty blocking position on the Kansas Line in case of a breakthrough by Communist forces during their attack on 25th Infantry Division outposts.

Elko could not survive if the enemy held both Vegas and Carson, the Turkish troops withdrew to the main line of resistance. By the time the withdrawal took place, the attacking Chinese had lost perhaps 3,000 men but showed no signs of breaking off the action.

In the meantime, General Clarke had ordered the 1st Marines to prepare for a possible counterattack. The regiment's infantry battalions and antitank and heavy mortar companies moved into position just south of the refurbished defenses of the old Kansas Line. In addition, the 1st Marine Division's reconnaissance company came under control of the 25th Infantry Division, replacing a company of that division's 14th Infantry in reserve along the east bank of the Imjin River.

The projected counterattack never occurred. Torrential rains on 30 May frustrated any ambitions the enemy may have had for trying to crack the main line of resistance. Moreover, a truce agreement that seemed on the verge of acceptance dissuaded the United Nations Command from making the obviously costly effort to recapture Carson, Vegas, and Elko.

The Truce in Jeopardy

After the bitter fighting on the outpost line at the end of May, a lull ensued on the I Corps front. The Chinese, however, did not abandon their attempt to improve their military position before a truce should go into effect and lunged instead at the South Korean II Corps, opposite the enemy-held town of Kumsong, attacking on 10 June and in six days of fighting forcing the defenders to pull back some 4,000 yards. A second blow drove the South Korean 20th Division back from the northern rim of the Punchbowl. By the time the fighting died down after 18 June, the South Koreans had suffered some 7,300 casualties, perhaps 600 more than the Chinese attackers. The enemy, moreover, had pushed the South Koreans back as far as 4,000 yards on segments of the frontlines totaling about 15,000 yards in width. The offensive of mid-June proved the most successful Chinese thrust since April and May 1951 when the enemy penetrated as deeply as 30 miles at some points along the United Nations line.

Even as the negotiators at Panmunjom were resolving the outstanding issues concerning the repatriation of prisoners, Syngman Rhee, the president of the Republic of Korea, became increasingly concerned about the impact of a ceasefire on the survival of his nation. His dream of one Korea unified under the government at Seoul was rapidly vanishing. He had little direct leverage on the Chinese government; his only hope—a slim one, indeed—was somehow to prod his war-weary American ally into turning its back on the Panmunjom negotiations and exerting renewed pressure on the battlefield. To achieve this unlikely end, President Rhee threatened to pull South Korean forces out of the United Nations Command. The threat, however unrealistic, brought a formal American offer to build up the South Korean armed forces and restore the nation's shattered economy, provided that the Seoul government accepted the settlement that was taking shape. Despite his weak bargaining position, the South Korean president ignored the offer and demanded the removal of Chinese forces from the Korean peninsula, along with a formal military alliance with the United States, the recently offered program of military and economic aid, and the stationing of American air and naval forces in South Korea.

When progress continued

The Offshore Islands

American Marines continued to provide essential help in defending the islands off the east and west coasts of the Korean peninsula that were garrisoned by their South Korean counterparts. The two defensive organizations, formerly the East Coast and West Coast Island Defense Commands, were redesignated task units effective 1 January 1953. The 2d Korean Marine Corps Regiment furnished the troops that manned these outposts under the direction of U.S. Marines.

Two battalions of South Korean Marines manned the West Coast Island Defense Task Unit, while the U.S. Marine Corps contributed some 17 officers and 100 enlisted men. The western task unit manned outposts on six islands in close proximity to the 38th Parallel, which had separated prewar South Korea from the Communist North. These islands, from north to south, were Sok-to, Cho-do, Paengyong-do, Taechong-do, Yongpyong-do, and Tokchok-to. An earlier attempt to occupy tiny Ho-do had ended in failure when the Chinese overwhelmed the garrison. The six surviving outposts served as bases for artillery, intelligence collection, the direction of guerrilla operations in North Korea, and training. At some of the

National Archives Photo (USN) 80-G-480993

Two members of the 1st Air and Naval Gunfire Liaison Company spot and correct gunfire laid down against Communist troops by the Dutch frigate Johann Maurits van Nassau.

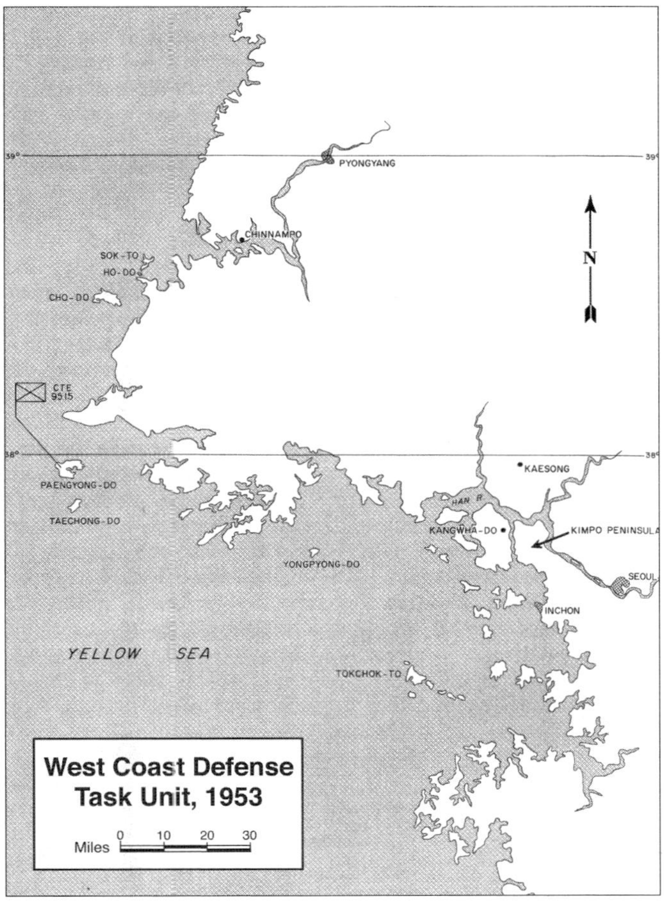

outposts, the South Koreans used their old-model radios to call down fire from artillery or naval guns against hostile batteries on the mainland. American Marines directed all this activity, although the language barrier hampered their role in training.

Late in 1952, Communist artillery batteries on the peninsula opened fire more frequently against United Nations warships as well as the western island outposts. For counterbattery missions from Cho-do, two 90mm guns supporting a force of South Korean guerrillas were shifted there from the island of Kanghwa-do at the mouth of the Han River. Enemy light aircraft contributed to the increased pressure on the islands, conducting nuisance raids against Cho-do, bombed as recently as October 1952, Sok-to, and Paengyong-do.

Besides using the two artillery pieces available at Cho-do, the guerrillas provided valuable information on enemy activity. The South Koreans reported an increasing number of Chinese junks offshore and identified major military units on the peninsula itself. As the threat of an assault against one or more of the western islands intensified, counterbattery fire from the outposts kept pace, only diminishing when the threat ebbed.

From time to time, American or British pilots, low on fuel or flying damaged aircraft, landed on the beach at Paengyong-do. Unfortunately, obtaining the needed fuel or aircraft parts proved difficult because of tangled and unresponsive lines of supply. Logistics difficulties also posed a threat to the large number of refugees that had found asylum during 1952 on the islands protected by the western task unit. As winter approached, concern mounted that they could not be fed by air or sea and would not survive the cold, but two supply-laden tank landing ships arrived before the seasonal storms began and eased the crisis.

By the spring of 1953, improving prospects for a ceasefire generally along the 38th Parallel raised doubts about the future control of two west-coast islands, Cho-do and Sok-to, which lay north of the demarcation line. Colonel Harry N. Shea, the task unit commander and his successor, Colonel Alexander B. Swenceski, carried out Operation Pandora, a plan for the evacuation of both islands. Artillery on Sok-to and Cho-do—each of which now had its own pair of 90mm guns—pounded enemy batteries along the coast, and warships joined in, including the battleship *New Jersey* (BB 62) with its 16-inch guns. Despite the avalanche of American firepower, the enemy stuck to his guns, firing some 1,800 rounds in June alone. By the end of June, Operation Pandora began the successful withdrawal first of the guerrillas and their families and then of the two island garrisons.

Whereas only two of the islands manned by the western task unit were located north of the 38th Parallel, the most important of those garrisoned by the East Coast Island Defense Unit—Tae-do, So-do, Sin-do, Mod-do, Ung-do, Hwangto-do, and Yo-do—lay just off the North Korean port of Wonsan, more than 100 miles north of the demarcation between the two Koreas. The unit also bore responsibility for two other potentially vulnerable outposts: Yang-do, off the town of Songjin, 150 miles northeast of Wonsan; and Nan-do, near the North Korean town of Kojo, some 40 miles south of Wonsan. A total of 1,270 South Korean Marines, 35 U.S. Marines, and 15 American sailors manned the defense unit. The individual garrisons varied from 300 at Yo-do, the largest of the island outposts and the site of unit headquarters and an airstrip, to the compact naval-gunfire spotting teams on the smallest of the islets off Wonsan.

Like the western islands, those off the east coast—especially the ones nearest Wonsan—endured frequent shelling from batteries on the peninsula. Because Wonsan had been a major port, the ability of the United Nations forces to blockade it served as an affront to the Chinese and North Koreans, as well as a disruption of their coastal supply line. Besides hammering the offshore outposts, hostile gunners fired upon United Nations warships, as they had off the opposite coast. At Wonsan, however, the enemy employed a more efficient fire-control technique than in the west. When United Nations gun crews on the eastern islands, or warships nearby, fired counterbattery missions against an enemy weapon, it fell silent and other artillery pieces took over the fire mission.

When winter gripped the east coast islands, temperatures dropped to 10 degrees below zero, Fahrenheit, and high winds disrupted the movement of supply ships. The Yo-do garrison, for example, survived on canned rations for a week, and for several days the defenders of Hwangto-do drank water from melted snow. When the weather improved, so did the determination of Chinese artillerymen, who fired more than 1,050 rounds against the island outposts in April 1953 and twice that number against blockading warships. In addition, United Nations sailors sighted 37 floating magnetic mines that month, although the devices caused no damage. This was the greatest number of sightings in any month since the previous summer, when mines sank the tug *Sarsi* (AT 111) off Hungnam and damaged the destroyer *Barton* (DD 722) off Wonsan.

As pressure mounted against the islands near Wonsan, Fleet Marine Force, Pacific, increased the size of its contingents at these outposts by some 40 percent, adding nine Marine officers and 44 enlisted Marines and sailors. A new unit commander arrived in the spring of 1953. After a year in charge, Lieutenant Colonel Robert D. Heinl, Jr., entrusted the organization to Lieutenant Colonel Hoyt U. Bookhart, Jr.

Scarcely had this change of command taken place, when preparations began for a withdrawal from the eastern islands after the signing of an armistice. On 11 June, the unit began retrenching by removing South Korean villagers, along with guerrillas and their families, from Yang-do, the farthest north and least defensible of the outposts, and withdrawing noncombatants from Yo-do, the site of the unit headquarters. Withdrawal of the other east coast outposts awaited the signing of the armistice.

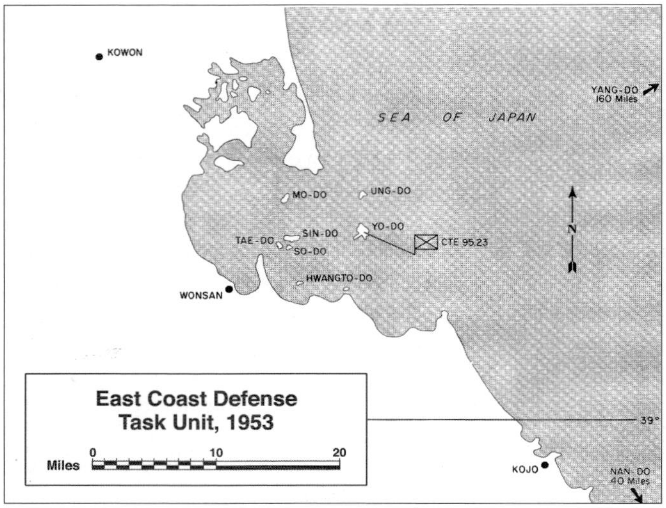

East Coast Defense Task Unit, 1953

toward a truce that he considered disastrous, President Rhee sought to disrupt the negotiations. On 18 June, he ordered the release of some 27,000 Communist prisoners who had refused repatriation, hoping to undo what the negotiators had accomplished in resolving the issue of repatriation. His gambit failed. The People's Republic of China launched additional limited attacks, perhaps in retaliation, but the government in Peiping had grown too weary of the long and bloody war to mount an all-out offensive. Similarly, the United States remained determined to end the Korean fighting, regardless of Rhee's objections. As an American special diplomatic mission made clear, South Korea could expect no more than a security treaty with the United States and a combination of military and economic aid that would include the presence of American troops.

On 24 June, six days after President Rhee tried to sabotage the truce negotiations, Chinese troops attacked the sector held by the South Korean II Corps, focusing on the South Korean 9th Division, which blunted the thrust after the loss of an outpost. On the following day, the enemy hit the South Korean 1st Division, to the right of the Commonwealth Division dug in at the eastern boundary of the sector held by General Clarke's I Corps. Despite fierce resistance, the Chinese overwhelmed three outposts manned by the South Korean 1st Division. Efforts to regain the lost ground halted on the 29th when General Clarke ended a series of gallant but unsuccessful counterattacks, repeating the decision made a month earlier during the fighting for Outposts Carson and Vegas elsewhere on the I Corps front that the cost in lives would be prohibitive.

As it had during the battles of late May, 1st Marine Division's the 4.5-inch rocket battery deployed eastward at the end of June to support the hard-pressed South Koreans dug in to the right of the I Corps line. The battery shifted about 20 miles closer to the action and fired some 25 missions that helped prevent a Chinese breakthrough. So grave was the threat on the right that General Clarke alerted the 7th Marines to stand by to reinforce the South Korean 1st Division, even though Eighth Army policy forbade American units from serving under South Korean command. The crisis abated, however, and the alert of the Marine regiment was canceled within 24 hours. The 1st Regimental Combat Team (less one battalion) of the Korean Marine Corps took over from the 7th Marines and relieved bloodied elements of the South Korean 1st Division.

Marine aviation took part in the fighting during June. Between the 14th and the 17th, when the Chinese made their greatest gains since 1951, American aircraft flew 8,359 sorties, 1,156 of them by Marines. The air war again intensified when the Chinese renewed their attacks. On the last day of June, for example, Marine airmen flew 301 sorties, which included 28 percent of the day's close air support and 24 percent of the interdiction missions.

Marines Return to the Bunkers

The abandonment of the least defensible of the offshore outposts took place as President Rhee was attempting to wrest stronger guarantees of future aid from his American ally, and time was approaching for the return of the 1st Marine Division to the frontline. The release of some 27,000 captured North Korean or Chinese prisoners of war, who sought to remain in the Republic of Korea, and escape attempts by still others,

A .50-caliber machine gun covers the field of fire across the Han River on the Kimpo Peninsula. Marines firing the weapon were members of the 1st Amphibian Tractor Battalion, which manned bunkers overlooking the river, and in mid-June had to deploy Company A to Ascom City to maintain order among the Chinese and North Korean prisoners.

Department of Defense Photo (USMC) A170806

had an impact on the Marines in I Corps reserve. The 1st Amphibian Tractor Battalion, for example, had to deploy one company to a prisoner of war compound at Ascom City to maintain order after escape attempts there.

By releasing the prisoners of war, the South Korean president caused a temporary suspension of the truce negotiations and persuaded the United States to agree formally to a mutual defense arrangement that maintained American forces in South Korea and provided economic and military aid. Since both the People's Republic of China and the United States were ready to end the fighting, the talks resumed at Panmunjom, where the delegations tried to obtain whatever advantage they could, although without sabotaging prospects for a ceasefire.

During the latter part of June, the 1st Marine Division completed its training and prepared to resume bunker warfare. The major elements that had been in reserve returned to the segment of the main line of resistance that the division formerly held, relieving the Army's 25th Infantry Division. By dawn 7 July, the 7th Marines took over on the right of the division's line and the 5th Marines on the left, while the 1st Marines formed the division reserve, protecting the bridges across the Imjin River and the Marine radar used to direct air strikes.

The tactical situation had changed for the worse since the Marines last occupied this sector. Chinese troops now controlled the three outposts—Carson, Vegas, and Elko—that blocked the best approach to Combat Outposts Berlin and East Berlin, now being taken over by the 7th Marines. As General Pate, the division's commander, immediately realized: "The loss of Outpost Vegas...placed Berlin and East Berlin in very precarious positions and negated their being supported by ground fire except from the MLR."

The Chinese tried to take advantage of any confusion resulting from the relief of the Army division by General Pate's Marines. On the evening of 7 July, Chinese mortars opened fire upon Outposts Berlin and East Berlin and the nearby portion of the main line of resistance, the area that Lieutenant Colonel Alexander Cereghino's 2d Battalion, 7th Marines, was taking over from Turkish soldiers attached to the 25th Infantry Division. By midnight, assault troops from the Chinese *407th Regiment, 136th Division,* advanced from the vicinity of Hill 190, a frequently used staging area, then moved along the ridgeline broken by Carson, Reno, and Vegas—all of them now in enemy hands—and pounced on Berlin and East Berlin.

At Berlin, Turkish soldiers stayed in place for a time after the Marines arrived, and a patrol dispatched from Lieutenant Colonel Cereghino's battalion to set up an ambush got no farther than the outpost when the enemy struck. The remaining Turks and the newly arrived patrol reinforced the Marines manning Berlin. Higher headquarters soon lost contact with both Berlin and the other embattled outpost.

Because Berlin and East Berlin lay no more than 325 yards from the main line of resistance, their capture could provide the enemy with a springboard for an attack designed to shatter the main defenses. As a result, Cereghino organized a provisional platoon from members of his battalion headquarters and sent the unit to reinforce the main line of resistance. Elements of Companies H and I, 3d Battalion, 7th Marines, came under Cereghino's operational control and prepared to counterattack if the enemy should break through.

Meanwhile, the mixed force of Marines and Turkish soldiers succeeded in clinging to Outpost Berlin. The Marines at East Berlin succumbed, however, to an overwhelming force that surged up a steep slope and seized the main trench despite stubborn resistance from the outpost itself and accurate fire from the main line of resistance and beyond. Supporting machine guns, mortars, and artillery—deadly though they were—could not save East Berlin.

A squad from Company F, 2d Battalion, 7th Marines, unsuccessfully counterattacked Outpost East Berlin at 0415 on 8 July, dispensing with the usual artillery barrage in the hope of achieving surprise. A second force of Marines from Company F moved out at about 0440 to reinforce the squad already committed to attacking East Berlin. Chinese artillery caught the reinforcements in the open and wounded 15 Marines, but the attempted counterattack continued for another hour until the men of Company F received orders to fall back so the 11th Marines could fire a time-on-target concentration against the enemy-held outpost.

The Chinese who overran East Berlin had advanced by way of Reno and Vegas, where additional forces were now gathering to exploit this early success. The 1st 4.5-inch Rocket Battery hammered the assembly areas and also the Chinese attacking Berlin and consolidating their hold on East Berlin. Artillerymen of the 2d Battalion, 11th Marines, fired a time-on-target concentration that shattered an enemy company as it was organizing on captured

Outpost Vegas to continue the attack. During the early morning's fighting, all four battalions of the 11th Marines took part in the firing, along with the seven U.S. Army and Turkish artillery battalions still in the area until the relief of the 25th Infantry Division was completed. These weapons matched their Chinese counterparts almost round for round, and Army and Marine Corps tanks joined the rocket battery and artillery battalions in battering the Chinese.

Despite the savage fighting that erupted at Outposts Berlin and East Berlin on the night of 7-8 July, the Marines succeeded in taking over from the 25th Infantry Division and its Turkish component. The 5th Marines and 1st Battalion, 7th Marines, effected a smooth transition, but Lieutenant Colonel Cereghino's 2d Battalion, 7th Marines, had to fight at Berlin and East Berlin. Not until 0630 did Cereghino obtain confirmation that East Berlin had fallen, and shortly afterward he learned that Berlin, some 300 yards west of the captured outpost, still survived. He promptly reinforced Berlin insofar as its compact size permitted, dispatching 18 additional Marines and roughly doubling the number of the outpost's American and Turkish defenders.

To recapture East Berlin would require a strong force of infantry supported by intense fire from mortars, tanks, and artillery. At 1000, taking advantage of an artillery and mortar barrage totaling perhaps 1,600 rounds, a reinforced platoon from Company G, 3d Battalion, 7th Marines, and another from that battalion's Company H—both companies now under Cereghino's operational control—launched the counterattack. The platoon from Company H led the way but encountered an accurate Chinese mortar barrage that pinned the Marines against the barbed wire protecting the main line of resistance and in 15 minutes reduced the force to about 20 men able to fight.

The platoon from Company G advanced through the battered unit and pressed home the counterattack. Shells fired by tanks and artillery exploded immediately in front of the infantrymen, enabling the assault force to reach the main trench on East Berlin and use grenades and small arms fire to kill, capture, or drive off the Chinese. At 1233, the platoon from

Pilots of Marine Fighter Squadron 311, from left, Capt John A. Ritchie, Capt Lenhrew E. Lovette, 1stLt Marvin E. Day, go over their upcoming combat mission step by step. Each must know exactly what to do.

National Archives Photo (USMC) 127-N-A346707

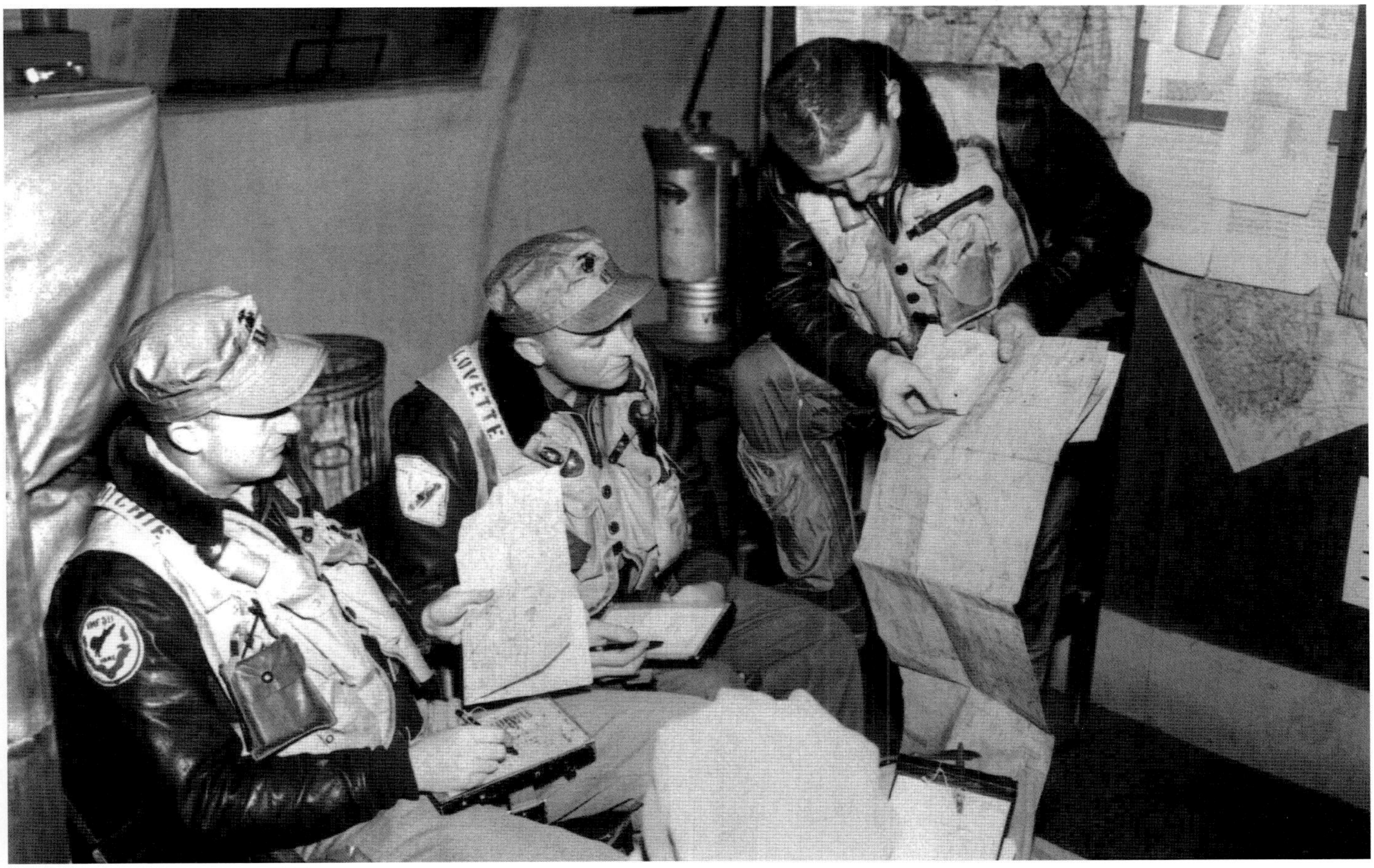

National Archives Photo (USMC) 127-N-A171052

Whether on the frontlines or in reserve Marines were paid monthly in Military Pay Certificates, a substitute for U.S. currency used to curb black market activities. With little to spend it on, most Marines saved their money or had allotments taken out.

Company I, reduced to fewer than two-dozen effectives, regained control of the outpost. Another platoon from the same company immediately moved forward to reinforce the survivors.

Throughout the fighting at Berlin and East Berlin, storms had disrupted the movement of supplies by sending the Imjin out of its banks and destroying a bridge. The bad weather also created mud that hampered movement on the battlefield and brought clouds that reduced visibility from the cockpits of supporting aircraft. At about noon on 8 July, however, four Marine F9Fs took advantage of ground-based Marine radar to attack targets a safe distance from East Berlin. Led by the commander officer of Marine Fighter Squadron 311, Lieutenant Colonel Bernard McShane, the jets dropped five tons of bombs on bunkers and troop concentrations.

The recapture of East Berlin enabled the last of the Turkish troops to withdraw, in effect completing the relief of the 25th Infantry Division. The 11th Marines, commanded after 5 July by Colonel Manly L. Curry, resumed its normal mission of direct support of the 1st Marine Division, as did the 1st Tank Battalion. By 13 July, the company of amphibian tractors that had been guarding prisoners of war at Ascom City rejoined the battalion under the control of the division. The Reconnaissance Company, the Kimpo Provisional Regiment, and the Korean Marine Corps regiment also returned to the division's control.

Once it returned to the main line of resistance, the 1st Marine Division again assumed operational control over VMO-6 and HMR-161. Cloud cover at first impeded the aerial observers from VMO-6, but they successfully directed four artillery fire missions on 8 July against targets behind enemy lines. On 10 July, Marine helicopters from HMR-161 delivered some 1,200 pounds of rations, water, and other cargo to Marine outposts.

Ten Days of Patrols

Regaining Outpost East Berlin on 8 July, which coincided with the resumption of truce negotiations at Panmunjom, did not end the Chinese pressure on the Marines. After dark on the 8th, Colonel Glenn C. Funk, who had assumed command of the 7th Marines on 27 March, moved a platoon from the regiment's 3d Battalion and four M-46 tanks into position to strengthen the main line of resistance. The tanks had just arrived at Hill 126, an outcropping just to the rear of the battle line, when the Marines heard the sound of trucks from beyond Chinese lines. From the hilltop, the M-46s directed 90mm fire against known Chinese positions, and the noise of truck motors ended. Chinese troops, who meanwhile had advanced from the assembly area on Vegas, probed Outpost Berlin and struck a stronger blow against East Berlin. Fighting raged for almost two hours before fire from mortars, artillery, and tanks forced the enemy to break off the action at about 0315 on the morning of 9 July.

After Lieutenant Colonel Cereghino's Marines ended this latest threat to East Berlin, the Chinese remained content to jab at the division rather than try for a knockout. Entire days might pass during which Marine aerial or ground observers and patrols saw few, if any, signs of Chinese. The enemy seemed to be improving his tunnels and bunkers instead of venturing out of them to mount an attack. The Marines still underwent sporadic shelling, but the bombardments did not approach in ferocity those of 8 and 9 July.

Mines for a time proved deadlier than artillery and mortars, as on 12 July when these weapons killed four Marines and wounded eight. At least one minefield contained a new type of Russian-designed weapon that could be detonated by pressure or with a trip wire. Most of the fields employed mines familiar to the Marines, types that may have been newly planted or perhaps had lain dormant under the frozen ground and become deadly when the weather grew warmer and the earth softer.

Although the enemy did not attack on the scale of 7-8 July, Chinese patrols repeatedly clashed all along the division's front with those sent out by the Marines. On the night of 12 July, for example, a 13-man patrol from the 5th Marines encountered a force of Chinese near Outpost Esther, and a combat patrol from the 7th Marines, looking for the enemy near Elko, engaged in an 18-minute firefight.

As the frequency of patrol actions increased, flooding again interfered with the supply effort. On the night of 14-15 July, the Imjin River reached a maximum depth of 26 feet. Only the solidly-built Freedom Bridge, carrying the road to Panmunjom across the swollen stream, could be used until the water subsided.

On the night of 16-17 July, patrols from the 5th Marines engaged in two firefights, suffering no casualties in the first, near Outpost Hedy, while killing three Chinese and wounding one. The regiment's second patrol of the night ran into an ambush near Hill 90. The Chinese proved more aggressive than in recent days, pinning down the patrol and unleashing a flurry of mortar and artillery fire that wounded every member of a unit sent to help break the ambush. Another group of reinforcements succeeded, however, in reaching the embattled patrol. After two hours of fighting and several attempts to isolate and capture individual Marines, the Chinese withdrew, having suffered 22 killed and wounded. When seven Marines failed to return to the main line of resistance, a platoon from the 5th Marines searched the battle site and recovered six bodies.

The third firefight of the night erupted just after midnight in the sector of the 7th Marines, when a 30-man patrol from Company A, 1st Battalion, was ambushed after it passed through a gate in the barbed wire northwest of Outpost Ava. Between 40 and 50 Chinese, supported by mortars, opened fire with grenades and small arms. After a 15-minute exchange of fire in which as many as 18 Chinese may have been killed or wounded, the ambush party vanished into the darkness. As the Marines from Company A returned through the gate, a head count revealed four men missing. A recovery squad crossed and recrossed the area until dawn drew near but found only three bodies. One Marine from Company A remained missing; three had been killed and 21 wounded.

The actions near Outpost Elko and in front of the Ava gate lent credence to Chinese propaganda. Since the 1st Marine Division returned to the main line of resistance, Chinese loudspeakers had

The never tiring doctors and corpsmen treat the wounded. At the forward aid stations patients are examined and their wounds dressed; few are discharged and most prepared for further evacuation.

National Archives Photo (USMC) 127-N-A173337

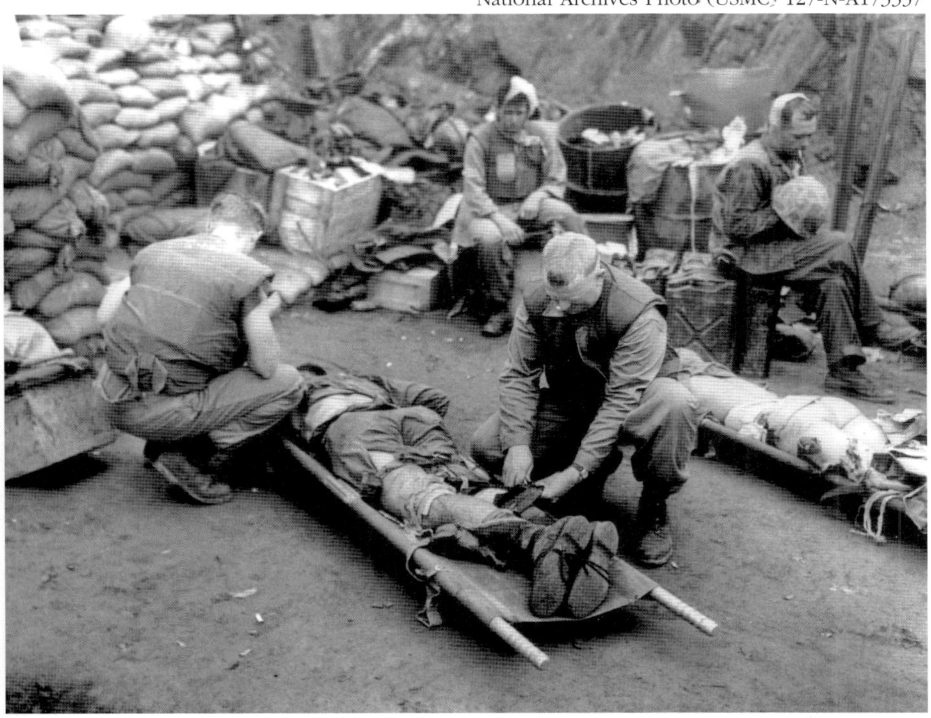

gone beyond the usual appeals to surrender, on at least one occasion warning of the fatal consequences of going on nighttime patrols. This threat, however, probably reflected a Chinese policy of maintaining overall military pressure after the resumption of truce talks rather than a specific effort to demoralize the Marines.

Whatever the purpose of the enemy's propaganda, the Marine patrols continued. On the night after the ambush of Company A, 1st Battalion, 7th Marines, a combat patrol from the regiment's Com-pany C advanced as far as the Ungok hills to silence a machine gun that had been harassing the main line of resistance and, after a successful 20-minute firefight, left a Marine Corps recruiting poster to mark the point of farthest advance. Meanwhile, the Korean Marines had four patrol contacts with the enemy, none lasting more than a few minutes.

The combat outposts like the Berlins, Esther, and Ava had become increasingly vulnerable. By mid-July, General Pate directed his staff to study the possibility of the 1st Marine Division's shifting from a linear defense—the continuous main line of resistance and the network of outlying combat outposts in front of it—to a system of mutually supporting defensive strongpoints that would result in greater depth and density. The Chinese attacks of 7 and 8 July on Berlin and East Berlin served as a catalyst for the study that General Pate launched. As the I Corps commander, General Clarke, later explained, these actions demonstrated that American minefields and barbed wire entanglements had channeled movement between the main line of resistance and the combat outpost lines into comparatively few routes that had become dangerously familiar to the enemy. As a result, Chinese mortars and artillery could savage the troops using these well-worn tracks to reinforce an embattled outpost, withdraw from one that had been overwhelmed, or counterattack to regain a lost position. Indeed, General Maxwell D. Taylor, in command of the Eighth Army since February 1953, agreed that the enemy could, if he chose to pay the price in blood and effort, overrun any of the existing outposts, and endorsed the concept that General Pate's staff was studying. The change in tactics, however, had not yet gone into effect when the Chinese next attacked the Marine positions, but the new assault forced 7th Marines to adopt, in a modified form, the principles of depth and density that the division commander was suggesting.

When the enemy again attacked, a ceasefire seemed imminent. President Rhee agreed on 11 July to accept American assurances of future support and enter into a truce. By the 19th, the negotiators at Panmunjom seemed to have resolved the last of the major issues. On this very date, however, the Chinese struck.

The Fighting Intensifies

Heavy downpours hampered frontline combat and grounded the 1st Marine Aircraft Wing for a total of 12 days early in July. Rain fell on 22 days that month, but the wing nevertheless reported 2,668 combat sorties, more than half of them flown in close support all along the United Nations line. The airmen supported their fellow Marines on the ground with some 250 missions, four-fifths of them using ground-based radar by night or day.

The weather improved after mid-month, enabling aerial activity to increase at a critical moment, for on the night of 19-20 July, the Chinese again assaulted Combat Outposts Berlin and East Berlin—now manned by the 3d Battalion, 7th Marines, which had relieved Lieutenant Colonel Cereghino's 2d Battalion—and also menaced Outposts Dagmar and Ingrid, held by elements of the 5th Marines. The positions of the 5th Marines held firm, thanks in part to accurate fire from the 11th Marines, but Berlin and East Berlin were in peril almost from the outset.

After a savage bombardment of both Berlins and nearby segments of the main line of resistance, Chinese troops at 2230 on the night of the 19th stormed the ridgeline where the two outposts were located, attacking East Berlin first and Berlin immediately afterward. Company I, 3d Battalion, 7th Marines, commanded by First Lieutenant Kenneth E. Turner, garrisoned both outposts, posting 37 Marines at East Berlin and 44 at Berlin. Mortars, machine guns, howitzers, and 90mm tank guns blasted the advancing Chinese in support of Company I. Despite the firepower massed against him, the enemy overran both outposts within three hours.

A duel between American and Chinese gunners continued after the fall of the two Berlins. The enemy fired some 3,000 rounds while overwhelming the outposts and trying to neutralize the nearby main line of resistance and the artillery batteries behind it. One Turkish and two Army artillery battalions joined three battalions of the 11th Marines—two of 105mm and one of 155mm howitzers—in responding to the Chinese bombardment, battering the assault force, its supporting mortars and howitzers, and the assembly areas used by reinforcements in exploiting the early suc-

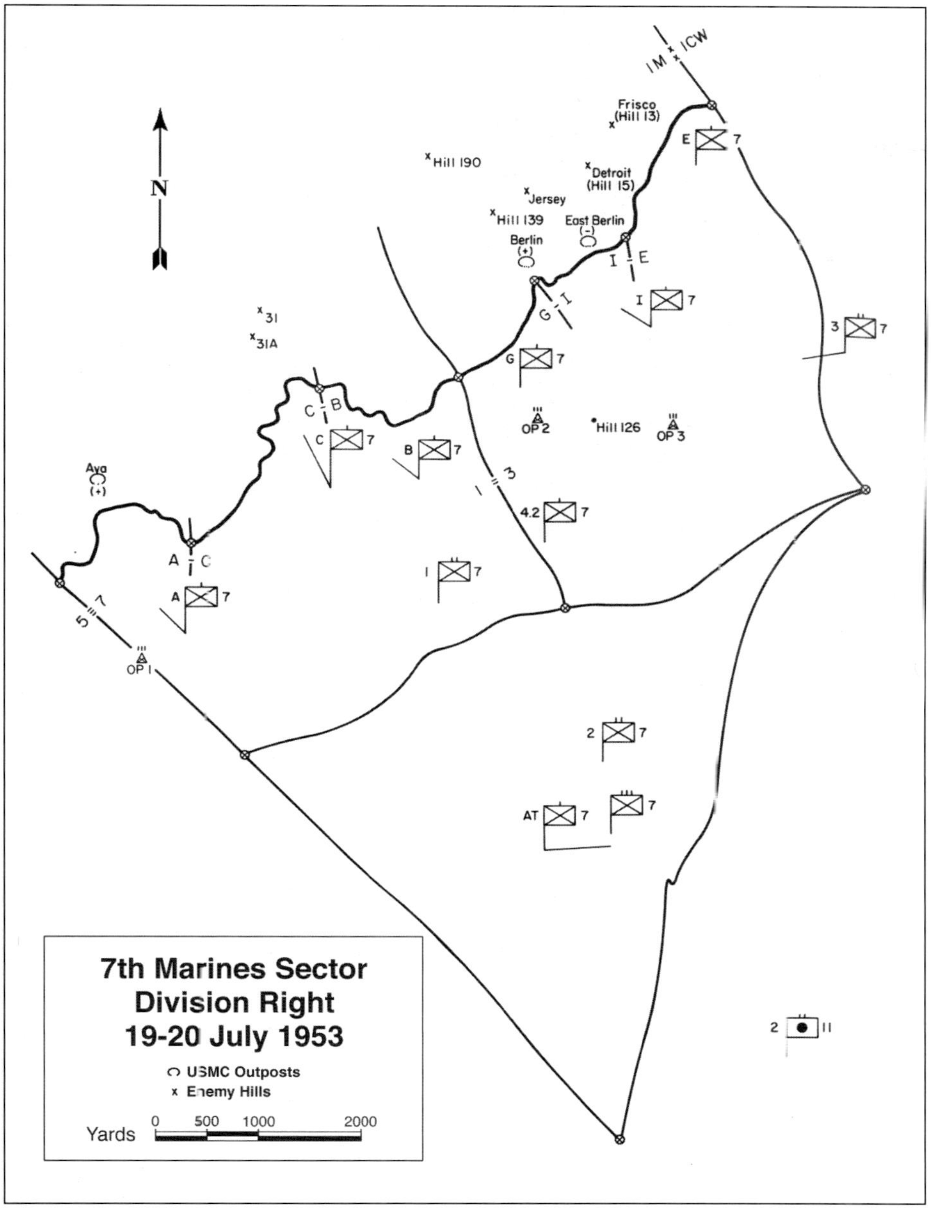

cess. Barrage and counterbarrage continued into the morning of 20 July; at 0520, for example, Chinese shells were exploding at the rate of one per second on the main line of resistance immediately behind Outposts Berlin and East Berlin.

Meanwhile, at 0400 Lieutenant Colonel Paul M. Jones, in command of the 3d Battalion, 7th Marines, alerted Companies D and E of the regiment's 2d Battalion, already under his operational control, to counterattack Berlin and East Berlin at 0730. Half an hour before the scheduled time, Jones received word to cancel the counterattack. Rather than restore the outpost line, General Pate shifted elements of the division reserve, the 1st Marines, to strengthen the main line of resistance in the event the enemy should try to exploit his capture of the two Berlins.

While Colonel Wallace Nelson's 1st Marines reinforced the main line of resistance, air power and artillery tried to neutralize the outposts the Chinese had captured. Since a ceasefire seemed only days away and any attempt to regain the lost ground would result in severe Marine casualties, there would be no counterattack to restore a position that seemed almost certain to be abandoned when a demilitarized zone took shape after the end of hostilities. Instead, air strikes and fire from tanks and artillery scourged the lost outposts to prevent Chinese from using them to mount an assault on the main defenses. Especially effective were attacks by Marine airmen against Berlin and East Berlin and bombardment by Army 8-inch and 240mm howitzers, adjusted by Marine aerial observers, which shattered bunkers and collapsed almost all the trenches on both enemy-held outposts.

Colonel Jones' 3d Battalion, 7th Marines, estimated that the deadly fighting on 19-20 July had killed perhaps 75 Chinese and wounded as many as 300, thus crippling an enemy battalion that had to be replaced by a fresh unit. The 7th Marines and attached units lost six killed, 118 wounded, and 56 missing, but 12 of the missing men survived as prisoners of war and returned in the general exchange when the fighting ended.

Once the enemy captured Berlin and East Berlin, the critical terrain feature on the right of the sector held by the 1st Marine Division became Hill 119, nicknamed Boulder City, the segment of the main line of resistance nearest the two lost outposts and therefore the likely objective of any deeper Chinese thrust. Company D, 2d Battalion, 7th Marines, (attached to the regiment's 3d Battalion) held Boulder City itself. Company E of the 2d Battalion, 7th Marines, (also attached to the 3d Battalion) joined Companies H and I of the 3d Battalion in defending the high ground extending from behind Boulder City—although within supporting distance—to Hill 111 at

Department of Defense Photo (USMC) A173925

Marines of the 2d Battalion, 1st Marines, board armored personnel carriers to be taken to the front. As a result of the critical tactical situation and number of casualties suffered during the Berlin operations, the battalion was positioned in the center of the regimental main line of resistance as the first step in the relief of the 7th Marines.

the boundary between the 1st Marine Division and the Commonwealth Division. The newly-arrived 2d Battalion, 1st Marines, under Lieutenant Colonel Frank A. Long, moved into position between the 3d Battalion, 7th Marines, on its right and Lieutenant Colonel Harry A. Hadd's 1st Battalion, 7th Marines, on its left. The 2d Battalion, 7th Marines, served as regimental reserve.

The introduction of Lieutenant Colonel Long's battalion, which came under control of the 7th Marines, served as the first step in a planned relief of the 7th Marines by the 1st Marines. For now, the newly arrived battalion added further depth and density to the main line of resistance, organizing Hill 126 and the other commanding heights in its sector. In effect, three

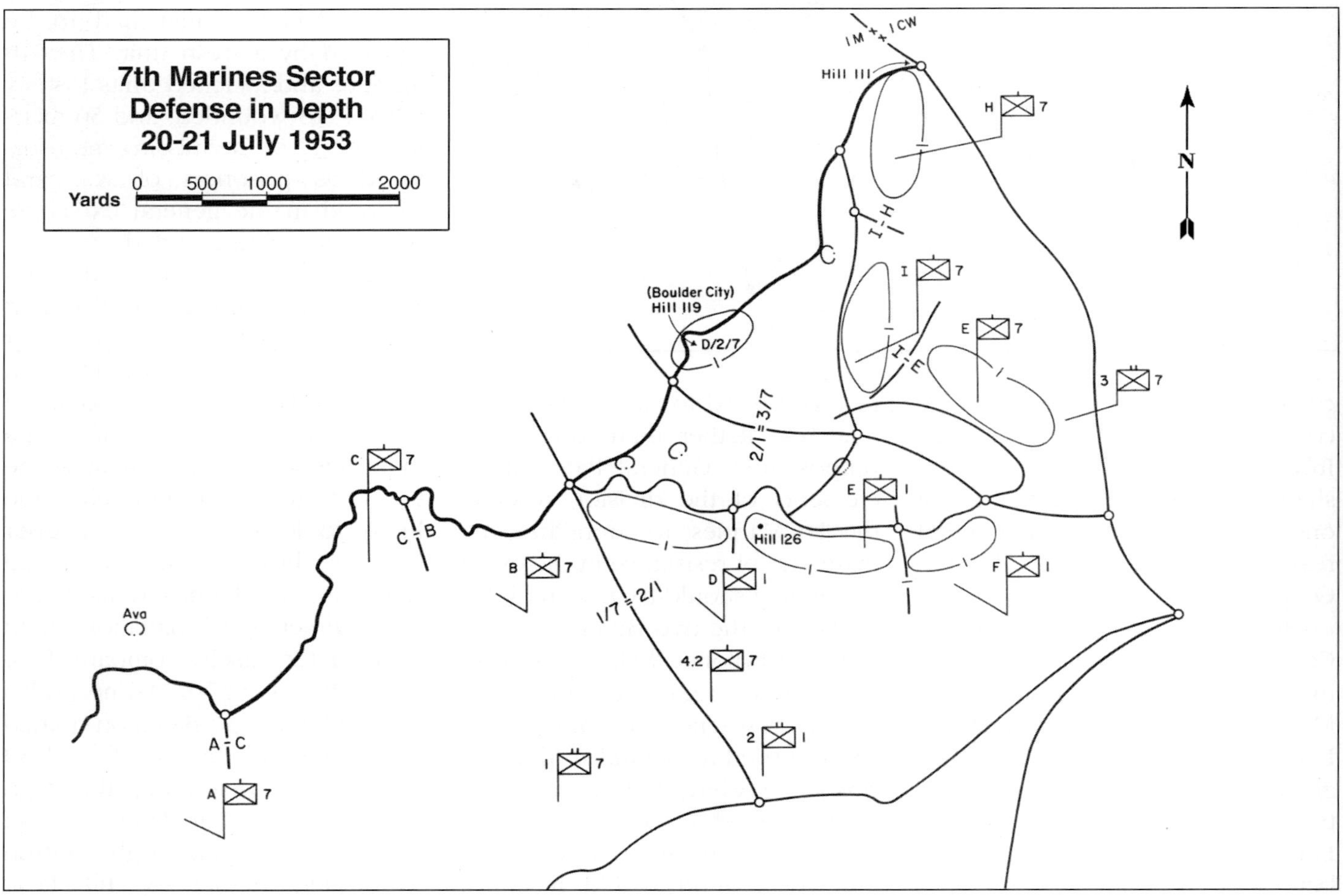

Department of Defense Photo (USMC) A173715

Men of the 1st Marines move toward the frontlines to reinforce other Marines fighting at Hill 119, known informally as Boulder City. Little did they know that they would see the last of the war's heavy fighting.

battalions, rather than the two previously defending the regimental area, formed a crescent of strongpoints designed to contain and defeat any offensive launched from Berlin and East Berlin.

In the sector held by the 7th Marines, Outpost Ava, manned by a squad from Company A of the regiment's 1st Battalion, survived on the far left, near the boundary between the 7th and 5th Marines. Boulder City, formerly a component of a continuous main line of resistance, now functioned as an outpost of the reconstituted defenses. By 22 July, Company G, 3d Battalion, 7th Marines, had taken over Boulder City, from the regiment's Company D, which reverted to the control of its parent battalion, the 2d, in reserve.

The Last Battle

Signs of an imminent Chinese attack multiplied as July drew to a close. The probable objectives seemed to include Outposts Hedy and Dagmar, but instead of attacking either in force, the enemy sent only a token force, wearing burlap camouflage, that appeared near Hedy on 21 July. The defenders opened fire, killing three of the Chinese, and the survivors fled.

Marine Fighter Squadrons 115 and 311, released by the Fifth Air Force to support the United Nations troops fighting in central and eastern Korea, joined Marine Attack Squadron 121 in pounding the Chinese threatening the 1st Marine Division. Recurring cloud cover produced frequent downpours that interfered with operations during the critical period of 21-23 July, but the three squadrons nevertheless flew more than 15 radar-directed missions that dropped some 33 tons of bombs.

As the threats to Outposts Hedy and Dagmar abated, Chinese forces menaced Boulder City, where Company G, 3d Battalion, 1st Marines, commanded by First Lieutenant Oral R. Swigart, Jr.,

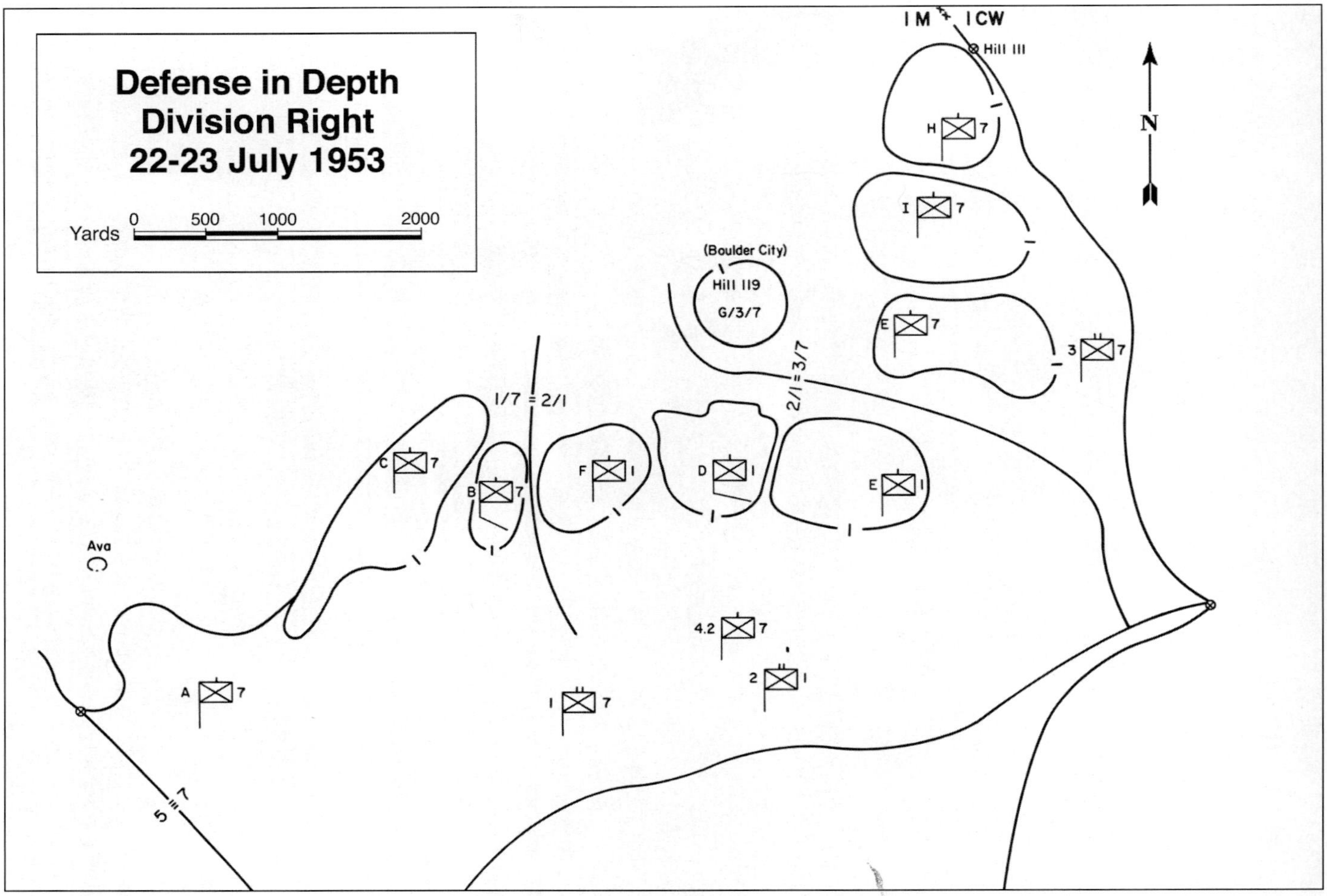

manned the defenses after relieving Company D, 2d Battalion, 7th Marines. On the evening of 24 July, hostile mortars and artillery began hammering Swigart's perimeter. Marine artillery and 4.5-inch rocket launchers immediately responded against targets that included a Chinese regiment massing behind Hill 139, northwest of enemy-held Outpost Berlin.

At 2030, Chinese troops began probing the right of the 1st Marine Division's line. After a powerful barrage by mortars and artillery, the assault force hit Hill 111 at the far right of the positions held by the 7th Marines, then shifted to Boulder City near the boundary between the 3d Battalion, 7th Marines, and the attached 2d Battalion, 1st Marines. As he had on 7 July, when he sought to capitalize on the Marine division's takeover of the lines of the 25th Infantry Division, the enemy sought to take advantage of the relief of the 7th Marines by the 1st Marines.

When the Chinese attack began, the 2d Battalion, 1st Marines, attached to the 7th Marines, had already taken over positions that included Boulder City. The 3d Battalion, 1st Marines, commanded by Lieutenant Colonel Roy D. Miller, was relieving the 3d Battalion, 7th Marines, as Company H took over Hill 111 and Company G defended the critical ground at Boulder City.

At about 1930 on 24 July the enemy attacked Hill 111 and soon cracked the perimeter now manned by Company H of Miller's battalion. For about 50 minutes, the Chinese clung to a salient on the hilltop, but then withdrew. After this flurry of action, apparently intended to divert attention from Boulder City, the enemy ignored Hill 111 until the morning of 25 July, when artillery fire battered the perimeter but no infantry assault followed.

The two Chinese battalions attacking on the Marine right had their greatest success at Boulder City, seizing a portion of the trenchline defended by Company G, 3d Battalion, 1st Marines. In an attempt to exploit this foothold, the enemy attacked the Berlin and East Berlin gates, passages through the wire that the Marines had used to supply and reinforce the two outposts before both were overwhelmed. Cloud cover prevented aerial observers from supporting the troops protecting the gates, and the Chinese managed to gain control of Berlin gate and

Second Lieutenant Raymond G. Murphy

Department of Defense Photo (USMC) A49767

An ardent athlete with a major in physical education, Murphy was born in Pueblo, Colorado, in 1930, and was commissioned in the Marine Corps Reserves in 1951. In Korea, he was awarded a Silver Star Medal for his actions on 22 November 1952 in assaulting an enemy strongpoint. Then his heroism, again as a platoon commander with Company A, 1st Battalion, 5th Marines, on 3 February 1953, resulted in a Medal of Honor with a citation, which read in part:

> Undeterred by the increasing intense enemy fire, he immediately located casualties as they fell and made several trips up and down the fire-swept hill to direct evacuation teams to the wounded, personally carrying many of the stricken Marines to safety. When reinforcements were needed by the assaulting elements, Second Lieutenant Murphy employed part of his unit as support and, during the ensuing battle, personally killed two of the enemy with his pistol. With all the wounded evacuated and the assaulting units beginning to disengage, he remained behind with a carbine to cover the movement of friendly forces off the hill and, though suffering intense pain from his previous wounds, seized an automatic rifle to provide more firepower when the enemy reappeared in the trenches.

After the war, he joined the Reserves and was discharged as a captain in 1959.

Hospital Corpsman Francis C. Hammond

Navy Medicine Photo Collection

Born in 1931, Hammond enlisted in the U.S. Navy in 1951. Assigned to the Marine Corps as a "Hospitalman," he gave his life at Sanae-dong, Korea, serving with Company C, 1st Battalion, 5th Marines, on 26 March 1953. His Medal of Honor citation reads, in part:

> Hospitalman Hammond's platoon was subjected to a murderous barrage of hostile mortar and artillery fire, followed by a vicious assault by onrushing enemy troops. Resolutely advancing through the veritable curtain of fire to aid his stricken comrades, Hospitalman Hammond moved among the stalwart garrison of Marines and, although critically wounded himself, valiantly continued to administer aid to the other wounded throughout an exhausting four-hour period. When the unit was ordered to withdraw, he skillfully directed the evacuation of casualties and remained in the fire-swept area to assist the corpsmen of the relieving unit [Company F, 2d Battalion, 5th Marines] until he was struck by a round of enemy mortar fire and fell, mortally wounded.

After the war, a school in his hometown of Alexandria, Virginia, a medical clinic at Camp Pendleton, California, and the Knox-class destroyer *Francis Hammond* (DE 1067) were named in his honor.

–Captain John C. Chapin, USMCR (Ret)

mount a second determined assault on the Boulder City perimeter. Hand-to-hand fighting raged all along the 700 yards of trench that Lieutenant Swigart's Marines still held. The company's ammunition ran low, and the plight of casualties became increasingly difficult as Chinese fire killed two of Boulder City's eight corpsmen and wounded most of the others. By midnight, Swigart's company could muster no more than half its earlier strength, but it still clung to the rear slope of Boulder City. In the words of one of Company G's Marines, "only a never-say-die resistance was keeping the enemy from seizing the remainder of the position."

Casualties had further eroded the strength of the Boulder City garrison, when Captain Louis J. Sartor, at 15 minutes after midnight on the morning of the 25th, led Company I, 3d Battalion, 1st Marines, toward the hill to reinforce Swigart's survivors. The Chinese intercepted and correctly interpreted the coded radio message ordering Sartor's Marines forward, thus obtaining information that enabled enemy artillery and mortars to wound or kill about a third of the reinforcements. Despite the deadly barrage, much of Company I reached Boulder City, joined forces with the remnants of Swigart's garrison, and took part in a counterattack led by Captain Sartor that recaptured the hill by 0330. Further reinforcements from Company E, 2d Battalion, 7th Marines, and Company E, 2d Battalion, 1st Marines, arrived by 0530 to consolidate the position. A few Chinese, however, continued to cling to positions on the slopes nearest their main line of resistance.

Since the enemy still controlled the approaches to Boulder City, he was able to mount another attack on that position at 0820, 25 July. Fire from Marine mortars and

Hospital Corpsman Third Class William R. Charette

A native of Ludington, Michigan, Charette was born in 1932 and enlisted in the U.S. Navy in 1951. For his actions during the early morning hours of 27 March 1953 in the Panmunjom Corridor, while attached to Company F, 2d Battalion, 7th Marines, he was recommended for and later received the Medal of Honor. His citation reads, in part:

> When an enemy grenade landed within a few feet of a Marine he was attending, he immediately threw himself upon the stricken man and absorbed the entire concussion of the deadly missile with his body. Although sustaining painful facial wounds, and undergoing shock from the intensity of the blast which ripped the helmet and medical aid kit from his person, Hospital Corpsman Third Class Charette resourcefully improvised emergency bandages by tearing off part of his clothing, and gallantly continued to administer medical aid to the wounded in his own unit and to those in adjacent platoon areas as well. . . . Moving to the side of another casualty who was suffering excruciating pain from a serious leg wound, Hospital Corpsman Third Class Charette stood upright in the trench line and exposed himself to a deadly hail of enemy fire in order to lend more effective aid to the victim and to alleviate his anguish while being removed to a position of safety.

Miraculously surviving his wounds, he rose to the rank of master chief hospital corpsman before retiring in 1977. A hospital facility at the Naval Medical Center, Portsmouth, Virginia, is named for Corpsman Charette.

Sergeant Daniel P. Matthews

Department of Defense Photo (USMC) A76499

Born in Van Nuys, California, in 1931, Matthews enlisted in the Marine Corps in 1951. After completing recruit training he was assigned to the 1st Battalion, 3d Marines, at Camp Pendleton, California. He sailed for Korea in January 1953, joining Company F, 2d Battalion, 7th Marines. On 28 March 1953 he was killed in a counterattack on Vegas Hill. His Medal of Honor citation reads, in part:

> Sergeant Matthews fearlessly advanced in the attack until his squad was pinned down by a murderous sweep of fire from an enemy machine gun located on the peak of the outpost. Observing that the deadly fire prevented a corpsman from removing a wounded man lying in an open area fully exposed to the brunt of the devastating gunfire, he worked his way to the base of the hostile machine-gun emplacement, leaped onto the rock fortification surrounding the gun and, taking the enemy by complete surprise, single-handedly charged the hostile emplacement with his rifle. Although severely wounded when the enemy brought a withering hail of fire to bear upon him, he gallantly continued his valiant one-man assault and, firing his rifle with deadly effectiveness, succeeded in killing two of the enemy, routing a third and completely silencing the enemy weapon, thereby enabling his comrades to evacuate the stricken Marine to a safe position. [He died of] his wounds before aid could reach him.

–Captain John C. Chapin, USMCR (Ret)

Staff Sergeant Ambrosio Guillen

Department of Defense Photo (USMC) 407003

Born in La Junta, Colorado, in 1929, Guillen grew up in El Paso, Texas; he enlisted in the Marine Corps in 1947. After sea duty and serving as a drill instructor at Marine Corps Recruit Depot, San Diego, he was assigned as a platoon sergeant with Company F, 2d Battalion, 7th Marines, in Korea. He died of wounds incurred near Songuch-on 25 July 1953. Ironically, the ceasefire was signed two days later. His Medal of Honor citation reads, in part:

> With his unit pinned down, when the outpost was attacked under cover of darkness by an estimated force of two enemy battalions supported by mortar and artillery fire, he deliberately exposed himself to the heavy barrage and attacks to direct his men in defending their positions and personally supervise the treatment and evacuation of the wounded. Inspired by his leadership, the platoon quickly rallied and engaged the enemy force in fierce hand-to-hand combat. Although critically wounded during the course of the battle, Staff Sergeant Guillen refused medical aid and continued to direct his men throughout the remainder of the engagement until the enemy was defeated and thrown into disorderly retreat.

A middle school in El Paso, Texas, is named for Staff Sergeant Guillen.

–Captain John C. Chapin, USMCR (Ret)

artillery, and from the 90mm weapons of 10 tanks dug in on the Marine positions, played the key role in breaking up the new assault, although the last of the attackers did not withdraw until afternoon. The M-46 tanks proved deadly against advancing Chinese troops, but also presented an inviting target for Chinese artillery and mortar crews, who directed some 2,200 rounds at the armored vehicles. Aircraft also helped repulse the 25 July attack on Boulder City, as when Panther jets flew nine missions, guided by Marine radar on the ground, against hostile positions threatening Boulder City and nearby defensive strongpoints.

Before midnight on 24 July, in an attack perhaps loosely coordinated with the thrust at Boulder City, Chinese forces hit the positions held by the 5th Marines. After probing the defenses of Outposts Dagmar and Esther, the enemy concentrated against the latter, manned by elements of Company H, 3d Battalion, 5th Marines. The Chinese tried to isolate Outpost Esther by shelling and patrolling the routes leading there from the main line of resistance and succeeded in overrunning outer portions of the perimeter. The defenders, commanded by Second Lieutenant William H. Bates, prevailed because of the skillful use of their own weapons, including flamethrowers and the support of mortars, machine guns, tanks, and the always-deadly artillery. The 3d

Aerial view of the pockmarked terrain in front of Boulder City taken from an HMR-161 helicopter. Although the monsoon rains of July limited normal support missions, when weather conditions cooperated planes of the 1st Marine Aircraft Wing worked from morning to sundown unleashing tons of ordnance on Chinese positions seen just beyond the Marine sector of the line.

National Archives Photo (USMC) 127-N-A173886

National Archives Photo (USMC) 127-N-A173809

A Marine holds a wounded buddy onto an armored personnel carrier being brought to the 3d Battalion, 1st Marines aid station during the heavy fighting in the Berlin sector.

Battalion, 11th Marines, fired 3,886 rounds against Chinese troops attacking Outpost Esther, and hostile gunners matched this volume of fire. The Marines suffered 12 killed and 98 wounded in the fighting that began at Dagmar and continued at Esther, while Chinese casualties may have totaled 195 killed and 250 wounded.

Dawn on 26 July brought a lull in these last battles. Chinese attempts to revive their attack by infiltrating reinforcements through the site of Outpost Berlin failed, thanks to accurate fire from Marine riflemen and machine gunners. The 1st Marines completed its relief of the 7th Marines at 1330. That night, the enemy probed Boulder City for the last time, sending a patrol from captured Outpost Berlin that failed to penetrate the defensive wire and shortly after midnight dispatching another platoon that prowled about before Marine fire repulsed it.

Although the last of the Chinese attacks seized Outposts Berlin and East Berlin, they failed to wrest Boulder City from its Marine defenders. Had the enemy captured Boulder City, he might have exploited it and seized the high ground to the south and east, from which he could have fired directly into the rear areas that sustained the 1st Marine Division in its positions beyond the Imjin River. In fighting the Chinese to a standstill during July 1953, the division suffered 1,611 casualties—killed, wounded, and missing—the most severe losses since October 1952 when savage fighting had raged at Outposts Carson, Reno, and Vegas, and on the Hook. Chinese losses during July 1953 may have exceeded 3,100.

The Final Patrols

During the last few nights of combat in July, Marines continued patrolling aggressively, even though a truce was fast approaching. *The Last Parallel*—a wartime memoir by Martin Russ, a corporal in Company A, 1st Battalion, 1st Marines, in July 1953—describes an action that took place just before the truce, as the 1st Marines was taking over from the 7th Marines. A Lieutenant from Company A, 7th Marines, led a 30-man combat patrol, made up of men from Companies A of both regiments, that set out from Outpost Ava near midnight and crossed a rice paddy on the way to raid Chinese positions on the Hill 104. A reconnaissance probe, which had preceded the raiding party, reported the presence of Chinese on the approaches to the objective.

The route to Hill 104 followed a trail flanked by waist-high rice, growth that not only impeded off-trail movement, but also might conceal a Chinese ambush. The soft muck in which the rice thrived provided poor support for some types of mines, thus reducing one threat to the patrol, but moving through the paddy would have snapped the brittle rice stalks and created noise to alert any lurking ambush party. Balancing less noise against greater danger from mines, the Marines decided to follow the trail.

Two members of the 3d Battalion, 7th Marines, quickly clean a semi-automatic M-1 carbine with grenade launcher after firing in support of a night patrol. A third Marine take a break in one of the sandbagged culverts, which provided overhead cover along the trenchline.

National Archives Photo (USMC) 127-N-A173671

The patrol's point man, as he approached the far end of the rice paddy, found a trip wire stretched across the trail and followed it to a Russian-built, antipersonnel mine attached to a stick thrust deep into the mud and aimed to scatter deadly fragments low across the surface of the trail. The patrol halted, bunching up somewhat as the members tried to see why the lieutenant was moving forward to confer with the point man. At this moment, Chinese soldiers crouching hidden in the rice opened fire from as close as 50 feet to the Marines.

The fusillade wounded the patrol leader and within 10 seconds killed or wounded nine of the first 10 men in the patrol. Dragged into the paddy by members of the ambush party, the lieutenant disappeared until he surfaced in the exchange of prisoners after hostilities ended. The point man proved luckier, however, hurling himself to the ground as soon has he realized what was happening, finding cover beside the trail, and escaping injury. From the rear of the patrol, the other survivors fired at the muzzle flashes of the Chinese weapons. Marine firepower prevailed, silencing the enemy after five or so minutes, although not before six members of the patrol had been killed, 14 wounded, and the wounded lieutenant carried off as a prisoner. The survivors regrouped and moved forward, retrieving as many

National Archives Photo (USN) 80-G-625731

Flanked on the left by Gen Otto P. Weyland, USAF, commanding general of Far East Air Forces, and on the right by VAdm Robert P. Briscoe, Commander, Naval Forces Far East, and VAdm Joesph J. Clark, Seventh Fleet commander, Gen Mark W. Clark, commander in chief of United Nations forces, countersigns the armistice agreement at Munsan-ni.

as they could of the dead and wounded. A half-dozen Marines covered the patrol's withdrawal, preventing the Chinese from encircling the group. When the men who had helped evacuate the wounded returned from the aid station and reinforced the firepower of the hastily formed screening force, the Chinese vanished into the night.

The Marines kept up their patrolling until the moment the truce took effect. On the night of 27 July, according to Martin Russ, his company sent out a patrol scheduled to return shortly before the fighting would end at 2200. As the Marines prepared to move out, Chinese mortars fired on Outpost Ava, through which the patrol staged, and nearby portions of the main line of resistance. The bombardment wounded five members of the patrol and two of the Marines defending the outpost.

The patrol returned as planned, and at 2200, Russ, who was not a member, watched from the main line of resistance as white star clusters and colored flares cast a pulsating light that set the shadows dancing in ravines and paddies and on hillsides, while the final shells fired in the 37-month war exploded harmlessly. As Russ described the scene:

> A beautiful moon hung low in the sky like a Chinese lantern. Men appeared along the trench, some of them had shed their flak jackets and helmets. The first sound we heard was the sound of shrill voices.... The Chinese were singing. A hundred yards or so down the trench, someone

National Archives Photo (USN) 80-G-626455

Marines take a well-deserved rest after word of the armistice is passed. Less than 24 hours after the signing of the armistice, Marine units began withdrawing from the Demilitarized Zone to their new battle positions.

was shouting the Marine Corps Hymn at the top of his lungs. Others joined in bellowing the words. All along the battle line, matches flared and cigarettes glowed, but no snipers peered through telescopic sights to fire at these targets. The war had ended.

After the Ceasefire

The ceasefire agreement, which was signed on the morning of 27 July and went into effect 12 hours later, required that both the United Nations forces and the Communist enemy withdraw from the most advanced positions held when the fighting ended. In effect, the abandoned area formed the trace of a Military Demarcation Line, as the opposing armies fell back 2,000 yards to organize new main battle positions, thus creating the 4,000-yard Demilitarized Zone between them. The Marines built as they destroyed, evacuating certain portions of the old main line of resistance, giving up some of its outposts, and dismantling fortifications simultaneously with their construction of the new line and its mutually supporting strongpoints.

A No-Fly Line supplemented the controls imposed by the Demilitarized Zone. Restrictions on the movement of aircraft applied throughout the Demilitarized Zone and in a corridor extending from the vicinity of Panmunjom to Kaesong. Only helicopters could fly beyond the No-Fly Line, provided they remained 500 yards from the Military Demarcation Line.

Establishing the Demilitarized Zone

The armistice document set a timetable for the creation of the Demilitarized Zone. Within 72 hours after the ceasefire went into effect, the combatants were to remove "all military forces, supplies, and equipment" and report the location of "demolitions, minefields, wire entanglements, and other hazards" capable of impeding the safe movement of the organizations that would oversee the armistice—the Military Armistice Commission, its Joint Observer Teams, and the Neutral Nations Supervisory Commission. For 45 days following the initial 72 hours, the parties to the ceasefire would salvage the materials still in the designated Demilitarized Zone, using only unarmed troops for the task. Meanwhile, during the 10 days after 27 July, Chinese or North Korean troops took over all offshore islands, east coast and west, that lay north of the 38th Parallel. The United Nations retained control of Taechong-do, Paengyong-do, Sochong-do, and Yongpyong-do, along with U-do, Tokchok-to, and Kangwha-do, all of them off the west coast and south of the demarcation line.

Since the agreement of July 27 called for a truce rather than peace—indeed, it sought to ensure a cessation of hostilities until "a final peaceful settlement is achieved"—the possibility existed that the fighting might resume. Despite the shower of flares and the singing that marked the moment the truce went into effect, some battle-tested Marines fully expected a Chinese attack in the hours after the ceasefire. As the skies lightened on the morning of 28 July, Chinese commanders obtained permission to recover the bodies of their men killed in the final assaults on Marine positions. The sight of "the enemy moving around within a stone's throw of our front lines" underscored the possibility of renewed attacks, but nothing happened, and as the morning wore on, it became obvious that the truce was holding—at least for the present.

Like the Chinese, the Marines used the first morning of the ceasefire to recover the bodies of men killed in recent days. Most of the

National Archives Photo (USMC) 127-N-A173878

Following the armistice, the commanding general of the 1st Marine Division, MajGen Randolph McC. Pate, his chief of staff, Col Lewis W. Walt, and the commanding officer of the 1st Marines, Col Wallace M. Nelson, survey the aftermath of the battle for Boulder City.

dead had fallen at Hill 111 and Boulder City. By the end of the day, all the bodies had been retrieved and were on the way to the rear.

In the three days immediately after the armistice took effect, some 50 companies of Marine infantry, both American and South Korean, began dismantling the old defenses, with the help of elements of the division's engineers. Working day and night—taking frequent breaks during the daytime heat, sleeping during the three hours after noon when the heat was most enervating, and using portable lighting to take advantage of the comparative cool of the night—the Marines and the Korean service troops helping them removed supplies and ammunition, tore apart bunkers, and stacked the timbers for shipment to the new battle line. Some of the work parties treated the bunkers like trenches and filled them with earth, which then had to be shoveled out to provide access to the salvageable timbers.

Dismantling the old battle line required the removal of some structures on elements of both the combat outpost line and the main line of resistance, while at the same time building the new main battle position and sealing off the Demilitarized Zone that the truce established between the contending armies. Some of the former Marine outposts like Bunker Hill, Esther, and Ava lay north of the Military Demarcation Line. As a result, the Marines could be sure of having access to them only during the 72 hours after the ceasefire began. Moreover, anything salvaged from Bunker Hill or nearby Outpost Hedy had to travel over a primitive road described as "particularly tortuous," which made the transfer "of first the ammunition and then the fortification materials a physical ordeal." Further complicating the dismantling of these distant outposts, a horde of reporters, photographers, and newsreel cameramen arrived, eager to record the activity of both the Marines and the Chinese soldiers who could be seen tearing down their own defenses on nearby ridgelines or hilltops.

Salvaging building materials proved to be hard work, whether tearing apart structures on the combat outpost line or on the main line of resistance. The picks, shovels, and steel pry bars available to the infantrymen could not remove timbers, measuring up to 12 inches square and secured by spikes 10 to 24 inches long, that formed the skeleton of bunkers measuring perhaps 12 by 20 feet. Wherever possible bulldozers bore the brunt, but heavy trucks fitted with power winches and even tow trucks helped out, as did medium tanks, their guns removed so they could enter the demilitarized area. The Marines found that the fastest method of dismantling a bunker was to uncover it, winch it out of its hole, and bounce it down the hill it had guarded.

Once the bunker had been disassembled, the Marines manhandled the timbers onto vehicles, usually several two-and-one-half-ton trucks, although heavier vehicles saw service during the first 72 hours until the ban took effect on disarmed tanks, tank retrievers, and other vehicles that fit the armistice agreement's imprecise definition of "military equipment." Fortunately, two-and-one-half-ton trucks could still be used over the next 45 days, although restricted to designated routes of access into the Demilitarized Zone.

As the dismantling of the abandoned defenses went ahead on schedule, the 1st Marine Division moved into its new positions, a transfer completed by the morning of 1 August. The main battle position, to the rear of the former line, consisted of a succession of strongpoints. From the division's right-hand boundary near the Samichon River, the new line formed a misshapen arc encom-

Marines begin the task of dismantling bunkers on the abandoned main line of resistance after the ceasefire went into effect on 27 July. Trenchlines were filled in, tank slots bulldozed under, and usable timber carried to salvage collection points.

National Archives Photo (USN) 80-G-626370

passing the town of Changdan on the north bank of the Imjin River and the bridges across that stream. The main battle position then crossed to the south bank of the Imjin, which it followed to the conflux of that river and the Han before shifting to the Han's south bank and continuing westward. The 5th Marines occupied the ground north of the Imjin, organizing strongpoints that in effect functioned as an outpost line for the rest of the division. The 7th Marines dug in south of the Imjin and the 1st Marines provided a reserve. The South Korean Marines formed the left of the line, south of the Imjin and along the Han.

From the end of the 72-hour period until the 45-day deadline of 13 September, the infantry companies north of the Imjin sent out each day work parties of from 25 to 100 men to finish the salvage effort begun immediately after the ceasefire took effect. By the time the task ended, the 5th Marines had retrieved some 12 tons of miscellaneous equipment, 2,000 miles of telephone line, 2,850 rolls of barbed wire and 340 of concertina wire, 19,000 pickets for use with the wire, 339,000 sand bags, and 150,000 linear feet of timber. Most of the salvaged items were incorporated on the positions being built by the regiment to defend the bulge north of the Imjin.

Department of Defense Photo (USMC)

Marine Sgt Richard J. Thompson checks in on his radio at one of the checkpoints throughout the 1st Marine Division's front while patrolling the buffer zone south of the line of demarcation.

Maintaining Order in the Demilitarized Zone

At the outset of its salvage operation, the 5th Marines marked the adjacent portion of the Demilitarized Zone and began controlling access to it. After constructing a so-called No-Pass Fence some 200 yards south of the near edge of the Demilitarized Zone, the regiment marked the fence with warning signs, engineer tape, and panels visible from the air. Next, the 5th Marines established 21 crossing stations into the Demilitarized Zone, each one manned by at least two Marines who denied access to anyone carrying weapons or lacking authorization. As the work of salvaging material drew to a close, the regiment closed the crossing stations it no longer needed.

Each person entering the Demilitarized Zone through the area held by the 5th Marines had to show a pass issued by the regiment. The salvage operation generated heavy traffic, especially in its earlier stages; indeed, vehicles passed through the crossing stations more than 3,500 times. After 13 September, when the salvage project ended, I Corps assumed responsibility for issuing passes.

Controlling access through its lines to the Demilitarized Zone and maintaining order there became continuing missions of the 1st Marine Division. The ceasefire

(Continued on page 596)

The Prisoners Return

Like the earlier exchange of sick and wounded prisoners, Operation Little Switch, the process of repatriation following the ceasefire took place within the sector held by the 1st Marine Division. Because the final exchange, appropriately named Big Switch, involved more than 10 times the number repatriated earlier, the medical facilities used in April could not meet the new challenge. A large Army warehouse at Munsan-ni, converted into a hospital by Marine engineers, replaced the old treatment center. A newly created administrative agency, the Munsan-ni Provisional Command, assumed overall responsibility for Big Switch, with the 1st Marine Division carrying out the actual processing.

When Operation Big Switch got underway on 5 August, a Marine receipt and control section, functioning as part of the provisional command, accepted the first group of prisoners at Panmunjom, checked them against the names on a roster the captors had submitted, and began sending them south along the Panmunjom Corridor. Returnees not in need of immediate medical care boarded ambulances for the drive to the Freedom Village complex at Munsan-ni. Those requiring prompt or extensive care were flown to Freedom Village in Marine helicopters; the most serious cases were then rushed either to hospital ships off Inchon or directly to hospitals in Japan.

Besides helping evacuate the sick or wounded, Marine helicopters played the principal role in an airlift designed to placate President Rhee, who believed that India's refusal to send troops to help defend his nation demonstrated support for North Korea and China. The United States sought to ensure his cooperation with Big Switch by promising that no Indian troops would set foot on South Korean soil. As a result, the Indian contingent involved in taking custody of the prisoners who refused repatriation flew by helicopter from ships anchored in Inchon harbor to a camp in the Demilitarized Zone.

Processing the Freed Prisoners

At Freedom Village—South Korea operated a Liberty Village for its returnees—the newly freed prisoners underwent a medical evaluation, brought their records up to date, received new uniforms, mail, newspapers, and magazines, and ate a light meal. The steak that many of them craved would have shocked weakened digestive systems and had to wait until they had readjusted to American fare. Those former captives judged healthy enough could meet with reporters at Freedom Village and answer questions about their experiences. After a brief period of recuperation, the former prisoners embarked for the United States.

Ambulances carrying sick and wounded United Nations prisoners of war arrive at Freedom Village. The seriously injured were transferred directly to hospital ships at Inchon or air-evacuated to Japan.

Department of Defense Photo (USMC) A174375

National Archives Photo (USMC) 127-N-A174381

For the benefit of the press and distinguished visitors, Marines maintained a map showing the progress of road convoys bringing former prisoners from Panmunjom to Freedom Village during Operation Big Switch.

During the voyage, or shortly after disembarking, they described for intelligence specialists the treatment they had received at the hands of the enemy.

The fifth former prisoner to arrive at Freedom Village on 5 August turned out to be a Marine, Private First Class Alfred P. Graham, captured in July 1951. Forced labor and malnutrition left him too weak to meet with reporters at Munsan-ni, but later, at a hospital in Japan, he would describe how he had routinely been compelled to carry firewood 11 miles for the stoves at his prison camp. Two other Marines reached Munsan-ni on the first day: Sergeant Robert J. Coffee, wounded and captured in November 1950, and Private First Class Pedron E. Aviles, knocked unconscious by a Chinese rifle butt in December 1952 and taken prisoner.

The stream of returning Americans continued until 6 September, apparently with scant regard for time in captivity, physical condition, rank, or duties. For example, Private First Class Richard D. Johnson, a machine gunner in the 5th Marines captured on 25 July 1953, returned on 24 August, two days before the repatriation of Captain Jesse V. Booker, an aviator and the first Marine taken prisoner, who was captured on 7 August 1950. Some of the returnees had been given up for dead, among them: First Lieutenant Paul L. Martelli, a fighter pilot officially listed as killed in action; First Lieutenant Robert J. O'Shea, an infantry officer whose name did not appear on any list of prisoners; and Private First Class Leonard E, Steege, believed by his buddies to have been killed in the fighting at Boulder City.

Operation Big Switch continued until 6 September and resulted in the release of 88,596 prisoners, 75,823 of them

Communist prisoners of war rip off their U.S. provided uniforms and toss them contemptuously to the ground. Shouting Communist slogans and hurling insults at United Nations forces, they put on a propaganda show for the benefit of world newsreel cameras.

National Archives Photo (USN) 80-G-626976

North Korean or Chinese. South Koreans, both soldiers and Marines, totaled 7,862, and those repatriated to other countries in the United Nations coalition numbered 4,911, 3,597 of them Americans, including 197 Marines.

Although the overwhelming majority of prisoners agreed to repatriation, more than 22,000 did not, entrusting themselves to the custody of the Indian troops overseeing the process on behalf of the five officially neutral nations. The Indians released 22,467 former prisoners; two-thirds of them Chinese, after a final attempt at persuasion allowed by the armistice agreement was made. Most of the Chinese were veterans of the Nationalist forces, who had been captured on the mainland and impressed into the service of the People's Republic. Once released, these veterans joined the Nationalists who had fled the mainland and established themselves on Taiwan and its satellite isles. The North Koreans disappeared into the populace of the South.

A total of 357 United Nations troops refused repatriation, 333 of them South Korean. Only 23 Americans—none of them a Marine—chose to stay behind, along with one British serviceman, a Royal Marine. Two of the American soldiers, swayed by the final efforts at persuasion, changed their minds at the last minute. Over the years, another 12 reconsidered their decision and returned to the United States.

Behind Barbed Wire

As had happened in previous wars, Americans captured in Korea endured mistreatment from their captors, whether inflicted deliberately or the result of callousness. In this war, unlike the earlier ones, prisoners served as pawns in an ideological contest in which the Chinese and North Koreans tried to convert them to Communism or, failing that, to force them to make statements that would further the Communist cause in its world-wide struggle against capitalism. The methods of conversion or coercion varied from unceasing lectures extolling Communism to threats and torture, with the harshest treatment meted out for acts of resistance. By using these techniques, the prison staffs sought a variety of objectives that included maintaining order, persuading prisoners to embrace Communism, obtaining military information, or extorting confessions to alleged war crimes, statements designed to turn worldwide public opinion against the United States. By 1952, the enemy was focusing in particular on forcing captured fliers of all the Services to confess to participating in germ warfare.

The vigor of the persuasion or punishment varied according to the camp, the whim of the guards, or the policy of the moment. In July 1951, for example, the Chinese,

MajGen Vernon E. Megee, center, Commanding General, 1st Marine Aircraft Wing, introduces former prisoner of war, Capt Gerald Fink, to MajGen Randolph McC. Pate, Commanding General, 1st Marine Division, at Freedom Village.

Department of Defense Photo (USMC) A174480

Department of Defense Photo (USN) 627709

Marine prisoner of war 1stLt Richard Bell is dusted with DDT prior to his departure for Inchon. Dusting with DDT was mandatory for all Marines leaving Korea for the United States.

who had assumed custody of the prisoners, announced a policy of leniency that offered organized athletic competition among the prisoners and promised better treatment in return for cooperation. The new policy aimed at winning over world opinion while converting "reactionaries"—those prisoners who resisted indoctrination—into "progressives," who did not.

The new leniency did not apply to Lieutenant Colonel William G. Thrash, an arch-reactionary. Accused of "Criminal Acts and Hostile Attitude," he spent eight months in solitary confinement. In January 1952, he was beaten, dragged outside, and exposed overnight to the deadly cold, an ordeal that nearly killed him.

To facilitate the indoctrination of the prisoners, a process that came to be called brainwashing, the enemy tried to shatter the chain-of-command. He separated officers from enlisted men and tried to place progressives in places of leadership. The reactionaries fought back by creating a network of their own to frustrate the Chinese tactics by restoring military discipline.

Lieutenant Colonel Thrash of the 1st Marine Aircraft Wing and Major John N. McLaughlin of the 1st Marine Division proved especially successful in creating solidarity among the prisoners. After recovering from the effects of his confinement, Thrash stirred up resistance to the indoctrination effort and to a related attempt to encourage the writing of letters that might serve as Chinese propaganda. He also warned that interrogators would try to pressure accused reactionaries into implicating fellow prisoners who shared reactionary views. McLaughlin created a chain-of-command using five veteran Marine Corps noncommissioned officers and, like Thrash, helped form the escape committees that took shape shortly before the armistice.

Resistance to the Chinese might be passive, essentially an internal rejection of Communist attempts at indoctrination, or active, with a range of actions that varied from symbolic to practical to defiant. Symbolic gestures included a celebration of the Marine Corps birthday on 10 November 1952, featuring a cake made from eggs, sugar, and flour stolen from the Chinese. A 22-inch crucifix, carved by Captain Gerald Fink with improvised tools and given the title "Christ in Barbed Wire," symbolized both hope and resistance. Captain Fink also fashioned practical implements, an artificial leg for an Air Force officer, Major Thomas D. Harrison, injured when his plane was shot down, and crude stethoscopes for the medical personnel among the prisoners.

Captain Fink's handiwork also played a part in an effort to keep track of as many of the prisoners as possible. The hollow portion of the artificial leg he had made contained a list of names, the dates of death for those known to have perished, and details of treatment while in captivity. The Chinese forced Harrison—a cousin of Lieutenant General William K. Harrison, Jr., the chief United Nations truce negotiator—to give them the leg when he was repatriated, but the information survived. The Air Force officer had a copy in the hollow handgrip of his crutch, and another prisoner carried a copy in a hollowed-out part of his cane.

Escape was the ultimate act of defiance, and also the most difficult. No underground existed in the Korean War, as there had been during World War II, to shepherd escaped prisoners or downed airmen through hostile territory to safety. Moreover, the average American could not blend as easily into the civilian populace of North Korea as he might have in Europe.

Despite the difficulty, captured Marines made several attempts to escape from various prison camps. Unfortunately, most attempts ended in failure. Sergeant Donald M. Griffith slipped away from a guard who had fallen asleep only to be recaptured when he asked a peasant family for food. Captain Byron H. Beswick, although badly burned when his aircraft was shot down, tried with four other prisoners to escape from a column on the march, but guards recaptured them all. While being held prior to transfer to a prison camp, Private First Class Graham, the first Marine repatriated in Operation Big Switch, joined another Marine in an escape attempt, but both were recaptured trying to obtain food. After one unsuccessful try, punished as usual by a period of solitary confinement, First Lieutenant Robert J. Gillette, accompanied for a time by a South African pilot, succeeded in remaining at large for 10 days. Lieutenant Colonel Thrash and Major McLaughlin, together with First Lieutenant Richard Bell, got beyond the barbed wire, only to be cornered and forced to try to sneak back into the compound; Bell, however, was caught and punished with solitary confinement. Captain Martelli, who was reported killed but literally returned from the dead as a result of Big Switch, escaped and evaded recapture for 10 days. The final attempt, the product of one of the recently formed escape committees, took place on 1 July 1953, but the escapees remained at large for only a short time. With a ceasefire, and presumably an exchange of prisoners, drawing nearer, escape planning was suspended.

On rare occasions, special circumstances enabled captured Marines to escape and rejoin the United Nations forces. In August 1950, Private First Class Richard E. Barnett, driving a jeep, took a wrong turn and was captured by North Korean soldiers, who confined him in a cellar. When his captors took him along on a night attack, he managed to lag behind, threw a rock to distract the guard nearest him, and bolted into the darkness. He avoided recapture and rejoined his embattled unit before it had reported him missing.

Other successful escapes took place in May 1951. On the 15th, four days after their capture, Corporal Harold L. Kidd and Private First Class Richard R. Grindle, both from Company B, 1st Battalion, 7th Marines, made their way to United Nations lines. Later that month, a group of 18 Marines and an Army interpreter assigned to the 1st Marine Division were forced to join Chinese troops near the 38th Parallel. As the prisoners approached the frontlines, the Army interpreter, Corporal Saburo Shimamura, reported to the senior Marine, First Lieutenant Frank E. Cold that their captors planned to release them near Marine lines. The Chinese issued safe-conduct passes, lending credence to the report, but rather than wait to see if the Chinese would do as they said, the group took advantage of the distraction caused by United Nations artillery registering nearby and fled into the hills. They evaded the patrols looking for them; spread out improvised signal panels, and caught the attention of an aerial observer who reported their location. On 25 May, the day after their escape, two Army tanks clattered up to the men and escorted them to the nearest United Nations position.

In the spring of 1951, the enemy actually did release two prisoners, presumably as an overture to the policy of leniency announced shortly afterward. Whatever the reason, Corporal William S. Blair and Private First Class Bernard W. Insco were turned loose near the frontlines. Captured on 24 April, they regained their freedom on 12 May.

At times, captured Marines sought to invent fictitious statements that would ease the pressure on them by creating an illusion of cooperation. Master Sergeant John T. Cain, an enlisted pilot, had the misfortune of being mistaken for a senior officer because of his age and military bearing, thus becoming a prized target for the interrogators. He tried to satisfy them with elaborate details about a non-existent logistics command in which he claimed to have served. The pressure continued, however, until he was taken to a hillside one day, blindfolded, and subjected to a mock execution. He spent 84 days in solitary confinement before the Chinese intelligence officers either gave up or realized their error.

Attempts to outsmart the enemy could end in tragic failure, as demonstrated by the fate of Colonel Frank H. Schwable, the chief of staff of the 1st Marine Aircraft Wing, who became a prisoner when his aircraft was shot down on a reconnaissance flight in July 1952. Deprived of sleep and medical care, subjected to relentless pressure to confess to war crimes, he tried to frustrate the enemy by making a statement so obviously false that it not only would fail to help the Chinese, but also demonstrate their use of coercion. This action, however, merely whetted the appetite of interrogators determined to exploit his rank and position. They rewrote the document to create an illusion of truth, and forced him to choose between endorsing the revised version or spending the rest of his life in prison. After he was freed as Operation Big Switch drew to an end, a Court of Inquiry—functioning like a civilian grand jury—recommended that he not face a court-martial for aiding the enemy because he had resisted to the best of his ability. The inquiry also found that his future in the Marine Corps was "seriously impaired" by his conduct as a prisoner, thus putting an end to a once-promising career. Only one Marine faced a trial, an enlisted man convicted of fraternizing with the enemy and dismissed from the Service.

The 221 Marines captured in Korea endured an unexpected ordeal. Prisoners in past wars had suffered malnutrition, forced labor, and other acts of cruelty, but never before had their captors tried systematically to coerce them into participating in a propaganda campaign. Despite the harsh treatment the Chinese meted out, 197 Marines survived captivity and returned in Operation Big Switch.

Five Marines received official recognition for their steadfast conduct and strong leadership while prisoners of war. Lieutenant Colonel Thrash received a gold star signifying his second Legion of Merit. Major McLaughlin also earned the Legion of Merit, as did Major Walter R. Harris, another reactionary, who set up a communications network in one camp by designating locations where messages could be hidden and picked up. Harris also took advantage of the policy of leniency by organizing Spanish-language classes as cover for providing information and encouragement to offset Chinese attempts at indoctrination. Captain John P. Flynn, who refused despite torture to confess to war crimes and encouraged others to resist, earned the Navy and Marine Corps Medal. Master Sergeant Cain, who did not yield to solitary confinement and threats of execution, received a Letter of Commendation with Ribbon.

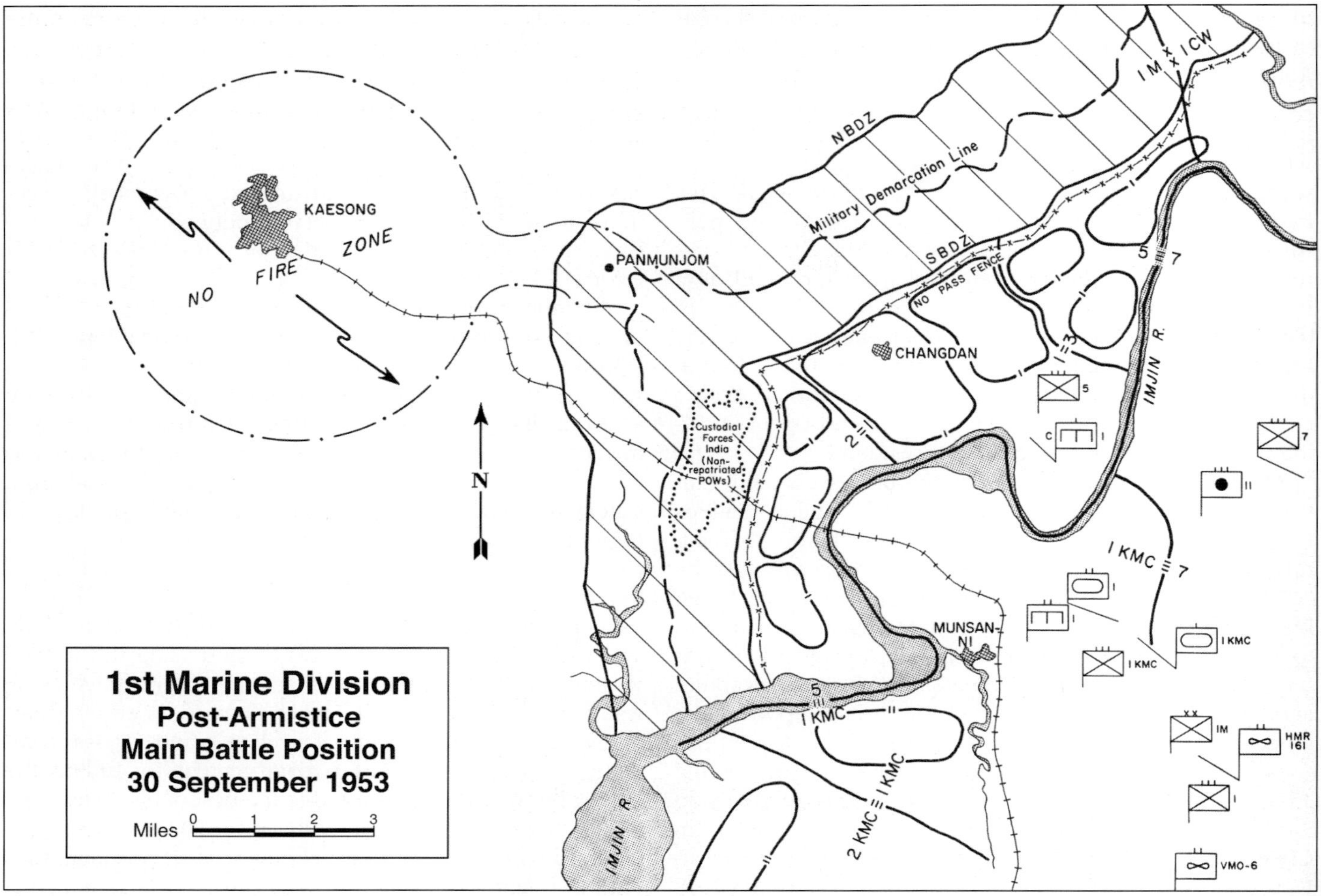

1st Marine Division Post-Armistice Main Battle Position 30 September 1953

(Continued from page 590)
agreement specified that each side, Communist and United Nations, maintain a force of 1,000 "civil police" to preserve the status of the demilitarized buffer. Since no civilian law enforcement agencies existed to provide this manpower, troops had to function as police. For maintaining the security of the Demilitarized Zone along its portion of the main battle position, the division on 4 September activated the 1st Provisional De-militarized Zone Police Company, drawing men from its infantry regiments. Attached to the 5th Marines, the police company, which initially numbered 104 men under Captain Samuel G. Goich, took over the crossing stations, manned observation posts to monitor the demilitarized Zone, and escorted members of the Military Armistice Commission, the Joint Observation Teams, and the Neutral Nations Supervisory Com-mission along with other persons authorized to enter the zone. According to a Marine Corps journalist, the average enlisted man assigned to the police company had to know "map reading on an officer level, first aid, radio, and understand the fine print of the ceasefire agreement like a striped-trouser diplomat."

Besides providing armed escorts—usually one-half-dozen Marines carrying rifles and pistols—and staffing fixed observation posts, the police company operated roving patrols. Traveling in radio-equipped jeeps these groups reported any unusual activity in the Demilitarized Zone. In case of a genuine emergency, a platoon, standing by as a mobile reserve, would respond.

Marine Demilitarized Zone police manning the observation posts monitored aerial activity as well as events on the ground. Besides keeping a record of all flights, they made sure that light reconnaissance planes had an appropriate clearance and that helicopters operating in front of the 5th Marines also had obtained permission for each flight. Helicopters responding to medical emergencies need not obtain specific approval from the ground commander.

The truce specified that prisoners of war who had refused repatriation to North Korea and China would enter the Demilitarized Zone, where their fellow countrymen, joined by Polish and Czech members of the Neutral Nations Supervisory Commission, could try to persuade them to return. The Communist persuasion teams required escorts from the police company, which therefore had to be tripled in size.

Main Battle Position

After the fighting ended, the 1st Marine Division formed the left of the line held by I Corps. The Commonwealth Division dug in on the right of the Marines then, in succession, the 1st Republic of Korea Division and the U.S. 7th Infantry Division. The Imjin River continued to challenge the ability of the Marines to move men and cargo, perhaps to a greater degree than before the armistice. A shallower and less defensible crescent of Marine-controlled ground lay north of the stream, which separated the supply center at Munsan-ni from the lines manned by the 5th Marines and forced the division to rely on bridges and fords to transport material and reinforcements across the Imjin. Because the salient was more vulnerable than before, the division would have to react faster and in greater strength to repulse any new attack there.

Machine gunners stand-by with loaded belts as riflemen take positions on the forward line. When the "Black Alert" siren sounded, Marines did not know whether it was "for real," but truce or not, they were always ready.

National Archives Photo (USMC) 127-N-A349038

The main battle position began at the No-Pass Fence, which served in effect as a trip wire to warn of an attack across the Demilitarized Zone, and consisted of successive lines of mutually supporting strongpoints, each stronger than the one to its front, that extended from the near edge of the Demilitarized Zone to the vicinity of the old Kansas and Wyoming Lines. Indeed, wherever possible, Marine engineers incorporated portions of these two lines in the new battle position. North of the Imjin, the 5th Marines manned the equivalent of an outpost line of resistance, in which firepower from large numbers of automatic weapons took the place of manpower. Within the northern salient, Colonel Tschirgi, the regimental commander, placed the 3d Battalion on the right, the 1st Battalion in the center, and the 2d Battalion on the left. South of the Imjin, the 7th Marines defended the right of the line, with the 1st Regiment of South Korean Marines in the center and the 1st Marines, the division reserve, manning a series of positions behind the South Koreans. The 11th Marines emplaced its howitzers behind the 7th Marines and prepared to fire in general support of the division.

In October, however, a major reshuffling took place south of the river. The ongoing exchange of prisoners—which brought to the Demilitarized Zone large numbers of North Koreans and Chinese who had refused repatriation, along with their former comrades who were trying to persuade the defectors to return—raised the

Department of Defense Photo (USMC) A365246

MajGen Randolph McC. Pate breaks ground for a new tuberculosis hospital donated and built by the 1st Marine Division. The completed hospital eventually bore a highly polished brass plaque, bearing a Marine Corps emblem and the words, "Built by the First Marine Division."

specter of another South Korean attempt to disrupt the settlement. To guard against this eventuality, troops from the reserve regiment, the 1st Marines, took over the sector manned by South Korean Marines.

Clearly, the strongpoints manned by the 5th Marines formed the most vulnerable portion of the new battle position. The regiment held a frontage of 36,000 yards, about three times the usual width during the Korean War. Under Colonel Rathvon McC. Tompkins, who assumed command on 2 August, the 5th Marines fortified hills that included wartime Outposts Marilyn and Kate, along with Boulder City and the Hook. Behind this arc of mutually supporting strongpoints, the regiment established bridgehead positions to protect the vital river crossings.

Maintaining an adequate supply of ammunition to defend the salient posed a problem because of the reliance on automatic weapons and the range at which they would have to open fire. Not only did the large number of machine guns devour .30-caliber ammunition faster than the usual mix of these weapons, rifles, and automatic rifles, the gunners would have to open fire sooner than normal if they were to close the gaps between outposts. As a result, the amount of ammunition at the outposts had to be increased and the reserve stocks moved forward from regimental to battalion dumps. These changes, however, could not solve the problem of dividing a finite supply of ammunition among the regiments and replenishing the 5th Marines by bridges or ferries vulnerable to

Courtesy of *Leatherneck* Magazine

Men of the 1st Battalion, 7th Marines, poke holes in a frozen Korean stream and ignore the icy waters to scrub their clothing before stowing it in their seabags for the trip back to the States.

flood, ice, or—if the fighting resumed—hostile fire.

The ceasefire provided an opportunity for training to hone the cutting edge of the Marine forces. Within the division, individuals and small units practiced their skills, and all but two of the infantry battalions took part in landing exercises. Besides remaining on alert in case the fighting resumed, Marine airmen also trained, sometimes in conjunction with Air Force squadrons. They practiced bombing and ground-controlled intercepts and supported an amphibious exercise at Tokchok-to Island involving the 1st Battalion, 7th Marines.

The division, as part of the Armed Forces Assistance to Korea program, also undertook 51 building projects—42 to them schools. As each school was completed it received kits containing instructional supplies and athletic gear. These extras were made possible solely by the donations of 1st Division Marines and matching funds from CARE, the Cooperative

National Archives Photo (USMC) 127-N-A366243

Two Marines from Company A, 1st Battalion, 1st Marines, show men of the U.S. Army's 24th Infantry Division their new positions near the Demilitarized Zone. The division's attitude about returning to Korea from Japan was summed up by two disillusioned privates: "Man, sixteen months in this forsaken place."

for American Remittances to Europe.

The post-armistice activity included athletics that provided a break from training and routine housekeeping. Athletic competition took place among Marine units and between organizations and bases. Indeed, some Marines stationed at Pyongtaek played on an Air Force softball team representing that airfield.

Routine housekeeping continued, however. Each day, for example, latrines had to be burned out, primarily to prevent disease but also to eliminate foul odors. A Marine poured a quart or so of gasoline into the privy, tossed in a lighted match, and stepped outside. Too much gasoline could produce spectacular results, as Corporal Lee Ballenger recalled. A buddy of his grew impatient when a succession of matches seemed to fizzle, poured gasoline directly from the can and was knocked down by a blast that splintered the wooden structure. Luckily, the flame did nothing worse than singe the Marine's hair and eyebrows.

The 1st Marine Division manned its portion of the main battle position for almost two years after the ceasefire. General Pate, in command at the time of the truce, was succeeded by Major General Robert H. Pepper (12 May 1954 to 22 July 1954), Major General Robert E. Hogaboom (23 July 1954 to 17 January 1955), and Major General Merrill B. Twining, who took over on 18 January 1955 and would bring the division home.

The first hint of the division's redeployment came from the

White House when a reporter "blandly asked the President if he could lend official credence to the 'authoritative sources' announcement that the First Division would be pulled out of Korea." The reporter gained a "grinning admission" from President Dwight D. Eisenhower that the Marines would be moved in the near future.

When the news hit Korea, it was greeted with mixed enthusiasm and skepticism—every Marine knew that "in the near future" could mean months or years. But soon clippings sent by mothers, wives, and girl friends began to arrive, lending credibility to the move. Marines who had been wary now became convinced. However, there was little spontaneous rejoicing; few celebrations or sayonara parties were planned.

Soon there was a gradual switch from hours spent on training to the packing and crating of gear. Most trucks, tanks, tents clothing, and other "common" gear would stay in Korea, while all Marine Corps "peculiar" gear would accompany the division to Camp Pendleton, California. Despite contemporary newspaper reports to the contrary, there was no massive swapping of troops from the 1st to the 3d Marine Division in Japan. The only exceptions were a Rocket Battery and an Armored Amphibian Company.

In mid-March 1955, the division turned its sector of the demarcation line over to the U.S. Army's 24th Infantry Division. Every Marine hoped that lugging water up steep, slimy hills, helmet baths, rations, sleeping bags, pot-belly stoves, long-handled skivves, parkas, DDT, and Mongolian "piss cutters" would become just memories.

The 1st Marine Aircraft Wing, now commanded by Brigadier General Samuel S. Jack, began withdrawing some of its components from South Korea in June 1956. The wing remained in the Far East, however. General Jack established his headquarters at Iwakuni Naval Air Station, Japan. Some of the wing's elements moved to Japan, but others continued to operate from bases in South Korea.

The 7th Marines rode trucks to the Munsan-ni railhead where they wait to board trains for Inchon. Divided into landing teams, the 7th Marines followed the 5th Marines, which was the first Marine unit to arrive in Korea.

Courtesy of *Leatherneck* Magazine

Courtesy of *Leatherneck* Magazine

Packed on board small landing craft like sardines, Marines were carried out to waiting Navy transports where they easily climb the cargo net with a full pack and rifle. They were homeward bound.

The War in Perspective

Marine participation in the Korean War began with the desperate defense of the Pusan Perimeter, continued when the 1st Marine Division spearheaded a daring amphibious assault at Inchon, and, after the Chinese intervention forced the United Nations to withdraw from the Yalu River to the vicinity of the 38th Parallel, ended in static warfare fought along a battle line that extended across the width of the Korean peninsula. A highly mobile amphibious assault force thus became tied to fixed positions, a condition that lasted from the spring of 1952 until the armistice in July of the following year. Indeed, by January 1953, defense overshadowed the offense, even though defensive operations required frequent and often bloody counterattacks to maintain the positions held by the United Nations. The Chinese succeeded in capturing and holding several outposts, but the 1st Marine Division clung tenaciously to its segment of the main line of resistance, helping force the enemy to agree to the ceasefire, with provisions for the exchange of prisoners of war, that had become the overriding American objective of the war.

Before North Korea invaded the South, American strategy focused on the defense of Western Europe against the Soviet Union and its satellites. The aggression in Asia presented an unexpected challenge, which the Truman administration decided to meet, although not at the expense of weakening the newly formed North Atlantic Treaty Organization. As a result, the United States fought a limited war in Korea, withholding nuclear weapons and relying with few exceptions on the weapons of World War II or improved versions of them.

In Korea, the Marines used to deadly effect the weapons they had, whether old or improved, massing artillery, supplementing the howitzers with rockets, and when necessary employing tanks as pillboxes on the battle line, but new weapons lay just over the horizon. The weight of the standard infantry weapons of World War II—for example, the M1 rifle weighing 9.5 pounds, and the Browning automatic rifle almost twice as heavy—aroused interest in lighter, fully automatic weapons like the sub-machine guns used by Chinese infantry. A lightweight automatic rifle, the M16, was therefore developed but did not become the standard weapon for Marine infantrymen until the Vietnam War. Although the steel helmet remained essentially unchanged since World War II, body armor had undergone improvement because of experience in Korea, as had the original antitank rocket launcher or Bazooka.

Infantry tactics changed to meet the evolving demands of the war on the ground. Fighting at night, especially patrol actions, received greater emphasis, for instance. Also, by the time of the battles for Berlin, East Berlin, and Boulder City in July 1953, the 1st Marine Division was committed to a defense in depth, made up of mutually supporting strongpoints, a principal followed in establishing the post-armistice main battle line.

Courtesy of *Leatherneck* Magazine

Col Raymond L. Murray, who commanded the 5th Marines during its first action in Korea almost five years before, greets the regiment's current commanding officer, Col Robert H. Ruud, upon his arrival at San Diego, California, in March 1955.

The construction of the bunkers and covered fighting positions that made up these strongpoints had improved greatly since the Marines took over their segment of the Jamestown Line in the spring of 1952.

The most dramatic changes took place in aviation. The new jet fighters, which had barely seen action against the enemy during World War II, proved their worth in support of the 1st Marine Division and also in aerial combat. Fourteen Marine pilots, flying aircraft of the 1st Marine Aircraft Wing downed 16 hostile airplanes, ranging from biplanes engaging in nighttime harassment to modern jets. Eleven other Marine pilots, flying North American F-86 jet fighters in U.S. Air Force squadrons, were credited with downing 21 Chinese jet fighters. Moreover, the helicopter showed tremendous potential in limited use during the Korean fighting; this potential, however, would not be realized

Marines of the 1st Division parade through downtown San Diego following their return from Korea. A hero's welcome was furnished the combat-clad veterans as they marched up Broadway to Balboa Park.

Marine Corps Historical Center Photo Collection

fully until the Vietnam conflict.

Because the Korean War had demonstrated the importance of integrated action between air and ground components, the Marine Corps in January 1953 established the 1st Provisional Air-Ground Task Force at Kaneohe in Hawaii. This unit, built around an infantry regiment and an aircraft group, stood ready for deployment throughout the Pacific.

In short, the Marine Corps had responded quickly and effectively to the peculiar demands of the Korean War without losing sight of its amphibious mission and its role as a force in readiness. To Marine eyes, the use of one of its amphibious divisions in an extended ground campaign seem-ed an aberration, a misapplication of resources, no matter how well that division had fought. Because of this commitment to amphibious operations, the 1st Marine Division participated in landing exercises whenever possible during the course of the fighting, activity that may well have reminded the enemy of the abiding threat of another assault like the victory at Inchon.

After the truce went into effect, amphibious training continued in the Far East for the 1st Marine Division, until its return of the United States. In August 1953, as the Demilitarized Zone took shape, the 3d Marine Division, teamed with two aircraft groups, arrived in Japan. A program of amphibious training, which included regimental landings at Iwo Jima and Okinawa, got underway while the 3d Marine Division was based in Japan and continued after its redeployment to Okinawa.

The continued existence of a force, normally one battalion, afloat with the Sixth Fleet in the Mediterranean provided a further demonstration of the abiding orientation of the Marine Corps toward amphibious warfare. Since 1949, a succession of battalions drawn from the 2d Marine Division at Camp Lejeune, North Carolina, had embarked with the Sixth Fleet. Clearly, the Korean fighting in 1952 and 1953 had not converted the Marine Corps from amphibious operations to extended ground warfare, from assault to attrition, or affected its dedication to the air-ground team.

A Douglas F3D is hauled on board ship after being ferried out by two amphibious trucks, as the 1st Marine Aircraft Wing transferred its headquarters and most of its aircraft from Korea to Iwakuni, Japan, in June 1956. Postwar plans called for the wing to occupy bases in both Korea and Japan.

National Archives Photo (USMC) 127-N-A349563

About the Author

Bernard C. Nalty, a member of the Marine Corps historical program from October 1956 to September 1961, collaborated with Henry I. Shaw, Jr., and Edwin T. Turnbladh on *Central Pacific Drive*, a volume of the *History of U. S. Marine Corps Operations in World War II.* He also completed more than 14 short historical studies, some of which appeared in *Leatherneck* magazine or the *Marine Corps Gazette.* He joined the history office of the Joint Chiefs of Staff in 1961, transferred in 1964 to the Air Force history program, and retired in 1994. Mr. Nalty has written or edited a number of publications, including *Blacks in the Military: Essential Documents, Strength for the Fight: A History of Black Americans in the Military, The Vietnam War, Tigers Over Asia, Air Power and the Fight for Khe Sanh,* and *Winged Shield, Winged Sword: A History of the U.S. Air Force.* In addition to contributing to this series on the Korean War by writing *Stalemate: U.S. Marines from Bunker Hill to the Hook*, he took part in the Marines in World War II commemorative series, completing two pamphlets, *Cape Gloucester: The Green Inferno* and *The Right to Fight: African-American Marines in World War II.*

Sources

The best account of operations by Marines during 1953, their role in establishing the Demilitarized Zone, and their eventual withdrawal from South Korea appears in *Operations in West Korea*, volume five of *U. S. Marine Corps Operations in Korea, 1950-1953* (Washington, D.C.: Historical Division, Headquarters, U.S. Marine Corps, 1972), by LtCol Pat Meid, USMCR, and Maj James M. Yingling, USMC.

The Marine Corps historical program also has dealt thoroughly with the treatment received by Marines held as prisoners of war, their reaction to deprivation and hostile pressure, and their repatriation. James Angus MacDonald has made extensive use of interviews with former prisoners in his *The Problems of U. S. Marine Corps Prisoners of War in Korea,* published by the History and Museums Division, Headquarters, U. S. Marine Corps, in 1988.

Walter G. Hermes discusses Marine Corps operations in his contribution to the official history of Army activity during the Korean War, *Truce Tent and Fighting Front* (Washington, D.C.: Office of the Chief of Military History, 1966), a volume that deals with negotiation as well as fighting.

Martin Russ, a wartime Marine Corps infantryman, has written the best known and most revealing memoir covering the final months of the fighting, *The Last Parallel: A Marine's War Journal* (New York: Rinehart, 1957).

Personal accounts by other Marines appear in *Korean Vignettes: Faces of War* (Portland, OR: Artwork Publica-tions, 1996), a compilation of narratives and photographs by 201 veterans of the Korean War, prepared by Arthur W. Wilson and Norman L. Stickbine.

Valuable insights and first-hand accounts are also available in *The Korean War: The Uncertain Victory; The Concluding Volume of an Oral History* (New York: Harcourt Brace Jovanovich, 1988), by Donald Knox with additional text by Alfred Coppel.

William H, Jantzen, a participant in one of the fiercer actions of 1953, has written a riveting account of that fight in "A Bad Night at Reno Block," in the March 1998 issue of *Leatherneck* magazine.

The *Marine Corps Gazette* has analyzed the tactics and lessons learned of the Korean War in three especially useful articles: Peter Braestrup's "Outpost Warfare" (November 1953) and "Back to the Trenches" (March 1955); and "Random Notes on Korea" (November 1955) by LtCol Roy A. Batterton. Also of use was the article by MSgt Paul Sarokin, "Going Home," in the May 1955 issue of *Leatherneck*.

The personal papers collection of the Marine Corps history program contains journals, photographs, letters, memoirs, and at least one academic paper, a master's thesis on outpost warfare by a Marine, Maj Norman W, Hicks. For events of the year 1953, the most valuable of these items were the submissions by Eldon D. Allen and Gen Vernon E. Megee.

WM
2

CORSAIRS TO PANTHERS
U.S. Marine Aviation in Korea

by Major General John P. Condon, USMC (Ret)
Supplemented by
Commander Peter B. Mersky, USNR (Ret)

The first major surprise of the post World War II years came into play when in late June 1950, the United States found itself responding in crisis fashion to the North Korean invasion of the new republic of South Korea, just four years and nine months after VJ-Day. The nation became involved in Korea as a result of the Cairo and Yalta conferences in which the United States and the Soviet Union agreed to the concept of a free and independent post-war Korea. Included in the agreement was a joint occupation of the country by the two powers, with the Soviets north of the 38th Parallel and the United States south. The concept of the occupation had a general objective of settling down Korea for a period so that it could learn to govern itself as a nation after many decades of Japanese rule. As the United States was painfully learning, however, it soon became apparent that what the Soviets said was one thing and what they intended was quite another with respect to a free, independent, and democratic Korea. When in 1948, they refused to participate in elections, supervised by the United Nations to form the first National Assembly, the hopes for a united Korea died. The Soviets formed a separate Communist state in their sector, the People's Democratic Republic of Korea. With the elections completed for the National Assembly in the south, the Republic of Korea (ROK) was established and the United States trusteeship in the country came to an end.

On 25 June 1950, the North Koreans attacked with nine well-equipped infantry divisions, spearheaded by one armored division equipped with Soviet-built T-34 tanks. The Republic of Korea's army had been in existence for just about a year and could only oppose the invasion with four lightly equipped divisions and one additional regiment. Needless to say, although there were some spirited but isolated small unit defensive actions, the Republic's forces were no match for the invaders. The North Koreans reached out with rapidly advancing armored columns, moving almost at will during the first four days. Seoul fell on 28 June, and at that time, the ROK army had 34,000 troops missing, although many of them later returned to their units. With the capture of Seoul, the invaders halted to regroup and those ROK forces, which were still intact, fell back through Suwon to set up some form of new defensive positions. The South Korean government had displaced to Taejon well to the south when the fall of Seoul became imminent. This state of near collapse was the basic situation faced by the United States and the United Nations in the opening week of the war. It was the first

AT LEFT: *After strafing enemy troops positions, a Vought F4U Corsair pilot hunts out a suitable target for his remaining napalm bomb.* Department of Defense Photo (USMC) A133540

A graduate of the open cockpit and silk scarf era of Marine Corps aviation, BGen Thomas J. Cushman saw service in Nicaragua, Haiti, and the Central Pacific before being named Assistant Wing Commander, 1st Marine Aircraft Wing, in June 1950. He commanded the wing's forward echelon, which provided air support for the 1st Provisional Marine Brigade at Pusan, and later served as Commanding General, Tactical Air Group (X Corps) during Inchon and the advance on Seoul.

Department of Defense Photo (USMC) A2108

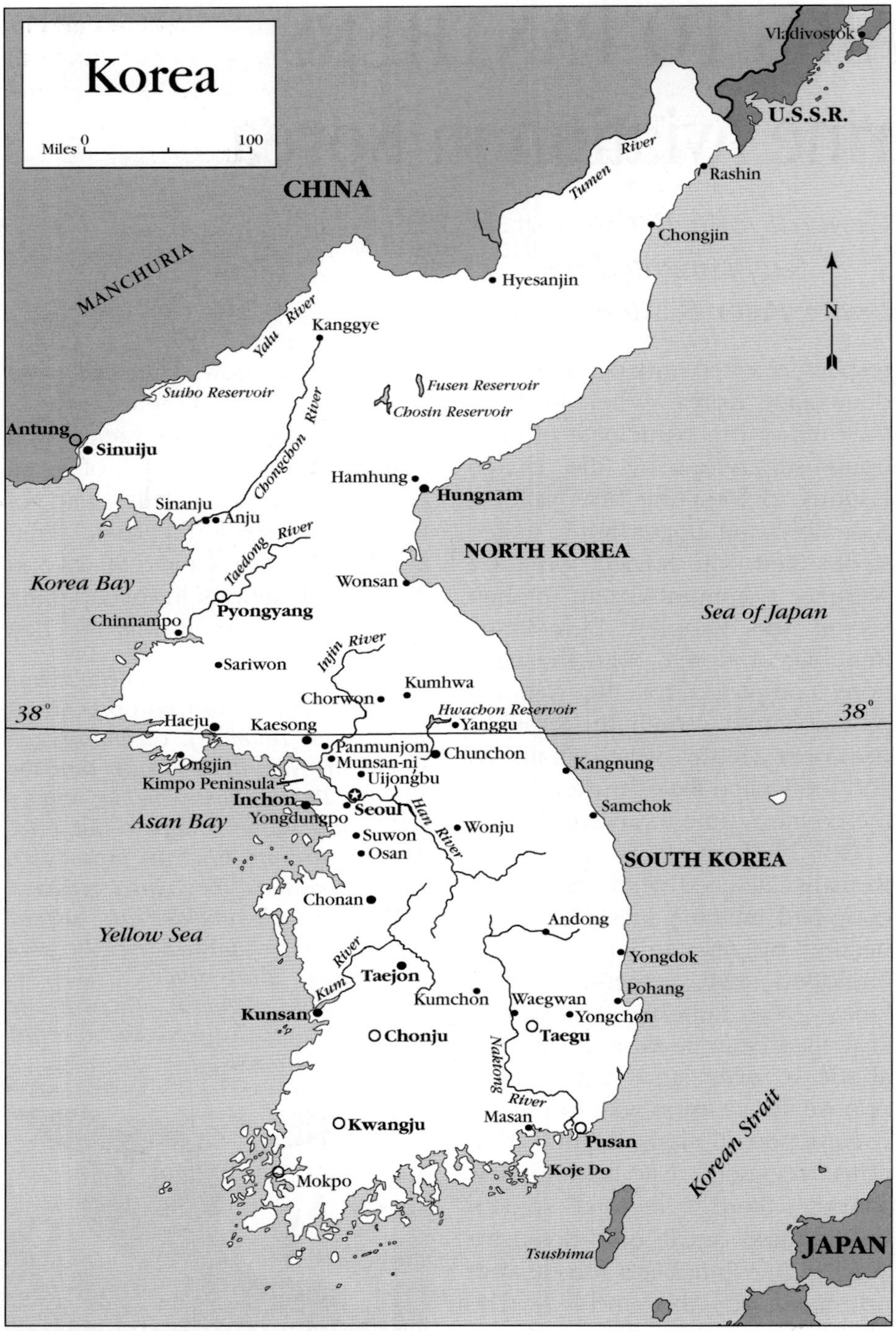

time that a Soviet-supported state was permitted to go as far as open warfare in their post-World War II depredations, and it constituted a definite showdown between the Communist and non-Communist worlds.

The United States responded to the invasion of South Korea both independently, and through strong support and leadership in a United Nations resolution condemning the breaking of world peace by the North Koreans. President Harry S. Truman gave General of the Army Douglas MacArthur, Commander in Chief, Far East, the go-ahead to send Army units into Korea from Japan and to take other actions in support of the shocked and shattered ROK forces. It is important to note that of the 56 respondents to the United Nations resolution, only three were opposed: the Soviet Union, Poland, and Czechoslovakia.

The United Nations participating pledges were substantial and included aircraft, naval vessels, medical supplies, field ambulances, foodstuffs, and strategic materials. In addition to the Army forces authorized by President Truman, a naval blockade of the entire Korean coast was ordered, and U.S. Air Force units based in Japan were authorized to bomb specific targets in North Korea. It is also important to note that these critical actions met with the wholehearted approval of the American people. Americans also applauded the strong stance of the United Nations, and they repeatedly expressed their thoroughgoing pride in the responses of their nation to the seriously deteriorating international situation.

Marine Brigade

In response to urgent requests for American reinforcements from the Far East Command, and as a result of unit offerings and proposals from the United States, the 1st Provisional Marine Brigade was activated on 7 July 1950. It was an air-ground team built around the 5th Marine Regiment and Marine Aircraft Group 33 (MAG-33), both based on the west coast at Camp Pendleton and Marine Corps Air Station, El Toro, respectively. Brigadier General Edward A. Craig, with Brigadier General Thomas J. Cushman, a renowned and experienced Marine aviator, assigned as his deputy commander, commanded the brigade.

The time and space factors in the activation and deployment of the brigade were, to say the least, something extraordinary. Activated on 7 July, the unit was given at the same time a sailing date five to seven days later. In looking back at this first of the post-World War II surprises, it is again important to fully understand what the radical demobilization steps had accomplished. It is impossible to list them all in this short account, but it will suffice to point out that rifle companies were at two platoons instead of three, infantry battalions at two rifle companies instead of three, and deep cuts in normal logistic back-ups of all types of "ready" supplies of everything from ammunition to field rations were common. It also must be emphasized that normally, after the cutbacks and reductions following World War II, the division-wing teams on both coasts would have been very hard-pressed to deploy one reinforced brigade of regiment-group-sized in 30 days, let alone the seven days granted in this case. A super performance is simply a classic understatement for the mount-out of the 1st Provisional Marine Brigade to Korea.

In late June 1950, Marine Fighter Squadron 214 was the only Corsair squadron operating from El Toro. Marine Fighter Squadron 323 was in the process of returning to the air station following several

Marine Corps Air Units and Primary Aircraft

Forward Echelon, 1st Marine Aircraft Wing (July–September 1950)

- Marine Aircraft Group 33
  - Headquarters Squadron 33
  - Service Squadron 33
  - Marine Fighter Squadron 214 (Vought F4U Corsair)
  - Marine Fighter Squadron 323 (Vought F4U Corsair)
  - Marine Night Fighter Squadron 513 (Grumman F7F Tigercat, Douglas F3D Skyknight)
  - Marine Tactical Air Control Squadron 2

1st Marine Aircraft Wing (September 1950 – July 1953)

- Headquarters Squadron 1
- Marine Wing Service Squadron 1
- Marine Wing Service Group 17
  - Headquarters Squadron 17
  - Marine Air Base Squadron 17
  - Marine Aircraft Repair Squadron 17
- Marine Aircraft Group 12
  - Headquarters Squadron 12 (Vought F4U Corsair, General Motors TBM Avenger)
  - Service Squadron 12
  - Marine Air Base Squadron 12
  - Marine Aircraft Maintenance Squadron 12
- Marine Aircraft Group 33
  - Headquarters Squadron 33 (Vought F4U Corsair, General Motors TBM Avenger)
  - Service Squadron 33
  - Marine Air Base Squadron 33
  - Marine Aircraft Maintenance Squadron 33
- Marine Fighter Squadron 115 (Grumman F9F Panther)
- Marine Attack Squadron 121 (Douglas AD Skyraider)
- Marine Fighter Squadron 212 (Redesignated Marine Attack Squadron 212 on 10 June 1952) (Vought F4U Corsair, Vought AU-1 Corsair)
- Marine Fighter Squadron 214 (Vought F4U Corsair)
- Marine Fighter Squadron 311 (Grumman F9F Panther)
- Marine Fighter Squadron 312 (Redesignated Marine Attack Squadron 312 on 1 March 1952) (Vought F4U Corsair)
- Marine Fighter Squadron 323 (Redesignated Marine Attack Squadron 323 on 30 June 1952) (Vought F4U Corsair, Vought AU-1 Corsair)
- Marine Attack Squadron 332 (Vought F4U Corsair)
- Marine Attack Squadron 251 (Douglas AD Skyraider)
- Marine Night-Fighter Squadron 513 (Vought F4U Corsair, Grumman F7F Tigercat, Douglas F3D Skyknight)
- Marine Night-Fighter Squadron 542 (Grumman F7F Tigercat, Douglas F3D Skyknight)
- Marine Transport Squadron 152 (Douglas R4D Skytrain)
- Marine Ground Control Squadron 1
- Marine Air Control Group 2
- Marine Tactical Air Control Squadron 2
- Marine Ground Control Intercept Squadron 1
- Marine Ground Control Intercept Squadron 3
- Marine Photographic Squadron 1 (McDonnell F2H-P Banshee)
- Marine Composite Squadron 1 (Douglas AD Skyraider)
- Marine Helicopter Transport Squadron 161 (Sikorsky HRS-1 Helicopter)
- Marine Observation Squadron 6 (Consolidated OY Sentinel, Sikorsky H03S Helicopter, Bell HTL Helicopter)
- 1st 90mm Antiaircraft Artillery Gun Battalion

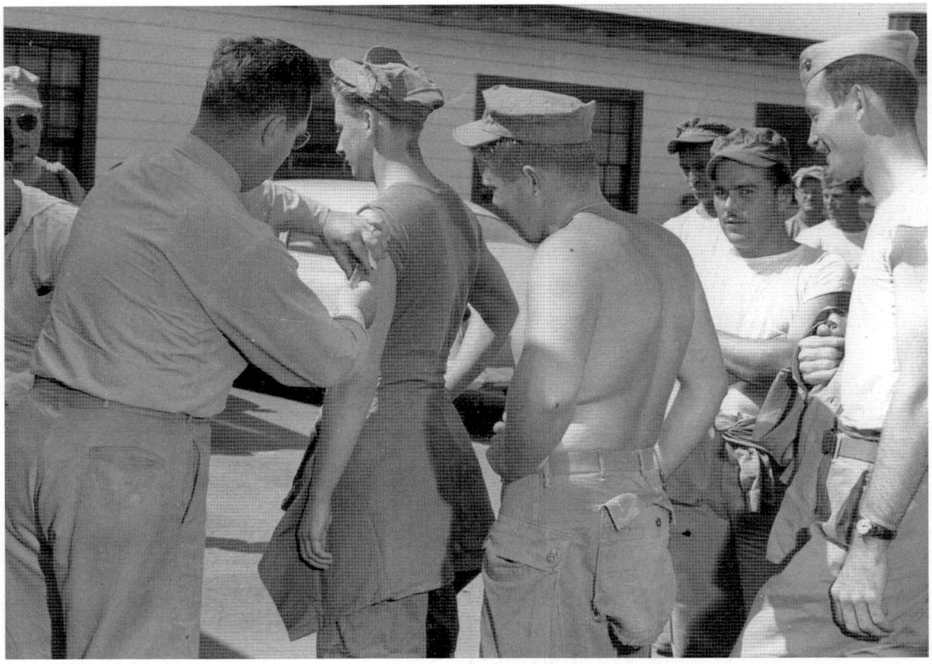

Department of Defense Photo (USMC) A130091

Marines of the wing's forward echelon receive their inoculations in early July at Marine Corps Air Station, El Toro, before leaving for Korea on board the escort carrier Badoeng Strait *(CVE 116) and transport* General A. E. Anderson *(AP 111).*

months of training at Camp Pendleton and on board the Essex-class carrier *Philippine Sea* (CV 47) off the California coast. Following its return, the aircraft and men of the squadron quickly prepared to deploy. The Black Sheep of Marine Fighter Squadron 214 likewise were in a high state of readiness, but had been "out of pocket" when the war broke out. The squadron was enroute to Hawaii on board the escort carrier *Badoeng Strait* (CVE 116), having been awarded the privilege of hosting the annual Naval Academy midshipman's cruise, when it received word of the North Korean invasion of South Korea. It was not long before the squadron's commanding officer, Major Robert P. Keller, was summoned to Headquarters Fleet Marine Force, Pacific, at Camp Smith. After flying off the carrier, Keller met with Colonel Victor H. Krulak, Lieutenant General Lemuel C. Shepherd, Jr.'s chief of staff. With a tone of dead seriousness only Krulak could project, he asked Keller: "Major, are you ready to go to war?" Keller, reflecting on the training and experience level of the squadron, assured him that the Black Sheep were ready. With no time to enjoy Hawaii, the midshipmen were offloaded and the carrier made a beeline back to California in anticipation of mobilization orders.

As can be readily imagined, Camp Pendleton and El Toro were twin scenes of mad confusion as Marines arrived hourly by train, bus, and plane, and "demothballed" equipment of all types arrived for marking and packing, literally at a rate measured in tons per hour. "Sleep on the boat" was the order of the day as the date of embarkation at San Diego and Long Beach for the first elements, 12 July, rapidly drew closer. By 14 July, all units were on board assigned shipping and underway westward.

At departure, the total strength of the brigade was 6,534. MAG-33 totaled 192 officers and 1,358 enlisted men, composed principally of the two fighter squadrons, VMF-214 and VMF-323, a night fighter squadron, VMF(N)-513, and an observation squadron, VMO-6. An important and historic component of VMO-6 was a detachment of four HO3S-1 Sikorsky helicopters, hurriedly assigned and moved to El Toro from the helicopter development squadron at Quantico, HMX-1. This was the first time that the United States Armed Services had actually deployed helicopters in a unit mounting out for combat service overseas, although a few had been

The Badoeng Strait *(CVE 116) was the carrier home from which the "Death Rattlers" of Marine Fighter Squadron 323 launched their initial Korean combat missions in August 1950.*

Marine Corps Historical Center Photo Collection

Courtesy of Cdr Peter B. Mersky, USNR (Ret)

An F4U Corsair of Marine Fighter Squadron 214 launches from the escort carrier Sicily *(CVE 118). In its second major war in five years, as the workhorse of Marine Corps aviation, the "U-Bird" was still considered a first-rate close air support aircraft.*

tried out in both the European and Pacific theaters at the end of World War II on an experimental basis. Aircraft strength at deployment added up to 60 Vought F4U Corsairs, eight Consolidated OY "Sentinels," and the four Sikorsky HO3S-1s.

By 16 July, the brigade commander and a key advance party took off by air for conferences and briefings at Honolulu and at the Far East Command in Tokyo. As these meetings progressed and the possibilities of immediate commitment on arrival of the main body came clearly into focus, an original plan to hold the brigade in Japan temporarily was abandoned. This was a result of the deteriorating position of the United Nations Command in Korea, which by the fourth week of the war had drawn into a perimeter-type defense of the port of Pusan at the southern tip of the peninsula On 2 August, the brigade debarked at Pusan and on 3 August at 0600, departed Pusan for the front by rail and 50 borrowed Army trucks. MAG-33 shipping had been directed to Kobe when the force reached far eastern waters, and debarkation began there on 31 July. The fighter squadrons were flown off the *Badoeng Strait* to Itami near Osaka, where they were checked for combat by the ground crews and hastily transported overland from Kobe. With just one refresher hop at Itami, VMF-214, now commanded by Lieutenant Colonel Walter E. Lischeid, landed on board the escort carrier *Sicily* (CVE 118) for operations on 3 August, and on 5 August, Major Arnold A. Lund's VMF-323 returned to the *Badoeng Strait* for the same purpose. VMF(N)-513, under the command of Major Joseph H. Reinburg, was assigned to the Fifth Air Force for control and began shore-based operations from Itazuke Airfield on the southern island of Kyushu. Its mission was to provide night "heckler" operations over the brigade and the Korean combat area generally, while the two carrier-based units would provide close air support. To furnish the essential communications and tactical links for close air support and general direct support to the brigade, on arrival at Kobe a tank landing ship was waiting to reembark Marine Tactical Air Control Squadron 2 (MTACS-2), led by Major Christian C. Lee, and the ground echelon of VMO-6, commanded by Major Vincent J. Gottschalk, for transport to Pusan. The aircraft of VMO-6 were readied at Kobe and Itami and ferried to Pusan by air. Thus the air-ground integrity of the brigade was held intact as it entered its first combat less than 30 days after activation, a truly remarkable achievement.

Pusan Perimeter Air Support

At the time of the commitment to action of the brigade in early August 1950, the United Nations defense had contracted to a perimeter around the southernmost port of Pusan. It was vital that the perimeter contract no more, since the port was the logistic link to a viable base position in support of a United Nations recovery on the peninsula. In bringing this desirable outcome to reality, the brigade became known variously as the "Fire Brigade," the "Marine Minutemen," and other into the breach sobriquets. In the process of their month-in-the-perimeter employment, the Marines were accorded the honor of restoring the confidence of United Nations troops through destruction of the myth that the North Koreans were somehow invincible. Marine aviation carried its portion of the brigade load in this restoration of pride and stature, once again relying on its ability to operate afloat as well as ashore. Like the deployments on board the fast carriers in World War II, the basing of VMFs -214 and -323 on board the escort carriers *Sicily* and *Badoeng Strait* once again showed the lasting wisdom of the long-standing commonality policies between naval and Marine aviation.

Sicily *Highlights from 1946 to 1951*

The commanding officer of VMF-214, LtCol Walter E. Lischeid, center, and the Sicily*'s captain, Capt John S. "Jimmy" Thach, seated right, listen intently as returning pilots report on the results of their mission.*

From *Sicily*, in the form of eight Corsairs, came the first Marine offensive action of the war. Led by Major Robert P. Keller, the squadron's executive officer, the eight VMF-214 Corsairs took off at 1630 on 3 August in a strike against Chinju and the Communist-held village of Sinban-ni. Using incendiary bombs, rockets, and numerous strafing runs it was a more than suitable and impressive greeting for the previously almost unopposed North Korean troops. On the following day, 21 additional sorties were flown to help relieve the pressure on the Eighth Army southern flank. These struck at bridges, railroads, and troop concentrations in the Chinju and Sachon areas. With -214 continuing the march from

On the afternoon of 3 August, the "Black Sheep" of VMF-214 made their first air strikes against North Korean positions from Chinju to Sachon. Earlier in the day the squadron's 24 planes landed on board the Sicily, *then cruising in the Tsushima Straits, following two days of field carrier landing practice and a short flight from Itami Air Force Base, Japan.*

Sicily *Highlights from 1946 to 1951*

Department of Defense Photo (USMC) A130914

Against a backdrop of rugged Korean terrain, an OY Sentinel light observation aircraft of Marine Observation Squadron 6, piloted by the squadron's commanding officer Maj Vincent J. Gottschalk, spots concentrations of North Koreans for Marine Corsairs to sear with napalm.

the deck of the *Sicily*, VMF-323 joined the fray from *Badoeng Strait* on 6 August with strikes west of Chinju along the Nam River, hitting large buildings and railroad lines with rockets and 500-pound bombs. Because the carriers were so close to the frontlines of the perimeter, the strikes could reach their targets in a matter of minutes at almost any point where support was requested. That the North Koreans realized something new had been added was apparent when on 11 August, -323 teamed up with North American F-51 Mustangs of the U.S. Air Force near Kosong in what became known as the "Kosong Turkey Shoot." In this action, the Corsairs hit a convoy of more than 100 vehicles of a North Korean motorized regiment, a mixed bag of jeeps, motorcycles, and troop-carrying trucks, stopping the convoy at both ends on the road. They got every one with the help of the F-51s. While hitting the jackpot in this manner was not an every-day occurrence, the daily sorties from the two carriers so conveniently nearby, began to climb in both number and effectiveness all along the length of the entire perimeter. MAG-33 aircraft were constantly orbiting on station over the frontline as the ground forces advanced, and communications within the air-ground team was steady from the Tactical Air Control Parties (TACP) with the battalions, all the way back to the brigade headquarters. The air support system, controlled by the active presence of Marine Tactical Air Control Squadron 2 and VMO-6 at brigade headquarters from 6 August on, worked to the wondrous amazement of the associated U.S. Army and other United Nations units.

The Fifth Air Force exercised overall control of tactical air operations in Korea, but Marine aviation units, as components of an integrated Fleet Marine Force, operated in support of the brigade as their highest priority, and in support of other United Nations units as a lower priority. The brigade control organization consisted of three battalion Tactical Air Control Parties and one regimental TACP, each consisting of one aviation officer, an experienced and fully qualified pilot, and six enlisted technicians. Each party was equipped with a radio jeep, portable radios, and remoting communications gear. In addition, there were the facilities and personnel of MTACS-2 at brigade headquarters, as well as the brigade air section of the staff, one officer and one enlisted. The air section was responsible for air planning, tactical control, and coordination of supporting aircraft. Lastly, but certainly of no lesser importance, there was also the brigade observation section consisting of the tactical air observer, three gunnery observers, and the light observation and rotary-wing aircraft of VMO-6. When supporting other United Nations forces, Marine air units operated under the Air Force-Army system for tactical air control.

The foregoing gives an abbreviated description of the brigade air support system, which operated very effectively through some of the most rugged fighting of the Korean War. The operations in the Pusan Perimeter basically were divided during the six-week period into three major actions. The first was the counterattack in the extreme southwest which ran approximately from 3 to 15 August, and was known as the Sachon offensive; the second was the First Naktong counteroffensive, from 16 to 19 August; and the third was the Second Naktong from 3 to 5 September. All three, rugged as they were, resulted in thorough defeats for the Communist forces but were not without cost to the brigade: 170 killed, 2 missing, and 730 wounded. Estimates showed that the brigade had inflicted almost 10,000 casualties in killed and wounded on the enemy units faced in the six weeks of its participation in the perimeter operation.

Throughout these three vital actions the morale and confidence of the United Nations forces facing the North Koreans was restored. Marine air units of the brigade carried their part of the rebuilding process on a daily, and nightly, basis. In addition, Major Vincent J. Gottschalk's VMO-6 established so many "firsts" with its helicopters during the period that it was obvious that a major tactical innovation was in the making. The new steed that Brigadier General Craig, his chief of staff, Colonel Edward W. Snedeker, and G-3, Lieutenant Colonel Joseph L. Stewart, had discovered clearly indicated that the helicopter was fully capable of working a revolution in command observation, inspection, and staff procedures.

Most importantly, it was apparent from action in the Pusan Perimeter that the Marine Corps air-ground team concept was a winner. The tight knit integration of close air support into the ground scheme of maneuver proved to be devastatingly effective. From 3 August to 14 September 1950, the two squadrons of MAG-33 on board the carriers (VMFs -214 and -323) and the shore-based night fighters of VMF(N)-513, flew 1,511 sorties, of which 995 were close air support missions in response to requests from engaged ground troops.

The strikes by Marine aircraft not only decimated the enemy's forces, but they rekindled the bond between air and ground that characterizes the Marine air-ground team. Ground Marines gained courage from seeing their fellow Marines in Corsairs swoop in to deliver ordnance oftentimes within 100 yards of the frontlines. The pilots became part of the fight on the ground and as a result gained a sense of pride and accomplishment in helping ground troops accomplish their mission. Captain John E. Barnett, one of the Corsair pilots summed up how aviators felt about their relationship with ground Marines: "With consummate conceit we doubted not that Marines were the best pilots supporting the best infantry, employing the best tactics; a brotherhood non-parallel. Pilots were in awe of the infantry, lavish praise from whom (regarding close air support) fed our ego."

To sum up the air component performance in the perimeter, the words of General Craig said it all: "The best close air support in the history of the Marine Corps. . . outstanding in its effectiveness." And from the envious viewpoint of an adjacent Army infantry regiment:

> The Marines on our left were a sight to behold. Not only was their equipment superior or equal to ours, but they had squadrons of air in direct support. They used it like artillery. It was 'Hey, Joe—This is Smitty—Knock the left off that ridge in front of Item Company.' They had it day and night. It came off nearby carriers, and not from Japan with only 15 minutes of fuel to accomplish the mission.

While there was literally no air opposition from the North Korean Air Force because it had been wiped out by the initial U.S. Air Force efforts at the beginning of the war, such accolades in addition to others were pleasant music to Marine aviation and to the brigade

For the first time in any war helicopters were used on the battlefront as liaison aircraft and in evacuating the wounded. Although underpowered and hard-pressed to carry more than a pilot, crew chief, and one evacuee, the Sikorsky H03S was relatively dependable.

Marine Corps Historical Center Photo Collection

as a whole. The performance of the brigade was a vital factor in stopping the invasion in August. The punishment meted out to the North Korean units was so severe that it set them up for the crushing defeat, which followed in September.

Inchon Landing

The North Korean invasion of South Korea occurred while Lieutenant General Lemuel C. Shepherd, Jr., was enroute to Hawaii to take over as Commanding General, Fleet Marine Force, Pacific. He cut short his trip on 25 June when he received word of the North Korean action and immediately proceeded to San Francisco and then directly to Hawaii. Following a rapid succession of conferences and briefings, he was off to Japan for meetings with the Far East Command. History was truly made in one of these meetings at Far East Command in which General MacArthur and General Shepherd were the major participants. In a sense, it was a reunion between the two because a few years before on New Britain, General Shepherd had been the assistant division commander of the 1st Marine Division when it was under the operational control of General MacArthur. It had long been a plan of MacArthur's that should a serious overrun of any part of his forces occur, he would attempt to recoup through the use of appropriate delay followed by an amphibious landing in the enemy rear. This was the primary subject to be discussed in the 10 July meeting.

The upshot of this historic conference was that following General Shepherd's assurance that the 1st Marine Division could be made available, MacArthur asked the

Department of Defense Photo (USMC) A1466

So great was his confidence in the Marine Corps Reserve that LtGen Lemuel C. Shepherd, Jr., took personal responsibility for promising Gen Douglas MacArthur that the 1st Marine Division with appropriate Marine air could be sent to Korea by 15 September for the landing at Inchon.

Joint Chiefs of Staff for it, with appropriate air in the form of the 1st Marine Aircraft Wing. As it was approved, almost immediately, it was "less the Brigade units" which would revert to the division and wing upon their arrival in the theater. General Shepherd knew full well that the under-strength division could hardly deploy the reinforced 5th Marines to the brigade, let alone field the balance of the division, but he had an abiding and deep faith in the loyalty and performance potential of the Marine Corps Reserve. The Reserve, ground and air, came through like the proverbial gang-busters, and in less than 60 days after receiving the initial orders, both wing and division made the landing at Inchon on 15 September, just 67 days after the 10 July conference in Tokyo.

The response of the Marine Corps Reserve was so much a key to the success not only of Inchon, but also to the firm establishment of the United Nations effort in Korea. By about 20 July, the exchanges between Far East Command and Washington had settled out that what would be deployed for Inchon would be a war-strength 1st Marine Division and 1st Marine Aircraft Wing. With the strengths of the two being only at 7,779 and 3,733 respectively, there was no way the war-strength manning levels could be reached and maintained without drawing heavily on both the ground and aviation organized Reserve contingents. Division war-strength ran about 25,000 and the wing about 9,500. On 19 July, President Truman authorized the mobilization of the Marine Corps Reserve and things began to move at a record pace. Minimum time warnings went out to all Reserve District Directors, and alerts were given to Camp Pendleton, El Toro, Camp Lejeune, and Cherry Point to expect literally thousands of reservists in a matter of days. The first reservists arrived at Camp Pendleton and El Toro on 31 July, and by utilizing some units and personnel of the 2d Marine Division and 2d Marine Aircraft Wing on the east coast, the 1st Marine Division and the 1st Marine Aircraft Wing were able to realistically consider their scheduled mount-out dates of 10-15 August. Because a very high percentage of the reservists were combat veterans of World War II, only about 10 percent required any form of basic indoctrination and training. This was a key factor, particularly in aviation, since the total time required from commencement of pilot training to operational status was a matter of some two years. To be called up so soon after World War II, was the theme of many a barracks-room and ready-room ballad that

1st Marine Aircraft Wing Leaders

Major General Field Harris

Speaking before a crowded gathering of the Wings Club at the Ambassador Hotel in Washington, D.C., in May of 1945, Major General Field Harris ended his address on Marine aviation in the scheme of National Defense with the words: "We are not an air force. We are a part of an air-ground team. I believe we will ever be a necessary part of our Nation's air-ground-sea team. As always, we will aspire to be a useful and helpful arm of the United States Marines." A little more than five years later, Harris' remarks would ring true. As Major General Oliver P. Smith, Commanding General, 1st Marine Division, wrote to Major General Harris, then commanding the 1st Marine Aircraft Wing in Korea, following the successful breakout from the Chosin Reservoir: "Never in its history has Marine aviation given more convincing proof of its indispensable value to the ground Marines. . . . A bond of understanding [between brother Marines on the ground and in the air] has been established that will never be broken."

Born in 1895 in Versailles, Kentucky, he received his wings at Pensacola in 1929. But before that he had 12 years of seasoning in the Marine Corps that included sea duty on board the *Nevada* and *Wyoming* and tours ashore with the 3d Provisional Brigade at Guantanamo, Cuba, and at Marine Barracks, Cavite, Philippines, and

Major General Field Harris

Department of Defense (USMC) A310952

Department of Defense (USMC) A30035

Major General Christian F. Schilt

the Office of the Judge Advocate in Washington.

After obtaining his gold wings, Harris served with a squadron of the West Coast Expeditionary Force in San Diego, followed by additional flight training and assignments at sea and on shore, including Egypt as assistant naval attaché. During World War II he was sent to the South Pacific where he served successively as Chief of Staff, Aircraft, Guadalcanal; Commander, Aircraft, North Solomons; and commander of air for the Green Island operation. Following the war, he became Director of Marine Aviation and in 1948 was given command of Aircraft, Fleet Marine Force, Atlantic, and a year later, Aircraft, Fleet Marine Force, Pacific, and 1st Marine Aircraft Wing at El Toro, California.

His Korean War service as Commanding General, 1st Marine Aircraft Wing was rewarded with both the Army's and Navy's Distinguished Service Medal. Harris returned to the United States in the summer of the 1951 and again became the commanding general of Aircraft, Fleet Marine Force, Atlantic. Upon his retirement in 1954 he was advanced to the rank of lieutenant general. He died in December 1967 at the age of 72.

Major General Christian F. Schilt

Major General Christian F. "Frank" Schilt, Major General Field Harris' replacement as Commanding General, 1st Marine Aircraft Wing, brought a vast amount of flying experience to his new post in Korea.

Born in Richland County, Illinois, in 1895, Schilt

entered the Marine Corps in June 1917 and served as an enlisted man in the Azores with the 1st Marine Aeronautical Company, a seaplane squadron assigned to anti-submarine patrol. In June 1919, on completion of flight training at Marine Flying Field, Miami, Florida, he received his wings and was commissioned a Marine second lieutenant, beginning a near 40-year career in Marine Corps aviation.

His initial assignments were to aviation units in Santo Domingo and Haiti, and in 1927, he was assigned to Nicaragua. As a first lieutenant in 1928, he received the Medal of Honor for his bravery and "almost superhuman skill" in flying out wounded Marines from Quilali. Schilt's career pattern during the interwar years consisted of a mix or school and flight assignments.

Prior to the United States entry into World War II, Colonel Schilt was assigned to the American Embassy in London as assistant naval attaché for air, and as such, traveled extensively in the war zones observing British air tactics in North Africa and the Middle East. During the war, he served as the 1st Marine Aircraft Wing's chief of staff at Guadalcanal, was later commanding officer of Marine Aircraft Group 11, and participated in the consolidation of the Southern Solomons and air defense of Peleliu and Okinawa.

In April 1952, on his return from Korea, Schilt became Deputy Commander, Fleet Marine Force, Pacific, and the next year he was given control of aircraft in the Pacific command. His last assignment was as Director of Aviation at Headquarters Marine Corps and upon his retirement in April 1957, he was advanced to four-star rank because of his combat decorations. General Schilt died in January 1987 at the age of 92.

Major General Clayton C. Jerome

MajGen Clayton C. Jerome, USMC

Major General Clayton C. Jerome

Like his predecessor, Major General Clayton C. Jerome had a distinguished flying career.

A native of Hutchinson, Kansas, born in 1901, he was commissioned a second lieutenant in 1922 upon graduation from the Naval Academy. After a year at Marine Barracks, Washington, D.C., he reported to Pensacola for flight training and received his naval aviator's wings in 1925. Foreign service in China, the Philippines, and Guam followed his first duty assignment at Naval Air Station, Marine Corps Base, San Diego.

In the mid-1930s, Jerome became naval attaché for air in Bogota, Columbia, and several other Latin and Central American republics. While serving as naval attaché he earned the Distinguished Flying Cross for his daring rescue of the survivors of a Venezuelan plane crash. Using an amphibious plane, he repeatedly flew over the treacherous jungles of Cuyuni in search of the wreck. After finding it he made two hazardous landings on the narrow Cuyuni River to rescue four survivors.

During World War II, he took part in the consolidation of the Northern Solomons and the Treasury-Bougainville operation as operations officer and later chief of staff to Commander, Aircraft, Northern Solomons. He was later Commander, Aircraft and Island Commander, Emirau, before serving with the U.S. Army in the Philippines. During the Luzon campaign, he commanded Mangalden Airfield and Marine aircraft groups at Dagupan, directing Marine air support for the Army ground operations.

Postwar duty included command of Marine Corps Air Station, Quantico; duty as Chief of Staff, Marine Corps Schools, Quantico; and simultaneous service as Director of Public Information, Recruiting, and Marine Corps History at Headquarters Marine Corps.

He was serving as Director of Aviation and Assistant Commandant for Air when reassigned as Commanding General, 1st Marine Aircraft Wing in Korea in April 1952.

In January 1953, Major General Jerome reported to Cherry Point, North Carolina, as Commanding General, 2d Marine Aircraft Wing, and Commander, Aircraft, Fleet Marine Force, Atlantic. Two years later he moved to El Toro, becoming the air commander for Fleet Marine Force, Pacific. Retiring in 1959 as a lieutenant general, he died in 1978 at the age of 77.

Major General Vernon E. Megee

A Marine aviator for more than 20 years, Major General Megee assumed command of the 1st Marine Aircraft Wing on 9 January 1953.

Born in Tulsa, Oklahoma, in 1900, he enlisted in the Marine Corps in 1919 after attending Oklahoma A&M College. Commissioned in 1922, Megee served in

1st MAW Historical Diary Photo Supplement, Apr53

Major General Vernon E. Megee

infantry, artillery, and expeditionary billets before undergoing pilot training in 1931. A year before, while quartermaster with the Aircraft Squadrons, 2d Marine Brigade, in Nicaragua, he earned the Navy and Marine Corps Medal for commendatory achievement while flying as an observer and machine gunner in an attack on a large force of Sandinista rebels.

Receiving his naval aviator's wings in 1932, Megee spent the pre-war decade as a flight instructor at Quantico; student at the Air Corps Tactical School, Maxwell Field, Alabama; and commander of a Marine fighter squadron. In 1940, Major Megee was assigned to the U.S. Naval Aviation Mission to Peru and spent the next three years as a special advisor to that government's Minister of Aviation.

During World War II, he was the first commander of an Air Support Control Unit, which was created specifically to provide close air support for ground troops. In combat operations at Iwo Jima, Megee was said to have told his pilots to "go in and scrape your bellies on the beach" in support of Marines on the ground. Later, at Okinawa, he commanded all Marine Corps Landing Force Air Support Control Units.

After promotion to brigadier general in 1949, Megee was named Chief of Staff, Fleet Marine Force, Atlantic, and after receiving his second star in 1951, he served as Commanding General, Aircraft, Fleet Marine Force, Pacific, prior to his assignment in 1953 to command the 1st Marine Aircraft Wing in Korea.

In 1956 he became the first Marine aviator to serve as Assistant Commandant of the Marine Corps and Chief of Staff. After having served as Commanding General, Fleet Marine Force, Pacific, he retired in 1959. In retirement, General Megee earned a master's degree from the University of Texas, Austin, and served as superintendent of the Marine Military Academy in Harlingen. He died in 1992 at the age of 91.

---

sustained both ground and aviation troops on the lighter side throughout the conflict. The "two-time losers," as they referred to themselves, put a lot of morale-building humor into Korea, but the factor of overriding importance was that they were well trained, experienced, and seasoned, ground and air.

Any discussion of Inchon must be considered incomplete if it fails to mention the difficult problem the site itself presented to the amphibious planners. First, the tidal variation at Inchon is one of the greatest ranges of rise and fall on the entire Korean coastline, east or west. Secondly, the approach channels to the landing sites essential to successful establishment of a major force ashore, were not only narrow and winding, but also were through extensive mud flats. The combination of these two factors alone meant that much of the unloading of heavy equipment would be over mud flats at low tide with the amphibious force ships on the bottom until the next tidal change. In order to accommodate to this problem somewhat and also meet the D-Day date of 15 September, and manage to negotiate the very narrow approach channels, it was essential to make the approach during daylight hours on the fall of the tide, thus deriving an assault H-Hour of late afternoon.

While the complexities of Inchon as a site were much discussed, with sides taken both at Far East Command and all the way back to Washington, General MacArthur held firm in his confidence in the amphibious experts of the Navy and Marine Corps. He believed that any other site or date would not yield the opportunity to quickly cut the North Korean supply lines to their forces in the south. MacArthur was right, and Inchon has achieved its place in history as the most audacious, daring, difficult, risky—and successful—amphibious landing, perhaps of all time.

Air Support Plan

Major General Field Harris, commanding general of the 1st

Department of Defense Photo (USMC) A29033

A Marine aviator since 1930 and a veteran of the Guadalcanal, Iwo Jima, and Okinawa campaigns during World War II, Col Frank G. Dailey led the bomb- and napalm-laden Corsairs of Marine Aircraft Group 33 from the Pusan Perimeter to the Chosin Reservoir.

Marine Aircraft Wing, arrived in Tokyo on 3 September, and immediately began to finalize the air support plans for the Inchon operation with Far East Command, the Navy, the Air Force, and the Army. Underlying the air plan was the decision that the sky over the objective area was to be divided between the air units of the Navy's Joint Task Force 7, and those of X Corps. X Corps had been assigned its own organic air under corps control in a manner reminiscent of the Tactical Air Force organization accorded X Army in the Okinawa operation. The command of X Corps tactical air was given to General Cushman who had been the brigade deputy commander to General Craig in the Pusan Perimeter. MAG-33, under Colonel Frank G. Dailey, was designated by the wing as Tactical Air Command X Corps, with principal units being VMFs -212 and -312, in addition to VMF(N)-542 and VMF(N)-513. Joint Task Force 7 counted on its fast carrier task force, Task Force 77, to gain air superiority in the area, as well as to furnish deep support and interdiction strikes. Close support for the landing was assigned to the task group including the two small carriers, *Sicily* and *Badoeng Strait*, still operating VMFs-214 and-323, which had supported the brigade so well in the Pusan Perimeter.

The lst Marine Aircraft Wing designated MTACS-2, which had controlled air support for the brigade, to function in that capacity for the landing and upon the establishment of X Corps ashore, to then continue to control for Tactical Air Command, X Corps.

Assault Phase Air Support

A primary and crucial objective in the Inchon landing was Wolmi-do Island, very close in to the main landing beaches of Inchon. Preparation of Wolmi-do began on 10 September with attacks by VMFs -214 and -323 with bombs, rockets, and napalm. The island was only about 1,000 yards wide and about the same dimension on the north-south axis, except that a long causeway extending to the south added another 1,000 yards to the length of the island. At the end of the causeway, a small circular islet with a lighthouse marked the entrance to the harbor. The main part of the island was dominated by a centrally situated piece of high ground known as Radio Hill. The Corsairs literally blackened the entire island with napalm to the extent that during the second day of attacks, the whole island appeared to be ablaze.

When the air strikes began, First Lieutenant John S. Perrin, a pilot with VMF-214, recalled that several North Korean military vehicles were flushed out. Evidently the enemy drivers believed that a moving target would be harder to hit. Perrin said that they got their Corsairs as low and slow as they could and literally chased the vehicles up and down streets and around corners in the island's small industrial sector. Eventually all the vehicles felt the wrath of the blue fighters.

While the two carriers were busy with replenishment at Sasebo on the third day of the pre-landing strikes, Task Force 77 took over the preparation effort with extensive bombing attacks, augmented by the Shore Bombardment Group of four cruisers and six destroyers, the latter closing to within 800 yards of the island. In five days of continuous pounding by this combined air and naval gunfire, Wolmi-do was one blasted piece of real estate as the 3d Battalion, 5th Marines, prepared to land at Green Beach on the morning of 15 September. Testifying to the effectiveness of the pre-landing preparation, Lieutenant Colonel Robert D. Taplett's battalion had completed their mopping-up operations by noon and its total casualties for the day were 17 wounded. In return, the battalion could count 136 prisoners, 108 enemy dead, and from interrogations of the prisoners, at least 150 more entombed in caves and emplacements throughout the island. During the afternoon of 15 September, from observation posts at the north tip of the island and at the top of Radio Hill, targets were picked out for special attention during the pre-H-Hour bombardment preparation for the landings at Red and Blue beaches at 1730. While the afternoon wore on, VMFs -214 and -323, in addition to three squadrons of Navy AD Skyraiders, alternately blasted Inchon, integrating their strikes with naval gunfire from 1430 right

Sketch by Cpl Ralph H. Schofield, USMCR

Marines charge ashore at Inchon on 15 September. After scaling the seawall, with the aid of ladders, they fan out rapidly to secure the beachhead as Corsairs of the 1st Marine Aircraft Wing blast enemy targets in support of the

up to H-Hour. In addition, Task Force 77 kept a continuous strike group of another 12 planes over the objective area to keep any movement of defensive forces toward the beaches at an absolute minimum. With this type and intensity of air and naval gunfire preparation, in addition to the support given the Red and Blue beach landings from Wolmi-do, plus the strong element of surprise carried by the Inchon assault, success of the operation was assured. In view of the very heavy element of risk involved with the hydrographic characteristics of the harbor and the many other departures from normal planning patterns for an amphibious assault of the magnitude of Inchon, a completely successful outcome was indeed welcome.

During the advance out of the beachhead, which commenced the day following the landing, the air support control system functioned precisely as previously described. On the first day of the advance toward Seoul, the obviously confused North Koreans learned even more about close air support and its effects than they had absorbed in the Pusan Perimeter a few weeks before. While the attack on D+1 had barely gotten underway, just five miles away from the advancing 5th and 1st Marines, six North Korean T-34 tanks were spotted in broad daylight rumbling along the Seoul-Inchon highway without escort of any kind, apparently ordered out to bust up the landing. An eight plane strike of VMF-214 hit the enemy armor near the village of Kansong-ni with napalm and rockets as 2d Battalion, 1st Marines, applauded from their positions less than two miles away. The Corsairs destroyed two of the T-34s and a third was damaged, but the North Korean crews abandoned some of the tanks and tried to take shelter in huts near the side of the road, which were promptly napalmed by the strike. This threw up large quantities of smoke and

Department of Defense Photo (USA) SC348504

A curious Marine passes three destroyed North Korean T-34 tanks five miles east of Inchon. The rocket-laden Corsairs of VMF-214 knocked out the tanks, part of a group of six ordered to break-up the landing.

led the pilots to believe that all six tanks had been destroyed, so they switched to other targets in the beachhead area.

Destruction of the tanks came at a price. Captain William F. Simpson, Jr., a pilot with VMF-214 was killed. Fellow pilot Captain Emmons S. Maloney recalled that Simpson "got so involved in it, hitting these tanks coming up, that, he almost flew straight into the tank. By the time he realized he was too low, it was too late to pull out."

Shortly after, 2d Battalion, 5th Marines, with a tank escort, came into Kansong-ni and as they were coming into the position, surprised three of the remaining T-34s, which were promptly destroyed by the escorting M-26 Pershing tanks. The close contacts between air and ground, as typified by this example, permitted the continuous and synergistic employment of the capabilities of the air-ground team during the advance to Seoul and beyond.

To briefly summarize all aspects of the Inchon landing, a quote from Rear Admiral James H. Doyle, the veteran amphibious group commander, does the job nicely:

The assault itself was successful only through the perfect teamwork that existed between the participating Naval and Marine elements. . . . Only the Unit-ed States Marines through their many years of specialized training in amphibious warfare, in conjunction with the Navy, had the requisite know-how to formulate the plans within the limited time available and execute those plans flawlessly without additional training or rehearsal.

Kimpo Airfield

One of the key objectives of the assault phase and the advance toward Seoul was the capture of Kimpo Airfield, the major air installation of the city, about seven miles west on the other side of the Han River from Seoul. While still in the relatively confined operating areas of the assault phase of the operation, the forces assigned could meet air support require-

In a destroyed hanger at Kimpo Airfield, Marines found one of several near-flyable North Korean Soviet-built aircraft. Captured by 2d Battalion, 5th Marines, Marines engineers quickly made the airfield operational with temporary repairs, ready to receive elements of MAG-33.

Department of Defense Photo (USMC) A3226

Gen Oliver P. Smith Collection, Marine Corps Research Center

Tactical Air Commander, X Corps, BGen Thomas J. Cushman, USMC, right, and his chief of staff, Col Kenneth H. Weir, USMC, meet with the commanding general of the Fifth Air Force, MajGen Earle R. Partridge, USAF, left, at Kimpo Airfield. While the wing headquarters remained in Japan, its task was to furnish administrative and logistical support to Cushman's command and MAG-33 during the Kimpo air operations.

On 19 September, Tactical Air Command X Corps, General Cushman, established his headquarters at Kimpo, and was quickly followed by Marine Ground Control Intercept Squadron 1, MTACS-2, and VMO-6. The first fighter squadron of MAG-33 to check in was Lieutenant Colonel Max J. Volcansek, Jr.'s VMF(N)-542 with five Grumman F7F Tigercats landing late in the afternoon of the 19th. They also flew the first combat mission from the field early the next morning when they destroyed two locomotives near Seoul. Corsairs of Lieutenant Colonel Richard W. Wyczawski's VMF-212 and Lieutenant Colonel J. Frank Cole's VMF-312 landed shortly after -542 and also got into action on the 20th.

During the transition of the squadrons assigned to MAG-33 ments. As the objective areas widened and expanded with the advance, however, it was essential to bring in more shore-based aviation to meet the demand quickly and with optimized dispatch on a constantly broadening front. The field was captured and declared secure in the mid-morning of 18 September. It was in such good shape after the assault that it was possible to almost immediately move in the first operating units. The first aircraft to land officially at Kimpo was an H03S helicopter of VMO-6, piloted by Captain Victor A. Armstrong, which brought in General Shepherd and his G-3, Colonel Victor H. Krulak, to confer with General Craig, who had just arrived by jeep. Later in the afternoon, Generals Harris and Cushman also arrived to make final plans for the deployment of the Marine squadrons from Japan and those that would fill out MAGs -33 and -12 for the follow-on operations.

Returning to the Sicily *after making the first landing at Kimpo Airfield, 1stLt John V. Haines points out the damaged section on his Corsair which caused the unscheduled landing to his squadron commander, LtCol Walter E. Lischeid. LtCol Lischeid would die six days later when his Corsair was shot down over the western suburbs of Seoul.*

National Archives Photo (USN) 80-G-420281

Department of Defense Photo (USMC) A130104

The Corsairs from VMF-312 take off from Kimpo Airfield as fast as they could be refueled and rearmed in around-the-clock attacks on retreating North Korean forces.

from MAG-12 in Japan, the operational burden of Marine air support was handled entirely by the two carrier-based Corsair squadrons, VMF-214 and VMF-323, now administratively assigned to MAG-12. Also supporting the displacement of the division-wing team into the Korean peninsula was Major Joseph H. Reinburg's VMF(N)-513, still operating from Itazuke Air Force Base in Japan.

The flexibility of Marine aviation in supporting a forward displacement of such magnitude with hardly a break in the continuity of operations is well illustrated in the rapid establishment of Tactical Air Command X Corps at Kimpo. Once again, the value of commonality between Marine and Naval aviation was effectively demonstrated in the coverage, without a break, of air support requirements of the 1st Marine Division, utilizing the two carrier-based squadrons. Regarding the capture of Kimpo, Lieutenant General George E. Stratemeyer, Commander, Far East Air Forces, had this to say to Major General Oliver P. Smith, the commanding general of the 1st Marine Division: "I want to take this opportunity of expressing my admiration and gratification for the manner in which elements of your Division recently captured Kimpo Airfield and so secured it as to make it available for use by Far East Air Forces and Marine Corps aircraft in shortest possible time."

Control of air support passed from the Amphibious Force Commander to MTACS-2 ashore on D+2 when the landing force commander (Major General Oliver P. Smith, Commanding General, 1st Marine Division) declared he was ready to assume control. Requests for close air support increased rapidly as the enemy recovered from the initial shock of the assault. For example, on 18-19 September, VMFs -323, -214, and -513 flew a total of 50 close support sorties, delivering napalm, rockets, and 500-pound bombs against troop concentrations in front of the 1st Marines, who were finding the going a bit tougher in the vicinity of Sosa on the Inchon-Seoul highway. In addition, -513 flew a total of 15 daylight close support missions during the period 17-19 September for Army units along the Pusan Perimeter, where the accompanying breakout to the north and west was being initiated.

With Kimpo in hand, the next major objective became the forced crossing of the Han River and the taking of essential key terrain from which to launch the assault on Seoul. MAG-33 and MAG-12 made their principal contributions to these major endeavors by steadily and rapidly increasing their strengths and capabilities at Kimpo, and through strikes against redeployment and reinforcing moves by the North Koreans attempting to improve the defenses of the city. Logistically, there was a vehicle shortage for the movement of aviation gasoline, ammunition, and oil from the port dumps at Inchon and Ascom City to Kimpo, but a timely offer from the Far East Air Force's Combat Cargo Command solved the problem. During the week of 18-24 September, the Command hauled a total of 1,545 tons of these vital aviation supplies in from Japan. Once again the theorem that the farther from Washington, the greater the inter-Service cooperation was proven, just as it was in the South Pacific a few years before. In addition to this air effort, about 1,450 tons were trucked to Kimpo from the port during the same period. Also, Marine Transport Squadron 152 flew in spare parts and urgently needed ground equipment from Japan or wherever it could be made available, practically around-the-clock.

The crossing of the Han was assigned to the 5th Marines in the vicinity of Haengju, and after an

Photo by Frank Noel, Associated Press

Marine amphibious tractors plow across the muddy Han River as the planes of Marine Aircraft Groups 12 and 33 provide close support for the 1st and 5th Marines in their assault toward the South Korean capital.

abortive attempt during the night of 19 September; the 3d Battalion accomplished it during daylight hours on the 20th. Four Corsairs of VMF-214 provided supporting fires against a key hill from which the North Koreans were directing accurate fire at the crossing tracked landing vehicles. As the assault on this hill continued, the Corsairs reported enemy in numbers hastily evacuating with strafing Marine aircraft in full pursuit. The three primary objectives were secured by mid-morning and the advance down the north bank of the river toward Seoul began immediately. The general plan was for the 5th Marines to continue the advance toward Seoul and to seize vantage points in support of the 1st Marines crossing at Yongdungpo. Yongdungpo, the industrial area of the city, was situated on the south bank of the Han on a large sandspit. The fighting on the north bank and in the attack on Yongdungpo both served notice to the division that it was going to be a "to the last man" defense of the city. By the 24th, after an extremely severe minute-to-minute three days of intensive battling, night and day, the 1st Marines was able to make the crossing and the battle for Seoul was underway.

From the 19th on, both MAGs -12 and -33 flew maximum effort schedules in close support of both the 1st and the 5th Marines in their assaults toward the city. Typical of the squadron performances during this period was a flight of five Corsairs led by Lieutenant Colonel Walter Lischeid of VMF-214, which effectively broke up a threatened counterattack on Hill 105 South, held by the 1st Battalion, 5th Marines. It was one of six close support missions flown by -214 on the 23d in the zone of the 5th Marines. As a counter to the air support rendered during the daylight hours, Marine artillery took over the complete support job after dark when the "closest" close support possibilities became somewhat diminished. When the terrain cooperates, this one-two counter to enemy counterattacks around the clock was most effective.

On the 24th, in front of Company F, 2d Battalion, 5th Marines, on the east slope of Hill 56, VMF-323 dropped 500-pound

Marines of the Second Platoon, Company G, 5th Marines, clean snipers out of a residential section of Seoul. Due to the confined nature of much of the fighting within the city, Marine close air support was used sparingly and at deeper distances from the advancing troops.

Department of Defense Photo (USMC) A3365

Flying Sergeants: Enlisted Marine Aviators

One area where the Marine Corps was probably alone among the aviation Services was the degree it used enlisted pilots, especially in combat. Enlisted pilots were not new. France in World War I and the Axis powers, Germany, Japan, and Italy, in World War II made considerable use of their enlisted aviators. The Royal Air Force would have been in even worse straits during the Battle of Britain in 1940 had it not been for its sergeant-pilots. For the most part, however, the United States required its pilots to be commissioned officers and, with few exceptions, that is the way it continues to be.

The Navy had instituted its Naval Aviation Pilot (NAP) designation in 1919 because of a pilot shortage. The Marines, too, authorized selection of enlisted members to become pilots and First Sergeant Benjamin Belcher was the first Marine NAP in 1923.

With the country's hurried and somewhat unexpected entry into World War II, the need for pilots transcended the niceties of rank and tradition. Therefore all the Services, at one time or another during the war, made use of enlisted pilots, sometimes elevating them to commissioned rank later. Marine ace Lieutenant Colonel Kenneth A. Walsh, who scored 21 kills and earned the Medal of Honor during the war, was an enlisted pilot until he was commissioned in 1942.

The Marine Corps probably had the largest number of noncommissioned aviators (131 in 1942), and not in second-line transport squadrons; many of these NAPs later flew helicopters and jets in very heavy action in Korea. Flying sergeants flew Corsairs and Tigercats at Pusan and Chosin, Panthers in close air support against the Chinese, and OYs on dangerous artillery-spotting missions.

Technical Sergeant Robert A. Hill accumulated 76 combat missions as an OY pilot, earning the moniker "Bulletproof" after coming home in planes that were more holes than aircraft. He received a Distinguished Flying Cross for evacuating wounded Marines near Chosin under heavy enemy fire. Marine NAPs piloted several of the R4D transports that also evacuated wounded from Hagaru-ri and Koto-ri during the Chosin breakout.

But the jet pilots were the glamour boys and NAPs were among the first Marine jet pilots, taking their training in Lockheed TO-1s along with their commissioned squadron mates. The training met some resistance from senior squadron commanders, a few of whom did not want enlisted pilots flying their new jets. NAPs were not allowed to train in jets until 1949. This provided a cadre of experienced and motivated personnel to draw upon during the action in Korea.

This somewhat confusing situation had the added facet that several now-enlisted NAPs had been commissioned lieutenants in World War II. However, after mustering out in 1945 and 1946, many of the former Corsair drivers regretted their decision to leave the active Marine Corps; several missed flying such powerful aircraft as the tough F4U. The Corps also found itself short of qualified aviators to fly its new jets and to man its remaining squadrons.

National Archives Photo (USN) 80G-428028

The flying sergeants of VMF-212 on board the light carrier Bataan *(CVL 29). Standing from left to right are: TSgt Gail Lane, MSgt John J. McMasters, MSgt Clyde B. Casebeer, and seated from left to right, MSgt Billy R. Green, MSgt Donald A. Ives, and MSgt Norman E. Payne, Jr.*

A program was developed whereby former Marine officer aviators could return as master sergeants (E-7 was the highest enlisted rating at the time), if they re-upped 90 days or less after leaving active duty. After the 90-day limit, the former aviator could rejoin as a technical sergeant, a grade below that of master sergeant.

When VMF-311 brought its F9F Panthers to Korea, several of its pilots were enlisted aviators. Master Sergeant Avery C. Snow was the first NAP to complete 100 combat missions in a jet. Snow had been a captain with Marine Torpedo Bomber Squadron 232 during World War II.

One specialized squadron that made heavy use of its NAPs was Marine Photographic Squadron 1 (VMJ-1), established on 25 February 1952, flying modified McDonnell F2H-2P Banshees with a long nose to accommodate several reconnaissance cameras. VMJ-1 established an enviable record in Korea. Several of its pilots, who were specially trained volunteers, were enlisted men who could double as lab technicians if the situation warranted. Squadron crews flew 5,025 sorties, shooting 793,012 feet of film, one-third of all United Nations photo reconnaissance output, and at times, 50 percent of all Far East Air Force intelligence missions. However, even with this outstanding record, the Banshee drivers of VMJ-1 could not respond to all requests, and as such, overall reconnaissance requirements suffered throughout the war, primarily because of a lack of assets—mainly planes, pilots, and trained photo interpreters. Real-time imagery for field commanders and their units was not available at times when it was most needed. This problem, although well known and accepted, especially by the ground units, continued through Vietnam, and even into the Gulf War.

Aerial photo reconnaissance is one of the most exacting and dangerous jobs in all military aviation. The "recce" pilot must be more than just a good pilot; that is just a base from which to start. He must be a crackerjack navigator and know his camera systems inside and out, their capabilities and their limitations. And he must be resourceful, as well as have an inexhaustible supply of courage. Sometimes these last qualities are all that enable him to bring the film home and successfully complete his mission.

Most jet reconnaissance aircraft were unarmed, relying upon their speed to get them home before being intercepted. During World War II, there were no specifically dedicated reconnaissance aircraft, merely modified fighters, which had cameras stuck in the most convenient space, sometimes behind the pilot in the cockpit, or below him in the belly. The F6F Hellcat and P-51 Mustang are examples of such modification. Usually, these aircraft retained most, if not all, their machine gun armament and could therefore fight their way to and from the target. During Korea, however, the dedicated photo-Banshees of VMJ-1 were toothless and needed

Five master sergeant NAPs of VMJ-1 pose by one of their Banshees. From left to right are: MSgt James R. Todd, MSgt Samuel W. Cooper, MSgt Lee R. Copland, MSgt Marvin D. Myers, and MSgt Lowell L. Truex. As a second lieutenant with VMF-224 during World War II, Truex shot down a Japanese "George" fighter during an engagement off Okinawa.

Courtesy of MSgt Lowell L. Truex

escorts. Sometimes another Banshee would go along, both as an escort and sometimes to ensure the coverage of the target with another camera. Air Force F-86s were sometimes called upon to shepherd the "recce" pilot. And sometimes, the photo pilot found himself alone.

In 1952, Master Sergeant Lowell T. Truex had made his photo runs against installations near the Yalu River, thinking that his F-86 escort would look out for any Communist fighters, which might try to come after him. However, as he looked around he found that the Sabres were nowhere to be seen, and he also spotted a gaggle of MiG-15s taking off across the river. Hurriedly, he finished his photo runs and ran for home. He found later, that the F-86s had been watching from above, had the MiGs in sight, and were ready to jump the Chinese fighters if they come after Truex. Recalling his time with VMJ-1, Truex said:

> My memories of the photo unit, which became a squadron during my tour, and all the plankowners, are good ones. We were completely self-contained and operated with field equipment from the well-point, water tank to the generators. The technicians were all superior guys, who worked with energy and diligence

While standing squadron watches, besides flying their regular missions, the NAPs were also required to work as division officers in the squadron photo lab and on the flight line. There was also the need to brief escort pilots, who were often Air Force F-86 pilots. The Sabre pilots did not always appreciate being briefed by an enlisted aviator, and occasionally made things difficult for their Marine compatriot. In Master Sergeant Truex's case, he had to submit to annoying identification exercises before he was allowed to conduct his brief. "I had to be verified," he recalled, "and they wouldn't let me into their briefing room without identification. But, I briefed the Air Force pilots precisely, even though some of them took a casual attitude about escorting a Marine reconnaissance pilot."

The haughty Sabre pilots tended to look down on the big, blue Banshee their Marine charges flew. How could it compare with their shiny, silver F-86s? As Master Sergeant Truex again recalled:

> They underestimated the Banshee's speed and climb, as well as the intensity our mission required. With our small J-34 engines and big tip tanks, our F2Hs had superior range. Although their F-86s looked good, and the Air Force *did* take care of us—and we certainly appreciated their presence—they usually bingoed before we were finished. We usually flew back alone.

Master Sergeant James R. Todd was VMJ-1's high-mission man, completing 101 photo missions before rotating home. Todd flew 51 reconnaissance missions in Banshees, 10 in F9F-2Ps, 23 in F7F-3Ps, 13 in F4U-5Ps, and 4 escort missions in F4U-4Bs. As he recalled: "The F4U-4B was used for armed escort only. The rest of the time, we relied on a thirty-eight pistol, a can of film and a lot of speed."

Like many of the enlisted aviators, Todd had been commissioned a second lieutenant in World War II, although he had just missed seeing combat service when the war ended, having spent much of his post-wing time as an instructor. He was mustered out in September 1946, but returned in November. He resigned his first lieutenant's commission, raised his hand as a private, then was immediately advanced to master sergeant and sent to El Toro and then to Pensacola. Arriving at the Florida air station, he joined other re-enlistees at the Naval School of Photography, where they learned the art of aerial reconnaissance. The training was to stand them in good stead in the coming years. By 1950, Todd and his friends had gained a lot of experience in Corsairs and Tigercats.

In September 1951, they were sent to Korea to supplement the meager photographic assets at K-3 (Pohang). At the time the Marines flew F7F-3Ps and F4U-5Ps. But Todd managed to check out in the F9F-2P, and thus, when VMJ-1 was commissioned the following February with brand-new McDonnell F2H-2P Banshees, he was a natural to slide into the new jet's cockpit. For a while, though, the squadron operated five different types: F7F-3P, F4U-5P, F4U-4B (for escort), F9F-2P, and F2H-2P. The props and Panthers remained until April 1952.

Several of VMJ-1's enlisted aviators also worked in the squadron's photo section, interpreting mission film. Although the squadron was administratively under MAG-33 and 1st Marine Aircraft Wing, it was the Air Force at K-14 (Kimpo) that tasked the targets, which was in keeping with the agreement with the Fifth Air Force. Occasionally, the 1st Marine Division could call in a requirement, but for the most part, Fifth Air Force called the shots.

Two MiGs near Chosin set on one of Todd's squadron mates, Master Sergeant Calvin R. Duke, who laid claim to being the oldest Marine NAP in Korea. In a dogfight that went from 10,000 feet to 30,000 feet, Duke outmaneuvered the Communist fighters and ran for home at 600 miles per hour.

Enlisted aviators were an integral part of the Marine Corps' capability. However, by Vietnam, there were only a few NAPs on active duty, and fewer still actually involved in flying duties. Some of these pioneers served with distinction throughout Vietnam. But by 1973, only four NAPs were still on active service with the Marines, and all four were simultaneously retired on 1 February 1973, closing a colorful era in naval aviation and Marine Corps history.

bombs only 100 yards from the attacking Marines, enabling them to seize the high ground. On the same day, to further illustrate the intensity of the air support effort, VMF-212 set a squadron record for the 1st Marine Aircraft Wing for combat operations by flying 12 missions and a total of 46 single-plane sorties. When refueling, rearming, and mechanical check times are considered, this became a rather remarkable achievement under the more or less "primitive" conditions of the first five days at Kimpo Airfield after moving in from Japan.

During the difficult and very heavy fighting in the city, there were many occasions where close air support could be called in with effect. But because of the confined nature of much of the action, the preponderance of air support was rendered at deeper distances from the advancing troops. VMO-6 helicopters and OYs rendered yeoman service in evacuating wounded, flying constant observation missions, and in providing helicopter communication, conference, and observation flights for the troop commanders.

By 28 September, the intensive fighting in the city was drawing to a close. The newly arrived 7th Marines joined the 1st and 5th Marines on the left after the assault on the city began on the 24th, and things began to move out with dispatch. By the 28th, the 5th Marines, according to plan, had been placed into division reserve and the 7th was preparing to push off in pursuit of the North Koreans fleeing the city toward Uijongbu, 10 miles to the north. On the 29th, the 1st Marines were to establish blocking positions about three miles east of the city and the 5th was assigned a similar mission to the northwest at Suyuhyon. These missions were carried out with relatively minor difficulty, as the North Korean resistance appeared to be collapsing.

The 7th Marines moved out for Uijongbu early on 1 October and ran into firm resistance about half way to the objective. The developing firefight exposed the enemy positions and enabled VMF-312 Corsairs to work them over heavily during the remainder of the day. The advance was renewed the second day and again -312 was busy around-the-clock in support of two battalions forcing their way through a tough defile on the main road, essential for tank and heavy vehicle passage. In addition to the close support missions, the Corsairs caught eight trucks in convoy and destroyed seven in one attack. It was a heavy day all around and two Corsairs were lost to North Korean antiaircraft fire, but one landed in friendly territory and the pilot was recovered in good shape. On the third day, with the progress that had been made to that point, one battalion was assigned to each side of the road to mop up while the third passed through on the road straight for Uijongbu. It soon became apparent that the enemy was in full flight, but the 7th Marines was in Uijongbu by afternoon. Establishment of the blocking position there marked the last large-scale fight of the Inchon-Seoul operation. The supply lines of the North Korean invading forces had been cut totally and the Inchon landing had crushed the North Korean army.

With the end of this phase of the war, U.S. Army and Republic of Korea Army units began to relieve the Marine forces of their area responsibilities. Division units were issued orders for movement to staging areas in Inchon and all were in place by 7 October to mount out for what looked like a follow-on amphibious assault on the east coast.

A few of the major highlights drawn from the operation will suffice to summarize the division-wing performance: (1) Expansion from a reduced peace strength to a reinforced war strength, less one regimental combat team, was completed in 15 days; (2) Movement of more than 15,000 personnel, organic heavy equipment, and partial resupply from San Diego to the Far East Command began in less than three weeks after the expansion order was issued; (3) Unloading, reembarkation, and combat loading for the Inchon landing was done at Kobe, Japan, in seven days, including two days lost to a typhoon in the Kobe area; (4) The 1st Provisional Marine Brigade was disengaged from active combat in the Pusan Perimeter at midnight on 5 September, moved to Pusan, and outloaded in combat shipping in less than seven days; (5) A successful assault landing was executed at Inchon on 15 September, under some of the most adverse hydrographic conditions in the history of amphibious operations; (6) The force beachhead line, approximately six miles from the landing beaches, was seized within 24 hours of the main landings; (7) Kimpo Airfield, one of the finest in the Far East, was captured 50 hours and 35 minutes after H-Hour; (8) The Han River was crossed, without major bridging equipment, and Seoul was seized 12 days after the Inchon landing; and (9) The effectiveness of the Marine air-ground team and close air support doctrine was reaffirmed with outstanding success.

Two more days of fighting remained for the squadrons of MAGs -12 and -33 at Kimpo after the relief of the ground units. During the 33-day period from 7

September to 9 October, the five squadrons flew a total of 2,774 sorties, most of them being in close support of infantry units. The accolades from all units supported under the Marine air support control system were many, and welcome, but one from the division artillery commander and fire support coordinator of the Army's 7th Infantry Division was particularly noteworthy. As Brigadier General Homer W. Kiefer said: "Allow me to reemphasize my appreciation for the outstanding air support received by this division. The Marine system of control. . . approaches the ideal and I firmly believe that a similar system should be adopted as standard for Army Divisions."

During the period of the Inchon-Seoul operation, 15 September-7 October, the lst Marine Division suffered losses of 415 dead, of whom 366 were killed in action and 49 died of wounds; six were declared missing in action and 2,029 were wounded in action, for a total of 2,450 battle casualties. The division took 6,492 North Koreans prisoner and the estimates of total casualties inflicted on the enemy added up to 13,666, most of whom were counted dead on the battlefield. These figures represent a ratio of better than 8 to 1, a thoroughly commendable performance considering the speed with which the air-ground team was put together and deployed.

Chosin Reservoir

Before the end of the Inchon-Seoul operation, when it became clear that the effect of the landing was a total rout of the North Koreans, the Commander in Chief, Far East, was formulating plans for the follow-up. With much prudence and caution emphasized from both Washington and the United Nations regarding the possible entry of either Soviet or Chinese Communist forces into Korea, it was decided that the United Nations Command could conduct pursuit operations beyond the 38th Parallel into North Korea. Concern over the possible outbreak of a general war remained strong, however, and the authority for General MacArthur to utilize his forces north of the 38th was burdened with several limitations. Briefly, there could be no entry of other than Republic of Korea forces if there was a clear indication of either Soviet or Chinese entry. Also, there could be no attack of any type against any portion of either Chinese or Soviet territory, including the use of Naval or Air forces. Further, only South Korean forces would be utilized in those provinces of North Korea bordering on the Soviet Union or Manchuria. It is interesting to note that in spite of these qualifications, on 29 September Secretary of Defense George C. Marshall included the following in a message to MacArthur: "We want you to feel unhampered tactically and strategically to proceed north of the 38th parallel." Coming so soon after the world-shaking experiences of World War II, there was justifiable cause for concern, but limitations and cautions over and above normal prudence certainly added much to the difficulties of the decisions faced by MacArthur as the move into North Korea was being executed.

Generally, the plan was for Eighth Army, commanded by Lieutenant General Walton H. Walker, USA, to advance along the Kaesong-Sariwon-Pyongyang axis; the II ROK Corps in the center along the Kumhwa-Yangdok-Sunchon axis; and the I ROK Corps up the east coast direct to Wonsan. The 1st Marine Division would make an assault landing at Wonsan and the Army's 7th Infantry Division would follow ashore in an administrative landing. After establishment ashore at Wonsan, X Corps, under the command of Major General Edward M. Almond, USA, would then advance west to Pyongyang, joining up with Eighth Army. The entire force would then advance north to two phase lines, the second being along the general line Songjin in the east, southwest to Chongju on the west coast. Only South Korean forces would advance beyond the second phase line, in keeping with MacArthur's restrictions. Eighth Army would cross the 38th Parallel on 15 October and the Wonsan landing was set for 20 October.

Wonsan Landing

As has been seen before in military operations, surprise comes in many different packages, and Wonsan indeed had its share. While the division was in the throes of feverishly meeting its tight combat loading schedules at Inchon, at 0815 on 10 October, the I ROK Corps in its rapid advance up the east coast entered Wonsan. By the next day they had completed mopping up the town and were guarding the airfield on Kalma Peninsula. MacArthur then zigzagged back and forth with the idea of a new assault objective at Hungnam, 50 miles north. However, by the time the harbor characteristics and the availability of both landing craft and ships for unloading at two ports were reviewed, the original plan for Wonsan was retained with D-Day still set for 20 October.

The next surprise was a combination of circumstances. First was the discovery that the harbor and approaches to Wonsan were thor-

Department of Defense Photo (USMC) A4452

Corsairs of VMF-312 at Wonsan Airfield are serviced for the continuing battle against Communist forces in northeast Korea. The squadron flew from Kimpo to Wonsan on 14 October, 12 days before landing craft brought elements of the 1st Marine Division ashore.

oughly sown with rather sophisticated Soviet mines of all kinds, from drifting contact types to magnetic ship-counting designs. It was determined that no entry for landing could possibly be made until the harbor was safely swept and the threat eliminated. The delays entailed in the sweeping combined with the early taking of Wonsan by the I ROK Corps to bring about an unusual reversal of the normal order in amphibious operations. When the assault elements of the division finally landed at Wonsan, they were welcomed ashore by the already well-established Marine aviation units.

Planning for 1st Marine Aircraft Wing operations in the northeast had kept up with the rapidly changing strategic situation. On 13 October, General Harris flew into Wonsan and on inspecting the condition of the field, decided to begin operations there immediately. With that decision, VMF-312 flew in from Kimpo on the 14th and wing transports flew in 210 personnel of the headquarters and VMF(N)-513 the same day. Two tank landing ships sailed from Kobe with equipment and personnel of MAG-12, and Far East Air Forces' Combat Cargo Command began flying in aviation fuel. Some bombs and rockets to "get 'em started" were flown in on the planes of VMF(N)-513. On the 16th, VMFs -214 and -323, still on the *Sicily* and the *Badoeng Strait*, began operations at Wonsan, covering the minesweeping activities until 27 October.

With the delay of almost 10 days before the beach landings could begin, the squadrons operating from Wonsan had to rely almost totally on air resupply for the period. The small amount of supplies that did arrive by ship arrived only by extreme effort and makeshift means. Edward S. John, the only second lieutenant in Marine Aircraft Group 12 at the time, was assigned the task of getting 55-gallon drums of aviation

One of the steady stream of Douglas R4D Skytrains that brought supplies to Marine fighter squadrons VMF-312 and VMF(N)-513 at Wonsan. The squadrons were totally dependent on airlift for all supplies during the 10 days it took to clear a lane through the Wonsan harbor minefields.

Department of Defense Photo (USMC) A130144

Department of Defense Photo (USMC) A130420

Speed was of the essence for these Marine airmen in rearming their Corsair for repeated strikes against Chinese and North Korean forces. Ground crewmen in the foreground mix a batch of deadly napalm while other Marines hastily change a tire.

fuel from a tank landing ship, floating clear of the minefield, ashore. In accomplishing the feat, the drums were manhandled into World War II vintage landing craft, now under Japanese operation, using Korean laborers. When close to the beach, the drums had to be manually lifted over the side, as the ramp had been welded shut. Once in the surf the drums were waded ashore through the icy water.

The squadrons also were faced with added difficulties resulting from few bomb carts, trucks, and refuelers. Consequently, the fuel trucks had to be loaded by hand from 55-gallon drums which had been rolled more than a mile from the dumps, also by hand. This slowed operations somewhat, but judicious planning and steady effort maintained a useful sortie rate. Armed reconnaissance flights were flown regularly which resulted in productive attacks on retreating North Korean troops. On the 24th, for example, a -312 flight surprised a column of about 800 North Koreans near Kojo, 39 miles south of Wonsan, and dispersed it with heavy losses.

With the change from an assault to an administrative landing at Wonsan, the 1st Marine Aircraft Wing was placed under the control of Far East Air Force, with delegation of that control to Fifth Air Force, north of the 38th Parallel. This required the daily operations schedule to be submitted to Fifth Air Force at Seoul by 1800 the previous day. Because of the distance involved and the poor communications that existed, it made it extremely difficult at best to get clearance back in time. This was resolved between General Harris and General Earl E. Partridge, Commanding General, Fifth Air Force, with permission for the former to plan and execute missions for X Corps in northeast Korea without waiting for Air Force clearance. Direction of support for X Corps was exercised for the wing by MAG-12 from 15 October to 9 November. Night operations did not begin until late in the month because of delay in getting runway lights at Wonsan, but -513 flew day missions along with -312 from the beginning. After the administrative landing on the 27th, the two carrier squadrons operated in similar fashion to the way they functioned at Inchon.

Generally, Marine aircraft reported to specified Tactical Air Control Parties at times given in the Fifth Air Force daily order, in response to previous requests by ground units for air support. Close air support requests, which were of a more urgent nature, were usually handled by aircraft on runway alert or by flights orbiting a specific point on stand-by status.

As at Inchon, Major Vincent Gottschalk's VMO-6 was under the operational control of the 1st Marine Division. Two helicopters were flown from Kimpo to Wonsan on the 23d and the rest of the squadron came in by tank landing ship on the 27th. A flight echelon of helicopters remained at Kimpo until early November at the request of Fifth Air Force, for evacuation of casualties of the 187th Airborne Regimental Combat Team in the Sukchon area.

After the landing of the initial elements of the division at Wonsan on the 26th, 1st Battalion, 1st Marines, was ordered to Kojo, southeast of Wonsan. The battalion occupied positions in the vicinity with the mission of protecting the I ROK Corps supply dump there in preparation for its displacement to the north. It developed that a sizeable remnant of North Korean troops was in the area and a series of significant actions took place toward the end of the month. It soon became clear that the "remnant" was actually a seasoned unit of experienced troops, and that possibly a major counterattack was in the offing in the Kojo area. However, as it evolved, most of these actions were confined to night attacks by smaller units than were at first suspected. Aviation supplied the need for emergency evacuation of wounded by helicopter, and although there was a significant loss in killed and wounded, there was no need to reinforce the battalion from Wonsan. Ironically, the South Korean supply dump had been essentially moved out before the attacks occurred and when the North Koreans were finally beaten off and dispersed, the battalion was ordered back to Wonsan. The final loss count was 23 killed, 47 wounded, and 4 missing. The battalion took 83 prisoners, with enemy casualties estimated at 250 killed, in addition to an undetermined number of wounded and a count of 165 enemy dead on the battlefield. The unit was back at Wonsan by 4 November.

1stMarDiv Historical Diary Photo Supplement, Nov50

In 1st Marine Division operations around Wonsan, evacuation of the wounded was accomplished by jeep ambulance, and in the case of more seriously wounded, by helicopter. The Sikorsky H03S and later Bell HTL helicopters attached to Marine Observation Squadron 6 were the cornerstones of the Korean War medical evacuation and rescue efforts.

Ordnance men operate a "belting" machine that inserted ammunition into a metal belt to be used in the 20mm cannons of the Corsair. The machine loaded the belts at a rate of 6,000 rounds per hour in comparison with the hand operation of 500 rounds per hour.

Department of Defense Photo (USMC) A130762

With the major changes in strategy that accompanied the collapse of the North Koreans, and the rapid advances of Eighth Army and the two Republic of Korea Army corps to the north, MacArthur issued new directions, which affected X Corps and the Marines. One was an order for the 1st Marine Division to "advance rapidly in zone to the Korean northern border." With the Eighth Army entering Pyongyang on 18 October, X Corps on the east coast was being left behind and the right flank of Eighth Army was becoming exposed. Hence the corps was under pressure to move north at the earliest. With the exception of a significant engagement of the 3d Battalion, 1st Marines, in the Majon-ni area near Wonsan, similar to the action of the 1st Battalion at Kojo, most of the security requirements laid on the 1st Marines in and around Wonsan had been met by early November.

At Majon-ni, the various actions were supported by Marine aviation in the normal manner during daylight hours, but most of the attacks on the perimeter defenses occurred at night. Evacuations of wounded were by helicopter primarily, and several airdrops of supplies were included in the general air support. Marine losses in these actions included 20 killed and 45 wounded. Enemy casualties were estimated at 525 killed and 1,395 prisoners were taken. More than 4,000 Korean refugees were screened at the roadblocks along the main supply route (MSR). The MSR itself was so precipitous, narrow, and difficult, the Marines lost 9 killed and 81 injured along one very tough stretch known as "ambush alley."

While the 1st Marines were busy at Kojo and Majon-ni, the 5th and 7th Marines had taken up their new assignments to the north. This meant that the division stretched a total of 130-road-miles from the 1st Marines in the south to the 7th in the north. It would be a gross understatement to only say that this complicated the delivery of the usual air support by the 1st Marine Aircraft Wing to its brothers on the ground. With the arrival of additional Army elements of X Corps in the area, however, it became possible to shorten lines somewhat. The division's command post was moved to Hungnam on 4 November, with the 5th and 7th Marines operating north. By 17 November, the 1st Marines were at Chigyong, 14 miles south of Hungnam, thus closing the stretch to less than 60 miles. This was a definite improvement, but Wonsan from the viewpoint of Marine aviation, was looking like a by-passed Japanese base from World War II. The concentration of the division north of Hungnam in its march to the Yalu River made the airfield at Yonpo increasingly attractive to the wing because it was in the center of the Hungnam-Hamhung area. This meant that response times for close air support would be considerably reduced for any actions that occurred to the north. Accordingly, on 6 November, MAG-33 was ordered to Yonpo from Japan, and by 10 November, was in operation there in time to receive VMF-212 from Wonsan. On the 15th, VMF-214 was ordered ashore from the *Sicily* and set up at Wonsan with MAG-12 supporting the squadron as best it could, bearing in mind that many pieces of vehicular equipment needed ashore are not required nor used on board carriers. Also, because of a shortage of shipping in the Far East, it took much longer to move essential shore-based equipment from where it was stored in Japan, in the case of both -214 and -212, to where it was needed in Korea. This meant that for a considerable period, bombs often had to be loaded by "muscle power," aircraft refueled by small hand "wobble" pumps from 55-gallon drums, weighing 450 pounds, and many other operational and maintenance factors that revert to the hard way when the equipment just is not there. It was just another throwback to the sustaining principle of Marine aviation of doing the best with what you have got because the job must get done—and in this case, again, it was done.

Ground Situation

At the time of the Wonsan landing, the Marines had been informed that X Corps would be a part of the dash north to the Yalu under the revised Far East Command/United Nations Command plans, and Eighth Army would be doing the same on the western side of the peninsula. The jump-off dates were set for 24 November for Eighth Army and the 27th for X Corps. There had been many sightings and identifications of Communist Chinese Forces (CCF) well below the Yalu as early as late October and in the first few days of November. Far East Command press releases, however, treated these sightings as being only "volunteers" to help the North Koreans resolve their problems. The usual sighting reports were invariably small groups in remote areas, but in some instances the sightings were characterized also by thousands of footprints and tracks in the snow. Furthermore, in a significant five-day battle from 4 to 9 November, the 7th Marines took 62 Chinese prisoners at Chinhung-ni, enroute to their objective at Hagaru-ri. The Chinese were interrogated and precisely identified as part of the

The outcome of the Chosin Reservoir campaign owed much to airdropped supplies by the wing's twin-engine R4D transports. Assisting the Air Force's Combat Cargo Command, Marine transports carried more than five million pounds of supplies to the front.

National Archives Photo (USMC) 127-N-A4841

124th Division, 42d Army, 13th Army Group, Fourth Field Army. Tokyo press releases dismissed these and other CCF contacts with the "volunteer" label and the plan remained in effect.

By the 27th, the 1st Marine Division was concentrated in the vicinity of the Chosin Reservoir, with the command post at Hagaru-ri, the 5th and 7th Marines at Yudam-ni, and the 1st Marines along the MSR with a battalion each at Chinhung-ni, Koto-ri, and Hagaru-ri. Colonel Homer L. Litzenberg, Jr., commanding the 7th Marines, while enroute from Hagaru-ri to Yudam-ni, had dropped off Company F at Toktong Pass to hold that critical point for any eventuality. On 25 November, Generals Smith and Almond conferred and the news was not good from the Eighth Army sector. The II ROK Corps had been overrun on Eighth Army's right and the Army itself was falling back before a wholesale CCF onslaught. In spite of these results, Almond ordered Smith to attack on the 27th as planned. At the time, 1st Marine Division intelligence had identified five more divisions from prisoner interrogations, and line-crossing agents had given firm indications of even more Chinese forces just to the immediate north.

On the morning of the 27th, the division began its attack from Yudam-ni on schedule but the lead regiment had only advanced about 2,000 yards when stiff resistance stopped it. On the night of the 27th, the CCF in great strength attacked all Marine positions from Yudam-ni to Koto-ri, including a division attack on Company F at Toktong Pass, and a strong assault of division-sized against the three-battalion task force of the 7th Infantry Division east of the reservoir. As the intelligence reports were gathered and analyzed, the results showed clearly that opposing the Marines and associated troops in the Chosin Reservoir area was the *9th Army Group, 3d Field Army*. This comprised a total of four corps-sized armies, a force that added to the five divisions already identified by the 1st Marine Division, totaled, by some estimates, almost 100,000 seasoned Chinese infantry troops. With the disposition of the division north of Hungnam and Hamhung, in addition to attached units of Royal Marines and assorted Army units totaling only 20,500 in all, the balance of the two forces favored the Chinese by better than 5 to 1. The fact that much of this was known to Far East Command and X Corps on the 25th, with Eighth Army estimating 200,000 CCF in front of them, cannot go unmentioned in connection with General Almond's order to General Smith to attack as planned two days later, the 27th.

Department of Defense Photo (USMC) A130488

Marines who used the "daisy-cutter" in the South Pacific rig them again for use against the Chinese in Korea. Attaching the bombs to the racks of the Corsair required delicate adjustments, often difficult in the sub-zero Korean winter weather.

The situation had changed so radically and so quickly that on the 28th, General MacArthur called Generals Walker and Almond to Tokyo for a lengthy conference. The result of these deliberations was a change of strategy. The previous plan for North Korea was abandoned and both the Eighth Army and X Corps were to pull back to a more defensible line to the south. General Smith had already decided to start moving without any further delay and ordered the 5th and 7th Marines to move back to Hagaru-ri from Yudam-ni, the first leg of what would be a 68-mile fight through thousands of enemy troops.

Air Situation

The 1st Marine Aircraft Wing's command post and attached Headquarters and Service units, in addition to five fighter squadrons, had moved to Yonpo from Wonsan and Japan by late in November. The sixth squadron, VMF-323, was still launching its maximum efforts from the *Badoeng Strait*. Rounding out the wing's combat lineup was VMO-6 with its OYs and H03S helicopters, operating mainly from Yonpo, but also from wherever else required. This was a crowd for Yonpo, especially when it is remembered that much of the ordnance and maintenance equipment of the squadrons was not available. Also included in the serious shortage category were both transportation generally, and provision of any form of heated space for bare-handed engine and aircraft engineering maintenance.

Cold weather maintenance proved difficult. Touching the metal surface of an aircraft parked on the flight line with bare skin would cause the skin to stick to it. Aircraft engines had to be started throughout the night to keep them from freezing. Tires on the planes would be frozen on the bottom in the morning and they would thump out for take off or slide along the snow and ice. Staff Sergeant Floyd P. Stocks, a plane captain with VMF-214 recalled the difficulties in accomplishing the simplest maintenance tasks, such as changing a spark plug in a F4U. "It isn't too bad removing the plugs, that can be done wearing gloves. Installing them is a different story. You can't start a plug wearing gloves, not enough clearance around the plug port. To change a sparkplug you have the old plug out and the new plug warm before you start. Wrap a new warm plug in a rag and hurry to the man standing by at the engine. That man pulls off his glove and gets the plug started. Once started he puts on his glove and completes the installation using a plug wrench."

Bombs, rockets, ammunition, and fuel were on hand at Yonpo, and with Marines to manhandle them all, the air part of the air-ground team was ready to do its job. The task it had to do was probably the heaviest responsibility ever placed on a supporting arm in relatively modern Marine Corps history. As Lieutenant General Leslie E. Brown, a Marine aviator who witnessed combat in three wars, recalled: "The Chosin Reservoir thing was the proudest I had ever been of Marine aviation...because those guys were just flying around the clock, everything that would start and move. And those ordnance kids out there dragging ass after loading 500-pound bombs for 20 hours. And aviation's mood and commitment to that division, my God it was total. There was nothing that would have kept them off those targets—nothing!"

From the time of the decision to fight their way south to the sea, Fifth Air Force had given the wing the sole mission of supporting the division and the rest of X Corps. Backup was provided by Task Force 77 aircraft for additional close support as required, and both the Navy and Fifth Air Force tactical squadrons attacked troop concentrations and interdicted approach routes all along the withdrawal fronts of Eighth Army and X Corps. The Combat Cargo Command was in constant support with requested airdrops of food and ammunition, and did a major job in aerial resupply of all types from basic supplies to bridge sections, as well as hazardous casualty evacuation from improvised landing strips at both Hagaru-ri and Koto-ri.

When reviewing the fighting withdrawal of the Marine air-ground team from the reservoir against these horrendous odds, and assessing the part Marine aviation played in the operation, it is important to remember the Tactical Air Control Party structure of the Marine air control system. Every strike against enemy positions

Marine Corsairs operating out of frozen Yonpo Airfield experienced a number of problems. The airstrip had to be continually cleared and sanded, aircraft had to be run every two hours during the night to keep the engine oil warm enough for morning takeoffs, and ordnance efficiency declined.

Department of Defense Photo (USMC) A130423

Sketch by Cpl Ralph H. Schofield, USMCR

Marine Corsairs hit enemy troop concentrations with rockets and napalm in support of Marines fighting around the Chosin Reservoir. However, approximately half of the Marine air missions were in support of South Korean and U.S. Army units.

along the route wherever the column was held up or pinned down, was under the direct control of an experienced Marine pilot on the ground in the column, known to the pilots in the air delivering the attack. Other methods had been tried repeatedly, but to put it colloquially, "there ain't no substitute for the TACP."

The Breakout

From the start of the 68-mile battle to the sea on 1 December to its completion at Hungnam on 12 December, so much happened on a daily basis that only shelves of books could tell the story in detail. It was one of the high watermarks for the Marine Corps, ground and air, cementing permanently a mutual understanding and appreciation between the two line branches of the Corps that would never be broken. It must be borne in mind that the same air support principles in almost every detail were followed in support of the division on its fight up to Hagaru-ri and Yudam-ni as were applied in supporting its fight back down to the sea.

Underlying the air support plan for the operation was the idea of having a flight over the key movement of the day at first light. This initial flight would be assigned to the forward air controller (FAC) of the unit most likely to be shortly in need of close air support. In turn, as soon as that flight had been called on to a target, another flight would be assigned to relieve it on station. This meant that response times from request to delivery on target could be reduced to the minimum. Naturally, the weather had to cooperate and communications had to stay on, but if minimum visibility and ceiling held so that positive delivery of weapons was possible, the targets were hit in minimum time. If the attack of the aircraft on station was not sufficient to eliminate that target, additional strength would be called in, either from Yonpo or from Task Force 77, or from time to time, by simply calling in any suitable aircraft in the area for a possible diversion from its assigned mission. The last possibility was usually handled by the Tactical Air Direction Center (TADC) of the air support system, or often by the tactical air coordinator airborne on the scene.

After dark each night, the column would be defended through unit assignments to key perimeters of defense. This was when they were most vulnerable to attacks by the Chinese. During daylight when Corsairs were on station, the Chinese could not mass their troops to mount such attacks because when they tried they would be immediately subjected to devastating air strikes with napalm, bombs, rockets, and overwhelming 20mm strafing. Not one enemy mass attack was delivered against the column during daylight hours. The night "heckler" missions over the column were effective in reducing enemy artillery, mortar, and heavy machine gun fires. But there was no way that they could do the things that were done in daylight controlled close air support, although the night controlled strikes against enemy positions revealed by their fires against the column were extremely effective as well. The general feeling in the column, however, was invariably one of relief with the arrival of daybreak.

The desire to have Marine aircraft overhead during daylight hours bears witness to the faith the Marines on the ground had in the potency and accuracy of Marine close air support. This was apparent to Captain William T. Witt, Jr., who led a flight of eight VMF-214 Corsairs that appeared over the Marine column one cold day as morning broke. As he checked in with the forward air controller on the ground he advised the controller that he had seen an enemy jeep heading north across the frozen reservoir and asked "if they wanted it shot up." The foot weary controller said: "Hell no, just shoot the driver."

The first leg of the fight south was from Yudam-ni to Hagaru-ri, a movement that would bring the 5th and 7th Marines together with elements of the lst Marines, division headquarters and command post. It was essential that Hagaru-ri be held because it gave the division its first chance to evacuate the seriously wounded by air. The evacuation was done from the hazardous but serviceable strip that had been hacked out of the frozen turf on a fairly level piece of ground near the town. Company D, 1st Engineer Battalion, accomplished this extraordinary job. Under fire much of the time, the work went on around-the-clock, under floodlights at night, and with flights from the two Marine night fighter squadrons orbiting overhead whenever possible. During the period from the first airstrip landing on 1 December to 6 December, the Combat Cargo Command's Douglas C-47 "Sky-trains," augmented by every Marine Douglas R4D in the area, flew out a total of 4,312 wounded, including 3,150 Marines, 1,137 Army personnel, and 25 Royal Marines. Until the Hagaru-ri strip became operational on 1 December, evacuation of the seriously wounded was limited as the only aircraft that could land at Yudam-ni, Hagaru-ri, and Koto-ri were the OYs and helicopters of VMO-6. For example, from 27 November to 1 December, VMO-6 evacuated a total of 152 casualties, including 109 from Yudam-ni, 36 from Hagaru-ri, and 7 from Koto-ri.

Casualties are helped on board a Marine R4D Skytrain at Hagaru-ri. From there, and later at Koto-ri to the south, more than 4,000 wounded men were snatched from death and flown to safety and hospitalization.

Department of Defense (USMC) A130281

Department of Defense Photo (USMC) A5439

Elements of the 7th Marines pause at the roadblock on the way to Koto-ri as Marine Corsairs napalm an abandoned U.S. Army engineer tent camp. The position had become a magnet for Chinese troops seeking food and shelter.

In the extreme cold and at the altitudes of the operation, these light aircraft had much less power and considerably reduced lift from normal conditions, but in spite of these handicaps, saved scores of lives.

The Yudam-ni to Hagaru-ri leg was completed by the afternoon of 4 December, with the first unit reaching Hagaru-ri in the early evening of the 3d. With most of the heavy action occurring on the 1st and 2d, wing aircraft flew more than 100 close support sorties both days, all in support of the division and the three Army battalions of the 7th Division, which were heavily hit east of the reservoir trying to withdraw to Hagaru-ri. The Marine FAC with the Army battalions, Captain Edward P. Stamford, directed saving strikes against the Chinese on 1 December, but during the night, they were overwhelmed and he was captured. However, the next day he managed to escape and made his way into Hagaru-ri. Of the three battalions, only a few hundred scattered troops survived to reach Hagaru-ri. On 4 and 5 December, wing aircraft continued the march with almost 300 sorties against enemy positions, vehicles, and troop concentrations throughout the reservoir area. But on 6 December, they resumed their primary role over the division as the second leg, Hagaru-ri to Koto-ri, began.

Air planning for the second leg drew heavily on the experience gained during the move from Yudam-ni. The FACs were again spotted along the column and with each flanking battalion, and were augmented with two airborne tactical air controllers who flew their Corsairs ahead and to each side of the advancing column. The addition of a four-engine R5D (C-54) transport configured to carry a complete TADC controlled all support aircraft as they reported on station, and assigned them to the various FACs or TACs, as appropriate for the missions requested. The system worked smoothly and made it possible for the column to keep moving on the road most of the time; even while the support aircraft were eliminating a hot spot. By evening of the 7th, the division rear guard was inside the perimeter of the 2d Battalion, 1st Marines, at Koto-ri. During the two days, Marine aircraft flew a total of 240 sorties in support of X Corps' withdrawal, with almost 60 percent of these being in support of the division, with the remainder being in support of other units. In addition, 245 sorties from Task Force 77 carriers and 83 from Fifth Air Force supported X Corps. The Navy sorties were almost entirely close support while the Air Force were mostly supply drops. The Koto-ri strip, although widened and lengthened, was not even as operable as the more or less "hairy" one at Hagaru-ri, but an additional 375 wounded were flown out. VMO-6, augmented by three TBMs on 7 December, also evacuated 163 more up to 10 December.

An enlisted squadron mechanic with VMF-214 noted the unpleasantness of unloading the TBMs in his diary: "Not only are the people seriously wounded, they are frozen too. This morning I helped with a Marine who never moved as we handled his stretcher. His head, framed by his parka, looked frozen and discolored. His breath fogging as it escaped his purple lips was the only sign of life. Between fingers on his right hand was a cold cigarette that had burned down between his fingers before going out. The flesh had burned but he had not noticed. His fingers were swollen and at places had ruptured now looking like a wiener that splits from heat."

With just one more leg to go, the epochal move was almost completed. But the third leg, Koto-ri to

National Archives Photo (USN) 80-G-425817

During the cold Korean winter it often took hours of scraping and chipping to clear several inches of ice and snow off the decks, catapults, arresting wires, and barriers of the Badoeng Strait *to permit flights operations. High winds, heavy seas, and freezing temperatures also hampered Marine carrier-based air mis-*

Chinhung-ni, was tough to contemplate because it included an extremely hazardous passage of a precipitous defile called Funchilin Pass, in addition to a blown bridge just three miles from Koto-ri that had to be made passable. The latter was the occasion of engineering conferences from Tokyo to Koto-ri, a test drop of a bridge section at Yonpo as an experiment, revision of parachutes and rigging, and finally the successful drops of the necessary material at Koto-ri.

The air and ground plans for the descent to Chinhung-ni amounted to essentially using the same coverage and column movement coordination that had been so successful on the first two legs, only this time there was one very effective addition. The 1st Battalion, 1st Marines, from its position in Chinhung-ni, would attack up the gorge and seize a dominating hill mass overlooking the major portion of the MSR. The battalion's attack was set for dawn on 8 December, simultaneous with the start of the attack south from Koto-ri. The night of 7-8 December brought a raging blizzard to the area, reducing visibility almost to zero and denying any air operations during most of the 8th. As a result, although both attacks jumped off on schedule, little progress was made from Koto-ri and the installation of the bridge sections was delayed. The one bright spot that day was the complete surprise achieved by the 1st Battalion, 1st Marines, in taking Hill 1081. Using the blizzard as cover, Captain Robert H. Barrow's Company A employed total silence and a double-envelopment maneuver by two of the company's three platoons with the third in frontal assault, to take an enemy strongpoint and command post, wiping out the entire garrison.

A General Motors TBM Avenger taxis out for takeoff. Largely flown by field-desk pilots on the wing and group staffs, the World War II torpedo bomber could fly out several litter patients and as many as nine ambulatory cases.

Department of Defense Photo (USMC) A131268

The night of the 8th saw the end of the weather problem and the clear skies and good visibility promised a full day for the 9th. From the break of day complete air coverage was over the MSR under the direction of the airborne TADC, the TACs, and the battalion FACs. The installation of the bridge was covered, and when it was in place, the column began its move down to Chinhung-ni on the plain below. It is interesting to note that the bridge was installed at the base of the penstocks of one of the several hydroelectric plants fed by the reservoir. (Eighteen months later in June 1952, two of these plants were totally destroyed by MAGs -12 and -33 in one attack, Chosin 3 by MAG -12 and Chosin 4 by MAG -33, the latter in one of the largest mass jet attacks of the war.)

The good weather continued on the 10th and the passage over the tortuous MSR was completed by nightfall. Early in the morning of the 11th, the truck movement from Chinhung-ni to Hungnam began, and by early afternoon, the last unit cleared the town. With the division loading out from Hungnam, the three shore-based fighter squadrons moved to Japan on the 14th, and by the 18th the last of the wing's equipment was flown out of Yonpo. Air coverage of the evacuation of Hungnam became the responsibility of the light carriers with the displacement of the wing. Under a gradual contraction of the perimeter, with the heavy support of the naval gunfire group, the movement and outloading were completed by the afternoon of the 24th.

The statistics of the outloading from Hungnam cannot go unmentioned. Included were 105,000 military personnel (Marine, Army, South Korean, and other United Nations units), 91,000 Korean refugees, 17,500 vehicles, and 350,000 tons of cargo in 193 shiploads by 109 ships. That would have been a treasure trove for the Chinese if it had not been for the leadership of General Smith who said that the division would bring its vehicles, equipment, and people out by the way they got in, by "attacking in a different direction."

Department of Defense Photo (USA) SC355021

As the last of the division's supplies and equipment were loaded on board U.S. Navy landing ships at Hungnam, the wing's remaining land-based fighter squadrons at Yonpo ended their air strikes and departed for Japan.

A few summary statistics serve to give an order of magnitude of the support 1st Marine Aircraft Wing rendered to the operation as a whole. From 26 October to 11 December, the TACPs of Marine, Army, and South Korean units controlled 3,703 sorties in 1,053 missions. Close air support missions accounted for 599 of the total (more than 50 percent), with 468 of these going to the 1st Marine Division, 8 to the 3d Infantry Division, 56 to the 7th Infantry Division, and 67 to the South Koreans. The balance of 454 missions were search and attack. On the logistics side, VMR-152, the wing's transport squadron, averaged a commitment of five R5Ds a day to the Combat Cargo Command during the operation, serving all units across the United Nations front. With its aircraft not committed to the Cargo Command, from 1 November to the completion of the Hungnam evacuation, -152 carried more than 5,000,000 pounds of supplies to the front and evacuated more than 4,000 casualties.

One other statistic for Marine aviation was its first jet squadron to see combat when VMF-311, under Lieutenant Colonel Neil R. McIntyre, operated at Yonpo for the last few days of the breakout. It is of interest to note that the tactical groups of 1st Marine Aircraft Wing, MAGs -12 and -33, were so constituted that just a year later MAG-33 was all jet and MAG-12 was the last of the props, for about a 50-50 split on the tactical strength of the wing.

On casualty statistics, the 1st Marine Aircraft Wing had eight pilots killed, four missing, and three wounded, while the division had 718 killed, 192 missing, and 3,485 wounded. The division also suffered a total of 7,338 non-battle casualties, most of which were induced by the severe cold in some form of frostbite or worse. The division estimated that about one third of these casualties returned to duty without requiring evacuation or additional hospitalization. Against these figures stands a post-action estimate of enemy losses at 37,500, with 15,000 killed and 7,500 wounded by the division, in addition to 10,000 killed and 5,000 wounded by the wing. In this case these estimates are based on enemy testimony regarding the heavy losses sustained by the Communists, and there is some verification in the fact that there was no determined attempt to interfere with the Hungnam evacuation.

In a letter from General Smith to General Harris on 20 December, Smith stated the sincere feeling of the division when he wrote:

> Without your support our task would have been infinitely more difficult and more costly. During the long reaches of the night and in the snow storms many a Marine prayed for the coming of day or clearing weather when he knew he would again hear the welcome roar of your planes as they dealt out destruction to the enemy. Even the presence of a night heckler was reassuring.
>
> Never in its history has Marine aviation given more convincing proof of its indispensable value to the ground Marines. . . . A bond of understanding has been established that will never be broken.

In any historical treatment of this epic fighting withdrawal, it is important to emphasize that there was total control of the air during the entire operation. Without that, not only would the action have been far more costly, but also it may have been impossible. It is well to keep firmly in mind that not one single enemy aircraft appeared in any form to register its objection.

Air Support: 1951-1953

After the breakout from the Chosin Reservoir and the evacuation from Hungnam, the Korean War went into a lengthy phase of extremely fierce fighting between the ground forces as the Eighth Army checked its withdrawal, south of Seoul. The line surged back and forth for months of intensive combat, in many ways reminiscent of World War I in France, with breakthroughs being followed by heavy counteroffensives, until it finally stabilized back at the same 38th Parallel where the conflict began in June 1950. In 1951, there were many moves of both the 1st Marine Division and elements of the 1st Marine Aircraft Wing. The basic thrust of the wing was to keep its units as close to the zone of action of the division as possible in order to reduce to the minimum the response time to requests for close air support. Coming under Fifth Air Force without any special agreements as to priority for X Corps, response times from some points of view often became ridiculous, measuring from several hours all the way to no response at all. The Joint Operations Center, manned by Eighth Army and Fifth Air Force, processed all requests for air support, promulgated a daily operations order, approved all emergency requests for air support, and generally controlled all air operations across the entire front. With the front stretching across the Korean peninsula, with a communications net that tied in many division and corps headquarters in addition to subordinate units, and many Air Force and other aviation commands, there was much room for error and very fertile ground for costly delays. Since such delays often could mean losses to enemy action, which might have been avoided, had close support been responsive and readily available, the Joint Operations Center was not highly regarded by Marines who had become used to the responsiveness of Marine air during the Chosin breakout, Inchon-Seoul campaign, and the Pusan Perim-eter. This was a difficult time for the wing because every time the Fifth Air Force was approached with a proposal to improve wing support of the division, the attempt ran head-on into the statement that there were 10 or more divisions on the main line of resistance and there was no reason why one should have more air support than the others. There is without question something to be said for that position. But on the other hand, it could never be sufficient to block all efforts to improve close air support response across the front by examining in detail the elements of different air control systems contributing to fast responsiveness.

Throughout the period from 1951 to mid-1953, there were various agreements between the wing and Fifth Air Force relative to the wing's support of 1st Marine Division. These covered emergency situations in the division sector, daily allocations of training close air support sorties, special concentrations for unusual efforts, and other special assignments of Marine air for Marine ground. While these were indeed helpful, they never succeeded in answering the guts of the Marine Corps question, which essentially was: "We developed the finest system of air support known and equipped ourselves accordingly; we brought it out here intact; why can't we use it?"

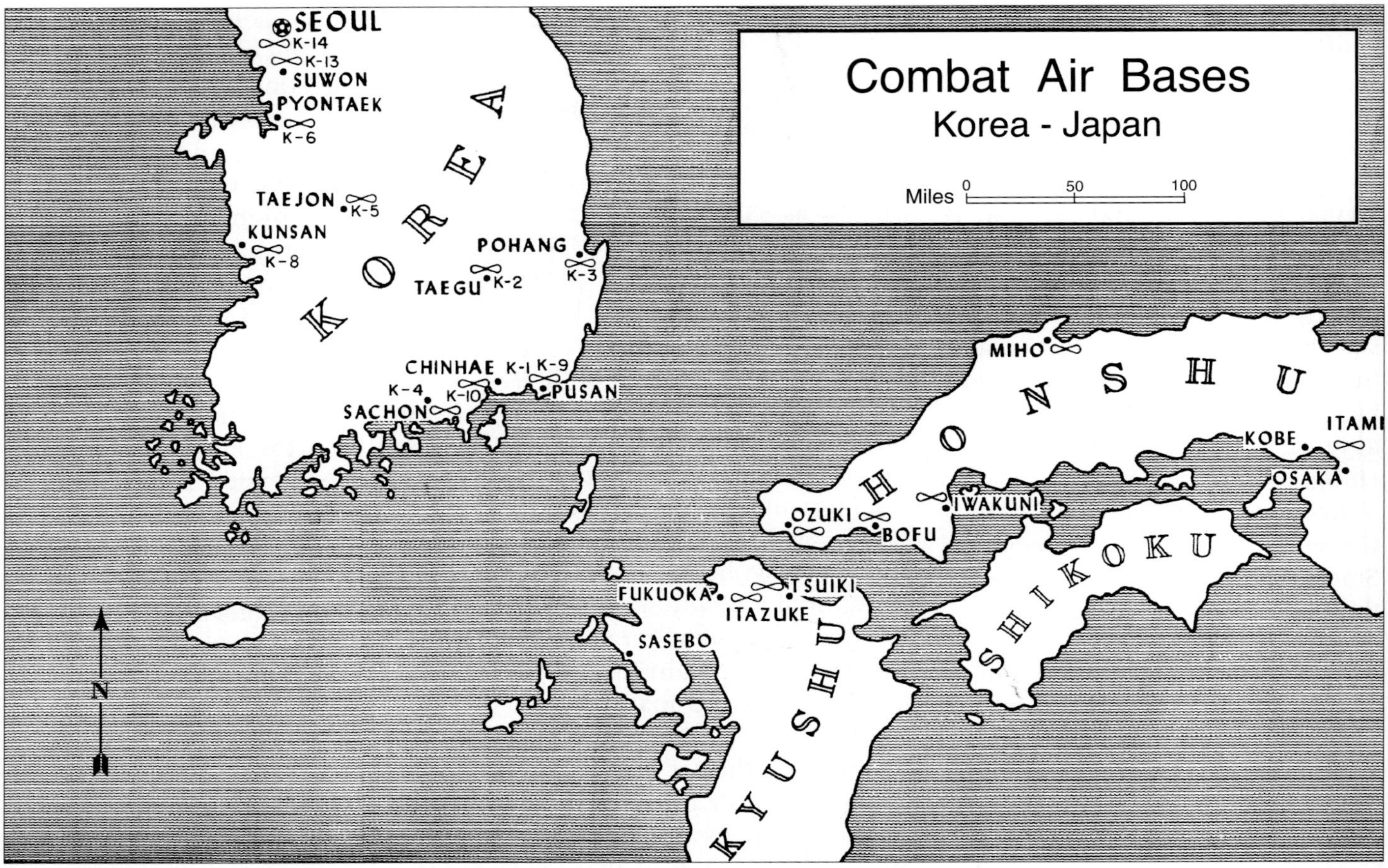

If the Army-Air Force Joint Operations Center system had been compared in that combat environment to the Marine system, and statistically evaluated with the objective of improved response to the needs of the ground forces, something more meaningful might have been accomplished. Instead, what improvements were tried did not seem to be tried all the way. What studies or assessments were made of possibilities such as putting qualified Air Force pilots into TACPs with Army battalions, seemed to receive too quick a dismissal. They were said to be impractical, or would undercut other standard Air Force missions such as interdiction and isolation of the battlefield. Since the air superiority mission was confined almost entirely to the vicinity of the Yalu River in this war, a good laboratory-type chance to examine the Joint Operations Center and Marine systems under the same microscope was lost, probably irretrievably. As if to prove the loss, the same basic questions were pondered, argued, and left unanswered a decade or more later in

The 1st Marine Aircraft Wing made a notable contribution in providing effective and speedy tactical air support. Simplified TACP control, request procedures, and fast radio system enabled wing pilots to reach the target area quickly and support troops on the ground successfully.

National Archives Photo (USN) 80-G-429965

Sketch by TSgt Tom Murray, USMC

Marine Ground Control Intercept Squadron 1 radio and radar van set-up atop Chon-san—the imposing 3,000-foot peak near Pusan. During the early years of the war, the squadron was hard-pressed to identify and control the hundreds of aircraft flying daily over Korea.

the puzzlement of the Vietnam War.

By early 1952, the stabilization of the front had settled in to the point where the fluctuations in the line were relatively local. These surges were measured in hundreds or thousands of yards at most, as compared to early 1951 where the breakthroughs were listed in tens of miles. The Eighth Army had become a field force of seasoned combat-wise veterans, and within limitations, was supported by a thoroughly professional Fifth Air Force. The wing, still tactically composed of MAG-33 at K-3 (Pohang) and K-8 (Kunsan) airfields, and MAG-12, newly established at K-6 (Pyontaek), was more or less settled down to the routines of stabilized warfare. Wing headquarters was at K-3, as was the Marine Air Control Group, which handled the air defense responsibilities of the southern Korea sector for wing. Air defense was not an over-exercised function in southern Korea, but the capability had to be in place, and it remained so throughout the remainder of the war. The control group's radars and communications equipment got plenty of exercise in the control and search aspects of all air traffic in the sector, and was a valuable asset of the wing, even though few if any "bogies" gave them air defense exercise in fact. MAG-33 was composed of VMFs -311 and -115, both with Grumman F9F Panther jets, and the wing's photographic squadron, VMJ-l, equipped with McDonnell F2H Banshee photo jets, the very latest Navy-Marine aerial photographic camera and photo processing equipment. All were at K-3 with accompanying Headquarters and Service Squadrons. At K-8, on the southwest side of the peninsula, MAG-33 also had VMF(N)-513 with Grumman F7F-3N's and Vought F4U-5Ns. In mid-1952, -513 received Douglas F3D Skyknights under Colonel Peter D. Lambrecht, the first jet night fighter unit of the wing. Colonel Lambrecht had trained the squadron in the United States as -542, moving in the new unit as -513, making MAG-33 entirely jet.

MAG-12 was the prop side of the house with VMAs -212, -323, and -312 equipped with the last of the Corsairs, and VMA-121 with Douglas AD Skyraiders. The AD was a very popular aircraft with ground Marines just like the Corsair, because of its great ordnance carrying capability. VMA-312, under the administrative control of MAG-12, and operating for short periods at K-6, maintained the wing's leg at sea and was based on board the carrier *Bataan* (CVL 29). The wing was supported on the air transport side by a detachment of VMR-152, in addition to its own organic R4Ds, and by Far East Air Force's Combat Cargo Command when required for major airlift. The rear echelon of the wing was at Itami, Japan, where it functioned as a supply base, a receiving station for incoming replacements, a facility for special aircraft maintenance efforts, and a center for periodic rest and recreation visits for combat personnel.

Operationally, the 1st Marine Aircraft Wing was in a unique position with respect to the Fifth Air Force because the air command treated the two MAGs in the same manner as they did their own organic wings. (Wing, in Air Force parlance, is practically identical to MAG in Marine talk.) This left the 1st Marine Aircraft Wing as kind of an additional command echelon between Fifth Air Force and the two MAGs which was absent in the line to all the other Air Force tactical wings. On bal-

Marine Corps Historical Center Photo Collection

Used as a night fighter during the early years of the war, the two-seat, twin-engine Grumman F7F Tigercat, with its distinctive nose-mounted radar and taller vertical tail, proved its capabilities time after time.

The Douglas AD Skyraider, one of the most versatile aircraft then in existence, was used on electronic countermeasure, night fighter, and attack missions. It could carry more than 5,000 pounds of ordnance in addition to its two wing-mounted 20mm cannon.

Department of Defense Photo (USMC) A133536

W.T. Larkins Collection, Naval Aviation History Office

The twin-engine Douglas F3D Skyknight jet night fighter gained the respect of many "former" members of the Chinese Air Force. With its state-of-the-art avionics, the big jet was soon tasked with escorting Air Force B-29s, which had been decimated by enemy MiGs.

The first Marine jet to see action in Korea, the Grumman F9F Panther compiled an enviable record in supporting United Nations forces. It speed however was offset by its relatively short endurance and poor service reliability.

Department of Defense Photo (USMC) A132958

Department of Defense Photo (USMC) A132423

Cardinal Francis J. Spellman visits the Korean orphanage at Pohang supported by the 1st Marine Aircraft Wing. To the Cardinal's left are: MajGen Christian F. Schilt, Commanding General, 1st Marine Aircraft Wing; Bishop Germain Mousset, head of the orphanage; and Col Carson A. Roberts, commanding officer of Marine Aircraft Group 33.

ance, the presence of the wing in the act was a definite plus of the most supportive kind for the two MAGs. For instance, the daily operation order for air operations came in to the two MAGs during the night and was popularly known as the "frag order," or simply, "the frag." The wing also received the frag at the same time by teletype and could check it over with MAG operations or even intercede with the Air Force if considered desirable. Relations between Fifth Air Force and wing were consistently good and although communications were somewhat hectic from time to time, the basic daily operational plans got through so that planned schedules could be met most of the time.

Maintenance of good command relations between the wing and the Fifth Air Force in the sometimes-difficult structure of the Korean War was a direct function of the personalities involved. Marine aviation was fortunate in this regard with a succession of wing commanders who not only gained the respect of their Air Force counterparts, but also did not permit doctrinal differences, which might occur from adversely affecting the mutuality of that respect. Relationships were very much aided also by the presence of a liaison colonel from the wing on duty at the Joint Operations Center, a post that smoothed many an operational problem before it could grow into something out of proportion. The teams of leaders of the 1st Marine Aircraft Wing and the Fifth Air Force were hard to match. Generals Field Harris–Earl E. Partridge, Christian F. Schilt–Frank E. Everest, Clayton C. Jerome– Glenn O. Barcus, and Vernon E. Megee–Glenn O. Barcus, constituted some of the most experienced and talented airmen the country had produced up to that time.

Operations of both MAGs generally ran to the same pattern throughout the war. Neither group was engaged in any except chance encounters with respect to air-to-air, and some of these brought an occasional startling result as when a Corsair shot down a Mikoyan-Gurevich MiG-15. However, since air combat was confined to the Yalu River area, the chance encounters were very infrequent. Considering the types of aircraft with which both groups were equipped, it is probably just as well that the Communists worked their MiGs largely in that confined sector. This left the usual frag order assignments to Marine aircraft mostly in the interdiction and close air support categories, with a lesser number in night interdiction and photo reconnaissance.

Interdiction as a category took a heavy percentage of the daily availability of aircraft because of the determination of the Air Force to show that by cutting the enemy's supply lines his ability to fight effectively at the front could be dried up. No one can deny the wisdom of this as a tenet. But in Korea at various stages of the war, it was conclusively shown that the North Koreans and the Chinese had an uncanny ability to fix roads, rails, and bridges in jury-rigged fashion with very little break in the flow of supplies. This was most evident at the main line of resistance where no drying up was noted. Because interdiction was not proving effective, any dissatisfaction stemmed from the low allocation of aircraft to close air support where air support was needed almost daily. To many, it seemed that having tried the emphasis on interdiction at the expense of close air support, pru-

Major-League Reservists

The Marine Corps Air Reserve, like other Reserve components of the United States military, had contracted after World War II. Unlike today's active organization, many reservists simply went inactive, remaining on the roles for call-up, but not drilling. Former SBD pilot, Guadalcanal veteran, and a greatly admired officer, Colonel Richard C. Mangrum (later lieutenant general) helped to establish a Aviation Reserve program, resulting in the Marine Corps Air Reserve Training Command that would be the nucleus of the "mobilizable" 4th Marine Aircraft Wing in 1962. By July 1948, there were 27 fighter-bomber squadrons, flying mostly F4U Corsairs, although VMF-321 at Naval Air Station Anacostia in Washington, D.C., flew Grumman F8Fs for a time, and eight ground control intercept squadrons. Major General Christian F. Schilt, who received the Medal of Honor for his service in Nicaragua, ran the revamped Air Reserves from his headquarters at Naval Air Station Glenview, Illinois.

When the North Koreans invaded South Korea, the Regular Marine forces were desperately below manning levels required to participate in a full-scale war halfway around the world. The Commandant, General Clifton B. Cates, requested a Reserve call-up. At the time, there were 30 Marine Corps Air Reserve squadrons and 12 Marine Ground Control Intercept Squadrons. These squadrons included 1,588 officers and 4,753 enlisted members. By late July 1950, Marines from three fighter and six ground control intercept squadrons had been mobilized—others followed. These participated in such early actions as the Inchon landing; 17 percent of the Marines involved were reservists.

The success of the United Nations operations in containing and ultimately pushing back the North Korean advance, prompted the Communist Chinese to enter the war in November and December 1950, creating an entirely new, and dangerous, situation. The well-documented Chosin breakout also resulted in a surge of applications to the Marine Corps Reserves from 877 in December 1950 to 3,477 in January 1951.

In January 1951, the Joint Chiefs of Staff authorized the Marine Corps to increase the number of its fighter squadrons from 18 to 21. Eight days later, nine fighter squadrons were ordered to report to duty. Six of these were mobilized as personnel, while three—VMFs -131, -251, and -451—were recalled as squadrons, thus preserving their squadron designations. Many of the recalled aviators and crewmen had seen sustained service in World War II. Their recall resulted from the small number of Marine aviators, Regular and Reserve, coming out of flight training between World War II and the first six months of the Korean War. Interestingly, few of the call-ups had experience in the new jet aircraft, a lack of knowledge that would not sit well with many Regular members of the squadrons that received the eager, but meagerly trained Reserve second lieutenants. As one reservist observed, without rancor: "The regulars had all the rank."

Major (later Lieutenant General) Thomas H. Miller, Jr., who served as operations officer and then executive officer of VMA-323, appreciated the recalled reservists. Remembering that the executive officer of the squadron, Major Max H. Harper, who was killed in action, was a reservist, Miller observed that although the Reserve aviators had to be brought up to speed on current tactics, they never complained and were always ready to do their part.

Miller was the eighth Marine to transition to jets and was looking forward to joining VMF-311 to fly Panthers. However, because he had flown Corsairs in World War II and was a senior squadron aviator, he was assigned to VMA-323 as a measure of support to the incoming Reserve aviators, most of who were assigned to the Corsair-equipped units in Korea. It was important, he observed, to show the Reserves that Regular Marines flew the old, but still-effective fighters, too.

The call-up affected people from all stations, from shopkeepers to accountants to baseball players. Two big-league players, Captain Gerald F. "Jerry" Coleman of the New York Yankees and Captain Theodore S. "Ted" Williams of the Boston Red Sox, were recalled at the same time, and even took their physicals at Jacksonville on the same day in May 1952. Another member of the 1952 Yankees, third baseman Robert W. "Bobby" Brown, was actually a physician, and upon his recall, served with an Army ground unit in Korea as battalion surgeon.

Capt Gerald F. "Jerry" Coleman poses in an F4U Corsair of VMA-323. Playing second base for the New York Yankees, the former World War II SBD dive-bomber pilot was recalled to duty for Korea.

Courtesy of Gerald F. Coleman

At 34, Williams was not a young man by either baseball or military standards when he was recalled to active duty in Korea in 1952. Of course, he was not alone in being recalled, but his visibility as a public figure made his case special. The star hitter took the event stoically. In an article, which appeared the August 1953 issue of *The American Weekly*, he said: "The recall wasn't exactly joyous news, but I tried to be philosophical about it. It was happening to a lot of fellows, I thought. I was no better than the rest."

Many in the press could not understand the need to recall "second hand warriors," as one reporter wrote somewhat unkindly. Most sports writers bemoaned the fact that Williams was really kind of old for a ball player as well as for a combat jet pilot.

However, the Boston outfielder reported for duty on 2 May 1952, received a checkout in Panthers with VMF-223 at Marine Corps Air Station Cherry Point, North Carolina, and was assigned to VMF-311 in Korea. His squadron mates got used to having a celebrity in their midst. Future astronaut and United States Senator John H. Glenn, then a major, was his flight leader for nearly half his missions.

On 16 February 1953, Williams was part of a 35-plane strike against Highway 1, south of Pyongyang, North Korea. As the aircraft from VMF-115 and VMF-311 dove on the target, Williams felt his plane shudder as he reached 5,000 feet. "Until that day I had never put a scratch on a plane in almost four years of military flying. But I really did it up good. I got hit just as I dropped my bombs on the target—a big Communist tank and infantry training school near Pyongyang. The hit knocked out my hydraulic and electrical systems and started a slow burn."

Unable to locate his flight leader for instructions and help, Williams was relieved to see another pilot, Lieutenant Lawrence R. Hawkins, slide into view. Hawkins gave his plane a once-over and told Williams that the F9F was leaking fluid (it turned out later to be hydraulic fluid). Joining up on the damaged Panther, Hawkins led Williams back to K-3 (Pohang), calling on the radio for a clear runway and crash crews. The baseball player was going to try to bring his plane back, instead of bailing out.

With most of his flight instruments gone, Williams was flying on instinct and the feel of the plane as he circled wide of the field, setting himself up for the approach.

It took a few, long minutes for the battered Panther to come down the final approach, but perhaps his athlete's instincts and control enabled Williams to do the job. The F9F finally crossed over the end of the runway, and slid along on its belly, as Williams flicked switches to prevent a fire. As the plane swerved to a stop, the shaken pilot blew off the canopy and jumped from his aircraft, a little worse for wear, but alive.

Later that month, after returning to El Toro, he wrote

Courtesy of Cdr Peter B. Mersky, USNR (Ret)

Capt Theodore S. Williams prepares for a mission in his VMF-311 Panther jet. Although in his mid-30s, Williams saw a lot of action, often as the wingman of another famous Marines aviator, Maj John H. Glenn.

a friend in Philadelphia describing the mission:

> No doubt you read about my very hairy experience. I am being called lucky by all the boys and with good cause. Some lucky bastard hit me with small arms and. . . started a fire. I had no radio, fuel pressure, no air speed, and I couldn't cut it off and slide on my belly. . . . Why the thing didn't really blow I don't know. My wingman was screaming for me to bail out, but of course, with the electrical equipment out, I didn't hear anything.

Williams received the Air Medal for bringing the plane back. He flew 38 missions before an old ear infection acted up, and he was eventually brought back to the States in June. After convalescing, Williams returned to the Boston Red Sox for the 1954 season, eventually retiring in 1960. Although obviously glad to come back to his team, his closing comment in the letter to his friend in Philadelphia indicates concern about the squadron mates he left behind: "We had quite a few boys hit lately. Some seem to think the bastards have a new computer to get the range. Hope Not."

Unlike Williams, who had spent his World War II duty as an instructor, Yankees second-baseman Coleman had seen his share of combat in the Philippines in 1945 as an SBD pilot with Marine Scout Bomber Squadron 341, the "Torrid Turtles," flying 57 missions in General Douglas MacArthur's campaign to wrest the archipelago from the Japanese.

Coleman had wanted gold wings right out of high school in 1942, when two young naval aviators strode into a class assembly to entice the male graduates with their snappy uniforms and flashy wings. He had signed up and eventually received his wings of gold. When Marine ace Captain Joseph J. "Joe" Foss appeared at his base, however, Coleman decided he would join the Marines. And he soon found himself dive-bombing the Japanese on Luzon.

Returning home, he went inactive and pursued a career in professional baseball. Before the war, Coleman had been a member of a semi-pro team in the San Francisco area, and he returned to it as a part of the Yankees farm system.

He joined the Yankees as a shortstop in 1948, but was moved to second base. Coleman exhibited gymnastic agility at the pivotal position, frequently taking to the air as he twisted to make a play at first base or third. His colorful manager, Casey Stengel, remarked: "Best man I ever saw on a double play. Once, I saw him make a throw while standin' on his head. He just goes 'whisht!' and he's got the feller at first." By 1950, the young starter had established himself as a dependable member of one of the game's most colorful teams. He had not flown since 1945.

As the situation in Korea deteriorated for the allies, the resulting call-up of Marine Reserve aviators finally reached Coleman. The 28-year-old second baseman, however, accepted the recall with patriotic understanding: "If my country needed me, I was ready. Besides, the highlight of my life had always been—even including baseball—flying for the Marines." After a refresher flight course, Coleman was assigned to the Death Rattlers of VMF-323, equipped with F4U-4 and AU-1 Corsairs.

Younger than Williams, whom he never encountered overseas, the second baseman had one or two close calls in Korea. He narrowly averted a collision with an Air Force F-86, which had been cleared from the opposite end of the same runway for a landing. Later, he experienced an engine failure while carrying a full bomb load. With no place to go, he continued his forward direction to a crash landing. Miraculously, the bombs did not detonate, but his Corsair flipped over, and the force jerked the straps of Coleman's flight helmet so tight that he nearly choked to death. Fortunately, a quick-thinking Navy corpsman reached him in time.

Coleman flew 63 missions from January to May 1953, adding another Distinguished Flying Cross and seven Air Medals to his World War II tally. With 120 total combat missions in two wars, he served out the remainder of his Korea tour as a forward air controller.

When the armistice was signed in July 1953, he got a call from the Yankees home office, asking if he could get an early release to hurry home for the rest of the season. At first the Marine Corps balked at expediting the captain's trip home. But when the Commandant intervened, it was amazing how quickly Coleman found himself on a Flying Tigers transport leaving Iwakuni bound for California.

Coleman had to settle for rejoining his team for the 1954 season, but he felt he never regained his game after returning from Korea. Retiring in 1959, he became a manager in the front office, indulged in several commercial ventures, and finally began announcing for the expansion team San Diego Padres in 1971, where he can still be found today.

The press occasionally quipped that the military was trying to form its own baseball club in Korea. However, the players never touched a bat or ball in their squadrons. In the privacy of the examination room, Dr. Robert "Bobby" Brown did try to show an injured soldier how to better his slide technique—all in the interests of morale.

According to the Assistant Secretary of the Navy for Air at the time, John F. Floberg, every third airplane that flew on a combat mission in Korea was flown by a Navy or Marine reservist. Of the total combat sorties conducted by the 1st Marine Aircraft Wing, Marine Air Reserves flew 48 percent.

---

dence and logic would have switched the preponderance of effort the other way, particularly where casualties were being taken which close air support missions might have helped reduce.

Other than in this doctrinal area, interdiction missions targeted supply dumps, troop concentrations, and vehicle convoys, as in the earlier days of the war. Day road reconnaissance missions became less productive as the months rolled by, and the Communists became very adept at the use of vehicle camouflage as they parked off the routes waiting nightfall. Flak became increasingly intense also and was invariably in place and active wherever a road or rail cut looked to the target analysts as if it might create a choke point leading to a supply break. The fact, however, that nothing moved except at night generally stated the effectiveness of day interdiction. But it was impossible to isolate the battlefield if the tactical air was only effective half of each day.

VMF(N)-513 carried the load for the 1st Marine Aircraft Wing with respect to night road reconnaissance, or "road recces" as they were known, using both F7F-3Ns and F4U-5Ns. Usually, they were assigned a specific section of road

National Archives Photo (USMC) 127-N-A347877

During a series of strike missions in June 1953, more than 68 Panther jets from VMFs –115 and –311 destroyed or damaged more than 230 enemy buildings using napalm and incendiary munitions.

and a time on station, coordinated with a flare plane which would sometimes be a wing R4D, at others another Tigercat or Corsair, or at still others an Air Force aircraft. A mission plan would be set up and briefed for all participants, and all intelligence available would be covered. At the agreed upon time, the flare plane would illuminate and the pilot of the attack plane would be in such a position that he could hopefully make maximum use of the light in delivering his ordnance, usually fragmentation bombs, napalm, and strafing. Here, as elsewhere, as the stabilized phase of the war continued, the Communists improved their use of organized light flak. Many planes were holed with hand-held weapons, indicating a policy of massed fires of all weapons when under air attack. In addition, a steadily increasing number of mobile twin 40mm mounts appeared on the roads, which added weight to the flak problem. The gradual improvement was effective to the point that in 1952, the F7F was taken off road recces because its twin-engine configuration was correlated with excessive losses without the protection of one big engine directly forward of the cockpit. The Corsair continued to fly road recces, but the Tigercat was used primarily for air-to-air intercepts at night from mid-1952. The F3D Skyknight, when it arrived in -513, was used for deep air-to-air patrolling and for night escort of B-29s, with the F7F for closer range patrols.

Close air support missions were of two types. The first, used the most, appeared in the frag as an assignment of a certain number of aircraft to report to a specific control point at a specific time, for use by that unit as required or speci-

Night MiG Killers

A Marine squadron that had both an unusual complement of aircraft and mission assignments was VMF(N)-513, the "Flying Nightmares." The squadron was on its way to the Pacific war zone when the Japanese surrendered, but it was an early arrival in Korea, operating Grumman's graceful twin-engine F7F Tigercat. Too late to see action in the Pacific, the F7F had languished, and it was not until the war in Korea that it was able to prove its worth.

Actually, a sister squadron, VMF(N)-542 had taken the first Tigercats over—by ship—and flew some of the first land-based Marine missions of the war, relinquishing the Grummans to -513 when it relieved -542.

The Flying Nightmares soon found their specialty in night interdiction, flying against Communist road supply traffic, much as their successors would do more than 10 years later and farther to the south in Vietnam, this time flying F-4 Phantoms.

Operating from several Air Force "K" fields, -513 quickly gained two other aircraft types—the F4U Corsair and the twin-jet F3D Skyknight. Thus, the squadron flew three frontline warplanes for the three years of its rotating assignment to the war zone.

The squadron accounted for hundreds of enemy vehicles and rolling stock during dangerous, sometimes fatal, interdiction strikes. Four Nightmare aviators were shot down and interned as prisoners of war.

Occasionally, Air Force C-47 flareships would illuminate strips of road for the low-flying Corsair pilots, a tricky business, but the high-intensity flares allowed the Marines to get down to within 200 to 500 feet of their targets.

Nightmare aviator First Lieutenant Harold E. Roland recounted how he prepared for a night interdiction flight in his Corsair:

> As soon as I was strapped in, I liked to put on my mask, select 100 percent oxygen and take a few deep breaths. It seemed to clear the vision. At the end of 4 1/2 hours at low altitude, 100 percent oxygen could suck the juices from your body, but the improved night vision was well worth it.
>
> We always took off away from the low mountains to the north. Turning slowly back over them, my F4U-5N labored under the napalm, belly tank, and eight loaded wing stations. I usually leveled off at 6,000 feet or 7,000 feet, using 1,650 rpm, trying to conserve fuel, cruising slowly at about 160 indicated.

The F4U pilots were expected to remain on station, within a quick call to attack another column of enemy trucks. Individual pilots would relieve another squadron mate as he exhausted his ordnance and ammunition.

VMF(N)-513 was also unique in that it scored aerial kills with all three types of the aircraft it operated. The Corsairs shot down one Yakovlev Yak-9 and one Polikarpov PO-2, while the F7Fs accounted for two PO-2s. The jet-powered F3Ds, black and sinister, with red markings, destroyed four MiG-15s, two PO-2s, and one other Communist jet fighter identified as a Yak-15, but sometimes as a later Yak-17.

Today, the squadron flies the AV-8B Harrier II, and although based at Marine Corps Air Station Yuma, Arizona, it is usually forward deployed in Japan. A detachment of VMA-513 Harriers flew combat operations during the 1991 Persian Gulf War.

Returning on the night he shot down a MiG-15, squadron commander LtCol Robert F. Conley greets SSgt Walter R. Connor. There was a second MiG, which was listed as a probable, hence SSgt Connor's two-fingered gesture.

Courtesy of Cdr Peter B. Mersky, USNR (Ret)

fied Applicable intelligence and coordinating information would be included most of the time, and ordnance would either be specified or assigned as a standard load. Depending on the target, if one was specified in the frag, flights of this type were usually of four aircraft but could often be as many as eight or twelve. The second type of close air support mission was known as strip alert. This concept was adapted usually to those fighter fields which were reasonably close to the main line of resistance, making it possible for the slower prop aircraft so assigned to reach any sector of the front from which a close support request was received, in minimum time. It was also used from fields farther back,

National Archives Photo (USN) 80-G-429631

Corsairs of Marine Fighter Squadron 312, based on the light carrier Bataan *(CVL 29), carry out a raid against several small North Korean boats suspected of being used to lay mines along the Korean coastline.*

primarily with jet aircraft, in order to conserve their fuel so that they could remain on station longer, time to reach any sector of the front not being as much of a factor as with prop aircraft. Ordnance loads for strip alert close air support could be specified or standard. Intelligence matters and coordinating data would usually be given while the aircraft were enroute. Strip alert aircraft were without exception under the "scramble" control of Joint Operations Center.

The same increasing antiaircraft capabilities of the Communists were found along the main line of resistance as elsewhere. In fact, stabilized warfare brought some weird and different tactics into play, which were somewhat reminiscent of the "Pistol Pete" days at Guadalcanal. Heavy antiaircraft artillery guns were sited close to

Department of Defense Photo (USMC) A168084

An entrenched Marine peers out over the lip of his bunker to observe an air strike against equally entrenched Communist soldiers on the western front in Korea.

the main line of resistance just out of friendly artillery range, and 37 and 40mm twins were a commonly encountered near the frontlines. Once the close air support flight checked in with the Tactical Air Control Party, the usual response was for the controller to bring the flight leader "on target" by having him make coached dummy runs.

When he had the target clearly spotted, he would mark it with a rocket or other weapon on another run, having alerted the orbiting flight to watch his mark. The flight would then make individual runs, in column and well spaced, invariably down the same flight path. While this was essential for accurate target identification, the whole process gradually told the enemy exactly who or what the target was, so that by the second or third run down the same slot, every enemy weapon not in the actual target was zeroed in on the next dive. The heavy antiaircraft artillery and automatic antiaircraft fire complicated the process because the flight, orbiting at 10,000 feet or so, now had other things to consider while watching the flight leader's dummy run and mark. In close air support, there is usually no way to change the direction of the actual attack run without subjecting friendly troops to inordinate danger of "shorts" or "overs."

The net effect stimulated more time on target coordinating tactics with the artillery, and also put more emphasis on the detailed briefing given by the forward air controller by radio to the flight. This measure served to reduce the number of dummy runs and marking runs required, while coordination with the artillery put airbursts into the area at precisely the right time to cut down on the massing of enemy weapons on each succeeding dive. These measures were effective counters to the increased antiaircraft capability of the enemy, without the sacrifice of any effectiveness in close air support delivery.

To attempt to fill the lack of Tactical Air Control Parties in the Army and other United Nations divisions, the Fifth Air Force used the North American T-6 training

A bird's-eye-view of Battery B, 1st 90mm Antiaircraft Artillery Gun Battalion's heavily sandbagged position north of Pusan. While the battalion's two 90mm batteries were centered on Pusan, its .50-caliber automatic weapons battery was stationed at K-3 (Pohang), the home base of MAG-33.

1st MAW Historical Diary Photo Supplement, Jul53

Department of Defense Photo (USN) 443503

Among the targets hit by Marine aircraft were the generating stations of hydroelectric plants along the Yalu River, which provided power to Communist-controlled manufacturing centers. The resultant blackout of the surrounding areas halted production of supplies needed by enemy forces.

aircraft which flew low over the frontlines and controlled air strikes in close support, in somewhat the same manner as was done by a forward air controller in the Tactical Air Control Party. Many of these controllers, known as "Mosquitos," were very capable in transmitting target information to strike aircraft and in identifying and marking targets. The Mosquito was an effective gap-filler, but with increased enemy antiaircraft fire, the effectiveness of the expedient fell off markedly.

In addition to interdiction and close air support missions, from time to time Fifth Air Force would lay on a maximum effort across the board when intelligence developed a new or important target. These missions would involve all Air Force wings, in addition to the two MAGs, and a heavy force from Task Force 77 carriers. Preliminary coordination and planning would

A Sikorsky HRS-1 helicopter picks up several Marines from a precarious frontline position. The helicopters of Marine Helicopter Transport Squadron 161 revolutionized frontline operations, bringing men and equipment into the battle zone and evacuating the wounded in minutes.

National Archives Photo (USMC) 127-N-A159962

Department of Defense Photo (USMC) A158624

Developed between 1946 and 1950, the MPQ-14 radar-controlled bombing equipment was employed by Marine Air Support Radar Team 1 to control night fighter sorties flown by day attack aircraft, achieving Marine aviation's primary goal of providing real 24-hour close air support, regardless of weather conditions.

usually be the subject of conferences at Joint Operations Center, to which the wing commanders (including the commanding officers of the Marine aircraft groups) would be summoned. When a non-scheduled wing commanders conference was called, it was a signal that a big one was in the offing. Examples of this type of targeting included the hydroelectric plant complex, long restricted and finally released in June 1952; intelligence indications of a high-level Communist conference in Pyongyang; or an important installation on the Yalu, just across from the MiG fields in Manchuria. These missions broke the routine of stabilized warfare and gave all units a chance to see what massing their aircraft could achieve—it was a good break from the usual flight-of-four routine.

While VMO-6 continued its support of the division through 1951-1953 with its OYs, OEs, and HO3Ss, the big news in helicopters was the arrival of Marine Transport Helicopter Squadron 161 on 31 August 1951. Commanded by Lieutenant Colonel George W. Herring, the first transport helicopter squadron was attached to the division and administratively supported by the wing in the pattern of VMO-6. Just two weeks later, the squadron executed the first resupply and casualty evacuation lift in just 2.5 hours, moving 19,000 pounds of cargo seven miles to the engaged 2d Battalion, 1st Marines, and evacuating 74 casualties. Called Operation Windmill, it was the first in a long and growing list of Marine Corps combat lifts. HMR-161 set standards on helicopter operations with troops, which are still in active use. The squadron was a leader in night and marginal weather operations, and pioneered many different movements of field equipment in combat for the division, quick tactical displacements which were previously impossible. A typical example was the preplanned emplacement of rocket launchers, which after a ripple discharge attracted immediate counterbattery fire. Lifting the launchers in by "chopper," and then immediately lifting them to another planned site after firing avoided an enemy response.

Another piece of Marine aviation equipment that was moved into the 1st Marine Division area early in 1951 was a radar bombing system that could direct aircraft to their proper release points at night or in bad weather. It was scaled down from an Air Force version mounted in large vans that was unsuitable for forward battlefield terrain, to a mobile configuration that could be used close to the frontlines. Designated the MPQ-14, the objective of the design was to provide close air support around-the-clock, regardless of the weather. While this ambitious goal was not attained, nevertheless the use of the MPQ-14 radar in Korea was an unqualified success in that it kept an "almost close" capability over the frontlines under conditions that previously had closed the door to air support. MPQ-14 air support was never as close and as positive as the close air support, but it was useful and continued to fill that type of need many years after Korea.

In practice, the MPQ controller would vector the aircraft to the release point and at the proper spot, would direct release by radio, and in later refinements, automatically. The aircraft would be in horizontal flight, and in effect it turned day fighters and day attack aircraft into all-weather horizontal bombers, without any major modification to the aircraft, ordnance, and communications systems. The work that was done

VMJ-1 Historical Diary Photo Supplement, Oct52

Crewmen load reconnaissance cameras on board one of Marine Photographic Squadron 1's MacDonnell F2H-2P Banshees. The squadron's wartime output of more than 793,000 feet of processed prints was equal to a continuous photographic strip six-and-one-half times around the earth at the equator.

Maj Marion B. Bowers, VMJ-1's executive officer, prepares to "light-off" his 550-mph F2H-2P twin-jet Banshee for another unarmed but escorted mission deep into North Korea to photograph enemy positions, airfields, powers plants, and other potential targets.

VMJ-1 Historical Diary Photo Supplement, Oct52

Who Were the Guys in the MiGs?

For decades, the public perception was the men in the cockpits on the other side were North Korean and Communist Chinese. While there were certainly pilots from these countries flying against allied aircraft, recent disclosures after the collapse of the Soviet Union in 1991 and subsequent release of previously classified files, point to a complete wing of MiG-15s flown exclusively by 'volunteered" Soviet aviators, many of whom had considerable combat experience in World War II. Several had sizeable kill scores against the Germans. Indeed, the leader of the wing, although he apparently did not actually fly MiGs in Korean combat, was Colonel, later Air Marshall, Ivan N. Kozhedub, with 62 kills on the Eastern Front, the top-scoring Allied ace of World War II.

The Soviets went to great lengths to disguise the true identities of their MiG drivers. They dressed the much larger Soviet aviators in Chinese flight suits, complete with red-topped boots, and tried (somewhat unsuccessfully) to teach them flying phrases in Chinese to use on the radio. But they could not hide the rapid-fire Russian American monitors and pilots heard once a major engagement had begun. The American Sabre and Panther pilots always suspected that the "honchos," the leaders of the so-called "bandit trains" that launched from the other side of the Yalu River, were actually Soviets.

While the MiG-15 was a match for the American F-86 Sabre jets, which several Marine Corps aviators flew during exchange tours with the Air Force, its pilots later described their cockpits as rather cramped with much less visibility compared to the Sabre. They flew without G-suits or hard helmets unlike their opposite numbers in the F-86s. MiG-15 pilots used the more traditional leather helmets and goggles—a kit used through the 1970s by North Vietnamese MiG-17 pilots.

The MiG's ejection seat required activating only one handle, whereas the Sabre pilot had to raise both arms of his seat to eject. While the Soviet arrangement might be advantageous if the pilot was hurt in one arm, it could also place him badly out of proper position when ejecting, and could result in major back injuries.

Yefim Gordon Archives

Soviet volunteer pilots inspect one of their MiG-15s in Korea. The MiG's small size shows up well, as does the bifurcated nose intake.

Korean service was hard, and decidedly inglorious for the Soviet crews, who remained largely anonymous for more than 40 years. Yet, it would seem that the top-scoring jet-mounted ace in the world is a Russian, Colonel Yevgeni Pepelyaev with 23 kills over United States Air Force F-86s and F-84 Thunderjets in Korea. He is closely followed by Captain Nikolai Sutyagin with 21 scores. The only other jet aces who approached these scores are two Israelis, with 17 and 15 kills, and American Air Force Captain Joseph "Mac" McConnell with 16 kills in F-86s. When McConnell was ordered home in May 1953, Marine Corps ace Major John F. Bolt, Jr., succeeded him as commander of Dog Flight, 39th Squadron, 51st Fighter Interceptor Wing.

with the MPQ-14 in Korea established confidence in its use and set procedures in its employment, which are still standard practice.

In the spring of 1952, MAG-33 acquired a new and special squadron, VMJ-1. A photo reconnaissance unit, the squadron was equipped with 10 McDonnell F2H-2P Banshees and the latest Navy-Marine camera configuration that made the aircraft by far the most efficient photo reconnaissance system in the Fifth Air Force. Not only were the side-looking and vertical cameras superior to anything else around, but also the squadron was equipped with its own organic field film processing equipment. The design of the Banshee photo equipment was the work of the photographic development section of the Bureau of Aeronautics of the Navy, the McDonnell Aircraft Corporation, and the Navy and Marine pilots assigned to the associate activities. Where the percentage of film exposed that after processing was readable had been no more than 30 percent, the comparable figure in VMJ-1 was more than 90 percent. This factor, along with other automated advances in the system, literally made the 10 Banshees, which comprised no

1stMAW Historical Diary Photo Supplement, Jul53

Maj John F. Bolt, Jr., while flying a North American F-86 Sabre jet with the Air Force's 51st Fighter Interceptor Wing shot down his sixth MiG-15 on 12 July 1953, becoming the Marine Corps' first jet ace. Bolt also achieved ace-status during World War II by downing six Japanese aircraft while flying with the Black Sheep of VMF-214.

more than 20 percent of the photo reconnaissance force available, carry upwards of 30 to 40 percent of the daily Air Force photo mission load.

The employment of the reconnaissance aircraft was interesting. Totally unarmed, almost all of its missions were flown unescorted at high altitude, except that often the pilot in the event of cloud obstruction would descend below a cloud deck to acquire his target if the area was not too hot. For the tougher targets, like Sinanju and Suiho on the Yalu, which were well within MiG range from across the Yalu, the Banshee was escorted by an ample flight of North American F-86 Sabre jets. There was an advantage, strange as it may seem, to the unescorted mission. A single Banshee at high altitude presented a very low profile to enemy antiaircraft radar and radar fighter direction equipment, compared to that of one photo plane with four or more fighter escorts in company. The unescorted missions penetrated all the way up the east coast to the Soviet border and at the extreme northeast end of the run, Vladivostok was clearly visible. Other missions would take the aircraft the length of the Manchurian border down the Yalu to the point where the range of the MiG dictated escort. If jumped when unescorted, the best defense against the MiG was a steep and very tight spiral to the deck or to the nearest heavy cloudbank.

The last highlight to mention was the system arranged between Fifth Air Force and 1st Marine Aircraft Wing which provided a few Marines, after they had finished their tours in MAG-33 jets, the experience of a few weeks temporary duty with the F-86 squadrons. Being very experienced jet pilots, they checked out quickly and were taken into the regular flights of the Air Force squadrons, some for as many as 50 or more missions against the MiG. From November 1951 to July 1953, these visitors shot down a total of 21 MiG-15s. At any given time, there was usually only one Marine on duty with each of the two F-86 wings. The high score and only Marine jet ace of the group was Major John F. Bolt with six, although Major John H. Glenn, getting three in July 1953, was closing in fast when the ceasefire was announced. It was a valuable program for Marine aviation, which was indebted to the Air Force for the experience; air-to-air experience being essentially denied because the straight-wing F9F was no match for the swept-wing MiG-15. With the Corsair, Tigercat, and Skyknight tolls added in, Marines shot down more than 37 Communist aircraft of all types during the Korean War.

The character of the Korean War for Marine aviation was light on air-to-air, heavy on air-to-ground,

Marine Pilots and Enemy Aircraft Downed

Date: Pilot	Squadron	Aircraft Flown	Aircraft Downed
21 Apr 51: 1stLt Harold D. Daigh	VMF-312	F4U-4	2 Yak-9
21 Apr 51: Capt Phillip C. DeLong	VMF-312	F4U-4	2 Yak-9
30 Jun 51: Capt Edwin B. Long/ WO Robert C. Buckingham	VMF(N)-513	F7F-3N	1 PO-2
12 Jul 51: Capt Donald L. Fenton	VMF(N)-513	F4U-5NL	1 PO-2
23 Sep 51: Maj Eugene A. Van Gundy/ MSgt Thomas H. Ullom	VMF(N)-513	F7F-3N	1 PO-2
4 Nov 51: Capt William F Guss	336 FIS (USAF)	F-86A	1 MiG-15
5 Mar 52: Capt Vincent J. Marzelo	16 FIS (USAF)*	F-86A	1 MiG-15
16 Mar 52: LtCol John S. Payne	336 FIS (USAF)	F-86A	1 MiG-15
7 Jun 52: 1stLt John W. Andre	VMF(N)-513	F4U-4NL	1 Yak-9
10 Sep 52: Capt Jesse G. Folmar	VMF-312	F4U-4	1 MiG-15
15 Sep 52: Maj Alexander J. Gillis	335 FIS (USAF)	F-86E	1 MiG-15
28 Sep 52: Maj Alexander J. Gillis	335 FIS (USAF)	F-86E2	2 MiG-15
3 Nov 52: Maj William T. Stratton, Jr./ MSgt Hans C. Hoglind	VMF(N)-513	F3D-2	1 Yak-15(17?)
8 Nov 52: Capt Oliver R. Davis WO Dramus F. Fessler	VMF(N)-513	F3D-2	1 MiG-15
10 Dec 52: 1stLt Joseph A. Corvi/ MSgt Dan R. George	VMF(N)-513	F3D-2	1 PO-2
12 Jan 53: Maj Elswin P. Dunn/ MSgt Lawrence J. Fortin	VMF(N)-513	F3D-2	1 MiG-15
20 Jan 53: Capt Robert Wade	16 FIS (USAF)	F-86E	1 MiG-15
28 Jan 53: Capt James R. Weaver/ MSgt Robert P. Becker	VMF(N)-513	F3D-2	1 MiG-15
31 Jan 53: LtCol Robert F. Conley/ MSgt James N. Scott	VMF(N)-513	F3D-2	1 MiG-15
7 Apr 53: Maj Robert Reed	39 FIS (USAF)	F-86F	1 MiG-15
12 Apr 53: Maj Robert Reed	39 FIS (USAF)	F-86F	1 MiG-15
16 May 53: Maj John F. Bolt	39 FIS (USAF)	F-86F	1 MiG-15
17 May 53: Capt Dewey F. Durnford	335 FIS (USAF)	F-86F	1/2 MiG-15
18 May 53: Capt Harvey L. Jensen	25 FIS (USAF)	F-86F	1 MiG-15
15 Jun 53: Maj George H. Linnemeier	VMC-1	AD-4	1 PO-2
22 Jun 53: Maj John F. Bolt	39 FIS (USAF)	F-86F	1 MiG-15
24 Jun 53: Maj John F. Bolt	39 FIS (USAF)	F-86F	1 MiG-15
30 Jun 53: Maj John F. Bolt	39 FIS (USAF)	F-86F	1 MiG-15
11 Jul 53: Maj John F. Bolt	39 FIS (USAF)	F-86F	2 MiG-15
12 Jul 53: Maj John H. Glenn	25 FIS (USAF)	F-86F	1 MiG-15
19 Jul 53: Maj John H. Glenn	25 FIS (USAF)	F-86F	1 MiG-15
20 Jul 53: Maj Thomas M. Sellers	336 FIS (USAF)	F-86F	2 MiG-15
22 Jul 53: Maj John H. Glenn	25 FIS (USAF)	F-86F	1 MiG-15

*** FIS (Fighter Interceptor Squadron)**

Department of Defense Photo (USMC) A348551

Future astronaut and United States Senator, Maj John H. Glenn smiles from the cockpit of his F-86 Sabre jet on his return from a flight over North Korea during which he shot down the first of three MiG-15s he would be credited with during the war.

and often primitive with respect to operating airfields. The part played by the enemy which directly affected Marine aviation, was the gradual and continuous buildup of his antiaircraft capability. The employment of heavy antiaircraft artillery in proximity to the front, the increased use of mobile automatic antiaircraft weapons of higher caliber, both at the front and on access routes, forced tactical changes but did not lessen the effectiveness of either close air support or interdiction missions. In addition, the time spent in advancing up the learning curve as changes occurred, are reflected in a summary of the aviation statistics for the war. These show that Marine aviation lost 258 killed (including 65 missing and presumed dead) and 174 wounded. A total of 436 aircraft were also lost in combat and in operational accidents. Of the 221 Marines captured during the three-year conflict, 31 were aviators.

Armistice and Aftermath

The possibility of a ceasefire and general armistice was a constant element in the Korean War from mid-1951. The peace talks gained more attention in early 1952 after a formal site was established at Panmunjom, with assigned United Nations, North Korean, and Communist Chinese negotiators in attendance at scheduled sessions. Marine aviation provided support for this aspect of the Korean War, and its aftermath. Aviation furnished several general officers, as did the ground Marine Corps, for the negotiating team, a shared assignment between all the United States Armed Services.

The 1st Marine Aircraft Wing post-armistice plan, a part of the Fifth Air Force strategy, was effective on 27 July 1953. Its basic objective was twofold: first, to carry out Fifth Air Force responsibilities as assigned; and second to maintain a high level of combat readiness in all units. The armistice delineated a "no-fly" barrier along a line just south of the United Nations southern boundary of the Demilitarized Zone, and day and night patrols of that barrier were missions assigned to the wing. The day missions were shared by the MAGs at K-3 and K-6, while the night patrols were flown by the F3Ds of VMF(N)-513 and the radar-configured ADs of Marine Composite Squadron l.

The armistice agreement created a set of administrative bottlenecks, with the limitation on airports of entry and departure to a total of six for South Korea. This meant that every aircraft entering, regardless of its ultimate destination, had to undergo a detailed inspection upon landing. Numerous forms were required to be filled out and untold reports rendered for each aircraft arriving in country or departing. When the personnel and unit reports were added to the list, it all became a formidable bureaucratic check on cheating with respect to the armistice agreement.

Because of the indeterminate nature and duration of the armistice, it was necessary to deploy additional Fleet Marine Forces to the Far East in order to maintain a posture of amphibious readiness in the area. Late in the summer of 1953, the 3d Marine Division arrived in Japan accompanied by MAGs -11 and -16, the latter a helicopter transport group equipped with Sikorsky HRS-2s. MAG-11, comprised of three F9F squadrons, was based at Atsugi, Japan, as was VMR-253, an additional transport squadron assigned to wing and flying the F4Q Fairchild Packet. MAG-16 was based at Hanshin Air Force Base with its two squadrons and service units.

Both in Korea and Japan, the

Aviator Prisoners of War

The long months of incarceration, torture, deprivation, and uncertainty made the prisoner of war experience a terrible ordeal. It was a harbinger of what the next generation of American prisoners of war would face barely a decade later in another Asian country.

While American treatment toward its prisoners of war in World War II was much more benevolent, it might be said that the stories told by returning prisoners from World War II Japanese and Korean War prison camps changed how we as a country looked at ourselves as warriors, and how we conducted ourselves regarding enemy soldiers we captured in future wars.

Certainly, the greatest change that resulted from the Korean War prisoners' collective experience was the institution of the Code of Conduct, which specifically outlined what an American serviceman would give his captors by way of information and how he would conduct himself.

The Code was at times quite nebulous and in its first test, in Vietnam, each American had to determine his own level of faith and endurance. The boundaries were defined in the Code, but as the years wore on, cut off from any contact with his government, and with only occasional meetings with his compatriots in the camps, each had to determine for himself how he could meet the requirements of his country. It was a trial of strength and courage far more terrible than the short-burst stress of aerial combat. Those who survived their internment in Southeast Asia could—in some measure—perhaps thank their predecessors in the cold mountain camps of Korea for bringing back information that helped them live. Of the 221 Marines captured during the Korean War, 31 were flight crewmen. Three died in captivity; one is presumed dead.

The first Marine aviator prisoner of war in Korea was Captain Jesse V. Booker of Headquarters Squadron 1. He was shot down on 7 August 1950 while flying a reconnaissance mission from carrier *Valley Forge* (CV 45). Captain Booker, who had shot down three Japanese aircraft in World War II, received several briefings on escape and evasion. He could be considered as well prepared as could be at this early stage of the war. After capture, he was beaten and tortured by his North Korean guards and was the only Marine pilot in enemy hands until April 1951.

Captain Paul L. Martelli was shot down on 3 April 1951 while flying Corsairs with VMF-323. As he attacked ground targets, his fighter's oil cooler was hit by small arms fire, and he soon had to bail out. His wingman initially reported that Captain Martelli had fallen from his F4U, and he was carried as killed in action.

Martelli was captured by Chinese troops, who took him to an interrogation center near Pyongyang. He

Department of Defense (USN) 628393

Maj Francis Bernardini, USMC, chats with returning prisoners of war Capt Jesse V. Booker, center, and 1stLt Richard Bell, right, at Freedom Village, Panmunjom, Korea. Booker and Bell were returned on 27 August 1953, the first Marine aviators to be sent back.

endured several painful sessions with a Major Pak, considered by many of the prisoners to be among the enemy's most sadistic "interviewers."

Captain Mercer R. Smith launched for an armed reconnaissance mission from K-3 (Pohang) on 1 May 1951. Flying F9F-2B Panthers with VMF-311, he and his wingman were at 6,000 feet when Captain Smith reported a fire in the cockpit. He climbed to 16,000 feet and ejected. At first, his wingman and the pilot of a rescue helicopter that arrived shortly afterward reported enemy troops standing over the body of the downed pilot, thereby giving rise to the belief that Captain Smith was dead. He initially was carried as killed in action, but was reported on the Communist 18 December 1951 list of prisoners of war.

The following day, Captain Byron H. Beswick, an F4U pilot with VMF-323, was part of a four-plane, close air support mission. It was his third mission of the day and the 135th of his tour. Small arms fire caught him during a strafing run, hitting a napalm tank, which did not ignite. However, his aircraft was hit soon afterward, catching fire, and forcing Captain Beswick to bail out. He suffered painful burns on his face, arms, hands, and right leg.

Communist troops captured him, placing him with a battalion of British prisoners of war, which fortunately included two doctors. Enduring long marches, Captain Beswick and his compatriots tried to escape, but were recaptured.

On 27 May 1951 while on an armed reconnaissance with two other aircraft, Captain Arthur Wagner, the pilot of an F4U-5N with VMF(N)-513 also was interned.

Captain Jack E. Perry of VMF-311 was the squadron-briefing officer and had to scrounge flights. By mid-June 1951 he had 80 missions. He knew about the danger of enemy flak sites in the Singosan Valley and scheduled himself for a mission against the traps on 18 June. However, the guns quickly found the range and hit his Panther's fuel tank. Captain Perry ejected and was captured by Chinese troops, who showed him bomb craters and their wounded soldiers as a result of American strikes.

Several other Marine aviators were shot down in subsequent months, mainly by antiaircraft guns. But VMF-311 lost a Panther to MiGs on 21 July 1951. First Lieutenant Richard Bell was part of a 16-plane strike in MiG Alley, the notorious area along the Yalu River in northwestern Korea. His division of three aircraft—a fourth F9F pilot had aborted the mission when his cockpit pressurization failed—flew their mission and were returning to base when no less than 15 MiG-15s appeared. The enemy fighters attacked the small American formation, whose pilots turned into the oncoming MiGs.

Unknown to his two other squadron mates, Lieutenant Bell, low on fuel, engaged the first MiGs, giving his fellow Marines the chance to escape. When his fuel was gone, Bell ejected from his powerless jet and was captured.

Other Marines were interned after leaving their crippled aircraft. On 30 July, Lieutenant Colonel Harry W. Reed, the commanding officer of VMF-312, was hit by another Corsair during an attack and bailed out. The other pilot, First Lieutenant Harold Hintz, was thought to have been killed when he apparently spun in. But subsequent prisoner of war debriefings revealed Hintz had died in captivity. Lieutenant Colonel Reed was captured and apparently hanged by the North Koreans because he had shot and killed four enemy soldiers during his capture.

Marine crews from nearly every squadron flying offensive missions in Korea were captured. VMF(N)-513's executive officer, Major Judson C. Richardson, Jr., was captured when his F4U-5N was shot down on a night armed-reconnaissance mission on 14 December 1951.

Lieutenant Colonel William G. Thrash was flying a TBM-3R as part of a strike with VMA-121. The old Grumman torpedo bombers, normally assigned to 1st Marine Aircraft Wing, flew as hacks—mainly short-range "taxis" and currency trainers, and occasionally carried observers. With two ground officers as passengers, Lieutenant Colonel Thrash accompanied the strike when his aircraft was hit by enemy flak. Thrash and the junior officer behind him were able to get out of the crippled Avenger, but the ground colonel could not open his canopy and died in the plane crash.

Four Marine aviators were shot down in May 1952: Major Walter R. Harris (VMF-323); First Lieutenant Milton H. Baugh (VMF-311); Captain John P. Flynn, Jr. (VMF[N]-513); and First Lieutenant Duke Williams, Jr. (VMF[N]-513).

Most prisoners of war of all Services and nationalities were subjected to periods of torture, starvation, and political indoctrination. The Chinese, in particular, were furious at the effort by the United Nations and took out their anger and frustration on many prisoners. The degree of interrogation and deprivation varied considerably, depending on requirements and how much inter-camp movement occurred in any particular period. Other prisoners were occasionally put in camps with newly captured forces.

Lieutenant Colonel Thrash became the senior officer in one camp, establishing rules of behavior that listed what tasks prisoners would do and not do. Thrash's policies eventually brought the wrath of the camp commander down on him, resulting in his removal and eight months of solitary confinement with constant interrogation and harassment.

The final Marine prisoner of war was actually captured after the armistice. Lieutenant Colonel (later Colonel) Herbert A. Peters was an experienced aviator with heavy combat experience in the Pacific, where he shot down four Japanese aircraft during service at Guadalcanal. On 5 February 1954, he took off in an OY light aircraft and became lost in a snowstorm among the mountains.

Circling, he saw a small landing strip through the clouds. He landed, but was immediately surrounded by North Korean soldiers, who held onto his small plane's wings so he could not take off. He languished in captivity at the airfield until August. No word of his internment had been sent, and his family and the Marine Corps had thought him missing, if not dead. His family was surprised and gratified to be notified of Peters' return in October 1954.

---

period was one of intensive training, including landing exercises, joint exercises with the U.S. Army and the U.S. Air Force, and a heavy concentration on bombing and gunnery. The principal bombing target for Korean-based squadrons was on the Naktong, where Marine pilots had done considerable bombing during the defense of the Pusan Perimeter. In addition an exchange program between Japan-based and Korean-based squadrons was established within the wing. The objective of the program was to familiarize new pilots to the area with flight conditions in Korea, just in case the ceasefire did not work out. There were many programs and competitions in athletics with one of the highlights being the winning of the Fifth Air Force and Far East Air Force softball championships by MAG-12 of K-6.

In June 1956, the wing moved its headquarters to Naval Air Station, Iwakuni, Japan, and control of the wing passed from Fifth Air Force to Commander in Chief, Pacific Fleet, in Hawaii, thus ending Marine Corps aviation's participation in the Korean War.

About the Authors

The main text of this pamphlet is derived from Major General John P. Condon's original draft of a history of Marine Corps aviation, an edited version of which appeared as *U.S. Marine Corps Aviation,* the fifth pamphlet of the series commemorating 75 years of Naval Aviation, published by the Deputy Chief of Naval Operations (Air Warfare) and Commander, Naval Air Systems Command in 1987.

Major General John Pomery Condon, Naval Academy Class of 1934, earned his wings as a naval aviator in 1937. On active duty from May 1934 to October 1962, he held command positions at the squadron, group, and wing levels. During World War II, he served with the Fighter Command at Guadalcanal and in the Northern Solomons and subsequently played a key role in training Marine Corps pilots for carrier operations. At Okinawa he commanded Marine Aircraft Group 14, and in Korea, Marine Aircraft Groups 33 and 12, the first group to fly jet aircraft in combat and the last to fly the Corsair against the enemy. As a general officer, he served with the U.S. European Command and commanded both the 1st and 3d Marine Aircraft Wings.

General Condon earned a Ph.D. at the University of California at Irvine and also studied at the U.S. Air Force's Air War College. He is the author of numerous essays and several works on Marine Corps aviation, the last, *Corsairs and Flattops: Marine Carrier Air Warfare, 1944-1945*, was published posthumously in 1998.

Commander Peter B. Mersky, USNR (Ret), provided supplemental materials. A graduate of the Rhode Island School of Design with a baccalaureate degree in illustration, Mersky was commissioned through the Navy's Aviation Officer Candidate School in 1968. Following active duty, he remained in the Naval Reserve and served two tours as an air intelligence officer with Light Photographic Squadron 306.

Before retiring from federal civil service, he was editor of *Approach*, the Navy's aviation safety magazine, published by the Naval Safety Center in Norfolk, Virginia. Commander Mersky has written several books on Navy and Marine Corps aviation, including *U.S. Marine Corps Aviation, 1912-Present* (3d Edition, 1997). He also authored two publications for the History and Museums Division: *A History of Marine Fighter Attack Squadron 321* and *Time of the Aces: Marine Pilots in the Solomons, 1942-1944*, a pamphlet in the World War II Commemorative Series.

Sources

The five volume official Marine Corps history of the Korean War provides the centerline for this account of Marine aviation in Korea: Lynn Montross and Capt Nicholas A. Canzona, USMC, *U.S. Marine Operations in Korea, 1950-1953: The Pusan Perimeter* (Washington, D.C.: Historical Branch, G-3 Division, HQMC, 1954); Lynn Montross and Capt Nicholas A. Canzona, USMC, *U.S. Marine Operations in Korea, 1950-1953: The Inchon-Seoul Operation* (Washington, D.C.: Historical Branch, G-3 Division, HQMC, 1955); Lynn Montross and Capt Nicholas A. Canzona, USMC, *U.S. Marine Operations in Korea, 1950-1953: The Chosin Reservoir Campaign* (Washington, D.C.: Historical Branch, G-3 Division, HQMC, 1957); Lynn Montross, Maj Hubard D. Kuokka, USMC, and Maj Norman W. Hicks, USMC, *U.S. Marine Operations in Korea, 1950-1953: The East-Central Front* (Washington, D.C.: Historical Branch, G-3 Division, HQMC, 1962); and LtCol Pat Meid, USMCR and Maj James M. Yingling, USMC, *U.S. Marine Operations in Korea, 1950-1953: Operations in West Korea* (Washington, D.C.: Historical Division, HQMC, 1972).

Other official accounts of use were Roy E. Appleman, *South to the Naktong, North to the Yalu* (Washington, D.C.: Office of the Chief of Military History, Department of the Army, 1961), and Ernest H. Giusti and Kenneth W. Condit, "Marine Air Over Inchon-Seoul," *Marine Corps Gazette*, June 1952; Ernest H. Giusti and Kenneth W. Condit, "Marine Air at the Chosin Reservoir," *Marine Corps Gazette*, July 1952; and Ernest H. Giusti and Kenneth W. Condit, "Marine Air Covers the Breakout," *Marine Corps Gazette,* August 1952.

Among useful secondary sources were BGen Edwin H. Simmons, USMC (Ret), *The United States Marines* (Annapolis, MD: Naval Institute, 1999); Andrew Greer, *The New Breed: The Story of the U.S. Marines in Korea* (New York: Harper Brothers, 1952); Richard P. Hallion, *The Naval Air War in Korea* (Baltimore, MD Nautical & Aviation Publishing Co., 1986); G. G. O'Rourke with E. T. Wooldridge, *Night Fighters Over Korea* (Annapolis, MD: Naval Institute, 1998); and Robert F. Dorr, Jon Lake, and Warren Thompson, *Korean War Aces* (London: Osprey, 1995).

Sources of great use were the oral histories, diaries, and memoirs of many of the participants. The most important of these were those of LtGen Robert P. Keller, LtCol John Perrin, LtCol John E. Barnett, LtCol Emmons S. Maloney, Col Edward S. John, LtCol William T. Witt, Jr., SgtMaj Floyd P. Stocks, LtGen Leslie E. Brown, MSgt James R. Todd, and MSgt Lowell T. Truex.

As is the tradition, members of the Marine Corps Historical Center's staff, especially Fred H. Allison, were fully supportive in the production of this pamphlet as were others: William T. Y'Blood and Sheldon A. Goldberg of the U.S. Air Force History Support Office; Hill Goodspeed of the Emil Buehler Naval Aviation Library, National Museum of Naval Aviation; and Warren Thompson, Joseph S. Rychetnik, Steven P. Albright, Steven D. Oltmann, Nicholas Williams, and James Winchester.

MARINES
HR-5

WHIRLYBIRDS
U.S. Marine Helicopters in Korea

by Lieutenant Colonel Ronald J. Brown, USMCR (Ret)

On Sunday, 25 June 1950, Communist North Korea unexpectedly invaded its southern neighbor, the American-backed Republic of Korea (ROK). The poorly equipped ROK Army was no match for the well prepared North Korean People's Army (NKPA) whose armored spearheads quickly thrust across the 38th Parallel. The stunned world helplessly looked on as the out-numbered and out-gunned South Koreans were quickly routed. With the fall of the capital city of Seoul imminent, President Harry S. Truman ordered General of the Army Douglas MacArthur, Commander in Chief, Far East, in Tokyo, to immediately pull all American nationals in South Korea out of harm's way. During the course of the resultant noncombatant evacuation operations an unmanned American transport plane was destroyed on the ground and a flight of U.S. Air Force aircraft were buzzed by a North Korean Air Force plane over the Yellow Sea without any shots being fired. On 27 July, an American combat air patrol protecting Kimpo Airfield near the South Korean capital actively engaged menacing North Korean planes and promptly downed three of the five Soviet-built Yak fighters. Soon thereafter American military forces operating under the auspices of the United Nations Command (UNC) were committed to thwart a Communist takeover of South Korea. Thus, only four years and nine months after V-J Day marked the end of World War II, the United States was once again involved in a shooting war in Asia.

At left: *A Sikorsky HRS-1 transport helicopter from HMR-161 sets down behind 1st Marine Division lines to pick up waiting Marines.* Department of Defense Photo (USMC) A159970

The United Nations issued a worldwide call to arms to halt Communist aggression in Korea, and America's armed forces began to mobilize. Marines were quick to respond. Within three weeks a hastily formed provisional Marine brigade departed California and headed for the embattled Far East. Among the aviation units on board the U.S. Navy task force steaming west was a helicopter detachment, the first rotary-wing aviation unit specifically formed for combat operations in the history of the Marine Corps. Although few realized it at the time, this small band of dedicated men and their primitive flying machines were about to radically change the face of military aviation. Arguably, the actions of these helicopter pilots in Korea made U.S. Marines the progenitors of vertical envelopment operations, as we know them today.

Helicopters in the Marine Corps

There is great irony in the fact that the Marine Corps was the last American military Service to receive helicopters, but was the first to formulate, test, and implement a doctrine for the use of rotary-wing aircraft as an integral element in air-ground combat operations. The concept of manned rotary-wing flight can be traced back to Leonardo da Vinci's Renaissance-era sketches, but more than four centuries passed before vertical takeoffs and landings by heavier-than-air craft became a reality. The Marines tested a rotary-wing aircraft in Nicaragua during the Banana Wars, but that experiment revealed the Pitcarin OP-1 autogiro was not ready for military use. Autogiros used rotary wings to remain aloft, but they did not use spinning blades to get airborne or to power the aircraft so autogiros were airplanes not helicopters. Some aviation enthusiasts, however, assert that the flight data accumulated and rotor technology developed for autogiros marked the beginning Marine Corps helicopter development. It was not until 1939 that the first practical American helicopter, aircraft de-signer Igor I. Sikorsky's VS-300, finally moved off the drawing board and into the air. The U.S. Army, Navy, and Coast Guard each acquired helicopters during World War II. The bulk of them were used for pilot training, but a few American-built helicopters participated in special combat operations in Burma and the Pacific. These early machines conducted noncombatant air-sea

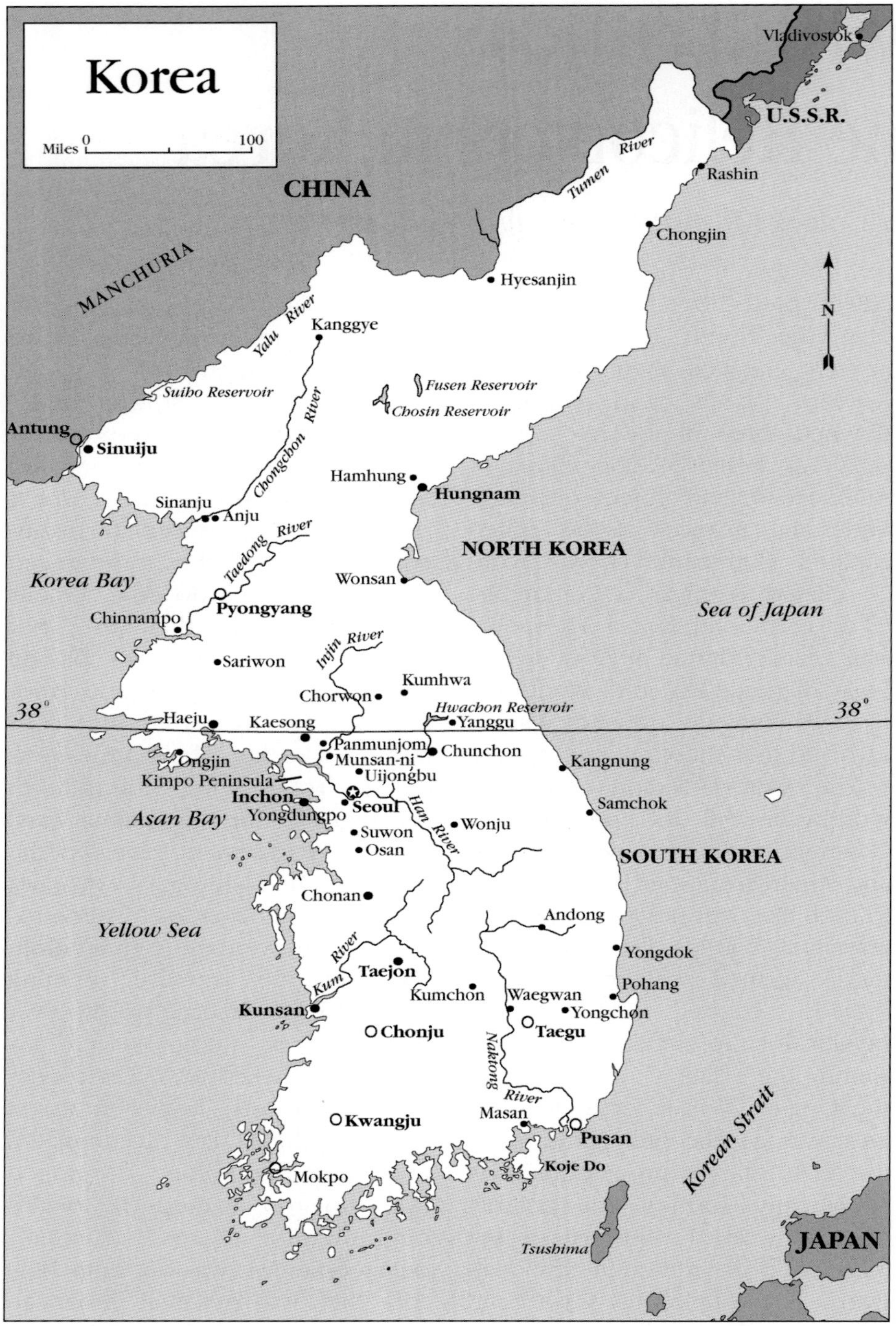

rescue, medical evacuation, and humanitarian missions during the war as well.

In 1946, the Marine Corps formed a special board headed by Major General Lemuel C. Shepherd, Jr., to study the impact of nuclear weapons on amphibious operations. In accordance with the recommendations made by the Shepherd Board in early 1947, Marine Corps Schools at Quantico, Virginia, began to formulate a new doctrine, eventually termed "vertical assault," which relied upon rotary-wing aircraft as an alternative to ship-to-shore movement by surface craft. The following year, Marine Corps Schools issued a mimeographed pamphlet entitled, "Amphibious Operations—Employment of Helicopters (Tentative)." This 52-page tome was the 31st school publication on amphibious operations, so it took the short title *"Phib-31."* Concurrently, the Marine Corps formed a developmental helicopter squadron to test the practicality of *Phib-31's* emerging theories. This formative unit, Colonel Edward C. Dyer's Marine Helicopter Squadron 1 (HMX-1), stood up in December 1947 and was collocated with Marine Corps Schools. The new squadron's primary missions were to develop techniques and tactics in conjunction with the ship-to-shore movement of assault troops in amphibious operations, and evaluate a small helicopter as replacement for fixed-wing observation airplanes. Among the officers as-signed to HMX-1 was the Marine Corps' first officially sanctioned helicopter pilot, Major Armond H. DeLalio, who learned to fly helicopters in 1944 and had overseen the training of the first Marine helicopter pilots as the operations officer of Navy Helicopter Develop-ment Squadron VX-3 at Lakehurst Naval Air Station, New Jersey.

In February 1948, the Marine Corps took delivery of its first helicopters when a pair of Sikorsky HO3S-1s arrived at Quantico. These four-seat aircraft featured a narrow "greenhouse" cabin, an overhead three-blade rotor system, and a long-tail housing that mounted a small vertical antitorque rotor. This basic outline bore such an uncanny resemblance to the Anisoptera subspecies of flying insects that the British dubbed their newly purchased Sikorsky helicopters "dragonflies." There was no Service or manufacturer's authorized nickname for the HO3S-1, but the most common unofficial American appellations of the day were "whirlybirds," "flying windmills,"

Pitcarin OP-1 Autogiro

The first rotary-winged aircraft used by naval aviation was not a helicopter. It was an autogiro, an airplane propelled by a normal front-mounted aircraft engine but kept aloft by rotating overhead wings, a phenomenon known as "autorotation." Although rather ungainly looking due their stubby upturned wings, large tails, and drooping rotors, autogiros took well to the air. Their ability to "land on a dime" made them favorites at air shows and an aggressive publicity campaign touted them as "flying autos, the transportation of the future." Autogiros, however, turned out to be neither a military nor a commercial success.

The aircraft itself was an odd compilation of a normal front-mounted aircraft engine used to generate thrust and three overhead free-spinning blades attached to a center-mounted tripod to provide lift. The fuselage included a pair of stubby wings that supported the landing gear and had a semi-standard elongated tail assembly. Typical of the day, it had an open cockpit.

Although a rotary-winged aircraft, the OP-1 was not a helicopter. The engine was used to start the rotors moving but was then disengaged and connected to the propeller to deliver thrust. A speed of about 30 miles per hour was needed to generate lift and maintained for controlled flight. The OP-1 could not hover, it required conventional engine power to take off and move forward in the air; the plane could, however, make a vertical landing. This unique feature made the OP-1 attractive to the military.

The specific autogiro model first tested by the Marine Corps was the OP-1 built by Harold F. Pitcarin, who would later found Eastern Airways. His company was a licensed subsidiary of a Spanish firm. All American autogiros were based upon designs formulated by Spanish nobleman Juan de la Cierva. His first successful flight was made near Madrid in 1923. More than 500 autogiros flew worldwide during the next decade. Although his airplanes never lived up to his high expectations, de la Cierva did develop rotor technology and recorded aerodynamic data later applied by helicopter designers Igor Sikorsky and Frank Piasecki.

The Navy purchased three Pitcarin autogiros for extensive field-testing and evaluation in 1931. The only carrier tests were conducted on 23 September of that year, but the OP-1's performance was virtually identical to that of carrier-borne biplanes then in use. The Marines took one OP-1 to Nicaragua to test it under combat conditions. Again, its performance was disappointing. The pilots of VJ-6M noted it lacked both payload and range. The only practical use they found was evaluation of potential landing areas. This was not enough reason to incorporate the OP-1 into the Marine inventory. Overall, the OP-1 was described as "an exasperating contraption," not fit for military use. Further trials of a wingless autogiro in 1935 revealed no improvement, so director of aviation Major Roy S. Geiger recommended against adoption of that aircraft type.

In the barnstorming days between the World Wars, autogiros proved to be the ultimate novelty attraction. Aviator Charles A. Lindbergh often put on demonstrations, aviatrix Amelia Earhart set an altitude record in one, and Secretary of the Navy Charles Francis Adams flew in an autogiro to join President Herbert C. Hoover at an isolated fishing camp in Virginia. The Royal Air Force actually used autogiros for convoy escort and observation during World War II, and the Soviet Union developed its own autogiro.

Although the OP-1 never became a mainstream Marine aircraft and was not a true helicopter, some aviation enthusiasts assert that the technology and data developed by de la Cierva was crucial for rotary-winged flight. They, therefore, make the case that the OP-1 should be considered the progenitor of today's helicopters.

Department of Defense Photo (USMC) 528139

and "pinwheels." The HO3S-1 had a cruising speed of less than 100 miles per hour, a range of about 80 miles, could lift about 1,000 pounds, and mounted simple instrumentation that limited the HO3S to clear weather and daylight operations. This very restricted flight envelope was acceptable because these first machines were to be used primarily for training and testing. They were, however, sometimes called upon for practical missions as well. In fact, the first operational use of a Marine helicopter occurred when a Quantico-based HO3S led a salvage party to an amphibious jeep mired in a nearby swamp.

The first Marine helicopter operational deployment occurred in May 1948 when five HMX-1 "pinwheels" flying off the escort carrier *Palau* (CVE 122) conducted 35

The Visionaries

The wake of the World War II, with its ominous specter of nuclear weapons, forced the Marine Corps to rethink existing amphibious doctrine. The conclusion was that previous methods of ship-to-shore movement were no longer sufficient to ensure a successful landing so alternative methods had to be developed. Several options looked promising, but the only one that stood the test of time and combat was vertical envelopment—the use of helicopters to move troops and supplies.

In 1946, Commandant Alexander A. Vandegrift—at the urging of Lieutenant General Roy S. Geiger, the "Gray Eagle" of Marine aviation who had just witnessed post-war nuclear tests—formed a special board culled from Marine Corps headquarters to study existing tactics and equipment then make recommendations for restructuring the Fleet Marine Force. Assistant Commandant Lemuel C. Shepherd, Jr., a graduate of Virginia Military Institute, who was arguably the Marines' most innovative division commander in the Pacific, headed the board. Shepherd was an excellent choice because he was both a traditionalist and a visionary who would later become Commandant. Other members of the board included Major General Field Harris, the director of Marine aviation, and Brigadier General Oliver P. Smith, the head of plans and operations division. All three men would be reunited in Korea in 1950 where they would put into practice the revolutionary doctrines they set in motion; Shepherd as the commanding general of Fleet Marine Force, Pacific, Harris as commanding general of the 1st Marine Aircraft Wing, and Smith as commanding general of the 1st Marine Division. Two colonels assigned to the board secretariat were particularly influential, Edward C. Dyer and Merrill B. Twining. Dyer, a Naval Academy graduate and decorated combat pilot, was master of all things aeronautical while Merrill Twining, a highly regarded staff officer, handled operational theory. Neither a formal member of the board nor its secretariat but keeping close tabs on what transpired was Brigadier General Gerald C. Thomas, Vandegrift's trusted chief of staff. Dyer eventually commanded the first Marine helicopter squadron and Thomas replaced Smith as 1st Marine Division commander in Korea.

Doctrinal development for vertical assault was done at Marine Corps Schools located at Quantico, Virginia. First, a board headed by Lieutenant Colonel Robert E. Hogaboom laid out what was needed in a document titled "Military Requirements for Ship-to-Shore Movement of Troops and Cargo." Even though no suitable aircraft were yet available, the thinkers at Quantico came up with new doctrine published as *Amphibious Training Manual 31,* "Amphibious Operations—Employment of Helicopters (Tentative)." One of the drivers of this project was Lieutenant Colonel Victor H. Krulak, a tough former paratrooper who had been wounded in the Pacific but was also known for his high intellect and an unsurpassed ability to get things done. He was a prolific writer and a demanding taskmaster who kept his finger on the pulse of several vital projects including helicopter development.

Despite the nearly unlimited future potential of helicopters for assault and support of landing forces, there was ingrained resistance to such a revolutionary concept. Most young pilots wanted to fly sleek jets and dogfight enemy aces, not manhandle temperamental aircraft to deliver troops and supplies; experienced fliers were comfortable with aircraft they already knew well and were reluctant to give up their trusted planes; and critics claimed helicopters were too slow and vulnerable. Twining took the lead in addressing these problems when he pointed out the Marine Corps had far more pilots than planes and noted that the wishes of the individual were always subservient to the needs of the Marine Corps. He also asserted that the speed and vulnerability of helicopters should not be properly compared to fixed-wing aircraft but to surface landing craft (helicopters were both faster and more agile than boats or amphibious tractors).

All early helicopter advocates were highly motivated and dedicated men. Their achievements and foresight kept the Marine Corps' reputation for innovation alive despite severe budgetary constraints and concurrent inter-Service unification battles. In fact, many of the men also played key roles in the "Chowder Society," whose behind-the-scenes work successfully protected Marine Corps interests during the bitter "unification battles" after the World War II.

National Archives Photo (USMC) 127-N-A130996

BGen Edward C. Dyer, here receiving the Legion of Merit for meritorious service as the 1st Marine Aircraft Wing's G-3 during the Inchon-Seoul campaign, was one of the most influential men involved in the adoption of the helicopter by the Marine Corps. A naval aviator, he helped to bring the concept to reality by formulating doctrine and then commanding HMX-1 at Quantico, Virginia.

flights to land 66 men and several hundred pounds of communications equipment at Camp Lejeune, North Carolina's Onslow Beach during amphibious command post exercise Packard II. As the year progressed, HMX-1's aircraft complement increased by six when the Marine Corps took delivery of two new types of helicopters, one Bell HTL-2 and five Piasecki HRP-1s. The Bell HTL, often called the "eggbeater," was a side-by-side two-seat trainer that could fly at about 85 miles per hour. It had two distinctive features, a rounded Plexiglas "fishbowl" cockpit canopy and a single overhead two-bladed rotor. This model had four landing wheels and a fabric-covered tail assembly, although later versions of the HTL mounted skids and left the tail structure bare. The larger Piasecki HRP-1 was a 10-place troop transport whose tandem-mounted rotors could push it along at about 100 miles per hour. The aircraft's unique bent fuselage (overlapping propeller radii meant the tail rotor had to be mounted higher than the forward rotor) gave it the nickname "Flying Banana." Unfortunately, it was a temperamental machine considered too fragile to be assigned to combat squadrons. The HRP-1 was instead relegated to use as a test bed and demonstration aircraft until a more capable transport helicopter could be procured.

During the next two years HMX-1 conducted numerous experiments, tests, exercises, demonstrations, and public appearances. Helicopters soon became crowd pleasers at air shows and were invariably the center of attention for dignitaries visiting Quantico. As a result of numerous tactical tests and performance evaluations, Colonel Dyer recommended that light helicopters should be added to Marine observation squadrons. Headquarters agreed, and it was

One of five Sikorsky HO3S-1s from HMX-1 prepares to land on the Palau *(CVE 122) during Operation Packard II in May 1948. This was the first test to determine the value of the helicopter in the movement of assault troops in an amphibious operation.*

Department of Defense Photo (USMC)

Department of Defense Photo (USMC) A55366

Piasecki HRP-1 "Flying Bananas" in action during a Basic School pre-graduation field problem at Quantico, Virginia. The HRP was the first Marine Corps transport helicopter, but technical constraints limited it to demonstration and training use and no HRPs saw action in Korea.

decided that an even mix of helicopters and airplanes should be adopted as soon as enough helicopters and trained personnel were available. Unfortunately, teething problems grounded each of the helicopter types at one time or another, and it was apparent more reliable aircraft with much greater lift capacity would be necessary to make vertical assault a true option in the future. Marine helicopter detachments participated in exercises Packard III (1949) and Packard IV (1950). This time period also featured many milestones. Among them were the first overseas deployment of a Marine helicopter pilot when Captain Wallace D. Blatt flew an HO3S-1 borrowed from the U.S. Navy during the American withdrawal from China in February 1949; the first unit deployment in support of a fleet exercise occurred in February 1950; and the largest single helicopter formation to that time took place when six HRPs, six HO3Ss, and one HTL flew by Quantico's reviewing stand in June 1950. By that time, Lieutenant Colonel John F. Carey, a Navy Cross holder who a dozen years later would lead the first Marine aviation unit sent to Vietnam, commanded HMX-1. The squadron mustered 23 officers and 89 enlisted men; its equipment list showed nine HRPs, six HO3Ss, and three HTLs. Since its inception the Marine helicopter program had garnered many laurels, but several vital items remained on the agenda—notably the creation of helicopter squad-rons for service with the Fleet Marine Force and the procurement of a combat-ready transport helicopter. This was the status of the Marine Corps helicopter program when the North Korean unexpectedly burst across the 38th Parallel.

Called to Action

The commitment of American combat troops to Korea on 30 June set off alarm bells throughout the Marine Corps. Although the official "word" had yet to be passed, within a few hours of the North Korean invasion most Marines surmised it would not be long before they would be on their way to war.

Marine Helicopter Squadron 1

Marine Helicopter Squadron 1 (HMX-1) is unique in the Marine Corps because it has several distinct missions and at least three different chains-of-command providing guidance and tasking.

HMX-1 was the first Marine rotary-wing squadron. It "stood up" at Marine Corps Airfield Quantico in Virginia on 1 December 1947 and has been located there ever since. Its activation was the first operational move that started a revolution in Marine aviation and tactical doctrine.

One interesting insight into the Marines' most unique aircraft squadron is the frequent misunderstanding of its official designation. Although HMX-1 was initially tasked to develop techniques and tactics in connection with the movement of assault troops by helicopter and to evaluate a small helicopter as an observation aircraft, the "X" does not designate "experimental" as is often inferred. The "Nighthawks" of HMX-1 do perform some developmental tasks, but their primary missions are to provide helicopter transportation for the President of the United States and to support Marine Corps Schools.

The squadron, initially manned by seven officers and three enlisted men, quickly grew and mustered 18 pilots and 81 enlisted men when the first helicopters, Sikorsky HO3S-1s, arrived. These first primitive machines carried only the pilot and up to three lightly armed troops, but they formed the basis for testing helicopter doctrine described in Marine Corps Schools operational manual *Phib-31*. Eventually, HMX-1 received a mix of early model helicopters with the addition of Piasecki HRP transports and Bell HTL trainers to test doctrine before the Korean War.

On 8 May 1948, HMX-1 pilots flew from Quantico to Norfolk, Virginia, to board the escort carrier *Palau* (CVE 122). The fly-on operation was described by HMX-1 commanding officer Colonel Edward C. Dyer as a "complete shambles [with] sailors running all over the place in mortal danger of walking into tail rotors, and the Marines were totally disorganized as well. It was complete bedlam, there was no organization and no real system [in place]." By the next day, however, the Navy and Marine Corps were using the same basic ship-board flight operations procedures practiced today—circular lines delineated danger areas as well as personnel staging areas and approach lanes. Five days later, the HO3S-1s delivered 66 men and several tons of equipment to Camp Lejeune, North Carolina's Onslow Beach during command post exercise Packard II.

The following year a similar exercise employed eight HRPs, three HO3Ss, and a single HTL. During Exercise Packard III, the HRP "Flying Banana" troop transports were carrier borne, the HTL was loaded on an LST for command and control, and the HO3Ss stayed ashore as rescue aircraft. The HRPs brought 230 troops and 14,000 pounds of cargo ashore even though choppy seas swamped several landing craft and seriously disrupted operational maneuvers. Many consider this superb performance to be the key factor in the acceptance of the helicopter as a viable ship-to-shore method, thus paving the way for the integration of rotary-wing aircraft into Marine aviation.

In 1957, HMX-1 acquired an unexpected mission—transporting the President of the United States. Helicopters were only considered for emergency situations until President Dwight D. Eisenhower used an HMX-1 Sikorsky HUS Sea Horse helicopter for transportation from his summer home on Narragansett Bay. After that, Marine helicopters were routinely used to move the President from the White House lawn to Andrews Air Force Base, the home of presidential plane "Air Force One." That transport mission became a permanent tasking in 1976 and continues to this day.

Currently mustering more than 700 personnel, HMX-1 is the largest Marine Corps helicopter squadron. It is divided into two sections. The "White" side flies two unique helicopters—both specially configured Sikorsky executive transports, the VH-3D Sea King and the VH-60N Seahawk. The "Green" side provides basic helicopter indoctrination training for ground troops, tests new concepts and equipment, and assists the Marine air weapons and tactics squadron. Unlike any other Marine squadron, HMX-1 answers to three distinct chains-of-command: the Marine Corps deputy chief of staff for air at Headquarters Marine Corps; the White House military office; and the operational test and evaluation force commander at Norfolk. Marine Helicopter Squadron 1 was not only the first such Marine unit, it also currently holds a unique place in naval aviation.

General MacArthur's formal request for a Marine regimental combat team and supporting aviation finally filtered through official channels on 2 July, and five days later the 1st Provisional Marine Brigade was activated. Brigadier General Edward A. Craig's 6,534-man unit included the 5th Marines as its ground combat element and the 1st Marine Aircraft Wing (Forward Echelon) as its aviation combat element.

Brigadier General Thomas J. Cushman, a veteran aviator who had commanded an aircraft wing in the Pacific during World War II, was "dual-hatted" as both the brigade deputy commander and the commander of the aviation component. The 1st Brigade's 1,358-man aviation element was built around Marine Aircraft Group 33 (MAG-33), which included three squadrons of propeller-driven Vought F4U Corsairs, two day fighter squadrons (VMF-214 and -323) and one night fighter squadron (VMF[N]-513). The remaining aviation units included headquarters, ground support, and air control personnel in addition to an observation squadron.

The observation squadron assigned to the 1st Marine Brigade was Marine Observation Squadron 6 (VMO-6) commanded by Major Vincent J. Gottschalk. Its mission was to conduct "tactical air reconnaissance, artillery spotting, and other flight operations within the capabilities of assigned aircraft in support of ground units." This last statement became a well-exercised elastic clause under the innovative guidance of Major Gottschalk, an engineering graduate of the University of Michigan who saw several years sea duty in the Pacific before earning his wings. In action, Gottschalk saw to it that practically any flying task in support of ground units, no matter how difficult or outrageous it initially seemed, fell within the capabilities of VMO-6 aircraft. He took command of VMO-6 on 3 July and was ordered to be ready for overseas deployment only four days later.

Capt Victor A. Armstrong, at the controls of a Sikorsky HO3S-1 helicopter, was the officer-in-charge of the VMO-6 helicopter section, the first Marine helicopter unit formed for combat duty. Holder of the Distinguished Flying Cross for actions in the Pacific during World War II, he would attain the rank of major general and serve as the deputy chief of staff for air.

Department of Defense Photo (USMC) A130162

Marine observation squadrons had been serving as indispensable components of Marine air-ground combat teams since the Banana Wars. Marine Observation Squadron 6 (then called VO-6M) was specifically formed for expeditionary duty in Nicaragua in 1928, but it was administratively transferred back to Quantico for duty as a training unit about six months later. Marine observation squadrons went by the wayside in 1933 and did not re-emerge until operations moved to the Western Pacific during World War II. There, flying small, nimble, high-wing, two-seat, single-engined Piper OE "Grasshoppers" and similar Stinson OY-1 "Sentinels" (often called Grass-hoppers as well), VMOs provided aerial reconnaissance and artillery-naval gunfire spotting as well as performing assorted utility duties while attached to various Marine divisions. Marine Observation Squadron 6 was reactivated in 1943, saw combat action on Okinawa in 1945, and participated in the post-war occupation of North China. Upon its return to the United States in 1947, the squadron flew in support of the 1st Marine Division located at Camp Pendleton, California. The aircraft of VMO-6 did occasional artillery spotting and sometimes supported ground maneuvers or performed administrative duties, but the main mission at Camp Pendleton was a practical one—spraying aerial insecticide. In early June 1950, VMO-6 was assigned to the 1st Marine Aircraft Wing stationed at nearby Marine Corps Air Station El Toro.

With the arrival of the first warning orders, both Camp Pendleton

Department of Defense Photo (USMC) A1280

A Sikorsky HO3S-1 helicopter transports a passenger from one ship to another while the convoy carrying the VMO-6 helicopter section is enroute to Korea. Marine "Whirlybirds" flying off the carrier Badoeng Strait *(CVE 116) were routinely used to deliver messages and personnel between ships.*

and Marine Corps Air Station El Toro became scenes of bedlam as people raced around to gather materials and units speedily absorbed new personnel. "Mothballed" weapons and equipment were hurriedly broken out of storage and readied for use. Trains and planes brought in personnel culled from posts and stations across the United States at all hours of the day and night. Arrivals were welcomed on board and sent to their new units as soon as the handshakes finished. Space was at a premium, as was time. Round-the-clock work schedules were instituted, and the unofficial order of the day became "sleep on the boat!"

Major Gottschalk was originally told to form a four-plane, four-officer, and 10-enlisted man detachment to accompany the 1st Brigade to Korea. Although this detachment was far smaller than a war-strength squadron, just finding enough airplanes was not an easy task. Gottschalk decided to take eight well-worn OYs to ensure that four of them would be flyable—the rest would become "hangar queens" until replacement parts or new aircraft were in the supply pipeline. While the search for planes and equipment got under way, Gottschalk's orders were modified on 7 July. The entire squadron would now be going and, in accord with earlier recommendations, the squadron aircraft mix would also include helicopters.

Eight officers and 30 enlisted men were pulled out of HMX-1 at Quantico, Virginia, with orders to move to the West Coast immediately. Captain (later Major General) Victor A. Armstrong was the officer-in-charge of the helicopter detachment. The other pilots included Captains George B. Farish and Eugene J. Pope, and First Lieutenants Arthur R. Bancroft, Lloyd J. Engelhardt, Robert A. Longstaff, Max N. Nebergall, and Gustave F. Lueddeke, Jr. The detachment's claim to historical fame was that this was the first permanent assignment of a Marine helicopter unit to the Fleet Marine Force. Contrary to some assertions, this detachment was neither the first Marine combat helicopter squadron nor was it the first U.S. helicopter detachment to see combat service—a helicopter element (later designated Flight F) from the U.S. Air Force 3d Air Rescue Squadron and carrier-based U.S. Navy helicopters assigned to Utility Helicopter Squadron 1 (HU-1) were already in action in Korea by the time VMO-6 arrived.

Armstrong's detachment made its way from Quantico to El Toro, California, leaving on 8 July and reporting for duty on the 10th. Upon arrival, helicopter detachment personnel were integrated into VMO-6, and Captain Armstrong was named that squadron's executive officer. Because only the personnel of the helicopter detachment transferred from HMX-1, aircraft had to be found. Six HO3S-1 helicopters were obtained from U.S. Navy sources (two each from Inyokern and Point Mugu, California, and two more from the overhaul and maintenance facility at San Diego). Only two days after reporting in, the helicopter detach-

Department of Defense Photo (USMC) A131099

Sikorsky HO3S-1

The Sikorsky HO3S-1 was the first helicopter assigned to the U.S. Marine Corps. The HO3S was the naval variant of Sikorky's model S-51 commercial helicopter. Despite its observation designation, the HO3S was actually a utility aircraft used for a variety of roles. Among the 46 conceptual uses initially listed by Marine Corps Schools were the ones most used in Korea: search and rescue; aerial reconnaissance; medical evacuation; and liaison. The U.S. Air Force flew the same aircraft as a search and rescue helicopter designated H-5F.

The HO3S was the lineal descendent of earlier Sikorsky designs, the initial HNS trainer and the first designated military observation helicopter (alternately known as the HO2S in naval service and the R-5A to the Army). The HO3S featured a more powerful engine that gave it added lift and an increased payload. During the immediate pre-war period, the HO3S proved to be an outstanding rescue craft that often utilized its winch to pull downed pilots out of the water. Likewise, the HO3S was an excellent observation platform for artillery spotting.

In Korea, its primary uses were as a liaison aircraft and as an aerial ambulance. A first-rate liaison aircraft with good range, the HO3S had a dependable engine, and was rugged enough that it required relatively little maintenance when compared to other rotary-wing aircraft of the day.

Even though the HO3S performed yeoman service at the Pusan Perimeter, it had significant shortfalls as a combat aircraft. The tricycle landing gear and its high center of gravity made the HO3S unstable on all but flat solid terrain; the aircraft could not accommodate interior stretcher loads; its lack of back-lit instrumentation precluded extended night and bad weather operations; and the high engine location made aircraft maintenance difficult. Another major drawback was that it required a great deal of strength and endurance to handle such a heavy aircraft for an extended period without servo-controls. In addition, the single main rotor and long tail assembly combined with a centrally located engine mount often required field expedient ballast adjustments to maintain in-flight stability, so it was not unusual for pilots to keep several sand bags or a seabag filled with rocks in the cabin.

Aircraft Data

Manufacturer: Sikorsky Division of United Aircraft Corporation

Power Plant: Pratt and Whitney R-985 AN-7 Wasp Jr., 9 cylinder, 450 horsepower, radial engine

Dimensions: Length, 57' 1/2"; height, 12' 11"; rotor, 48' composite construction blade

Performance: Cruising speed, 85 mph; range, 260 miles

Lift: Pilot plus two passengers or about 500 pounds of cargo (excluding fuel)

ment moved to San Diego to board ship.

The crowded escort carrier *Badoeng Strait* (CVE 116) carrying 60 Corsairs, 8 OY Sentinels, and 6 Marine helicopters along with their aircrews sailed for the Far East on 14 July. Enroute helicopters were used for inter-ship supply delivery, mail runs, and personnel transfers. The 1st Marine Brigade was originally slated for a temporary layover in Japan where cargo could be sorted out then combat loaded and some rudimentary amphibious training would be conducted before the Marines entered the combat zone. That was the plan until the situation in Korea became so grave that the 5th Marines was ordered to go directly to the beleaguered South Korean port city of Pusan. The aviation element was still slated to land in Japan, however, so the ships carrying the aviation component split off and headed for the Japanese port city of Kobe.

As the ships of Navy Task Group 53.7 plowed through the Pacific, Brigadier General Craig and his operations officer Lieutenant Colonel Joseph L. Stewart flew to Korea to attend a series of command conferences. On 30 July, they learned that upon landing the Marines would be attached to a U.S. Army task force assigned to shore up the crumbling southwest flank of the United Nations defense lines. Colonel Stewart called the aviation advance party command post in Japan to warn that combat action was imminent and requested that VMO-6 and Marine Tactical Air Control Squadron 2 (MTACS-2) be sent on to Korea as quickly as possible. This emergency phone call confirmed that the situation in Korea was desperate. Accordingly, when the *Badeong Strait* made landfall on the evening of 31 July 1950, Major Gottschalk received word to begin operations at first light the next morning.

Marine Observation Squadron 6's airplanes and helicopters went ashore on 1 August. The next day the Marine air elements scattered to the four winds. The day fighter squadrons boarded a pair of escort carriers and then sailed for the combat zone; the night fighter squadron joined an Air Force all-weather squadron at Itazuke Air Base on Kyushu; VMO-6 ground crews and their equipment "transshipped" to a tank landing ship (LST) for transportation to Korea; and headquarters personnel moved to Itami Air Base near Osaka on the island of Honshu.

Helicopters Enter Combat

From Kobe, the helicopters of VMO-6 proceeded to Itami where two helicopters were assigned to MAG-33 headquarters. They would be held in Japan to provide liaison services between the widely scattered aviation units and, at the same time, be available as emergency replacements if needed. The other four HO3S-1s proceeded to Korea. They made their way from Itami to Iwakuni Air Base where they stayed overnight. After a detailed situation brief and a hasty final maintenance inspection at Ashiya Air Base on northern Kyushu on the morning of 2 August, the helicopters made the hop across the Tsushima Sraits. They landed at an airfield near Pusan, the logistics keystone of the United Nations defensive perimeter.

The outlook in Korea was not good when they arrived. The hard-pressed United Nations Command was struggling to hold onto a 60-by-90-mile area of southeast Korea known as the Pusan Perimeter. The North Korean drive south was slowing, but the outcome of the battle for the Korean peninsula was far from certain when the 1st Provisional Marine Brigade was welcomed on board by Eighth Army commander Lieutenant General Walton H. Walker, USA.

At the Pusan Perimeter, the Marine brigade acquitted itself well and showcased the combat effectiveness of the Marine air-ground team. The Marines were used as a "fire brigade" moving from place to place to stamp out enemy threats. They spearheaded the first U.N. offensive in Korea, and then twice threw back NKPA penetrations of the U.N. defensive lines. Marine air hit the enemy when Corsairs swept out of the sky on the same day that the ground element was coming ashore at Pusan harbor. The brigade then consolidated at a temporary assembly area near Changwon before mounting the first sustained United Nations offensive of the war. The initial ground action occurred in the vicinity of Chindong-ni from 6 to 9 August. From there the Marines pressed south to Kosong before turning north to the Changchon Pass after wiping out an enemy motorized regiment during the Kosong "Turkey Shoot." On 13 August, as they neared Sachon, the Marines were abruptly ordered back to Masan to prepare to seal off an enemy penetration across the Naktong River. Hard fighting at Red Slash Hill and carefully coordinated supporting arms fires threw the North Koreans back. While recuperating at an area dubbed the Masan "Bean Patch," the Marines had to return to the Naktong bulge to repulse the enemy one more time. Finally, on 5 September, the Marines pulled out of the line and returned to Pusan so they could mount out to lead MacArthur's amphibious turning movement at Inchon. Throughout the campaign, the hard-working HO3S-1s of VMO-6 performed a wide variety of tasks and were so indispensable that Marine and Army commanders were soon demanding more helicopters.

Upon its arrival at Pusan on 2 August, the VMO-6's forward echelon was temporarily billeted in a South Korean schoolhouse located about 10 miles west of the port until the squadron support element caught up and a more permanent, and less crowded, site could be occupied. The rear party, which sailed from Kobe on board a Japanese-manned landing ship, actually arrived at Pusan on 4 August but could not move out for two more days due to the lack of transportation. Squadron supplies and equipment were laboriously loaded (there was no cargo handling machinery at hand) onto the dock then reloaded onto a train for shipment west to Chinhae on 6 August. Chinhae was a South Korean naval base, as well as the future home of the Korean Marine Corps, located only a short hop across the bay from Masan. The site of a former Japanese ammunition depot with an airstrip, it was selected because it was close to the action, had a 2,600-foot grass and concrete runway (already being used by a combined US-ROK Air Force training squadron), and included a pair of completed hangars with a third under construction. There were enough Quonset huts to house the men, provide adequate office space, and warehouse supplies. This facility would be VMO-6's home field and base of operations until the 1st Provisional Marine Brigade was dissolved in early September.

In Korea, VMO-6 would be under the operational control of the brigade but under the administrative control of the wing. This meant that the brigade, and later the division, commander through his air section would assign daily missions while the aircraft wing would provide supplies and personnel administration. Unfortunately, the helicopters, which belonged partially to both, but not fully to the ground or aviation commanders, seemed to be neither fish nor fowl. To use Major Gottschalk's words to describe this awkward command and control system: "Observation squadrons were the stepchildren of Marine aviation." This theoretical dichotomy, however, in no way diminished the practical use of helicopters. They soon proved their worth in combat and, in fact, became so indispensable that virtually every ground commander recommended additional helicopters be made immediately available by the time the Marines departed the Pusan Perimeter.

The commanding officer of VMO-6 holds a pre-mission pilot brief during the early stages of the Korean War. From left to right are Capt George B. Farish, 1stLt Eugene P. Millette, Capt Victor A. Armstrong, 1stLt Lloyd J. Engelhardt, Maj Vincent J. Gottschalk, Capt Alfred F. McCaleb, Jr., 2dLt Edgar F. Gaudette, Jr., 1stLt Gustave F. Lueddeke, Jr., and enlisted pilot TSgt Robert A. Hill.

Department of Defense Photo (USMC) A1991

The hard-working Marine helicopters were used for a wide variety of missions that taxed them to the limit during the month of August 1950. The most common uses were for command and control, aerial reconnaissance, medical evacuation, and combat search and rescue; however, they also spotted artillery fire, dispensed emergency supplies, lifted individuals to remote outposts, and provided high-speed communications wire laying services as well. An operational pattern soon emerged. Each morning the two duty helicopter pilots would fly to General Craig's command post where they would report to Major James N. Cupp, the brigade's air officer, for tasking. At about noon, these two helicopters would be relieved on station by the other two. This aircraft rotation ensured adequate pilot rest and gave ground crews time for daily maintenance work. In addition, an ad

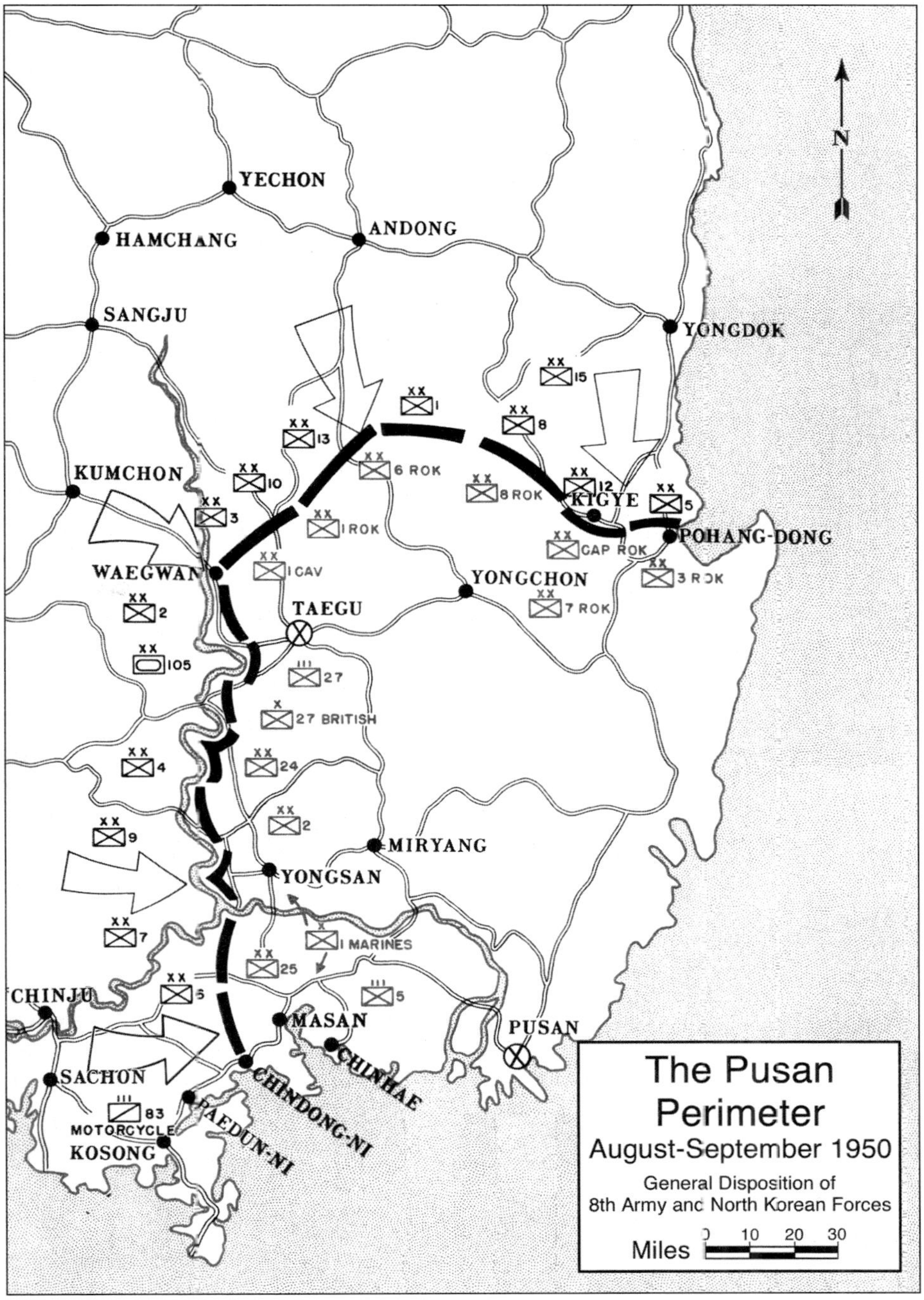

hoc control system evolved whereby the helicopter pilots would check in and out with the MTACS-2 air control section on their way to and from assigned missions. As air traffic control squadron commander Major Elton Mueller explained:

> We maintained the same positive radio contact with the helicopters that we did with all the other aircraft operating with us. The division air officer, however, controlled the helicopters. When they went out on a mission, they would fly by our operating site, give us a call—a radio check—on our reporting-in-and-out net. . . . In this manner [we] knew when [they] went out on a mission [and] they would tell us what type of mission they were going on, i.e. whether they were going on a reconnaissance, whether they had any rank on board, whether they were carrying the commanding general out to one of the units, or whether they were going out on an evacuation mission. . . . Since we had communications facilities and the air officer [did not] we could. . . keep [him] abreast of the situation.

Korea was a difficult arena of operations due to its rugged terrain, weather extremes, and poorly developed infrastructure as aerial observer Second Lieutenant Patrick G. Sivert recalled: "It was hot and dusty, the road network was very poor, and the country very mountainous. There was no apparent pattern of any sort to the mountains. . . no particular ranges or draws, compartments, or corridors." The Marines were first greeted by sweltering heat and choking dust, but within a few months bitter cold and heavy snow brought south by the so-called "Siberian Express" would create vastly different operational challenges. The already difficult topography was exacerbated by the lack of modern hard surface roads as well as poor overland communications links. River valleys provided the only flat space suitable for roadways, but they were susceptible to flash flooding. The lack of reliable telephone communications was also a problem because the short-ranged infantry radios of the day did not function well when out of the line-of-sight. The cumulative result of these disparate problems made Korea an operational nightmare. Luckily, helicopters provided the ideal technological fix. They were unrestrained by the terrain, could act as radio relays or lay wire at high-speed, and easily flew over

Department of Defense Photo (USMC) A130163

On 3 August 1950, 1stLt Gustave F. Lueddeke, Jr., flew the first command liaison mission in Korea. In addition to ferrying commanders around, he also logged numerous medical evacuations and flew rescue missions behind enemy lines.

traffic jams or roadless wilderness.

According to Major Gottschalk, the use of HO3S helicopters at Pusan for command liaison work had the greatest tactical value. General Craig faced many unusual command circumstances due to the emergency situation in Korea. Hurried planning, reliance upon oral orders, incomplete intelligence, poor communications, and inadequate maps all plagued the brigade staff. Craig turned to the helicopter to help solve his problems. While stationed on Guam in 1949, he became acquainted with helicopters when he borrowed a carrier-based Navy HO3S-1 to make command visits and observe field training, and Craig immediately put this experience to use in Korea. On the morning of 3 August 1950, he and his operations officer, Colonel Stewart, climbed into First Lieutenant Gus-tave Lueddeke's waiting HO3S, beginning the first Marine helicopter flight in an active combat zone. Craig and Stewart were airborne almost all of that day. The initial leg took them from Pusan 30 miles west to the brigade staging area at Changwon. Along the way, Lueddeke set down amid some Korean huts to allow Craig to confer with a battalion commander leading the convoy to its new assembly area. After a few minutes on the ground, Craig continued his journey to the actual site selected to become his forward command post. Next, he flew back to Masan to meet with the Eighth Army commander and the commanding general of the U.S. Army task force slated to carry out the first United Nations offensive in Korea. On the way home, Craig stopped three times to inform small unit troop leaders about the upcoming operation. Although this trip seems routine by modern standards, that was certainly not true in 1950. Marine Corps historian Lynn Montross noted the uniqueness of this feat and its impact on the future: "Only a helicopter could have made this itinerary possible in a period of a few hours. A fixed-wing plane could not have landed in such unlikely spots, and a jeep could not have covered the same route before nightfall over narrow, twisting roads choked with Army and Marine vehicles." He further

A Korean rice paddy serves as a makeshift-landing pad for a Marine HO3S-1 helicopter. The air panels laid out in the foreground mark the landing area and indicate wind direction.

Department of Defense Photo (USMC) A131089

National Archives Photo (USMC) 127-N-A1385

The leaders of the 1st Provisional Marine Brigade, BGen Edward A. Craig, left, and his deputy, BGen Thomas J. Cushman, right, wait in the shade of a Sikorsky HO3S-1. The commanding officer of VMO-6 felt that command visits were missions of the most tactical value during the fighting at the Pusan Perimeter.

opined: "A general and his staff could now make direct . . . contact with operations at the front as had never been possible before [and this] enabled a commander to keep in personal touch with his forward units since the helicopter could land virtually anywhere without asking favors of the terrain."

General Craig also said: "Time was always pressing. Fortunately . . . helicopters . . . were always available for observation, communications, and control. . . . Without them I do not believe we would have had the success we did."

In addition to command and control, a second valuable tactical use for helicopters was visual reconnaissance A major problem during the attack toward Sachon was a scarcity of tactical maps, compounded by the fact that the only maps readily available were inaccurate ones created by Japanese cartographers sometime before World War II. Villages were misnamed and misplaced, many roads were either not shown or were incorrectly plotted, there were no contour lines to accurately depict terrain features, and the complex grid system was too confusing to be of much value. Although no one at Quantico had predicted that helicopters might have to replace maps for navigation, this is exactly what happened in Korea. Small unit commanders often used helicopters to reconnoiter their routes of advance or to locate good ground for defensive positions. On the march helicopters shadowed ground movements and provided over-the-horizon flank security. In addition, HO3Ss were used to direct artillery fire, a task made difficult for ground observers due to the poor maps and hilly terrain that frequently masked targets.

Another ground support duty, one that had received much play at Quantico, was aerial wire laying. A helicopter flying nap-of-the-earth could put down communications wire at the rate of about a mile per minute, far faster than a ground party could do it. The heavy and cumbersome spools presented no problem for a helicopter, whereas ground-based wire layers were severely limited as to how much wire they could carry and which terrain they could cross. An additional bonus was that by flying over tree lines or narrow defiles, helicopters could keep the wire overhead where it was not subject to destruction by tank treads or artillery bursts. Today, wire laying seems like a small thing but, in the days before needed two-way radio reliability, land line communications was vital for command and control.

Two missions of marginal tactical value had a significant impact upon morale, aerial medical evacuations and airborne search and rescue. Helicopter evacuations, reported Major Gottschalk: "exert a very positive effect on ground troops since they know their chances of survival are tremendously in-creased. . . . A unit cut off by land [could still] have its wounded evacuated [and] it helped units by relieving them of the necessity of caring for them [thus] freeing more men for fighting. The use of helicopters for rescue of downed pilots [was] also important in bolstering [air crew] morale."

On 4 August, Marine helicopters performed their first aerial medical evacuation when a Marine wounded by an accidental weapon discharge was flown from Changwon to the naval hospital train at Masan. The next day helicopters were called out to deliver water and rations to an infantry platoon sent to a nearby hilltop to check out reports of an enemy observation party located there. "Whirlybirds" were used because they could deliver the cargo in a matter of minutes where it would have taken a carrying party hours to bring up in the rugged terrain and intense heat. Five Marines suffering severe heat exhaustion and in need of advanced medical attention were taken out by helicopter.

On 8 August, the squadron conducted a night helicopter evacuation—another first. This was a daring feat because the HO3S did not have proper instrumentation for night operations. Disregarding these limitations, Captain Armstrong flew off into the fading light to pick up a critically wounded man and the regimental surgeon of the 5th Marines. The nearly blind helicopter was guided back by the light of flares and came to earth amid the glow of headlights. This dramatic flight was the first of more than 1,000 night evacuations conducted in Korea.

The first of many Marine helicopter medical evacuations occurred when VMO-6 helicopters lifted several severe heat casualties to safety. "Whirlybirds" were often used because ground transportation could not traverse the rugged terrain and stretcher-bearer evacuation would take too long.

Department of Defense Photo (USMC) A2855

National Archives Photo (USN) 80-G-420545

The Sikorsky HO3S-1 was a civilian model helicopter acquired for use as an observation aircraft. Unfortunately, the aircraft was poorly configured for medical evacuations, which often required Marines and Navy Corpsmen to lift patients into the aircraft from odd angles.

As helicopter pilot Captain Norman G. Ewers later recalled:

> Normally, helicopter evacuation missions [were] performed on orders from the division air officer who relay[ed] the requests from the medical officers of the battalions or regiments. Helicopters [were] used to evacuate only those who [were] critically wounded and require[d] immediate hospital treatment. The helicopter [made] it possible not only to get the man to the hospital much more quickly, but it [provided] a much easier ride than travel by roads over rough terrain [and] this smoother ride . . . prevent[ed] hemorrhages.

Medical evacuations were flown without regard for difficult circumstances. The pilots took off in all kinds of weather, without the benefit of proper instrumentation or homing devices, and often disregarded enemy fire in the landing zones. A tribute to the helicopter pilots of VMO-6 was rendered by a ground officer: "The flying of evacuation helicopters from jury rigged and inadequate landing sites was nothing short of miraculous. . . . The pilots of the observation squadron received far less credit than they deserved. They used to fly at night [into] frontline landing strips where I had trouble walking." Frontline medical officers likewise credited the flying skills and bravery of the medical evacuation pilots for saving many lives. The mortality rate in Korea fell to a new low of only two percent, less than half the rate of World War II and far below the nearly 50 percent rate prior to the American Civil War, due in large part to the rapid evacuation of seriously wounded and the immediate availability of helicopter-provided whole blood at forward medical stations.

Unfortunately, the HO3S-1 was a civilian model aircraft adopted for use as a military machine; it was not designed to be a flying ambulance and, thus, poorly configured to be used as such. Marine ground crews in Korea quickly modified the HO3Ss to carry stretcher cases. The starboard observation window was removed and straps secured the stretcher in flight, but still a wounded man's legs protruded from the cabin. This was a minor annoyance that summer, but during cold-weather operations several cases of frostbitten feet and lower legs caused by the severe airborne wind chill were recorded. In addition, the wounded man most often had to be loaded into the helicopter from a position above the heads of the stretcher-bearers, a ponderous and awkward process. Inside the cabin, the pilot had to make quick ballast adjustments to ensure proper trim on the way home. Another problem was the HO3S-1's high profile and unstable tricycle landing gear; at least one HO3S tipped over while idling on rough ground. Although all agreed that the HO3S was invaluable in emergencies, there was room for mechanical improvement. This was handled in two ways. First, requests for immediate deployment of an off-the-shelf medical evacuation helicopter, the Bell HTL trainer, were sent up the chain-of-command. Second, Sikorsky Aircraft made design modifications to its newest observation helicopter, the developmental model S-52, which reached the fleet as the HO5S.

One mission of mercy for which the HO3S was perfectly suited was the rescue of downed pilots.

Helicopters were virtually the only means by which a downed pilot could be snatched from behind enemy lines and returned safely home within hours. The HO3S's side-mounted winch was an ideal tool for pulling an unfortunate aviator from the chilly waters off the Korean coast. The pilot or his crewman located the downed man and then the helicopter hovered overhead while the stricken man was lifted to safety. Lieutenant Lueddeke made the first of these rescues on 10 August while conducting a ground reconnaissance with the brigade commander on board. Second Lieutenant Doyle H. Cole's Corsair was struck by ground fire during a strafing run. Cole was unable to make it back to the *Badoeng Strait*, so his plane plunged into the water. Luckily, he was able get out and inflate his life raft before the plane sank. Lueddeke's helicopter quickly rushed toward the sinking plane to affect an airborne rescue. General Craig winched the soaked pilot up into the helicopter as Lueddeke hovered over the wreckage. Once safely inside the grinning pilot slapped his benefactor on the back with the words "Thanks, Mac" before he noticed the general's rank insignia and was able to render proper honors. The unperturbed senior officer simply replied: "Glad to be of service, Lieutenant."

Not every rescue had such a happy ending. Later that same day Lieutenant Lueddeke was sent to rescue another VMF-323 pilot. This time the downed flyer was Captain Vivian M. Moses whose plane had been hit by antiaircraft fire in enemy territory. Lueddeke skillfully negotiated a low-level approach behind enemy lines to pick up the stranded pilot and returned him to Chinhae for an overnight stay. The next morning, Moses returned to his ship where he promptly volunteered to fly another combat mission. Ironically, he was shot down once again before the helicopter that delivered him returned to action. His plane crashed into a rice paddy and flipped over when it struck the dike. Captain Moses was knocked unconscious as he fell from the plane and drowned before helicopter pilot Captain Eugene J. Pope could save his life. Sadly, Vivian Moses became the first Marine pilot to die in combat in Korea.

Marines refuel a VMO-6 helicopter in a rice paddy during the fighting in the Pusan Perimeter. When a helicopter could not make it back to base, 55-gallon drums of fuel and a supply of oil had to be trucked out to the make-shift landing pad.

Marine Corps Historical Center Photo Collection

On 7 August, the first Marine helicopter came under fire when the commanding general's HO3S-1 was caught in an enemy artillery barrage. Luckily, the plane emerged undamaged after dropping General Craig off. The first combat damage to a Marine helicopter occurred a week later when an HO3S-1 lost its windshield while evad-ing enemy antiaircraft fire. No "whirlybirds" were lost to enemy fire during the 580 missions flown by the helicopter section of VMO-6 during the fighting at Pusan.

On 12 August, the Marine advance toward Sachon was abruptly halted due to a breakthrough that penetrated the U.N. lines near Miryang on the Naktong River. The situation was so critical that a battalion of the 5th Marines was immediately ordered north to counterattack. Once again, the helicopter proved invaluable as a liaison vehicle. The battalion commander and the brigade operations officer mounted First Lieutenant Robert Longstaff's HO3S-1 to rendezvous with a U.S. Army representative. They flew to the appointed place but could not locate their man. Luckily, they were able to orbit the area until they found a reconnaissance unit, which was able to contact their division headquarters. The Marines were told that instead of joining the Army unit as planned they should instead "look the situation

Early Naval Helicopters

The first U.S. Navy experience with rotary-wing aircraft was not a good one. The Pitcarin OP-1 autogiro, an airplane not a true helicopter, had been tested and found wanting during the era between the World Wars. It was not until Igor Sikorsky introduced his VS-316 model helicopter on 13 January 1942 that vertical takeoff and landing aircraft became feasible. Sikorsky had earlier flown the first practical American helicopter, the VS-300, but that machine was only a test bed. The follow-on VS-316, designated the XR-4 by the U.S. Army, had a two-seat side-by-side enclosed cabin. A 200 horsepower Warner R-550-3 engine that ran a single overhead main rotor and a smaller anti-torque rotor on the tail powered the aircraft. The XR-4 prototype could hit a top speed of around 85 miles per hour, cruised at about 70 miles per hour, and had a range of about 130. In July 1942, the Navy tested its first one; an R-4 transferred from the Army and then promptly redesignated HNS-1 by the Bureau of Aeronautics. Two more were requisitioned from Army stocks in March 1943. The new helicopter was a success, and 22 more were procured for use as trainers beginning on 16 October 1943. The HNS-1 served as the primary naval aviation helicopter trainer until the Bell HTL-series replaced it.

Several other early helicopters (the Platt LePage R-1 and the Kellet R-2 and R-3) produced by other manufacturers were considered but not selected. All was not lost, however, because a bright young Kellet engineer, Frank Piasecki, would later develop tandem-rotor helicopters that would become a mainstay of naval aviation. The Bell Aircraft Company was too busy turning out jets to enter the initial helicopter competition, but that corporation's mathematician and engineer Arthur M. Young would soon revolutionize light helicopter design.

Sikorsky Aircraft produced 133 HNS helicopters; the Navy accepted 23, the Army kept 58, and the British Royal Air Force got 52. The first shipboard helicopter trials were conducted by America's first certified military helicopter pilot, Army Captain Hollingworth "Frank" Gregory. He put his HNS through its paces by repeatedly landing and taking off from the tanker *Bunker Hill* operating in Long Island Sound on 7 May 1943. Coast Guard Lieutenant Commander Frank A. Erickson flew the initial naval service helicopter mercy mission when he delivered two cases of blood plasma to a hospital at Sandy Hook on the New Jersey shore. Doctors credited Erickson's timely arrival with saving several lives. Other rescue missions aiding both civilian and military personnel in the New York area soon followed. The U.S. Army and the Office of Strategic Services both used helicopters for special combat missions in Asia during World War II.

The Navy was satisfied enough with the HNS to order an additional 150 helicopters from Sikorsky, 100 HOS-1s (designated R-6A by the USAAF) and 50 HO2S-1s (Army designation R-5A) before the end of the war. The HOS-1 was more compact, more powerful, and more maneuverable than its HNS predecessor. It mounted a single overhead main rotor, and was powered by a 240 hp Franklin O-405-9 engine. Three XHOS-1s were requested for testing from Army R-6A stocks in late 1942 and were accepted by the U.S. Coast Guard, which was by then running Navy helicopter training at New York's Floyd Bennett Field in March 1944. After the war a second batch of 36 HOS-1s were assigned to the Navy helicopter development squadron (VX-3) after passing acceptance tests. The Navy also took two HO2S-1 (Army R-5A) test models in December 1945, but opted to place an order for slightly modified S-51 commercial models (designated HO3S-1) which became the standard Navy, Marine, and Coast Guard light utility helicopters in 1947.

When the Coast Guard returned to the Treasury Department from the Navy Department on 28 December 1945, the U.S. Navy took over helicopter training and development. Marine helicopter pilots learned their trade with VX-3 before moving on to HMX-1 at Quantico, Virginia, prior to the Korean War.

over and do what [they] thought proper [to] ensure the safety of the 159th Field Artillery." The Marines had neither detailed maps of the area nor locating coordinates, so they took to the air to conduct a visual reconnaissance and, hopefully, find the lost Army artillerymen. This was done, and the Marines returned to meet the rescue convoy on the road. After giving an estimate of the situation and further instructions, the two Marines returned to the Army position to prepare for the rescue column's arrival. Concurrently, a helicopter piloted by Lieutenant Lueddeke carrying the artillery regimental commander and his operations officer located several survivors of an overrun artillery battery. They dropped a note of encouragement then led a relief party to the spot. During this excursion, Lueddeke's HO3S-1 came under small arms fire and had to "buck and jerk" its way out of the area using maneuvers not found in the pilot's manual. Only helicopters could have provided such assistance. Ground transportation would have been unable to find the misplaced units in a timely manner, while a light observation plane could not have moved back and forth between the supported and supporting units with such speed and efficiency. The helicopters of VMO-6 saved the day.

National Archives Photo (USMC) 127-N-A1990

Air-sea rescue was an important mission flown by VMO-6 with the first such rescue made in August. Here, Capt Eugene J. Pope, at the controls of his HO3S-1 helicopter, is congratulated by his still-wet fellow VMO-6 observation pilot Capt Alfred F. McCaleb, Jr.

Two HO3S-1 helicopters, two pilots, and five mechanics assigned to headquarters squadron in Japan were released from that duty and joined VMO-6 at Chinhae on 15 August, just in time for one of the biggest battles for the Pusan Perimeter. The 5th Marines had been pulled back from Sachon, hurriedly replenished, and then marched north to seal off the NKPA penetration near Miryang. Helicopters were used for visual reconnaissance of the battle area, conducted liaison visits, scouted the routes of advance, screened the flanks, spotted artillery fires, brought in supplies, and evacuated casualties as the Marines were twice called on to throw the North Koreans back across the Naktong River. During that time the helicopter pilots began to perfect evasive maneuvers that allowed them to dodge enemy ground fire. It also became obvious that the frail looking helicopters were tougher than previously thought. Several were hit by enemy small arms fire but kept on flying, and others survived some very hard landings in rough country. As General Lemuel Shepherd later noted about the toughness of helicopters: "I saw [them] come in with a dozen bullet holes [but] unless they are hit in a vital part, they continue to fly." Still, the helicopters carried no armor or weapons so they were used in supporting roles except for emergency evacuations or deep search and rescue missions. The best tactic for those risky missions was to get in and out as quickly as possible while flying nap-of-the-earth using terrain to mask ingress and egress routes.

The Marine defense of the Pusan Perimeter ended with the arrival of other elements of the 1st Marine Division and the remainder

Airfield Designations

United Nations Command airfields in Korea used letter-number designators: the letter "K" indicated a major air base; the letter "X" indicated an auxiliary landing strip; and the letter "A" indicated that the airfield adjoined a U.S. Army base or headquarters. During the Korean War, Marine helicopter squadrons operated from the following locations:

Marine Observation Squadron 6
Chinhae (K-10)
Kimpo (K-14)
Wonsan (K-25)
Yonpo (K-27)
Masan (unnamed)
Pohang (K-3)
Wonju (K-38)
Hongchon (K-47)
Kwandae-ri (unnamed)
Sonjong-ni (unnamed)
Sohung-ni (X-77)
Sinchon (X-83)
Tonggo-ri (A-9)

Marine Transport Squadron 161
Pusan (K-1)
Kangnung (K-18)
Chodo-ri (X-83)
Yongpo-ni (A-17)
Taejong-ni (A-33)

of Major General Field Harris' 1st Marine Aircraft Wing from California in preparation for the landing at Inchon. By late August, the helicopter detachment had logged 580 sorties and 348 flight hours, conducted 35 medical evacuations, and flew 85 aerial reconnaissance missions. Throughout that time helicopter availability was 100 percent. In his final report Major Gottschalk attributed this remarkable accomplishment to two factors: the excellent facilities at Chinhae and the ground support crew's professionalism, skill, and willingness to put in long hours. This was no small achievement because helicopters required a great deal more effort; more spare parts, more man-hours, and more sophisticated tools and work spaces than did the OY Sentinels. On the other hand, Gottschalk also noted that larger transport helicopters could have provided much needed services such as troop lifts, resupply, and command liaison, which were beyond the capabilities of the HO3S-1.

General Craig, the first Marine commander to use a helicopter as a command and control aircraft later wrote:

> Helicopters are a godsend. . . . The mountainous terrain of Korea presents a difficult problem for security. . . . [Transport] helicopters would be ideal to [quickly] post patrols and outguards on high, dominating terrain which would [normally]

A VMO-6 helicopter lands near the artillery positions of the 1st Battalion, 11th Marines, along the Naktong River. The HO3S was designated as an "observation" platform but was actually used as a light utility aircraft in Korea.
Department of Defense Photo (USMC) A2204

National Archives Photo (USMC) 127-N-A130052

MajGen Field Harris, left, commanding general of the 1st Marine Aircraft Wing, MajGen Oliver P. Smith, commanding the 1st Marine Division, and BGen Thomas J. Cushman, assistant wing commander, meet in Tokyo, Japan, a week before the landing at Inchon.

> take hours to climb. . . . [More helicopters] would . . . insure the earlier defeat of the enemy. They should be made available for use at the earliest possible date.

He also noted other Service interest in rotary-wing aircraft by stating: "The Army is enthusiastic over our ideas of employment of this type of aircraft and is going ahead with the idea of employing them on a large scale." Like Major Gottschalk, Craig also recommended that a transport helicopter squadron be formed and sent to Korea as quickly as possible. The Director of Marine Corps Aviation, Brigadier General Clayton C. Jerome, made the case for additional helicopters in a memorandum to the Deputy Chief of Naval Operations (Air):

> There are no superlatives adequate to describe the general reaction to the helicopter. Almost any individual questioned could offer some personal story to emphasize the valuable part played by [the] HO3S planes. . . . There is no doubt the enthusiasm voiced . . . is entirely warranted. . . . No effort should be spared to get helicopters—larger than the HO3S if possible—to the theater at once, and on a priority higher than any other weapon. [We need] helicopters, more helicopters, and more helicopters.

The Inchon-Seoul Campaign

On 9 September, VMO-6 was placed under the operational control of the 1st Marine Division, commanded by Major General Oliver P. "O. P." Smith, and under the administrative control of the 1st Marine Aircraft Wing. The Marines' next mission was destined to become a military classic—the amphibious assault at Inchon, a battle that dramatically reversed the course of the Korean conflict. U.S. Army X Corps, spearheaded by the 1st Marine Division, launched a difficult daylong amphibious landing then rapidly moved inland to secure the supply depot at Ascom City and Kimpo Airfield. The campaign culminated with the retaking of the South Korean capital of Seoul. This seizure cut the enemy's main supply routes and left the NKPA forces in the south isolated. By the time the lead elements of X Corps in the north and Eighth Army coming up from the Pusan Perimeter linked up the NKPA was in full flight. That once awesome fighting force had been completely routed and was headed for the dubious safety of North Korea.

To prepare for the Inchon landing, Major Gottschalk divided his squadron into forward and rear echelons. The forward echelon, 10 officers, 48 enlisted men, and 8 helicopters, loaded on board Japanese-manned LST Q079 at Chinhae. During the voyage, the Marines and Japanese crew shared mess facilities. Luckily, detachment commander Captain Victor Armstrong spoke fluent Japanese—he had resided in Japan for 15 years before the outbreak of World War II. Four officers and 43 enlisted men remained behind to safeguard squadron property at Chinhae.

Once ashore the Marine helicopter detachment picked up right where it left off, but on a much larger scale. The main missions remained command and liaison, aerial evacuation of seriously wounded, combat search and rescue of downed fliers, and visual reconnaissance. Although the number of HO3S helicopters had doubled since August, the demands for their time continued to increase.

Major General Smith, the 1st Marine Division commander and a former member of the Shepherd Board in 1946, quickly became a helicopter advocate. "The helicopter was of inestimable value to the division commander and his staff in keeping personal contact with subordinate units in a minimum of transit time," he asserted. Generals Smith and Craig, now assistant division commander, depended upon helicopters to visit the front on a daily basis and unit commanders scouted proposed routes of advance, although emergency medical evacuations were given priority over liaison and reconnaissance. With as few as only four helicopters operational, however, command and liaison visits were often interrupted when the commander's helicopter was diverted for emergency missions. When critically wounded men needed a ride the generals and colonels either used alternative transportation or waited until their "chopper" returned. The list of dignitaries using helicopter transport during September 1950 included Fleet Marine Force, Pacific, commander Lieutenant General Lemuel Shepherd, Commandant Clifton B. Cates, and X Corps commander, U.S. Army Major General Edward M. Almond. At Inchon, just as at Pusan, the most often heard complaint about helicopters was that there were not enough of them.

Although Marine helicopters played no combat role on the first day at Inchon, Navy helicopters did spot naval gunfire during the preliminary bombardment. On 16 September (D+1), Marine helicopters entered the fray flying 14 missions. The landing ship-based Marine "whirlybirds" flew reconnaissance and artillery spotting missions over Wolmi-do Island, and First Lieutenant Max Nebergall pulled a Navy pilot out of the drink. On the afternoon of 17 September, ground Marines captured Kimpo Airfield, the largest

"Whirlybird" pilots in Korea were famous for their daring feats while rescuing downed flyers and evacuating seriously wounded men; among the very best were 1stLts Robert A. Longstaff and Gustave F. Lueddeke, Jr. of VMO-6. Tragically, the Marine Corps lost two of it most promising pioneer helicopter pilots when Longstaff was killed in action at the Chosin Reservoir and Lueddeke succumbed to poliomyelitis not long after returning from Korea.

National Archives Photo (USMC) 127-N-A130403

Department of Defense Photo (USMC) A130249

Capt Victor A. Armstrong, right, proudly displays the cake sent from the carrier Philippine Sea *(CV 47) as Maj Vincent J. Gottschalk, VMO-6's commanding officer, looks on. Capt Armstrong made a daring behind-the-lines rescue of a Navy pilot shot down near Seoul and the cake was sent ashore as a mark of appreciation.*

airfield in Korea, virtually intact. The first U.S. aircraft to land there was Captain Armstrong's HO3S, which arrived at mid-morning on 18 September as Marines searched for the remnants of the previous night's NKPA counterattack force. Armstrong carried two early proponents of Marine helicopter operations, Lieutenant General Shepherd and his operations officer Colonel Victor H. Krulak.

On 19 September, the 1st Marine Division moved its command post from Inchon to Oeoso-ri. The next day VMO-6 moved to nearby Kimpo, which thereafter served as the squadron's base of operations until the subsequent move north. The final phase of the Inchon turning movement—the recapture of Seoul—was about to begin, and helicopters proved to be particularly valuable when terrain obstacles separated elements of the division during the drive to retake the capital. The general operational pattern was for one helicopter to be earmarked for each regimental commander in addition to one each for the division commander and his assistant commander. The regimental helicopters were primarily used for reconnaissance and medical evacuations, the division commander's for liaison, and the assistant division commander's for reconnaissance; any unassigned helicopters underwent maintenance while standing by for emergency evacuations or combat search and rescue.

The major obstacle on the way to Seoul was the Han River. Brigadier General Craig used his helicopter to locate a suitable crossing area, scout key terrain, and survey the road approaches to the South Korean capital. Although few enemy soldiers actually showed themselves, Captain Armstrong, Craig's pilot, had to dodge scattered small arms fire along the way. As a result of his aerial reconnaissance, Craig recommended that the 5th Marines move across the Han at an abandoned ferry site near Haengju and then seize the high ground overlooking Seoul.

Just as before, combat search and rescue was an important additional duty for the helicopters of VMO-6. On 21 September 1950, the squadron received word that a pilot had gone down behind enemy lines and was jammed inside his cockpit. Anticipating a difficult extraction, First Lieutenant Arthur R. Bancroft loaded his plane captain on board then took off to make the rescue. The area was "hot," so friendly planes maintained a rescue combat air patrol to strafe any enemy who showed their heads. Bancroft set his HO3S down and remained at the controls while the helicopter idled with its rotor blades slowly turning. The crew chief could not free the encased pilot alone, so Bancroft had to leave the aircraft to assist.

Who was the First Marine Helicopter Pilot?

There is some dispute about who the first Marine Corps helicopter pilot actually was. According to Marine lore that honor goes to fighter ace and famed test pilot Marion E. Carl, but the official records of the naval service identify Major Armond H. DeLalio as Marine helicopter pilot number one, and Marion Carl himself proclaimed that Desmond E. Canavan was probably the first Marine to fly a helicopter.

According to the Marine Corps' official history, *Marines and Helicopters, 1962-1973,* "Major General Marion E. Carl is generally credited with being the first Marine to learn how to fly a helicopter in July 1945 [but] it was not until some years later that he was officially designated [as such]." In his autobiography, *Pushing The Envelope* (Annapolis, MD: Naval Institute Press, 1994), Carl relates that he learned how to fly a Sikorsky HNS (R-4) while a test pilot stationed at the Naval Air Test Center, Patuxent River, Maryland. He was given about three hours of instruction before he soloed. In that same memoir, however, he states that fellow Marine Desmond Canavan was flying helicopters in late 1944. Carl's claim that he was helicopter pilot number one rests upon the fact that he was the first Marine to log the 40 hours required for certification even though he never applied for such certification. Neither Carl nor Canovan appear on the naval service helicopter pilot certification list prior to June 1950.

Marine Corps Historian Lynn Montross, the recognized authority on early Marine helicopter operations, lists Navy Cross holder Armond DeLalio as having flown U.S. Navy helicopters at New York's Floyd Bennett Field then under the auspices of the U.S. Coast Guard in 1944. He is officially recognized as the first Marine certified as a helicopter pilot, achieving that honor on 8 August 1946. DeLalio was the operations officer for Navy helicopter squadron VX-3 at that time. He was killed during a test flight in 1952 when a rocket-assisted takeoff pod malfunctioned causing his HRS helicopter to catch fire and then crash.

The Navy register of early helicopter pilots lists 250 qualifiers prior to the onset of the Korean War in June 1950; 33 are Marines, including three enlisted naval aviation pilots (the famous "Flying Sergeants" of the Marine Corps).

While who should be recognized as the true "Gray Eagle" of Marine helicopter aviation remains murky, there is little doubt about the specific incident that started the Marine Corps helicopter program. That event occurred at Quantico, Virginia, in 1946 and was described by helicopter pioneer Edward C. Dyer:

> One day Marion Carl, a test pilot at Patuxent, flew a helicopter to Marine Corps Schools to demonstrate it to the students. . . . He hoisted [Lieutenant Colonel Victor H.] Brute Krulak . . . about 15 feet [off the ground] and pulled him into the cockpit. [Lieutenant Colonel Merrill B.] Twining and I were standing by the window and watching and I said 'Bill, let's . . . quit fooling around.' He said 'OK! . . . He wrote the theory . . . principles . . . background . . . reasoning . . . and I wrote [an implementation] program."

Marine Corps Historical Center Photo Collection

LtCol Armond H. DeLalio, recipient of the Navy Cross for heroism as a pilot with Marine Scout-Bomber Squadron 241 during the battle of Midway and a Marine Corps helicopter pioneer, was honored in 1965 when an elementary school at Camp Lejeune, North Carolina, was dedicated in his name.

Marion Carl recalled that he specifically selected Lieutenant Colonel Krulak because his small stature and lightweight could be accommodated by the limited room and lift capability of his HOS-1 helicopter. Krulak thereafter became a helicopter devotee.

While the two Marines busily freed the trapped pilot, the helicopter's collective friction device worked loose and the plane tipped on its side where the beating rotors destroyed the aircraft. Luckily, Lieutenant Robert Longstaff was able to pick up the grounded trio although his overloaded HO3S staggered under the excessive weight until it reached friendly lines. Bancroft then promptly mounted another helicopter to rescue a second Navy flier before the day ended.

Two days later, Captain Armstrong recorded the longest search and rescue operation yet by a VMO-6 helicopter when he flew nearly 100 miles behind enemy lines to rescue a downed Navy pilot. On the return flight, he ran out fuel over friendly territory, temporarily put down, refueled, and then landed at Kimpo after dark using a flashlight to illuminate his control panel. The rescued pilot turned out to be a squadron commander from the carrier *Philippine Sea* (CV 47). The next day, VMO-6 received a large layer cake, compliments of the U.S. Navy as a reward for Armstrong's fine work. Conversely, Lieutenant Longstaff flew the shortest rescue mission of the war picking up a pair of Marines from a Grumman F7F Tigercat that crashed after taking off from Kimpo. That mission on the 25th took less than six minutes. The pilot was Lieutenant Colonel Max J. Volcansek, Jr., of Marine Night Fighter Squadron 542, one of three squadron commanders to go down that day.

A more dramatic rescue also occurred on 25 September. A Navy helicopter "on loan" to the Marines suffered battle damage during a deep rescue mission and was forced to put down near the Han River. Word that an American aircrew was down in enemy territory did not reach the division air officer until about 2100—after sunset. Captain Armstrong took off despite the fact that the HO3S had neither proper instrumentation nor landing lights for limited visibility flying. Armstrong needed both arms and both feet to control the helicopter, so he held a flashlight between his knees to illuminate the unlit instrument panel. He spotted the downed aircraft in the glow of light cast from the burning city of Seoul and set down on a nearby sandbar. The crew, a Navy pilot and a Marine enlisted man, swam to Armstrong's waiting helicopter for a safe ride home. He once again had to rely upon makeshift lights upon arrival at the landing zone.

Thus far in Korea, VMO-6 had lost helicopters to operational incidents but had suffered no fatalities. Tragically, this string of luck came to an end on 29 September. A VMO-6 Sentinel was shot down about five miles north of Seoul. Reports indicated the aerial observer was killed in the crash, but the pilot was able get out. First Lieutenants Lloyd Engelhardt and Arthur Bancroft, both of who previously had logged deep search and rescue missions, were at the division command post when the call for help came in. Both immediately volunteered to go, but Major James Cupp, the division air officer, ordered them to wait until more detailed information became available. A few minutes later they learned that the OY went down beyond the Marine frontlines near Uijongbu, an unsecured area teeming with enemy and known to be infested with antiaircraft guns. Bancroft, who won a coin flip to decide who would make the rescue, took the lead with Engelhardt trailing by about a half mile. They found the crash site, but as Bancroft's helicopter began to settle it was hit by enemy fire and disintegrated in a fireball. Engel-hardt called for fighter planes to survey the area. They reported Bancroft had been killed, and there was no sign of the downed pilot. First Lieutenant Arthur R. Bancroft thus became the first Marine helicopter pilot to die in action.

Helicopters became crucial for command liaison. The rugged terrain, a major river, and wide dispersal of fighting units made control difficult. Helicopter mobility made it possible for commanders to scout approach routes, identify key terrain, attend conferences in the rear, and then quickly thereafter meet subordinate commanders face-to-face. On 28 September, Major General Smith coordinated the defense of Seoul as he visited each of his three regimental command posts: the 1st Marines at Seoul's Duk Soo Palace; the 5th Marines at the Seoul Women's University; and the recently arrived 7th Marines on the city's western outskirts. The 1st and 5th Marines were to defend in place while the 7th attacked toward Uijongbu. On 3 October, Armstrong flew Commandant Cates on an aerial survey of the Inchon-Seoul area and a frontline inspection tour highlighted by observation of an attack by the 7th Marines on the 4th. This was the final ground combat action of the campaign, although Marine helicopters continued to fly deep rescue and medical evacuation missions from Kimpo throughout the rest of October. Lieutenant Engelhardt rescued a Marine pilot near Chunchon on 3 October and then plucked an Air Force pilot up at Sibyon-ni on the 5th.

When the Inchon-Seoul campaign was officially declared over at noon on 7 October 1950, VMO-

U.S. Naval Aviation Designations

During the Korean conflict, the Navy Bureau of Aeronautics used designation systems that conveyed a lot of information about its squadrons and aircraft in a concise manner.

Squadron Designations:

The Bureau recognized three aircraft squadron types: lighter than air (Z); heavier than air (V); and helicopter (H). In addition, Marine aircraft squadrons were identified by the insertion of the letter "M" between the aircraft type and the squadron function. In general, a three letter prefix followed by up to three numbers was used to identify individual Marine aircraft squadrons. The first letter (a "V" or "H") identified the primary aircraft type used by the squadron, the second letter ("M") identified it as a Marine aviation unit, and the third ("O" indicating observation and "R" for transportation) identified the squadron's primary mission; the numbers in the suffix sometimes identified the squadron's unit affiliation and always noted its precedence order.

Thus, VMO-6 was the sixth heavier-than-air Marine observation squadron formed. The single digit indicated that the squadron was not specifically affiliated with a particular aircraft wing (observation squadrons were attached to ground units). On the other hand, HMR-161 was the first Marine helicopter transport squadron assigned to the 1st Marine Aircraft Wing (the first "1" indicating initial assignment to the wing, numbers above "6" were used for non-fixed wing aircraft, and the last "1" signifying it was the first squadron formed).

Aircraft Designations

Individual aircraft designations used a similar identification system. The Bureau of Aeronautics gave each naval aircraft a mixed letter and number designation. Except for experimental or prototype helicopters, the first letter was an "H" indicating rotary-wing status; the second letter indicated its primary purpose ("O" for observation, "R" for transport, or "T" for trainer); a number (except in the case of the first model) indicated the manufacturer's sequence for producing that specific aircraft type; the next letter identified the manufacturer ("L" for Bell, "P" for Piasecki, or "S" for Sikorsky); and the number following a dash indicated a sequential modification of that aircraft model.

Thus, the HO3S-1 was Sikorsky Aircraft's third model observation helicopter with one modification; the HRP was Piasecki's first transport helicopter; the HTL-4 was the fourth modification to Bell Aircraft's original trainer helicopter; the HO5S was Sikorsky's fifth observation model; and the HRS-1 was Sikorsky's first transport helicopter.

The Bureau's system was a good one that remained in use for four decades, but there were a few problems. First, aircraft were often used for roles other than those assigned. For example, the HO3S-1 was actually a utility aircraft that during field service performed many tasks other than observation, a task that actually became a seldom-used secondary mission in Korea. Second, the proliferation of missions and manufacturers as time passed led to confusing duplication of letters ("T" was variously used to indicated torpedo, trainer, and transport aircraft). Third, lack of inter-Service consistency produced confusion (the Navy HO3S-1 was an H-5F to the Air Force and Army). The naval aircraft designation system was replaced by a joint aircraft designation system in 1962, but the Bureau's squadron designation system remains in effect.

6 helicopters had flown 643 missions, evacuated 139 seriously wounded men, and rescued 12 airmen from behind enemy lines or out of the water.

The success of VMO-6's fledgling helicopter detachment had wide-ranging effects that spread well beyond the theater of operations and impacted more than just the Marine Corps. In the United States, military dogmatists and civilian pundits complained long and loud about lack of inter-Service unity in Korea. However, in the words of Major General John P. Condon, an expert in joint operations and an experienced air group commander in Korea: "The farther from Washington, the less inter-Service differences came into play." This dictate was borne out by Marine helicopter operations in late October. On the 21st, Captain Gene W. Morrison made a series of flights to evacuate eight seriously wounded Army paratroopers from Sukchon to Pyongyang in his HO3S. Three days later, Captain Wallace D. Blatt, who had provided helicopter coverage for the withdrawal of U.S. forces from China, and First Lieutenant Charles C. Ward flew deep into enemy territory to rescue a pair of Air Force pilots down near Koto-ri, more than 100 miles inland from their temporary base at Wonsan Harbor. These were only a few of many times Marine helicopters rescued or aided other American servicemen in Korea. Although both the U.S. Navy and Air Force were flying helicopters in Korea, the Marine success with rotary-wing operations at Pusan and Inchon prodded the Air Force to attach helicopter units specifically earmarked for medical evacuation to Army field hospitals. Likewise, a clamor for organic transport and observation helicopters arose from U.S. Army commanders. The utility and practicality of helicopters in combat zones had been firmly established by the Marines of VMO-6 in less than three months.

The Chosin Reservoir

General MacArthur's successful turning movement at Inchon drastically changed the course of the Korean War. Thereafter, the NKPA was a broken machine with its scattered remnants headed for the protection of North Korea's hinterlands or a safe haven inside China. MacArthur, sensing a chance to end the conflict by trapping the remaining North Korean forces, sent his United Nations Command speeding north beyond the 38th Parallel in a race for the Yalu River despite warnings not to do so.

MacArthur split his forces to hasten the pursuit. He ordered the Eighth Army forward in the west and opted to use X Corps, including the 1st Marine Division, for an amphibious landing at Wonsan in northeast Korea. Once again, VMO-6 split into forward and rear elements. The advance party (4 officers and 70 enlisted men known as the "surface" echelon) embarked on board LST 1123 and then sailed for Wonsan on 13 October. Most pilots, all VMO-6 aircraft, and a skeleton ground-support crew remained at Kimpo. Fifth Air Force specifically tasked the Marine helicopters with supporting a U.S. Army parachute drop near Pyongyang, but the Marines also would conduct combat search and rescue as needed. This "flight" echelon was composed of 17 officers and 19 enlisted men with Captain Armstrong as officer-in-charge. The stay-behind element was to continue operations from Kimpo until ramp space at Wonsan became available. Included in the helicopter flight echelon were several newly arrived pilots and replacement aircraft ferried in from the United States on board the aircraft carrier *Leyte* (CV 32). The new aircraft were welcome additions that made nine Marine HO3S helicopters available.

United Nations ground forces pressed forward against only token resistance. A South Korean division occupied Wonsan in early October, but the amphibious task force carrying VMO-6 had to mark time sailing up and down the east coast until the harbor could be cleared of mines. Consequently, members of VMO-6's stay-behind echelon actually set down in North Korea before the advance party. On 23 October, Captain Blatt and Lieutenant Ward flew north from Kimpo to Wonsan. The airfield served as the squadron's home base from then until VMO-6 moved to Yonpo on 3 November. The embarked surface echelon finally got ashore on the 25th, and the flight echelon completed its movement to Wonsan three days later.

Capt Wallace D. Blatt, a helicopter pilot assigned to VMO-6 in Korea, had been a Marine multi-engined transport pilot at Guam and Okinawa during World War II. He learned to fly helicopters immediately after the war and was the first Marine helicopter pilot deployed overseas when he flew a borrowed Navy HO3S-1 during the occupation of North China.

National Archives Photo (USMC) A130580

Immediately after landing, the 1st Marine Division began operations. One regiment occupied Wonsan and manned two battalion-sized outposts (Majon-ni to the west and Kojo to the south) while two regiments proceeded about 50 miles north to the port of Hungnam and the railway junction at Hamhung before moving out toward the Chosin Reservoir some 78 miles farther inland. Although intelligence estimates indicated there would be little resistance and X Corps commander, Major General Edward M. Almond, wanted a rapid inland movement, the enemy had other ideas. A night attack at Kojo caught the Americans by surprise and cut the main supply route while unexpectedly strong NKPA forces encircled the Majon-ni outpost. With no overland routes open, helicopters became the only reliable link with both outposts.

The 1st Marine Division was alerted that the Kojo garrison was under attack in the early morning hours of 28 October. Emergency requests for medical assistance, specifically aerial evacuation helicopters and a hospital receiving ship in addition to ground reinforcements, were quickly acted upon. Six HO3S helicopters were dispatched. As Captain Gene Morrison later recalled, the situation was desperate enough that he never shut his engine down after arriving at Wonsan on his ferry flight from Kimpo. Instead, he received a hurried cockpit brief and was on his way to Kojo without ever leaving the aircraft. Captains Blatt and Morrison, and

Department of Defense Photo (USMC) A134641

The hospital ship Repose *(AH 17) at anchor in Inchon Harbor. Note the helicopter-landing pad mounted on the ship's stern; this configuration became standard on all hospital ships during the course of the Korean War.*

Lieutenants Engelhardt, Lueddeke, and Ward, collectively flew 17 seriously wounded men from Kojo to the hospital ship *Repose* (AH 17) at Wonsan Harbor. Captain George B. Farish provided airborne search and rescue. During a search on 29 October, he spotted the word "HELP" spelled out in straw about a mile northeast of Tongchon. As Farish trolled the area, a lone figure emerged from cover and then began waving. Farish shouted: "Hey Mac, looking for a ride?" He then plucked up the first of several lost Marines he brought in that day. During several of the rescues Farish left his helicopter to assist badly wounded men to the idling aircraft. Unfortunately, his daring attempt to rescue a Navy pilot under fire late in the day came to naught when it was discovered the man was already dead.

Helicopters played an important role at Majon-ni, a vital road junction located in a Y-shaped valley about 25 miles west of Wonsan.

Capt Gene W. Morrison, a helicopter pilot with VMO-6, was one of the first Marine "Whirlybirds" to arrive in northern Korea from Kimpo Airfield to support the Chosin Reservoir campaign. At Yonpo Airfield, he was immediately diverted to help evacuate serious wounded Marines from Kojo to the hospital ship Repose *(AH 17) in Wonsan Harbor.*

National Archives Photo (USMC) 127-N-A130604

The village was occupied without resistance on 28 October, but within a week the garrison was completely surrounded and the vulnerable main supply route became known as "Ambush Alley." Radio communications between Majon-ni and Wonsan was uncertain because intervening high ground and intermittent atmospheric interference allowed an open window of only a few hours each day, so the only reliable communications links were messages carried in and out by helicopter or OY pilots. For the most part, the Majon-ni strong point was supplied by airdrop and casualty evacuation was by helicopter from 2 November until the siege lifted.

The Chosin Reservoir campaign tested the endurance of the "whirlybirds" and the skill of their pilots and the fortitude of their ground crews like no other period before Chinese anti-aircraft fire began to light up the clouded skies of northeast Korea. The via-

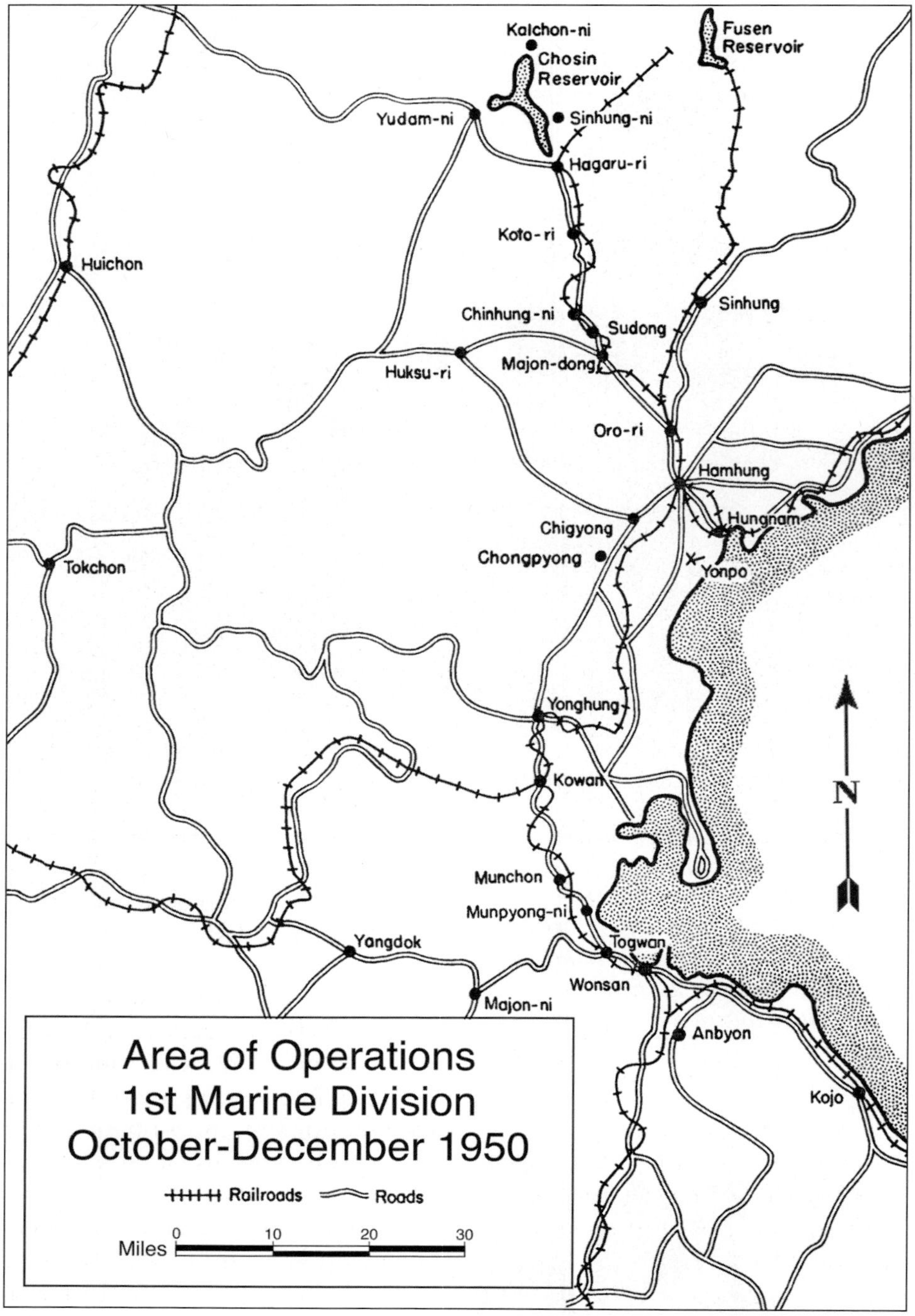

bility of extended helicopter operations at high altitude and in difficult weather conditions was at that time still conjectural. It was believed that helicopters might not be able to operate safely at any point beyond Chinhung-ni at the mouth of the Funcilin Pass, about two-thirds of the way to the Chosin Reservoir, due to the thin air at that altitude. The effect of prolonged cold weather on helicopter operations was also a source of concern. This issue came to the fore when Captain Eugene Pope had to return his HO3S after only four minutes aloft because the collective and cyclic controls were too stiff to adequately control flight. Ground crews subsequently switched to light weight lubricants and tried to either hangar or cover all aircraft when not in use. These measures compensated for, but did not completely alleviate, cold weather-induced problems. Re-duced lift in low temperatures at high altitude and flight in windy conditions made flying in the mountainous terrain hazardous, but there was no choice when emergencies occurred. It also became apparent that ground-effect hovers would not be possible in the foreseeable future. An additional problem was the ungainly configuration of the HO3S-1, which required stretcher cases to extend outside the cabin. Sub-freezing temperatures and extreme airborne wind chill factors put already wounded men at risk for frostbite while enroute to safety. Thus, the already limited flight envelope of the HO3S-1 was further restricted by terrain and weather.

On 2 November, the 1st Marine Division began its ascent toward the reservoir following a helicopter reconnaissance of the Sudong Valley. No enemy troops were located from the sky, but ground units were soon mixing it up with the first Chinese Com-munist units yet encountered. General Smith ignored the advice of the X Corps commander to speed it up and instead moved his division steadily ahead along a single-lane road, keeping all units tied in and establishing strong points along the way. His foresight and prudence likely saved the 1st Marine Division from annihilation when the Chinese sprang their trap a few weeks later.

Helicopters scouted hill-masked flanks, reconnoitered the roadway, laid communications wire, provided radio relays, and brought in crucial small items in addition to their by-then normal jobs of command liaison and medical evacuation. Despite increasingly poor weather, First Lieutenant Ward flew 115 miles from Yonpo to Songjin to rescue an Air Force airborne for-

Department of Defense Photo (USMC) A4912

Aerial evacuation of wounded and severely frostbitten Marines and soldiers from Hagaru-ri saved numerous lives. From late November to mid-December, Maj Vincent J. Gottschalk's squadron conducted a total of 191 helicopter evacuations out of a total of 1,544 flights.

ward air controller whose plane had gone down near the Chosin Reservoir on 5 No-vember. The HO3S was badly buffeted by crosswinds and strained to bite into the chilly thin air. Three days later, Captain Pope's helicopter was blown out of the sky by turbulent winds while on a resupply run. The helicopter was a wreck, but Pope escaped without serious injury. Lieutenant Ward arrived to take him out but was beset by a temperamental starter, so both pilots spent the night at a ground command post.

By 26 November, the 1st Marine Division was dangerously spread out. Little active resistance had yet been encountered, but veteran commanders were leery that things might be going too well. The Marines had moved upward through the snow-covered Funchilin Pass over the main supply route, a treacherous, icy, winding, narrow, dirt road. General Smith wisely established a series of outposts along the way; a regimental supply base at Koto-ri just north of the Funchilin Pass, an airstrip and division headquarters at Hagaru-ri on the southern tip of the reservoir, a company-sized outpost guarded the Toktong Pass from Fox Hill, and a jump off point manned by two regiments at Yudam-ni on the western tip of the reservoir.

Conditions were terrible. Swirling snow and sub-zero temperatures were the result of the winds, which blew down from Manchuria. It would be hard to imagine more difficult flying conditions for helicopter operations. The bitterly cold, short days and lack of repair facilities hampered helicopter maintenance. It was under these dire circumstances that the mettle of VMO-6's helicopter section was truly tested.

Beginning on the night of 27 November, the advance elements of the 1st Marine Division became heavily engaged at Yudam-ni and Hagaru-ri. The fierce fighting at the Chosin Reservoir required an all-hands effort by VMO-6 when more than six Chinese divisions tried to overrun two Marine regiments and cut the main supply route at several points. Helicopter pilots Blatt and Morrison both reported enemy roadblocks between Koto-ri and Hagaru-ri, the first official confirmation that the 1st Marine Division was surrounded. The Marines' abortive advance was about to become a breakout an epic of modern warfare during which the Marines "attacked in a different direction" bringing out most of their equipment and all of their wounded.

Several helicopters moved forward to Hagaru-ri to save flight time on 28 November, and all available aircraft flew from dawn until dusk each day for the next week-and-a-half. General Smith often used helicopters to visit his scattered units during that time. The helicopters of VMO-6 logged 40 sorties (1 reconnaissance, 16 transport, and 23 medical evacuations) in 73.7 flight hours on the 29th. Fifty seriously injured men were flown out and numerous vital supplies (particularly radio batteries and medicine) were brought in; General Smith visited the forward command posts, and a large group of enemy was spotted by helicopter that day. Captain Farish's HO3S was hit several times as he delivered supplies to an isolated rifle company perched atop Fox Hill. As he later related: "They ran me off." Farish limped back to Hagaru-ri and safely landed under covering fire by the Marines in the perimeter, but his aircraft was operationally grounded due to damage to the main rotor transmission. The next day, Lieutenant Engelhardt's HO3S was hit while delivering vital radio batteries to Fox Hill. A bullet just missed the pilot, and the helicopter was so damaged that it had to return to base for emergency repairs after carrying out one wounded Marine. Forty-three other casualties were successfully brought out that day as well by the HO3S-1s of VMO-6. Helicopters carried out 50 wounded and brought in medical supplies, gasoline, radio batteries, and tank parts in almost 60 hours of flight time on 30 November.

National Archives Photo (USN) 80-G-420288

Helicopter pilot 1stLt Robert A. Longstaff poses in front of his HO3S-1 helicopter decked out in flight gear after delivering a downed Navy airman to the Sicily *(CVE 118). Longstaff was the second Marine helicopter pilot killed in action when his aircraft was hit by antiaircraft fire at the Chosin Reservoir.*

The Marines consolidated at Hagaru-ri, broke out of the Chinese trap at Koto-ri, and moved back toward the sea by way of an air dropped Treadway portable bridge that spanned the Funchilin Pass. Throughout the ordeal at the reservoir, helicopters were the only dependable means of physical contact between scattered units. They provided liaison, reconnaissance, and medical evacuation; whenever a "whirlybird" flew a medical supply mission, ammunition and radio batteries were part of the incoming load. These operations were not without cost. On 3 December, First Lieutenant Long-staff was killed at Toktong Pass when his helicopter was brought down by enemy fire while trying to rescue a critically wounded man. Captain Blatt played a role in a daring but ultimately tragic event. After several frustrating hours trying to start his frozen helicopter, Blatt was finally able to get his aircraft to crank up just as an emergency rescue mission came in. Blatt took off but then returned when the covering air patrol told him an ax and fire extinguisher

Although operating at the extreme edge of their performance envelope, the Marine HO3S-1s of VMO-6 provided reliable service at the Chosin Reservoir. During the most critical period the squadron's helicopters and OYs provided the only physical contact between units separated by enemy action.

National Archives Photo (USMC) 127-N-A5398

would be needed to free the trapped pilot, Ensign Jesse L. Brown, USN. Arriving at the scene, Blatt joined Navy Lieutenant Junior Grade Thomas J. Hudner, Jr., who had purposely crash-landed his plane in order to assist Brown. Despite their best efforts, they could not extract the mortally wounded man before he died. The saddened men had to return empty handed, but Hudner later received the Medal of Honor for his unselfish actions to rescue the United States Navy's first African-American combat pilot.

After the 1st Marine Division departed Koto-ri for Hungnam on 6 December, VMO-6 moved back to Yonpo. During the ensuing voyage from Hungnam some of the squadron's helicopters were earmarked to conduct emergency rescues during carrier-borne air operations. On 12 December, the first elements of the squadron (including two helicopters) began to back load on board LST Q082 for immediate transportation to Hungnam, which would then be followed by a seaborne redeployment to Pusan. Seven helicopters remained behind until additional ship spaces could be found. On 17 December, three HO3Ss flew from their temporary home on the beach at Wonsan to the battleship *Missouri* (BB 63) and then each transshipped to three different carriers (the *Leyte* [CV 32], *Princeton* [CV 37], and *Philippine Sea* [CV 47]) for duty as standby plane guards, a fourth HO3S served the heavy cruiser *St. Paul* (CA 73). Three additional helicopters embarked on board the *Missouri* late in the day. Enroute, Lieutenant Colonel Richard W. Wyczawski, commander of Marine Fighter Squadron 212, was charged with overseeing the movement of VMO-6's "lost sheep" during the voyage to Pusan. They were successively located on board their various ships and then gathered together on board the light carrier *Bataan* (CVL 29) as the convoy sailed south. Unfortunately, three helicopters were damaged enroute by high winds and heavy seas. The four operable "whirlybirds" flew off their host carriers to Masan on 26 December 1950. The others were off loaded at Pusan Harbor and underwent repairs.

The return to Masan closed the books on the Chosin Reservoir campaign. During the movement north and the ensuing breakout between 28 October and 15 December, Marine helicopters flew 64 reconnaissance, 421 transport, 191 medical evacuation, 60 utility, and 11 search and rescue missions; more than 200 wounded men were flown out, most of whom would have died without speedy medical assistance. All of this, of course, could not have been possible without the outstanding support of the tireless ground crews aided by Mr. Harold Nachlin, the much-respected civilian technical representative from Sikorsky Aircraft. As impressive as these achievements were, however, the Chosin campaign once again pointed out the inadequacy of the HO3S as a military aircraft. A more effective medical evacuation platform was desperately needed, as was a viable transport helicopter. Fortunately, each of these was in the pipeline and would soon see combat service.

Pohang to the Punchbowl

The unexpected Communist Winter Offensive initiated the longest retreat in American military history. While X Corps pulled back from northeast Korea, the Eighth Army fell back more than 600 miles before halting south of Seoul. During the next eight months the U.S. Marines would rest and rebuild at Masan, chase elusive North Korean guerrillas near Pohang, lead the United Nations Command drive up central Korea from Wonju to the Hwachon Reservoir, survive the last major Chinese offensive of the war, then once again claw their way north to a rugged mountain area just north of the 38th Parallel where the U.N. lines would remain until the end of the war. Throughout those U.N. counteroffensives the helicopters of VMO-6 continued to provide outstanding support.

While the Marines in Korea were slogging their way back from Chosin, several Bell HTL helicopters arrived in Japan. The HTL was a two-seat, single-engine aircraft that was already familiar to every helicopter pilot because they had learned to fly helicopters using Bell-made trainers at Lakehurst and Quantico. These small "fishbowls" (so called due to their prominent plexiglass bubble canopies) mounted two evacuation pods, one on each side of the fuselage. This handy configuration made the Bells much better adapted for medical evacuation than the venerable Sikorskys. Unfortunately, their relatively underpowered engines were unsuited for high-altitude, cold-weather operations, so they were kept in reserve until the Marines returned to Pusan. Most of the older HTL-3s were assigned to headquarters or maintenance squadrons while all of the newer HTL-4s went to VMO-6. The plan was to gradually replace the HO3S-1s as HTL-4s became available. On 28 December 1950, three HTL-4s, two HTL-3s, and another HO3S-1 joined the ranks of VMO-6. First Lieutenant John L. Scott flew the first operational mission with an HTL-4 on 2 January 1951. As the New Year dawned, VMO-6 mustered 13 helicopters and nine

VMO-6 Historical Diary Photo Supplement, Nov52

Bell HTL

Thanks to the opening credits of the long-running television series "M*A*S*H," a helicopter delivering wounded men to a field hospital remains one of the most enduring images of the Korean conflict. The aircraft featured on that show was a Bell Model-47, the same type flown by the Marines under the designation HTL and by the Army and Air Force as the H-13.

The Model-47 first flew in 1946, was granted the first ever U.S. commercial helicopter license in 1947, and remained in production for almost 30 years. Military versions saw extensive service in both Korea and Vietnam, and several generations of naval aviation helicopter pilots learned to fly using HTLs. Early model HTL-2 trainers used at Lakehurst Naval Air Station, New Jersey, mounted wheels instead of skids and were covered in fabric when the first Marine trainees learned to fly rotary-wing aircraft. The Chief of Naval Operations designated the HTL as the prospective observation helicopter in 1949. The press of combat operations in Korea, particularly the need for a more suitable aerial medical evacuation platform than the HO3S, led to a massive influx of HTL-4s to Marine Observation Squadron 6 at the end of 1950.

The unique technical feature of all Bell helicopters was a two-bladed rotor and stabilizer system that reduced flying weight without harming performance, and the unique visual feature of the HTL was its clear Plexiglas "goldfish bowl" cabin canopy that allowed all-round vision. The HTL-4's squat configuration and skids allowed it to land in rough terrain while the inclusion of two exterior stretcher pods made it the preferred aircraft for field evacuations of seriously wounded men. Unfortunately, it had an unreliable engine and a notoriously weak electrical system that together required inordinate maintenance time while its limited fuel supply severely reduced the helicopter's combat radius.

Several generations of naval aviators learned to fly using HTL trainers, and the Bureau of Aeronautics eventually purchased more than 200 HTLs, the last of which were still regularly flying more than two decades after the first one took to the air. Advanced versions of the HTL developed into the UH-1 Huey and AH-1 Cobra, the utility and attack helicopters that arm today's Fleet Marine Forces.

Aircraft Data

Manufacturer: Bell Aircraft Company
Power Plant: 200 hp Franklin O-335-5
Dimensions: Length, 41'5"; height, 9' 2"; rotor, 35' two blade with stabilizer
Performance: Cruising speed, 60 mph; ceiling, range, 150 miles
Lift: Pilot plus two passengers or two externally mounted stretchers

National Archives Photo (USMC) 127-N-A130600

Capt George B. Farish, a helicopter pilot with VMO-6, stands by his Bell HTL. Farish, by late 1950, had participated in more than 100 combat missions and was responsible for the evacuation of more than 55 seriously wounded United Nations troops.

OY observation aircraft. An influx of fresh faces was a welcome sight as well because, according to Captain Gene Morrison, "the old hands . . . were . . . pretty tired" after six-months of grueling combat duty. Just as with the ground units, a significant personnel change was underway. The Regulars were giving way to recalled reservists. By the end of January 1951, the number of Reserve pilots in VMO-6 equaled the number of Regulars.

The 1st Marine Division spent a month recuperating throughout the uneventful respite at the Masan Bean Patch. During that time, VMO-6 operated from an airstrip near the waterfront. A maintenance detachment including four officers and 11 enlisted men moved from Korea to Itami Air Base in Japan to prepare the growing fleet of arriving helicopters for combat service. Most helicopter missions at Masan were utility and liaison flights, although occasional aerial reconnaissance and familiarization flights were also made. Concurrently, plans were being formulated for the Marines to move about 70 miles northeast to secure the X Corps eastern flank by conducting antiguerrilla operations near the coastal village of Pohang.

Helicopters proved invaluable for liaison work even before the 1st Marine Division moved to Pohang. Unfortunately, poor weather often hampered flying conditions. General Smith had several hair-raising encounters en-route to planning conferences, but he always arrived on time. Two HO3S-1s were tossed about by high winds as they carried General Smith's forward command group to meet with the new Eighth Army commander, Lieutenant General Matthew B. Ridgway, USA, on 30 December 1950 at Kyongju. They made it on time despite the harrowing flight conditions. On 8 January 1951, General Smith was summoned to a commander's conference at Taegu. Dense fog grounded all fixed-wing aircraft, so Smith boarded Lieutenant Lueddeke's HO3S for the flight. Lueddeke followed some dimly visible railroad tracks at about 400 feet, twice having to suddenly swerve to avoid mountainsides along the way. Once, the visibility was so reduced that Lueddeke had to put the plane down in a rice

One of the little noted, but important missions performed by VMO-6 helicopters was laying telephone wire between frontline positions. Here, a squadron ground crewman loads wire spools onto a HTL-4 flown by Capt James R. O'Moore.

Department of Defense Photo (USMC) A131086

paddy; Smith lit his pipe and made small talk while waiting to resume his journey. Not long thereafter, the pair took to the air once again; this time using roadside telephone posts to guide them.

In early 1951, the 1st Marine Division rooted out remnants of a North Korean division that had infiltrated the region surrounding Pohang and threatened X Corps headquarters at Taegu. Dubbed the "Pohang Guerrilla Hunt," the campaign sought to secure this area as it held the only usable port on Korea's southeastern coast, the main supply route for east-central Korea, and three vital airfields.

The VMO-6 ground support elements moved from Masan to Pohang by air, truck convoy, and ship beginning on 13 January 1951. The move was complete by 16 February. Pohang's mountainous and forested terrain hid the enemy who quickly broke up into small groups when the Marines arrived. The solution was saturation patrolling. The Marines sent out fire-team and squad-sized patrols operating from platoon- and company-bases to flush out enemy stragglers. Helicopters were used for observation, reconnaissance, laying wire, command and control, medical evacuations, re-supply of isolated small units, and transportation of fire teams to remote hilltops. The guerrillas were driven underground by relentless Marine pressure, but not decisively defeated. In the words of the official history: "In retrospect, had [a full] squadron of helicopters been available . . . its quick lift . . . increased mobility and surveillance would have made quite a difference in the conduct of action." Unrealized at the time, the use of helicopters at Pohang was actually a foretaste of the methods that would be used by the U.S. Marines and Army on a much larger scale in Vietnam more than a decade later.

National Archives Photo (USMC) 127-N-A131826

Capt Clarence W. Parkins points out the spot where he was forced to crash land his helicopter in the water during a test flight. Parkins later became VMO-6's acting commander.

The most notable helicopter incident of the guerrilla hunt occurred when First Lieutenant John Scott flew the first night medical evacuation by a Bell helicopter. There were several other nerve-wracking experiences as well. On 27 January, for example, an HTL-4 flown by Captain Harold G. McRay caught a skid on a low-strung cable and crashed while attempting to takeoff from Andong. The aircraft was wrecked but neither the pilot nor his passenger, Brigadier General Lewis B. "Chesty" Puller, who had been "frocked" to this rank the night before, were injured.

The helicopters of VMO-6 evacuated 59 men, most from the 7th Marines at Topyong-dong, between 25 and 31 January. Helicopter evacuations directly to hospital ships became routine operations. The advantages of this time-saving and life-saving method were enumerated by Captain John W. McElroy, USNR, the commander of the hospital ship *Consolation* (AH 15): "tests . . . conclusively proved the superiority of [helicopters for] embarking and evacuating patients to and from the ship. There was less handling in that patients were moved directly from airstrip to ship in one short hop, thereby eliminating . . . long and rough stages by boat and ambulance [and] 'choppers' [could] operate when seas were too rough for boat handling." When the *Consolation* returned stateside for an overhaul in July, a helicopter-landing platform designed by Marine Major Stanley V. Titterud was added and Marine pilots instructed the ship's company in proper landing procedures. Upon return her to Korean waters, a pair of Sikorsky H-19 (U.S. Air Force designation for the HRS) search and rescue helicopters were stationed permanently on board the *Consolation* to carry out medical evacuation flights. U.S. Army aircraft eventually replaced these Air Force helicopters. Operations became so smooth that it was not unusual for a litter case to be off the helicopter and on the way to the emergency room within a minute or less. Eventually, all hospital ships were similarly outfitted with landing platforms. There is no definitive tally as to how many seriously wounded men were saved due to the swift treatment afforded by the helicopters of all Services, but most estimates reach well into the hundreds.

On 1 February, Captain Gene Morrison made a daring night landing on the deck of the *Consolation*. The next day a similar evacuation flight to the *Consolation* almost ended in tragedy when a delirious patient became so violent that Captain Clarence W. Parkins had to make an unscheduled landing so he and the corpsman on board could subdue and bind the man. Parkins then resumed the mercy flight.

From Pohang, the Marines were tapped to lead IX Corps up the center of the peninsula during a series of limited objective attacks, Operations Killer, Ripper, and Rugged, collectively called the "Ridgway Offensives." These successive attacks, which began in late February and continued throughout March and April, gradually pushed the Communists out of the Som River Valley and back above the Hwachon Reservoir. During that time, VMO-6 followed in trail of the advance, successively moving forward from Pohang to Chung-ju, Wonju, Hongchon, and Chungchon, only to move back again when the Chinese mounted their spring offensives.

The Marines jumped off on 21 February, but traffic congestion delayed the arrival of Marine assault troops and hampered command and control. Luckily, General Smith had the use of a helicopter and was able to communicate directly with his subordinates and be present to observe the initial attack. In the words of Marine Corps historian Lynn Montross: "Only the helicopter . . . enabled General Smith to solve his time and space problems prior to Operation Killer. The division was required to move 150 miles by road and rail from Pohang to the objective area near Wonju in central Korea, with only one road being available for the last 30 miles."

Three days later, Marine General Smith was hurriedly summoned to the IX Corps advanced command post to take command after the commanding general died of a heart attack. This battlefield promotion, however, was only temporary until a more senior Army general arrived. Smith commandeered a Marine helicopter to use during his time at IX Corps. As he later explained: "at the Corps level the helicopter was even more essential for command purposes than at the division level."

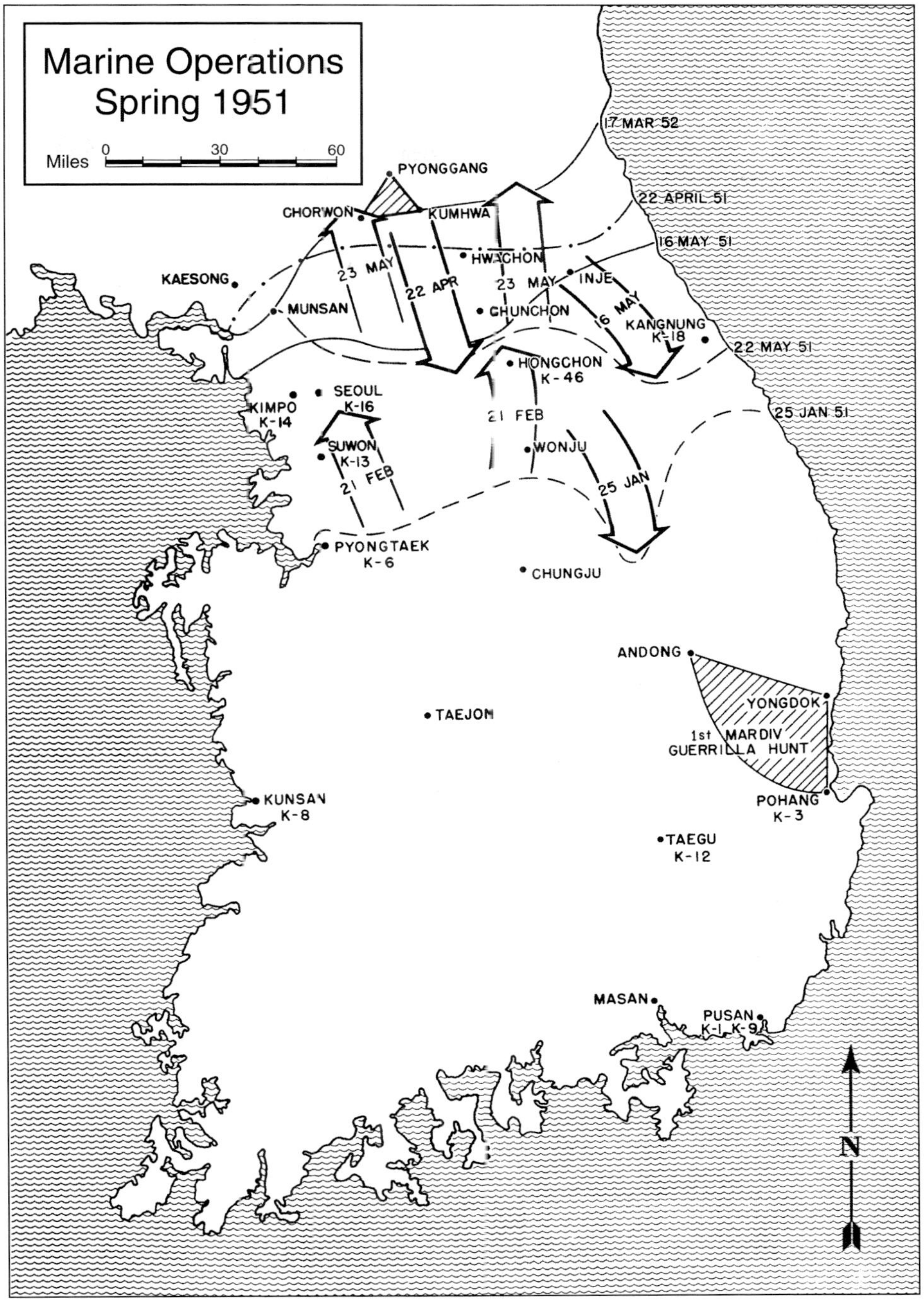

Just as before, although not an official task for observation squadrons, combat search and rescue missions remained a high priority. Captain Morrison picked up a Marine fighter pilot downed near Song-gol on 12 March. On 27 March, two Marine helicopters flown by Captain Norman C. Ewers and First Lieutenant Robert A. Strong were called out to conduct a search and rescue mission for an Air Force C-119 Flying Boxcar that had gone down behind enemy lines. They found the site, set down, picked up three injured crewmen, and recovered the body of a fourth airman. The impact of helicopters on operations in Korea was such that by that time this daring mission that would once have garnered stateside headlines, had become routine.

Between 1 January and 30 March, VMO-6 evacuated 539 wounded Marines (60 in January, 99 in February, and 370 in March). The helicopter section was extremely fortunate; it lost only two aircraft (General Puller's HTL-4 and an HO3S-1 lost to a takeoff incident on 12 March) and suffered no one killed in action. Unfortunately, the month of April was a tough one; three helicopters would be lost during heavy fighting.

April began with a command change for VMO-6. Major Gottschalk departed on the last day of March and the officer-in-charge helicopter section, Captain Clarence W. Parkins, became the acting squadron commander until the arrival of Major David W. McFarland who would command the squadron for the next six months. The squadron at that time numbered 28 officers and 125 enlisted men with nine OY observation aircraft, five HO3S-1s, and six HTL-4s.

The 13th of April was a busy day for helicopter search and rescue. First, Captain James R. O'Moore and Technical Sergeant Philip K. Mackert took off to search for a lost aircraft with the help of a flight of Marine Corsairs. They were unable to locate that pilot and one of the Corsair escorts was shot down. O'Moore set his HO3S down, then he and Mackert rushed over to try to save the pilot but it was too late. Later that day, Captain Valdemar Schmidt, Jr.'s HO3S-1 was brought down by enemy fire during a rescue mission about 20 miles behind enemy lines. Several hits from small arms fire caused a loss of power and control as the helicopter made its final approach. He crash landed in hilly terrain and his

As commanding officer of VMO-6, Maj David W. McFarland initiated night aerial observation flights by OY planes. Instead of the intended improvement in Marine artillery accuracy, the mere presence of an OY overhead would often silence enemy artillery.

National Archives Photo (USMC) 127-N-A131464

aircraft rolled over upon impact. Schmidt suffered only minor injuries, but his passenger, Corporal Robert Sarvia wrenched his leg, cut his hand, and went into shock. American aircraft circling above kept the enemy at bay with strafing runs until helicopter pilot Captain Frank E. Wilson arrived on the scene. Wilson picked up the two Marines in addition to the Air Force pilot they had came after and then made his precarious way back in the dark, flying an overloaded helicopter without navigational aids. Jeeps, trucks, and flares lit the field for Wilson's returning aircraft.

Not every mission had a happy ending. Sometimes, despite great effort on the part of helicopter pilots, a rescue could not be made. On 14 April, Captain Gene Morrison made three attempts to pick up a downed pilot, but his HO3S was turned away by enemy fire each time. Captain Norman Ewers then tried, but he took so many hits he had to return to base empty handed as well. Plans were made to rig a stretcher to lift the pilot out the next morning, but inclement weather intervened. When OY aircraft flying over the target area could not locate the man, the helicopter rescue was scrubbed.

On the night of 22 April, the Chinese mounted their long expected Fifth Phase Offensive. When a South Korean unit on the Marines' left flank broke and ran, the 1st Marine Division pulled back and formed a semi-circle on the high ground to defend several vital river crossings. The bitter fighting, collectively known as the battle of Horseshoe Ridge, was marked by fierce hand-to-hand combat and several last ditch defensive stands by isolated units that equaled the combat intensity at the Naktong bulge or the Chosin Reservoir. The division suffered about 500 casualties in three days fighting.

The last days of April found the helicopters of VMO-6 busily evacuating wounded men from dawn until dusk in an all-hands effort until the Marines reached the No Name Line. At about 0600 on the 23d, all helicopters were airborne and most continued operations throughout the day with 36 individual flights made (15 by HO3S-1s and 21 by HTL-4s). Fifty wounded Marines were evacuated. Captain Dwain L. Redalen logged 18 evacuations in almost 10 hours of flying; First Lieutenant George A. Eaton was a close second with 16 men brought out. The next day an HTL-4 was lost to enemy fire when First Lieutenant Robert E. Mathewson was shot out of the sky as he attempted a medical evacuation. Enemy fire hit the engine, instrument pedestal, and tail sections rendering Mathewson's aircraft uncontrollable as he hovered over the air panels set out to mark the landing zone. Mathewson crash-landed but was uninjured. Lieutenant John Scott, who set a record with 18 evacuations in one day, tried to fly

Capt Norman G. Ewers receives word that a helicopter is needed in the forward area for a reconnaissance mission. During daylight hours VMO-6 pilots stood by with elements of the 1st Marine Division, maintaining constant contact with tactical air controllers by field telephone.

National Archives Photo (USMC) 127-N-A131065

National Archives Photo (USMC) 127-N-A132015

1stLt Joseph C. Gardiner, left, an HO3S-1 pilot with VMO-6, is awarded a Navy Commendation Ribbon and a pair of gold stars denoting second and third awards of the Air Medal for combat actions during the Inchon-Seoul campaign. Marine helicopters played an important role in the drive inland by providing transportation, medical evacuation, and visual reconnaissance for the 1st Marine Division.

in despite the danger, but was waved off by Mathewson who then picked up a rifle and temporarily joined the infantry. His crippled aircraft was destroyed by demolitions before the Marines departed. Thirty-two helicopter missions were flown, and about another 50 seriously wounded were evacuated by Mathewson's fellow pilots.

The United Nations Command briefly regrouped behind the No Name Line, repelled a second Communist offensive, then once again set off north—this time heading the Kansas Line along the 38th Parallel. Non-stop fighting had exhausted the enemy and his forces were seriously depleted after suffering grievous losses in the recently concluded spring offensive. The desperation of the enemy was evident as unprecedented numbers of them began to surrender. This time it was the Communists who were "bugging out." By the end of June, the United Nations Command was once again about to enter North Korea. At that point, the Communists called for a cessation of offensive actions as a prelude to peace talks. The United Nations accepted this condition, and the fighting forces of both sides temporarily settled down along a line not far from the original pre-war border between the two Koreas.

During August, VMO-6 operated from Songjong until the 28th, then moved to Sohung. The month saw several rescue missions. First Lieutenant Joseph C. Gardiner, Jr., picked up a downed Marine fighter pilot on 12 August. On 28 August, Major Kenneth C. Smedley used his HTL-4 to pull two communications men stranded on a small island in the middle of a rapidly rising river out of harm's way. That same day, Captain Frank E. Wilson lost control of his HTL-4 when a crewman jumped out of the hovering aircraft during an attempted rescue. Captain Frank G. Parks was credited with saving several lives by delivering whole blood in darkness on 29 August despite the fact his helicopter had no lit instrumentation, no landing lights, and no homing locator.

When peace talks broke down in September, Lieutenant General James A. Van Fleet, USA, commander of the Eighth Army since mid-April, mounted a series of limited attacks intended as much to pressure the Communists back to the peace table as to secure dominating terrain just north of the Kansas Line. The Marine sector featured a volcanic depression known as the Punchbowl. Its capture was a bloody three-week slugfest fought over nearly impassable roadless mountain terrain, so helicopters were much in demand. Marine pilots were at risk as they courageously defied enemy fire on their missions of mercy. The HO3 and HTL helicopters delivered small loads of medicine, ammunition, and radio batteries to the front and then brought out 541 severely wounded men. Another frequent mission was the delivery of whole blood to forward-deployed Medical Companies A and E of the 1st Medical Battalion.

On 16 September the light helicopters of VMO-6 evacuated 85 men. First Lieutenant Joseph Gardiner led the pack with 17 medical evacuations. Major Edward L. Barker's HTL was hit by enemy artillery as he tried to lift out a pair of wounded Marines. He escaped without injury, but one of his passengers succumbed to his wounds before reaching medical sanctuary. The following day, Captain William G. Carter's HTL-4 crashed while conducting an emergency medical evacuation. Ground personnel attempting to assist the landing on rough terrain grabbed the helicopter's skids but inadvertently tipped the aircraft causing it to crash. The aircraft was lost and the pilot suffered non-threatening injuries. Captain Gilbert R. Templeton's HO3S-1 was hit by enemy fire during a resupply mission on 21 September; Templeton was able to return to base for repairs, but the mission had to be scrubbed. Major Kenneth C. Smedley, the squadron's executive officer, crashed when his HO3S-1 lost hover and set down hard on uneven ground. When the plane began to slip over the steep cliff, Smedley had to intentionally roll the helicopter on its side to stop its descent. Neither he nor his passenger was injured, but the helicopter was wrecked.

The fighting for the Punchbowl lasted until late September. After that, both sides settled down and began to dig in. The capture of the Punchbowl marked the last major offensive action by the Marines in Korea.

As the first year of the Korean War came to a close there could be little doubt that the helicopter was the most important tactical innovation to date. The plucky little aircraft had proven themselves adaptable, versatile, and survivable. The ability of the helicopter to traverse difficult terrain, to land in tight spots, and to rapidly scout unfamiliar territory made it the preferred mode of transportation for generals and colonels; downed pilots could look forward to being hoisted out of the freezing water or grabbed up from behind enemy lines with a certainty never before experienced; and almost 2,000 men had been lifted to hospitals with in a few hours of being wounded, a factor that greatly increased survival rates. There was little doubt the helicopter was here to stay, but thus far in the war the "whirlybirds" had not yet been used for their proposed main missions and original raison d'etre: vertical en-velopment and assault support. This was due to the inadequate lift of the machines currently available, but that was about to change as the war entered its second year.

Arrival of HMR-161

Marine Transport Helicopter Squadron 161 (HMR-161) was the first transport helicopter squadron in history. It was also the first full helicopter squadron committed to combat. Mounted in brand new Sikorsky HRS-1 helicopters, HMR-161 arrived in Korea in early September 1951 and was soon testing new operational methods under actual combat conditions, a little more than one year after Brigadier General Edward Craig's original recommendation that such a squadron be sent into combat. The squadron's arrival at that particular juncture in the war was fortuitous because the 1st Marine Division, then slogging its way north against stubborn Communist resistance in the mountains of east-central Korea, was led by two early and very influential proponents of helicopters—division commander Major General Gerald C. Thomas and his chief of staff Colonel Victor H. Krulak. Both Marines were plank holders in the helicopter program; from Washington, D.C., and Quantico, Virginia, they pushed for adoption of rotary-winged aircraft and created a test-bed squadron immediately after the war. Krulak helped write initial helicopter doctrine and drew up many of the first operational plans used by HMX-1, while Thomas pushed for expanded helicopter development at Head-quarters Marine Corps in the immediate post-war period, then gained practical experience in their use at Quantico after his return from China in the late 1940s. Both men were known throughout the Corps as innovators and visionaries, but they also garnered reputations for thorough planning and meticulous execution of those plans. In retrospect, it was clear that HMR-161 and the 1st Marine Division formed a perfect match.

Plans to create transport helicopter squadrons had been on the board well before the outbreak of the Korean War. In fact, early post-war planners envisioned a Marine helicopter aircraft wing comprising 10 squadrons with 24 helicopters each. The proposed machines should be able to carry 15-20 men or 4,000 pounds of cargo. This was no small order because that number of aircraft just about equaled the entire American helicopter production to that time and no existing helicopter could come close to lifting the specified number of troops or amount cargo. The main sticking points were lack of funds, a ceiling on aircraft procurement, and—most importantly—lack of a suitable aircraft. The demands of the Korean War loosened up funding and virtually eliminated aircraft procurement restrictions. Thus, the only remaining roadblock became the machines themselves.

Long-range plans in the late 1940s called for the creation of up to six transport helicopter squadrons by the mid-1950s. This leisurely pace was driven as much by technology as by anything else. The Marines wanted a reliable, high-performance, heavy-lift helicopter to carry cohesive tactical units ashore from escort carriers and then rapidly build up supplies within the beachhead. The problem was the machines of the day were too limited in range, lift, and

National Archives Photo (USN) 80-G-433339

Sikorsky HRS-1

The HRS transport helicopter was the military version of the Sikorsky S-55 commercial aircraft. It featured the familiar Sikorsky design signatures, a single overhead main rotor and a small anti-torque rotor on the tail boom. Although many of its components were simply enlarged versions of similar ones found in the HO3S, the HRS did not look much like the Marines' earliest observation helicopter. It was much larger, its cargo space included seats for eight passengers, the two-seat cockpit was located high on the fuselage and set farther back than the HO3S, and the engine was mounted low on the front of the aircraft rather than high amidships. Although initially selected as only an interim model until a larger heavy-lift helicopter became available, the Navy Department eventually purchased 235 variants of the S-55. The U.S. Army and Air Force flew similar models as H-19s, and the Coast Guard variant was the HO4S-3G.

The Marine Corps turned to the Sikorsky S-55 after its first choice, the Piasecki H-16, outgrew the ability to operate from small escort carriers—foreseen as the transport helicopter's primary mission. The Navy was already looking at one version of the S-55; an antisubmarine variant designated the HO4S. There was no obvious external difference between the HRS and the HO4S. This was because the main difference was each respective aircraft's mission. The Marine transport helicopter did away with mine detection equipment but mounted troop seats and had self-sealing fuel tanks. The most innovative feature of the S-55 was its engine placement. It was set low in the helicopter's nose. A drive shaft ran up through the back of the cockpit to provide power to the three-bladed overhead main rotor. The engine placement made it easy to reach, cutting maintenance time. That configuration also eliminated critical center-of-gravity problems that plagued both the HO3S and the HTL. The HRS also mounted a drop hook to carry external loads under the cabin. The main shortfalls of the HRS were that the machine was underpowered and mechanical failures required them to be grounded on several occasions. No Marine HRSs were lost to enemy fire, but several crashed while hovering and at least two went down in mid-air due to engine failure.

The HRS was a great step forward, but it was not the transport helicopter Marine planners envisioned. They wanted an aircraft that could carry 15 or more men to ensure unit integrity during assaults and generating enough lift to carry most division equipment. The main problem with the HRS was lifting power. Although rated for eight passengers, in the harsh reality of the Korean mountains the HRS could only carry about six men—only four if they were fully combat loaded. Both Igor Sikorsky and Frank Piasecki worked feverishly to deliver a more capable aircraft, but that advance would have to wait until the development of a practical turbine helicopter engine.

The first batch of Marine HRS-1s included 60 machines and the second order of HRS-2s mustered 91, the final version (HRS-3) included 89 more. Only the first two variants saw action in Korea, but some HRS-3s were still in the Marine inventory when their designation was changed to the CH-19E in accordance with the Department of Defense unified designation system in 1962.

Aircraft Data

Manufacturer: Sikorsky Aircraft Division of United Aircraft Corporation
Type: Transport helicopter
Accommodation: Ten-places (two crew and eight passengers)
Power Plant: One 600 hp Pratt & Whitney R-1340-57
Cruising speed: 80 mph
Payload: 1,050 pounds

avionics. Frank Piasecki's tandem rotor helicopters seemed to offer the best potential. However, the development of an improved version of the Flying Banana was taking too long, and its projected size was not compatible with escort carrier deck space. The Marines, therefore, reluctantly opted to go with an interim transport helicopter until a more capable aircraft became a reality. The machine they chose was a variant of the Sikorsky model S-55, which was already in naval service as the HO4S. The HO4S featured the standard Sikorsky frame: a single overhead rotor with a tail-mounted anti-torque rotor. Many of its components were little more than larger versions of those of the HO3S, but a front-mounted engine greatly enhanced ease of maintenance and in-flight stability. Luckily, the antisubmarine warfare HO4S helicopter required only minor modifications to meet Marine Corps requirements. A Marine assault transport helicopter, designated the HRS, was created by eliminating the antisubmarine warfare suites and then adding self-sealing fuel tanks and placing troop seats in the cargo bay. An initial order for 40 HRS-1s was sent to Sikorsky Aircraft in July 1950. The "interim" tag, however, may have been premature. Every U.S. Armed Service and many of our allies eventually used the S-55 (designated H-19 by the Army and Air Force), and 235 HO4S/HRS variants entered naval service over the next decade.

On 15 January 1951, the first Marine transport squadron was formed at Marine Corps Air Station El Toro. The unit tentatively was designated HMR-1 ("H" for helicopter, "M" for Marine, "R" for transport, and "1" for first), but that name was changed before the squadron became operational. The new squadron was given the prefix "1" because it would be assigned to the 1st Marine Aircraft Wing; the middle number "6" was adopted because the highest fixed-wing designator to that time had been "5"; and the last "1" indicated it was the first squadron formed, thus the new squadron became HMR-161. The commanding officer was Lieutenant Colonel George W. Herring, the former executive officer of HMX-1. A mix of regulars and reservists populated the new transport helicopter squadron. Most of the pilots, like the squadron's executive officer Major William P. Mitchell, had been fixed-wing pilots in the Pacific. Lieutenant Colonel Herring, however, had received the Navy Cross as a Marine raider before receiving his wings. While the mix of regular and reserve pilots was about equal, most of the squadron's enlisted personnel were reservists. The squadron trained at the Navy's former lighter-than-air base located at

In July 1951, Marine Helicopter Transport Squadron 161 staged a helicopter demonstration for the press at Camp Pendleton, California. Its purpose was to show how helicopters would be used in modern warfare as envisioned by the Marine Corps.

Marine Corps Historical Center Photo Collection

Marine Corps Historical Center Photo Collection

A Marine Sikorsky HRS-1 transport helicopter is loaded on board the escort carrier Sitkoh Bay *(CVE 86) at San Diego, California, for the journey to Korea. The arrival of HMR-161 and the HRS-1 would mark a new era in Marine airborne support to ground troops.*

Tustin, California, not far from Camp Pendleton while waiting for its new helicopters. The squadron gradually built up to its full strength of 43 officers and 244 enlisted men flying 15 HRS-1 helicopters before receiving orders to prepare to ship out for Korea in July 1951.

The squadron embarked at San Diego on 16 August with the helicopters and aircrews on board the escort carrier *Sitkoh Bay* (CVE 86) and the equipment and a working party on board the civilian-manned cargo ship *Great Falls*. The squadron arrived at Pusan on 2 September. In Korea, HMR-161 came under the administrative control of the 1st Marine Aircraft Wing and the operational control of the 1st Marine Division, the same command and control arrangements used by VMO-6. Four days after landing, HMR-161 moved from airfield K-1 (Pusan East) to airfield K-18 (Kangnung Airdrome) in central Korea. From there, the advance echelon moved by truck and air to X-83 at Chodo-ri, an auxiliary airstrip not far from the division headquarters, already hosting VMO-6. A rear echelon remained at K-18 to conduct advanced maintenance and make complex repairs.

The fact that HMR-161 was even in Korea was at least partially due the efforts of Major General Thomas and Colonel Krulak who actively pushed to speed the pace of getting transport helicopters into the combat zone. Thomas and Krulak were well aware of the technical limitations of the HRS-1 and the demands of Korea's difficult weather and rugged terrain, so they began testing its abilities slowly. The initial helicopter operations were modest ones to test the waters, carefully conducted with little risk. First came a couple of resupply efforts well shielded from enemy observation and direct fire. Next came small-scale troop lifts, eventually increasing to battalion-sized movements. Tactical innovations were also on the agenda: counter-guerrilla activities; a night assault; and rapid movement of rocket batteries. It was not long before a division of labor emerged. The smaller aircraft of VMO-6 concentrated on medical evacuations, reconnaissance, observation, and liaison work, while HMR-161 conducted aerial resupply, moved troops, and experimented with vertical envelopment. Although the HRS could do everything its smaller kin could, medical evacuations and combat search and rescue were secondary missions for HMR-161. This was possible because of the static nature of the fighting. In fact, the combat situation eventually became stable enough that it was possible to increase emphasis on amphibious training even though the squadron remained in the combat zone, a factor that lent elements of realism and urgency to the helicopter training program that were probably not present at Quantico, Virginia, or Onslow Beach, North Carolina. The stunning success in Korea of helicopters used for assault support silenced critics and converted skeptics. In the words of historian

National Archives Photo (USMC) 127-N-A156740

LtCol George W. Herring, right, commanding officer of HMR-161, is welcomed to Korea by LtCol Edward V. Finn, the 1st Marine Division's air officer. LtCol Herring commanded the world's first transport helicopter squadron used in combat.

Lynn Montross, with the introduction of HMR-161 to Korea "a new era of military transport had dawned."

The first order of business was to conduct familiarization flights so the pilots could become accustomed to the terrain and get a feel for the tactical and operational conditions at the front. The veteran pilots of VMO-6's helicopter element indoctrinated the new men of HMR-161 in flying conditions and combat procedures. Also during this time various potential landing zones and flight routes were identified. While the pilots were busy flying, selected members of the shore party battalion became familiar with helicopter landing and loading procedures while planners met to prepare for the squadron's first combat operation. General Thomas wisely decided to use a series of cautious activities until both the helicopter crews and ground units got up to speed, he then pushed an aggressive agenda featuring a wide variety missions that became progressively more complex and that thoroughly tested existing operational procedures and new theories for helicopter employment.

The initial combat operation by HMR-161 took place only two weeks after its arrival. It was dubbed Operation Windmill to honor the HRS's unofficial nickname, "Flying Windmill." Mindful of the chaotic experiences of the first Packard exercise at Camp Lejeune, North Carolina, and well aware of the dictates of *Phib-31*, Krulak and Thomas ensured the new transport helicopters would

MajGen Gerald C. Thomas, center, in command of the 1st Marine Division, discusses plans for using the new 10-place Sikorsky HRS-1 helicopters with the leaders of HMR-161, LtCol George W. Herring, the commanding officer, right, and his executive officer, Maj William P. Mitchell. Thomas was instrumental both in bringing helicopters into the Marine Corps and getting the first Marine helicopter transport squadron to Korea in 1951.

National Archives Photo (USMC) 127-N-132027

Department of Defense Photo (USMC) A156768

Vital supplies are transferred from a truck to a helicopter for delivery to frontline troops. Helicopters often offered the only practical way to supply positions in the trackless mountain terrain near the Punchbowl.

be carefully integrated into a Marine air-ground combat team, not just used as a "nice-to-have" aviation adjunct as was sometimes the case with VMO-6's light helicopters. One of the first steps in this process was to train elements of the 1st Shore Party Battalion for helicopter operations. Shore parties had been formed during World War II to handle supplies coming ashore by landing craft. The logical extension of this mission to landing zones as well as landing beaches eventually led to the formation of specially trained helicopter support teams. In addition, the energetic division chief of staff, Colonel Krulak, held a series of planning conferences with the 1st Marine Division staff even before HMR-161 was in Korea to draw up tentative standard operating procedures. Ground units needed to learn the intricacies of helicopter movement and their leaders were encouraged to apply the unique capabilities of helicopters in tactical situations. Before HMR-161 left Korea, its helicopters had performed virtually every mission envisioned under operational conditions. The squadron's main functions, however, were to test the practicality of vertical envelopment and to practice assault support by ferrying troops and delivering supplies to units in the field. The latter was the most exercised mission while on the East-Central Front. After moving to western Korea in 1952, emphasis eventually shifted to vertical envelopment using a continuing series of amphibious exercises. These exercises and combat operations were the foundation of the sophisticated airmobile tactics and techniques still used by the U.S. Army and Marine Corps of today.

In September 1951, Marines were clearing the enemy from a series of ridges around an extinct volcano called the Punchbowl. The ground battalion commander, Lieutenant Colonel Franklin B. Nihart described the difficult tactical situation:

> We were attacking from Hill 673 toward Hill 749 . . . our supply and evacuation route was four miles of mountainous foot trails. The only way to keep supplies moving . . . was by using Korean Service Corps porters. . . . [They] could not keep up with the logistical demands imposed by heavy casualties and high ammunition expenditure [so] HMR-161 was called in to fill the . . . gap.

On 12 September, the first combat helicopter support team—a platoon from 1st Shore Party Battalion—attended briefings about proper loading techniques and learned how to transmit landing signals to incoming aircraft. The next morning was devoted to arranging supplies into 800-pound bundles. The first flight consisting of four helicopters made its way about seven miles and then deposited the shore party landing point section to enlarge and improve the landing zone, direct landing operations using hand signals, unload arriving helicopters and collect cargo bundles, establish supply dumps, and load battle casualties. In mid-afternoon, seven HRS-1s began lifting off with cargo loads suspended from belly hooks. The ingress and egress routes followed a deep valley that masked the helicopters from direct enemy observation. A restrictive fire plan was in effect to avoid friendly fire. The landing zone was marked with fluorescent panels, but the first incoming aircraft could only place two of their four wheels on the landing platform, which was situated on the reverse slope of a steep hill. The first supply helicopter dropped its sling at 1610 and then picked up seven battle casualties (two stretcher cases and five walking wounded). Operation Windmill I comprised 28 flights that delivered 18,848 pounds of supplies and evacuated 74 seriously wounded men. The elapsed time was two hours and 40 minutes with a total of 14.1 flight hours logged. Lieutenant Colonel Nihart's final evaluation of HMR-161's first combat action in support of his battalion was that "they . . . performed admirably."

Continued fierce fighting in the vicinity of the Punchbowl, particularly for an outcropping dubbed "the Rock," led to the second transport helicopter assault support mission. Spurred on the unquestioned success of Windmill I as well as the need for heavy fortification materials such as sand bags, timber, barbed wire, and land mines, it was decided to conduct a follow-on aerial supply operation, Windmill II. The need to move bulky fortification materials to a nearly inaccessible position drove operational planning. Sand bags, barbed wire, land mines, and timber were all too cumbersome and heavy to be moved forward by Korean laborers so General Thomas turned to his rotary-winged "mule train" for the second time in a week. The formal request was made on the morning of 19 September, approved before noon, and underway before nightfall. Ten HRS-1s delivered more than 12,000 pounds of cargo using 16 flights in about one hour. The same operational procedures for Windmill I were used: an advance helicopter support team was inserted to operate the landing zone; the helicopters used covered and concealed routes; and material was delivered using sling-loaded bundles for speed and ease of handling. The major difference was the rapid planning process, this time preparations took only a few hours instead of several days. Once again, the helicopters of HMR-161 did within a few hours what would have taken the trail-bound South Korean porters several days.

With the ability of HMR-161 to deliver supplies fully established, the next evolution was to lift human cargo. This was Operation Summit. The mission was for the 1st Marine Division reconnaissance company to replace a South Korean unit occupying Hill 884, a key observation post located atop a rugged mountain. It was estimated that it would take a Marine rifle company about 15 hours to scale the roadless heights with all resupply thereafter accomplished either by foot or by helicopter. General Thomas decided instead to mount the first combat helicopter troop lift in history.

Once again, careful planning and preparation were the hallmarks of this operation. Lieutenant Colonel Herring and Major Mitchell coordinated their tactical plans with Major Ephraim Kirby-Smith (the ground unit commander) and worked out the loading plans with First Lieutenant Richard

Marine transport helicopter "HR-10," one of 15 HRS-1s assigned to HMR-161, lands to deliver supplies near the Punchbowl. Detachments from the 1st Air Delivery Platoon loaded the supplies while helicopter support teams from the 1st Shore Party Battalion controlled the unloading at forward area landing zones.

Department of Defense Photo (USMC) A131993

National Archives Photo (USMC) 127-N-A156727

A Sikorsky HRS-1 transport helicopter delivers supplies using "sling loading" techniques. Sling loading employed prepackaged materials that were carried in nets, lifted by a powered winch, and dropped by a remotely controlled hook that allowed helicopters to rapidly deliver vital supplies without landing.

C. Higgs, representing the division embarkation section. Aerial reconnaissance indicated the landing was going to be a tight squeeze. The only two available spots were located some 300 feet below the topographical crest about a football field length apart, and each was less than 50-feet square with a sheer drop on two sides. Terrain limited each landing zone to one arrival at a time. Operational planning was based upon the dictates of *Phib-31* and practical experience during the Windmill operations. The landing force would consist of a reinforced reconnaissance company. Helicopter support teams from the 1st Shore Party Battalion would control loading and unloading. Landing serials were compiled and rehearsals began on 20 September. H-hour was slated for 1000 the next day.

Several problems soon became apparent. First was the number of troops each helicopter could carry. The HRS-1 was rated to carry eight combat-loaded troops but practical experience in California and Korea quickly showed this figure to be overly optimistic. The actual safe load was six men carrying only small arms and personal equipment. A second problem was weather. The threat posed by high winds and the possibility of limited visibility or rain influenced operations. An additional problem in the mountainous region was reliable radio communications. The solution was to earmark one helicopter as a radio relay aircraft, the first use of a helicopter for air-to-air command and control during ground operations.

Operation Summit was delayed on the morning of 21 September by dense ground fog. Finally, about a half-hour later than expected, the first wave of three helicopters at landing field X-83 departed for the 14-mile run to Hill 884. They approached their objective flying low along a streambed between the ridgelines and then hovered over Hill 884's reverse slope. A security element went hand-over-hand down knotted ropes and then fanned out. Next in were two landing site preparation teams. About 40 minutes later, idling helicopters at X-83 received word to begin loading. Each carried five riflemen. Two hundred and twenty four men, including a heavy machine gun platoon, and almost 18,000 pounds of equipment were brought in using 12 helicopters requiring a total elapsed time of about four hours. The bulk of the equipment and supplies were delivered by suspended cargo nets, which had to be released on top of the mountain because the hillsides were so steep. This took place in full view of the enemy, but no helicopters were hit by enemy fire. The final touch was the airborne laying of two telephone lines in about a quarter hour from Hill 884 to the ground command post more than eight miles away. This would have been a daylong task for a wire party on foot. The event was headline news in the States, and congratulations from higher headquarters poured in: General Shepherd noted Operation Summit was "a bright new chapter in the employment of helicopters"; and X Corps commander Major General Clovis E. Byers claimed: "Your imaginative experiment with this kind of helicopter is certain to be of lasting value to all the Services."

Holding the Minnesota Line

In late September 1951, the United Nations Command once again halted offensive operations. All across the trans-peninsular frontline troops began digging in. Soon, a series of interconnected trenchlines reminiscent of World War I extended from the Sea of Japan in the east to the Yellow Sea in the west, and the U.S. Eighth Army was prohibited from launching new attacks. Ground activities were limited to conducting daily foot patrols, mounting tank-infantry raids, manning small outposts, and setting up nightly ambushes. The overriding tactical concern was a penetration of the main line of resistance by the Communists. The war in Korea had once again entered a new phase; but, unlike the others, this one would last from the fall of 1951 until the ceasefire almost two years later.

The 1st Marine Division was assigned 22,800 yards of front along the northern edge of the Punchbowl with orders "to organize, construct, and defend" the Minnesota Line. Much of the main line of resistance ran through roadless mountains, and the re-serve regiment was located almost 17 miles to the rear. With his manpower stretched to the limit and terrain and distance limiting rapid overland reaction by reserve

An aerial view of auxiliary airfield X-83 located near Chodo-ri, not far from 1st Marine Division headquarters. Both helicopter squadrons (VMO-6 and HMR-161) shared this forward airfield situated near the Punchbowl.

HMR-161 Historical Diary Photo Supplement, Nov-Dec51

National Archives Photo (USMC) 127-N-A132540

All veteran World War II fighter pilots, the officers of Marine Helicopter Transport Squadron 161 pose with their commanding officer, LtCol George W. Herring, kneeling in front left, and the squadron's executive officer, Maj William P. Mitchell.

forces, General Thomas turned to HMR-161 to help solve his time and distance problems. He decided to test-lift a single rifle company. In addition, since most Chinese attacks occurred under cover of darkness, this helicopter lift would take place at night.

Once again careful planning and detailed rehearsals were conducted. Fortunately, the HRS-1, unlike the light utility helicopters of VMO-6, had flight attitude instruments, albeit not the sophisticated instrumentation found on fixed-wing aircraft. Daylight reconnaissance of the operational area, daytime practice inserts, and night indoctrination flights were conducted. The helicopter embarkation zone was a dry riverbed southeast of Hill 702, and the landing zone was located near the northwest rim of the Punchbowl. The straight-line five-mile ingress route, however, actually became a 13-mile round trip due to tactical considerations.

A daylight rehearsal on the morning of 27 September got Operation Blackbird off to an inauspicious start. Six helicopters lifted more than 200 men into a 50-by-100-foot area cleared by a provisional helicopter support team. This practice lift took about two hours. During the march out, however, a rifleman detonated an antipersonnel mine, and subsequent investigation revealed that the proposed route to the main line of resistance was seeded with unmarked mines. It was decided to change the ground scheme of maneuver but to keep the helicopter landing zone the same.

At 1930 on the 27th, Operation Blackbird, the first night combat helicopter troop lift in history, got underway. Departing at three-minute intervals as they shuttled between the departure and arrival landing zones, each aircraft carried five riflemen. Different altitudes were used for ingress and egress to avoid collisions, and running lights were switched on for two minutes as aircraft neared the landing zones. Only two hours and 20 minutes were required to lift all 223 men, a movement that would have required at least nine hours by foot.

Unfortunately, there were many problems. Rotor wash blew out many of the flare pots that illuminated the embarkation area, battery-powered lanterns in the landing zone were inadequate, windshield glare temporarily blinded the pilots, artillery flashes distracted the pilots as they wormed their way through the high mountain ridges, and many in-bound pilots needed radio assistance to find the landing zone. As the squadron's after action report candidly stated: "Night lifts are feasible with present equipment [but they] should be limited to movements within friendly territory." Although the operation was a marginal success that affirmed the possibility of emergency night reinforcement and intermittent night indoctrination flights continued, Operation Blackbird was the only major night helicopter troop lift conducted in Korea.

The next day, HMR-161 lost its first helicopter to an operational mishap. A dozen light helicopters (HO3S-1s and HTLs) had been previously lost to enemy fire and operational mishaps, but this was the first HRS to go down. The HRS-1 piloted by Major Charles E. Cornwell and First Lieutenant Frederick D. Adams came into the landing zone too low. The helicopter struck the ground, bounced into the air, canted on its side losing all lift, and then careened to the ground and caught fire. The flaming helicopter was a total wreck, but both pilots escaped without injury.

HMR-161 pilots and crewmen attend a pre-mission brief for Operation Blackbird, the first major night helicopter troop lift. Careful planning and rehearsals were conducted to test the feasibility of night helicopter operations; after action reports noted it was possible in an emergency, but Blackbird was the only night helilift actually carried out.

HMR-161 Historical Diary Photo Supplement, Nov-Dec51

The ability to rapidly move a single rifle company had been established by Operations Summit and Blackbird, so Thomas and Krulak were eager to see if the same principles could be applied to a larger lift. On 9 October, a warning order for Operation Bumble-bee, the lift of an entire rifle battalion, was issued. Second Lieu-tenant Clifford V. Brokaw III, at that time an assistant operations officer with the 7th Marines, recalled that the genesis of the operation actually occurred much earlier when Colonel Krulak inquired if helicopters could support a frontal attack. Colonel Herman Nickerson, the regimental commander replied with a firm "no!" While in reserve, however, the regiment was tasked to prepare an amphibious contingency plan including a helicopter lift. Then, at Krulak's insistence, that plan was adapted to provide for the heliborne relief of a rifle battalion on the main line of resistance. Well aware that such a major helicopter event would become headline news, the division public relations officer asked what the operation was going to be called. Sergeant Roger Hanks, a former University of Texas football player, mindful of the many vociferous skeptics who questioned the viability of helicopters for combat duty, quickly piped up: "Bumblebee because supposedly they can't fly either."

Colonel Krulak headed the planning group that included Lieutenant Colonel Herring and Major Mitchell from HMR-161, Lieutenant Colonel Harry W. Edwards, the rifle battalion commander, and Lieutenant Colonel George G. Pafford, the shore party battalion commander. Bumblebee was planned as if it were an amphibious operation. Assignment and loading tables were carefully constructed, detailed arrival and departure schedules were prepared, and helicopter loading and unloading serials were established with each person assigned a specific spot in the helicopter, and order of embarkation and debarkation charts were distributed. Lieutenant Brokaw recalled that this time eight troops, carrying only small arms and limited ammunition loads, were squeezed in and only one pilot flew each helicopter to test if such "surge loading" was practical in an emergency. Familiarization classes and rehearsals were held on 10 October.

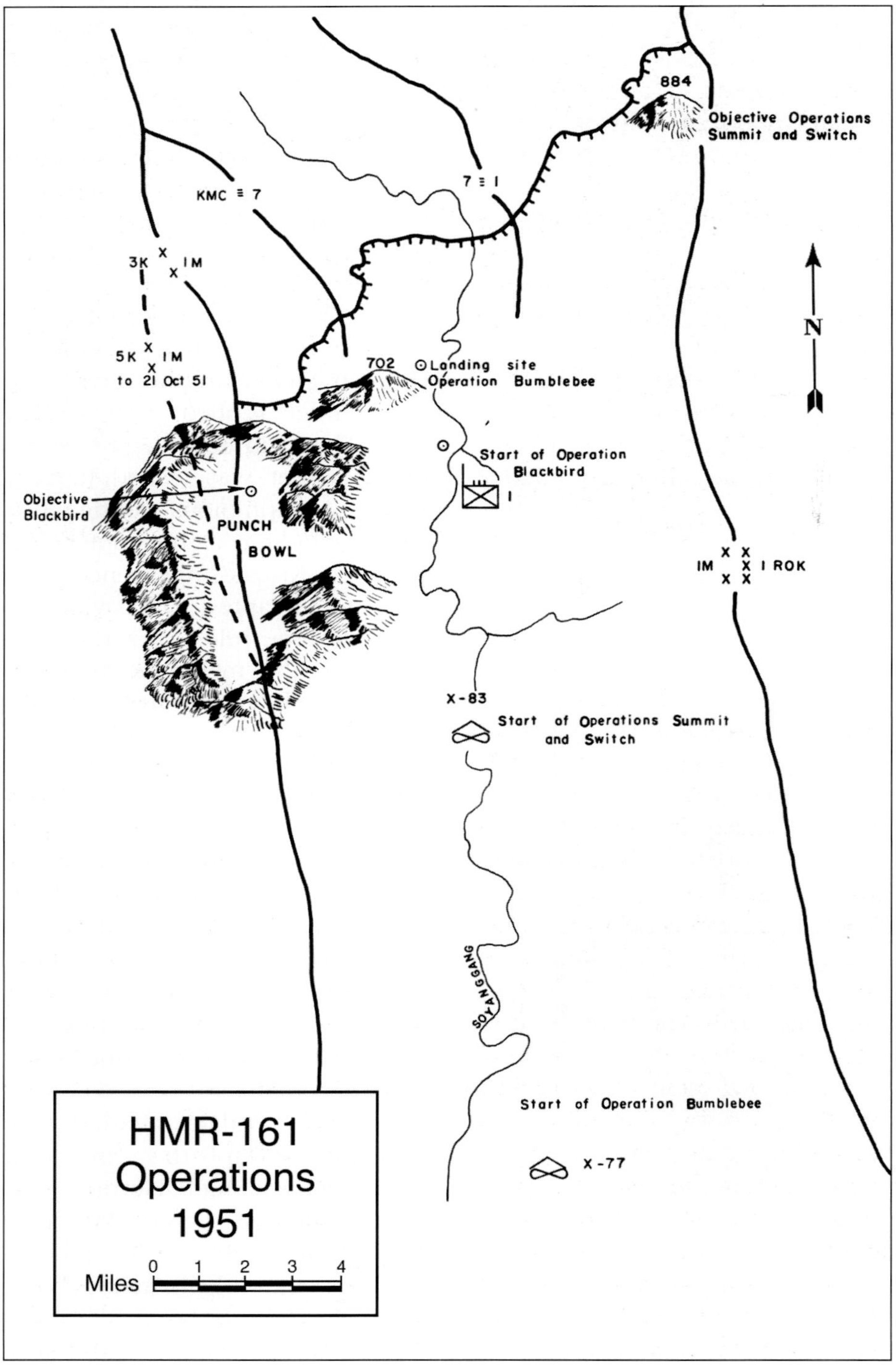

Operation Bumblebee kicked off at 1000 on 11 October. Twelve HRS-1 helicopters, working at about 30 second intervals and flying nap of the earth 15-mile routes, carried 958 passengers and more than 11 tons of supplies from airfield X-77 to Hill 702 using 156 individual flights in a total elapsed time of a little more than six hours. Two debarkation zones, Red and White, were used. In each, passenger manifests were used to control loading. The men moved from an assembly area to the "standby" box to the "ready" box and then into the helicopter. If any serial was short, additional passengers were summoned from a nearby "casual" area. At the offload spots shore party personnel "vigorously assisted the passengers by grasping their arms and starting them away from the aircraft." The first man out was the team leader and the last man out checked to see if any gear was adrift. Guides furnished by the ground units hurried the debarking men on their way to keep the landing zones clear for the oncoming waves. Bumblebee made the stateside headlines, but more importantly for the Marine Corps it was a giant step toward turning vertical envelopment theory into reality.

Four days later, the helicopters of HMR-161 again demonstrated their flexibility by mounting Operation Wedge, a short notice lift of 10 tons of ammunition and the evacuation of two dozen seriously wounded South Korean soldiers. Upon learning that a Republic of Korea unit was surrounded and in need of ammunition and medical assistance, Major Mitchell led six HRS-1 helicopters to the rescue. Captains Albert A. Black and James T. Cotton each made four flights into the embat-

HMR-161 Historical Diary Photo Supplement, Nov-Dec51

Troops load on board an HRS-1 at Airfield X-77 during Operation Bumblebee in October 1951. The Bumblebee troop lift was actually made to test contingency plans in case Chinese Communist forces cracked the Minnesota Line.

tled landing zone. At the end of this ad hoc operation IX Corps commander, Major General Claude F. Feren-baugh, USA, personally thanked each pilot for his effort in support of an allied nation.

While the main focus of effort was defense of the main line of resistance, several incidents behind the lines led to the use of HMR-161 helicopters for antiguerrilla activities. The first of these, Operation Bushbeater, used helicopter-borne teams to sweep the Soyang River Valley in late October. Unfortunately, the uneven terrain and lack of emergency power combined to make this operation the most costly in Korea in terms of aircraft lost. Three HRS-1s went down on 22 October while trying to insert ground units using knotted ropes for debarkation due to rough terrain. The pilots had difficulty maintaining station at the specified landing site. It was virtually impossible to hover above the ridge because inconsistent wind conditions sometimes caused the sudden loss of ground effect. When an aircraft is near the surface a thick layer of air builds up between the rotor and the surface. This cushion is known as "ground effect," and it creates additional lift. The loss of ground effect requires quick action by the pilot, who must either add power or go into motion before the helicopter plummets. Most pilots were able to avert a crash by gaining forward speed, making an abrupt turn, or diving into the valley. Unfortunately, three helicopters were unable to take such actions and crashed; two were lost and the third badly damaged its tail rotor; fortunately, only one man was injured. The follow-on salvage operation became another pioneering effort. Supervised by Major Edwin E. Shifflett, and led by Technical Sergeant Thomas M. McAuliffe, Marine working parties were able to dismantle the injured aircraft so all usable parts and one airframe could be recovered. Major Mitchell used his HRS as a "flying crane" to lift out an entire fuselage secured by ropes and harnessed to his cargo hook. Despite the initial setback, the operation continued when more suitable sites were used. Forty insertions were made and more than 200 men landed. Post-crash investigators determined that similar operations should continue but only after a careful study of the proposed terrain and evaluation of existing atmospheric conditions.

Several other heliborne anti-guerrilla operations followed. Operation Rabbit Hunt used helicopters for systematic patrols of the vast wilderness area behind the main line of resistance. This operation was not unlike those mounted by the 1st Marine Division to control the An Hoa Basin southwest of Da Nang in the Republic of Vietnam 15 years later. Operation Houseburner was mounted on 26 October to deprive enemy irregulars hiding behind friendly lines of much-needed shelter as winter approached. Two helicopters each carried four-man destruction teams armed with demolitions, flamethrowers, and incendiary grenades. Initially, one ship provided cover while the other hovered and sprayed the target area with a flammable mixture prior to dropping incendiary grenades. Later, both helicopters landed and let the destruction teams do their work from the ground. Operation Houseburner II used four helicopters to destroy 113 dwellings on the last day of the month. This action also featured the first extended firefight between helicopters and ground troops when an airborne automatic rifle team engaged an enemy position. Although the helicopter itself was not armed, this incident was probably the forerunner of the helicopter gunship.

A helicopter-borne Marine destroys a potential enemy guerrilla hideout during Operation Rabbit Hunt. In addition to troops on the ground, Marine demolition crews on board helicopters sprayed gasoline on huts serving as enemy cover and then set them ablaze with phosphorus grenades.

HMR-161 Historical Diary Photo Supplement, Nov-Dec51

Operation Switch, the relief and replacement of a full regiment at the frontlines, was the largest helicopter effort so far. On 11 November, nearly 2,000 combat loaded troops swapped positions between Hill 884 (unofficially dubbed "Mount Helicopter" because so many helicopter lifts took place there) and airfield X-83 in about 10 hours. Standard operating procedures included a three-plane flight that dropped off the advanced helicopter support team to supervise operations at the landing zone, departure teams controlled operations at X-83, and naval gunfire kept enemy heads down during flight operations.

Operation Farewell on 19 December saw the rotation of one battalion for another and marked the last flight by HMR-161 commanding officer Lieutenant Colonel George Herring. After that flight, he departed Korea to assume duties as commanding officer of HMX-1 at Quantico. His replacement was that unit's previous commander, Colonel Keith B. McCutcheon. The holder of an advanced degree in aeronautical engineering, McCutcheon had been a proponent and pioneer of Marine close air support during World War II before learning to fly helicopters. Major Mitchell remained as squadron executive officer.

After only two months in the combat, HMR-161 had logged more than 1,200 flight hours comprising more than 1,000 sorties to deliver 150 tons of supplies and carry out 192 medical evacuations. The "flying windmills" of HMR-161 participated in morale building as

HMR-161 Historical Diary Photo Supplement, Nov-Dec51

Col Keith B. McCutcheon, left, shakes the hand LtCol George W. Herring, the departing commanding officer of HMR-161. The squadron's last troop lift under Herring was dubbed "Operation Farewell" in his honor.

well as tactical operations: they delivered large cakes so the front-line Marines could celebrate the Marine Corps birthday on 10 November; on Thanksgiving they brought turkey dinners to the front; a heavy snow storm interrupted plans for Christmas feasts, but the arrival of several United Service Organization entertainers around the new year helped raise morale. The New Year also saw implementation of an additional duty that would last until the end of the war. After ice destroyed a bridge spanning the Imjin River, one HRS-1 and its crew were dispatched on a weeklong rotation to the United Nations Command advanced headquarters at Munsanni to ferry United Nations peace delegates to and from Panmunjom.

Colonel McCutcheon's first full month as squadron commander was the most ambitious helicopter-borne effort thus far; HMR-161 flew the most missions (820) in a single month so far and logged the most combat missions (506) in a single month during the entire war. Three major efforts were launched in January 1952—Muletrain, Changie-Changie, and Mouse Trap. Each exercised a different capability. Muletrain and Changie-Changie were assault support (helicopter-borne resupply and troop transport), while Mouse Trap was an exploration of tactical vertical envelopment. Operation Muletrain (named for a popular song of the day) called for the complete supply of a battalion located on the main line of resistance for one week. The destination was once again Hill 884. McCutcheon's squadron used a "flying crane" technique developed by Major Charles E. Cornwell whereby the HRS-1s mounted underslung nets carrying about 850 pounds and controlled from the cockpit to deliver cargo rather than pallets as had been previously done. Tentage, stoves, rations, fuel oil, and ammunition comprised the various loads. Four helicopters, operating on a rotating basis, were so effective that they actually flew in more cargo than could be handled by the shore party during the first week of January; 219 loads equaling 150,730 pounds were ferried about 10 miles from the supply dump to Mount Helicopter.

Operation Changie-Changie (pidgin Korean-English meaning "exchange") was a troop lift that

National Archives Photo (USMC) 127-N-A159212

Marines in Korea for the first time are moved into frontline positions the "modern way." Instead of climbing the steep trails, and spending hours to reach the ridges' crest, helicopters airlift troops in a matter of minutes.

began on 10 January. The essential difference between this troop movement and previous ones was that this time the helicopters flew into company-sized positions located within a few hundred yards of the frontline. In addition, the 35-man 1st Air Delivery Platoon took over helicopter ground support operations, relieving the hard-pressed 1st Shore Party Battalion of that duty. This realignment of missions was more in line with each unit's stated missions; First Lieutenant William A. Reavis' 1st Air Delivery Platoon was thereafter tasked "to prepare and deliver supplies by air, whether by parachute, air freight, or helicopter." Operation Mouse Trap, conducted from 14 to 17 January, tested the ability of Marines to launch a counter-guerrilla reaction operation on short notice. The squadron was not notified until just after midnight to be prepared to mount a two-company lift by mid-morning on the 14th. the operation went off with only minor difficulties and was so smooth that three similar lifts were made by the 17th.

For the most part, HMR-161 ceded deep search and rescue operations to VMO-6. The most notable such mission occurred in early February when the Eighth Army-Fifth Air Force Joint Operations Center requested help to bring back a fighter pilot and helicopter crew downed in enemy territory. Two previous attempts had been turned away by the time Major Mitchell's HRS-1 departed X-83 for airfield K-50 where it would pick up fighter escorts. Diverted enroute, the helicopter landed on the cruiser *Rochester* (CA 124) for a pre-flight brief before setting out. Fighter planes strafed the valley and surrounding ridgelines as the helicopter neared the crash site, but no activity was spotted so Mitchell reluctantly aborted the mission. The techniques used on this mission became standard operating procedure even though the rescue attempt had come up empty.

February 1952 was a harbinger of trouble on the horizon. Another relief in place, Operation Rotate, was successfully conducted on 24th. That same day, however,

Hot chow is served at the HMR-161 forward operating base near the Punchbowl; after finishing their meal, Marines go through the wash line. Living and working conditions were primitive, but the hard-working ground crews carried on.

Department of Defense Photo (USMC) A133622

HMR-161 Historical Diary Photo Supplement, Jan52

The Marine metal shop was located at K-18 airdrome near Kangnung in central Korea. This major maintenance facility served HMR-161 which was flying from the forward strip X-83 at Chodo-ri, behind the Minnesota Line.

were installed in mid-March.

After the end of the fighting at the Punchbowl, VMO-6 continued to support the 1st Marine Division flying from Sinchon in the X Corps sector of the East-Central Front. Indicative of the changing roles for light utility helicopters, the squadron listed four HO3S-1s and four HTL-4s in October 1951, but only one HO3S-1 remained on the rolls by March 1952 while the number of HTL-4s had increased to 10. The wisdom of combining helicopters and fixed-wing aircraft within observation squadrons was confirmed by combat experience. A well-defined division of labor between the fixed-wing airplanes and helicopters of VMO-6 had evolved since the early days of the war. The nimble OY were best suited for reconnaissance, artillery spotting, and airborne control of close air support while the helicopter niche combined combat search and rescue and medical evacuation. Transportation and administrative flights were divided

Captain John R. Irwin was enroute from Seoul to X-83 when he encountered severe vibrations. After putting down to locate the trouble, he was amazed to discover the broken remnants of his tail assembly lying in the snow. Four days later, Captain Calvin G. Alston's HRS-1 began to buck and jerk without warning. Suspecting he had been hit by enemy fire, Alston set down to inspect the damage. Like Irwin, he quickly found that a broken tail assembly was the culprit. Similar accidents outside the combat zone prompting the Chief of Naval Operations to ground all HRS-type aircraft until the problem could be isolated, analyzed, and corrected. The squadron was not able to resume normal operations until after new tail assemblies for each aircraft

Sikorsky representative, Louis Plotkin, left, explains the intricacies of an HRS-1 aircraft engine to HMR-161 pilots. Representatives of the Bridgeport, Connecticut, company accompanied the squadron to Korea not only to assist with technical problems, but also to report on how the aircraft performed in combat.

HMR-161 Historical Diary Photo Supplement, Nov-Dec51

HMR-161 Historical Diary Photo Supplement, Jul52

In March 1952, the 1st Marine Division moved from the Minnesota Line located in central Korea to the Jamestown Line in western Korea. HMR-161 likewise relocated its forward base to A-17 (Yongpu-ni), while the rear echelon's maintenance facility was moved to A-33 not far from the massive supply base at Ascom City outside the capital of Seoul.

about equally between fixed-wing and rotary-wing aircraft. Unfortunately, hopes for all-helicopter observation squadrons still were considered impractical. As time passed, HTL-4s gradually replaced the aging HO3s-1s, and by February 1952 the one remaining Sikorsky was no longer flying combat missions. The helicopter section's priorities gradually changed to reflect the new tactical situation as well. Positional warfare placed more emphasis on ground support and administrative missions while deep combat search and rescue had become the bailiwick of Navy and Air Force helicopter detachments. In September 1951, medical evacuation and combat search and rescue had been at the top of the list, but by March 1952 the new priorities were: evacuation of wounded; reconnaissance and observation; liaison and transportation; administrative and resupply flights; and combat search and rescue, in that order. The vulnerability of helicopters was an early concern, but this proved not to be the case as few helicopters were lost and the number coming back with bullet holes became all too common to merit special mention.

Between October 1951 and March 1952, the helicopter section logged 2,253 total flights (1,277 combat and 976 non-combat missions), including 637 medical evacuations to deliver 1,096 seriously wounded men. Most transportation flights involved bringing distinguished visitors to the front. Among them was Dr. Charles Mayo of the famed Mayo Clinic who visited units of the 1st Medical Battalion. Liaison flights included transportation of the Commandant of the Marine Corps, and Fleet Marine Force, Pacific, and IX Corps commanders. The bulk of the administrative and resupply flights went for medical support; the delivery of fresh whole blood or plasma, medicine, and medical records. After the frontlines stabilized, very few search and rescue missions were called for. Only seven such missions were flown between 1 October 1951 and 15 March 1952.

The helicopter section's only combat casualty during that time occurred when Captain David T. Gooden's HTL-4 was shot down as it wandered past friendly lines during a medical evacuation mission on 7 February. Neither the pilot nor the helicopter could be recovered due to their location behind enemy lines.

Defending the Jamestown Line

With both sides roughly equal in manpower and firepower on the ground, the frontlines remained unchanged during the winter of 1951-1952. In March, the United Nations Command decided to realign its forces. The 1st Marine Division moved from its positions along the Minnesota Line on the East-Central Front to the Jamestown Line astride the Pyongyang-Seoul corridor on the western flank. This move initiated the so-called "outpost war" which lasted from March 1952 until July 1953 during which no significant changes of territory occurred. The major actions of the outpost war included those at "Bunker Hill" in August 1952, a temporary incursion of the main line of resistance at the "Hook" in October 1952, tough fighting for positions "Berlin" and "East Berlin" in early 1953, the "Nevada Cities" (Outposts Reno, Carson, and Vegas) battles in March 1953, and the last fight at "Boulder City" just before the armistice in July 1953.

Although the generally flat terrain of western Korea simplified logistical challenges, the Jamestown Line was no tactical bargain. Terrain and diplomatic conditions prohibited defense in depth and severely hampered the ability of Marine commanders to maneuver or commit reserve forces in case of a Communist breakthrough. The 35-mile Marine sector was the longest defensive zone held by any Eighth Army division. The low-rolling hills on the Marine side of the line were dominated by the high ground on the far side of no-man's-land held by the Com-munists. To make matters worse, the Imjin River, with only four crossing points, ran behind the main line of resistance. Major waterways separated the line at two points, and a diplomatic demilitarized "neutral corridor" from Munsan-ni to Panmunjom divided the defensive zone. The United Kingdom's 1st Commonwealth Division anchored the Marine flank on the northeast at the Samichon River, where the 38th Parallel crossed the Jamestown Line. From there the main line of resistance generally traced the Imjin for about 10 miles until it intersected that river; the main line then followed the south bank of the Imjin estuary to where the Han River joined the Imjin, and across the Han an isolated defense sector was located on the Kimpo Peninsula. The main line of resistance was extremely vulnerable and had to be protected by a series of combat outposts scattered throughout no-man's-land. The scrub-covered, low-lying areas that predominated the Marine sector were subject of year-round enemy observation and flooding each spring. Overall, the Jamestown Line was a tactician's nightmare.

In late March, the 1st Marine Division moved 180 overland miles from the Punchbowl to Munsan-ni, an urban rail junction located near the Imjin River about 30 miles from Seoul. Lieutenant Colonel William T. Herring's VMO-6 and Colonel McCutcheon's HMR-161 began displacing from Sinchon in mid-March and had completed their respective moves by the end of the month. Each took up residence at separate landing fields near the 1st Marine Division's command post. The VMO-6 airstrip (A-9) was located in the

His predecessor, Maj William G. MacLean, right, welcomes LtCol William T. Herring on board as the new commanding officer of VMO-6. A graduate of the Naval Academy, Herring served as the commanding officer of Marine Fighter Squadron 111 and operations officer of the 4th Marine Aircraft Wing during World War II.

National Archives Photo (USMC) 127-N-A133045

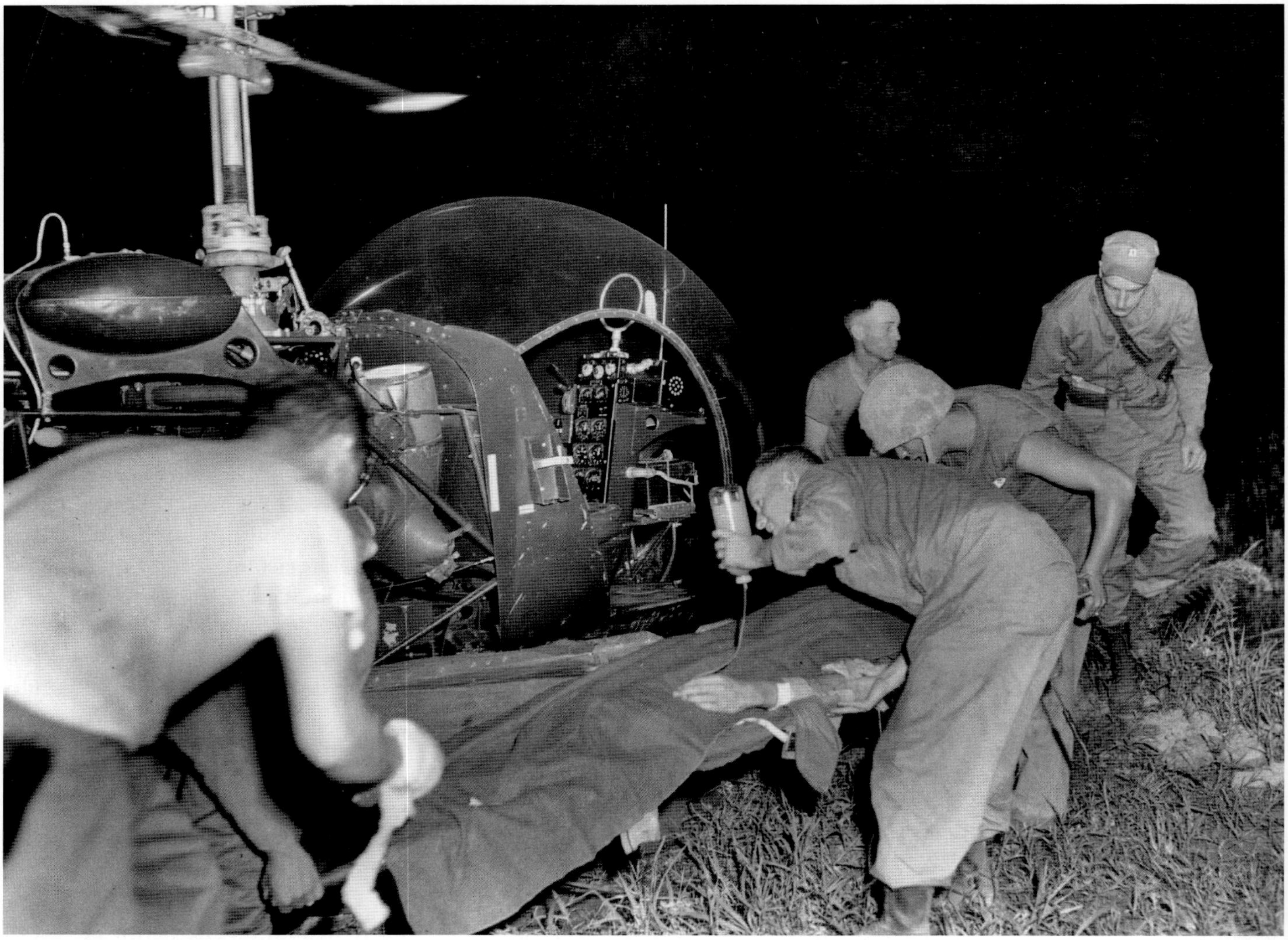

Department of Defense Photo (USMC) A134463

A badly wounded Marine receives life-sustaining plasma and will be flown to an advance medical care facility in the dark. Night evacuations were hazardous affairs because early helicopters lacked instrumentation and back lighting.

village of Tonggo-ri about three miles south of the division command post. The airfield was quickly named Bancroft Filed to honor the first Marine helicopter pilot killed in action. HMR-161's forward flight echelon was located at Yongpu-ni's A-17, while its rear echelon including advanced maintenance personnel was at airfield A-33 (Taejong-ni, a well-developed airdrome that served the massive Eighth Army supply base known as Ascom City, which should not be confused with airfield K-5 located at Taejon in south-central Korea).

For the most part, VMO-6 continued flying missions as before with medical evacuation as its number one priority. During this time the squadron's executive officer, Major William G. MacLean, Jr., developed a plan to station evacuation helicopters, crews, and maintenance personnel at the command post of the centrally located, frontline infantry regiment on weeklong rotations. This "forward evacuation echelon" was on-call around-the-clock and could reach any part of the frontline within a few minutes, cutting evacuation time in half thereby keeping severely injured men within what the surgeons called "golden minute"—the period during which immediate treatment could save a man's life. These operations began in June, and they included the first routinely scheduled night evacuations. The normal forward evacuation echelon complement was five officers, nine enlisted men, and two helicopters. In addition, close liaison with American and other allied nations' medical stations and hospital ships was maintained. Other missions performed by VMO-6's helicopter section were liaison flights and visual reconnaissance. The former usually brought important visitors to the front while the latter flew commanders along the main line of resistance and offered high altitude glimpses into enemy territory.

Major General John T. Selden,

the commander of the 1st Marine Division, required that a strong defensive line be established but was still nervous because it would be difficult to quickly reinforce the Jamestown Line. Accordingly, two existing "fallback" lines, Wyoming and Kansas, were strengthened, and a series of rapid deployment exercises by the division reserve regiment were planned. Primary among them were those conducted by HMR-161 in the spring and summer of 1952.

The first test of the ability to move across water obstacles was Operation Pronto. On 5 April, a 662-man battalion and about 10,000 pounds of supplies were transported from Munsan across the Han River to the Kimpo Peninsula. Ironically, Operation Pronto was both the longest distance and the shortest notice helicopter-borne troop lift so far. Colonel McCutcheon was not notified until about 0210 in the morning, yet, the first helicopter lifted off only about three-and-a half hours later. The initial wave carried specialists from the 1st Air Delivery Platoon to two landing zones. Thereafter, nine helicopters (seven of them manned by pilots fresh from the United States) were used. The hour-long round trips were almost 60 miles because of flight restrictions in the vicinity of the Panmunjom "neutral corridor." The squadron logged 99 flights in more than 115 flight hours with a total elapsed time of 14 hours, an all-time high. After the operation, Colonel McCutcheon noted: "This airlift . . . proved that a Marine transport helicopter squadron can successfully operate as an 'on call' tactical tool." The operation was carried out with only minimal liaison between flight and ground units and virtually none of the detailed planning previously employed. Helicopter operations, which only a few months earlier made front-page news in the United States, had by that time, become routine.

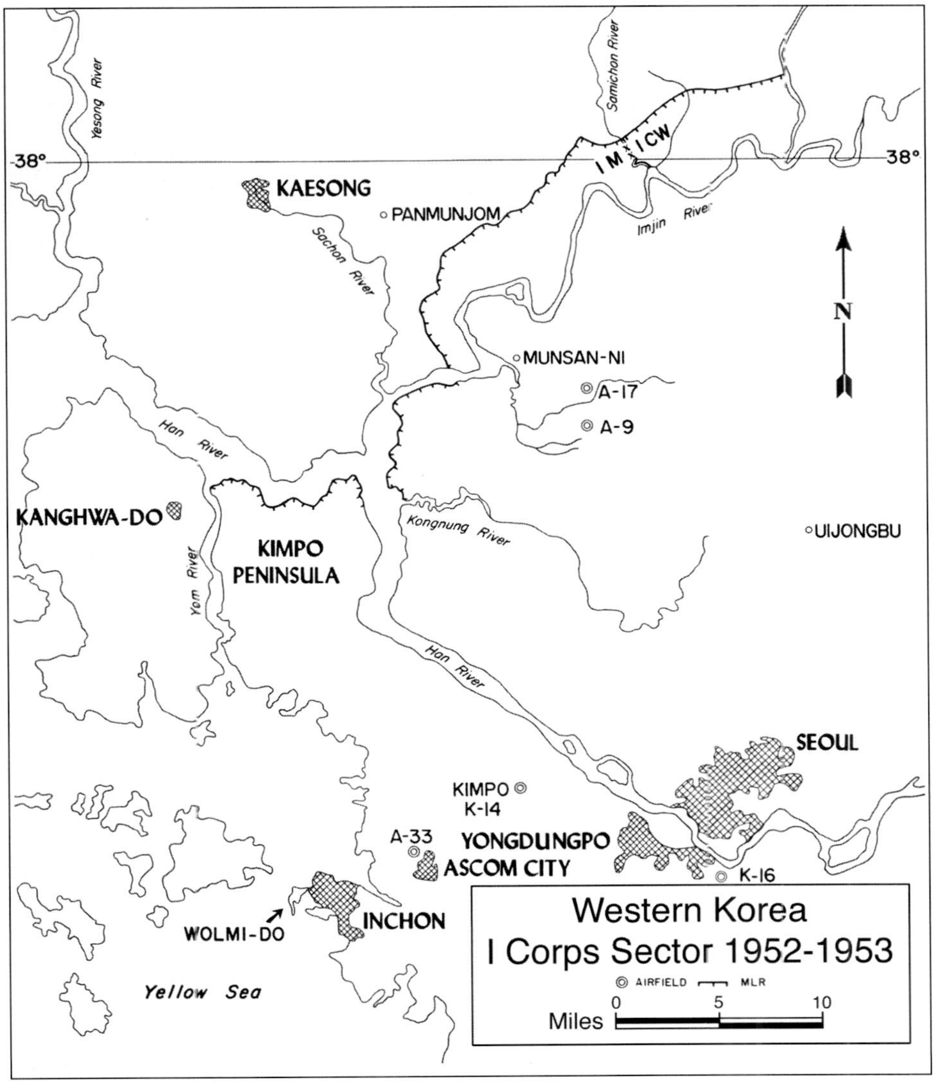

An intermittent series of troop lift exercises were interspersed with several tactical and logistical operations over the next year and a half. Pronto was promptly followed by two-day Operation Leapfrog, the helicopter-borne exchange of one South Korean Marine battalion for another on 18 and 19 April. Operation Circus, the lift of a U.S. battalion closed out the month. All operations, except for emergencies, were discontinued on 27 April after the Chief of Naval Operations grounded all HRS-1s due to structural problems. By the middle of May, HMR-161's helicopters were back in action. Two more short notice troop lifts, Operation Butterfly and Ever Ready, were conducted in June and a third, Operation Nebraska, took place in November. The last such exercise was Operation Crossover II held the following spring.

Although combat search and rescue was not a primary mission of HMR-161, one dramatic episode occurred in late May. Two helicopters were dispatched from A-17 to look for a downed Navy pilot near Hapsu, North Korea. During the airborne search, the HRS-1 flown by Major Dwain L. Lengel and Captain Eugene V. Pointer

National Archives Photo (USMC) 127-N-A134367

LtCol John F. Carey, left, bids farewell to his predecessor as commanding officer of HMR-161, Col Keith B. McCutcheon; both men had previously commanded HMX-1 at Quantico. McCutcheon made his reputation as a close air support advocate in the Pacific and eventually commanded all Marines in Vietnam; later he was slated to become the first active duty Marine aviator to receive a fourth star but illness intervened.

with crew chief Technical Sergeant Carlyle E.J. Gricks on board lost flight control due a combination of low speed and high altitude. Flying low to the earth, the helicopter was unable to gain enough power to maneuver around a stump. The aircraft crashed and no one was seriously injured, but the would-be rescuers now needed to be rescued. The crew quickly moved to a pick up location but had to wait almost two days due to bad weather. Squadron mates Captain Robert J. Lesak, First Lieutenant Wallace Wessel, and Technical Sergeant Elmer DuBrey flew the rescue mission. It was a complicated pick-up procedure. In order to keep from repeating the previous crash, Captain Lesak had to keep his aircraft in motion to stay aloft. This required the downed crew to grab a trailing rope ladder as the HRS passed overhead. Fortunately, the rescue was a success, if not a frustrating one.

One of the primary purposes of rushing HMR-161 to Korea was to test vertical envelopment concepts as they applied to amphibious operations under wartime conditions. This was not possible at first due to geographic restrictions when the Marines manned the East-Central Front. The move to western Korea brought the Marines close to the sea, but initial operational tempo and subsequent grounding of the HRS-1s delayed the opportunity until June. At that time as series of Marine landing exercises (MarLExs) were held at a rate of about two each month for the remainder of 1952 with two more held the following year. Although there were minor variations in each MarLEx, they generally followed a similar pattern: a detachment of a half dozen HRS-1s from HMR-161 would lift one battalion of the division reserve from one small island to a larger one during a simulated amphibious assault. The purpose of these exercises was twofold. First, tactics and techniques were perfected with each passing exercise as lessons were learned and assimilated. Second, aircraft and ground personnel became familiar with the standard operating procedures for helicopter-borne operations. The main problems were the lack of an escort carrier and helicopters. Untested Marine amphibious doctrine envisioned individual transport helicopter squadrons and rifle battalions embarked on board escort carriers during the movement to the amphibious objective area. Once there, the helicopters would conduct one portion of the ship-to-shore movement then be on call to deliver supplies and evacuate casualties until the beachhead was secured and operations could safely move ashore. The trouble was that the Navy did not have enough carriers or crews to implement this policy, and no escort carrier was readily available to support most MarLEx operations. In addition, the demands on HMR-161 prohibited the entire squadron from participating in the exercises.

MarLEx I was held on 10 and 11 June with its announced purpose to gain experience in vertical envelopment as part of an amphibious operation. Because no escort carrier was available, the island of Sung Bong-do about 40 miles southwest of Inchon would stand in for the missing ship. Nearby Tokchok-to, a five-mile-long island with two broad sandy beaches located about six miles southeast, was selected as the objective. As was standard practice,

An HRS-1 helicopter of HMR-161 approaches the escort carrier Sicily *(CVE 118). The Marine transport squadron experimented with the newly developed concept of vertical assault in the many landing exercises conducted during the Korean War.*

National Archives Photo (USMC) 127-N-A134628

1st Air Delivery Platoon Historical Diary Photo Supplement, Feb53

Col Harry N. Shea, right, commanding officer of the 11th Marines, briefs from left, 1stLt Donald L. Seller, Commanding Officer, 1st Air Delivery Platoon, 1stLt William B. Fleming, Executive Officer, 1st 4.5-inch Rocket Battery, Capt Edwin T. Carlton, Commanding Officer, 1st 4.5-inch Rocket Battery, and LtCol John F. Carey, Commanding Officer, HMR-161, prior to a fire mission in support of Korean Marines.

helicopter support teams descended from hovering helicopters using rope ladders to prepare landing zones. Seven aircraft delivered 236 fully equipped troops the first day and another 236 the following day. Unfortunately, the exercise did not go well. Communications were unsatisfactory, and the time required to land troops by helicopter was "too great in comparison to the time needed to land troops . . . by boat." It was decided to use a closer island the next time to reduce the strain on the helicopters and the time in the air. MarLEx II was held later that month. This time Soya-do, two miles from Tokchok-to, was used as the simulated carrier. Four instead of seven aircraft were used to lift 235 men, and the exercise was deemed a success.

The Marines finally got to train with an actual aircraft carrier in September. The escort carrier *Sicily* (CVE 118) was available to support MarLEx VII. This exercise offered the most realistic test of amphibious doctrine as envisioned by planners at Quantico. On 1 and 2 September, the bulk of HMR-161's12 HRS helicopters were used to lift 964 troops from the carrier deck to Landing Zones Able and Baker on Tokchok-to. Five more MarLEx operations were held in 1952. They were followed by a six-month break, then two more amphibious exercises were held in the summer of 1953 before the ceasefire took effect.

On 30 July, HMR-161 received a request to launch a humanitarian effort in support of the U.S. Army and South Koreans. More than 600 American soldiers and about 150 Korea civilians had been stranded by flash flooding of the Pukkan River. Six Marine helicopters flew to the rescue. This spur-of-the-moment evacuation was made without written orders or advanced scheduling. The squadron mounted 182 flights over about three hours. The squadron after action report noted: 'The average load was five men and gear [but we lifted] as many as nine small children complete with dogs and chickens . . . in one trip . . . . The Army and Air Force . . . marveled at the expeditious way our helicopters carried out the operation."

The month of August saw a change of command when Lieutenant Colonel John F. Carey, yet another veteran of HMX-1, replaced Colonel McCutcheon as commanding officer of HMR-161. September was the busiest month of the war for HMR-161. The squadron flew 1,195 missions. Included in that total were the largest aerial supply operation thus far, the first of many regularly scheduled helicopter-borne troop rotations, the only amphibious exercise supported by an aircraft carrier, and the tactical lift of a rocket artillery battery.

The first of two large logistical support operations took place from 22 to 26 September. Operation Haylift was designed to completely support a frontline regiment for five days. Included in the loads were rations, water, ammunition, fortification material, and fuel. These supplies were carried internally or suspended below the helicopter frame in wire baskets and cargo nets. The distance from loading zone to landing was about 20 miles, depending upon which loading zone was used. The 1st Service Battalion supplied Loading Zone Able while the 1st Ordnance Battalion did so at Loading Zone Baker; air delivery platoon personnel supervised loading operations and shore party personnel unloaded the incoming aircraft. More than 350,000 pounds of cargo and 75 passengers were lifted despite rainy weather. This effort tripled the output of Opera-tion Muletrain, the previously biggest logistical operation. Op-eration Haylift was summed up in a single sentence in the squad-ron report: "No unusual problems were encountered and the operation progressed smoothly and continuously throughout."

The use of helicopters to rotate troops between the rear and the front had become routine by the summer of 1952. So much so that the 1st Marine Division initiated regularly scheduled replacement operations intended "to effect the relief of a unit on the MLR and return the relieved unit to a rear area as expeditiously as possible," using the codename "Silent Redline." Silent Redlines were conducted at the rate of about one per month during the rest of the year, but were only intermittently used the following year due to tactical considerations (the 1st Marine Division was either off the line or heavily engaged). The first of these began on 11 September with the lift of a Korean Marine battalion. Because these operations were carried out under enemy observation, if they came under direct fire squadron aircraft were directed to seek landing spots in defilade and maintain communications while the ground troops debarked and sought the best defensive terrain. Ten aircraft, each carrying six men or five men and a crew-served weapon, transported 1,618 troops in an overall time of six-and-one-half hours during Silent Redline I.

By the summer of 1952, the strategy in Korea had developed into positional warfare and artillery began to dominate tactical thinking. Unfortunately, the Chi-nese actually began to outgun the Americans as a result of massive Soviet aid that furnished excellent weapons and plenty of ammunition, and in western Korea the enemy controlled the Taedok Mountain spur which gave them superior observation of the United Nations lines. The Marines countered by adopting mobile artillery tactics using multiple gun positions. One innovative solution came about as the result of a cooperative effort between the pilots of HMR-161 and the ar-tillerymen of the 11th Marines. A particularly valuable weapon was the towed multiple rocket launcher. The problem was that these mobile

Two rocket battery crewmen prepare the launcher for action as an air deliver platoon signalman assigned by HMR-161 to direct incoming cargo-carrying aircraft to the landing site loads a rocket round into the tube.

1st Air Delivery Platoon Historical Diary Photo Supplement, Feb53

rocker launchers were vulnerable because their back blast kicked up dust and debris that was visible from the Communist side. The rapid insertion of light artillery into defiladed positions followed by a rapid withdrawal, however, would allow the Marines to land, set up, fire a barrage, and then leave before enemy counterbattery fire could pinpoint the target. Experiments at Quantico, Virginia, and Camp Lejeune, North Carolina, proved the feasibility of lifting a 4.5-inch rocket launcher along with a skeleton crew and a small amount of ammunition in a single load. On 19 August, HMR-161 put this theory into practice during Operation Ripple. Rehearsals in Korea tested new delivery methods using a variety of external hooks and release mechanisms, but there was some trouble during the initial lift of the 1st 4.5-inch Rocket Battery. The problems were solved that night and additional operations the following day went much more smoothly. Colonel Carey was able to recommend that helicopters were suitable for rocket launcher transportation, and Operation Ripple was followed by several more similar tactical operations over the next few months. These were the only operations in which helicopters were directly responsible for putting rounds on the target. As such, they were the distant forerunners of the "fire base" concept that became a tactical mainstay in Vietnam.

HMR-161 Historical Diary Photo Supplement, Dec52

Transport squadron's "HR-69" helicopter is decked out as Santa Claus to deliver toys and food on Christmas Day 1952. Although the visionaries at Quantico, Virginia, before the Marines received helicopters foresaw almost every possible use for rotary-winged aircraft and made up a potential task list, it is doubtful that this humanitarian mission appeared on that list.

Beginning in October, one helicopter and a standby crew were assigned to Marine Aircraft Group 12 on a rotating basis to provide air-sea rescue and administrative transportation. As a result of heavy fighting at the Hook the squadron logged the most medical evacuations that month as well, 365. During the month, the squadron also began receiving Sikorsky HRS-2 helicopters. Although a newer model, the HRS-2 offered no significant increase in performance because it used the same engine as the HRS-1; the main differences were that the HRS-2 was about a foot shorter and a few inches closer to the ground. Operation Nebraska, conducted on the 13th, tested the ability of HMR-161 to move troops from one phase line to another. Ten helicopters lifted one rifle battalion and a heavy mortar platoon (820 men) in only two-and-one-half hours comprising 169 individual flights.

Activities in December included Operations Crossover, the movement of a reserve rifle battalion from the Wyoming Line to the Kansas Line; Silent Redline III, the by-then standard helicopter-borne rotation of a frontline battalion by one from the reserve area; MarLEx XII-52, the last amphibious exercise of the year; and Operation Santa Claus during which helicopter HR-69 was made up to look like jolly old Saint Nicholas as it delivered toys and food to about 100 orphans adopted by the squadron. Also during the month, Colonel Carey flew out to the newly arrived Danish hospital ship *Jutlandia* to test its helicopter-landing platform and to familiarize the crew with helicopter landing procedures. The cold, damp weather and fog continued to interfere with flight operations, but new hangars eased the maintenance burden to some degree. Although few of the shivering mechanics would have believed it at the time, conditions on the Jamestown Line were far superior to those encountered on the East-Central Front the previous year.

January 1953 witnessed the use of HRS helicopters as "flying squad cars" as they carried members of the 1st Military Police Battalion searching for Communist infiltrators. On the 23d, fire was exchanged be-tween the airborne Military Police and guerrillas on the ground. Three enemy troops were killed while the helicopter suffered only minor gunfire damage.

February hosted the largest helicopter supply operation in Korea, Operation Haylift II. While Haylift I the previous September had supported one frontline regiment for five days, Haylift II was twice as ambitious. This time, two frontline regiments would receive helicopter-transported Class I (rations), III (fuel), IV (construction materials), and V (ammunition) supplies for five days, from 23 to 27 February. The planning and execution of Haylift II was similar to its forerunner, but on a much larger scale. And, this one would take place in much more difficult weather conditions. One hundred and thirty tons per day were required to support both regiments, but this total was actually exceeded on the first day. The unloading time per load was less than one minute. On 25 February, HMR-161 brought in more than 200 tons, a record. By the third day, the supply build-up had actually surpassed the ability of the ground logisticians to cope with it. Fortuitously, emergency requests for ammunition by other units lessened the backlog. Ground fog on the last two days slowed operations. Still, the final results were impressive, 1,612,406 pounds lifted without the loss of crew or aircraft. February also saw records set for the number of combat hours (765), total flight hours (1,275.5), combat flights (575), and total flights (1,183), and the gross lift of more than two million pounds that month was the largest of the entire war for HMR-161.

Sadly, that month also ended HMR-161's streak of not losing a man. On the 12th, a three-plane flight departed A-33 for Pusan to rendezvous with a carrier that was to take them to Japan. Along the way the HRS-1 carrying Captain Allen W. Ruggles and Technical Sergeant Joe L. Brand, Jr., became separated and crashed into the sea about 25 miles south of Pusan. The cause was believed to have been mechanical failure, but this was never confirmed because there were no survivors and wreckage was never located.

A helicopter from HMR-161 prepares to lift bundled supplies from the 1st Air Delivery Platoon area to frontline troops. The air delivery platoon signalman holds the hook while the other Marine holds the net in a hook-up position.

1st Air Delivery Platoon Historical Diary Photo Supplement, Feb53

On 15 March, Colonel Owen A. Chambers took over HMR-161. Ten days later, a second HRS-1 went down with three crewmembers on board. Major Doil R. Stitzel was making a test hop out of Ascom City with mechanics Master Sergeant Gilbert N. Caudle, Jr., and Sergeant Richard L. Parsell when their aircraft suddenly lost power, crashed, and burned. All three men were lost.

Only two major operations were conducted that spring, both were troop lifts from the reserve area to the Jamestown Line, Operation Crossover II and Silent Redline VI. On 27 March, all HRS-2 helicopters with more than 200 flight hours were grounded because of rotor blade problems. This was a precautionary move due to stateside incidents, and no HRS-2s in Korea were lost to this cause.

Beginning on 26 April, HMR-161 participated in Operation Little Switch, the six-day exchange of prisoners of war. The United Nations released 6,670 North Korean and Chinese prisoners while the Communists returned only 684 captives, including 15 Marines and three Navy corpsmen. From the middle of the month, HMR-161 provided transportation from Freedom Village near Munsan-ni to Panmunjom for various international delegates and

American negotiators. During the actual prisoner exchange, helicopters stood by to transport the seriously ill or wounded Marines from Panmunjom to one of three hospital ships, the *Consolation* (AH 15), the *Haven* (AH 12), or the *Jutlandia,* riding at anchor in Inchon harbor. Four Marines had to be evacuated.

In May the 1st Marine Division came off the line for the first time since the Masan interlude ended in January 1951. While this represented a relief from the rigors of combat, it was not exactly a time of rest and relaxation. The division staff ordered HMR-161 to get busy on the first Marine landing exercise of the New Year (MarLEx I-53). On 13 May, after careful planning and

VMO-6 Historical Diary Photo Supplement, Nov52

Sikorsky HO5S

The HO5S helicopter, developed from Sikorsky's S-52 design begun in 1948, was the purpose-built replacement observation helicopter for the HO3S. The S-52 was first conceived as a compact two place machine, but it eventually incorporated recommendations from the fighting front under the designation S-52-2. The HO5S was more compact than its predecessor and featured several new design features to overcome technical problems identified in the HO3S. Forty-eight HO5S-1s were ordered for the Marine Corps in 1951 and accession began in January 1952.

Although its theoretical performance statistics appear only marginally better than its predecessor, the HO5S was actually a much-improved aircraft that addressed many of the HO3S's shortcomings. The HO5S was the first U.S. helicopter fitted with all-metal blades, could mount two stretchers internally, and was much more stable on the ground due to its low center of gravity and four-wheel landing gear. The most unique practical innovation was a hinged, two-piece, forward-mounted observation bubble. Opening the left seat side of the bubble allowed access to the cabin interior for two stretcher-borne patients. In addition, the HO5S could carry three combat-loaded men over short distances.

By the time of the armistice in 1953, almost all VMO-6 helicopters were HO5Ss. Unfortunately, plans to replace light airplanes with HO5S helicopters in Marine observation squadrons had to be put on hold due to performance problems and structural defects that came to the fore in Korea. It was decided that the Marine Corps needed a machine that offered better stability and easier in-flight control in addition to a more powerful engine. Thus, instead of becoming the backbone of Marine observation squadrons, the HO5S was actually replaced by the Kaman HOK beginning in 1954; the later aircraft remained in operational service for the next decade until was it in turn replaced by the Bell UH-1 Iroquois ("Huey"), which remains the designated Marine observation and utility helicopter to this day. Marine observation squadrons were equipped with fixed-wing airplanes after light helicopter squadrons were created during the Vietnam-era.

Aircraft Data

Manufacturer: Sikorsky Division of United Aircraft Corporation

Power Plant: 245 hp Franklin O-425-1 engine

Dimensions: Length, 27' 5"; height, 8'8"; rotor, three 33' metal blades

Performance: Cruising speed, 96 mph

Lift: Pilot and three passengers or two internal stretchers

Department of Defense Photo (USMC) A169730

A seriously wounded Marine near the Jamestown Line is loaded on board a Sikorsky HO5S-1 helicopter from VMO-6. By 1953, the HO5S-1, which was designed to remedy shortfalls of the HO3S-1, had become the Marines' primary medical evacuation aircraft.

rehearsals, HMR-161 brought a battalion landing team to Yongdong-ni, a beach area southwest of Seoul. This exercise was followed in June by a special helicopter assault demonstration as part of the rehearsal for MarLEx II-53. Similar to the previous amphibious exercise in scope and purpose, MarLEx II-53 actually turned out to be the last major amphibious exercise during the Korean War. The squadron re-turned to the lines on 10 July and thereafter continued routine operations delivering supplies to various outposts and transporting mail and personnel until the ceasefire was declared on 27 July 1953.

The Korean Conflict had finally ended, but HMR-161 was not yet homeward bound. Immediately after the guns cooled off, HMR-161 would support a massive prisoner of war exchange and then enter a period of "watchful waiting" before returning stateside.

The intervening 16 months between the move to the Jamestown Line in March 1952 and the armistice in July 1953 were busy ones for the helicopter section of Marine Observation Squadron 6. The last of the venerable HO3S-1s (Bureau of Aeronautics number 124343) departed VMO-6 in April 1952. Bell HTLs carried the load throughout the spring until the arrival of replacement helicopters that summer. The first of the new Sikorsky HO5S-1 helicopters arrived in July. This new machine, the first helicopter equipped with all metal rotor blades, was a three-seat utility aircraft that mounted a three-bladed overhead main rotor and a two-bladed tail rotor. Powered by a 245-horsepower Franklin engine, it could carry a 750-pound load at a cruising speed of 96 miles per hour. The most

The Innovators

The first rotary-winged flight machines were children's toys believed to have been developed in China. Just after the end of World War I a rotary-winged airplane, the autogiro, was developed and gained some popularity during the Jazz Age. True helicopter technology, however, did not really take off until just before World War II. Pre-war helicopter enthusiasts in France, Italy, Spain, and Germany spread their gospel throughout Europe and on to the United States. Of the early American designers, three stand out: Igor I. Sikorsky, Frank N. Piasecki, and Arthur M. Young. Each of these men left an indelible mark on U.S. helicopter development, and their legacy lives on in the aircraft used by current Marine aviators.

IGOR I. SIKORSKY

America's preeminent helicopter advocate, pioneer, and designer was a Russian émigré who moved to the United States to escape communism. He did not invent, nor was he the first to fly, a helicopter. He did, however, formulate a solution to movement stability for rotary-wing flight that has since evolved into the most popular modern helicopter configuration—a single, large, horizontal, overhead rotor stabilized by a small, vertical, anti-torque tail rotor with forward movement controlled by varying the main rotor's pitch while using the tail rotor to determine direction. Igor Ivanovich Sikorsky developed a wide variety of helicopters that became versatile aircraft equally suited for both commercial and military use, aircraft able to perform unique tasks on land, at sea, and in the air. After World War II, Sikorsky worked closely with the U.S. Marine Corps to adapt his helicopters to military use, a symbiotic relationship between manufacturer and user that carried on even after his death. Today, Igor Sikorsky is rightfully considered the "Father of American helicopters."

Although long interested in rotary-wing flight, Sikorsky actually first gained fame for his multiple-engine aircraft designs. Born in Kiev, Russia, on 25 May 1889, his interest in, and aptitude for, aeronautical engineering became evident early in his life. He began experimenting with flying machines and the principles of aerodynamics prior to entering the Russian Naval Academy. After three years, Sikorsky left St. Petersburg to study in Paris and then returned to his homeland to attend Kiev Polytechnic Institute. He returned to Paris a second time to learn more about rotary-winged flight.

Inventor Igor I. Sikorsky, the father of American helicopters visits HMX-1 at Marine Corps Air Station Quantico, Virginia. In the background is an HO3S-1 helicopter, one of the first two "Whirlybirds" assigned to the U.S. Marine Corps.

National Archives Photo (USMC) 127-N-A322389

He built his first helicopter, a wooden box mounting two horizontal propellers powered by a 25-horsepower motorcycle engine, upon his return to Kiev in 1909. He could never get this machine to fly and concluded that the technology of the day was not adequate, but he also remained convinced that in time rotary-winged aircraft would surpass fixed-wing airplanes as flying machines. Sikorsky continued his experiments using engine-powered sleighs until he turned to designing multiple-engine airplanes. In 1913, he designed and built the world's first four-engine airplane. Thus, he embarked upon a new career path for the next 40 years.

Driven from Russia by the Bolshevik Revolution, a nearly penniless Igor Sikorsky fled to the United States by way of France. In America he eked out a living teaching mathematics and consulting part time. Among his projects was a proposed, but never adopted, trimotor bomber for the U.S. Army. After much hardship, he was able to live the American dream when he converted a Long Island, New York chicken farm into the Sikorsky Aero Engineering Company in 1923. Six years later, the company joined Boeing, Pratt & Whitney, and Chance Vought in forming the United Aircraft and Transportation Corporation. These humble beginnings comprised the genesis of one of America's most successful aviation enterprises. Sikorsky's first successful U.S. design—an all-metal, twin-engine transport, the S-29A—established his reputation for building aircraft noted for their ability to withstand hard landings on rough surfaces, poor weather conditions, and continuous operations with only rudimentary maintenance. By far his most successful airplane to that time was his eight-passenger, high-wing, twin-engine S-38 amphibian. Although designed for commercial use, 16 variants bearing Bureau of Aeronautics "RS" designations were purchased by the U.S. Navy, some of which saw service with the Marine Corps. Sikorsky next turned to large, long-range, four-engine, transoceanic passenger planes. His successive S-40,-41, and-42 models gained fame as the "American Clippers," large flying boats that plied their trade for Pan-American Airways in the Carribean and across the Pacific Ocean. Economic troubles forced the shutdown of United Aircraft's Sikorsky Division in 1938, but this setback fortuitously once again whetted his long-standing interest in rotary-winged aircraft.

Heartened by technological progress and spurred into action by recent European developments—notably Germany's spectacular public exhibitions of Heinrich Focke's Fa-61—Sikorsky went back to developing helicopters. By 1939 he had created the VS-300, an ungainly looking contraption consisting of a sprawling bare metal frame mounting a single main rotor for lift and a small-tail rotor for control. Although this "flying bedspring" was not aesthetically pleasing and performed more like a bucking bronco than a steady workhorse, it became the first practical American helicopter after its initial free flight on 13 May 1940. Not long thereafter, the U.S. military became interested in helicopter development. In early 1942, Sikorsky won an unofficial competition by producing the VS-316. This two-place, single-main rotor helicopter was given the military designation R-4 (R-1,-2, and-3 were competing designs by other manufacturers). It was soon followed by improved versions labeled R-5 and R-6. By the end of the war more than 400 Sikorsky helicopters had been built. The U.S. Navy procured its first Sikorsky helicopter, an Army R-4 given the designation HNS upon transfer in October 1943. Sikorsky-built helicopters have been a mainstay of naval aviation ever since. Marines currently fly the Sikorsky three-engine CH-53E heavy-lift transport helicopter, one of the largest helicopters in the world.

FRANK N. PIASECKI

Frank Piasecki, the son of an immigrant Polish tailor born in 1919, was considered the "wonder boy" of early helicopter development. By his 21st birthday he already held degrees in mechanical engineering from the University of Pennsylvania and aeronautical engineering from New York University. He began working as a mechanic for Kellet Autogyro while a teenager then became a designer with Platt-LePage after college before branching off on his own. Piasecki developed the second successful American helicopter using castoff auto parts and an outboard motor. He endeared himself to Marine helicopter proponents with his theories of how tandem rotors could support very large or heavy loads, an innovation that promised to make ship-to-shore movement of complete units and bulky equipment when other machines of the day could lift only a pilot and one or two others. Piasecki co-founded P.V. Engineering Forum, a consortium of aircraft designers interested in rotary-wing flight and was the driving force behind that firm's most successful project, the PV-3. The PV-3 was a large, elongated, bent fuselage, tandem rotor transport helicopter; the first of a series nicknamed "Flying Bananas." The PV-3 was unique because the Flying Banana was rated for eight passengers as well as a crew of two. After its first flight in 1945, the Navy purchased 22 PV-3s (designated HRPs by the Bureau of Aeronautics). The HRPs quickly established the practicality of tandem rotors for heavy lift, and orders for improved models quickly poured in. Piasecki's notable early success was the famous H-21 Workhorse, which was used by the Air Force as a rescue craft and by the Army ("Shawnees") to haul troops and cargo. The P.V. Engineering Forum became the Piasecki Helicopter Corporation in 1947, then a division of Vertol Aircraft, which in turn became a division of Boeing Aircraft. Venerable Boeing-Vertol CH-46 Sea Knight assault helicopters, lineal descendants of the first Flying Banana have been the backbone of Marine helicopter aviation for more than four decades and continue to serve with the fleet to this day.

ARTHUR M. YOUNG

The brilliant, but somewhat eccentric, scion of a wealthy Pennsylvania family, Arthur M. Young invented a rotor stabilizer bar that allowed two-bladed rotors to power light utility helicopters. His invention enabled Bell Helicopter Corporation to produce the two most prolific helicopter models in history, each of which remained in production for more than 30 years.

Young began developing his idea while employed by Lawrence D. Bell's aircraft company, the same firm that produced the first U.S.-built jet (the P-59 Bell Airacomet) and the first supersonic aircraft (the X-1 rocket plane). After 15 years of building models and researching rotary-winged flight, Young perfected his revolutionary new concept. He knew that Sikorsky's tail rotor concept eliminated torque, but he wanted to improve flight stability and reduce weight. His solution was a small counter-weighted stabilizer bar linked directly to the rotor that functioned like a flywheel, a device that kept the rotor blades independent from the movement of the fuselage. In 1941, he assigned his patents to Bell Aircraft with an agreement to oversee the production of a few prototype Model-30 helicopters. The first of these rolled out at Gardenville, New York, in December 1942, and then made its first untethered flight the following June. The second prototype looked like an automobile with its fully enclosed cabin and four wheels. That aircraft was the first helicopter used to transport a doctor on an emergency call, and it also rescued a pair of fishermen stranded on an ice floe in 1945. The third prototype featured an advanced instrument panel, a bare metal tubular tail boom, and a distinctive Plexiglas bubble canopy.

Building on the lessons learned while improving the early models, Young next developed the first full production Bell helicopter labeled the Model-47. This machine, first flown on 8 December 1945, was the first helicopter certified for sale by the Civil Aeronautics Administration. It was quickly adopted as a training aircraft by the military under the Army designation H-13 and the Navy designation HTL. The Navy Department purchased 10 HTL-1s for evaluation in 1947. A dozen HTL-2s followed in 1949, with nine HTL-3s the next year. The HTL-4 was virtually identical to the HTL-3 except for some internal mechanical improvements. Eventually, nine variants of the HTL saw naval service, and the Navy purchased more than 200 of them between 1947 and 1958. The Model-47 was so successful that the last HTLs were not stricken from the Marine Corps flight line until 1962, and H-13s were still in service with the U.S. Army well into the Vietnam War.

Modern-day Marines fly two descendants of the HTL, the Bell UH-1N Huey Twin utility helicopter and the heavily armed AH-1W Super Cobra attack helicopter. Both have rendered yeoman duty thus far and are slated to continue naval service for the foreseeable future.

---

unique feature of the aircraft was a removable forward canopy that allowed access for two stretchers inside the cabin. This latter feature protected injured passengers from the elements while enroute to advanced medical care, a significant improvement over both the HO3S and HTL models. The HO5S also possessed superior flight characteristics that made it a good reconnaissance and observation aircraft. Unfortu-nately, its underpowered engine and some structural defects limited the aircraft's performance. By the end of July, VMO-6 mustered eight HO5S-1s in addition to nine HTL-4s. Plans called for the HO5S to completely replace the HTLs as soon as possible. The number of HTLs steadily declined as time passed until only one HTL-4 remained when the ceasefire was declared a little over a year later.

Throughout the remainder of its tour, VMO-6 had a reputation for being a "happy ship." In the words of commanding officer Major Wallace J. Slappey, Jr.: "Morale was extremely high. . . . The squadron was loaded with gung-ho personnel. Pilots were actually stealing flights from one another. . . . The engineering department was outstanding, working round the clock . . . . Every man pulled his weight by simply knowing what needed to be done and doing it willingly."

From April 1952 until the armistice in July 1953 the helicopter section of VMO-6 averaged about 600 missions per month, usually flying out between 200 and 300 wounded. More than 1,000 missions were mounted in two different months during the summer of 1952 with the single month record of 721 non-combat missions flown in September. Squadron records for combat missions (375) and medical evacuations (428) were achieved in October 1952 during intense fighting at the Hook when the Communists made their only successful, albeit temporary, penetration of the Jamestown Line. In May 1953, the helicopter section was reorganized into three echelons: a liaison and medical evacuation flight assigned to the 1st Marine Division command post; the squadron headquarters and most aircraft located at airfield A-9; and a maintenance crew in addition to any "down" aircraft stationed at Ascom City (A-33). In May and June, HO5S-1 helicopters from VMO-6 served as standby plane guards in support of the MarLEx I and II amphibious exercises.

Two HTLs and five HO5S-1s suffered major damage due operational mishaps or crashed due to mechanical failures during the squadron's stay in western Korea, and all HO5S-1s were grounded in July 1953 because of numerous

Aviator and Aircraft Losses in Korea, 1950-1953

Crews Killed

1stLt Arthur R. Bancroft
TSgt Joe L. Brand, Jr.
MSgt Gilbert N. Caudle, Jr.
Capt David T. Gooden
1stLt Robert A. Longstaff
1stLt Charles B. Marino
Sgt Richard L. Parsell
Capt Allen W. Ruggles
Maj Doil R. Stitzel

Helicopter Losses

Date	Unit	Type	BuAer No	Cause
12 Sep 50	VMO-6	HO3S-1	122514	Operational mishap
25 Sep 50	HU-1 (USN)*	HO3S-1	122720	Enemy fire
29 Sep 50	VMO-6	HO3S-1	Unknown	Operational mishap
9 Nov 50	VMO-6	HO3S-1	Unknown	Operational mishap
3 Dec 50	VMO-6	HO3S-1	Unknown	Enemy fire
27 Jan 51	VMO-6	HTL-4	Unknown	Operational Mishap
12 Mar 51	VMO-6	HO3S-1	122518	Operational mishap
13 Apr 51	VMO-6	HO3S-1	122517	Enemy fire
19 Apr 51	VMO-6	HTL-4	128638	Operational mishap
24 Apr 51	VMO-6	HTL-4	128632	Enemy fire
25 Aug 51	MAG-33	HTL-3	124566	Operational mishap
28 Aug 51	VMO-6	HTL-4	128633	Operational mishap
17 Sep 51	VMO-6	HTL-4	128902	Operational mishap
22 Sep 51	VMO-6	HO3S-1	124342	Operational mishap
28 Sep 51	HMR-161	HRS-1	127802	Operational mishap
22 Oct 51	HMR-161	HRS-1	127789	Operational mishap
22 Oct 51	HMR-161	HRS-1	127792	Operational mishap
12 Jan 52	MAMS-12	HO3S-1	122528	Mechanical failure
21 Jan 52	HMR-161	HRS-1	127797	Operational mishap
23 Jan 52	VMO-6	HTL-4	122521	Operational mishap
7 Feb 52	VMO-6	HTL-4	128892	Enemy fire
1 Mar 52	HMR-161	HRS-1	Unknown	Structural failure
14 Mar 52	VMO-6	HTL-4	128625	Operational mishap
17 Mar 52	VMO-6	HTL-4	128887	Operational mishap
27 May 52	HMR-161	HRS-1	127784	Operational mishap
12 Feb 53	HMR-161	HRS-1	127798	Mechanical failure
25 Mar 53	HMR-161	HRS-1	127822	Mechanical failure
18 JuL 53	VMO-6	HO5S-1	130112	Enemy fire

* U.S. Navy helicopter "on loan" to VMO-6 with one Navy pilot and one Marine crewman on board.

stateside tail boom failures. The only combat loss occurred on 18 July 1953 when an HO5S-1 piloted by First Lieutenant Charles B. Marino was hit by enemy antiaircraft fire while on an artillery spotting mission. The helicopter lost control and crashed killing both the pilot and the artillery observer. This was the last helicopter-related combat casualty in Korea.

Ceasefire

At 1000 in the morning on 27 July 1953, the United Nations and Communist delegations sat down inside Panmunjom's "Peace Pagoda" to sign the formal ceasefire agreement that would bring an end to the fighting in Korea. The deed was done in only a few minutes, and the guns fell silent a half-day later, at 2000 that evening. It was, however, an uneasy peace. Neither side fully trusted the other. The fighting had stopped, but few believed the war was really over.

The Marines did not stand down and were not going home to march in any victory parades as they had in 1945. Instead, the 1st Marine Division was ordered to organize post-armistice battle positions and to establish a "no pass" line south of the Demilitarized Zone. The Marines were also charged with assisting in the final prisoner exchange of the war, Operation Big Switch. This would be a high-profile undertaking that would be conducted under the watchful eyes of the international press. Anticipating many of the former prisoners would need medical assistance, helicopters from HMR-161 stood by to carry litter patients or those too weak to travel by ambulance to the U.S. Army 11th Evacuation Hospital at Free-dom Village near Munsan-ni. Seriously injured men were taken directly to the hospital ships by helicopter or were air evacuated to Japan by fixed-wing transport planes.

Even at this late date, Marine helicopters in Korea were called upon to perform another mission never dreamed of by the early planners at Quantico. This time the HRSs of HMR-161 provided the best solution to a tangled diplomatic knot. The Marines were responsible for the safety of non-repatriated enemy prisoners, Chinese and North Koreans, who did not want to return home and would instead be placed in the custody of a neutral country, India. The problem was that Syng-man Rhee, the president of the Republic of South Korea, refused permission for Indian troops to enter his country. In the words of General Mark W. Clark, USA, the United Nations field commander: "We had to go to great lengths to live up to our pledge . . . that no Indian troops would set foot on South Korean soil. Therefore, we set up an airlift operation, which carried more than 6,000 Indians from the decks of our carriers off Inchon by helicopter to the De-militarized Zone. It was a major undertaking which just about wore out our helicopter fleet."

Marine Helicopter Transport Squadron 161 carried on in Korea for almost two more years. Its HRS-2s and-3s transported cargo, personnel, and medical evacuees until orders to prepare to leave Korea arrived in late February 1955. The squadron moved from A-17 to Ascom City and the helicopters flew to Iwakuni, Japan, to prepare for the sea journey. On 12 March, HMR-162 officially as-sumed responsibility for supporting the 1st Marine Division in Korea. By that time part of the squadron had already departed on board the amphibious cargo ship *Seminole* (AKA 104) and the remaining personnel, helicopters, and gear were stowed on board the aircraft carrier *Wasp* (CV 18) when it bid goodbye to the Far East and sailed for Marine Corps Air Station Kaneohe, Hawaii, on 26 March 1955.

Following the ceasefire, VMO-6's helicopter section continued to provide liaison, observation, and medical evacuation for the 1st Marine Division. The squadron also supported training exercises. The last HTL-4 departed in August 1953, and all HO5S-1s were back in action by October. The squadron reached a helicopter milestone of note when Major John T. Dunlavy flew VMO-6's 55,000th flight hour in Korea during an HO5S-1 test hop on 14 May 1954. The squadron began standing down on 4 Feb-ruary 1955, and finally departed Korea when four separate increments sailed from Inchon for San Diego in April 1955.

Contributions

The final accounting showed nine Marine pilots and aircrew men lost their lives during helicopter operations in Korea, four due to enemy fire. Helicopters proved to be generally more resilient and far less vulnerable to enemy fire than most thought possible prior to the test of combat—only six (all from VMO-6) of more than two dozen helicopters destroyed during the war were shot down while an uncounted number suffered some damage at the hands of the enemy but returned to base for repairs. The helicopter section of VMO-6 flew 22,367 missions including 7,067 medical evacuations in 35 months of combat flying. During its time in the combat zone, HMR-161 logged 19,639 flights (4,928 combat and 14,711 non-combat), transported 60,046 people, evacuated 2,748 seriously wounded, and offloaded

7,554,336 pounds of cargo.

Marine Observation Squadron 6 was awarded a individual U.S. Presidential Unit Citation and shared two others as an organic component of senior commands. In addition, the squadron received a Navy Unit Commendation, an Army Distinguished Unit Citation, and three Korean Presidential Unit Citations for its actions in Korea. Marine Helicopter Transport Squadron 161 was recognized for its participation as a component of commands that were awarded one U.S. Presidential Unit Citation, a Navy Unit Commendation, and one Korean Presidential Unit Citation.

Some notable early Marine helicopter pilots met mixed fates after their combat service. First Lieutenant Gustave Lueddeke succumbed to poliomyelitis not long after returning to HMX-1 at Quantico, Virginia. Major Armond Delalio was killed during a test flight when his specially configured HRS caught fire and crashed at Patuxent River Naval Air Station, Maryland. First Lieutenant Lloyd Engelhardt and Captain Gene Morrison each commanded Marine Medium Helicopter Squadron 161 as lieutenant colonels in the 1960s. Morrison, in fact, got to put into practice the helicopter combat tactics and techniques he pioneered in Korea when he led the squadron during its deployment to Vietnam in 1965. Brigadier General Edward C. Dyer and Colonel Keith B. Mc-Cutcheon both sat on the influential Hogaboom Board that restructured the Fleet Marine Force in 1956. The board recommended that all Marine divisional equipment be air transportable and entire assault battalion landing teams be helilifted ashore to secure beachheads using vertical assault techniques. Captain Victor Armstrong and

HMR-161 Historical Diary Photo Supplement, Nov-Dec51

HMR-161 conducted its first aerial medical evacuation on 13 September 1951. Although medical evacuation was a secondary mission, the squadron carried more than 2,000 seriously wounded men to various locations for advanced medical care. Its sister helicopters of VMO-6 evacuated more than 7,000 during 35 months of combat flying.

Colonel Mc-Cutcheon both rose to the highest aviation post in the Marine Corps. McCutcheon was the director of aviation on the eve of the Vietnam era and then later both he and Armstrong held the post of deputy chief of staff (air) as major generals—McCutcheon from 1966 to 1970 and Armstrong in 1975. Lieutenant General McCutcheon was actually slated to become the first Marine aviator to wear four stars on active duty until he was tragically felled by cancer immediately after commanding Marine forces in Vietnam.

Much like that first Marine HO3S that guided the rescue party to the mired amphibious jeep in the marsh at Quantico in 1948, VMO-6 and HMR-161 led the way for helicopters in the other Services. The United States Army owes a salute to the Marines for conceptualizing and testing the principles of modern airmobile warfare. The Army had long been interested in rotary-winged aircraft and actually used some primitive helicopters during World War II. The Marine Corps, however, pioneered doctrine, employed full helicopter units in combat, and developed hands-on tactical concepts in Korea. *Phib-31*, written at Quantico, Virginia, before the Marines even had a helicopter squadron, is arguably the forerunner of today's airmobile doctrine. According to Air Force historian Robert F. Futrell: "Army officers were [so] impressed by the utility of Marine helicopters in Korea [that] General Ridgway asked the

Department of the Army to provide four Army helicopter transport battalions, each with 280 helicopters." His request was significantly scaled down (to only two companies), but within a decade the Army went on to create an airmobile division whose assault elements could be helilifted into combat. The Navy and the Air Force took their cues from VMO-6 whose light utility helicopters performed search and rescue, medical evacuation, liaison, and reconnaissance—missions that closely paralleled the needs of those Services.

Today the legacy of those early helicopter pioneers of HMX-1, VMO-6, and HMR-161 lives on within the Marine Corps as well. Marine skeptics were silenced by helicopter performance in combat, and helicopters thereafter became a full partner in naval aviation rather than the "stepchildren" they had previously been. It is a tribute to the dedication, bravery, and skill of Marine helicopter air and ground crews in Korea that helicopters are vital components of the modern Marine air-ground team. Current Marine helicopter pilots are mounted in the direct descendants of those simple rotary-winged machines that traversed the Korean skies from 1950 to 1953: Bell UH-1 "Hueys" and AH-1 Sea Cobras were sired by the HTL "eggbeaters," the tandem-rotor Boeing-Vertol CH-46 Sea Knights are advanced developments of Frank Piasecki's HRP "flying banana," and the massive Sikorsky CH-53 Sea Stallions evolved from the much smaller HO3S-1 "pinwheels." Currently, the tilt-rotor Boeing MV-22 Osprey is making true the vision of designer Frank Piasecki about the future of rotary-winged flight voiced a half century earlier: "The most dramatic progress will be increased speed of vertical-lift aircraft. This will come from two directions: helicopter designers will add speed to their machines; conversely, airplane designers will add vertical-lift capabilities to their high-speed aircraft. The result will be a blending of flight into machines fully capable of both helicopter flight as we know it and high-speed flight."

While we cannot be certain exactly what the future holds, we can safely state that vertical assault and rotary-winged assault support will remain mainstays of Marine Corps doctrine well into the 21st century. With this in mind, we should always remember this is due to the achievements of the Korean "whirlybirds" that led the way.

Col Keith B. McCutcheon became one of the most versatile and best-known Marine aviators during his career. He was an innovator and theoretician as well as a doer, and, like his hero MajGen Roy S. Geiger, he commanded both air and ground units in combat.

National Archives Photo (USMC) 127-N-A132705

About the Author

Lieutenant Colonel Ronald J. Brown, USMCR (Ret), is a freelance writer, a high school football coach, and an educational consultant. The author of several official histories *(A Brief History of the 14th Marines, With Marines in Operation Provide Comfort,* and *With Marine Forces Afloat in Desert Shield and Desert Storm),* he was also a contributing essayist for the best-selling book, *The Marines,* and was the sole author of *A Few Good Men: The Fighting Fifth Marines.* After almost four years active duty from 1968 to 1971, Brown returned to teaching high school for the next three decades; intermittently, he served as an activated reservist traveling to Korea among other places. He is a combat veteran of both the Vietnam and Persian Gulf conflicts. He spent 20 years as a reservist with Mobilization Training Unit DC 7, the Reserve unit that supports the History and Museums Division. Lieutenant Colonel Brown commanded the training unit before retiring from the Marine Corps Reserve in 1996. He is the author of an earlier pamphlet in this series, *Counteroffensive: U.S. Marines from Pohang to No Name Line.*

Sources

The most important sources used in preparing the pamphlet are the various official Service histories and several monographs produced by the Marine Corps as well as primary documents and oral history interviews held by the Marine Corps Historical Center located in Washington, D.C. Selected secondary works were also used to provide context and technical information.

The starting point for this study of Marine Corps helicopter operations in Korea was the five volume *History of U.S. Marine Operations in Korea, 1950-1953* including: Lynn Montross and Capt Nicholas A. Canzona, *The Pusan Perimeter,* v. 1 (Washington, D.C.: Historical Branch, G-3 Division, HQMC, 1954); Lynn Montross and Capt Nicholas A. Canzona, *The Inchon-Seoul Campaign,* v. 2 (Washington, D.C.: Historical Branch, G-3 Division, HQMC, 1955); Lynn Montross and Capt Nicholas A. Canzona, *The Chosin Reservoir Campaign,* v. 3 (Washington, D.C.: Historical Branch, G-3 Division, HQMC, 1957); Lynn Montross, Maj Hubard D. Kuokka, and Major Norman W. Hicks, *The East-Central Front,* v. 4 (Washington, D.C.: Historical Branch, G-3 Division, HQMC, 1962); LtCol Pat Meid and Maj James M. Yingling, *Operations in West Korea,* v. 5 (Washing-ton, D.C.: Historical Division, HQMC, 1972). Other official histories consulted were: Robert F. Futrell, *The United States Air Force in Korea, 1950-1953* (Washing-ton, D.C.: Office of Air Force History, 1983); and James A. Field, Jr., *History of United States Naval Operations: Korea* (Washington, D.C.: Navy History Division, 1962).

Monographs used included: Capt John C. Chapin, *Fire Brigade: U.S. Marines in the Pusan Perimeter* (Washington, D.C.: History and Museums Division, HQMC, 2000); LtCol Kenneth J. Clifford, *Progress and Purpose: A Developmental History of the U.S. Marine Corps* (Washington, D.C.: History and Museums Division, HQMC, 1973); MajGen John P. Condon, *Corsairs To Panthers: U.S. Marine Aviation in Korea* (Washington, D.C.: History and Museums Division, HQMC, 2002); LtCol Eugene W. Rawlins, *Marines and Helicopters, 1946-1962* (Washington, D.C.: History and Museums Division, HQMC, 1976); LtCol Gary W. Parker, *A History of Marine Medium Helicopter Squadron 161* (Washington, D.C.: History and Museums Division, HQMC, 1978); and LtCol Gary W. Parker and Maj Frank M. Batha, Jr., *A History of Marine Observation Squadron Six* (Washington, D.C.: History and Museums Division, HQMC, 1982).

Other books reviewed included: Cdr Malcolm W. Cagle and Cdr Frank A. Manson, *The Sea War in Korea* (Annapolis: U.S. Naval Institute, 1957); Col Robert D. Heinl, *Soldiers of the Sea: The U.S. Marine Corps, 1775-1962* (Annapolis: U.S. Naval Institute, 1962); Robert Jackson, *Air War Korea* (Osceola, WI: Motorbooks International, 1998); Allan R. Millett, *Semper Fidelis: The History of the United States Marine Corps* (New York: Macmillan, 1980); Lynn Mon-tross, *Cavalry of the Sky* (New York: Harper and Brothers, 1954); Wayne Mutza, *H-13 Sioux* (Carrollton, TX: Squadron/Signal Books, 1995); Edwin H. Simmons, *The United States Marines: The First Two Hundred Years* (New York: Viking, 1974); Gordon Swanborough and Peter M. Bowers, *United States Navy Aircraft Since 1911* (London: Putnam Aeronautical Books, 1976); and Warren R. Young, *The Helicopters* (Alexandria, VA: Time-Life Books, 1982).

History and Museums Division oral history transcripts provided observations by: LtCol Clifford V. Brokaw III; Capt Norman G. Ewers; Maj Vincent J. Gottschalk; Capt Andrew L. McVicars; Capt Gene W. Morrison; Maj Elton Mueller; Capt Clarence W. Parkins; 1stLt John L. Scott; and 2dLt Patrick G. Sivert.